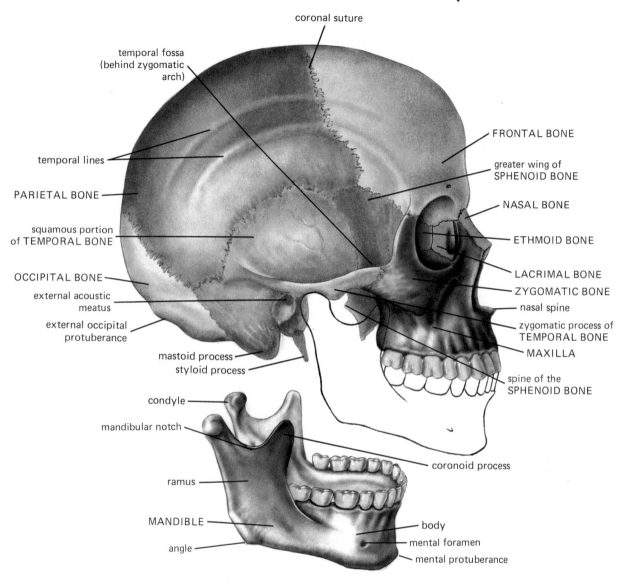

Right lateral view of the skull and the mandible.
(See Fig. 9-11 on page 146 of the text.)

HUMAN ANATOMY AND PHYSIOLOGY
Structure and Function

HUMAN ANATOMY AND PHYSIOLOGY
Structure and Function

Second Edition

Dorothy S. Luciano, Ph.D.
Formerly of The University of Michigan

Arthur J. Vander, M.D.
Professor of Physiology
The University of Michigan

James H. Sherman, Ph.D.
Associate Professor of Physiology
The University of Michigan

McGraw-Hill Book Company
New York   St. Louis   San Francisco   Auckland   Bogotá   Hamburg
Johannesburg   London   Madrid   Mexico   Montreal   New Delhi
Panama   Paris   São Paulo   Singapore   Sydney   Tokyo   Toronto

HUMAN ANATOMY AND PHYSIOLOGY: Structure and Function

1234567890DOWDOW89876543

ISBN 0-07-038962-4

See pages 817–819 for illustration credits not already found in legends. Copyrights included on this page by reference.

Library of Congress Cataloging in Publication Data

Luciano, Dorothy S.
  Human anatomy and physiology.

    Rev. ed. of: Human function and structure. 1978.
    Includes index.
    1. Human physiology. 2. Anatomy, Human.
I. Vander, Arthur J., date   . II. Sherman,
James H., date   . III. Title.
QP34.5.L83 1983 612   82-13986
ISBN 0-07-038962-4

Cover Photograph Credit
The Bathers, a series of six sculptures by Pablo Picasso. Photograph courtesy of the National Trust for Historic Preservation, Nelson A. Rockefeller Collection.

This book was set in Times Roman by Ruttle, Shaw & Wetherill, Inc. The editors were David T. Horvath, Sibyl Golden, and Stephen Wagley; the designer was Robin Hessel; the production supervisor was Charles Hess. New anatomical illustrations were prepared by Judith Glick. Remaining new drawings were done by J & R Services, Inc.
R. R. Donnelley & Sons Company was printer and binder.

To Horace W. Davenport
with respect and affection

CONTENTS

PREFACE

The primary purpose of this book remains what it was in the first edition: to present the fundamental facts and concepts of human anatomy and physiology. It is intended for introductory courses in anatomy and physiology taken by students of biology, nursing, and allied health. It is suitable for these students regardless of their science backgrounds: Although a previous introductory course in either biology or chemistry would be useful, it is not essential, since relevant background is presented where necessary.

The first edition of this book was written in response to a large number of requests from teachers of combined anatomy and physiology courses who expressed a desire for a book with the approach to physiology presented in our *Human Physiology: The Mechanisms of Body Function,* but with extensive anatomy as well. This edition maintains the same approach—an attempt at a balanced integrated presentation of the two disciplines.

In this edition, material, particularly in the anatomy sections, is extensively rewritten for greater clarity and completeness; new anatomical illustrations are added; the material is organized in parts and chapters of more manageable size; the chapter order is rearranged so it is more compatible with commonly used teaching schedules; the design is changed so the organization of the material is more readily apparent and easily understood; the format is more open so the book is less forbidding; many of the illustrations are reworked for increased crispness and clarity; and review exercises and sets of

key words are included with each chapter to help guide the students' assessment of their study needs. New terms are highlighted in boldface or italics; the more important terms are in boldface. As with the previous edition, a separate Study Guide (by Kent M. Van De Graaff and R. Ward Rhees of Brigham Young University) and an Instructor's Manual (by John P. Harley of Eastern Kentucky University) keyed to the basic text are available.

Authors of anatomy texts meet special problems in terminology. By the beginning of this century, there were already some 50,000 terms for the 5000 to 6000 structures in the body, an average of ten terms per structure. After considering the problem of what names to use, we have chosen to follow the official Nomina Anatomica terminology. In cases where older terms are still commonly used, we have included them in parentheses.

The physiology in this book is more extensive than that in most other combined anatomy and physiology texts, emphasizing as it does mechanisms and control systems. It stresses that the body's various coordinated functions—circulation and respiration, for example—result from the precise control and integration of specialized cellular activities, serve to maintain relatively constant the internal composition of the body, and can be described in terms of control systems similar to those designed by engineers. Consequently, the book progresses from the structure and function of cells and tissues in Part 1 to Parts 2 through 5 on how the structures and activities of these discrete tissues are integrated to achieve overall form and function of the organ systems of the body.

The presentation of material is flexible enough that the book can be used in several ways. A course which has basic biology as a prerequisite could omit Part 1 completely. If gross anatomy is to be the major anatomical focus, the chapters of Part 1 could be skimmed; alternatively, time could be saved by eliminating from Part 3 whole topics such as consciousness and behavior or from Part 5 the chapter on the immune system. A longer course could use the book in its complete form. The anatomy is presented at two levels of complexity: The text and illustrations and their accompanying descriptive material are at a general introductory level, and the extensive tables serve as more detailed sources of information.

We would like to thank those colleagues who reviewed all or part of this revision for their criticisms and suggestions: Barton Bergquist, University of Northern Iowa; William D. Chapple, University of Connecticut; Robert D. Everson, William Paterson College; Ruth Lanier Hays, Clemson University; R. T. Justus, Arizona State University; Patricia A. Murski, St. Francis Hospital School of Nursing; Robert Resau, Essex Community College; Christopher Schatte, Colorado State University; and Marjorie F. Sparkman, School of Nursing, Florida State University.

Several anatomists have provided valuable assistance to us. We would like to thank Sara Winans and Ray Kahn who helped us to develop our overall approach to anatomy. There is no way that we can adequately express our gratitude to Thomas Oelrich, who was our major anatomy consultant. Dr. Oelrich graciously devoted an immense amount of time and effort to going over anatomical illustrations and tables; merely having him to work with was a source of great reassurance to us. Any errors in these materials almost certainly represent our failure to follow his advice. We were delighted to have the opportunity to work with Sibyl Golden, our editor at McGraw-Hill, and to benefit from her patience, attention to detail, and balanced wisdom. Once again, Peggy Rogers did a superb job of typing the manuscript.

Dorothy S. Luciano
Arthur J. Vander
James H. Sherman

PART

1

BASIC
PRINCIPLES

CHAPTER

ORGANIZATION
OF THE BODY

This book presents the fundamental facts and concepts of two basic biological sciences as they apply to the human body: **anatomy,** which deals with the structure of the body, and **physiology,** which deals with the functions of the body, the mechanisms by which the various anatomical structures in the body operate. Parts 1 and 2 are devoted to an analysis of the discrete components of the body, beginning with basic cellular biology and concluding with the anatomy and physiology unique to individual tissues. The remaining parts describe how the structures and activities of these discrete tissues are integrated to achieve overall form and function in the body.

A SOCIETY OF CELLS

Cells, microscopic units of living matter, are the simplest units of biological structure and function into which an organism can be divided and still retain the characteristics associated with life. Each human organism begins as a single cell, the fertilized *ovum*. This single cell divides to form two cells, each of which in turn divides, resulting in four cells, and so on as repeated divisions increase the number of cells in the developing organism. If cell multiplication were the only event occurring, the end result would be a spherical mass of identical cells. However, other processes that shape the organism and form the various anatomical structures of the body are also occurring during this time.

In the fully developed body all cells are not identical in structure or function. Some cells, for example, have developed the ability to generate forces that produce movement. These are the muscle cells. Others become specialized in their ability to generate and transmit electrical signals. These are called nerve cells. Altogether there are about 200 different types of cells that can be identified in terms of differences in their structure and specialized functions. These specialized cells are arranged in various combinations to form **organs**—the heart, liver, kidneys, pancreas, etc.—each specialized cell contributing to the overall functioning of a particular organ. A collection of organs whose functions are interrelated form an **organ system.** For example, the kidneys, the bladder, and the tubes leading from the kidneys to the bladder and from the bladder to the exterior constitute the urinary system.

In essence, then, the human body can be viewed as a complex society of cells of many different types that are structurally and functionally combined and interrelated in a variety of ways to carry on the functions essential to the survival of the organism as a whole. Yet the fact remains that individual cells still constitute the basic units of this society and that almost all these cells individually exhibit the fundamental activities common to all forms of life. Indeed, many of the body's different cell types can be removed from the body and maintained in test tubes as free-living cells.

There is a definite paradox in this analysis. If each individual cell performs the fundamental activities required for its own survival, what contributions do the different organ systems make? How can we refer to a system's functions as being "essential to the survival of the organism as a whole" when each individual cell of the organism seems to be capable of performing its own fundamental activities? The resolution of this paradox is found in the isolation of most of the cells of a multicellular organism from the environment surrounding the body (**external environment**).

An ameba and a human liver cell both obtain most of their required energy by the breakdown of certain organic nutrients. The chemical reactions involved in this intracellular process are remarkably similar in the two types of cells and involve the utilization of oxygen and the production of carbon dioxide. The ameba picks up required oxygen directly from its environment and eliminates the carbon dioxide into it. But how can the liver cell obtain its oxygen and eliminate the carbon dioxide when, unlike the ameba, it is not in direct contact with the external environment? Supplying oxygen to the liver is the function both of the respiratory system, which takes up oxygen from the environment, and of the circulatory system, which distributes oxygen to all parts of the body. Conversely, the circulatory system carries the carbon dioxide generated by the liver cells and all the other cells of the body to the lungs, which eliminate it to the exterior. Similarly, the digestive and circulatory systems, working together, make nutrients from the external environment available to all the body's cells. Wastes, other than carbon dioxide, are carried by the circulatory system from the cells that produced them to the kidneys and liver, which excrete them from the body. Thus, the overall effect of the activities of organ systems is to create *within* the body the environment required for all cells to function.

Clearly, the society of cells that constitutes the human body bears many striking similarities to a society of people (although the analogy must not be pushed too far). Each person in a complex society must perform an individual set of fundamental activities (eating, excreting, sleeping, etc.) that are virtually the same for all persons. In addition, because the complex organization of a society makes it virtually impossible for individuals within the society to raise their own food, arrange for the disposal of their wastes, and so on, each individual participates in the performance of one of these supply-and-disposal operations required for the survival of all. A specialized activity, therefore, becomes an *additional* part of each person's daily routine, but it never allows him or her to cease or to reduce the performance of the fundamental activities required for survival.

THE INTERNAL ENVIRONMENT

A cell is a very fragile chemical machine. Large fluctuations in the physical and chemical properties of the fluid medium surrounding it can disrupt the regulated flow of chemical reactions maintaining the life of the cell. Seawater, whose temperature and chemical composition do not change rapidly (because of its large volume), provided the stable environment for the first living cells, which appeared on earth about 3 billion years ago. These single, free-living cells obtained nutrients from and excreted wastes directly into the external environment, seawater. Thus, life at this early stage depended upon chemical exchanges between two fluid environments, the fluid surrounding the cell and the fluid within the cell.

A loose association of independent cells into small clusters was the first step in the evolution of multicellular organisms. Only the cells at the surfaces of such clusters were in immediate contact with the seawater. Within the cluster, cells were surrounded by other cells and by **extracellular fluid** that had been trapped between the cells. This extracellular fluid provided the immediate environment for the interior cells of the cluster.

Even though organisms have grown in size and complexity from these simple clusters, they all have within them a thin layer of extracellular fluid that bathes each of the cells. In other words, the environment in which the trillions of cells in the body live is not the *external environment* surrounding the total organism but is the fluid that immediately surrounds each cell. It is this extracellular fluid, or **internal environment,** from which a cell receives nutrients and into which it excretes wastes.

A multicellular organism can survive only as long as it is able to maintain the composition of its internal environment in a state compatible with the survival of its individual cells. The French physiologist Claude Bernard first clearly described, in 1857, the central importance of the extracellular fluid of the body: "It is the fixity of the internal environment which is the condition of free and independent life. . . . All the vital mechanisms, however varied they may be, have only one object, that of preserving constant the conditions of life in the internal environment." This concept of an internal environment and the necessity for maintaining its composition relatively constant is the single most important unifying concept to be kept in mind while attempting to unravel and understand the structure and function of the human body.

Body-Fluid Compartments

Multicellular organisms obtain nutrients and oxygen from the external environment and deliver them to their cells by way of the extracellular fluid. Likewise, waste products of metabolism produced by cells must travel the reverse route through the extracellular fluid to the surface of the organism where they are released into the external environment. The extracellular fluid of the internal environment thus provides the medium by which substances are exchanged between cells and the external environment. The extracellular fluid is divided into two compartments: the **plasma,** which is the fluid surrounding cells in the blood vessels, and the **interstitial fluid,** which surrounds all the other cells, i.e., all the cells except the blood cells (Fig. 1-1).

To generalize one step further, the body can be viewed as containing three fluid compartments: (1) blood plasma, (2) interstitial fluid, and (3) **intracellular fluid,** the fluid inside cells (Fig. 1-2). The major molecular component of all three compartments is *water,* which accounts for about 60 percent of the body weight, or 42 L, in an average-sized individual. Two-thirds of the total body water (28 L) is inside the cells: the intracellular fluid. The remaining one-third (14 L) comprises the two extracellular fluids: 80 percent as interstitial fluid (11 L) and 20 percent as plasma (3 L).

The above figures for the volumes of fluids in the various body compartments are those of a 70-kg

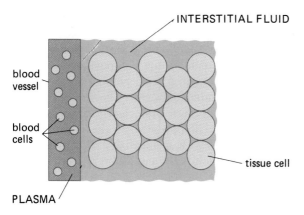

FIG. 1-1 The extracellular fluid (*internal environment*) is in two compartments: plasma and interstitial fluid.

man. Obviously, the absolute amount or size of any component will be different in different individuals. Some of these values may vary directly with body weight and can be expressed as a percentage of the body weight, whereas others depend on age, sex, and state of health as well as body weight. The values for a healthy, 70-kg (154-lb), 21-year-old male have been used for years as representative of an average individual because a majority of the data collected over the years on normal healthy people have come from measurements made on students attending medical school. The 21-year-old, 70-kg male represents the average body size, sex, and age of the medical students from which the data were obtained. Measurements on females of the same

age and corrected for differences in body weight give values of a similar magnitude, although slight sex-specific differences do exist. Thus, females have a slightly lower total-body-water content than males of the same weight because female bodies are composed of a higher percentage of fatty tissue, which contains less water than other tissues. Unless otherwise stated, quantitative values given in this text refer to a healthy 70-kg, 21-year-old male.

UNITS OF STRUCTURAL ORGANIZATION

The cells of the body are combined to form a hierarchy of structural organization. Individual specialized cells are arranged into tissues, which are combined to form organs, which are linked together to form organ systems (Fig. 1-3).

Specialized Cell Types

Each cell functions at two levels: (1) It performs those basic processes essential for its own survival—release and utilization of chemical energy; synthesis of molecules necessary for the maintenance of its structure and function; and exchange of nutrients, salt, and water with the extracellular fluid. (2) It performs one or more specialized functions that, in cooperation with other specialized cells, maintain the composition of the internal environment in which all the cells must live. Rather than being a jack-of-all trades, as with free-living,

FIG. 1-2 Fluid compartments of the body. Volumes are for the "average" 70-kg man.

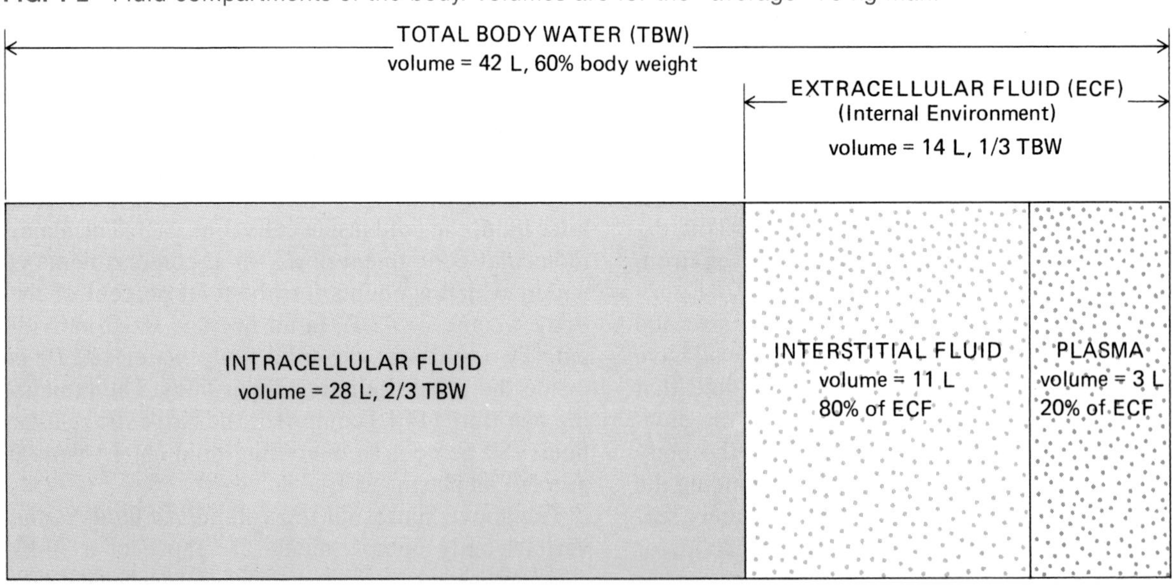

TOTAL BODY WATER (TBW) volume = 42 L, 60% body weight		
	EXTRACELLULAR FLUID (ECF) (Internal Environment) volume = 14 L, 1/3 TBW	
INTRACELLULAR FLUID volume = 28 L, 2/3 TBW	INTERSTITIAL FLUID volume = 11 L 80% of ECF	PLASMA volume = 3 L 20% of ECF

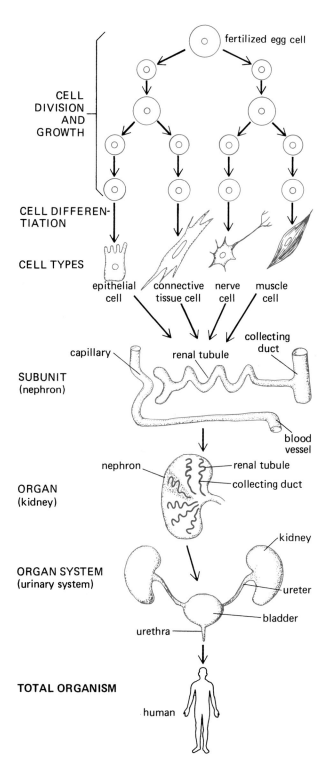

FIG. 1-3 Levels of organization, using the urinary system as an example.

single-celled organisms, the cells of a multicellular organism became specialists in performing one or more particular jobs, leaving other jobs to other types of specialized cells. However, each of the functions of the specialized cell types is the result of an augmented development of one or more of the essential functions carried out by all cells.

Many of the 200 different types of cells perform similar but not identical functions. When cells are classified according to the general type of function they perform, four broad categories emerge: (1) **muscle cells,** specialized for the production of forces which produce movement; (2) **nerve cells,** specialized for initiation and conduction of electric signals over long distances; (3) **epithelial cells,** specialized for the selective secretion and absorption of organic molecules and ions; and (4) **connective-tissue cells,** specialized for the formation and secretion of various types of extracellular connective and supporting elements. In each of these functional categories there are a variety of cell types that perform variations of the general type of specialized function. For example, there are three different types of muscle cells—skeletal, cardiac, and smooth muscle cells—all of which generate forces and produce movements but which differ from each other in shape, mechanisms controlling their contractile activity, and location in the various body organs.

Tissues

Most of the specialized cells are associated with other cells of a similar kind, forming multicellular aggregates known as **tissues.** Just as there are four general categories of cells, there are, corresponding to these cell types, four general categories of tissues: **muscle tissue, nerve tissue, connective tissue,** and **epithelial tissue.** It should be noted that the term *tissue* is frequently used in several different ways. It is formally defined as described above, an aggregate of a single type of specialized cell. However, it is commonly used to denote the general cellular fabric of any given organ or structure, for example, kidney tissue, lung tissue, etc., which are in fact usually composed of all four specialized cell types.

The general term **parenchyma** is used to refer to the tissues of an organ which perform the *specialized functions* associated with that organ, while the term **stroma** refers to the *supporting structures,* such as connective tissues, nerves, and blood vessels, of the organ.

Histology is the branch of microscopic anatomy

that deals with the structure of tissues and their arrangement in the various organs; **cytology** is concerned with the structure of individual cells.

Organs

The organs of the body are composed of the four kinds of tissues arranged in various proportions and patterns: sheets, tubes, layers, bundles, strips, etc. For example, the kidney (Fig. 1-3) consists largely of (1) a series of small tubules, each composed of a single layer of epithelial cells; (2) blood vessels, whose walls consist of an epithelial lining and varying quantities of smooth muscle and connective tissue; (3) nerve fibers with endings near the muscle and epithelial cells; and (4) a loose network of connective-tissue elements that are interspersed throughout the kidney and that form an enclosing connective-tissue capsule. The structural compo-

nents of many organs are organized into small, similar subunits, each performing the function of the organ. For example, the kidneys' 2 million functional units, the nephrons, are the tubules with their closely associated blood vessels. The total production of urine by the kidney consists of the sum of the amounts formed by the individual nephrons.

Organ Systems

There are ten major organ systems in the body (Table 1-1). Nine of them (the reproductive system is the exception) contribute to the maintenance of specific aspects of the internal environment. The structures and functions of each of the organ systems are understandable in terms of the problems that confronted multicellular organisms as their growth in size increased the separation of the majority of their cells from direct contact with the external environment. For example, the evolution

TABLE 1-1   ORGAN SYSTEMS OF THE BODY

System	Major Organs or Tissues	Primary Function(s)
Circulatory	Heart, blood vessels, blood	Rapid bulk flow of blood throughout the body's tissues
Respiratory	Nose, throat, larynx, trachea, bronchi, bronchioles, lungs	Exchange of carbon dioxide and oxygen and regulation of hydrogen-ion concentration
Digestive	Mouth, pharynx, esophagus, stomach, intestines, salivary glands, pancreas, liver, gallbladder	Digestion, absorption, and processing of nutrients
Urinary	Kidneys, ureters, bladder, urethra	Regulation of plasma composition through excretion of organic wastes, salts, and water
Musculo-skeletal	Cartilage, bone, ligaments, tendons, joints, skeletal muscle, bone marrow	Support, protection, and movement of the body; blood cell formation
Immune	White blood cells, lymph vessels and nodes, spleen, thymus, and other lymphatic tissues	Defense against foreign invaders; return of extracellular fluid to blood and the formation of white blood cells
Nervous	Brain, spinal cord, peripheral nerves and ganglia, special sense organs	Regulation and coordination of many activities in the body; detection of changes in the internal and external environments; states of consciousness
Endocrine	All glands secreting hormones: Pituitary, thyroid, parathyroid, adrenal, pancreas, testes, ovaries, intestinal glands, kidneys, hypothalamus, thymus, pineal	Regulation and coordination of many activities in the body
Reproductive	Male: Testes, penis, and associated ducts and glands	Production of sperm; transfer of sperm to female
	Female: Ovaries, uterine tubes, uterus, vagina, mammary glands	Production of egg cells; provision of a nutritive environment for the developing embryo
Integumentary	Skin	Protection against injury and dehydration; defense against foreign invaders; regulation of temperature

of a rapid-transit **circulatory system** allowed an increase in body size by rapidly distributing molecules to the cells once the molecules had entered the body from the external environment across the body's surface. Moreover, the increase in body size presented a problem related to the rate at which molecules could be moved across the body's surface. As an organism increases in size, the volume of its body increases more rapidly than the surface area; i.e., the ratio of surface area to volume decreases. Eventually a size is reached at which the rate of exchange of molecules across a unit area of the body's surface would not be sufficient to supply the metabolic requirements of the large volume of underlying tissue, in spite of a rapid circulatory system. The solution to this problem was either to extensively fold the body's surface or to form highly coiled small-diameter tubes that, although "inside" the body, remain in direct contact with the external environment. Both of these solutions increase the surface area in contact with the external environment without significantly increasing the volume of the organism.

Such specialized regions of the body's surface evolved over millions of years into those organ systems involved with the exchange of molecules be-

tween the external and internal environments: the **respiratory system,** which exchanges oxygen and carbon dioxide between the air and blood; the **digestive system,** which is the site of entry for nutrients, salts, and water from the external environment; and the **urinary system,** which excretes organic waste products, salts, and water into the external environment. Both the respiratory and digestive systems achieve a large surface area by an extensive folding of the layers of cells lining the organs of these systems, whereas the large surface area of the urinary system results from numerous small tubules that make up the kidneys. As these organ systems have evolved, their surface areas have increased, become internalized, and become connected to the exterior of the body. The surface area of each provides an interface between the external environment and the blood plasma, which, in turn, is linked to the interstitial fluid and, thus, to the body's cells through exchanges across the capillaries (Fig. 1-4).

The **musculoskeletal system** supports and moves the body; in addition, it provides the delicate tissues of the brain and spinal cord, lungs, heart, and great vessels with bony protection. Such a system makes it possible for an organism to defend itself against

FIG. 1-4 Exchanges of matter occur between the external environment and the circulatory system via the digestive, respiratory, and urinary systems.

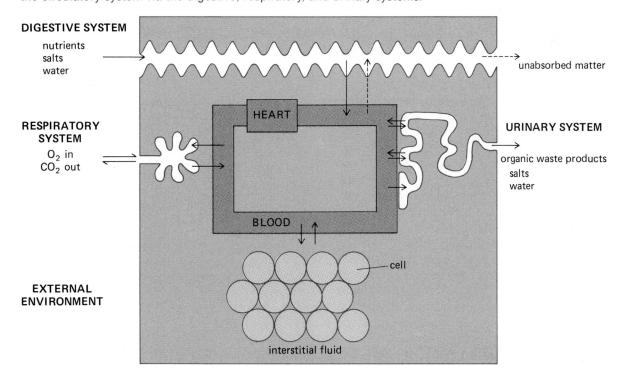

large predatory organisms, but there are also microscopic invaders, the bacteria and viruses, that can injure the body from within. Defense against the microscopic invaders is provided by the **immune system** and the **integumentary system,** or skin.

As the complexity of organisms increased, survival depended upon the coordination and regulation of the multiple activities of the body so that a stable internal environment could be maintained. Leon Fredricq, in 1885, amplifying the concepts of Claude Bernard, noted: ''The living being is an agency of such a sort that each disturbing influence induces by itself the calling forth of compensatory activity to neutralize or repair the disturbance. The higher in the scale of living beings, the more numerous, the more perfect, and the more complicated do these regulatory agencies become. They tend to free the organism completely from the unfavorable influence and changes occurring in the environment.'' Two organ systems play the major role in this regulation and coordination: the **nervous system** and the **endocrine system.** The nervous system makes use of electric signals conducted along the surfaces of nerve cells; the endocrine system makes use of chemicals, *hormones,* secreted into the blood. These two organ systems provide the major routes for linking the activities of cells in one part of the body with those in another. In addition, the nervous system transmits to the interior of the body information received at its surface by specialized *receptor cells* of the eyes, ears, nose, and skin and is responsible for the various states and contents of consciousness.

Adequate maintenance of the composition of the internal environment depends on the proper functioning of each of these organ systems, and the failure of any one of them can be fatal. In the case of a few organs, such as those of the immune and endocrine systems, a major malfunction may not directly kill the organism but will severely limit the types of external environment in which it can survive. For example, in the absence of an immune system, an organism can survive only in an environment free of microscopic organisms.

The one remaining organ system, the **reproductive system,** does not regulate the composition of the internal environment. It can be removed without affecting the organism's ability to survive. It is, however, obviously essential for the survival of the species. The reproductive organs produce the primary sex cells, the *spermatozoa* in the male and

TABLE 1-2 PREFIXES AND SUFFIXES

PREFIXES

a-, ab-	away from, outside of, deviating from
a-, an-	without, lacking, not
ad-	to, toward
ante-	before
anti-	against
auto-	self
bi-	two
bi-, bio-	life (biology)
brady-	slow
chrom-, chromo-	color, pigment
circum-	around
contra-	opposed, against
de-	remove, decrease
di-	two
dia-	through
dys-	bad, difficult
ect-, ecto-	on, without, on the outside
end-, endo-	in, within
ep-, epi-	upon, above, among
erythr-, erythro-	red
eu-	good, well (euphoria)
ex-	out, away from
hemi-	half
hetero-	varied, unlike, different
hist-, histo-	tissue
homoeo-, homoio-, homeo-	similar, like
hyper-	above, beyond, excess
hyp-, hypo-	below, under, deficient
in-	into, not
infra-	below, under, within
inter-	between, among, mutual
intra-	within, inside
ipse-	same
is-, iso-	equal, like
juxta-	next to
leuc-, leuco-, leuk-, leuko-	white
macr-, macro-	large, long
mal-	bad
meg-, mega-, megal-, megalo-	large, great, enlarged
mes-, meso-	medium, middle
micr-, micro-	small, minute, undersized
mon-, mono-	one, single
morph-, morpho-	shape, form
necr-, necro-	dead
noci-	injurious, pain
olig-, oligo-	scant, sparse, deficiency

TABLE 1-2 continued	

PREFIXES continued

par-, para-	beside, near, closely resembling, beyond
peri-	around
phag-, phago-	eat
poli-, polio-	gray
poly-	many
post-	behind, after
pre-	before, in front of
pro-	before, giving rise to
retro-	backward, behind
scler-, sclero-	hard
semi-	half
sub-	below
super-, supra-	above, excess
syn-, sym-	with, binding together
tachy-	swift, accelerated
trans-	across
ultra-	beyond, excessive
uni-	one

SUFFIXES

-algia	pain
-ase	enzyme
-ectomy	cut out, surgically remove
-emia, -aemia	blood
-gen, -gene	producing
-graph	writing
-itis	inflammation
-lysis	dissolving, destruction, separation
-meter	measure
-oid	like, similar to
-ole	small
-opia, -opy	vision
-osis	a condition, a process
-plegia, -plexy	paralysis, stroke
-pnea, pnoea	respiration
-rrhage	to burst forth, excessive discharge
-rrhea, -rrhoea	flow
-soma, -some	body
-stomy	make an opening
-tome, -tomy	cutting
-trope	to turn
-trophe	nourishment
-ule, -ulus	small
-uria	urine

TABLE 1-2 continued	

COMBINING FORMS

aden-, adeno-	gland
amyl-, amylo-	starch
angi-, angio-	vessel
ano-	anus
arthr-, arthro-	joint
bili-	bile
brachi-, brachio-	arm
branchi-, bronchio-	bronchus, trachea
cardi-, cardia-, cardio-	heart
cephal-, cephalo-	head
cerebr-, cerebri-, cerebro-	brain
chol-, chole-, cholo-	bile
chondr-, chondri-, chondro-	cartilage
cor-, core-, coro-	eye
corpus	body
cost-, costi-, costo-	rib
cyt-, cyto-	cell
derm-, derma-, dermo-	skin
enter-, entero-	intestine, intestinal
gastr-, gastro-	stomach
gloss-, glosso-	tongue
gluc-, gluco-, glyc-, glyco-	glucose, sugar
hem-, hema-, hemo-, haem-, haema-, haemo-	blood
hepat-, hepato-	liver
hydr-, hydro-	water
hyster-, hystero-	uterus
ili-, ilio-	ilium
my-, myo-	muscle
myel-, myelo-	marrow
nephr-, nephros-	kidney
neur-, neuro-	nerve
ophthalm-, ophthalmo-	eye
ost-, oste-, osteo-	bone
ot-, oto-	ear
ovi-, ovo-	egg
path-, patho-	disease
phleb-, phlebo-	vein
pneum-, pneumo-	air, lungs
psych-, psycho-	mind
pulmo-	lung
ren-, reni-, reno-	kidney
sacr-, sacro-	flesh
therm-, thermo-	heat
thromb-, thrombo-	clot
ur-, uro-	urine, tail
vas-, vasi-, vaso-	vessel

the *ova* in the female. In the formation of a new individual, multicellular organisms develop from a single cell, the fertilized ovum, or *zygote*. Reproductive organs in the female provide a special environment for the stages of embryonic and fetal development that follow fertilization.

STRUCTURAL ORGANIZATION

Terminology

The task of learning—and remembering—many of the terms in anatomy and physiology can be made easier if you know some of the key parts of words that are found in these terms. In fact, some terms consist entirely of such parts (e.g., apnea, bronchitis, cardiovascular). The prefixes, suffixes, and combining forms listed in Table 1-2 are commonly used in anatomy and physiology. The effort required to become familiar with these terms will be repaid many times over in the ease with which you will be able to learn and remember new vocabulary.

In the **standard anatomical position** (Fig. 1-5), the body is erect with the feet together, arms hanging at the sides, palms of the hands facing forward, and thumbs pointing away from the body. In order to describe the location of a particular structure, a number of directional terms are used (Table 1-3). All directions, including the various movements of the limbs, are described relative to this standard anatomical position.

The terms **posterior** and **dorsal** mean toward the back, and **anterior** and **ventral** mean toward the front. The **superior,** or **cranial,** direction is toward the head, and the **inferior,** or **caudal,** direction is toward the feet.

One other set of directions is necessary to locate a position across the width of the body. The **midsagittal plane** divides the body symmetrically into right and left halves; directions away from it are **lateral,** and directions toward it are **medial.** For example, the eyes are lateral to the nose, and the nose is medial to each eye.

Two other primary planes of the body are the **coronal plane,** parallel to the anterior and posterior

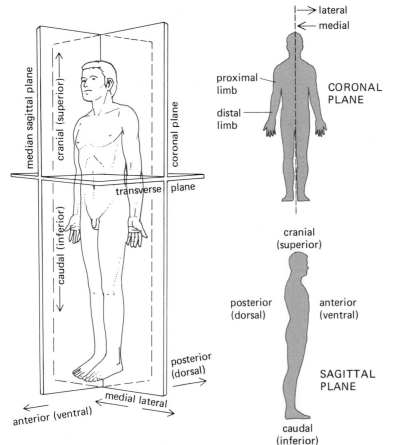

FIG. 1-5 Anatomical directions and planes.

surfaces of the body, and the **transverse plane,** perpendicular to the coronal and sagittal planes (Fig. 1-5). Note that a transverse plane can be placed through the body at any point along the cranial-caudal axis, passing through the head, the chest, or the legs, depending upon its location. Numerous sagittal and coronal planes can also be passed through various segments of the body in their respective planar orientations.

The terms used thus far refer to directions relative to the standard anatomical position. The two terms **proximal** and **distal** have a more generalized meaning, referring to directions toward (proximal) or away from (distal) the origin of a particular structure. For example, the hands and feet are located at the distal ends of the limbs; the elbows and knees are located proximal to the hands and feet, respectively, but remain distal to the shoulder and hip.

The anatomical terminology used in referring to various regions of the body's surface is illustrated in Fig. 1-6, along with the more common terms.

The Body Cavities

Most of the body's organs are in two large cavities, the **dorsal** and **ventral cavities** (Fig. 1-7), each of which has smaller subdivisions.

The dorsal cavity consists of the connected **cranial cavity** in the skull and the smaller **vertebral canal** that runs through the vertebral bones. The bones surrounding these cavities protect the delicate tissues of the brain, located in the cranial cavity, and the spinal cord, extending from the base of the brain through the vertebral canal. The dorsal cavity, surrounded on all sides by solid bone, has a fixed volume. Any abnormal growth of tissue or accumulation of fluid within the dorsal cavity exerts pressure on the brain or spinal cord that can affect their functioning.

In contrast to the dorsal cavity, the ventral cavity can vary in capacity and shape, depending on its contents and the muscular activity of the surrounding walls. The ventral cavity is divided into two chambers, the **thoracic cavity** and the **abdominal cavity,** by a sheet of muscle, the *diaphragm* (Fig. 1-7).

The thoracic, or chest, cavity is surrounded by a protective rib cage. The diaphragm forms its floor. The major organs located in the thoracic cavity are the heart and lungs (Fig. 1-8), each enclosed in its own separate chamber surrounded by a membranous lining. The heart is suspended in a fluid-filled sac, the **pericardial cavity.** On either side of the pericardial cavity are the two lungs, each of which is enclosed in a separate chamber, a **pleural cavity** (Fig. 1-9). The region of the thoracic cavity located between the pericardial cavity and the two pleural cavities is the **mediastinum.** The major blood

TABLE 1-3 TERMS THAT INDICATE THE RELATIVE LOCATION OF PARTS OF THE BODY		
Term	**Definition**	**Example**
Anterior (ventral)	Closer to (or at) the front of the body	The nose is on the anterior surface of the head.
Posterior (dorsal)	Closer to (or at) the back of the body	The backbone (vertebral column) is posterior to the breastbone (sternum).
Lateral	Farther from the central axis of the body	The ears are lateral to the nose.
Medial	Closer to (or at) the central axis of the body	The neck is medial to the shoulders.
Inferior (caudal)	Closer to the feet or lower end of the body	The chin is at the inferior border of the face.
Superior (cranial)	Closer to the head or upper end of the body	The neck is superior to the heart.
Deep	Away from the surface	The heart is deep to the skin.
Superficial	Closer to (or at) the surface of the body	The skin is superficial to the muscles.
Distal	Farther from the trunk or central axis of the body	The hand is distal to the elbow.
Proximal	Closer to the trunk or midline	The shoulder is proximal to the hand.

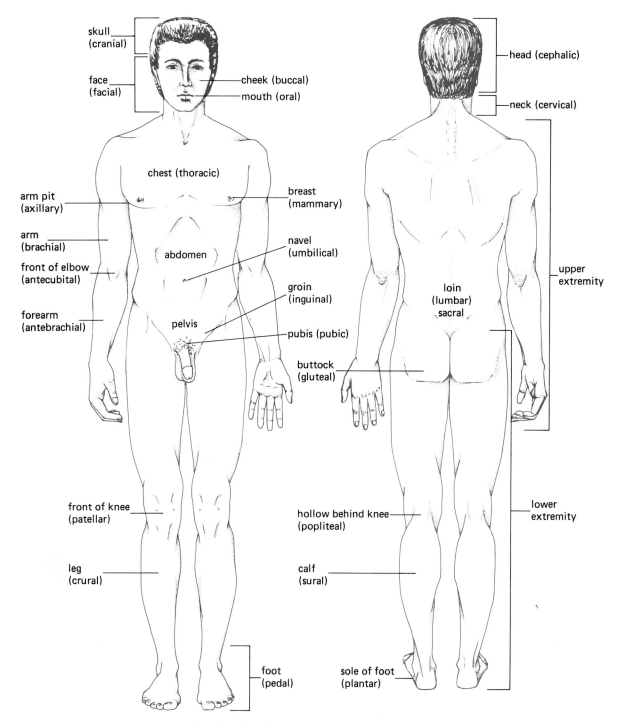

FIG. 1-6 Anatomical regions of the body surface.

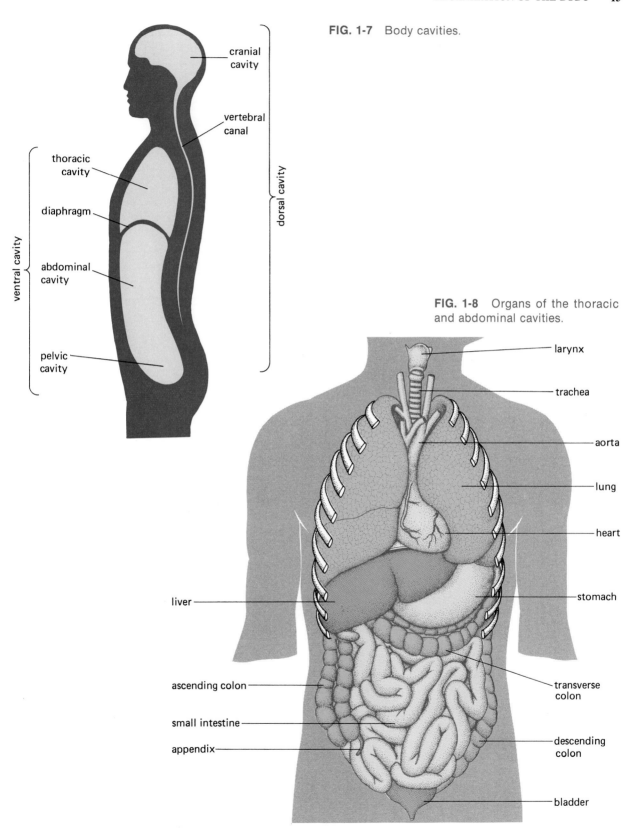

FIG. 1-7 Body cavities.

cranial cavity

vertebral canal

dorsal cavity

thoracic cavity

diaphragm

abdominal cavity

ventral cavity

pelvic cavity

FIG. 1-8 Organs of the thoracic and abdominal cavities.

larynx

trachea

aorta

lung

heart

stomach

liver

ascending colon

transverse colon

small intestine

appendix

descending colon

bladder

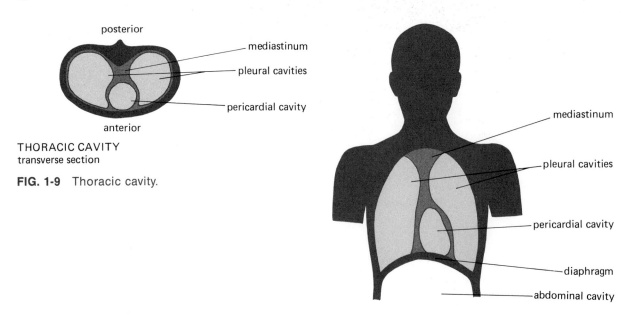

FIG. 1-9 Thoracic cavity.

THORACIC CAVITY
transverse section

THORACIC CAVITY
coronal section

FIG. 1-10 Cross section of the trunk at the level of (A) the thoracic cavity and (B) the abdominal cavity.

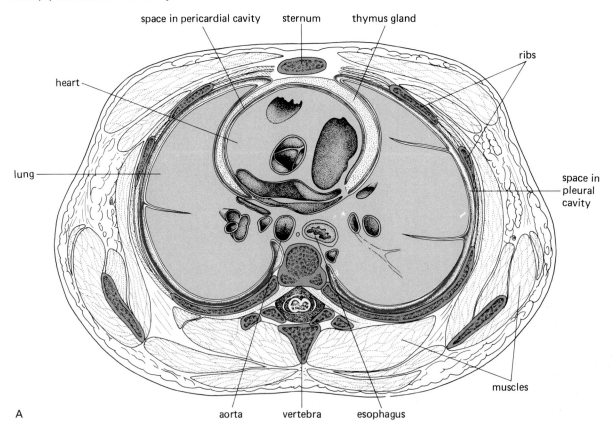

vessels to the lungs and heart, the trachea, and the esophagus pass through the mediastinum; the thymus gland is also located there.

The abdominal cavity, which forms the second portion of the ventral cavity, is the largest cavity. This cavity contains the organs most often referred to as the viscera (Fig. 1-8), although the term *viscera* actually refers to any large organ in either the dorsal or ventral cavities. The abdominal cavity has arbitrarily been divided into an upper **abdominal cavity** and a lower region, the **pelvic cavity,** which is partially protected by the surrounding pelvic bones.

The body cavities are lined by membranes that secrete a watery fluid. These are known as **parietal membranes** when they line the outer wall of the body cavity and **visceral membranes** when they cover the surfaces of the organs in the cavity. The parietal and visceral membranes are separated from each other in most cases by only a thin film of watery serous fluid. The actual "free space" in a given cavity is therefore very small, as can be seen in the cross sections at the thoracic and abdominal levels in Fig. 1-10A and B.

For the purpose of specifying the location of organs within the abdominal cavity, this region is divided into four **quadrants**—*right upper, left upper, right lower,* and *left lower* (Fig. 1-11). The imaginary lines that form the boundaries of these four quadrants intersect at the umbilicus. The body surface overlying the abdominal cavity is divided into nine **regions**—the *right* and *left hypochondriac* regions, the *right* and *left lumbar* regions, the *right* and *left iliac* regions, and the *epigastric, umbilical,* and *hypogastric* regions (Fig. 1-12). The two horizontal lines that define these regions occur at the level of the ninth ribs and iliac crests (the top of the hip, or pelvic, bones) respectively; the two vertical lines pass through the midpoints of the right and left inguinal ligaments (which mark the separation of the abdominal wall and the legs).

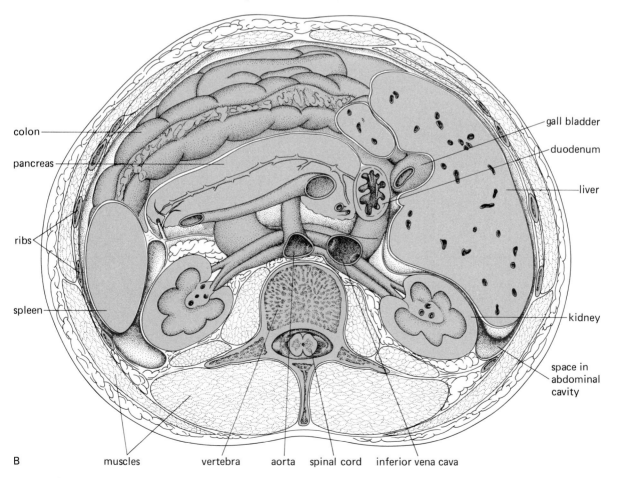

colon

pancreas

ribs

spleen

gall bladder

duodenum

liver

kidney

space in abdominal cavity

B   muscles   vertebra   aorta   spinal cord   inferior vena cava

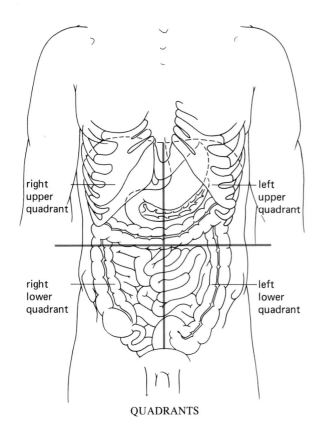

QUADRANTS

FIG. 1-11 (at left) Quadrants of the abdominal cavity.

The upper abdominal cavity is the least protected by bony structures of any of the body's cavities. Located between the rib cage of the thoracic cavity and the pelvic bones of the pelvic cavity, its walls are composed of layers of muscle and connective tissue but no bony elements except for segments of the vertebral column at the back. The major organs in the upper abdominal cavity are the liver, stomach, pancreas, spleen, and intestines.

The pelvic cavity contains the bladder as well as portions of the intestines; the latter terminate in the short segment of the rectum leading to the anus. Also located in the pelvic region are the reproductive organs—the vagina, uterus, uterine tubes, and ovaries in the female, and the prostate, seminal vesicles, and portions of the ductus deferens in the male. The gonads of the male, the testes, are outside the pelvic cavity in a separate scrotal sac suspended from the lower abdomen.

The major organs not located in one of the body cavities are the special sense organs associated with the eyes, ears, nose, and mouth and the glands of the neck, the thyroid and parathyroid glands.

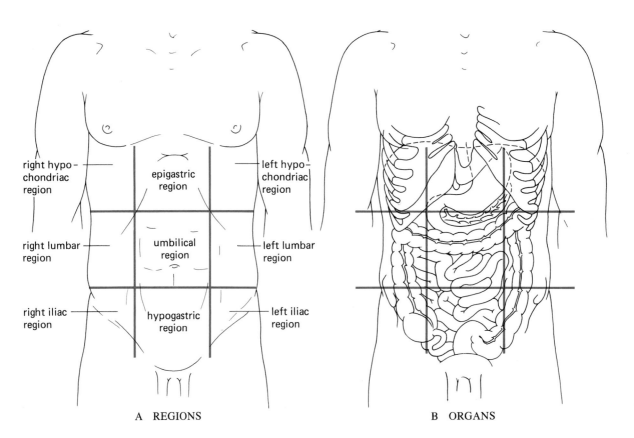

A REGIONS

right hypo-chondriac region · epigastric region · left hypo-chondriac region
right lumbar region · umbilical region · left lumbar region
right iliac region · hypogastric region · left iliac region

B ORGANS

KEY TERMS

anatomy	**external environment**	**standard anatomical position**
physiology	**intracellular fluid**	**cranial cavity**
cell	**extracellular fluid**	**vertebral canal**
tissue	**internal environment**	**thoracic cavity**
organ	**plasma**	**abdominal cavity**
organ system	**interstitial fluid**	

REVIEW EXERCISES

1 Describe the overall relation between the activities of the organ systems in the body and the survival of individual cells.

2 Distinguish between the external and internal environments.

3 Describe the relation between the internal environment and the survival of individual cells in a multicellular organism.

4 List the three fluid compartments of the body and the percent of the total body water located in each compartment. Which of these fluids comprise the internal environment?

5 Describe the two general categories of functions performed by all cells in a multicellular organism.

6 List the four categories of specialized cells found in the body and the type of specialized function performed by representative cells in each category.

7 Define the terms tissue, histology, and cytology.

8 Distinguish the following levels of structural organization in the body: tissue, functional unit, organ, and organ system.

9 List the 10 organ systems in the body and identify the major functions of each.

10 Describe the standard anatomical position.

11 Describe the orientation of the sagittal, coronal, and transverse planes of the body.

12 Define the following directions in relation to the body: posterior, anterior, cranial, caudal, proximal, distal, medial, and lateral.

13 Identify the following regions on the surface of the body: axillary, antecubital, inguinal, gluteal, cranial, brachial, femoral, lumbar, cervical, sacral, and patellar.

14 Describe the location of the subdivisions of the following body cavities and identify the organs located within each of them: dorsal cavity, thoracic cavity, and abdominal cavity.

15 Identify the location of the four quadrants and nine regions of the abdominal cavity and its overlying body surface.

FIG. 1-12 (at left) (A) Regions of the abdominal surface, and (B) the internal organs underlying these regions.

CHAPTER

2

CELL STRUCTURE AND CHEMICAL COMPOSITION

CELL STRUCTURE

Microscopic Observations of Cells
Cell Organelles

CHEMICAL COMPOSITION OF CELLS

Atoms and Molecules
Water
Organic Molecules

CELL STRUCTURE

One of the unifying generalizations of biology is that certain fundamental activities that represent the minimal requirements for maintaining the life of a cell are common to all cells. A human liver cell and an ameba are remarkably similar in their means of exchanging materials with their immediate environments, obtaining energy from organic nutrients, synthesizing complex molecules, and duplicating themselves. This is not to say that there are no significant differences between an ameba and a liver cell or between a muscle and a nerve cell. These differences in cell function, however, represent a specialized development of one or more of the general properties common to all the trillions of cells in the human body. These specializations in cell function occurred during the evolution of single cells into the coordinated society of cells which make up the human body.

This chapter and Chaps. 3 to 5 describe those elements of cell structure and function that most cells in the body have in common. Subsequent chapters will describe the properties of specialized cells and their organization into tissues and organs.

Microscopic Observations of Cells

When a living cell is examined under a light microscope, numerous granules and particles, some of which appear to be long filamentous structures, can be seen. Many of these oscillate back and forth, as if floating in a fluid medium. This medium is water, which accounts for about 80 percent of the cell's weight. Such an abundance of water, which is highly transparent to light, makes a cell almost invisible under an ordinary light microscope, and many devices are used to make the cellular structure more visible. One is the application to the cell of dyes that combine with various cell structures, staining them so that they become visible.

Even with staining techniques, however, only a portion of the cell's structure can be seen through a light microscope. The smallest object the naked eye can resolve is about 0.1 mm in diameter. A light microscope can magnify an object about 2000 times, producing an image of a 10-μm (Table 2-1) cell that is about 20 mm in diameter. But even at this magnification many cell structures are too small to be seen. For example, the outer membrane surrounding the cell is about 7.5 nm thick; after a magnifi-

TABLE 2-1   ENGLISH AND METRIC UNITS

	English	Metric
Length	1 foot = 0.305 meter 1 inch = 2.54 centimeters	1 meter (m) = 39.37 inches 1 centimeter (cm) = 1/100 meter 1 millimeter (mm) = 1/1000 meter 1 micrometer (μm) = 1/1000 millimeter 1 nanometer (nm) = 1/1000 micrometer *(1 angstrom (Å) = 1/10 nanometer)
Mass	†1 pound = 433.59 grams 1 ounce = 27.1 grams	1 kilogram (kg) = 1,000 grams = 2.2 pounds 1 gram (g) = 0.037 ounce 1 milligram (mg) = 1/1000 gram 1 microgram (μg) = 1/1000 milligram 1 nanogram (ng) = 1/1000 microgram 1 picogram (pg) = 1/1000 nanogram
Volume	1 gallon = 3.785 liters 1 quart = 0.946 liter	1 liter (L) = 1000 cubic centimeters = 0.264 gallon 1 liter = 1.057 quarts 1 milliliter (mL) = 1/1000 liter 1 microliter (μL) = 1/1000 milliliter

*The angstrom unit of length is not a true metric unit but has been included because of its frequent use, until recently, in the measurement of molecular dimensions.
†A pound is a unit of force not mass. When we write 1 kg = 2.2 pounds, this means that one kilogram of *mass* will have a *weight* under standard conditions of gravity at the Earth's surface of 2.2 pounds *force*.

cation of 2000 times it would be only 0.015 nm thick and still invisible under the light microscope. The great diversity of structures within the cell began to be discovered only after the development of the *electron microscope* in the late 1940s.

Cell Organelles

Although a few cells are essentially spherical in shape and some are long and cylindrical, the majority of cells are box-shaped as a result of being surrounded by other cells.

The interior of the cell, far from being a semiuniform medium, is structurally divided into a number of compartments known as **cell organelles** (little organs), which have different chemical compositions and carry out specific functions. Figure 2-1 is a diagrammatic view of a cell, its organelles, and its association with adjacent cells.

The outer surface of a cell is covered by a very thin structure known as the **plasma membrane.** It is this membrane that separates the intracellular from the extracellular fluid. The plasma membrane, acting as a selective barrier, regulates the flow of molecules into and out of the cell. This membrane, as well as the membranes that surround the cell organelles, is very flexible, much like a piece of cloth. It can be bent and folded, but it cannot be stretched without being torn. The plasma membrane is often invaginated (folded) into the cell, forming clefts, or extended from the cell surface in fingerlike projections (Fig. 2-1).

The interior of a cell is composed of two large regions: the **nucleus,** a spherical or oval body generally near the center of the cell, and the **cytoplasm,** the region outside the nucleus. Most of the cell organelles are in the cytoplasm, suspended in the intracellular fluid known as the **cytosol.**

The nucleus (Fig. 2-2) is the largest structure in the cell. It is surrounded by a barrier, the **nuclear envelope,** which consists of two membranes separated by a small space. At regular intervals along the surface of the nuclear envelope, the two nuclear membranes become joined, forming the rims of circular openings known as **nuclear pores.** These pores provide access to and from the nucleus for large molecules that otherwise could not cross the membranes of the nuclear envelope. The interior of the nucleus consists of granular and filamentous elements in various states of aggregation. The most

prominent structure in the nucleus is the densely staining **nucleolus,** a highly coiled filamentous structure associated with numerous granules but not surrounded by a membrane. Fibrous threads, known as **chromatin,** are distributed throughout the remainder of the nucleus. These chromatin threads, which carry genetic information from parent to offspring and from parent cell to daughter cell, are coiled to a greater or less degree, producing a variation in the granular density of the nucleus. All human cells contain a single nucleus during some stage of their life cycle. However, a few specialized types of cells, such as the red blood cell, lose their nucleus during the process of cell differentiation, while other cells formed by the fusion of many single-nucleated cells come to have more than one nucleus, as is the case with skeletal muscle cells.

The most extensive cytoplasmic cell organelle is the system of membranes that forms the **endoplasmic reticulum** (Fig. 2-3). Like the nuclear envelope, the endoplasmic reticulum consists of two opposing membranes separated by a small space. These membranes form a series of relatively flat sheets that are distributed throughout the cytoplasm and interconnect with each other. The small space between the membranes of the reticulum appears to be continuous throughout this membranous network and with the space between the two nuclear membranes. Two types of endoplasmic reticulum can be distinguished: *granular* (rough-surfaced) and *agranular* (smooth-surfaced) endoplasmic reticulum. The granular endoplasmic reticulum has small particles, **ribosomes,** attached to its membrane surface (see Figs. 2-1 and 2-5). Other ribosomal particles, *free ribosomes,* occur in the cytoplasm as free particles not attached to membranes. The agranular endoplasmic reticulum (Fig. 2-3) has no ribosomal particles on its surface and is more fragmented in appearance, being less likely to exist as extended sheets of membranes. Both granular and agranular endoplasmic reticulum can exist in the same cell, but the relative amounts of the two types vary in different cells and even in the same cell with changes in cell activity.

The ribosomes are the sites at which proteins are synthesized. (Proteins are large molecules that perform a multitude of structural and functional roles in living organisms. Their structure and functions will be discussed later in this and the following chapter.) When the ribosomes are attached to the

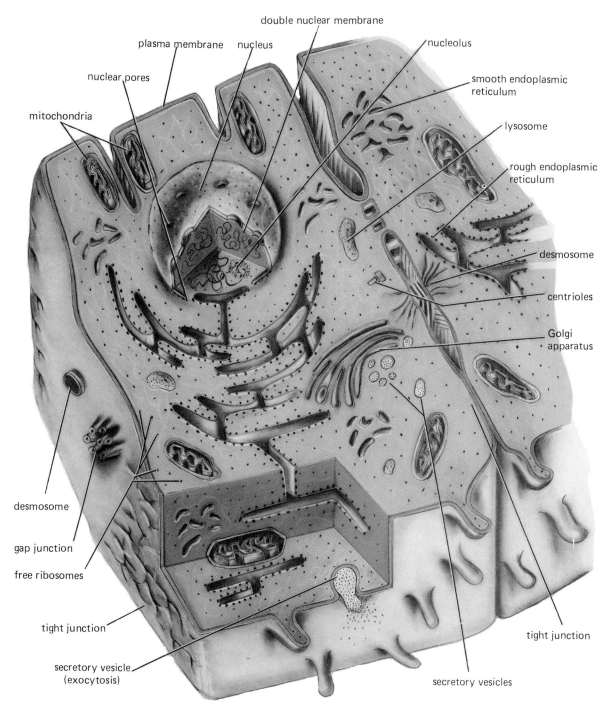

FIG. 2-1 Diagram of the structures and organelles found in most animal cells.

nuclear envelope
(double membrane)   nuclear pore   nucleolus

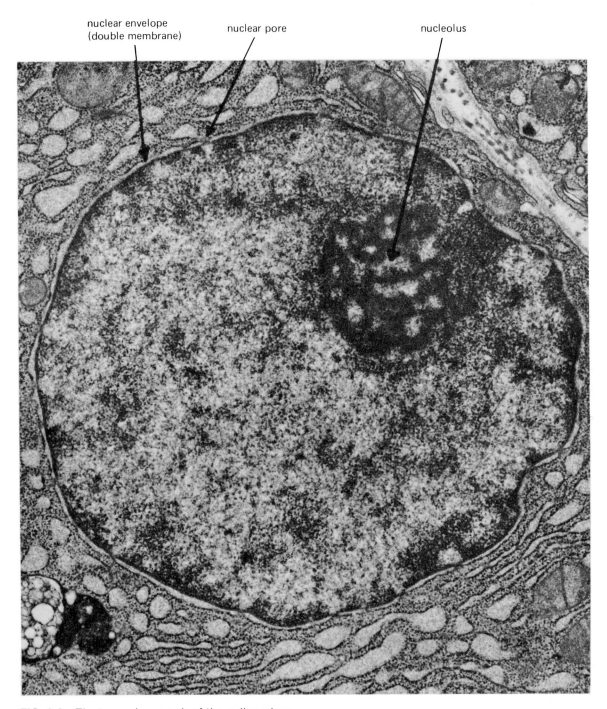

FIG. 2-2 Electron micrograph of the cell nucleus.

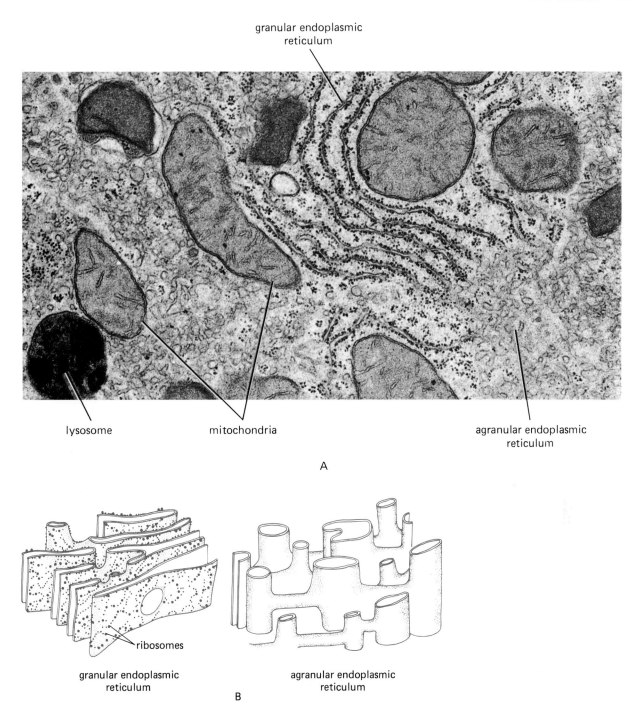

granular endoplasmic
reticulum

lysosome   mitochondria   agranular endoplasmic
reticulum

A

ribosomes

granular endoplasmic   agranular endoplasmic
reticulum   reticulum

B

FIG. 2-3 (A) Electron micrograph of a portion of the cytoplasm of a liver
cell, showing both granular and agranular endoplasmic reticulum, mitochondria, and lysosomes. (B) Three-dimensional shapes of the two types of
endoplasmic reticulum.

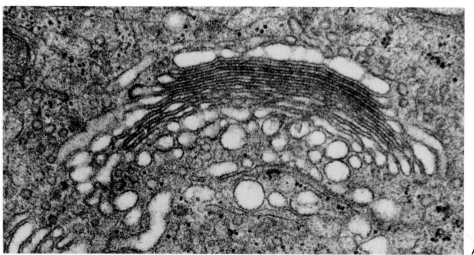

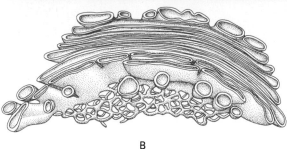

B

FIG. 2-4 (A) Electron micrograph and (B) three-dimensional shape of the Golgi apparatus.

endoplasmic reticulum, the synthesized protein is released into the lumen (space) between the reticulum membranes. From this point the protein is eventually released to the exterior of the cell in the process of protein secretion. Proteins synthesized on free ribosomes are released into the cytosol for internal use by the cell.

The **Golgi apparatus** (Fig. 2-4) is named in honor of Camillio Golgi, who first described this cell organelle. It consists of a series of closely opposed, flattened membranous sacs that are slightly curved, forming a cup-shaped organelle. Associated with this organelle, particularly near its concave surface, are a number of membrane-enclosed vesicles. These vesicles, which are produced by the Golgi apparatus, move to the periphery of the cell where they fuse with the plasma membrane and empty their contents to the outside of the cell during certain types of cell secretion. The vesicles are therefore referred to as either **secretory granules** (when they contain densely staining material) or **secretory vesicles** (when enclosing less dense material). Most cells have a single Golgi apparatus located near the nucleus, although some cells may have several.

The **mitochondria** (Greek *mitos,* thread; *chondros,* granule) are the cell organelles that provide the major source of chemical energy utilized by cells in performing their various functions. These organelles, which are usually rod- or oval-shaped (Fig. 2-5), are surrounded by two membranes, an inner membrane and an outer membrane. The outer membrane is smooth, whereas the inner membrane is folded into sheets or tubules, known as *cristae,* that extend into the inner space *(matrix)* of the mitochondrion. The mitochondria are found scattered throughout the cytoplasm. Large numbers of them are present in cells that utilize large amounts of energy; for example, a single liver cell may contain 1000 mitochondria. Less active cells contain fewer mitochondria.

Lysosomes (Fig. 2-3) are small, spherical or oval bodies, surrounded by a single membrane that encloses a densely staining, granular matrix. Lysosomes often cannot be distinguished structurally from the secretory granules derived from the Golgi apparatus and are in fact probably formed in the Golgi region of the cell. In contrast to the secretory vesicles that export material from the cell, the ly-

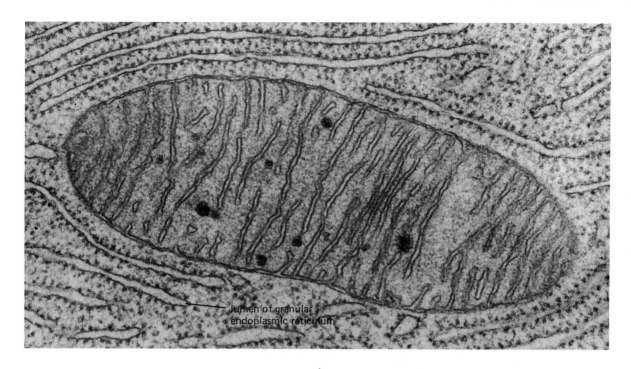

lumen of granular
endoplasmic reticulum

A

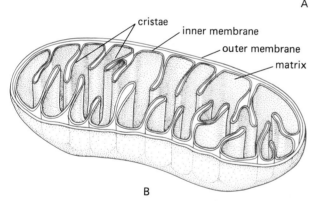

cristae
inner membrane
outer membrane
matrix

B

FIG. 2-5 (A) Electron micrograph of a mitochondrion surrounded by an extensive array of granular endoplasmic reticulum. (B) Three-dimensional shape of a mitochondrion.

In addition to these membranous organelles, several other structures are found in the cytoplasm. Most cells contain rodlike **filaments** and hollow **microtubules.** The microtubules appear to be more rigid than the filaments and may provide structural support (a *cytoskeleton*) for maintaining various cell shapes. Both filaments and microtubules have been implicated in cell processes that involve movement, whether it be movement of whole cells, the movement of organelles within cells, or the specialized movements of cell division. The **centrioles** (Fig. 2-6), microtubular structures involved in the process of cell division, consist of two small cylindrical bodies generally located near the nucleus in the region that contains the Golgi apparatus. Finally, the least specialized of the cytoplasmic structures are the fat droplets and various granules that are aggregates or crystals of particular chemical compounds.

Table 2-2 summarizes the major cell structures and their functions. Later chapters will describe the mechanisms by which the cell structures perform these functions.

sosomes function intracellularly to break down various complex structures, such as bacteria and cellular debris, that have been engulfed by the cell or, in some cases, to break down other intracellular organelles that have been damaged and are no longer functioning normally. The lysosomes are thus a highly specialized intracellular digestive system.

The cytoplasmic organelles we have described thus far—the endoplasmic reticulum, Golgi apparatus, secretory granules, mitochondria, and lysosomes—all have one structural element in common: They are surrounded by membranes. Membranes are the major structural elements in cells, and we will discuss their structure and function in Chap. 4.

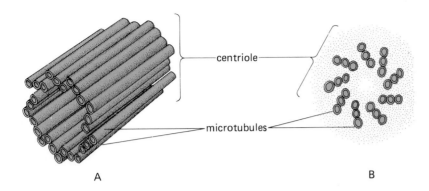

FIG. 2-6 (A) The microtubular structure of a single centriole. (B) Cross section of a centriole.

TABLE 2-2   CELL STRUCTURES*

Structure	Number per Cell	Structural Organization	Function
PLASMA MEMBRANE	1 at cell surface	7.5 nm thick. Composed of a phospholipid bilayer and protein.	Selective barrier to movement of ions and molecules into and out of cell.
NUCLEUS			
Nuclear envelope	1 surrounds nucleus	Two opposed membranes separating small space. Nuclear pores, 50–70 nm diameter.	Barrier to movement of most molecules. Messenger RNA passes to cytoplasm through pores.
Chromatin	46 strands per human cell nucleus	Coiled threads composed of DNA and protein.	DNA stores genetic information determining amino acid sequence of cell proteins. Chromatin condenses into chromosomes at time of cell division.
Nucleolus	Usually 1	Coiled filamentous structure associated with granules. Not surrounded by membrane.	Site of ribosomal RNA synthesis.
CYTOPLASM A Membrane-bound cell organelles			
Endoplasmic reticulum (ER)	1 interconnected cell organelle	Two opposed membranes separating a space continuous throughout organelle and interconnecting with space of nuclear envelope.	
Granular ER		Ribosomal particles bound to ER membrane.	Synthesis of proteins to be secreted from cell.
Agranular ER		Smooth membrane; no ribosomes.	Fatty acid and steroid synthesis. Calcium storage and release in muscle cells.
Golgi apparatus	Usually 1, near nucleus	Cup-shaped series of closely opposed membranous sacs and vesicles.	Concentration and modification of protein prior to secretion.

TABLE 2-2   CELL STRUCTURES continued

Structure	Number per Cell	Structural Organization	Function
Secretory vesicles (granules)	Many	Membrane-bound sacs containing concentrated solution of proteins.	Protein secretion.
Mitochondria	Many	Rod- or oval-shaped bodies surrounded by two membranes. Inner membrane folds into inner matrix, forming cristae.	Major site of ATP production, oxygen utilization, and CO_2 formation. Contains enzymes of Krebs cycle and oxidative phosphorylation.
Lysosomes	Several	Densely staining oval body surrounded by membrane, containing hydrolytic enzymes.	Digestive organelle, specialized for breakdown of engulfed bacteria and damaged cell organelles.
B Nonmembranous structures			
Ribosomes	Many	20 nm–diameter particles composed of RNA and protein.	Site at which amino acids are assembled into proteins.
Free ribosomes		Not bound to membranes.	Site at which proteins to be used intracellularly are assembled.
Bound ribosomes		Bound to membranes of granular endoplasmic reticulum.	Site at which proteins to be secreted from cell are assembled.
Filaments	Many	5–15 nm–diameter protein threads of variable length.	Cell movements, especially in muscle cells. Also used to provide structural support at cell junctions.
Microtubules	Many	25 nm–diameter protein tubules with 15 nm–diameter hollow core.	Maintenance of cell shape (cytoskeleton). Associated with movements of cilia, flagella, and mitotic spindle.
Centrioles	2, near nucleus	Two small cylindrical bodies composed of nine sets of three fused microtubules.	Formation of spindle apparatus at poles of cell during cell division. Also associated with formation and movement of cilia.
Granules	Few to many	Aggregates or crystals of chemical substances.	Storage of specialized end products of metabolism. Glycogen granules most common.
Fat droplets	Few	Spherical globule of triacylglycerol.	Storage of fat.

*Much of the information in this table will be described in later chapters.

CHEMICAL COMPOSITION OF CELLS

Atoms and molecules are the chemical units of cell structure and function just as the cells are the structural and functional units of multicellular organisms. In this section we describe the structures of the major classes of chemicals in the body and certain characteristics of atoms that underlie molecular structure in general. The specific functions and reactions of the various classes of molecules will be discussed in subsequent chapters. Thus, this section is, in essence, an expanded glossary of chemical terms and structures, and like a glossary it should be consulted according to need.

Atoms and Molecules

Atoms are the smallest units of matter that singly or in combination with other atoms form all substances. Over 100 different types of atoms have been identified. Matter composed of identical types of atoms is known as a **chemical element.** For example, oxygen, carbon, and iron are pure chemical elements, each of which is composed of many small units of matter, i.e., oxygen atoms, carbon atoms, and iron atoms, respectively. Each atom is approximately spherical, the smallest (hydrogen) being about 0.1 nm in diameter.

A one- or two-letter symbol is used as a shorthand identification of each chemical element, for example, O for oxygen, C for carbon, and Fe for iron (Table 2-3). Hydrogen, being the smallest atom, is the lightest atom. Carbon is a larger atom and is about 12 times heavier than a hydrogen atom. The **atomic weight** of an atom is a relative weight that indicates the weight of an atom relative to other atoms. Thus the atomic weight of carbon is 12, which indicates that this atom is 12 times heavier than the smallest atom, hydrogen, which has an atomic weight of 1.

Only 24 of the chemical elements are known to play essential roles in the human body (Table 2-3). In fact, just four elements, hydrogen, carbon, nitrogen, and oxygen, account for 96 percent of the body weight and about 99 percent of the atoms in the body. These four *major elements* combine to form water and organic molecules. Most of the seven *mineral elements* (Table 2-3) are found either in crystallized solid structures, such as teeth and bone, or as charged particles, *ions,* that are dis-

TABLE 2-3 ESSENTIAL ELEMENTS IN THE BODY

Symbol	Element	Atomic Weight
Major elements: 99.3% total atoms		
H	Hydrogen	1
O	Oxygen	16
C	Carbon	12
N	Nitrogen	14
Major minerals: 0.7% total atoms		
Ca	Calcium	40
P	Phosphorus	31
K (Latin, *kalium*)	Potassium	39
S	Sulfur	32
Na (Latin, *natrium*)	Sodium	23
Cl	Chlorine	35
Mg	Magnesium	24
Trace elements: less than 0.01% total atoms		
Fe (Latin, *ferrum*)	Iron	56
I	Iodine	127
Cu (Latin, *cuprum*)	Copper	64
Zn	Zinc	65
Mn	Manganese	55
Co	Cobalt	59
Cr	Chromium	52
Se	Selenium	79
Mo	Molybdenum	96
F	Fluorine	19
Sn (Latin, *stannum*)	Tin	119
Si	Silicon	28
V	Vanadium	51

solved in the body fluids. The 13 *trace elements* are present in very small quantities but perform specific chemical functions essential for the normal growth and functions of the body.

Atoms are linked together by **chemical bonds** to form **molecules.** Some pairs of atoms are linked by not one but two chemical bonds, forming what is known as a *double bond.* One type of chemical bond, a **covalent bond,** is formed by the electric forces associated with the sharing of two negatively charged subatomic particles, called *electrons,* one provided by each atom. In some cases, rather than sharing electrons to form a neutral covalent chemical bond, one atom will completely transfer an electron to another atom, forming two separate electrically charged atoms known as **ions.** Ions have a net positive charge if the electrically neutral atom has lost an electron, as in the case of the sodium (Na^+) and potassium (K^+) ions, or a net negative charge if the atom has gained an electron, as in the case of the chloride (Cl^-) ion.

The formation of ions as a result of gaining or losing electrons is known as **ionization.** Mineral elements are not the only atoms that can form ions. The most commonly encountered groupings of atoms that undergo ionization in molecules are the **carboxyl group** (R—COOH) and the **amino group** (R—NH$_2$), where R stands for the remainder of the molecule. The hydrogen-oxygen bond in the carboxyl group ionizes to form a carboxyl ion and a hydrogen ion:

$$\text{R—COOH} \rightleftharpoons \text{R—COO}^- + \text{H}^+$$

| Carboyxl group | Carboxyl ion | Hydrogen ion |

The amino group is able to combine with a hydrogen ion to form an ionized amino group:

$$\text{R—NH}_2 + \text{H}^+ \rightleftharpoons \text{R—NH}_3{}^+$$

| Amino group | Ionized amino group |

The ionization of carboxyl and amino groups provides some molecules with a net electric charge.

Note that hydrogen ions were involved in the ionization of both carboxyl and amino groups. The **hydrogen ion** plays a central role in the chemistry of the body as a result of its participation in ion formation. The **acidity** of a solution is a measure of the concentration of hydrogen ions present; the higher the hydrogen-ion concentration, the greater the acidity. Molecules that release hydrogen ions are known as **acids.** For example, acetic acid (CH_3COOH) is an acid since ionization of its carboxyl group releases a hydrogen ion (and an acetate ion, CH_3COO^-). The ionized amino group, $R—NH_3{}^+$, is also an acid, since it can release a hydrogen ion (to form the neutral R—NH$_2$).

In addition to forming neutral chemical bonds and ions, atoms can interact to form **polarized chemical bonds.** Such polarized bonds are formed when the negative electrons that form the chemical bond reside closer to one atom (which becomes slightly negative) than to the other atom (which becomes slightly positive). The atoms remain bonded together but the chemical bond is electrically polarized. The most commonly encountered polarized chemical bond occurs between hydrogen and oxy-

gen in the **hydroxyl group** (R—OH), in which the oxygen atom is slightly negative and the hydrogen atom slightly positive. Polarized chemical bonds play important roles in determining the solubility of molecules and their interactions with other molecules.

In summary, atoms chemically interact through the sharing or transfer of electrons to form chemical bonds or ions. The resulting molecules may be completely uncharged or may have regions that are electrically polarized or completely ionized.

Water

Water is the most abundant molecule in the body; 99 out of every 100 molecules is a water molecule, and 60 percent of the total body weight is water. As a first approximation, cells can be viewed as water-filled sacs in which various ions and molecules are dissolved.

The primary role of water in the body is to provide a fluid medium in which chemical reactions can occur. The two chemical bonds between hydrogen and oxygen in the water molecule (H—O—H) are polarized bonds, with the oxygen atom being slightly negative and the hydrogens slightly positive. Water molecules are electrically attracted to each other through these polarized bonds (Fig. 2-7). A molecule that dissolves in a liquid is a **solute,** and the liquid in which it dissolves is the **solvent.** Water is the solvent for the solutes in the body. The high solubility of many types of molecules in water is a result of the interactions between the electrically polarized bonds of the water molecules and the charged regions of the dissolved molecules. For example, table salt ($NaCl$) dissolves in water because the polar water molecules are attracted to the charged sodium and chloride ions, forming clusters of water molecules around each ion. These water molecules decrease the electric attraction between the sodium and chloride ions that had held them together in the crystalline state (Fig. 2-7). Thus, in order for a molecule to dissolve in water, its molecular structure must contain a certain number of polar or ionized groups that can associate with the polar water molecules. Conversely, molecules that have few, if any, polar or ionized groups are insoluble in water. Most molecules found in the body are soluble in water, with the exception of one general class of compounds known as **lipids.** The chem-

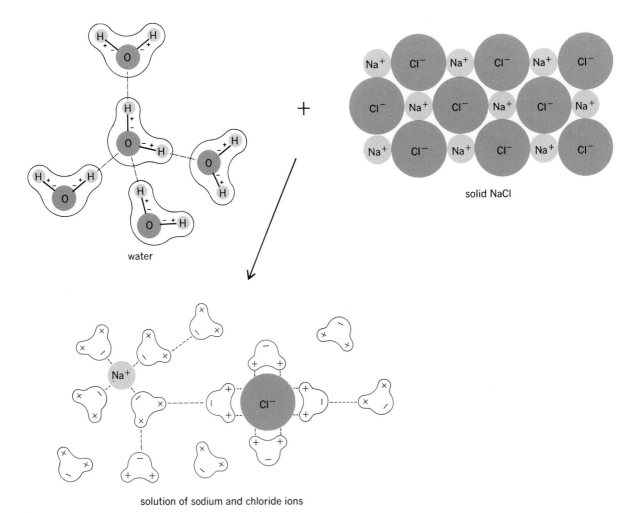

water

+

solid NaCl

solution of sodium and chloride ions

FIG. 2-7 The ability of water to dissolve sodium chloride crystals depends upon the electric attraction between the polar chemical bonds of the water molecules and the charged sodium and chloride ions. (Each positive region in a water molecule represents a hydrogen atom; each negative region, an oxygen atom.)

ical bonds of lipid molecules are minimally polarized or ionized and thus will not interact with water.

Some molecules contain polar or ionized groups at one end of the molecule and a nonpolar region at the opposite end; such molecules are called **amphipathic.** When mixed with water, amphipathic molecules form spherical clusters known as **micelles** in which the polar regions of the molecules are located at the surface of the micelle where they are attracted to the surrounding water molecules, and the nonpolar ends are oriented toward the center of the micelle. As we shall see, this orientation of amphipathic molecules in aqueous (water) environ-

ments plays an important role in the structure of cell membranes and the transport of nonpolar molecules in the blood.

Since many of the substances in the body are dissolved in water and their chemical interactions depend upon the number of molecules of various kinds that are present in solution, we need to define units that will describe the amount of a particular substance present in solution.

Concentration is defined as the amount of material present per unit of volume. The standard unit of volume in the metric system is a *liter* (L). Smaller units are the *milliliter* (mL, 0.001 liter) and the *mi-*

croliter (μL, 0.001 milliliter) (Table 2-1). One way of expressing the amount of a chemical compound is by its mass in *grams* (g); its concentration then is expressed as the number of grams of compound present in a liter of solution.

A comparison of the concentrations of two different types of compounds on the basis of the number of grams per liter of solution does not, however, directly indicate how many molecules of each compound are present. For example, 10 g of compound X, whose molecules are heavier than those of compound Y, will contain fewer molecules than 10 g of compound Y. Thus, we need a unit that describes the *number* of molecules of a substance present in solution. The unit used to express the number of particles of a particular substance is known as a **mol.** The number of mols of a particular substance that are present is given by the following equation:

$$\text{Number of mols} = \frac{\text{weight of substance in grams}}{\text{molecular weight of substance}}$$

(If the substance is not a molecule, but a pure chemical element, the weight of the element in grams is divided by the element's atomic weight.)

The **molecular weight** is equal to the sum of the atomic weights of all the atoms in the molecule. For example, as you can see from the atomic weights of the elements listed in Table 2-3, glucose ($C_6H_{12}O_6$) has a molecular weight of 180 (6-C × 12 + 12-H × 1 + 6-O × 16 = 180). Thus, 180 g of glucose equals 1 mol of glucose. Water (H_2O) has a molecular weight of 18 (2-H × 1 + 1-O × 16 = 18); therefore, 18 g of water is equal to 1 mol of water. Note that since one molecule of glucose is 10 times heavier than one molecule of water, 180 g of glucose contains the same number of molecules as 18 g of water. To generalize, 1 mol of any substance contains the same number of molecules; (the actual number of molecules in 1 mol is 6 × 10^{23}). Clearly, 1 L of solution that contains 180 g of glucose will have a glucose concentration of 1 mol per liter; this is called a *1 molar glucose solution.* Such a solution will contain the same number of solute molecules as a 1 molar solution of any other type of solute.

Organic Molecules

The chemistry of living organisms is centered on the chemistry of the *carbon atom.* The carbon atom is able to form four separate bonds with other atoms, such as oxygen, nitrogen, or hydrogen, but most importantly, with other atoms of carbon. The ability of the carbon atom to form chemical bonds with other carbon atoms permits the formation of an unlimited variety of molecules by the linkage of more and more carbon atoms. When carbon atoms are linked together, the molecule grows in three dimensions. Although we draw diagrammatic structures of molecules on flat sheets of paper, these molecules have a three-dimensional shape. Within limits, these molecules are flexible and can change their shape without breaking their bonds because the bonds linking the carbon atoms behave like an axle about which the atoms can rotate.

Most of the thousands of **organic molecules** (those that contain carbon) in the body can be classified into four groups: carbohydrates, lipids, proteins, and nucleic acids (Table 2-4).

Proteins   The term **protein** comes from the Greek *proteios* (of the first rank), which aptly describes their importance. Proteins, which account for about 50 percent of the organic material in the body, are components of most of the body structures and play critical roles in almost all the chemical transformations that occur in the body. Proteins are extremely large molecules, formed by the linking together of much simpler molecular subunits, the **amino acids.** There are 20 different amino acids, all of which have one structural feature in common: The terminal carbon atom is attached to both an amino group (—NH_2) and a carboxyl group (—COOH) (Fig. 2-8). The amino group of one amino acid reacts with the carboxyl group of the next amino acid, splitting off a molecule of water and forming a chemical bond, known as a **peptide bond,** between the two amino acids. Thus, proteins

TABLE 2-4 CHEMICAL COMPOSITION OF THE BODY	
Category	Percent of Body Weight
Water	60
Proteins	17
Lipids	15
Minerals (Na^+, K^+, Cl^-, Ca^{2+}, Mg^{2+}, etc.)	5
Carbohydrates	2
Nucleic acids	1

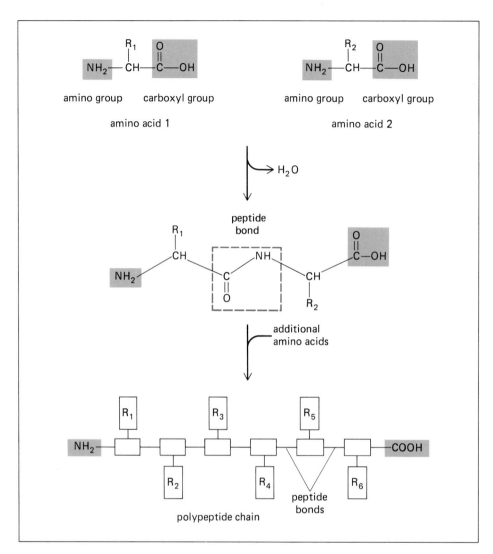

FIG. 2-8 Primary structure of a protein molecule: a sequence of amino acids linked together by peptide bonds between the carboxyl group of one amino acid and the amino group of the next amino acid in the sequence to form a polypeptide chain.

are linear sequences of amino acids (i.e., **polypeptide chains**) linked together by peptide bonds. The remainder of the amino acid, referred to as the amino acid *side chain,* has a different structure for each of the 20 amino acids. Some of these side chains contain ionized and polarized chemical bonds, whereas others are electrically neutral.

The sequence of amino acids along a polypeptide chain constitutes the *primary structure* of the protein molecule. There are two variables in this primary structure: (1) the total number of amino acids in the sequence and (2) the specific type of amino acid at each position along the sequence. If the

number of amino acids in the polypeptide chain is less than about 50, the molecule is known as a **peptide;** if the sequence is more than 50 amino acids long, the polypeptide is known as a protein. Since there are 20 different amino acids, there are 20 different types of side chains that can branch off the protein backbone, although all 20 amino acids need not be present in any given protein.

Thus, starting with 20 different amino acids, an almost unlimited variety of protein molecules can be constructed by simply rearranging the sequence and altering the total number of amino acids in the sequence. If we consider a peptide consisting of a sequence of three amino acids and assume that only two different types of amino acids, A and B, occur in the peptide, then $2 \times 2 \times 2 = 8$ different molecules can be formed, each having a different amino acid sequence. If any of the 20 different amino acids normally found in proteins occupy any of the three different positions in this peptide the number of possible molecules that can be formed becomes $20 \times 20 \times 20 = 8000$. If we consider peptides that are six amino acids long, we find that $20^6 = 64,000,000$ different molecules can be formed. But even this number is infinitesimal when we consider that many proteins have sequences of several hundred amino acids. We shall see examples later of how changing a single amino acid in a sequence of several hundred can completely alter the function of a protein molecule.

Thus far we have described the structure of a protein as a linear sequence of amino acids, which would produce a molecular structure analogous to a long piece of rope. Since segments of the protein chain can rotate about their chemical bonds, a variety of three-dimensional shapes can be formed, just as a long piece of flexible rope can be twisted into many types of configurations (Fig. 2-9). The three-dimensional shape of any given protein (its *conformation*) is primarily determined by the types and locations of the polarized and ionized amino acid side chains, since these will electrically interact with each other, causing the protein chain to fold into a specific shape. Thus, some proteins are rod-shaped, and others are coiled into almost spherical globules. The three-dimensional shape of a protein plays an important role in its ability to interact with other molecules having a complementary shape.

Lipids   The second major category of organic molecules is the **lipids** (Greek *lipos,* fat), which make up about 40 percent of the organic matter in the body. The lipids are defined primarily by a single property: their insolubility in water. The lipids are insoluble because their chemical structure, composed mostly of carbon and hydrogen atoms, contains few polar chemical bonds or ionized groups that can interact with polar water molecules. The lipids can be divided into three subclasses on the basis of their chemical structure: the triacylglycerols, the phospholipids, and the steroids.

The **triacylglycerols** constitute the largest subclass of lipids in the body and are generally referred

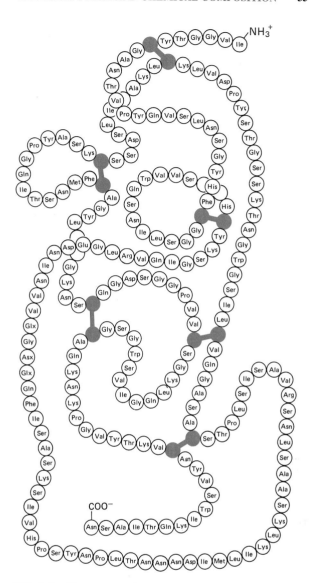

FIG. 2-9 Diagrammatic representation of a sequence of amino acids (circles) that form the polypeptide chain of the protein trypsin. The bars cross-linking various segments of the polypeptide chain represent chemical bonds between the side chains of particular amino acids.

to simply as *fat*. The primary function of fat is to provide a store of chemical fuel that can be used by cells to provide energy. Two chemical units make up the structure of this lipid: **fatty acids** and **glycerol.** Fatty acids consist of a sequence of carbon atoms, most commonly 16 or 18 carbons, combined with hydrogen and have an acidic carboxyl group at one end (Fig. 2-10). Although the carboxyl group can ionize and thus associate with water, the long

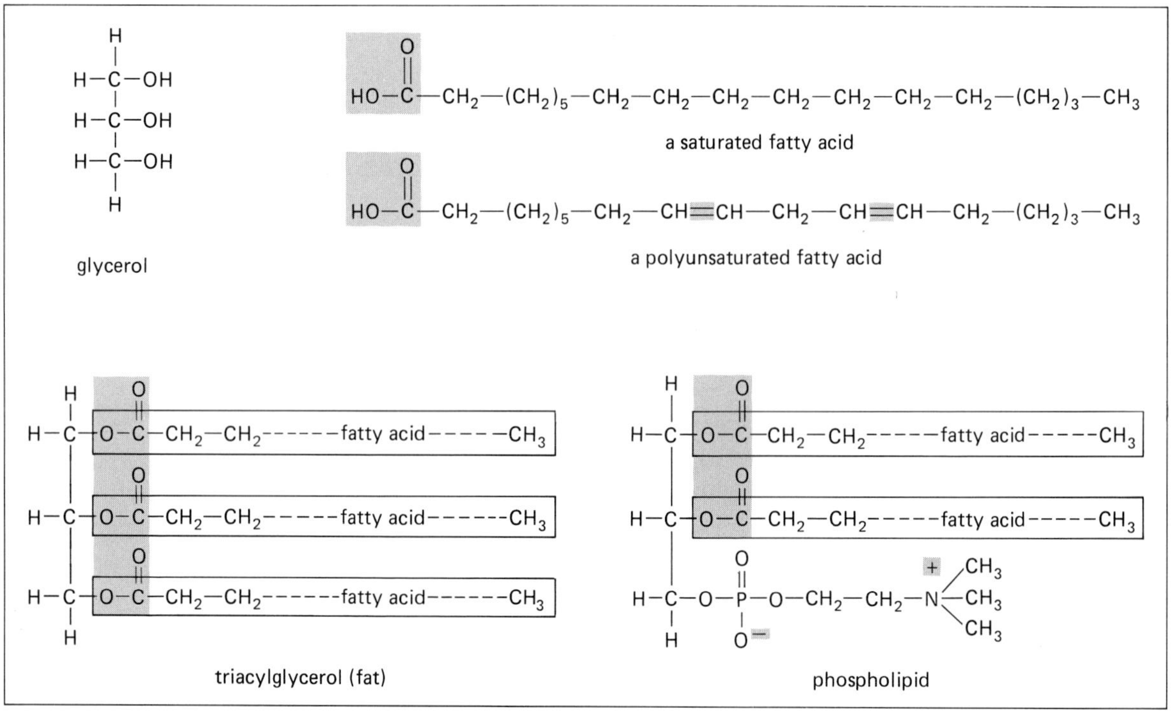

FIG. 2-10 Glycerol and fatty acids are the major subunits that combine to form triacylglycerol (fat) and phospholipids.

hydrocarbon chain, which constitutes the bulk of the fatty acid, cannot because it contains no polar or ionized groups. Some fatty acids contain one or more double bonds (two chemical bonds linking the same two atoms) at various positions in their hydrocarbon chain (Fig. 2-10) and are said to be *unsaturated* (not saturated with hydrogen atoms). If more than one double bond is present they are known as *polyunsaturated fatty acids*. There are many different fatty acids in the body, varying in number of carbon atoms (length of the fatty acid chain) and number and location of double bonds.

The second unit of a triacylglycerol is glycerol (Fig. 2-10), a three-carbon molecule that is soluble in water because of the three polar hydroxyl groups it contains. Thus, glycerol, by itself, is not a lipid and, in fact, belongs to the category of organic molecules known as carbohydrates. A molecule of fat is formed by linking the carboxyl group of three fatty acids to the three hydroxyl groups of glycerol. The result is a molecule that no longer contains any chemical groups that can interact with water and is therefore insoluble (Fig. 2-10).

The second subclass of lipids, the **phospholipids,** are similar in structure to fat (Fig. 2-10). However, these molecules contain only two fatty acids attached to glycerol, the third carbon on glycerol being linked to a phosphate group ($—PO_4^-$) to which, in turn, is usually attached a small polar or ionized nitrogen-containing compound. Thus, unlike the totally nonpolar fat molecule, a phospholipid contains several charged or polar groups. The combination of a polar region at one end of the molecule and nonpolar fatty acid chains at the other end makes these molecules amphipathic. Most of the phospholipids in the body are incorporated into the structure of cell membranes and are responsible for many of the membranes' special properties.

The third class of lipids, the **steroids,** is quite different structurally from the fats and phospholipids (Fig. 2-11). Four interconnected rings of carbon atoms form the basic structure of all steroids. Attached to this ring structure are small chemical groups or short hydrocarbon chains. Few polar groups are present, and thus the steroids are insoluble in water. Cholesterol is a steroid, as are several

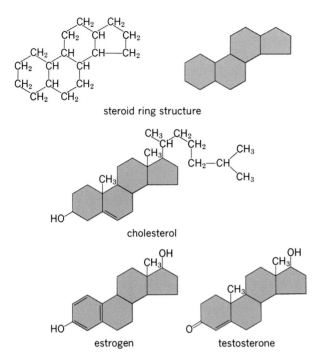

steroid ring structure

cholesterol

estrogen   testosterone

FIG. 2-11 Steroids, a subclass of lipids, are characterized by four interconnected rings of carbon atoms. Examples of specific steroids include cholesterol and the female and male sex hormones, estrogen and testosterone.

of the hormones, such as the male and female sex hormones, testosterone and estrogen.

Carbohydrates   The third category of organic molecules is the **carbohydrates.** These molecules account for only a small percentage of body weight, but they play a central role in the chemical reactions that provide cells with energy. Carbohydrates are water-soluble molecules composed of carbon, hydrogen, and oxygen in the proportions represented by the general formula $C_n(H_2O)_n$, where n is any whole number. The oxygen in carbohydrates is found primarily in the form of hydroxyl groups attached to the carbon atoms. These polar hydroxyl groups account for the water solubility of the carbohydrates.

The simplest carbohydrates are the **monosaccharides** (Greek *sakcharin,* sugar), most of which consist of five or six carbon atoms formed into a ring, closed by an atom of oxygen (Fig. 2-12). Hydroxyl groups and hydrogen atoms attached to the carbon atoms extend above or below the plane of the ring.

Different sugars may have the same number and type of atoms but differ in the orientation of their individual hydroxyl groups above or below the plane of the ring. The most important monosaccharide in the body is the six-carbon sugar **glucose** (Fig. 2-12). The chemical breakdown of glucose into carbon dioxide and water provides much of the energy utilized by cells.

Two sugars can be linked together to form a **disaccharide.** Common table sugar, sucrose, is a disaccharide composed of two sugars, glucose and fructose. When many sugars are linked together, the resulting molecule is known as a **polysaccharide.** Two similar polysaccharides, **starch** and **glycogen,** composed entirely of glucose, are found in the plant and animal kingdoms, respectively. Compared with its almost unlimited capacity for the storage of fat, the human body has only a limited ability to store carbohydrate, mainly in the form of glycogen. Plants, in contrast, store fuel mainly as starch whose structure is similar to glycogen, differing only in the number of branched chains and the length of these chains. Starch provides the high carbohydrate content of most foods of plant origin.

Nucleic Acids   The **nucleic acids** contribute very little to the body's weight, but these molecules are the largest and most specialized of all. They are responsible for the storage of genetic information and its passage from parent to offspring and from cell to cell. It is the nucleic acids that determine whether one is a man or a mouse, or whether a cell is a muscle cell or a nerve cell.

There are two types of nucleic acids: **deoxyribonucleic acid** (DNA) and **ribonucleic acid** (RNA). The

FIG. 2-12 Difference between two sugars, glucose and galactose, depends upon the position of the hydroxyl group on the fourth carbon atom. It is below the plane of the ring in glucose and above the ring in galactose.

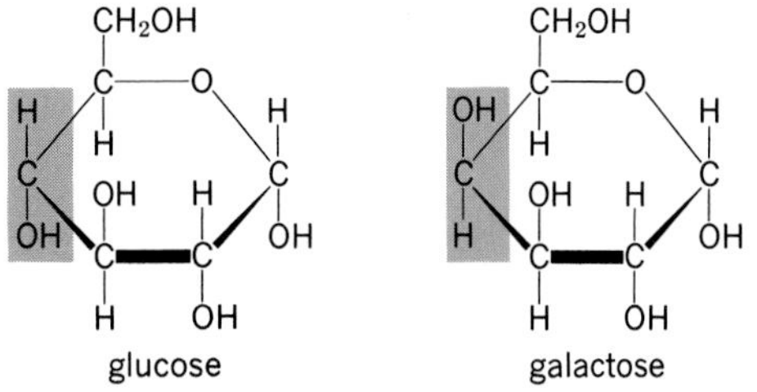

glucose   galactose

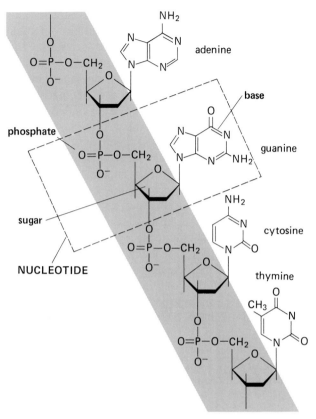

FIG. 2-13 Phosphate-sugar bonds (color) link nucleotides in sequence to form nucleic acids.

tide is linked to the sugar of the adjacent nucleotide to form a chain with the bases sticking out to the side of the phosphate-sugar backbone (Fig. 2-13).

A DNA molecule consists of not one but two chains of nucleotides coiled around each other in the form of a *double helix* (Fig. 2-14). Polar groups along the base on one chain are attracted to those of a base on the second chain to link the two chains of DNA. The structure of the four bases is such that adenine (A) always pairs with thymine (T) and guanine (G) with cytosine (C). As we shall see in the next chapter, this specificity in base pairings

FIG. 2-14 Base pairings between two nucleotide chains form the double-helical structure of DNA.

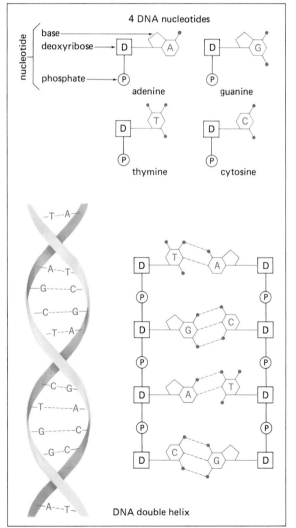

DNA molecules store genetic information coded within their structures, whereas RNA molecules are involved in the decoding of this information into a form that can be utilized by a cell, specifically, into instructions for protein synthesis.

Both types of nucleic acids are *polymers* (i.e., they are composed of linear sequences of repeating subunits). Each subunit, known as a **nucleotide,** has three components: a *phosphate group,* a *sugar,* and a ring of carbon and nitrogen atoms known as a *base* (Fig. 2-13). The nucleotides in DNA contain the sugar deoxyribose, thus the name deoxyribonucleic acid. Four different nucleotides are present in DNA, corresponding to the four different bases that may be attached to deoxyribose. The four bases are divided into two classes: (1) the *purine bases,* adenine and guanine, which have two rings of nitrogen and carbon atoms; and (2) the *pyrimidine bases,* cytosine and thymine, which have only a single ring. The phosphate group of one nucleo-

provides the mechanism for duplicating and transferring genetic information.

The general structure of RNA molecules differs in only a few respects from that of DNA: (1) RNA consists of a single, rather than a double chain of nucleotides. (2) The sugar in each nucleotide is ribose rather than deoxyribose. (3) The base thymine in DNA is replaced by the base uracil (U). Although RNA contains only a single chain of nucleotides, portions of this chain may bend back upon itself to undergo base pairings with other nucleotides in the same chain.

KEY TERMS

cell organelles
nucleus
cytoplasm
chemical bonds
ions
molar concentration
proteins

amino acids
lipids
triacylglycerols
phospholipids
steroids
fatty acids
carbohydrates

monosaccharides
polysaccharides
nucleic acids
nucleotides
deoxyribonucleic acid (DNA)
ribonucleic acid (RNA)

REVIEW EXERCISES

1   Describe some of the limits of using a light microscope to study cell structure.
2   Define and describe the location of the following: plasma membrane, nucleus, cytoplasm, and cytosol.
3   Describe the general structure of the following nuclear components: nuclear envelope, nucleolus, chromatin, and nuclear pores.
4   Describe the general structure of the following cytoplasmic cell organelles: rough and smooth endoplasmic reticulum, ribosomes, Golgi apparatus, mitochondria, and lysosomes.
5   Describe the types of structures that make up the cytoskeleton of a cell.
6   Define the following terms: atom, chemical element, and atomic weight.
7   Identify the four most abundant chemical elements in the body.
8   Describe the difference between an atom and a molecule.
9   Define the terms covalent chemical bond and ion.
10 Identify the electrical charge (positive or negative) that is present on the ionized amino group; on a carboxyl ion.
11 Identify the substance whose concentration in a solution determines the acidity of the solution.
12 Define and give an example of a polarized chemical bond.

13 Identify the most abundant molecule in the body.
14 Identify the types of chemical bonds that are present in a water molecule.
15 Define the terms solute and solvent.
16 Describe the characteristics of molecules that are soluble in water. Describe the characteristics of molecules that are insoluble in water.
17 Describe the characteristics of an amphipathic molecule.
18 Explain why knowing the molar concentration of a substance provides more information than just knowing the number of grams of the substance that are present in the solution.
19 Describe the subunits of protein structure.
20 Describe the two variables that determine the primary structure of a protein molecule.
21 Define the difference between a peptide and a protein.
22 Describe the factors that determine the conformation of a protein molecule.
23 Describe the general characteristics of the class of organic molecules known as lipids.
24 Describe the structure of the molecules that are commonly referred to as fat.
25 Describe the differences between the structure of a molecule of a triacylglycerol and a phospholipid molecule.

26   Describe the general structure of steroid molecules.

27   Identify the type of chemical groups that are most abundant in carbohydrate molecules.

28   Match the following carbohydrates–glucose, sucrose, and glycogen–with the general subclasses of carbohydrates to which they belong–monosaccharides, disaccharides, or polysaccharides.

29   Identify the subunits of nucleic acid polymers.

30   Describe the basic differences between DNA and RNA molecules.

31   Describe the distinguishing characteristics of the chemical structures of carbohydrates, lipids, proteins, and nucleic acids.

CHAPTER

PROPERTIES OF PROTEINS: DNA AND CELL METABOLISM

The outstanding accomplishment of twentieth-century biology has been the elucidation of the chemical basis of heredity and its relationship to protein synthesis. Whether an organism is a man or a mouse, has blue eyes or black, has light skin or dark is directly determined by the proteins it possesses. Moreover, within an individual organism, muscle cells differ from nerve cells or epithelial cells or any other cell types solely because they have different proteins. This is true not only because proteins function as structural elements in cells but also because they regulate the rates at which almost all cellular chemical reactions occur.

Crucial for an understanding of protein functions is the fact that each type of protein consists of a unique sequence of amino acids that confers upon it a unique three-dimensional shape. In Section A of this chapter we describe how these shapes underlie protein function.

So preeminent is the role of proteins in cellular function that the primary hereditary information passed from cell to cell is a set of specifications for the amino acid sequences of all proteins produced by the cell. This information is coded into the molecular structure of DNA. Sections B through D describe the nature of this code, how it is translated into protein synthesis, and how genetic information is duplicated and passed from cell to cell during the process of cell division. Section E describes the role of proteins in regulating the chemical reactions that occur in cells—cell metabolism.

SECTION A: PROTEIN BINDING SITES

CHARACTERISTICS OF PROTEIN BINDING SITES

Proteins differ from other classes of molecules in their extraordinary ability to bind organic molecules and ions selectively. This binding may be so specific that one type of protein may bind only one type of organic molecule and no other. Such selectivity allows a protein to "identify" (by binding) the presence of one particular type of molecule in a solution containing hundreds of different kinds of molecules.

A **ligand** is any molecule or ion that is bound to the surface of another molecule, such as a protein, by forces other than covalent chemical bonds.

These forces are either electrical attractions between oppositely charged (or polarized) groups on the two molecules or very weak attractions between the electric fields surrounding electrically neutral atoms in them. On the surface of a protein, the chemical groups of only a small region, known as a **binding site,** interact with any given ligand. A single protein may contain several different binding sites, each specific for different ligand types. The reversible binding of a ligand (L) to a protein binding site (P_b) to form a bound complex ($P_b \cdot L$) is written:

$$L + P_b \rightleftharpoons P_b \cdot L$$

The dot indicates the absence of a covalent bond between P_b and L.

A protein binding site with a specific shape can bind only those ligands having a complementary shape. Thus, the **chemical specificity** of a protein binding site is determined by the shape of the protein molecule (Fig. 3-1).

FIG. 3-1 Complementary shapes of ligand and protein binding site determine the chemical specificity of binding.

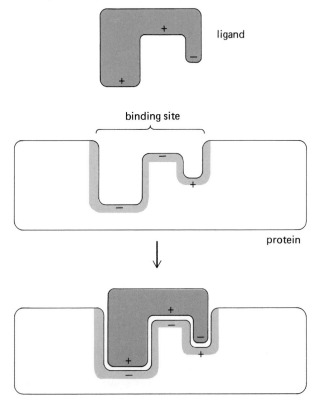

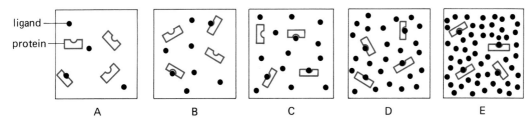

FIG. 3-2 Increase in ligand concentration increases the number of binding sites that are occupied (percent saturation). At 100 percent saturation, all the binding sites have been occupied and further increases in concentration of ligand will not increase the amount bound.

Specificity depends upon shape, whereas the strength of binding (**affinity**) depends upon both the number of interacting loci on the ligand and binding site and the strength of the electrical attraction at each of the loci.

At any one time, a single binding site on a protein is either occupied by a ligand or it is not. In a solution containing a number of identical proteins and complementary ligands, some of the sites may be occupied by ligands and others not. When all are occupied, the binding sites are said to be fully **saturated**—there are no unoccupied sites available to bind additional ligands. When half of the available sites are occupied, the system is 50 percent saturated, and so on (Fig. 3-2). The probability that a given binding site will be occupied by a ligand depends upon two factors: (1) the concentration of free ligand in the solution and (2) the affinity of the binding site for the ligand.

FUNCTIONS OF PROTEIN BINDING SITES

Having defined the physical characteristics of ligand binding to protein binding sites, we now ask what roles such binding reactions play in cell function. A detailed answer to this question is given in the remaining chapters of this book, since practically every cell function and its regulation is mediated by specific interactions between ligands and their corresponding binding sites on proteins. In this section, a few examples will suffice to illustrate general principles.

Enzymes

Substances that accelerate chemical reactions but do not themselves undergo any net rearrangement of their chemical structure during the reaction are known as **catalysts.** Proteins that catalyze chemical reactions in the body are called **enzymes.** The property of enzymes that enables them to accelerate only certain chemical reactions is the specificity of the enzyme's binding sites. The ligands that bind to these binding sites are known as **substrates,** and the molecules that are ultimately formed in the chemical reaction catalyzed by the enzyme are known as **products.**

Practically every chemical reaction in the body is catalyzed by a specific enzyme. Without enzymes these chemical reactions would proceed at a negligible rate. Therefore, the types of chemical reactions that occur in a particular cell depend upon which enzymes are present in that cell. Furthermore, particular types of reactions are localized to particular regions of the cell because the specific enzymes required for their catalysis are in those regions. For example, the chemical reactions that occur in the mitochondria are different from those that occur in the nucleus or in the endoplasmic reticulum because each of these cell organelles contains a different set of enzymes.

Certain enzymes are present in almost all cells, whether they be muscle, nerve, etc., and to that extent these cells are all similar (although not identical since the concentrations of these common enzymes may differ from cell to cell). In contrast, other enzymes may be absent from certain cells so that the reactions they catalyze occur so slowly in those cells as to be negligible. The crucial point is that cells differ from each other almost entirely because they contain different enzymes. Moreover, a given cell may undergo dramatic changes in its activity solely as a result of a change in its rate of enzyme production. Again we see that a cell is what it is because of the instructions it receives concerning the types of proteins (enzymes, in this case) it synthesizes and the rates at which it synthesizes them.

Detection of Chemical Messengers

In addition to holding molecules together and catalyzing the chemical reactions in cells, protein binding sites play a key role in the communications systems of the body. Chemical signals that are transmitted between cells or between one area of a cell and another are detected by specific binding sites for the chemical messengers. For example, all the cells are exposed to the same hormones (chemical messengers secreted by glands into the blood) reaching them by way of the circulatory system, but only those cells that contain binding sites for a particular hormone will respond to it. Thus, hormone binding sites provide the basis for the selective action of hormones.

There are many other types of chemical signals that depend on specific binding sites. For example, the activity of nerve cells depends on chemical agents that bind to specific binding sites on them. To take another example, chemical signals within a cell influence the activity of enzymes by binding to them at specific protein binding sites.

REGULATION OF THE CHARACTERISTICS OF BINDING SITES

Because a cell is what its binding sites make it, molecular control mechanisms center upon alteration of the number or effectiveness of these sites. One mechanism for achieving either more or fewer sites is to control the synthesis of the proteins that contain them. This process is discussed later in this chapter. The other mechanisms, the subject of this section, operate not by altering the number of proteins but rather by changing the characteristics of their binding sites.

There are a number of nonspecific factors in the environment of a protein that have the potential for altering the characteristics of its binding sites. Two of the most important are temperature and acidity, but normally these factors are not critical because they are maintained relatively constant in the body's fluids. (Moreover, should large changes occur, as in disease, marked alterations of binding sites can occur and produce life-threatening consequences because of the resulting alterations in the conformation of protein binding sites and the biological activities that are controlled by these sites).

Allosteric Regulation

When a ligand binds to the surface of a protein, the electrical forces between the ligand and the protein may produce a change in the shape of the protein in the region of the binding site. This may in turn alter the electrical interactions between other regions of the protein, thereby producing a change in the total conformation of the folded polypeptide chain, as illustrated in Fig. 3-3. Such changes in protein shape may cause new binding sites to appear at regions of the molecule where none existed previously, or they may modify the shapes of already existing binding sites. In the latter case, the change in shape may alter the chemical specificity of the site or its affinity for ligand. The alteration of one binding site as a result of the binding of a ligand to a second site on the same protein is known as **allosteric** (other site) regulation. The chemical messenger whose binding at one site (the regulator site) regulates the activity of another binding site on the same protein is known as a **modulator.**

FIG. 3-3 Allosteric modulation. Binding of a modulator ligand at the regulatory site alters the conformation of a second binding site.

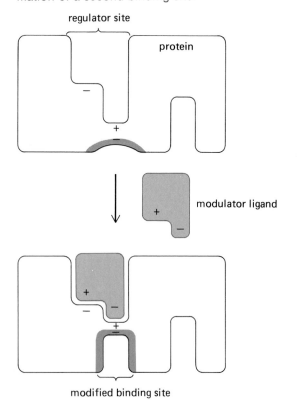

Covalent Modulation

Allosteric modulation involves changing the shape of protein molecules but not their chemical composition. In contrast, another means of altering the configuration of binding sites, **covalent modulation,** is to link various chemical groups (primarily phosphate) covalently to the amino acid side chains of the protein and, for example, convert a slightly polar side chain into a highly ionized one. The altered charge on the side chain interacts with other charged groups in the protein, altering the shape of the polypeptide chain and thereby altering its binding sites. The addition or removal of such chemical groups is normally mediated by specific enzymes.

SECTION B: GENETIC INFORMATION AND PROTEIN SYNTHESIS

The transmission of hereditary information from cell to cell is the function of deoxyribonucleic acid (DNA), which contains instructions for the synthesis of proteins coded into its molecular structure. The portion of DNA that contains the information required to determine the amino acid sequence of

FIG. 3-4 The genetic information in a cell is transcribed from DNA to RNA in the nucleus, followed by the translation of the RNA information into protein synthesis in the cytoplasm.

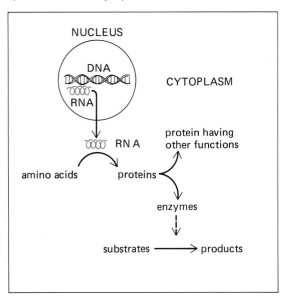

a single polypeptide chain is known as a **gene,** the unit of hereditary information. A single DNA molecule contains many genes.

Although DNA contains the information necessary for the synthesis of specific proteins, DNA does not itself participate directly in the assembly of proteins. The DNA of a cell is in the nucleus, whereas most protein synthesis occurs in the cytoplasm. The transfer of information from the DNA genes to the site of protein synthesis is the function of RNA (ribonucleic acid), synthesized on the surface of the DNA. This mechanism of expressing genetic information occurs in all living organisms and has led to what is now called the "central dogma" of molecular biology: In all organisms, genetic information flows from DNA to RNA and then to protein:

$$DNA \longrightarrow RNA \longrightarrow PROTEIN$$

Figure 3-4 summarizes the general pathway by which the information stored in DNA influences cell activity via the process of protein synthesis.

DNA AND THE GENETIC CODE

As described in Chap. 2, DNA consists of two polynucleotide chains coiled around each other to form a double helix. Each chain is a sequence of nucleotides joined by phosphate-sugar linkages. Each nucleotide contains one of four different bases—adenine (A), guanine (G), cytosine (C), or thymine (T). Each of these bases is specifically paired, A to T and G to C, with a base on the corresponding polynucleotide chain of the double helix (refer to Fig. 2-14). Thus, both chains contain a precisely ordered sequence of bases, one chain being complementary to the other. It is the sequence of these bases along the polynucleotide chains of the helix that provides the code that ultimately specifies the sequence of amino acids in proteins.

Since there are 20 different amino acids in proteins but only four different bases, a single base cannot be the code word for a single amino acid. Rather, a specific sequence of three bases, for example, CGG or ACT, serves as the code word, or **codon,** corresponding to a specific amino acid. Thus, each codon specifies a particular amino acid, and a sequence of codons along a single chain of the DNA double helix (a gene) specifies the se-

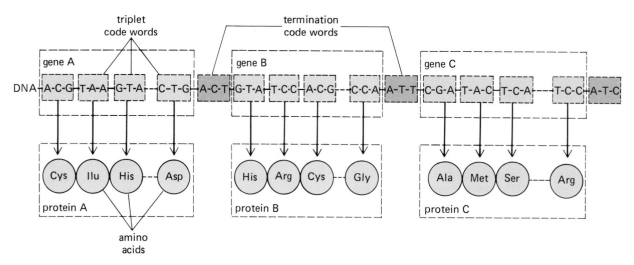

FIG. 3-5 Relation between the sequence of triplet code words (codons) in genes and the amino acid sequences of proteins that correspond to them. The names of the individual amino acids are abbreviated. Note that all three genes are part of the same DNA, which may contain hundreds of genes in a linear sequence.

quence of amino acids in the protein. A single molecule of DNA contains the genetic information corresponding to many different proteins arranged as a sequence of genes. Some of the three-base codons, rather than specifying a particular amino acid, act as punctuation marks, termination code words that indicate the points along the DNA chain where a gene begins and ends (Fig. 3-5).

Almost all of a cell's DNA is in the nucleus, where it exists as long, coiled threads. The DNA molecules are too large to pass through the nuclear pores into the cytoplasm. Since the actual assembly of a protein from individual amino acids occurs in the cytoplasm, a message containing the information specifying the sequence of amino acids must be carried from DNA to the cytoplasm. This message is in the form of RNA known as **messenger RNA** (mRNA). The transfer of genetic information from DNA into protein thus occurs in two stages: first, the genetic message is passed from DNA to mRNA (**transcription**); second, the message in mRNA is used to direct the assembly of the proper sequence of amino acids to form a protein (**translation**).

Messenger RNA is synthesized in the nucleus on the surface of DNA by the linking together of the appropriate sequence of nucleotides. Recall from Chap. 2 that RNA contains three of the same bases

as DNA but that its fourth base is uracil rather than thymine. The nucleotide sequence in mRNA is determined by base pairings between free nucleotides and corresponding nucleotides in one of the two chains of a DNA double helix. Thus, a free adenine nucleotide pairs with thymine in the DNA chain, C with G, U with A, and G with C, creating a sequence of bases in mRNA that is the mirror image of the base sequence in DNA. For example, when a codon sequence of bases in DNA is ATC, the corresponding codon sequence in mRNA is UAG.

PROTEIN ASSEMBLY

Once a molecule of mRNA has been synthesized, it leaves the nucleus through the nuclear pores and enters the cytoplasm where it becomes attached to a **ribosome,** composed of both protein and RNA. Free amino acids cannot bind directly to their corresponding codon in mRNA. Each amino acid must first be attached to a second type of RNA known as **transfer RNA** (tRNA), which contains in its nucleotide sequence a specific three-base sequence, known as an **anticodon,** that is able to base-pair to its mirror-image codon in a mRNA. There are several bases in tRNA that are not found in other RNAs, such as the base inosine (I). There is a different tRNA for each of the 20 different amino

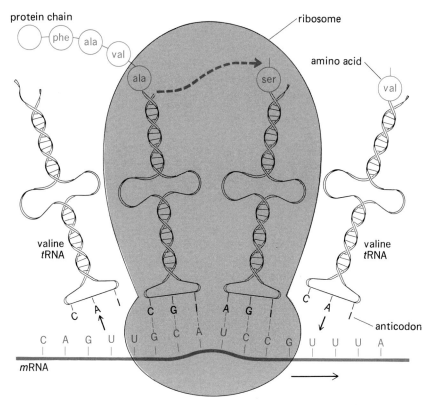

protein chain

ribosome

amino acid

FIG. 3-6 Sequence of events during the synthesis of a protein on the surface of a ribosome. The three-base anticodon of tRNA carrying one amino acid, e.g., valine (val), binds to the corresponding codon of mRNA and transfers its amino acid to the growing polypeptide chain. As the ribosome moves along the strand of mRNA, successive amino acids are added to the polypeptide chain.

acids, as well as a different cytoplasmic enzyme specific for catalyzing the linkage of each type of amino acid with its corresponding tRNA. Once a given amino acid has been attached to a tRNA containing the anticodon for that particular amino acid, the tRNA–amino acid complex base-pairs with the appropriate codon in mRNA on the surface of the ribosome (Fig. 3-6). This recognition system allows the amino acids to be linked in a sequence determined by the codon sequence in mRNA.

We have now identified all the major ingredients required for protein synthesis. A mRNA carrying the codon sequence for a particular protein becomes attached to a ribosomal particle. Each ribosome also has, in addition to a binding site for mRNA, two binding sites for tRNA. One site holds the tRNA already attached to a growing chain of amino acids, and a second site holds the tRNA that contains the next amino acid to be added to the sequence (Fig. 3-6). Ribosomal enzymes catalyze the formation of a peptide bond between the last amino acid attached at one site and the amino acid linked to its tRNA at the adjacent site. The ribo-

some now moves a distance of one codon down the mRNA, releasing one tRNA and binding the next tRNA–amino acid complex. This process is repeated as each amino acid is added in succession to the growing peptide chain. When the ribosome reaches the end of the mRNA or the end of a coded sequence for a single protein, the completed protein is released from the surface of the ribosome. It takes about 1 min to synthesize a protein containing 100 amino acids, that is, 1 or 2 sec to add each amino acid. Table 3-1 summarizes the sequence of events leading from DNA to the completed synthesis of a protein.

SECTION C: REPLICATION OF GENETIC INFORMATION

The growth of a multicellular organism and the propagation of the species from generation to generation depend on the fundamental property of *cell division*. When a cell divides, identical copies of

TABLE 3-1   SEQUENCE OF EVENTS LEADING FROM DNA TO PROTEIN SYNTHESIS

Transcription

1 The two strands of the DNA double helix separate in the region of the gene to be transcribed.
2 Free nucleotides base-pair with the nucleotide bases in DNA.
3 The nucleotides paired with one strand of DNA are linked by an enzyme to form mRNA containing a sequence of bases complementary to the DNA base sequence.

Translation

4 mRNA passes from the nucleus to the cytoplasm where one end of the mRNA binds to a ribosome.
5 Free amino acids are linked to their corresponding tRNAs by enzymes in the cytoplasm.
6 An amino acid–tRNA complex binds to a site on the ribosome and the three base anticodons in tRNA pair with the corresponding codons in mRNA.
7 Each amino acid is then transferred from its tRNA to the growing peptide chain, which is attached to the adjacent tRNA.
8 The tRNA freed of its amino acid is released from the ribosome.
9 mRNA moves one condon step along the ribosome.
10 Steps 6 to 9 are repeated over and over.
11 The completed protein chain is released from the ribosome when the termination codon in mRNA is reached.

the genetic information stored in DNA must be formed and passed on to the daughter cells. As cells become highly specialized through the process of differentiation, they often lose their capacity for division. Such specialized cells can be replaced only by the division of less specialized, undifferentiated cells, followed by differentiation into the specialized cell type, or, as is the case with nerve cells, they cannot be replaced at all.

REPLICATION OF DNA

The initial event leading to cell division is the replication of the DNA followed by the distribution of these replicates to each of the two new daughter cells. DNA replication (Fig. 3-7) begins with the separation of the two base-paired strands of the DNA double helix. The exposed bases of the two

FIG. 3-7 Replication of DNA begins with the separation of the two strands of the DNA double helix, followed by the pairing of the bases of free nucleotides with the exposed bases of the old DNA strands, the process gives rise to two identical double-helical molecules of DNA, each of which contains one old (black) and one new (color) nucleotide strand.

NUCLEOTIDE

base
deoxyribose
phosphate
pyrophosphate

strands can now act as templates that bind the corresponding bases of free nucleotides. Enzymes catalyze the formation of bonds between successive free nucleotides to form the new strands. The result is the formation of two identical molecules of DNA, each containing one strand of nucleotides that was formerly present in the original DNA and one new strand formed by the nucleotides that have base-paired with the old strand.

Mistakes, although rare, can occur during the process of DNA replication. Any alteration in the genetic message carried by DNA is known as a **mutation.** If, for example, a nucleotide containing the base guanine (G) is inserted at a point normally occupied by adenine (A) in the codon sequence CAT, the three-base codon at that site will be altered to CGT. Since the new codon may specify a different amino acid from that originally coded, the protein synthesized by instructions from the mutated gene may have a different chemical activity in the cell. A number of factors in the environment increase the rate at which genes mutate. These factors, known as **mutagens,** include certain chemical substances and various forms of ionizing radiation, such as x-rays and atomic radiation. Most of these agents cause the breakage of chemical bonds, which allows the substitution of new bases in the DNA sequence.

Mutation is the mechanism that underlies the evolution of living organisms. If an organism carrying a mutant gene is able to perform some function more efficiently than an organism lacking such a gene, it has a better chance of surviving and passing the mutant gene on to its descendants (this is the principle of *natural selection*). Although some mutations may enable a cell to function more efficiently in a given environment, the majority of mutations lead to modifications in protein structure that result in less effective functioning or even the death of the cell carrying the mutant gene. Medical science is beginning to identify a number of human diseases that are the result of abnormal enzymes produced by mutant genes. Such diseases have been termed *inborn errors of metabolism.*

CELL DIVISION

Rapidly growing cells divide about once every 24 hr, whereas other cells may go for weeks or months before undergoing division, and highly specialized cells may not divide at all. Although the general morphologic changes in cell structure that occur during division have been studied for the past hundred years, little is known about the chemical events that initiate and regulate this process.

A reproductively active cell passes through three periods: interphase, mitosis, and cytokinesis. The first period, **interphase,** is by far the longest and represents the time between cell divisions. Most of the cells in the body are in the interphase state, and the major bodily functions are carried out by cells in interphase. Once the chemical events that initiate cell division have begun at the end of interphase, the physical process of splitting the cell in two lasts only about 1 to 2 hr. By this time one very important event related to cell division has occurred, namely, the replication of DNA, a process that may last 10 to 12 hr.

The next phase, **mitosis** (Greek *mitos,* thread), involves the division of the cell nucleus with the passage of the replicated chromosomes to the two daughter cells. Mitosis is then followed by **cytokinesis,** the division of the cytoplasm of the whole cell into two daughter cells, each containing a replica of the DNA of the parent cell.

The interphase nucleus contains 46 long chromatin threads composed of DNA and protein. Half of these chromatin threads were derived from the female parent and half from the male parent at the time of conception, and the genetic information they contain has been passed on from cell to cell at each subsequent cell division. Each of the 23 chromatin threads from one parent carries a complete set of genes. Thus, each cell contains two genes for every protein in the cell, one gene from each parent.

Mitosis

Mitosis begins at the end of interphase when the DNA in the 46 chromatin threads replicates, but the duplicate copies of each chromatin thread (called *chromatids*) remain joined together at one point, the *centromere*. These duplicated chromatin threads start to coil up, forming 46 highly condensed **chromosomes** (Fig. 3-8).

While the chromatin threads are condensing in the nucleus, other events are taking place in the cytoplasm in preparation for cell division. The two microtubular centrioles, located just outside the nu-

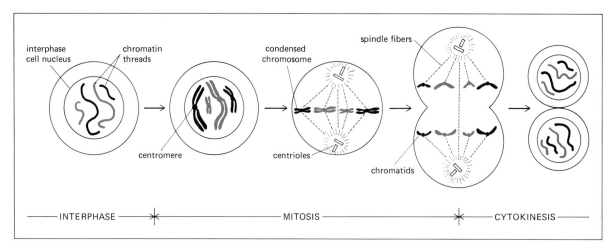

FIG. 3-8 Sequence of events during cell division. (Only 4 of the 46 chromosomes in a human cell are illustrated.)

cleus, replicate to form four centrioles. The two sets of centrioles then begin migrating to opposite sides of the nucleus. During this migration, a number of new microtubules are formed, the **spindle apparatus,** which link the two sets of centrioles together.

By the time the centrioles have completed their migration, the chromosomes have condensed, and the nuclear envelope begins to disintegrate. At this point, the centromere of the chromosomes attaches to the fibers of the spindle apparatus, causing the chromosomes to become aligned in a plane midway between the two sets of centrioles. The centromeres now split, each part holding one of the two duplicates of each chromosome, and move toward the opposing centrioles along the spindle fibers. This separation of the chromatids provides each daughter cell with an identical set of DNA molecules. A special type of nuclear division, known as *meiosis,* occurs in the germ cells which give rise to ova and sperm and will be described in the chapter on reproduction.

Cytokinesis

The movement of the chromatids to the opposite poles of the cell marks the end of mitosis, and the cell now undergoes **cytokinesis.** The surface of the cell begins to constrict along a plane perpendicular to the spindle apparatus. This constriction continues until the cell has been pinched in half, forming two separate daughter cells, The condensed chro-

matids in each of the daughter nuclei now uncoil, forming the extended interphase chromatin threads. The spindle apparatus disintegrates, and the nuclear envelope re-forms.

Most of the cell organelles are distributed randomly between the two daughter cells during cytokinesis. As the daughter cells grow, new membrane material is synthesized, probably by the endoplasmic reticulum, to form the membranes of the nuclear envelope, the Golgi apparatus, and lysosomes. The mitochondria, on the other hand, appear to be able to duplicate themselves. Small amounts of DNA are found in these organelles, and morphologic evidence suggests that new mitochondria are formed by the growth of a membranous partition across an old mitochondrion, separating it into two new organelles.

RECOMBINANT DNA

In 1972 a bacterial enzyme known as a *restriction enzyme,* which splits DNA into a number of fragments in a unique manner, was discovered. Rather than splitting both strands of the double helix at the same site, this enzyme cuts the two strands at different locations, resulting in fragments that have exposed segments of single-stranded DNA at each end (Fig. 3-9). The enzyme acts only at locations that have a particular sequence of nucleotides, and the resulting single-stranded ends therefore have exposed sets of complementary bases. These ends

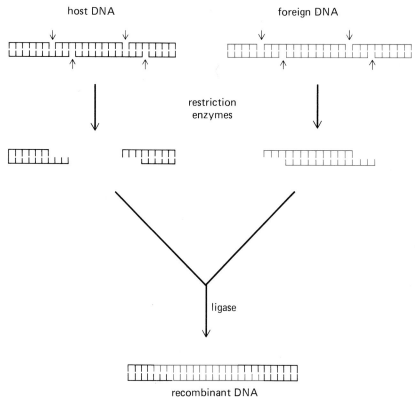

FIG. 3-9 Formation of recombinant DNA by the use of bacterial enzymes.

can undergo complementary base pairings with the ends of other DNA fragments produced in the same way. Therefore, when DNA from two different organisms is treated with restriction enzyme, the resulting fragments can be recombined through base pairings between their single stranded ends (the ends of the fragments are covalently linked by another enzyme known as a *ligase*) to form a new molecule of DNA containing a mixture of genes from the two organisms. When this **recombinant DNA** is then inserted into a living cell, both sets of genes may be transcribed into mRNA and translated into proteins.

The potential benefits and hazards of this technique are many. For example, splicing the gene that codes for human insulin onto a bacterial DNA and inserting it into a bacterium allows the organism to synthesize insulin. This hormone is then extracted from the bacterium and used to treat diabetic patients who are unable to synthesize this hormone. On the other hand, it would be equally possible, theoretically, to insert a gene that codes for a toxic protein into a harmless bacterium that commonly inhabits the human gastrointestinal tract and

thereby produce a widespread epidemic. Only the future will tell whether actual benefits will outweigh the hazards of being able to manipulate the genetic constitution of organisms.

SECTION D: TISSUE DEVELOPMENT

MECHANISMS OF TISSUE DEVELOPMENT

Cell Differentiation

The process by which a cell is transformed into a specialized cell type is **cell differentiation.** The general sequence of development is thus one of cell multiplication, followed by differentiation into the various types of specialized cells. Cells that have not undergone differentiation are known as *undifferentiated,* or *stem,* cells. Although most cells become differentiated during the period of embryonic and fetal development, some undifferentiated cells remain after birth and multiply in the various organs where, upon receiving an appropriate stimulus, they differentiate into a specialized cell type.

Beginning with the division of the fertilized egg, identical copies of the genetic information obtained from the male and female parent at the time of fertilization are distributed to each of the daughter cells of the growing embryo. Thus each cell contains an identical set of genes. How then is it possible for one cell to differentiate into a muscle cell that synthesizes muscle proteins, while another cell containing the same set of genes differentiates into a nerve cell that synthesizes the proteins characteristic of nerve cells? We must conclude that not all the genes in a given cell transcribe their genetic information into protein synthesis; otherwise, all cells would synthesize the same proteins and presumably would have the same characteristics. Thus, to understand how cells differentiate, we must discover the mechanisms that allow some of the genes in one type of cell to be "turned on" (i.e., form mRNA and the corresponding proteins), whereas in another type of cell these particular genes are "turned off" and another set "turned on."

In addition to those genes that are turned on or off in a particular type of differentiated cell, there are other genes that are active in all cells. Even these genes may be more active in one type of cell than another, in the sense that they may produce larger quantities of a particular enzyme in one cell than in another. Thus there is a quantitative as well as a qualitative control of gene expression associated with cell differentiation.

In some cases the process of cell differentiation appears to be programmed by chemical events in the cell and independent of influences from the extracellular environment, whereas in other cases cells may remain in an undifferentiated state until they receive a specific chemical signal from an extracellular source. It remains for future biologists to work out the detailed mechanisms that control these many aspects of genetic expression.

As a result of differential gene expression, cells acquire differences in their locomotion, adhesiveness, shape, and ability to divide. These differences account, in large part, for the patterns of cell organization into tissues and organs.

Cell Adhesion

Most of the cells in the body adhere to neighboring cells or to an extracellular fibrous matrix produced by connective-tissue cells. If cells did not adhere to each other, the body would collapse like a bag of dry sand. At least three classes of adhesive mechanisms hold cells together: (1) electrostatic attraction between charged surfaces of the cells, (2) specific proteins that cross-link the surfaces of cells, and (3) specialized cell junctions.

The weakest and least specific type of cell adhesion is electrostatic attraction. The phospholipids and proteins on the surface plasma membranes, along with various sugar residues that are attached to these molecules, contain ionized groups, primarily negatively charged carboxyl groups. Thus the surface of cells has a negative charge built into the structure of the plasma membrane. (This fixed electric charge should not be confused with the separation of sodium and potassium ions across plasma membranes that is responsible for membrane potentials.) Two negatively charged surfaces, when brought together, should repel each other, since like electric charges repel; however, two negatively charged surfaces can be linked if a positively charged material is placed between them. The calcium ion (Ca^{2+}) appears to play this role in some cases, since the removal of calcium from the extracellular medium causes some cells to separate from each other.

Proteins also play an essential role in linking certain cells. Proteins known as **aggregation factors,** when added to a population of separated cells, cause the cells to bind to each other. Furthermore, such proteins appear to be specific for the types of cells they will bind together, since different cells have different binding sites to which the aggregation factor can become attached. The strength of cell adhesion can be varied by varying the number of binding sites and thus the number of cross links that can be formed. Again, it appears to be the expression of certain genetic elements during cell differentiation that determines the types of aggregation factors that are formed and the number and types of binding sites that are present on a given cell's surface.

The third mechanism of binding cells together involves the specialized junctions that will be discussed in Chap. 4: *tight junctions, gap junctions,* and *desmosomes.* In addition to holding cells together, these junctions perform a number of other functions associated with the flow of molecules through the extracellular space between cells and from one cell to another.

Cell locomotion and the specificity of cell adhe-

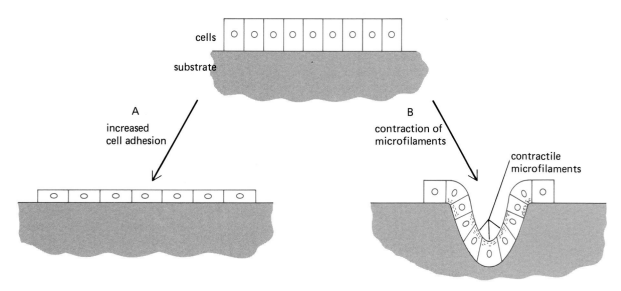

FIG. 3-10   Morphologic shaping of a layer of cuboidal cells as a result of (A) increased adhesion between the layer of cells and the underlying cells, which results in flattening of cells; and (B) contractile microfilaments at the base and top of the cell that produce conical-shaped cells that cause invaginations of the cell layer into the underlying tissue.

sion can explain a great deal of the patterned cellular organization that occurs during embryonic development. For example, when the cells of an embryo are dissociated and placed in tissue culture, the interactions of the cells can be followed over a period of days. Within 1 or 2 days the cells arrange themselves in various specific patterns with most of the cells of a similar cell type bound together in clusters. Aggregates of different cell types are arranged so that a cluster of one cell type surrounds a cluster of another type. This sorting out of cell types in tissue culture is a result of the specificity of cell adhesion as cells of various types randomly collide with each other as they move over the surface of the glass. Weakly adhering cells will break their contacts in favor of forming stronger adhesions until a population of strongly adhering cells comes to occupy the center of the cluster surrounded by more weakly adhering cells.

Cell Shape

Cell differentiation is usually accompanied by a change in the shape of the cell. Most differentiated cells can be readily identified because of their char-

acteristic shapes. Nerve cells have long branched processes, muscle cells are long cylinders, some cells are thin and flat, and others are cuboidal. Two general factors are responsible for the shape of a cell: (1) the attachments of the cell to other cells and extracellular materials, and (2) structural elements in the cell or associated with the plasma membrane.

Most cells, when suspended in a fluid medium, tend to become spherical; when they become attached to a surface, however, they become flattened, often with an irregular cell outline. When cells are surrounded by neighboring cells, the close association may give the cells a polyhedral shape. The more characteristic shapes of cells, however, are determined by structural elements in the cell itself. Microfilaments and microtubules appear to play a major role in maintaining specific cell shapes. The microtubules, which act like relatively stiff rods, are usually found in abundance in regions of the cell where processes extend from the cell surface, for example, the long cylindrical processes of nerve cells. Microfilaments, which appear to have a contractile function, are often found in bands underlying the plasma membrane and may even be attached to the membrane. Contraction of these filaments may alter the shape of the cell surface.

Figure 3-10 provides two examples of how changes in cell shape in turn contribute to the morphologic shaping of tissues and organs. A layer of cuboidal cells may become flattened, thus extending the layer of cells. Such flattening may arise because of a change in the adhesive properties of the cell surface for the underlying layer of cells, which leads to a greater area of surface contact between the two cell layers. The second example illustrates the role of microfilaments arranged in bands just below the plasma membrane at the two ends of the cells. Contraction of these microfilaments can produce a purse-string type of action, converting the cuboidal cell into a more cone-shaped cell, with the result that the layer of cells invaginates into a pocket.

Cell Growth

Obviously the whole process of development could not occur without the multiplication of cells. Like other cellular processes, cell division is regulated and does not occur at the same rate in all cells. Since the body does not continue to increase in size indefinitely, cells must either stop dividing or be destroyed as rapidly as they are replaced. Actually, both processes occur in the adult. Many types of cells, such as nerve and muscle cells, lose the ability to divide following differentiation; other cell types, including most epithelial, glandular, and connective-tissue cells, retain this ability. In general, cells that develop highly specialized internal structures lose the ability to divide; an example is muscle cells with their highly organized array of contractile filaments. In contrast, cells specialized for the secretion or absorption of materials use structures normally present in all cells and retain the ability to divide.

Growth of the embryo is the result of two processes: (1) an increase in the number of cells, **hyperplasia,** and (2) an increase in the size of individual cells, **hypertrophy.** By the time of birth many of the cells have differentiated into cell types that are no longer capable of division. This means that at birth the body contains all the cells of certain types that it will ever have; yet from birth to the adult state there is a considerable increase in the size of these tissues. Thus, this increase in size is due to the hypertrophy of individual cells and not to an increase in their number.

Both hyperplasia and hypertrophy are regulated processes, as can be illustrated by a few examples. The cells of the liver are capable of cell division; yet the liver stops growing when it reaches the adult size. However, if a large section of the liver is removed from an adult, the remaining liver tissue undergoes hyperplasia until the lost tissue has been replaced and the liver has returned to its original size. Muscle provides an example of the control of tissue hypertrophy. If a muscle is repeatedly exercised, its size increases as a result of an increase in the diameter of its individual muscle cells, the latter due to the synthesis of new contractile filaments. There is no increase in cell number because the adult muscle cell is incapable of division. If the exercise that leads to muscle hypertrophy is discontinued, the size of the muscle cells decreases as the contractile filaments are broken down.

The general concept that overall size depends upon the number of cells as well as the size of individual cells can also be applied to the development of the functional units in various organs. In some organs, the number of functional units is fixed by the time of birth. Throughout the remainder of life, the number of units will not increase but the size of individual units may increase, both by cell division and hypertrophy of the cells composing the unit. If a portion of the organ is damaged, the remaining portion may be able to maintain the organ's function by increasing the size of the remaining functional units. In other organ systems, particularly glands, the number of functional units is not fixed at birth, and new units can differentiate and become organized in the adult body.

CANCER

Now that we have examined some of the factors involved in the normal growth and organization of tissues, it is instructive to examine briefly the abnormal growth of cells that occurs in **cancer,** the uncontrolled growth of cells. Although the causes of cancer at the molecular level are still unknown, the weight of evidence points to an alteration in the mechanisms controlling the genetic elements of the cell associated with its motility, adhesion, shape, and division.

It is not the ability to divide or the rate of cell division that characterizes a cancer cell. Cancer arises when a cell no longer responds to the various regulating mechanisms that control its growth, with the result that the cancer cell undergoes unlimited growth, primarily by hyperplasia, and forms a growing mass of cells known as a **tumor.** Viruses, chemical agents, and radiation are all capable of inducing the transformation from a normal to a cancerous cell. All these agents have in common the ability to alter the genetic elements contained in DNA. Any cell in the body has the potential of being transformed into a cancer cell.

Cancer cells are less adhesive than corresponding normal cells. As a result, they tend to break away from the initial site of the growing tumor and travel through the bloodstream to become lodged in other tissues, where they grow into secondary tumors. This process of the spreading of cancer cells is known as **metastasis.** Because of this tendency to metastasize, it is important that the growth of cancerous cells be detected before they can spread to other tissues. At an early stage, the small nodule of growing cancer cells can often be completely removed by surgery, a process that is practically impossible once the cancer has spread to multiple organs in the body.

Cancer cells not only show alterations in their ability to adhere to other cells, they also lack *contact inhibition,* the process whereby normal cells stop moving when they come in contact with other cells. In tissue culture, cancer cells move over the tops of other cells and form multiple layers. This lack of contact inhibition increases the ability of cancer cells to invade surrounding tissues during the tumor's growth and metastasis.

SECTION E: CELL METABOLISM

Metabolism (Greek: change) refers to all chemical reactions that occur in cells or living organisms. Virtually all the molecules in most cells undergo a continuous transformation as some molecules are broken down while others of the same type are being synthesized. Chemically, no person is the same at noon as at 8 o'clock in the morning, because during this short period much of the body's structure has been torn apart and replaced with newly synthesized molecules. These transformations occur within a relatively short time compared with the total life-span of the cell. Therefore, the human body is in a dynamic chemical state. Those chemical reactions that result in the fragmentation of a molecule into smaller and smaller parts are **degradative** or **catabolic reactions,** and those reactions that put molecular fragments together to form larger molecules are **synthetic** or **anabolic reactions.** Accompanying most catabolic reactions there is a release of chemical energy; this in turn is used by the cell for the synthesis of new molecules and the performance of other energy-requiring functions such as the contraction of muscle.

ENERGY AND CHEMICAL REACTIONS

Chemical Energy

Energy can most simply be defined as the ability of a physical or chemical system to undergo change. Accordingly, energy is measured in terms of the magnitude of change that occurs when various forces are applied to matter. The *law of the conservation of energy* states that energy can be transferred from one system to another but cannot be created or destroyed. All physical and chemical change is the result of the transfer of energy from one system to another.

Since energy can neither be created nor destroyed, the difference in energy content between reactant molecules and product molecules must equal the amount of energy added or released during the reaction. The released chemical energy appears mainly in the form of heat, i.e., as increased molecular motion. Heat energy is measured in units known as **calories** (cal), one calorie being the amount of heat energy required to raise the temperature of one gram of water one degree Celsius. Energies associated with most chemical reactions are of the order of several thousand calories and are given as kilocalories (1 kcal = 1000 calories).

The formation of 1 mol of water from hydrogen and oxygen releases 68 kcal of heat energy; the full reaction is

$$\text{H---H} + \text{O} \rightleftharpoons \text{H---O---H} + 68 \text{ kcal/mol}$$

In other words, the chemical energy stored in 1 mol of water molecules is less by 68 kcal than the energy originally present in hydrogen and oxygen. Any

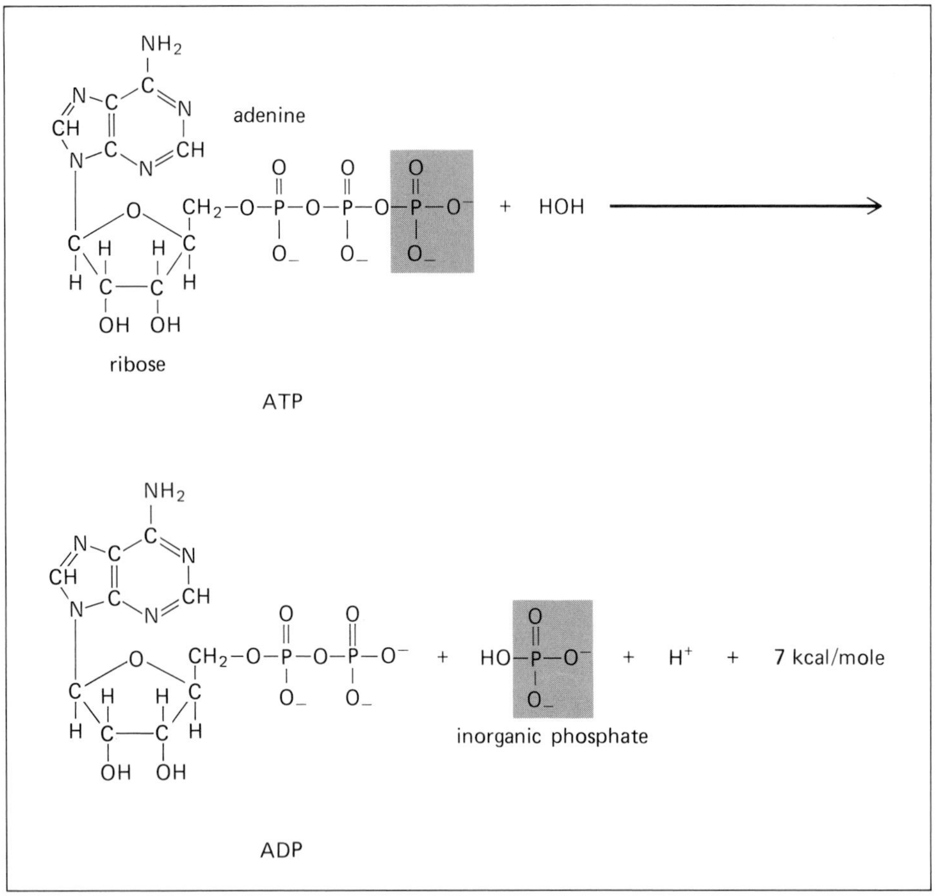

FIG. 3-11 Chemical structure of ATP. Its breakdown to ADP and inorganic phosphate is accompanied by the release of 7 kcal/mol of energy.

chemical reaction is, at least in theory, a **reversible chemical reaction** (the double arrow in the equation on p. 55 indicating the forward and reverse reaction), provided energy equal to the amount released when a reaction proceeds in one direction is put back into the reaction when it goes in the opposite direction.

ATP

When energy must be added to a chemical reaction, where does the energy come from? One source is heat energy. During the collisions that occur between moving molecules, the heat energy can be transferred into the energy of a chemical bond. However, for those reactions requiring a large input of energy, this would require a much higher temperature than the body temperature of 37°C (98.6°F). Therefore, in the body, the energy for

high-energy-requiring reactions must be obtained by coupling the reaction to a second reaction that releases a large amount of energy.

During catabolism of carbohydrates, lipids, and proteins, chemical energy associated with these molecules is released. About half of this energy appears directly as heat, and the remainder is coupled through a special set of chemical reactions to the synthesis of **adenosine triphosphate** (ATP). ATP (Fig. 3-11) has three components: *adenine, ribose* (a sugar), and three *phosphate* groups. ATP is synthesized by the addition of inorganic phosphate (P_i) to the terminal phosphate of ADP (*adenosine diphosphate*); 7 kcal/mol is required to form this bond. Conversely, 7 kcal/mol of energy is released when ATP breaks down into ADP and inorganic phosphate.

$$ATP + H_2O \rightleftharpoons ADP + P_i + 7 \text{ kcal/mol}$$

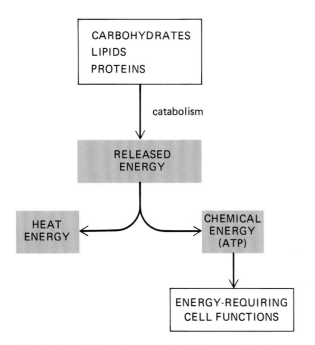

FIG. 3-12 ATP is the chemical intermediate through which about half of the energy released during the catabolism of carbohydrates, fats, and proteins flows to the energy-requiring processes in cells. The other half of the energy released appears as heat energy.

This energy is used universally by living organisms to carry out energy-requiring cellular functions, whether it be the contraction of muscle or the synthesis of a new molecule. The function of ATP is to transfer the energy released from the breakdown of carbohydrates, lipids, and proteins to the many different energy-requiring processes in the cell (Fig. 3-12).

Enzymes

As mentioned in Section A, chemical substances that accelerate the rate of a chemical reaction without themselves being chemically altered by the overall reaction are known as catalysts. The molecules that act as chemical catalysts in the body are the proteins known as enzymes. Although all enzymes are proteins, not all proteins act as enzymes.

The molecules acted upon by an enzyme are substrates and the molecules resulting from the reaction are products.

$$A + B \xrightarrow{\hspace{2cm}} C + D$$
Substrates   **Enzyme**   Products

After an enzyme has reacted with its substrates and released its products, it can combine with new substrate molecules and repeat the reaction. Thus, enzymes are not used up during the reaction and can be used over and over again. Therefore, only small quantities of a particular enzyme are required to transform large amounts of substrate into product.

When a substrate (S) binds to a binding site on an enzyme (E), a temporary enzyme-substrate complex (ES) is formed, which then breaks down into the enzyme and product (P) molecules.

$$E \; + \; S \rightleftharpoons ES \rightleftharpoons E \; + \; P$$
Enzyme   Substrate   Enzyme-   Enzyme   Product
substrate
complex

The binding of the substrate to the enzyme weakens particular chemical bonds in the substrate and is responsible for the accelerated rate at which the reaction occurs in the presence of an enzyme.

The characteristics of protein binding sites were discussed earlier. Thus, enzymes have the properties of: (1) *chemical specificity*—a given enzyme will combine with only certain types of substrate molecules whose shape is complementary to the shape of the enzyme's binding site; (2) *saturation*—the rate of an enzyme-catalyzed reaction will increase with increasing substrate concentration until all of the binding sites on the enzymes are occupied (100 percent saturation), and further increases in substrate concentration will not increase the rate of the reaction; (3) *regulation*—an enzyme's binding site can be altered by allosteric and covalent modulation, providing mechanisms for the regulation of enzyme activity.

There are almost as many different types of enzymes in a cell as there are chemical reactions. Each enzyme acts on only one or two chemical bonds in a particular type of molecule. The products of one enzyme-mediated reaction can become the substrates for another enzyme reaction in a sequence. By a sequence of small changes catalyzed by enzymes, a molecule can be transformed into a totally different chemical structure. The sequence of enzyme-catalyzed reactions leading from an initial major substrate to a major end product is a **metabolic pathway.** Thus, the total metabolism of the cell can be analyzed in terms of a number of separate and interrelated metabolic pathways.

Some types of enzymes are associated with particular cell organelles, either bound into the structure of the membrane surrounding the organelle or free in the internal compartment of the organelle. Therefore, the enzymes associated with different cell organelles serve to segregate different metabolic pathways to different regions of the cell.

SOURCES OF METABOLIC ENERGY

Carbohydrate Catabolism

The catabolism of glucose to carbon dioxide and water in the presence of molecular oxygen releases 686 kcal of energy per mol of glucose.

$$C_6H_{12}O_6 + 6O_2 \longrightarrow 6CO_2 + 6H_2O + 686 \text{ kcal/mol}$$

This overall reaction involves 18 separate enzyme-mediated steps, which are grouped into two sequences based upon the location of the enzymes, the requirement for oxygen, and the amount of ATP formed. The first ten reactions constitute the metabolic pathway known as **glycolysis** (Fig. 3-13), and the last eight are referred to as the **Krebs cycle** (Fig. 3-14) in honor of the biochemist Hans Krebs. Both glycolysis and the Krebs cycle illustrate the general principle applicable to all metabolic pathways, that each step in the metabolic sequence results in only a small modification of chemical structure, generally the breakage and formation of only one or two chemical bonds.

   Glycolysis   Glycolysis is mediated by soluble enzymes in the cytoplasm and not associated with any particular cell organelle. Furthermore, no oxygen is used during these reactions. A net synthesis of 2 mol of ATP per mol of glucose occurs during glycolysis. Since the synthesis of 1 mol of ATP "traps" 7 kcal of energy, a total of only 14 kcal (2

FIG. 3-13 End products formed from the breakdown of glucose through the glycolytic pathway.

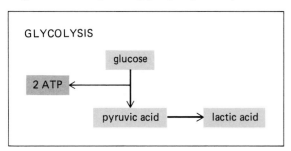

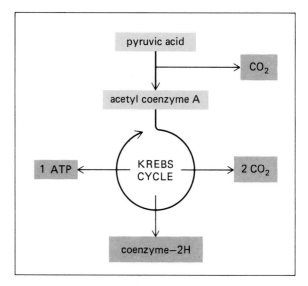

FIG. 3-14 Inputs and outputs of the Krebs cycle reactions, which occur in mitochondria.

percent) out of the 686 kcal of potential energy available from glucose has been transferred to ATP.

   The sequence of chemical reactions that constitute glycolysis is shown in detail in Fig. 3-15. Each of the 10 reactions is catalyzed by a separate enzyme. Note that at each step only small changes occur in the rearrangement of substrates into product molecules. The reaction occurring at step six (Fig. 3-15) is essential for maintaining the flow of glucose through the glycolytic pathway. In this reaction two atoms of hydrogen are transferred from 3-phosphoglyceraldehyde to NAD (nicotinamide adenine dinucleotide) to form $NADH_2$. NAD participates in many chemical reactions in cells, where it transfers hydrogen from one chemical reaction to another (several other molecules in the cell perform a similar role and all of these will be referred to simply as *coenzyme* molecules). If $NADH_2$ were unable to transfer its hydrogens to another molecule to re-form NAD, the supply of NAD in the cell would become depleted, and glucose breakdown, which requires NAD at step six, would come to a halt. As we shall see in the next section, the major substrate to which $NADH_2$ transfers its hydrogen is oxygen. However, in the absence of oxygen, $NADH_2$ can transfer its hydrogens to pyruvic acid to form **lactic acid** and regenerate NAD that can now return to reaction six and maintain the breakdown of glucose and generation of ATP through the glycolytic pathway. Thus, in the absence of oxygen

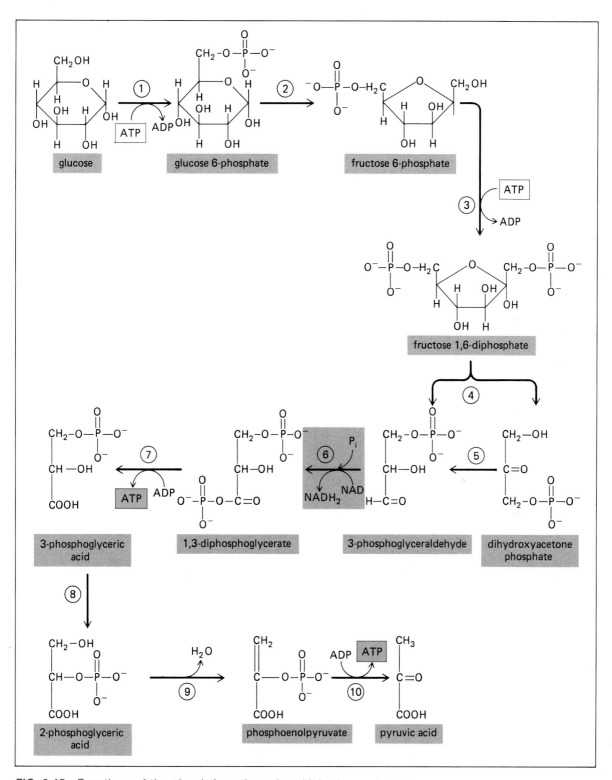

FIG. 3-15 Reactions of the glycolytic pathway by which glucose is broken down into two molecules of pyruvic acid with the net formation of two molecules of ATP.

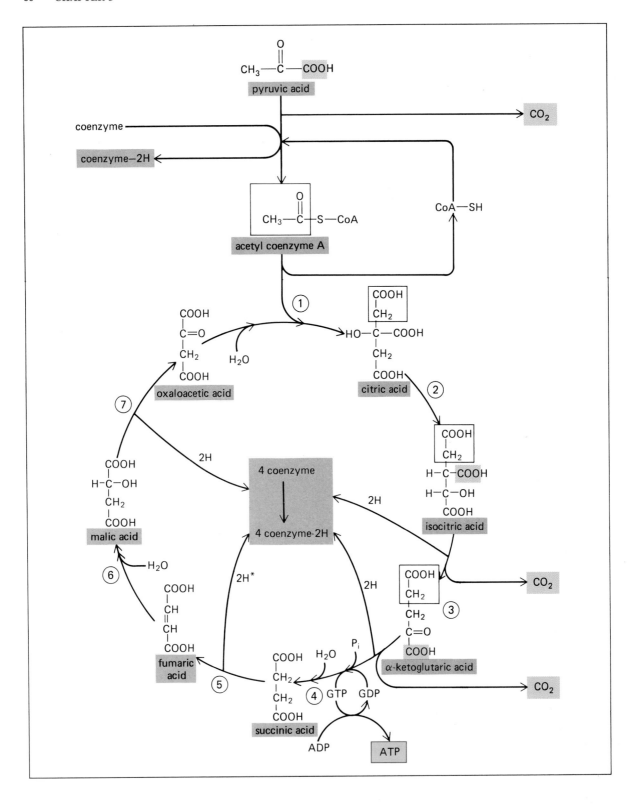

FIG. 3-16 Krebs cycle reactions that break down a molecule of pyruvic acid into three molecules of carbon dioxide and form five molecules of coenzyme-2H. One molecule of ATP is formed during reaction 4.

(**anaerobic conditions**), lactic acid becomes the end product of glycolysis because it provides the mechanism for regenerating NAD. In contrast, in the presence of oxygen (**aerobic conditions**), the end product of glycolysis is pyruvic acid because NAD can be regenerated by reacting with oxygen. Although all cells contain the enzymes of the glycolytic pathway and can thus generate limited amounts of ATP in the absence of oxygen, the small amounts of ATP formed are not sufficient to meet the large energy demands of most cells.

Krebs Cycle   The details of the sequence of chemical reactions that occur in the Krebs cycle are shown in Fig. 3-16. As with glycolysis, only small changes in the structure of the molecules occur at any one step in the sequence. The main products of this cycle of reactions are coenzyme-2H molecules, formed by the transfer of hydrogen from various substrates to coenzyme molecules, and carbon dioxide. As we shall see the coenzyme-2H molecules will be processed by the oxidative phosphorylation pathway leading to the formation of ATP.

The Krebs-cycle reactions convert the carbons in a molecule of pyruvic acid into three molecules of carbon dioxide and in the process form five molecules of coenzyme-2H. The removal of the carboxyl group from pyruvic acid leaves a two-carbon molecule, *acetate,* which combines with *coenzyme A*

(CoA) to form **acetyl coenzyme A.** Acetyl coenzyme A is one of the main sources of carbon fragments used as a starting material to synthesize a variety of organic molecules. In the Krebs cycle, acetyl coenzyme A transfers its two-carbon acetate to the four-carbon molecule, oxaloacetic acid, to form the six-carbon molecule of citric acid. In the remainder of the Krebs cycle, two of the carboxyl groups of citric acid are eventually converted into carbon dioxide, leaving at the end of the cycle the starting material, oxaloacetic acid, which can begin the cycle again by combining with a new molecule of acetyl coenzyme A. In contrast to the enzymes for glycolysis, the enzymes of the Krebs cycle are in the mitochondria.

Oxidative Phosphorylation   Embedded in the inner membrane of the mitochondria are the **cytochrome** enzymes that mediate the chemical process known as oxidative phosphorylation: the reaction between hydrogen and oxygen that releases energy that is coupled to the synthesis of ATP. The oxygen for the reaction is molecular oxygen taken in during the process of breathing, and the hydrogen, derived from the breakdown of carbohydrates, fats, and proteins, enters the reaction in the form of coenzyme-2H.

$$\text{Coenzyme-2H} + \tfrac{1}{2}O_2 \longrightarrow$$
$$\text{coenzyme} + H_2O + 52 \text{ kcal/mol}$$

Almost 50 percent of the energy released in the reaction is coupled to ATP synthesis.

The overall reaction between coenzyme-2H and oxygen proceeds in a series of reactions, each of which releases a small amount of energy and is

FIG. 3-17 Energy is coupled to the formation of ATP at three points in the cytochrome chain during oxidative phosphorylation. Some coenzyme-2H enters this chain beyond the point at which the first ATP is formed, which gives rise to two rather than three ATP for each pair of hydrogen atoms that react with oxygen to form water.

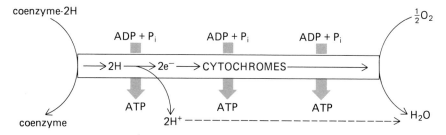

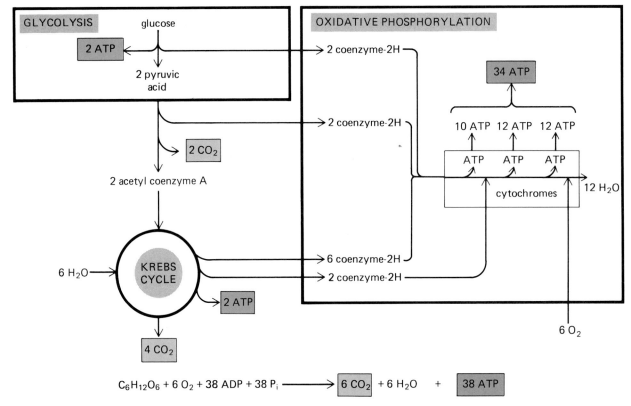

FIG. 3-18 Pathways of aerobic glucose catabolism and their linkage to ATP formation.

mediated by one of the cytochrome enzymes. Coenzyme-2H donates two electrons to the first cytochrome in the enzyme chain. These electrons are then passed on to successive cytochromes until, at the end of the sequence, they are added to oxygen, which then combines with the two hydrogen ions to form a molecule of water. At three separate points in the cytochrome chain, the energy released during the transfer of electrons from one cytochrome to the next is coupled to the phosphorylation of ADP to produce ATP (Fig. 3-17).

We can now calculate the amount of ATP formed during the breakdown of one molecule of glucose into carbon dioxide and water in the presence of oxygen (Fig. 3-18). Two coenzyme-2H ($NADH_2$) are formed during the glycolysis of glucose and 10 molecules are formed during the Krebs-cycle reactions, giving a total of 12 coenzyme-2H. The transfer of the hydrogens in these 12 molecules to oxygen generates 34 ATP by oxidative phosphorylation. In addition, 2 ATP were formed by the glycolytic pathway and 2 ATP by the Krebs cycle,

giving a total of 38 ATP from the catabolism of one molecule of glucose in the presence of oxygen. This represents about 39 percent ($38 \times 7 = 266$ kcal) of the total energy (686 kcal) released during glucose breakdown. The remaining 61 percent of the energy appears as heat. Since 36 out of the total 38 ATP were formed in the mitochondria, the majority of the cells' ATP production (95 percent) is formed in this cell organelle and is dependent on the presence of oxygen.

Carbohydrate Storage Although the major role of carbohydrates is to provide energy for ATP formation, a small amount of carbohydrate is stored as a reserve supply of fuel. The small amount that is stored, about 430 g in a 70-kg man, is in the form of the branched-chain polysaccharide known as *glycogen,* which is similar in structure to starch, the primary storage form of carbohydrate in plants. Most of the glycogen is stored in skeletal muscles and the liver, with smaller amounts in most other tissues.

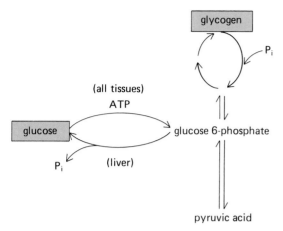

FIG. 3-19 Pathways for the formation and breakdown of glycogen.

Glycogen is synthesized from glucose by the pathway illustrated in Fig. 3-19, the enzymes of which are in the cytosol. The first step, the transfer of phosphate to glucose from ATP, forming glucose 6-phosphate, is the same as the first step in glycolysis (thus, glucose 6-phosphate can either be broken down to pyruvic acid or incorporated into glycogen). Note that (as indicated by the bowed arrows in the figure) a different enzyme is required to split glucose 6-phosphate from glycogen than to incorporate it into glycogen. The existence of different enzymes provides a mechanism for regulating the flow of glucose to and from glycogen. These enzymes are allosteric enzymes whose activity can be stimulated or inhibited by modulator molecules. When an excess of glucose is available to a liver or muscle cell, the enzymes in the glycogen-synthetic pathway are activated and the enzyme breaking down glycogen is simultaneously inhibited, leading to the net storage of glucose in the form of glycogen. When less glucose is available, the reverse combination of enzyme stimulation and inhibition occurs and there is a net breakdown of glycogen.

Most cells, including skeletal muscle cells, do not have the enzyme that catalyzes the removal of phosphate from glucose 6-phosphate to form free glucose, which can cross the plasma membrane. Thus, the glucose that is stored as glycogen in these cells cannot be released to provide glucose for other cells. The major function of the relatively large quantities of glycogen in muscle cells is to provide an immediately available source of energy (by way of glycogen → glucose 6-phosphate → glycolysis)

for the cells themselves during periods of intense contractile activity, when their consumption of glucose may greatly exceed the rate at which glucose is delivered to them by the blood.

Liver cells, on the other hand, do contain the enzyme that converts glucose 6-phosphate to glucose, thus allowing phosphorylated glucose intermediates formed from glycogen breakdown or other pathways to be converted into free glucose and released from the cells into the blood. Because of this ability to release glucose, the liver plays a very important role in maintaining blood glucose concentration.

The sources of carbohydrates in the body are the carbohydrates in food or the carbohydrates synthesized from certain types of amino acids found in proteins (Fig. 3-20). Carbohydrates can be catabolized to provide ATP or can be stored in the form of glycogen. Carbohydrates can also be used to synthesize fat. In this case the chemical energy in the carbohydrate molecule is stored in the molecules of fat. This energy can be made available to a cell by the catabolism of fat.

Fat Catabolism

The metabolic pathway for the breakdown of fat into carbon dioxide and water feeds into the metabolic pathway for glucose breakdown. Fat catabolism begins with an enzyme that splits off the three fatty acids linked to glycerol in triacylglycerol. Glycerol is a three-carbon carbohydrate that can be converted into one of the three-carbon intermediates in the glycolytic pathway, from which point it can be metabolized through the rest of the glyco-

FIG. 3-20 Carbohydrates in food or formed from proteins can be catabolized to provide energy, stored as glycogen, or converted into fat.

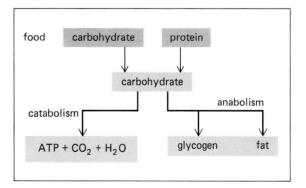

lytic pathway and on into the Krebs cycle just like a molecule of glucose. The breakdown of the fatty acids proceeds through a pathway in which the two carbons at the carboxyl end of the fatty acids are split off (by enzymes in the mitochondria) and transferred to coenzyme A, and hydrogen is transferred from the fatty acid to coenzyme. This reaction is repeated until the entire fatty acid is converted, two carbons at a time, into acetyl coenzyme A, which enters the Krebs cycle to undergo further breakdown to carbon dioxide. Likewise, the coenzyme-2H formed during fatty acid breakdown can donate its hydrogens to the cytochrome system, giving rise to ATP synthesis by the oxidative phosphorylation pathway (Fig. 3-21).

Since most ATP formed from the breakdown of fatty acids results from the hydrogens donated to

FIG. 3-21 Pathway for the catabolism of fat. Carbohydrates and proteins in food can be converted into fat.

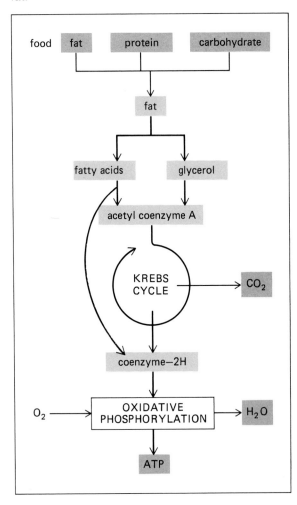

TABLE 3-2	FUEL CONTENT OF A 70-kg INDIVIDUAL			
	Total Body Content	Energy per Gram	Total Body Energy Content	
	(kg)	(kcal/g)	(kcal)	(%)
Triacylglycerols	15.6	9	140,000	78
Proteins	9.5	4	38,000	21
Carbohydrates	0.5	4	2,000	1

the cytochrome system (the remainder coming from ATP formed in the Krebs cycle), the production of ATP by fatty acid catabolism requires oxygen. Each 18-carbon fatty acid forms 147 molecules of ATP. The three fatty acids from one molecule of fat can thus form 441 molecules of ATP; adding the 22 ATP formed from glycerol gives a total of 463 ATP formed from each molecule of fat. Taking the molecular weights into account (842 for a molecule of fat containing three 18-carbon fatty acids and 180 for glucose), one can calculate that, on a per gram basis, fat catabolism provides slightly more than twice as much ATP as does glucose. Therefore, fat is a much more efficient means of storing fuel than is carbohydrate; indeed, fat accounts for most of the fuel stored in the body (Table 3-2). This fat comes not only from the fat in foods but also from the conversion of carbohydrates into fat, as well as the conversion of certain amino acids into fat (Fig. 3-21).

Protein Catabolism

The major difference in the composition of proteins, compared with carbohydrates and fats, is the presence of nitrogen in the amino groups of the amino acids. Once this nitrogen has been removed, the remainder of the molecule (a keto acid) can be metabolized via the Krebs-cycle reactions (Fig. 3-22). For example, removal of the amino group from the amino acid alanine yields pyruvic acid, the end product of glycolysis.

$$\underset{\text{Alanine}}{CH_3-CH-COOH} + H_2O + NAD \longrightarrow$$
$$\qquad\qquad |$$
$$\qquad\quad NH_2$$

$$\qquad\qquad\qquad O$$
$$\qquad\qquad\qquad ||$$
$$\underset{\text{Pyruvic acid}}{CH_3-C-COOH} + \underset{\text{Ammonia}}{NH_3} + NADH_2$$

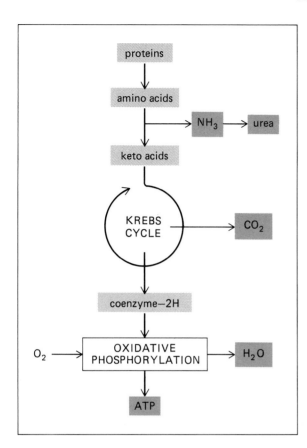

FIG. 3-22 Pathway for the catabolism of proteins.

The initial step in protein catabolism is the breakdown of the protein into its subunit amino acids, followed by the removal of the amino group as ammonia, NH_3. This ammonia is combined with carbon dioxide in a series of reactions which occur in the cells of the liver to form **urea,** which is then excreted in the urine.

$$2NH_3 + CO_2 \longrightarrow NH_2-\overset{\overset{\displaystyle O}{\|}}{C}-NH_2 + H_2O$$
$$\text{Ammonia} \qquad\qquad\qquad \text{Urea}$$

Interconversion of Carbohydrate, Fat, and Protein

Having discussed the metabolism of the three major classes of organic molecules, we can now briefly review how each class is related to the others and to the process of synthesizing ATP. Figure 3-23 shows the major pathways and the relations of the common intermediates. All three classes of molecules give rise to intermediates that can enter the Krebs cycle, and thus all three can be used for the synthesis of ATP by oxidative phosphorylation. Hydrogen and oxygen provide the substrates for

this oxidative phosphorylation. Some of these hydrogens are derived from the Krebs-cycle reactions directly; others may come from glycolysis or the breakdown of fatty acids and amino acids.

Glucose can be converted into fat or amino acids by way of the common intermediates such as pyruvic acid, α-ketoglutaric acid, and acetyl CoA. Similarly, amino acids can be converted into glucose and fat. There is, however, one major restriction on the interconversion of carbohydrates, fats, and proteins: Fatty acids cannot be used to synthesize net amounts of glucose or amino acids. On the other hand, glucose and amino acids can be broken down to acetyl coenzyme A and used to synthesize fatty acids. Metabolism is, therefore, a highly integrated process in which all classes of molecules can be used, if necessary, to provide energy for the cell through ATP synthesis and in which each class of molecule can, with the exception of fatty acids, provide the raw materials required to synthesize members of other classes.

The total metabolism of the body depends upon the supply and distribution of nutrients to the cells where they are metabolized. A cell's metabolic activity may be primarily anabolic following the ingestion of a meal and primarily catabolic during a period of fasting. Under one set of conditions carbohydrate catabolism may provide the major source of a cell's ATP, whereas under other conditions fatty acid catabolism may predominate. Some of the organs, notably the liver and fat tissue, have become specialized in their ability to store and release various nutrients into the blood. The various mechanisms regulating these aspects of total body metabolism will be the subject of a later chapter.

ESSENTIAL NUTRIENTS

There are many substances that are required for normal or optimal body function but that are synthesized by the body either not at all or in amounts inadequate to maintain health. They are known as *essential nutrients.* Because they are all excreted or catabolized at some finite rate, a continuous new supply must be provided by the diet. Approximately 50 in number, they are water, 8 amino acids, several unsaturated fatty acids, approximately 20

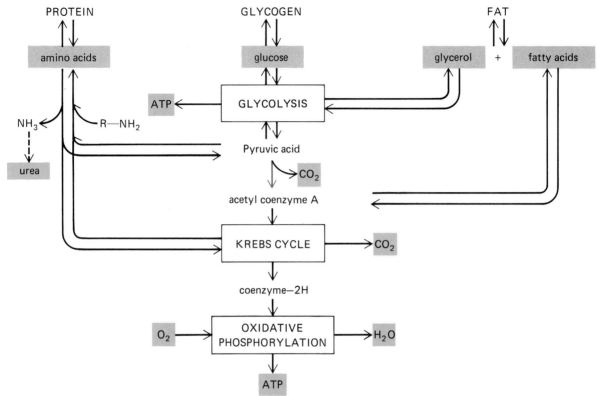

FIG. 3-23 Interrelations between the pathways for the metabolism of carbohydrates, fats, and proteins.

vitamins, and a similar number of inorganic minerals. *Vitamins* are distinguished from other essential organic nutrients in that only trace amounts are required; they are classified as water-soluble (vitamin C and the B complex vitamins) or fat-soluble (vitamins A, D, E, and K).

It should be reemphasized that the term essential nutrient is reserved for substances that fulfill *two* criteria: They not only must be essential for good health but must not be synthesized by the body in adequate amounts. Thus, glucose, although "essential" for normal metabolism, is not classified as an essential nutrient because the body normally can synthesize all it needs.

METABOLIC REGULATION

The rates at which molecules pass through metabolic pathways vary and depend on two primary factors: (1) the concentration of the substrates and (2) the activity of the enzymes mediating the reactions, both of which are subject to physiologic control.

Consider a simple reversible reaction in which an enzyme catalyzes the splitting of substrate A into products B and C.

$$A \overset{\text{enzyme}}{\rightleftharpoons} B + C$$

In order for the reaction to occur, a molecule of substrate must become bound to the active site of the enzyme as a result of random collisions between the substrate and enzyme. The probability of a collision is increased if there are more molecules of substrate in the solution. Therefore, the rate of the reaction will be increased as the substrate concentration increases. This is an example of the principle of **mass action,** which states that increasing the concentration of any molecule in a chemical reaction will cause the reaction to proceed in a direction that tends to decrease the concentration of that molecule. Increasing the concentration of A causes the reaction to proceed toward the formation of B and C; increasing the concentration of either B or C will cause the reaction to proceed toward the formation of A. This principle of mass action is often the determining factor governing the direction in which reactions proceed along a reversible metabolic pathway. However, there is an upper limit to the rate at which an enzyme-mediated reaction can be increased by increasing substrate concentration. This point is reached when all the binding sites on the enzyme molecules are occupied by substrate; the enzyme is then *saturated* with substrate and the reaction proceeds at a maximal rate.

Regulation of Enzyme Activity

By **enzyme activity** we mean the maximum rate at which an enzyme-mediated reaction proceeds when the enzyme is saturated with substrate. One way to vary the activity of an enzyme is to vary its concentration; the more enzyme that is present, the more product will be formed in a given time. Since all enzymes are proteins, the concentration of an enzyme depends on the rates of protein synthesis and degradation. For most enzymes, these rates remain essentially constant, and thus the concentration of the enzyme remains constant. In contrast, other enzymes undergo marked changes in their rates of synthesis (and therefore their concentrations) under various conditions. Some of these enzymes are present in a cell only when their substrate is also present. The substrate, in this case, appears to interact with the gene in DNA that corresponds to the enzyme that acts upon the substrate. This interaction leads to the transcription of

mRNA for that gene and the subsequent synthesis of the enzyme by cytoplasmic ribosomes. This process is called **enzyme induction.** In other cases (**enzyme repression),** product molecules inhibit the synthesis of the enzyme that forms the products by inhibiting the transcription of the gene corresponding to the enzyme into mRNA. The concentration of the enzyme decreases when the products of the reaction it catalyzes increase. Enzyme repression allows a cell to turn off the synthesis of an end product when its concentration rises above a certain level.

Since it takes time to synthesize and degrade a protein, the processes of enzyme induction and repression cannot be used to regulate minute-to-minute changes in the flow of substrates through various metabolic pathways. However, it is possible to alter the activity of already synthesized enzymes by means of allosteric and covalent modulation. For example, the end products of a metabolic pathway often act as modulator molecules that inhibit the activities of allosteric enzymes in the same pathway, a form of regulation known as **feedback inhibition.** In some cases the end products of one metabolic pathway are the modulator molecules that interact with the enzymes in a completely different pathway. Such interaction allows various metabolic pathways in a cell to be coordinated. Figure 3-24 summarizes the various factors we have discussed that influence the rates of enzyme-mediated reactions in a cell.

FIG. 3-24 Factors that affect the rate of enzyme-mediated reactions. The arrows indicate the sites at which the various factors affect the reaction rate.

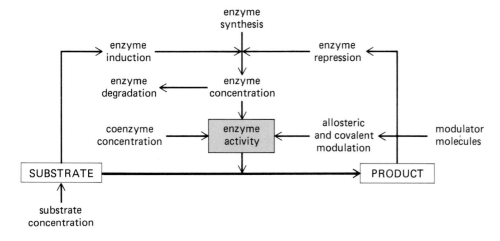

Interacting Metabolic Pathways

The flow of molecules through a metabolic pathway is influenced not only by the activity of the individual enzymes in the pathway but also by the characteristics of certain key reactions in the sequence. Consider the following arbitrary metabolic pathway from A to F that consists of five steps:

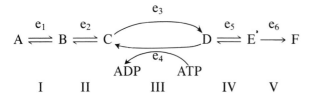

Three of the reactions (I, II, and IV) are reversible; a single enzyme mediates the flow of metabolites in either direction. During these reversible reactions, only small amounts of energy are released or must be added to cause the reaction to occur in the opposite direction. This small amount of energy can be obtained from the collisions between molecules. Reaction V is irreversible because it is associated with the release of a large amount of energy. Without a large source of energy to put back into the reaction, F cannot be converted back into E. One of the most important examples of such an irreversible reaction is the formation of carbon dioxide and acetyl coenzyme A from pyruvic acid, the reaction which links the end of glycolysis to the beginning of the Krebs cycle. Once acetyl coenzyme A has been formed, it cannot be converted directly back into pyruvic acid.

Reaction III represents a reaction that is usually present at one or two steps in a metabolic pathway and is often the major control point regulating the flow of metabolites through the pathway. The con-version of C to D is an irreversible reaction similar to reaction V, and if e_3 were the only enzyme present, D could not be converted directly back into C. However, D can be converted into C by a second enzyme e_4 that couples the reaction to a source of energy: the breakdown of ATP. By controlling the activities of the two enzymes e_3 and e_4, the direction of flow through the pathway can be regulated so that it occurs from C to D under one set of conditions and in the reverse direction under other conditions.

When a series of reactions are linked in a metabolic pathway, the rate at which A is converted into F depends on the activity of the one enzyme in the pathway that has the lowest enzyme activity, since the overall rate cannot be faster than the reaction proceeding at the slowest rate, the so-called **rate-limiting reaction.** By regulating the activity of the enzyme mediating the rate-limiting step, the overall rate of flow of metabolites through the pathway can be controlled. The rate-limiting reaction in most pathways is mediated by an allosteric enzyme whose activity can be altered by modulator molecules. For example, ATP and its products, ADP and P_i, are modulator molecules able to activate or inhibit various allosteric enzymes. If a cell is utilizing large amounts of energy, the rate of ATP breakdown to ADP and P_i will be rapid, leading to a fall in ATP concentration and a rise in the concentrations of ADP and P_i. The changes in concentrations of these molecules allosterically activate or inhibit various enzymes so that the overall rates of glucose and fat catabolism are increased, thereby providing the energy necessary to form more ATP. Likewise, when the cell uses less energy, the rise in ATP and the fall in ADP and P_i inhibit these catabolic enzymes and activate the anabolic ones, leading to a net storage of fuel as glycogen and fat.

KEY TERMS

protein binding sites	**chromosome**	**Krebs cycle**
saturation	**cell differentiation**	**oxidative phosphorylation**
allosteric regulation	**metabolism**	**glycogen**
gene	**adenosine triphosphate (ATP)**	**urea**
genetic code	**enzyme**	
messenger RNA (mRNA)	**glycolysis**	
ribosome	**pyruvic acid**	
mitosis	**lactic acid**	

REVIEW EXERCISES

Section A: Protein Binding Sites

1 Define the term ligand.
2 Describe the characteristics of a protein binding site which determine the chemical specificity of the binding site.
3 Define the term saturation.
4 Define the terms catalyst, enzyme, substrate, and product.
5 Describe the role of proteins in detecting chemical messengers.
6 Describe the process of allosteric and covalent modulation of protein binding sites.

Section B: Genetic Information and Protein Synthesis

1 Define the term gene.
2 State the "central dogma" of molecular biology.
3 Describe the relation between codons, DNA, and proteins.
4 Describe the function of messenger RNA.
5 Describe the role of transfer RNA in protein synthesis.
6 Describe the sequence of events that take place on a ribosome during the synthesis of a protein molecule.

Section C: Replication of Genetic Information

1 Describe the general mechanism of DNA replication.
2 Define the term mutation.
3 Define the terms interphase, mitosis, and cytokinesis.
4 Describe the relation between chromatin, chromatids, and chromosomes.
5 Describe the events associated with the separation of the chromosomes during mitosis.

Section D: Tissue Development

1 Define the term cell differentiation.
2 Describe how protein synthesis is related to cell differentiation.
3 Describe the adhesive mechanisms that hold cells together.
4 Define the terms hyperplasia and hypertrophy.
5 Describe some of the characteristics that distinguish a cancer cell from a normal cell.

Section E: Cell Metabolism

1 Distinguish between catabolic reactions and anabolic reactions.
2 Describe the overall function of ATP in cells.
3 Define the general term metabolic pathway.
4 Describe the following characteristics of the metabolic pathway known as glycolysis:
   a the location in a cell of the enzymes for this metabolic pathway
   b the major substrates that enter this pathway
   c the end products formed by this pathway
   d the net amount of ATP formed per glucose molecule entering the pathway
5 Define the terms aerobic and anaerobic.
6 Describe the following characteristics of the Krebs cycle pathway:
   a the location in a cell of the enzymes for this metabolic pathway
   b the major substrates that enter this pathway
   c the end products formed by this pathway
   d the net amount of ATP formed by this pathway during the catabolism of one glucose molecule
7 Describe the following characteristics of oxidative phosphorylation:
   a the location in a cell of the enzymes for this metabolic pathway
   b the major substrates that enter this pathway
   c the end products formed by this pathway
   d the net amount of ATP formed by this pathway during the catabolism of one glucose molecule
8 Describe the metabolic pathway leading to the formation and breakdown of glycogen. Explain how this pathway differs in liver and muscle.
9 Explain why it is more efficient for the body to store fat rather than carbohydrate as a reserve fuel.
10 Describe the relation between ammonia, urea, and protein catabolism.
11 Identify the one major restriction governing the interconversion of carbohydrates, fats, and proteins.
12 Describe the process of enzyme induction, enzyme repression, and feedback inhibition.
13 Summarize the various factors that can influence the rate of an enzyme reaction.
14 Define the rate-limiting reaction of a metabolic pathway.

CHAPTER

CELL MEMBRANES: STRUCTURE AND FUNCTION

The contents of a cell are separated from the surrounding extracellular medium by the **plasma membrane,** a very thin structure composed of lipids and protein. In most cells membranes of similar structure surround various cell organelles—mitochondria, endoplasmic reticulum, lysosomes, the Golgi apparatus, and the nucleus—and divide the intracellular fluid into compartments. The term *membrane* refers nonspecifically to any of these cell membranes, whereas plasma membrane denotes specifically the membrane at the surface of a cell. If all the membranes in 1 g of liver were unfolded and pieced together, they would cover an area of approximately 30 m^2 (250 ft^2).

The contributions of these membranes to cell function fall into two general categories: (1) They provide barriers to the movements of molecules and ions between the various compartments in the cell and between the cell and the extracellular fluid; (2) they provide a scaffolding to which various cell components are anchored. For example, the outer surface of the plasma membrane contains a variety of binding sites that function as "recognition sites" (receptors) for hormones and other chemical messengers. Thus the plasma membrane serves as a signal-receiving device for chemical signals that regulate the cell's activity. Surface binding sites also participate in the arrangement of cells into tissues during development by providing sites for specific cell-to-cell adhesions.

MEMBRANE STRUCTURE

Membranes are composed of lipids and proteins in about equal proportions by weight. The lipids form the general barrier to movements of molecules through the membrane; the proteins provide selective means for the transfer of certain molecules through the barrier and also constitute the binding sites and enzymes associated with the membrane.

The majority of the membrane lipids are phospholipids. As described in Chap. 2, a phospholipid is an amphipathic molecule that has a charged region at one end (due to the presence of a negatively charged phosphate group); the remainder of the molecule, consisting of two long-chain fatty acids, is electrically neutral. When phospholipids are mixed with water, they associate with each other to form bimolecular layers in which the polar ends are

positioned at the surfaces of the bilayer because of their electrical attraction to water molecules; the nonpolar fatty acid chains are perpendicular to the surfaces and form a nonpolar region in the bilayer.

The phospholipids in cell membranes are organized into a bimolecular layer similar to those that form spontaneously in phospholipid-water mixtures, their polar ends at the inner and outer surfaces of the membrane. The individual phospholipids in the membrane have considerable freedom of movement because the fatty acid chains are flexible and can wiggle back and forth, and the entire molecule can move laterally. Thus, the lipid phase of the membrane is more like a fluid than a rigid crystalline matrix. This fluidity makes the entire membrane quite flexible; it can easily be bent and folded, although horizontal stretching may rupture its loose phospholipid associations. The phospholipid bilayer is an effective barrier to the transmembrane movements of polar and charged molecules because these molecules are unable to associate with the nonpolar (fatty acid) layer of the membrane.

The membrane is an asymmetric structure; the types of proteins on the outer surface are quite different from those on the inner surface. Two classes of proteins, peripheral proteins and integral proteins, are associated with cell membranes. **Peripheral proteins** are water-soluble and are associated with the polar and ionized surfaces of the membrane. In contrast, the **integral proteins** are insoluble in water because the polypeptide chain is folded in such a way that certain regions contain mainly ionized and polar amino acid side chains, whereas the remaining portion of the protein has primarily nonpolar side chains. This creates an amphipathic molecule that, like the amphipathic phospholipids, orients so that its polar regions are at the surface of the membrane and the nonpolar region is in the middle. In some cases, the integral proteins are on only one side of the membrane; in other cases, the protein has a polar region at each end, separated by a nonpolar region, allowing it to span the entire membrane (Fig. 4-1).

Some of the integral proteins that span the membrane seem to be organized in clusters to form **pores,** aqueous channels through which small water-soluble molecules and ions can pass and thereby bypass the nonpolar lipid regions of the membrane. As we shall see, they also provide spe-

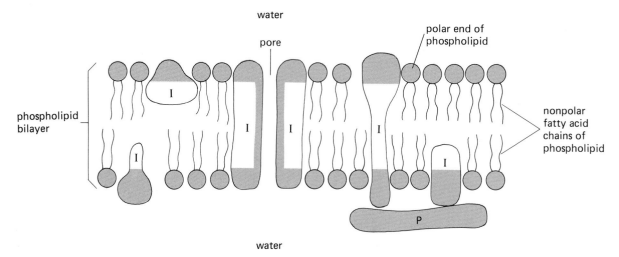

FIG. 4-1 Arrangement of integral (I) and peripheral (P) membrane proteins in association with bimolecular layer of phospholipids. Shaded areas at the membrane surface indicate the polar regions of the proteins and phospholipids.

cialized mechanisms for the movement of certain substances across the membrane. Integral proteins that span the membrane may also participate in transmitting signals across the membrane.

Membranes are quite thin, varying from about 6 to 10 nm in thickness, depending on the amount and types of protein associated with each lipid surface. Most of the proteins as well as the lipids are free to move in the plane of the membrane independently of each other; thus, the membrane proteins appear to float in the lipid bilayer. This dynamic membrane model is known as the **fluid mosaic model** (Fig. 4-2).

FIG. 4-2 Fluid mosaic model of membrane structure.

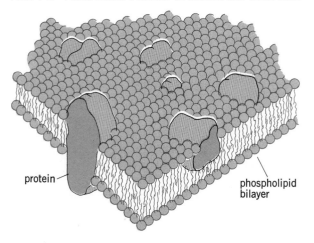

DIFFUSION

The movement of molecules within a cell, across cell membranes, or from one area of the body to another is essential for the normal functioning of the body. When molecules must be moved over long distances, the contraction of muscles provides the required forces. This type of movement, known as *bulk flow* (Fig. 4-3), is the mechanism that produces blood circulation and the flow of air into and out of the lungs.

FIG. 4-3 Bulk flow processes are required to move molecules rapidly over long distances (blood circulation.) Random molecular motion (diffusion) can produce rapid molecular movements over short distances between the blood and tissue cells.

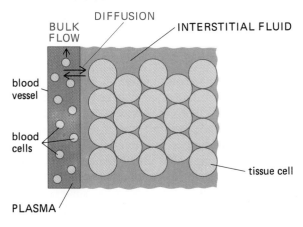

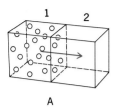
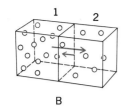
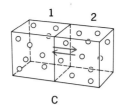

FIG. 4-4 Diffusion of glucose between two compartments of equal volume. (A) Initial conditions: No glucose is present in compartment 2. (B) Some glucose molecules have moved into compartment 2, and some of these are moving at random back into compartment 1. (C) Diffusion equilibrium has been reached, the flux of glucose between the two compartments being equal in the two directions.

Diffusion, unlike bulk flow, is the movement of molecules from one location to another by random molecular motion. The individual molecules in a solution are in a state of continuous random movement, bouncing off each other like rubber balls, each collision altering the direction of molecular movement. This frenzied molecular motion is the physical equivalent of heat; the hotter an object becomes, the faster its molecules move. It is this random motion of individual molecules that is responsible for diffusion. Unlike bulk flow, in diffusion there is no external force pushing each individual molecule from one region to another.

Figure 4-4 illustrates the characteristics of the diffusion process. A solution containing glucose is separated from pure water by an imaginary boundary across which glucose can freely move. Initially, as a result of their random motion, some of the glucose molecules move across the boundary into compartment 2. The amount of material crossing a surface in a unit of time is known as a **flux.** The magnitude of this initial one-way flux of glucose depends on the glucose concentration in compartment 1; the higher the glucose concentration, the greater the number of glucose molecules randomly moving in the direction of compartment 2 at any instant. As the concentration of glucose in compartment 2 increases, some of the glucose will randomly move back into compartment 1. The magnitude of this reverse, one-way flux depends on the concentration of glucose in compartment 2. Therefore, the **net flux,** F (the difference between the two one-way fluxes) of glucose from compartment 1 into compartment 2 depends on the difference in the glucose concentration between the two compartments, $\Delta C = C_1 - C_2$ (often referred to as the **concentration gradient**). Accordingly, the net flux of molecules across a boundary can be written in the form of the following equation:

$$F = k_D A \Delta C$$

where k_D is the **diffusion coefficient,** a number that reflects the molecular weight of the diffusing molecule (large molecules move more slowly and thus have a smaller k_D) and the temperature of the solution (the higher the temperature, the faster molecules move, producing a larger k_D), and A is the surface area separating the two compartments (the greater the surface area, the greater the magnitude of the net flux).

The concentration gradient, ΔC, determines the direction of the net flux. The net flux always proceeds from a region of high concentration to a region of lower concentration. When the concentrations of the two compartments become equal, the system is said to be in *diffusion equilibrium;* there is no concentration gradient ($\Delta C = 0$) and thus no net flux.

Although individual molecules travel at very high speeds, the number of collisions they undergo prevents them from traveling very far in a straight line. Thus, diffusion can lead to the rapid movement of molecules over short distances but is a very slow process for transferring molecules over long distances. For example, the diffusion of glucose from a blood vessel to a point one cell diameter away would take only about 3.5 sec to reach 90 percent of the glucose concentration in the blood, whereas it would take about 11 years for this glucose concentration to be reached 10 cm (about 4 in) away from the vessel. Fortunately, all the cells in the body are within a very short distance of a blood vessel so that nutrients and products of cell metabolism can diffuse rapidly between the cells and the blood. Because of the slowness of diffusion over long distances, bulk-flow systems, such as the circulatory system, evolved in multicellular organisms to move molecules rapidly over long distances.

MEMBRANE PERMEABILITY

The plasma membrane separates the extracellular chemical environment from the intracellular fluid. All exchanges of material between the cell and its environment must occur across this structure. The plasma membrane acts as a selective barrier that prevents certain molecules from entering or leaving the cell while it allows others to diffuse freely across its surface. The rates at which molecules diffuse through the plasma membrane are a thousand to a million times slower than the rates at which these same molecules diffuse through water, indicating that the plasma membrane acts as a partial barrier to diffusion.

Oxygen, carbon dioxide, water, lactic acid, urea, and steroid hormones, are examples of molecules that cross plasma membranes by diffusion. The net flux, F, of a solute crossing a membrane by diffusion can be described by the same equation that was used to describe the diffusion of a solute through water. The difference in concentration (ΔC) between the inside of the cell (C_i) and the outside (C_o) and the membrane surface area (A_m), determine the magnitude of the net flux across the membrane.

$$F = k_p A_m \Delta C$$

The proportionality constant k_p is known as the **permeability constant.** The numerical value of this constant depends not only on the molecular weight of the penetrating molecule and the temperature of the medium but also upon the thickness and chemical composition of the plasma membrane. The net flux can be altered by a change in either the concentration gradient across the membrane or the permeability properties of the plasma membrane, i.e., a change in the value of the permeability constant.

Most of the molecules that cross the membrane rapidly by diffusion have one chemical property in common: They are composed almost entirely of nonpolar, uncharged chemical groups. These nonpolar molecules cross the membrane by dissolving in the fatty acid layers. As would be predicted, the permeability of the membrane to various molecules decreases in proportion to the number of polar and charged groups in their molecular structure. Highly charged molecules may be unable to diffuse through the membrane at all and are either excluded from entering the cell or trapped within it. Many of the phosphorylated and ionized intermediates of the various metabolic pathways (such as glucose 6-phosphate) fall into this category, as well as most proteins.

Although most of the molecules that are able to diffuse rapidly across the plasma membrane are nonpolar, there are a few important exceptions. Water, which is highly polar, diffuses across most membranes, as do the ions Na^+, K^+, and Cl^-. The distinguishing characteristic of this group is their small size, which has led to the hypothesis that the plasma membrane contains a number of small holes through which these small charged ions and polar molecules can pass. Larger molecules are excluded from these *membrane pores*. These channels consist of integral proteins that extend all the way through the membrane (Fig. 4-2).

Membrane pores are too small to be seen with the electron microscope, but their size has been estimated from the largest charged particle that can diffuse through the membrane. They appear to be about 0.8 nm in diameter (about three times the size of a water molecule), and they occupy less than 1 percent of the membrane surface area.

Different membranes have different permeabilities to the same molecule, and in some membranes the permeability may be altered during certain states of activity. Differences in permeability must result from differences in the two pathways for diffusion through the membrane—the lipid pathway or the pores.

MEDIATED-TRANSPORT SYSTEMS

Some of the major cell metabolites, including glucose and the amino acids, cannot cross cell membranes by diffusion, since they are polar or charged and too large to pass through pores. Yet these essential molecules do enter cells; they cross cell membranes by special **mediated-transport** processes that involve chemical interactions between the transported molecule and proteins in the cell membrane. The properties of mediated-transport systems result from the fact that there are specific protein binding sites in the membrane that combine with the transported molecule. These protein binding sites have been called **carriers** because it was once thought that the proteins actually "carried" the bound solute molecules across the membrane. However, we now know that, although protein molecules can move laterally in the plane of the membrane, they cannot readily move across it from one side to the other. The mechanism by which a carrier protein enables a solute to pass through the membrane is still unknown, but is presumed to involve a change in the shape of the carrier brought about by the solute's binding to it. The term carrier has been retained despite the fact that we now know that these proteins do not actually move across the membrane.

The carrier model accounts for a number of the properties of mediated-transport systems. Carriers have chemical specificity; i.e., they transport only those solutes that are able to bind to the binding

FIG. 4-5 The net flux of molecules diffusing across a plasma membrane increases in proportion to the extracellular concentration, whereas the net flux of molecules entering by mediated-transport systems reaches a maximal value that does not increase with increasing concentration. This maximal mediated-transport flux corresponds to the saturation of all the available carrier molecules by the transported solute.

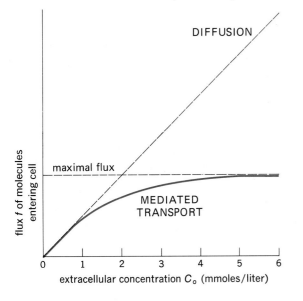

site on the carrier. Also, the saturation of the carriers in a mediated-transport system, reached at a solute concentration where all the carriers are combined with solute, places an upper limit on the rate of solute movement through the membrane. Any further increase in the concentration will produce no increase in the flux (Fig. 4-5). Contrast this to diffusion in which the flux will continue to increase in proportion to the concentration. One way to alter the maximal rate of mediated transport is to add or remove carriers from the membrane, and in fact some hormones influence cell activity through the regulation of the synthesis of carrier molecules.

The behavior of carrier-mediated transport systems can be divided into two classes: **facilitated-diffusion systems,** which lead to equal concentrations of solute on both sides of the membrane, and **active-transport systems,** which can move solutes "uphill" from a low-concentration region on one side of a membrane to a high-concentration region on the other side.

Facilitated Diffusion

In a facilitated-diffusion system, the carrier undergoes identical binding reactions with the transported solute at both membrane surfaces. Therefore, a net flux of solute across the membrane occurs only when a concentration gradient exists for that solute. If the solute concentrations on the two sides of the membrane are equal, the carriers will move equal amounts of solute into and out of the cell. In other words, facilitated diffusion is similar to diffusion in two respects: Net transport occurs only from high to low concentrations, and it continues until equal concentrations of the solute have been reached on the two sides of a membrane. Accordingly, the real function of such a system is to facilitate the movement through the membrane of molecules that are unable to diffuse across because they are polar and too large to pass through the membrane pores. For example, glucose crosses most cell membranes by facilitated diffusion.

Active Transport

Whereas a facilitated-diffusion system moves molecules only from high to low concentrations, active-transport systems are able to move solutes from a lower to a higher concentration. This is analogous to pumping water uphill, whereas diffusion and facilitated diffusion are analogous to the flow of water downhill. In order for a net movement of molecules to occur against a concentration gradient, energy must be expended, just as energy is required to pump water uphill. Accordingly, active-transport systems are often referred to as "pumps." There are a number of different active-transport systems in cell membranes, some of which pump molecules into the cell whereas others pump molecules out.

Movement against a concentration gradient is possible because of the asymmetry of the affinity of the carrier binding site for the solute on the inner and outer surfaces of the membrane, which allows the carrier to bind solute to different degrees on the two sides. Energy, usually in the form of ATP, is coupled to active-transport systems at the inner surface by a reaction that modifies the structure of the carrier. Even when the extracellular and intracellular concentrations of solute are equal, more solute will combine with the high-affinity carriers, which are on one side (let us say, the extracellular side), than with low-affinity carriers, which are on the other side. More solute will therefore move into the cell than will move out; i.e., there will be a net flux of solute into the cell even though the solute concentrations on the two sides of the membrane are initially equal. As the intracellular concentration continues to increase, more of the low-affinity carrier sites will become occupied. The flux out of the cell will be increased until an intracellular concentration is reached at which the flux into the cell is equal to the flux out. At this point a steady state will have been reached at which no further change in concentrations will occur. For example, the concentrations of amino acids are 2 to 20 times higher inside cells than in the blood, as a result of their active transport across the plasma membrane.

Since we have already seen that Na^+ and K^+ can diffuse through channels in the membrane, one might expect these ions to reach diffusion equilibrium across the membrane. However, the concentration of K^+ in the extracellular fluid is about 5 mmol/L, and the concentration of K^+ in cells is about 150 mmol/L, 30 times higher. Given this high concentration gradient and the presence of membrane pores, there is a net diffusion of potassium out of cells. However, simultaneous with this leak of potassium there is the active transport of potassium into the cell, so that the rate of net diffusion

out is equal to the rate of active transport in, and the large concentration gradient is maintained. If energy, in the form of ATP, were not being supplied to the ion pump, it would stop and the large intracellular potassium concentration would decrease as potassium diffused out of the cell.

The situation is reversed for sodium. The concentration of sodium ions in the blood is about 145 mmol/L, whereas the intracellular concentration is quite low, about 15 mmol/L. There is a continuous net diffusion of sodium into the cell through the sodium channels but an equal active transport out by the sodium pump. Here again, energy is continuously expended by the cell to operate the sodium pump and prevent the concentration of sodium in the cell from rising.

This unequal distribution of Na^+ and K^+ ions is found across all plasma membranes, and a considerable portion of the cells' metabolic energy is used to maintain these ion pumps. As we shall see in later chapters, these ion concentration gradients lead to the electric activity associated with nerve and muscle cells. In addition, some cells contain active chloride, calcium, and hydrogen-ion pumps.

FIG. 4-6 Summary of the three pathways by which molecules can cross cell membranes: (1) diffusion through the lipid matrix of the membrane, (2) diffusion through a pore, and (3) carrier-mediated transport.

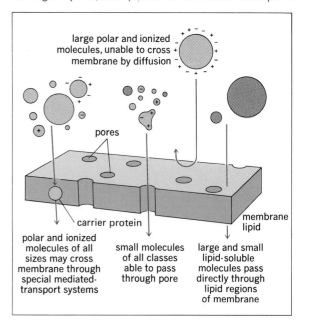

Figure 4-6 summarizes the three major routes by which molecules may cross cell membranes: (1) by diffusion through the lipid regions of the membrane if the molecule is nonpolar; (2) by diffusion through membrane pores if the molecule, though polar, is small; (3) by a carrier-mediated transport system if carriers specific for the solute exist in the membrane.

OSMOSIS

Water, which makes up the bulk of the fluid inside and outside the cells, is a small, polar molecule about 0.3 nm in diameter that crosses most plasma membranes very rapidly by diffusion. A net movement of water across a plasma membrane occurs whenever there is a difference between the water concentrations on the two sides of the membrane; this is known as **osmosis.** If water happens to be the only molecule in the solutions able to move through the membrane, osmosis into or out of the cell will lead to a change in cell volume as the cell gains or loses water.

How can the concentration of water in a solution be altered? By the addition of solute. If a solute, such as glucose, is dissolved in water, the concentration of water in the resulting solution will be less than that of pure water. Each molecule of glucose that is added to the solution adds an element of volume that cannot be occupied by water. The more solute that is added, the less water there will be in a given volume of solution and thus the water concentration will be lowered (Fig. 4-7).

The degree to which the concentration of water is decreased by the addition of solute depends upon the *number* of solute molecules that are added to

the solution and not upon the chemical nature of the solute. The total solute concentration, irrespective of the type of solutes that may be present, is known as the **osmolarity** of the solution. The *higher* the osmolarity (total solute concentration) of a solution, the *lower* will be its water concentration.* Normally, water is in diffusion equilibrium across the plasma membrane; thus the concentration of water inside and outside of cells is the same. Note that this implies that the solute concentration (osmolarity) of the intracellular and extracellular solutions is also the same. If the osmolarity of the extracellular solution were to be increased or decreased by adding or removing solute, this would alter the water concentration in the extracellular fluid and produce a water concentration gradient across the plasma membrane, which would lead to the net diffusion of water into or out of the cell—osmosis.

Note that in order to produce a water concentration gradient across a membrane there must also be a difference in solute concentration across the same membrane. If the solutes are able to diffuse across the membrane they will reach diffusion equilibrium, at which point they will have the same concentration on the two sides of the membrane and thus will not produce a difference in water concentration. Therefore, it is the concentration of *nonpenetrating solutes* on the two sides of a membrane that determines the net concentration gradient for water diffusion during osmosis.

The total solute concentration of the extracellular fluid is about 300 mmol/L. About 85 percent of the solutes in the extracellular fluid are sodium and chloride ions, which can diffuse into the cell through channels in the plasma membrane. However, as we have seen, the membrane contains an active-transport system for pumping sodium ions out of the cell, and the result is that there is no net movement of sodium across the membrane. It is as if sodium ions were unable to cross the membrane, since for every sodium ion entering the cell another

FIG. 4-7 Decrease in water concentration as a result of the addition of solute.

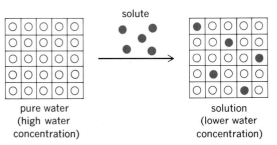

pure water
(high water
concentration)

solute

solution
(lower water
concentration)

---

* We have chosen to describe osmosis in terms of the diffusion of water from high to low water concentration (low to high osmolarity). Osmosis can also be described in terms of the osmotic pressure of a solution which, like osmolarity, depends on the total solute concentration and is thus inversely related to the water concentration. Thus, osmosis proceeds from a region of lower to higher osmotic pressure.

sodium ion is returned to the outside by the active-transport system. Since chloride ions follow the sodium because of the electric attraction between them (sodium is a positive ion and chloride a negative one), both sodium and chloride ions act as if they were nonpenetrating solutes. Any change in the sodium chloride concentration outside the cell alters the concentration of extracellular water; water diffuses across the plasma membrane, and the cell changes volume.

Inside the cell the major solutes are potassium ions and a number of organic molecules. Most of the latter are unable to cross the membrane, partly because they contain polar or ionized groups (which limit their diffusion through the lipid portions of the membrane) and partly because they are too large to pass through the pores. Potassium ions can leak out of the cell, but potassium is actively transported back in, and the net effect is as if potassium could not cross the membrane. Thus, sodium chloride outside the cell and potassium and organic solutes inside the cell represent the major effective nonpenetrating solutes determining the water concentrations on the two sides of the membrane.

If a cell is placed in a solution of sodium chloride that has a total solute concentration of 300 mmol/L (150 mmol of Na^+ plus 150 mmol of Cl^-), equivalent to the osmolarity of the extracellular fluids, the cell will neither swell nor shrink. Such a solution is said to be **isotonic.** Any solution that contains the same concentration of nonpenetrating solutes as the extracellular fluid is an isotonic solution. Isotonic saline (NaCl) is often used as the basic solution in which drugs or other substances that are to be injected into the body are dissolved. Since this solution is isotonic, it will not cause osmotic

swelling or shrinking of the cells at the site of injection.

A solution that contains a lower concentration of nonpenetrating solutes than the extracellular fluid is said to be **hypotonic.** The water concentration in a hypotonic solution is greater than the water concentration within cells and thus water will diffuse into the cell and cause the cell to swell (Fig. 4-8). Injection of a drug dissolved in pure water, or other hypotonic solution would lead to swelling and possible rupture of the cells at the site of injection. Solutions that contain a higher concentration of nonpenetrating solutes than the extracellular fluid are known as **hypertonic** solutions, which cause cells to shrink as water diffuses out of the cell (Fig. 4-8). As we shall see, mechanisms have evolved for maintaining the osmolarity of the extracellular fluid nearly constant by regulating the rates at which the kidneys excrete salt and water in the urine.

ENDOCYTOSIS AND EXOCYTOSIS

Earlier in this chapter we described the various pathways by which molecules pass through plasma membranes. There is an additional pathway that does not actually require the molecule to cross the structural matrix of the membrane. When living cells are observed under a light microscope, small regions of the plasma membrane can be seen to invaginate into the cell. These invaginations then become pinched off; when this happens, small, intracellular membrane-bound vesicles that enclose a small volume of extracellular fluid are formed. This process is known as **endocytosis** (Fig. 4-9). A similar process in the reverse direction, known as **exocytosis,** occurs when membrane-bound vesicles

FIG. 4-8 Changes in cell volume that result from osmosis when a cell is placed in a hypotonic, isotonic, or hypertonic extracellular solution of sodium chloride.

FIG. 4-9 Endocytosis and exocytosis.

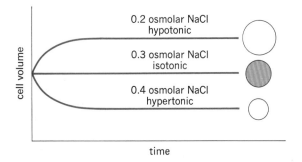

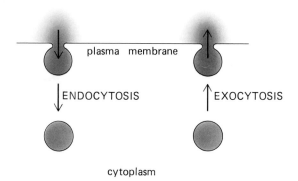

in the cytoplasm fuse with the plasma membrane and release their contents to the outside of the cell. The molecular mechanisms that produce endocytosis and exocytosis are still unclear, although they require metabolic energy and appear to involve contractile proteins associated with the plasma membrane.

Endocytosis

Several varieties of endocytosis can be identified. When the vesicle encloses a small volume of extracellular fluid, the process is known as *fluid endocytosis*. The composition of the vesicle contents is the same as that of the extracellular fluid. In other cases, specific molecules bind to sites on the membrane and are carried into the cell when the membrane invaginates. This is known as *adsorptive endocytosis*. Both fluid and adsorptive endocytosis are often referred to as **pinocytosis** (meaning cell drinking). A third type of endocytosis occurs when large multimolecular particles, such as bacteria, are engulfed by the plasma membrane and enter the cell. In this case, little extracellular fluid is enclosed within the vesicle. This process is known as **phagocytosis** (meaning cell eating).

FIG. 4-10 Fate of endocytotic vesicles. Pathway 1 transfers extracellular materials from one side of the cell to the other. Pathway 2 leads to fusion with lysosomes and the digestion of the vesicle contents.

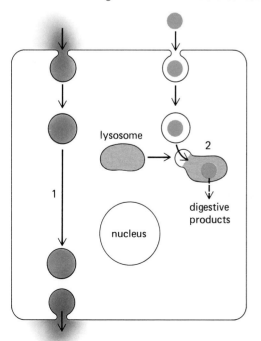

What is the fate of endocytotic vesicles once they have entered the cell? In some cases the vesicle passes through the cytoplasm and fuses with the opposite plasma membrane, and thus releases its contents to the extracellular space on the opposite side of the cell via exocytosis (pathway 1 in Fig. 4-10). This provides a pathway for transferring large molecules, which cannot cross plasma membranes by other means, from one side of a layer of cells to the other. However, in most cells endocytotic vesicles do not cross the cell but rather fuse with the membranes of lysosomes (pathway 2 in Fig. 4-10), the cell organelles that contain the digestive enzymes that break down large molecules such as protein, polysaccharides, and nucleic acids. The fusion of the vesicle with the lysosomal membrane exposes the contents of the vesicle to these digestive enzymes. The endocytosis of bacteria and their digestion in the lysosomes is one of the body's major defense mechanisms against microorganisms and will be discussed in a later chapter.

Obviously, the process of endocytosis removes a small portion of the membrane from the surface of the cell. If this membrane were not replaced, the surface area of the cell would decrease. In cells that have a large amount of endocytotic activity, more than 100 percent of the plasma membrane may be internalized in an hour, yet the cell volume and membrane surface area remain constant because the membrane is replaced at about the same rate that it is removed. Much of this replacement seems to occur through the fusion of intracellular vesicles with the membrane via exocytosis.

Exocytosis and Protein Secretion

Exocytosis performs two functions for cells: (1) It provides a way to replace portions of the plasma membrane that have been removed by endocytosis or to add new membrane during cell growth; (2) it provides a route by which certain types of molecules that are synthesized by cells can be released into the extracellular fluid. Many of the substances secreted by certain cells are unable to cross plasma membranes because they are large polar molecules (proteins, for example). By being enclosed in secretory vesicles they can be released from the cell by exocytosis.

The packaging of proteins for secretion into vesicles creates special problems for the cell because of the complex enzymatic machinery associated with protein synthesis and the inability of these

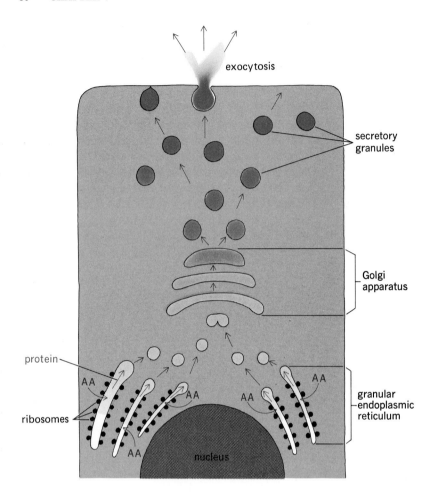

exocytosis

secretory
granules

Golgi
apparatus

protein

AA   AA   AA   AA

ribosomes

granular
endoplasmic
reticulum

AA

nucleus

FIG. 4-11 Pathway that leads to the release of protein from a secretory cell by exocytosis. (A signifies free amino acids that are assembled into proteins on ribosomes.)

large molecules to pass through membranes. As described in Chap. 3, ribosomes are the sites at which amino acids are assembled into proteins. Proteins that are to be secreted from the cell are synthesized on ribosomes attached to the surface of the endoplasmic reticulum. As the protein is assembled, it passes through the ribosome and the underlying membrane of the endoplasmic reticulum and enters the lumen of the reticulum (Fig. 4-11). This is the only stage along its route to the outside of the cell at which the protein passes through a membrane. Portions of the ends of the reticulum then pinch off and form small vesicles that contain the newly synthesized protein. These vesicles migrate to the region of the Golgi apparatus, where they fuse with the Golgi membranes. In the Golgi apparatus, the protein solution becomes progressively more concentrated as a result of fluid removal from the vesicles. In addition, carbohydrate groups are linked to many of the proteins by en-

zymes in the Golgi apparatus. Following their concentration and enzymatic modification in the Golgi apparatus, the small vesicles that contain the protein are known as *secretory* (zymogen) *granules*. They have pinched off from the stack of Golgi membranes and migrate toward the plasma membrane, where they can be released by exocytosis in response to the proper stimulus. The movement of secretory granules to a particular end of the cell may be guided by the network of cellular microtubules and contractile filaments.

The endoplasmic reticulum, in addition to se-

FIG. 4-12 Schematic diagram of three types of specialized cell junctions: (A) desmosome, (B) tight junction, and (C) gap junction. (D) Electron micrograph of two intestinal epithelial cells joined by a tight junction near the luminal surface and a desmosome a short distance below the tight junction.

A

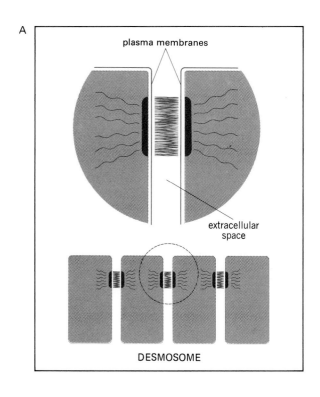

plasma membranes

extracellular space

DESMOSOME

B

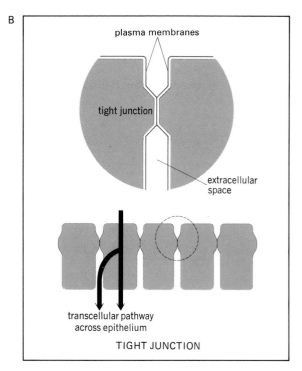

plasma membranes

tight junction

extracellular space

transcellular pathway across epithelium

TIGHT JUNCTION

C

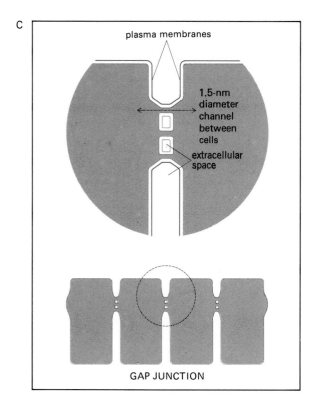

plasma membranes

1.5-nm diameter channel between cells

extracellular space

GAP JUNCTION

D

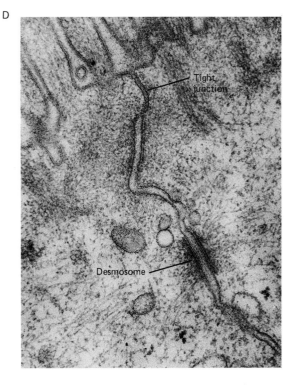

Tight junction

Desmosome

questering newly synthesized secretory proteins, also appears to be the site at which membranes are formed. The enzymes for the synthesis of membrane phospholipids are associated with the endoplasmic reticulum, and membrane proteins, following their synthesis on the ribosomes, can be added to the phospholipid bilayer of the reticulum. The newly synthesized membrane then passes to the Golgi apparatus and ultimately to the plasma membrane by a series of vesicle-membrane fusions.

MEMBRANE JUNCTIONS

In addition to providing a barrier to the movements of molecules between the intracellular and extracellular media, plasma membranes are involved in interactions between cells to form organized tissues. Some cells, particularly those of the blood, do not associate with other cells but remain suspended individually in a fluid (the blood plasma). Most cells, however, are packaged into tissues and organs and are not free to move around the body. Even these cells, however, are not packaged so tightly that the adjacent cell surfaces are in direct contact with each other. There usually exists a space of at least 20 nm between adjacent cells. This space is filled with extracellular fluid and provides the pathway for the diffusion of substances within the tissue.

The electron microscope has revealed that a variety of cells are joined by specialized junctions. One type, a **desmosome** (Greek *desmos,* binding), is illustrated in Fig. 4-12A and D. A desmosome consists of two opposed membranes that remain separated by about 20 nm but show a dense accumulation of matter at each membrane surface and between the two membranes. In addition, fibers extend from the inner surface of the desmosome into the cytoplasm and appear to be linked to other desmosomes of the cell. The desmosome holds adjacent cells together in areas that are subject to considerable stretching, such as in the skin and heart muscle. Desmosomes are usually disk-shaped and thus could be likened to rivets or spotwelds between cells.

Another type of membrane junction, the **tight junction,** is found between epithelial cells. Epithelial layers usually separate two compartments that have differing chemical compositions. For example, the intestinal epithelium lies between the lumen of the intestinal tract that contains the products of food digestion and the blood vessels that pass beneath the epithelial layer. These epithelial layers generally mediate the passage of molecules between the two compartments. A tight junction (Fig. 4-12B and D) is an actual fusing of the two adjacent plasma membranes so that there is no space between adjacent cells at that point. This type of fusion extends around the circumference of the cell and greatly reduces (but does not eliminate) the passage of molecules in between the cells. Therefore, for the most part, in order to cross the epithelium, a molecule must first cross the plasma membrane of an epithelial cell, pass through the cytoplasm, and exit through the membrane on the opposite side. Thus, the tight junction, in addition to helping to hold cells together, seals the passageway in between adjacent cells. Figure 4-12D shows both a tight junction and a desmosome near the luminal border between two epithelial cells.

When a molecule to which the plasma membrane is impermeable is injected into a cell through a micropipette inserted into the cell, the molecule will usually remain in the cell and not pass into adjacent cells. When this experiment is performed on certain types of cells, however, the marker molecule is found in the cells adjacent to the injected cell but not in the extracellular medium, which suggests that there is a direct channel that links the cytoplasms of the two cells. The electron microscope has revealed a structure in these cells known as a **gap junction** (Fig. 4-12C). At the gap junction the two opposing plasma membranes come within 2 to 4 nm of each other. Small channels about 1.5 nm in diameter extend across this gap and link the cytoplasm of the two cells. The small diameter of these channels limits the size of molecules that can pass between the connected cells to small molecules and ions, such as sodium and potassium, and excludes the exchange of large protein molecules. A variety of cell types possess gap junctions, including the muscle cells of the heart and smooth muscle cells. As we shall see, the gap junctions in these cells play an important role in the transmission of electrical signals between adjacent cells. In other cases, gap junctions are thought to coordinate the activities of adjacent cells by allowing chemical messengers to move from one cell to the next.

KEY TERMS

plasma membrane	**membrane pores**	**endocytosis**
fluid mosaic model	**mediated transport**	**exocytosis**
bulk flow	**carrier proteins**	**protein secretion**
diffusion	**facilitated diffusion**	**desmosome**
flux	**active transport**	**tight junction**
concentration gradient	**osmosis**	**gap junction**
membrane permeability	**osmolarity**	

REVIEW EXERCISES

1 List the general contributions of membranes to cell function.
2 Describe the arrangement of phospholipid molecules in cell membranes.
3 Define and identify the location of integral and peripheral membrane proteins.
4 Describe properties of membrane structure that account for the fact that membranes are the flexible structures that can easily be bent and folded.
5 Summarize the main structural characteristics of the fluid mosaic model of membrane structure.
6 Define and distinguish between the two methods of moving molecules in the body, bulk flow and diffusion.
7 Define the terms flux, concentration gradient, and diffusion coefficient.
8 Describe the relation between the concentration gradient of a given substance and
   a the direction in which the net diffusional flux will occur.
   b the magnitude of the net diffusional flux.
9 Describe the value of the concentration gradient and net diffusional flux when a system has reached diffusion equilibrium.
10 Describe the properties of membrane structure that prevent large, charged molecules from diffusing across the membrane.
11 Describe the properties of membranes that allow small charged molecules, such as the ions Na^+, K^+, and Cl^-, to diffuse across membranes.
12 Describe the function of a carrier protein in mediating the movements of some molecules across membranes.
13 Describe the property of carrier molecules that is responsible for the observation that there is an upper limit to the flux of molecules that can cross a membrane by mediated transport. Describe a change in the membrane composition that could produce a change in this limiting maximal flux.

14 Describe and compare the characteristics of the two forms of mediated transport: facilitated diffusion and active transport.
15 Identify some of the molecules or ions that can cross plasma membranes by:
   a faciliated diffusion
   b active transport
16 Describe how the concentration of water in a solution can be increased or decreased.
17 Identify the most abundant solutes in the extracellular and intracellular fluids which are the main determinents of the water concentrations on the two sides of cell plasma membranes.
18 Define the terms isotonic solution, hypertonic solution, and hypotonic solution.
19 A cell placed in a hypertonic solution of sodium chloride will: (*a*) shrink, (*b*) swell, (*c*) undergo no volume change. Explain.
20 Describe and distinguish between the processes of endocytosis and exocytosis.
21 Define the terms pinocytosis and phagocytosis.
22 Describe the relation between the process of endocytosis and the functions of the cell organelles known as lysosomes.
23 Beginning at a ribosome, summarize the pathway that leads to the release (secretion) of a protein from a cell into the extracellular fluid.
24 Identify the site(s) in a cell where the proteins and phospholipids are synthesized and assembled into membranes.
25 Explain how a newly synthesized membrane gets inserted into the plasma membrane at the cell surface.
26 Describe the structural and functional characteristics of the three types of cell to cell junctions:
   a desmosomes
   b tight junctions
   c gap junctions

CHAPTER

5

ELECTRICAL PROPERTIES OF CELLS

BASIC PRINCIPLES OF ELECTRICITY

All chemical bonds are basically electric because they involve exchanging or sharing negatively charged electrons between atoms. Some molecules have no net electric charge because they contain equal numbers of electrons and protons. However, many molecules, as we have seen, have a net electric charge due to components such as the negative carboxyl group, $RCOO^-$, and the positive amino group, RNH_3^+. Moreover, inorganic substances such as sodium, potassium, and chloride (Na^+, K^+, and Cl^-) are present as charged ions. With the exception of water, the major chemical components of the extracellular fluid are, as we have seen, sodium and chloride ions, whereas the intracellular fluid contains high concentrations of potassium ions and organic molecules (particularly proteins and phosphate compounds) that contain ionized groups. Since the environment of the cell contains many charged particles, it is not surprising to discover that electric phenomena resulting from the interaction of these charged particles play a significant role in cell function.

When positive and negative charges are separated, an electric force draws the opposite charges together. Like charges repel each other; thus, positive charge repels positive charge and negative charge repels negative charge. The amount of force acting between electric charges increases when the charged particles are moved closer together and with increasing quantity of charge. When oppositely charged particles come together as a result of the attracting force between them, energy that can be used to perform work is released. Conversely, to separate oppositely charged particles, energy must be added to overcome the attractive forces between the particles. Thus, when electric charges are separated, they have the "potential" of doing work if they are allowed to come together again. The potential of a separated electric charge to do work when it is moving from one point in a system to another is called **electric potential.** One, therefore, refers to the potential difference between two points. The units of electric potential are **volts** (V). The greater the potential difference (voltage) between two points, the greater the work done by an electric charge moving between those two points. Since the total amount of charge that can be separated in most biological systems is very small, the potential differences are small and are measured in millivolts (mV) (1 mV = 0.001 V).

The movement of electric charge is known as **current.** If electric charge is separated between two points, there is a potential difference between these points. The electric force of attraction between the separated charges tends to make charges flow, producing a current. The amount of charge that does move, the current, depends upon the nature of the material lying between the separated charges. The higher the **resistance** of the material, the lower is the amount of current flow for any given voltage. Some materials, like glass and rubber, have such high electric resistance that the amount of current flow through them, even when high voltages are applied, is very small. Such materials, known as *insulators,* are used to prevent the flow of current. Thus, the rubber insulation around electrical wires prevents the flow of current from the wire to areas outside it. Materials having low resistance to current flow are known as *conductors.* Pure water is a relatively poor conductor because it contains very few charged particles. But when sodium chloride is added to the water, the solution becomes a relatively good conductor with a low resistance because the sodium and chloride ions provide charges that can carry the current. The water compartments inside and outside the cells contain numerous charged particles (ions) that are able to move between areas of charge separation. Lipids contain very few charged groups and thus have a high electric resistance. The lipid components of the cell membrane provide a region of high electric resistance that separates two water compartments of low resistance.

MEMBRANE POTENTIALS

The presence of an electric potential across a plasma membrane can be determined by inserting a very small electrode (an electrical conductor) into the cell, another into the extracellular fluid, and connecting the two to a voltmeter. In this manner it has been found that all cells under resting conditions have a membrane potential oriented so that the inside of the cell is negatively charged with respect to the outside. This potential is the **resting membrane potential;** its magnitude varies from 5 to 100 mV, depending upon the type of cell and its chemical environment. As we shall see, the resting

TABLE 5-1	DISTRIBUTION OF MAJOR IONS ACROSS THE PLASMA MEMBRANE OF A NERVE CELL	
Ion	Extracellular Concentration, mmol/L of Water	Intracellular Concentration, mmol/L of Water
Na^+	150	15
Cl^-	110	10
K^+	5	150

potential of some cells, for example nerve and muscle cells, changes in response to various types of stimuli.

The concentrations of the major ions in the fluid bathing cells (the extracellular fluid) is approximately that listed in Table 5-1. As we have mentioned before, the ionic composition of the intracellular fluid is entirely different. There are many other ions, such as Mg^{2+}, Ca^{2+}, HCO_3^-, HPO_4^{2-}, and SO_4^{2-}, in both fluid compartments; but sodium, potassium, and sometimes chloride, which are the ions present in greatest concentrations, play the most important roles in the generation of the resting membrane potential.

Diffusion Potentials

The separation of electric charge across the plasma membrane, which results in the resting membrane potential, is produced mainly by the diffusion of ions across the membrane. Thus, the resting membrane potential is a **diffusion potential,** as are most of the other electric potentials we will encounter in association with various types of cell activity.

In order to understand how the diffusion of ions across a membrane can lead to the separation of electric charge, consider the situation illustrated in Fig. 5-1, which depicts two solutions separated by a membrane. On one side of the membrane there is a 0.15 M solution of sodium chloride; the other compartment contains a 0.15 M solution of potassium chloride. Initially there is no separation of electric charge across the membrane. There are just as many sodium ions (Na^+) as chloride ions (Cl^-) in compartment 1 and thus there is no net charge in this compartment. The same is true of compartment 2, where the total number of potassium ions (K^+) is equal to the number of chloride ions.

What will happen if the membrane that separates these two solutions is permeable to potassium but

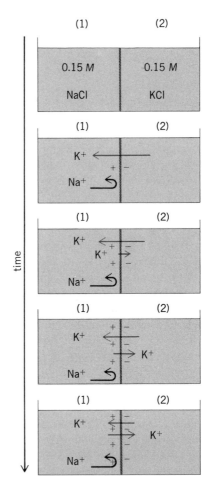

FIG. 5-1 Generation of a diffusion potential across a membrane permeable only to potassium ions. An equilibrium is reached when there is zero net diffusion of potassium across the membrane, as seen in the bottom part of the figure.

does not allow either sodium or chloride to cross? Potassium ions will diffuse from their high concentration in compartment 2 into compartment 1, which had no potassium initially. As each potassium ion crosses the membrane, it adds one net positive charge to compartment 1 and leaves behind one net negative charge in compartment 2 in the form of a chloride ion, which is unable to cross the membrane. Thus, compartment 1 will develop a net positive charge and compartment 2 will become negatively charged. Electric charge is being carried across the membrane by the diffusion of potassium ions, and thus a diffusion potential is generated.

Will potassium ions continue to diffuse across the membrane, adding more and more positive charge to compartment 1, until the concentration of potassium ions in compartment 1 becomes equal to the concentration of potassium ions in compartment 2? The answer is no because there are actually two forces acting on the movement of potassium across the membrane. One force is the concentration gradient that leads to the net movement of potassium from high to low concentration. If this were the only force acting on potassium, diffusion would continue until the concentrations of potassium in the two compartments became equal. But there is a second force acting on the movement of potassium, namely the electric force created by the net separation of charge across the membrane. Recall that oppositely charged particles attract each other. Thus, the positively charged ions in compartment 1 will be attracted to the negative charges in compartment 2. Since potassium ions are free to move, this electric attraction will pull some of the potassium ions from compartment 1 into compartment 2, a movement that is opposite to the net diffusion produced by the concentration gradient. Because like charges repel each other, the positive charge that is developing in compartment 1 will also oppose the diffusion of positive potassium ions into compartment 1.

Consider the net effect of these two forces, the concentration gradient and the electrical gradient, on the movement of potassium ions across the membrane. Initially the only force acting to produce potassium net movement is the concentration gradient. As soon as a few potassium ions have crossed the membrane, an electric gradient will be produced that will decrease the one-way flux of potassium into compartment 1 and increase the one-way flux of potassium back into compartment 2. The overall result will be a decrease in the net flux of potassium across the membrane. The electric gradient will continue to increase in magnitude as long as there is a net diffusion of potassium into compartment 1. Eventually, however, the electric gradient will reach a magnitude where the decreasing one-way flux of potassium out of compartment 2 will become equal to the increasing one-way flux of potassium into compartment 2. At this membrane potential there will be no net flux of potassium across the membrane and no further change in the concentrations of potassium on the two sides of the membrane. This potential, at which there is zero net diffusion of an ion across a membrane, is known as the ion's **equilibrium potential** and is illustrated in the bottom part of Fig. 5-1. The equilibrium poten-

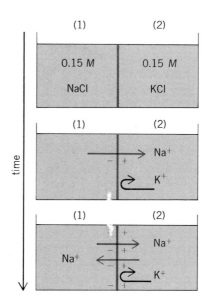

FIG. 5-2 Generation of a diffusion potential across a membrane permeable only to sodium ions. An equilibrium is reached when there is zero net diffusion of sodium across the membrane, as seen in the bottom part of the figure.

tial is reached after a very small number of the total potassium ions in compartment 2 have moved into compartment 1. There is thus very little change in the concentration of potassium in compartment 2— only about 0.001 percent. Even though there is still a large concentration gradient across the membrane, there is no net diffusion of potassium across the membrane because the concentration's driving force has been exactly balanced by the electric force of the equilibrium potential.

It can be seen that the value of the equilibrium potential for any ion depends upon the concentration gradient for that ion across the membrane. If the concentrations in the two compartments were equal, the concentration force would be zero and the electric potential required to oppose it would also be zero. The larger the concentration gradient, the larger is the equilibrium potential. Using potassium concentrations typical for nerve cells and extracellular fluid, the equilibrium potential for potassium is close to 90 mV, the inside of the cell being negative with respect to the outside.

If the membrane separating compartments 1 and 2 is replaced with one permeable only to sodium, the initial net flow of the positively charged sodium will be from compartment 1 to 2 and compartment 2 will become positive (Fig. 5-2). This movement of sodium down its concentration gradient is opposed by the electric force generated by that movement. A sodium equilibrium potential will be established with compartment 2 positive with respect to compartment 1, at which point net movement will cease. The equilibrium potential for sodium based on the sodium concentrations found across most nerve plasma membranes is about 60 mV, inside positive. The equilibrium potential for each ion species is different because the concentration gradients are different.

The Resting Cell Membrane Potential

It is not difficult to move from these hypothetical experiments to a nerve cell at rest where (1) the potassium concentration is much greater inside the cell than out and the sodium concentration gradient is the opposite and (2) the plasma membrane is some 50 to 75 times more permeable to potassium than to sodium. Given these characteristics, it should be evident that a diffusion potential will be generated across the membrane, largely because of the movement of potassium out of the cell, so that the inside of the cell becomes negative with respect to the outside. The experimentally measured membrane potential of −70 mV is not, however, equal to the potassium equilibrium potential of −90 mV because the membrane is also permeable to sodium and some sodium continually diffuses into the cell, carrying a small amount of positive charge. Note that at a resting membrane potential of −70 mV there will be both a concentration gradient and an electric gradient operating to move sodium into the cell. However, because of the low membrane permeability to sodium the magnitude of the net sodium flux is low. Since neither sodium nor potassium is at their equilibrium potential, there will be a net flux of sodium into the cell and of potassium out of the cell at the resting membrane potential. The reason the resting potential does not change, in spite of all this charge movement across the membrane, is that the net diffusion of sodium into the cell is equal to the net diffusion of potassium out of the cell; thus, there is zero net movement of positive charge across the membrane.

When more than one ion species can diffuse across the membrane, the membrane permeability properties as well as the concentration gradient of each species must be considered when accounting for the membrane potential. If the membrane is impermeable to a given ion species, no ion of that species can cross the membrane and contribute to a diffusion potential, regardless of the electric and concentration gradients that may exist. For a given concentration gradient, the greater the membrane permeability to an ion species, the greater the influence that ion species will have on the membrane potential. Since the resting membrane is much more permeable to potassium than to sodium, the resting membrane potential is much closer to the potassium equilibrium potential than to that of sodium.

If there is net movement of sodium into and potassium out of the cell, why do the concentration gradients not run down? The reason is that active-transport mechanisms in the membrane utilize energy derived from cellular metabolism to pump the sodium back out of the cell and the potassium back in. Actually, the pumping of these ions is linked because they are both transported by the same carrier in the membrane. If the linkage were to result in the exchange of sodium and potassium on a one-to-one basis, the pump would not directly separate charge. Whenever the exchange is not one-to-one, the pump will directly separate charge and is known as an **electrogenic pump.** Thus, a membrane potential can result both from diffusion (as we have been describing) and from the operation of an electrogenic pump. In most cells (but by no means all), the electrogenic contribution of the pump to the membrane potential is quite small. However, the pump always makes an essential indirect contribution to the membrane potential, because it maintains the concentration gradients down which ions can diffuse to establish a membrane potential.

SIGNALS OF THE NERVOUS SYSTEM: GRADED POTENTIALS AND ACTION POTENTIALS

Changes in the membrane potential from its resting level can convey meaningful information to a cell. Nerve and muscle cells in particular use such voltage changes as signals in receiving, integrating, and transmitting information. These signals occur in two forms: **graded potentials** and **action potentials.** Graded potentials are extremely important in signaling over short distances along the plasma membrane; action potentials, on the other hand, are the

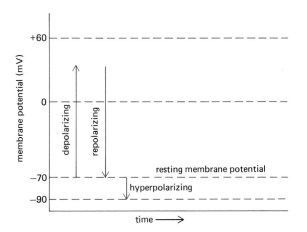

FIG. 5-3 Depolarizing, repolarizing, and hyperpolarizing changes in membrane potential.

long-distance signals of nerve and muscle membranes.

Recall that the word *potential* means a voltage difference between two points. Adjectives such as *membrane* potential, *resting* potential, *diffusion* potential, *graded* potential, etc., define the conditions under which the potential is measured or the way it developed. *Depolarize, hyperpolarize,* and *repolarize* will be used to describe changes in membrane potential. The membrane is said to be **depolarized** when the membrane potential is less negative than the resting membrane potential, i.e., closer to zero; it is **hyperpolarized** when the potential is more negative than the resting level. When the membrane potential is changing so that it moves toward or even above zero, it is **depolarizing;** when it moves away from zero back toward its resting level, it is **repolarizing** (Fig. 5-3). When it becomes more negative than its resting level, it is **hyperpolarizing.**

Graded Potentials

Graded potentials are local changes in membrane potential in either a depolarizing or hyperpolarizing direction (Fig. 5-4, part 1). They are usually produced by some specific change in the cell's environment acting on the plasma membrane. They are

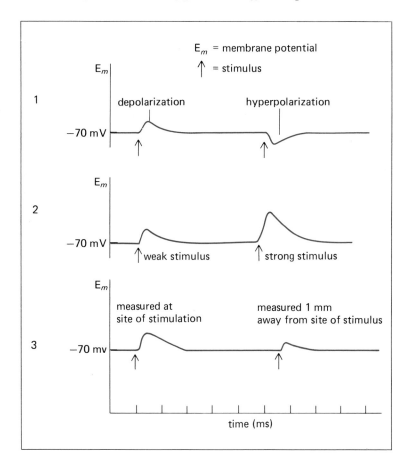

FIG. 5-4 Graded potentials: (1) can be depolarizing or hyperpolarizing, (2) vary in magnitude depending on stimulus strength, (3) decrease in magnitude with distance from the site of stimulus.

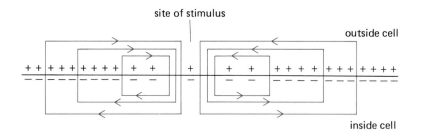

site of stimulus

outside cell

inside cell

FIG. 5-5 Ion current flow in the cytoplasm of a nerve fiber and in the extracellular fluid surrounding it. Membrane is slightly depolarized (less negative inside) at the site of stimulus.

called graded potentials because the magnitude of the potential change is variable and related to the magnitude of the external event (see Fig. 5-4, part 2).

Graded potentials cause current to flow in the region around the potential change; the greater the potential change, the greater the current flow. By convention the direction in which positive ions move is designated the direction of current flow, although negatively charged particles simultaneoulsy move in the opposite direction. The intracellular and extracellular fluids are fairly good conductors, and current flows through them whenever a voltage difference occurs. When a stimulus causes a graded potential that slightly depolarizes a membrane, that area of the membrane has an electric potential different from that of adjacent areas. Because unlike charges attract and like charges repel, inside the cell current flows away from the activated membrane region through the intracellular fluid, and outside the cell current flows toward the activated region through the extracellular fluid. This local current flow removes positive charges from the outside of the membrane adjacent to the site of depolarization and adds positive charge to the inside of the membrane in this region producing depolarization of the adjacent membrane (see Fig. 5-5).

These local currents almost completely die out within a few millimeters of their point of origin. For this reason, local current flow is *decremental;* i.e., its magnitude decreases with increasing distance. The resulting change in membrane potential, therefore, also decreases with distance from the site at which the graded potential was initiated (Fig. 5-4, part 3).

Because of decremental conduction, graded potentials (and the current flows they generate) can function as signals only over very short distances. Nevertheless, graded potentials play very important roles in information processing, as we shall see in Chap. 13.

Action Potentials

Action potentials are very different from graded potentials. They are rapid alterations in membrane potential that may last only 1 msec. During this time the membrane changes from -70 to $+30$ mV and then returns to its original value (Fig. 5-6). Only a few types of cells (nerve, muscle, and some gland cells) have plasma membranes that are capable of producing action potentials. These membranes are known as **excitable membranes,** and their ability to generate action potentials is known as **excitability.**

Ionic Basis of the Action Potential   We have seen that the magnitude of the resting membrane potential depends upon ion concentration gradients and membrane permeabilities to ions, particularly sodium and potassium. This situation is true for the period of the action potential as well. Thus, the action potential must result from a transient change in either the concentration gradients or the membrane permeabilities. The latter is the case. In the resting state the membrane is 50 to 75 times more

FIG. 5-6 Changes in membrane potential during an action potential. (The sodium equilibrium potential is $+60$ mV; the one for potassium is -90 mV.)

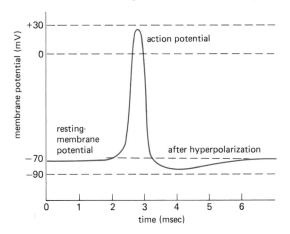

permeable to potassium than to sodium ions. Thus, the magnitude and polarity of the resting potential are due almost entirely to the movement of potassium ions out of the cell. During an action potential, however, the permeability of the membrane to sodium and potassium ions is markedly altered. In the depolarizing phase of the action potential the membrane permeability to sodium ions increases several hundredfold and sodium ions rush into the cell, whereas there is little change in the potassium permeability of the membrane. During this period more positive charge enters the cell in the form of sodium ions than leaves in the form of potassium ions. Thus the membrane potential decreases and eventually reverses its polarity, becoming positive on the inside and negative on the outside of the membrane. In this phase the membrane potential approaches but does not quite reach the sodium equilibrium potential.

Action potentials in nerve cells last about 1 msec (0.001 sec). What causes the membrane to return so rapidly to its resting level? The answer to this question is twofold: (1) The increased sodium permeability (*sodium activation*) is rapidly turned off (*sodium inactivation*), and (2) the membrane permeability to potassium increases over its resting level. The timing of these two events can be seen in Fig. 5-7. The decrease in sodium permeability alone would restore the potential to its resting level. However, the entire process in speeded up by the simultaneous increase in potassium permeability, which causes more potassium ions to move out of the cell down their concentration gradient. These two events, decreased sodium permeability and increased potassium permeability, rapidly return the membrane potential to its resting level. In fact, during the time when the potassium permeability is greater than normal, there is generally a small hyperpolarizing overshoot of the membrane potential (*after-hyperpolarization,* Fig. 5-7).

In our description it may have seemed as though the sodium and potassium fluxes across the membrane involved large numbers of ions. Actually, only about 1 out of every 100,000 potassium ions in the cell diffuses out to charge the membrane potential to its resting value, and very few sodium ions enter the cell during an action potential. Thus, there is almost no change in the concentration gradients during an action potential. Yet if the tiny number of additional ions crossing the membrane with each action potential were not eventually moved back across the membrane, the concentration gradients of sodium and potassium across the membrane would gradually disappear and action potentials could no longer be generated. As might be expected, an accumulation of sodium and loss of potassium are prevented by the continuous action of the membrane active-transport system for sodium and potassium. This restoration occurs mainly after the action potential is over. The number of ions that cross the membrane during an action potential is so small, however, that the pump need not keep up with the action-potential fluxes, and hundreds of action potentials can occur even if the pump is stopped experimentally. Thus, the pump plays no direct role in the generation of the action potential.

Mechanism of Permeability Changes   The cause of the permeability changes that underlie action potentials has been elucidated by experiments in which membrane permeability and ion fluxes are measured as the membrane potential is changed. The change in membrane potential is accomplished by electrodes that add positive charge to the outside of the membrane while simultaneously removing positive charge from the inside of the cell, thereby causing the membrane potential to become hyperpolarized. If the electrodes are reversed, they will remove positive charge from the outside while adding it to the inside, thereby depolarizing the cell membrane.

FIG. 5-7 Changes in the membrane permeability to sodium (P_{Na}) and potassium (P_K) ions during an action potential.

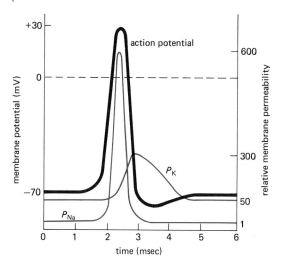

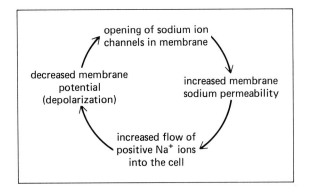

FIG. 5-8 Positive-feedback relation between membrane depolarization and increased sodium permeability, which leads to the rapid depolarizing phase of the action potential.

Measurement of membrane-permeability changes during these experiments revealed that the permeability to sodium was altered whenever the membrane potential was changed; specifically, hyperpolarization of the membrane caused a decrease in sodium permeability whereas depolarization caused an increase in sodium permeability. In light of our previous discussion of the ionic basis of membrane potentials, it is very easy to confuse the cause-and-effect relationships of the statement just made. Earlier we pointed out that an increase in sodium permeability *causes* membrane depolarization; now we are saying that depolarization *causes* an increase in sodium permeability. Combining these two distinct causal relationships yields the positive-feedback cycle (Fig. 5-8) responsible for the rising phase of the action-potential spike: Depolarization alters the plasma membrane structure so that its permeability to sodium increases. Because of increased sodium permeability, sodium diffuses into the cell. This addition of positive charge to the cell further depolarizes the membrane, which, in turn, produces a still greater increase in sodium permeability, which, in turn, causes. . . . And so on.

The sodium ions move through membrane channels that behave like "gates," such that the channels are open under some conditions but closed under others. The channels are thought to consist of assemblies of proteins that stretch through the membrane. The gating functions are probably achieved by changes in the shape of these proteins. For example, the gates to the channels could be closed when the proteins are arranged so that parts of the molecules effectively block the channel. Un-

der the influence of a reduced voltage across the membrane (depolarization) a rearrangement of the proteins could occur and open the channel. Sodium inactivation may be another change in protein structure, blocking the channels again at the peak of the action potential and thus keeping them closed even if the membrane remains depolarized.

Excitable membranes contain these voltage-sensitive ion channels, while other types of membranes, which do not generate action potentials, have ion channels that do not open or close in response to changes in membrane potential.

Threshold   Not all depolarizations trigger the positive-feedback relationship that leads to an action potential. Action potentials occur only when the membrane is depolarized enough (and, therefore, sodium permeability is increased enough) so that sodium entry exceeds potassium efflux. In other words, an action potential occurs only when the net movement of positive charge is *inward*. The membrane potential at which this happens is called the **threshold potential,** and stimuli strong enough to depolarize the membrane to this level are called *threshold stimuli* (Fig. 5-9). The threshold potential of most excitable membranes is 5 to 15 mV more depolarized than the resting membrane potential. Thus, if the resting potential of a neuron is -70 mV, the threshold potential may be -60 mV. In order to initiate an action potential in such a membrane, the potential must be depolarized by at least 10 mV. At depolarizations less than threshold, potassium movement still dominates and the positive-feedback cycle cannot get started despite the increase in sodium entry. In such cases, the membrane returns to its resting level as soon as the stimulus is removed and no action potentials are generated. These weak depolarizations are *subthreshold potentials,* and the stimuli that cause them are *subthreshold stimuli.*

Stimuli of more than threshold magnitude (*suprathreshold stimuli*) elicit action potentials, but as can be seen in Fig. 5-9, the action potentials resulting from such stimuli have exactly the same magnitude as those caused by threshold stimuli. This is explained by the fact that, once threshold is reached, membrane events are no longer dependent upon stimulus strength. The depolarization continues to become an action potential because the positive-feedback cycle is operating. An action po-

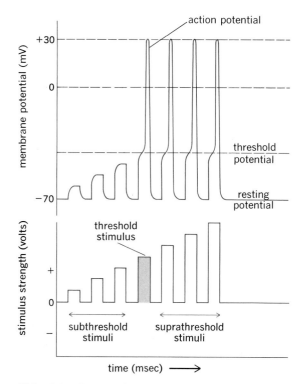

FIG. 5-9 Decreasing membrane potential with increasing strength of depolarizing stimulus. When the membrane potential reaches the threshold potential, action potentials are generated. Increasing the stimulus strength above threshold level does not alter the magnitude of the action potential response.

tential either occurs as determined by the electrochemical conditions across the membrane or it does not occur at all. Another way of saying this is that action potentials are *all or none.*

Because of the all-or-none nature of the action-potential response, a single action potential cannot convey any information about the magnitude of the stimulus that initiated it, since a threshold-strength stimulus and one of twice threshold strength give the same response. Since one function of nerve cells is to transmit information, one may ask how a system operating according to an all-or-none principle can convey information about the strength of a stimulus. The answer, as we shall see in Chap. 13, depends upon the number of action potentials transmitted per unit time, i.e., the frequency of action potentials, and not upon their size. It also depends upon the number of nerve cells activated.

Refractory Periods   How soon after firing an action potential can an excitable membrane be stimulated to fire a second one? If we apply a threshold-strength stimulus to a membrane and then stimulate the membrane with a second threshold-strength stimulus at various time intervals following the first, the membrane does not always respond to the second stimulus. Even though identical stimuli are applied, the membrane appears unresponsive for a certain time. The membrane during this period is said to be **refractory** to a second stimulus.

Instead of applying the second stimulus at threshold strength, if we increase it to suprathreshold levels, we can distinguish two separate refractory periods associated with an action potential During the action potential a second stimulus will not produce a second action potential no matter how strong it is. The membrane is said to be in its **absolute refractory period.** Following the absolute refractory period there is an interval during which a second action potential can be produced but only if the stimulus strength is considerably greater than the usual threshold level. This is known as the **relative refractory period** and can last some 10 to 15 msec or longer.

The mechanisms responsible for the refractory periods are related to the membrane mechanisms that alter the sodium and potassium permeability. The absolute refractory period corresponds roughly with the period of sodium permeability changes, and the relative refractory period corresponds roughly with the period of increased potassium permeability. Following an action potential, time is required to return the voltage-sensitive ion channels to their original resting state.

The refractory periods limit the number of action potentials that can be produced by an excitable membrane in a given period of time. Recordings made from nerve cells in the intact organism that are responding to physiologic stimuli indicate that most nerve cells respond at frequencies up to 100 action potentials per second although some may produce much higher frequencies for brief periods.

Action-Potential Propagation   In describing the generation of action potentials, we used stimulating electrodes to bring the membrane to threshold. In the body the threshold stimulus is provided by the graded potentials described earlier. How this occurs will be presented later.

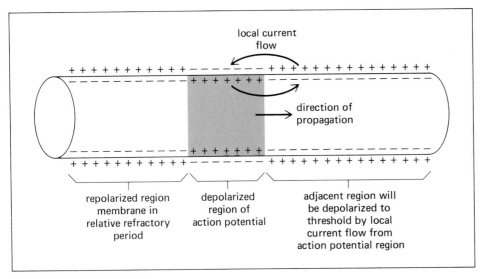

FIG. 5-10   Propagation of an action potential along a plasma membrane.

Once generated, how does the action potential travel from one point to another? One particular action potential does not itself travel along the membrane; rather, each action potential triggers, by local current flow, a *new* action potential at an adjacent area of membrane. The local current flow is great enough to depolarize the adjacent membrane site to just past its threshold potential, the sodium positive feedback cycle at this membrane site takes over, and an action potential occurs there. Once the adjacent site is depolarized to threshold, the action potential generated there is solely dependent upon the ion gradients and membrane permeability properties at the new site. Since these factors are usually identical to those involved in the generation of the initial action potential, the new action potential is virtually identical to the old. This is an important point because it means that no distortion occurs as the signal passes along the membrane; the action potential that arrives at the end of the membrane is identical to the initial one.

Meanwhile, at the original membrane site, sodium inactivation is occurring and potassium permeability is increasing so that the membrane is repolarizing. These processes repeat themselves until the end of the membrane is reached: A wave of depolarization and repolarization travels smoothly along the membrane as each active area in turn stimulates by local current flow the area immediately ahead of it (Fig. 5-10).

TABLE 5-2 THE DIFFERENCES BETWEEN GRADED POTENTIALS AND ACTION POTENTIALS	
Graded Potentials	**Action Potentials**
1 Graded response; amplitude reflects conditions of the initiating event.	All-or-none response; once membrane is depolarized to threshold, amplitude is independent of initiating event.
2 Graded response; can be summed.	All-or-none response; cannot be summed.
3 Has no threshold.	Has a threshold that is usually 10–15 mV depolarized relative to the resting potential.
4 Has no refractory period.	Has a refractory period.
5 Is conducted decrementally; i.e., amplitude decreases with distance.	Is conducted without decrement; the amplitude is constant.
6 Duration varies with initiating conditions.	Duration is constant.
7 Can be a depolarization or hyperpolarization.	Is a depolarization (with overshoot).
8 Initiated by stimulus, neurotransmitter, or spontaneously.	Initiated by membrane depolarization.

Local current flow occurs wherever there is an electric gradient; however, the membrane areas that have undergone an action potential are refractory and cannot undergo another. Thus the only direction of action-potential propagation is away from the stimulation site.

The action potentials in skeletal muscle cells are initiated near the middle of these cylindrical cells and propagate from this region toward the two ends, but in most nerve-cell membranes action potentials are normally initiated at one end of the cell and propagate toward the other end of the cell. This unidirectional propagation of action potentials in nerve cells is determined by the stimulus location rather than an intrinsic inability to conduct in the opposite direction.

The velocity with which an action potential is propagated along a membrane depends upon the cell diameter and whether or not the membrane is covered by special insulating coats, as occurs in many nerve cells (Chap. 13). The larger the cell diameter, the faster the action-potential propagation, because a large cell offers less resistance to local current flow; therefore, adjacent regions of the membrane are brought to threshold faster.

Some of the differences between graded potentials and action potentials are summarized in Table 5-2.

KEY TERMS

electrical potential	hyperpolarize	voltage-sensitive ion channels
current	graded potentials	threshold potential
resting membrane potential	local currents	all-or-none principle
diffusion potential	decremental conduction	refractory period
equilibrium potential	action potential	action potential propagation
depolarize	excitable membrane	

REVIEW EXERCISES

1 Define the term electrical potential. Name the units in which electrical potentials are measured.
2 Identify the polarity (extracellular side positive or negative) of the membrane potential of all cells under resting conditions.
3 Describe the process by which the diffusion of ions across a membrane can result in a membrane potential (diffusion potential).
4 Identify the two "forces" that determine the magnitude of the net flux of ion diffusion across a membrane.
5 Describe the flux of potassium across a membrane if the membrane potential is equal to the potassium equilibrium potential.
6 Describe the movements of sodium and potassium ions across a plasma membrane that are responsible for producing the resting membrane potential.
7 Explain why the resting membrane potential is closer to the potassium equilibrium potential than to the sodium equilibrium potential.
8 Describe the role of sodium and potassium active transport across the plasma membrane in producing a membrane potential.
9 Describe the local current flows around a region of a membrane that has been depolarized.
10 Describe the decremental conduction of a graded membrane potential.
11 Define the term excitability. List the types of cells that have excitable membranes.
12 Describe the changes in membrane potential that occur during an action potential.
13 Identify the change in membrane permeability that is responsible for the depolarizing phase of an action potential.
14 Identify the two factors that are responsible for the rapid repolarization of an action potential from its peak to the resting membrane potential.
15 Describe the change in the membrane permeability to sodium ions that occurs in excitable membranes when the membrane is depolarized.
16 Define the terms threshold potential, subthreshold stimuli, and suprathreshold stimuli.
17 Explain what is meant by the statement that action potentials are "all or none" potentials.
18 Define the terms absolute refractory period and relative refractory period.
19 Describe how local current flows lead to the propagation of an action potential along an excitable membrane.
20 Describe the relation between the diameter of an excitable cell and the velocity at which action potentials are propagated along the plasma membrane.
21 Summarize the differences and similarities between graded potentials and action potentials.

EPITHELIUM, CONNECTIVE TISSUE, AND SKIN

The four categories of specialized tissue—epithelium, connective tissue, nerve, and muscle—that are arranged in various proportions to form the body's organs, arise during embryonic growth and development by the process of cell differentiation. This chapter will examine the structural and functional characteristics of epithelial and connective tissues as well as skin, which is largely a composite of these two tissues.

EPITHELIUM

The external and internal surfaces of the body are covered by layers of epithelial cells arranged in flat or tubular sheets to form boundaries between various compartments. Just as the plasma membrane of a cell provides a selective barrier to the flow of molecules between the cell and the extracellular fluid, the epithelial cells covering or lining the body's surfaces are selective barriers that regulate the movement of molecules across these surfaces. For example, the epithelial cells of the skin form a barrier at the body's surface; the epithelial lining of the lungs, gastrointestinal tract, and kidneys regulates the movement of molecules between the blood and the external environment; the epithelial cells lining the smallest blood vessels (capillaries) form the barrier between the blood and the interstitial fluid. In addition, almost all glands (a cell or group of cells that secrete a substance) are composed of epithelial tissue.

The specialized functions performed by epithelial cells are: *absorption,* the movement of molecules from the external environment into the internal environment (interstitial fluid and plasma); *excretion,* the movement of molecules from the internal environment to the external environment; *secretion,* the release of specific molecules into the extracellular medium (Fig. 6-1). Each of these processes is associated with the movement of molecules from one compartment to another and, thus, the specialized functions of epithelial cells are primarily dependent on the permeability and transport characteristics of their plasma membranes.

FIG. 6-1 Epithelial surfaces and their relation to excretion, absorption, and secretion.

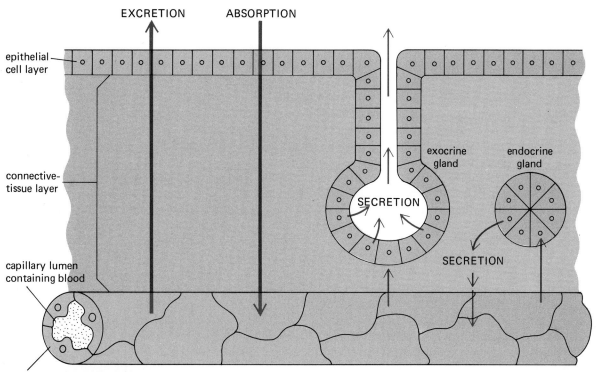

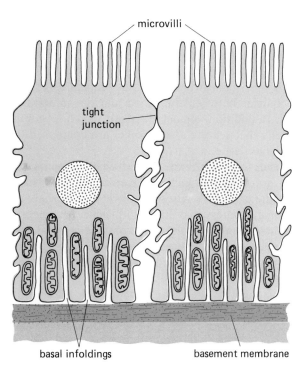

microvilli

tight junction

basal infoldings

basement membrane

FIG. 6-2 Extensive foldings of the plasma membrane are characteristic features of epithelial cells at absorptive and secretory surfaces.

Types of Epithelia

Epithelial cells are organized into sheets of one or more layers of cells. The *apical* surface of the epithelial sheet is exposed to the aqueous or gaseous medium at the outer surface of the epithelium, and the *basal* region rests upon the underlying tissues (Fig. 6-2). At the boundary between the basal region and the underlying tissues is an extracellular structure known as the **basement membrane.** It consists of a matrix of polysaccharides (secreted by the overlying epithelial cells), in which are embedded fine filaments formed by the underlying connective-tissue cells. The basement membrane forms a continuous layer along the basal surface of the epithelium. In some types of epithelia the basement membrane may be quite thin, whereas in others it is thick and may offer considerable resistance to the diffusion of large molecules, such as proteins. All blood vessels are below the basement membrane and do not penetrate into the epithelial layers. Nerve fibers, however, may penetrate the basement membrane and terminate near the plasma membranes of the epithelial cells.

Epithelial cells are classified according to the shape of the cells and the number of layers of cells constituting the epithelial surface. Epithelia that consist of a single layer of cells are **simple epithelia** (Fig. 6-3). The cells may be thin and flat, termed *squamous;* roughly symmetric in shape, *cuboidal;* or tall and thin, *columnar.* Simple epithelia are usually found at surfaces that move large numbers of molecules across the epithelial layer; examples are the columnar epithelium lining the kidney tubules and the intestines. A single layer of squamous epithelium lines the inside of the heart and blood vessels and is known as *endothelium;* a similar layer of squamous epithelium *(mesothelium)* is found in the lining of the pleural, peritoneal, and pericardial cavities. One form of simple epithelium known as *pseudostratified* appears to be composed of several layers but is, in fact, a single layer of cells sitting on the basement membrane and having unequal heights, some of which do not reach all the way to the apical surface (Fig. 6-3).

Epithelia composed of more than one layer of cells are **stratified epithelia** (Fig. 6-4). Such multilayered epithelia are found where the surface is subject to mechanical stress; examples are the skin and various ducts that link parts of the body to the external environment. The shapes of the cells in the various layers of stratified epithelia may differ, but the type of stratified epithelium is designated as squamous, cuboidal, or columnar on the basis of the shape of the cells at the apical surface. Thus, *stratified squamous epithelia* have thin, flat cells at their apical surfaces although the cells in the underlying layers may be cuboidal or columnar.

In regions such as the urinary bladder, where the epithelium is subject to very large degrees of stretch, a special type of stratified epithelium, known as *transitional* epithelium, is found. As transitional epithelial membranes stretch, the cuboidal superficial cells flatten out and resemble squamous epithelium; thus, there is a considerable increase in surface area without rupturing the cells.

Cell junctions (Chap. 4) between the epithelial cells play important roles in maintaining the mechanical integrity of the surface layers and forming a barrier to the extracellular movement of molecules between the apical and basal surfaces. Desmosomes, which provide regions of high mechanical strength between cells, are quite numerous between the cells in the epithelial layers of the skin, a tissue that is subject to considerable mechanical stress.

Epithelial cells divide at a fairly rapid rate, and some epithelial linings (for example, the luminal surface of the intestinal tract) are entirely replaced every few days. In those epithelia composed of several cell layers, the cells at the basal surface provide the stem cells that undergo division; the new cells then migrate toward the apical surface. Since most epithelial layers are subject to varying degrees of mechanical and chemical stress, the continual replacement of damaged cells is essential to maintaining the integrity of the epithelial boundary.

Epithelial Membranes   The term *epithelial membrane* is often used to denote the combination of an epithelial-cell layer and its underlying connective-tissue layer. Thus, it refers to a multicellular structure located at the surface of an organ, not to the thin lipoprotein plasma membrane at the surface of individual cells. There are two types of epithelial membranes in the body: (1) mucous membranes and (2) serous membranes.

Mucous membranes line the surfaces of structures that are connected to the outside of the body; ex-

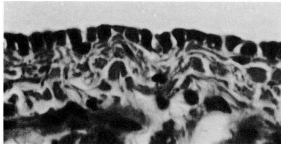
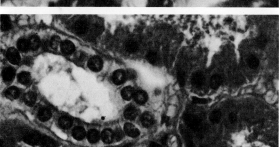
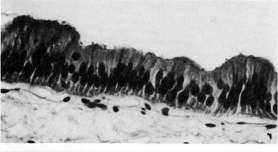
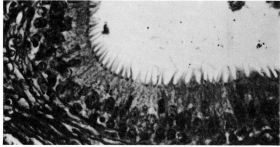

FIG. 6-3   Types of simple epithelia.

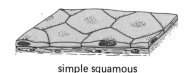

simple squamous

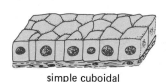

simple cuboidal

simple columnar

pseudostratified columnar (ciliated)

amples are the lining of the mouth and gastrointestinal tract, the airways leading to the lungs, and the passages from the reproductive and urinary systems to the exterior. The epithelial cells at the surface of a mucous membrane secrete **mucus,** a mixture of proteins and polysaccharides having a thick, viscous consistency that moistens and lubricates the epithelial surface.

Serous membranes line the body cavities that are not connected to the exterior of the body; these include the abdominal, thoracic, and pericardial cavities. These membranes secrete a fluid that contains salts and proteins, similar in composition to blood serum. These secretions moisten and lubricate the surfaces of the body cavities, allowing the

organs within these cavities to move smoothly relative to each other as, for example, during inflation of the lungs, when the expanding lung surface moves relative to the surface of the thoracic and pericardial cavities.

Epithelial Surfaces   A characteristic of epithelial cells found at boundaries across which large amounts of material are actively transported is the extensive folding of the plasma membrane (Fig. 6-2). At the apical surface of the cell, numerous fingerlike projections, **microvilli,** extend into the surrounding medium. The microvilli may be short and irregularly spaced, or fairly long and densely packed, giving a brushlike appearance known as a

FIG. 6-4 Types of stratified epithelia.

stratified cuboidal

stratified squamous

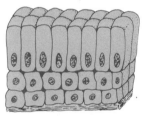
stratified columnar

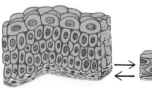

transitional epithelium
(unstretched)

transitional epithelium
(stretched)

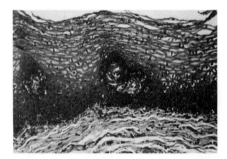

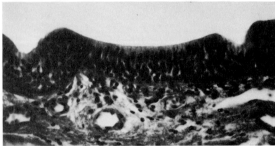

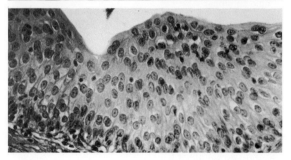

brushborder to the surface of the cell. Foldings of the plasma membrane may also occur on the lateral and basal surfaces of the cell; in the basal region they often form extensive infoldings deep into the cytoplasm of the cell. By increasing the surface area of the cell these microvilli and infoldings increase the diffusion of molecules across the membrane because this flux is proportional to the surface area. Moreover, carrier-mediated transport systems are located in these plasma membranes and the greater the membrane surface area, the greater the number of carrier proteins, although the types and concentrations of carrier molecules may differ

Fig. 6-5 Cilia extending from the surface of an epithelial cell. The microtubular substructure of a single cilium is shown in cross section at top.

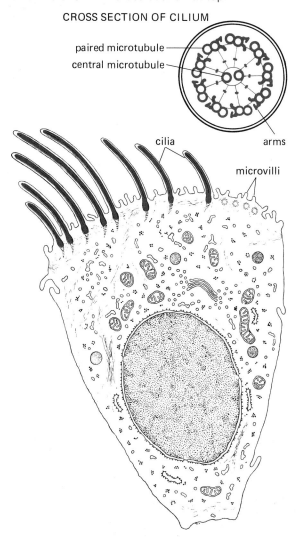

CROSS SECTION OF CILIUM

paired microtubule

central microtubule

cilia

arms

microvilli

at the apical and basal regions of the cell. Numerous mitochondria are often found associated with the basal infoldings of the membrane, where they supply ATP to the active ion transport mechanisms concentrated in this region of the cell.

A further modification of the cell surface occurs in certain types of epithelia that have motile, hairlike cilia extending from their apical surface. In cross section a **cilium** consists of nine sets of paired microtubules surrounding a central pair and resembles the microtubular structure of the cell's centrioles from which the cilia are derived (Fig. 6-5). Changes in the shapes of proteins associated with the microtubules generate forces that bend the cilia. The beat of a cilium consists of a power stroke, during which the cilium behaves like a relatively stiff rod, and a return stroke, when it resembles more a piece of flexible rope. Cycles of ciliary beating thus provide a unidirectional propulsive force at the cell surface. In single-cell organisms, such as a paramecium, the beating cilia propel the cell through the water. In the case of ciliated epithelia, the cell remains anchored to the epithelial surface and the beating cilia propel any overlying material at the surface along the epithelial layer in the direction corresponding to the power stroke of the ciliary beat.

Glands

Glands are formed during embryonic development by the invagination of the epithelial surfaces. Many of the glands remain connected by ducts to the epithelial surface from which they were formed and are known as **exocrine glands.** The secretions of the exocrine glands pass along the ducts and are discharged into the lumen of the organ, or, in the case of skin glands, onto the surface of the skin. Thus, the functions of the secreted substances are carried out at the external surface of the epithelial layer onto which they are discharged.

A second group of glands, known as **endocrine glands,** have lost the ducts that connect them to the epithelial surface from which they were formed. The molecules secreted by these glands are released into the interstitial fluid surrounding the gland cells, diffuse into the blood, and are carried by the blood to other cells in the body. The molecules synthesized and secreted by the endocrine glands are **hormones.** Hormones are blood-borne chemical messengers that, with the nervous system, coordinate the activities of different cells in the body. The general properties of hormones and their mechanisms of action will be described in Chap. 18.

In practical usage, the term *ductless gland* has come to be synonymous with endocrine gland. However, it should be noted that there are cells that secrete nonhormonal, organic substances into the blood; for example, fat cells secrete fatty acids and glycerol into the blood, and the liver secretes glucose, amino acids, fats, and proteins. These substances serve as nutrients for other cells or perform special functions in the blood but they do not act as hormones.

Exocrine Glands Most exocrine glands are multicellular. However, one of the few types of unicellular gland is the *goblet cells,* which secrete mucus; many goblet cells are interspersed throughout the

epithelia of mucous membranes. The nature of the duct and terminal portion of an exocrine gland determines its classification (Fig. 6-6). If the duct passes directly to the surface, the gland is *simple;* if the duct branches, the gland is *compound.* If the terminal portions are tubular in shape, the gland is *tubular;* if they are saclike, the glands are *acinar* (alveolar), and the sacs themselves are *acini* (alveoli). If the terminal portions of a compound gland are both tubular and acinar, the gland is *tubuloacinar.* In a tubular gland, the secretory cells may occur throughout the tubule or be restricted to its terminal portion. In an acinar gland, the primary secretion is produced by the cells of the acini, but the cells lining the duct may alter the secreted material by additional secretion or reabsorption during the material's passage from acinus to surface. The gland cells are usually separated from the surround-

FIG. 6-6 Classification of tubular and acinar glands.

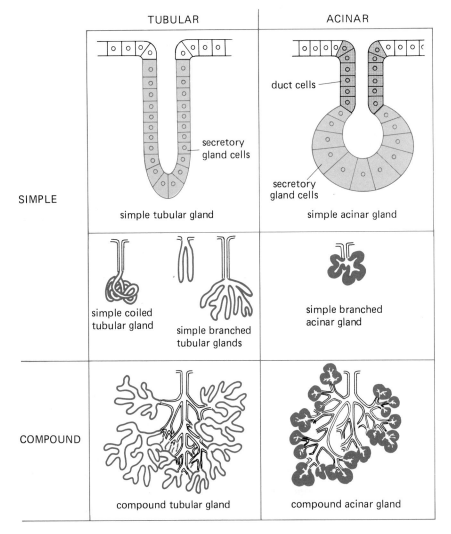

ing connective tissue by a basement membrane.

Endocrine Glands   As described above, endocrine glands lose their connection with surface epithelium during embryonic development. In most cases, the separated clusters of cells lose their original lumen as well and form compact masses of cells penetrated by blood vessels and connective tissue. Certain endocrine cells (for example, those that produce the gastrointestinal hormones) remain scattered throughout the epithelial lining from which they were derived and are not aggregated into specific multicellular glandular structures. Others constitute distinct organs that have migrated considerable distances from their epithelium of origin.

Mixed Glands   Some organs consist of both exocrine and endocrine glands. In most cases (the pancreas, for example), the cells constituting the different gland types are quite distinct.

Gland Secretion   The secretion of salts by exocrine glands is accomplished by ion-transport systems in the plasma membranes of these cells. Various ions are delivered to the cells by the blood and are actively transported into the lumen of the gland where they reach higher concentrations than are present in the blood or cytoplasm of the cell. Water then diffuses into the lumen because of the high osmolarity of the salt solution there. This primary secretion may be modified as it flows along the excretory ducts as a result of the secretion and reabsorption of ions and water by the cells that line these ducts.

Some of the organic molecules secreted by exocrine cells are merely transported from the blood into the lumen of the gland in a manner similar to that described above for ions. However, many organic molecules secreted by exocrine cells and all the hormones secreted by endocrine cells are synthesized in the cells themselves.

Gland cells differ in the manner in which they release products synthesized by the cell (Fig. 6-7).

FIG. 6-7 Merocrine and apocrine discharge of secretory products.

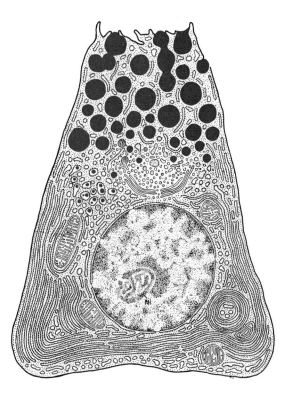

merocrine secretion

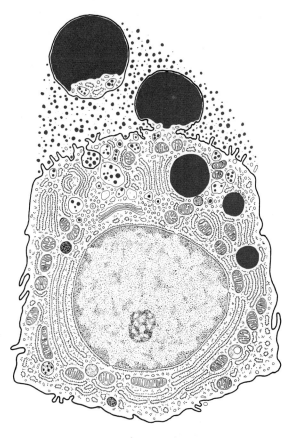

apocrine secretion

The type of secretion in which product is released by exocytosis and the gland cell remains intact during release is known as **merocrine secretion.** During **apocrine secretion,** in contrast, the secretory droplets accumulate at the apical end of the cell and, at the time of discharge, the entire top portion of the cell breaks away, releasing the secretory product and some of the cytoplasm. The remaining membrane of the ruptured cell reseals, and the cell repeats the process after synthesizing and accumulating more secretory product. The secretion of lipid droplets by the cells of mammary glands is an example of apocrine secretion. A third way of releasing secretory products, **holocrine secretion,** involves the accumulation of secretory products within the cell followed by the complete disintegration of the cell, thereby releasing the entire cell contents and at the same time destroying the cell. The secretion of oils by the sebaceous glands in the skin occurs by this mechanism.

CONNECTIVE TISSUE

Structure of Connective Tissue

Of the four types of specialized tissues, connective tissue is the most diverse in its function and structure. Unlike nerve, muscle, and epithelia, connective tissue has a considerable amount of extracellular material between widely spaced connective-tissue cells. It ranges from the loose meshwork of cells and fibers underlying most epithelial layers to the solid structure of bone, and it includes cell types

as diverse as fat-storing (adipose) cells and red and white blood cells. Blood cells, cartilage, and bone are described in separate chapters.

As its name implies, a major function of connective tissue is to connect, anchor, and support the other tissues of the body. The actual structures that provide this physical support are in the extracellular spaces surrounding the connective-tissue cells. These supporting elements are secreted by the connective cells into the extracellular medium where they form a matrix that consists of various types of fibers embedded in a ground substance. This matrix may vary in consistency from a semifluid gel to the solid crystalline structure of bone. The three characteristic features of most connective tissues are thus (1) fibers, (2) ground substance, and (3) cells.

Connective-tissue Fibers   The extracellular fibers of connective tissue are formed from protein molecules synthesized by the connective-tissue cells. **Fibroblasts** are the major source of the proteins that form these extracellular fibers; however, some other types of cells, such as smooth muscle cells, can synthesize them. The proteins are synthesized by the ribosomes of the rough endoplasmic reticulum and released at the fibroblast surface by exocytosis. Once outside the cell, they aggregate to form filaments (Fig. 6-8). This occurs spontaneously because of the inherent ability of the molecules to bind together. However, the rate of assembly and the thickness of the fibers formed are also influenced by the surrounding chemical environment created by the fibroblasts. The fibers may

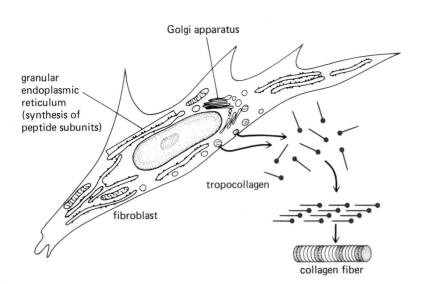

FIG. 6-8   Formation of collagen fibers in the extracellular matrix from tropocollagen subunits synthesized and secreted by fibroblasts.

Golgi apparatus

granular endoplasmic reticulum (synthesis of peptide subunits)

tropocollagen

fibroblast

collagen fiber

slowly undergo further structural modifications that are associated with an increase in the number of chemical bonds that cross-link the proteins in the filaments. These modifications contribute to the progressive changes in the flexibility and elasticity of many regions in the body associated with the process of aging.

Connective-tissue fibers are classified as collagenous, elastic, or reticular. **Collagenous** fibers are composed of the protein *collagen,* which is secreted by fibroblasts as the rod-shaped molecule *tropocollagen.* The tropocollagen rods form a parallel organization in which each rod overlaps about one-quarter of the adjacent rods, giving rise to a staggered arrangement that is manifested as a banded appearance in the fibers (Fig. 6-9). Collagen fibers range in diameter from 1 to 20 μm and may be many centimeters long; they do not branch. The physical properties of collagen fibers resemble those of a piece of rope; they are flexible and have a high tensile strength that resists stretching.

Elastic fibers are formed from the fibrous protein *elastin*. These fibers, which are thinner than the collagen fibers, do not have a periodic banding pattern. They have numerous branches that interconnect with other elastic fibers, producing a loose, fibrous network. Like rubber bands, they can be stretched and upon release return to their original length. It is the presence of these fibers in the connective-tissue layers of various structures, such as the walls of the blood vessels, that give these structures their elastic behavior. Aging is accompanied by a progressive decrease in the number of elastic fibers and an increase in the number of collagen fibers in connective tissues, which leads to certain forms of hardening of arteries.

Reticular fibers are very fine, highly branched fibers. They show the same periodic banding pattern as collagen and may actually represent an early stage in the formation of the larger mature collagen fibers. They can be identified in histologic preparations by their ability to bind silver stains; collagen and elastic fibers do not bind these stains.

Ground Substance   In addition to secreting fibrous proteins, connective-tissue cells release several types of polysaccharides and proteins that form a nonfibrous, homogeneous extracellular matrix known as **ground substance.** These substances convert the extracellular fluid from a liquid into the semisolid state of a gel. The physical consistency of the extracellular medium varies considerably, depending on the amount of ground substance, the number and type of fibers embedded in it, and the possible deposition of various inorganic salts to form a more solid, crystallized extracellular material such as bone.

The structural polysaccharides that form the

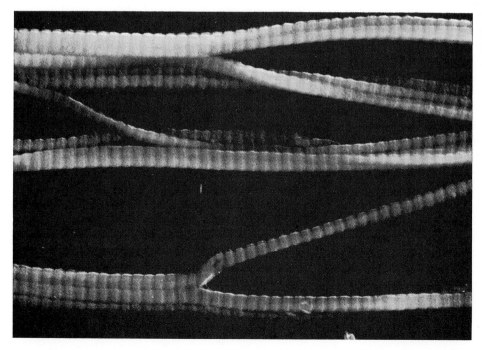

FIG. 6-9 Electron micrograph showing the banded structure of collagen fibers.

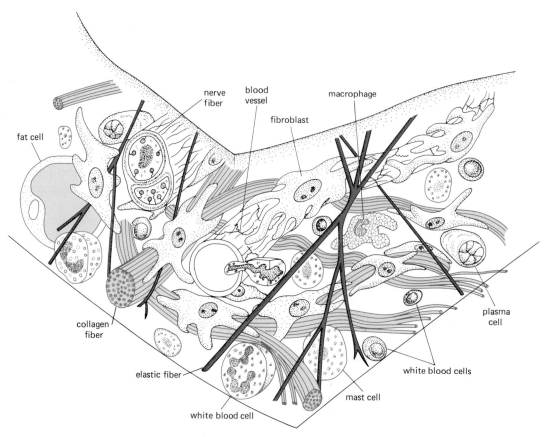

FIG. 6-10 Assortment of cells and fibers dispersed throughout loose connective tissue.

ground substance are known as *mucopolysaccharides;* they contain sugar residues that have both amino and carboxyl groups as part of their structure. The proteins of the ground substance contain similar sugar residues attached to some of their amino acid side chains, and these proteins are known as *mucoproteins*. Mucoproteins and mucopolysaccharides form most of the structure of basement membranes.

Mesenchyme Embryonic connective tissue, or **mesenchyme,** consists of a loose arrangement of stellate-shaped cells surrounded by large extracellular spaces filled with interstitial fluid and a meshwork of extracellular fibers. Mesenchyme fills in the spaces between the developing organs of the embryo. It is from these mesenchymal cells that all the connective cells of the adult body are ultimately derived.

Types of Connective Tissue Proper

Loose (Areolar) Connective Tissue Loose connective tissue contains large extracellular spaces composed of a loose network of collagenous, elastic, and reticular fibers embedded in a gel-like ground substance (Figs. 6-10 and 6-11A). It serves as a ''filler'' throughout the body's structures and also forms the connective-tissue layer underlying most epithelia. Reticular fibers are particularly prominent in the loose connective tissue of the spleen and lymph nodes where numerous white blood cells lie in the interfiber spaces. A number of different cell types are distributed throughout this loose network, the majority of which are the fibroblasts that synthesize and secrete the fibers and ground substance of the extracellular matrix. Variable numbers of adipose-tissue cells may be present in loose connective tissue as well as a number of

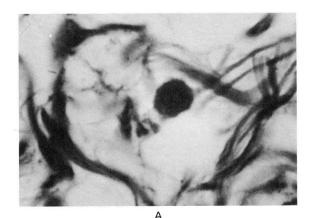

A

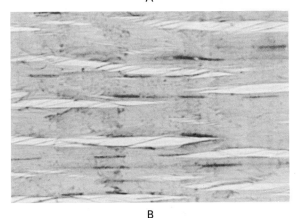

B

FIG. 6-11 (A) Loose connective tissue. (B) Dense connective tissue.

contents of the secretory granules include *heparin,* a substance that inhibits the coagulation of blood, and *histamine,* a substance that has powerful effects upon the smooth muscles and glands in the regions surrounding the connective tissues. These effects will be described in later chapters.

Dense Connective Tissue   This type of connective tissue is distinguished from loose connective tissue by the greater abundance of fibers, particularly large collagen fibers, in its extracellular matrix. Fewer cells are found between the fibers than in

FIG. 6-12 (A) Adipose (fat) tissue cell. A thin rim of cytoplasm surrounds the large lipid droplet that is composed predominantly of triacylglycerol molecules. (B) Photomicrograph of adipose tissue.

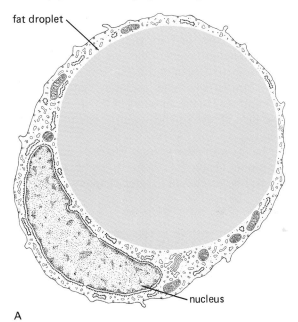

fat droplet

nucleus

A

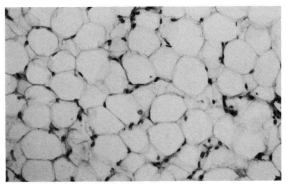

B

specialized cells—macrophages, plasma cells, mast cells, and some white blood cells—that participate in defense mechanisms.

Macrophages (or *histiocytes*) are cells that are able to engulf and digest bacteria and other particulate matter that may invade the connective-tissue layers. The macrophages are known as "fixed" when they remain anchored to one position and as "free" or wandering if they migrate by ameboid movement throughout the connective-tissue layers. **Plasma cells** synthesize and secrete antibodies, proteins that play an important role in the defense mechanisms of the body and whose properties will be described in Chap. 28. These cells are found primarily in the lymph nodes but also occur in smaller numbers throughout the loose connective tissues. **Mast cells** contain numerous dense secretory granules that are released into the surrounding medium in response to various local stimuli. The

loose connective tissue, and these are mostly fibroblasts (Fig. 6-11B). Dense connective tissue forms the capsules, which enclose most internal organs, and the partitions or septa, which subdivide many organs into subunits. A particularly dense form of connective tissue occurs in the tendons and ligaments that link muscles to bones and bones to each other, respectively. These fibrous structures are composed almost entirely of compact bundles of parallel collagen fibers.

Adipose Tissue   This tissue consists of large numbers of adipose cells in a loose or reticular network of extracellular fibers. These specialized cells synthesize and store large quantities of triacylglycerols, which accumulate to form large droplets of fat surrounded by a thin layer of cytoplasm so that the cells look like a ring with the nuclear region corresponding to the setting in the ring (Fig. 6-12). Adipose-tissue cells are found throughout the loose connective tissues and are especially numerous in the subcutaneous layers of the skin where they provide a protective cushion for the overlying skin as well as forming a layer of insulating material at the body's surface. Large amounts of fat are also found around the kidneys and heart and as padding around the joints.

We have examined several types of connective tissue proper (Table 6-1). Other types of connective tissue will be discussed in later chapters.

TABLE 6-1   CONNECTIVE-TISSUE TYPES
I Mesenchyme
II Adult connective tissue
A Connective tissue proper
1 Loose (areolar)
2 Dense
3 Reticular
4 Adipose
B Cartilage
1 Hyaline
2 Elastic
3 Fibrous
C Bone
1 Spongy (cancellous)
2 Compact
D Dentin
E Hemopoietic tissue, blood and lymphoid cells

SKIN

The multiplicity of structure and function of epithelium and connective tissue can be illustrated by examining the largest organ in the body, the *skin*. The skin, or **integument,** is composed of two layers: an outer epithelial layer, the **epidermis,** and an underlying connective-tissue layer, the **dermis,** that also contains numerous nerve fibers and blood vessels.

Functions of Skin

In lower animals, many of the exchanges of molecules between the external and internal environments occur across the skin. In human beings and other higher organisms, these exchanges occur mainly across epithelial surfaces that have invaginated from the surface of the body into the interior (forming the lungs and intestinal tract, etc.), and the skin forms a barrier that prevents most substances from entering or leaving the body. For example, potentially harmful bacteria are barred from entering, and loss of water by evaporation from the moist surfaces of the underlying tissues is greatly diminished. The skin also protects tissues from damage by the ultraviolet radiation present in sunlight. The toughness of this flexible outer covering also protects internal organs from physical damage that would result from frictional contacts between the body and objects in the external environment.

Much of our awareness of the external world results from stimulation of the nerve fibers in the skin by various environmental stimuli, leading ultimately to the sensations of touch, pressure, and temperature.

The skin also assists in the regulation of body temperature, since heat, released during the metabolism of cells, is carried by the blood to the skin, from which heat loss can be regulated by controlling the amount of blood flow to the skin. The secretion of sweat by glands of the skin also assists in temperature regulation by carrying heat away from the body in association with the process of water evaporation.

Although the skin is not a major excretory organ, small amounts of water, salts, and a variety of organic molecules are secreted onto the surface of the skin by the sweat and sebaceous (oil) glands. During periods of intense physical exercise, large quan-

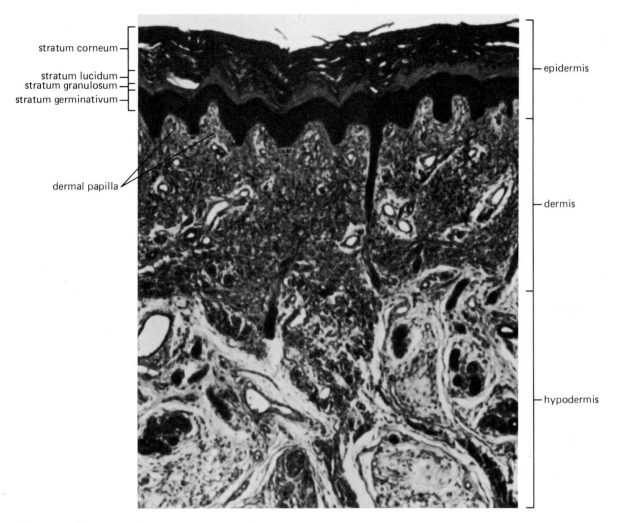

stratum corneum —
stratum lucidum —
stratum granulosum —
stratum germinativum —

dermal papilla

— epidermis

— dermis

— hypodermis

FIG. 6-13 Skin, showing the four layers of the epidermis which overlie the dermis and hypodermis.

tities of salt and water may be lost through sweating. If sustained for a long period of time, such losses may significantly alter the composition of the internal environment.

The skin performs a number of metabolic functions, including the formation of vitamin D_3 in the presence of sunlight. Further modifications of vitamin D_3, as we shall see, must occur in the liver and kidney before it is able to function in the body.

Damage to large areas of the skin can be life-threatening. Excessive amounts of water and salt pass through the damaged area and are lost from the body; infection becomes a major problem because bacteria readily penetrate the damaged surface.

Structure of Skin

Epidermis The outer surface of the skin consists of a layer of stratified squamous epithelium that varies from 30 to 50 cells thick. Proceeding from the innermost to the outermost zone, four major zones of epidermis (Fig. 6-13) can be distinguished: (1) stratum germinativum, (2) stratum granulosum, (3) stratum lucidum, (4) stratum corneum.

The **stratum germinativum,** composed of 8 to 10 layers of cells, is separated from the underlying dermis by a basement membrane. The mitotic activity of the cuboidal cells that rest upon this membrane continually replaces the cells in the upper epidermal layers. It takes about 14 days for a cell to move from the stratum germinativum to the stratum corneum, where it may reside for another 30 days before being shed from the outer surface of the skin. In the upper layers of the stratum germinativum the cells are irregularly shaped and connected by multiple desmosomes. As the cells continue to be pushed toward the surface of the epidermis, they become flattened, and many granules of the fibrous protein *keratin,* synthesized on the ribosomes, begin to appear. These flattened cells containing granules form the **stratum granulosum.**

Scattered throughout the stratum germinativum and the stratum granulosum are pigment-synthesizing cells known as **melanocytes.** These cells synthesize a black pigment, *melanin,* that accumulates in dense, membrane-bound vesicles in the melanocytes. These vesicles are then secreted and taken up by the surrounding epithelial cells by phagocytosis (endocytosis). The number of melanocytes relative to the number of epithelial cells is approximately the same in all skin colors. The wide range of skin colors found among the human population is determined by the numbers, size, and distribution of the melanin granules. The pink color of Caucasian skin is due to the small amount of melanin pigment, which incompletely masks the blood vessels in the underlying dermis. Ultraviolet radiation can damage chromosomes and kill cells. However, the ultraviolet radiation in sunlight can also stimulate the synthesis of melanin. The result is an increase in the melanin content and a darkening or tanning of the skin. In a highly pigmented skin, a large portion of the ultraviolet light is absorbed by the melanin and thus does not reach the underlying tissues. As a result of genetic mutation, some individuals are unable to synthesize melanin. Such individuals, known as *albinos,* have no pigment in their skin, although melanocytes are present. This lack of pigment makes the underlying tissues in the albino's skin particularly sensitive to the damaging effects of ultraviolet light.

FIG. 6-14 Organization of epidermis in region of fingernail.

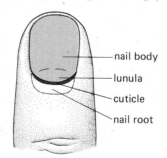

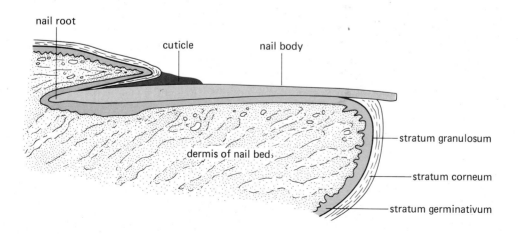

The **stratum lucidum** is a thin, clear area between the stratum granulosum and the stratum corneum that is most pronounced in the thick epidermal layer on the palms of the hands and soles of the feet. By the time the outwardly migrating cells enter the stratum lucidum, they show signs of disintegration and have lost their nuclei and organelles. The cells in this layer are filled with a transparent substance that appears to be a precursor of keratin.

The outermost layer of the epidermis is the **stratum corneum,** or horny layer. It is composed of nonnucleated disintegrated flat, squamous cells that are filled with keratin fibers and contain very little water. It is this compact, fibrous layer of dead cells that forms the protective barrier of the skin.

At the tips of the fingers and toes, the epidermis forms the horny plates of the nails (Fig. 6-14), which, like the stratum corneum of the skin, are composed of a dense layer of dead, flat cells filled with keratin. Below a thin fold of skin, the **cuticle,** in the region of the **lunula** (the white, half moon-shaped area at the base of the nail) is the growing portion of the nail, which contains the proliferating cells of the stratum germinativum. As the cells accumulate keratin, they become flattened, die, and form the hard compact structure of the nail plate, which grows at a rate of about 0.5 mm per week. Nails appear pink because underlying blood vessels can be seen through the translucent, keratinized layer of the nail. Unlike superficial layers of the skin, the keratinized surface of the nail is not shed; the nail will continue to grow until trimmed or broken.

Dermis   The epidermis overlies the connective-tissue layer of skin, the *dermis* (Figs. 6-13 and 6-15), which is from 1 to 3 mm thick. The outermost region is the papillary layer; the remaining portion, which occupies about 80 percent of the thickness of the dermis, is the reticular layer.

The *papillary layer* consists of loose connective tissue, containing many fine collagen and elastic fibers and numerous fibroblasts, but few fat cells. This layer is folded into ridges, or papillae, that extend into the epidermis and produce ridges on the surface of the skin. These are especially noticeable on the palms of the hands and fingers where they produce the patterns of the fingerprints and enable the hand to form a firmer grip on objects than would be possible if the surface of the skin

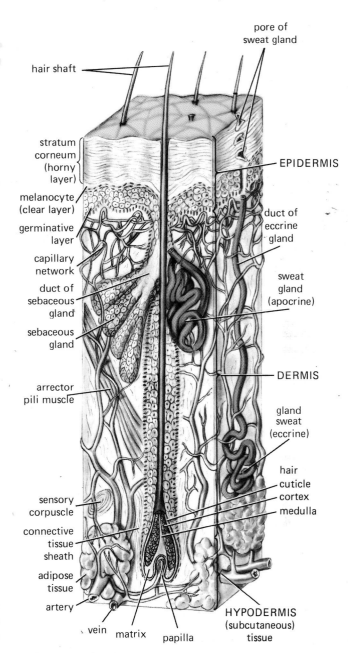

FIG. 6-15   Structure of hair follicle located in the dermal layer of skin.

were smooth. Numerous small blood vessels extend into the papillary folds and supply nutrients to the overlying epidermis, which is avascular. The papillary layer also contains numerous nerve fibers and associated structures that respond to stimuli applied to the surface of the skin.

There is no distinct boundary between the papillary layer and the *reticular layer,* which contains denser connective tissue with fewer fibroblasts but many bundles of thick collagen fibers. It is this region that is responsible for the toughness of leather formed by processing animal skins.

The reticular layer borders on the underlying subcutaneous tissue known as the **hypodermis,** which is technically not part of the skin. This region is composed of loose connective tissue and contains an abundance of fat cells. In addition to providing a store of fat that can be metabolized by the body, it provides a layer of insulation, which prevents heat loss through the skin from the underlying organs of the body. The various amounts of adipose tissue that underlie the skin smooth out the angular surfaces of the body.

Hair   One of the characteristics of mammals, including human beings, is the presence of hair on most of the body's surface except for the palms of the hands, soles of the feet, and lips. In furry mammals, this hairy surface provides an extra layer of insulation, but human beings are not sufficiently hairy to benefit from this insulating property. The hair on the top of the head provides protection from the rays of the sun, and the eyelashes partially protect the eyes from various airborne particles. Likewise, the hairs in the nostrils and ears help to prevent particles from entering these passages.

Hair is formed by epidermal cells that invaginate into the underlying dermal layer and thus form a **hair follicle** (Fig. 6-15). Hair consists of keratin that is synthesized by cells in the *papilla* at the base of the hair's shaft. Similar to the process of forming the stratum corneum and the nails, the cells in the papilla divide, synthesize keratin, flatten, and die, and thus leave behind compact layers of keratin that form the shaft of the hair. Hair does not continue to grow indefinitely; after it reaches a certain length, the cells in the papilla stop dividing and the cells at the base of the hair become completely keratinized. Subsequently, the papillary cells begin to divide again and form a new hair that forces the old hair out of the follicle.

Hair color is determined by the amount of melanin that is present. Blond hair does not contain melanin, nor does gray hair, which has many air spaces between the fibrous layers of keratin that give the hair a gray or white appearance.

Attached to the middle region of the hair follicle is a small, smooth muscle, the **arrector pili,** which is anchored to the connective tissue of the dermis. The hair follicles invaginate into the dermis at an angle, so that when the arrector pili contracts it moves the hair to a more vertical position, causing the hair to "stand on end." At the same time it causes an indentation of the skin around the hair, which produces a "goose pimple." In furry animals, this erection of the body hair produces a thicker layer of insulation.

Skin Glands   The skin contains several types of exocrine glands that are formed by invagination of the epithelial cells into the dermis. The secretory portions of these glands are connected to the surface of the skin by ducts that pass through the epidermis or empty into a hair follicle.

The **sebaceous (oil) glands** have a branched acinar structure; their ducts empty into hair follicles, from which point the secretion reaches the surface of the skin along the hair shaft (Fig. 6-15). These glands secrete a mixture of various lipids known as *sebum,* which lubricates the surface of the skin. The secretory process in these glands is holocrine, which involves the complete disintegration of the cell at the time of release.

The most common type of sweat gland is the **eccrine gland,** which secretes water and salts. The secretory unit of the gland consists of a highly coiled tube that lies in the dermis; it sends a long, relatively straight duct to the surface of the skin (Fig. 6-15). The secretory activity of these glands is regulated by the nervous system via nerve fibers that end in the secretory region of the gland.

A second type of sweat gland is found in the armpits, nipples, anus, and pubic regions. These sweat glands are known as odoriferous, or **apocrine glands.** They are considerably larger than the eccrine glands, and their ducts usually empty into hair follicles. In addition to water and salts, the secretions of these glands contain a variety of organic molecules. Although these secretions are initially odorless, the action of bacteria on the skin's surface decomposes the organic molecules, which produces the distinctive body odors.

KEY TERMS

epithelium
basement membrane
mucous membranes
serous membranes
cilia
exocrine glands
endocrine glands

fibroblasts
collagen fibers
elastic fibers
ground substance
mucopolysaccharides
loose connective tissue
dense connective tissue

adipose tissue
epidermis
dermis
hypodermis
sebaceous glands
eccrine glands

REVIEW EXERCISES

1   Identify the four categories of specialized tissues in the body.
2   Describe the specialized functions associated with epithelial tissues.
3   Define the apical and basal surfaces of an epithelial layer of cells.
4   Describe the location and structure of the basement membrane of an epithelial layer of cells.
5   Describe the following types of epithelia: simple, squamous, cuboidal, and columnar.
6   Distinguish between stratified, pseudostratified, and transitional epithelia.
7   Distinguish between mucous membranes and serous membranes.
8   Describe the following structures found at the surfaces of epithelial cells: microvilli, brushborder, and cilia.
9   Describe the differences in function and structure between an exocrine gland and an endocrine gland.
10 Describe the structure of the following types of exocrine glands: simple, compound, tubular, acinar, and tubuloacinar.
11 Describe the following types of glandular secretion processes: merocrine secretion, apocrine secretion, and holocrine secretion.
12 Describe the general characteristics common to most connective tissues.
13 Describe the following types of connective tissue fibers: collagenous fibers, elastic fibers, and reticular fibers.
14 Describe the composition of the ground substance of connective tissue.
15 Describe the structure of the following types of connective tissue: mesenchyme, loose connective tissue, dense connective tissue, and adipose tissue.
16 Describe the functional characteristics of the following types of cells found in connective tissues: fibroblasts, macrophages, plasma cells, and mast cells.
17 List the functions of the skin.
18 Describe the cellular structure of the epidermis of the skin.
19 Define the role of the melanocytes in the epidermis of the skin.
20 Describe the structure of fingernails and toenails.
21 Describe the cellular structure of the dermis layer of the skin.
22 Describe the growth of hair.
23 Describe the process responsible for the formation of a "goose pimple."
24 Describe the characteristics of the following skin glands: sebaceous glands, eccrine glands, and apocrine glands.

CHAPTER

7

HOMEOSTASIS AND CONTROL SYSTEMS

HOMEOSTASIS

As described in Chap. 1, Claude Bernard was the first to recognize the central importance of maintaining a stable internal environment (extracellular fluid). This concept was further elaborated and supported by the American physiologist W. B. Cannon, who emphasized that such stability could be achieved only through the operation of carefully coordinated physiologic processes that he termed **homeostatic.** The activites of tissues and organs must be regulated and integrated with each other in such a way that any change in the internal environment automatically initiates a reaction to minimize the change. **Homeostasis** denotes the stable conditions that result from these compensating regulatory responses. Some changes in the composition of the internal environment do occur, of course, but the fluctuations are minimal and are kept within narrow limits through the multiple coordinated homeostatic processes.

Concepts of regulation and relative constancy have already been introduced in the context of a single cell. In Chap. 3 we described how metabolic pathways within a cell are regulated by the principle of mass action and by changes in enzyme activity so as to maintain the concentrations of the various metabolites. We also described there how the control of protein synthesis is regulated by the genetic apparatus. Each individual cell exhibits some degree of self-regulation, but the existence of a multitude of different cells organized into specialized tissues, which are further combined to form organs, obviously imposes the need for overall regulatory mechanisms to coordinate and integrate the activities of all cells. For this, intercellular communication over relatively long distances is essential. Such communication is accomplished by means of *nerves* and the blood-borne chemical messengers known as *hormones.*

The mechanisms by which these two communications systems operate are the subject of other chapters, but their overall role and the basic characteristics of homeostatic processes can be appreciated only in terms of **control systems,** to which we now turn.

GENERAL CHARACTERISTICS OF CONTROL SYSTEMS

We shall define a **homeostatic system** as a control system that consists of a collection of interconnected components that function to keep a physical or chemical parameter of the body relatively constant. Let us first analyze a nonbiological control system (Fig. 7-1), one designed to maintain the temperature of a water bath at approximately 30°C despite fluctuations in room temperature from 25 to 10°C. Since the water temperature is always to be higher than the room temperature, there is a continuous loss of heat to the room from the water. Moreover, the lower the room temperature, the greater is this heat loss. Accordingly, the water must be continuously heated in order to offset the loss, and the degree of heating must be altered whenever the room temperature changes. This adjustment of heat input to heat loss so that water temperature remains approximately 30°C is the job of the control system.

The first component of the system is known as a *sensor.* This particular sensor is sensitive to temperature, and it generates an electric current the magnitude of which increases as water temperature decreases. The current flows through wire B into the control box C constructed in such a way that the amount of current flowing out of the box in wire D is directly proportional to that entering along B. The current from the box is fed to the heating unit E, the water bath. Its activity and therefore the amount of heat it produces per unit time is directly proportional to the strength of the signal to it, i.e., the current flow in D.

We fill the bath with water at room temperature, close the switches, and allow the system to operate

FIG. 7-1 Components of a control system for regulating the temperature of a water bath.

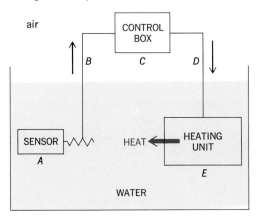

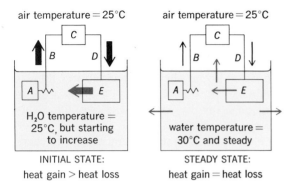

air temperature = 25°C   air temperature = 25°C

INITIAL STATE: STEADY STATE:
heat gain > heat loss   heat gain = heat loss

FIG. 7-2 Effects of filling a water bath that has an operating point of 30°C with water at room temperature and allowing the system to operate. Shown are the initial state, i.e., just after the water was added, and the ultimate steady state achieved. Note the difference in output of the sensor and the heating unit in the two states.

(Fig. 7-2). Current generated in the sensor *A* controls the output of the heating unit by way of the control box. Because of the initially low water temperature, the magnitude of current flow from *A* is large and the heating unit is running full blast. This heats the water rapidly, but as the water temperature rises, two opposing events occur: (1) More heat is lost from the bath to the room, and (2) the signal from *A* decreases and results in a decreased input to the heating unit, thereby decreasing the amount of heat it produces. The system ultimately stabilizes at a particular water temperature when heat loss to the room exactly equals heat gain from the unit. At this point, the system is said to be in a **steady state:** input equals output, and the temperature remains steady. (Actually there is always some oscillation around the steady-state temperature because of the time required for the heating unit to heat up or cool down.)

The steady-state temperature is determined, in large part, by the characteristics of the sensor and the control box *C,* since these components are what determine the output of the heating unit. If we had chosen a heat-sensitive sensor that generates only half as much current at any given temperature as *A* does, the input to the control box would always be less, the output from the control box would be less, and the heating unit would always be generating less heat. Therefore, the steady-state temperature of the bath would be lower. Similarly, by altering the transforming function, i.e., the relationship between current in and current out, of the control box, we alter the steady-state temperature ultimately reached. In any control system, the actual steady state, or so-called **operating point** of the system, depends upon the characteristics of the individual components of the system. Our components were chosen to achieve an operating point of 30°C when the air temperature is 25°C.

Most important is that this type of system resists any changes from the operating point; thus, the control system in our example will automatically prevent any large deviation of the water temperature from the steady state established. Suppose that after the steady state (30°C) has been reached, the room temperature is suddenly lowered (and kept low) so that the loss of heat from water to room is increased. This loss unbalances heat loss and heat gain, and the water temperature falls. But the decrease in water temperature immediately increases the current generated in *A*. More current flows out of the control box *C,* the heating unit increases its activity, and the water temperature rises back toward its original value. A new steady state will be reached when heat loss once again equals heat gain, both of which having been increased by a proportionate amount. What is the new steady-state temperature at this point? If the system is extremely sensitive to change, the temperature will be only very slightly below what it was before the room temperature was lowered *but it cannot be precisely the same.* Compensation is incomplete because the new steady state depends upon the maintenance of an increased heat production to balance the increased heat loss, and this increased heat production is a result of the increased signal coming from *A*. The reason *A* generates more current is the slightly lower water temperature. If the temperature actually returned completely to 30°C, the signal from *A* would return to normal, heat production would return to its value for that temperature, and the water temperature would immediately decrease as heat loss again became greater than heat gain. (Recall that the room temperature is being maintained lower than the bath temperature.) Thus, control systems of this type cannot absolutely prevent changes from occurring in the physical or chemical variable being regulated, but they do keep such changes within very narrow limits, depending upon the sensitivity of the system. A crucial generalization emerges: The operating point for a particular control system is not a single number but rather a narrow range of values (say 29 to 31°C in our example); the precise value within that range existing at any time is dependent on conditions in the external environment (the room temperature in our example).

We can now summarize the basic characteristics of a control system in general terms. There must be a component that is sensitive to the variable being regulated and that changes its signal rate as the variable changes. There must be a continuous flow of information from this sensor to an integrating component box, from which, in turn, the so-called command signal flows to an apparatus that responds to the signal by altering its rate of output (heat, in our example).

There remains one more concept that we have described but not named: **feedback.** The ultimate effects of the change in output by the system must somehow be made known (fed back) to the sensor that initiated the sequence of events. Our example illustrated perhaps the commonest form of feedback: When the water temperature is reduced, the sensor *A* detects this change and relays the information to the control box, which in turn signals the heating unit to increase its output. Sensor *A* is "informed" of this change in output by the resulting rise in water temperature; accordingly, its current generation is again altered, which, in turn, results in an alteration of input to the control box and thereby the heating unit. The water temperature acts as a continuous link between the sensor and the heating unit (without feedback, the sensor's signal would be unrelated to the heating-unit output and the system would be unable to maintain constancy). This type of feedback, in which a change in the variable being regulated brings about responses that minimize the change, is known as **negative feedback.** It clearly leads to stability of a system and is crucial to the efficient operation of homeostatic mechanisms.

Note that a major characteristic of negative-feedback control systems is that they *restore* the regulated variable toward its operating point after its initial displacement, but they cannot *prevent* the initial displacement. Suppose, however, that we add another component to our temperature-control system of Fig. 7-2, namely, a thermosensor on the *outside* of the water bath that can detect changes in room temperature. Now, when the room temperature is lowered, as in our example, this information is immediately relayed to the control box that causes the heating unit to increase its output. In this manner, additional heat can be supplied to the water bath *before* the water temperature begins to fall. Thus, the use of the external sensor permits the system to *anticipate* a pending fall in water temperature and begin to take action to counteract the change before it occurs. This provides **feedforward** information that has the net effect of minimizing fluctuations in the level of the variable that is being regulated. We shall see that the body frequently makes use of feedforward control in conjunction with negative-feedback systems.

There is, however, another type of feedback known as **positive feedback** in which an initial disturbance in a system sets off a train of events that increases the disturbance ever further. Thus, positive feedback rapidly displaces a system *away* from its steady-state operating point, often explosively. Because it does not lead to stability it is encountered much less frequently than are negative-feedback systems. However, several important positive-feedback relationships occur in the body, blood clotting being an example.

COMPONENTS OF LIVING CONTROL SYSTEMS

Reflexes

Homeostatic control systems in living organisms manifest virtually the same characterstics as those just described although some of the terminology is different. Many belong to the general category of stimulus-response sequences known as reflexes. A **stimulus** is a detectable change in the environment, such as a change in temperature, potassium concentration, pressure, etc., and a reflex is the sequence of events elicited by a stimulus. We may be aware of only the final event in the sequence, the **reflex response** (pulling one's hand away from a hot stove, for example). Indeed, the entire reflex, including the response, often occurs without any awareness on the part of the person. The pathway that mediates the reflex is known as the **reflex arc,** and its components are shown in Fig. 7-3; note that they are completely analogous to those of Fig. 7-1.

A **receptor** is the component that detects the environmental changes (it is identical to the sensor of

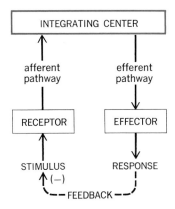

FIG. 7-3 General components of a reflex arc with negative feedback. The response of such a system has the effect of counteracting or eliminating the original stimulus. This phenomenon of negative feedback is indicated by the minus sign in the feedback loop.

the previous section). The stimulus acts upon the receptor to alter the signal emitted by the receptor, and this signal is the information relayed to the control box, or **integrating center.** The pathway between the receptor and the integrating center is known as the **afferent pathway.**

The integrating center usually receives input from many receptors, some of which may be responding to quite different types of stimuli. Thus, the output of the integrating center reflects the net effect of the total afferent input; i.e., it represents an integration of numerous and frequently conflicting bits of information.

The output of the integrating center is relayed to the last component of the system, the device whose change in activity constitutes the overall response of the system. This component, known as an **effector,** is analogous to the heating unit of our previous example. The information going from the integrating center to the effector is like a command directing the effector to alter its activity. The pathway along which this information travels is known as the **efferent pathway.**

As a result of the effector's response, the original stimulus (environmental change) that triggered this entire sequence of events may be counteracted (at least in part); the counteracting of the stimulus by the effector's response constitutes the negative feedback.

In summary, the components of a typical reflex arc are:

1 Receptor
2 Afferent pathway
3 Integrating center
4 Efferent pathway
5 Effector

To illustrate, let us again take a thermoregulatory system, this time one of the several reflexes that maintain internal body temperature relatively constant. The receptors are the endings of certain nerve cells in the body that generate electric signals at a rate determined by their temperature. This information is relayed by the nerve fibers (the afferent pathway) to a specific part of the brain (which acts as integrating center) that in turn influences, via a chain of nerve cells, the rate of firing of those nerve cells that stimulate skeletal muscle to contract. These latter nerve fibers are the efferent pathway, and the muscles they stimulate are the effectors. The amount of muscle contraction is altered so as to raise or lower heat production (heat is produced during muscle contraction). Thus, when external temperature decreases, thereby enhancing heat loss from the body, this reflex automatically increases heat production by increasing muscle contraction—familiar to us as shivering—so that heat loss and gain can remain equal and body temperature relatively unchanged.

Traditionally, the term reflex is restricted to situations in which the first four of the components listed above are all parts of the nervous system, as in the thermoregulatory reflex just described. However, present usage is not so restrictive and recognizes that the principles are essentially the same when a hormone, rather than a nerve fiber, serves as the afferent or (more commonly) efferent pathway, or when a hormone-secreting gland serves as integrating center. Thus, in our example above, the integrating brain centers not only trigger increased activity of nerve fibers to skeletal muscle but also cause release of several hormones that travel by the blood to cells and increase the cells' heat production by altering the rates of various heat-producing chemical reactions. These hormones, therefore, also serve as an efferent pathway in thermoregulatory reflexes.

Accordingly, in our use of the term *reflex* we include hormones as reflex components, so that afferent or efferent information can be carried by either nerve fibers or hormones. In any case, two different components must serve as afferent and efferent pathways; thus the input to, or output from, the integrating center may both be neural, but they must be two different nerve fibers. Of course, they can also both be hormonal, or one neural and the other hormonal. Moreover, depending on the specific nature of the reflex, the integrating center may reside in either the nervous system or an endocrine gland.

Finally, we must identify the effectors, the cells whose outputs constitute the ultimate responses of the reflexes. Actually, most cells of the body act as effectors in that their activity is subject to control by nerves or hormones. Muscles and glands, however, constitute the major effectors of biological control systems.

Thus, most biological control systems function to keep a physical or chemical parameter of the body relatively constant. One may analyze any such system by answering a series of questions: (1) What is the parameter (blood glucose, body temperature, blood pressure, etc.) that is being maintained constant in the face of changing conditions? (2) Where are the receptors that detect changes in the state of this parameter? (3) Where is the integrating center to which these receptors send information and from which information is sent out to the effectors, and what is the nature of these afferent and efferent pathways? (4) What are the effectors and how do they alter their activities so as to restore the regulated parameter toward normal (i.e., the operating point of the system)?

Local Homeostatic Responses

Besides reflexes, another group of biological responses is of immense importance for homeostasis. We shall call them **local responses.** Local responses are initiated by a change in the external or internal environment, i.e., a stimulus, which acts upon cells in the immediate vicinity of the stimulus, thus inducing an alteration of cell activity with the net effect of counteracting the stimulus. Thus, a local response is, like a reflex, a sequence of events that proceeds from stimulus to response. Unlike a reflex, however, the entire sequence occurs only in the area of the stimulus; no hormones or nerves are involved.

Two examples should help clarify the nature and significance of local responses: (1) Damage to an area of skin causes the release of certain chemicals from cells in the damaged area that help the local defense against further damage; and (2) an exercising muscle liberates into the extracellular fluid chemicals that act locally to dilate the blood vessels in the area, thereby permitting the required inflow of additional blood to the muscle. The great significance of such local responses is that they provide individual areas of the body with mechanisms for local self-regulation.

Chemicals as Intercellular Messengers in Homeostatic Systems

It should be evident that the *sine qua non* of reflexes is the ability of cells to communicate with one another, i.e., the capacity of one cell to alter the activity of another. When a hormone is involved in a reflex, it is clear that the communication between cells, i.e., between endocrine gland cell and effector, is accomplished by a chemical agent, the hormone (the blood, of course, acting as the delivery service). What has not been said, however, is that virtually all nerve cells also communicate with each other or with effectors by means of chemical agents known as **neurotransmitters.** Thus, one nerve cell (neuron) alters the activity of the next nerve cell in a reflex chain by releasing from its ending a neurotransmitter that diffuses across the very narrow space separating the two nerve cells and acts upon the second, altering its activity. Similarly, neurotransmitters released from the ends of the nerve cells that go to effectors constitute the immediate signal, or input, to the effector cells.

The detailed physiology of these chemical transmitters and of neuron-neuron or neuron-effector communication will be described in Chaps. 11 and 13. We mention them here to emphasize that chemical mediators, whether they be secreted by endocrine gland cells or released from nerve cell endings, constitute the ultimate messages by which one cell signals another to alter its activity.

This is true not only for reflexes but for local responses (as the examples above illustrate). The chemical messengers involved in local responses are referred to as **paracrine agents** (or **paracrines**) to distinguish them from the hormones. Many paracrines are synthesized by local cells and released, given the appropriate stimulus, into the extracellular fluid surrounding the cell; there, they exert their particular effects on nearby cells. Perhaps the best known example of such an agent is **histamine,** which is produced and stored in several cell types and plays important roles in the body's responses to injury (Chap. 28). Other paracrines are not released by local cells but are generated locally from inactive precursors circulating in the plasma; this generation is initiated by a change in the local environment.

As we shall see, a given chemical messenger may be synthesized by a number of different cell types so that a single molecular type may serve as a neurotransmitter (released from neuron terminals), as a hormone (released from endocrine gland cells), and as a paracrine agent. For example, epinephrine functions both as a hormone and as a neurotransmitter, and estrogen is both a hormone and a paracrine agent.

Prostaglandins   In recent years, an enormous amount of work has been devoted to the study of one particular group of chemical messengers, the family of unsaturated fatty acids that contain a five-membered (cyclopentane) ring and are known collectively as **prostaglandins.** Prostaglandins were originally discovered in semen but they are now known to be synthesized in most, possibly all, organs in the body. They are divided into main groups, each designated by the abbreviation PG and an additional letter—PGA, PGE, PGF, etc.—on the basis of the configuration of their cyclopentane ring. Within each group, further subdivision occurs according to the number of double bonds in the side chains. Thus, PGE_2 (Fig. 7-4) has two double bonds, which is denoted by the subscript in its name. At present, at least 14 different prostaglandins as well as a number of closely related substances called *thromboxanes* have been isolated from tissues; they are all synthesized from *arachidonic acid* and other essential unsaturated fatty acids.

The prostaglandins are not stored in tissues to any great extent but are synthesized and released immediately in response to a wide variety of stimuli. They then act locally in the tissues in which they are formed and are quickly metabolized to inactive forms. Accordingly, the prostaglandins fit best into the category of paracrine agents (whether blood-borne prostaglandins have any important effects on sites distant from their tissues of origin remains unsettled).

The prostaglandins and thromboxanes exert a bewildering array of effects, and it is proving a difficult task to elucidate their precise physiological roles. They almost certainly are important in blood clotting, regulation of smooth muscle tone, modulation of neurotransmitter release and action, multiple processes in the reproductive system, control of hormone secretion, and the body's defenses against injury and infection. Clearly, this widely distributed and all-encompassing system of chemical messengers is of enormous importance, and agents such as aspirin that inhibit one or more of the enzymes involved in their synthesis may cause many alterations of body function.

Receptors

The first step in the action of any chemical messenger on its target cell is its chemical combination with certain specific molecules of that cell; these molecules are known as **receptors.** This term can be the source of confusion because the same word is used to denote the "sensors" described earlier in this chapter (stimulation of which initiates reflexes). You should keep in mind that "receptor" has two totally distinct meanings, but the context in which it is used usually makes it quite clear which meaning is intended.

What is the nature of these receptors with which chemical messengers combine? Present evidence suggests that most, if not all, are segments of proteins in either the cell's plasma membrane or cytoplasm (in the general language of Chap. 3, the

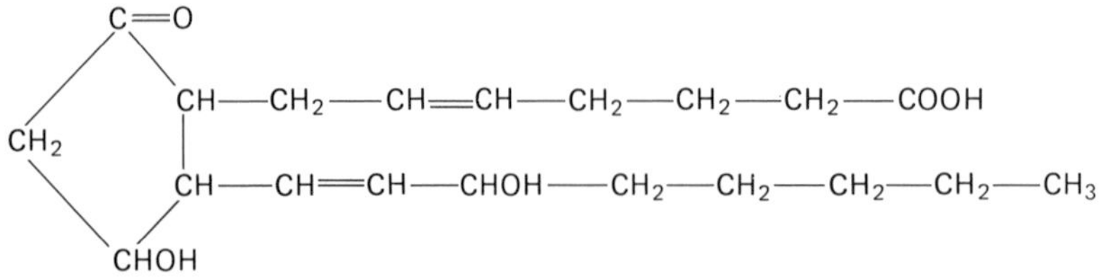

FIG. 7-4 Structure of prostaglandin PGE_2.

messenger is a *ligand* and the receptor a *binding site*). It is the combination of chemical messenger and receptor that initiates the cellular events that lead to the cell's response. This response takes the ultimate form of a change in membrane structure, permeability, or transport, in the rate at which a particular substance is synthesized or secreted by the cell, or in the rate or strength of muscle contraction.

One extremely important characteristic of chemical mediation that is understandable in terms of receptors is specificity. A chemical messenger—hormone, neurotransmitter, or paracrine—influences only certain cells and not others. The explanation is that the membranes or cytoplasm of different cell types differ in the types of receptors they contain. Accordingly, only certain cell types, frequently just one, possess the precise protein receptor required for combination with a given chemical messenger. Here is another example of how protein structure confers specificity upon a biological process.

We have pointed out that receptors may be in the cell's plasma membrane or cytoplasm. In fact, the former location is much more common, the major important exception being the cytoplasmic receptors for steroid hormones. Thus, the cell plasma membrane not only maintains the physical and chemical integrity of the cell but also serves as a signal-receiving device for the detection of chemical signals that regulate the cell's activity. A single cell usually contains many different receptor types, each capable of combining with only one chemical messenger.

It should be emphasized that receptors are, themselves, subject to physiological regulation. The number of receptors on a cell and the affinity of the receptors for their specific messenger can be increased or decreased, at least in certain systems. Such alterations provide an additional mechanism for feedback control.

How does the messenger-receptor combination elicit the cell's responses? In some cases, the alteration in protein structure produced by the combination of messenger with receptor may be the immediate cause of the altered function observed. For example, if the receptor were a membrane-bound protein involved in membrane permeability to sodium, the alteration of its structure by the chemical messenger might alter the passive flux of sodium

into the cell. However, in most cases, the path from messenger-receptor combination to cell response is far more complicated and requires the participation of "second messengers."

SECOND MESSENGERS

Second messengers are substances that, as a result of the original messenger-receptor combination, alter the cell's enzymes, membranes, contractile proteins, etc. (i.e., they trigger the cell's overall response). They may be generated in the cell or may enter its cytosol either across the plasma membrane or by release from cell organelles. In other words, the only function of the "first messenger"—hormone, neurotransmitter, or paracrine—is to combine with the receptors and bring a second messenger into play; the latter is then responsible for eliciting the cell's response. One of the most exciting developments in physiology has been the identification and elucidation of some of these second messengers and the recognition that the same ones may operate in many areas of the body and in different types of cells. The two best understood second messengers at present are calcium and cyclic AMP.

Calcium

In many cells, the combination of a first chemical messenger with a membrane receptor leads to an increase in cytosolic calcium, a second messenger that triggers, in turn, the cell's overall response. The cytosolic concentration of calcium is extremely low in virtually all cells because of the active transport of calcium either out of the cell across the plasma membrane or into cellular organelles, particularly the endoplasmic reticulum and mitochondria. Because of the low cytosolic calcium and the fact that the inside of the cell is negatively charged relative to the outside, there always exists a large electrochemical gradient for the net diffusion of calcium *into* the cell. In the unstimulated state, however, this net inward movement is exactly counterbalanced by net active transport of calcium *out* of the cell (or into organelles). However, should the membrane permeability to calcium increase, the net inward diffusion would increase and a transient elevation of cytosolic calcium would occur. This is precisely what happens in certain cells as a result

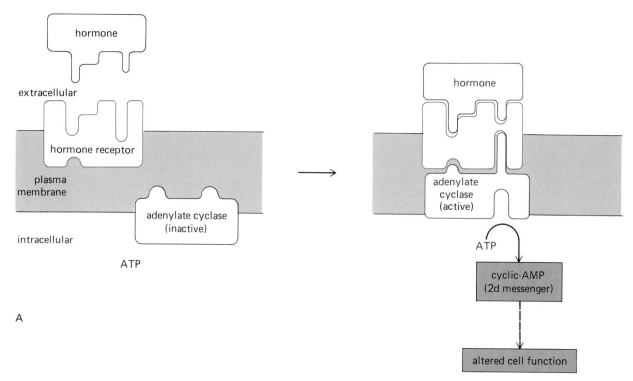

FIG. 7-5 Generation of cyclic AMP. (A) The first messenger, in this case a hormone, combines with a specific membrane receptor, which changes its configuration and permits it to activate adenylate cyclase. (The allosteric nature of this activation pictured in the figure is hypothetical.) (B) Structures of ATP, cyclic AMP, and AMP, the latter a result of enzymatic alteration of cyclic AMP.

of membrane changes induced by the combination of first chemical messenger and membrane receptor.

In other cells, notably muscle, the source of at least a fraction of the increased cytosolic calcium triggered by messenger-receptor combination is not an influx of extracellular calcium but rather a release of calcium from cell organelles into the cytosol. The end result is, however, the same: The increased calcium acts as a second messenger to elicit the response.

Calcium acts as a second messenger in many different cells, including muscle, nerve, and glands. Its role as a second messenger is frequently expressed by the term *coupler*. Thus calcium is an *excitation-contraction coupler* in muscle, meaning that it carries the "message" from the "excited" membrane to the contractile apparatus inside the muscle. Similarly, it is termed an *excitation-secretion coupler* in those cells in which it is the second

messenger between the membrane events and the secretory apparatus.

Just how an increased cytosolic calcium concentration leads to altered cell function probably differs from cell type to cell type, but a common initial event seems to be combination of the additional calcium with a protein inside the cell. This combination then triggers a sequence of chemical alterations that lead ultimately to the cell's response (contraction, secretion, etc.). In many cells, the specific protein to which the additional calcium ions bind is known as **calmodulin.** The calcium-calmodulin complex in turns binds to other cellular proteins, increasing or decreasing their activity.

Cyclic AMP

Most nonsteroid first messengers combine with receptors on the outer surface of the plasma membrane. The next step is identical for a large number

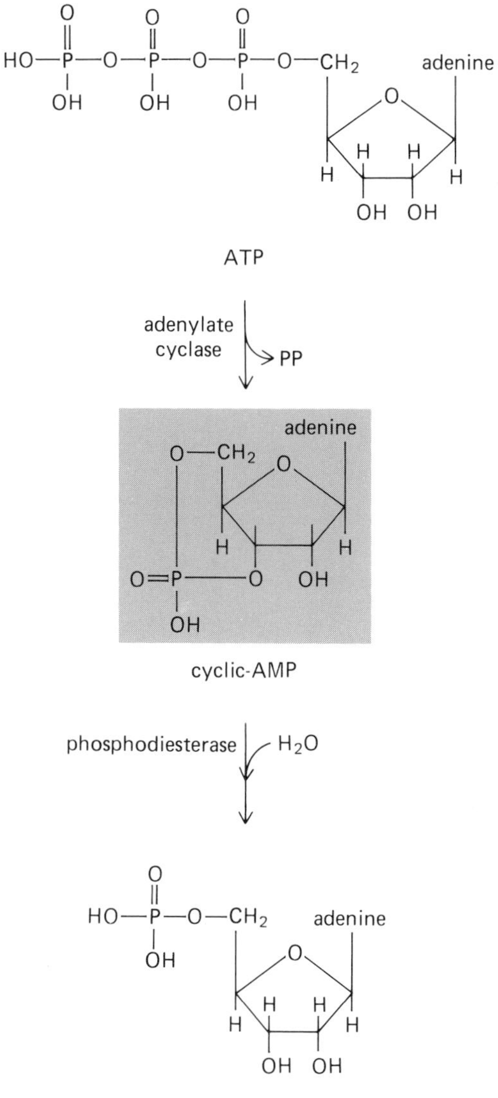

ATP

adenylate
cyclase → PP

adenine

cyclic-AMP

phosphodiesterase / H₂O

AMP

B

of these messengers: The receptor molecule, when combined with the messenger, is able to bind to a protein on the inner surface of the membrane (Fig. 7-5). This latter protein is an enzyme, **adenylate cyclase,** which becomes activated by the binding of receptor to it. The activated adenylate cyclase then catalyzes the transformation of cell ATP to another molecule known as **cyclic 3′,5′-adenosine mono-phosphate,** or simply **cyclic AMP (cAMP),** which then acts in the cell as a second messenger to trigger the intracellular sequence of biochemical events that lead ultimately to the cell's overall response. The action of cAMP is terminated by its breakdown to noncyclic AMP, a reaction catalyzed by the enzyme phosphodiesterase.

The organic molecule **cyclic AMP** was first thought to be a second messenger only for hormones. However, it is now recognized to serve this function in the responses to certain neurotransmitters and paracrine agents as well.

Of particular interest is the mechanism of action of certain first messengers, the cell responses to which are opposite to those produced by other first messengers known to generate cAMP. For example, the actions of the hormone insulin are opposite to those of epinephrine, the latter known to be mediated by cAMP, and this has led to the suggestion that insulin acts by inhibiting adenylate cyclase and thereby lowering intracellular cAMP. An alternative explanation invokes a role for a second cyclic nucleotide, **cyclic 3′,5′-guanosine monophosphate (cGMP),** present in many cells. This theory proposes that, since cGMP exerts actions opposed to those of cAMP, hormones like insulin may cause the generation of cGMP in their target cells. Thus, according to this theory, cells may be subject to bidirectional control by opposing second messengers.

KEY TERMS

homeostasis	receptor (detector, sensor)	neurotransmitters
steady state	afferent pathway	hormones
negative feedback	integrating center	paracrines
positive feedback	efferent pathway	receptors
stimulus	effector	second messenger
reflex	local homeostatic responses	cyclic AMP

REVIEW EXERCISES

1 Describe the components of a biological control system.
2 Identify these general components in any reflex.
3 State which cell types constitute the afferent pathways, efferent pathways, and effectors of reflexes.
4 Contrast negative-feedback and positive-feedback systems.
5 Contrast local homeostatic responses with reflexes.
6 State the different categories of intercellular chemical messengers.
7 Describe the types of prostaglandin, their precursor, and their general function.
8 State the initial event in the action of all chemical messengers.
9 Describe the locations and characteristics of receptors (with which chemical messengers combine).
10 Describe how messenger-receptor combinations elicit the cell's response to that messenger.
11 Describe the sequence of events in which cyclic AMP serves as second messenger.
12 State the factors influencing free cytosolic calcium concentration and its role as a second messenger.
13 Define the following terms: homeostatic control system, operating point, prostaglandins, first messenger, adenylate cyclase, cyclic AMP.

PART

2

SUPPORT AND MOVEMENT

CHAPTER

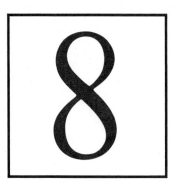

SKELETAL TISSUE: CARTILAGE AND BONE

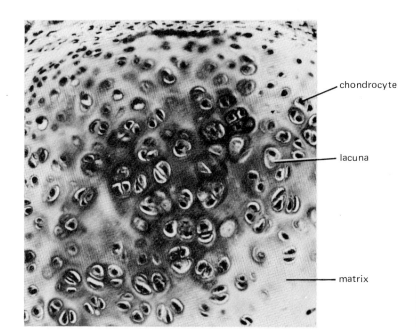

chondrocyte

lacuna

matrix

FIG. 8-1 Cartilage. Each chondrocyte lies in its own lacuna, which is encased by the surrounding matrix.

CARTILAGE

Cartilage is a firm but flexible connective tissue that provides to various structures a rigidity intermediate between that of the soft tissues and bone. **Chondroblasts,** which arise from undifferentiated cells in the mesenchyme, form the extracellular matrix of cartilage. That matrix is composed of bundles of collagen fibers embedded in a ground substance that consists of acid mucopolysaccharides. As this matrix is laid down around each cell, the chondroblasts become further and further separated from each other, and each cell comes to exist in a small cavity, or **lacuna,** enclosed by the surrounding matrix. These isolated cells scattered throughout the matrix of fully developed mature cartilage are now called **chondrocytes** (Fig. 8-1). Surrounding most cartilaginous tissue is a sheath of dense connective tissue, the **perichondrium,** composed of fibroblasts and bundles of collagen fibers. The fibroblasts in this sheath can differentiate to form chondroblasts, which then deposit new layers of matrix along the surface of the cartilage.

Most cartilage has no blood vessels; thus, nutrients can reach the chondrocytes only by diffusing through the matrix from blood vessels in the perichondrium. Since diffusion cannot deliver nutrients rapidly over long distances, the survival of the chondrocytes depends upon their proximity to the perichondral surface. As a result, cartilaginous structures are relatively thin plates or sheets.

There are three types of cartilage: (1) hyaline, (2) fibrous, and (3) elastic. **Hyaline cartilage** is the most abundant form. Its matrix of collagen fibers and mucopolysaccharides is translucent, and the individual collagen fibers cannot be seen with the light microscope because of their small size and the density of the surrounding mucopolysaccharides. This type of cartilage is found in the nose, in supporting rings surrounding the large airways to the lungs, between the ends of the ribs and the sternum (breast bone), and on the surfaces of many bones in the region of joints. Hyaline cartilage is also found in regions of bone growth, as will be described in the next section.

The structure of **fibrous cartilage** is intermediate between hyaline cartilage and dense connective tissue, and its collagen fibers are large and quite visible. Fibrous cartilage forms the intervertebral disks of the spinal column and is present at the ends of tendons and ligaments where they attach to bones.

Elastic cartilage is similar to hyaline cartilage but contains an additional elastic material in its matrix, which consists of a loose meshwork of branched thin fibers and granules embedded in an amorphous material. The elasticity of the matrix material gives this type of cartilage greater flexibility than is found in other forms. It is found in the external ear and auditory tube.

BONE

Bone is the special type of connective tissue that forms the solid matrix of **bones,** which are technically organs composed of bone tissue, other forms of connective tissues, nerves, and blood vessels. Bones form the internal supporting and protecting structures of the body's skeleton and provide the attachments for the muscles that move the skeleton. In addition to these mechanical functions, the central cavity of some bones contains the connective tissue, **hematopoietic tissue,** which forms a variety of blood cells. Bone also performs an important role in the regulation of the calcium and phosphate concentrations in the internal environment through the controlled deposition of these salts in the bone matrix or their release. Living bone is a dynamic structure that is continually resorbed, re-formed, and remodeled.

The basic features of a typical long bone can be seen in Fig. 8-2. The long, central shaft is the **diaphysis,** and the two ends are the **epiphyses.** The portion of each epiphysis that contacts other bones at a joint is the **articular surface;** it is covered with hyaline cartilage. The rest of the bone is covered by a tightly adhering sheath of connective tissue, the **periosteum.**

Histology of Bone

A longitudinal section through a long bone (Fig. 8-2) reveals a central **marrow cavity** that contains **yellow marrow,** which is composed largely of fat. The dense, ivorylike surface layer of the bone (the **cortex**) is compact bone. Below this layer is **cancellous** (or **spongy**) **bone,** which has a latticelike structure that contains many small cavities that merge with the large central cavity. The **trabeculae,** or spicules, of spongy bone (Fig. 8-3) are arranged like struts to resist forces that tend to compress the bone or put tension on it. The spaces between the trabeculae are filled with **red marrow,** which consists mainly of hematopoietic tissue (its color due to the presence of numerous red blood cells and their pigmented precursors), some macrophages, and fat cells. Although red marrow occurs throughout the skeleton at birth, most of it is gradually replaced by inactive yellow marrow so that, in the adult, blood cell formation occurs mainly in certain regions of the skull, clavicles, ribs, vertebrae, ster-

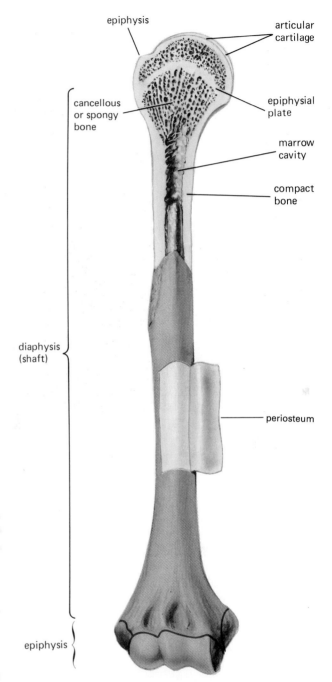

FIG. 8-2 Structure of a typical long bone. Surface regions and interior anatomy in a longitudinal section.

num, and pelvis. Active bone marrow constitutes approximately 5 percent of the total adult body weight.

The developing blood cells in the red marrow are supported in a loose meshwork formed by *adven-*

epiphysis   trabeculae

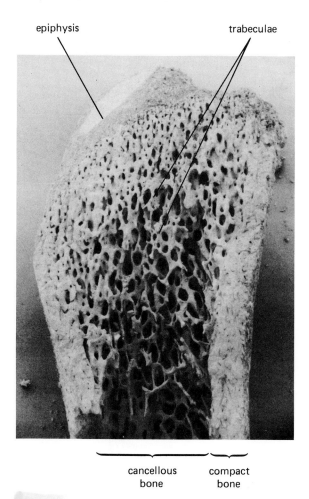

cancellous   compact
bone   bone

FIG. 8-3 Scanning electron micrograph of a section through the end of a long bone from which the organic constituents were removed, thus exposing the calcified bone matrix.

titial, or *reticular, cells.* The newly formed blood cells enter the circulation through the walls of *vascular sinuses,* which connect the blood vessels that supply the marrow with those that drain it.

The long bones are subject to the greatest bending and twisting forces in the central region of the diaphysis, and here the compact bone is relatively thick and the spongy bone almost nonexistent. Spongy bone occurs mainly in the epiphyses and near the ends of the diaphysis where the bone is subject to the greatest weight-bearing stresses; here the shell of compact bone is relatively thin.

The tubular shape characteristic of most long bones provides strength yet keeps the structure relatively light. The tensile strength of bone, i.e., its resistance to being stretched under applied forces, is comparable to that of cast iron, yet it is only a third the weight of iron. In its flexibility, bone resembles steel more than cast iron; it has about half the flexibility of steel.

As we shall see in Chap. 9, there are bone types other than long bones. Their structure is generally similar to that of the epiphyses of long bones, i.e., primarily cancellous bone that underlies a thin layer of compact bone and has no marrow cavity. As with long bones, the entire bone, except for the surfaces involved in joints, is encased in periosteum.

The composition of the matrix that surrounds the bone-forming cells is basically similar in both compact and cancellous bone. The organic components are collagen fibers embedded in an amorphous mucopolysaccharide ground substance. Bound to these fibers are needlelike crystals known as *hydroxyapatite,* which consists of inorganic minerals, mostly calcium phosphate in proportions given by the formula $Ca_{10}(PO_4)_6(OH)_2$, calcium carbonate, and sodium salts. These minerals, which compose 65 percent of the bone's weight, give bone its hard consistency.

The basic unit of compact bone is a **haversian system,** or *osteon.* Each system contains at its center a small channel, a **haversian canal,** which runs essentially parallel to the long axis of the bone and contains one or two small blood vessels and nerve fibers. Surrounding each canal are successive layers of mineralized matrix **(lamellae)** that, in a cross section of the bone, appear as a concentric series of rings (Fig. 8-4). Scattered along the boundaries of each lamella are small cavities known as **lacunae,** each of which contains a single bone cell, or **osteocyte.** Extending radially from the lacunae are numerous small channels, **canaliculi,** that link the lacunae of adjacent rings and ultimately extend to the haversian canal. Projections from the cell extend from the osteocytes into the canaliculi, and it is through these channels that nutrients diffuse to the osteocyte from the blood vessels in the haversian canal.

The relatively thin, interconnected trabeculae of cancellous bone do not contain typical haversian systems, although the tissue is still arranged in lacunae-containing lamellae. In cancellous bone nutrients diffuse to the osteocytes through the canaliculi from blood vessels that pass through the bone marrow and hollow regions of the spongy bone.

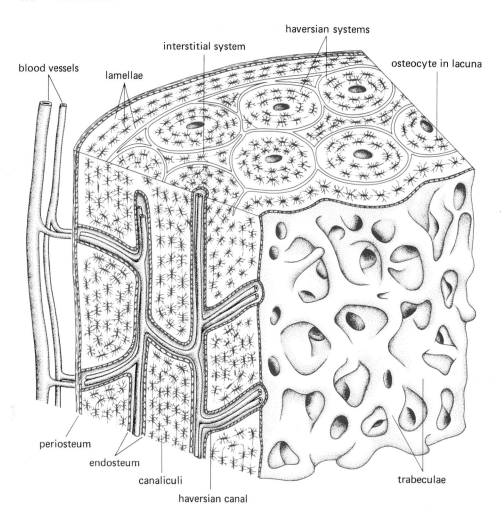

blood vessels

lamellae

interstitial system

haversian systems

osteocyte in lacuna

periosteum

endosteum

canaliculi

haversian canal

trabeculae

FIG. 8-4 Arrangement of haversian systems in compact bone.

Table 8-1 presents a comparison of cartilage and bone.

Bone Development and Repair

Calcified bone is formed by the process known as **ossification.** Although it occurs in two different ways (intramembranous bone formation and endochondral bone formation), the resulting structures of the calcified bone are similar. **Intramembranous bone formation** involves direct calcification of the organic bone matrix in regions of mesenchyme where there is no preexisting cartilage. **Endochondral bone formation** involves the replacement of a cartilaginous structure by bone.

Intramembranous Bone Formation Intramembranous bone formation begins in the embryo with

TABLE 8-1   COMPARISON OF CARTILAGE AND BONE
Similarities
1   Both consist of cells in lacunae and an organic intercellular substance called the matrix.
2   The matrix consists of collagen fibers embedded in an amorphous component.
3   Both develope from mesenchyme.
Differences
1   The ratio of fibers to amorphous substance in the matrix is much higher in bone than in cartilage.
2   The cells of bone are osteocytes; those of cartilage are chondrocytes.
3   The connective-tissue of bone is called periosteum; that of cartilage, perichondrium.
4   Bone is much stronger than cartilage.
5   Bone is a vascular tissue, but cartilage is not; thus bone tissue, while appearing solid, is in reality permeated with capillaries and other vessels enclosed in haversian canals.

the differentiation of mesenchymal cells into bone-forming cells, the **osteoblasts.** These osteoblasts synthesize and secrete collagen-fiber precursors and an organic ground substance, and thus surround themselves with a nonmineralized matrix. As calcium salts are laid down in the newly formed matrix and the matrix becomes mineralized, small spicules (trabeculae) of bone are formed, each surrounded by a single layer of osteoblasts. Some of the osteoblasts, however, become trapped in lacunae in the mineralized matrix and become osteocytes. Adjacent spicules fuse at various points to form the spongelike structure of cancellous bone. The matrix becomes mineralized almost immediately and further bone growth can occur only at the surface of the bone, i.e., by apposition to already existing bone. As the cells covering the trabeculae continue to multiply and form new bone, the trabeculae become thicker and the spaces between them become smaller. When the amount of bone exceeds the amount of space between bone, the tissue is known as compact bone.

Mesenchymal tissue trapped in the spaces of the spongy bone eventually gives rise to the hematopoietic tissues of the bone marrow. The layer of mesenchymal tissue that surrounds the developing bone becomes the periosteum, from which further osteoblasts arise and add new concentric layers of ossified bone to the growing network of spicules. The layer of osteoblasts that line the haversian canal and marrow cavity is the **endosteum.**

Endochondral Bone Formation   Endochondral bone formation begins with the clumping together of mesenchyme cells into a group roughly the shape of the bone-to-be. The cells at the center of the packed mesenchyme then begin to differentiate into chondroblasts, which deposit a cartilage matrix in the shape of the developing bone (Fig. 8-5A). The mesenchyme cells immediately surrounding the cartilaginous core then transform into a perichondrium, the inner cells of which continue to form cartilage (Fig. 8-5B). The cartilaginous bone-to-be continues to grow in length and width by cell division at the surface *(appositional growth)* and within the cartilage core *(interstitial growth)*, most of the interstitial growth occurring near the ends of the bone.

As the number of blood vessels in the developing embryo increases, the cells that surround the bone-to-be differentiate into osteoblasts rather than chondroblasts and the perichondrium becomes periosteum; this occurs initially around the midregion of the bone shaft. At the inner surface of the periosteum, adjacent to the cartilaginous bone, osteoblasts begin to lay down a mineralized matrix in a

FIG. 8-5 Stages in endochondral bone formation.

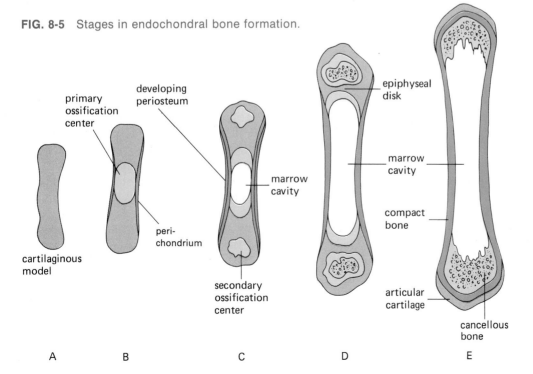

primary ossification center

developing periosteum

peri-chondrium

marrow cavity

secondary ossification center

cartilaginous model

epiphyseal disk

marrow cavity

compact bone

articular cartilage

cancellous bone

A   B   C   D   E

manner resembling the process of intramembranous bone formation. As the shaft of the cartilaginous bone widens and its surface becomes mineralized, the cells at the core no longer receive adequate nutrition and begin to die. The lacunae coalesce, and the spaces of the disintegrating cartilage are invaded by bone-forming cells and capillaries from the periosteum. These form the primary *center of ossification,* which spreads out to replace most of the cartilage shaft (Fig. 8-5B). The first bone formed in this area is cancellous bone. When this stage is completed, the developing bone consists of a mineralized shaft with two large masses of cartilage at the epiphyses.

The next stage involves the invasion of the epiphyses by osteoblasts to form *secondary centers of ossification* at the two ends of the bone (the ossification center in the shaft is the primary center). The formation of the secondary centers begins in different bones sometime between birth and the first 3 years of life. Between the primary ossification center and the two secondary centers at the epiphyses are layers of growing cartilage (Fig. 8-5C–D). In these regions, known as the **epiphyseal disks** or **plates,** the growth of the long bones occurs by interstitial growth of the cartilage and its replacement by mineralized bone.

Four zones exist in the epiphyseal plate of growing bone (Fig. 8-6). *Resting cartilage* constitutes the zone nearest the epiphyseal aspect of the bone and contains chondrocytes distributed at random throughout its matrix. Next is the zone of *chondrocyte multiplication,* which is the major area for new cartilage deposition; in this zone the cells become arranged in rows parallel to the long axis of the bone. In the third zone, the zone of *chondrocyte maturation,* the cells become enlarged and show signs of beginning disintegration. In the last zone (nearest the shaft of the bone), the zone of *cartilage ossification,* chondrocytes die and disintegrate, and the region is invaded by bone-forming cells and capillaries. By this continuous process of cartilaginous growth at the epiphyseal side of the plate and replacement by mineralized bone at the diaphyseal side, the long bones are able to elongate during the years of growth. Eventually, the epiphyseal plates become completely replaced by bone tissue (**epiphyseal closure**) and further lengthening of the bone ceases. This occurs at various ages in the different bones but is usually complete in all the long bones by the age of 20 years.

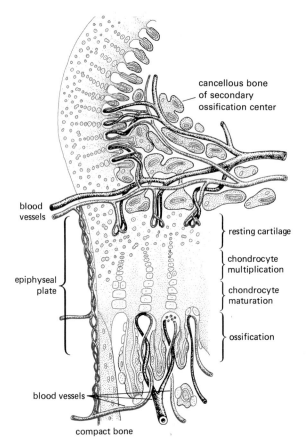

FIG. 8-6 Schematic representation of active cartilage growth and transformation into bone in the epiphyseal plate. Note that the bone is vascularized, whereas the cartilage is not.

Mature Bone and Bone Remodeling The formation of new bone tissue and bone mineralization does not cease with closure of the epiphyseal plates. The deposition of newly mineralized bone occurs throughout life, although normally at a rate balanced by the removal of old bone; thus, there is a continual resorption and reforming of bone but no bone elongation. This removal and deposition of bone can occur only at the surface of the bone, either under the periosteum or along the edge of the marrow cavity. As long as the rates of bone deposition and removal at these two locations remain equal, the thickness of the compact bone will remain constant. In old age the rates of bone removal exceed those of bone deposition and the bones become thinner and more easily broken.

By a combination of the addition of bone at one site on the surface of a preexisting bone and the resorption of bone at another site the shape of the bone can undergo **remodeling.** In general, the remodeling occurs in response to tensile and compressive forces on the bone. Thus a bone subjected to unusual strain will gradually undergo resorption and rebuilding of its structure to best fit the new forces acting upon it. The resorption of bone is accomplished by **osteoclast cells** that appear on the bone surfaces. These cells are also responsible for the removal of trabeculae in the center of the growing bone to form the marrow cavity.

A number of hormones are known to influence bone growth, epiphyseal closure, and the rates at which calcium and phosphate are exchanged between the bone matrix and the blood. These hormones will be described later, primarily in Chap. 24.

There are a large number of diseases in which abnormal bone metabolism is due not to an inherent defect in the bone cells but is secondary to a metabolic disturbance that originates elsewhere in the body. For example, **osteosclerosis**—increased density of calcified bone—occurs in patients with certain tumors, lead poisoning, and deficient parathyroid gland function. **Osteomalacia** and **rickets**—inadequate mineralization per unit matrix—are caused mainly by a lack of the active form of vitamin D. **Osteoporosis**—a decrease in the amount of bone below that necessary for adequate mechanical support but with no change in the microscopic appearance or mineralization-matrix ratio—occurs commonly in postmenopausal women and elderly men.

Fractures and Their Repair   A fracture is any break in the continuity of a bone. A **simple,** or **closed fracture** is one in which the bone is broken in two parts or fragments but the skin or mucous membrane overlying the bone is intact. The periosteum is usually torn, the bone fragments have moved out of their proper alignment, and blood vessels in the haversian systems, periosteum, and adjacent tissues have been damaged. When both the bone and the overlying skin or mucous membrane are broken, the fracture is called **compound,** or **open** (Fig. 8-7). A **compression fracture** results in

FIG. 8-7 Various bone fractures.

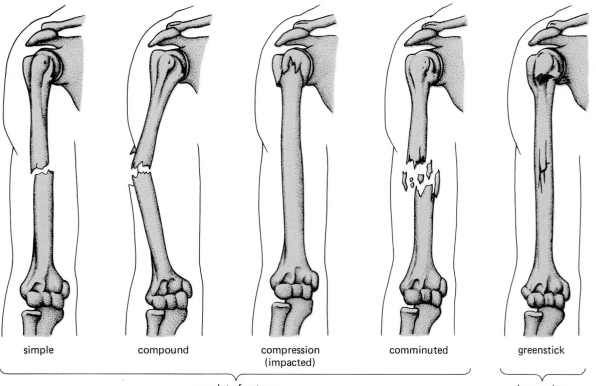

simple   compound   compression
                                  (impacted)   comminuted   greenstick

complete fractures   incomplete
                                                          fracture

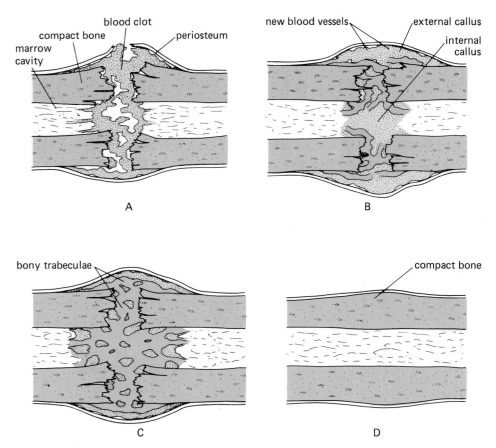

FIG. 8-8 (A) In a fracture the periosteum is usually torn, blood vessels damaged, and bone fragments separated. (B) Rapid division of bone- and cartilage-forming cells in the region of the break forms a thickened band that is composed of an internal and external callus. (C) Osteoblasts form trabeculae, which adhere to existing bone and extend across the break. (D) The break is bridged by compact bone, and the contour of the now intact bone is remodeled.

decreased length or width of a bone, whereas a **comminuted fracture** contains more than two fragments of bone.

If the fracture crosses the entire width of a bone, it is a **complete fracture;** if it does not, it is **incomplete,** or **partial.** Thus, a "greenstick" fracture, in which the surface opposite to the break is still intact, is a type of incomplete fracture.

The first changes in the repair of the fracture occur in the periosteum adjacent to the break. There, the primitive bone-forming (osteogenic) cells differentiate into chondroblasts when there is little blood supply, or into osteoblasts when there is better vascularization, i.e., when the cells are accompanied by newly formed capillaries. The rapid division of the osteogenic cells causes a thickened

band (the **callus**) to form around the broken edge of the bone. As the band thickens, it bulges across the break and eventually meets the band from the opposite fragment and fuses with it. Fusion of the two fragments is aided by the formation of an internal callus as osteoblasts along the edge of the marrow cavity and those lining haversian canals form trabeculae, which extend across the region of the break (Fig. 8-8). Thus, the two fragments come to be joined first by trabeculae, which adhere to the underlying bone, and finally by compact bone. The site of the healing fracture is remodeled as osteoblasts lay down new bone and osteoclasts remove the damaged and temporary bone. Often, the fracture site cannot be felt or seen on x-ray after healing is complete.

KEY TERMS

cartilage
lacuna
chondrocyte
perichondrium
bone
hematopoietic tissue
diaphysis

epiphysis
periosteum
marrow cavity
compact bone
cancellous or spongy bone
marrow
haversian system

osteocyte
intramembranous bone formation
endochondral bone formation
epiphyseal disk
epiphyseal closure
osteoclast

REVIEW EXERCISES

1   Draw a section of hyaline cartilage; label a chondrocyte, a lacuna, the perichondrium, and the matrix.
2   Distinguish between the structure and location of hyaline, fibrous, and elastic cartilage.
3   Draw a typical long bone; label the diaphysis, epiphyses, shaft, and articular surfaces.
4   Draw a diagram of a microscopic cross section of compact bone; label a haversian system, haversian canal, lamella, lacuna, canaliculus, and the periosteum.
5   Describe the processes of intramembranous and endochondral bone formation.
6   Distinguish between closed and open fractures; between simple and compound fractures; between complete and incomplete fractures.
7   Describe the stages of fracture repair.

CHAPTER

9

THE SKELETAL SYSTEM

TYPES OF BONES

Bones vary greatly in size and shape, but despite this diversity they can be divided into five general categories. **Long bones** (Fig. 9-1A) occur in the upper and lower limbs. As described in Chap. 8, they are tube-shaped and have a long shaft, expanded epiphyses and contain a central marrow cavity. The femur, the large thigh bone, is an example of a long bone. The **short bones** (Fig. 9-1B) such as the bones of the wrists or ankles, are similar in shape to the long bones but do not develop epiphyseal centers for bone growth; instead they depend upon appositional growth at the ends of the bone. The small, rounded **sesamoid bones** (Fig. 9-1C) of which the patellas (knee caps) are the most familiar example, are almost always near joints or embedded in tendons where the tendons make a sharp bend around a bony surface. With the obvious exception of the patella, the sesamoid bones are usually associated with joints in the hands and feet. The **flat bones** (Fig. 9-1D) include the bones of the upper portion of the skull, the ribs, and the shoulder blades. They are formed of an inner layer of trabecular bone (the *diploë*) between two layers of compact bone. Bones such as the vertebrae and some of the bones of the face, which fit none of the above categories, are classified as **irregular** (Fig. 9-1E).

The surfaces of bones are generally characterized by structural variations such as grooves, protrusions, holes, and shallow depressions. The grooves and holes provide passageways for nerves and blood vessels; the large protrusions at the ends of bones form parts of joints; and shallow depressions and ridges often serve as the attachment points for muscles.

The terms used to describe such surface markings are as follows:

Protrusions (processes):
Condyle: A relatively large, rounded protuberance that forms articulations
Crest: A prominent ridge or border
Epicondyle: A prominence above or on a condyle
Head: A rounded projection separated from the main part of the bone by a constricted region (the *neck*)
Line: A low ridge
Spine: A sharp, pointed process
Trochanter: A large, blunt process found only on the femur
Tubercle: A small, rounded projection
Tuberosity: A large, rounded process, often with a roughened surface

Flat surfaces and depressions:
Facet: A flat or shallow surface for articulations
Fissure: A groove or narrow cleft passage
Foramen: A hole or perforation through a bone
Fossa: A depression or groove
Fovea: A small, shallow depression
Groove or **sulcus:** A furrow that accommodates a blood vessel, nerve, tendon, or other soft structure
Meatus: A tube-like passageway or canal within a bone
Neck: The narrow section of bone between the head and the shaft
Notch: A deep indentation in the edge of a bone
Sinus: An air-filled cavity in a bone
Sulcus: See groove

FIG. 9-1 Five categories of bone: (A) long, (B) short, (C) sesamoid, (D) flat, and (E) irregular.

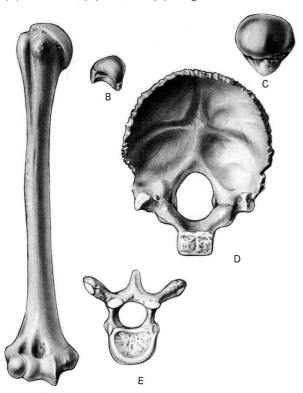

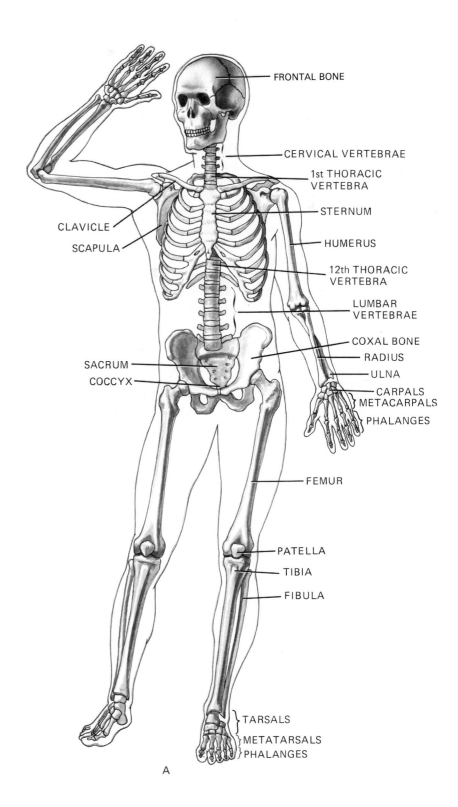

FRONTAL BONE

CERVICAL VERTEBRAE

1st THORACIC
VERTEBRA

STERNUM

HUMERUS

12th THORACIC
VERTEBRA

LUMBAR
VERTEBRAE

CLAVICLE

SCAPULA

COXAL BONE

RADIUS

SACRUM

ULNA

COCCYX

CARPALS
METACARPALS

PHALANGES

FEMUR

PATELLA

TIBIA

FIBULA

TARSALS

METATARSALS

PHALANGES

FIG. 9-2 The adult human
skeleton: (A) anterior, and
(B) posterior views.

A

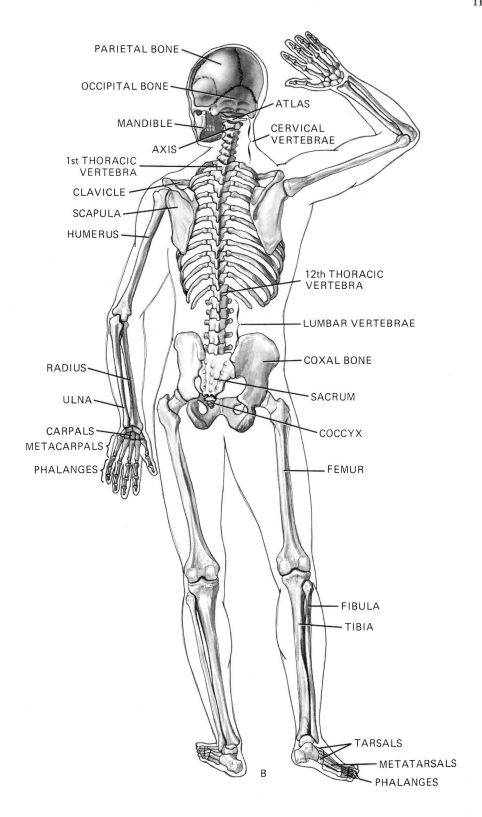

PARIETAL BONE

OCCIPITAL BONE

MANDIBLE

AXIS

1st THORACIC
VERTEBRA

CLAVICLE

SCAPULA

HUMERUS

RADIUS

ULNA

CARPALS
METACARPALS

PHALANGES

ATLAS

CERVICAL
VERTEBRAE

12th THORACIC
VERTEBRA

LUMBAR VERTEBRAE

COXAL BONE

SACRUM

COCCYX

FEMUR

FIBULA

TIBIA

TARSALS

METATARSALS

PHALANGES

B

TABLE 9-1	BONES OF THE HUMAN SKELETON

	Number of Bones
AXIAL SKELETON	**80**
Skull	**29**
Cranium (brainbox) 15	
Parietal 2	
Temporal 2	
Frontal 1	
Occipital 1	
Sphenoid 1	
Ethmoid 1	
Nasal 2	
Lacrimal 2	
Inferior nasal concha 2	
Vomer 1	
Face 7	
Maxilla 2	
Zygomatic 2	
Palatine 2	
Mandible 1	
Others 7	
Auditory ossicles:	
Malleus 2	
Incus 2	
Stapes 2	
Hyoid 1	
Vertebral column	**26**
Cervical vertebra 7	
Thoracic vertebra 12	
Lumbar vertebra 5	
Sacral vertebra (5 fused) 1	
Coccygeal veretbra (usually 4 fused)	
Rib cage	**25**
Rib 24	
Sternum (3 fused) 1	

	Number of Bones
APPENDICULAR SKELETON	**126**
Upper limb	**64**
Pectoral girdle 4	
Clavicle 2	
Scapula 2	
Arm 60	
Humerus 2	
Radius 2	
Ulna 2	
Carpal 16	
Metacarpal 10	
Phalanges 28	
Lower limb	**62**
Pelvic girdle 2	
Coxal bone (3 fused) 2	
Leg 60	
Femur 2	
Patella 2	
Tibia 2	
Fibula 2	
Tarsal 14	
Metatarsal 10	
Phalanges 28	
TOTAL	**206**

In this chapter, these surface markings will, for the most part, be merely identified; in the subsequent chapters on articulations (joints) and muscles, their specific functions will be given.

The adult skeleton (Fig. 9-2A and B) consists of approximately 206 bones (Table 9-1). We say "approximately" because extra bones are rather frequently encountered, particularly in the hands and feet.

The skeleton is divided into two basic parts: the **axial skeleton,** which consists of the head and trunk (skull, vertebral column, ribs, and sternum or breastbone); and the **appendicular skeleton,** which consists of the bones of the upper and lower limbs and of the pectoral (shoulder) and pelvic (hip) girdles by which the limbs are attached to the axial skeleton.

THE AXIAL SKELETON

The Skull

The skull is a complex of bones adapted to support and protect the brain and several of the special senses, to isolate the cerebral blood vessels from extracranial variations in pressure, and to get and process food. To serve these functions the skull has two regions: a "brainbox," called the **cranium,** and a **facial skeleton.**

Many of the bones of the skull are formed in whole or in part by intramembranous bone formation, whereas other skull bones are formed by endochondral bone formation. Even though the earliest centers of ossification appear during the sixth

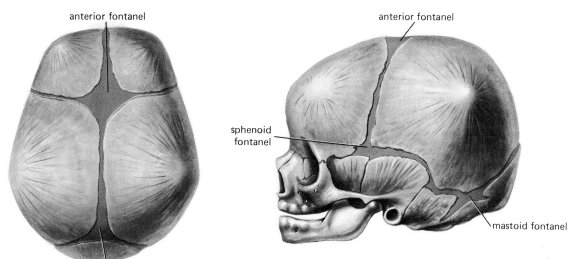

anterior fontanel

anterior fontanel

sphenoid
fontanel

mastoid fontanel

posterior fontanel

FIG. 9-3 Skull of a newborn infant showing the fontanels.

and seventh months of fetal life, ossification is not complete at the time of birth. The bones forming the sides and roof of the skull are still joined by membrane while some at the base of the skull are united by cartilage, both of which allow further growth of the skull. The largest areas of membrane at birth are the **anterior, sphenoid, mastoid,** and **posterior fontanels** (Fig. 9-3). The mastoid and sphe-

noid fontanels lie deep to muscle, but the anterior and posterior fontanels lie directly under the scalp where they can be felt as the "soft spots" of an infant's head.

Of the combined cranial and facial portions of the skull of an adult, only one, the **mandible** or lower jaw bone, is movable, whereas the other 21 bones are firmly bound together. The skull contains five

FIG. 9-4 Top of the calvaria of the skull.

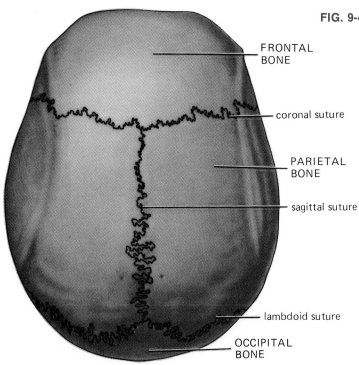

FRONTAL
BONE

coronal suture

PARIETAL
BONE

sagittal suture

lambdoid suture

OCCIPITAL
BONE

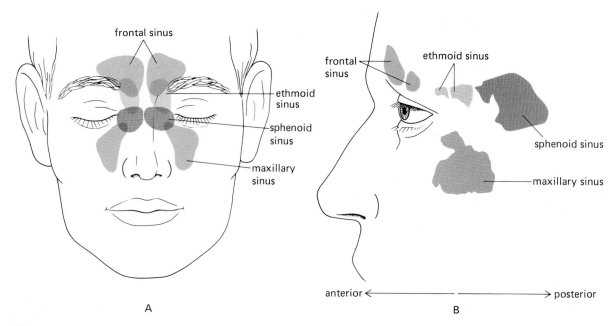

FIG. 9-5 (A) Frontal and (B) lateral view of the paranasal sinuses, which are in the frontal, ethmoid, sphenoid, and maxillary bones.

large cavities, four of which (the paired **nasal cavities** and **orbits,** or eye sockets) open to the outside. The largest cavity, the **cranial cavity,** which contains the brain, is closed except where the spinal cord joins the brain. The roof of the cranial cavity is the **calvaria.**

The anterior part of the calvaria is formed by the **frontal bone,** which passes back over the top of the skull as far as the *coronal suture* (Fig. 9-4) where it meets the right and left **parietal bones.** These two bones meet in the midline at the top of the skull forming the *sagittal suture* and extend back to meet the unpaired **occipital bone** at the *lambdoid suture*.

Most of the bones of the cranium consist of two layers of compact bone (the *laminae*) separated by a layer of spongy bone (the *diploë*) and relatively large veins. In some of the bones, the diploë is in part replaced by air-filled cavities, the **paranasal sinuses** (Fig. 9-5), which open into the nasal cavity. Cavities in the temporal bones, the *mastoid cells,* may contain air or bone marrow.

Anterior View Figure 9-6 shows the anterior bones of the face. The two **maxillae** contain the sockets for the teeth of the upper jaw. They also form a large part of the opening of the nasal cavities

in the skull (the *piriform aperture*) and part of the orbits. Below the orbit, an *infraorbital foramen* pierces the maxilla, and thus allows passage of part of the maxillary branch of the trigeminal nerve (Vth cranial nerve), which supplies the cheek, upper lip, and part of the nose. The *frontal process* of the maxilla passes upward between the orbit and nose to form part of the lateral wall of the nasal cavity, and the *zygomatic process* extends to meet the zygomatic bone. The body of the maxilla contains, as mentioned earlier, the large maxillary sinuses.

Nasal Cavities The nasal skeleton includes the bony and cartilaginous support of the external nose and the walls of the nasal cavity. The bridge of the nose is formed by the paired nasal bones and the frontal processes of the two maxillae, and the nasal bones, with the maxillae, complete the piriform aperture (Fig. 9-6). In the midline of the lower border of the piriform aperture is the *anterior nasal spine*. The nasal **septum** is the partition that separates the nasal cavity into two irregular spaces that open anteriorly via the piriform aperture and posteriorly by paired **choanae.** Each space communicates with the paranasal sinuses, which were mentioned earlier.

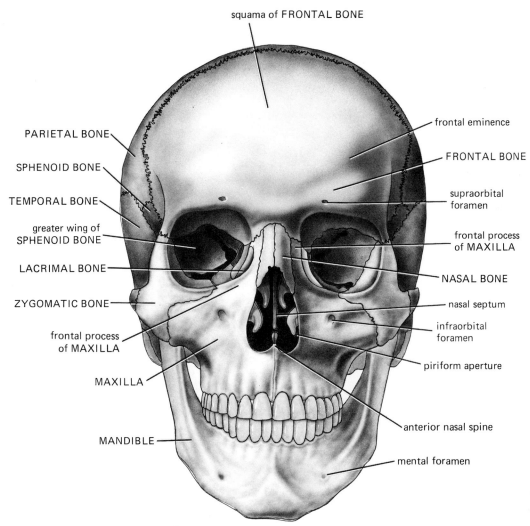

squama of FRONTAL BONE

frontal eminence

PARIETAL BONE

FRONTAL BONE

SPHENOID BONE

supraorbital
foramen

TEMPORAL BONE

frontal process
of MAXILLA

greater wing of
SPHENOID BONE

NASAL BONE

LACRIMAL BONE

nasal septum

ZYGOMATIC BONE

infraorbital
foramen

frontal process
of MAXILLA

piriform aperture

MAXILLA

MANDIBLE

anterior nasal spine

mental foramen

FIG. 9-6 The skull, anterior view.

The roof of each half of the nasal cavity (Fig. 9-7) is formed by the nasal and frontal bones in front, and by the cribriform plate of the ethmoid bone (seen better from inside the skull, Fig. 9-15A) and the sphenoid. The posterior part of the cavity roof consists of the tip of the upward projecting vomer (not shown in Fig. 9-7 but visible in Figs. 9-8, 9-13, and 9-14), the body of the sphenoid, and a portion of the palatine bone.

The floor of the nasal cavity is formed mainly by the palatine process of the maxilla and the horizontal plates of the palatine bones, the same bones that form the hard palate of the roof of the mouth.

The septum of the nose (Fig. 9-8) is formed mainly by the perpendicular plate of the ethmoid bone (Fig. 9-9) and the vomer. The anterior border of the bony septum has a notch that in life is filled in by the **septal cartilage** of the nose, and the posterior border (between the two choanae) is formed by the posterior edge of the vomer. The septum is usually deflected from the midline.

The lateral wall of the nasal cavity (Fig. 9-7) is formed by the maxilla, lacrimal bone, ethmoid bone, and the inferior nasal concha. The three **conchae** (*superior, middle,* and *inferior*), which project medially into the nasal cavity, form ridges along the lateral wall.

The choanae are bounded by the vomer, medial

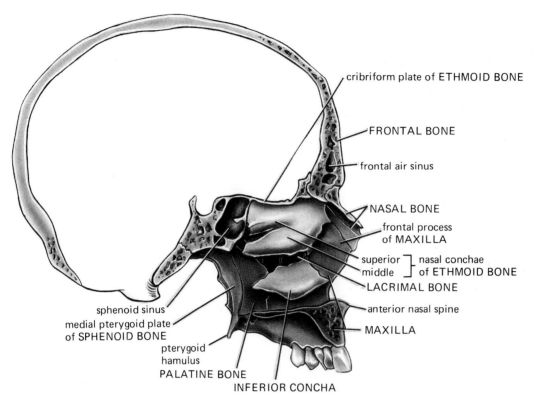

cribriform plate of ETHMOID BONE

FRONTAL BONE

frontal air sinus

NASAL BONE

frontal process
of MAXILLA

superior ⎤ nasal conchae
middle ⎦ of ETHMOID BONE

LACRIMAL BONE

anterior nasal spine

MAXILLA

sphenoid sinus

medial pterygoid plate
of SPHENOID BONE

pterygoid
hamulus

PALATINE BONE

INFERIOR CONCHA

FIG. 9-7 Left lateral wall of the nasal cavity.

FIG. 9-8 Bony and cartilaginous structures of (A) the
external nose and (B) the nasal septum.

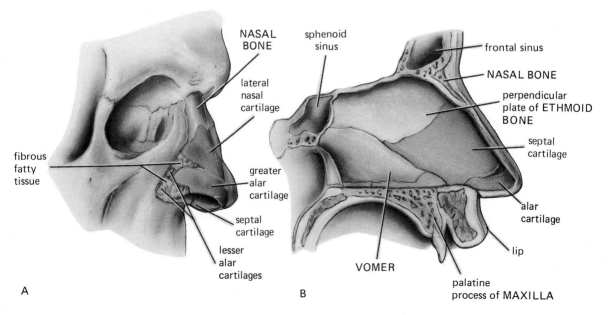

NASAL
BONE

sphenoid
sinus

frontal sinus

NASAL BONE

lateral
nasal
cartilage

perpendicular
plate of ETHMOID
BONE

septal
cartilage

fibrous
fatty
tissue

greater
alar
cartilage

alar
cartilage

septal
cartilage

lip

lesser
alar
cartilages

VOMER

palatine
process of MAXILLA

A

B

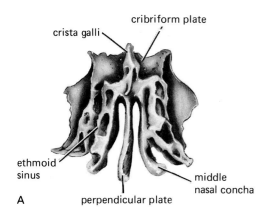

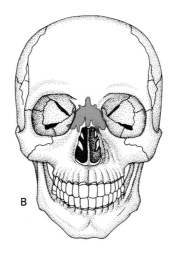

FIG. 9-9 (A) The ethmoid bone. (B) Its location in the skull.

pterygoid plate of the sphenoid bone, and the horizontal plate of the palatine bones.

The Orbit The **orbit,** which largely surrounds and protects the eyeball and the nerves, muscles, and vessels associated with it, lies below the anterior cranial fossa, a large, flat depression in the interior of the skull that supports the anterior portions of the brain. The roof of the orbit is composed chiefly of the frontal bone (Fig. 9-10); the floor is formed by orbital plate of the maxilla and a small portion of the zygomatic bone. The medial wall is formed by the lacrimal bone and the orbital plate of the ethmoid bone; the lateral wall by the frontal

process of the zygomatic bone and the greater wing of the sphenoid bone. At the back of the orbit is a somewhat triangular opening, the *superior orbital fissure* and, medial to this, the rounded *optic canal,* through which passes the optic (IId cranial) nerve.

Lateral View We now turn to a discussion of the bones seen on a lateral view of the skull (Fig. 9-11). The **frontal bone,** as mentioned previously, forms a large part of the roof of the orbit. In addition to this *orbital portion,* the frontal bone consists of a *squama,* which forms the forehead and contains the paired frontal sinuses, and a *nasal portion.* Just lateral to the orbit the zygomatic process of the

FIG. 9-10 (A) The orbit. (B) Its location in the skull.

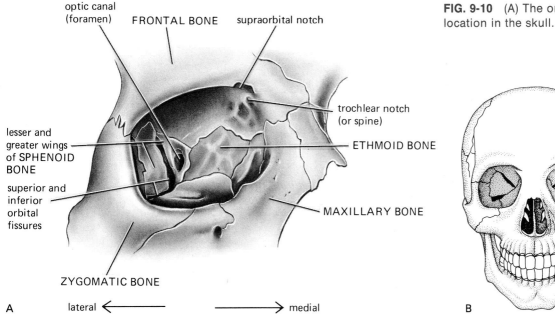

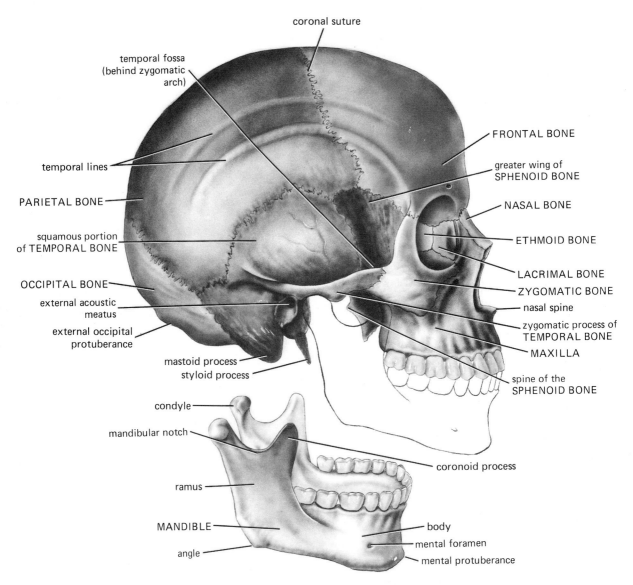

FIG. 9-11 The right lateral view of the skull and the mandible.

frontal bone dips down to meet the zygomatic bone and complete the lateral wall of the orbit.

The **mandible** consists of a *body,* which bears the sockets for the teeth of the lower jaw, and a more vertical process, the *ramus,* which receives the attachments for the major muscles of the jaw and articulates with the temporal bone. The ramus and body of the mandible meet at the *angle.* The body of the mandible bears the *mental protuberance* and contains the paired *mental foramina,* through which pass branches of the mandibular nerve to supply the lower lip and chin. The ramus ends in two processes, an anterior *coronoid process,* which provides for muscle attachments and the posterior *condylar process,* which helps form the *temporomandibular joint;* between the two processes is the *mandibular notch.* On the inner surface of the ramus is the opening to the *mandibular canal* (not shown in Fig. 9-11) through which pass the nerves and vessels to the roots of the teeth. A small bony projection, the *lingula,* partially overlaps the opening and serves as a landmark for injections of the nerve.

In the lateral view of the skull (Fig. 9-11), most of the bones seen are those of the cranium, although

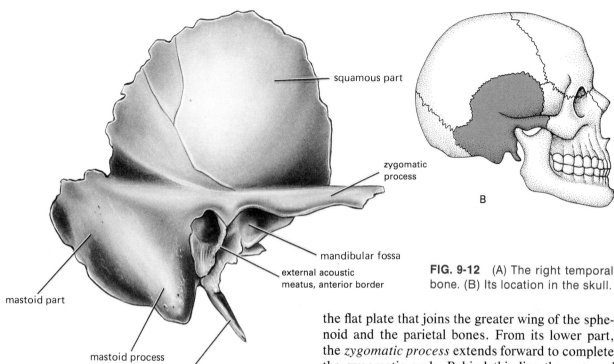

squamous part

zygomatic process

mandibular fossa

external acoustic meatus, anterior border

mastoid part

mastoid process

styloid process

A

B

FIG. 9-12   (A) The right temporal bone. (B) Its location in the skull.

the maxilla and the **zygomatic bone,** which forms the lateral rim of the orbit and the bony prominence of the cheek, can also be seen. The zygomatic bone sends a *frontal process* upward along the side of the orbit to meet the frontal bone and a *temporal process* horizontally backward to meet a process from the temporal bone; the temporal and zygomatic processes together form the *zygomatic arch.*

The greater wing and spine of the **sphenoid bone** are visible in Fig. 9-11. This irregularly shaped bone is situated at the base of the skull and, except for its greater wing, is best discussed later. The *greater wing* projects laterally from the body of the bone and is perforated by the *foramen rotundum,* which transmits a branch of the trigeminal nerve. The region of joining of the frontal, parietal, sphenoid, and temporal bones is the *pterion.* With the temporal bone, the greater wing of the sphenoid forms a portion of the side of the skull, deep to the zygomatic arch, known as the *temporal fossa.*

The **temporal bone** (Figs. 9-11 and 9-12) lies immediately behind the greater wing of the sphenoid. It forms the lower part of the lateral skull, contributes to the base, and houses the middle and inner ears. The *squamous part* of the temporal bone is

the flat plate that joins the greater wing of the sphenoid and the parietal bones. From its lower part, the *zygomatic process* extends forward to complete the zygomatic arch. Behind this lies the *external acoustic (auditory) meatus,* or ear canal, and behind the external meatus is the *mastoid process,* which is actually part of the *petrous* (or dense) *portion* of the temporal bone. Most of the petrous portion is associated with the base of the skull and will be discussed later. Projecting downward from the temporal bone in front of the mastoid process is the slender *styloid process.*

Base of the Skull   The **occipital bone** is the chief bone of the posterior part of the base of the skull (Fig. 9-13). Although the parts of the occipital bone are not distinguished by discrete landmarks, there are basilar and lateral parts as well as a *squama,* which is the portion that rises up to articulate with the two parietal bones at the *lambdoidal suture.* Sometimes the highest part of the squama persists as a separate *interparietal bone.* The *external occipital protuberance* projects from the posterior surface of the squama, and it may continue as an *external occipital crest* that extends downward, from which are *superior* and *inferior nuchal lines.* The occipital bone surrounds the *foramen magnum,* which is the opening through which the spinal cord passes as it joins the brain. Adjacent to the foramen magnum are the *occipital condyles,* which articulate with the *atlas,* or first cervical vertebra. Behind

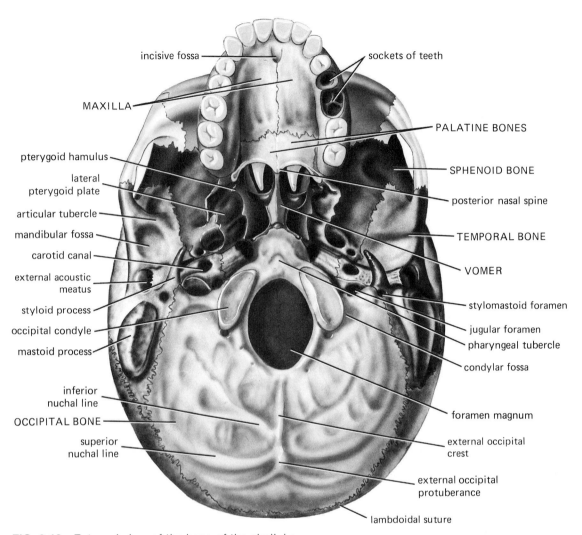

incisive fossa

sockets of teeth

MAXILLA

PALATINE BONES

pterygoid hamulus

SPHENOID BONE

lateral
pterygoid plate

posterior nasal spine

articular tubercle

TEMPORAL BONE

mandibular fossa

VOMER

carotid canal

external acoustic
meatus

stylomastoid foramen

styloid process

jugular foramen

occipital condyle

pharyngeal tubercle

mastoid process

condylar fossa

inferior
nuchal line

OCCIPITAL BONE

foramen magnum

superior
nuchal line

external occipital
crest

external occipital
protuberance

lambdoidal suture

FIG. 9-13 External view of the base of the skull, i.e., the skull is viewed upside down.

each condyle is a *condylar fossa,* and hidden under each condyle is a *hypoglossal canal* through which passes the hypoglossal, or XIIth cranial, nerve. Anteriorly on the base of the skull, the occipital bone fuses with the body of the sphenoid bone.

The *petrous portion* of the **temporal bone** (Figs. 9-12 and 9-13) is lateral to the occipital bone on the base of the skull, and its *mastoid* and *styloid processes* can be seen clearly. Just behind the base of the styloid process is the *stylomastoid foramen,* through which passes the facial, or VIIth cranial, nerve. Medial to the styloid process is the *jugular fossa,* which opens through the *jugular foramen* to the inside of the skull. Immediately to the front of the jugular foramen is the *carotid canal.* Anterior

to the external acoustic meatus is the *mandibular fossa,* which receives the condyle of the mandible in formation of the temporomandibular joint.

Anterior to the base of the occipital bone is the **sphenoid bone** (Figs. 9-13, 9-14, and 9-15), the greater wing of which was discussed previously. The sphenoid forms a major part of the floor of the skull. Its *pterygoid processes* (Fig. 9-14), which project downward between the greater wing and body, each consist of a *medial plate,* which articulates with the palatine bone, and a *lateral plate.* The *body* of the sphenoid, which contains an air-filled sinus, bears on its upper surface (Fig. 9-15) the *sella turcica,* which houses the pituitary gland. Posterior to the sella is the *dorsum sellae.* The *lesser wings* of the sphenoid, best seen from inside the skull (Fig. 9-15), are perforated by the optic

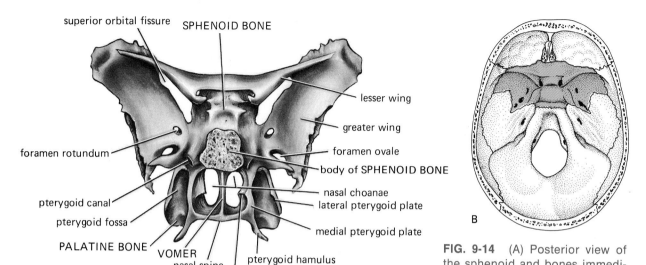

superior orbital fissure SPHENOID BONE

lesser wing

greater wing

foramen rotundum

foramen ovale

body of SPHENOID BONE

nasal choanae

pterygoid canal

lateral pterygoid plate

pterygoid fossa

medial pterygoid plate

PALATINE BONE

VOMER

nasal spine

pterygoid hamulus

INFERIOR NASAL CONCHA

A

B

FIG. 9-14 (A) Posterior view of the sphenoid and bones immediately below it. (B) Its location in the skull.

FIG. 9-15 The top of the skull has been removed, exposing the inside of the base of the skull and the three cranial fossae.

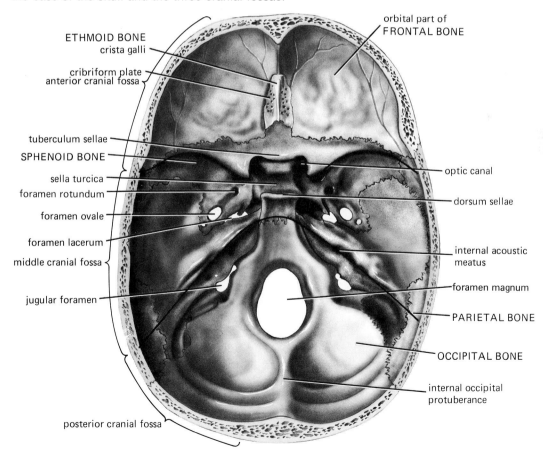

orbital part of FRONTAL BONE

ETHMOID BONE

crista galli

cribriform plate

anterior cranial fossa

tuberculum sellae

SPHENOID BONE

optic canal

sella turcica

foramen rotundum

dorsum sellae

foramen ovale

foramen lacerum

internal acoustic meatus

middle cranial fossa

jugular foramen

foramen magnum

PARIETAL BONE

OCCIPITAL BONE

internal occipital protuberance

posterior cranial fossa

canal, through which pass the optic (IId cranial) nerves from the eye. Slitlike openings between the greater and lesser wings form the *superior orbital fissures,* which allow passage of the ophthalmic branch of the trigeminal (Vth cranial) nerve and the three cranial nerves that supply the muscles of the eyeball; the oculomotor (IIId cranial), trochlear (IVth cranial), and abducens (VIth cranial) nerves. The *foramen rotundum,* which transmits the maxillary branch of the trigeminal nerve, and the *foramen ovale,* which transmits the mandibular branch of the same nerve, open through the greater wing.

Interior of the Skull   When the top of the skull and the brain are removed to expose the floor of the skull (Fig. 9-15), one can see three distinct shallow depressions—the anterior, middle, and posterior cranial fossae. The floor of the **anterior fossa** forms the roof of the eye sockets (orbits) and nasal cavity and houses the frontal lobes of the brain. The frontal bone forms most of the orbital roofs, but a narrow strip of perforated bone, the *cribriform plate* of the ethmoid bone, forms the roof of the nasal cavity. The olfactory (Ist cranial) nerve fibers pass through these holes on their way from the nose to the brain. The **middle cranial fossa** has a narrow central region but expands on each side. The central part, formed by the body of the sphenoid bone, contains the sella turcica. The lateral parts of the middle fossa are formed by the greater wings of the sphenoid bone and petrous parts of the temporal bones. The floor of the **posterior cranial fossa** is formed to a large extent by the occipital bone, which contains the foramen magnum. The posterior fossa accommodates the cerebellum, pons, and medulla of the brain.

The Hyoid Bone   The **hyoid bone** lies in the anterior part of the neck superior to the "Adam's apple." It is suspended from the styloid process of each temporal bone by a slender *stylohyoid ligament,* and it is connected by a broad *thyrohyoid membrane* to the thyroid cartilage of the larynx. The hyoid is a U-shaped bone (Fig. 9-16) that consists of a central region, the *body,* and two pairs of processes, the *greater* and *lesser horns.*

The Vertebral Column

The **vertebral column,** often called the *spinal column* or simply the *spine,* in the adult typically con-

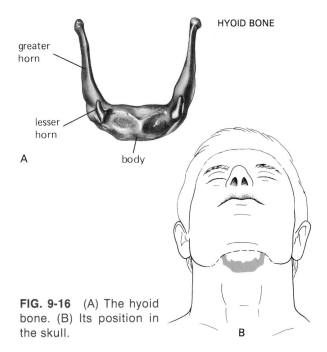

FIG. 9-16   (A) The hyoid bone. (B) Its position in the skull.

sists of 26 vertebrae (sing., vertebra). The first seven, located in the neck, are the **cervical vertebrae;** the next twelve are in the thorax and form the **thoracic vertebrae;** the next five form the **lumbar vertebrae** (Fig. 9-17). These 24 bones are classed together as the *movable vertebrae.* The five sacral vertebrae become fused together in the adult to form one bone, the **sacrum,** and the four terminal rudimentary (i.e., not fully developed) bones of the spine tend to become fused to form the **coccyx.** The lower vertebrae thus lose their mobility and are known as the *fused vertebrae.*

The vertebral column of an early embryo is C-shaped, but it gradually straightens during growth except for the curves in the thoracic and sacral portions, where the original curves are not completely lost. The curve is reversed, however, in the cervical portion of the vertebral column, particularly as the infant learns to hold up its head, and in the lumbar portion as the child learns to sit, stand, and walk (Fig. 9-16B). Abnormal curvatures can occur, such as a pronounced thoracic curve (*kyphosis,* or humpback), an exaggerated lumbar curve (*lordosis*), or a lateral curve (*scoliosis*).

The movable vertebrae vary in size and other characteristics from one region of the spine to another and even within a region, but they all share one fundamental plan (Fig. 9-18A). Each consists of two basic parts: a ventral *body* and a dorsal *arch.* The body supports the weight of the trunk, head,

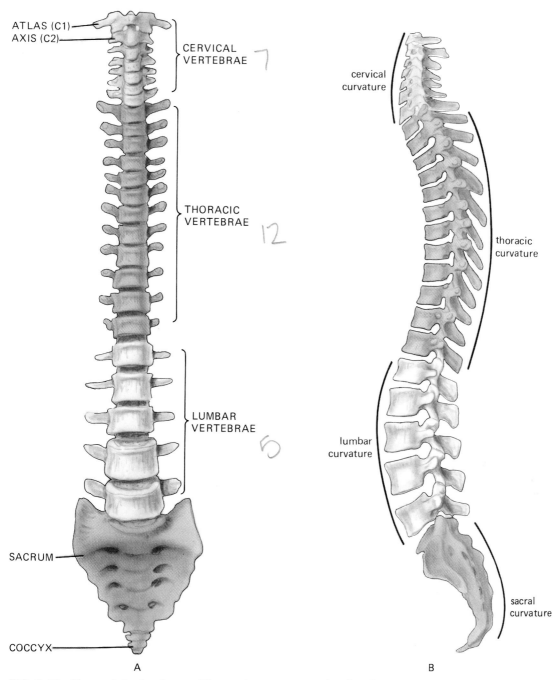

FIG. 9-17 The vertebral column: (A) anterior and (B) lateral views.

and arms, whereas the arch encloses a space, the *vertebral foramen,* in which the spinal cord with its associated coverings and blood vessels is situated. The arch is formed by lateral *pedicles* through which it joins the body, and *lamina.* Where the two lamina fuse together a midline *spinous process* occurs; it is these "spines" that can be felt or seen protruding along the midline of the back. Extending somewhat to the sides of the spinous process are pairs of *superior* and *inferior articular processes* and *transverse processes* (Figs. 9-18A and B).

When the vertebrae are in place in the column, the inferior articular processes, which bend downward, meet the upward jutting superior processes

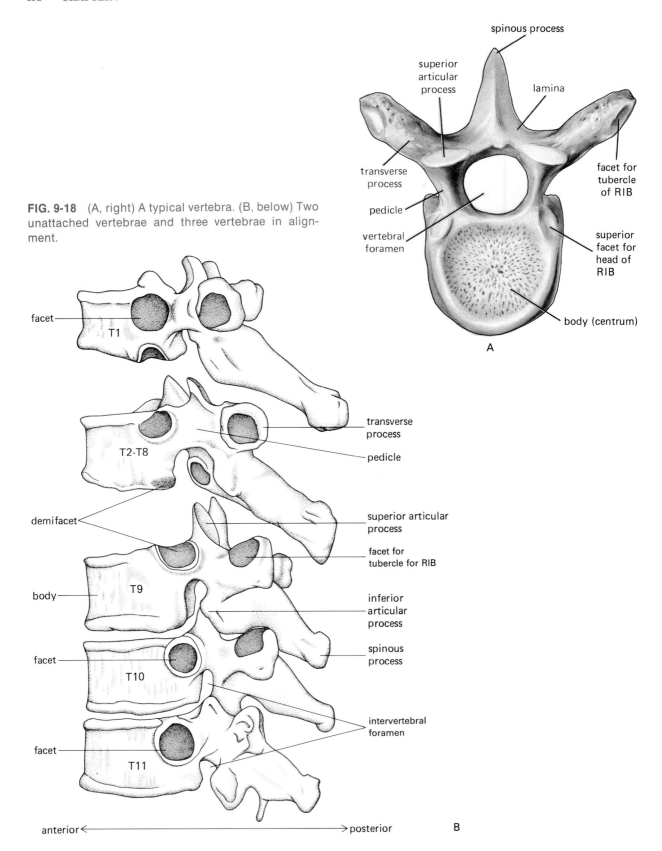

FIG. 9-18 (A, right) A typical vertebra. (B, below) Two unattached vertebrae and three vertebrae in alignment.

Labels in A:
- spinous process
- superior articular process
- lamina
- transverse process
- facet for tubercle of RIB
- pedicle
- vertebral foramen
- superior facet for head of RIB
- body (centrum)

A

Labels in B:
- facet — T1
- T2-T8
- transverse process
- pedicle
- demifacet
- superior articular process
- facet for tubercle for RIB
- body — T9
- inferior articular process
- facet — T10
- spinous process
- facet — T11
- intervertebral foramen
- anterior ←→ posterior

B

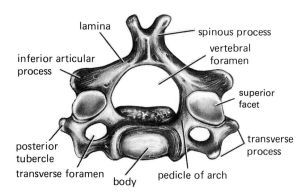

FIG. 9-19 A typical cervical vertebra, superior view.

of the vertebra just below. Between these joined processes and the main column of the vertebral bodies and discs is an opening, the *intervertebral foramen* (Fig. 9-18B), through which the peripheral nerves pass.

Cervical vertebrae are distinguished by their small size and by the presence of paired *transverse foramina* adjacent to the vertebral body (Fig. 9-19). In the upper five or six cervical vertebrae these openings transmit the vertebral arteries, which help supply the brain with blood. The spinous process

varies: It is absent on the first cervical vertebra; it is bifed, or forked, in the third to sixth vertebrae; and it is much longer in the sixth and seventh vertebrae than in the other cervical vertebrae.

The first two cervical vertebrae are decidedly different. The **atlas** (Fig. 9-20) is the first cervical vertebra. It received its common name because it literally supports the "globe" of the head. It is unusual in that it lacks a vertebral body and consists chiefly of an anterior and posterior arch. The space of the missing body in the atlas is filled in by an upward projecting *dens* from the body of the second cervical vertebra, the **axis.** The dens forms, in fact, a pivot around which the atlas and head rotate. The condyles of the occipital bone of the skull sit on the facets of the two lateral masses of the atlas, rocking in the facets during nodding movements of the head.

The **thoracic vertebrae** articulate with the ribs and therefore bear articular facets (or parts of them) for at least one rib. The head of the correspondingly numbered rib articulates with the upper end of the body of the vertebra, and the tubercle of that rib articulates with the transverse process. (See also

FIG. 9-20 (A) The first two cervical vertebrae: C1, the atlas, and C2, the axis. The superior articular facets of the atlas articulate with the occipital condyles of the skull. (B) Superior view of the atlas. (C) Superior view of the axis.

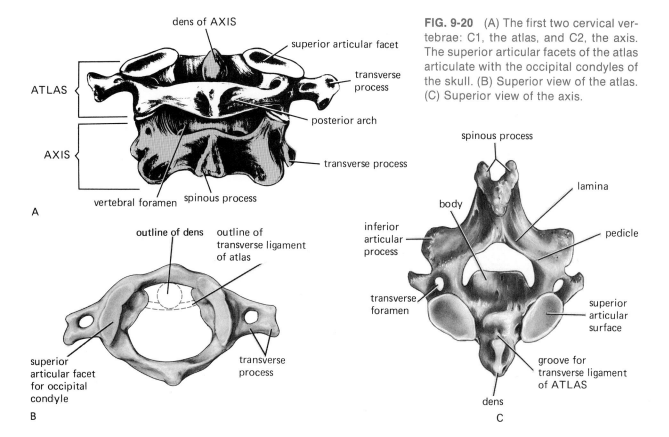

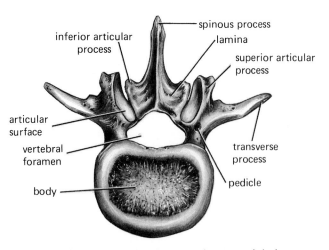

FIG. 9-21 A typical lumbar vertebra, cranial view.

Fig. 9-25C.) The spinous processes of the middle thoracic vertebrae are the longest. The typical vertebra in Fig. 9-18A is a thoracic vertebra.

The **lumbar vertebrae** are large and heavy (Fig. 9-21). The lamina of a lumbar vertebral arch is shallower, or thinner than the body; thus between any two adjacent lamina there is a gap bridged only by ligaments. It is through these gaps that *lumbar punctures* (introduction of a needle into the subarachnoid space of the spinal cord coverings to sample the cerebrospinal fluid) are performed.

The **sacrum** (Fig. 9-22) is a large triangular bone; the wide upper part articulates with the fifth lumbar vertebra and the narrow lower end with the coccyx. The characteristics of the sacrum reflect the fusion of the bodies, processes, and laminae of the individual sacral vertebrae. The wide upper surface, or *base,* of the sacrum has on its pelvic surface the *promontory* and the pelvic *sacral foramina,* which open into the sacral portion of the vertebral canal and transmit the ventral branches of the first four sacral nerves. On its dorsal surface (Fig. 9-22B) are: the articular surfaces for the fifth lumbar vertebra; the *auricular surface,* which articulates with the ilium of the hip bone; and the *median sacral crest,* which represents the fused spinous processes of the sacral vertebrae. The *dorsal sacral foramina* allow exit of the dorsal roots of the sacral nerves. The lower portion of the sacrum bears the *sacral hiatus,* a defect leading to the sacral canal and covered only by ligaments. Adjacent to the hiatus, *sacral cornu* present articular surfaces for the first segment of the coccyx. Typically the sacra of males

FIG. 9-22 (A) Anterior and (B) Posterior views of the sacrum and coccyx.

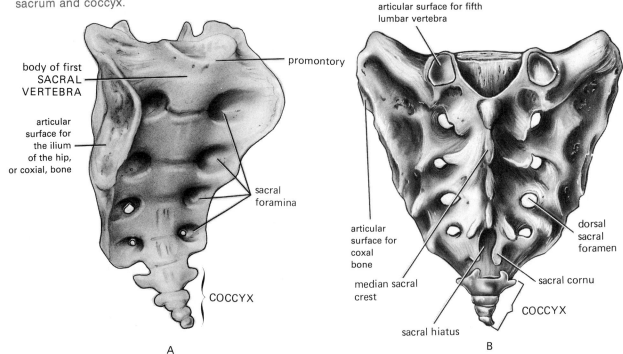

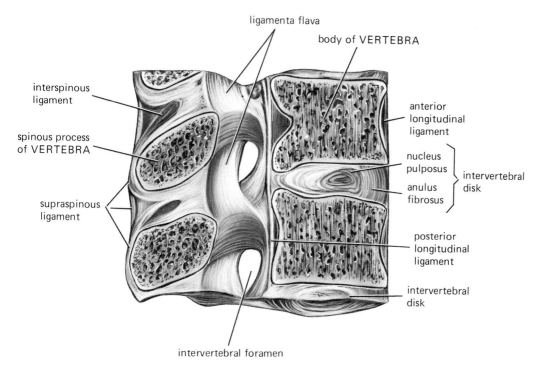

FIG. 9-23   Sagittal section of a portion of the vertebral column showing the intervertebral discs and ligaments.

and females differ; that of the female is shorter and wider and has a deeper concavity than that of the male.

The **coccyx** (Fig. 9-22) is a small triangular bone that consists of from three to five rudimentary vertebrae, the second, third, and fourth coccygeal vertebrae being simply tiny nodules of bone. All the coccygeal vertebrae may be fused together, but the first often remains separate. Lacking a central canal, the coccygeal vertebrae are solid.

The upper and lower surfaces of adjacent vertebral bodies are strongly bound to each other by fibrocartilage **intervertebral disks** (Fig. 9-23), which lend a certain degree of flexibility to the complete vertebral column and act as shock absorbers; they become temporarily flatter and broader and bulge from between the vertebrae when they are compressed. This is facilitated by the presence of a semifluid fibrogelatinous mass (the *nucleus pulposus*) within each disk. However, sudden heavy strains can cause the nucleus to break through the ring of cartilaginous fibers (the *anulus fibrosus*) surrounding it; this is called a *ruptured disk*. Disks occur between all the vertebrae including the space between the last lumbar and first sacral vertebrae;

they account for one-fourth of the total length of the spinal column.

Anterior and *posterior longitudinal ligaments* join the bodies of adjacent vertebrae while strong *ligamenta flava* join the laminae (Fig. 9-23). *Supraspinous ligaments* join the spinous processes below the seventh cervical vertebra, but above that level they give way to the *ligamentum nuchae,* which is a triangular, sheetlike ligament that extends up to attach to the skull.

The Thorax

The skeleton of the chest, which comprises the **sternum** (or breastbone), 12 pairs of **ribs,** and the 12 **thoracic vertebrae,** forms the bony part of the **thoracic cage** (Fig. 9-24). It protects the heart and lungs and provides attachments for muscles of the upper arms, back, and abdomen as well as the thorax itself. Moreover, the red marrow of the ribs and sternum is one of the main sites of red blood cell formation in the adult.

The **sternum** has three parts: the *manubrium, body,* and *xiphoid process.* On each side of the manubrium is a *clavicular notch* for articulation

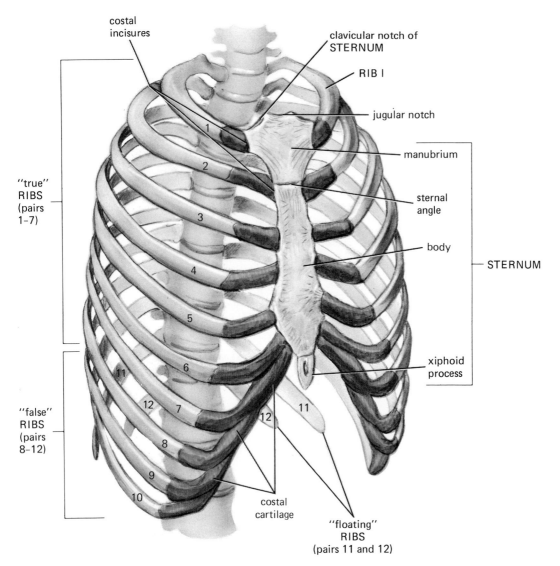

FIG. 9-24 The skeleton of the thoracic cage. The costal cartilages are shown in color.

with the clavicle, between which is the *jugular notch*. Below the clavicular notch is the first of the *costal incisures* for articulation with the first rib. The first seven pairs of ribs articulate with the lateral borders of the sternum.

There are 12 pairs of **ribs** numbered from above downward; the first pair lies just beneath the clavicles. The main parts of a typical rib are the *anterior end, shaft, head, neck,* and *tubercle* (Fig. 9-25). The head and tubercle, which are at the posterior, or vertebral, end of the rib, bear facets for articulation with the vertebrae, the head attaching

to the body of a vertebra, and the neck and tubercle attaching to the transverse process (Fig. 9-25). *Costal cartilages*, pieces of hyaline cartilage, lie between the anterior ends of the ribs and the sternum (Fig. 9-24) and contribute significantly to the mobility and elasticity of the walls of the thorax. The costal cartilages of the first seven pairs of ribs attach to the sternum, and these are called the *true ribs*. The cartilages of the 8th, 9th, and 10th pairs attach to the cartilages of the ribs immediately above them, and those of the 11th and 12th pairs do not attach to anything (thus, these last ribs are

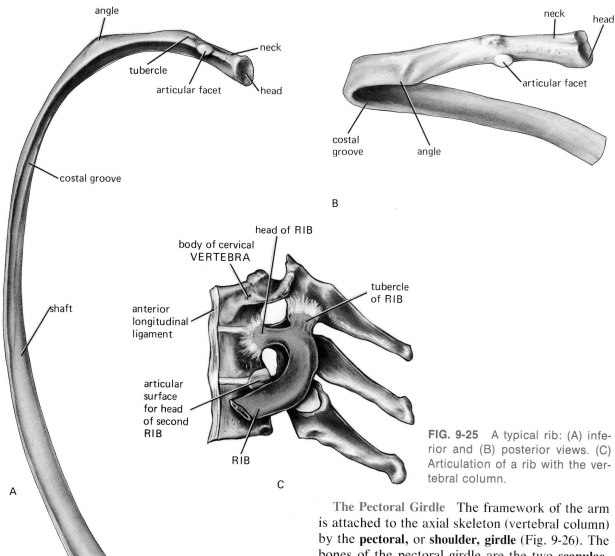

FIG. 9-25 A typical rib: (A) inferior and (B) posterior views. (C) Articulation of a rib with the vertebral column.

called *floating ribs*). The last five pairs of ribs (pairs 8 through 12), including the floating ribs, do not attach to the sternum and are called *false ribs*.

THE APPENDICULAR SKELETON

Upper Limbs

The upper members of the appendicular skeleton consist of the shoulders, arms, forearms, wrists, and hands.

The Pectoral Girdle The framework of the arm is attached to the axial skeleton (vertebral column) by the **pectoral,** or **shoulder, girdle** (Fig. 9-26). The bones of the pectoral girdle are the two **scapulae,** or shoulder blades, on the posterior wall of the thoracic cage, and the two **clavicles,** or collar bones. The single point of bony contact between the upper limb and the trunk is at the *sternoclavicular joint,* a small and mobile articulation between the medial end of each clavicle and the manubrium of the sternum. There is no direct articulation between the skeleton of the upper limb and the vertebral column. The lateral end of each clavicle articulates with the *acromion process* of a scapula. The pectoral girdle articulates with the arms at freely movable ball-and-socket joints between the *glenoid cavity* of each scapula and *head* of the humerus.

The two **clavicles,** or collar bones, extend almost horizontally across the upper part of the thorax as

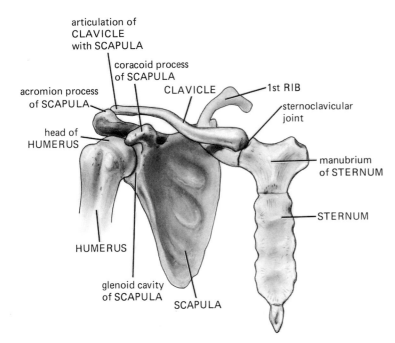

articulation of
CLAVICLE
with SCAPULA

coracoid process
of SCAPULA

acromion process
of SCAPULA

CLAVICLE

1st RIB

sternoclavicular
joint

head of
HUMERUS

manubrium
of STERNUM

STERNUM

HUMERUS

glenoid cavity
of SCAPULA

SCAPULA

FIG. 9-26   The pectoral, or shoulder, girdle.

far as the shoulder. They lie just under the skin and
can be easily felt. They are an important part of the
shoulder girdle because they serve as struts for the
upper limbs by propping the shoulders out from the
chest so the arms can swing freely. They are also
the attachment site of many neck muscles. The
lateral end of each clavicle is flattened and bears an
acromial articular surface for articulation with the

scapula; the medial end is enlarged and bears an
articulate surface for the *clavicular notch* on the
manubrium portion of the sternum. The clavicle is
an "S"-shaped bone. The curves enable the bone
to withstand the shocks it transmits from the shoulder to the axial skeleton; nevertheless, the clavicle

FIG. 9-27   The right scapula: (A) dorsal and (B)
lateral views. In (B) the dorsal surface of the
bone is facing left, and the costal (i.e., toward
the ribs) surface is facing right.

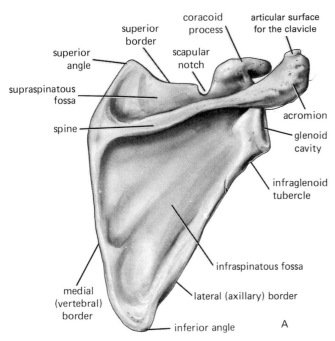

coracoid
process

superior
border

articular surface
for the clavicle

superior
angle

scapular
notch

supraspinatous
fossa

spine

acromion

glenoid
cavity

infraglenoid
tubercle

infraspinatous fossa

medial
(vertebral)
border

lateral (axillary) border

inferior angle

A

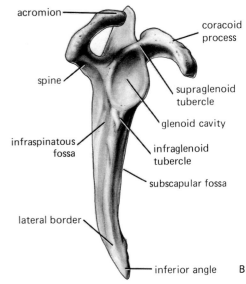

acromion

coracoid
process

spine

supraglenoid
tubercle

glenoid cavity

infraspinatous
fossa

infraglenoid
tubercle

subscapular fossa

lateral border

inferior angle

B

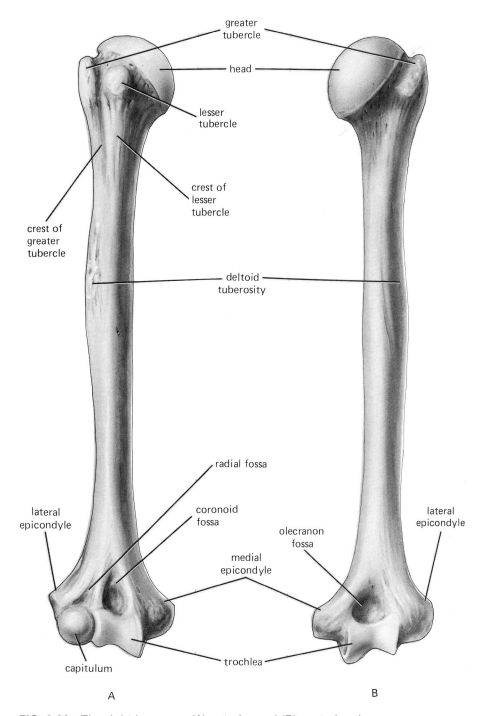

greater
tubercle

head

lesser
tubercle

crest of
lesser
tubercle

crest of
greater
tubercle

deltoid
tuberosity

radial fossa

coronoid
fossa

lateral
epicondyle

olecranon
fossa

lateral
epicondyle

medial
epicondyle

capitulum

trochlea

A

B

FIG. 9-28 The right humerus: (A) anterior and (B) posterior views.

is the most frequently fractured bone in the body.

The **scapula** (Fig. 9-27) is a large, flattened bone that lies against the posterior wall of the thorax, from the level of the second to the seventh ribs. It is attached to the trunk by means of its articulation with the clavicle and by muscles. The *spine* of the scapula projects from the dorsal surface and extends from the medial (vertebral) edge of the bone to the *acromion* at the point of the shoulder where the acromion projects forward to articulate with the clavicle. The superior border of the scapula is marked by a *scapular notch*, lateral to which is the *coracoid process*. A shallow depression of the lateral surface of the bone, the *glenoid cavity*, forms the shoulder joint with the head of the humerus. On the upper surface of the cavity is the *supraglenoid tubercle*, and below the cavity is the *infraglenoid tubercle*. The *body* of the bone is thin and even translucent, but along its edges and in the region of the glenoid fossa, acromion, and coracoid process the bone is weighty.

The Arm   The **humerus** (Fig. 9-28) is the longest and heaviest bone of the upper limb. The expanded upper end of the bone has a rounded *head*, which articulates with the glenoid cavity of the scapula, and greater and lesser tubercles. The *lesser tubercle* is a small, rough projection on the medial side of the front of the humerus, the *crest of the lesser tubercle* extending below it. The larger *greater tubercle* is on the lateral side of the bone at the top of the *crest of the greater tubercle*. The lower end of the humerus is more irregular, bearing articular surfaces for the radius and ulna. The lateral and medial borders of the shaft, or *body*, of the bone end in thickened *lateral* and *medial epicondyles*, which are separated by the condyle. The *condyle* consists of articular surfaces (the lateral *capitulum* and more medial *trochlea*) and concavities. Above the trochlea are the *coronoid fossa* on the front of the bone and the *olecranon fossa* on the back. The *radial fossa* lies above the capitulum.

The Forearm   The **radius** and **ulna** are the two bones of the forearm (Fig. 9-29); the radius is on the lateral (thumb) side and the ulna on the medial side. The two bones articulate with each other at the upper and lower ends of the forearm and are

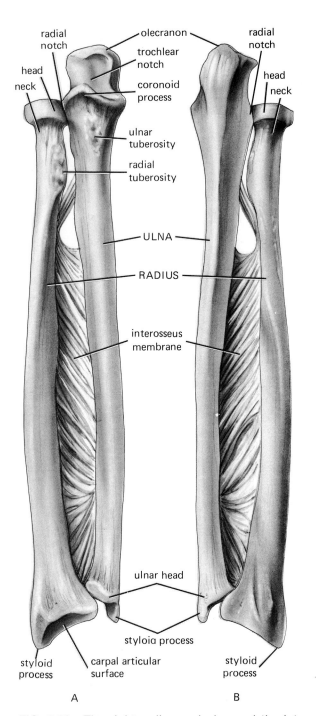

FIG. 9-29   The right radius and ulna and the interconnecting interosseous membrane: (A) anterior and (B) posterior views.

connected throughout their length by a flexible connective-tissue *interosseous membrane*.

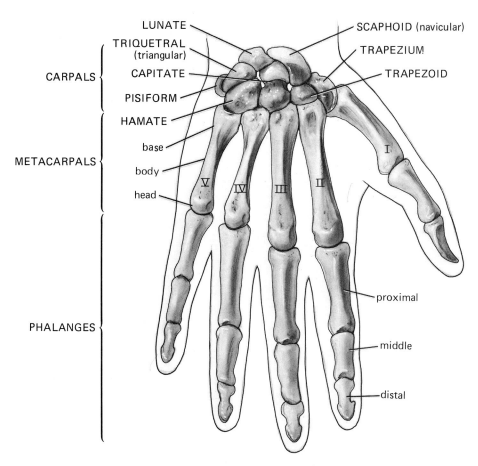

CARPALS {
LUNATE
TRIQUETRAL (triangular)
CAPITATE
PISIFORM
HAMATE

METACARPALS {
base
body
head

PHALANGES {

SCAPHOID (navicular)
TRAPEZIUM
TRAPEZOID

I
II
III
IV
V

proximal
middle
distal

FIG. 9-30   Dorsal view of the bones of the right wrist and hand.

The **radius** articulates with the capitulum of the humerus (and *radial notch* of the ulna) at its upper end (the *head*) and with the wrist (and ulna) at its lower end. Immediately below the head of the radius is the narrowed *neck*, and a little below that is a raised projection, the *radial tuberosity*, to which the biceps muscle is attached. The radius is better developed at its lower end and plays a larger role in the articulations of the wrist than in the elbow joint, while the ulna is just the opposite.

The upper end of the **ulna** bears *olecranon* and *coronoid processes*, which act like stationary jaws to grasp the trochlea of the humerus in the *trochlear notch*. The lower end of both radius and ulna extends as a *styloid process*.

The Hand   The skeleton of the hand has three parts: bones of the wrist (**carpals**), palm (**metacar-**

pals), and fingers (**phalanges**) (Fig. 9-30). The eight carpal bones are arranged in two rows, each with four bones. Except for the *pisiform bone*, all the carpal bones articulate with more than one of the adjacent bones. At the wrist the *scaphoid, lunate,* and *triquetral bones* articulate with the lower end of the radius and with the articular disc of the ulna to form the *radiocarpal*, or *wrist, joint;* they also articulate with each other. There are five metacarpal bones in each hand; their *bases* articulate with the outer row of carpal bones and their *heads* with the proximal phalanges, and thus form the knuckles. Each of the four fingers contain three phalanges, designated as *proximal, middle,* and *distal*. The thumb has only two phalanges; the middle phalanx is absent.

Lower Limbs

The skeleton of the lower limb is in general analogous to that of the upper limb but differs in several important respects. (1) The girdle of the lower limb (the **pelvic girdle**) is attached firmly to the vertebral column, a difference that can be directly correlated with the weight-bearing function of the lower limb. (2) In accordance with this weight-bearing function, the hip joint is structurally much stronger than the shoulder joint; thus, for example, the cavity, or *acetabulum,* of the hip bone is far deeper than the comparable glenoid cavity of the shoulder, and the thigh bone, the **femur,** is the longest and strongest bone of the body. (3) The expansion of the lower end of the femur and the leg bones (**tibia** and **fibula**) with which it articulates are adapted for weight bearing; the stresses of anterioposterior and lateral movements of the knee joint are sustained not directly by bony elements of the skeleton but rather by ligaments and muscles associated with the joint.

Of the two bones in the leg, the tibia supports most of the weight, and transmits it from the femur to the ankle, whereas the fibula serves chiefly to help stabilize the knee. The posterior bones of the foot, the **tarsal bones,** correspond to the carpal bones of the wrist, although the tarsal bones are larger and arranged differently. The other bones of the foot are the **metatarsals,** equivalent to the metacarpals of the hand, and the **phalanges** of the toes.

The Pelvic Girdle   The **pelvic girdle** is formed of three paired bones (ischium, ilium, and pubis) that fuse with each other during development to form a single bone on each side, the **coxal** (*os coxae*), or **hip, bone** (formerly known as the "innominate bone"). The two coxal bones are firmly articulated to each other anteriorly at the *symphysis pubis,* and posteriorly each is fused to the sacrum at a *sacroiliac joint.*

The ilium forms the upper, larger part of the coxal bone, the ischium forms the lower, posterior part of the bone, and the pubis forms its anterior medial portion (Fig. 9-31). All three of the component bones contribute to the *acetabulum,* which is a deep, cup-shaped cavity on the lateral surface of the coxal bone. The acetabulum faces laterally, downward, and forward; its inferior wall is interrupted by the *acetabular notch.* Below the acetabulum is a large opening, the *obturator foramen,* which is closed by the *obturator membrane* except for a small region at the top of the foramen through which nerves and vessels pass.

The inner, somewhat concave surface of the **ilium** is the *iliac fossa,* behind which is a thickened region which ends in the *auricular surface* (so called because of its resemblance to an ear), which articulates with the sacrum. Above the auricular surface is the *iliac tuberosity,* which receives the attachments of ligaments that stretch between the sacrum and ilium. The *iliac crest,* which forms the upper thickened border of the ilium, ends posteriorly at the *posterior superior iliac spine,* and anteriorly at the *anterior superior iliac spine.* Below these are *posterior* and *anterior inferior iliac spines.* Farther down along the posterior edge of the ilium is the *greater sciatic notch.* The *arcuate line* passes between the border of the auricular surface and the tubercle and crest of the pubic bone. The lateral (external) surface of the ilium is marked by the *interior, anterior,* and *posterior gluteal lines.*

The **ischium** bears a large posterior inferior swelling, the *ischial tuberosity,* and a smaller, more pointed *ischial spine.* The tuberosity and spine are separated by the *lesser sciatic notch;* the *greater sciatic notch* lies above the ischiac spine between the spine of the ischium and the posterior inferior spine of the ilium.

The **pubis** has on its medial surface a *symphyseal surface* at which it is bound to the pubis on the other side by an interpubic disc. The thickened anterior part of the pubis is the *pubic crest,* which ends laterally at the *pubic tubercle.* This tubercle is the beginning of a raised line, the *pecten,* which is continuous with the arcuate line of the ilium.

The **pelvis** (Fig. 9-32) is divided into major and minor parts by the *terminal line,* that passes obliquely through the sacral promontory, the arcuate lines of the two iliac bones, and the pecten lines of the two pubic bones (Fig. 9-32B). The terminal line marks the *pelvic brim* and outlines the *superior pelvic aperture,* or *pelvic inlet.* Above it is the *major,* or *greater, pelvis,* which is related to the lower abdominal cavity. The *minor,* or *lesser, pelvis* is below the terminal line and forms the *true pelvis.* The *inferior pelvic aperture,* or *pelvic outlet,* is closed by muscles that form a floor for the rounded basin of the pelvis and support the viscera that open to the outside through it. These muscles form the *pelvic diaphragm.*

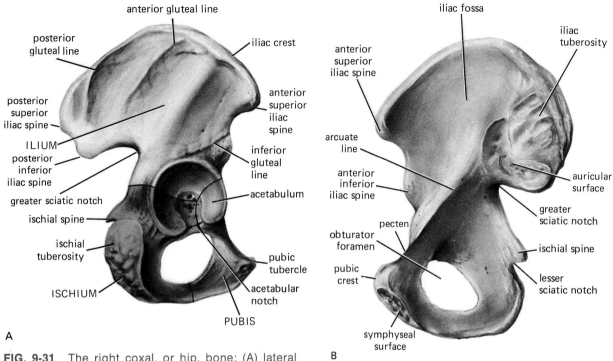

FIG. 9-31 The right coxal, or hip, bone: (A) lateral (external) and (B) medial views.

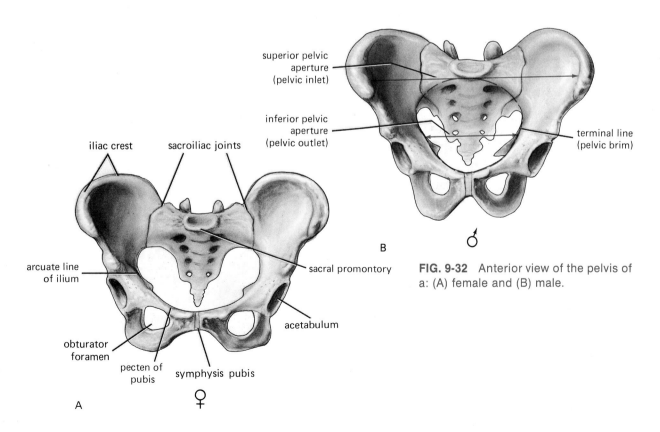

FIG. 9-32 Anterior view of the pelvis of a: (A) female and (B) male.

The female pelvis is different from that of the male (Fig. 9-32) in ways that relate to childbearing and the changes in the body weight and center of gravity that must be supported in upright posture and locomotion during pregnancy. The main difference is an enlargement of the pelvic outlet; thus, the angle at which the two pubic bones meet is wider in the female, the ischial tuberosities are flattened outward, and the symphysis pubis is shallower. As a consequence, the iliac fossa is deeper in the male pelvis, and the cavity of the true pelvis is broader and shallower in the female. Moreover, the sacrum in the female is broader and less curved. Other differences are the triangularity and smaller size of the obturator foramen, the less prominent markings for muscle and ligament attachments, the smaller size and more forward attitude of the acetabulae, and the greater delicacy of the bones in the female pelvis (Table 9-2).

The Thigh The **femur** (Fig. 9-33), the bone of the thigh, articulates at its upper end with the coxal bone and at its lower end with the patella and the tibia. Its upper end has a *head* and *neck* that are at an angle to the long axis of the bone and thus allow greater mobility of the hip joint but impose unusual strain, since the weight the bone must bear is transmitted along a curve. The ball-like head is covered with cartilage except for a depression, the *fovea capitis,* where a ligament attaches. The *greater trochanter* projects from the bone at the junction of the neck and shaft and has on its medial side a *trochanteric fossa.* The *lesser trochanter* is behind and medial to the greater trochanter, the two trochanters being joined by an *intertrochanteric crest* on the posterior surface of the bone and by an *intertrochanteric line* on the anterior surface. Also on the posterior surface is a *gluteal tuberosity,* sometimes called a third trochanter.

The shaft, or *body,* of the femur is marked along its posterior surface by the *linea aspera,* which divides at its lower end to form the *popliteal surface.* The two rounded *condyles* at the distal end of the bone are separated on the posterior surface of the bone by a deep *intercondylar fossa,* but on the anterior surface they blend together across the *patellar articular surface.* On the sides of the condyles are the roughened *medial* and *lateral epicondyles.*

The Leg The **tibia** is the more medial and stronger of the two bones of the leg (Fig. 9-34). Its upper end has two prominent masses, the *medial*

TABLE 9-2	SOME STRUCTURAL DIFFERENCES BETWEEN THE MALE AND THE FEMALE PELVIS	
	Male	Female
Bones	Larger, heavier, rougher	Smaller, lighter, smoother
Sacrum	Narrower, more curved	Broader, less curved
Ilium	Less vertical	More vertical
Iliac fossa	Deeper	Shallower
Greater sciatic notch	Deeper, narrower	Shallower, wider
Sciatic spines	More projecting	Less Projecting
Symphysis pubis	Narrower, pointed	Wider, more rounded
Obturator foramen	Oval	Triangular
Superior aperture	More heart-shaped	More oval
Major pelvis	Wider	Narrower
Minor pelvis	Deeper, narrower	Shallower, wider
Capacity of minor pelvis	Less	Greater
Pubic angle (between inferior rami of pubic bones)	Narrower, pointed	Wider, more rounded

and *lateral condyles,* which articulate with the corresponding condyles of the femur to form the knee joint. (The *patella,* a small, sesamoid bone also associated with the knee joint, will be discussed in Chapter 10.) The condyles are separated by an *intercondylar eminence.* The lower end of the tibia is smaller and bears on its medial surface a large *medial malleolus.* The shaft of the tibia lies just under the skin and, over most of its length, forms a sharp crest, the shin, which begins above at the *tibial tuberosity.* The posterior surface of the tibia is marked by the *line of the soleus muscle,* or *soleal line.*

The **fibula** (Fig. 9-34), the lateral bone of the leg, is more slender than the tibia and serves mainly for muscle attachments and to complete the lateral part

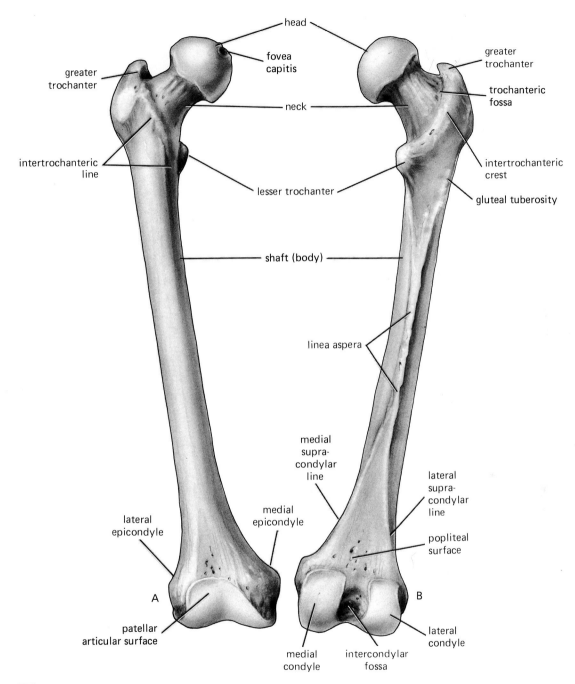

FIG. 9-33   The right femur: (A) anterior and (B) posterior views.

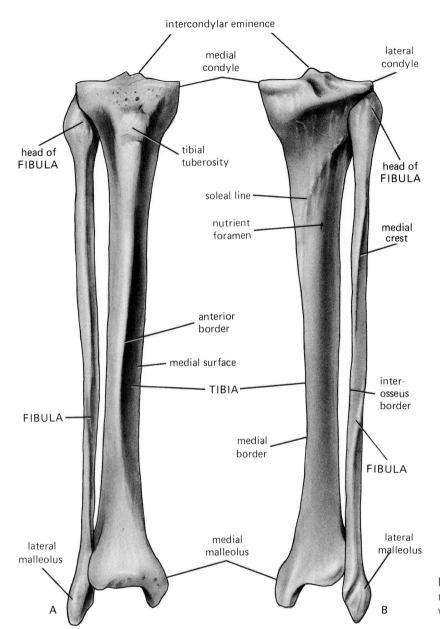

intercondylar eminence

medial
condyle

lateral
condyle

head of
FIBULA

tibial
tuberosity

head of
FIBULA

soleal line

nutrient
foramen

medial
crest

anterior
border

medial surface

TIBIA

inter-
osseus
border

FIBULA

medial
border

FIBULA

lateral
malleolus

medial
malleolus

lateral
malleolus

A

B

FIG. 9-34   The right tibia and fibula: (A) anterior and (B) posterior
views.

of the ankle joint. The bone consists of a head, shaft, and lower end; the *head* articulates with the tibia at the inferior surface of the lateral condyle, and the lower end articulates with the talus bone in the ankle and again with the tibia. The posterior surface of the shaft is divided by a *medial crest.* The shafts of the tibia and fibula are connected by an *interosseous membrane,* and the lower ends of the two bones are connected by a special type of joint, the *tibiofibular syndesmosis,* which is to be

discussed in Chap. 10. The interosseous membrane and syndesmosis allow only very slight movements of the bones relative to each other, a distinct dif-ference between these two bones and the radius and ulna of the forearm.

The Foot   The skeleton of the foot (Fig. 9-35) has three divisions: the **tarsal** bones of the ankle, the **metatarsal** bones of the foot proper, and the **phalanges** of the toes. The seven tarsal bones con-

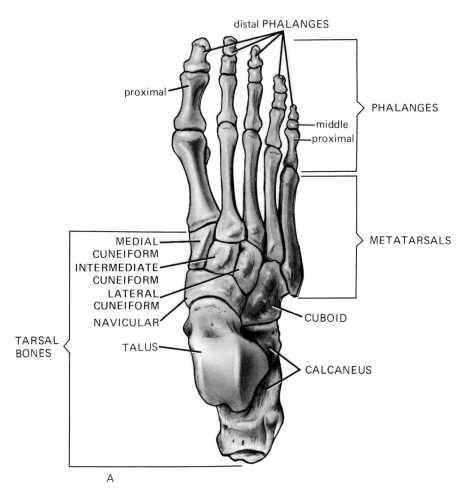

distal PHALANGES

proximal

PHALANGES

middle
proximal

METATARSALS

MEDIAL
CUNEIFORM
INTERMEDIATE
CUNEIFORM
LATERAL
CUNEIFORM
NAVICULAR

CUBOID

TALUS

CALCANEUS

TARSAL
BONES

A

FIG. 9-35 The bones of the right foot: (A) dorsal and (B) lateral views.

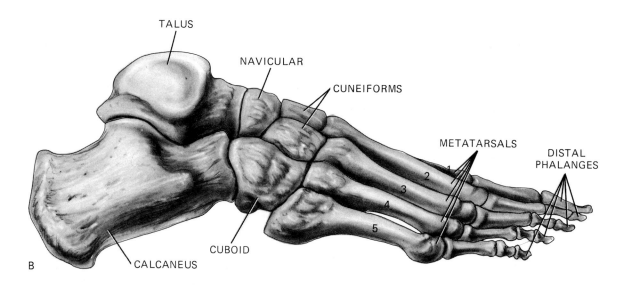

TALUS

NAVICULAR

CUNEIFORMS

METATARSALS

DISTAL
PHALANGES

1
2
3
4
5

CUBOID

CALCANEUS

B

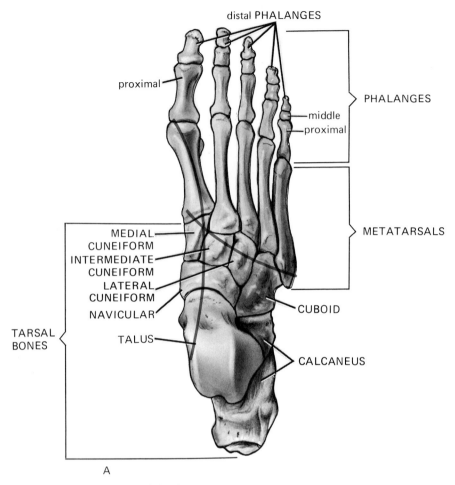

distal PHALANGES

proximal

middle
proximal

PHALANGES

METATARSALS

MEDIAL
CUNEIFORM
INTERMEDIATE
CUNEIFORM
LATERAL
CUNEIFORM
NAVICULAR

TARSAL
BONES

TALUS

CUBOID

CALCANEUS

A

FIG. 9-36 The arches of the foot.

stitute the skeleton of the heel *(calx)* and posterior part of the foot. Like the carpal bones of the wrist, they are arranged in two rows, although the proximal row (i.e., that nearer the heel) is irregular, with the *talus* lying above the *calcaneus,* and the *navicular* bone lying between the talus and bones of the second row. This arrangement allows the body weight to be transmitted from the talus to the other tarsal and metatarsal bones over the arched form of the foot rather than directly downward. The calcaneus is the largest of the tarsal bones; it projects posteriorly to form the ''heel bone'' and receives the attachments of the calf muscles via the *calcaneal,* or *Achilles tendon.* Five metatarsal bones connect the tarsal bones with the two phalanges in the great toe and three in each of the other four toes. Unlike the thumb, which is rotated even in its

resting position, the great toe is in the same plane as the other toes with its flexor surface facing the ground.

Although the foot is constructed on the same general plan as the hand, it differs in that muscles and ligaments hold its bones in two arches: a *longitudinal arch,* which springs between the calcaneal bone and the heads of the metatarsals, and a *transverse arch,* formed by the proximal parts of the metatarsals and the distal parts of the tarsal bones (Fig. 9-36). Both arches increase the strength and elasticity of the foot. The medial part of the foot, where the longitudinal arch is the highest, has greater elasticity than the lateral part. It provides support during walking and jumping, whereas the lateral part provides support during standing.

KEY TERMS

condyle	**facet**	**foramen**
crest	**fovea**	**meatus**
head	**sulcus**	**cranium**
tubercle	**fossa**	**orbit**
tuberosity	**neck**	**suture**
trochanter	**sinus**	
spine	**fissure**	

REVIEW EXERCISES

1   List the five basic types of bones; give a description and example of each.
2   Classify the bones of the skeleton according to the five basic types.
3   Distinguish between the axial and appendicular skeletons.
4   Identify the bones of the skull.
5   List the four regions of the spinal column and describe a typical vertebra from each region.
6   Describe the components of the pectoral girdle; describe their relationships to each other and to the spinal column.
7   Describe the components of the pelvic girdle; describe their relationships to each other and to the spinal column.
8   Identify the bones of the upper and lower limbs.
9   Explain how the upper and lower limbs are connected to the spinal column.

CHAPTER

ARTICULATIONS

An **articulation,** or **joint,** is defined as the meeting place of two or more bones. Some bones meet at movable joints, others at only slightly movable joints, and still others at immovable joints. The bones involved in an articulation do not actually touch each other but are separated by various types of tissue.

TYPES OF JOINTS

Joints are classified in three groups—fibrous joints, cartilaginous joints, and synovial joints. Classification depends on the type of connecting tissue between the articulating bones, which, in turn determines the type and range of movement between them.

Fibrous Joints

At **fibrous joints,** the two bones are held firmly together by fibrous connective tissue so that little, if any, movement occurs between them. Fibrous joints are further subdivided, depending upon the actual form of the joint. These subdivisions include sutures and syndesmoses, which are distinguished from each other, in part, by the length of the connecting fibers. **Sutures** (Fig. 10-1A), which have short connections, occur between the flat bones in the cranial portions of the skull where the bones, each one growing from its own ossification center, contact each other and trap fibrous connective tissue between them. In **syndesmoses** (Fig. 10-1B), the fibers of the connective tissue in the joint are longer and more plentiful and some movement is possible. The joint between the lower ends of the

FIG. 10-1   Types of fibrous joints: (A) suture, (B) syndesmosis, and (C) gomphosis.

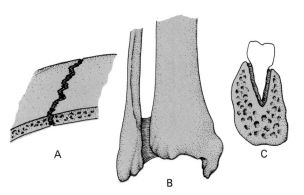

A

B

C

tibia and fibula is a syndesmois. In still another type of fibrous joint, the **gomphosis** (Fig. 10-1C), a conical process is inserted into a socket-like portion. In all three types of fibrous joints the connective-tissue fibers blend with the periosteum of the bones.

Cartilaginous Joints

Cartilaginous joints are symphyses, which retain a permanent cartilage or modified cartilage between the articulating bones, and synchondroses, in most of which cartilage appears temporarily and is eventually replaced by bone. In **symphyses,** a layer of cartilage is retained on the surface of each of the articulating bones; the cartilages of the two bones are then joined by fibrous connective tissue or a fibrocartilage pad plus strong ligaments. Together these can be compressed or displaced and therefore offer some movement. Vertebrae are joined by symphyses, and the cartilaginous discs between them give considerable flexibility to the spinal column. Another example of a symphysis is the joint between the pubic bones, the symphysis pubis (Fig. 10-2) **Synchondroses** typically allow no movement. The most common type of synchrondrosis is in growing bone, in which an epiphyseal cartilage separates the bone of the shaft from that of the epiphysis. After growth is complete, this region ossifies. A second example of a synchrondrosis occurs between the sternum and first rib.

Synovial Joints

Synovial joints work on an entirely different principle from the fibrous and cartilaginous joints and have a mechanism that allows free movement. The major parts of the articulating surfaces of the bones involved are separated by a cavity; the union of the bones is effected by a fibrous **articular capsule (joint capsule)** (Fig. 10-3) and often by **accessory ligaments** as well. These ligaments may be inside or outside the capsule, and they may be separate or applied to it. They serve to limit the types of movement and help hold the articulating bones in place.

The bony surfaces involved in the joint are covered by a thin layer of hyaline **articular cartilage,** and it is actually the two cartilaginous surfaces that slide past each other during movement. Their movement is facilitated by viscous **synovial fluid,** which lubricates the cartilages, cushions shocks,

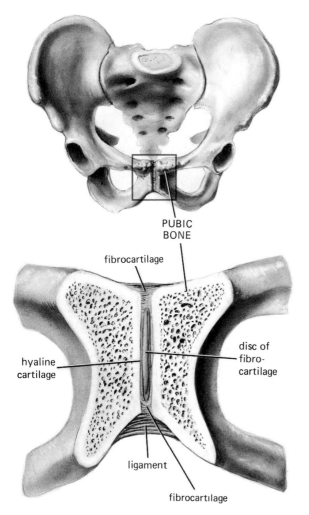

PUBIC
BONE

fibrocartilage

hyaline
cartilage

disc of
fibro-
cartilage

ligament

fibrocartilage

FIG. 10-2 Section through the pubic region of the coxal bone to show the structure of a symphysis.

and provides a nutrient source. This fluid is formed by a sheet of specialized tissue, the **synovial membrane.** The membrane lines the entire joint capsule and thus forms a cavity (the **articular,** or **joint, cavity**) that contains the synovial fluid. The cavity extends along the sides of the bones under the capsule and ends as the capsule fibers attach to the bone. In some synovial joints a pad of fibrous tissue or fibrocartilage lies in the joint cavity (Fig. 10-3B). Synovial joints are called *simple* if only two bones are involved and *composite* if more then two bones contribute to the joint.

The movement possible at synovial joints is considerable but basically four types of movement can be distinguished: gliding movements, angular movements, circumduction, and rotation. **Gliding movements,** the simplest, occur when one joint surface slides over another. This is the only type of movement possible at some of the joints of the wrist and ankle, but many other joint movements have a gliding component in addition to one or more other movement types. There are two categories of **angular movements.** One occurs with a decrease or increase of the angle between adjoining bones, as with *flexion* (bending) or its opposite, *extension* (straightening), which occurs, for example, at the elbow. A second category of angular movements includes *abduction* (movement away from the midline of the body) or its opposite, *adduction* (movement toward the body midline), as when the arm is extended away from or brought closer to the body wall. **Circumduction,** which most commonly involves the shoulder and hip joints, occurs when the distal end of a long bone is moved in a circle as the proximal end is held relatively stationary, the bone seeming to pass along the walls of an imaginary

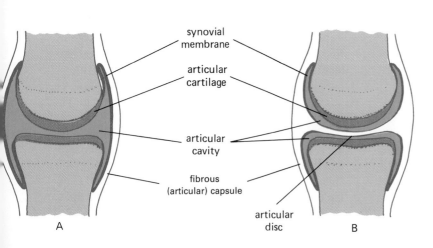

synovial
membrane

articular
cartilage

articular
cavity

fibrous
(articular) capsule

articular
disc

A   B

FIG. 10-3 Diagrammatic sections through (A) a simple synovial joint and (B) a synovial joint with an articular disc.

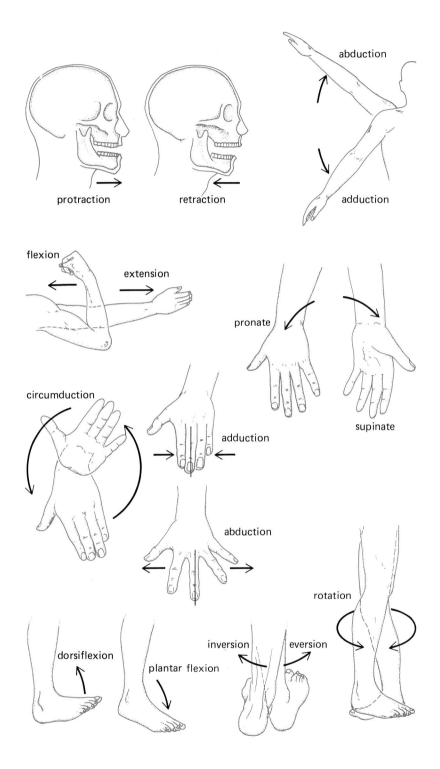

FIG. 10-4 Types of movements possible at various synovial joints.

TABLE 10-1 MOVEMENTS AT SYNOVIAL JOINTS

Movement	Definition	Movement	Definition
Flexion	A movement that decreases the angle between two bones, bending a joint	Circumduction	Movement of a bone in a circular direction so the distal end scribes a circle while the proximal end remains stationary, as in "winding up" to throw a ball
Extension	A movement that increases the angle between two bones, straightening a joint		
Dorsiflexion	Flexion of the foot at the ankle, the foot and toes are turned upward as in standing on the heel	Inversion	Turning inward, movement of the foot at the ankle joint so the sole faces inward
Plantar flexion	Extension of the foot at the ankle, the foot and toes are turned downward toward the sole of the foot	Eversion	Turning outward, movement of the foot at the ankle joint so the sole faces outward
Hyperextension	Continuation of extension beyond the anatomical position, as in bending the head backward	Protraction	Movement of the clavicle (collar bone) or mandible (lower jaw bone) forward on a plane parallel to the ground
Abduction	Movement of a bone away from the midline of the body or body part as in raising the arm or spreading the fingers	Retraction	Movement of the clavicle or mandible backward on a plane parallel to the ground
Adduction	Movement of a bone toward the midline of the body or part		
Rotation	Movement of a bone around its own axis as in moving the head to indicate "no" or turning the palm of the hand up and then down	Supination	Rotation of forearm so that the palm faces forward or upward, movement of the whole body so that the face and abdomen are upward
		Pronation	Rotation of the forearm so that the palm faces backward or downward, movement of the whole body so the face and abdomen are downward

cone-shaped space. **Rotation** occurs as a bone twists about its longitudinal axis. The amazing variety of movements possible are best discussed in the context of specific joints, but the general terms can be understood best by consulting Table 10-1 and Fig. 10-4.

The classification of synovial joints is extremely varied; recent terminology is determined by the shapes of the articulating surfaces.

The **spheroid,** or **cotyloid,** (commonly called the *ball-and-socket*) **joint** offers the greatest degree of movement by allowing relatively free movements in all directions, including rotation of the bone around its long axis, unless the movements are restricted by ligaments or muscles. In this type of joint, a rounded head on one bone fits into a cuplike depression (cotyloid means cup) in a second bone (Fig. 10-5A), such as occurs at the hip and shoulder joints.

Condylar joints, such as the joints between the metacarpals and phalanges in the hands, are modified ball-and-socket joints that differ by having ellipsoid, shallow sockets rather than round, deep ones (Fig. 10-5B). Such joints allow movement in two planes at right angles to each other. Thus, such joints permit flexion-extension, abduction-adduction, and circumduction, but prevent rotation.

Ginglymus, or **hinge, joints,** which occur in the elbow and between the phalanges of the fingers, allow movement around a single axis at right angles to the bones and permit only flexion-extension (Fig. 10-5C).

A **trochoid,** or **pivot, joint,** in which a projection from one bone is encircled by a ring on a second bone, allows rotation. The joint between the first and second cervical vertebrae is of this type.

Plane joints provide a limited amount of sliding between two almost flat surfaces (Fig. 10-5D). Plane

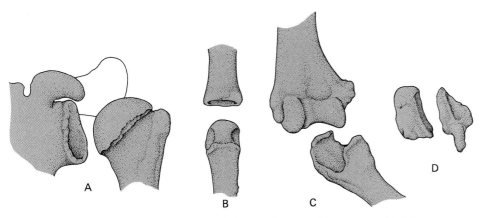

FIG. 10-5 Some examples of synovial joints: (A) spheroid, or cotyloid, joint at the shoulder; (B) condylar joint at the metacarpalphalangeal joint, or knuckle; (C) ginglymus, or hinge, joint at the elbow; (D) plane joint between two carpal bones at the wrist.

joints occur, for example, between the carpal bones (of the wrist).

A joint that consists of two opposed curved surfaces, the surface on one bone being concave, the other convex, is called a **sellar,** or **saddle, joint.** This joint allows some flexion and extension, although less than a ball-and-socket joint, and limited rota-

tion. The joint between the trapezium (one of the carpal bones) and metacarpal of the thumb is of this type.

The **joint capsule** consists of two layers: an outer fibrous capsule and an inner synovial membrane. The fibrous capsule is continuous with the periosteum of the bones that are involved in the joint (see

FIG. 10-6 (A) Diagram of a bursa and synovial sheath. (B) Diagram of a tendon sheath. (C) Section through the elbow joint.

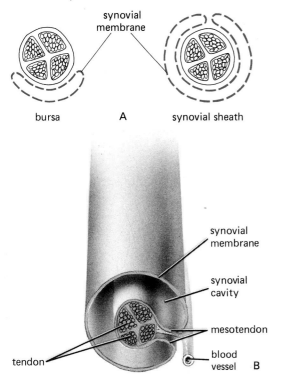

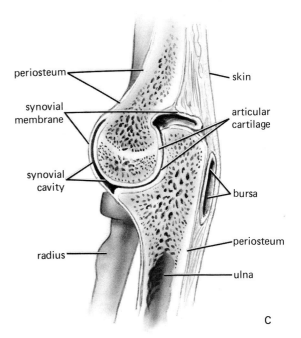

TABLE 10-2   CLASSIFICATIONS OF JOINTS

I. Recent Classification (according to structure)

Type	Description	Example
Fibrous		
Sutures	Small amounts of fibrous tissue separate the bones	Sutures between the bones in the cranial region of the skull
Syndesmosis	Longer fibers bind the bones together	Syndesmosis between the distal ends of the tibia and fibula
Gomphosis	The bones, one shaped like a peg and the other containing a hole, are held together by small amounts of fibrous tissue	A tooth in its socket
Cartilaginous		
Symphysis	A fibrocartilaginous disc forms the binding unit between the two bones	Pubic symphysis
Synchrondrosis	Hyaline cartilage separates two bones or parts of a single bone	Epiphyseal plate between the diaphysis and epiphysis of a growing bone
Synovial		
Plane (gliding)	The surfaces of the opposing bones are almost flat	Between two carpal bones
Spheroid (cotyloid, ball-and-socket)	A rounded head of one bone fits into a concavity on another	Shoulder and hip
Condylar	A modified ball-and-socket in which the socket is shallow and ellipsoidal and the ball is not truly round	The metacarpophalangeal joint at the knuckles
Ellipsoidal	Similar to condylar except that both socket and ball have ellipsoidal shape	The radiocarpal joint

Fig. 10-3). It consists of long collagen fibers and is relatively inelastic and thus contributes to the stability of the joint.

Ligaments, which are connective-tissue cords that join two bones, and **tendons,** which are connective-tissue bundles that connect skeletal muscle to bone, are often associated with joints. Ligaments are sometimes embedded in the joint capsule itself, where they appear as a thickened band on the capsule, or they may be separated from the capsule by a tissue space. Friction can be a problem as the tissues of a moving joint rub against each other—against skin, muscle, and bone, for example. And just as a fluid-filled, synovial membrane–lined cavity (the joint cavity) reduces friction between the opposed bones at a point, similar cavities protect other areas of the joint where friction may be a problem. The cavities are called **bursae** (sing., bursa) when they are flattened, and they are **syn-** **ovial sheaths** when they are tubelike (Fig. 10-6A). A typical synovial sheath is the **tendon sheath** (Fig. 10-6B). Bursae, associated with most joints, can be seen in Fig. 10-6C, a section through the elbow joint in which the bursa and other structures are identified. Synovial cavities normally contain only a thin film of fluid, but when inflamed (as in bursitis, inflammation of a bursa) they can swell.

Other Classifications

In other terminology, a *synarthrosis* is any form of articulation that is incapable of any appreciable motion (such as the sutures of the skull). An *amphiarthrosis* is an articulation that permits little motion, the opposed bones being connected by a disc of fibrocartilage or a ligament, such as a symphysis and syndesmosis. Thus, neither synarthroses nor amphiarthroses contain joint cavities. *Diarthroses*

TABLE 10-2 *(continued)*		

I. Recent Classification (according to structure)

Type	Description	Example
Trochoid (pivot)	A pivot joint in which one bone forms a pin and the other forms a circle in which the pin rotates	The middle joint between the atlas and axis (C1 and C2 vertebrae)
Sellar (saddle)	Each bone has both concave and convex curves like those of a saddle; the curves of the two bones are reciprocal	The joint between the trapezium of the wrist and the metacarpal of the thumb
Ginglymus (hinge)	A spool-like surface on one bone fits into a concave surface on another to allow movement in only one plane	The elbow

II. Older Classification (according to movements)

Synarthroses (immovable joints)
  Suture
  Gomphosis
  Synchondrosis
Amphiarthroses (slightly movable joints)
  Symphysis
  Syndesmosis
Diarthroses (freely movable joints)
  Plane (gliding)
  Spheroidal (ball-and-socket)
  Condylar (modified ball-and-socket)
  Ellipsoidal
  Trochoid (pivot)
  Sellar (saddle)
  Ginglymus (hinge)

are joints that are freely movable, a category that includes most joints of the body. These joints contain a joint cavity partially lined with synovial membrane. The joint classifications are listed in Table 10-2.

DISORDERS OF JOINTS

The most common injuries to joints are dislocations and sprains. A **dislocation** is the actual displacement of the articulating surfaces of the bones; realignment may be difficult if the surrounding connecting tissues have been damaged or if the muscles attached to the bones are in spasm. In a **sprain,** the bones are not displaced, but the ligaments attached to the bones are stretched, torn, or otherwise damaged. (A **strain** is a stretched or torn muscle.) **Arthritis** is defined as inflammation of a joint. Any of the joint tissues may be involved, and the results

are pain, swelling, and impaired movement. Arthritis can be caused by a large number of diseases; the cause is sometimes known (as in gouty arthritis, the result of accumulation of uric acid crystals in the joints), but in most cases the cause remains unclear.

EXAMPLES OF SPECIFIC JOINTS

The Shoulder

The roughly spherical head of the humerus and shallow glenoid cavity of the scapula form the **shoulder joint,** which is of the ball-and-socket type (Fig. 10-7A). Both articular surfaces are covered by a layer of hyaline articular cartilage. The shallowness of the glenoid cavity allows considerable range of motion but little security, and the joint is supported by a ring of muscles and ligaments (the so-

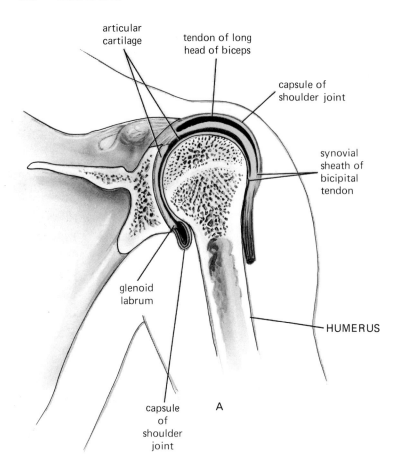

articular
cartilage

tendon of long
head of biceps

capsule of
shoulder joint

synovial
sheath of
bicipital
tendon

glenoid
labrum

HUMERUS

capsule
of
shoulder
joint

A

FIG. 10-7 (A) A section through the shoulder joint. Bone is indicated by gray outlined with color, cartilage is shaded in light color, and the synovial membranes are shaded in dark color. (B) Ligaments of the shoulder joint between the clavicle, scapula, and humerus.

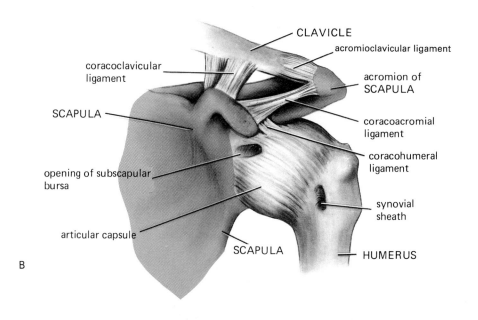

CLAVICLE

acromioclavicular ligament

coracoclavicular
ligament

acromion of
SCAPULA

SCAPULA

coracoacromial
ligament

coracohumeral
ligament

opening of subscapular
bursa

synovial
sheath

articular capsule

SCAPULA

HUMERUS

B

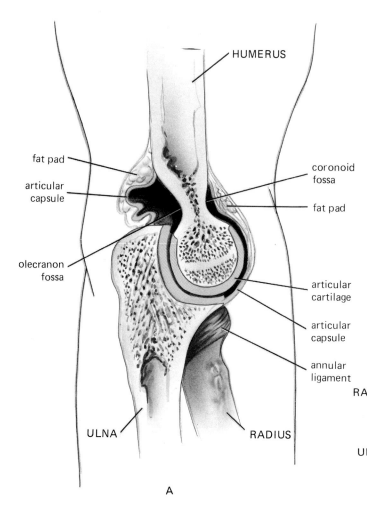

fat pad

articular capsule

olecranon fossa

ULNA

coronoid fossa

fat pad

articular cartilage

articular capsule

annular ligament

RADIUS

HUMERUS

A

FIG. 10-8 (A) A section through the elbow. Bone is indicated by gray outlined with color, cartilage is shaded in light color, and the synovial membranes are shaded in dark color. (B) Lateral view of the elbow showing the ligaments associated with the joint.

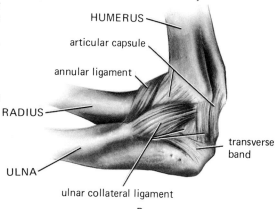

HUMERUS

articular capsule

annular ligament

RADIUS

ULNA

transverse band

ulnar collateral ligament

B

called *musculotendinous,* or *rotator, cuff)* that surrounds it, and a connective-tissue ring, the *glenoid labrum,* that surrounds the fossa and deepens it. The ligamentous structures of the acromial end of the clavicle and the shoulder joint are shown in Fig. 10-7B.

The Elbow

The **elbow** includes three articulations. One is between the capitulum of the humerus and the head of the radius *(humeroradial articulation).* Another is the hinge-like articulation between the trochlea of the humerus and the trochlear notch of the ulna (the *humeroulnar articulation).* The third is between the edge of the head of the radius and the radial notch of the ulna and the annular ligament (the *proximal radioulnar joint).* Thus, it is a composite joint. It acts as a hinge, the radius and ulna

swinging on the capitulum and trochlea of the humerus during flexion and extension of the joint. There is also a slight degree of rotation of the forearm during these movements. The three joints are enclosed in a single joint capsule and have a common synovial cavity (Fig. 10-8A). The ligament that forms almost a complete circle around the heads of the radius and ulna is the *annular ligament.* This and other ligaments of the elbow are shown in Fig. 10-8B.

The Hip

The head of the femur fits into the shallow cup-like fossa of the acetabulum of the coxal bone to form the **hip joint** (Fig. 10-9A). The articular surfaces of both bones are almost completely covered with cartilage except at the point of attachment of the ligaments. Associated with the joint is a fibrocartilage

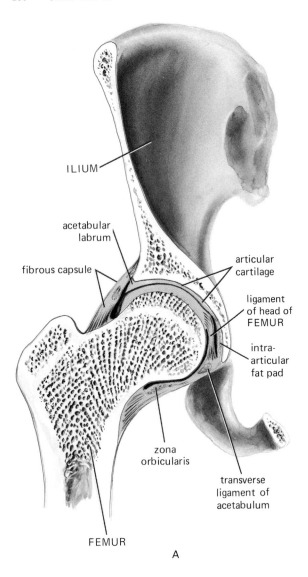

ring, the *acetabular labrum,* around the rim of the acetabulum, which increases the depth of the acetabulum; in fact, the depth is so great that the head of the femur cannot be pulled from the acetabulum without tearing the labrum. The *transverse acetabular ligament* continues the labrum across the acetabular notch. The joint capsule attaches to the labrum on the coxal bone and to the greater trochanter, lesser trochanter, and intertrochanteric line of the femur; thus, it extends a considerable distance along the neck of the femur. The fibrous membrane of the capsule blends with the labrum and periosteum of the femur at its attachments.

The capsule contains four thickenings, the *iliofemoral, ischiofemoral,* and *pubofemoral ligaments* (Fig. 10-9B and C) and the *zona orbicularis* (Fig. 10-9A), which circles around the joint. The iliofemoral ligament is the most important ligament of the hip joint and functions to resist hyperextension and internal rotation of the joint. The *ligament of the head* lies within the cavity of the joint and passes from the acetabulum to the fovea of the head of the femur. The joint is of the ball-and-socket type and is capable of flexion-extension, adduction-abduction, and circumduction (a combination of the previous four) as well as lateral and medial rotation.

FIG. 10-9   (A) Section through the hip joint. Bone is indicated by gray outlined in color, cartilage is shaded in light color, and the synovial membranes are shaded in dark color. (B) Anterior, and (C) posterior views of the ligaments associated with the hip joint.

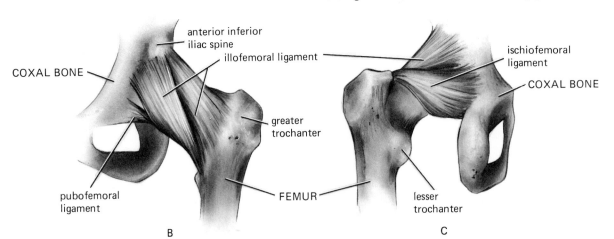

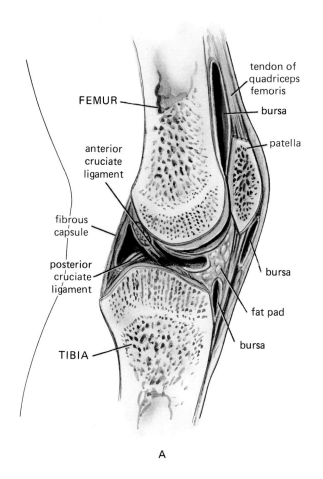

A

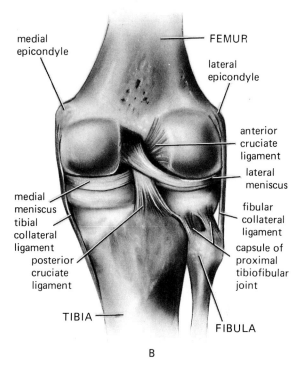

B

FIG. 10-10 (A) Section through the knee joint. Bone is indicated by gray outlined with color, cartilage is shaded in light color, and the synovial membranes are shaded in dark color. (B) Posterior view of the right knee showing the ligaments associated with the joint.

The Knee

The **knee** (Fig. 10-10A), the largest joint in the body, is composed of articulations between the femur, tibia, and patella. (The *patella* is a sesamoid, i.e., more or less oval, seed-shaped bone in front of the knee that shares a common joint capsule with it.) The joint capsule of the knee is complicated, being partially subdivided by wedge-shaped fibrocartilage plates, the *medial* and *lateral menisci* (Fig. 10-10B) and by folds of the synovial membrane.

The support of the body on the opposed ends of two long bones is not basically stable, but the stability is improved by strong ligaments and a strong capsule reinforced by tendons (Fig. 10-10B). The knee is strengthened laterally and medially by the *tibial* and *fibular collateral ligaments,* and posterially by the *oblique popliteal ligament,* and the *arcuate ligament. Anterior* and *posterior cruciate ligaments* pass between the tibia and condyles of the femur, the anterior ligament passing to the lateral condyle and the posterior ligament passing slightly medially, crossing behind the anterior one. The tibial collateral ligament in particular, but the cruciates as well, exert a steadying effect on the knee throughout its full range of motion.

The knee is functionally a hinge joint, but it also allows a small amount of rotation, particularly when in the flexed position. Full extension of the knee involves a screw-type action, the lateral condyle of the femur sliding forwards and the medial condyle backwards, the collateral ligaments tightening to hold the bones against one another. As the knee is flexed, the obliquely oriented popliteus muscle rotates the femur laterally and draws the lateral condyle of the femur backward, "unlocking" the joint.

KEY TERMS

articulation
joint
fibrous joint
suture
syndesmosis
cartilaginous joint
symphysis
synchondrosis

synovial joint
articular capsule
synovial fluid
synovial membrane
flexion
extension
abduction

adduction
circumduction
rotation
ligament
tendon
bursa
synovial sheath

REVIEW EXERCISES

1 Differentiate between fibrous, cartilaginous, and synovial joints; draw an example of each.
2 Differentiate between a suture and syndesmosis; give an example of each.
3 Differentiate between a symphysis and synchondrosis; give an example of each.
4 Draw a diagram of a synovial joint, label the joint capsule, articular cartilage, accessory ligament, synovial membrane, and articular cavity.
5 List the advantages and disadvantages of the various joint types (spheroid, condylar, ginglymus, trochoid, plane, and sellar).
6 Differentiate between a ligament and tendon.
7 Explain the function served by bursae, synovial sheaths, and tendon sheaths.

8 Differentiate between a dislocation, strain, and sprain.
9 Describe the shoulder joint; explain the advantages and disadvantages of the somewhat shallow ball-and-socket joint at this location.
10 Describe the elbow joint, including the humeroradial, humeroulnar, and proximal radioulnar articulations and the role of the capitulum and trochlea of the humerus.
11 Describe the hip joint; explain the advantages and disadvantages of the deep ball-and-socket joint at this location.
12 Describe the knee joint, including the relation of the patella to the femur and tibia and the role of the ligaments.

CHAPTER

MUSCLE TISSUE

Three types of muscle cells can be identified on the basis of structure and contractile properties: (1) skeletal muscle, (2) smooth muscle, and (3) cardiac muscle. Most **skeletal muscle,** as the name implies, is attached to bones, and its contraction is responsible for the movements of parts of the skeleton. The contraction of skeletal muscle is controlled by a division of the nervous system known as the somatic nervous system (see Chap. 14) and is under voluntary control. The movements produced by skeletal muscle are primarily involved with interactions between the body and the external environment.

Smooth muscle surrounds such hollow organs and tubes as the stomach, intestinal tract, urinary bladder, uterus, blood vessels, and the air passages to the lungs. The contraction of this smooth muscle either may propel the luminal contents out of or through the hollow organs, or it may regulate the flow of the contents through tubes by changing their diameters without itself initiating the propulsion. Smooth muscle is also found as single cells distributed throughout organs (for example, the spleen) and in small groups of cells attached to the hairs in the skin or the iris of the eye. Smooth muscle contraction is controlled by factors intrinsic to the muscle itself, by the autonomic nervous system, and by hormones; therefore, it is not normally under direct conscious control.

The third type of muscle, **cardiac muscle,** is the muscle of the heart, and its contraction propels blood through the circulatory system. Like smooth muscle, it is regulated by intrinsic factors, the autonomic nervous system, and hormones.

Although there are significant differences in the structure, contractile properties, and control mechanisms of the three types of muscle, the force-generating mechanism is similar in all of them. The properties of skeletal muscle and the molecular events associated with its force generation will be described first, followed by smooth muscle with emphasis on the ways in which the latter differs from skeletal muscle. Cardiac muscle, which combines some of the properties of both skeletal and smooth muscle, will be described in Chap. 20 in the context of its role in the circulation.

STRUCTURE OF SKELETAL MUSCLE

Skeletal muscle accounts for 40 to 45 percent of the total body weight. Figure 11-1 illustrates the various levels of skeletal muscle structural organization that will be discussed in this section. An individual **muscle,** such as the biceps or gastrocnemius, is made up of many cells, called **muscle fibers.** Each fiber is cylindrical, having a diameter of 10 to 100 μm and a length of up to 300,000 μm (about 1 ft). In contrast to most cell types, each fiber contains hundreds of nuclei, which are located just beneath the cell membrane. The fibers, each surrounded by a loose connective-tissue **endomysium,** are organized into various-sized bundles, or **fascicles.** These in turn are surrounded by a dense connective-tissue sheath known as the **perimysium.** The entire muscle, which is composed of many fascicles, is surrounded by still another connective-tissue sheath, the **epimysium.** Blood vessels and nerves follow these connective-tissue sheaths into the muscle interior.

Over 600 muscles can be identified in the human body. Some of them are very small, consisting of only a few hundred fibers; larger muscles may contain several hundred thousand fibers. In some muscles the individual fibers are as long as the muscle itself; but most fibers are shorter than the total muscle, and their ends are attached to the connective-tissue network interlacing the muscle fibers.

Generally each end of a muscle is attached to a bone by collagen-fiber bundles known as **tendons,** which have great strength but no active contractile properties. The collagen fibers of the perimysium and epimysium are continuous with those in the tendons, and together they act as a structural framework to which the muscle fibers and bone are attached. The forces generated by the contracting muscles are transmitted by the connective tissue and tendons to the bones. The transmission of force from muscle to bone is like a number of people pulling on a rope, each person corresponding to a single muscle fiber and the rope to the connective tissue and tendons.

Subcellular Organization of Skeletal Muscle Fibers

Figure 11-2 shows a section through a skeletal muscle as seen with a light microscope. In between the individual muscle fibers are capillary blood vessels that contain red blood cells. The most striking feature of the muscle fibers is the series of transverse light and dark bands that form a regular pattern along the fiber. Both skeletal and cardiac muscle fibers have this characteristic banding and are known as **striated muscles;** smooth muscle cells

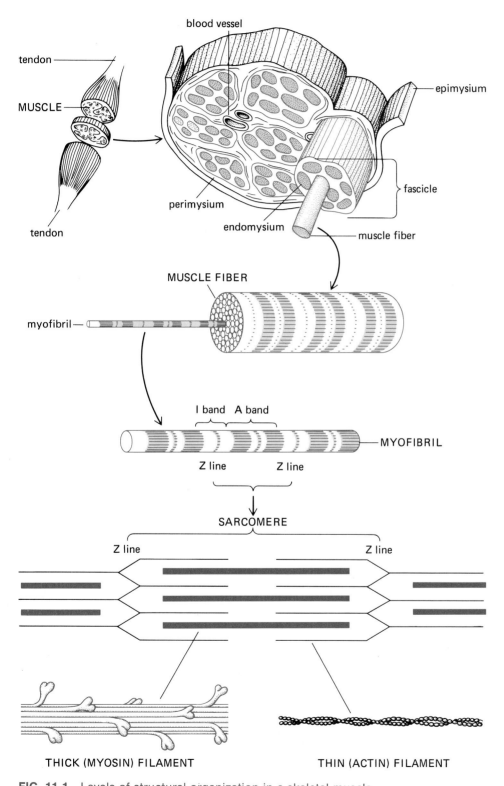

FIG. 11-1   Levels of structural organization in a skeletal muscle.

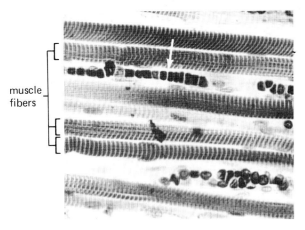

FIG. 11-2 Photomicrograph of skeletal muscle fibers. Arrow indicates capillary blood vessel containing blood cells.

show no banding pattern. Although the banding pattern appears to be continuous across a single fiber, these bands are actually confined in the cytoplasm of the fiber to independent cylindrical elements, known as **myofibrils** (Fig. 11-1). Each myofibril is about 1 to 2 μm in diameter and continues through the length of the muscle fiber. Myofibrils occupy about 80 percent of the fiber volume and vary in number from several hundred to several thousand, depending on the fiber diameter.

Viewed with the electron microscope, the structures responsible for the banding patterns become evident. The myofibrils consist of smaller **filaments** (Figs. 11-1 and 11-3), which form a regular repeating pattern along the length of the fibril. One unit of this pattern is known as a **sarcomere.** It is the functional unit of the contractile system in muscle, and the events occurring in a sarcomere are duplicated in the other sarcomeres along the myofibrils.

Each sarcomere contains two types of filaments: thick and thin. The **thick filaments,** 12 to 18 nm in diameter, are located in the central region of the sarcomere where their orderly parallel arrangement gives rise to the dark bands, known as *A bands,* that are seen in striated muscles. These thick filaments contain the protein known as **myosin.** The **thin filaments,** 5 to 8 nm in diameter, contain the protein **actin** and are attached at either end of the sarcomere to a structure known as the *Z line.* Two successive Z lines define the limits of one sarcomere. The Z lines consist of short elements that interconnect the thin filaments from two adjoining sarcomeres and thus provide an anchoring point for

the thin filaments. The thin filaments extend from the Z lines toward the center of the sarcomere where they overlap with the thick filaments.

A third band, the *I band* (Fig. 11-3) represents the region between the ends of the A bands of two adjoining sarcomeres. This band contains that portion of the thin filaments that do not overlap with the thick filaments and is bisected by the Z line. Because it contains only thin filaments it usually appears as a light band separating the dark A bands. A fourth band, the *H zone* appears as a thin, lighter band in the center of the A band and corresponds to the space between the ends of the thin filaments. Only thick filaments are found in the H-zone region. Finally, a thin dark band can be seen in the center of the H zone; this is the *M line* and is produced by cross linkages between the thick filaments, which keep the thick filaments in parallel alignment. Thus, neither the thick nor the thin filaments are free floating; each is linked either to Z lines, in the case of the thin filaments, or to M lines, in the case of the thick filaments.

In the region of overlapping thick and thin filaments in the A band, the gap between thick and thin filaments is bridged by projections, **cross bridges,** which are portions of myosin molecules extending from the thick filaments and making contact with the actin molecules on the adjacent thin filaments. As we shall see, it is these cross bridges that are the sites of force generation in muscle cells.

MOLECULAR MECHANISMS OF CONTRACTION

Sliding-Filament Model

When a skeletal muscle fiber shortens during contraction, the width of the A band in a sarcomere remains constant (Fig. 11-4). Since this width corresponds to the length of the thick filament, the filament must not have changed length during shortening. In contrast, the width of the I band and H

FIG. 11-3 Filament organization of myofibrils. (A) Numerous myofibrils in a single skeletal muscle fiber. Arrows indicate mitochondria located between the myofibrils. (B) High magnification of a single sarcomere within a single myofibril; arrow indicates end of thick filament. (C) Arrangement of thick and thin filaments that produces the striated banding patterns in myofibrils.

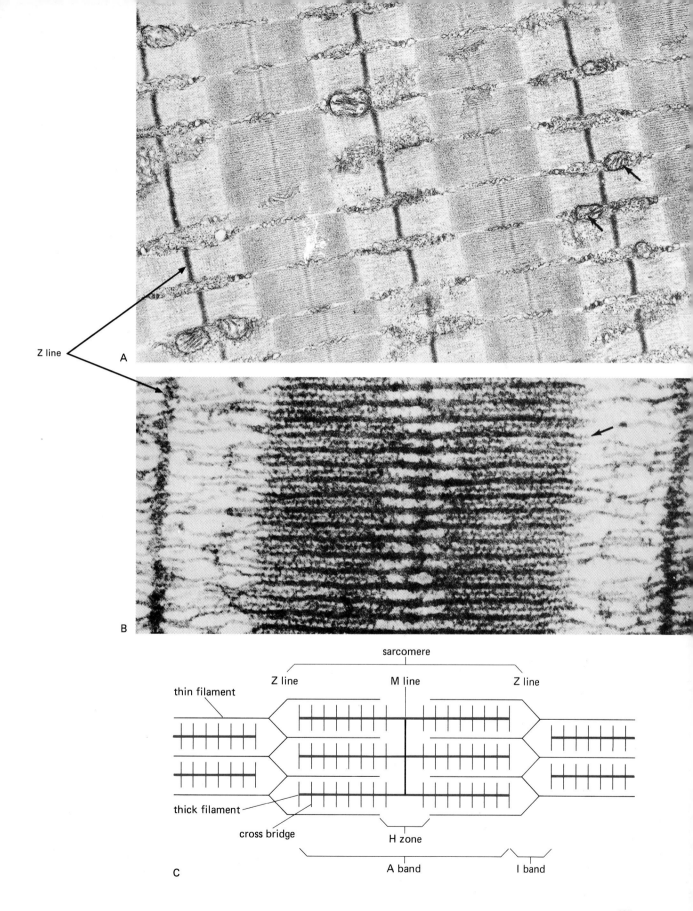

Z line

A

Z line

B

sarcomere

Z line M line Z line

thin filament

thick filament

cross bridge

H zone

A band I band

C

187

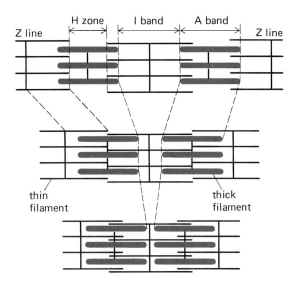

FIG. 11-4 Changes in filament alignment and banding pattern during shortening as thin filaments slide past thick filaments. (Cross bridges are not illustrated in this diagram.)

zone decreases during shortening as a result of the movement of the thin filaments past the thick filaments. Thus, each H zone decreases as the ends of the thin filaments from opposing ends of its sarcomere approach each other. The I band also decreases in width for the same reason, i.e., as more and more of the thin filament length comes to overlap the thick filaments. Thus, the lengths of neither the thick filaments nor the thin filaments change during shortening; rather the two sets of filaments slide past each other. These observations led to the **sliding-filament model** of muscle contraction.

What produces the movement of these filaments? The answer is the cross bridges that extend from the surface of the thick filaments and make contact with the adjacent thin filaments. When a cross bridge is activated, it moves in an arc parallel to the long axis of the thick filament, much like an oar on a boat. If, at this time, the cross bridge is attached to a thin filament, this swiveling motion slides the thin filament toward the center of the A band, thereby producing shortening of the sarcomere. One stroke of a cross bridge produces only a small displacement of a thin filament relative to a thick, but the cross bridges undergo many cycles during a single contraction, each cycle requiring attachment of the bridge to the thin filament, angular movement of the bridge while attached, de-

THIN FILAMENT

FIG. 11-5 Two helical chains of globular actin proteins form the primary structure of the thin filaments.

tachment from the thin filament, reattachment at a new location, and repetition of the cycle. Each cross bridge undergoes its own independent cycle of movement, so that at any one instant during contraction only about 50 percent of the bridges are attached to the thin filaments, while the others are at intermediate stages of the cycle.

Actin, Myosin, and ATP   The generation of force by a cross bridge involves the interaction of two proteins, actin and myosin, and a source of chemical energy, ATP. Globular-shaped molecules of actin are arranged in two chains helically intertwined to form the thin filaments (Fig. 11-5). Myosin is a much larger molecule and has a globular end attached to a long tail. Approximately 200 myosin molecules constitute a single thick filament. The tails of the molecules lie along the axis of the filament and the globular heads extend out to the sides, forming the cross bridges. Each globular head contains a binding site able to bind to a complementary site on an actin molecule. The myosin molecules in the two halves of each thick filament are oriented in opposite directions, such that all their tail ends are directed toward the center of the filament (Fig. 11-6). The power strokes of the cross bridges in the

FIG. 11-6 Orientation of myosin molecules in a thick filament, with the globular ends of the myosin molecules forming the cross bridges.

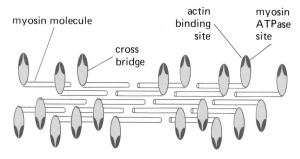

THICK FILAMENT

two halves of each thick filament are directed toward the center (M line), thereby moving the thin filaments of the sarcomere toward the center of the sarcomere during shortening.

In addition to the binding site for actin, the globular end of myosin, that portion of the myosin molecule that forms the cross bridge, contains a separate enzymatic site that catalyzes the breakdown of ATP to ADP and inorganic phosphate, releasing the chemical energy stored in ATP. The actual splitting of ATP occurs on the myosin molecule before it attaches to actin, but the ADP and inorganic phosphate generated remain bound to myosin. The chemical energy released at the time of ATP splitting is transferred to myosin (M), producing a high-energy form of myosin (M*)

$$M \cdot ATP \longrightarrow M^* \cdot ADP \cdot P_i$$
$$\text{(ATP breakdown)}$$

The subsequent binding of this high-energy form of myosin to actin (A) triggers the discharge of the energy stored in the myosin cross bridge, with the

resultant production of the force-causing movement of the cross bridge (the ADP and P_i are released from myosin at this time):

$$A + M^* \cdot ADP \cdot P_i \rightarrow A \cdot M^* \cdot ADP \cdot P_i \rightarrow A \cdot M + ADP + P_i$$
$$\text{(actin binding)} \qquad \text{(bridge movement)}$$

This sequence of energy storage and release by myosin is analogous to a mousetrap: Energy is stored in the trap when the spring is cocked (ATP splitting) and released when the trap is sprung (binding to actin). This is, of course, only an analogy, and the actual changes in the shape of the myosin molecule that accompany energy storage and release have yet to be determined.

During contraction, as we have seen, each cross bridge undergoes many cycles of attachment, movement, and dissociation from a thin filament, each cycle being accompanied by the splitting of one molecule of ATP. However, the myosin cross bridge binds very firmly to actin and this linkage must be broken at the end of each bridge cycle; the binding of a new molecule of ATP to

FIG. 11-7 Chemical and mechanical changes during the four stages of a single cross-bridge cycle. Start reading the figure in the lower left.

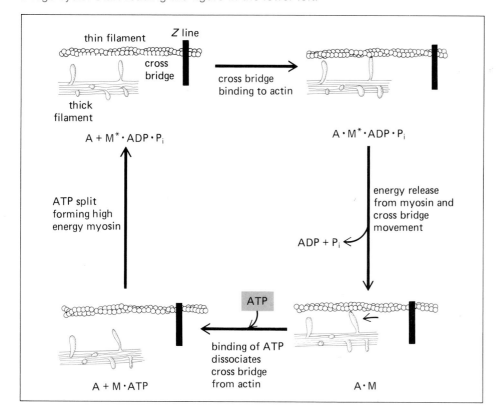

myosin is responsible for breaking this link:

$$A \cdot M + ATP \longrightarrow A + M \cdot ATP$$
(A·M dissociation)

Thus, upon binding (but not splitting) a molecule of ATP, myosin dissociates from actin. The free myosin bridge then splits its bound ATP, thereby reforming the high energy state of myosin, which can now reattach to a new site on the actin filament, and so on. Thus, ATP performs two distinct roles in the cross-bridge cycle: (1) The energy released from the splitting of ATP provides the energy for cross-bridge movement; and (2) the binding (not splitting) of ATP to myosin breaks the link between actin and myosin at the end of a cross-bridge cycle, allowing it to be repeated. Figure 11-7 provides a summary of these chemical and mechanical changes that occur during one cross-bridge cycle.

The importance of ATP in dissociating actin and myosin at the end of a bridge cycle is illustrated by the phenomenon of **rigor mortis** (death rigor), in which the muscles become very stiff 3 to 12 hr after death. Rigor mortis results directly from the loss of ATP in the dead muscle cells. In the absence of ATP the myosin cross bridges are able to bind to actin but the bond between them is not broken. The thick and thin filaments become cross-linked to each other, producing the rigid condition of the dead muscle. In contrast, in the living muscle at rest, the myosin bridges are not bound to actin and the filaments readily slide past each other when the muscle is passively stretched.

Regulator Proteins and Calcium

Since a muscle fiber contains all the ingredients necessary for cross-bridge cycling—actin, myosin, and ATP—the question arises: Why are muscles not in a continuous state of contractile activity? The reason is that in a resting, relaxed muscle fiber, the cross bridges are unable to bind to actin and initiate the cross-bridge cycle that leads to contraction. This inhibition of cross-bridge binding is caused by two proteins, **troponin** and **tropomyosin**, which are bound to the thin filaments (Fig. 11-8). Tropomyosin molecules are arranged end-to-end

FIG. 11-8 In the absence of calcium, tropomyosin blocks the cross-bridge binding sites on actin. Binding of calcium to troponin moves the tropomyosin to one side, exposing the binding sites, allowing myosin cross bridges to bind to actin.

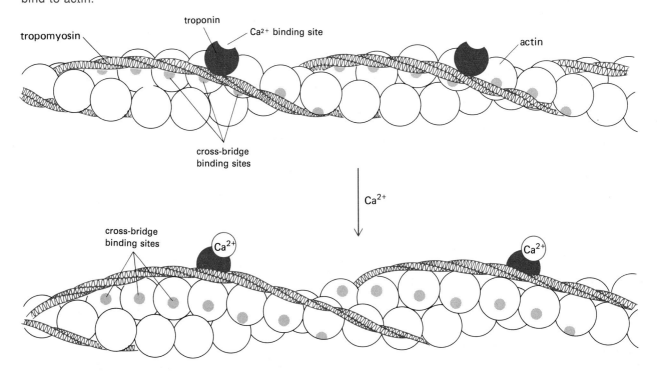

along the chains of actin, so that they partially cover the myosin binding sites on the actin molecules, thereby preventing them from binding to the myosin cross bridges. Each tropomyosin molecule is held in this blocking position by a molecule of troponin, itself bound to both tropomyosin and actin.

Having accounted for the mechanism that prevents cross-bridge activity and thus keeps a muscle fiber turned "off," we can now ask: What turns the muscle fiber "on," i.e., allows cross-bridge activity to proceed? In order for the cross bridges to bind to actin, the tropomyosin molecules must be moved away from their blocking position. This occurs when calcium ions bind to a specific site on troponin. The binding produces a change in the shape of troponin such that it pulls the tropomyosin bound to it to one side, uncovering the cross-bridge binding sites on actin (Fig. 11-8). Conversely, removal of calcium from troponin reverses the process and tropomyosin moves back into its blocking position so that cross-bridge activity ceases. Note that the activating effect of calcium on contraction is not directly on either actin or myosin, but is mediated through the two regulator proteins, troponin and tropomyosin.

Thus, the availability of calcium ions to the troponin binding sites determines whether a muscle fiber is turned on or off. We now describe how this availability is coupled to the electrical events occurring in the muscle plasma membrane.

Excitation-Contraction Coupling

Excitation-contraction coupling refers to the process by which an action potential in the plasma membrane of the muscle fiber triggers off the sequence of events that lead to cross-bridge activity (via increased availability of calcium). The plasma membrane of a skeletal muscle fiber is an excitable membrane capable of generating and propagating action potentials by mechanisms described in Chap. 5. A skeletal muscle's action potential lasts 1 to 2 msec and is completed before any signs of mechanical activity begin (Fig. 11-9). The period between the action potential and the beginning of mechanical activity lasts several milliseconds and is known as the **latent period;** it is during this period that the events of excitation-contraction coupling occur. Once begun, the mechanical activity following a single action potential may last 100 msec or more. Note that the mechanical activity far outlasts the duration of the action potential; thus, the electrical activity in the plasma membrane does not directly act upon the contractile proteins but initiates a process which continues to activate the contractile apparatus long after the electrical activity in the membrane has ceased.

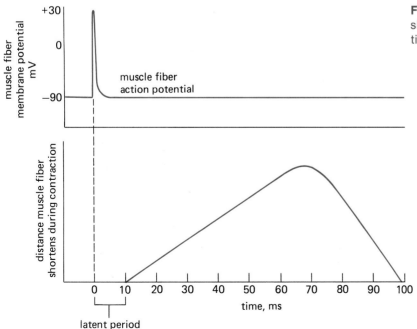

FIG. 11-9 Time relations between a skeletal muscle fiber action potential and mechanical contraction.

The link between membrane electrical activity and cross-bridge activity is provided by calcium ions. In a resting muscle fiber the concentration of free calcium in the cytosol surrounding the thick and thin filaments is very low, calcium is not bound to troponin, and thus cross-bridge activity is blocked by tropomyosin. Following an action potential there is a rapid increase in the calcium con-centration surrounding the contractile filaments, calcium binds to troponin, removing the blocking effect of tropomyosin and thereby initiating con-traction.

Sarcoplasmic Reticulum   In skeletal muscle, the calcium ions that initiate cross-bridge activity are released, following membrane excitation, from the

FIG. 11-10   (A) Diagrammatic representation of the geometrical relationships between the membranes of the sarcoplasmic reticulum, the transverse tu-bules, and the myofibrils. (B) Three-dimensional view of transverse tubules and sarcoplasmic reticulum in human skeletal muscle.

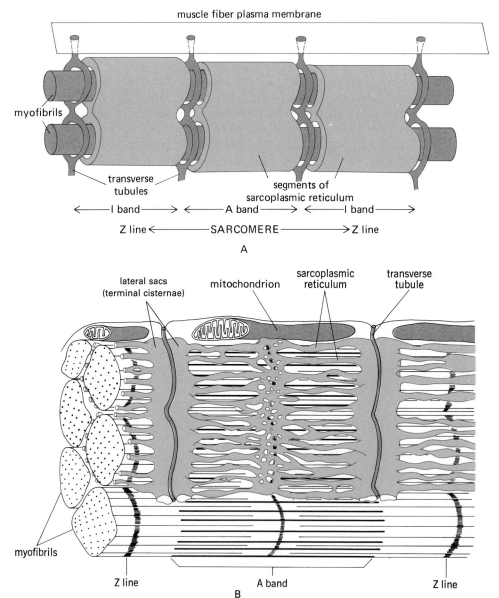

muscle fiber plasma membrane

myofibrils

transverse tubules

segments of sarcoplasmic reticulum

←——I band——→ ←——A band——→ ←——I band——→

Z line ←————SARCOMERE————→ Z line

A

lateral sacs (terminal cisternae)

mitochondrion

sarcoplasmic reticulum

transverse tubule

myofibrils

Z line

A band

Z line

B

lumen of the cell organelle known as the **sarcoplasmic reticulum.** The sarcoplasmic reticulum in muscle is homologous to the endoplasmic reticulum found in most cells. The sarcoplasmic reticulum forms a sleeve-like structure around each of the myofibrils (Fig. 11-10). One segment of the sarcoplasmic reticulum surrounds the region of the A band, while an identical but separate segment surrounds the I-band region. Each of these segments of the sarcoplasmic reticulum possesses two enlarged regions at each end, the **lateral sacs,** which are connected to each other by a series of smaller tubular elements. The lateral sacs store the calcium to be released following membrane excitation. Associated with the sarcoplasmic reticulum is a tubular structure, the **transverse tubule** (t tubule) at the level of each A-I junction, which passes between the adjacent segments of the sarcoplasmic reticulum and eventually joins the plasma membrane of the muscle fiber. The lumen of the t tubule is continuous with the extracellular medium surrounding the muscle fiber.

The membrane of the t tubule, like the plasma membrane, is able to propagate action potentials. Thus, an action potential, once initiated in the plasma membrane, is rapidly conducted over the surface of the muscle fiber and into the interior of the fiber along the t tubules. As the action potential in the t tubule passes the lateral sacs of the sarcoplasmic reticulum, it triggers, by an unknown mechanism, the release of calcium from the lateral sacs. Since the reticulum surrounds the myofibrils, the released calcium has only a short distance to diffuse to reach troponin where its binding initiates contraction.

How is contraction, once initiated by the release of calcium from the sarcoplasmic reticulum, turned off? As mentioned above, the contraction is turned off by removing the calcium from troponin, and this is achieved by once again lowering the calcium concentration in the region of the troponin binding sites. The membrane of the sarcoplasmic reticulum contains a carrier-mediated active transport system that pumps calcium ions from the cytosol into the lumen of the reticulum. Calcium is rapidly released from the reticulum upon arrival of an action potential, but the pumping of the released calcium back into the reticulum requires a much longer time. Therefore, the contractile activity continues for some time after the action potential.

To reiterate, just as contraction results from the release of calcium ions stored in the sarcoplasmic reticulum, so relaxation occurs as calcium is pumped back into the reticulum (Fig. 11-11). ATP is required to provide the energy for the calcium pump, and this is the third major role of ATP in the mechanics of muscle contraction, the others providing the energy for cross-bridge movement and inducing the dissociation of actin and myosin at the end of each cross-bridge cycle.

The Neuromuscular Junction   We have just seen that an action potential in the muscle fiber membrane is the signal that leads to mechanical activity by triggering the release of calcium. What are the mechanisms that lead to the initiation of action potentials in muscle-fiber membranes? There are three answers to this question, depending on the type of muscle that is being considered: (1) stimulation by a nerve fiber; (2) stimulation by hormones and local chemical agents; and (3) spontaneous electrical activity within the membrane itself. Stimulation by nerve fibers is the only mechanism by which skeletal muscles are normally excited, whereas all three mechanisms are involved in initiating excitation in smooth and cardiac muscle. (Mechanisms 2 and 3 will be described in the section on smooth muscle.)

The nerve cells that innervate skeletal muscle fibers are known as **motor neurons.** Extending from each *cell body* (the region of the neuron that contains the cell nucleus and most of the cell organelles) is a long cylindrical process known as an *axon.* The cell bodies of the motor neurons are located in the central nervous system (the brain and spinal cord), whereas the axon of each motor neuron extends from the central nervous system and makes contact with specific skeletal muscle fibers. Action potentials are propagated at high velocities along the plasma membrane of the axon from the cell body to skeletal muscle fibers. The high velocity with which these action potentials are propagated provides the mechanism for the rapid coordination of skeletal muscle activity by the central nervous system. As the axon from a single neuron approaches the muscle, it divides into many branches, each of which forms a single junction with a muscle fiber (Fig. 11-12). Thus, each motor neuron is connected through its branching axon to several muscle fibers (although each motor neuron innervates many muscle fibers, each muscle fiber is

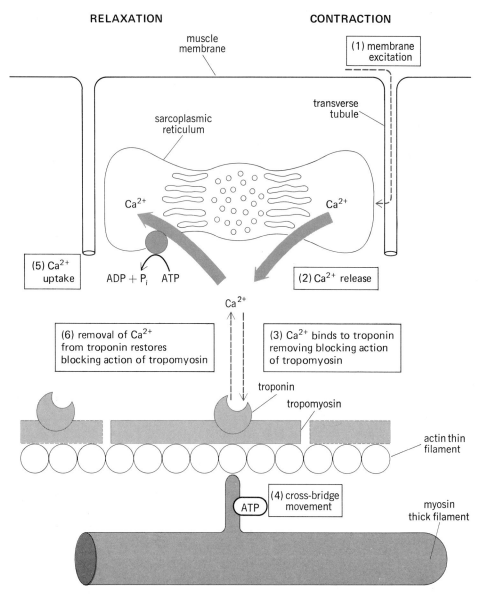

RELAXATION

CONTRACTION

muscle membrane

(1) membrane excitation

transverse tubule

sarcoplasmic reticulum

Ca^{2+}

Ca^{2+}

(5) Ca^{2+} uptake

$ADP + P_i$ ATP

(2) Ca^{2+} release

Ca^{2+}

(6) removal of Ca^{2+} from troponin restores blocking action of tropomyosin

(3) Ca^{2+} binds to troponin removing blocking action of tropomyosin

troponin

tropomyosin

actin thin filament

(4) cross-bridge movement

ATP

myosin thick filament

FIG. 11-11   Role of calcium in muscle excitation-contraction coupling.

innervated by only a single motor neuron). The motor neuron plus the muscle fibers it innervates are a **motor unit.** When a motor neuron fires an action potential, all the muscle fibers in that motor unit are activated.

As a branch of the motor axon approaches the muscle surface, it divides into a fine terminal arborization that lies in grooves on the muscle fiber surface. The region of the muscle membrane that lies directly under the terminal portion of the axon

has special properties and is known as the **motor end plate.** The entire junction, including the axon terminal and motor end plate, is known as a **neuromuscular junction** (Fig. 11-13).

The terminal ends of the motor axon contain membrane-bound vesicles that contain the chemical transmitter **acetylcholine** (ACh). When an action potential in the motor axon arrives at the neuromuscular junction, it depolarizes the nerve membrane and triggers release of ACh into the extracellular

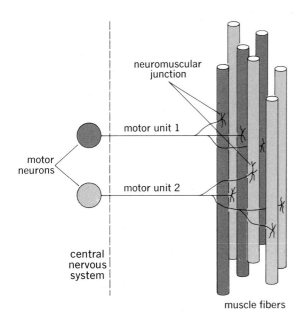

FIG. 11-12 Muscle fibers associated with two motor neurons, forming two motor units.

cleft that separates the nerve and muscle membranes. The released acetylcholine diffuses across the cleft and combines with receptor sites on the

FIG. 11-13 Neuromuscular junction.

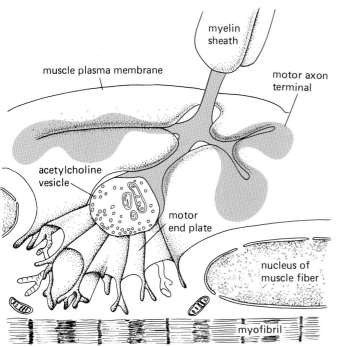

motor-end-plate membrane. The binding of ACh to these sites increases the permeability of the membrane to sodium and potassium ions, producing a depolarization of the motor end plate, known as the **end-plate potential** (EPP). Current flows locally between the end-plate regions and the muscle plasma membrane adjacent to it, depolarizing the adjacent membrane to threshold, initiating an action potential. The action potential thus initiated in the muscle fiber's plasma membrane is then propagated over the surface of the muscle fiber by the same mechanism described for the propagation of action potentials along axon membranes (see Chap. 5). The EPP lasts for only a short period of time because ACh is rapidly broken down by an enzyme, acetylcholinesterase, located at the end plate.

Because each EPP produces an action potential in a muscle fiber, every action potential in a motor neuron produces an action potential in the muscle fibers of its motor unit. Thus, there is a one-to-one transmission of activity from the motor neuron to the muscle fiber.

There are many ways in which events at the neuromuscular junction can be modified by disease or drugs. For example, the deadly South American Indian arrowhead poison, curare, is strongly bound to the acetylcholine receptor site, but it does not change membrane permeability, nor is it destroyed by acetylcholinesterase. When a receptor site is occupied by curare, acetylcholine cannot bind to the receptor; therefore, although the motor nerves still conduct normal action potentials and release acetylcholine, there is no resulting EPP or contraction. Since the skeletal muscles responsible for breathing movements depend upon neuromuscular transmission to initiate their contraction, death comes from asphyxiation. Curare and similar drugs are used in small amounts to prevent muscular contractions during certain types of surgical procedures while the patient's breathing is maintained by artificial ventilation.

Neuromuscular transmission can also be blocked by inhibition of acetylcholinesterase. Some organophosphates, which are the main ingredients in certain pesticides and nerve gases (the latter developed for biological warfare), inhibit this enzyme; acetylcholine is not destroyed and its prolonged action maintains depolarization of the motor end plate. The failure of repolarization prevents new action potentials from being initiated; thus the mus-

TABLE 11-1   SEQUENCE OF EVENTS BETWEEN A MOTOR NEURON ACTION POTENTIAL AND CONTRACTION OF A SKELETAL MUSCLE FIBER

1   An action potential is initiated and propagates in the axon of a motor neuron.

2   The action potential causes the release of acetylcholine from the axon terminals at the neuromuscular junction.

3   Acetylcholine diffuses from axon terminals to motor-end-plate membrane.

4   Acetylcholine binds to receptor sites on the motor-end-plate membrane.

5   Bound acetylcholine increases the permeability of the motor end plate to sodium and potassium ions, producing an end-plate potential (EPP).

6   By local current flow the EPP depolarizes the muscle membrane to its threshold potential and thus generates an action potential that then propagates over the surface of the muscle membrane.

7   The action potential propagates from the surface into the muscle fiber along the transverse tubules.

8   Depolarization of transverse tubules leads to the release of calcium ions from the lateral sacs of the sarcoplasmic reticulum that surrounds the myofibrils.

9   These calcium ions bind to troponin on the thin filaments, causing tropomyosin to move away from its blocking position covering the cross-bridge binding sites on actin.

10   The energized myosin cross bridges on the thick filaments bind to actin: $A + M^* {\cdot} ADP {\cdot} P_i \rightarrow A {\cdot} M^* {\cdot} ADP {\cdot} P_i$

11   This binding triggers the release of energy stored in myosin, producing an angular movement of the cross bridge: $A {\cdot} M^* {\cdot} ADP + P_i \rightarrow A {\cdot} M + ADP \times P_i$

12   ATP binds to myosin, breaking the linkage between actin and myosin, thereby allowing the cross bridge to dissociate from actin: $A {\cdot} M + ATP \rightarrow A + M {\cdot} ATP$

13   The ATP bound to myosin is split, transferring energy to the myosin cross bridge and readying it for another cycle: $M {\cdot} ATP \rightarrow M^* {\cdot} ADP {\cdot} P_i$

14   The cross bridges repeat the cycle (10 to 13), leading to the movement of the thin filaments past the thick filaments. These cycles of cross-bridge movement continue as long as calcium remains bound to troponin.

15   The concentration of calcium ions around the myofibrils decreases as calcium is actively transported into the sarcoplasmic reticulum by a membrane pump that uses energy derived from the splitting of ATP.

16   Removal of calcium ions from troponin restores the blocking action of tropomyosin, the cross-bridge cycle ceases, and the fiber relaxes.

---

cle does not contract in response to further nerve stimulation, and the result is skeletal muscle paralysis and death from asphyxiation.

A third group of substances, such as botulinus toxin, produced by the bacterium *Clostridium botulinum,* blocks the release of acetylcholine from the nerve terminals, and thus prevents excitation of the muscle membrane. Botulinus toxin is responsible for one type of food poisoning and is one of the most deadly poisons known. Less than 0.0001 mg is sufficient to kill a human being, and 500 g could kill the entire human population.

One form of neuromuscular disease, known as *myasthenia gravis,* which is associated with skeletal muscle weakness and fatigue, is due to decreased numbers of ACh receptor sites at the motor end plates. The release of ACh from the nerve terminals is normal but the magnitude of the muscle EPP is markedly reduced because of the decreased number of receptor sites. The destruction of the ACh receptors in this disease is brought about by the body's own defense mechanisms gone awry, as will be described in Chap. 28.

The sequence of events leading to contraction of a skeletal muscle fiber is summarized in Table 11-1.

MECHANICS OF MUSCLE CONTRACTION

Contraction refers to the active process of generating a force within a muscle. This force, generated by the cross bridges, is exerted parallel to the muscle fiber. The force exerted by a contracting muscle on an object is known as the muscle **tension,** and the force exerted by the object on the muscle is the **load.** Thus, muscle tension and load are opposing forces. To move a load the muscle tension must be greater than the load.

When a muscle shortens and moves a load, the muscle contraction is said to be **isotonic** (constant tension) since the load on the muscle remains con-

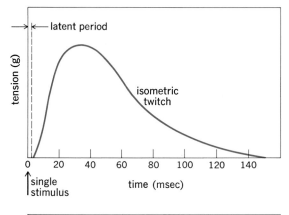

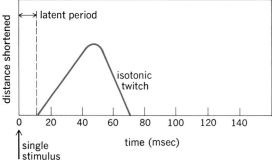

FIG. 11-14 Isometric and isotonic skeletal muscle twitches following a single action potential.

stant and equal to the muscle tension throughout most of the period of shortening. In contrast, when a muscle develops tension but does not shorten, the contraction is said to be **isometric** (constant length). Such contractions occur when the muscle supports a load in a fixed position or attempts to move a load that is greater than the tension developed by the muscle.

The electrical and chemical events occurring in the muscle fibers are the same in both isotonic and isometric contraction, i.e., the cross bridges are activated and exert force on the thin filaments. During an isotonic contraction the thin filaments move past the thick filaments, causing the muscle to shorten, whereas during isometric contractions the cross bridges still exert a force on the thin filaments but there is no overall muscle shortening.

Single Twitch

The mechanical response of a muscle to a *single* action potential is known as a **twitch.** Figure 11-14 shows the main features of an isometric and an isotonic twitch. As mentioned earlier, there is an interval of a few milliseconds, the latent period, before contraction begins. The time interval from the stimulus to the peak contractile response is the *contraction time*. Not all skeletal muscles contract at the same rate. Some "fast" fibers have contraction times as short as 10 msec, whereas slower fibers may take 100 msec or longer. The time from peak contraction until the contraction has decreased to zero is known as the *relaxation time*.

One can see from Fig. 11-14 that the duration of the isotonic twitch is considerably shorter than that of an isometric twitch, whereas the latent period is considerably longer. Once shortening begins, in an isotonic twitch, it proceeds at a constant velocity over about 70 percent of the total distance shortened, as can be seen from the straight-line relation between distance shortened and time. The slope of this line indicates the velocity of shortening. Both the velocity of shortening and the duration of an isotonic twitch depend upon the magnitude of the load being lifted (Fig. 11-15). At heavier loads the latent period lasts longer but the velocity of shortening, the duration of the twitch, and the distance shortened all decrease. During the latent period of an isotonic contraction, the cross bridges begin to develop force but actual shortening does not begin until the muscle tension becomes equal to the load; therefore, the heavier the load, the longer the latent period. The maximum velocity of shortening occurs when there is no load on the muscle. As the load on a muscle is increased, eventually a load will be reached that the muscle is unable to lift, the velocity of shortening will be zero, and the contraction becomes isometric.

FIG. 11-15 Change in the isotonic twitch response of a muscle fiber with different loads.

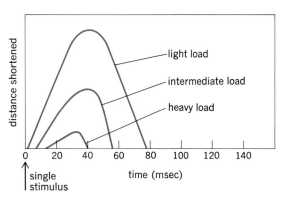

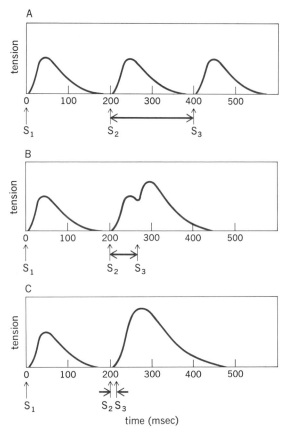

FIG. 11-16 Summation of isometric contractions produced by shortening the time between stimulus S_2 and S_3.

Summation of Contractions

Since a muscle action potential lasts 1 to 2 msec, whereas the mechanical response to it (single twitch) may last for 100 msec, it is possible for a second action potential to be initiated during a period of mechanical activity. Figure 11-16 illustrates the isometric contractions of a muscle in response to three successive stimuli. In Fig. 11-16A the isometric twitch following the first stimulus S_1 lasts 150 msec. The second stimulus S_2, applied to the muscle 200 msec after S_1 when the muscle has completely relaxed, causes a second identical twitch. In Fig. 11-16B the interval between S_1 and S_2 remains 200 msec, but a third stimulus is applied 60 msec after S_2, when the mechanical response resulting from S_2 is beginning to decrease. Stimulus S_3 induces a contractile response whose peak tension is greater than that produced by S_2. In Fig. 11-16C the interval between S_2 and S_3 is further reduced to 10 msec and the resulting peak tension is even greater; the mechanical response to S_3 is a continuation of the mechanical response already induced by S_2.

The increase in the mechanical response of a muscle to action potentials occurring in rapid succession is known as **summation.** The greater the frequency of stimulation, the greater is the intensity of the mechanical response (summation) until a frequency is reached beyond which the response no longer increases (Fig. 11-17). The maximal response to high-frequency stimulation is known as a **tetanus.** This is the greatest tension the muscle can develop and is generally about three to four times greater than the isometric twitch tension produced by a single stimulus. A muscle contracting isotonically can also undergo summation and tetanus, repetitive stimulation leading in this case to an increase in the distance the muscle shortens.

FIG. 11-17 Isometric contractions produced by multiple stimuli of (B) 10 per second and (C) 100 stimuli per second as compared with a single twitch (A).

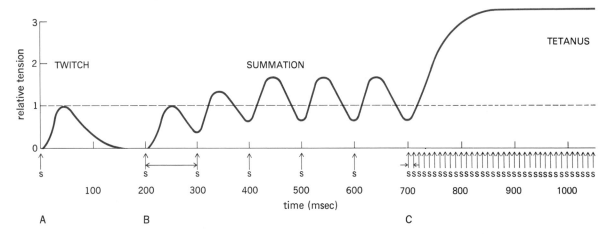

Just as the contraction time of different muscle fibers varies considerably, so does the stimulus frequency that will produce a tetanus. Frequencies of about 30 per second may produce a tetanus in "slow" fibers, whereas frequencies of 100 per second or more are necessary in very rapidly contracting fibers.

Some muscle fibers can maintain a tetanic contraction for extended periods of time, whereas in others the muscle tension declines even though a tetanic frequency of stimulation is maintained. This drop in tension with prolonged stimulation is known as **muscle fatigue.**

What is the mechanism responsible for the increase in the mechanical response of a muscle fiber to repetitive stimulation? The explanation involves the passive elastic properties of the muscle, which cause a dissociation between the *internal* tension generated by the cross bridges and the *external* manifestation of this activity—shortening or tension development by the muscle fiber. Let us first look once more at the internal events. In skeletal muscle fibers, the amount of calcium released from the sarcoplasmic reticulum by a single action potential is sufficient to produce nearly complete saturation

of all troponin sites so that all of the cross bridges are turned "on" and the cross bridges exert their maximal force. (This is probably not the case in cardiac and smooth muscle, as we shall see). This internal tension generated by the filaments as a result of cross-bridge cycling at any instant is known as the **active state** of the muscle fiber. The number of active cross bridges exerting tension on the actin filaments declines as calcium is pumped back into the sarcoplasmic reticulum. Thus, the active state (internal tension) first increases following stimulation and the release of calcium and then declines as calcium is removed. As can be seen from Fig. 11-18A, the time course of the active state is markedly different from that of the external tension being exerted on a load during an isometric twitch. What causes this dissociation?

Tension is transmitted from the cross bridges through the thick and thin filaments, across the Z lines, and eventually through the extracellular connective tissue and tendons to the load on the muscle fiber. All these structures have a certain amount of elasticity and are collectively known as the **series elastic component.** The series elastic component has the properties of a spring placed between the force-

FIG. 11-18 (A) Time course of the active state tension produced by the contractile component (actin and myosin) in comparison with the development of external tension during an isometric twitch. (B) The tension in the contractile component (active state) stretches the series elastic component. It is only the tension in the series elastic component that is transmitted to the exterior of the muscle as external tension.

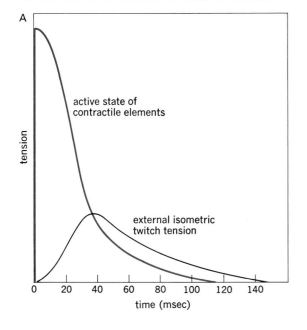

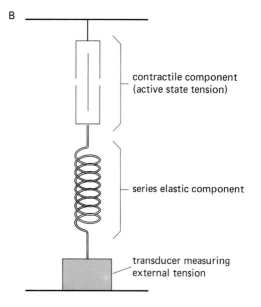

generating cross bridges and the external load. The force generated by the cross bridges stretches this spring (the series elastic component), which in turn transmits its tension to the external load (Fig. 11-18B). The tension in the series elastic component is therefore the external tension and depends on the extent to which the series elastic component has been stretched by cross-bridge activity; the greater the stretch, the greater the tension in the series elastic component and the greater the tension transmitted to the load on the muscle fiber. Here is the crucial point: It takes time to stretch the series elastic component. In a single twitch, the tension in the series elastic component is rising as it is being stretched but at the same time the cross-bridge tension that is stretching the series elastic component is beginning to decline (Fig. 11-18A). Therefore, the peak tension developed by the stretched series elastic component (external twitch tension) is never as great as the maximal cross bridge (active state) tension.

When a muscle is stimulated repetitively, the cross bridges remain active for a longer period of time (i.e., the maximal level of the active state is prolonged) because the repetitive release of calcium from the sarcoplasmic reticulum maintains the required free calcium concentration. Therefore, because more time is available to stretch the series elastic component, it is stretched farther, and the tension transmitted to the load is greater, i.e., summation occurs. In a tetanus, the continuous maintenance of the active state at its maximal level allows sufficient time to fully stretch the series elastic component so that the external tension becomes equal to the force exerted by the cross bridges. To summarize, the increase in external tension (summation and tetanus) that accompanies repeated stimulation is the result of an increase in the length of time the cross bridges remain active, which in turn allows more time to stretch the series elastic component and increase the amount of tension transmitted to the load on the muscle.

Length-Tension Relationship

One of the classic observations in muscle physiology is the relationship between a muscle's length and the tension it develops. A muscle fiber can be passively stretched to various lengths, and the magnitude of the isometric tetanic tension (maximal tension) produced by stimulation measured at each

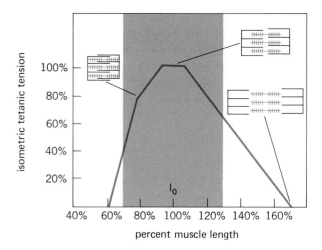

FIG. 11-19 Variation in isometric tetanus tension with muscle fiber length. Maximal tension is produced at muscle length $l_0 = 100$ percent. The shaded band represents the range of length changes (from 70 to 130 percent) that can occur in the body while the muscles are attached to bones.

length (Fig. 11-19). The length at which the muscle develops the greatest tension is termed l_0. When the muscle is set at a length 60 percent of l_0, it develops no tension when stimulated. As the length of the muscle is increased, the developed tension rises to a maximum at l_0 and further lengthening of the muscle causes a drop in tension. When the muscle is stretched to 175 percent of l_0 or beyond, it again develops no tension.

This relationship can be explained in terms of the sliding-filament model. Stretching a muscle changes the amount of overlap between the thick and thin filaments in the myofibrils. Stretching a muscle to 175 percent of l_0 pulls the thick and thin filaments so far apart that there is no overlap between the two, and since there is no overlap there can be no cross-bridge interaction and no tension developed. A larger and larger region of the thick filaments overlaps with thin filaments as the muscle length is decreased from 175 percent l_0 to l_0; the tension developed increases in proportion to the increased number of interacting cross bridges in the overlap region. At l_0 there is a maximal overlap of thick and thin filaments, and tension is maximal. At lengths less than l_0 two mechanical factors lead to decreasing tension: (1) The thin filaments in the two halves of the sarcomere begin to overlap each other, interfering with cross-bridge binding and decreasing the total number of active cross bridges; and (2) the thick filaments become compressed against the two Z lines and resist further shortening of the sarcomeres. In addition to these mechanical factors, at muscle lengths below about 80 percent of l_0 the transmission of action potentials along the t tubules and its triggering of calcium release from the sarcoplasmic reticulum becomes disrupted, preventing full activation of the cross bridges.

In the body, where muscles are attached to bones, the relaxed length of the muscle is very nearly l_0 and thus is optimal for force generation. The total range of length changes that a skeletal muscle can undergo while still attached to bone is limited to about a 30 percent increase or decrease of its resting length and is often much less. Therefore, the length-tension relation contributes, to some extent, to variations in skeletal muscle force in the body, but as we shall see, it plays a more important role in determining the force generated by cardiac muscle. The length of cardiac muscle fibers is not limited by their attachment to bones and thus they can undergo much larger changes in length, leading to corresponding changes in their capacity to develop tension.

CONTROL OF MUSCLE TENSION

The total tension a muscle can develop depends upon two factors: (1) the amount of tension developed by each contracting fiber, which as we have seen depends upon the frequency of action potentials, the length of the muscle fiber, and fatigue; and (2) the number of muscle fibers in the muscle that are contracting at any given time.

The number of muscle fibers contracting at any time depends upon both the number of motor neurons activated and the number of muscle fibers associated with each of them. In other words, one way to alter the total tension a muscle develops is to vary the number of motor units that are activated. An increase in the number of motor neurons that are actively discharging action potentials to a given muscle is known as **recruitment.**

The number of muscle fibers associated with a single motor unit varies considerably in different types of muscles. For example, in an eye muscle, one motor neuron innervates only about 13 muscle fibers while a single motor unit in the large calf muscle of the leg contains about 1700 muscle fibers.

If a muscle is composed of small motor units, the total tension produced by the muscle can be increased in small steps by the recruitment of additional motor units, whereas if the motor units are large, big jumps in tension occur as each additional motor unit is recruited. Thus, finer control of muscle tension can be achieved in muscles with small motor units.

MUSCLE ENERGY METABOLISM

As we have seen, ATP performs three major functions in muscle contraction and relaxation: (1) The energy released from ATP splitting is directly coupled to the movement of the cross bridges; (2) ATP binding to myosin breaks the link between the cross bridges and actin, allowing the cross bridges to operate cyclically; and (3) energy released from ATP splitting is utilized by the sarcoplasmic reticulum to reaccumulate calcium ions, producing relaxation. Without ATP a muscle fiber would be unable to generate force and movement.

Recall that the total quantity of ATP in cells is relatively small (Chap. 3). When a muscle is stimulated, the splitting of ATP by the cycling cross bridges greatly increases the rate of ATP breakdown. The ATP content of a muscle fiber would decline and force-generation would soon cease if ATP were not rapidly resynthesized by the rephosphorylation of ADP. In order to maintain muscle tension, the rate of ATP formation must keep pace with the rate of ATP breakdown. There are three sources of energy that can be used to form ATP from ADP (Fig. 11-20): (1) creatine phosphate, (2) glycolysis, and (3) oxidative phosphorylation. They will be discussed below.

Creatine phosphate (CP) provides the most rapid means of forming ATP in the muscle cell. It contains energy and phosphate, both of which can be transferred to a molecule of ADP to form ATP and creatine (C):

$$CP + ADP \rightleftharpoons C + ATP$$

Energy is stored as creatine phosphate by the reversal of this reaction in resting muscle. Resting fibers build up a concentration of creatine phosphate approximately five times that of ATP. When the ATP level begins to fall at the beginning of contraction and ADP levels rise, mass action favors

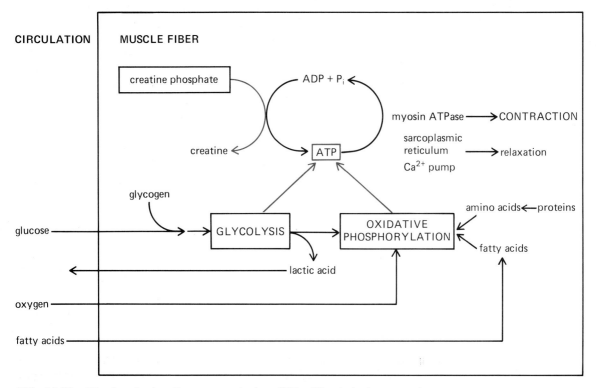

FIG. 11-20   Biochemical pathways producing ATP utilized during muscle contraction.

the formation of ATP from creatine phosphate. This reaction is so efficient that the actual concentration of ATP in the cell changes very little at the start of contraction, whereas the concentration of creatine phosphate falls rapidly.

If contractile activity is to be maintained for any period of time, the muscle must be able to derive ATP from sources other than the limited creatine phosphate stores. At moderate levels of muscle activity (moderate rates of ATP breakdown) most of the additional ATP can be formed by the process of **oxidative phosphorylation** in the mitochondria, with fatty acids and to a lesser extent carbohydrates, providing the predominant source of nutrient (see Chap. 26). However, during very intense exercise, when the breakdown of ATP is very rapid, the cell's ability to replace ATP by oxidative phosphorylation is limited by a number of factors: (1) the delivery of oxygen to the muscle, (2) the availability of nutrients, and (3) the rates at which the enzymes in the metabolic pathways can process these nutrients. Any of these may become rate-limiting for oxidative phosphorylation under appropriate con-

ditions, and the rate at which oxidative phosphorylation can produce ATP may therefore become inadequate to keep pace with the rapid rate of ATP breakdown.

Accordingly, when the level of exercise exceeds about 50 percent of maximum (50 percent of the maximal rate of ATP breakdown), **glycolysis** begins to contribute an increasingly significant fraction of the total ATP produced by the muscle. The glycolytic pathway, although producing only small quantities of ATP from each molecule of glucose metabolized, can operate at a high rate. Not only can glycolysis produce ATP rapidly, but it can proceed in the absence of oxygen. The lactic acid formed during glycolysis diffuses out of muscle into the blood.

Glycolysis, however, has the disadvantage of requiring very large quantities of glucose to produce relatively small amounts of ATP. The ability of muscle to store glucose (in the form of glycogen) provides a certain degree of independence from externally supplied glucose, and during intense exercise the glycogen content of the muscle falls.

In summary, creatine phosphate provides a very rapid mechanism for replacing ATP at the onset of contraction until the metabolic pathways can adjust to the increased demand for ATP production. During mild exercise, the metabolism of fatty acid (and, to a lesser extent, glucose) by oxidative phosphorylation provides most of the additional ATP needed. As the intensity of exercise increases, more and more of the ATP is supplied by glycolysis of internal glycogen stores. **Fatigue** (the decline in the force of muscle contraction that occurs with prolonged stimulation) is not simply a depletion of the muscle ATP content, since there is very little decline in ATP concentration in fatigued muscle as compared to resting muscle. The specific molecular mechanisms responsible for fatigue are unknown.

MUSCLE FIBER DIFFERENTIATION AND GROWTH

Each muscle fiber is formed during embryological development by the fusion of a number of small, mononucleated cells known as *myoblasts* to form a single long, cylindrical, multinucleated fiber. It is only after the fusion of the myoblasts that the muscle fibers begin to form actin and myosin filaments and become capable of contracting. Once the myoblasts have fused, the resulting muscle fibers no longer have the capacity for cell division but do retain the ability to grow in length and diameter. This stage of muscle differentiation is completed around the time of birth. Therefore, in an adult, if skeletal muscle fibers are destroyed, they cannot be replaced by division of other existing, differentiated fibers. There may occur some formation of new fibers from undifferentiated cells in the adult, but the major compensation for a loss of muscle tissue occurs through the increased growth in the size (hypertrophy) of the remaining differentiated fibers.

Following the fusion of the myoblasts in the embryo into muscle fibers, motor neurons send axon processes into the muscle, and thus form neuromuscular junctions and bring the muscle under the control of the central nervous system. From this point on there is a very critical dependence of the muscle fiber on its motor neuron, not only to provide a means of initiating contraction but also for its continued survival and development. If the nerve fibers to a muscle are severed or otherwise destroyed, the denervated muscle fibers become progressively smaller and their content of actin and myosin decreases. This is known as *denervation atrophy* (a muscle can also atrophy with its nerve supply intact if it is not used for a long period of time, as when a broken arm or leg is immobilized in a cast, and this is known as *disuse atrophy*). In contrast to the decrease in muscle mass that results from a lack of regular neural stimulation, increased amounts of neural activity, such as accompany repeated exercise, may produce a considerable hypertrophy of muscle fibers as well as other changes in their chemical composition. Action potentials in nerve fibers appear to release chemical substances that influence the biochemical activities of the muscle fiber, but the identity of these tropic agents is unknown.

TYPES OF SKELETAL MUSCLE FIBERS

Skeletal muscle fibers differ in their speed of contraction and capacity to split ATP. This is because all myosin molecules are not identical, some having a higher ATPase activity than others.

A second major difference between skeletal muscle fibers is the type of enzymatic machinery available for synthesizing ATP. Some fibers contain numerous mitochondria and thus have a high capacity for oxidative phosphorylation; the activity of the glycolytic enzymes in these fibers is relatively low. Therefore, most of the ATP produced by such fibers is dependent upon a supply of oxygen, and they are surrounded by numerous capillaries. These high-oxidative fibers also contain a protein, **myoglobin,** which is similar to hemoglobin, found in red blood cells. Myoglobin binds oxygen and increases the rate of oxygen diffusion into the muscle cell, as well as providing a small store of oxygen in the fiber. Fibers containing large amounts of myoglobin have a dark red color that distinguishes them from the paler fibers that lack appreciable amounts of myoglobin.

In contrast to these high-oxidative fibers, other fibers have few mitochondria but a very high capacity for glycolysis and a large store of glycogen. These fibers are specialized for the production of ATP by glycolysis in the absence of oxygen. Corresponding to their low requirement for

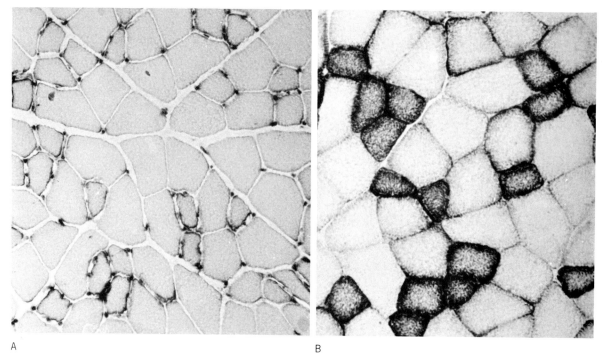

FIG. 11-21 Cross sections of a skeletal muscle containing different types of muscle fibers. (A) The capillaries surrounding the muscle fibers have been stained. Note the large number of capillaries surrounding the small-diameter fibers. (B) Dark staining indicates the presence of oxidative enzymes in the small-diameter fibers.

oxygen, they are surrounded by relatively few capillaries, contain little myoglobin, and have few mitochondria.

Thus, skeletal muscle fibers can be distinguished on the basis of differences in both their ATP-splitting activity and their ATP-synthesizing characteristics.

Some muscles may contain predominantly one type of fiber, but most muscles have a mixture of several types interspersed with each other (Fig. 11-21). All the muscle fibers belonging to a single motor unit are of the same type. Different muscles may contain different proportions of fiber types so that the contractile properties of whole muscles, compared with single muscle fibers, show a range of contraction speeds and fatigability.

The significance of having distinct muscle-fiber types is, in part, that skeletal muscles are called upon to perform various functions. The muscles that support the weight of the body (the postural muscles of the back and legs) must be able to maintain their activity for long periods of time without fatigue. The muscles in the arms may be called

upon to produce large amounts of tension rapidly, as in the lifting of heavy objects. The leg muscles perform the movements of walking and running in addition to supporting the body's weight. Moreover, the activities that a given skeletal muscle may be called upon to perform also vary.

Different types of exercise alter the proportion of fiber types in skeletal muscle. Individuals who perform regular exercises to improve muscle performance must be careful to choose a type of exercise that is compatible with the type of activity they ultimately wish to perform. Thus lifting weights will not improve the endurance of a long-distance runner, and jogging will not produce the increased strength desired by a weight lifter. As we shall see in later chapters, endurance exercise produces changes not only in the skeletal muscles but also in the respiratory and circulatory systems, changes that improve the delivery of oxygen and nutrients to the muscle fibers.

It should be reemphasized that changes in the proportion of fiber types with different types of exercise occur without a change in the total number of fibers. The molecular mechanisms by which changes in the pattern of neural activity influence metabolism and myosin composition are not known.

SMOOTH MUSCLE

Smooth muscle, like skeletal muscle, uses cross-bridge movements between actin and myosin filaments to generate force and calcium ions to control cross-bridge activity. However, the structural organization of the contractile filaments, the process of excitation-contraction coupling, and the time-course of contraction are quite different in the two types of muscle. Furthermore, there is considerable diversity in the properties of smooth muscle in different organs, especially with respect to the mechanism of excitation-contraction coupling. Nevertheless, two anatomical characteristics are common to

all smooth muscles: They lack the cross-striated banding pattern found in skeletal and cardiac fibers (thus the name "smooth muscle"); and the nerves to them are derived from the autonomic division of the nervous system (to be discussed in Chap. 14) rather than the somatic division (thus, smooth muscle is not normally under direct voluntary control).

Unlike skeletal muscle fibers, the precursor cells of smooth muscle fibers do not fuse during embryological development. Each differentiated smooth muscle fiber is spindle-shaped and contains a single nucleus in its central portion. The diameter of these small fibers ranges from 2 to 10 μm, compared to a range of 10 to 100 μm for skeletal muscle fibers. In most hollow organs, the smooth muscle fibers in the walls of the organ are arranged in bundles organized into two layers—an outer longitudinal layer and an inner circular layer. In blood vessels, bundles of muscle fibers are arranged in a circular or helical fashion around the vessel wall.

FIG. 11-22 Electron micrograph of portions of three smooth muscle fibers (insert). Higher magnification of thick filaments with projections (arrows) suggestive of cross bridges connecting adjacent thin filaments.

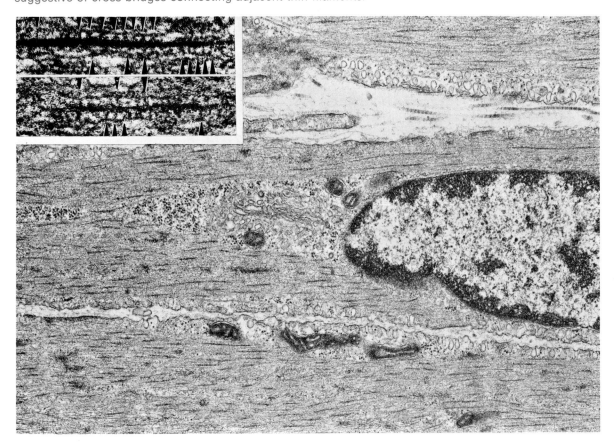

The cytoplasm of smooth muscle fibers is filled with filaments oriented approximately parallel to the long axis of the fiber (Fig. 11-22). Three types of filaments are present: thick myosin-containing filaments, thin actin-containing filaments, and intermediate-sized filaments (the latter filaments do not seem to play a role in the active generation of force and probably function as an elastic framework or cytoskeleton to maintain the shape of the cell). There are no Z or M lines to cross-link filaments of similar type, nor are there sarcomeres or any other regular alignment of the filaments; this accounts for the absence of a banding pattern in smooth muscle fibers. However, the actin filaments are anchored, either to the plasma membrane or to cytoplasmic structures (*dense bodies*). The amount of myosin in smooth muscle is only about one-third that in skeletal muscle, whereas the actin content can be as much as twice that of skeletal muscle.

Excitation-Contraction Coupling

Changes in the calcium concentration around the filaments controls the contractile activity of smooth muscle fibers, as it does in skeletal muscle. However, there are significant differences between the two types of muscle as regards the way in which calcium exerts its effects on cross-bridge activity and the mechanisms by which its intracellular concentration is controlled.

Smooth muscle cells lack troponin, the protein that is the site of calcium regulation of cross-bridge activity in skeletal muscle. In smooth muscle, calcium regulates cross-bridge activity by influencing myosin ATPase activity via **calmodulin,** a calcium-binding protein (which mediates cell responses to changes in intracellular calcium in many types of cells). Once activated by calcium, the rate at which ATP is split is much slower than in skeletal muscle, and this accounts for the relatively slow speed of smooth muscle contraction. A single twitch may last several seconds in smooth muscle but only a fraction of a second in skeletal muscle.

There are two sources of the increased free intracellular calcium that triggers contraction in smooth muscle: Some is released from the sarcoplasmic reticulum and some enters the cell across the plasma membrane. Different types of smooth muscles rely more on one source of calcium than the other, but most make use of both sources to some extent. Relaxation is brought about by the

TABLE 11-2	INPUTS THAT INFLUENCE SMOOTH MUSCLE CONTRACTILE ACTIVITY
1	Spontaneous electrical activity in the smooth muscle plasma membrane
2	Neurotransmitters released by autonomic neurons
3	Hormones
4	Locally induced changes in the chemical composition of the extracellular fluid that surrounds the smooth muscle fibers (paracrines, acidity, oxygen, osmolarity, ion concentrations)
5	Rapid stretching of the smooth muscle

removal of calcium from the cytosol through the action of calcium pumps either in the membrane of the sarcoplasmic reticulum or in the plasma membrane.

In a sense we have approached the question of excitation-contraction coupling backward by first describing the "coupling" (changes in free cytosolic calcium), but now we must ask what constitutes the "excitation" that elicits the changes in calcium.

Contractile activity can be altered in smooth muscle, in response to a variety of inputs, summarized in Table 11-2. (This multiplicity of controls is in contrast to skeletal muscle, in which contractile activity is dependent on a single input—the somatic neurons to the muscle.) All of these inputs have the ultimate effect of altering cytosolic calcium.

Spontaneous Electrical Activity Some types of smooth muscle fibers generate action potentials spontaneously in the absence of any neural or hormonal input. The membrane in fibers exhibiting such spontaneous activity is unstable and does not maintain a constant potential but gradually depolarizes until it reaches the threshold potential and an action potential is initiated. Following repolarization, the membrane again begins to depolarize (Fig. 11-23), leading to a sequence of action potentials, the frequency being determined by the rate at which the membrane depolarizes and the nearness of the average membrane potential to threshold. The spontaneous depolarization to threshold is known as the **pacemaker potential.** (As will be described later, certain cardiac muscle fibers and some neurons in the central nervous system also have pacemaker potentials and can spontaneously generate action potentials.)

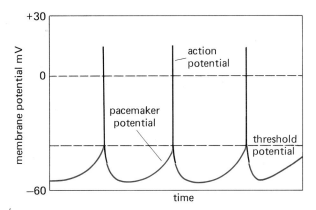

FIG. 11-23 Generation of action potentials in a smooth muscle fiber resulting from spontaneous depolarizations of the membrane potential (pacemaker potentials).

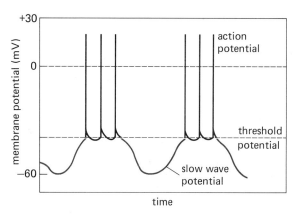

FIG. 11-24 Slow wave oscillations in membrane potential trigger bursts of action potentials in smooth muscle plasma membrane.

In addition to these pacemaker potentials, some types of smooth muscle have even slower oscillations of membrane potential, with cycle times of seconds or even minutes. These **slow wave potentials** arise from spontaneous cyclical changes in the rates at which ions are transported across the plasma membrane by electrogenic pumps (Chap. 5). These pacemaker potentials set off bursts of action potentials that occur at the peaks of the slow wave potentials, i.e., at the membrane potential closest to threshold (Fig. 11-24). It should be noted that pacemaker potentials and slow wave potentials do not occur in all types of smooth muscle cells, nor are slow wave potentials present in all cells that show spontaneous pacemaker activity.

What is the significance of these bursts of action potentials? A twitch produced by a single action potential in smooth muscle is very small compared to a skeletal muscle twitch. Therefore, in order for significant changes in tension to occur, bursts of action potentials leading to summation of mechanical activity and to tetanus are required.

Nerves and Hormones Smooth muscles are sensitive to a variety of hormones and many are innervated by the autonomic division of the nervous system. Unlike skeletal muscle, smooth muscle does not have a single neuromuscular junction for each fiber or a specialized motor-end-plate region. Rather, as the nerve fiber enters the region of smooth muscle tissue, it divides into numerous branches, each branch containing a series of swollen regions or varicosities. Each varicosity contains numerous vesicles filled with neurotransmitter, which is released as an action potential conducted along the nerve fiber passes the varicosity. The concentration of released neurotransmitter at the surface of the muscle fiber's membrane is dependent on the distance between the varicosity and the muscle surface, since the neurotransmitter becomes diluted as it diffuses away from its site of release.

Whereas some hormones and neurotransmitters depolarize smooth muscle membranes, others produce a hyperpolarization that leads to inhibition of contractile activity. Thus, in contrast to skeletal muscle, which receives only excitatory input from its motor neurons, smooth muscle can be either excited or inhibited by neural activity (and by circulating hormones). Moreover, a given neurotransmitter or hormone may produce opposite effects in different smooth muscle tissues. For example, the neurotransmitter norepinephrine depolarizes certain types of vascular smooth muscle, thereby initiating contraction of the muscle; whereas in intestinal smooth muscle, it inhibits spontaneous activity and produces a relaxation of the muscle. Thus, the nature of the response depends on events initiated by the binding of the chemical agent to receptor sites in the membrane and is not an inherent property of the agent itself.

Local Chemical Factors Local factors, including paracrines, acidity, oxygen concentration, osmolarity, and the ion composition of the extracellular fluid, may produce changes in the smooth muscle membrane potential and/or alter the permeability of the membrane to calcium and the release of calcium from the sarcoplasmic reticulum by mechanisms independent of potential changes. This responsiveness to local stimuli provides a direct means for altering smooth muscle tension in accord with changes in the metabolic activity of the surrounding tissues. Such responses play an important role in the local regulation of the internal environment in a tissue undergoing moment-to-moment fluctuations in activity. In contrast, neural and hormonal inputs to smooth muscle generally regulate contractile activity so as to control the internal environment of the body as a whole.

Classification of Smooth Muscle

The great diversity of smooth muscle characteristics has made it difficult to classify the different types found in various organs. One of the most general classifications divides smooth muscles into two groups based on the excitability characteristics of the plasma membranes and on the conduction of electrical activity from fiber to fiber by way of gap junctions (see Chap. 4) between adjacent fibers: (1) **single-unit smooth muscles** that are capable of propagating action potentials from fiber to fiber and may manifest spontaneous action potentials; and (2) **multiunit smooth muscles** that exhibit little, if any, propagation of electrical activity from fiber to fiber and whose contractile activity is primarily regulated by neural input. The smooth muscles of the intestinal tract, the uterus, and small-diameter blood vessels are predominantly *single-unit* smooth muscles. The smooth muscles in the large airways to the lungs, in large arteries, and those attached to the hairs in the skin are predominantly *multiunit* smooth muscles.

SUMMARY

Table 11-3 compares some of the properties of the different types of muscle. Cardiac muscle has been included for completeness, although its properties have not been discussed in this chapter (see Chap. 20). In brief, cardiac muscle has a filament organization very similar to that of skeletal muscle, with myofibrils, sarcomeres, transverse tubules, and a well-developed sarcoplasmic reticulum. Like single-unit smooth muscle, some cardiac muscle fibers act as pacemakers to generate action potentials spontaneously; the action potentials are then propagated throughout the heart by way of gap junctions between the cardiac fibers.

TABLE 11-3 CHARACTERISTICS OF MUSCLE FIBERS

Characteristic	Skeletal Muscle	Smooth Muscle Single-Unit	Multiunit	Cardiac Muscle
Thick and thin filaments	yes	yes	yes	yes
Sarcomeres—banding pattern	yes	no	no	yes
Transverse tubules	yes	no	no	yes
Sarcoplasmic reticulum (SR)	+ + +	+	+	+ +
Gap junctions between fibers	no	yes	few	yes
Source of activating calcium	SR	SR & extra-cellular	SR & extra-cellular	SR & extra-cellular
Site of calcium regulation	troponin (thin filaments)	myosin (thick filaments)	myosin (thick filaments)	troponin (thin filaments)
Speed of contraction	fast-slow	very slow	very slow	slow
Spontaneous production of action potentials by pacemakers	no	yes	no	yes
Slow wave potentials may be present	no	yes	no	no
Effect of nerve stimulation	excitation	excitation or inhibition	excitation or inhibition	excitation or inhibition
Physiological effects of hormones on excitability and contraction	no	yes	yes	yes

KEY TERMS

muscle fibers	troponin and tropomyosin	isotonic and isometric contractions
myofibrils	excitation-contraction coupling	twitch and tetanus contractions
sarcomere	sarcoplasmic reticulum	muscle fatigue
thick and thin filaments	transverse tubule	length-tension relation
actin and myosin	motor unit	creatine phosphate
cross bridges	neuromuscular junction	single-unit and multiunit smooth
sliding-filament model	end-plate potential	muscle

REVIEW EXERCISES

1 Identify the three types of muscle tissue in the body; give examples of the location of each and the types of input that initiate contraction.

2 Describe the organization of skeletal muscle fibers within a muscle and their attachment to bones.

3 Describe the arrangement of the thick and thin filaments and their associated connections within a sarcomere that give rise to the striated-muscle banding patterns known as the A band, I band, Z line, M line, and H zone.

4 Describe the movements of the thick and thin filaments during sarcomere shortening.

5 Describe the arrangement of actin and myosin molecules in the thin and thick filaments, respectively.

6 Summarize the mechanical and chemical events involving actin, myosin, and ATP that occur during one cross-bridge cycle.

7 Describe the role of tropomyosin, troponin, and calcium ions in regulating cross-bridge activity.

8 Describe the sequence of events that occur during excitation-contraction coupling in skeletal muscle fibers.

9 Describe the role of the sarcoplasmic reticulum in skeletal muscle contraction and relaxation.

10 Define the term *motor unit*.

11 Describe the transmission of electrical activity from a motor neuron to a muscle fiber at a neuromuscular junction.

12 Describe several ways in which the events at a neuromuscular junction can be altered by disease and drugs.

13 Describe the difference between an isometric and an isotonic contraction.

14 Describe the effects during an isotonic twitch of increasing the load on: (a) the latent period, (b) the contraction time, and (c) the shortening velocity.

15 Define the terms *summation* and *tetanus*.

16 Explain the mechanism responsible for summation and tetanus in terms of the active state and the series elastic component.

17 Describe the mechanism responsible for the changes in the active isometric tension that occur when a muscle is passively stretched to different muscle lengths prior to stimulation.

18 List the factors that determine the total tension developed by a skeletal muscle composed of many muscle fibers.

19 Describe the role of creatine phosphate in muscle energy metabolism.

20 List the three processes by which ADP can be phosphorylated to form ATP during muscle contraction.

21 Describe the differences between adult muscle fibers and embryonic myoblasts.

22 Compare the effects on muscle fiber size of (a) long periods of decreased nerve activity to the muscle and (b) long periods of increased nerve activity to the muscle.

23 Compare the structure of smooth muscle with that of skeletal muscle.

24 Compare the mechanism of calcium activation of cross-bridge activity in smooth and skeletal muscle.

25 Identify the two sources of calcium ions that contribute to the activation of smooth muscle contraction.

26 List the various types of input signals to smooth muscle that can alter the intracellular calcium concentration and thus influence contractile activity.

27 Define the terms *pacemaker potential* and *slow wave potential*.

28 Define and compare the characteristics of single-unit and multiunit smooth muscle.

CHAPTER

THE MUSCULAR SYSTEM

MUSCLES: ATTACHMENTS AND FIBER ARRANGEMENTS

The contraction of skeletal muscles causes the movement of bones at joints or, as in the face, the movement of skin. The two ends of a skeletal muscle are usually attached via tendons to the bones that meet at the joint. In other cases, they attach to mucous membrane (as do the muscles of the tongue); to sheets of connective tissue known as *fascia;* or, in the circular bands of sphincters and the several orbicularis muscles, to other muscles. Attachment via tendons, a *fibrous attachment,* allows the force of the muscle contraction to be concentrated on a small region of bone. Other types of attachment, the *fleshy attachments,* spread the muscle's force over a wider area. Regardless, the attachment is always via a blending of the connective-tissue elements of the muscle with those of the structure to which the muscle is attached.

Contraction of a sphincter-type muscle narrows an orifice. When the attachment is to bone, contraction of the muscle produces relatively more movement of one bone than of the other. The muscle attachment to the bone of lesser movement is called the muscle's **origin;** that to the bone of greater movement is the **insertion.**

The force that a muscle generates via its attachment depends on the number and thickness of its fibers, while the *maximum range* (i.e., amount of shortening) depends on the fibers' lengths. The fiber arrangement within the muscle is also important. Fascicles in a muscle may be *parallel, oblique,* or *spiral* relative to the muscle's line of pull (Fig. 12-1). Muscles with parallel fascicles may be flat, short, and rectangular or straplike. Straplike muscles, which are long and thin, are particularly useful in situations requiring a large range of motion and little power, as in stretching. When the fascicles are

FIG. 12-1 Types of muscle based on general form and arrangement of fascicles. Note that not all muscles have demonstrable tendons of origin and insertion.

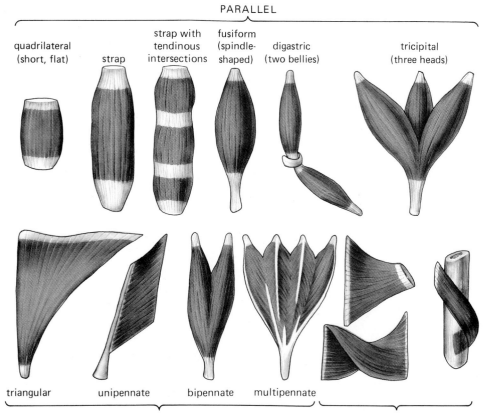

PARALLEL

quadrilateral (short, flat)   strap   strap with tendinous intersections   fusiform (spindle-shaped)   digastric (two bellies)   tricipital (three heads)

triangular   unipennate   bipennate   multipennate

OBLIQUE   SPIRAL

oblique to the muscle's line of pull, the muscles are classed as triangular or pennate (featherlike). Pennate muscles frequently have a short range of motion, but the large number of fibers can generate considerable force. Spiral muscles apply rotational forces to the bones to which they are attached. Many muscles have components that fit more than one of these categories.

MECHANICS OF MOVEMENTS AT JOINTS

A contracting muscle exerts a force on bones through its connecting tendons. The effect brought about by this contraction is the muscle's **action.** When the force is great enough, the bone moves as the muscle shortens. A contracting muscle exerts only a pulling force; as the muscle shortens, the bones attached to it are pulled toward each other. **Flexion** of a limb is its bending or movement toward the body and **extension** is straightening or movement away from the body. These motions require at least two separate muscles, one to cause flexion and the other, extension. From Fig. 12-2 it can be seen how contraction of the biceps causes flexion

of the forearm and contraction of the triceps causes its extension. Both muscles exert a pulling force upon the forearm when they contract, but they pull in opposite directions. Groups of muscles which produce oppositely directed movements of a joint are known as **antagonists.** Other sets of antagonistic muscles are required to cause side-to-side movements or rotation at a joint.

In some muscles, contraction under different conditions leads to different types of joint movement. For example, in Fig. 12-3 contraction of the large muscle in the calf of the leg (the gastrocnemius) normally acts on the heel to cause plantar flexion, i.e., to raise the heel, shifting the body

FIG. 12-3 Flexion of the leg or extension of the foot follows contraction of the gastrocnemius muscle, depending upon the activity of the quadriceps femoris muscle.

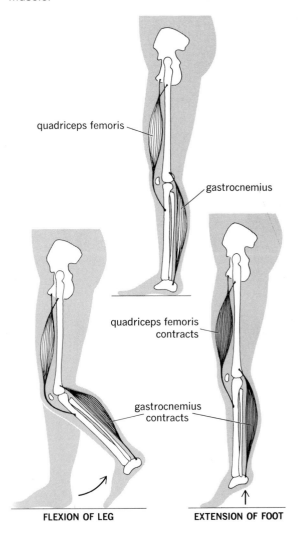

FIG. 12-2 Antagonistic muscles for flexion and extension of the forearm.

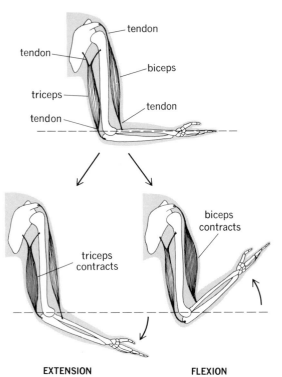

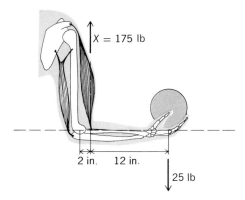

FIG. 12-4 Mechanical equilibrium of forces acting on a forearm supporting a 25-lb load.

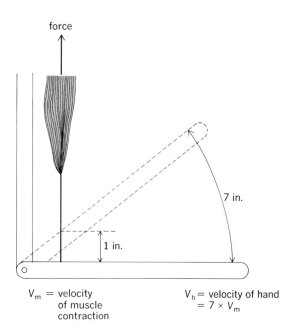

V_m = velocity of muscle contraction V_h = velocity of hand = 7 × V_m

FIG. 12-5 Small movements of the biceps muscle are amplified by the lever system of the arm, producing large movements of the hand.

weight to the toes, as in walking. When one is standing quietly and the heel is relatively firmly stabilized, the gastrocnemius acts to prevent the forward bending at the ankle that would occur whenever the weight of the body is centered ahead of the ankle joint. Moreover, contraction of the gastrocnemius simultaneously with the quadriceps femoris (which causes extension of the lower leg) prevents the knee from bending, leaving only the ankle capable of moving; the foot is extended, and the body rises on tiptoe.

The arrangements of the muscles, bones, and joints in the body form lever systems, as can be seen, for example, with the flexion of the forearm by the biceps muscle (Fig. 12-4). The biceps exerts an upward pulling force on the forearm about 2 in away from the elbow. A 25-lb weight held in the hand exerts a downward force of 25 lb about 14 in from the elbow. According to the laws of physics, in a lever system a rigid body (the forearm) is in mechanical equilibrium (not accelerating) if the product of the downward force (25 lb) and its distance from the elbow (14 in) is equal to the product of the upward force exerted by the muscle (X) and its distance from the elbow (2 in). From this relationship one can calculate the force the biceps must exert on the forearm to support a 25-lb load: 25 × 14 = 2X, thus X = 175 lb. Such a system is working at a mechanical disadvantage since the force exerted by the muscle is considerably greater than the load it is supporting.

Contracting muscles exert much more force on the bones than the physical weight of the body does. During quiet standing, the forces exerted by the muscles on the hip joint are six times greater

than the forces applied by the body's weight. In powerful contraction, the forces exerted by the muscles are even larger. In athletes with very powerful muscles, the large forces exerted by contracting muscles under conditions of maximum exertion sometimes tear the tendon away from the muscle or bone, or, in rare cases, break the bone.

The mechanical disadvantage under which most muscles operate is offset by increased maneuverability. In Fig. 12-5, when the biceps shortens 1 in, the hand moves through a distance of 7 in. Since the muscle shortens 1 in in the same amount of time that the hand moves 7 in, the velocity at which the hand moves is seven times faster than the rate of muscle contraction. The lever system of the arm amplifies the movements of the muscle. Short, relatively slow movements of the muscle produce longer and faster movements of the hand. Thus, a pitcher can throw a baseball at 100 mi/hr even though his muscles shorten at only a fraction of this velocity. Skeletal muscles shorten at the rate of about 5 to 10 muscle lengths per second. Thus, the longer the muscle, the faster is its velocity of shortening.

This completes our analysis of the general characteristics of the musculoskeletal system. We will now describe the specific muscles of the body.

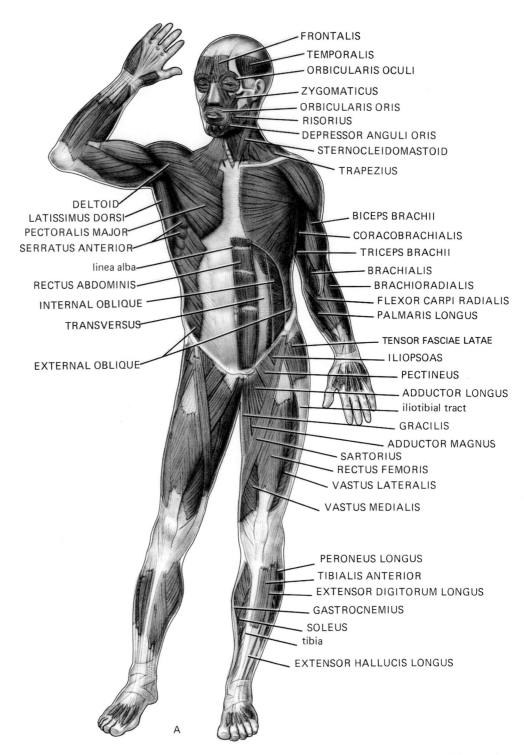

FRONTALIS
TEMPORALIS
ORBICULARIS OCULI
ZYGOMATICUS
ORBICULARIS ORIS
RISORIUS
DEPRESSOR ANGULI ORIS
STERNOCLEIDOMASTOID
TRAPEZIUS

BICEPS BRACHII
CORACOBRACHIALIS
TRICEPS BRACHII
BRACHIALIS
BRACHIORADIALIS
FLEXOR CARPI RADIALIS
PALMARIS LONGUS

TENSOR FASCIAE LATAE
ILIOPSOAS
PECTINEUS
ADDUCTOR LONGUS
iliotibial tract
GRACILIS
ADDUCTOR MAGNUS
SARTORIUS
RECTUS FEMORIS
VASTUS LATERALIS
VASTUS MEDIALIS

PERONEUS LONGUS
TIBIALIS ANTERIOR
EXTENSOR DIGITORUM LONGUS
GASTROCNEMIUS
SOLEUS
tibia
EXTENSOR HALLUCIS LONGUS

DELTOID
LATISSIMUS DORSI
PECTORALIS MAJOR
SERRATUS ANTERIOR
linea alba
RECTUS ABDOMINIS
INTERNAL OBLIQUE
TRANSVERSUS
EXTERNAL OBLIQUE

A

FIG. 12-6 Major muscles of the body: (A) anterior view and (B) posterior view.

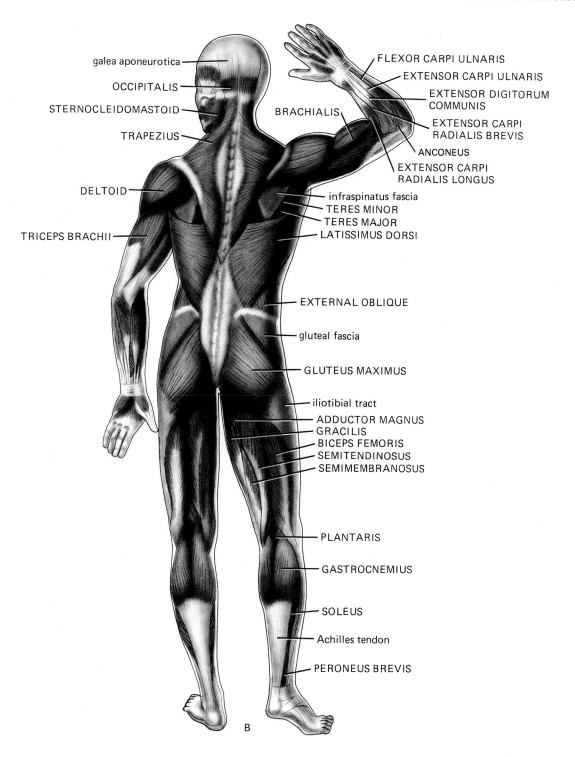

galea aponeurotica

OCCIPITALIS

STERNOCLEIDOMASTOID

TRAPEZIUS

DELTOID

TRICEPS BRACHII

BRACHIALIS

FLEXOR CARPI ULNARIS

EXTENSOR CARPI ULNARIS

EXTENSOR DIGITORUM COMMUNIS

EXTENSOR CARPI RADIALIS BREVIS

ANCONEUS

EXTENSOR CARPI RADIALIS LONGUS

infraspinatus fascia

TERES MINOR

TERES MAJOR

LATISSIMUS DORSI

EXTERNAL OBLIQUE

gluteal fascia

GLUTEUS MAXIMUS

iliotibial tract

ADDUCTOR MAGNUS

GRACILIS

BICEPS FEMORIS

SEMITENDINOSUS

SEMIMEMBRANOSUS

PLANTARIS

GASTROCNEMIUS

SOLEUS

Achilles tendon

PERONEUS BREVIS

B

SKELETAL MUSCLES OF THE BODY

Some of the major muscles of the body are illustrated in Fig. 12-6. Their number and names may seem formidable, but the task of learning them is simplified because the shapes, sizes, locations, attachments, and actions of the muscles are often used in naming them. Thus, the *trapezius* (trapezoidal in form) is named for its shape, the *gluteus maximus* (*gloutos,* buttock; *maximus,* largest) for its size, the *interossei* (*inter,* between; *ossa,* bones) for their location, the *coracobrachialis* (which passes between the coracoid process of the scapula and the arm, or brachium) for its attachments, and

FIG. 12-7 Muscles of the face, anterior view. The superficial muscles are shown on the left half of the figure, the deep muscles on the right.

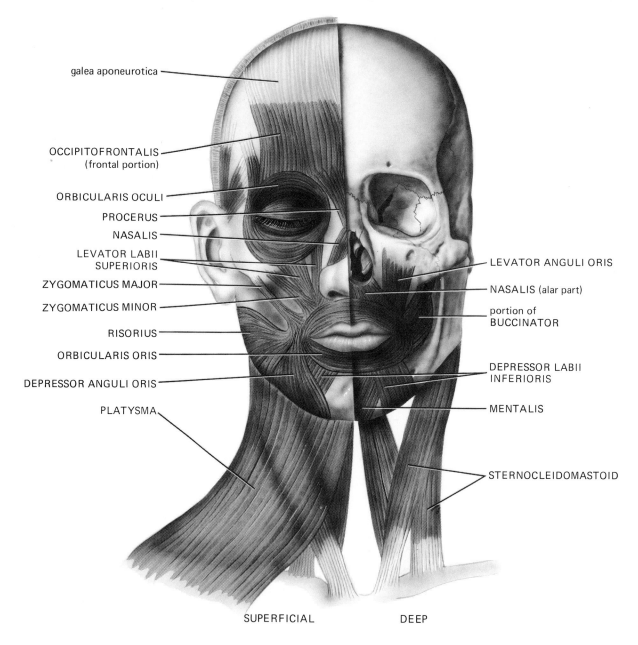

galea aponeurotica

OCCIPITOFRONTALIS
(frontal portion)

ORBICULARIS OCULI

PROCERUS

NASALIS

LEVATOR LABII
SUPERIORIS

ZYGOMATICUS MAJOR

ZYGOMATICUS MINOR

RISORIUS

ORBICULARIS ORIS

DEPRESSOR ANGULI ORIS

PLATYSMA

LEVATOR ANGULI ORIS

NASALIS (alar part)

portion of
BUCCINATOR

DEPRESSOR LABII
INFERIORIS

MENTALIS

STERNOCLEIDOMASTOID

SUPERFICIAL   DEEP

the *levator scapulae* (which raises the scapula) for its action. In addition, muscles are often named for more than one of their attributes; for example, the *flexor digitorum superficialis* is named for its action (flexing the fingers; *digitis,* finger) and its location (above the deeper lying *flexor digitorum profundus; profundus,* deep). Thus, knowing the reason behind a particular muscle's name will often help in learning about that muscle, and these reasons are generally indicated in the tables throughout this chapter. Moreover, we remind you that the markings of the bones, which were described and illustrated in seemingly endless detail in Chap. 9, serve mostly as the origins and insertions of muscles, and we encourage you to refer to that chapter where relevant in your study of the muscles.

Finally, in any discussion of the muscles, the material can be organized in one of two ways: by *location* or by *function.* A grouping by location allows easier visualization of the specific muscles, and we have organized the major muscle tables in this way. It is sometimes useful, however, to know which muscles act on a given joint, and we have included tables of those muscles involved in move-

ments of the shoulder, elbow, wrist, hip, knee, and ankle.

Muscles of the Head

Many of the muscles of the face (Figs. 12-7 and 12-8) serve facial expression, one of the most revealing outward signs of inner emotional states. These muscles differ from most skeletal muscles in that while their origin is most frequently on bone, their insertion is on skin and its underlying connective tissue. The facial muscles are generally thin and flat and are usually named for their actions. In fact, if you study the muscles' positions in Fig. 12-7, their actions become apparent and (usually) their names logical. The muscles of the face are listed in Table 12-1 according to their location.

Other muscles move the eyes and control the intricate lip and cheek movements associated with speech, eating, and drinking. The muscles that control the eyeball will be listed in Table 15-1, and those of the tongue and pharynx in Figs. 12-10C and 25-14. The muscle of the scalp (**occipitofrontalis,** or **epicranius**) contains two frontal and two

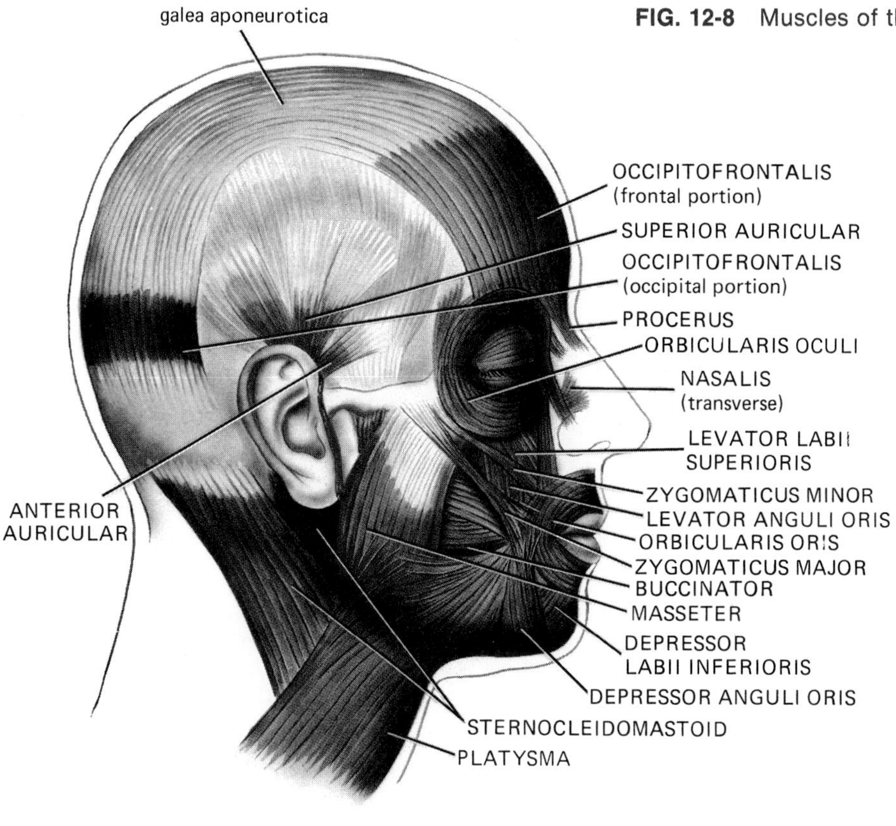

FIG. 12-8 Muscles of the face, lateral view.

galea aponeurotica

OCCIPITOFRONTALIS
(frontal portion)
SUPERIOR AURICULAR
OCCIPITOFRONTALIS
(occipital portion)
PROCERUS
ORBICULARIS OCULI
NASALIS
(transverse)
LEVATOR LABII
SUPERIORIS
ZYGOMATICUS MINOR
LEVATOR ANGULI ORIS
ORBICULARIS ORIS
ZYGOMATICUS MAJOR
BUCCINATOR
MASSETER
DEPRESSOR
LABII INFERIORIS
DEPRESSOR ANGULI ORIS
STERNOCLEIDOMASTOID
PLATYSMA

ANTERIOR
AURICULAR

TABLE 12-1 MUSCLES OF THE FACE

Muscle	Origin	Insertion	Action and Comments
Muscles of the Scalp (Epicranius) (*epi*, on; *cranium*, skull) Occipitofrontalis (*occiput*, back of head; *frontal*, forehead)			
Occipital parts (Fig. 12-8)	Superior nuchal line of the occipital bone	Galea aponeurotica	Draws scalp backward
Frontal parts (Figs. 12-7 and 12-8)	Galea aponeurotica	Skin of the eyebrows, and the root of the nose	Elevates eyebrows, wrinkles forehead, draws scalp forward
Auricular (*auricle*, ear)			
Anterior (Fig. 12-8)	Temporal fascia	Auricle of ear	
Superior (Fig. 12-8)	Temporal fascia	Auricle of ear	
Posterior	Mastoid process of temporal bone	Auricle of ear	
Muscles of the Eyelids Orbicularis oculi (*orbis*, round; *oculus*, eye) (Figs. 12-7 and 12-8)	Frontal, maxillary, and zygomatic bones and medial palpebral ligament	Skin around eye, lateral palpebral ligament	An elliptical muscle that occupies the eyelids, surrounds the orbit, and spreads onto the temporal region and cheek; closes eyelids
Corrugator supercilii (*corruga*, wrinkle together; *supercilium*, eyebrow)	Brow ridge of frontal bone	Skin of eyebrow	Draws eyebrows together; forms vertical wrinkles in forehead above nose
Muscles of the Nose Procerus (Figs. 12-7 and 12-8)	Lower part of nasal bone	Skin between the eyebrows	Forms horizontal wrinkles across bridge of nose
Nasalis (*nasus*, nose) (Figs. 12-7 and 12-8)	Maxilla next to incisor and canine teeth	Bridge and side of nose	Has transverse and alar parts; widens anterior nasal aperture, especially in deep inspiration
Depressor septi (*depressor*, lowers; *septum*, divides nasal cavity into two chambers)	Maxilla	Septum of nose	Draws septum downward

occipital parts in the forehead and occipital regions, respectively. They are joined by a strong membranous sheet, the *galea aponeurotica,* which is separated from the bone of the calvaria by loose connective tissue so that the scalp is movable. Sustained, involuntary contraction of the epicranius muscles accounts for most of the pain of tension headaches, probably the most common form of headache encountered. The shape of the mouth and position of the lips are controlled by elevators and retractors of the upper lip, depressors of the lower lip, and a sphincter, the **orbicularis oris,** which surrounds the mouth opening.

TABLE 12-1   MUSCLES OF THE FACE (*Continued*)

Muscle	Origin	Insertion	Action and Comments
Muscles of the Mouth			
Levator labii superioris (*levator*, raises; *labium*, lip; *superioris*, upper) (Figs. 12-7 and 12-8)	Maxilla	Upper lip	Raises upper lip and turns it outward
Zygomaticus (*zygomatic*, bone in cheek)			
Minor (Figs. 12-7 and 12-8)	Zygomatic bone	Upper lip	Elevates upper lip
Major (Figs. 12-7 and 12-8)	Zygomatic arch	Muscles and skin at angle of mouth	Draws angle of mouth upward and laterally
Levator anguli oris (*angulis*, corner; *oris*, mouth) (Figs. 12-7 and 12-8)	Canine fossa of maxilla	Muscles at angle of mouth	Raises angle of mouth
Mentalis (*mentum*, chin) (Fig. 12-7)	Incisor fossa of mandible	Skin of chin	Raises and protrudes lower lip, wrinkles skin of chin
Depressor labii inferioris (*inferioris*, lower) (Figs. 12-7 and 12-8)	Mandible	Skin and muscles of lower lip	Draws lower lip downward and a little laterally
Depressor anguli oris (Figs. 12-7 and 12-8)	Mandible	Muscles at angle of mouth	Draws angle of mouth downward and laterally
Buccinator (*bucca*, cheek) (Figs. 12-7, 12-8, and 12-10)	Mandible and maxilla in region near molars	Muscles at angle of mouth	Compresses cheeks against teeth; provides a stable lateral wall to oral cavity for pressure in speech, sucking, and mastication
Orbicularis oris (Figs. 12-7 and 12-8)	Maxilla and muscle fibers surrounding mouth	Fibers encircle mouth; some attach to skin and muscles at angle of mouth	Closes lips, presses lips against teeth, protrudes lips, and shapes lips in speech
Risorius (*risor*, laughter) (Figs. 12-7 and 12-8)	Fascia of masseter muscle	Skin at angle of mouth	Retracts angle of mouth
Levator labii superioris alaequae nasi (*alae*, wing) (Figs. 12-7 and 12-8)	Frontal process of maxilla	Skin of lip and ala of nose	Draws upper lip upward and widens nostril

Muscles of the Jaw   The muscles of the jaw attach to the mandible (lower jaw) and are known as the muscles of mastication (chewing), or mandibular muscles (Fig. 12-9 and Table 12-2). The muscles that pass between the mandible and hyoid bone (the *suprahoid muscles*, discussed under Muscles of the Neck) may assist in movements of the jaw but are not commonly included as muscles of mastication.

The **masseter** is largely covered by the parotid salivary gland posteriorly and by the facial muscles anteriorly, and it overlies the ramus of the mandible. The large fan-shaped **temporal muscle** covers much of the side of the head, and its posterior fibers maintain the resting position of the closed mouth. The *masseter, temporal,* and *medial pterygoid muscles* are powerful jaw closers, accounting for the

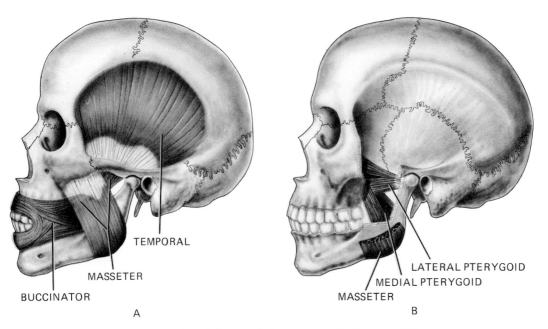

FIG. 12-9 Muscles of mastication, left lateral view. In part B, the mandible and zygomatic arch have been cut and the masseter and temporal muscles partially or completely removed to show the underlying pterygoid muscles.

strength of the bite, and the same three muscles assist in lateral movements which, when combined with closure of the jaw, produce the grinding movements associated with chewing. During grinding, the tongue positions the food on the teeth, and the **buccinator muscle** helps keep the food there.

Muscles of the Neck and Trunk

Muscles of the Neck   The neck consists of an anterior region, the **cervix** (thus the vertebrae in the neck are *cervical vertebrae),* and a posterior region, the **nucha.** The cervix portion itself is further di-

TABLE 12-2 MUSCLES OF THE JAW			
Muscle	**Origin**	**Insertion**	**Action and Comments**
Masseter (*maseter,* chewer) (Fig. 12-9)	Zygomatic arch	Angle of mandible	Elevates mandible, closing jaw; small effect in lateral movements to same side or protrusion
Temporal (*temporal,* temples) (Figs. 12-8 and 12-9)	Temporal fossa	Coronoid process and ramus of mandible	Elevates mandible, closing jaw; draws mandible backward after protrusion, assists in lateral movements to same side
Lateral pterygoid (*lateral,* further from midline; *pterygion,* wing) (Fig. 12-9)	Greater wing of the sphenoid bone and lateral pterygoid plate	Tissues of temporomandibular joint and neck of mandible	Assists in opening mouth, protrusion of jaw, and lateral movements to the opposite side
Medial pterygoid (*medial,* toward the midline) (Fig. 12-9)	Pterygoid plate, palatine bone, and maxilla	Ramus and medial surface of angle of mandible	Assists elevation and protrusion of mandible and lateral movements to the opposite side

220

TABLE 12-3   MUSCLES OF THE NECK (*Continued on following page*)

Muscle	Origin	Insertion	Action and Comments
Superficial and Lateral Cervical Muscles			
Platysma (*platysma*, platelike) (Figs. 12-7 and 12-8)	Fascia covering pectoralis major and deltoid muscles	Body of mandible and skin and subcutaneous tissue of lower part of face	Wrinkles skin in neck, assists in lowering mandible, draws down lower lip and angle of mouth
Trapezius (Described with scapular muscles; see Table 12-9)			
Sternocleidomastoid (*sternum*, breastbone; *cleidal*, pertaining to the clavicle; *mastoid*, pertaining to mastoid process of the temporal bone) (Figs. 12-10A, B and 12-18)	Manubrium of sternum and clavicle	Mastoid process of temporal bone	*Singly:* tilts head toward shoulder of same side, rotates face to opposite side; *together:* draw head forward, flex cervical part of vertebral column; if head is fixed, they help raise the thorax in forced inspiration
Suprahyoid Muscles (*supra,* above, *hyoid,* hyoid bone)			
Digastric (*di,* two; *gaster,* belly) (Fig. 12-10A, B)	Mastoid portion of temporal bone, symphysis of mandible	Hyoid bone, occasionally mandible	Depresses mandible, can elevate hyoid bone
Stylohyoid (*stylo,* stake or pole, pertaining to styloid process of temporal bone) (Fig. 12-10A, B)	Styloid process of temporal bone	Hyoid bone	Elevates the hyoid bone and draws it back, elongating the floor of the mouth; can stabilize hyoid bone for tongue muscle action
Mylohyoid (*mylo,* pertaining to molar teeth) (Fig. 12-10A, C)	Mylohyoid line on mandible	Hyoid bone and median raphe	Elevates floor of mouth in the first phase of swallowing; elevates hyoid bone or depresses mandible
Geniohyoid (*geneion,* chin) (Fig. 12-10C)	Inferior mental spine on mandible	Body of hyoid bone	Elevates hyoid bone and draws it forward; depresses mandible when hyoid bone is fixed
Infrahyoid Muscles (*infra,* below)			As a group, these are antagonists to the suprahyoid muscles
Sternohyoid (Fig. 12-10A, B)	Manubrium of the sternum	Body of hyoid bone	Depresses hyoid bone after it has been raised, as in swallowing
Sternothyroid (*thyroid,* thyroid cartilage, Adam's apple) (Fig. 12-10B)	Manubrium of sternum and cartilage of first rib	Lamina of thyroid cartilage	Draws larynx downward after it has been elevated, as in swallowing
Thyrohyoid (*thyro,* pertaining to thyroid cartilage) (Fig. 12-10A, B)	Lamina of the thyroid cartilage	Body of hyoid bone	Depresses hyoid bone or raises larynx
Omohyoid (*omos,* shoulder) (Fig. 12-10A, B)	Upper border of scapula	Body of hyoid bone	Depresses hyoid bone after it has been elevated

TABLE 12-3   MUSCLES OF THE NECK (*Continued*)

Muscle	Origin	Insertion	Action and Comments
Anterior Vertebral Muscles			
Longus colli (*longus*, long; *collum*, neck)			
Lower oblique part	Bodies of T1–3 vertebrae	Transverse processes of C5–6 vertebrae	Bend neck forward, flex it laterally, and rotate it to the opposite side
Upper oblique part	Transverse processes of C3–5 vertebrae	Anterior arch of atlas	
Vertical part	Bodies of C5–7 and T1–3 vertebrae	Bodies of C2–4 vertebrae	
Longus capitis (*caput*, head)	Transverse processes of C3–6 vertebrae	Basilar part of occipital bone	Flexes the head
Rectus capitis anterior (*rectus*, straight; *anterior*, before, on the front part of the body)	Lateral mass of atlas	Basilar part of occipital bone	Flexes the head
Rectus capitis lateralis (*lateralis*, at the side)	Transverse process of atlas	Jugular process of occipital bone	Bends the head to the same side
Lateral Vertebral Muscles			
Anterior scalene (*skalenus*, uneven) (Fig. 12-10A)	Transverse processes of C3–6 vertebrae	First rib	When first rib is stabilized, bends cervical portion of vertebral column forward and laterally and rotates it in the opposite direction; when cervical vertebrae are stabilized, assists elevation of the first rib in respiration
Middle scalene (Fig. 12-10A)	Transverse processes of C2–7 vertebrae	First rib	When first rib is stabilized, bends cervical part of vertebral column to same side; when cervical vertebrae are stabilized, helps raise first rib in respiration

vided into *anterior* and *posterior triangles,* and the muscles we discuss initially, the *supra-* and *infrahyoid muscles,* are in the anterior one. However, the muscles of the neck (Fig. 12-10 and Table 12-3) are often continuous with those of the head, trunk, or upper limb, and a clear demarcation of "neck muscles" is not possible.

The muscles that connect the hyoid bone with the mandible are the **suprahyoid muscles** (Fig. 12-10 and Table 12-3). They are classified as muscles of the neck despite the fact that the **mylohyoid** actually

forms the floor of the mouth, filling much of the space between the two sides of the body of the mandible. The submandibular salivary gland lies in the triangle formed by the two bellies of the **digastric muscle** and the lower edge of the mandible, and it extends upward deep to the mandible. The suprahyoid muscles raise the hyoid bone, tongue, and floor of the mouth. While elevating the hyoid bone, they elevate the larynx as well because it is attached to the hyoid by ligaments; this action is important in swallowing because it helps prevent food from

TABLE 12-3 MUSCLES OF THE NECK (Continued)

Muscle	Origin	Insertion	Action and Comments
Posterior scalene (*posterior,* towards the back of the body)	Transverse processes of C4–6 vertebrae	Second rib	Bends lower part of the cervical portion of the vertebral column to the same side if second rib is fixed; when cervical vertebrae are stabilized, helps elevate the second rib in respiration
Intrinsic Muscles of the Larynx			
Cricothyroid (*crico,* pertaining to the cricoid cartilage) (Fig. 12-10C)	Outer surface of the cricoid cartilage	Inferior horn and lower border of the thyroid cartilage	Regulates tension of vocal cords
Cricoarytenoid (*arytenoid,* pertaining to the arytenoid cartilage)			
Posterior (Fig. 23-7)	Lamina of the cricoid cartilage	Arytenoid cartilage	Opens the glottis, separates the vocal folds, increases tension on the vocal folds
Lateral (Fig. 23-7)	Arch of the cricoid cartilage	Arytenoid cartilage	Closes the glottis, approximates the vocal processes
Arytenoid Transverse (Fig. 23-7)	Arytenoid cartilage	The opposite arytenoid cartilage	An unpaired muscle; bridges the space between the two arytenoid cartilages, approximating them and thus closing the glottis
Oblique	Arytenoid cartilage	The opposite arytenoid cartilage	Acts as a sphincter to the inlet of the larynx
Thyroarytenoid (Fig. 23-7)	Angle of the thyroid cartilage and the cricothyroid ligament	Arytenoid cartilage	Shortens and relaxes the vocal ligaments; the lower, deeper fibers of this form a bundle known as the *vocalis*

passing into the lungs. If, however, the *infrahyoid muscles* (those below the hyoid) are activated at the same time as the suprahyoid, the hyoid bone is steadied to form a firm base for the tongue, and the suprahyoid muscles then assist in opening the mouth. The suprahyoid muscles are also active in singing high notes, when they move the hyoid and, therefore, the larynx, up. (In contrast, the infrahyoid muscles are active when low notes are sung.)

The **infrahyoid muscles** are often referred to as the "strap muscles." They, with one of the suprahyoid muscles (the *geniohyoid*), represent continuations into the neck of the muscle mass that, in the abdomen, forms the *rectus abdominis*. The infrahyoid muscles lower the larynx and hyoid bone and, as mentioned previously, when acting with the suprahyoids, make firm the position of the hyoid.

The anterior and posterior triangles are separated by the **sternocleidomastoid muscle,** which passes obliquely upward from its origins on the sternum and clavicle to insert behind the ear on the mastoid process of the temporal bone. Lying in part beneath the sternocleidomastoid and in part in the posterior triangle are the three **scalene muscles** (anterior, middle, and posterior) and behind them a muscle of the shoulder, the **levator scapulae.**

Closely bound to the front of the vertebral bodies and transverse processes of the cervical vertebrae are the complex **longus colli** and **longus capitis mus-**

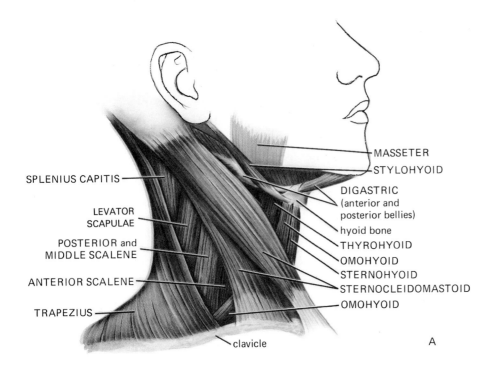

SPLENIUS CAPITIS

LEVATOR
SCAPULAE

POSTERIOR and
MIDDLE SCALENE

ANTERIOR SCALENE

TRAPEZIUS

clavicle

MASSETER
STYLOHYOID

DIGASTRIC
(anterior and
posterior bellies)

hyoid bone
THYROHYOID
OMOHYOID
STERNOHYOID
STERNOCLEIDOMASTOID
OMOHYOID

A

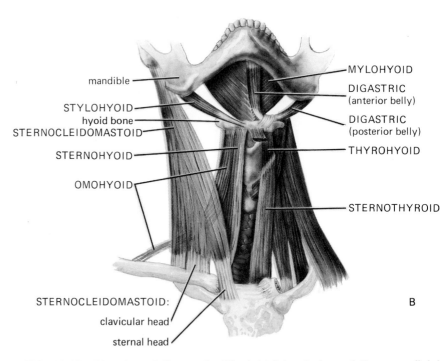

mandible

STYLOHYOID
hyoid bone
STERNOCLEIDOMASTOID

STERNOHYOID

OMOHYOID

MYLOHYOID
DIGASTRIC
(anterior belly)

DIGASTRIC
(posterior belly)

THYROHYOID

STERNOTHYROID

STERNOCLEIDOMASTOID:

clavicular head

sternal head

B

FIG. 12-10   Muscles of the neck: (A) right lateral view of the superficial muscles; (B) an anterior view showing the supra- and infrahyoid muscles (the clavicle, sternohyoid, and sternocleidomastoid have been cut on the right side); (C) right lateral view showing the muscles of the pharynx (parts of overlying bones and muscles have been removed).

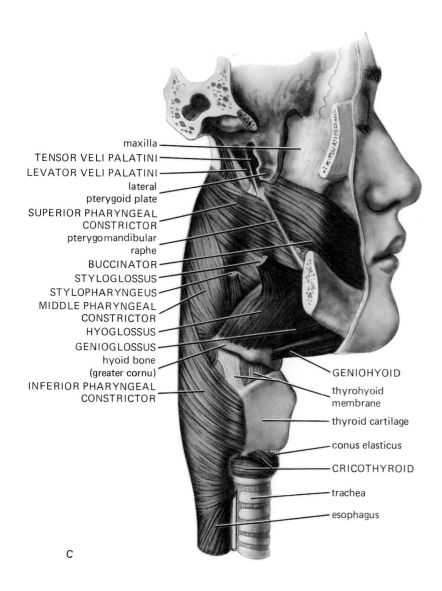

maxilla
TENSOR VELI PALATINI
LEVATOR VELI PALATINI
lateral
pterygoid plate
SUPERIOR PHARYNGEAL
CONSTRICTOR
pterygomandibular
raphe
BUCCINATOR
STYLOGLOSSUS
STYLOPHARYNGEUS
MIDDLE PHARYNGEAL
CONSTRICTOR
HYOGLOSSUS
GENIOGLOSSUS
hyoid bone
(greater cornu)
INFERIOR PHARYNGEAL
CONSTRICTOR

GENIOHYOID
thyrohyoid
membrane
thyroid cartilage
conus elasticus
CRICOTHYROID
trachea
esophagus

C

cles (not shown). They overlap, the longus colli being the lower muscle and the longus capitis the upper. They are primarily flexors of the head and neck. Also not shown are the deep anterior vertebral muscles that pass from one vertebra to another or from the vertebral column to the skull. These muscles flex the head on the neck or, contracting on only one side, tip the head toward that side.

Muscles of the Back   The true muscles of the back are largely covered by the muscles of the shoulder that extend to the vertebral column and partly by muscles of the ribs (the **serratus posterior muscles**). The fascia connective-tissue sheet covering the deep muscles in the upper part of the thorax is continuous with that in the neck and is generally quite thin, but in the lower back (the *thoracolumbar fascia*) it is much thicker and consists of not only fascia but also **aponeuroses** (very broad, flat tendons that, because of their thinness and breadth, resemble fascia).

In general, the deep muscles of the back (Figs. 12-11 and 12-12 and Table 12-4) are particularly complex because while they technically run only from one vertebra to the next, there is a great deal of fusion of these short segments to form long muscles. These muscles, moreover, also split into a

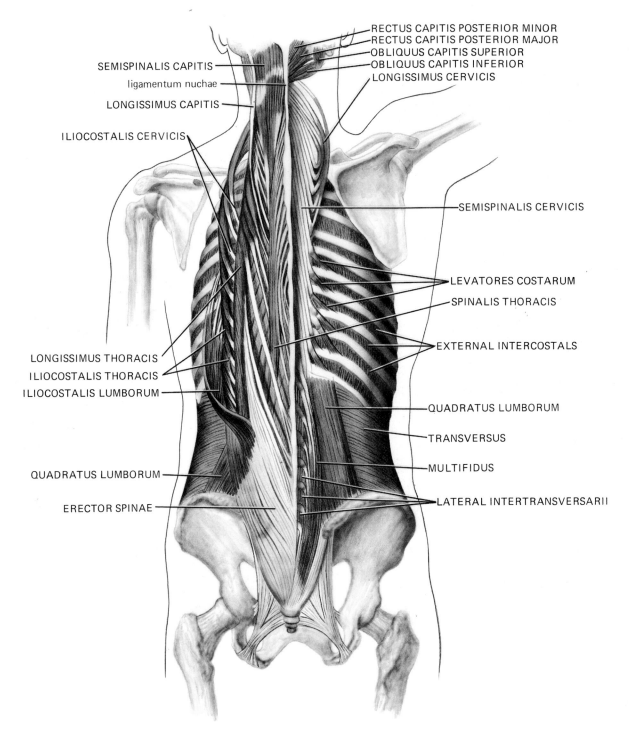

FIG. 12-11   Deep muscles of the back. The semispinalis capitis and erector spinae and some of its upward extensions have been removed on the right. The longissimus cervicis has been displaced laterally.

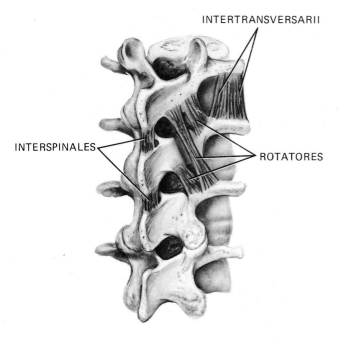

INTERTRANSVERSARII

INTERSPINALES

ROTATORES

FIG. 12-12 Section of the vertebral column showing the short muscles of the back and their relation to the vertebrae.

number of superimposed layers. Furthermore, these muscles (except for the short, very deep-lying ones) have multiple origins and insertions, which in turn results in muscles of varying lengths.

One of the major muscles of the back is the **erector spinae,** which lies in the angles between the spine and transverse processes of the vertebrae. As it ascends through upper lumbar regions, it breaks into three columns, the lateral *iliocostalis,* an intermediate *longissimus,* and a medial *spinalis muscle.* When the muscles on both sides act together, they straighten or extend the vertebral column and hold it upright; acting singly, they twist the column. The **suboccipital muscles** (Table 12-5) extend the head at the atlantooccipital joints and rotate the head and atlas on the axis. The muscles of the thorax connect

TABLE 12-4   DEEP MUSCLES OF THE BACK (*Continued on following page*)

Muscle	Origin	Insertion	Action and Comments
Splenius (*splenion*, bandage)			
Splenius capitis (*caput*, head) (Figs. 12-10A and 12-18)	Ligamentum nuchae, spine of C7 and T1–4 vertebrae	Mastoid process of temporal bone	Acting together: draw head backward; separately: draw head to one side and slightly rotate it, turning the face to the same side
Splenius cervicis (*cervix*, neck)	Spines of T3–6 vertebrae	Transverse processes of C1–3 vertebrae	
Iliocostalis (*ilium*, pertaining to the flank, or ilium)			
Iliocostalis lumborum (*lumbus*, pertaining to the loin) (Fig. 12-11)	Iliac crest and sacrospinal aponeurosis	Lower six or seven ribs	Extend the vertebral column and flex it laterally
Iliocostalis thoracis (*thorax*, chest) (Fig. 12-11)	Lower seven ribs	Upper seven ribs and transverse process of C7 vertebra	
Iliocostalis cervicis (Fig. 12-11)	3–6 ribs	Transverse processes of C4–6 vertebrae	

TABLE 12-4   DEEP MUSCLES OF THE BACK (*Continued*)

Muscle	Origin	Insertion	Action and Comments
Longissimus (*longissimus*, longest) Longissimus thoracis (Fig. 12-11)	Transverse processes of the L1–2 and T6–12 vertebrae and the thoracolumbar fascia as far as the crest of the ilium and sacrum	Transverse processes of thoracic and lumbar vertebrae and lower borders of ribs	Bends vertebral column backward and laterally
Longissimus cervicis (Fig. 12-11)	Transverse processes of T1–5 vertebrae	Transverse processes of C2–6 vertebrae	Bends vertebral column backward and laterally
Longissimus capitis (Fig. 12-11)	Transverse processes of T1–5 and articular processes of C4–7 vertebrae	Mastoid process of the temporal bone	Extends the head and turns the face toward the same side
Spinalis thoracis (*spinalis*, sharp process, pertaining to the spinal column) (Fig. 12-11)	Spines of T11–12 and L1–2 vertebrae	Spines of upper thoracic vertebrae	Extends the vertebral column; blends with longissimus thoracis and semispinalis thoracis
Semispinalis (*semi*, half) Semispinalis thoracis	Transverse processes of T6–10 vertebrae	Spines of T1–4 and C6–7 vertebrae	Extends thoracic portion of vertebral column and rotates it toward the opposite side
Semispinalis cervicis (Fig. 12-11)	Transverse processes of T1–6 vertebrae	Spines of C2–5 vertebrae	Extends cervical portion of vertebral column and rotates it toward the opposite side
Semispinalis capitis (Fig. 12-11)	Transverse processes of T1–7 and C7 vertebrae and the articular processes of C4–6	Occipital bone between superior and inferior nuchal line	Extends the head and turns the face slightly toward the opposite side
Multifidus (*multi*, many; *findere*, to split; *multifidus*, split into many parts) (Fig. 12-11)	Processes of vertebrae from C4 to sacrum	Spinous processes of vertebrae two to four segments above origin	Extends vertebral column and rotates it toward opposite side
Rotatores (*rotare*, to turn) (Fig. 12-12)	Processes of vertebrae from C4 to the sacrum	Spinous process of vertebra adjacent to or second from origin	Extends vertebral column and rotates it toward opposite side

TABLE 12-4   DEEP MUSCLES OF THE BACK (*Continued*)

Muscle	Origin	Insertion	Action and Comments
Interspinales (minor) (*inter*, between; *spines*, sharp projections) (Fig. 12-12)	Spinous processes of C and L vertebrae	Spinous process of vertebra adjacent to origin	Extends vertebral column; steadies vertebrae during movement of the vertebral column as a whole
Intertransversarii (minor) (*transverse*, pertaining to the transverse projections of the vertebrae) (Fig. 12-12)	Transverse processes of C and L vertebrae	Transverse process of adjacent vertebra	Flexes vertebral column laterally; steadies vertebrae during movement of the vertebral column as a whole

TABLE 12-5   SUBOCCIPITAL MUSCLES

Muscle	Origin	Insertion	Action and Comments
Rectus capitis posterior (*rectus*, straight; *caput*, head; *posterior*, toward the back)			
Major (*major*, larger) (Fig. 12-11)	Spine of the axis	Inferior nuchal line of the occipital bone	Extends the head and turns the face toward the same side
Minor (*minor*, smaller) (Fig. 12-11)	Posterior arch of the atlas	Below the inferior nuchal line of the occipital bone	Extends the head
Obliquus capitis (*oblique*, slanting)			
Inferior (*inferior*, lower, more caudal) (Fig. 12-11)	Spine and lamina of the axis	Transverse process of the atlas	Turns the face toward the same side
Superior (*superior*, higher, more rostral) (Fig. 12-11)	Lower four ribs	Occipital bone between the superior and inferior nuchal lines	Bends the head backward and to the same side

the ribs to each other or to the vertebrae. They are all involved in movements of the ribs and are active during respiration.

Deep to these muscles are small ones that run between spinous processes or transverse processes of adjacent vertebrae (Fig. 12-12). Short muscles (*multifidus, rotatores, interspinales,* and *intertrans-* *versarii*) function for the most part as postural muscles. They control the movement of the vertebrae relative to one another so segments of the column can be stabilized during movements of the whole column. In this way, the short muscles of the back ensure the efficient action of the long muscles.

TABLE 12-6 MUSCLES OF THE THORAX

Muscle	Origin	Insertion	Action and Comments
Intercostals (*inter,* between; *costa,* rib)			Thin layers of muscle and tendon which occupy the spaces between the ribs; external intercostals are more superficial than internal intercostals
External (*external,* closer to the surface) (Figs. 12-11 and 12-13)	Lower border of one rib	Upper border of the rib below the rib of origin	Eleven pairs; possibly elevate the ribs
Internal (*internal,* deeper) (Fig. 12-13)	Costal groove of one rib	Upper border of the rib below the rib of origin	Eleven pairs; oriented obliquely with their fibers at right angles to those of the external intercostals; probably depress the ribs; maintain integrity of the intercostal space and reduce the space between the ribs in respiration
Subcostals (*sub,* below)	Inner surface of one rib	Inner surface of the second or third rib below the rib of origin	Their fibers are parallel to the internal intercostals; like them they probably depress the ribs; well developed only in lower thorax
Transversus thoracis (*transversus,* crosswise; *thorax,* chest)	Body of the sternum, the xiphoid process and costal cartilages of lower ribs	Costal cartilages of the second to sixth ribs	Draws down the costal cartilage to which it is attached
Levatores costarum (*levare,* to raise) (Fig. 12-11)	Transverse process of C7 and T1–11 vertebrae	Upper edge of the rib below the vertebrae of origin	Elevate the ribs, but their role in respiration is disputed; also rotate and flex vertebral column
Serratus posterior (*serra,* saw; *posterior,* toward the back) Superior (*superior,* higher, more rostral)	Ligamentum nuchae, spines of C7 and T1–3 vertebrae, and supraspinous ligament	Second through fifth ribs	Can elevate ribs, but their role in respiration is not clear
Inferior (*inferior,* lower, more caudal) (Fig. 12-18)	Spines of T11–12 and L1–3 vertebrae and the supraspinous ligament	Lower four ribs	Draws lower ribs downward and backward, but possibly not in respiration; stabilizes lower ribs in respiration
Diaphragm (*dia,* across; *phragma,* fence)	Xiphoid process, cartilages and lower six ribs on each side, the lumbocostal arches (arcuate ligaments), and the lumbar vertebrae	Central tendon of the diaphragm	Dome-shaped sheet that separates the thoracic and abdominal cavities; principal muscle of inspiration, the contracting fibers draw the central tendon downward

Muscles of the Thorax   The actual muscles of the thoracic wall are the **intercostals,** the **subcostal,** and the **transversus thoracis,** although many other muscles attach to the ribs, including some from the upper arm, some of the long muscles of the back, the *serratus posterior muscles,* the *levatores costarum,* muscles of the abdomen, and the *diaphragm* (Table 12-6). Muscles of the thorax are discussed with muscles of the back, abdomen, and shoulder girdle.

Muscles of the Abdominal Wall   The anterolateral abdominal muscle group consists of three broad, flat sheets (the **internal oblique,** the **external oblique,** and the **transversus**) and a strap-like muscle (the **rectus abdominis**), which lies on each side of the midline. (Figs. 12-11, 12-13, and Table 12-7). Two smaller muscles (the *cremaster* and *pyramidalis*),

which suspend the testes and tense the abdominal fascia, lie in the inguinal (groin) region. The paired muscles of the posterior abdominal wall are the **quadratus lumborum muscles** (Fig. 12-11).

The aponeuroses of the anterolateral muscles meet at the anterior midline where the tendon fibers interlace to form a line, the *linea alba,* which extends from the sternum to the symphysis pubis. The *inguinal ligament* is formed from the lower border of the aponeurosis of the external oblique muscle. It passes between the anterior superior iliac spine and the pubic tubercle and marks the separation between the abdominal wall and the leg. Above and parallel to the inguinal ligament is the inguinal canal, a slanted opening in the lower abdominal wall, which gives passage in the male to the spermatic cord.

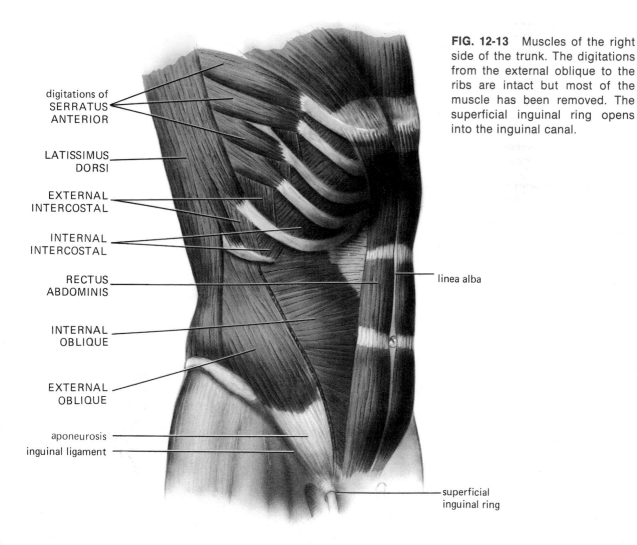

digitations of
SERRATUS
ANTERIOR

LATISSIMUS
DORSI

EXTERNAL
INTERCOSTAL

INTERNAL
INTERCOSTAL

RECTUS
ABDOMINIS

INTERNAL
OBLIQUE

EXTERNAL
OBLIQUE

aponeurosis

inguinal ligament

linea alba

superficial
inguinal ring

FIG. 12-13   Muscles of the right side of the trunk. The digitations from the external oblique to the ribs are intact but most of the muscle has been removed. The superficial inguinal ring opens into the inguinal canal.

TABLE 12-7   MUSCLES OF THE ABDOMEN

Muscle	Origin	Insertion	Action and Comments
External oblique (*m. obliquus externus abdominis*) (*external*, closer to the surface; *oblique*, slanting) (Figs. 12-13 and 12-18)	Lower eight ribs	Iliac crest and spine; pubic tubercle and crest; interdigitates with muscle of opposite side (linea alba)	Rotates and flexes vertebral column; tenses abdominal wall
Internal oblique (*m. obliquus internus abdominis*) (*internal*, deeper) (Figs. 12-13 and 12-18)	Inguinal ligament, iliac crest, and thoracolumbar fascia	Lower three or four ribs, aponeurosis; pubic crest; interdigitates with same muscle of opposite side (linea alba)	Rotates and flexes vertebral column; tenses abdominal wall
Cremaster (*kremasthai*, to suspend)	Inguinal ligament, internal oblique and transversus muscles	Pubic tubercle	Pulls the testes upward; although its fibers are striated, it is usually not under voluntary control
Transversus (*transversus*, crosswise) (Fig. 12-11)	Inguinal ligament, iliac crest, and the thoracolumbar fascia, ribs 7–12	Interdigitates with muscle of opposite side (linea alba), conjoint tendon to pubis	Supports abdominal viscera
Rectus abdominis (*rectus*, straight; *abdominis*, pertaining to the abdomen) (Fig. 12-13)	Xiphoid process and fifth to seventh costal cartilage	Pubic crest and symphysis pubis	Flexes vertebral column, tenses abdominal wall
Quadratus lumborum (*quadratus*, square; *lumbus*, pertaining to the loin) (Fig. 12-11)	Iliac crest, transverse processes of L1–4 vertebrae	Twelfth rib, transverse processes of L1–4 vertebrae	Stabilizes the last rib, steadies the origin of the diaphragm; may flex vertebral column if pelvis is fixed

Muscles of the Pelvis and Perineum   The lower part of the pelvis (the *inferior pelvic aperture* or *pelvic outlet,* see Fig. 9-32) is closed by a sheet of muscles, the **pelvic diaphragm,** which forms most of the rounded bottom of the pelvic basin. The muscles that form the pelvic diaphragm are the **coccygeus** and the **levator ani** (Fig. 12-14). They extend from one side of the pelvis to the other, attaching to various viscera, such as the terminal portion of the large intestine (the rectum) and the tube leading from the urinary bladder out of the body (the urethra), that pass through the pelvic diaphragm. The levator ani is important in voluntary control of micturition (urination) and, in the female, support of the uterus. Two pairs of muscles, the *obturator internus* and the *piriformis* muscles, are associated with the pelvic outlet but are properly listed with the muscles of the lower limb.

Below the levator ani, and therefore, not part of the pelvic diaphragm, are the **perineal muscles** (Figs. 12-15 and 12-16, and Table 12-8), which are associated with the anal canal and the urinary and reproductive tracts. The **perineum** is that region at the base of the trunk and outlet of the pelvis bounded on the outside of the body by the scrotum in the male and the mons veneris in the female in

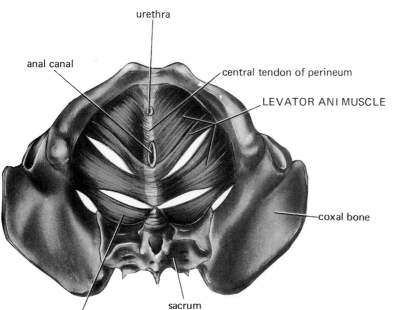

urethra

anal canal

central tendon of perineum

LEVATOR ANI MUSCLE

coxal bone

COCCYGEUS MUSCLE

sacrum

FIG. 12-14 The pelvic diaphragm of the male viewed from below.

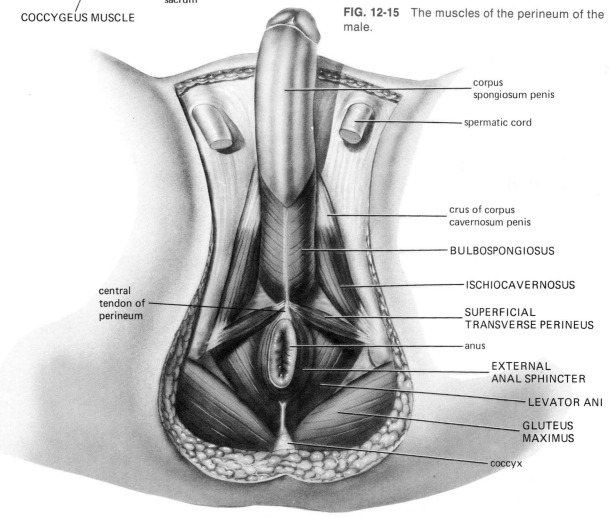

FIG. 12-15 The muscles of the perineum of the male.

corpus spongiosum penis

spermatic cord

crus of corpus cavernosum penis

BULBOSPONGIOSUS

ISCHIOCAVERNOSUS

SUPERFICIAL TRANSVERSE PERINEUS

anus

EXTERNAL ANAL SPHINCTER

LEVATOR ANI

GLUTEUS MAXIMUS

coccyx

central tendon of perineum

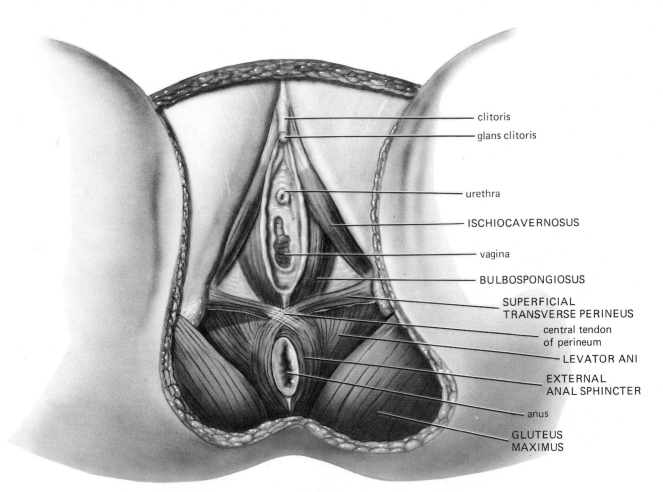

FIG. 12-16 The muscles of the perineum of the female.

TABLE 12-8	MUSCLES OF THE PELVIS AND PERINEUM (*Continued on following page*)

Muscle	Origin	Insertion	Action and Comments
Muscles of the Pelvis Piriformis (See Table 12-19) Obturator internus (See Table 12-19)			
Levator ani (*levare,* to raise; *ani,* pertaining to the anus) (Figs. 12-14, 12-15, and 12-16)	Dorsal surface of the pubis	Spine of the ischium	Forms great part of floor of the pelvic cavity; constricts the lower end of the rectum and, in the female, of the vagina
Coccygeus (*coccygeal,* pertaining to the coccyx) (Figs. 12-14 and 12-15)	Spine of the ischium and supraspinous ligament	Margin of the coccyx and side of the sacrum	Draws the coccyx forward and raises the pelvic floor
External anal sphincter (*sphincter ani externus*) (*external,* closer to the surface; *sphinkter,* ringlike binder) (Figs. 12-15 and 12-16)	Anteriorly, ends over the bulbospongiosus muscle	Posteriorly, ends in subcutaneous tissue over the coccyx	The *superficial portion* lies just under the skin of the anus, the *deep portion* circles the anal canal and internal anal sphincter; composed of skeletal muscle tissue

TABLE 12-8   MUSCLES OF THE PELVIS AND PERINEUM (*Continued*)

Muscle	Origin	Insertion	Action and Comments
Male Urogenital Region (*uro,* pertaining to the urinary system; *genital,* pertaining to the reproductive organs)			
Transverse perineus (*transversus,* crosswise; *perineal,* pertaining to the perineum) Superficial (*superficial,* closer to the surface) (Fig. 12-15)	Tuberosity of the ischium	Perineal body	Fixes central tendon, thereby stabilizing muscles inserting on it
Deep	Fascia over ischial ramus	Perineal body	Compresses urethra
Bulbospongiosus (*bulbus,* bulb, pertaining to the bulb of the penis; *spongia,* sponge) (Fig. 12-15)	Median raphe and perineal body	Fascia of erectile tissue of the penis	Compresses urethra and can stop urination; aids erection by compressing veins draining penis
Ischiocavernosus (*ischion,* ischium, hip; *caverna,* hollow place) (Fig. 12-15)	Tuberosity of the ischium	Aponeurosis of bulb of penis	Aids erection by compressing bulb of penis
Sphincter urethrae (*urethrae,* pertaining to the urethra)	Perineal ligament and fascia of pudendal vessels; inferior pubic ramus	Central tendinous raphe	Compresses urethra, essential for the voluntary control of urination
Female Urogenital Region Transverse perineus Superficial (Fig. 12-16)	Tuberosity of the ischium	Perineal body	Stabilizies central tendon and muscles that anchor upon it
Deep	Fascia over the ramus of the ischium	Joins muscle of opposite side, some fibers join wall of vagina	Compresses urethra and vagina
Bulbospongiosus (Fig. 12-16)	Perineal body	Blends with external anal sphincter	Surrounds the vaginal opening and acts as a sphincter of vaginal opening
Ischiocavernosus (Fig. 12-16)	Tuberosity and ramus of the ischium	Aponeurosis of clitoris	Compresses veins draining clitoris, thereby aiding erection
Sphincter urethrae	Transverse perineal ligament	Joins muscle of opposite side or wall of vagina	Compresses urethra and vagina, essential for the voluntary control of urination

front, by the medial surface of the thighs laterally, and by the buttocks behind. The perineum is below the outlet of the pelvic girdle and therefore contains not only the perineal muscles but also much of the genitalia; on the exterior, it contains the external genitalia with the openings of the urinary, reproductive, and intestinal tracts. Several of the perineal muscles come together in the midline between the urogenital and anal regions at a fibromuscular mass, the *central tendon of the perineum.*

Muscles of the Shoulder Girdle and Upper Limb

The anterolateral muscles of the thoracic wall and those in the superficial region of the back are included here because of their relation to the arm and shoulder girdle. The parts of the anterolateral thorax involved in movements of the shoulder girdle and arm include the *pectoral region,* whose contour, at least in the male, is formed by the large **pectoralis major** muscle. The pectoral region is further divided into a *mammary* (breast) *region,* an *axillary* (armpit) *region,* and an *infraclavicular* (below the clavicle, or collarbone) *region.* On the back of the body, the muscles involved in the shoulder extend from the occipital region of the head to the sacral level of the vertebral column. They include muscles in the region of the shoulder blade (the *scapular region*) as well as the regions above and below it (*suprascapular* and *infrascapular regions,* respectively) and the *lumbar region.*

The upper limb itself is divided into several regions including the *acromion* at the tip of the shoulder (overlying the acromion of the scapula); the *axilla,* or armpit; the *brachium,* or arm (upper arm);

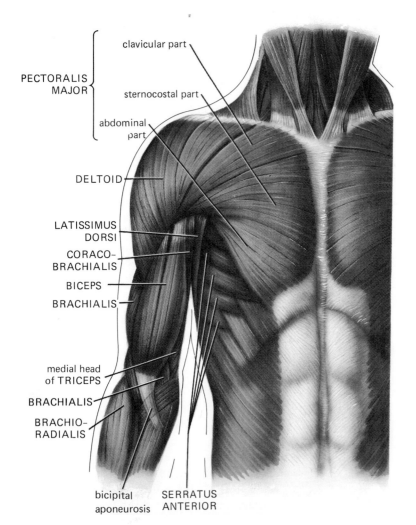

FIG. 12-17 Superficial muscles of the front of the chest.

PECTORALIS MAJOR
clavicular part
sternocostal part
abdominal part
DELTOID
LATISSIMUS DORSI
CORACO-BRACHIALIS
BICEPS
BRACHIALIS
medial head of TRICEPS
BRACHIALIS
BRACHIO-RADIALIS
bicipital aponeurosis
SERRATUS ANTERIOR

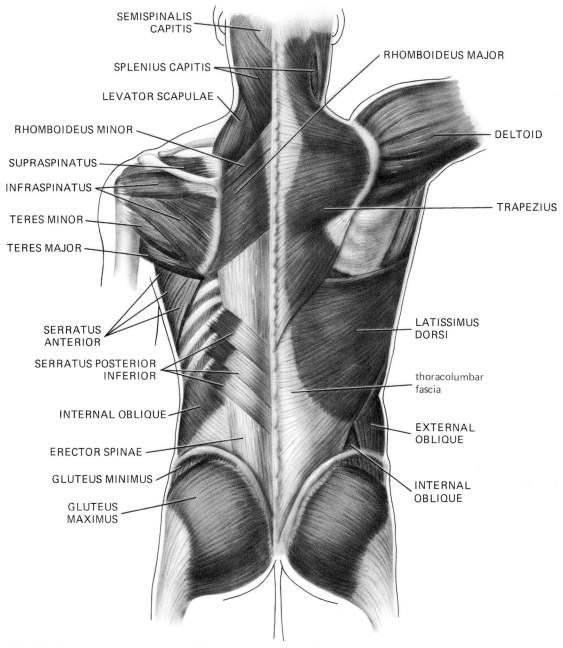

FIG. 12-18 Muscles of the back of the trunk and neck. The superficial muscles have been removed on the left to expose deeper muscle layers.

the *cubitus,* or elbow; the *antebrachium,* or forearm; the *carpus,* or wrist; and the *manus,* or hand. The *digits,* or fingers, are numbered, digit I being the thumb, or *pollex.*

The muscles of the upper limb can be divided into four basic groups: (1) the pectoral muscles, (2) the muscles of the (upper) arm, (3) the muscles of the forearm, and (4) the muscles of the hand.

Muscles of the Shoulder and Shoulder Girdle The pectoral, or shoulder, muscles insert on the shoulder girdle and have their bellies on the trunk of the body. Muscles connecting the pectoral girdle to the vertebral column (the **trapezius, latissimus dorsi, rhomboideus major** and **minor,** and **levator scapulae,** Figs. 12-17 and 12-18 and Tables 12-9 through 12-12) are located superficially in the back and arise

TABLE 12-9   MUSCLES THAT CONNECT THE UPPER LIMB AND VERTEBRAL COLUMN

Muscle	Origin	Insertion	Action and Comments
Trapezius (*trapezion,* trapezoid-shaped) (Figs. 12-10A and 12-18)	Superior nuchal line of the occipital bone, the occipital protuberance, the ligamentum nuchae, the spinous processes of C7 and T1–12 vertebrae, and the corresponding supraspinous ligaments	Clavicle, the acromion and spine of the scapula	Steadies the scapula and controls its position and movement during use of the arm; acting with other muscles, it elevates, rotates, or retracts the scapula; when the shoulder is fixed, it draws the head backward and laterally
Latissimus dorsi (*latissimus,* widest; *dorsi,* pertaining to the back) (Figs. 12-13, 12-17, 12-18, and 12-19)	Spines of T6–12, lumbar, and sacral vertebrae, the supraspinous ligaments, crest of the ilium, and the lower ribs	Intertubercular sulcus of the humerus	Active in adduction, extension, and medial rotation of the humerus; aids in depressing the arm against resistance and backward swinging of the arm; pulls trunk upward and forward during climbing
Rhomboideus (*rhombus,* kite; *eidos,* shape) (Fig. 12-18) Major (*major,* bigger) (Fig. 12-18)	Spines of T2–5 vertebrae and the supraspinous ligaments	Medial border of the scapula	Help control position and movement of the scapula during use of the arm; acting with other muscles, retract and rotate the scapula and brace back the shoulder
Minor (*minor,* smaller) (Fig. 12-18)	Ligamentum nuchae and spines of C7 and T1 vertebrae	Medial border of scapula at the spine	
Levator scapulae (*levare,* to raise; *scapulae,* pertaining to the scapula) (Fig. 12-18)	Transverse processes of the atlas, axis, and C3–4 vertebrae	Medial border of scapula above the spine	Helps control position and movement of the scapula during use of the arm; when the cervical part of the vertebral column is fixed, helps elevate the scapula; when the shoulder is fixed, bends the neck to the same side

TABLE 12-10   MUSCLES THAT CONNECT THE UPPER LIMB AND THORACIC WALL

Muscle	Origin	Insertion	Action and Comments
Pectoralis (*pectoral,* pertaining to the breast or chest)			
Major (*major,* bigger) (Fig. 12-17)	Clavicle, anterior surface of the sternum, cartilages of the true ribs, and aponeurosis of the external oblique muscle	Crest of greater tubercle of the humerus	Adduction and medial rotation of the humerus; when the arm is extended, draws it forward and medially; in climbing, draws trunk forward and upward
Minor (*minor,* smaller) (Fig. 12-19A)	3d–5th ribs and intercostal fascia	Coracoid process of the scapula	Aids in drawing scapula forward around the chest wall and rotating scapula to depress point of shoulder
Subclavius (*sub,* beneath; *clavius,* pertaining to the clavicle) (Fig. 12-19A)	Junction of first rib and its costal cartilage	Clavicle	Pulls point of shoulder down and forward and steadies the clavicle during movement of the shoulder
Serratus anterior (*serra,* saw; *anterior,* toward the front) (Figs. 12-13, 12-17, 12-18, and 12-19A)	Upper nine ribs and intercostal fascia	Medial border of the scapula	Helps draw scapula forward in reaching and pushing movements; with the trapezius, rotates the scapula to elevate the point of the shoulder

TABLE 12-11   MUSCLES OF THE SHOULDER (*Continued on following page*)

Muscle	Origin	Insertion	Action and Comments
Deltoid (*deltoideus,* triangular shape) (Figs. 12-17, 12-18, and 12-19B)	Clavicle, acromion and spine of the scapula	Shaft of the humerus	Aids in drawing arm forward or backward and rotating it medially or laterally, depending upon which of its fibers are active; principal abductor of the humerus
Subscapularis (*sub,* beneath; *scapularis,* pertaining to the scapula) (Fig. 12-19A)	Subscapular fossa of scapula	Lesser tubercle of humerus	Steadies head of humerus, rotates arm medially
Supraspinatus (*supra,* above; *spinatus,* sharp projection, spine) (Figs. 12-18 and 12-19B)	Supraspinous fossa of scapula	Greater tubercle of the humerus	Steadies head of humerus, initiates abduction of the arm
Infraspinatus (*infra,* below) (Figs. 12-18 and 12-19B)	Infraspinous fossa of scapula	Greater tubercle of the humerus	Steadies head of humerus, rotates arm laterally

TABLE 12-11   MUSCLES OF THE SHOULDER (Continued)

Muscle	Origin	Insertion	Action and Comments
Teres (*teres,* long and round, cylindri- cal)			
Minor (*minor,* smaller) (Figs. 12-18 and 12-19B)	Dorsal surface of scapula	Greater tubercle of the humerus	Steadies head of humerus, rotates arm laterally
Major (*major,* larger) (Figs. 12-18 and 12-19A, B)	Inferior angle of scapula	Crest of the lesser tubercle of the humerus	Draws humerus medially and back- ward and rotates it medially

TABLE 12-12   MUSCLES THAT ACT ON THE ARM AT THE SHOULDER

Action	Muscles	Action	Muscles
Flexors		**External Rotators**	
Chief	Deltoid (anterior part) Pectoralis major (clavicular part) Coracobrachialis	Chief	Infraspinatus Teres minor
Accessory	Biceps (?)	Accessory	Deltoid (posterior part)
Extensors		**Internal Rotators**	
Chief	Latissimus dorsi Deltoid (posterior part) Pectoralis major (sternal part; from flexion only)	Chief	Subscapularis
Accessory	Teres major Triceps (long head)	Accessory	Teres major Deltoid (anterior part) Pectoralis major Latissimus dorsi Supraspinatus (?)
Abductors			
Chief	Deltoid Supraspinatus		
Accessory	Biceps (long head)		
Adductors			
Chief	Pectoralis major Latissimus dorsi Teres major		
Accessory	Deltoid (posterior part) Coracobrachialis Triceps (long head)		

from the vertebrae and ligamentum nuchae. (The **ligamentum nuchae,** seen in Fig. 12-11, is a connective-tissue sheet in the upper back and back of the neck. It provides sites for muscle attachments that cannot be accommodated by the shortened spines of the cervical vertebrae.) Deep muscles that pass between the scapula and humerus (**deltoid, supra-** and **infraspinatus, teres major** and **minor,** and the **subscapularis**) are illustrated in Figs. 12-18 and 12-19 and are listed in Table 12-11.

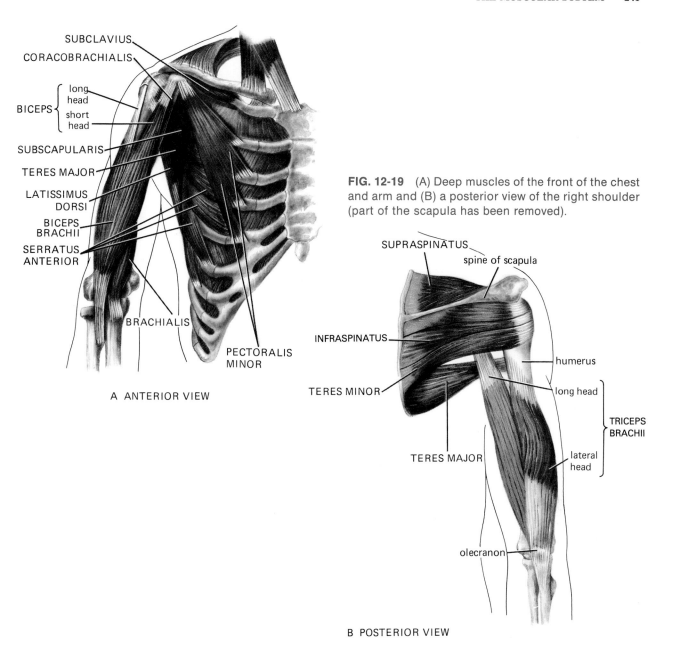

FIG. 12-19 (A) Deep muscles of the front of the chest and arm and (B) a posterior view of the right shoulder (part of the scapula has been removed).

A ANTERIOR VIEW

B POSTERIOR VIEW

Muscles of the (Upper) Arm The muscles of the upper arm (Fig. 12-17 and Table 12-13) have their bellies (i.e., their fleshy parts) in the arm and move the forearm at the elbow, the humerus at the shoulder, or both. There are three anterior muscles of the arm: the *coracobrachialis, biceps brachii,* and *brachialis*. The **coracobrachialis** crosses only the shoulder joint, the **biceps brachii** crosses both the shoulder and elbow, and the **brachialis** crosses only the elbow. The biceps brachii, as its name suggests,

has two heads, the upper ends of which are under the deltoid muscle.

The **triceps brachii** is the important posterior muscle of the arm. The three heads of the triceps arise from the scapula (the long head) and the shaft of the humerus (the lateral and medial heads).

Muscles of the Forearm The muscles of the forearm (Table 12-14) have their bellies in the forearm and act at the elbow (Table 12-15), wrist (Table 12-16), or on the fingers or thumb. The flexor muscles

TABLE 12-13 MUSCLES OF THE UPPER ARM

Muscle	Origin	Insertion	Action and Comments
Coracobrachialis (*coracoid*, pertaining to the coracoid process of the scapula; *brachium*, upper arm) (Figs. 12-17 and 12-19A)	Coracoid process of scapula	Shaft of the humerus	Draws arm forward and medially
Biceps brachii (*bi*, two; *caput*, head; *biceps*, having two heads) (Figs. 12-17 and 12-19A)	*Short head:* coracoid process of scapula; *long head:* supraglenoid tubercle	Tuberosity of radius	A powerful supinator of the arm, flexes elbow joint, stabilizes head of humerus
Brachialis (Figs. 12-17 and 12-19A)	Lower half of the humerus and intermuscular septa	Tuberosity and coronoid process of the ulna	Flexes the elbow joint
Triceps (*tri*, three; *triceps*, having three heads) (Figs. 12-17 and 12-19B)	*Long head:* infraglenoid tubercle of scapula; *lateral head:* shaft of the humerus; *medial head:* shaft of the humerus	Olecranon process of ulna	Extends the arm at the elbow joint; slightly extends and adducts humerus and stabilizes head of humerus

TABLE 12-14 MUSCLES OF THE FOREARM (*Continued on following page*)

Muscle	Origin	Insertion	Action and Comments
Superficial Flexor Muscles			
Pronator teres (*pronate*, turn the palm downward or posterior; *teres*, cylindrical) (Figs. 12-20A and 12-21)	*Humeral head:* medial epicondyle of humerus; *ulnar head:* coronoid process of the ulna	Shaft of radius	Rotates the radius upon the ulna turning the palm backward (i.e., it pronates the forearm)
Flexor carpi radialis (*flex*, to bend; *carpi*, pertaining to the wrist; *radialis*, pertaining to the radius) (Fig. 12-20 and 12-21)	Medial epicondyle of humerus	Base of second and third metacarpal bones	Aids in flexing wrist and abducting the hand
Palmaris longus (*palmar*, pertaining to the palm; *longus*, long) (Fig. 12-20A)	Medial epicondyle of humerus	Palmar aponeurosis	Flexes the wrist and tenses palmar fascia
Flexor carpi ulnaris (*ulnar*, pertaining to the ulna) (Figs. 12-20A, B and 12-21)	Medial epicondyle of the humerus and the olecranon process of the ulna	Pisiform, hamate, and fifth metacarpal bones	Aids in flexing the wrist and adducting the hand

TABLE 12-14   MUSCLES OF THE FOREARM (*Continued*)

Muscle	Origin	Insertion	Action and Comments
Flexor digitorum superficialis (*digitus,* finger, or toe; *superficialis,* closer to the surface) (Figs. 12-20A and 12-21)	Medial epicondyle of the humerus, medial ulna, anterior border of the radius	Middle phalanges of digits 2 to 5	Flexes the middle and then the proximal phalanges and flexes the wrist
Deep Flexor Muscles			
Flexor digitorum profundus (*profundus,* deep) Fig. 12-20B)	Upper three-fourths of the ulna and adjacent interosseus membrane	Distal phalanges of digits 2 to 5	Flexes distal phalanges and aids in flexing the wrist
Flexor pollicis longus (*pollex,* thumb) (Figs. 12-20A, B and 12-21)	Radius, adjacent interosseus membrane	Distal phalanx of the thumb	Flexes the phalanges of the thumb
Pronator quadratus (*quadratus,* square) (Figs. 12-20B and 12-21)	Distal end of shaft of ulna	Distal end of radius	Pronates the forearm, opposes separation of the lower ends of the radius and ulna
Superficial Extensor Muscles			
Brachioradialis (Figs. 12-17, 12-19A, and 12-20B)	Lateral supracondylar ridge of the humerus	Styloid process of radius	Flexes the elbow joint
Extensor carpi radialis (*extensor,* straightens, opens an angle)			
Longus (Figs. 12-20A, B, and 12-21)	Lateral supracondylar ridge of humerus	Base of the second metacarpal bone	Aids in extending the wrist and abducting the hand
Brevis (*brevis,* short, brief) (Fig. 12-20A)	Lateral epicondyle of humerus	Base of second and third metacarpal bone	Aids in extending the wrist and abducting the hand
Extensor digitorum (Fig. 12-21)	Lateral epicondyle of the humerus	Phalanges of second to fifth fingers	Extends fingers, hands, and forearm
Extensor digiti minimi (*minimi,* smallest)	Common extensor tendon and intermuscular septa	Phalanges of little finger	Extends the little finger and wrist
Extensor carpi ulnaris (Fig. 12-21)	Lateral epicondyle of the humerus, posterior border of ulna	Base of fifth metacarpal bone	Aids in extending and adducting the wrist
Anconeus (*ankon,* elbow)	Lateral epicondyle of the humerus	Olecranon process and shaft of the ulna	Extends elbow joint

TABLE 12-14 MUSCLES OF THE FOREARM (*Continued*)

Muscle	Origin	Insertion	Action and Comments
Deep Extensor Muscles			
Supinator (*supinate*, to turn the palm upward or anterior) (Fig. 12-20B and 12-21)	Lateral epicondyle of the humerus and ligaments of the elbow joint and from supinator crest of ulna	Upper shaft of radius	Supinates hand and forearm
Abductor pollicis longus (*ab*, from; *ducere*, to draw or direct; *abduct*, to draw away from the midline of the body or an adjacent part) (Fig. 12-20B and 12-21)	Shaft of the radius and ulna and the interosseous membrane	First metacarpal bone	Abducts and extends the thumb
Extensor pollicis brevis	Posterior surface of the radius and the interosseous membrane	Base of the proximal phalanx of the thumb	Extends proximal phalanx of the thumb and helps extend metacarpal bone
Extensor pollicis longus (Fig. 12-21)	Shaft of the ulna and the interosseous membrane	Base of the distal phalanx of the thumb	Extends the metacarpal and proximal and distal phalanges of the thumb; rotates extended thumb laterally
Extensor indicis (*indicis*, index finger)	Posterior surface of the ulna and interosseus membrane	Extensor expansion of index finger	Extends the index finger, aids in extending the wrist

TABLE 12-15 MUSCLES THAT ACT ON THE FOREARM AT THE ELBOW

Action	Muscles
Flexors	
Chief	Biceps
	Brachialis
	Brachioradialis
Accessory	Pronator teres
	Radial muscles of the extensor group
Pronators	
Chief	Pronator teres
	Pronator quadratus
Accessory	Flexor carpi radialis
	Palmaris longus
	Brachioradialis (?)
Supinators	
Chief	Supinator
	Biceps
Accessory	Extensor carpi radialis longus
	Brachioradialis (?)

(Fig. 12-20A and B) generally lie on the anterior side of the forearm, whereas the extensors are on the posterior surface. The origins of these muscles are on the sides of the elbow (their bulk would interfere with the joint's motion if they were on the front or back). A common tendon of origin of several of the superficial flexors attaches to the medial epicondyle of the humerus, whereas the lateral epicondyle and supracondylar ridge are important origins for the superficial extensors. The origins of the deep flexors and extensors are from the radius and ulna and the interosseous membrane.

The muscles of the forearm are enclosed in a sheath of fascia, which is reinforced on the dorsal side of the wrist by connective-tissue fibers that run between the radius and the ulna, triquetral bone, and pisiform bone; this thickening forms the *extensor retinaculum*. A similar reinforcement, the *palmar carpal ligament*, occurs on the anterior side of the wrist. Fig. 12-21 shows more complete dissections of some of the forearm muscles.

TABLE 12-16	MUSCLES THAT ACT ON THE HAND AT THE WRIST
Action	**Muscles**
Flexors	
Chief	Flexor carpi radialis
	Flexor carpi ulnaris
	Palmaris longus
Accessory	Abductor pollicis longus
	Flexor digitorum superficialis
	Flexor digitorum profundus
	Flexor pollicis longus
Extensors	
Chief	Extensor carpi radialis longus
	Extensor carpi radialis brevis
	Extensor carpi ulnaris
Accessory	Extensor digitorum
	Extensor indicis
	Extensor digiti minimi
	Extensor pollicis longus
Radial abductors	Abductor pollicis longus
	Extensor pollicis brevis
Ulnar abductors	Extensor carpi ulnaris
	Flexor carpi ulnaris

FIG. 12-20 Muscles on the forearm, anterior view: (A) superficial and (B) deep muscles (see p. 246).

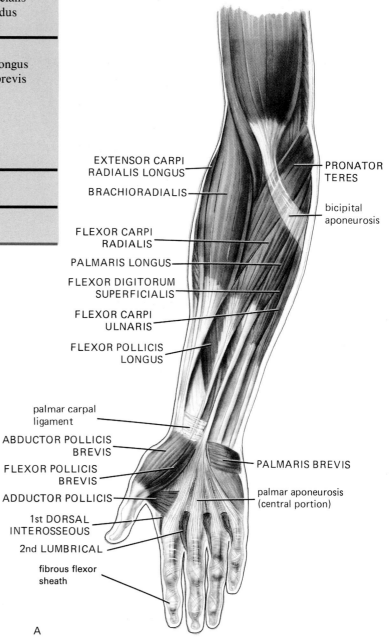

EXTENSOR CARPI RADIALIS LONGUS

PRONATOR TERES

BRACHIORADIALIS

bicipital aponeurosis

FLEXOR CARPI RADIALIS

PALMARIS LONGUS

FLEXOR DIGITORUM SUPERFICIALIS

FLEXOR CARPI ULNARIS

FLEXOR POLLICIS LONGUS

palmar carpal ligament

ABDUCTOR POLLICIS BREVIS

PALMARIS BREVIS

FLEXOR POLLICIS BREVIS

palmar aponeurosis (central portion)

ADDUCTOR POLLICIS

1st DORSAL INTEROSSEOUS

2nd LUMBRICAL

fibrous flexor sheath

A

FIG. 12-20B Deep muscles of the forearm, anterior view.

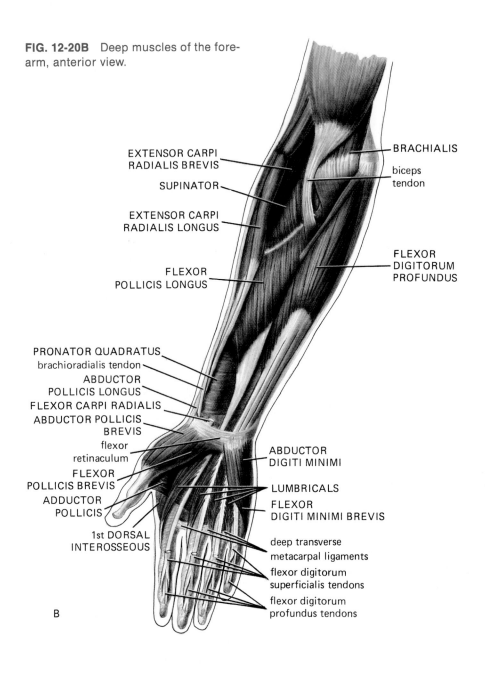

EXTENSOR CARPI
RADIALIS BREVIS

SUPINATOR

EXTENSOR CARPI
RADIALIS LONGUS

FLEXOR
POLLICIS LONGUS

BRACHIALIS

biceps
tendon

FLEXOR
DIGITORUM
PROFUNDUS

PRONATOR QUADRATUS
brachioradialis tendon
ABDUCTOR
POLLICIS LONGUS
FLEXOR CARPI RADIALIS
ABDUCTOR POLLICIS
BREVIS
flexor
retinaculum
FLEXOR
POLLICIS BREVIS
ADDUCTOR
POLLICIS
1st DORSAL
INTEROSSEOUS

ABDUCTOR
DIGITI MINIMI

LUMBRICALS

FLEXOR
DIGITI MINIMI BREVIS

deep transverse
metacarpal ligaments

flexor digitorum
superficialis tendons

flexor digitorum
profundus tendons

B

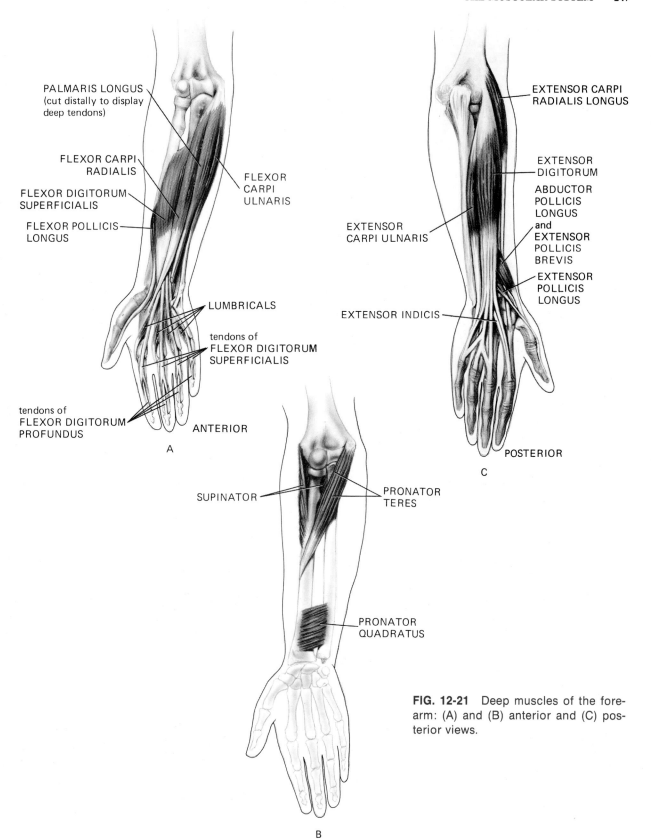

PALMARIS LONGUS
(cut distally to display
deep tendons)

FLEXOR CARPI
RADIALIS

FLEXOR DIGITORUM
SUPERFICIALIS

FLEXOR POLLICIS
LONGUS

FLEXOR
CARPI
ULNARIS

LUMBRICALS

tendons of
FLEXOR DIGITORUM
SUPERFICIALIS

tendons of
FLEXOR DIGITORUM
PROFUNDUS

ANTERIOR

A

EXTENSOR CARPI
RADIALIS LONGUS

EXTENSOR
DIGITORUM

ABDUCTOR
POLLICIS
LONGUS
and
EXTENSOR
POLLICIS
BREVIS

EXTENSOR
CARPI ULNARIS

EXTENSOR
POLLICIS
LONGUS

EXTENSOR INDICIS

POSTERIOR

C

SUPINATOR

PRONATOR
TERES

PRONATOR
QUADRATUS

B

FIG. 12-21 Deep muscles of the forearm: (A) and (B) anterior and (C) posterior views.

TABLE 12-17   MUSCLES OF THE HAND

Muscle	Origin	Insertion	Action and Comments
Abductor pollicis brevis (*ab*, from; *ducere*, to draw or direct; *abduct*, to draw away from the midline of the body or an adjacent part; *pollex*, thumb; *brevis*, short, brief) (Fig. 12-20A, B)	Flexor retinaculum, tubercles of scaphoid bone and trapezium	Base of the proximal phalanx of the thumb	Abducts thumb and rotates it medially
Opponens pollicis (*opponens*, opposing)	Tubercle of the trapezium and the flexor retinaculum	Palmar surface of the metacarpal bone of the thumb	Flexes metacarpal bone of thumb and rotates it medially, bringing the tip of the thumb in contact with the palmar surface of the fingers
Flexor pollicis brevis (*flex*, to bend) (Fig. 12-20A, B)	Flexor retinaculum, tubercle of the trapezium	Base of the proximal phalanx of the thumb	Flexes the proximal phalanx of the thumb and flexes the metacarpal bone and rotates it medially
Adductor pollicis (*ad*, to; *adduct*, to draw toward a median line) (Fig. 12-20A, B)	Capitate bone, bases of the second and third metacarpal bones	Base of the proximal phalanx of the thumb	Approximates the thumb to the palm of the hand
Palmaris brevis (*palmar*, pertaining to the palm)	Flexor retinaculum and palmar aponeurosis	Skin of the ulnar side of the hand	Wrinkles the skin on the ulnar side of the palm of the hand and deepens the hollow of the palm
Abductor digiti minimi (*digitus*, finger or toe; *minimi*, smallest) (Fig. 12-20B)	Pisiform bone, tendon of flexor carpi ulnaris	Base of the proximal phalanx of the little finger	Abducts the little finger away from the fourth
Flexor digiti minimi brevis (Fig. 12-20B)	Hamate bone and flexor retinaculum	Base of the proximal phalanx of the little finger	Flexes the little finger
Opponens digiti minimi	Hamate bone and flexor retinaculum	Fifth metacarpal bone	Draws the fifth metacarpal bone forward and rotates it laterally
Lumbricals (*lumbricus*, earthworm; thus, earthworm-shaped) (Figs. 12-20A, B and 12-21)	Tendons of the flexor digitorum profundis	Extensor expansion covering the dorsal surface of each finger	Flex digits at joint between metacarpals and phalanges; extend terminal phalanges
Dorsal and palmar interossei (*dorsal*, back, of the hand; *inter*, between; *ossa*, bones) (Fig. 12-20A, B)	Metacarpal bones	Proximal phalanges and extensor expansion	Abduct and adduct digits II to V, aid in extending terminal phalanges

Muscles of the Hand   In the hand, all the muscles lie on the palmar side, and the muscles considered here are those that lie entirely within the hand, i.e., the **intrinsic muscles of the hand** (Figs. 12-20A and B and Table 12-17). Three of the muscles of the thumb (the **abductor pollicis brevis, flexor pollicis brevis,** and **opponens pollicis**) form a rounded contour (the *thenar eminence*) next to the thumb; the muscles of the little finger (the *"digiti minimi"*) form the *hypothenar eminence*. The **lumbricals** and the two sets of **interossei** (dorsal and palmar) fill in the spaces between the metacarpal bones.

Muscles of the Pelvic Girdle and Lower Limb

The regions of the lower limb are the *gluteal region,* or buttock; the *thigh,* or femoral region; the *genu,* or knee, (with its posterior, *popliteal* and anterior, *patellar,* surfaces); the leg; and the foot (*pes*). The borders of the leg are medial (or tibial) and lateral (or fibular).

Unlike the articulation of the shoulder girdle to the vertebral column, the pelvic girdle is firmly attached to the sacrum, and muscles comparable to those that attach the shoulder girdle to the vertebral column are sparse. Most of the muscles that do attach to the pelvic girdle move the leg or they are abdominal, pelvic, or perineal muscles.

Muscles of the Gluteal Region   The large muscle of the buttock, the **gluteus maximus,** covers most of the other structures. It arises from the ilium, sacrum, and coccyx, and inserts on the *iliotibial tract,* a particularly dense part of the *fascia lata* (see below) along the lateral portion of the thigh (Fig. 12-22A). The gluteus maximus has been cut away in Fig. 12-22B, and the muscles beneath it can be seen. The muscles of the gluteal region are listed in Table 12-18.

FIG. 12-22 Posterior muscles of the hip and thigh. In (B) the gluteus maximus has been cut away to make visible the underlying muscles.

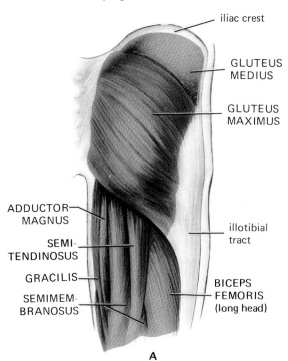

A

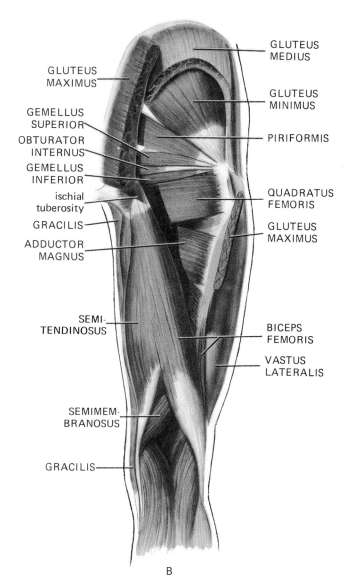

B

TABLE 12-18 MUSCLES OF THE GLUTEAL REGION

Muscle	Origin	Insertion	Action and Comments
Gluteus (*gloutos*, buttock)			Extensor and abductors of the hip joint
Maximus (*maximus*, largest) (Fig. 12-22)	Posterior gluteal line of the ilium and area above and behind it, lower part of the sacrum, side of the coccyx, and sacrotuberous ligament	Iliotibial tract of the fascia lata and gluteal tuberosity of the femur	When pelvic girdle is stabilized, extends the flexed thigh; when the hip joint is stabilized, maintains upright posture; lateral rotator of thigh; active in raising trunk after stooping; steadies the femur on the tibia; largest muscle mass in the body
Medius (*medius*, middle) (Fig. 12-22)	Outer surface of the ilium and adjacent fascia	Greater trochanter of the femur	When pelvic girdle is stabilized, abducts thigh and rotates it medially, aids in maintaining upright posture during stepping; stabilizes pelvis when one foot is raised
Minimus (*minimus*, smallest) (Fig. 12-22)	Outer surface of the ilium	Greater trochanter of the femur	When pelvic girdle is stabilized, abducts the thigh and rotates it medially; aids in maintaining upright posture during stepping; stabilizes pelvis when one foot is raised
Piriformis (*pirum*, pear; *forma*, shape) (Fig. 12-22)	Pelvic surface of the sacrum	Greater trochanter of the femur	Rotates the extended thigh laterally; abducts the flexed thigh
Obturator internus (*obturator*, closed; *internus*, inside)	Wall of obturator foramen of coxal bone	Greater trochanter of the femur	Rotates extended thigh laterally; abducts flexed thigh
Gemellus (*gemelli*, twins)			
Superior (*superior*, higher, more rostral) (Fig. 12-22)	Spine of the ischium	Trochanter fossa with the tendon of the obturator internus	Rotates the extended thigh laterally; abducts the flexed thigh
Inferior (*inferior*, lower, more caudal) (Fig. 12-22)	Tuberosity of the ischium	Trochanter fossa with the tendon of the obturator internus	Rotates the extended thigh laterally; abducts the flexed thigh
Quadratus femoris (*quadratus*, square; *femora*, thigh) (Fig. 12-22)	Tuberosity of the ischium	Trochanteric crest of the femur	Rotates thigh laterally

Muscles of the Thigh The muscles of the thigh (Figs. 12-22, 12-23, and 12-24) may be conveniently divided into three groups that are separated by inward extensions from the *fascia lata,* a sheet of fascia that is fused with the fascia of the external abdominal oblique muscle and forms an almost stocking-like covering of the tissues of the thigh.

Muscles of the anterior group (see Table 12-19) occupy the entire anterior half of the thigh and almost completely surround the femur. They powerfully extend the leg at the knee and, in a weaker action, flex the thigh at the hip. The major flexor of the thigh is the **iliopsoas** (*iliacus* and *psoas major,* Table 12-20). Of the anterior group, the *rectus fe-*

FIG. 12-23 Anterior muscles of the right hip and thigh.

FIG. 12-24 Specific muscles of the right thigh, posterior view and anterior views.

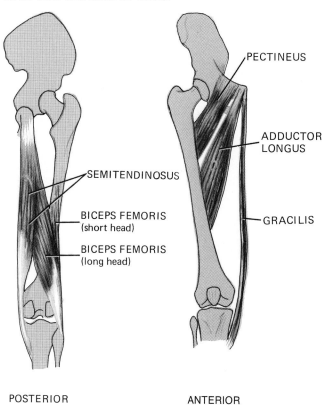

PSOAS MAJOR

anterior superior iliac spine

inguinal ligament

TENSOR FASCIAE LATAE

ILIACUS

PSOAS MAJOR

PECTINEUS

ADDUCTOR LONGUS

RECTUS FEMORIS

GRACILIS

SARTORIUS

VASTUS LATERALIS

VASTUS MEDIALIS

PECTINEUS

ADDUCTOR LONGUS

SEMITENDINOSUS

BICEPS FEMORIS (short head)

BICEPS FEMORIS (long head)

GRACILIS

POSTERIOR

ANTERIOR

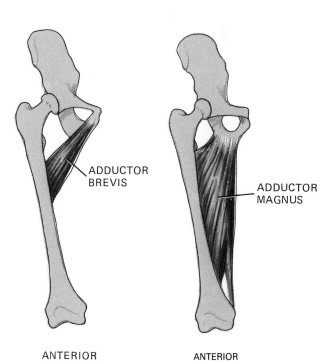

ADDUCTOR BREVIS

ADDUCTOR MAGNUS

ANTERIOR

ANTERIOR

TABLE 12-19   MUSCLES OF THE THIGH (*Continued on following page*)

Muscle	Origin	Insertion	Action and Comments
Anterior Femoral Muscles (*anterior,* toward or on the front of the body; *femora,* thigh)			
Tensor fasciae latae (*tensor,* to make tense; *fasciae latae,* the wide, flat connective-tissue sheaths of the muscles of the thigh) (Fig. 12-23)	Outer lip of the iliac crest, anterior superior iliac spine, and fascia lata	Iliotibial part of fascia lata, lateral femoral condyle and femur via lateral septum	Extends the knee with lateral rotation of the leg; when the knee is stabilized, steadies the head of the femur in the pelvis and the condyles of the femur on the tibia and thus helps maintain upright posture
Sartorius (*sartor,* tailor, refers to the cross-legged way in which tailors sit) (Figs. 12-22, 12-23, and 12-25B)	Anterior superior iliac spine and notch	Medial surface of the tibia	The longest muscle in the body; assists in flexing the leg on the thigh and the thigh on the pelvis; helps abduct the thigh and rotate it laterally
Psoas (*psoa,* muscle of the loin)			Flexes thigh upon the pelvis
Major (*major,* larger) (Fig. 12-23)	Transverse processes of the lumbar vertebrae; the vertebral bodies and intervertebral discs	Inserts with the fibers of the iliacus on the lesser trochanter of the femur	Acts with the iliacus (see below)
Minor (*minor,* smaller)	Bodies of T12 and L1 vertebrae and the disc between them	Iliopectinal eminence and iliac fascia	Weak flexor of the trunk
Iliacus (*iliacus,* pertaining to the ilium) (Fig. 12-23)	Upper two-thirds of the concavity of the iliac fossa, inner lip of the iliac crest, lateral part of the sacrum, and adjacent ligaments	Tendon of psoas major and femur below lesser trochanter	Assists in flexion of the thigh; when acting with psoas major and the femur is stabilized, bends the trunk and pelvis forward, as in raising the trunk from a lying to a sitting posture
Quadriceps femoris (*quadra,* four; *caput,* head; *quadriceps,* four-headed) (Figs. 12-22, 12-23, and 12-24)	(See individual entries below)	The tendons of the four parts of the quadriceps unite to form a single tendon that attaches to the base of the patella and tubercle of the tibia	The great extensor muscle of the leg; covers almost all the front and sides of the femur; can be divided into four parts

TABLE 12-19   MUSCLES OF THE THIGH (*Continued on following page*)

Muscle	Origin	Insertion	Action and Comments
Rectus femoris (*rectus,* straight, parallel to the midline) (Fig. 12-23)	Arises by two heads: from the inferior iliac spine and from the groove above the acetabulum and capsule of the hip joint	Base of the patella	Extends leg at the knee; also helps flex the thigh on the pelvis; can flex hip and extend knee simultaneously
Vastus lateralis (*vastus,* great, vast; *lateralis,* at the side) (Fig. 12-23)	Intertrochanteric line, borders of the greater trochanter, gluteal tuberosity, and linea aspera of the femur	Lateral border of the patella and the quadriceps femoris tendon	Extends leg at knee; stabilizes knee joint
Vastus medialis (*medialis,* medial) (Figs. 12-23 and 12-25B)	Intertrochanteric line, spiral line, linea aspera, and medial supracondylar line of the femur, tendons of the adductor longus and magnus	Medial border of the patella and the quadriceps femoris tendon	Extends leg at knee; stabilizes patella and knee joint
Vastus intermedius (*intermedius,* between, in the middle)	Upper two-thirds of the femoral shaft and the lateral intermuscular septum	Quadriceps femoris tendon, lateral border of the patella, and lateral condyle of the tibia	Extends leg at knee
Articularis genu (*articulatio,* articulation, or joint; *genu,* knee)	Lower part of the shaft of the femur	Upper part of the synovial membrane of the knee joint	Pulls the synovial membrane of the knee joint upward during extension of the leg
Medial Femoral Muscles			
Gracilis (*gracile,* slender, delicate) (Figs. 12-22, 12-23, 12-24, and 12-25B)	Lower half of the body and inferior ramus of the pubis and ramus of the ischium	Upper part of the medial surface of the tibia	Flexes the leg and rotates it medially; may also adduct the thigh
Pectineus (*pectus,* pubes, pertaining to the pubic bone) (Figs. 12-23 and 12-24)	Pecten of the pubis and adjacent fascia	Femur between the lesser trochanter and linea aspera	Adducts the thigh and flexes it on the pelvis

TABLE 12-19 MUSCLES OF THE THIGH (*Continued*)

Muscle	Origin	Insertion	Action and Comments
Adductor (*ad*, to; *ducere*, to draw or direct; *adduct*, to draw toward a median line)			Adducts thigh; aids in control of gait and posture
Longus (*longus*, long) (Figs. 12-23 and 12-24)	Front of the body of the pubis	Linea aspera of the middle third of the femur	Probably also a medial rotator of the thigh
Brevis (*brevis*, short, brief) (Fig. 12-24)	Body and inferior ramus of the pubis	Femur between the lesser trochanter and linea aspera	Probably also a medial rotator of the thigh
Magnus (*magnus*, large) (Fig. 12-22 and 12-24)	Inferior ramus of the pubis, ramus of the ischium and ischial tuberosity	Gluteal tuberosity, linea aspera, medial supracondylar line, and the medial condyle of the femur	Probably also a medial rotator of the thigh
Obturator externus (*obturator*, closed; *externus*, superficial)	Ramus of the pubis and ischium on medial side of obturator foramen and obturator membrane	Trochanteric fossa of the femur	Rotates thigh laterally
Posterior Femoral Muscles			
Biceps femoris (*bi*, two; *biceps*, having two heads) (Figs. 12-22 and 12-24)	*Long head:* ischial tuberosity and sarcotuberous ligament; *short head:* linea aspera, lateral supracondylar line, and lateral intermuscular septum	Head of the fibula, fibular collateral ligament, and lateral condyle of the tibia	When the hip joint is stabilized, flexes the leg on the thigh; when the knee is stabilized, extends the hip joint, drawing the trunk upright after stooping; rotates semiflexed knee laterally; rotates thigh laterally when hip is extended; the tendon of the biceps femoris is called the *lateral hamstring*
Semitendinosus (*semi*, half; *tendinous*, composed in part of tendons) (Figs. 12-22, 12-24, and 12-25B)	Ischial tuberosity and adjacent aponeurosis	Upper part of the medial surface of the tibia	When the hip joint is stabilized, flexes the leg on the thigh; when the knee is stabilized, extends the hip joint and draws trunk upright; rotates thigh medially when hip is extended
Semimembranosus (*membranous*, composed in part of membranes)	Ischial tuberosity	Posterior border of the medial condyle of tibia	When the hip joint is stabilized, flexes the leg on the thigh; when the knee is stabilized, extends the hip joint and draws the trunk upright; rotates thigh medially when hip is extended

TABLE 12-20   MUSCLES THAT ACT ON THE THIGH AT THE HIP

Action	Muscles	Action	Muscles
Flexors		**Abductors**	
Chief	Iliopsoas	Chief	Gluteus medius
	Pectineus		Gluteus minimus
	Tensor fasciae latae	Accessory	Tensor fasciae latae
	Adductor brevis		Sartorius
	Sartorius		Piriformis
Accessory	Adductor longus	**Medial Rotators**	
	Adductor magnus (anterior part)	Chief	Gluteus minimus
	Gracilis (?)		Tensor fasciae latae
	Gluteus minimus	Accessory	Gracilis
Extensors			Adductor brevis
Chief	Gluteus maximus		Adductor longus
	Adductor magnus (posterior part)		Adductor magnus (anterior part)
Accessory	Semimembranosus		Adductor magnus (posterior part)
	Semitendinosus		Pectineus
	Biceps femoris (long lead)		Semitendinosus
	Gluteus medius		Semimembranosus
	Piriformis	**Lateral Rotators**	
Adductors		Chief	Gluteus maximus
Chief	Adductor brevis		Iliopsoas
	Adductor longus		Piriformis
	Adductor magnus		Obturator externus
	Gracilis		Obturator internus
Accessory	Gluteus maximus		Superior gemellus
	Pectineus		Inferior gemellus
	Obturator externus		Quadratus femoris
	Iliopsoas	Accessory	Gluteus medius
	''Hamstrings''		Sartorius
			Biceps femoris (long head)

moris and the three muscles of the *vastus group* act on the knee through a common tendon and are considered as one muscle, the **quadriceps femoris.**

The medial group includes the muscles that lie in the gluteal region and abduct and laterally rotate the thigh as well as the thigh adductors.

The posterior group includes the thigh flexors (**biceps femoris, semitendinosus,** and **semimembranosus**), which span the hip and knee joint and integrate extension of the hip with flexion of the knee and are often grouped together as the *"hamstring muscles."*

The actions of the muscles at the hip and the knee are listed in Tables 12-20 and 12-21.

TABLE 12-21   MUSCLES THAT ACT ON THE KNEE

Action	Muscles
Extensors	
Chief	Quadriceps femoris
Accessory	Gluteus maximus
	Gastrocnemius
	Soleus
Flexors	
Chief	Semimembranosus
	Semitendinosus
	Biceps femoris (short head)
	Gracilis
	Sartorius
Accessory	Popliteus
	Gastrocnemius
	Biceps femoris (long head)

Muscles of the Leg   A comparison between the forearm and leg can be made if the forearm is pronated and the hand held in dorsiflexion. Thus, the tibial side of the leg corresponds to the radial side of the forearm, and the fibular side of the leg corresponds to the ulnar side of the forearm. With the arm in this position, the anterior surface of both leg and forearm contains extensor muscles and the posterior surface contains flexors.

The fascia lata of the thigh continues into the leg as the *crural fascia*. Intermuscular septa, which pass from the fascia deep into the leg, and the interosseous membrane separate the muscles into anterior, lateral, and posterior compartments (Table 12-22). The large superficial muscles of the posterior group (**gastrocnemius, plantaris,** and **soleus**) form the calf of the leg (Fig. 12-25). Their contraction produces plantar flexion of the foot; their size is directly related to human beings' upright posture and method of walking. The crural fascia is reinforced in the ankle region by transverse and oblique fibers that form thick bands, or *retinacula* (Fig. 12-25). These hold the tendons that cross the ankle close to the bone and prevent their bowing out when under tension.

The actions of the leg muscles at the ankle are listed in Table 12-23.

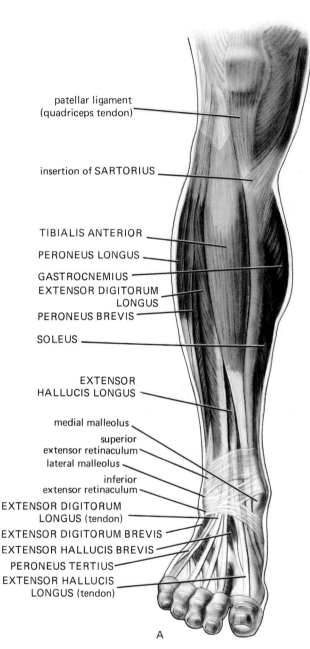

FIG. 12-25 Superficial muscles of the right leg: (A) anterior, (B) medial, and (C) posterior views.

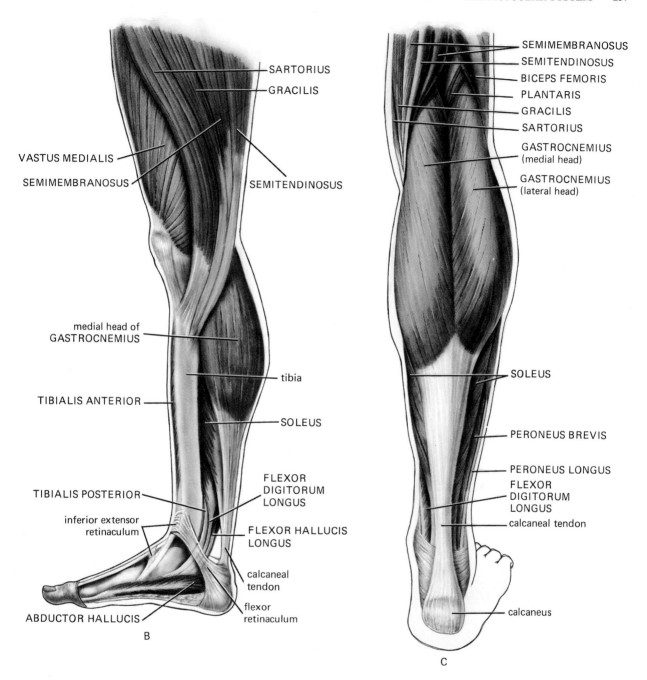

SARTORIUS

GRACILIS

VASTUS MEDIALIS

SEMIMEMBRANOSUS

SEMITENDINOSUS

medial head of
GASTROCNEMIUS

tibia

TIBIALIS ANTERIOR

SOLEUS

TIBIALIS POSTERIOR

FLEXOR
DIGITORUM
LONGUS

inferior extensor
retinaculum

FLEXOR HALLUCIS
LONGUS

calcaneal
tendon

flexor
retinaculum

ABDUCTOR HALLUCIS

B

SEMIMEMBRANOSUS

SEMITENDINOSUS

BICEPS FEMORIS

PLANTARIS

GRACILIS

SARTORIUS

GASTROCNEMIUS
(medial head)

GASTROCNEMIUS
(lateral head)

SOLEUS

PERONEUS BREVIS

PERONEUS LONGUS

FLEXOR
DIGITORUM
LONGUS

calcaneal tendon

calcaneus

C

TABLE 12-22   MUSCLES OF THE LEG (*Continued on next page*)

Muscle	Origin	Insertion	Action and Comments
Anterior Muscles			
Tibialis anterior (*tibial*, pertaining to the tibia; *anterior*, at or to the front) (Fig. 12-25A, B)	Lateral condyle and upper half of shaft of tibia, interosseous membrane, and crural fascia	Medial cuneiform and base of the first metatarsal bone	Dorsiflexes and inverts the foot; if ankle is stabilized, can draw the body forward
Extensor hallucis longus (*extensor*, straightens, or opens an angle; *hallux*, the great toe; *longus*, long) (Fig. 12-25A)	Medial surface of the fibula and interosseous membrane	Base of the distal phalanx of the great toe	Extends the phalanges of the great toe and dorsiflexes the foot
Extensor digitorum longus (*digit*, a toe or finger) (Fig. 12-25A)	Lateral condyle of the tibia, medial surface of the fibula, interosseous membrane, and crural fascia	Divides into four slips on the dorsum of the foot; joined by tendons of the extensor digitorum brevis to form *dorsal digital expansion;* attaches to middle and distal phalanx	Extends the toes and aids in dorsiflexing the foot
Peroneus tertius (*perone*, the fibula; *tertius*, third in order) (Fig. 12-25A)	Medial surface of the fibula, interosseous membrane, and intermuscular septum	Base of the fifth metatarsal bone	Dorsiflexes the foot, acting as part of the extensor digitorum longus
Lateral Muscles			
Peroneus longus (Fig. 12-25A, C)	Head and lateral surface of the fibula, crural fascia, and intermuscular septum	Base of the first metatarsal bone and the medial cuneiform	Everts and plantar flexes the foot, maintains the concavity of the foot in the early phase of stepping and in tiptoeing
Peroneus brevis (*brevis*, short, brief) (Fig. 12-25C)	Lateral surface of the fibula and middle two-thirds of the intermuscular septum	Base of the fith metatarsal bone	Participates in eversion of the foot; may help steady the leg on the foot and prevent overinversion of the foot

TABLE 12-22 MUSCLES OF THE LEG (*Continued*)

Muscle	Origin	Insertion	Action and Comments
Posterior Muscles			
Gastrocnemius (*gaster,* stomach, belly; *kneme,* leg) (Fig. 12-25A, B, C)	Arises by two heads from the condyles of the femur and adjacent parts of the joint capsule	Posterior surface of the calcaneus as part of the calcaneal (Achilles) tendon	Forms the belly of the calf of the leg; one of the chief plantar flexors of the foot; flexes the knee; provides propelling force in walking, running, and leaping
Soleus (*soleus,* pertaining to the sole of the foot) (Fig. 12-25A, B, C)	Back of the head and posterior surface of the fibula and medial border of the fibula, and a fibrous band between the tibia and fibula	Posterior surface of the calcaneus as part of the calcaneal tendon	An important plantar flexor of the foot, steadies leg on foot for postural stability
Plantaris (*planta,* sole of the foot) (Fig. 12-25C)	Lateral supracondylar line of the femur	Posterior surface of the calcaneus as part of the calcaneal tendon	Accessory to the gastrocnemius
Popliteus (*poples,* ham, pertaining to the posterior surface of the knee)	Lateral condyle of the femur and a popliteal ligament	Posterior surface of the tibia and the tendinous expansion covering it	If the tibia is stabilized, it rotates the femur laterally; "unlocks" the fully extended knee at the beginning of flexion
Flexor hallucis longus (*flexor,* to bend) (Fig. 12-25B)	Posterior surface of the fibula, interosseous membrane, and intermuscular septum	Base of the distal phalanx of the great toe	Flexes the phalanges of the great toe when the foot is off the ground and plantar flexes the foot; when the foot is on the ground, aids in stabilizing the metatarsal bones and ball of the foot. Is the push-off muscle in walking
Flexor digitorum longus (Fig. 12-25B, C)	Posterior surface of the tibia and adjacent fascia	Forms long flexor tendons of the sole of the foot; ends on bases of distal phalanges	When the foot is off the ground, flexes phalanges of toes as plantar flexes foot; when the foot is on the ground, aids in stabilizing the metatarsal bones and ball of the foot
Tibialis posterior (*posterior,* at or toward the back) (Fig. 12-25B)	Posterior surface of the tibia and interosseous membrane and posterior surface of the fibula, adjacent fascia, intermuscular septum	Tubercle of the navicular, medial, and intermediate cuneiform, and bases of the 2–4 metatarsals	Principal invertor of the foot, affects degree of flattening of foot during walking

TABLE 12-23	MUSCLES THAT ACT ON THE FOOT AT THE ANKLE

Action	Muscles
Plantar Flexors	
Chief	Triceps surae (gastrocnemius-soleus)
Accessory	Plantaris
	Flexor hallucis longus
	Flexor digitorum longus
	Peroneus longus
	Peroneus brevis (?)
	Tibialis posterior (?)
Dorsiflexors	
Chief	Tibialis anterior
	Extensor digitorum longus
	Peroneus tertius
Accessory	Extensor hallucis longus
Invertors and Adductors (Supinators)	
Chief	Tibialis posterior
	Tibialis anterior
Accessory	Flexor hallucis longus
	Flexor digitorum longus
	Triceps surae (gastrocnemius-soleus)
	Extensor hallucis longus
Evertors and Abductors (Pronators)	
Chief	Peroneus longus
	Peroneus brevis
	Peroneus tertius
	Extensor digitorum longus (lateral part)

Muscles of the Foot   Muscles of the foot (Fig. 12-26 and Table 12-24) often parallel those of the hand, particularly on the plantar surface (sole) where they are similar in name and number. However, the term *hallucis* (*hallux,* great toe) is substituted for the term *pollicis* (*pollex,* thumb). Also, the muscles of the foot serve quite different purposes than those of the hand because the foot is adapted for stability, bearing weight, and providing propulsion for walking, running, and jumping, whereas the hand is adapted for grasping.

The muscles on the plantar surface of the foot are covered by a tough thickening of the fascia, the *plantar aponeurosis.* The plantar aponeurosis begins at the calcaneus bone, spreads over the sole of the foot, and splits into five forks as it approaches the toes. It helps sustain the arched form of the foot and gives origin to some of the intrinsic muscles (muscles that lie entirely within the foot). The plantar muscles may be divided into three groups: medial, lateral, and intermediate. The medial group contains the abductor (**abductor hallucis**) and flexor (**flexor hallucis brevis**) of the great toe; the lateral group contains the abductor and flexor of the small toe, i.e., the "digiti minimi." The central group, which is deep to the plantar aponeurosis, contains the lumbricals, interossei, and short flexors of the toes. The dorsal surface of the foot has few muscles.

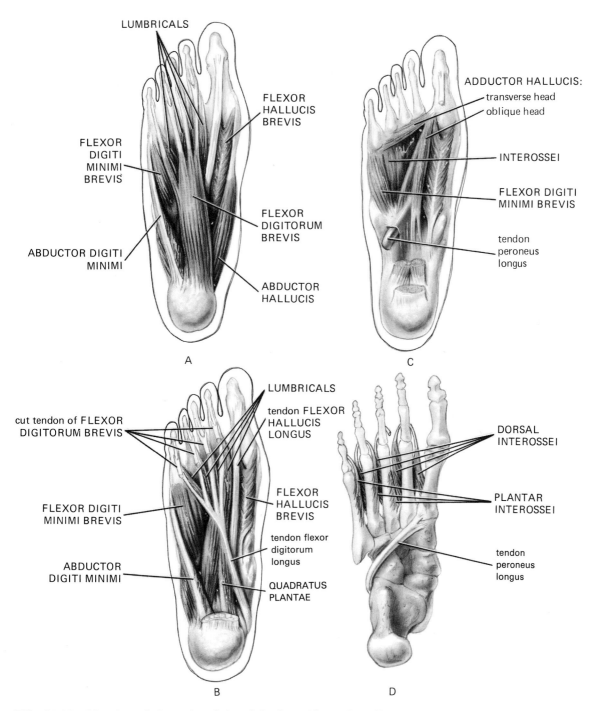

FIG. 12-26 Muscles of the sole of the right foot. Views A to D present successively deeper layers.

TABLE 12-24   MUSCLES OF THE FOOT (*Continued on following page*)

Muscle	Origin	Insertion	Action and Comments
Dorsal Muscles of the Foot (*dorsal,* pertaining to the upper surface of the foot)			
Extensor digitorum brevis (*extensor,* straightens, or opens an angle; *digit,* toe or finger; *brevis,* short, brief) (Fig. 12-25A)	Calcaneus, interosseous ligament, and extensor retinaculum	Base of the proximal phalanx of the great toe, tendons of extensor digitorum longus	Aids in extension of the phalanges of the middle three toes
Plantar Muscles of the Foot (*planta,* sole of the foot) Abductor hallucis (*ab,* from; *ducere,* to draw or lead; *abduct,* to draw away from the body or from an adjacent part; *hallux,* the great toe) (Figs. 12-25B and 12-26A, C)	Calcaneal tuberosity, flexor retinaculum, plantar aponeurosis, and adjacent intermuscular septum	Base of the proximal phalanx of the great toe	Aids in maintaining the concavity of the foot
Flexor digitorum brevis (*flex,* to bend, or make an angle smaller) (Fig. 12-26A)	Calcaneal tuberosity, plantar aponeurosis, adjacent intermuscular septum	Shaft of the intermediate phalanges	Aids in maintaining the concavity of the foot
Abductor digiti minimi (*minimi,* smallest) (Fig. 12-26A, B)	Calcaneal tuberosity, plantar aponeurosis, adjacent intermuscular septum)	Base of the proximal phalanx of the fifth toe	Aids in maintaining the concavity of the foot
Lumbricals (*lumbricus,* earthworm; thus, earthworm-shaped) (Fig. 12-26A, B)	Tendons of flexor digitorum longus	Proximal phalanges	Four small muscles that flex the proximal phalanges
Flexor hallucis brevis (Fig. 12-26A, B, C)	Cuboid and lateral cuneiform bones, tendon of tibialis posterior	Base of the proximal phalanx of the great toe	Flexes the proximal phalanx of the great toe
Adductor hallucis (*ad,* to, or toward; *adduct,* to draw toward a median line) (Fig. 12-26C)	*Oblique head:* bases of 2–4 metatarsal bones; *transverse head:* plantar ligaments of 3–5 toes	Sesamoid bone and base of first phalanx of great toe	Adducts the great toe and aids in maintaining the arches of the foot

TABLE 12-24   MUSCLES OF THE FOOT (*Continued*)

Muscle	Origin	Insertion	Action and Comments
Flexor digiti minimi brevis (Figs. 12-26A, B, C)	Base of the fifth metatarsal and adjacent muscle sheaths	Base of the proximal phalanx of the fifth toe	Flexes the proximal phalanx of the small toe
Dorsal interossei (*inter,* between; *ossa,* bones) (Fig. 12-26D)	Four muscles, each arising by two heads from the sides of the adjacent metatarsal bones	Bases of the proximal phalanges	Abduct the toes
Plantar interossei (Fig. 12-26D)	Three muscles, each arising from the base and side of the 3–5 metatarsal bones	Base of the proximal phalanges of the same toe	Adduct the toes

KEY TERMS

fascia   force
origin   range
insertion   muscle action

REVIEW EXERCISES

1 Differentiate between fibrous and fleshy muscle attachments.
2 Describe the various arrangements of fascicles in muscles; explain the advantages of each.
3 Explain how the muscles produce movement; use the terms *origin* and *insertion* in your discussion.
4 Explain the principles of a lever system; explain the advantages and disadvantages of levers in body movement.
5 Explain the role of agonists, antagonists, and synergists in starting, regulating, and stopping body movements.
6 Explain why it is easier to lift a heavy weight held close to your body than to lift a similar weight at arm's length.
7 Give two examples each of muscles named according to their function, location, size, shape, and bony attachments.
8 Name the muscles involved in smiling, frowning, looking surprised, showing your teeth, and squinting. Describe the action, origin, and insertion of each muscle.

9 List the muscles involved in chewing, giving the action, origin, and insertion of each.
10 Name the muscles that raise the chin and the bones to which each is attached. Repeat the exercise for the muscles that lower the chin and those that turn the head to the side.
11 Describe the various movements of the spinal column; list the muscles involved in these movements.
12 List the muscles that compress the anterior abdominal wall.
13 Name the muscles that raise the shoulders and that draw the shoulders down and back. Name the bones to which these muscles are attached.
14 List the muscles that form the pelvic diaphragm and state their function.
15 List the chief muscles that flex the arm at the shoulder, the forearm at the elbow, the hand at the wrist, and the fingers. Name the bones to which these muscles are attached.
16 Repeat exercise 15 using the extensor muscles.
17 Name the chief muscles that draw the arm to the side and that lift the arm from the side. Name the

bones to which each of these muscles are attached.

18 List the chief muscles that rotate the arm outward; inward. Name the bones to which these muscles are attached.

19 Name the muscles that pronate and supinate the forearm and list the bones to which each muscle is attached.

20 Name the muscles that swing the thumb toward the radius, and that swing the little finger toward the ulna; list the origins and insertions of each muscle.

21 In general terms, compare the size, function, and range of motion of the muscles of the upper and lower limb.

22 Name the chief muscles that flex the thigh at the hip; that extend it. Name the bones to which each muscle is attached.

23 List the chief muscles that extend the leg outward from the midline; that draw it back to the midline. Name the bones to which each muscle is attached.

24 Name the chief muscles that rotate the leg outward at the hip; that rotate it inward. Name the bones to which each muscle is attached.

25 Name the muscles that flex and extend the knee and the bones to which each is attached.

26 Name the chief muscles that act on the foot; give the action of each. Name the bones to which each muscle is attached.

27 Describe the effects of wearing high-heeled shoes on the muscles of the leg. Which muscles are stretched; which are allowed to shorten?

28 Name the muscles that stabilize the knee in standing.

PART

3

CONTROL SYSTEMS
OF THE BODY

CHAPTER

13

NEURAL
TISSUE

The nervous system consists of the brain and spinal cord and the nerves that pass between these two structures and the receptors, muscles, or glands that they innervate. With the other great communications system, the endocrine system, the nervous system is responsible for regulating many internal functions of the body as well as for coordinating the activities we know collectively as "human behavior," activities that include not only easily observed acts such as smiling or reading, for example, but also less apparent behaviors such as feeling angry or being motivated, having an idea or remembering a long past event. Moreover, most neurobiologists consider that the entire range of mental functions engaged in by the brain (those functions we attribute to the "mind") are related to the activities of the nervous system.

FUNCTIONAL ANATOMY OF NEURONS

The basic unit of the nervous system is the individual nerve cell, or **neuron.** Only about 10 percent of the cells in the nervous system are neurons; the remainder are **glial cells,** which sustain the neurons metabolically, support them physically, and help regulate the ionic concentrations in the extracellular space. (The glia do not, however, branch as extensively as the neurons do, so that neurons occupy 50 percent of the volume of the central nervous system.)

Nerve cells occur in a wide variety of sizes and shapes; nevertheless, they can be considered as consisting of three basic parts (Fig. 13-1): (1) the dendrites and cell body, (2) the axon, and (3) the axon terminals. The **dendrites** form a series of highly branched cell outgrowths connected to the **cell body** and may be looked upon as an extension of its membrane. The dendrites and cell body are the site of most of the specialized junctions where signals are received from other neurons. As in other types of cells, the cell body contains the nucleus and many of the organelles involved in metabolic processes and is responsible for maintaining the metabolism of the neuron and for its growth and repair.

The **axon,** or **nerve fiber,** is a single process extending from the cell body. It is long (sometimes more than a meter) in neurons that connect with distant parts of the nervous system or with muscles

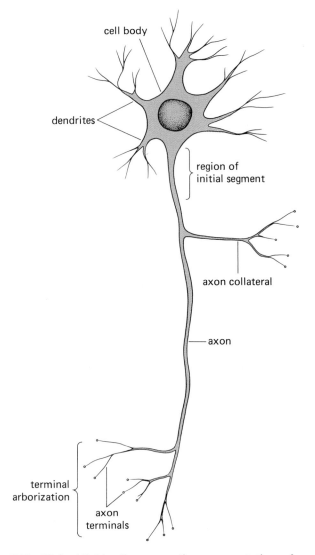

FIG. 13-1 Highly diagrammatic representation of a neuron.

and glands; such cells function as *projection neurons*. Axons are short or even nonexistent in *local circuit neurons,* which connect only with cells in their immediate vicinity. The first portion of the axon plus the part of the cell body where the axon is joined is known as the **initial segment.** The axon may give off branches called **collaterals** along its course, and near the end it undergoes considerable branching into numerous **axon terminals,** which are responsible for transmitting signals from the neuron to the cells contacted by the axon terminals.

The axons of some (but not all) neurons are cov-

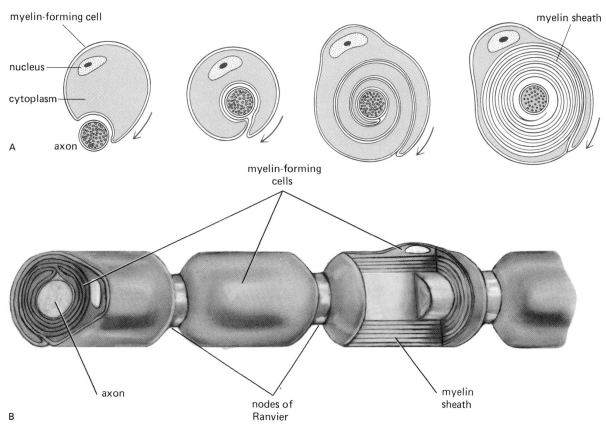

myelin-forming cell

nucleus

cytoplasm

A   axon

myelin sheath

myelin-forming
cells

axon

nodes of
Ranvier

myelin
sheath

B

FIG. 13-2   (A) Myelin-forming cells (Schwann cells in the peripheral nervous system and oligodendroglia in the central nervous system) wrap around the axon, trailing successive layers of their cell membrane. (B) They are separated by a small space, a node of Ranvier.

ered by a fatty material, **myelin,** which electrically insulates the membrane, making it more difficult for local currents to flow between intra- and extracellular fluid compartments (Fig. 13-2).(You may find it helpful at this time to review the material on the electrical properties of cells in Chap. 5.) Action potentials do not occur along the sections of membrane protected by myelin. They occur only where the myelin coating is interrupted and the membrane is exposed to the extracellular fluid. These interruptions, called the **nodes of Ranvier,** occur at regular intervals along the axon. Thus the action potential appears to jump from one node to the next as it propagates along a myelinated fiber, and for this reason this method of propagation is called **saltatory conduction,** from the Latin *saltare,* to leap. This saltatory conduction in myelinated axons results in a more rapid conduction velocity than oc-

curs in an unmyelinated axon of the same diameter. The velocity of action-potential propagation in large myelinated fibers can approach 120 m/sec (250 mi/hr).

Anatomical Classification of Neurons

As we have mentioned, neurons occur in many different shapes (Fig. 13-3). **Multipolar cells** are similar to the typical neuron described above in which one long axon and multiple short, highly branched dendrites extend from the cell body. **Bipolar cells** have, as their name suggests, only two processes, which extend from opposite ends of the cell body. One process (the axon) transmits away from the cell body, and the other (nominally a dendrite) transmits toward it. **Unipolar cells** have only a single short process connected to the cell body; this pro-

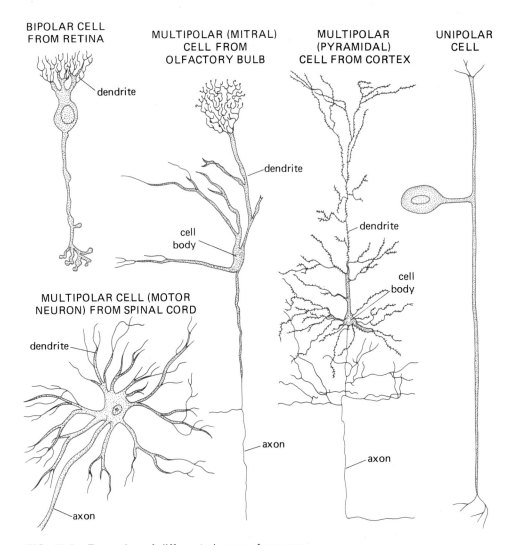

BIPOLAR CELL
FROM RETINA

dendrite

MULTIPOLAR (MITRAL)
CELL FROM
OLFACTORY BULB

dendrite

cell
body

MULTIPOLAR
(PYRAMIDAL)
CELL FROM CORTEX

dendrite

cell
body

axon

UNIPOLAR
CELL

MULTIPOLAR CELL (MOTOR
NEURON) FROM SPINAL CORD

dendrite

axon

axon

FIG. 13-3 Examples of different shapes of neurons.

cess divides into two processes that extend in opposite directions. One, the *central process,* enters the brain or spinal cord; the other, the *peripheral process,* extends out to the periphery. In unipolar cells, it is hard to know what to call these two processes. For example, the peripheral process carries action potentials from receptors in the periphery toward the cell body and is therefore functionally a dendrite; yet it looks like an axon and is, in fact, often called an axon, or an axon-dendrite. Moreover, in unipolar cells the action potential travels directly from one process to another without first passing through the cell body.

Functional Classification of Neurons

The many different shapes of neurons depend on their location and function. Yet, regardless of their shape, neurons can be divided into three functional classes: afferent neurons, efferent neurons, and interneurons (Fig. 13-4). Afferent and efferent neurons lie largely outside the skull or vertebral column, and interneurons lie within the central nervous system. At their peripheral endings **afferent neurons** have *receptors,* which, in response to various physical or chemical changes in their environment, cause action potentials to be generated

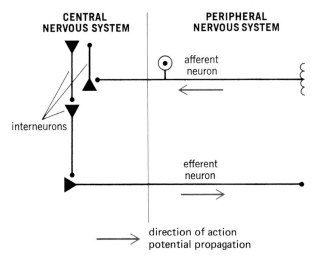

CENTRAL NERVOUS SYSTEM

PERIPHERAL NERVOUS SYSTEM

afferent neuron

interneurons

efferent neuron

direction of action potential propagation

FIG. 13-4 Three classes of neurons. Note that the interneurons are entirely within the central nervous system.

in the afferent neuron. The afferent neurons carry information from the receptors *into* the brain or spinal cord. After transmission to the central nervous system, some of this afferent information may be perceived as a conscious sensation. **Efferent neurons** transmit the final integrated information *from* the central nervous system out to the effector organs (muscles or glands). Efferent neurons that innervate skeletal muscle are also called *motor neurons*. The **interneurons,** which both originate and terminate within the central nervous system, account for 99 percent of all nerve cells. The number of interneurons in the pathway between afferent and efferent neurons varies according to the complexity of the action. One type of basic reflex, the stretch reflex, has no interneurons (the afferent neurons synapse directly upon the efferent neurons), whereas stimuli that invoke memory or language may involve millions of interneurons.

Information is relayed along neurons in the form of action potentials, which are initiated physiologically by the three types of graded potentials: receptor, synaptic, and pacemaker potentials. The frequency of action potentials can be altered to signal different types of information. The mechanisms of action-potential initiation at receptors and synapses and the determinants of action-potential frequency are the subject of this chapter. The spontaneous generation of action potentials by pacemaker potentials in certain neurons occurs in the absence of any identifiable external stimulus and

thus appears to be an inherent property of these neurons. Such neurons are thought to play important roles in rhythmic events such as breathing.

RECEPTORS

Information about the external world and internal environment exists in different energy forms—pressure, temperature gradients, light, sound waves, etc.—but only receptors can deal with these energy forms. The rest of the nervous system can extract meaning only from action potentials or, over very short distances, from graded potentials. Regardless of its original energy form, information from peripheral receptors must be translated into the language of action potentials.

The devices that do this are receptors. These are either specialized peripheral endings of afferent neurons or separate cells whose activity affects the afferent neurons. In the former case, the afferent neuron is activated directly when the stimulus impinges on the specialized membrane of the neuron's receptor ending (Fig. 13-5A); in the latter case, the separate receptor cell contains the specialized membrane that is activated by the stimulus (Fig. 13-5B), and upon stimulation, the receptor cell releases a chemical transmitter. This transmitter diffuses across the extracellular cleft separating the receptor cell from the afferent neuron and activates specific sites on the afferent neuron. In both cases, the result is an electrical change (a graded potential) in the afferent neuron that leads to the generation of action potentials in it.

FIG. 13-5 The sensitive membrane that responds to a stimulus is either (A) an ending of the afferent neuron itself or (B) on a separate cell adjacent to the afferent neuron.

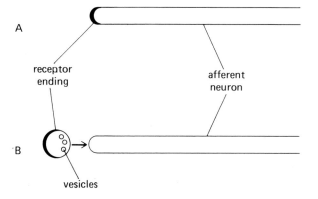

A

receptor ending

afferent neuron

B

vesicles

There are several types of receptors, each of which is specific; i.e., each responds much more readily to one form of energy than to others, although virtually all receptors can be activated by several different forms of energy if the intensity is sufficient. (For example, the receptors of the eye normally respond to light, but they can be activated by intense mechanical stimuli like a poke in the eye.) Usually much more energy is required to excite a receptor by energy forms to which it is not specific. On the other hand, most receptors are exquisitely sensitive to their specific energy forms. For example, olfactory receptors can respond to as few as three or four odorous molecules in the inspired air, and visual receptors can respond to a single photon, the smallest known quantity of light.

Receptor Activation: The Receptor Potential

The basic mechanism of receptor activation, believed to be the same for all types of receptors, is that the stimulus acts on a specialized receptor membrane to increase its permeability and thereby produce a graded potential, called a **receptor potential** (sometimes called a *generator potential*). The membrane mechanisms that generate the potential seem to be the same regardless of whether the receptor membrane is on the afferent neuron itself or on a separate receptor cell and the afferent neuron is activated by transmitters diffusing to it from the receptor cell.

In describing here the general mechanisms for receptor activation we will use the simple example in which the receptor is the peripheral ending of the afferent neuron itself and responds to pressure or mechanical deformation (Fig. 13-6A and B). In subsequent chapters, we shall describe receptors that respond to stimuli such as light and sound.

In the mechanoreceptor, stimuli such as pressure may bend, stretch, or press upon the receptor membrane, and somehow, perhaps by opening up channels in the membrane, increase its permeability to certain ions. Because of the increased permeability, ions move across the membrane down their electric and concentration gradients. Although these ion movements have not been worked out as clearly as those associated with action potentials, the permeability increases in mechanoreceptors appear to be nonselective and to apply to all small ions.

Remember from Chap. 5 that the intracellular

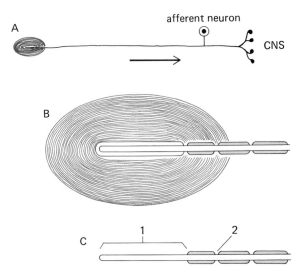

FIG. 13-6 (A) An afferent neuron with a mechanoreceptor (pacinian corpuscle) ending. (B) A pacinian corpuscle showing the nerve-ending modification by cellular structures. (C) The naked nerve ending of the same mechanoreceptor. The receptor potential arises at the nerve ending, 1, and the action potential arises at the first node in the myelin sheath, 2.

fluid of all nerve cells has a higher concentration of potassium and a lower concentration of sodium than the extracellular fluid and that the inside of a resting neuron is about 70 mV negative with respect to the outside. The effect of a nonselective increase in membrane permeability at a stimulated receptor is a net outward diffusion of a small number of potassium ions and the simultaneous movement in of a larger number of sodium ions. The result is net movement of positive charge into the cell, leading to a decrease in membrane potential (depolarization). The movement of sodium into the nerve fiber thus plays the major role in producing the receptor potential.

In the receptor region (Fig. 13-6C, area 1) the cell membrane has a very high threshold, so that the receptor potential cannot depolarize this region enough to cause an action potential. Instead, this depolarization (i.e., the receptor potential) is conducted by local current flow a short distance from the nerve ending to a region where the membrane's threshold is lower (Fig. 13-6C, area 2). In myelinated afferent neurons, this is usually at the first node of the myelin sheath. As is true of all graded potentials, the magnitude of the receptor potential decreases with distance from its site of origin but, if the amount of depolarization that reaches the first node is large enough to bring the membrane there to threshold, action potentials are initiated. The action potentials then propagate along the nerve fiber; the only function of the receptor potential is to trigger action potentials.

Receptor potentials, being graded potentials, are not all-or-none; their amplitude and duration vary with stimulus strength and other variables to be discussed shortly. A change in the amplitude of the receptor potential causes, via local current flow, a similar (but smaller) change in the degree of depolarization at the first node. As long as the first node is depolarized to threshold, action potentials continue to fire and propagate along the afferent neuron. The greater the depolarization of the membrane, the greater the frequency of action potentials; therefore, the amplitude and duration of the receptor potential determine action-potential frequency in the afferent neuron. In fact, it is much more common for stimuli to cause trains, or bursts, of action potentials than single ones.

Amplitude of the Receptor Potential

Since the amplitude and duration of the receptor potential determine the number of action potentials initiated in the afferent neuron, the factors that control these parameters are important. They include the stimulus intensity, rate of change of stimulus intensity, summation of successive receptor potentials, and adaptation. Note that receptor-potential amplitude determines action-potential *frequency,* i.e., number of action potentials fired per unit time; it does *not* determine action-potential magnitude. Since the action potential is all-or-none, its amplitude is always the same regardless of the magnitude of the stimulus.

Intensity and Velocity of Stimulus Application Receptor potentials become larger with greater intensity of the stimulus (Fig. 13-7), possibly because the permeability changes increase and the transmembrane ion movements are greater. The amplitude of the receptor potential also rises with a greater rate of change of stimulus application, i.e., the faster the stimulus is applied, the larger the receptor potential.

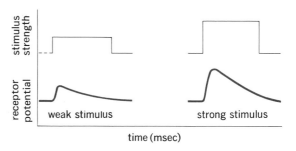

FIG. 13-7 A weak stimulus produces a small receptor potential. The amplitude of the receptor potential increases as stimulus intensity increases.

Summation of Receptor Potentials Another way of varying the amplitude of the receptor potential is by adding two or more together. This is possible because they are graded and because they may last as long as the stimulus is applied. If the receptor membrane is stimulated again before the receptor potential from a preceding stimulus has died away, the two potentials sum and make a larger depolarization.

Adaptation Adaptation is a decrease in frequency of action potentials in the afferent neuron despite a constant stimulus magnitude. The actual reasons for adaptation vary in different receptor types. Adaptation of an afferent neuron in response to the constant stimulation of its mechanoreceptor ending can be seen in Fig. 13-8. Some receptors adapt completely so that in spite of a constantly maintained stimulus, the generation of action potentials stops. In some extreme cases, the receptors fire only once at the onset of the stimulus. In contrast to these rapidly adapting receptors, slowly adapting types merely drop from an initial high action-potential frequency to a lower level, which is then maintained for the duration of the stimulus.

Summary Because the receptor potential is a graded potential, its magnitude can vary with stimulus intensity, rate of change of stimulus intensity, summation, and adaptation. The amplitude of the receptor potential in turn determines the frequency of the action potentials in the afferent neuron.

FIG. 13-8 Action potentials in a single afferent nerve fiber showing adaptation.

stimulus on   stimulus off

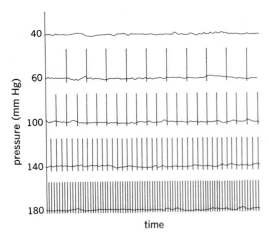

FIG. 13-9 Action potentials recorded from a blood-pressure receptor afferent nerve fiber as the receptor was subjected to pressures of different magnitudes.

Intensity Coding

We are certainly aware of different stimulus intensities. How is information about stimulus strength relayed by action potentials of constant amplitude? One way is related to the frequency of action potentials; increased stimulus strength means a larger receptor potential and higher frequency of firing of action potentials. A record of an experiment in which increased stimulus intensity is reflected in increased action-potential frequency in a single afferent nerve fiber is shown in Fig. 13-9.

In addition to the increased frequency of firing in a single neuron, similar receptors on the endings of other afferent neurons are also activated as stimulus strength increases, because stronger stimuli usually affect a larger area. For example, when one touches a surface lightly with a finger, the area of skin in contact with the surface is small and only receptors in that area of skin are stimulated; pressing down firmly increases the area of skin stimulated. This "calling in" of receptors on additional nerve cells is known as **recruitment.** These generalizations are true of virtually all afferent systems: Increased stimulus intensity is signaled both by an increased firing rate of action potentials in a single nerve fiber and by recruitment of additional receptors on other afferent neurons.

SYNAPSES

A **synapse** is an anatomically specialized junction between two neurons where the electrical activity

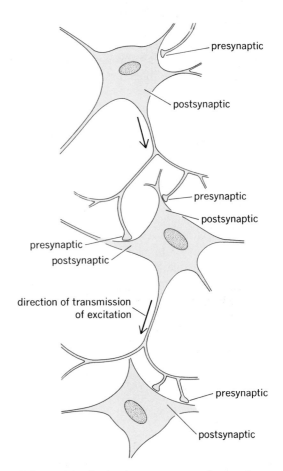

FIG. 13-10 A single neuron can be postsynaptic to one group of cells and presynaptic to another.

in one neuron influences the activity of the second. Most synapses occur between the axon terminals of one neuron and the cell body or dendrites of a second. In certain areas, however, synapses also occur between dendrite and dendrite, dendrite and cell body, or axon and axon. The neurons that conduct information toward synapses are called **presynaptic neurons,** and those that conduct information away are **postsynaptic neurons.** Figure 13-10 shows how, in a multineuronal pathway, a single neuron can be postsynaptic to one group of cells and, at the same time, presynaptic to another.

A postsynaptic neuron may have thousands of synaptic junctions on the surface of its dendrites or cell body so that information from hundreds or even thousands of presynaptic nerve cells converges upon it (Fig. 13-11). A single motor neuron in the spinal cord probably receives some 15,000 synaptic endings, and it has been calculated that certain neurons in the brain receive more than 100,000.

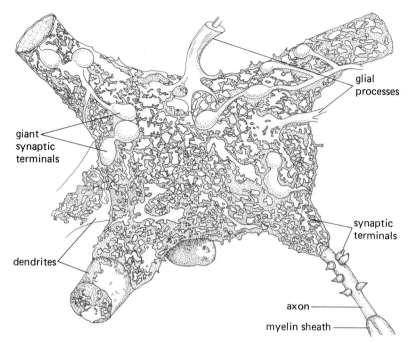

giant
synaptic
terminals

glial
processes

synaptic
terminals

dendrites

axon

myelin sheath

FIG. 13-11 Synaptic endings on a nerve cell body. (There are also some glial processes.)

Each activated synapse produces a brief, small electric signal, either excitatory or inhibitory, in the postsynaptic cell. The picture we are left with is one of thousands of synapses from many different presynaptic cells *converging* upon a single postsynaptic cell. The level of excitability of this cell at any moment, i.e., how close the membrane potential is to threshold, depends upon the number of synapses active at any one time and how many are excitatory or inhibitory. If the postsynaptic neuron reaches threshold it will generate action potentials.

FIG. 13-12 Convergence of neural input from many neurons onto a single neuron and divergence of output from a single neuron onto many others (here only a few).

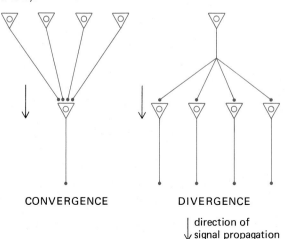

CONVERGENCE   DIVERGENCE

↓ direction of
↓ signal propagation

These are transmitted out along its axon to the terminal branches, which *diverge* to influence the excitability of many other cells. Figure 13-12 demonstrates the neuronal relationships of convergence and divergence.

In this manner, postsynaptic neurons function as neural **integrators;** i.e., their output reflects the sum of all the incoming bits of information arriving in the form of excitatory and inhibitory synaptic inputs.

Functional Anatomy

Synapses are of two types: *electrical* and *chemical*. At electrical synapses, the local currents that result from action potentials in the presynaptic neuron flow through gap junctions (see Chap. 4) directly into the postsynaptic cell. Chemical synapses are much more common in the mammalian nervous system, and it is this type of synapse that is discussed in the following sections.

Figure 13-13 shows the anatomy of a single chemical synaptic junction. The axon terminal of the presynaptic neuron ends in a slight swelling, the **synaptic terminal.** A narrow extracellular space, the **synaptic cleft,** separating the pre- and postsynaptic neurons, prevents direct propagation of the action potential from the presynaptic neuron to the postsynaptic cell.

Information is transmitted across the synaptic cleft by means of a chemical **neurotransmitter** sub-

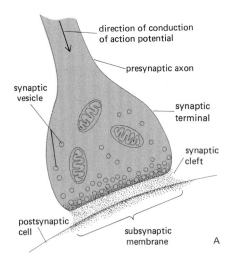

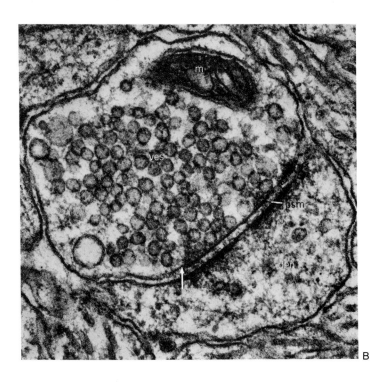

FIG. 13-13 (A) Diagram of synapse. (B) Electron micrograph of synapse. Synaptic terminal contains vesicles (ves) and a mitochondrion (m), and dendrite (den) shows thickening of postsynaptic membrane (psm). Arrow marks beginning of synaptic cleft.

stance stored in small, membrane-enclosed vesicles in the synaptic terminal. When an action potential in the presynaptic neuron reaches the end of the axon and depolarizes the terminal, small quantities of the neurotransmitter are released from the synaptic terminal into the synaptic cleft. The link between membrane depolarization and neurotransmitter release is calcium. Depolarization of the synaptic terminal causes an increase in the permeability of the terminal membrane to calcium ions. Calcium enters the presynaptic terminal during the action potentials and causes some of the vesicles to fuse with the cell membrane and liberate their contents into the synaptic cleft. Once released from the vesicles, the transmitter molecules diffuse across the cleft and bind to receptor sites on the membrane of the postsynaptic cell lying right under the synaptic terminal (**subsynaptic membrane**). The combination of the transmitter with the receptor site causes changes in the permeability of the subsynaptic membrane and thereby in the membrane potential of the postsynaptic cell.

Because all the transmitter is stored on one side of the synaptic cleft and all the receptor sites are on the other side, chemical synapses operate in only one direction. Because of this one-way conduction across synapses, action potentials are transmitted along a given multineuronal pathway in only one direction.

There is a delay between excitation of the synaptic terminal and membrane-potential changes in the postsynaptic cell; it lasts less than a thousandth of a second and is called the **synaptic delay.** The delay is caused by the mechanism that releases transmitter substance from the synaptic terminal, since the time required for the transmitter to diffuse across the synaptic cleft is negligible. The postsynaptic permeability change is then terminated when the transmitter is removed from the subsynaptic membrane. At different synapses this may be caused by (1) its chemical transformation into an ineffective substance, (2) diffusion away from the receptor site, or (3) reuptake by the synaptic terminal or, in some cases, by nearby glial cells.

Excitatory Synapse   The two kinds of chemical synapses, excitatory and inhibitory, are differentiated by the effects that the neurotransmitter has on the postsynaptic cell. Whether the effect of the transmitter is inhibitory or excitatory depends on the structure of the subsynaptic membrane. An excitatory synapse, when activated, increases the likelihood that the membrane potential of the postsynaptic cell will reach threshold and generate ac-

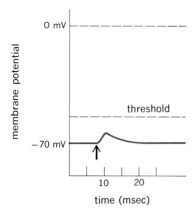

FIG. 13-14   Excitatory postsynaptic potential (EPSP). Stimulation of the presynaptic neuron is marked by the arrow. Note the short synaptic delay before the postsynaptic cell responds.

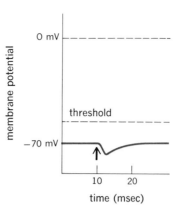

FIG. 13-15   Inhibitory postsynaptic potential (IPSP). Stimulation of the presynaptic neuron is marked by the arrow. Note the short synaptic delay.

tion potentials. Here the effect of the neurotransmitter-receptor combination is to increase the permeability of the subsynaptic membrane to sodium and potassium ions so that they are free to move according to the electric and chemical forces acting upon them. Thus, at the subsynaptic membrane of an excitatory synapse there occurs the simultaneous movement of a relatively small number of potassium ions out of the cell and a larger number of sodium ions into the cell. The *net* movement of positive ions is into the neuron, which slightly depolarizes the postsynaptic cell. This potential change, called the **excitatory postsynaptic potential (EPSP)**, brings the membrane closer to threshold (Fig. 13-14). The EPSP, like the receptor potential, is a graded potential that spreads decrementally, by local current flow; its only function is to help trigger action potentials.

Inhibitory Synapse   Activation of an inhibitory synapse produces changes in the postsynaptic cell that lessen the likelihood that the cell will generate an action potential. At inhibitory synapses the combination of the neurotransmitter molecules with the receptor sites on the subsynaptic membrane also changes the permeability of the membrane, but only the permeabilities to potassium or chloride ions are increased; sodium permeability is not affected. At some inhibitory synapses the transmitter acts to increase potassium permeability; at others (e.g., motor neurons) it increases chloride permeability; and at some it may increase both. Earlier it was

noted that if a cell membrane were permeable only to potassium ions the resting membrane potential would equal the potassium equilibrium potential; i.e., the resting membrane potential would be -90 mV instead of -70 mV. The increased potassium permeability at an activated inhibitory synapse makes the postsynaptic cell more like the hypothetical cell that is permeable only to potassium ions. Consequently, the membrane potential becomes closer to the potassium equilibrium potential. This increased negativity (hyperpolarization) is an **inhibitory postsynaptic potential (IPSP)** (Fig. 13-15). Thus, when an inhibitory synapse is activated, the postsynaptic neuron's membrane potential is moved farther away from the threshold level. In many cells the chloride equilibrium potential is also more negative (e.g., -80 mV) than the resting potential so that an increase in chloride permeability would have a similar hyperpolarizing effect. Even in cells in which the equilibrium potential for chloride is very close to the resting membrane potential, a rise in chloride-ion permeability will still lessen the likelihood that the cell will reach threshold, the reason being that it increases the tendency of the membrane to stay at the resting potential.

Activation of the Postsynaptic Cell: Neural Integration

A feature that makes postsynaptic integration possible is that, in most neurons, one excitatory synaptic event is not enough by itself to change the membrane potential of the postsynaptic neuron

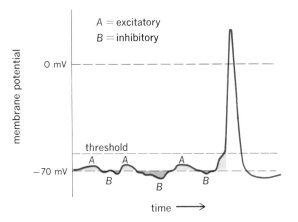

FIG. 13-16 Intracellular recording from a postsynaptic cell during episodes when (A) excitatory synaptic activity dominates and the cell is facilitated and (B) inhibitory synaptic activity predominates.

from its resting level to threshold; e.g., a single EPSP in a motor neuron is estimated to be only 0.5 mV whereas changes of up to 25 mV are necessary to depolarize the membrane from its resting level to threshold. Since a single synaptic event does not bring the postsynaptic membrane to its threshold level, an action potential can be initiated only by the combined effects of many synapses. Of the thousands of synapses on any one neuron, probably hundreds are active simultaneously (or at least close enough in time that the effects can summate), and the membrane potential of the neuron at any one moment is the result of all the synaptic activity affecting it at that time. There is a general depolarization of the membrane toward threshold when excitatory synaptic activity predominates (this is known as **facilitation**) and a hyperpolarization when inhibition predominates (Fig. 13-16).

Let us perform a simplified experiment to see how two EPSPs, two IPSPs, or an EPSP plus an IPSP interact (Fig. 13-17). Let us assume that there are three synaptic inputs to the postsynaptic cell; *A* and *B* are excitatory and *C* is an inhibitory synapse. There are stimulators on the axons to *A, B,* and *C* so that each of the three inputs can be activated individually. A very fine electrode is placed in the cell body of the postsynaptic neuron and connected to record the membrane potential. In part I of the experiment we shall test the interaction of two EPSPs by stimulating *A* and then, a short while later, stimulating *A* again. Part I of Fig. 13-17 shows that no interaction occurs between the two EPSPs. The reason is that the change in membrane potential associated with an EPSP is fairly short-lived. Within a few milliseconds (by the time axon *A* is restimulated) the postsynaptic cell has returned to its resting condition. In part II of the experiment, axon *A* is restimulated before the first EPSP has died away, and so-called **temporal summation** occurs. In part III, axons *A* and *B* are stimulated simultaneously and the two EPSPs that result also summate in the postsynaptic neuron; this phenomenon is called **spatial summation.** The summation of EPSPs can bring the membrane to its threshold so that an action potential is initiated. So far we have tested only the patterns of interaction of excitatory synapses. What happens if an excitatory and inhibitory synapse are activated so that their effects occur at the postsynaptic cell simultaneously? Since EPSPs and IPSPs are due to local currents flowing in opposite directions, they tend to cancel each other, and the resulting potential change is smaller (Fig. 13-17, part IV). Inhibitory potentials can also show temporal and spatial summation.

As described above, the subsynaptic membrane is depolarized at an activated excitatory synapse

FIG. 13-17   Interaction of EPSPs and IPSPs at the postsynaptic neuron.

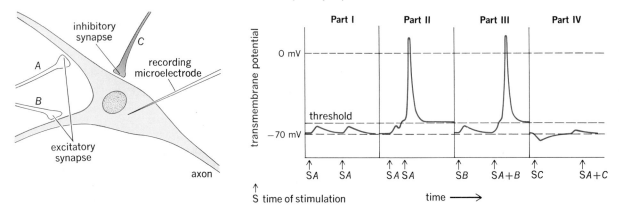

and hyperpolarized at an activated inhibitory synapse. By the mechanisms of local current flow described in Chap. 5, current flows through the cytoplasm *away from* an excitatory synapse, whereas at an inhibitory synapse the current flow through the cytoplasm is *toward* the synapse. Thus, the entire cell body, including the initial segment, becomes slightly depolarized during activation of excitatory synapses and slightly hyperpolarized during activation of inhibitory synapses.

In the above examples we referred to the threshold of the postsynaptic neuron. However, the fact is that different parts of a neuron have different thresholds. The cell body and larger dendritic branches reach threshold when their membrane is depolarized about 25 mV from the resting level, but in many cells the initial segment has a threshold that is much closer to the resting potential. In cells whose initial-segment threshold is lower than that of their dendrites and cell body, the initial segment is activated first whenever enough EPSPs summate, and the action potential that originates there is propagated down the axon.

Individual postsynaptic potentials last much longer than do action potentials. In the event that the initial segment is still depolarized above threshold after an action potential has been fired and the refractory period is over, a second action potential will occur. In fact, the greater the summed postsynaptic depolarization, the greater is the number of action potentials fired (up to the limit imposed by the duration of the absolute refractory period). Neuronal responses are almost always in the form of so-called bursts or trains of action potentials.

This discussion has no doubt left the impression that one neuron can influence another only at a synapse. This view requires qualification. We have emphasized how local currents influence the membrane of the cell in which they are generated (as, for example, in action-potential propagation and synaptic excitation). Under certain circumstances, local currents in one neuron may directly affect the membrane potential of other nearby neurons. This phenomenon is particularly important in areas of the central nervous system that contain a high density of unmyelinated neuronal processes.

Neurotransmitters

As described above, presynaptic neurons influence postsynaptic neurons by means of chemical trans-

TABLE 13-1 SOME CHEMICALS KNOWN OR PRESUMED TO BE NEUROTRANSMITTERS

Dopamine ⎫
Norepinephrine ⎬ ← Catecholamines
Epinephrine ⎭
Acetylcholine
Serotonin
Gamma-aminobutyric acid
Glycine
Aspartic acid
Glutamic acid
Enkephalins
Endorphins
Substance P
Somatostatin
Histamine
Adenosine

mitters. There are many different suspected synaptic transmitters used by different types of cells in the nervous system (Table 13-1). All synaptic terminals from a given cell probably liberate the same transmitter. It must be emphasized that many of these same molecules are found in nonneural tissue and function as hormones and paracrine agents. In other words, evolution has selected the same chemical messengers for use in widely differing circumstances.

Modification of Synaptic Transmission by Drugs and Disease

The great majority of drugs that act upon the nervous system do so by altering synaptic mechanisms, and all the synaptic mechanisms labeled in Fig. 13-18 are vulnerable. Drugs that, upon binding to the receptor, produce a response similar to that caused by the transmitter released at the synapse are called **agonists.** On the other hand, agents that bind to the receptor but are unable to activate it are **antagonists.** By occupying the receptors, antagonists prevent binding by the neurotransmitter.

Disease processes, too, often affect synaptic mechanisms. For example, the toxin produced by the tetanus bacillus acts at inhibitory synapses on neurons that supply skeletal muscles, presumably by blocking the receptors. This eliminates inhibitory input to the neurons and permits unchecked

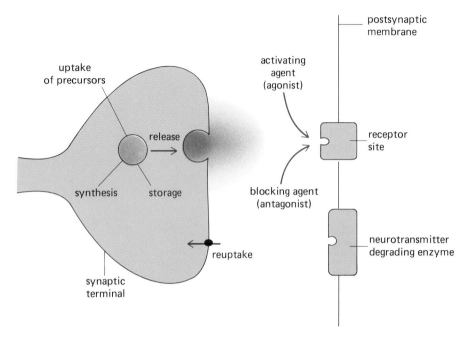

FIG. 13-18   Sites of drug action at synapses.

influence of the excitatory inputs, leading to muscle spasticity and seizures; the spasms of the jaw muscles that appear early in the disease are responsible for the common name, lockjaw.

Presynaptic Inhibition

As discussed above, the activation of an inhibitory synapse hyperpolarizes the postsynaptic cell, thereby decreasing the effect of all excitatory inputs to that cell. A second type of inhibitory action, called **presynaptic inhibition,** provides a means by which only certain excitatory inputs are depressed.

Presynaptic inhibition works by affecting the transmission at a single excitatory synapse. The structures that underly presynaptic inhibition consist of a synaptic junction between the terminal of the inhibitory fiber (*A* in Fig. 13-19) and the synaptic

FIG. 13-19   Presynaptic inhibition.

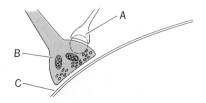

terminal of the excitatory neuron (neuron *B*). When activated, this synapse reduces the amount of the transmitter released by neuron *B*. Therefore, the size of the EPSP in the postsynaptic cell (neuron *C*) will be less influenced by input from neuron *B*.

NEUROMODULATORS

The concept of neuromodulation is new and still not clearly delineated. **Neuromodulators** do not, themselves, function as specific neurotransmitters but rather amplify or dampen neuronal activity by altering neurons directly or by influencing the effectiveness of a neurotransmitter. They might do the latter by affecting the neurotransmitter's synthesis, release, interaction with receptor sites, reuptake, or metabolism. For example, the hormone cortisol, by influencing the steady-state levels of the rate-limiting enzyme in the synthesis of norepinephrine, modulates the activity of norepinephrine-mediated synapses.

From this example, it is clear that neuromodulators may be hormones, i.e., substances that are secreted by endocrine glands into the blood via which they reach their distant target cells (Chap. 18). However, not all hormones act as neuromodulators, nor are all neuromodulators hormones.

TABLE 13-2 SOME POSSIBLE NEUROMODULATORS
Histamine
Prostaglandins
Enkephalins
Endorphins
Substance P
Somatostatin
Aldosterone
Cortisol
Estrogens
Testosterone
Thyroid hormone
Gastrin
Vasoactive intestinal peptide (VIP)
Adrenocorticotropic hormone (ACTH)
Thyrotropin-releasing hormone (TRH)
Gonadotropin-releasing hormone (GnRH)
Vasopressin (antidiuretic hormone, ADH)
Angiotensin
Oxytocin

The others are paracrine agents that are synthesized and released locally. In this last regard we do not know whether it is possible for a neuron to secrete both neurotransmitters and neuromodulators. It is also unclear whether a given substance may act both as a neurotransmitter in one area of the nervous system and as a neuromodulator in the same or different area.

Table 13-2 lists some of those substances presently thought to function as neuromodulators. Trying to memorize such a table is of no value. We have presented it only to emphasize a particular point: Comparison of this table with Table 13-1 (neurotransmitters) and Table 18-1 (hormones) emphasizes the overlapping functions of neuromodulators. In this regard, each of the neuromodulators shown in the table will be encountered in a nonneuromodulator context elsewhere in the book; i.e., all presently known neuromodulators also exert nonneural effects. All this emphasizes one of the most exciting developments in physiology: A given chemical messenger may be synthesized in differing sites and may serve as a neurotransmitter, paracrine agent, or hormone. The brain and gastrointestinal tract, in particular, seem to produce many of the same peptide messengers.

NEUROEFFECTOR COMMUNICATION

Efferent neurons innervate muscle or gland cells. The junction between nerve fibers and skeletal muscle cells, the **neuromuscular junction,** was described in Chap. 11. The junctions between nerve fibers and smooth and cardiac muscle cells or gland cells are given the more general term, **neuroeffector junctions.** Such junctions are basically like synapses, but there are important structural and functional differences. As the nerve fibers approach the effector cells, the axons form discrete swellings, or *varicosities,* that give the axon a beaded appearance. Chemical transmitter substance, which is stored in vesicles in the varicosities, is released during action potentials in the axon. The extracellular space, analogous to the synaptic cleft, separating the varicosities from nearby effector cells may be as small as 15 to 20 nm, but cells as far as 1 to 2 μm may also respond to transmitter released from the varicosity. In cases where the extracellular cleft is small, there may be a specialized membrane, analogous to the subsynaptic membrane, on the effector cell, but where the cleft is wide there may be no membrane specialization at all. Thus, the absence of a synapse-like structure does not indicate the absence of neural influence on a given effector cell. Neuroeffector junctions can be effective providing that the chemical transmitter can diffuse to the cell from the varicosity and that appropriate receptor sites exist on the effector cell. The major known neuroeffector transmitters are acetylcholine and norepinephrine.

PATTERNS OF NEURAL ACTIVITY

In order to illustrate some of the complexity exhibited by neural control systems and to review the neural mechanisms presented in the previous sections of this chapter, we will examine one of the simpler reflexes—the flexion reflex.

The Flexion Reflex

The sequence of events elicited by a painful stimulus to the toe leads to withdrawal of the foot from the source of injury: This is an example of the flexion, or withdrawal, reflex (Fig. 13-20). Receptors in the toe are stimulated and, via receptor potentials, transform the energy of the stimulus into

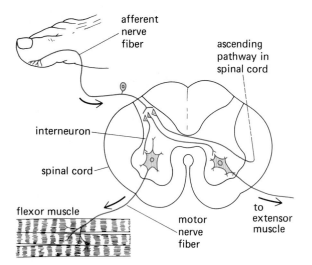

afferent
nerve
fiber

ascending
pathway in
spinal cord

interneuron

spinal cord

flexor muscle

motor
nerve
fiber

to
extensor
muscle

FIG. 13-20 Flexion, or withdrawal, reflex pathway. Arrows indicate direction of action potential propagation.

the electrochemical energy of action potentials in afferent neurons. Information about the intensity of the stimulus is coded both by the frequency of action potentials in single afferent fibers and by the number of afferent fibers activated. The afferent fibers branch after entering the spinal cord, each branch terminating at a synaptic junction with another neuron. In the flexion reflex, the second neurons in the pathway are interneurons. Because interneurons are interspersed between the afferent and efferent limbs of the reflex arc, the flexion reflex is one of the very large class of **polysynaptic** reflexes.

The afferent-neuron branches serve different functions. Some branches synapse with interneurons whose processes carry the information to the brain; it is only after the information transmitted in these pathways reaches the brain that the conscious correlate of the stimulus, i.e., the sensation of pain, is experienced. Other branches synapse with interneurons that, in turn, synapse upon the efferent neurons innervating flexor muscles. These muscles, when activated, cause flexion (bending) of the ankle and withdrawal of the foot from the stimulus. If the

stimulus is very intense and the afferent discharge of very high frequency, the number of motor neurons that fire action potentials is large, and muscles of other joints are activated so that the knee and thigh also flex.

Still other branches of the afferent neurons activate interneurons that inhibit the motor neurons of antagonistic muscles whose activity would oppose the reflex flexion. This inhibition of antagonistic motor neurons is called **reciprocal innervation.** In this case, the motor neurons innervating the foot and leg extensor muscles (which straighten the ankle and leg) are inhibited. Other interneurons excite extensor and inhibit flexor motor neurons that control movement of the opposite leg so that as the injured leg is flexed away from the stimulus, the opposite leg is extended more strongly to support the added share of the body's weight. This involvement of the opposite limb is called the **crossed-extensor reflex.**

The neural networks activated in the flexion reflex are among the simplest in humans, yet the reflex response is purposeful and coordinated. The foot flexion away from the stimulus removes the receptors from the stimulus source, and further damage is prevented (an example of negative feedback). The purposefulness of the reflex response is independent of the sensation of pain, for the withdrawal occurs before the sensation of pain is experienced. In fact, the reflex response can still be elicited after the spinal cord pathways that would normally transmit the afferent information up to the brain have been severed. An important characteristic of this type of control system is that the observed response varies with stimulus strength (between threshold and maximal stimulus strength). In general, low levels of stimulation lead to responses localized to the area of stimulus application whereas higher levels of stimulation cause more widespread responses. Thus, to take another example, the hotter an object, the more intense is the response to touching it. First one simply withdraws the finger; if it is hotter, the wrist and arm are pulled away; if hotter still, one jumps away with appropriate exclamations.

KEY TERMS

neuron	motor neuron	convergence
dendrite	receptor	divergence
axon	receptor potential	EPSP
nerve fiber	summation	IPSP
myelin	recruitment	facilitation
afferent neuron	synapse	neurotransmitter
efferent neuron	presynaptic	neuroeffector junction
interneuron	postsynaptic	polysynaptic reflex

REVIEW EXERCISES

1　Differentiate between neurons and glia in terms of numbers, function, and volume in the nervous system.

2　Describe the three basic parts of a neuron.

3　Distinguish between projection neurons and local circuit neurons.

4　Draw and label a neuron; include cell body, dendrites, initial segment, axon, axon collaterals, and axon terminals.

5　Describe the structure and function of myelin.

6　Draw a multipolar, bipolar, and unipolar neuron; where possible, label the axon, dendrites, and cell body.

7　Describe the location and function of afferent neurons, efferent neurons, and interneurons.

8　Describe the formation and function of a receptor potential.

9　Describe the functions of receptors.

10　Explain the relationship between receptor-potential amplitude and duration and action-potential frequency.

11　Describe how the coding of stimulus intensity is accomplished by variations in firing frequency and recruitment.

12　Differentiate between presynaptic and postsynaptic neurons.

13　Draw a diagram showing the difference between convergence and divergence.

14　Draw a synapse; label the synaptic terminal, synaptic cleft, subsynaptic membrane, and synaptic vesicles.

15　List three mechanisms by which transmitter is removed from the synaptic cleft.

16　Differentiate between excitatory and inhibitory synapses; describe the results of each on the resting membrane potential of the postsynaptic cell.

17　Differentiate between temporal and spatial summation and describe their relationship to facilitation, excitation, or inhibition of the postsynaptic cell.

18　Describe the postsynaptic cell as integrator using the concepts in exercises 16 and 17.

19　Differentiate between neurotransmitters and neuromodulators.

20　Describe presynaptic inhibition and explain how it differs from postsynaptic inhibition.

21　Define neuroeffector transmission and give two examples.

22　Describe the flexion reflex, including a discussion of polysynaptic reflexes, reciprocal innervation, and the crossed-extensor reflex.

CHAPTER

STRUCTURE OF
THE NERVOUS
SYSTEM

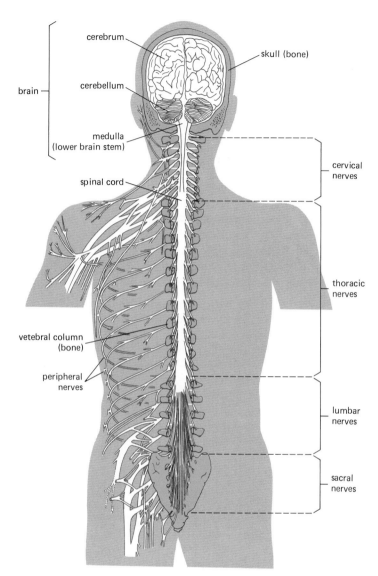

cerebrum

skull (bone)

cerebellum

brain

medulla
(lower brain stem)

cervical
nerves

spinal cord

thoracic
nerves

vetebral column
(bone)

peripheral
nerves

lumbar
nerves

sacral
nerves

FIG. 14-1   Nervous system viewed from behind. The back of the skull and backs of the
bony vertebrae have been removed to expose
the brain and spinal cord. The cranial nerves
are not shown.

The various parts of the nervous system are interconnected, but for convenience they can be divided
into the **central nervous system,** composed of the
brain and spinal cord, and the **peripheral nervous
system,** consisting of the nerves extending from the
brain and spinal cord (Fig. 14-1). The anatomy of
the nervous system is extremely complex, and its
basic features can best be approached in terms of
its embryologic development.

Three layers of cells, the *ectoderm, mesoderm,*
and *entoderm,* give rise to all the tissues and organs
of the body. The central nervous system develops
from a column of ectoderm cells on the dorsal surface of the embryo. As the cells divide, increasing
in number, a **neural groove** forms along the length

of the column, with a ridge along each side (Fig.
14-2). The groove dips deeper and deeper into the
mesoderm layer of the embryo and eventually seals
over at the top, forming the **neural tube.** The head,
or rostral, end of the neural tube becomes the brain,
changing during the fourth and fifth weeks of development from a simple to a complex tube with
five distinct regions, or *vesicles,* each corresponding to one of the five basic regions of the adult brain
(Fig. 14-3). The caudal (tailward) part of the neural
tube develops into the spinal cord. The hollow cavities within the neural tube are filled with fluid, the
cerebrospinal fluid. As the brain develops, the
shapes of the cavities change but they remain connected with each other through narrow channels

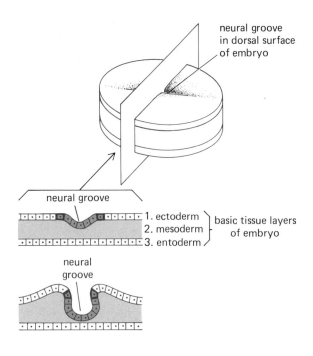

neural groove
in dorsal surface
of embryo

neural groove

1. ectoderm
2. mesoderm   } basic tissue layers of embryo
3. entoderm

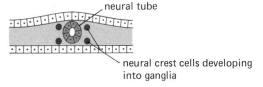

neural
groove

neural tube

neural crest cells developing
into ganglia

FIG. 14-2 Successive early stages in the development of the nervous system.

(*aqueducts*); these interconnected fluid-filled cavities form the **ventricular system** of the mature central nervous system.

The caudal end of the primitive neural tube is lined with cells that divide many times to form the billions of neurons of the spinal cord. This cell division leads to the formation of the central **gray matter,** which is composed of neuronal cell bodies (and neuroglia). Clusters of neurons in the gray matter are called **nuclei,** or *cell columns*. Processes (primarily axons) grow out from these cell bodies to form a layer on the outer surface of the tube; this is the early **white matter** (Fig. 14-4).

Some of the cells in the neural tube send axons out to the primitive muscles and glands, becoming one group of efferent neurons of the peripheral nervous system. Cells from two small columns, the

FIG. 14-3 Schematic dorsal (top row) and lateral (bottom row) views of the developing and adult brains. The hollow center of the neural tube becomes the ventricular system in the adult brain. The terms that identify the various brain regions will be discussed later.

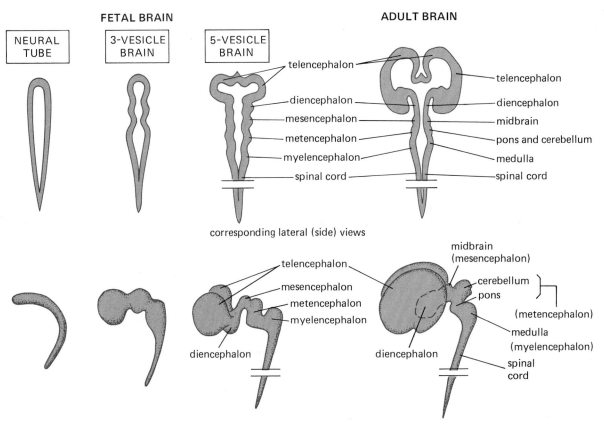

FETAL BRAIN   ADULT BRAIN

NEURAL TUBE   3-VESICLE BRAIN   5-VESICLE BRAIN

telencephalon   telencephalon
diencephalon   diencephalon
mesencephalon   midbrain
metencephalon   pons and cerebellum
myelencephalon   medulla
spinal cord   spinal cord

corresponding lateral (side) views

telencephalon   midbrain (mesencephalon)
mesencephalon   cerebellum
metencephalon   pons
myelencephalon   (metencephalon)
                           medulla (myelencephalon)
diencephalon   diencephalon   spinal cord

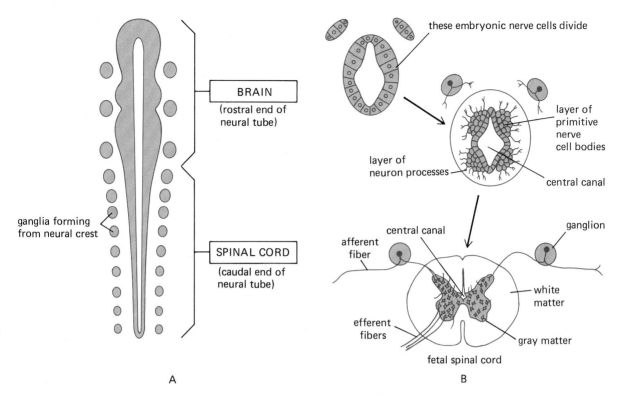

FIG. 14-4 Early stages in the formation of the spinal cord and afferent and efferent neurons. (A) Coronal section through the developing nervous system. (B) Cross sections cut at the spinal cord region in (A).

neural crest columns, become the afferent neurons. These cells send out a process that branches, one branch migrating outward with the peripheral nerve to connect with or become a sensory receptor, the other migrating centrally to enter the brain or spinal cord where it synapses with other neurons. The important result is that the cell bodies of all afferent neurons lie outside the central nervous system. These cell bodies of the afferent neurons form the **dorsal root ganglia,** known also as *sensory ganglia* at the level of the spinal cord. (A **ganglion** is any cluster of neuron cell bodies *outside* the central nervous system; such clusters *inside* the brain or spinal cord are called *nuclei,* or singular, *nucleus.*) Other neural crest cells migrate away from the original columns and differentiate into a second type of ganglia, those for the autonomic neurons. (The autonomic nervous system will be described later in this chapter.)

SUPPORTING STRUCTURES OF THE NERVOUS SYSTEM

The supporting structures of the nervous system develop in the embryo from the mesoderm tissue that immediately surrounds the neural tube. The outer layer (the skull and vertebral column) is bony. Between the bone and the nervous tissue are three membranes, or **meninges.**

Skull

Certain bones of the skull are adapted to support and protect the brain and several of the special sense organs and to isolate the cerebral blood vessels from extracranial variations in pressure. These bones form the **cranium;** they are shown in Figs. 9-4, 11, and 15. When the top of the skull and the brain are removed to expose the floor of the cranial cavity, the *anterior, middle,* and *posterior cranial fossae* are exposed (Fig. 9-15); these support the base of the brain. As mentioned in Chap. 9, the posterior cranial fossa, which accommodates the parts of the brain known as the cerebellum, pons, and medulla, is perforated by a large opening, the *foramen magnum,* which encircles the central nervous system at the transition between the medulla and spinal cord. Blood vessels and connective-tissue structures also pass through it.

Vertebral Column

The spinal cord is surrounded in its entire length by the bony vertebral column, which was discussed in

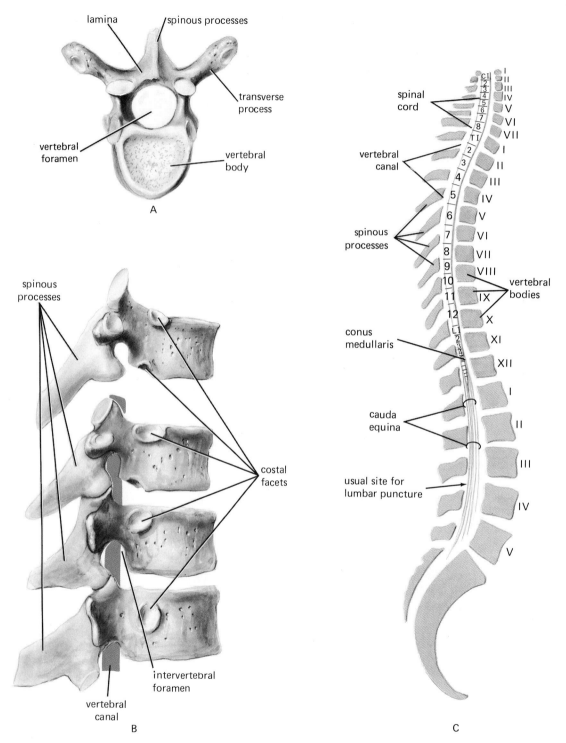

FIG. 14-5 (A) Typical vertebra (superior view). (B) Location of the vertebral canal relative to several vertebrae (lateral view). (C) Position of the spinal cord within the vertebral canal. Posterior is toward the left in B and C.

detail in Chap. 9. We are concerned here only with its relationship to the spinal cord. The spinal cord extends from the point where it joins the brain at the foramen magnum down to the region of the second lumbar vertebra, a length of about 45 cm (18 in). It lies in the *vertebral canal,* a passage through the successive *vertebral foramens* of the interconnected vertebrae (Fig. 14-5). Rather than ending abruptly, the lower part of the spinal cord narrows, forming the *conus medullaris.*

Meninges

Within the skull and vertebral column, respectively, the brain and spinal cord are enclosed in three membranous coverings, or **meninges** (Fig. 14-6): the **dura mater** (next to the bone), the **arachnoid** (in the middle), and the **pia mater,** which is next to the nervous tissue. There is a space, the **subarachnoid space,** between the pia and arachnoid, which is filled with cerebrospinal fluid. The pia follows the contours of the nervous tissue more closely than the arachnoid does, and in some places the two layers are quite widely separated. These spaces, called **cisterns,** contain large pools of cerebrospinal fluid; they are not shown in Fig. 14-6. A **subdural space** separates the dura and arachnoid, but it is so narrow that it is better termed a potential space.

The spinal cord is anchored to the dura by the *dentate ligaments,* which are narrow, fibrous bands lying on each side of the spinal cord (Fig. 14-6).

FIG. 14-6 Cross section through the spinal cord and its membranes.

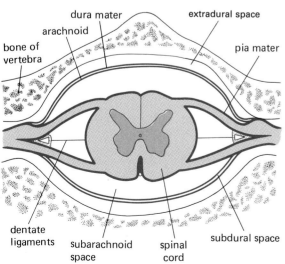

They extend from the pia mater, along which they form a continuous band running the length of the spinal cord, out to the dura mater where they are attached at intervals between the exiting nerves.

As the spinal cord ends, the pia mater becomes continuous with a connective-tissue thread, the **filum terminale,** which descends in the midst of a bundle of nerve roots passing from the spinal cord down through the vertebral canal to their point of exit to the periphery. These nerve roots resemble a horsetail and have, therefore, been named the **cauda equina** (Latin *cauda,* tail; *equus,* horse) (Fig. 14-5C). The filum terminale descends past the point of exit of the nerve roots and finally becomes closely associated with the dura and with it attaches to the back of the coccyx.

The subarachnoid space surrounding the filum is fairly wide, and samples of cerebrospinal fluid may be obtained from it by passing a needle through the midline of the back, between the spines of the third and fourth (or fourth and fifth) lumbar vertebrae, and through the dura mater into the subarachnoid space, care being taken to avoid damaging the nerve roots of the cauda equina. This procedure is called a *lumbar, or spinal, puncture.*

The dura that surrounds the spinal cord is attached to bone in relatively few places. Over most of its surface it is separated from the periosteum and ligaments of the vertebral canal by an **extradural** (or *epidural*) **space** (Fig. 14-6), which contains a small amount of fatty tissue and a few blood vessels.

The meninges surrounding the brain (*cerebral meninges*) are continuous with those around the spinal cord. The cerebral dura is attached directly to the bone of the skull; there is no extradural space outside the cerebral dura. It contains the large blood vessels (**venous sinuses**) that drain the blood from the brain. At the openings, or foramina, in the skull, the dura follows the edges of the bone, passing from the interior of the skull to its outer surface, where it blends with the periosteum of the skull and the connective-tissue coat of the peripheral nerves. (At the foramen magnum, it is continuous with the spinal dura.)

GLIA

Only 10 percent or so of the cells in the central nervous system are neurons; the remainder are **glial**

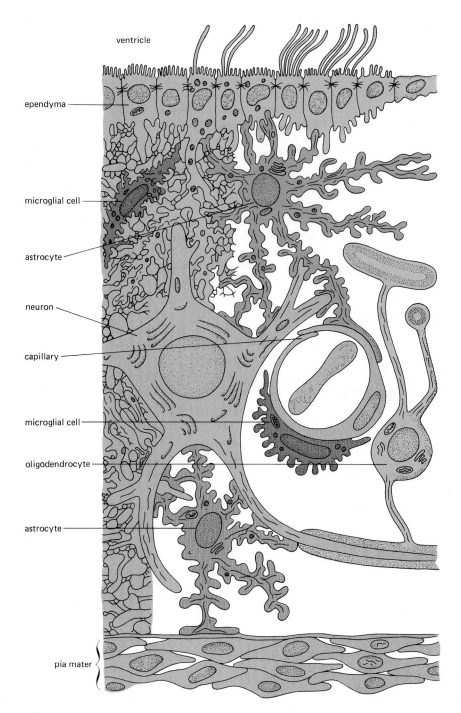

ventricle

ependyma

microglial cell

astrocyte

neuron

capillary

microglial cell

oligodendrocyte

astrocyte

pia mater

FIG. 14-7 The four types of neuroglia: astrocytes, oligodendrocytes, microglia, and ependymal cells.

cells, or **neuroglia** (Fig. 14-7). Glial cells do not give action-potential-like responses, and except for the few cases mentioned below, their precise functional relation to neurons and their role in the physiology of the central nervous system are unknown. Glial cell bodies are generally smaller than the cell bodies of neurons; the processes are shorter, although they may be numerous and highly branched.

Processes of one type of glial cell, the **astrocyte,** are tightly packed around the cell bodies and dendritic processes of neurons, sometimes forming a particularly tight covering over the regions of synapses. Astrocytes may affect the composition of the extracellular fluid in their immediate vicinity by providing a membrane across which ions released from active neurons can distribute, and it has been suggested that they play a role in the developing brain by guiding neurons as they migrate to their final adult locations and by directing the growing neuronal processes. They may affect the composition of the extracellular fluid as a whole by regulating the passage of some substances between the blood and the extracellular fluid of the central nervous system.

Oligodendrocytes, a second type of neuroglia, are sometimes found around nerve cell bodies, but they also occur along bundles of axons in the brain and spinal cord. Some of the oligodendrocytes surrounding nerve fibers in the central nervous system form the myelin coating of the axons; the functions of the others are not known. **Microglia,** a type of small glial cell, are numerous only in disease states in which they possibly act as scavengers. **Ependymal cells,** the final type of glial cells, line the brain ventricles. There are no glial cells in the peripheral nervous system, but the Schwann cells and satellite cells in the ganglia there are similar to glia and perform some of the glial functions, e.g., myelin formation.

CENTRAL NERVOUS SYSTEM

Spinal Cord

The spinal cord is a slender cylinder about as big around as the little finger. Figure 14-8 shows the basic division of the internal structures of the cord. The central butterfly-shaped area is the **gray matter;** it is filled with interneurons, the cell bodies and dendrites of efferent neurons, the entering fibers of afferent neurons, and glial cells. The two halves of the "butterfly" are joined by a central **gray commissure.** Each "wing" has a **dorsal** and **ventral horn** and, in some segments of the cord, a **lateral horn** as well. The cell bodies of neurons in the gray matter are often clustered together with other neurons having similar functions into groups called **nuclei,** which can be visualized as columns that extend the length of the spinal cord. The gray region is surrounded by **white matter,** which consists largely of bundles of myelinated nerve fibers (**tracts** and **pathways**) running longitudinally through the cord, some descending to convey information from the brain to the spinal cord (or from upper to lower

FIG. 14-8 Diagrammatic cross section of the spinal cord.

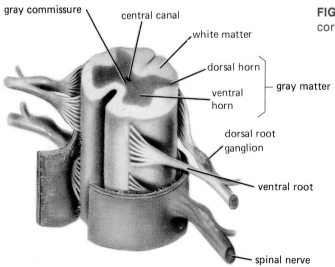

gray commissure
central canal
white matter
dorsal horn
ventral horn
gray matter
dorsal root ganglion
ventral root
spinal nerve

TABLE 14-1   MAJOR ASCENDING AND DESCENDING TRACTS OF THE SPINAL CORD (See also Fig. 14-9)

Tract	Comments
Ascending Tracts	
Fasciculus gracilis and fasciculus cuneatus	Also called the *posterior columns*. Tracts consist of the axons of afferent neurons that have entered via the dorsal root of the same side of the cord, and tracts end in the medulla (nuclei gracilis and cuneatus). Information resulting from stimulation of nerve fibers in these tracts contributes to an awareness of the position of the body in space (proprioception); discrimination between two points stimulated on the skin simultaneously; ability to recognize an object by its size, texture, and shape (stereognosis); weight discrimination; and vibration. (See Fig. 15-7B.)
Spinocerebellar tracts	Beginning with neurons in the dorsal horn of the spinal cord that receive proprioceptive information from afferent fibers, two tracts ascend to the cortex of the cerebellum: the *dorsal* (or posterior) *spinocerebellar* and the *ventral* (or anterior) *spinocerebellar*. The dorsal tract receives information from the side of the body on which the tract lies, and the ventral tract receives information from the same and opposite sides. The dorsal tract enters the cerebellum via the *inferior cerebellar peduncle,* the ventral tract via the *superior cerebellar peduncle.* Part of the information carried via these tracts signals the instantaneous progress of movements and the activity of muscles.
Anterior spinothalamic tract	This is the pathway of simple touch (concerned with the sense of light touch, light pressure, and a crude sense of tactile localization, and probably also tickling). The tract begins with neurons in the dorsal horn of the spinal cord that have received input from afferent neurons activated by receptors in the skin and hair follicles. Fibers from the dorsal horn neurons cross to the opposite side of the spinal cord and ascend to the thalamus, where they synapse with other neurons that relay the information to cerebral cortex.
Lateral spinothalamic tract	This tract conveys information about pain and temperature. Axons from spinal cord neurons, which have been activated by afferent neurons, cross to the opposite side of the cord and ascend to the thalamus. The information is then relayed by other neurons to the cerebral cortex. (See Fig. 15-7A.)
Descending Tracts	
Corticospinal tract	Also called the *pyramidal tract*. Descends from the cerebral cortex via the internal capsule, cerebral peduncle, and pons pyramid to the medulla, where most fibers cross over to descend in the opposite side of the spinal cord as the *lateral corticospinal tract*. The remaining uncrossed fibers descend in the *ventral corticospinal tract*. Fibers in these pathways are also known as the *upper motor neurons* (as opposed to the peripheral motor neurons, the somatic efferent fibers, that are known as the lower motor neurons). (See Fig. 16-8.)
Vestibulospinal, olivospinal, tectospinal, and rubrospinal pathways	These tracts are known collectively as the *extrapyramidal tracts*. They begin at various locations in the brainstem; the particular nucleus of origin is indicated by the name of the pathway (e.g., *vestibular nuclei* in the medulla, *olivary nucleus* in the medulla, the *tectum* or roof of the midbrain, and the *red nucleus*—rubro means red—in the midbrain). They descend in the spinal cord in discrete columns to influence motor neurons, especially those that control muscle fibers generally used in the unconscious postural support of parts of the body.

levels of the cord), others ascending to transmit in the opposite direction. The fiber bundles are so organized that a given tract or pathway generally contains fibers that transmit one type of information. For example, the fibers that transmit information from light-touch receptors in the skin travel together in one pathway. Some of these spinal-cord pathways are illustrated in Fig. 14-9 and listed in Table 14-1.

As shown in Fig. 14-8, groups of afferent fibers

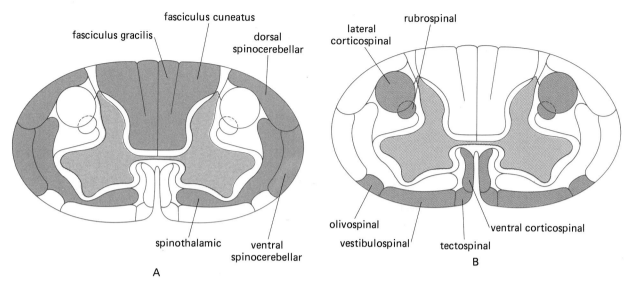

FIG. 14-9 Major tracts of the spinal cord. Ascending tracts are shaded and labeled in A and descending tracts in B.

enter the spinal cord via the **dorsal roots,** which contain the **dorsal root ganglia** (the cell bodies of the afferent neurons). Efferent fibers leave the spinal cord via the **ventral roots.** Shortly after leaving the cord, the ventral and dorsal roots from the same level combine to form a pair of **spinal nerves,** one on each side of the spinal cord. (The spinal nerves are discussed with the peripheral nervous system.)

Brain

The adult brain is composed of six subdivisions: the *telencephalon, diencephalon, midbrain, pons, medulla oblongata,* and *cerebellum* (Fig. 14-3 and Table 14-2). The midbrain, pons, and medulla oblongata together form the **brainstem;** the telencephalon (cerebrum) and diencephalon together constitute the **forebrain.**[1]

Brainstem The brainstem (Fig. 14-10A) is literally the stalk of the brain, through which pass all the nerve fibers that relay signals of afferent input and efferent output between the spinal cord and higher brain centers. In addition, the brainstem contains the cell bodies of neurons whose axons go out to the periphery to innervate the muscles and glands of the head, the heart, and the smooth muscles and

glands of most thoracic and abdominal viscera. The brainstem also receives many afferent fibers from the head and visceral cavities via the cranial nerves. In contrast to the distinct white and gray areas of the spinal cord, the tracts and nuclei of the brainstem are intermingled.

The **medulla oblongata** is the section of the brainstem continuous with the spinal cord below and the pons above. Its junction with the cord reflects a gradual change from the external tracts and internal columns of nuclei that exist at the upper levels of the cord. Efferent axons emerging from the medulla via cranial nerves VIII, IX, X, XI, and XII control

[1] Some neuroanatomists classify the diencephalon with the brainstem rather than with the forebrain.

TABLE 14-2 DIVISIONS OF THE CENTRAL NERVOUS SYSTEM		
Brain	Telencephalon; cerebral hemispheres (includes cerebral cortex and basal ganglia)	Forebrain
	Diencephalon (includes thalamus and hypothalamus)	
	Midbrain	Brainstem
	Pons	
	Medula oblongata	
	Cerebellum	
Spinal cord		

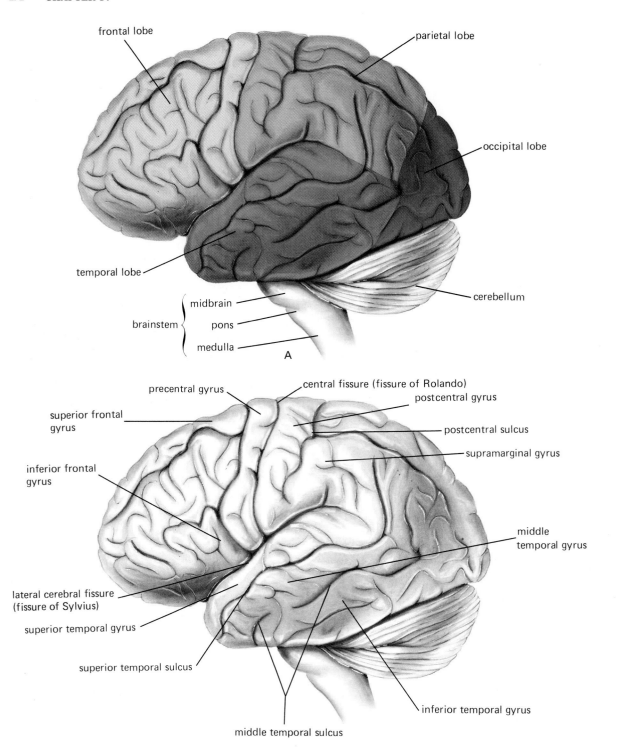

A

frontal lobe

parietal lobe

occipital lobe

temporal lobe

cerebellum

brainstem { midbrain
pons
medulla

B

precentral gyrus

central fissure (fissure of Rolando)

postcentral gyrus

superior frontal gyrus

postcentral sulcus

supramarginal gyrus

inferior frontal gyrus

middle temporal gyrus

lateral cerebral fissure (fissure of Sylvius)

superior temporal gyrus

superior temporal sulcus

inferior temporal gyrus

middle temporal sulcus

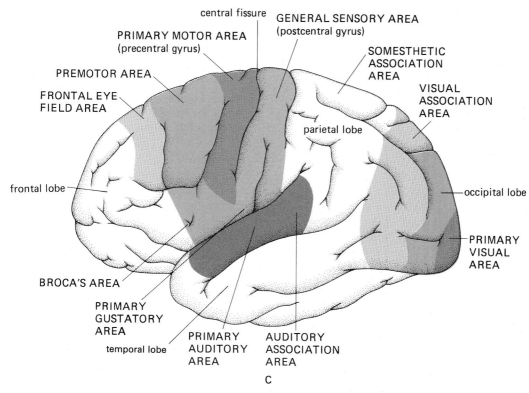

central fissure
GENERAL SENSORY AREA
(postcentral gyrus)
PRIMARY MOTOR AREA
(precentral gyrus)
SOMESTHETIC
ASSOCIATION
AREA
PREMOTOR AREA
VISUAL
ASSOCIATION
AREA
FRONTAL EYE
FIELD AREA
parietal lobe
frontal lobe
occipital lobe
PRIMARY
VISUAL
AREA
BROCA'S AREA
PRIMARY
GUSTATORY
AREA
temporal lobe
PRIMARY
AUDITORY
AREA
AUDITORY
ASSOCIATION
AREA
C

FIG. 14-10   (A) Regions of the surface of the adult human brain. (B) View of the lateral surface of the brain. (C) Functional areas of the cerebral cortex.

areas of the mouth, throat, neck, thorax, and abdomen.

The **pons** is both wider and thicker than the medulla and is easily distinguished by a band of fibers running across its ventral surface (see Fig. 14-27). These fibers converge at each side of the pons into bundles called the *middle cerebellar peduncles,* one of the three pairs of fiber bundles that carry information between the brainstem and cerebellum. The afferent and efferent components of cranial nerves V, VI, and VII connecting with the pons are from the head.

The **midbrain** is a relatively short part of the brainstem and is somewhat constricted in comparison with the pons. It is traversed by a huge number of axons that subserve the corticospinal and spinocortical pathways. It contains major nuclei associated with eye movements and hearing.

Running through the entire brainstem is a core of tissue called the **reticular formation,** which is composed of a diffuse collection of small, many-branched neurons. The neurons of the reticular formation receive and integrate information from

many afferent pathways as well as from many other regions of the brain. Some reticular-formation neurons are clustered together, forming certain of the brainstem nuclei and "centers," such as the cardiovascular, respiratory, swallowing, and vomiting centers. The output of the reticular formation can be divided functionally into descending and ascending systems. The descending components influence efferent neurons in the cranial and spinal nerves and frequently afferent neurons as well; the ascending components affect such things as wakefulness and the direction of attention to specific events.

Cerebellum   The **cerebellum** (Figs. 14-1 and 14-10A) is chiefly involved with skeletal muscle functions; it helps to maintain balance and provide smooth, directed movements (Chap. 16). It consists of paired **cerebellar hemispheres** joined by a median **vermis.** The surface of the hemispheres is thrown into folds called *folia.* Parts of the cerebellum are connected with each other by association fibers and with other parts of the brain and spinal cord by projection fibers. (*Association fibers* connect parts within a given brain structure, whereas *projection fibers* connect different structures. Association and projection fibers are found not only in the cerebellum but in other parts of the brain as well.) The many projection fibers of the cerebellum are grouped into three large bundles, or **peduncles,** on each side; the cerebellum is connected to the midbrain by the *superior cerebellar peduncles,* to the pons by the *middle cerebellar peduncles,* and to the medulla by the *inferior cerebellar peduncles.*

Forebrain   The large part of the brain that remains when the brainstem and cerebellum have been excluded is the **forebrain.** It consists of a central core, the **diencephalon,** and right and left **cerebral hemispheres** (the **cerebrum**). The surface of the cerebral hemisphere is extensively folded into ridges, called **gyri** (singular, gyrus), and grooves called **sulci** (singular, sulcus). Particularly deep sulci are often known as **fissures.** All the sulci, gyri, and fissures are named, but only the more prominent are labeled in Fig. 14-10B. The elaborate folding of the hemisphere surface allows a greater surface to

fit within the confines of the skull. The hemispheres are connected to each other by fiber bundles known as **commissures,** the *corpus callosum* (Fig. 14-11) being the largest. Areas within a single hemisphere are connected to each other by association fibers (Fig. 14-12).

The outer portion of the cerebral hemispheres, the **cerebral cortex,** is a cellular shell about 3 mm thick covering the entire surface of the cerebrum. It forms the outer rim of the brain cross section in Fig. 14-13. Popular opinion calls the cortex the ''site of the mind and the intellect.'' Scientific opinion considers it to be an integrating area necessary for the bringing together of basic afferent information into complex perceptual images and ultimate refinement of control over all efferent systems. The cortex is divided into several parts, or lobes (*frontal, parietal, occipital,* and *temporal*) (Fig. 14-10A). The general functions of various regions of the cortex are indicated in Fig. 14-10C, and the more detailed functions will be discussed in later chapters. The cortex is an area of gray matter, so called because of the predominance of cell bodies. There are also cell bodies in the subcortical nuclei, but in

FIG. 14-11   View of a midsagittal section of the adult human brain.

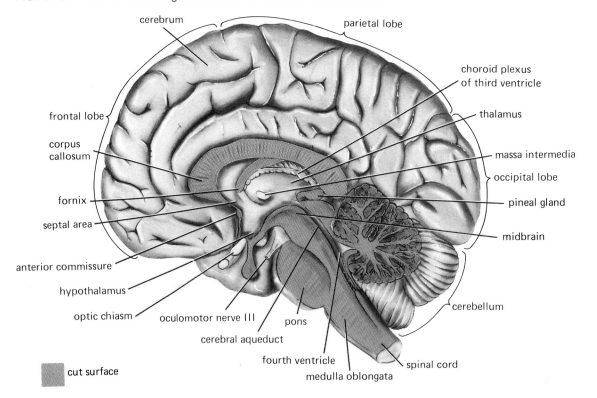

cut surface

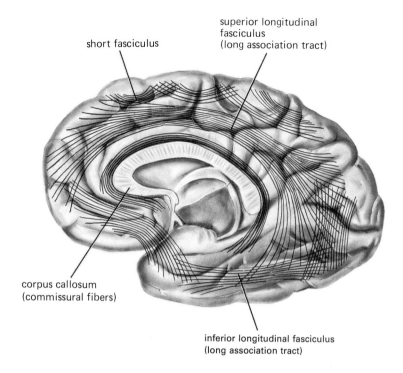

short fasciculus

superior longitudinal fasciculus (long association tract)

corpus callosum (commissural fibers)

inferior longitudinal fasciculus (long association tract)

FIG. 14-12 Association pathways of the cerebral hemispheres.

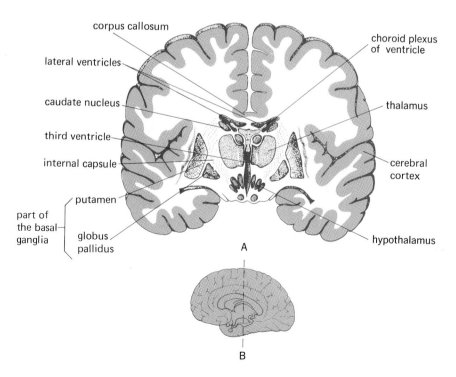

corpus callosum

lateral ventricles

caudate nucleus

third ventricle

internal capsule

putamen

part of the basal ganglia

globus pallidus

choroid plexus of ventricle

thalamus

cerebral cortex

hypothalamus

A

B

FIG. 14-13 Coronal section of the brain. The dashed line AB indicates the location of the section. The caudate nucleus is part of the basal ganglia.

other parts of the forebrain, nerve-fiber tracts predominate, distinguishable as white matter by their whitish myelin coating.

The **subcortical nuclei** form areas of gray matter that lie, as the name suggests, under the surface of the cortex and contribute to the coordination of muscle movements. Fig. 14-13 shows part of the **basal ganglia,** an example of the subcortical nuclei.

The **diencephalon,** a part of the forebrain, is rostral to the brainstem (Fig. 14-3) and is hidden beneath the cerebral hemispheres (Fig. 14-11). It is divided by a thin, vertical, fluid-filled space, the *third ventricle*. The largest part of the diencephalon is the **thalamus** (Fig. 14-13), which comprises two large egg-shaped masses of gray matter, one on each side of the third ventricle. Each is divided into several major parts and each part into distinct nuclei, which are the way stations and important integrating centers for all sensory input (except smell) on its way to the cortex. In addition to their connections with the cortex, the thalamic nuclei have connections with each other, with neighboring nonthalamic masses of gray matter, and with the long ascending and descending paths in the brainstem and spinal cord. Through its connections with the hypothalamus, the thalamus is involved in a wide range of activities that involve hormones, smooth muscle, and glands. The thalamus also contains a central core, which is part of the reticular system. The **hypothalamus,** a tiny region whose volume is about 5 to 6 cm^3, lies below the thalamus (Fig. 14-13) and is also part of the diencephalon. It is responsible for the integration of many basic behavioral patterns that involve correlation of neural and endocrine function. Indeed, the hypothalamus appears to be the single most important control area for regulation of the internal environment. It is also one of the brain areas associated with emotions; stimulation of some hypothalamic areas leads to behavior interpreted as rewarding or pleasurable, and stimulation of other areas is associated with unpleasant feelings. Neurons of the hypothalamus are also affected by a variety of hormones and other circulating chemicals. The bands of white matter separating the basal ganglia from the nuclei of the diencephalon are the **internal capsules** (Fig. 14-13), radiations of fibers passing between the cortex and other parts of the central nervous system.

The **limbic system** is not a single brain region but is an interconnected group of brain structures within the cerebrum, including portions of the frontal-lobe cortex, temporal lobe, thalamus, and hypothalamus, as well as the circuitous neuron pathways connecting these parts (Fig. 14-14). The limbic system is concerned with emotional behavior and learning. Besides being connected with each other,

FIG. 14-14 Structures of the limbic system are shaded in this partially transparent view of the brain.

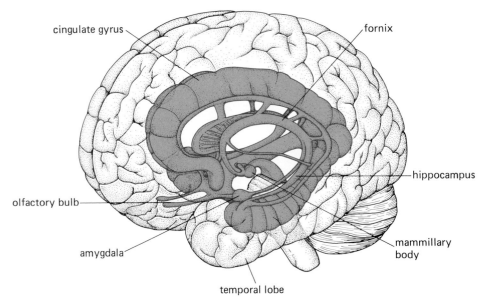

cingulate gyrus

fornix

hippocampus

olfactory bulb

mammillary body

amygdala

temporal lobe

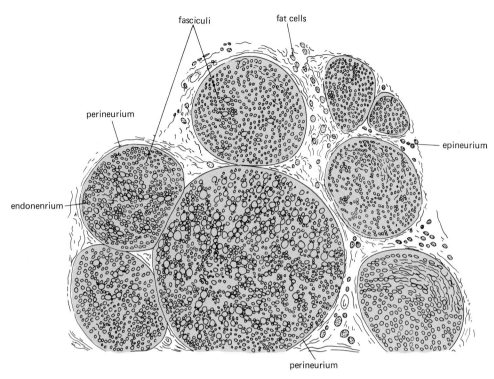

FIG. 14-15 Cross section of a peripheral nerve showing the connective-tissue sheaths of the nerves and nerve fibers.

the parts of the limbic system have connections with many other parts of the central nervous system. For example, it is likely that information from all the different afferent modalities can influence activity within the limbic system, whereas activity of the limbic system can result in a wide variety of automatic responses and body movements. This should not be surprising since emotional feelings are accompanied by automatic responses such as sweating, blushing, and heartrate changes, and by somatic responses such as laughing and sobbing.

PERIPHERAL NERVOUS SYSTEM

Nerve processes (axons) in the peripheral nervous system are grouped into bundles called **nerves,** whereas the individual processes (axons or dendrites) are **nerve fibers.** The peripheral nervous system consists of 43 pairs of nerves; 12 pairs connect with the brain and are called the **cranial nerves,** and 31 pairs connect with the spinal cord as the **spinal nerves.** (There are no "nerves" in the central ner-

vous system; recall that the bundles there are called *tracts* or *pathways*.)

The individual fibers in each nerve are grouped into **fasciculi;** a single fasciculus may contain relatively few fibers or hundreds of them (Fig. 14-15). The size and total number of fasciculi in a nerve vary greatly from one nerve to another and along the length of any one nerve. The fasciculi are bound together into a nerve by a connective-tissue sheath, the **epineurium,** which, along with arteries, veins, lymphatics, and fat cells, fills in the spaces between the fasciculi. A similar but less fibrous connective-tissue sheath, the **perineurium,** directly surrounds each fasciculus. The perineurium serves not only the protective and supportive functions of the other connective-tissue coverings; it also performs barrier functions, isolating to some extent the enclosed nerve fibers from the outside environment. The individual nerve fibers are covered by a delicate connective-tissue **endoneurium.**

Each peripheral nerve fiber is surrounded by **Schwann cells,** much as the central fibers are surrounded by oligodendroglia. The Schwann cells

axons

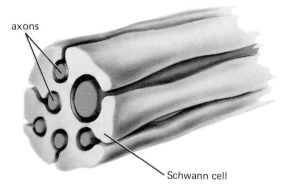

Schwann cell

FIG. 14-16 Schwann cell enclosing several unmyelinated nerve fibers.

have been likened to elongated beads strung on a slender thread, each bead representing a Schwann cell and the thread representing the axon. The narrow spaces between the beads represent the nodes of Ranvier, described earlier in Chap. 13. The Schwann cells lie beneath the endoneurium, separating it from the nerve fiber itself. Some of the axons of the peripheral nervous system are wrapped in layers of Schwann-cell membranes, which form the fiber's myelin coat. Other fibers are unmyelinated and are simply tucked into invaginations of the Schwann cell, one Schwann cell often enclosing several unmyelinated fibers (Fig. 14-16).

The individual nerve fibers in a nerve may be processes of either afferent or efferent neurons; accordingly they may be classified as belonging to the *afferent* or *efferent division* of the peripheral nervous system. All the spinal nerves and some of the cranial nerves contain processes of both afferent and efferent neurons.

One convenient way the divisions of the peripheral nervous system can be classified is presented in Table 14-3; each division will be discussed in turn.

TABLE 14-3	DIVISIONS OF THE PERIPHERAL NERVOUS SYSTEM

I Afferent
II Efferent
  a Somatic
  b Autonomic
    1 Sympathetic
    2 Parasympathetic

Peripheral Nervous System: Afferent Division

Afferent neurons convey information from receptors in the periphery to the central nervous system. Regardless of whether the initial stimulus is a prick of the skin, a stretch of skeletal muscle, a distention of the intestine, or a loud sound, the action potentials relaying information to the central nervous system travel along neurons that are structurally very similar. The neurons described in the receptor section of Chap. 13 are typical afferent neurons.

The cell bodies of afferent neurons are in structures called *ganglia* that are outside but close to the brain or spinal cord. From the region of the cell body, one long process extends away from the ganglion to innervate the receptors; commonly, the process branches several times as it nears its destination, each branch containing or innervating one receptor. A second process passes from the cell body into the central nervous system where it branches; these branches terminate in synaptic junctions onto other neurons.

Afferent neurons are sometimes called **primary afferents,** or *first-order neurons,* because they are the first cells in the synaptically linked chains of neurons that handle incoming information and carry it to higher centers in the brain. They are also frequently called sensory neurons, but we hesitate to use this term because it implies that the information transmitted by these neurons is destined to reach consciousness, and this is not always true. For example, we have no conscious awareness of our blood pressure even though we have receptors sensitive to this variable.

Peripheral Nervous System: Efferent Division

The efferent division is more complicated than the afferent and is subdivided into a **somatic nervous system** and an **autonomic nervous system.** These terms are somewhat unfortunate because they conjure up additional "nervous systems" distinct from the central and peripheral nervous systems. Keep in mind that the terms refer simply to the *efferent* divisions of the peripheral nervous system. Although separating the somatic and autonomic divisions is justified by many anatomical and physiological differences, the simplest distinction is that the somatic nervous system innervates skeletal muscle and the autonomic nervous system innervates smooth and cardiac muscle and glands. Other differences are listed in Table 14-4.

TABLE 14-4   DIFFERENCES BETWEEN SOMATIC AND AUTONOMIC NERVOUS SYSTEMS

Somatic Nervous System
1   Fibers do not synapse once they have left the central nervous system.
2   Innervates skeletal muscle.
3   Always leads to excitation of the muscle fibers.

Autonomic Nervous System
1   Fibers synapse once in ganglia after they have left the central nervous system.
2   Innervates smooth or cardiac muscles or gland cells.
3   Can lead to excitation or inhibition of the effector cells.

Somatic Nervous System   The somatic division of the peripheral nervous system is made up of all the fibers going from the central nervous system to skeletal muscle cells. The cell bodies of these neurons are in groups in the brain or spinal cord. Their large-diameter, myelinated axons leave the central nervous system and pass directly, i.e., without any synapses, to skeletal muscle cells. The transmitter substance released by these neurons is acetylcholine. Because activity of somatic efferent neurons causes contraction of the innervated skeletal muscle cells, these neurons are often called *motor neurons*. (Thus, it was motor neurons that were discussed in Chap. 11.) Motor neurons can be activated by local reflex mechanisms, as in the flexion reflex, or they can be activated by pathways that descend from higher brain centers, but in either case, their activation always leads to *contraction* of skeletal muscle cells; there are no *inhibitory* somatic motor neurons.

Autonomic Nervous System   Fibers of the autonomic division of the peripheral nervous system innervate cardiac muscle, smooth muscle, and glands. Anatomic and physiologic differences in the autonomic nervous system are the basis for its further subdivision into **sympathetic** and **parasympathetic** components. The cell bodies of the first neurons in the two subdivisions are located in different areas of the central nervous system, and their fibers leave at different levels, the sympathetic from the thoracic and lumbar regions of the spinal cord, and the parasympathetic from the brain and the sacral portion of the spinal cord (Fig. 14-17). Thus, the

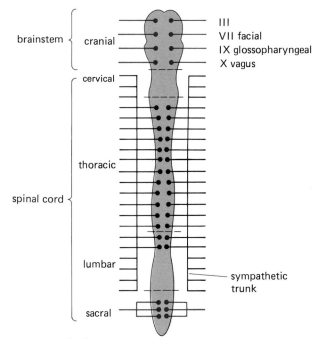

● parasympathetic
● sympathetic

FIG. 14-17 Origins of the sympathetic and parasympathetic divisions of the autonomic nervous system.

sympathetic division is also called the *thoracolumbar division,* and the parasympathetic the *craniosacral division.*

The fibers of the autonomic nervous system synapse once after they have left the central nervous system and before they arrive at the effector cells (Fig. 14-18). These synapses outside the central nervous system occur in cell clusters called ganglia. The two divisions of the autonomic nervous system differ with respect to the locations of their ganglia. Many of the sympathetic ganglia lie close to the spinal cord, forming the paired chains of ganglia known as the **sympathetic trunks,** and others lie halfway between the spinal cord and innervated organ. In contrast, the parasympathetic ganglia lie in the walls of the effector organ. The fibers passing between the central nervous system and the sympathetic or parasympathetic ganglia are the **preganglionic** autonomic fibers; those passing between the ganglia and the effector organ are the **postganglionic** fibers.

Each preganglionic fiber synapses with more than one neuron in the ganglion, and each ganglionic neuron receives synapses from more than one pre-

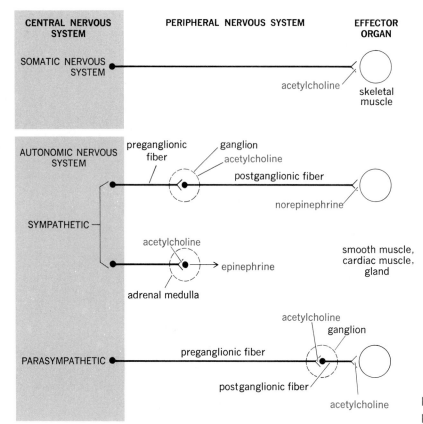

FIG. 14-18 Efferent divisions of the peripheral nervous system.

ganglionic fiber. Thus, there is not merely a simple relay of information through the autonomic ganglia, and integration and modification of signals can occur outside the central nervous system. This is a significant difference between the autonomic and somatic systems because in the somatic nervous system all signal processing must take place in the central nervous system.

In both sympathetic and parasympathetic divisions, the chemical transmitter for the ganglionic synapse between pre- and postganglionic fibers is acetylcholine. The chemical transmitter in the parasympathetic division between the postganglionic fiber and the effector cell is also acetylcholine. (Fibers that release acetylcholine are called **cholinergic fibers.**) The transmitter in the sympathetic division between the postganglionic fiber and the effector cell is norepinephrine (Fig. 14-18).[2] Because early experiments on the function of the sympathetic division were done using epinephrine rather than the closely related norepinephrine, and because the epinephrine was called by its British name, adren-

[2] There are a few exceptions to this statement that will be identified where appropriate.

aline, fibers that release norepinephrine came to be called **adrenergic fibers.**

Many drugs stimulate or inhibit various components of the autonomic nervous system. An important site of action for certain of them is the receptor for the neurotransmitters. Indeed, study of these drugs has revealed that there are several different types of receptors for each transmitter. For example, acetylcholine receptors on the postsynaptic neurons of all autonomic ganglia also respond to the drug nicotine and are therefore *nicotinic receptors.* In contrast, the acetylcholine receptors on smooth muscle and gland cells are stimulated not by nicotine but by the mushroom poison, muscarine; they are called *muscarinic receptors.* Similarly, adrenergic receptors (i.e., receptors for epinephrine and norepinephrine) form two classes distinguished by the specific drugs that stimulate or block them. They are called *alpha-adrenergic* and *beta-adrenergic receptors.*

One "ganglion" in the sympathetic division never developed long postganglionic fibers; instead, upon activation of its preganglionic nerves, the cells of this "ganglion" discharge their transmitters into

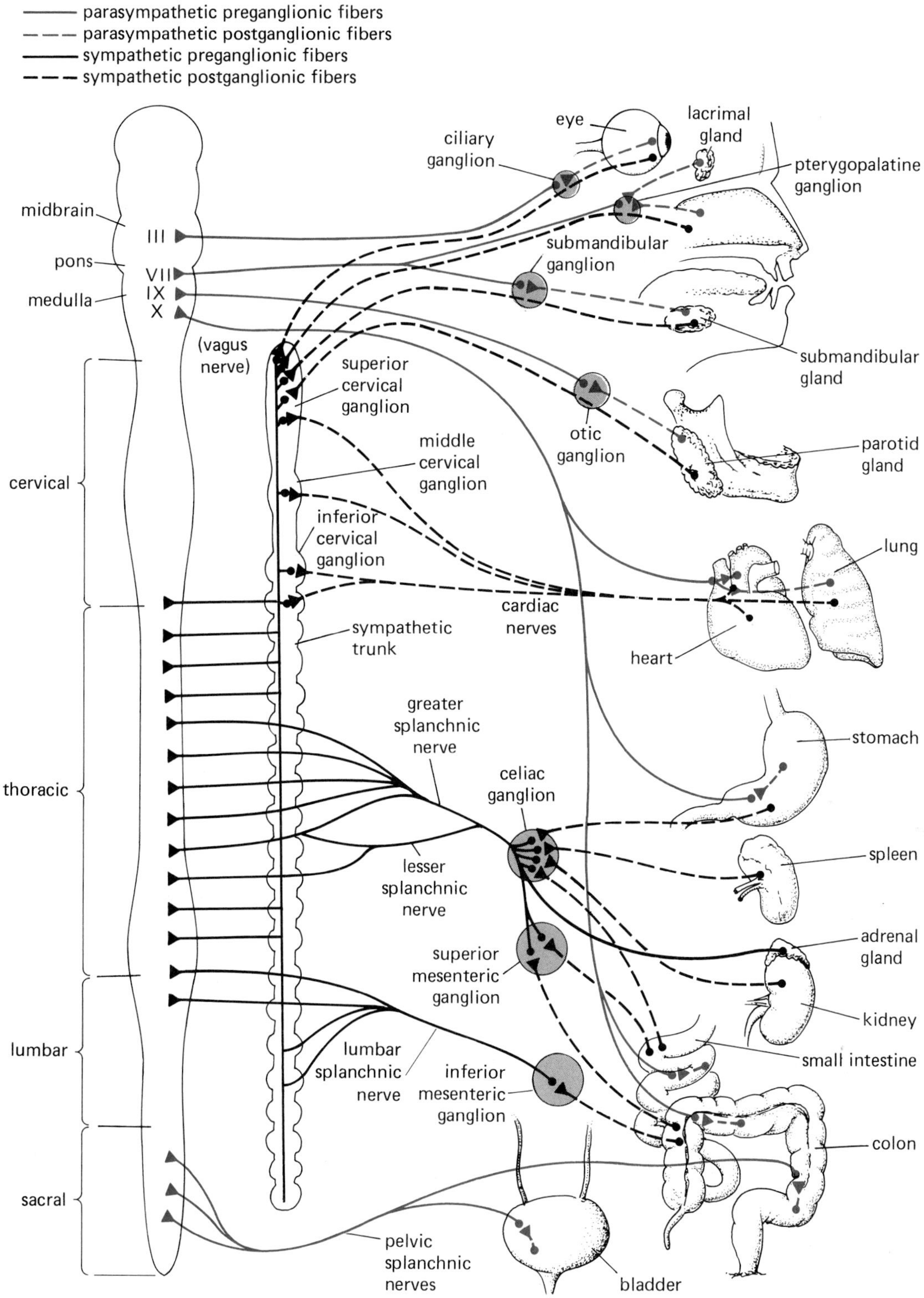

FIG. 14-19 Diagrammatic summary of autonomic nervous system. Only one of the two sympathetic trunks is shown. Not shown are the fibers passing to the glands and smooth muscle cells in the body walls.

TABLE 14-5   SOME EFFECTS OF AUTONOMIC NERVOUS SYSTEM ACTIVITY*

Effectors	Sympathetic Nervous System	Parasympathetic Nervous System
Eye		
Muscles of the iris	Contracts radial muscle (widens pupil)	Contracts sphincter muscle (makes pupil smaller)
Ciliary muscle	Relaxation (tightens suspensory ligaments, thus flattening lens for far vision)	Contraction (relaxes ligament, allowing lens to become more convex for near vision)
Heart		
S-A node	Increases heart rate	Decreases heart rate
Atria	Increases contractility	Decreases contractility
A-V node	Increases conduction velocity	Decreases conduction velocity
Ventricles	Increases contractility	———
Arterioles		
Coronary	Constriction	Dilation
Skin and mucous membrane	Constriction	———
Skeletal muscle	Constriction or dilation	———
Abdominal viscera and kidneys	Constriction	———
Salivary glands	Constriction	———
Penis or clitoris	Constriction	Dilation (causes erection)
Veins	Constriction	———
Lung		
Bronchial muscle	Relaxation	Contraction
Broncial glands	Inhibits secretion	Stimulates secretion
Salivary glands	Stimulates secretion	Stimulates secretion

the blood stream. This "ganglion," called the **adrenal medulla,** is therefore an endocrine gland (Chap. 18). It releases a mixture of about 80 percent epinephrine and 20 percent norepinephrine. These substances, in this case more properly called hormones rather than neurotransmitters, are transported via the blood to receptor sites on effector cells sensitive to them. The receptor sites may be the same ones that sit beneath the terminals of sympathetic postganglionic neurons and are normally activated by the transmitter delivered directly to them, or they may be other, noninnervated sites that are activated only by the circulating epinephrine or norepinephrine.

The heart and many glands and smooth muscles are innervated by both sympathetic and parasympathetic nerve fibers, i.e., they receive **dual innervation** (Fig. 14-19). (Do not confuse this with reciprocal innervation, which involves pairs of muscles with opposite actions; when one muscle is activated, its antagonist is inhibited.) Whatever one

division does to the effector organ, the other division frequently (but not always) does just the opposite (Table 14-5). For example, action potentials arriving over the sympathetic nerves to the heart increase the heart rate, whereas action potentials arriving over the parasympathetic fibers decrease it. In the intestine, activation of the sympathetic fibers reduces contraction of the smooth muscle in the intestinal wall, whereas the parasympathetics increase contraction. Dual innervation with fibers of opposite action provides for a very fine degree of control over the effector organ, for it is like equipping a car with both an accelerator and a brake. With only an accelerator, one could slow the car simply by decreasing the pressure on the accelerator, but the combined effects of releasing the accelerator and applying the brake provide faster and more accurate control. To prevent the sympathetic and parasympathetic divisions' opposing effects from conflicting with each other, the two divisions are usually activated reciprocally; i.e., as

TABLE 14-5   SOME EFFECTS OF AUTONOMIC NERVOUS SYSTEM ACTIVITY (Continued)

Effectors	Sympathetic Nervous System	Parasympathetic Nervous System
Stomach		
Motility	Decreases	Increases
Sphincters	Contraction	Relaxation
Secretion	Possibly inhibition	Stimulation
Gallbladder and ducts	Relaxation	Contraction
Liver	Glycogenolysis, gluconeogenesis	———
Pancreas		
Exocrine glands	Decreased secretion	Stimulates secretion
Endocrine glands (islets)	Inhibits insulin secretion, stimulates glucagon secretion	Stimulates insulin secretion
Fat cells	Stimulates lipid breakdown	———
Urinary bladder	Relaxation	Contraction
Uterus	Pregnant: contraction Nonpregnant: relaxation	Variable
Reproductive tract (male)	Ejaculation	———
Skin		
Muscles causing hair to stand erect	Contraction	———
Sweat glands	Stimulates secretion	Stimulates secretion
Lacimal glands	———	Stimulates secretion

*Specific effectors will be discussed in later chapters.
Table adapted from Louis S. Goodman and Alfred Gilman, *The Pharmacological Basis of Therapeutics*, 5th ed., Macmillan, New York, 1975.

the activity of one division is enhanced, the activity of the other is depressed. Skeletal muscle cells, in contrast, do not receive dual innervation; they are activated by somatic motor neurons, which are only excitatory.

Because glands, smooth muscles, and the heart participate as effectors in almost all bodily functions, it follows that the autonomic nervous system has an extremely widespread and important role in the homeostatic control of the internal environment. It exerts a wide array of effects (Table 14-5) that can be very difficult to remember. Some common denominators are that, in general, the sympathetic division helps the body to cope with challenges from the outside environment, and the parasympathetic seems to be more responsible for internal housekeeping, such as digestion, defecation, and urination. The sympathetic division is utilized in situations that involve stress or strong emotions such as fear or rage, whereas the parasympathetic division is most active during recovery or at rest. The sympathetic division provides the responses to a situation leading to "fight or flight." For example, the sympathetic division increases blood flow to exercising muscles and sustains blood pressure in case of severe blood loss. It decreases activity of the gastrointestinal tract, increases the metabolic production of energy, and increases sweating, changes that provide energy utilization most appropriate to the emergency. These and many other functions of the sympathetic (and parasympathetic) division will be discussed in relevant places throughout the book.

Autonomic responses usually occur without conscious control or awareness as though they were indeed autonomous (in fact, the autonomic nervous system has been called the involuntary nervous system). However, it is wrong to assume that this need always be the case; it has been shown that discrete visceral and glandular responses can be learned. For example, to avoid an electric shock, a rat can learn to selectively increase or decrease its heart rate and a rabbit can learn to constrict the vessels in one ear while dilating those in the other. The implications of such voluntary control of autonomic functions in human medicine are enormous.

These experiments are also important in showing that small segments of the autonomic response can be regulated independently; thus, overall autonomic responses, made up of many small components, are quite variable. Rather than being gross, undiscerning discharges, they are finely tailored to the specific demands of any given situation.

This completes our general description of the afferent and efferent divisions of the peripheral nervous system. We now turn to the specific ways in which these components are organized in the spinal and cranial nerves which constitute the peripheral nervous system.

Spinal Nerves

Shortly after leaving the spinal cord, the ventral root (carrying efferent fibers) and the dorsal root (carrying afferent fibers) from the same side at the same level combine to form a **spinal nerve;** thus, all spinal nerves contain afferent and efferent fibers. The 31 pairs of spinal nerves are grouped as follows: 8 cervical, 12 thoracic, 5 lumbar, 5 sacral, and 1 coccygeal, the individual nerves being commonly named for the intervertebral foramina through which they pass as they leave the vertebral canal. Thus, the *first cervical nerve* leaves between the occipital bone of the skull and the first cervical vertebra (the atlas), the *second cervical nerve* leaves between the atlas and the second cervical vertebra, the *eighth cervical nerve* leaves between the seventh cervical and the first thoracic vertebrae, etc.

Shortly after emerging from the intervertebral foramen, each spinal nerve divides into a **dorsal** and a **ventral ramus,** or **branch** (Fig. 14-20). The dorsal branches supply the muscles and skin of the back; the ventral branches supply the limbs and front and sides of the trunk. Figure 14-21 shows the segmental distribution of the spinal nerves as they travel to the periphery. The dorsal branches remain separate; i.e., they do not join the dorsal branches of

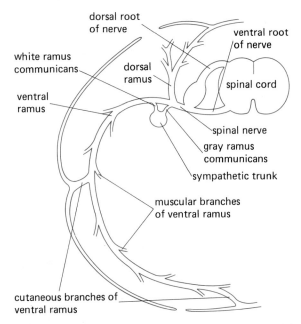

FIG. 14-20 A typical (thoracic) spinal nerve and its branches. The white and gray rami communicantes (singular, ramus communicans) will be described later in this chapter.

FIG. 14-21 Segmental distribution of the spinal nerves. C = cervical, T = thoracic, L = lumbar, and S = sacral.

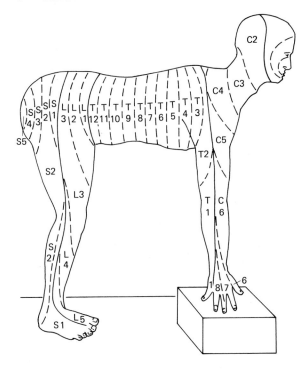

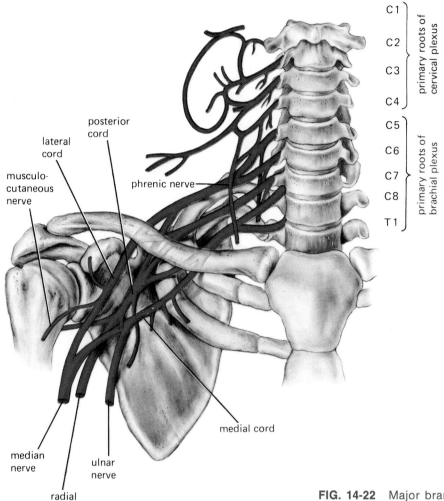

C1 ⎫
C2 ⎬ primary roots of
C3 ⎪ cervical plexus
C4 ⎭

C5 ⎫
C6 ⎪
C7 ⎬ primary roots of
C8 ⎪ brachial plexus
T1 ⎭

posterior
cord

lateral
cord

musculo-
cutaneous
nerve

phrenic nerve

median
nerve

ulnar
nerve

radial
nerve

medial cord

FIG. 14-22   Major branches of the cervical and brachial plexuses.

other nerves. The ventral branches from the thoracic region also remain separate, but the ventral branches from other regions of the spinal cord intermingle with fibers from neighboring nerves and form **nerve plexuses.** Therefore, a single peripheral nerve leaving a plexus may contain fibers from several different levels of the spinal cord. The four great plexuses, the *cervical, brachial, lumbar,* and *sacral plexuses,* can be seen in Fig. 14-1; their compositions are shown more clearly in Figs. 14-22 and 14-24.

The **cervical plexus** (Fig. 14-22) is formed mainly from the ventral rami of the second, third, and fourth cervical nerves. (Note that we are speaking of the ventral and dorsal rami, or branches, of the spinal nerves, not the ventral and dorsal roots.) An important muscular branch (i.e., branch to the muscles) that rises in part from this plexus is the **phrenic nerve,** which innervates the respiratory diaphragm, although the largest branches that leave the cervical plexus are cutaneous ones, i.e., branches to the skin.

The **brachial plexus** (Fig. 14-22) lies in the lower part of the neck and in the axillary region and is formed from the ventral rami of the fifth, sixth, seventh, and eighth cervical nerves and part of the first thoracic. The fibers from the various spinal nerves that enter the plexus unite to form **trunks,** and the fibers of the trunks are then sorted out into **anterior** and **posterior divisions,** which innervate an-

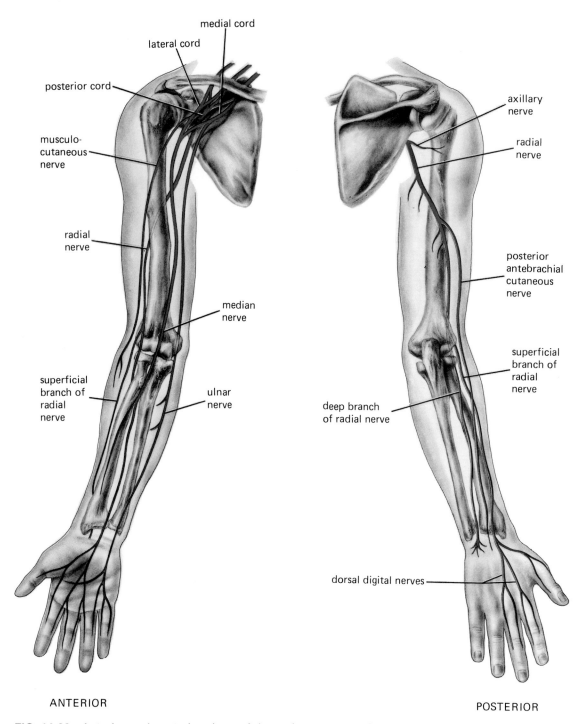

ANTERIOR

POSTERIOR

FIG. 14-23 Anterior and posterior views of the major nerves to the arm.

terior and posterior muscles, respectively. Although fibers of the two divisions are kept separate from each other, fibers within a division are further mixed, forming **cords,** and it is primarily the three cords from the brachial plexus that are seen in Fig. 14-22, the lateral and medial cords representing the anterior division of the plexus and the posterior cord representing the posterior divisions. The nerves, which emerge from the three cords, are shown in Fig. 14-23.

Of the four large nerves that enter the leg, two arise from the lumbar plexus and two from the sacral plexus. (The lumbar and sacral plexuses are some-

times spoken of as a single entity, the **lumbosacral plexus.**) The **lumbar plexus** (Fig. 14-24) is formed from the first, second, third, and part of the fourth lumbar nerves; it gives rise to two major branches, the *obturator nerve,* which innervates the adductor (i.e., the anteriomedial) muscles of the front of the thigh, and the *femoral nerve,* which is distributed to other muscles on the front of the thigh.

The **sacral plexus** (Fig. 14-24) is usually formed by the ventral rami of the fourth and fifth lumbar and the first three or four sacral nerves. The two large nerves that arise from this plexus are at first bound together to form the *sciatic nerve,* but they

FIG. 14-24   Major branches of the lumbar and sacral plexuses. The fourth lumbar nerve usually contributes to both these plexuses.

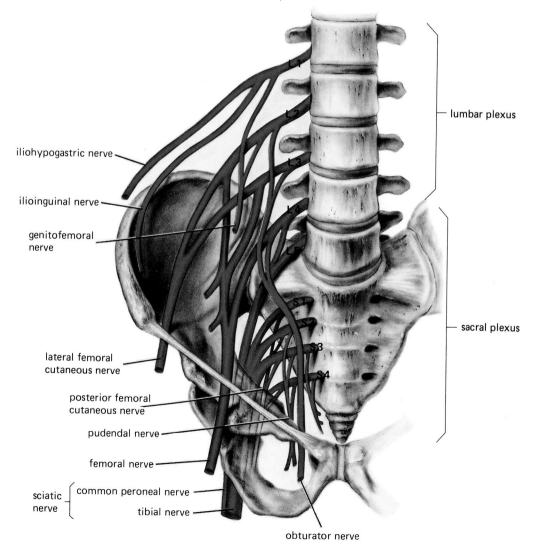

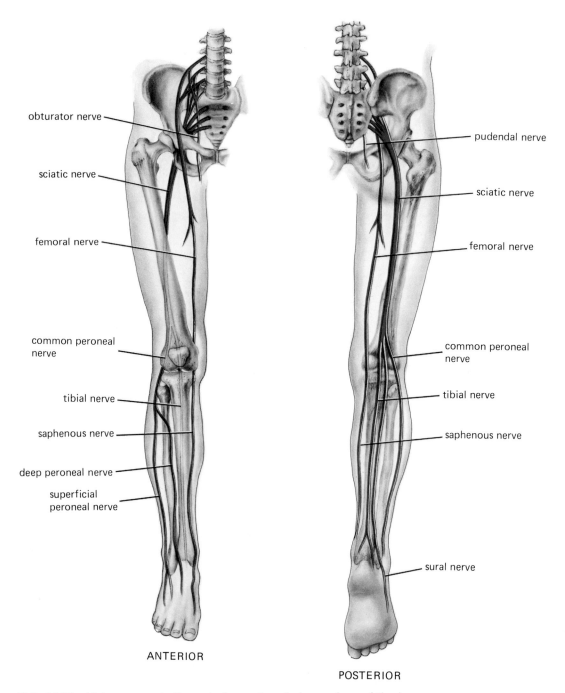

obturator nerve

sciatic nerve

femoral nerve

common peroneal
nerve

tibial nerve

saphenous nerve

deep peroneal nerve

superficial
peroneal nerve

pudendal nerve

sciatic nerve

femoral nerve

common peroneal
nerve

tibial nerve

saphenous nerve

sural nerve

ANTERIOR

POSTERIOR

FIG. 14-25   Major nerves to the anterior and posterior surface of the leg.

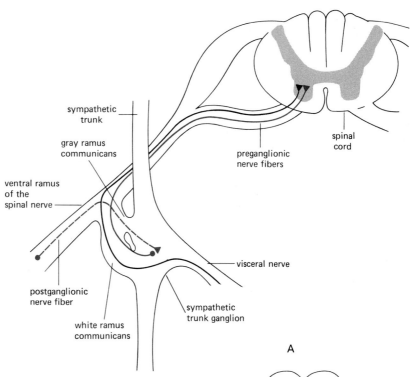

FIG. 14-26 (A) Relationship between the sympathetic trunk and spinal cord. (B) Various courses that preganglionic sympathetic fibers (solid lines) may take through the sympathetic trunk. Dashed lines represent postganglionic fibers. The numbers in the figure correspond to the numbers in the text discussion.

eventually separate into the *common peroneal* and *tibial nerves*. While the muscles of the buttock are innervated by several other branches of the sacral plexus, the posterior muscles of the thigh and all the muscles below the knee are innervated by the common peroneal and tibial nerves. The peripheral ramifications of the lumbar and sacral plexuses are shown in Fig. 14-25.

Autonomic Component of the Spinal Nerves   A potentially confusing aspect of spinal-nerve anatomy is the relationship of the nerves to the sympathetic trunks. Recall that the sympathetic preganglionic fibers (whose cell bodies of origin are in the lateral horn of the spinal cord) leave the cord via the ventral roots. The sympathetic fibers then join the spinal nerve and enter its ventral ramus. After a short distance, they leave this ramus by way of a branch called the *white ramus communicans*, which carries them to the sympathetic trunk (Figs. 14-20 and 14-26A). The white ramus also contains the afferent fibers that innervate the viscera.

The **sympathetic trunks** (or *sympathetic chains* as they are sometimes called) are paired, each consisting of a series of interconnected ganglia that extend from the first cervical vertebra to the tip of the sacrum. The number of ganglia does not exactly

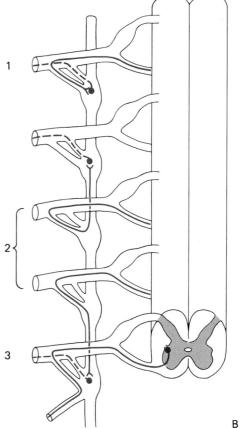

match the number of spinal segments, there typically being 3 cervical ganglia, instead of 8, 11 thoracic ganglia, instead of 12, and a variable number of ganglia in the lumbar and sacral regions. But since sympathetic preganglionic fibers leave the spinal cord typically only through the 12 thoracic and first 2 lumbar vertebrae, only these spinal nerves have white rami communicantes. The ganglia in the upper, cervical part of the trunks and in the lumbosacral regions (where there are no white rami) receive their preganglionic input from fibers that ascend or descend in the trunks themselves.

On entering the sympathetic chain, a preganglionic fiber may pursue one of three courses (Fig. 14-26B): (1) It may synapse immediately with postganglionic neurons of the ganglion it first enters. (2) It may travel up or down the trunk and synapse with postganglionic neurons in ganglia belonging to more superior or inferior segments. This explains how it is that the ganglia above or below the thoracolumbar segments having no white rami communicantes receive preganglionic input. (3) It may not synapse in the sympathetic trunk at all but may pass directly through the trunk into the *visceral nerves* to reach the farther removed ganglia (or the adrenal medulla) where it synapses with postganglionic neurons.

How do the axons of postganglionic neurons leave the sympathetic trunk? The axons of neurons that have synapsed at some point in the sympathetic trunks leave them by way of the *gray rami communicantes* (Fig. 14-26A) to return to the spinal nerves with which they then travel to their destinations in the glands and smooth muscle of the body wall. Unlike the limited number of white rami communicantes, every spinal nerve receives a gray ramus from the sympathetic trunk; thus, the peripheral branches of all spinal nerves contain sympathetic fibers. (As shown in Fig. 14-26A, an anatomic curiosity is that the gray rami enter the spinal nerve at a point closer to the intervertebral foramen than the site of exit of the white rami; thus these sympathetic fibers form a loop as they pass from the spinal cord to the periphery.)

The cervical plexuses receive sympathetic fibers from the *superior cervical ganglion* of the sympathetic trunk, and the brachial plexuses receive their sympathetics from the *middle* and *inferior cervical ganglia* of the sympathetic trunk (Fig. 14-19). (Sometimes the inferior cervical and first thoracic

ganglia are combined to form a single ganglion known as the *stellate ganglion.*)

As mentioned above, not all preganglionic sympathetic fibers synapse in the sympathetic trunk; many merely pass through it (Fig. 14-26B, part 3) and leave to form **visceral,** or **splanchnic, nerves.** These nerves also contain afferent fibers that come from the viscera as well as a very small number of postganglionic sympathetic fibers. The fibers of the splanchnic nerves form complex nerve nets, or plexuses, such as the **celiac,** or **solar, plexus** in the region of the aorta. The plexuses contain ganglia in which the preganglionic fibers synapse; the postganglionic fibers that arise from the ganglia tend to follow blood vessels to organs they will innervate. The sympathetic ganglia outside the sympathetic trunk are collectively called the **prevertebral ganglia.** Several prevertebral ganglia, e.g., the celiac ganglion and superior and inferior mesenteric ganglia, can be seen in Fig. 14-19. Thus in general (but certainly not always), the sympathetics that innervate smooth muscles in the body wall or limbs synapse in the sympathetic trunks, whereas the sympathetics that innervate viscera synapse in prevertebral ganglia.

The preganglionic neurons of the sacral components of the parasympathetic nervous system lie in the sacral portion of the spinal cord, and their axons leave the cord as part of the ventral root of a sacral nerve. They usually leave the nerve with which they are associated and end in widely scattered ganglia close to or even in the walls of the organs they innervate. In general, the parasympathetics do not travel in the same nerves as do the sympathetics, and they do not rejoin the spinal nerves. Unlike the sympathetic system, which is distributed to all major parts of the body, the parasympathetic fibers have a limited distribution.

Cranial Nerves

The 12 pairs of cranial nerves are listed in Table 14-6; the origins of all but the first (the olfactory nerve, which joins the olfactory bulb) are shown in Fig. 14-27. The nerves pass from the brain out to the periphery through holes, or foramina, in the skull; these were mentioned in Chap. 9.

Not all the cranial nerves have both afferent and efferent components, and not all of those with efferent components have both autonomic and so-

TABLE 14-6 CRANIAL NERVES (MAJOR COMPONENTS)

No.	Name	Components	Peripheral Distribution	Function
I	Olfactory	Afferent	Mucous membrane of olfactory region of nose	Smell
II	Optic	Afferent	Retina of eye	Vision
III	Oculomotor	Efferent	Muscles (superior, inferior, and medial rectus; inferior oblique; and levator palpebrae)	Eye movements, raise upper eyelid
			Pupillary constrictor and ciliary muscles of eye	Regulation of pupil size, accommodation of lens
IV	Trochlear	Efferent	Superior oblique muscle	Eye movements
V	Trigeminal	Afferent	Face, greater part of scalp, teeth, mouth, nasal cavity	
		Efferent	Muscles (medial and lateral pterygoids, masseter, temporalis)	Chewing movements
VI	Abducens	Efferent	Lateral rectus muscle	Eye movements
VII	Facial	Afferent	Anterior part of tongue and soft palate	Taste
		Efferent	Muscles of face, scalp, and outer ear	Facial expressions
			Submandibular and sublingual salivary glands, lacrimal glands, glands of nasal and palatine mucosa	Secretions
VIII	Vestibulocochlear	Afferent	Hair cells of organ of Corti, semicircular canals, maculae, and saccule	Hearing, balance, change of rate of motion of head
IX	Glossopharyngeal	Afferent	Back of tongue, pharynx, carotid sinus, and body	Taste and other sensations of tongue, changes in levels of blood pressure and gases
		Efferent	Stylopharyngeus muscle	Swallowing movements
			Parotid salivary gland	Secretions
X	Vagus	Afferent	Taste buds on epiglottis; larynx, trachea, pharynx, heart, lungs, esophagus, small intestine, part of colon	Pain, stretching, changes in levels of blood pressure and gases, taste
		Efferent	Smooth muscle of bronchi, esophagus, stomach, and intestines; heart; striated muscles of pharyngeal constrictors and intrinsic muscles of larynx	Movement and secretion by the organs supplied
XI	Accessory	Efferent	Muscles of larynx, trapezius, sternocleidomastoid	Shoulder movements, turning head, voice production
XII	Hypoglossal	Efferent	Muscles of tongue	Tongue movements

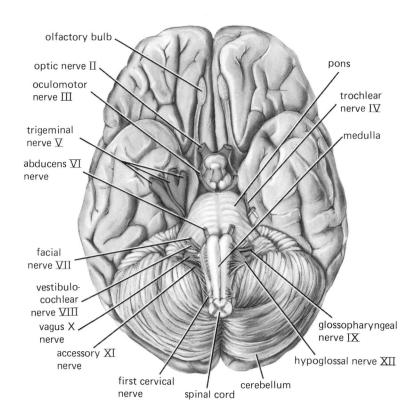

olfactory bulb

optic nerve II

oculomotor
nerve III

trigeminal
nerve V

abducens VI
nerve

facial
nerve VII

vestibulo-
cochlear
nerve VIII

vagus X
nerve

accessory XI
nerve

first cervical
nerve

spinal cord

cerebellum

pons

trochlear
nerve IV

medulla

glossopharyngeal
nerve IX

hypoglossal nerve XII

FIG. 14-27 View of the base of the brain showing the cranial nerves except for the olfactory nerve (I). The olfactory nerves (not shown) synapse with neurons in the olfactory bulbs.

matic parts. Efferent fibers arise in the brainstem from cell bodies contained in the nucleus of that nerve. For example, in the brainstem (from which most of the cranial nerves arise) there are nuclei of the oculomotor, trochlear, abducens, and facial nerves.

Afferent fibers arise outside the brain from cell bodies situated either in ganglia that occur along the nerve as it passes to the periphery or in a sense organ such as the eye, ear, or nose. The central processes of the afferent neurons enter the brain and synapse in a given nucleus of termination.

Autonomic Component of the Cranial Nerves Autonomic fibers occur only in cranial nerves III, VII, IX, and X and belong to the parasympathetic division of the autonomic nervous system. The cranial parasympathetic ganglia occur in four pairs: the *ciliary ganglia,* which innervate certain smooth muscles in the eye; the *submandibular* and *otic ganglia,* which innervate the salivary glands; and the *pterygopalatine ganglia,* which innervate mucous membranes and blood vessels in the nose (see Fig. 14-19). The parasympathetic ganglia that innervate thoracic and abdominal viscera, like the

ganglia of the head, lie close to or in the walls of the organs they innervate and are called, collectively, **enteric ganglia.**

BLOOD SUPPLY, BLOOD-BRAIN BARRIER PHENOMENA, AND CEREBROSPINAL FLUID

Glucose is the only substrate that can usually be metabolized sufficiently rapidly by the brain to supply its energy requirements, and energy from glucose is generally transferred during its oxidative breakdown to high-energy ATP molecules. The glycogen stores of the brain are negligible; thus the brain is completely dependent upon a continuous blood supply of glucose and oxygen. Although the adult brain is only 2 percent of body weight, it receives 15 percent of the total blood supply at rest to support the high oxygen utilization. If the oxygen supply is cut off for 4 to 5 min or if the glucose supply is cut off for 10 to 15 min, brain damage will occur. In fact, the most common cause of brain damage is stoppage of the blood supply (a *stroke*). The cells in the region deprived of nutrients cease to function and die.

The central nervous system receives a rich blood supply (see Chap. 21), but the exchange of substances between the blood and neurons in the central nervous system is handled differently from the somewhat unrestricted movement of substances from capillaries in other organs. A complex group of **blood-brain barrier** mechanisms closely controls both the kinds of substances that enter the extracellular space of brain and the rate at which they enter. The barrier probably comprises both anatomical structures and physiological transport systems that handle different classes of substances in different ways. The blood-brain barrier mechanisms precisely regulate the chemical composition of the extracellular space of the brain and prevent harmful substances from reaching neural tissue. However, certain areas of the brain lack a blood-brain barrier and can be readily influenced by blood-borne substances.

The central nervous system is perfused not only by its blood supply but by a second fluid, the **cerebrospinal fluid.** This clear fluid surrounds the outer surface of the brain and spinal cord so that the central nervous system literally floats in a cushion of cerebrospinal fluid. Since the brain is a soft and delicate tissue about the consistency of Jello, it is thus protected from sudden and jarring movements of the head. Cerebrospinal fluid also fills the four ventricles in the brain. The first and second ventricles (called the *lateral ventricles;* Fig. 14-28) are deep inside the two cerebral hemispheres. They meet in the midline and connect there with the *third ventricle,* which dips down vertically between the two halves of the diencephalon. The third ventricle, in turn, connects through a narrow channel, the *cerebral aqueduct,* with the *fourth ventricle,* an expanded space on the dorsal surface of the pons and part of the medulla. The fourth ventricle contracts at its posterior end to form the fine *central canal,* which runs through the caudal part of the medulla and on through the full length of the spinal cord. The central canal is usually occluded in the adult.

Cerebrospinal fluid is formed in the ventricles by highly vascular tissues, the **choroid plexuses.** A blood-brain barrier is present between the capillaries of the choroid plexus and the cerebrospinal fluid. Consistent with the barrier mechanisms, cerebrospinal fluid is a selective secretion, not a simple filtrate of plasma. For example, protein, potassium, and calcium concentrations are lower in cerebrospinal fluid than in a filtrate of plasma, whereas sodium and chloride are higher. The mechanisms responsible for this selective formation of spinal fluid are not known.

As the cerebrospinal fluid flows from its origin at the choroid plexuses in the ventricles, substances diffuse between it and the extracellular space of brain tissue because the walls of the ventricles are permeable to most substances. This exchange across the ventricular walls allows nutrients to enter brain tissue from cerebrospinal fluid and the end products of brain metabolism to leave. (However, the major sites of nutrient and end-product exchange are the cerebral capillaries.) The cerebrospinal fluid moves from its origin, back through the interconnected ventricular system to the brainstem, where it passes through three openings in the roof of the fourth ventricle (a single *median* and two *lateral apertures*) out to the subarachnoid space on the surface of the brain. This is a continuation of the cerebrospinal fluid-filled subarachnoid space surrounding the spinal cord.

Aided by circulatory, respiratory, and postural pressure changes, the cerebrospinal fluid finally flows to the top of the outer surface of the brain, where it enters the venous sinuses through one-way valves. If the path of flow is obstructed at any point between its site of formation and its final reabsorption into the vascular system, cerebrospinal fluid builds up, causing *hydrocephalus,* or "water on the brain." The pressure against which cerebrospinal fluid continues to be secreted is quite high—high enough to damage the brain so that mental retardation accompanies severe, untreated cases.

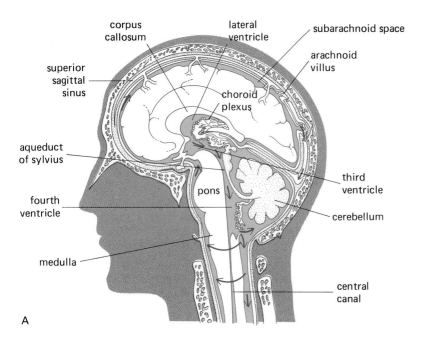

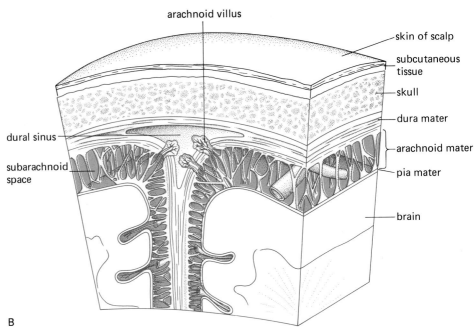

FIG. 14-28   (A) The ventricular system of the brain and the distribution of cerebrospinal fluid. (The second lateral ventricle is not shown.) Cerebrospinal fluid is formed in the ventricles (much of it by the choroid plexuses), passes to the subarachnoid space outside the brain, and then through the arachnoid villi which are valve-like structures, into large veins (sinuses) of the head. (B) A detailed view of the subarachnoid space, arachnoid villi, and large vein, or sinus, in the dura.

KEY TERMS

gray matter
white matter
ganglion
meninges
tract
spinal nerve
cerebellum

brainstem
forebrain
cerebrum
cerebral hemisphere
lobe
somatic nervous system
autonomic nervous system

sympathetic
parasympathetic
plexus
visceral nerve
blood-brain barrier
cerebrospinal fluid
cerebral ventricle

REVIEW EXERCISES

1 Distinguish between the central and peripheral nervous systems.

2 Describe the embryonic development of the nervous system, including a discussion of the neural groove, neural tube, vesicles, fluid-filled cavities, gray and white matter, and the neural crest.

3 Name the two bony and three membranous supporting structures of the central nervous system.

4 Draw a cross section through the spinal cord and bony vertebrae surrounding it (similar to Fig. 14-6); label the three meninges and the subdural, subarachnoid, and epidural spaces.

5 Draw a cross section through the spinal cord; label the gray and white matter; the gray commissure; the dorsal, ventral, and lateral horns; the dorsal and ventral roots; the dorsal root ganglion; and spinal nerve.

6 List the six subdivisions of the brain; indicate which are parts of the brainstem and which are parts of the forebrain.

7 Differentiate between the brainstem, cerebellum, and forebrain by describing their general location and function.

8 Describe the structures that constitute the brainstem; include discussions of the medulla oblongata, pons, cerebellar peduncles, midbrain, and reticular formation.

9 Describe the structural features of the cerebellum.

10 Describe the structural features of the forebrain; include discussions of the cerebral hemispheres, corpus callosum, cerebral cortex, subcortical nuclei, thalamus, and hypothalamus.

11 Draw a lateral view of the brain; indicate the locations of the frontal, parietal, occipital, and temporal lobes.

12 Describe the surface of the cerebral hemispheres; use the terms *gyri, sulci,* and *fissures.*

13 Describe the limbic system.

14 Describe the peripheral nervous system; differentiate between cranial and spinal nerves.

15 Differentiate between nerve and nerve fiber.

16 Describe the relationship between a Schwann cell and peripheral nerve fibers.

17 Describe the anatomy and function of neurons in the afferent division of the peripheral nervous system.

18 Differentiate between the autonomic and somatic divisions of the peripheral nervous system.

19 Compare the anatomy and neurotransmitters in the sympathetic and parasympathetic divisions of the autonomic nervous system.

20 Distinguish between dual innervation and reciprocal innervation; give two examples of dual innervation.

21 Compare the functions of the sympathetic and parasympathetic divisions of the autonomic nervous system; use Table 14-5 as a guide.

22 List the five classes of spinal nerves and give the number of pairs of nerves in each class.

23 Name and describe the four major nerve plexuses.

24 Draw a spinal nerve; show the relationship between the sympathetic fibers and the sympathetic trunk (as in Fig. 14-26A); label the preganglionic fibers, white ramus communicans, sympathetic trunk, and gray ramus communicans.

25 Describe the three different courses a sympathetic preganglionic fiber may take after it enters the sympathetic trunk.

26 Describe the prevertebral ganglia; use the terms *sympathetic preganglionic fiber, visceral nerve,* and *mesenteric ganglia* in your discussion.

27 Name the twelve cranial nerves and state their functions.

28 Describe the autonomic component of the cranial nerves; use the terms *parasympathetic preganglionic fibers* and *enteric ganglia* in your discussion.

29 Describe the function of the blood-brain barrier, the cerebral ventricles, and the cerebrospinal fluid.

CHAPTER

$$\boxed{15}$$

SENSORY FUNCTIONS

A human being's awareness of the world is determined by the physiological mechanisms involved in the processing of afferent information, the initial steps of which are the conversion of stimulus energy into action potentials in nerve fibers. This code represents information from the external world even though, as is frequently the case with symbols, it differs vastly from what it represents. Afferent information may or may not have a conscious correlate; i.e., it may or may not give rise to a conscious awareness of the physical world. Afferent information that does have a conscious correlate is called **sensory information,** and the conscious experience of objects and events of the external world that we acquire from the neural processing of sensory information is called **perception.**

Intuitively, it might seem that sensory systems operate like familiar electrical communications equipment, but this is true only up to a point. As an example, let us compare telephone transmission with our auditory sensory system. The telephone changes sound waves into electric impulses, which are then transmitted along wires to the receiver; thus far the analogy holds. (Of course, the mechanisms by which electric currents and action potentials are transmitted are quite different but this does not affect our argument.) The telephone then changes the coded electric impulses *back into sound waves.* Here is the crucial difference, for our brain does not physically translate the code into sound; rather the coded information itself or some correlate of it is what we perceive as sound. At present there is absolutely no understanding of how coded action potentials or composites of them are perceived as conscious sensations.

BASIC CHARACTERISTICS OF SENSORY CODING

It is worthwhile restating the fact that all information transmitted by the nervous system over distances greater than a few millimeters is signaled in the form of action potentials traveling over discrete neural pathways. Several different kinds of information must be relayed by this code: stimulus modality, intensity, and localization.

Stimulus Modality

As described in Chap. 13, different receptors have different sensitivities; i.e., each receptor type responds more readily to one form of energy than to

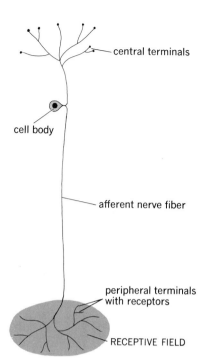

FIG. 15-1 Sensory unit and receptive field.

others. These different forms of stimulus energy are called **modalities.** Therefore, the type of receptor activated by a stimulus constitutes the first step in the coding of different types (modalities) of stimuli. The afferent nerve fibers from these receptors and at least some of the ascending spinal cord and brain pathways activated by them retain the same specificity in that they carry information that pertains to only one sensory modality. Thus, there are specific pathways ("labeled lines," as it were) for the different modalities. In tracing these pathways we shall begin at the receptor.

A single afferent neuron plus all the receptors it innervates make up a **sensory unit.** In a few cases the afferent neuron innervates a single receptor, but generally the peripheral end of an afferent neuron divides into many fine branches, each terminating at a receptor (Fig. 15-1). All the receptors of a sensory unit are preferentially sensitive to the same stimulus modality. The **receptive field** of a neuron is that area which, if stimulated, leads to activity in the neuron.

The central processes of the afferent neurons terminate in the central nervous system, *diverging* to terminate on several (or many) interneurons (Fig. 15-2A) and *converging* so that the processes of many afferent neurons terminate upon a single interneuron (Fig. 15-2B). Many of the parallel chains of interneurons are grouped together to form the

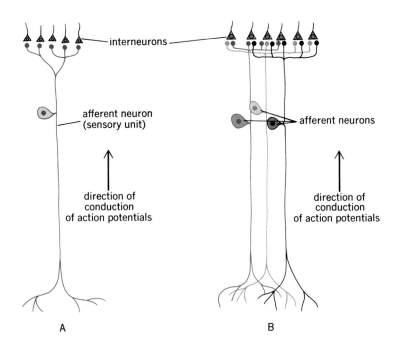

interneurons

afferent neuron
(sensory unit)

afferent neurons

direction of
conduction
of action potentials

direction of
conduction
of action potentials

A

B

FIG. 15-2 (A) Divergence of afferent neuron (sensory unit) terminals. (B) Convergence of input from several afferent neurons onto single interneurons.

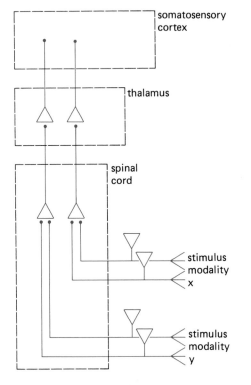

somatosensory
cortex

thalamus

spinal
cord

stimulus
modality
x

stimulus
modality
y

A SPECIFIC ASCENDING PATHWAY

FIG. 15-3 Diagrammatic representation of (A) specific and (B) nonspecific ascending pathways.

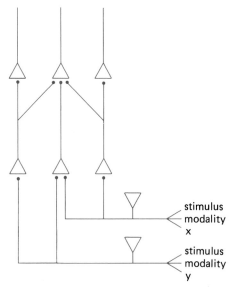

stimulus
modality
x

stimulus
modality
y

B NONSPECIFIC ASCENDING PATHWAY

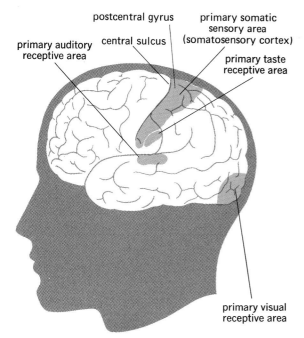

FIG. 15-4 Primary sensory areas of the cerebral cortex. The central sulcus is also called the central fissure or fissure of Rolando.

primary auditory receptive area

postcentral gyrus

central sulcus

primary somatic sensory area (somatosensory cortex)

primary taste receptive area

primary visual receptive area

specific ascending pathways of the central nervous system. Each chain consists of three to five synaptically connected neuronal links. Several sensory units may converge upon a given chain of neurons, but all these sensory units respond to the same stimulus modality so that specificity is maintained (Fig. 15-3A). For example, certain brainstem interneurons that form links in the specific pathways fire action potentials only when hair-follicle receptors are stimulated, others only when vibratory stimuli are applied to pacinian corpuscles, and still others when a specific joint is moved.

The specific pathways (except for the olfactory pathways) pass to the brainstem and thalamus, the final neurons in the pathways going to different areas of the cerebral cortex (Fig. 15-4). The fibers transmitting the **somatic sensory modalities** (touch, temperature, etc.) synapse at cortical levels in **somatosensory cortex,** a strip of cortex which lies on the postcentral gyrus in the parietal lobe just behind the junction between the parietal and frontal lobes. (The word *somatic* in this context refers to the framework or outer walls of the body, including skin, skeletal muscle, tendons, and joints, as opposed to the viscera, i.e., the organs in the thoracic and abdominal cavities. Somatic also includes the

eye and ear but not taste and smell, which are visceral.)

The specific pathways that originate in receptors of the taste buds, after synapsing in brainstem and thalamus, probably pass to cortical areas adjacent to the face region of the somatosensory strip. The specific pathways from the ears and eyes do not pass to the somatosensory cortex but go instead to other primary cortical receiving areas (Fig. 15-4). The pathways that subserve olfaction are different from all the others in that they do not pass through thalamus and have no representation in cerebral cortex; they pass instead into parts of the limbic system (see Fig. 14-14).

To reiterate, stimulus modality is indicated by the specific sensitivity of individual receptors and the pathways that convey the information to the primary sensory areas of the brain. However, these specific pathways just described are not the only ascending pathways. In contrast to the specific pathways, neurons in **nonspecific pathways** (Fig. 15-3B) are each activated by sensory units of several different modalities and therefore signal only general information about the level of excitability; i.e., they indicate that *something* is happening, usually without specifying just what (or where). A given cell in the nonspecific pathways may respond, for example, to maintained skin pressure, heating, cooling, skin stretch, and other stimuli applied to afferent nerves from deep tissues. Such cells, which respond to different kinds of stimuli, are called **polymodal.** The nonspecific pathways feed into the brainstem reticular formation and regions of the thalamus and cerebral cortex that are not highly discriminative.

Stimulus Intensity

The second kind of information contained in the action-potential code indicates the intensity of the stimulus. As described in Chap. 13, one important mechanism for signaling intensity is the number of sensory units activated; generally the greater the intensity of the stimulus, the greater is the number of sensory units activated. A second mechanism is the frequency at which any given sensory unit fires; an example is illustrated in Fig. 15-5, which shows the impulses in a single afferent fiber as a temperature receptor in the skin is gradually cooled from 34 to 26°C (normal body temperature is close to

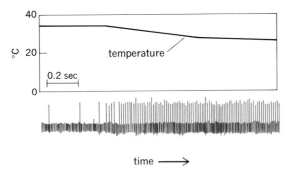

FIG. 15-5 Single cold receptor signals a drop in temperature from 34 to 26°C with an increase in firing rate of action potentials in the afferent nerve fiber.

37°C). An increased action-potential frequency correlates with increased stimulus intensity (and vice versa) for afferent pathways leading to the sensory experiences not only of temperature, but of touch, limb position (joint extension of flexion), taste, sound (loudness), and light (brightness).

Stimulus Localization

A third type of information to be signaled is the location of the stimulus. Since only sensory units from a restricted area converge upon any one interneuron of the specific pathways, the pathway that begins with that particular interneuron transmits information about that restricted area. Thus, the specific pathway indicates stimulus location as well as stimulus modality.

The precision with which a stimulus can be localized and differentiated from an adjacent stimulus depends both on the size of the receptive field covered by a single afferent neuron (the sizes of the receptive fields of individual sensory units may vary considerably, e.g., from 2 to 200 mm² in skin) and on the amount of overlap of nearby receptive fields. For example, the ability to discriminate between two adjacent mechanical stimuli to the skin is greatest on the thumb, fingers, lips, nose, and cheeks, where the sensory units are small and overlap considerably. The localization of visceral sensations is less precise than that of somatic stimuli because there are fewer afferent fibers and each has a larger receptive field.

Control of Afferent Information

All incoming afferent information is subject to extensive control before it reaches higher levels of the central nervous system. Much of it is reduced or even abolished by inhibition from other neurons. Some of the neural elements mediating these controls are collaterals from afferent neurons, interneurons in the local vicinity, and descending pathways from higher regions, particularly the reticular formation and cerebral cortex. The inhibitory controls are exerted via synapses mainly upon two sites: (1) axon terminals of the afferent neurons and (2) the second-order neurons, i.e., the interneurons directly activated by the afferent neurons. The afferent terminals are influenced by presynaptic inhibition (Chap. 13), whereas the second-order cells are inhibited by both pre- and postsynaptic mechanisms.

In some cases afferent input is continually, i.e., tonically, inhibited to some degree. This provides the flexibility either of removing the inhibition (**disinhibition**) to allow a greater degree of signal transmission or of increasing the inhibition to block the signal more completely. The thin myelinated and unmyelinated fibers that play an important role in pain seem to be under such tonic influence.

Afferent information can also be modified by central facilitation as well as by inhibition. Both methods increase the contrast between "wanted" and "unwanted" information, thereby increasing the effectiveness of selected pathways and focusing sensory-processing mechanisms on "important" messages.

In some afferent systems, the control is organized in such a way that the stronger inputs are enhanced and the weaker inputs of adjacent sensory units simultaneously inhibited. Such **lateral inhibition** can be demonstrated in the following way. While pressing the tip of a pencil against the finger with one's eyes closed, one can localize the pencil point quite precisely, even though the region around the pencil tip is also indented and mechanoreceptors within this entire area are activated; this is because the information from the peripheral region is removed by mechanisms of lateral inhibition. Lateral inhibition occurs in the pathways of most sensory modalities but is utilized to the greatest degree in the pathways that provide the most accurate localization. For example, a stimulus can be localized very precisely in the pathways relaying information about hair deflection, whereas stimuli activating temperature receptors, whose pathways lack lateral inhibition, are localized only poorly.

Lateral inhibition also occurs between pathways

that relay information about different sensory modalities. One example may explain why we rub an area to stop a pain there: Afferent fibers whose receptors are activated by mechanical stimuli can to a certain extent inhibit the output of afferent fibers whose receptors are stimulated by irritating or painful stimuli; therefore, a mechanical stimulus (rubbing) can inhibit an irritating or painful stimulus.

With these general principles as background, we now turn to the specific sensory systems, their receptor mechanisms, and the way the resulting neural signals are processed by the central nervous system.

THE SENSORY SYSTEMS

Receptors are classified in various ways. For example, they may be classified according to their modality (*thermoreceptors,* which sense changes in temperature; *mechanoreceptors,* which are sensitive to pressure changes; *chemoreceptors,* which are sensitive to chemical changes in the surrounding fluid; etc.). Alternatively, receptors may be classified according to their locations in the body and their functions in sensory awareness. **Exteroceptors** are located in the skin and organs of the special senses (that is, the eyes, ears, nose, etc.) and respond to stimuli from the external environment. **Proprioceptors** respond to stimuli in the deeper tissues, particularly the joints, muscles, and tendons, and, with some of the exteroceptors, provide information about the body's position in space. **Interoceptors,** which are generally sensitive to stretch, include the receptors in the walls of the viscera and blood vessels. The afferent neurons having interoceptors at their peripheral ends are **visceral afferent neurons;** those having exteroceptors or proprioceptors are called **somatic afferent neurons.** In this book, however, we shall group them together and refer simply to **afferent neurons.**

Somatic Sensation

The different kinds of somatic receptors respond to mechanical stimulation of the skin or hairs and underlying tissues, rotation or bending of joints, temperature changes, and possibly some chemical changes. Their activation gives rise to the sensations of touch, pressure, heat, cold, the awareness of the position and movement of the parts of the body, or pain. Recalling that by receptor we mean

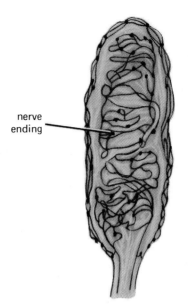

nerve ending

FIG. 15-6   Meissner's corpuscle.

the specialized peripheral ending of an afferent nerve fiber or the specialized receptor cell associated with it, we can say that each of these sensations is probably associated with a specific type of receptor; i.e., there are distinct receptors for heat, cold, touch, pressure, joint position, and pain.

The somatic receptors are classified as either **free nerve endings** or **encapsulated endings.** Free nerve endings, which have neither myelin nor Schwann-cell coatings (although they may be part of an afferent neuron that is myelinated), divide and ramify through many tissues of the body, including the epidermis and connective tissue of the skin. These endings are often implicated in the transduction of painful or thermal stimuli. Encapsulated endings have specialized nonneural cells associated with them, and as the name implies, the nerve terminal and its surrounding cells are enclosed in a capsule. One type of encapsulated ending, the pacinian corpuscle, was introduced in Chap. 13. Another is the **Meissner corpuscle** (Fig. 15-6), in which the nerve terminal winds back and forth between extensions of a specialized lamellar cell in the core of the corpuscle. Like the pacinian corpuscle, the Meissner corpuscle is presumed to be a rapidly adapting mechanoreceptor. Other encapsulated endings exist, but their functions are often not clear.

After entering the central nervous system, the afferent fibers from the somatic receptors synapse onto interneurons that form the specific pathways that go to somatosensory cortex via the brainstem and thalamus. The specific pathways cross from the side of afferent-neuron entry to the opposite side of the central nervous system in the spinal cord or brainstem; thus the sensory pathways from receptors on the left side of the body go to the somatosensory cortex of the right cerebral hemisphere and vice versa (Fig. 15-7). In the somatosensory cortex, the endings of the individual fibers of the specific somatic pathways are grouped according to the location of the receptors. The pathways that originate in the foot terminate nearest the longitudinal dividing line between the two cerebral hemispheres. Passing laterally over the surface of the brain, one finds the terminations of the pathways from leg, trunk, arm, hand, face, tongue, throat, and viscera (Fig. 15-8). The parts with the greatest sensitivity (fingers, thumb, and lips) are represented by the largest areas of somatosensory cortex. The sensory strip of one cerebral hemisphere is duplicated in the opposite hemisphere.

Pain   A stimulus that causes or is on the verge of causing tissue damage often elicits a sensation of pain and a reflex escape or withdrawal response as well as the gamut of physiological changes mediated by the sympathetic nervous system similar to those elicited during fear, rage, and aggression. These changes usually include faster heart rate, higher blood pressure, greater secretion of epinephrine into the bloodstream, increased blood sugar, less gastric secretion and motility, decreased blood flow to the viscera and skin, dilated pupils, and sweating. In addition, a painful stimulus elicits the experience of an emotional component of fear, anxiety, and sense of unpleasantness as well as information about the stimulus's location, intensity, and duration. And probably more than any other type of sensation, the experience of pain can be altered by past experiences, suggestion, emotions (particularly anxiety), and the simultaneous activation of other sensory modalities.

To reiterate, the stimuli that give rise to pain result in a sensory experience *plus* a reaction to it, the reaction including the emotional response (anxiety, fear) and behavioral response (withdrawal or other defensive behavior). Both the sensation and the reaction to the sensation must be present for tissue-damaging stimuli to cause suffering. The sensation of pain can be dissociated from the emotional and behavioral reactive component by drugs, e.g.,

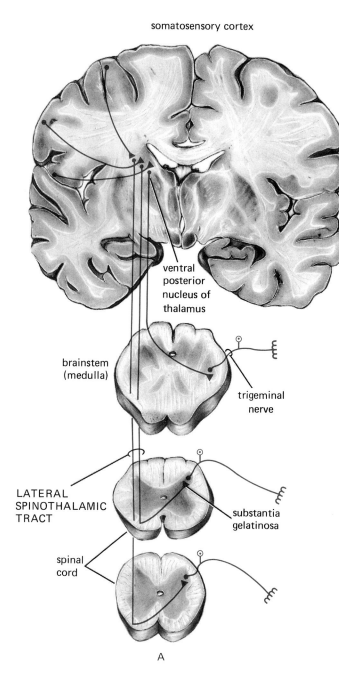

FIG. 15-7   Examples of ascending pathways. (A) The lateral spinothalamic tract, which conveys pain and temperature information. (B) The posterior columns (fasciculus gracilis and fasciculus cuneatus) and spinocerebellar tracts, which convey proprioceptive information. Axons in the posterior columns are the central processes of afferent (first order) neurons; they terminate in the nuclei gracilis and cuneatus.

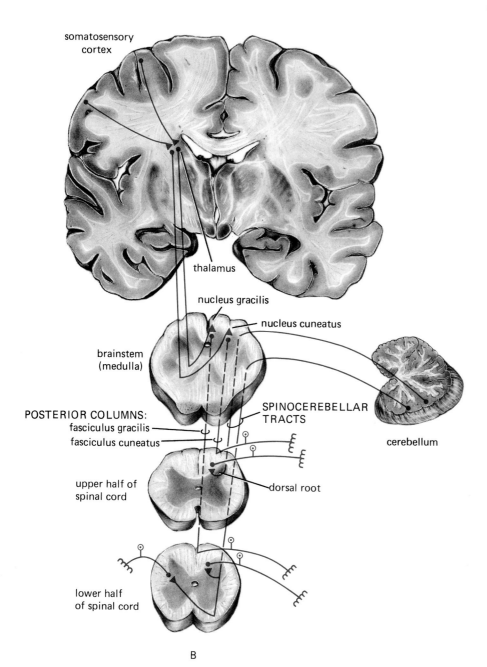

somatosensory
cortex

thalamus

nucleus gracilis

nucleus cuneatus

brainstem
(medulla)

POSTERIOR COLUMNS:
fasciculus gracilis
fasciculus cuneatus

SPINOCEREBELLAR
TRACTS

cerebellum

upper half of
spinal cord

dorsal root

lower half
of spinal cord

B

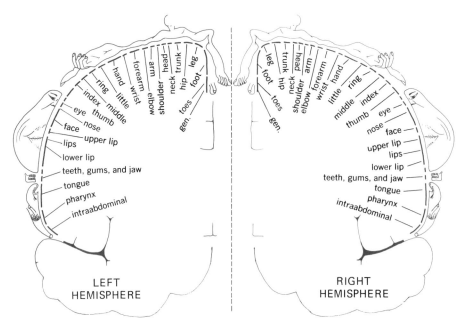

FIG. 15-8 Location of pathway terminations for different parts of the body in the somatosensory cortex. The left half of the body is represented on the right hemisphere of the brain, and the right half of the body is represented on the left cerebral hemisphere.

morphine, or by selective brain operations that interrupt pathways that connect the frontal lobe of the cerebrum with other parts of the brain. When the reactive component is no longer associated with the sensation, pain is felt, but it is not necessarily disagreeable; the patient does not mind as much. Thus, pain relief can be obtained even though the perception of painful stimuli is not reduced.

The receptors whose stimulation gives rise to pain are at the ends of certain small unmyelinated or lightly myelinated afferent neurons. These receptors, known as **nociceptors,** all appear the same when viewed with an electron microscope, yet different endings respond preferentially to harmful mechanical, chemical, or thermal stimuli. It is hypothesized that a chemical is released from the damaged tissue and depolarizes the nearby nerve endings, initiating action potentials in the afferent nerve fibers. The chemical has not been identified, although there are many candidates.

The primary afferents coming from nociceptors synapse onto two different types of interneurons after entering the central nervous system (one chemical thought to be released at this synapse is the neurotransmitter known as *substance P*) such that information about pain is transmitted to higher centers via the specific ascending pathways for pain

(the **lateral spinothalamic tracts,** Fig. 15-7A) and via nonspecific pathways. It is hypothesized that the specific pathways, which go to the thalamus and cerebral cortex and convey information about where, when, and how strongly the stimulus was applied, convey information about the sharp, localized aspect of pain. The nonspecific pathways, which go to the brainstem reticular formation and a part of the thalamus different from that supplied by the specific pathways, are believed to convey information about the aspect of pain that is duller, longer lasting, and less well localized.

At some of the synaptic links in the pain pathways the enkephalin and endorphin neurotransmitters are thought to be involved, and morphine, which activates the receptor sites for these transmitters, is thought to relieve pain by its ability to activate this inhibitory system. Since the dull, long-term, poorly localized type of pain responds more to morphine than does the sharp, well-localized type, this inhibitory pathway presumably affects mainly the nonspecific pathways.

Vision

The Eye The eyes are embedded in a fatty cushion within the orbits of the skull. The exposed sur-

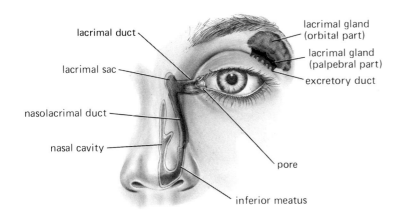

lacrimal duct

lacrimal sac

nasolacrimal duct

nasal cavity

lacrimal gland (orbital part)

lacrimal gland (palpebral part)

excretory duct

pore

inferior meatus

FIG. 15-9 Lacrimal apparatus, which consists of the lacrimal gland and its excretory ducts and the lacrimal ducts, lacrimal sac, and nasolacrimal duct.

face of the eyeball is kept moist by a continual flow of fluid (tears) from the **lacrimal gland,** located in the superiolateral corner of the orbit (Fig. 15-9). The tears drain via **excretory ducts** onto the surface of the eyeball and flow across the eye to its medial corner where they pass via **pores** into the **lacrimal ducts, lacrimal sac,** and **nasolacrimal duct,** which, in turn, empties into the nasal cavity. The lacrimal fluid not only keeps the eyeball moist and clean but also protects it against infection by means of antibacterial chemicals contained in it. The lacrimal glands are innervated by autonomic neurons that stimulate the increased secretion associated with emotions or irritation of the eyes. The eyeballs are also protected by a thin layer of epithelium, the **conjunctiva,** which covers part of their anterior surface.

Each eye consists of a three-layered membranous sac divided into three compartments: the **anterior** and **posterior chambers** which contain the **aqueous humor** (the posterior chamber contains the **lens** as well); and the **vitreous chamber,** which contains the **vitreous body** (Fig. 15-10A, B).

The external layer of the eye is the opaque **sclera** in the posterior part of the eye and the **cornea** in the anterior part (Fig. 15-10A, B). The transparent cornea, which projects as a flattened dome from the front of the sclera, has neither blood nor lymph vessels, but it has numerous free nerve endings. The highly vascular middle layer from the back of the eye forward is made up of the **choroid, ciliary body,** and **iris.** The iris, which is behind the cornea and in front of the lens, is a circular diaphragm with a central opening, the **pupil.** Light can enter the eye only through the pupil because the iris itself is pig-

mented and light cannot pass through it. The iris exerts important control over the amount of light entering the eye. Near the border of the pupil the iris contains a ring of smooth muscle cells, the **pupillary sphincter,** innervated by the parasympathetic division of the autonomic nervous system. Activation of the sphincter muscle decreases the diameter of the pupil and decreases the amount of light that enters the eye. Radially arranged smooth muscle fibers in the iris, the **pupillary dilator muscles,** are innervated by sympathetic nerve fibers; when activated, they enlarge the pupil.

The ciliary body supports the lens via **suspensory,** or **zonular, ligaments.** It also contains the **ciliary muscles,** which change the shape of the lens for near or far vision, and it secretes the aqueous humor that fills the anterior and posterior chambers. The aqueous humor provides a path for the metabolites of the avascular lens and cornea, and it also maintains pressure in the eyeball so that the latter's dimensions remain constant. It is drained from the anterior chamber via the **canal of Schlemm** (Fig. 15-10B). An increased aqueous-humor pressure, often caused by obstruction of the canal of Schlemm, is associated with the condition *glaucoma*. The inner, neural layer of the eye is the **retina,** which contains the receptors (**rods** and **cones**) for light. The cup formed by the sclera, choroid, and retina is filled by the transparent vitreous body.

Light   The receptors of the eye are sensitive to only that tiny portion of the vast spectrum of electromagnetic radiation that we call light (Fig. 15-11). Radiant energy is described in terms of wavelengths and frequencies. The **wavelength** is the distance be-

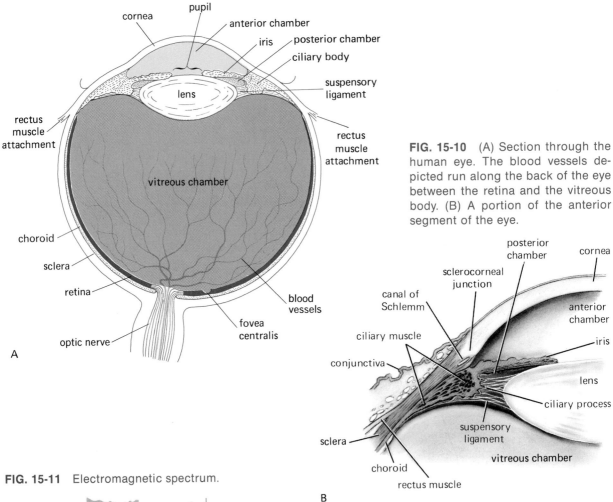

FIG. 15-10 (A) Section through the human eye. The blood vessels depicted run along the back of the eye between the retina and the vitreous body. (B) A portion of the anterior segment of the eye.

FIG. 15-11 Electromagnetic spectrum.

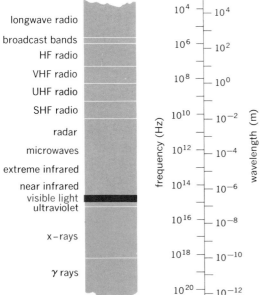

tween two successive wave peaks of the electromagnetic radiation (Fig. 15-12) and varies from several kilometers at the top of the spectrum to minute fractions of a millimeter at the bottom end. Those wavelengths capable of stimulating the receptors of the eye (the visible spectrum) are between 400 and 700 nm (nanometers), and light of different wavelengths in this band is associated with different color sensations.

The Optics of Vision   Light can be represented most simply by a ray or line drawn in the direction in which the wave is traveling. Light waves are propagated in all directions from every point of a visible object. These divergent light waves must pass through an optical system that focuses them back into a point before an accurate image of the

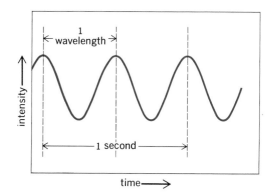

FIG. 15-12 Properties of a wave. The frequency of this wave is 2 cycles per second (cps or Hz).

object is achieved. In the eye itself, the image of the object being viewed must be focused upon the **retina,** where the light-sensitive receptor cells of the eye are located.

The **lens** and **cornea** of the eye are the optical systems that focus the image of the object upon the retina. At a boundary between two substances, such as the cornea of the eye and the air outside it, the rays are bent so that they travel in a new direc-

tion. The degree of bending depends in part upon the angle at which the light enters the second medium. The cornea plays the larger role in focusing light rays because the rays are bent more in passing from air into the cornea than in passing into and out of the lens or other transparent structures of the eye.

The surface of the cornea is curved so that light rays coming from a single point source hit the cornea at different angles and are bent different amounts, but all in such a way that they are directed to a point after emerging from the lens (Fig. 15-13A). When the object being viewed has more than one dimension (Fig. 15-13B), the image on the retina is upside down relative to the original light source and it is also reversed left to right.

The shape of the cornea and lens and the length of the eyeball determine the point where light rays reconverge. Moreover, light rays from objects close to the eye strike the cornea at greater angles (are more divergent) and have to be bent more in order to reconverge on the retina. Although the cornea performs the greater part quantitatively of focusing the visual image on the retina, all adjustments for

FIG. 15-13 Refraction (bending) of light by the lens system of the eye. (A) The focusing of light rays from a single point. (B) The focusing of light rays from more than one point to form the image of the object.

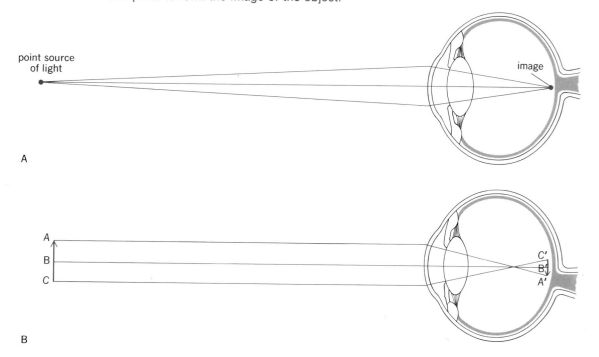

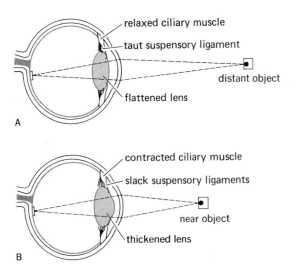

A

B

FIG. 15-14 Accommodation for distant and near vision by the pliable lens. (A) The lens is stretched for distant vision so that it adds the minimum amount of focusing power. (B) The lens thickens for near vision to provide greater focusing power.

distance are made by changing the shape of the lens. Such changes are called **accommodation.** The shape of the lens is controlled by the ciliary muscles and the tension they apply to the suspensory ligaments of the lens. The lens is flattened when distant objects are to be focused upon the retina and allowed to assume a more spherical shape to provide additional bending of the light rays when near objects are viewed (Fig. 15-14). The ciliary muscles are controlled by parasympathetic nerve fibers.

Cells are added to the lens throughout life but only to the outer surface. This means that cells at the center of the lens are both the oldest and the farthest away from the nutrient fluid that bathes the outside of the lens (if capillaries ran through the lens, they would interfere with its transparency). These central cells age and die first, and with death they become stiff, so that accommodation of the lens for near and far vision becomes more difficult.

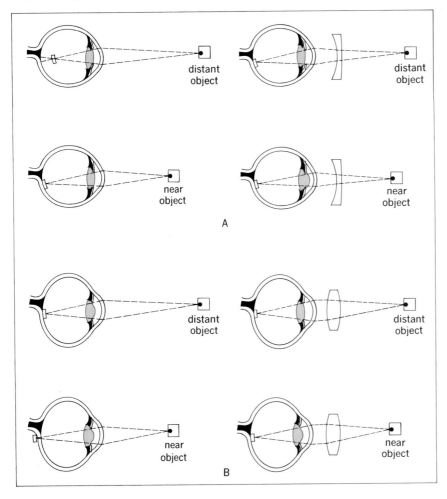

A

B

FIG. 15-15 (A) In the nearsighted eye, light rays from a distant source are focused in front of the retina. A concave lens placed before the eye bends the light rays out sufficiently to move the focused image back onto the retina. When near objects are viewed through concave lenses, the eye accommodates to focus the image on the retina. (B) The farsighted eye must accommodate to focus the image of distant objects upon the retina. (The normal eye views distant objects with a flat, stretched lens.) The accommodating power of the lens of the eye is sufficient for distant objects, and these objects are seen clearly. The lens cannot accommodate enough to keep images of near objects focused on the retina, and they are blurred. A convex lens converges light rays before they enter the eye and allows the eye's lens to work in a normal manner.

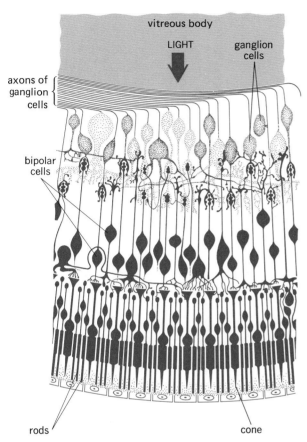

axons of ganglion cells

bipolar cells

rods

vitreous body

LIGHT

ganglion cells

cone

FIG. 15-16 Human retina. Light entering the eye must pass through the fibers and cells of the retina before reaching the sensitive tips of the rods and cones. After leaving the eyeball, the axons of the ganglion cells form the optic nerve.

This impairment is known as *presbyopia* and is one reason why many people who never needed glasses before start wearing them in middle age.

Cells of the lens can also become opaque so that detailed vision is impaired; this is known as *cataract.* The defective lens can usually be removed surgically from persons suffering from cataract, and with the addition of compensating eyeglasses, contact lenses, or even an implanted artificial lens, effective vision can be restored although the ability to accommodate will be lost.

Defects in vision occur if the eyeball is too long in relation to the lens size, for then the images of near objects fall on the retina but the images of far objects come into focus at a point in front of the retina (Fig. 15-15A). This is a *nearsighted,* or *myopic,* eye, which is unable to see distant objects

clearly. If the eye is too short for the lens, distant objects come into focus on the retina while near objects come into focus behind it (Fig. 15-15B); this eye is *farsighted,* or *hyperopic,* and near vision is poor. Figure 15-15 also shows the corrections for near- and farsighted vision achieved by lenses. Defects in vision also occur where the lens or cornea does not have a smoothly spherical surface. The improperly shaped eyeball or irregularities in the cornea (*astigmatism*) or lens can usually be compensated for by eyeglasses.

As mentioned above, the amount of light entering the eye is controlled by the iris, the color being of no importance as long as the tissue is sufficiently opaque to prevent the passage of light. The pupillary sphincter muscle of the iris by reflex contracts in bright light, decreasing the diameter of the pupil; this not only reduces the amount of light entering the eye but also directs the light to the central and most optically accurate part of the lens. Conversely, the sphincter relaxes in dim light, when maximal sensitivity is needed.

Receptor Cells and the Retina   The receptor cells in the retina (the photoreceptors) are called **rods** and **cones** because of their microscopic appearance (Fig. 15-16). Photoreceptors contain light-sensitive molecules called **photopigments** that absorb light. There are four different photopigments in the retina, each of which is made up of a protein (**opsin**) bound to a **chromophore** molecule. The chromophore is always **retinal** (a slight variant of vitamin A) but the opsin differs in each of the four types. This difference causes each of the four photopigments to absorb light most effectively at a different part of the visual spectrum. For example, one photopigment absorbs wavelengths in the range of red light better than those in blue or green ranges whereas another absorbs green light better than red or blue.

The events leading to receptor activation have been most studied for the rods. The rod-shaped ending of these receptor cells contains many stacked, flattened membrane discs parallel to the light-receiving surface of the retina. The disc membranes contain a high concentration of the photopigment **rhodopsin.** The initial change in molecular shape, caused by the absorption of light, leads to an unstable molecule that spontaneously undergoes further changes, the result being the dissociation of

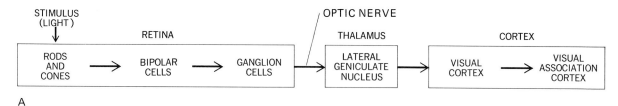

A

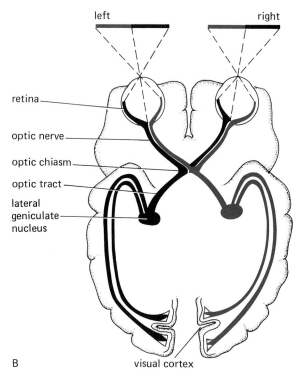

B

FIG. 15-17 (A) Diagrammatic representation of the cells in the visual pathway. (B) The visual pathway.

the chromophore from the opsin. A substance (possibly calcium) is released and causes a decrease in the sodium permeability of the plasma membrane, which produces a receptor potential. Since the change is a *decrease* in sodium permeability, the receptor potential in this case is a hyperpolarization. Thus the receptors are relatively depolarized in the dark and hyperpolarized in the light, and transmitter release, which occurs with depolarization, increases in the dark and is reduced by illumination. Needless to say, this is still an effective signaling mechanism. After dissociation of the photopigment, rhodopsin is regenerated by reactions that do not depend on light but are enzyme-mediated.

The receptor cells (whether rods or cones) synapse upon **bipolar cells** in the retina (Fig. 15-17A), which in turn synapse upon **ganglion cells.** Generally, cone receptor cells have relatively direct lines to the brain; i.e., there is little convergence along the neural pathway. This relative lack of convergence provides precise information about the area of the retina that was stimulated, and cone visual acuity is very high. Because cones are concentrated in the center of the retina, it is that part that we use for finely detailed vision. However, the lack of convergence in the cone visual pathways offers little opportunity for the summation of subthreshold events to fire action potentials in the ganglion cells, and high levels of illumination are needed to activate these pathways. Thus, the cone pathways are useful only in ''daytime'' vision. In contrast, there is much convergence in the rod visual pathways, and although acuity is poor, opportunities for spatial and temporal summation are good. Therefore, a relatively low-intensity light stimulus that would cause only a subthreshold response in a cone ganglion cell can cause an action potential in a rod ganglion cell. Thus, the difference in acuity and light sensitivity between rod and cone vision is a result, at least in part, of the anatomical wiring patterns of the retina.

These differences explain why objects in a darkened theater are indistinct; with such low illumination the cones do not generate effective signals, so that all vision is supplied through the more sensitive but less accurate rod vision. Moreover, rods contain only one type of photopigment (rhodopsin) and, as we shall see later, cannot give rise to color vision; to perceive color at least two types of receptors with different photopigments must be activated. Thus, objects in a darkened theater appear in shades of gray.

The sensitivity of the eye is decreased under conditions of bright illumination (**light adaptation**) and improves after being in the dark for some time (**dark adaptation**). The state of adaptation is partly related to the amount of intact rhodopsin; thus visual sensitivity improves as more rhodopsin is regenerated.

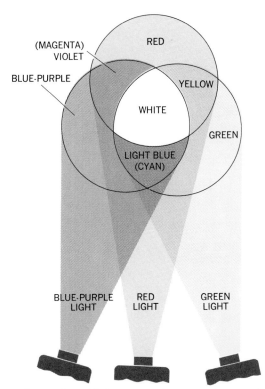

excite the photopigment of the retina most sensitive to them. Light perceived as white is a mixture of all wavelengths, and black is the absence of all light. Sensation of any color can be obtained by the appropriate mixture of three lights, red, blue, and green (Fig. 15-18). Light and pigments are properties of the physical world, but color exists only as a sensation in the mind of the beholder.

Color vision begins with the activation of the photopigments in the cone receptor cells. Human retinas have three kinds of cones, which contain either yellow-, green-, or blue-sensitive photopigments, responding optimally to light of 570-, 535-, and 445-nm wavelengths, respectively. (The 570 pigment sensitivity extends far enough to sense the long wavelengths, which correspond to red, and this pigment is sometimes called the red photopigment.) Although each type of cone is excited most effectively by light of one particular wavelength, it responds to other wavelengths as well; thus, for any given wavelength, the three cone types are excited to different degrees (Fig. 15-19). For example, in response to a light of 535-nm wavelength, the green cones respond maximally, the red cones less, and the blue cones hardly at all. Our sensation of color depends upon the ratios of these three cone outputs and their comparison by higher-order cells in the visual system.

The fact that there are three different kinds of cone cells explains the various types of *color blindness*. Most people (over 90 percent of the male population and over 99 percent of the female population) have normal color vision; i.e., their color

FIG. 15-18 All known colors can be produced by different combinations of the three primary wavelengths of light, those giving rise to the color sensations of red, blue, and green.

Adaptation is mainly determined, however, by a neural mechanism that alters the "gain" of the system so that in the dark-adapted condition activation of a rhodopsin molecule produces a larger electric signal than in the light-adapted state.

The axons of the ganglion cells form the **optic nerve** (cranial nerve II), which passes from the retina to the brain (Fig. 15-17A). The optic nerves from the two eyes meet at the **optic chiasm** (optic crossing) where some of the fibers cross over to the opposite side of the brain. This partial crossover provides both cerebral hemispheres with input from both eyes (Fig. 15-17B).

Color Vision Light is the source of all colors. Pigments, such as those mixed by a painter, serve only to reflect, absorb, or transmit different wavelengths of light, and the nature of the pigments determines how light of different wavelengths will react. For example, an object appears red because the shorter wavelengths are absorbed by the material and the longer wavelengths are reflected to

FIG. 15-19 Sensitivities of the three photopigments found in human cones. One pigment is maximally sensitive to blue (445 nm), another to green (535 nm), and the third to yellow (570 nm). The yellow pigment sensitivity extends far enough to sense red (about 650 nm) and is, in fact, called the red photopigment.

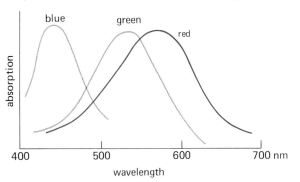

vision is determined by the differential activity of the three types of cones. Most color-blind (or better, color-defective) people appear to lack one of the three photopigments, and their color vision is therefore formed by the differential activity of the remaining two types of cones. For example, people with green-defective vision see as if they have only red- and blue-sensitive cones.

The pathways for color vision follow those described earlier (Fig. 15-17A); i.e., the cones synapse upon bipolar cells, and the bipolar cells synapse upon ganglion cells, etc.

Eye Movement   The cones are most concentrated in a specialized area of the retina known as the **fovea,** and images focused there are seen with the greatest acuity. In order to get the visual image focused on the fovea and keep it there, the six **extraocular muscles** (Fig. 15-20) attached to each eyeball are utilized. Of these six muscles, the **medial** and **lateral recti** move the eyeball in one axis only; each of the other four muscles moves it in all three axes. Together these muscles are capable of rotating the eyeball in any direction (Table 15-1). The four recti pass from their attachment at the back of the orbit to the sclera, their positions of attachment on the eye being designated by their names. The **superior oblique** also originates from the back of the orbit, but before attaching to the eye it passes through a cartilaginous loop, the *trochlea,* which allows a change in the direction of the muscle's pull much as a pulley changes the direction of pull by a rope. The **inferior oblique** originates on the nasal side of the orbit and passes under the eyeball to attach to its lateral surface.

The lateral and medial recti are antagonists, one muscle relaxing as the other contracts. They can

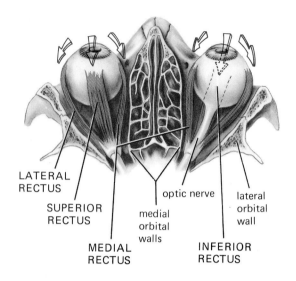

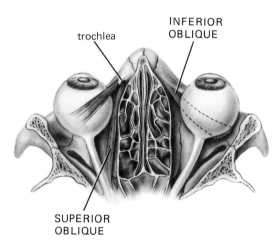

FIG. 15-20 Extraocular muscles.

either sweep the two eyes horizontally, both eyes turning to the right or left simultaneously, or they can turn the eyes medially (**convergence**) for near

TABLE 15-1 EXTRAOCULAR MUSCLES AND THEIR ACTIONS ON THE EYE		
Muscles	**Actions**	**Innervation**
Superior rectus	Elevates, adducts, rotates medially	Oculomotor (cranial III)
Inferior rectus	Depresses, adducts, rotates laterally	Oculomotor (cranial III)
Superior oblique	Depresses, abducts, rotates medially	Trochlear (cranial IV)
Inferior oblique	Elevates, abducts, rotates laterally	Oculomotor (cranial III)
Medial rectus	Adducts	Oculomotor (cranial III)
Lateral rectus	Abducts	Abducens (cranial VI)

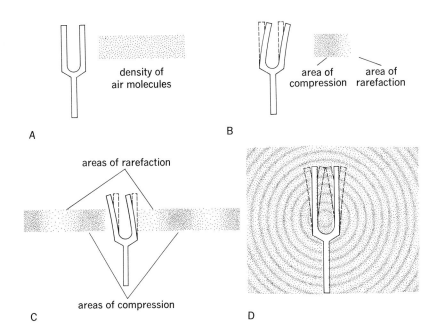

density of
air molecules

A

area of
compression

area of
rarefaction

B

areas of rarefaction

areas of compression

C

D

FIG. 15-21 Formation of sound waves. (See text for further details.)

vision or laterally (**divergence**) for distant vision. They cannot raise or lower the eyes; these movements are done by the inferior and superior recti and the two oblique muscles (Table 15-1).

The extraocular muscles perform two basic types of movements, *fast* and *slow*. The fast movements include **saccades,** small jerking movements that rapidly bring the eye from one fixation point to another to allow search of the visual field. In addition, saccades move the visual image over the receptors, thereby preventing adaptation. In fact, if saccades are prevented, all color and most detail fade away in a matter of seconds. Saccades also occur during certain periods of sleep when the eyes are closed, and perhaps they are associated with "watching" the visual imagery of dreams. Saccades are among the fastest movements in the body.

Slow eye movements are involved in tracking visual objects as they move throughout the visual field and during compensation for movements of the head. If the head is moved to the left, the eyes must be moved an equal distance to the right if the image of a stationary visual object is to remain focused on the fovea; if the head moves up, the eyes must move down. These compensating movements obtain their information about the movement of the head from the semicircular canals of the vestibular system, which will be described shortly.

Control systems for the other slow movements require the continual feedback of visual information about the moving object.

Hearing

Sound energy is transmitted through air by a movement of air molecules. When there are no air molecules, as in a vacuum, there can be no sound. The disturbance of air molecules that makes up a sound wave consists of regions of *compression,* in which the air molecules are close together and the pressure is high, alternating with areas of *rarefaction,* where the molecules are farther apart and the pressure is lower. Anything capable of creating such disturbances can serve as a sound source. A tuning fork at rest emits no sound (Fig. 15-21A), but if struck sharply it vibrates at a single fixed frequency and gives rise to a pure tone. As the arms of the tuning fork move, they push air molecules ahead of them, creating a zone of compression, and leave behind them a zone of rarefaction (Fig. 15-21B). As they move in the opposite direction, they again create pressure waves of compression and rarefaction (Fig. 15-21C).

The molecules in an area of compression, pushed together by the vibrating prong of the tuning fork, bump into the molecules ahead of them, push them

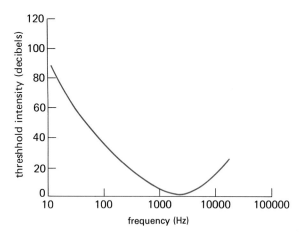

FIG. 15-22 Human audibility curve. The threshold of hearing varies with sound frequency. The conversational voice of an average male is about 120 Hz, that of an average female about 250 Hz.

together, and create a new region of compression (Fig. 15-21D). Individual molecules travel only short distances, but the disturbance passed from one molecule to another can travel many miles; and it is by these disturbances (*sound waves*) that sound energy is transmitted. The sound dies out only when so much of the original sound energy has been dissipated that one sound wave can no longer disturb the air molecules around it. The tone emitted by the tuning fork is said to be *pure* because the waves of rarefaction and compression are regularly spaced. The waves of speech and many other common sounds are not regularly spaced but are complex waves made up of many frequencies of vibration.

The sounds heard most keenly by human ears are those from sources that vibrate at frequencies between 1000 and 4000 Hz, or cycles per second (Fig. 15-22), but the entire range of frequencies audible to human beings extends from 20 to 20,000 Hz. The *frequency* of vibration of the sound source is related to the pitch we hear; the faster the vibration, the higher the pitch. We can also detect loudness and tonal quality, or timbre, of a sound. The difference between the packing (or pressure) of air molecules in a zone of compression and a zone of rarefaction, i.e., the *amplitude* of the sound wave, is related to the loudness of the sound that we hear. The number of sound frequencies in addition to the fundamental tone, i.e., the lack of *purity* of the sound wave, is related to the timbre of the sound. We can distinguish some 400,000 different sounds; we can distinguish the note A played on a piano from the same note played on a violin, and we can identify voices heard over the telephone. We can also selectively *not* hear sounds, tuning out the babel of a party to concentrate on a single voice.

The first step in hearing is usually the entrance of pressure waves into the *ear canal* (**external acoustic meatus**) (Fig. 15-23A). The waves reverberate from the side and end of the ear canal, filling it with the continuous vibrations of pressure waves. The **tympanic membrane** (eardrum) is stretched across the end of the ear canal. The air molecules, under higher pressure during a wave of compression, push against the membrane, causing it to bow inward. The distance the membrane moves, although always very small, is a function of the force and velocity with which the air molecules hit it and is therefore related to the loudness of the sound. During the following wave of rarefaction, the membrane returns to its original position. The exquisitely sensitive tympanic membrane responds to all the varying pressures of the sound waves, vibrating slowly in response to low-frequency sounds and rapidly in response to high tones. It is sensitive to pressures to which the most delicate touch receptors of the skin are totally insensitive.

The tympanic membrane separates the ear canal from the **middle-ear** (or **tympanic**) **cavity** (Fig. 15-23A). The pressures in these two air-filled chambers are normally equal. The ear canal is at atmospheric pressure, but the middle ear is exposed to atmospheric pressure only through the **auditory** (*eustachian*) **tube,** which connects the middle ear to the pharynx. The slitlike ending of this tube in the pharynx is normally closed, but during yawning, swallowing, or sneezing, when muscle movements of the pharynx open the entire passage, the pressure in the middle ear equilibrates with atmospheric pressure. A difference in pressure can be produced with sudden changes in altitude, as in an elevator or airplane, when the pressure outside the ear changes while the pressure within the middle ear remains constant because of the closed auditory tube. This difference distorts the tympanic membrane and causes pain.

The second step in hearing is the transmission of sound energy from the tympanic membrane, through the middle ear cavity, and then to the fluid-filled chambers of the **inner ear.** Because the liquid in the inner ear is more difficult to move than air,

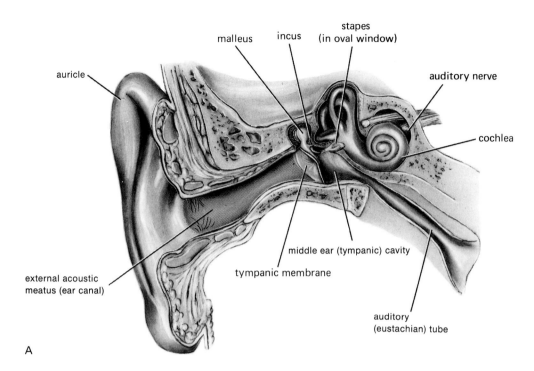

malleus incus stapes (in oval window)

auricle

auditory nerve

cochlea

middle ear (tympanic) cavity

tympanic membrane

external acoustic meatus (ear canal)

auditory (eustachian) tube

A

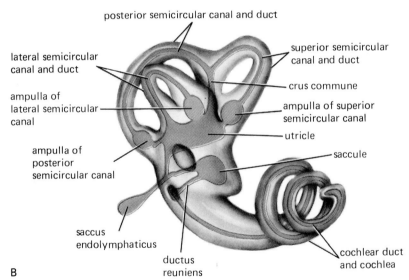

posterior semicircular canal and duct

lateral semicircular canal and duct

superior semicircular canal and duct

crus commune

ampulla of lateral semicircular canal

ampulla of superior semicircular canal

utricle

ampulla of posterior semicircular canal

saccule

saccus endolymphaticus

ductus reuniens

cochlear duct and cochlea

B

FIG. 15-23 (A) Anatomy of the human ear. The malleus, incus, and stapes are the middle ear bones. (B) The membranous labyrinth within the bony labyrinth. The saccus endolymphaticus is a blind pouch of the membranous labyrinth.

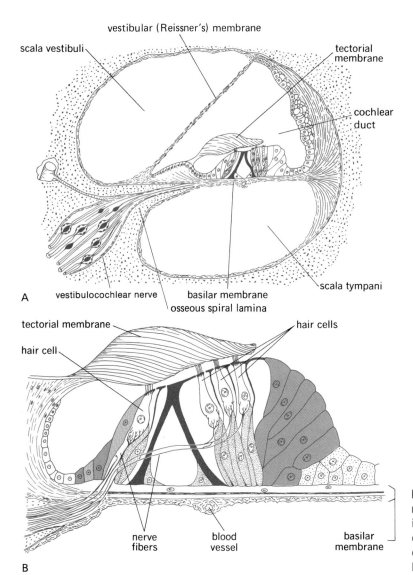

vestibular (Reissner's) membrane

scala vestibuli

tectorial membrane

cochlear duct

A

vestibulocochlear nerve

basilar membrane

osseous spiral lamina

scala tympani

tectorial membrane

hair cells

hair cell

nerve fibers

blood vessel

basilar membrane

B

FIG. 15-24 (A) Cross section of the membranes and compartments of the inner ear. (B) Detailed view of the spiral organ of Corti with its hair cells and other structures upon the basilar membrane.

the pressure transmitted to the inner ear must be amplified. This is achieved by a movable chain of three small middle-ear bones (the **auditory ossicles: the incus,** or anvil; the **malleus,** or hammer; and the **stapes,** or stirrup). These small bones couple the tympanic membrane to the membrane-covered opening (the **vestibular,** or **oval, window**) that separates the middle and inner ear.

The *total* force on the tympanic membrane is transferred to the oval window, but because the oval window is so much smaller, the *force per unit area* (i.e., pressure) is increased 15 to 20 times. Additional advantage is gained through the lever

action of the three middle-ear bones. The amount of energy transmitted to the inner ear can be modified by the contraction of two small muscles in the middle ear that alter the tension of the tympanic membrane (the **tensor tympani muscle**) and the position of the third middle-ear bone (stapes) in the oval window (the **stapedius muscle**). These muscles protect the delicate receptor apparatus from sustained intense sound stimuli and possibly aid intent listening over certain frequency ranges.

Thus far, the entire system has been concerned with the transmission of the sound energy into the inner ear, where the receptors are located. The inner ear consists of two parts: a series of cavities (the **bony labyrinth**) in the petrous portion of the temporal bone and a series of communicating membranous sacs (the **membranous labyrinth**) contained within the bony labyrinth (Fig. 15-23B). The bony labyrinth is made up of the vestibule, semicircular canals, and cochlea and is filled with a clear fluid (**perilymph**) in which the membranous labyrinth is suspended. The membranous labyrinth is filled with **endolymph.**

The **cochlea,** which is that part of the bony labyrinth concerned with hearing, contains a portion of the membranous labyrinth known as the **cochlear duct** (Fig. 15-23B) the base of which is the **basilar membrane** (Fig. 15-24A). The cochlear duct divides the cochlea lengthwise into two passageways, or scalae: the **scala tympani,** which is on the basilar membrane side of the cochlear duct, and the **scala vestibuli,** which is above the duct and separated from it by the thin **vestibular membrane** (Fig. 15-24A). The two scalae are connected with each other only at the tip of the cochlea where the cochlear duct ends. The cochlea spirals for 2¾ turns around a bony central pillar (the *modiolus*). A shelf of bone (the *osseous spiral lamina*) projects from the modiolus and is continued by the basilar membrane, which extends from the osseous spiral lamina to the opposite side of the cochlea.

As the pressure wave pushes in on the tympanic membrane, the chain of bones rocks the footplate of the stapes against the membrane that covers the oval window, causing it to bow into the scala vestibuli compartment of the cochlea (Fig. 15-25) and create there a wave of pressure. The wall of the scala vestibuli is largely bone, but there are two paths by which the pressure waves can be dissipated. One path is to the end of the scala vestibuli, where the waves pass around the end of the cochlear duct, into the scala tympani, and back to a second membrane-covered window, the **round window,** which they bow out into the middle-ear cavity. However, most of the pressure waves do not follow this route but are transmitted to the cochlear duct and thereby to the basilar membrane, which is deflected into the scala tympani.

The pattern by which the basilar membrane is deflected is important because this membrane contains the spiral **organ of Corti** with its sensitive receptor cells that transform sound energy, i.e., the pressure wave, into action potentials. At the end of the cochlea closest to the middle-ear cavity, the basilar membrane is narrow and relatively stiff, but it becomes wider and more elastic as it extends throughout the length of the cochlear spiral. The stiff end nearest the middle-ear cavity vibrates immediately in response to the pressure changes transmitted to the scala vestibuli, but the responses of the more distant parts are slower. Thus, with each change in pressure in the inner ear, a wave of vibrations travels down the basilar membrane (Fig. 15-26).

The region of maximal displacement of the basilar membrane varies with the frequency of vibration of the sound source. The properties of the membrane nearest the oval window and middle ear are such that this region resonates best with high-pitched tones and undergoes the greatest amplitude of vibration when high-pitched tones are heard. The vibration of the basilar membrane in response to high-frequency sound waves soon dies out once it is past this region. Lower tones also cause the basilar membrane to vibrate near the middle-ear cavity, but the vibration wave travels out along the membrane for greater distances. The most distant regions of the basilar membrane vibrate maximally in response to low tones. Thus the frequencies of the incoming sound waves are in effect sorted out along the length of the basilar membrane (Fig. 15-27).

Vibration of the basilar membrane serves to stimulate the receptor cells (**hair cells**) of the organ of Corti (Fig. 15-24B), which ride upon the membrane (the greatest stimulation occurs wherever displacement of the basilar membrane is greatest). Some of the fine hairs on the top of the receptor cells are in contact with the overhanging **tectorial membrane,** which projects inward from the side of the cochlea. As the basilar membrane is displaced by pressure waves, the hair cells move in relation to the tectorial membrane and the fluid surrounding the hairs and, consequently, the hairs are displaced. The resulting mechanical deformation of the hair cells generates a receptor potential, possibly by opening channels for ion flow across the hair-cell membrane; the receptor potential in turn causes the release of a chemical transmitter from the hair cell. The transmitter diffuses across an extracellular gap and activates the afferent neuron receptor sites that underlie the hair cell, and the neuron is depolarized. The nerve fibers from the organ of Corti form the *cochlear branch* of the **vestibulocochlear nerve** (cranial nerve VIII).

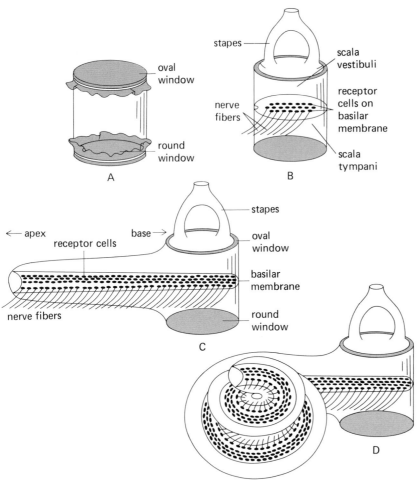

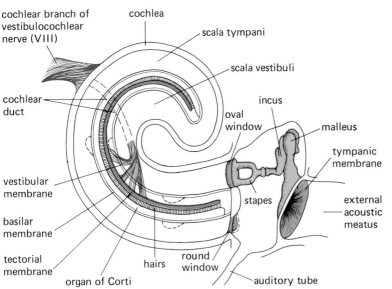

FIG. 15-25 Basic plan of the cochlea. (A) The cochlea is represented as a fluid-filled container with an elastic membrane (the oval and round window membranes) covering each end. The rigid wall of the container corresponds to the bony walls of the cochlea. (B) A driving piston (the stapes) and an elastic partition (the basilar membrane) are added. Sensory hair cells are placed on the basilar membrane. The nerve fibers from these cells will form the cochlear nerve. A hole is placed in one end of the basilar membrane. (C) The wall of the cochlea near the hole is extended and the basilar membrane is elongated. (D) The elongated cochlea is coiled along its length. The base of the cochlear duct, i.e., the basilar membrane, is shown, but for simplicity, the rest of the cochlear duct (or scala media), which lies between the scala vestibuli and the tympanic membrane, has not been shown. (E) The basic plan of the cochlea relative to the actual anatomy.

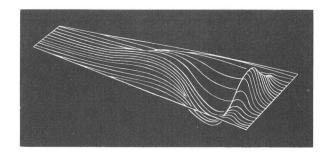

The hair cells are easily damaged by exposure to high-intensity noises such as the typical live amplified rock music concerts, engines of jet planes, and revved-up motorcycles. The damaged sensory hairs form giant, abnormal hair structures or are lost altogether, and in cases of long exposure to loud sounds, some hair cells and their supporting cells completely degenerate (Fig. 15-28). Much lesser noise levels also cause damage if exposure is chronic.

In normal hearing, the depolarizations formed in the endings of the afferent neurons trigger bursts of action potentials that are transmitted into the central nervous system. The greater the energy of the sound wave (loudness), the greater is the movement of the basilar membrane, the greater the depolarization of the afferent neuron, and the greater the frequency of action potentials in it. Nerve pathways from different parts of the basilar membrane are connected to specific sites along the strip of auditory cortex in much the same way that signals from different regions of the body are represented at different sites in somatosensory cortex; it is as though the basilar membrane were unrolled and spread along a strip of cortex. Each neuron along the auditory pathways responds at threshold levels of sound intensity (loudness) to a very small range of sound frequencies, its so-called *best frequency.* Different neurons have different best frequencies, depending on the location of their receptive fields in the basilar membrane.

Vestibular System

Changes in both the motion and position of the head are detected by mechanoreceptors that are part of the **vestibular apparatus,** parts of the membranous

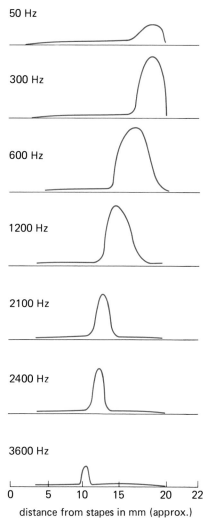

FIG. 15-27 Point along the basilar membrane where the traveling wave peaks is different with different sound frequencies. The region of maximal displacement of the basilar membrane occurs near the end of the membrane for low-pitched (low-frequency) tones and near the oval window and middle ear for high-pitched tones.

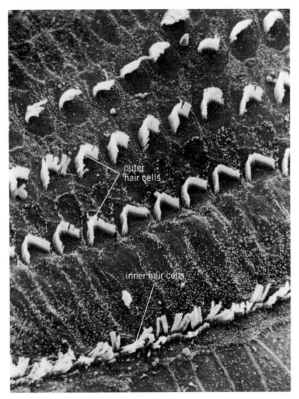

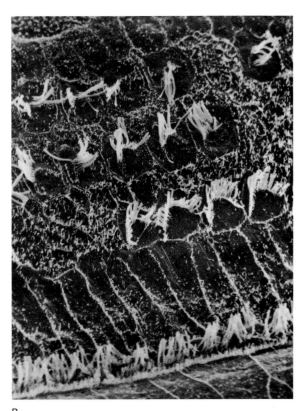

A   B

FIG. 15-28 Injury to the inner ear by intense noise. (A) Normal organ of Corti (guinea pig) showing the three rows of outer hair cells and single row of inner hair cells. (B) Injured organ of Corti after 24-hr exposure to noise levels typical of very loud rock music (2000 Hz–octave band at 120 dB). Several outer hair cells are missing, and the cilia of others no longer form the orderly pattern of the normal ear. Note also the increased number and size of small villi on the cell surfaces.

labyrinth of the inner ear filled with fluid (endolymph) and lying in the bony labyrinth tunnels in the temporal bone of the skull, one on each side of the head. The vestibular apparatus consists of three **semicircular canals** and a slight bulge for the **utricle** and **saccule** (Fig. 15-29A).

The three **semicircular canals** are arranged at right angles to each other (Fig. 15-29B). The mechanoreceptors of the semicircular canals are hair cells, the hairs of which are closely ensheathed by a gelatinous mass, the **cupula,** which extends into the channel of the semicircular canal at the **ampulla,** a slight bulge in the wall of each canal (Fig. 15-30). The connective-tissue ridge that supports the hair cells and cupula is the **crista** (or *crista ampullaris*). Whenever the head is moved, the bony tunnel wall

of the semicircular canal, its enclosed membranous sac, and the attached bodies of the hair cells all, of course, turn with it. The endolymph fluid that fills the membranous sac, however, is not attached to the skull. Because of inertia, the fluid tends to retain its original position, i.e., to be "left behind," and pushes against the gelatinous mass, bending the hairs within it and thereby stimulating the hair cells (Fig. 15-31). The speed and magnitude of the movement of the head determine the way in which the hairs are bent and the hair cells stimulated. As the inertia is overcome and the endolymph fluid begins to move at the same rate as the rest of the head, the hairs slowly return to their resting position. For this reason, the hair cells are stimulated only during *changes* in the rate of motion, i.e., during acceler-

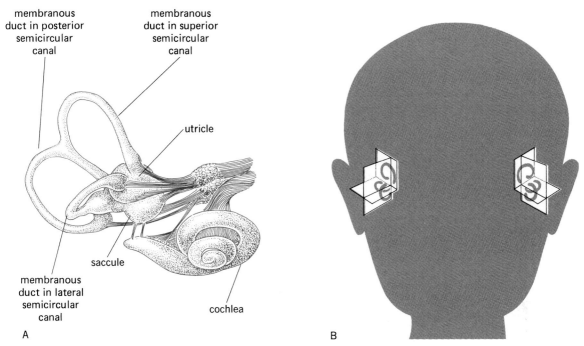

membranous duct in posterior semicircular canal

membranous duct in superior semicircular canal

utricle

saccule

membranous duct in lateral semicircular canal

cochlea

A

B

FIG. 15-29   (A) Vestibular system. (B) Relationship of the two sets of semicircular canals.

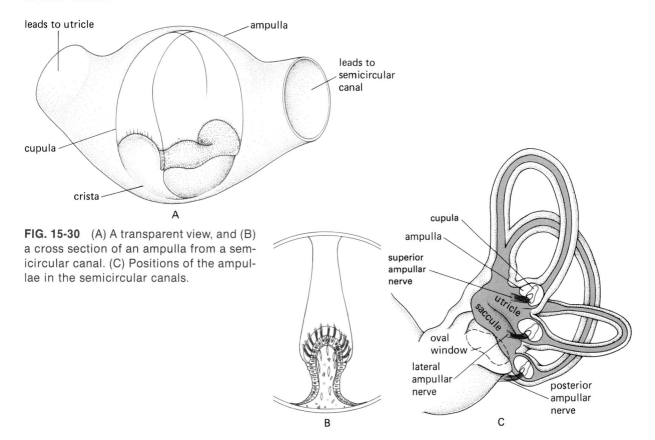

leads to utricle

ampulla

leads to semicircular canal

cupula

crista

A

FIG. 15-30   (A) A transparent view, and (B) a cross section of an ampulla from a semicircular canal. (C) Positions of the ampullae in the semicircular canals.

B

cupula

ampulla

superior ampullar nerve

utricle

saccule

oval window

lateral ampullar nerve

posterior ampullar nerve

C

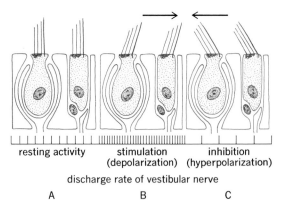

FIG. 15-31 Relation between position of hairs and activity in afferent neurons. Movement of the hairs in one direction (B) increases the action potential frequency in the afferent nerve fiber activated by the hair cell, whereas movement in the opposite direction (C) decreases the rate relative to the resting state (A).

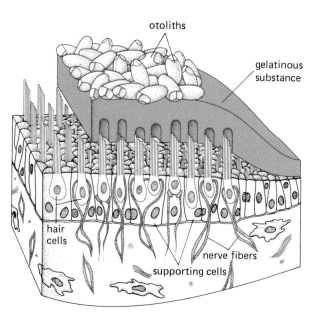

FIG. 15-32 Diagram of a section of an otolith organ such as occurs in the utricle and saccule.

ation of the head. In contrast, during motion at a constant speed, stimulation of the hair cells ceases.

The hair cells are functionally connected by chemically mediated synapses to the afferent nerve fibers that underlie them. Some transmitter is released from the hair cells in the absence of stimulation so that even when the head is motionless, the afferent neurons fire at a relatively low rate. Thus, the vestibular receptors can signal information by either increasing or decreasing the frequency of action potentials in the afferent-nerve fiber.

Whereas the semicircular canals signal the rate of change of motion of the head, the **utricle** and **saccule** contain receptors that provide information about the position of the head relative to the direction of the forces of gravity and about any linear acceleration of the head. The receptor cells here, too, are mechanoreceptors sensitive to the displacement of projecting hairs. The hair cells of the utricle and saccule are collected into groups from which the hairs protrude into a gelatinous substance (Fig. 15-32). Tiny calcium carbonate stones, or **otoliths,** are embedded in the gelatinous covering of the hair cells, making it heavier than the surrounding endolymph, or attached to the ends of some of the hairs. When the head is tipped, the gelatinous otolith material changes its position, pulled by gravitational forces to the lowest point in the utricle or saccule. The shearing forces of the gelatinous otolith substance against the hair cells bend the hairs and stimulate the receptor cells.

The information from the vestibular apparatus is

used for two purposes. One is to control the muscles that move the eyes so that, in spite of changes in the position of the head, the eyes remain fixed on the same point. As the head is turned to the left, for example, impulses from the vestibular nuclei in the brainstem activate the ocular muscles which turn the eyes to the right and inhibit their antagonists; the eyes therefore turn toward the right and remain fixed on the point of interest.

The second use of vestibular information is in reflex mechanisms for maintaining upright posture. In monkeys, cats, and dogs, for example, the vestibular apparatus plays a definite role in the postural fixation of the head, orientation of the animal in space, and reflexes that accompany locomotion. However, in human beings very few postural reflexes are known to depend primarily on vestibular input despite the fact that the vestibular organs are sometimes called the sense organs of balance.

Chemical Senses

Receptors sensitive to certain chemicals in the environment are **chemoreceptors.** Some of these respond to chemical changes in the internal environment (e.g., the oxygen and hydrogen-ion receptors in certain of the large blood vessels); others respond to external chemical changes, and in this category are the receptors for taste and smell. Taste and

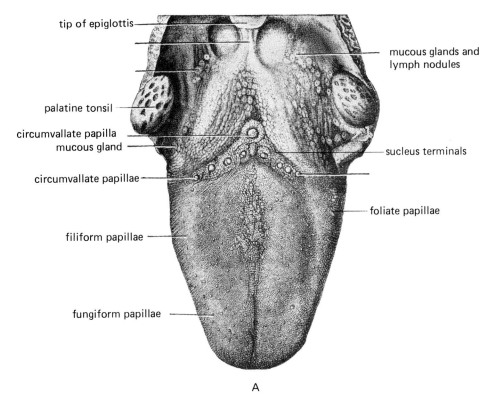

tip of epiglottis

mucous glands and lymph nodules

palatine tonsil

circumvallate papilla
mucous gland

sucleus terminals

circumvallate papillae

foliate papillae

filiform papillae

fungiform papillae

A

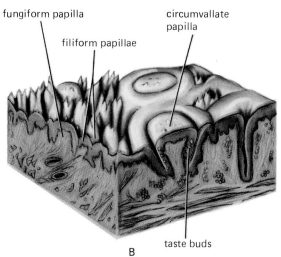

fungiform papilla

filiform papillae

circumvallate papilla

taste buds

B

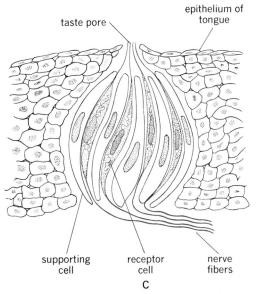

taste pore

epithelium of tongue

supporting cell

receptor cell

nerve fibers

C

smell affect a person's appetite, flow of saliva and gastric secretions, and the avoidance of harmful substances.

Taste The specialized receptor organs for the sense of taste are the 10,000 or so **taste buds,** some of which are located in the walls of **papillae,** small elevations on the tongue that are visible to the naked eye (Fig. 15-33A, B). Four types of papillae are distinguished: **vallate** (also called *circumval-*

FIG. 15-33 (A) Dorsal surface of the tongue indicating major structures. (B) Cross section of the tongue surface showing three of the four types of papillae. (C) Structure and innervation of a taste bud.

late), **fungiform, filiform,** and **foliate.** Most of the taste buds are scattered over the surface of the tongue, roof of the mouth, pharynx, and larynx. In the taste buds the receptor cells are arranged like segments of an orange so that the multifolded upper surfaces of the receptor cells extend into a small pore at the surface of the taste bud, where they are bathed by the fluids of the mouth (Fig. 15-33C).

Taste sensations are traditionally divided into four basic groups: sweet, sour, salt, and bitter, but different types of taste buds or receptor cells that would support this specificity have not been identified. In fact, a single receptor cell can respond in varying degrees to many different chemical substances that fall into more than one of the basic categories. The mechanisms by which the taste receptors are stimulated are not known.

Afferent nerve fibers enter the buds to end on the receptor cells, separated from them by a cleft across which transmitters released from the receptors diffuse. One nerve fiber may innervate several receptor cells, and one receptor may be innervated by several different neurons. There is clearly no one-to-one relationship by which each receptor cell has a direct line into the central nervous system. The frequency of action potentials in single nerve fibers increases in response to increasing concentrations of the chemical stimulant; therefore, frequency signals the *quantity.* But what signals the *quality*? How can we distinguish so many different taste sensations when the receptor cells lack specificity both in terms of the kind of chemical to which they respond and the way in which they are connected to the brain?

The afferent fibers involved in taste show different firing patterns in response to different substances; e.g., one fiber may fire very rapidly when the stimulatory substance is salt but only sporadically when it is sugar, and another fiber may have just the opposite reaction. Therefore, awareness of the specific taste of a substance probably depends upon the pattern of firing within a group of neurons rather than that in a specific neuron. Identification of the substance is also aided by information about its temperature and texture transmitted to the central nervous system from other receptors on the tongue and surface of the oral cavity. The odor of the substance clearly helps, too, as is attested to by the common experience that food lacks taste when one has a stuffy head cold.

Smell The olfactory receptors that give rise to the sense of smell lie in a small patch of mucus-secreting membrane (the **olfactory mucosa**) in the upper part of the nasal cavity (Fig. 15-34). The olfactory receptors are not separate cells but are specialized areas of the afferent neurons. The cell body of these neurons bears a knob from which several long ciliary processes extend out to the surface of the olfactory mucosa (Fig. 15-34). The knob and cilia of the afferent neurons contain the receptor sites, and the axons project to the brain as the **olfactory nerve** (cranial nerve I).

Before an odorous substance can be detected, it must release molecules that diffuse into the air and pass into the nose to the region of the olfactory mucosa, dissolve in the layer of mucus covering the

FIG. 15-34 Location and structure of the olfactory receptors.

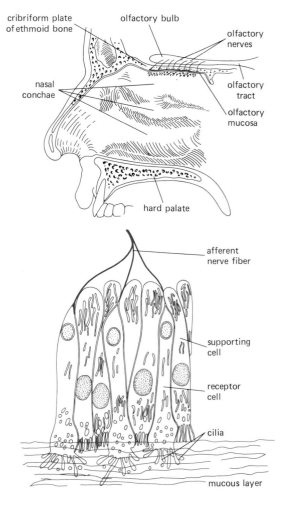

receptors, establish some sort of relation with the receptors, and depolarize the membrane enough to initiate action potentials in the afferent-nerve fiber. Upon stimulation of the olfactory mucosa with odorous substances, receptor potentials can be recorded that change with stimulus quality and intensity.

The physiological basis for discrimination between the tens of thousands of different odor qualities is speculative. There are no apparent differences in receptor cells, at least on a microscopic level, but meaningful differences between receptors are thought to exist at molecular levels.

The molecules of the odor substance combine with most types of receptor sites, but the "fit" varies. A good contact occurs when the odor-producing molecule and receptor site's size, shape, and polarity match; in such a case the resulting depolarization of the receptor cell is large. In cases of poorer fit, the receptor cell still depolarizes but to a smaller degree. All depolarizations occurring together in one cell are summed, forming a receptor potential, which determines the firing rate in the afferent-nerve fiber. According to this theory, receptor cells with different receptor-site populations respond to the same odor substance with different firing rates. Thus, it is the simultaneous yet differential stimulation of many receptor cells that provides the basis for odor discrimination.

It is known that olfactory discrimination depends only partially upon the action-potential pattern generated in the different afferent neurons. It also varies with attentiveness, state of the olfactory mucosa (acuity decreases when the mucosa is congested, as in a head cold), hunger (sensitivity is greater in hungry subjects), sex (women in general have keener olfactory sensitivities than men), and smoking (decreased sensitivity has been repeatedly associated with smoking). And just as the awareness of the taste of an object is aided by other senses, so the knowledge of the odor of a substance is aided by the stimulation of other receptors. This is responsible for the description of odors as pungent, acrid, cool, or irritating.

FURTHER PERCEPTUAL PROCESSING

Our actual perception of the events around us often involves areas of brain other than the primary sen-

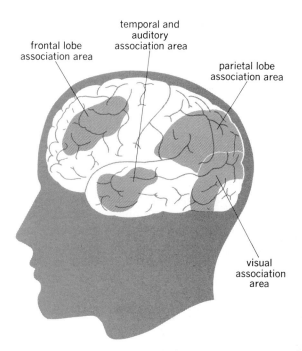

FIG. 15-35 Areas of association cortex.

sory cortex. Information from the primary sensory areas achieves further elaboration through the neural activity in **cortical association areas.** These brain areas lie outside the classic primary cortical sensory or motor areas but are connected to them by association fibers (Fig. 15-35). Although it has not often been possible to elucidate the specific roles performed by the association areas, they are acknowledged to be of the greatest importance in the maintenance of higher mental activities. For example, if areas of primary visual cortex are stimulated (when the brain surface is exposed under local anesthesia during neurosurgical procedures and the patient is awake), the patient "sees" a flash of light. Upon stimulation of the association areas surrounding visual cortex, the patient reports seeing more elaborate visual sensations such as "brilliant colored balloons floating around in an infinite sky." Upon stimulation of association areas still farther from primary visual cortex, the patient might report visual memories that seem to be reenacted before the patient's eyes.

Another example of the embellishment of the sensory experience provided by association areas can be found in persons who have undergone removal of parts of cortex because of tumors or accidents. A person who has no primary visual cortex is blind;

if a chair is placed in his path, he walks into it because he does not see it. In contrast, a person who has a functional primary visual cortex but has no visual association areas sees the chair and therefore does not walk into it, but he is not able to say that the object in his path is a chair or to explain its function. Similarly, a patient with damaged auditory association areas may hear a spoken word but not comprehend its meaning. A patient with damaged parietal association areas can feel a small, cool object in his hand but know neither that it is called a key nor that it is used to open locks.

Further perceptual processing involves arousal, attention, learning, memory, language, and emotions, as well as comparing the information presented via one sensory modality with that of another. For example, we may hear a growling dog, but our perception of the actual event taking place varies markedly, depending upon whether our visual system detects that the sound source is an angry animal or a loudspeaker.

We put great trust in our sensory-perceptual processes despite the inevitable modifications we know to exist. Some factors known to distort our perceptions of the real world are as follows.

1 Afferent information is distorted by receptor mechanisms and by processing along afferent pathways, e.g., by adaptation.
2 Such factors as emotions, personality, and social background can influence perceptions so that two people can witness the same events and yet perceive them differently.
3 Not all information that enters the central nervous system gives rise to conscious sensations. Actually, this is a very good thing because many unwanted signals, generated by the extreme sensitivity of our receptors, are canceled out. The afferent systems are very sensitive. Under ideal conditions the rods of the eye can detect the flame of a candle 17 mi away. The hair cells of the ear can detect vibrations of an amplitude much lower than that caused by the flow of blood through the vascular system and can even detect molecules in random motion bumping against the tympanic membrane. Olfactory receptors respond to the presence of only four to eight odorous molecules. It is possible to detect one action potential generated by a pacinian corpuscle. If no mechanisms existed to select, re-

strain, and organize the barrage of impulses from the periphery, life would be unbearable. Information in some receptors' afferent pathways is not canceled out; it simply does not give rise to a conscious sensation. For example, stretch receptors in the muscles detect changes in the length of the muscles, but activation of these receptors does not lead to a conscious sense of anything. Similarly, stretch receptors in the walls of the carotid sinus effectively monitor both absolute blood pressure and its rate of change, but people have no conscious awareness of their blood pressure.
4 We lack suitable receptors for many energy forms. For example, we can have no direct information about radiation and radio or television waves until they are converted to an energy form to which we are sensitive. Many regions of the body are insensitive to touch, pressure, and pain because they lack the appropriate receptors. The brain itself has no pain or pressure receptors, and brain operations can be performed painlessly on patients who are still awake, provided that the cut edges of the sensitive brain coverings are infused with local anesthetic.

However, the most dramatic examples of a clear difference between the real world and our perceptual world can be found in illusions and drug- and disease-induced hallucinations, when whole worlds can be created.

Any sense organ can give false information; e.g., pressure on the closed eye is perceived as light in darkness, and electrical stimulation of the afferent nerve fibers from any sense organ produces the sensory experience normally arising from the activation of that receptor. Why do such illusions appear? The two bits of false information just mentioned arise because most afferent fibers transmit information about one modality, and action potentials in a given afferent pathway going to a certain area of the brain signal the kind of information normally carried in that pathway.

In conclusion, the two processes of transmitting data through the nervous system and interpreting it cannot be separated. Information is processed at each synaptic level of the afferent pathways. There is no one point along the afferent pathways or one particular level of the central nervous system below which activity cannot be a conscious sensation and above which it is a recognizable, definable sensory experience. Perception has many levels, and it seems that the many separate stages are arranged in a hierarchy, with the more complex states receiving input only after it is processed by the more elementary systems. Every synapse along the afferent pathways adds an element of organization and contributes to the sensory experience.

KEY TERMS

sensory information	specific pathways	receptive field
perception	nonspecific pathways	exteroceptor
stimulus modality	somatosensory cortex	interoceptor
stimulus intensity	somatic sensory modalities	proprioceptor
stimulus localization	sensory unit	association cortex

REVIEW EXERCISES

1 Differentiate between afferent information and sensory information.
2 Describe the general plan and function of specific and nonspecific ascending pathways.
3 Distinguish between sensory unit and receptive field.
4 Describe two ways of coding stimulus intensity.
5 Describe the effect of the size and degree of overlap of receptive fields on stimulus localization.
6 Describe ways in which stimulus localization is enhanced by lateral inhibition.
7 Differentiate between exteroceptors, proprioceptors, and interoceptors; give two examples of each.
8 Describe the general ascending pathway for somatic sensation.
9 On a lateral view of the brain (such as Fig. 14-10B), locate the somatosensory cortex.
10 Explain how the perception of painful stimuli differs from other modalities.
11 Diagram the lacrimal apparatus; label the lacrimal gland, excretory ducts, lacrimal ducts, lacrimal sac, and nasolacrimal duct.
12 Draw a cross section of an eye (such as Fig. 15-10A); label the anterior, posterior, and vitreous chambers; lens; sclera; cornea; ciliary body; pupil; iris; suspensory ligament; ciliary muscles; retina; optic nerve; choroid; fovea; and conjunctiva.
13 Describe in general terms that part of the electromagnetic spectrum that stimulates the receptors of the eye.
14 Describe how the lens focuses an image on the retina in viewing near and distant objects.
15 Describe presbyopia, cataract, myopia, and hyperopia, and explain the defects responsible for each.
16 Differentiate between the functioning of the rods and cones, and describe the role of each in visual acuity, dark adaptation, and color vision.

17 Name the six extraocular muscles and the movement controlled by each.
18 Describe the two basic types of eye movement and name the adaptive value of each.
19 Describe the relation between the frequency, amplitude, and purity of a sound wave and the sound perceived by the listener.
20 Draw a section of the ear (such as Fig. 15-23A); label the ear canal, tympanic membrane, middle-ear cavity, auditory tube, inner ear, auditory ossicles, oval window, bony and membranous labyrinths, cochlea, and auditory nerve.
21 Draw a cross section through the inner ear (such as Fig. 15-24A); label the cochlear duct, basilar membrane, scala tympani, scala vestibuli, vestibular membrane, osseous spiral lamina, tectorial membrane, and hair cells of the organ of Corti.
22 Describe the transmission of sound from the atmosphere to the oval window, including the role played by each structure in the outer and middle ear.
23 Describe the role of the oval window, basilar membrane, scala vestibuli, scala tympani, perilymph, round window, and hair cells in the reception of sound.
24 Describe the basic structural elements of the vestibular system and their role in the detection of motion and position in space.
25 Describe the receptor mechanisms for taste and smell.
26 Describe the role of the association cortex in perception.
27 List and describe four factors known to distort our perception.
28 Differentiate between sensory input and perception.

CHAPTER

$$\boxed{16}$$

MOTOR FUNCTIONS OF THE NERVOUS SYSTEM

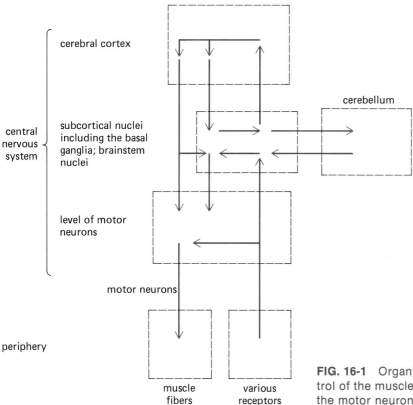

FIG. 16-1 Organization of the motor system. All control of the muscles by the motor system is exerted via the motor neurons.

The execution of a coordinated movement is a complicated process. Consider reaching to pick up an object. The fingers are extended and then flexed, the degree of extension depending upon the size of the object to be grasped, and the force of flexion depending upon the weight and consistency of the object. Simultaneously, the wrist, elbow, and shoulder are extended, and the trunk is inclined forward, the exact movements depending upon the distance of the object and the direction in which it lies. The shoulder must be stabilized to support the weight first of the arm and then the additional weight of the object. Upright posture must be maintained in spite of the body's continually shifting center of gravity.

The building blocks for this simple action–as for all movements–are active **motor units,** each comprising one motor neuron together with all the skeletal muscle cells it innervates (Chap. 11). Thus, anything that affects the movement of skeletal muscle does so by means of synaptic input to the motor neurons. Neural inputs from many sources converge upon the motor neurons to control their ac-

tivity, and the precision of coordinated muscle movement depends upon the *balance* of their influence. If for example, an inhibitory system is damaged and its input to the motor neurons is lessened, the still-normal excitatory input will be unopposed; the motor neurons will fire excessively and the muscle will be hyperactive. In cases of abnormal synaptic input, the motor units do not operate with the same degree of coordination or under the wide range of conditions typical of a normal motor system. No one source of input to the motor neuron is essential for movement, but to provide the precision and speed of normally coordinated movements, the balanced input from all sources is necessary.

Each of the myriad coordinated body movements is characterized by a set of motor-unit activities occurring over space and time, and the interrelating systems that converge upon the motor neurons to control their activity are the subject of this chapter. Figure 16-1 is a summary diagram of the motor system pathways that control the motor units. In the remainder of this chapter, we will describe each of these motor system components in greater detail.

First, a general **command** such as "pick up sweater" or "write signature" or "answer telephone" is developed, but it is not known where such a command is generated or how the neurons involved in this function are activated. Regions of association cortex and cerebellum show the earliest signs of coordinated electrical activity preceding and related to motor movements, and if these areas do not themselves house the "command neurons," they are certainly important early coordinating centers. Certain neurons in these brain areas, as well as in the basal ganglia, continually receive information from receptors in the muscles and joints, skin, vestibular apparatus, and eyes. These brain areas, therefore, have information about the starting position of the body part that is to be moved. Integrating this information with that from the "command neurons," they produce an initial **program** of motor unit activation needed to perform the desired movement. However, this overall program is then influenced by program subsets contributed by other brain regions; such subset programs exist because of inherent genetically determined connections between neurons and the connections formed later as a result of learning.

The complete program arrived at is relayed via nuclei in the thalamus to the motor cortex where it results in changes in the firing patterns of the motor-cortex cells. Action potentials in the axons of these cells then descend to alter the activity of the appropriate motor units; via axon collaterals, they also inform the basal ganglia and cerebellum of the signals that are being sent. This starts the process over again; the basal ganglia and cerebellum compare the new directives with ongoing information about the limb position, and compute adjustments, which are then sent back to the motor cortex. Thus, the basal ganglia and cerebellum receive a constant stream of information from the cortex about what actions are supposed to be taking place, while they simultaneously receive reports from the periphery about the actions that actually are taking place. Any discrepancies between the intended and actual movements are detected, and program corrections are sent to the motor cortex. This loop from the motor cortex to the cerebellum and basal ganglia and back again continues throughout the course of the movement. However, although activation of this loop takes only 1/50 sec, very rapid movements

do not provide enough time for continual correction, and the entire course of the action in such movements is completely preprogrammed. The cerebellum deals more with these rapid movements, whereas the basal ganglia are more involved with slow, continuous movements.

We have acknowledged that we know very little concerning the location or activation of the command neurons. Note that our system as described also does not take into account the mechanisms by which the "decision" to make a particular movement is reached. What neural events actually occur in the brain to cause one to "decide" to pick up an object in the first place? Presently we have no insight on this question.

Given such a model, it is difficult to use the word **voluntary** with any real precision, but we shall use it to refer to those actions that are characterized as follows.

1 These are actions we think about. The movement is accompanied by a conscious awareness of what we are doing and why we are doing it rather than the feeling that it "just happened," a feeling that often accompanies reflex responses.
2 Our attention is directed toward the action or its purpose.
3 The actions are the result of learning. Actions known to have disagreeable consequences are less likely to be performed voluntarily.

In the previous example of reaching to pick up an object, the activation of some of the motor units, such as those actually involved in grasping the object, can be classified clearly as voluntary; but most of the muscle activity associated with the act is initiated without any conscious, deliberate effort. In fact, almost all motor behavior involves both conscious and unconscious components, and the distinction between the two cannot be made easily.

Even a highly conscious act such as threading a needle involves the unconscious postural support of the hand and arm and inhibiton of antagonistic muscles (those muscles whose activity would oppose the intended action, in this case the finger extensor muscles that straighten the fingers). Moreover, unconscious basic reflexes such as dropping a hot object can be influenced by conscious effort. If the hot object is something that took a great deal of time and effort to prepare, one probably would not drop it but would try to inhibit the reflex, holding on to the object until it could be put down safely. Most motor behavior is neither purely voluntary nor purely involuntary but falls at some point on a spectrum between these two extremes. But even this statement is of little help because patterned muscle movements shift along the spectrum according to the frequency with which they are performed.

For example, when a person first learns to drive a car with standard transmission, stopping is a fairly complicated process that involves the accelerator, clutch, and brake. The sequence and force of the various operations depend upon the speed of the car, and their correct implementation requires a great deal of conscious attention. With practice the same actions become automatic. If a child darts in front of the car of an experienced driver, the driver does not have to think about the situation and decide to remove the foot from the accelerator and depress the brake and clutch. Upon seeing the child, the driver immediately and automatically stops the car. A complicated pattern of muscle movements is shifted from the highly conscious end of the spectrum over toward the involuntary end by the process of learning. Such a shift presumably occurs because relevant programs have been established, presumably by altering the synaptic connections between neurons (see Chap. 17). During early stages of learning, movements are slow and clumsy and are characterized by hesitant exploration; such movements depend greatly upon sensory feedback for guidance. With repetition of the movement, programming occurs (i.e., learning takes place). Thus, with later attempts at the movement, the altered synaptic connections are utilized, the dependence on sensory feedback is decreased, and the movement is carried out with greater precision and efficiency.

Whether activated voluntarily or involuntarily, individual motor units are frequently called upon to serve many different functions. For example, one demand upon the muscles of the limbs, trunk, and neck is made by postural mechanisms. These muscles must support the weight of the body against gravity, control the position of the head and different parts of the body relative to each other to maintain equilibrium, and regain stable, upright posture after accidental or intentional shifts in postion. Su-perimposed upon these basic postural requirements are the muscle movements associated with locomotion. For these purposes, the muscles must be capable of transporting the body from one place to another under the coordinated commands of neural mechanisms for alternate stepping movements and shifting the center of gravity. And added to the requirements of posture and locomotion can be more highly skilled movements such as those of a ballet dancer or hockey player. The motor units are activated and the sometimes conflicting demands are settled, usually without any conscious, deliberate effort.

We now turn to an analysis of the individual components of the motor-control-system model, beginning with local control mechanisms because their activity serves as a base upon which the descending pathways frequently exert their influence.

LOCAL CONTROL OF MOTOR NEURONS

Much of the synaptic input to the motor neurons arises from neurons at the same level of the central nervous system as the motor neurons. Indeed, some of these neurons are activated by receptors in the very muscles controlled by the motor neurons, in other nearby muscles, and in tendons associated with the muscles. These receptors monitor muscle length and tension and pass this information via afferent nerve fibers into the central nervous system. This input forms the afferent component of purely local reflexes that provide negative-feedback control over muscle length and tension. In addition, the information is transmitted to cerebral cortex, cerebellum, and the basal ganglia where it can be integrated with input from other types of receptors.

Length-Monitoring Systems and the Stretch Reflex

Embedded in skeletal muscle are stretch receptors that are made up of afferent-nerve endings wrapped around modified muscle cells, both partially enclosed in a fibrous capsule. The entire structure is called a **muscle spindle.** The modified muscle fibers in the spindle are known as **spindle** (or **intrafusal**) **fibers;** the typical skeletal muscle cells outside the spindle are the **skeletomotor** (or **extrafusal**) **fibers** (Fig. 16-2). The nerve fibers that innervate the two muscle types, the alpha and gamma efferent fibers, will be discussed later.

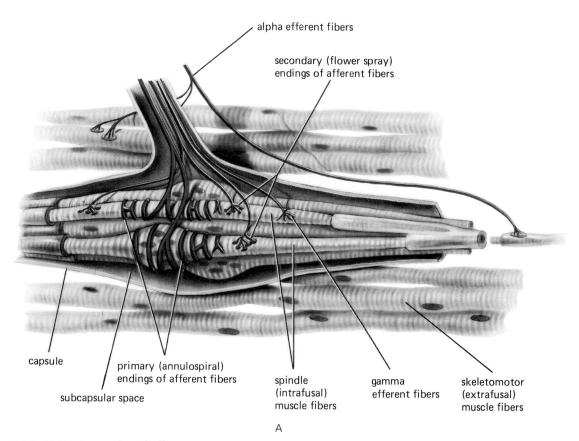

FIG. 16-2   A muscle spindle.

FIG. 16-3   (A) Passive stretch of the skeletomotor muscle fibers activates the spindle stretch receptors and causes an increased rate of firing in the afferent nerve. (B) Contraction of the skeletomotor fibers removes tension on the spindle stretch receptors and lowers the rate of firing in the afferent nerve. Arrows in (A) and (B) indicate the direction of force on the muscle spindles.

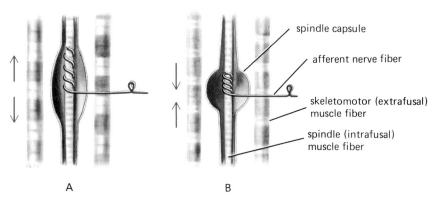

The muscle spindles are parallel to the muscle such that passive stretch of the muscle pulls on the spindle fibers, stretching them and activating their receptors (Fig. 16-3A); the greater the stretch, the faster the rate of firing. In contrast, contraction of the skeletomotor fibers and the resultant shortening of the muscle release tension on the muscle spindle (Fig. 16-3B) and slow down the rate of firing of the stretch receptor.

There are different kinds of spindle receptors, one responding to the magnitude of the stretch, another to both the absolute magnitude of the stretch and the speed with which it occurs. The importance of the kind of information relayed by the first receptor is apparent: It tells the central nervous system about the length of the muscle. However, by indicating the rate of change of the muscle length as well, the second type of receptor allows the central nervous system to anticipate the magnitude of the stretch. If the rate of stretch is increasing very rapidly, the stretch itself cannot stop immediately, and an additional change in length can be predicted. Although the two kinds of stretch receptors are separate entities, they will be referred to collectively as the **muscle-spindle stretch receptors.**

When the afferent neurons from the muscle spindles enter the central nervous system, they divide into branches that can take several different paths. One group of terminals (A in Fig. 16-4) directly forms excitatory synapses upon the motor neurons going back to the muscle that was stretched, thereby completing a reflex arc known as the **stretch reflex.** This reflex is probably most familiar in the form of the knee jerk, which is tested as part of routine medical examinations. The physician taps on the patellar tendon, which stretches over the knee and connects muscles in the thigh to a bone in the foreleg. As the tendon is depressed, the muscles to which it is attached are stretched, and the receptors in the muscle spindles are activated. Information about the change in length of the muscles is fed back to the motor neurons that control the same muscles. The motor units are excited, the thigh muscles shorten, and the patient's foreleg is raised to give the familiar knee jerk. The proper performance of the knee jerk tells the physician that the afferent limb of the reflex, the balance of synaptic input to the motor neuron, the motor neuron itself, the neuromuscular junction, and the muscle

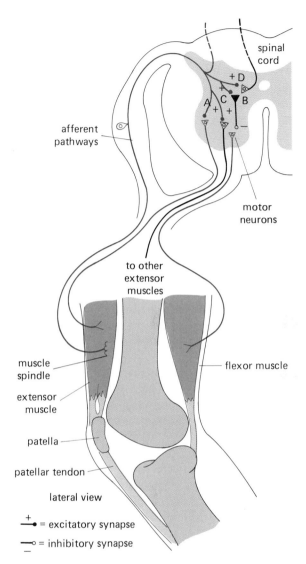

FIG. 16-4 Terminals of the afferent fiber from the muscle spindle involved in the knee jerk.

are functioning normally. In usual circumstances, of course, the stretch receptors are activated neither simultaneously nor so strongly, and the response is not a jerk. Rather, the stretch reflex maintains resting tension and smooths motions.

Because the group of afferent terminals mediating the stretch reflex synapses directly with the motor neurons without the interposition of any interneurons, the stretch reflex is called **monosynaptic.** Stretch reflexes are the only known monosynaptic motor reflex arcs in people; all other reflex arcs are **polysynaptic,** having at least one interneuron (and usually many) between the afferent and efferent pathways.

A second group of afferent terminals (B in Fig. 16-4) ends on interneurons that, when excited, inhibit the motor neurons that control antagonistic muscles whose contraction would interfere with the reflex response. For example, the normal response to the knee-jerk reflex is straightening of the knee to extend the foreleg. The antagonists to these extensor muscles are a group of flexor muscles that, when activated, draw the foreleg back and up against the thigh. If both opposing groups of muscles are activated simultaneously, the knee joint is immobilized and the leg becomes a stiff pillar. This is certainly what is required in some situations, but if the foreleg is to be extended from a flexed position, the motor neurons that activate the flexor muscles must be inhibited as the motor neurons that control the extensor muscles are activated. The excitation of one muscle and the simultaneous inhibition of its antagonistic muscle is called **reciprocal innervation.**

A third group of terminals (C in Fig. 16-4) ends on interneurons that, when excited, activate **synergistic muscles,** i.e., muscles whose contraction assists the intended motion. For example, in the knee jerk, interneurons facilitate motor neurons that control other leg extensor muscles.

A fourth group of afferent terminals (D in Fig. 16-4) synapses with interneurons that convey information about the muscle length to areas of the brain (particularly cerebellum and basal ganglia) that deal with coordination of muscle movement. Although the muscle stretch receptors initiate activity in pathways eventually reaching cerebral cortex, the information relayed by these action potentials does not have a strong conscious correlate; rather the conscious awareness of the position of a limb or joint comes more from the joint, ligament, and skin receptors.

Alpha-Gamma Coactivation Because the muscle spindles are parallel to the large skeletomotor muscle fibers, stretch on the spindles is removed when the skeletomotor fibers contract. If the spindle stretch receptors were permitted to go slack at this time, they would stop firing action potentials and this important afferent information would be lost. To prevent this, the spindle muscle fibers themselves are frequently made to contract during the

shortening of the skeletomotor fibers and thus maintain tension in the spindle and firing in the receptors. The spindle fibers are not large and strong enough to shorten whole muscle and move joints; their sole job is to produce tension on the spindle stretch receptors.

The muscle fibers in the spindles shorten in response to motor neuron activity (Fig. 16-5), but the motor neurons that activate the spindle fibers are usually not the same motor neurons that activate the skeletomotor muscle fibers. The motor neurons that control the skeletomotor muscle fibers are larger and are classified as **alpha motor neurons;** the smaller neurons whose axons innervate the spindle fibers are known as the **gamma motor neurons** (Fig. 16-2). The latter neurons are activated primarily by synaptic input from descending pathways. The overall route—descending pathway, gamma motor neuron, spindle muscle fiber, stretch receptor and afferent neuron, alpha motor neuron—is known as the **gamma loop** (Fig. 16-6). Because it is only the alpha motor neurons that activate skeletomotor fibers, i.e., the muscle fibers that actually do the work of moving joints, these motor neurons con-

FIG. 16-5 As the two striated ends of a spindle fiber contract in response to gamma motor neuron activation, they pull on the center of the fiber and stretch the receptor, which is located there.

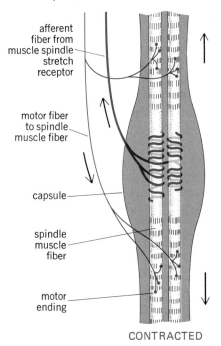

afferent fiber from muscle spindle stretch receptor

motor fiber to spindle muscle fiber

capsule

spindle muscle fiber

motor ending

CONTRACTED

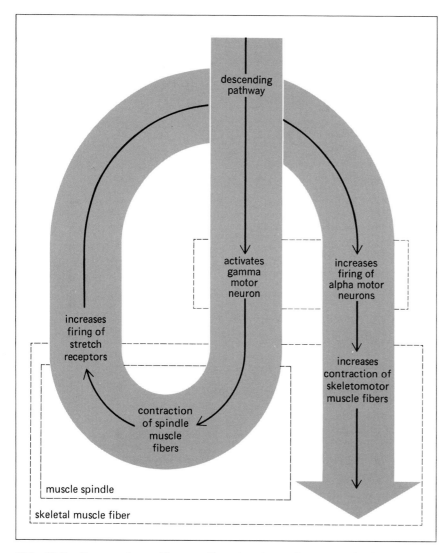

FIG. 16-6   Gamma loop. The small rectangle at the center indicates those events that occur in the brainstem or spinal cord; the large rectangle indicates the events that occur in the muscle.

stitute the **final common pathway** of the motor system.

In many voluntary and involuntary movements alpha and gamma motor neurons are **coactivated,** i.e., fired at almost the same time. To understand the usefulness of alpha-gamma coactivation, consider picking up a book whose weight is unknown. The position of the involved joints is determined by the length of the muscles. Suppose that the initially programmed strength of alpha motor-unit firing is not sufficient to lift the book. The skeletomotor muscle fibers will be unable to shorten but the spin-

dle fibers, activated simultaneously by the gamma motor neurons, will shorten and the spindle receptors will be stretched. By way of the stretch reflex, the excitatory synaptic input to the alpha motor units will increase, causing summation of contraction, recruitment of additional motor units, and greater muscle tension. Thus coactivation and the gamma loop provide a mechanism by which motor commands and muscle performance can be compared at the local level and compensation brought about on the spot. Moreover, information that the spindles are longer than expected (i.e., that the skeletomotor fibers have not shortened enough) will be transmitted to those higher brain centers involved in programming and controlling motor behavior so that they can alter their output as well.

Alpha-gamma coactivation works the other way too. If the initial program caused too intense alpha motor-unit activity, the book would be lifted too rapidly. The faster-than-expected shortening of the spindle fibers would remove tension from the spindles, stopping the receptors' firing. This would by reflex remove a component of excitatory input from the alpha motor neurons, automatically slowing the muscle movement to a more desirable rate. Thus, coactivation of alpha and gamma motor neurons can lead to fine degrees of regulation of muscle activity.

Tension-Monitoring Systems

A second component of the local motor-control apparatus monitors tension rather than length. The receptors employed in this system are the **Golgi tendon organs,** located in the tendon near its junction with the muscle. Endings of afferent-nerve fibers are wrapped around collagen bundles of the tendon, which are slightly bowed in the resting state. When the skeletomotor fibers of the attached muscle contract, they pull on the tendon, straightening the collagen bundles and distorting the receptor endings of the afferent nerves. The receptors fire in proportion to the increasing force or tension generated by the contracting muscle. The afferent neuron's activity causes inhibitory postsynaptic potentials in the motor neurons of the contracting muscle (Fig. 16-7).

Some of the Golgi tendon organs have high thresholds and respond only when the tension is very great. These high-threshold receptors may function as safety valves, inhibiting the muscle when the force it generates is great enough to damage the tendon or bone. The remainder of the Golgi tendon organs have lower thresholds, comparable to the receptors of the muscle spindle, and they supply the motor control systems with continuous information about the tension generated. This information is necessary for effective movement because a given input to a group of motor neurons does not always provide the same amount of tension. The tension developed by a contracting muscle depends on the muscle length and the degree of muscle fatigue, as well as the number of activated

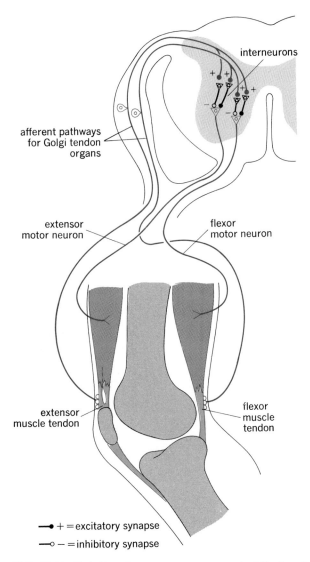

FIG. 16-7 Golgi tendon organ component of the local control system.

motor neurons and the rate at which they are firing (Chap. 11). Because one set of inputs to the motor neurons can lead to a large number of different tensions, feedback of information is necessary to inform the motor control systems of the tension actually achieved.

DESCENDING PATHWAYS AND THE BRAIN CENTERS THAT CONTROL THEM

The cerebral cortex, basal ganglia and other subcortical nuclei, brainstem nuclei, and cerebellum influence the motor neurons by descending pathways. There are three mechanisms by which these pathways alter the balance of synaptic input converging upon the alpha motor neurons.

1 By synapsing directly upon the alpha motor neurons themselves. This has the advantage of speed and specificity.
2 By synapsing on the gamma motor neurons, which, via the gamma loop, influence the alpha motor neurons. This pathway and number 1 above usually operate together (alpha-gamma coactivation). As described earlier, this has the advantage of maintaining output from the stretch receptors and providing a means for local on-the-spot compensation.
3 By synapsing on interneurons, often the same ones subserving the local reflexes. Although this route is not as fast as directly influencing the motor neurons, it has the advantage of the coordination built into the interneuron network as described earlier (e.g., recruitment of synergistic muscles, reciprocal innervation).

The degree to which each of these three mechanisms is employed varies, depending upon the nature of the descending pathway, of which there are two major categories: the **corticospinal pathway** and the **multineuronal pathways.**

Corticospinal Pathway

The fibers of the **corticospinal pathway,** as the name implies, have their cell bodies in the cerebral cortex. The axons of these cortical neurons pass without any additional synapsing to end in the immediate vicinity of the motor neurons (Fig. 16-8). The axons of these cortical neurons pass through the internal capsule of the cerebrum, through the pons, and into the medulla where about two-thirds of the fibers cross the midline and turn to descend on the opposite side of the spinal cord as the **lateral corticospinal tract.** Thus, the skeletal muscles on the left side of the body are controlled largely by neurons in the right half of the brain, and vice versa. The fibers that do not cross in the medulla pass straight into the spinal cord, forming the **ventral corticospinal tract.** (Most of these fibers eventually cross at their level of termination in the spinal cord.) The group of fibers that innervate muscles of the eye, face, tongue, and throat branch away from

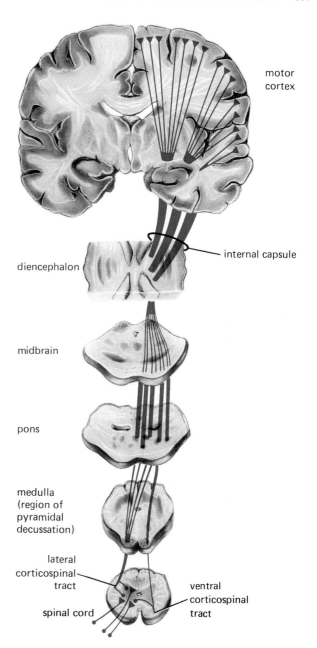

FIG. 16-8 Diagram of the corticospinal pathway. Most of the fibers cross in the medulla to descend in the opposite side of the spinal cord.

these descending pathways in the brainstem as the **corticobulbar pathways** to contact motor neurons whose axons travel out with the cranial nerves. The corticospinal pathway is also called the **pyramidal tract** or **pyramidal system,** perhaps because of its shape in some parts of the brain or because it was formerly thought to arise solely from the giant pyramidal neurons of the cortex.

The corticospinal pathway is the major mediator of fine, intricate movements. However, it is not the sole mediator of these movements, since surgical section of this pathway in people does not completely eliminate such movements, although it does make them weaker, slower, and less well coordinated. Clearly, the multineuronal pathways (to be described next) must also contribute to the performance of delicate movements.

The fibers of the corticospinal pathway end in all three of the ways described, i.e., on alpha motor neurons, gamma motor neurons, and interneurons. In addition they end presynaptically on the terminals of afferent neurons and, through collateral branches, on neurons of the ascending afferent pathways. The overall effect of their input to afferent systems is to limit the area of skin, muscle, or joints allowed to influence the cortical neurons, thereby sharpening the focus of the afferent signal and improving the contrast between important and unimportant information. The collaterals also convey the information that a certain motor command is being delivered and possibly give rise to the sense of effort. Because of this descending (motor) control over ascending (sensory) information, there is clearly no real functional separation of these two systems.

Multineuronal Pathways

Some of the neurons of cerebral cortex do not go directly to the region of the motor neurons; rather, they form the first link in **multineuronal pathways** that pass through the basal ganglia, several other subcortical nuclei, and the brainstem and form multiple synapses along their course. At each successive neuron, the information carried by the pathways is altered according to the balance of excitatory and inhibitory input to them. Some of these chains descend to the level of the motor neurons, ending there either on interneurons or gamma motor neurons; thus, the motor neurons (or the interneurons which directly influence them) receive input from both the corticospinal and multineuronal pathways. Other multineuronal pathways do not descend to the level of the motor neurons, but rather, at some point, loop back to early way stations, including cerebral cortex. These loops constitute the paths by which the programs and program corrections described in our original model modify the activity of motor cortex, including the

neurons of the corticospinal pathway. These multineuronal pathways and the structures that they connect are often called the **extrapyramidal system** to distinguish them from the corticospinal (pyramidal) pathways.

Whereas the corticospinal neurons have greater influence over the motor neurons that control muscles in the more distal ends of the limbs, e.g., the fingers and hands, the multineuronal pathways have more effect on the muscles of the trunk and more proximal parts of the limbs. However, it is wrong to imagine a complete separation of function between these two pathways, for the distinctions between them are not at all clear cut; moreover, some areas of motor cortex give rise to neurons of both pathways. All movements, whether automatic or voluntary, require the continual coordinated interaction of both pathways.

Cerebral Cortex

Many areas of the cortex give rise to the two types of descending pathways described above, but a large number of the fibers come from the posterior part of the frontal lobe, which is therefore called the **motor cortex** (Fig. 16-9). The function of the neurons in the motor cortex varies with position in the cortex. As one starts at the top of the brain and

FIG. 16-9 Regions of the motor cortex.

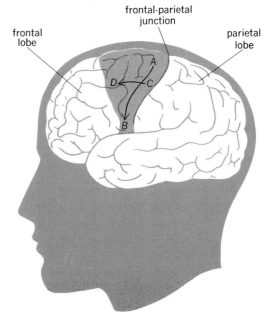

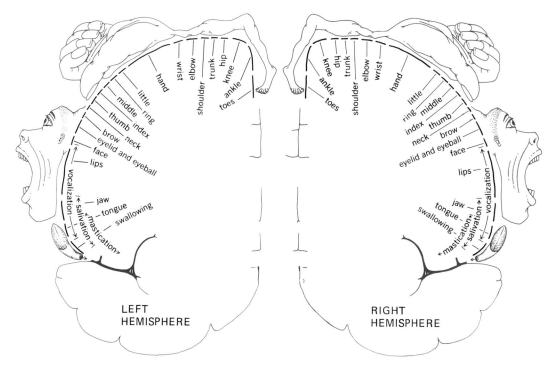

FIG. 16-10 Arrangement of the motor cortex.

moves down along the side (A to B in Fig. 16-9), the cortical neurons that affect movements of the toes and feet are at the top of the brain, followed (as one moves laterally along the surface of the brain) by neurons that control leg, trunk, arm, hand, fingers, neck, and face. The size of each of the individual body parts in Fig. 16-10 is proportional to the amount of cortex devoted to its control; clearly, the cortical areas that represent hand and face are the largest. The great number of cortical neurons for innervation of the hand and face is one of the factors responsible for the fine degree of motor control that can be exerted over those parts.

The neurons also contribute to different pathways as one explores from the back of the motor cortex forward (C to D in Fig. 16-9). Those in the back portion, i.e., closest to the junction of the frontal and parietal lobes, mainly contribute to the corticospinal pathway. Moving anterior (forward) in the motor cortex, this zone of neurons gradually blends into the group that forms the first link in the multineuronal pathways. However, none of these cortical neurons functions as an isolated unit. Rather, they are interconnected so that those cortical neurons that control motor units that have

related functions fire together. The more delicate the function of any skeletal muscle, the more nearly one-to-one is the relationship between cortical neurons and the muscle's motor units.

Colonies of cells that have similar functions are arranged in columns perpendicular to the surface of motor cortex, each column containing hundreds of neurons. The neurons in the columns function together.

Throughout this description we have presented the cerebral cortex as the origin of the descending pathways to motor neurons, but the neurons of cerebral cortex manifest no spontaneous generation of action potentials. Rather, their firing pattern, like that of almost all neurons, is determined solely by the balance of synaptic activity impinging upon them. In other words, as we emphasized in the earlier description of the overall model of motor system control, the motor cortex is not the prime initiator of movement; it is simply a tremendously important relay station. What then are the inputs that drive the neurons of motor cortex? Again as emphasized in our model, where the "command" neurons are located is not known, but regions of association cortex and cerebellum have been implicated. For example, electrodes placed on the skull to record the electrical activity of the brain during the "decision making period" pick up over wide areas of association cortex distinctive activity patterns (the so-called **readiness potential**) about 800 msec (0.8 sec) before the movement begins. Only afterwards, 50 to 60 msec prior to the movement, does sharper electrical activity (the **motor potential**) appear over the motor-cortex region of the hemisphere that controls the involved muscles.

Another source of important input to the motor cortex is the pathways from the receptors listed earlier in the chapter. Yet such input does not generally account for the actual initiation of movement. For example, if the neurons of the motor cortex are stimulated in conscious patients during surgical exposure of the cortex, the patient moves but not in a purposefully organized way. The movement made depends upon the part of the motor cortex stimulated. The patient is aware of the movement but says, "You made me do that," recognizing that it is not a voluntary movement. Stimulation of parts of cortex in the parietal lobe sometimes causes patients to say that they want to make a certain movement, but the movement does not occur.

Finally, the input to the cortex neurons not only must be directed toward some purpose but must select from a variety of ways of achieving that purpose since learned motor programs can be executed in many ways with many sets of muscles. For a simple example, a rat trained to press a lever will do so quite consistently but, depending upon its original position in the cage, its movements can be quite varied. If the rat is to the left of the lever, it moves to the right; if it is to the right, it moves to the left. If its paw is on the floor, it raises the paw; if its paw is above the lever, it lowers the paw. The only consistent act is pressing the lever. The movements performed to achieve the purpose are variable and seem almost inconsequential. Yet, although it is the end result that matters, only the intervening acts or movements can be programmed by the nervous system. How a given program is selected is not known.

Subcortial Nuclei (Including the Basal Ganglia) and Brainstem Nuclei

As mentioned above, the multineuronal pathways that descend from cerebral cortex synapse in the basal ganglia, other subcortical nuclei, and brainstem nuclei; collateral branches from axons in the corticospinal pathway also end in these regions. These nuclei and the pathways that interconnect them serve to correlate fine, detailed voluntary movements with the appropriate postural mechanisms upon which these movements are superimposed. In addition, the basal ganglia play a special role in the control of slow, smooth, voluntary movements.

Perhaps the roles these regions play in the control of muscle activity can be better understood by learning what happens when they are damaged by accident or disease. Such conditions usually involve one or more of the following.

1 There can be an increase in muscle tone, i.e., a resistance to stretch, so that movements become stiffer.
2 Unwanted, purposeless, uncontrolled movements are often present. The latter can take the form of rapid contractions of opposing muscle groups (*tremor*); writhing, continuous swings between two positions (*athetosis,* e.g., hyperextension of the fingers with the palm facing upward alternating with clenching the fingers with the palm facing downward); or sudden, random, coordinated movements of the distal part of the arms or legs that are relatively mild (*chorea*) or more violent (*ballismus*).
3 There is usually a decrease or loss of associated movements, such as arm-swings during walking or the facial movements that normally give vitality and emotional expression to one's appearance.

One particular subcortical nucleus is the *substantia nigra* (black substance), which gets its name from the deep pigment in its cells. It is this group of neurons, whose fibers synapse in the basal ganglia, that is implicated in the disease *parkinsonism,* characterized by tremor, rigidity, and a delay in the initiation of movement. The neurons of the substantia nigra liberate the transmitter dopamine at their axon terminals in the basal ganglia. When they degenerate, as they do in parkinsonism, the amount of dopamine delivered to the postsynaptic cells is reduced.

Cerebellum

The **cerebellum** sits on top of the brainstem, as seen in Fig. 14-10A. It consists of two *cerebellar hemi-*

spheres joined by a median *vermis;* its surface is marked by numerous parallel fissures. The *cerebellar cortex* is a convoluted shell of gray matter over a core of nerve fibers that ascend to and descend from it. Purkinje-cell neurons of the cerebellar cortex provide the sole output pathway from the cortex. They synapse in deep nuclei in the white matter at the base of the cerebellum where final program adjustments are made before the cerebellar output is transmitted to specific nuclei in the brainstem and diencephalon. The neural connections between the cerebellum and the rest of the brain pass through the three pairs of cerebellar peduncles (Chap. 14).

The cerebellum does not initiate movement but acts by influencing other regions of the brain responsible for motor activity. Destruction of the cerebellum does not cause the loss of any specific movement; instead it is associated with a general inadequacy of that movement.

In cases of severe cerebellar damage, the combined difficulties of poor balance and unsteady movements may become so great that the person is incapable of walking or even standing alone. Since speech depends on the intricately timed coordination of many muscle movements, cerebellar damage is accompanied by speech disturbances. The person with cerebellar damage shows no evidence whatsoever of sensory or intellectual deficits.

From this discussion, you may deduce (correctly) that the cerebellum is involved in the control of muscles used in maintaining steady posture and effecting coordinated, detailed movements. As stated earlier, the cerebellum receives input both from cortex and subcortical centers with information about what the muscles *should* be doing and from many afferent systems with information about what the muscles *are* doing. If there is a discrepancy between the two, an error signal is sent from cerebellum to the cortex and subcortical centers where new commands are initiated to decrease the discrepancy and smooth the motion.

The cerebellum plays a special role in the control of rapid movements (and thus serves as the counterpart to the basal ganglia, which influence slow movements). Rapid movements are largely "preprogrammed" in their entirety, rather than being modified during their course as slower movements are. Such preprogramming requires calculation of the time needed for the movement (taking into account the particular amount of muscle force nec-

essary in any special situation) and integration of these data with information about the moved structure's initial position and final location. The cerebellum performs this function and, like the basal ganglia, projects the information to the motor cortex.

The afferent inputs to the cerebellum come from the vestibular system, eyes, ears, skin, muscles, joints, and tendons, i.e., from the major receptors affected by movement. Inputs from receptors in a single small area of the body end in the same region of the cerebellum as do the inputs from the higher brain centers that control the motor units in that same area. Thus information from the muscles, tendons, and skin of the arm arrive at the same area of cerebellar cortex as the motor commands for "arm" from the cerebral cortex. This permits the cerebellum to compare motor commands with muscle performance.

This completes our analysis of the components of motor control systems. We will now analyze the interactions of these components in two situations: maintenance of upright posture and walking.

MAINTENANCE OF UPRIGHT POSTURE AND BALANCE

The skeleton that supports the body is a system of long bones and a many-jointed spine that cannot stand alone against the forces of gravity. Even when held together with ligaments and covered with flesh, it cannot stand erect unless there is coordinated muscular activity. This applies not only to the support of the body as a whole but also to the fixation of segments of the body on adjoining segments, e.g., the support of the head.

Added to the problem of supporting one's own weight against gravity is that of maintaining equilibrium. A person is a very tall structure balanced on a relatively small base, and the center of gravity is quite high, being situated just above the pelvis. (The center of gravity is the point in an object at which all the downward forces caused by gravity seem to be localized.) For stability, the center of gravity must be kept within the small area determined by the vertical projection of the base of the body (feet) in contact with the surface upon which a person stands. Yet human beings are almost always in motion, swaying back and forth and side to side even when standing still. Clearly, they often operate under conditions of unstable equilibrium and would be toppled easily by physical forces in the environment if their equilibrium were not protected by reflex postural mechanisms.

The maintenance of posture and balance is accomplished by means of complex counteracting reflexes, all the components of which we have met previously. The efferent arc of the reflexes is, of course, the alpha motor neurons to the skeletal muscles. The major coordinating centers are, as usual, the basal ganglia, brainstem nuclei, and reticular formation, all of which influence the motor neurons mainly via the descending multineuronal pathways. What is the source of afferent input? One might predict the existence of "center-of-gravity" receptors but, in fact, no such receptors exist. Rather, information about the location of the center of gravity is given by the integration of all the afferent signals from muscles, joints, skin, vestibular system, and eyes. This integration provides the coordinating centers with a "map" of the position of the whole body in space.

In such a system it is extremely difficult to assign a certain percentage of importance to any one afferent system, even when the exact conditions are specified. However, it does seem that in people, under conditions that permit vision, visual information is probably most important. Yet, so influential are the other inputs and so adaptable is the overall system that a blind person maintains balance quite well, with only slight loss of precision. Moreover, with changing circumstances, the dependence on different afferent inputs may change considerably.

Let us take the vestibular input as an example. Despite the fact that the vestibular organs are called the sense organs of balance, persons whose vestibular mechanisms have been destroyed may have very little disability in everyday life (one such person was even able to ride a motorcycle). Such persons are not seriously handicapped as long as their visual system, joint position receptors, and cutaneous receptors are functioning. However, they do have difficulty walking in darkness over uneven ground or walking down stairs, where they cannot see a point immediately in front of their feet for visual reference. Thus, in a normal person vestibular input must increase in importance under such circumstances. Finally, vestibular information provides the only clue to orientation with respect to gravity when one is swimming under water, where visual, skin, and joint input is inappropriate.

Skin receptors sensing contact of the body with other surfaces also play a role in regulation of body posture, as can be shown by the following test. Walking in the dark, one's gait is halting and uncertain; stability and confidence are greatly improved by the simple act of running a fingertip along a wall. The fingertip certainly provides no physical support, but it adds significant afferent information to the coordinating centers in the subcortical centers and cerebellum.

WALKING

Postural fixation of the body is intimately related to the problems of locomotion and maintenance of equilibrium in the face of movement. In fact, the mechanisms for posture and movement are the same, and a disturbance of one is almost always associated with a disturbance of the other. In considering locomotion, the need for some structure or mechanism capable of carrying the body along is added to the basic requirement of antigravity support of the body. In walking, the human body is balanced on the very small base provided by one foot. The weight of the body is supported on each leg alternately, and to accomplish this the body moves from side to side in such a way that the center of gravity is alternately poised over the right and then the left leg. Only when the center of gravity is shifted over the right leg can the left foot be raised from the ground and advanced. As the left foot is lifted, the trunk of the body sways to the right to counterbalance the weight of the left leg (Fig. 16-11). It must be apparent that strict and delicate control of the center of gravity is essential to permit these movements without loss of equilibrium.

The stimulus necessary to trigger stepping is a slight forward tilt of the body. When a person takes a step, the weight of the body is shifted to one foot, and the opposite foot is lifted from the ground. The body is allowed to fall forward and loses its equilibrium until it is caught on the leg that has swung forward. During this process, the center of gravity has moved both sideways and forward from its original position over one leg to a similar position over the other leg. This action is repeated rhythmically,

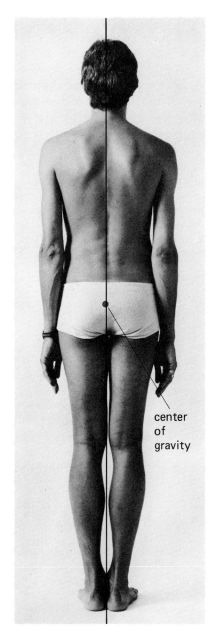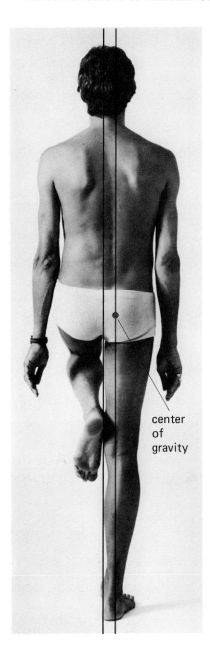

FIG. 16-11 Postural changes with stepping. (A) Normal standing posture. The line from the center of gravity falls directly between the two feet. (B) As the left foot is raised, the whole body leans toward the right so that the center of gravity shifts above the right foot.

and its continuation depends on both components of the shift in the center of gravity, the forward shift, which causes the fall forward, and the sideways shift, which allows one foot to be lifted and advanced. Thus, there are four necessary components for locomotion: *antigravity support of the body, stepping, control of the center of gravity* to provide equilibrium, and *a means of acquiring forward motion*. All four of these components must be present simultaneously and continuously. It would obviously be futile, for example, to apply forward motion without an adequate stepping mechanism.

KEY TERMS

motor unit	gamma motor neuron	motor cortex
program	coactivation	subcortical nuclei
local control	final common pathway	pyramidal system
stretch reflex	Golgi tendon organ	extrapyramidal system
reciprocal innervation	corticospinal pathway	basal ganglia
synergistic	multineuronal pathways	center of gravity
alpha motor neuron		

REVIEW EXERCISES

1 Define motor unit.
2 Differentiate between voluntary and involuntary motor activities.
3 Describe the role of learning in the development of involuntary responses.
4 Draw a muscle spindle; label the spindle capsule, intrafusal muscle fiber, muscle-spindle stretch receptors, afferent nerve fibers, and gamma efferent fibers.
5 Diagram the components of the stretch reflex.
6 Discuss each component of the stretch reflex as an element of a negative-feedback control system for muscle length.
7 Describe the role of reciprocal innervation and the activation of synergistic muscles in the knee jerk.
8 Differentiate between the intrafusal and extrafusal muscle fibers and between the alpha and gamma motor neurons.
9 Diagram the gamma loop; label the alpha and gamma motor neurons, intrafusal and extrafusal muscle fibers, muscle spindle, and final common pathway.

10 Define alpha-gamma coactivation and describe its significance.
11 Describe a Golgi tendon organ and state its function.
12 Describe the three ways in which descending pathways can modify the activity of alpha motor neurons.
13 Describe the corticospinal pathways; use the terms *lateral* and *ventral corticospinal tract, internal capsule,* and *pyramidal system.*
14 Differentiate between the corticospinal pathway and the multineuronal pathways in both anatomical and functional terms.
15 Describe the interactions of the motor cortex and the subcortical centers in relation to coordination of muscle movement.
16 Explain the interaction of cortical input and general afferent input in the functioning of the cerebellum.
17 List and describe the four components necessary for locomotion.

CHAPTER

CONSCIOUSNESS
AND BEHAVIOR

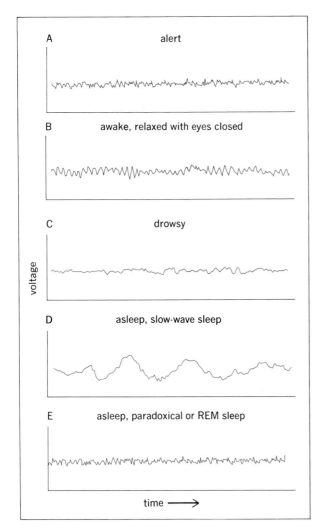

A   alert

B   awake, relaxed with eyes closed

C   drowsy

D   asleep, slow-wave sleep

E   asleep, paradoxical or REM sleep

voltage

time ⟶

FIG. 17-1 EEG patterns corresponding to various states of consciousness.

The term **consciousness** includes two distinct concepts, **states of consciousness** and **conscious experience**. The second concept refers to those things of which a person is aware—thoughts, feelings, perceptions, ideas, dreams, reasoning—during any of the states of consciousness. In contrast, a person's state of consciousness, i.e., whether awake, asleep, drowsy, etc., is defined both by behavior, covering the spectrum from coma to maximum attentiveness, and by the pattern of brain activity that can be recorded electrically, usually as the electric-potential difference between two points on the scalp. This record is the **electroencephalogram (EEG)** (Fig. 17-1).

STATES OF CONSCIOUSNESS

Electroencephalogram

The EEG is such an important tool in identifying the different states of consciousness that we begin with it. The wavelike pattern of the EEG is about 100 times smaller than the amplitude of an action potential, and the frequency of the waves may vary from 3 to 22 Hz, the patterns being distinguished from one another by both their frequencies and amplitudes. Correlated with changes in EEG pattern are changes in behavior that span the entire normal range from attentive alertness to sleep.

Each EEG pattern is the result of varying degrees of intermittent synchronization of membrane potential changes in groups of neurons located in the cerebral cortex. These potential changes are thought to be synaptic potentials (and other graded potential changes) rather than action potentials. The involved synapses manifest more or less synchronous electrical changes because of intimately coordinated activities of the thalamus and cortex. Small areas of thalamus act as pacemakers, or rhythm generators; each area projects to and influences the activity of a specific small region of the cortex.[1] The cells in a given small thalamic unit are interconnected and they tend to fire together. Moreover, the individual units are affected by both excitatory and inhibitory feedback loops from other thalamic units as well as by loops with the cells in the cortex to which the thalamic units project. These loops lead to the synchronization of large areas of cortex, and it is these bursts of cortical activity that are recorded as the EEG. Cortical neurons other than those synchronized at any given moment also have fluctuating membrane potentials caused by synaptic input, but their activity is not synchronized and, therefore, the potentials tend to cancel each other out. Periodically, new groups of neurons are synchronized.

We have been speaking thus far mainly of the highly synchronized EEG waves characteristic of slow-wave sleep (Fig. 17-1D). The EEG pattern becomes less synchronized during arousal, and it becomes completely desynchronized with full alertness. This breakdown in synchrony occurs as re-

[1]Although other regions of the brain can affect cortical synchrony, these thalamocortical interactions form the basis of the EEG-generating mechanism.

gions of the midline thalamus are activated. Their input to cortex causes, in turn, greater activation of the excitatory feedback loops that pass from cortex back to thalamus, thereby counteracting the IPSPs that are largely responsible for strong synchronization of the pacemakers.

The EEG is a useful clinical tool because the normal patterns are altered over brain areas that are diseased or damaged. It is also useful in defining states of consciousness, but it is not known what function, if any, this electric activity serves in the brain's task of information processing. We do not know whether these electric waves actually influence brain activity or whether they are merely epiphenomena. (An epiphenomenon is a phenomenon that occurs with an event but is not causally related to it, for example, the sound of a baseball bat striking a ball—the sound results from the impact but does not influence how far the ball will travel.)

The Waking State

Behaviorally, the waking state is far from homogeneous, comprising the infinite variety of things one can be doing. The prominent EEG wave pattern of an awake relaxed adult whose eyes are closed is a slow oscillation of 8 to 12 Hz, known as the **alpha rhythm** (Fig. 17-1B) (each region of the brain has a characteristic alpha rhythm). The alpha rhythm is associated with decreased levels of attention, and when alpha rhythms are being generated, subjects commonly report that they feel relaxed and happy. A high degree of alpha rhythm is also associated with meditational states. However, people who normally experience high numbers of alpha episodes have not been shown to be psychologically different from others with lower levels, and the relation between brain-wave activity and subjective mood is obscure. People have been trained to increase the amount of alpha brain rhythms by providing a feedback signal such as a tone whenever alpha rhythm appears in their EEG.

When people are attentive to an external stimulus (or are thinking hard about something), the alpha rhythm is replaced by lower, faster oscillations (Fig. 17-1A). This transformation is known as **EEG arousal** and is associated with the act of attending to stimuli rather than with the perception itself; for example, if people open their eyes in a completely dark room and try to see, EEG arousal occurs.

With decreasing attention to repeated stimuli, the EEG pattern reverts to the alpha rhythm.

Sleep

Although adults spend about one-third of their time sleeping, we know little of the functions served by it. We do know that sleep is an active process and not a mere absence of wakefulness. Moreover, it is not a single simple phenomenon; there are distinct states of sleep characterized by different EEG and behavior patterns.

The EEG pattern changes profoundly in sleep. As a person becomes drowsy, the alpha rhythm is gradually replaced (Fig. 17-1C), and as sleep deepens, the EEG waves become slower and larger (Fig. 17-1D). This **slow-wave sleep** is periodically interrupted by episodes of **paradoxical sleep,** during which the subject still seems asleep but has an EEG pattern similar to that of EEG arousal, i.e., an awake alert person (Fig. 17-1E).

It is difficult to tell precisely when a person passes from drowsiness into slow-wave sleep, for in slow-wave sleep there is considerable tonus in postural muscles and only a small change in cardiovascular or respiratory activity. The sleeper can be awakened fairly easily and if awakened, rarely reports dreaming. Slow-wave sleep has a characteristic kind of conscious experience described by subjects as "thoughts" rather than "dreams." The thoughts are more plausible and conceptual; they are more concerned with recent events of everyday life and more like waking-state thoughts than are true dreams.

At the onset of paradoxical sleep, there is an abrupt and complete inhibition of tone in the postural muscles, although periodic episodes of twitching of the facial muscles and limbs and rapid eye movements behind the closed lids occur. (Paradoxical sleep is therefore also called *rapid-eye-movement* or **REM sleep.**) Respiration and heart rate are irregular, and blood pressure may go up or down. When awakened during paradoxical sleep, 80 to 90 percent of the time subjects report that they have been dreaming.

Continuous recordings show that the two states of sleep follow a regular 30- to 90-min cycle, each episode of paradoxical sleep lasting 10 to 15 min. Thus, slow-wave sleep constitutes about 80 percent of the total sleeping time in adults, and paradoxical sleep about 20 percent. The time spent in paradoxical sleep increases toward the end of an undisturbed night. Normally it is not possible to pass directly from the waking state to an episode of paradoxical sleep; it is entered only after at least 30 min of slow-wave sleep.

What is the functional significance of sleep, i.e., what happens to the brain during sleep? The brain, as a whole, does not rest during sleep and there is no generalized inhibition of activity of cerebral neurons; however, there is a reorganization of neuronal activity, some individual neurons being less active during sleep than during waking and others showing just the opposite pattern. The total blood flow and oxygen consumption of the brain, signs of its metabolic activity, do not decrease in sleep.

Although sleep is not a period of generalized rest for the whole brain, it may represent a period of rest for certain specific elements, during which they can replenish substrates necessary for their generation of action potentials. Yet, when isolated neural tissue is exposed to extreme rates of stimulation far exceeding those occurring under physiological circumstances, neurons recover within a period of minutes. Alternatively, it has been suggested that the functional significance of sleep lies not in short-term recovery but in the relatively long-term chemical and structural changes that the brain must undergo to make learning and memory possible.

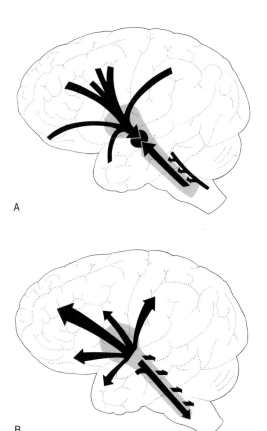

FIG. 17-2 (A) Convergence of descending, local, and ascending influences upon the reticular formation (shaded area). (B) Projections from the reticular formation to the spinal cord, brainstem, and cerebellum.

Neural Substrates of States of Consciousness

The prevailing state of consciousness is mainly the result of the interplay between three neuronal systems, one causing arousal and the other two sleep. All three are parts of the **reticular formation** (see Chap. 14). The reticular formation proper lies in the central core of the brainstem in the midst of the neural pathways that ascend and descend between the brain and spinal cord, but neurons in the midline thalamus also function as an extension of this brainstem system. Neurons of the reticular formation are influenced by and influence virtually all areas of the central nervous system (Fig. 17-2).

The reticular formation is not homogeneous, and discrete areas frequently have specific functions, several of which we have already discussed: It helps to coordinate skeletal muscle activity (Chap. 16); it contains the primary cardiovascular and respiratory control centers (Chaps. 22 and 23); it monitors the huge number of messages ascending and descending through the central nervous system (Chaps. 14 and 15). In this chapter we are concerned with its role in determining states of consciousness.

Pathways for the Waking State, Arousal, and Attention   In 1934, it was discovered that after the cerebrum is surgically isolated from the spinal cord and lower three-fourths of the brainstem, the EEG loses the wave patterns typical of an awake animal, indicating that some neural structures in the separated brainstem or spinal cord are essential for the maintenance of a waking EEG. The relevant neural structures actually lie in the reticular formation and form part of the **reticular activating system (RAS),** destruction of which produces coma and the EEG characteristic of the sleeping state.

Single neurons in the reticular formation may be activated by any afferent modality—a flash of light, a ringing bell, a touch on the skin. As they pass from the brainstem into the central core of the cerebrum, fibers of the RAS activate the **diffuse thalamic projection system.** Fibers of these thalamic neurons synapse in the cortex, but unlike the specific thalamic projections described in Chap. 15, they are not involved in the transmission of information about specific sensory modalities; rather, they maintain the behavioral characteristics of the awake state.

The RAS is crucial not only for maintenance of a waking state but for arousal and attention. The important generalization to keep in mind during this analysis is that human beings are conscious of a stimulus only when the nervous system is oriented and appropriately receptive toward it, and it is the neurons of the reticular activating system that arouse the brain and facilitate information reception by the appropriate neural structures. However, the sensitivity of this system is selective. A mother may awaken instantly at her baby's faintest whimper, whereas she can sleep peacefully through the roar of a jet plane passing overhead.

Since attention is directed only to meaningful stimuli, it seems that at some point in the nervous system there must be a comparison of present stimuli with those that have gone before to answer questions such as: "Is this new information?" "Is it relevant in terms of past events?" The neural outcome of this questioning biases (or fails to bias) the relevant cortical neurons so that attention is paid to the stimulus or it is ignored. In other words, a "yes" answer to either of the above questions results in increased alertness and a "no" answer leads to a progressive decrease in response (**habituation**). For example, when a loud bell is sounded for the first time, it may evoke increased alertness (or a startle response) in the animal; but after several ringings, the animal makes progressively less response and eventually may ignore the bell altogether. Subsequent stimulus of another modality or the same stimulus at a different intensity restores the original response (**dishabituation**). Habituation is not due to receptor fatigue or adaptation.

Not unexpectedly, the neuronal mechanisms that "select" the response are thought to be located in the thalamus and reticular formation, a hypothesis supported in part by the finding there of *"novelty detectors,"* which will be described shortly.

The attention-focusing mechanism just described is automatic, and it is suggested that in humans (and probably other higher animals) there exists a second, "voluntary," attention-directing mechanism that involves temporal and frontal cortex. This mechanism would permit one to "choose" to pay attention to selected stimuli, but it still requires much of the organizational function of the reticular formation.

It should be evident from this description that both the cortex and reticular formation are required for sustained wakefulness and normal arousal and attention-focusing. In addition, certain areas of the limbic system (see Chap. 14) are implicated in the EEG and behavioral aspects of the waking state. For example, during increased alertness a portion of the limbic system known as the *hippocampus* manifests low-frequency, high-amplitude waves, the so-called *theta rhythm* (or hippocampal "arousal"); note that the theta rhythm is essentially opposite to the EEG pattern of the cerebral cortex during arousal. Injury to the hippocampus results in faulty mechanisms for increased alertness as well as to the faulty learning described in a subsequent section.

The Sleep Centers   The control of sleep is exerted by two neuronal systems that oppose the tonic activity of the RAS. The two neuronal clusters, one in the central core of the brainstem, the other in the pons, are also part of the reticular formation (Fig. 17-3). The sleep-wake cycle is thought to occur because the brainstem-core neurons tonically release the transmitter serotonin; when serotonin levels become high enough, the neurons of RAS are inhibited. This results in the loss of awake conscious behavior and its EEG manifestations and the replacement of these by the behavior and EEG characteristics of slow-wave sleep.

These brainstem-core neurons also facilitate the sleep center of the pons, whose activity induces paradoxical sleep. Pathways ascending from this paradoxical-sleep center of the pons establish the low-voltage, fast EEG pattern and activate the muscles of the eyes; descending pathways inhibit motor-neuronal activity, which results in loss of muscle tone.

Of great importance is the fact that, while instituting the EEG and behavioral manifestations of paradoxical sleep, the neurons of the sleep center

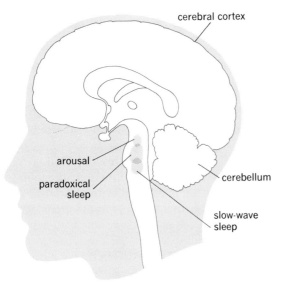

FIG. 17-3 Brainstem structures involved in arousal, paradoxical sleep, and slow-wave sleep.

in the pons also influence the central core of the brainstem in a true feedback fashion. We have said that the release of serotonin by these central core neurons causes slow-wave sleep and facilitates the paradoxical-sleep center. It is thought that the paradoxical-sleep-center neurons feed back and stimulate the *uptake* of serotonin by the endings of the brainstem-core neurons. This decrease in free (i.e., extracellular) serotonin concentration lessens the inhibition on the RAS and permits a return of the awake state. Waking continues until sleep is again triggered by the release of sufficient serotonin to inhibit the RAS. Thus, the cycling of the sleeping and waking states of consciousness is probably due, at least in part, to the slow accumulation and dissipation of chemical transmitters.

As presented above, sleep is basically the result of the cyclical inhibition of the RAS by brainstem-core neurons. However, other brain regions also affect sleep. Many of them are in the limbic system, brain regions associated with emotions. This provides a neural basis for the common experience that thoughts, worries, fears, etc., can interfere with sleep.

In summary, two opposing systems exist in the brain, one a sleep-producing system and the other an arousal system; the two are connected in such a way that activity in one suppresses activity in the other, and vice versa. It is the interaction of these two systems that produces the sleep-waking cycles. However, the periodic inhibition of the arousal systems (which would normally cause sleep) can be overridden by input from afferent pathways or other brain centers so that RAS activity is maintained sufficiently high to keep one awake or interrupt sleep. In fact, the waking mechanisms seem to be more easily activated than those that cause sleep. An example familiar to all parents is that it is much easier to arouse a sleeping child than to get an alert, attentive child to sleep.

CONSCIOUS EXPERIENCE

All subjective experiences are popularly attributed to the workings of the mind. This word conjures up the image of a nonneural ''me,'' a phantom interposed between afferent and efferent impulses, with the implication that mind is something more than neuronal activity. The truth of the matter is that physiologists and psychologists have absolutely no idea of the mechanisms that give rise to conscious experience. Nor are there even any scientifically meaningful hypotheses concerning the problem.

Conscious experiences are difficult to investigate because they can be known only by verbal report. Such studies lack scientific objectivity and must be limited to people. In an attempt to bypass these difficulties scientists have studied the behavioral correlates of mental phenomena in other animals. For example, a rat deprived of water performs certain actions to obtain it. These actions are the behavioral correlates of thirst. But it must be emphasized that we do not know whether the rat consciously experiences thirst; this can only be inferred from the fact that human beings are conscious of thirst under the same conditions.

However, one crucial question that cannot be investigated in experimental animals is whether conscious experiences actually influence behavior. Although, intuitively, it might seem absurd to question this, the fact is that the answer is crucial for the development of one's concept of humanity. It is possible that conscious experience is an epiphenomenon.

Consider the following sequence of events: The ringing of a telephone reminds a student that he had promised to call his mother; he finishes the page he has been reading and makes the call. What causes him to do so? The epiphenomenon view holds that the conscious awareness accompanies but does not influence the passage of information from afferent to motor pathways. Thus, in this view, behavior occurs automatically in response to a stimulus, memory stores supplying the direct link between afferent and efferent activity. In contrast, the processing of the afferent information (the sound of the bell), acting through memory stores, could result in the conscious awareness of his promise, which, in turn, leads to the relevant activity in motor pathways descending from the cortex. There is no way of choosing between these two views at present.

Is there a specific brain area in which conscious experience resides? The answer can be gleaned from conscious persons undergoing neurosurgical procedures and from persons who have accidental or disease-inflicted damage to parts of the brain; but the answer is still far from clear. We have mentioned that it is possible that conscious experience may be just another aspect of the neural activity in those brain centers that receive and process afferent information. Or, on the other hand, conscious experience may depend upon the transmission of the processed afferent information to special parts of the brain whose function is to "release" the contents of conscious experience. The only available clues as to how or where this might be done have been gained from inference. For example, with evolution comes (we assume) greater complexity of conscious experience. Since the human brain is distinguished anatomically from that of other mammals by a greatly increased volume of cerebral cortex, this is a logical place to look for the seat of conscious experience.

The cortex has been stimulated when patients on the operating table are fully alert and the brain is exposed. If a certain area of association cortex in temporal lobe is stimulated, a subject may report one of two types of changes in her conscious experience. Either she is aware of a sudden change in her interpretation of the present situation, i.e., what she is seeing or hearing suddenly becomes familiar or strange or frightening or coming closer or going away, or she has a sudden flashback or awareness of an earlier experience. Although she is still aware of where she is, an earlier experience comes to her and repeats itself in the same order and detail as the original experience. It may have been a particular occasion when she was listening to music. If

asked to do so, she can hum an accompaniment to the music. If in the past she thought the music beautiful, she thinks so again.

During such electric stimulation, visual or auditory experiences are recalled only if the patient was attentive to them when they originally occurred. Other experiences that are also part of conscious experience have never been produced by such stimulation. No one has ever reported periods when he or she was trying to make a decision or solve a problem or add up a row of figures. It is also interesting that no brain area other than temporal association cortex (see Fig. 15-35) has been found from which complete memories have been activated. However, one cannot assume that association cortex is the site of the stream of consciousness, since stimulation of these cortical neurons causes the propagation of action potentials to many other parts of the brain.

It is best to take a different view of the matter; for, in the words of one famous neurosurgeon, "Consciousness is not something to be localized in space." It is a function of the integrated action of the brain. Sensations and perceptions form part of conscious experience, and yet there is no one point along the ascending pathways or one particular level of the central nervous system below which activity cannot be a conscious sensation and above which it is a recognizable, definable sensory experience. Every synapse along the ascending pathways adds an element of meaning and contributes to the sensory experience.

On the other hand, consciousness and, we presume, the accompanying conscious experiences, are inevitably lost when the function of regions deeper in the cerebrum or the nerve fibers passing to association cortex from the reticular formation are interrupted by injury. Although the matter is far from settled, evidence suggests that neuronal systems in the reticular formation of the brainstem and regions deep in the cerebrum are involved in brain mechanisms necessary for perceptual awareness. It has been suggested that this system of fibers has widespread interactions with various areas of the cortex and somehow determines which of these functional areas is to gain temporary dominance in the on-going stream of the conscious experience.

The concept we want to leave as an answer to the question of where the conscious experience resides is perhaps best presented in the following analogy. In an attempt to say which part of a car is responsible for its controlled movement down a highway, one cannot specify the wheels or axle or engine or gasoline. The final performance of an automobile is achieved only through the coordinated interaction of many components. In a similar way, the conscious experience is the result of the coordinated interaction of *many* areas of the nervous system. One neuronal system would be incapable of creating a conscious experience without the effective interaction of many others.

MOTIVATION AND EMOTION

Motivation

Motivation is presently undefinable in neurophysiological terms, but it can be defined in behavioral terms as the processes responsible for the goal-directed quality of behavior. Much of this behavior is clearly related to *homeostasis,* i.e., the maintenance of a stable internal environment, an example being putting on a sweater when one is cold. In such homeostatic goal-directed behavior, specific bodily needs are being satisfied, the word "needs" having a physicochemical correlate. Thus, in our example the correlate of need is a drop in body temperature, and the correlate of need satisfaction is return of the body temperature to normal. The neurophysiologic integration of much homeostatic goal-directed behavior will be discussed subsequently (thirst and drinking, Chap. 24; food intake and temperature regulation, Chap. 26; reproduction, Chap. 27).

However, many kinds of motivated behavior, e.g., the selection of a particular sweater on the basis of style, have little if any apparent relation to homeostasis. Clearly, much of human behavior fits this latter category. Nonetheless, the generalization that motivated behavior is induced by needs and is sustained until the needs are satisfied is a useful one despite the inability to understand most needs in physicochemical terms.

A concept inseparable from motivation is that of **reward** and **punishment,** rewards being things that organisms work for or things that strengthen behavior leading to them, and punishments being the opposite. They are related to motivation in that rewards may be said to satisfy needs. Many psychologists believe that rewards and punishments constitute the incentives for learning. Because virtually all behavior is shaped by learning, reward and punishment become crucial factors in directing behavior. Although some rewards and punishments have conscious correlates, many do not. Accordingly, much of human behavior is influenced by factors (rewards and punishments) of which the individual is unaware.

We have thus far described motivation without regard to its neural correlates. Nothing is known of the mechanisms that underlie the subjective components of this phenomenon, nor is it known how rewards and punishments influence learning and behavior. Present knowledge is limited to recognition of some of the brain areas (and their interconnecting pathways) that are important in motivated behavior.

It should not be surprising that the brain area most important for the integration of motivated behavior related to homeostasis is the hypothalamus, since it contains the integrating centers for thirst, food intake, temperature regulation, and many others. Much information concerning the reinforcing effects of rewards and punishments on hypothalamic function has been obtained through **self-stimulation** experiments, in which an unanesthetized experimental animal regulates the rate at which electric stimuli are delivered through electrodes previously implanted in discrete brain areas. The animal is placed in a box that contains a lever he can press. If no stimulus is delivered to the animal's brain when the bar is pressed, he usually presses it occasionally at random. However, if a stimulus is delivered to the brain as a result of the bar press, a different behavior can result, depending upon the location of the electrodes. If the animal increases his bar-pressing rate above control, the electric stimulus is, by definition, rewarding; if he decreases it, the stimulus is punishing. Thus, the rate of bar pressing is taken to be a measure of the effectiveness of the reward (or punishment). Bar pressing that results in self-stimulation of the sensory and motor systems produces response rates not significantly different from the control rate. Brain stimulation through electrodes implanted in certain areas of hypothalamus serves as a positive reward. Animals with electrodes in these areas bar-press to stimulate their brains from 500 to 5000 times per hour. In fact, electric stimulation of some areas of hypothalamus is more rewarding than external rewards; e.g., hungry rats often ignore available food for the sake of electrically stimulating their brains.

This rewarding effect of self-stimulation is not found in all areas of the hypothalamus but is most closely associated with those areas that normally mediate highly motivated behavior, e.g., feeding, drinking, and sexual behavior. Consistent with this is the fact that the animal's rate of self-stimulation in some areas increases when he is deprived of food; in other areas, it is decreased by castration and restored by administration of sex hormones. Thus, it appears that neurons controlling homeostatic goal-directed behavior are themselves intimately involved in the reinforcing effects of reward and punishment.

Emotion

Related to motivation are the complex phenomena of **emotion.** Scientists are presently trying to understand the operation of the chain of events leading from the perception of an emotionally toned stimulus, i.e., the subjective feeling of fear, love, anger, joy, anxiety, hope, etc., to the complex display of emotional behavior, and they are beginning to find some answers. Most experiments point to the involvement of the limbic system, which has been studied in experimental animals, using electric stimulation of specific areas within it. The physiological results of these procedures vary markedly but justify the concept that three distinct neural systems mediate the various emotional behaviors. Of course, in these experiments there was no way to assess the subjective emotional feelings of the animals; instead they were observed for behaviors that usually are associated with emotions in human beings. As different areas of the limbic system were electrically stimulated in awake animals three types of behavior resulted.

During stimulation of one particular area the animal actively approaches a situation as though expecting a reward. Stimulation of a second area causes the animal to stop the behavior he is performing, as though he knew it would lead to punishment. Stimulation of a third area of the limbic system causes the animal to arch its back, puff out its tail, hiss, snarl, bare its claws and teeth, flatten its ears, and strike. Simultaneously, its heart rate, blood pressure, respiration, salivation, and concentrations of plasma epinephrine and fatty acids all increase. Clearly, this behavior typifies that of an enraged or threatened animal. In fact, an animal's behavior can be changed from quiet to savage or from savage to docile simply by electrically stimulating different areas of the limbic system.

Limbic areas have also been stimulated in awake human beings undergoing neurosurgery. These patients, relaxed and comfortable in the experimental situation, report vague feelings of fear or anxiety during periods of stimulation to certain areas even though they are not told when the current is on. Stimulation of other areas induces pleasurable sensations that the subjects find difficult to define precisely.

Surgical damage to parts of the limbic system in experimental animals is another commonly used tool; it leads to a great variety of changes in behavior, particularly that associated with emotion. Destruction of a nucleus in the tip of the temporal lobe produces docility in an otherwise savage animal, whereas surgical damage to an area deep in the brain produces vicious rage in a tame animal; and the rage caused by this lesion can be counteracted by a lesion in the tip of the temporal lobe. A rage response can also be caused by destruction of part of the hypothalamus. Lesioned animals sometimes manifest bizarre sexual behavior in which they attempt to mate with animals of other species; females frequently assume male positions and attempt to mount other animals.

Self-stimulation experiments have shown the presence of reward and punishment responses in various parts of the limbic system. When the electrodes are in certain midline areas, the animal presses the bar once and never goes back, indicating that stimulation of these brain areas has a punishing effect. These are the same brain areas that, when stimulated, give rise to behavioral activity that signifies avoidance, rage, or escape. In contrast, self-stimulation of other limbic areas has a strong rewarding effect.

Stimulation of certain hypothalamic areas (like the stimulation of other limbic structures described above) elicits behavior that *seems* to have a strong subjective emotional component; yet if the hypothalamus is isolated from the other portions of the

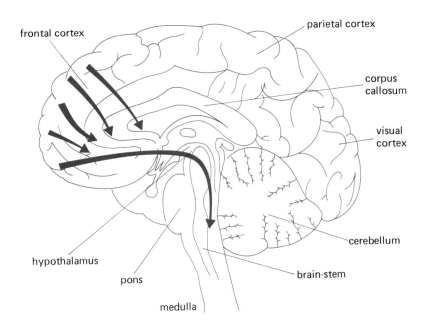

frontal cortex

parietal cortex

corpus callosum

visual cortex

cerebellum

hypothalamus

pons

brain-stem

medulla

FIG. 17-4 Location of the reward-system pathways in the human brain. These correspond to the areas of rat brain that yield the highest bar-pressing rate in self-stimulation experiments.

limbic system, the emotional component of the behavior is lacking. For example, stimulation of a cat's hypothalamus under such conditions can cause enraged, aggressive behavior complete with attack directed at any available object, but as soon as the stimulation ends, the animal immediately reverts to its usual friendly behavior. It seems as though the actions lacked emotionality and purpose, representing only the motor component of the behavior. We use the word "purpose" but could have said that, except for the experimentally induced stimulus, the behavior lacked "motivation."

The subjective aspects, or feelings, that make up part of an emotional experience possibly involve the cortex, particularly cortex of the frontal lobes, which is implicated because changes in emotional states frequently occur following damage there. These alterations in mood and character are described as fear, aggressiveness, depression, rage, euphoria, irritability, or apathy. There are indications that frontal regions may exert inhibitory influences upon the hypothalamus and other areas of the limbic system. There may be facilitatory frontal regions as well. Anatomical connections between frontal cortex and hypothalamus exist to support the suggested interrelationship of these two areas in motivated and emotional behaviors. Excitatory and inhibitory influences from the limbic system and possibly from nonlimbic areas of cortex are of great importance in determining and patterning the

level of excitability of hypothalamic and brainstem neurons. How activity in the limbic system is initiated and influences other brain areas is still poorly understood.

Chemical Mediators for Emotion-Motivation

The pathways that underlie the brain reward systems and motivation (Fig. 17-4) are, as are other neural pathways, made up of synaptically interconnected neurons, and the neurotransmitters and the many drugs that affect the synapses in these pathways are fairly well-defined. A role for norepinephrine and dopamine (both of which belong, along with epinephrine, to the family of *catecholamines*) is suggested by several kinds of evidence. First, the anatomical sites that give high rates of self-stimulation contain catecholamine-mediated pathways. Second, rates of self-stimulation by experimental animals are altered by drugs that affect transmission at catecholamine-mediated synapses; for example, drugs (such as *amphetamine*) that increase synaptic activity in the catecholamine pathways, increase self-stimulation rates. As mentioned below, these drugs also elevate mood in humans. Conversely, drugs (such as chlorpromazine, an antipsychotic agent) that lower activity in the catecholamine pathways decrease self-stimulation and depress mood. (The fact that catecholamines correlate both with the brain reward-system and moods is further evidence for a neural relation between motivation and emotion.)

The catecholamines are also implicated in the pathways that subserve learning. This association is not unexpected since we have just stated that they are involved in the neural systems that underly reward and punishment. As we have mentioned, many psychologists believe that rewards and punishments constitute the incentives for learning.

PSYCHOACTIVE DRUGS, TOLERANCE, AND ADDICTION

Psychoactive drugs such as amphetamines are used in a deliberate attempt to elevate mood (*euphorigens*) and produce unusual states of consciousness ranging from meditative states to hallucinations. All these drugs seem to act on neuronal membranes or synaptic mechanisms in one way or another (neurotransmitter release or reuptake, membrane permeability, etc.).

Actually, psychoactive drugs are often chemical forms related to neurotransmitters such as serotonin, dopamine, and norepinephrine; moreover, it is possible that the endorphins and enkephalins, in conjunction with their effects in the pain-inhibiting pathways (Chap. 15), serve as natural euphorigens. Several of the naturally occurring euphorigens have been implicated in diseases such as schizophrenia.

In general, there seems to be a tendency for the neurotransmitter-receptor interaction to stay at some baseline state and to regain that state if moved from it. Perhaps this balance is upset by some malfunction in disease states such as schizophrenia, the depressions, and the manias. In any case, it may be this tendency to return to a baseline that underlies the dual phenomena of tolerance and dependence (addiction) to administered psychoactive drugs.

Tolerance to a drug occurs when increasing doses of that drug are required to achieve effects that initially occurred in response to a smaller dose, i.e., it takes more drug to do the same job. One cause of tolerance is that the presence of the drug may stimulate the manufacture (especially, in the liver) of those enzymes that degrade it. As drug concentrations increase so do the concentrations of the degratory enzymes so that more drug must be administered to produce the initial effect.

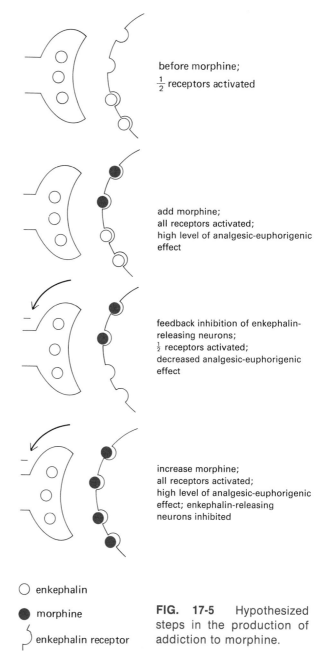

before morphine; $\frac{1}{2}$ receptors activated

add morphine; all receptors activated; high level of analgesic-euphorigenic effect

feedback inhibition of enkephalin-releasing neurons; $\frac{1}{2}$ receptors activated; decreased analgesic-euphorigenic effect

increase morphine; all receptors activated; high level of analgesic-euphorigenic effect; enkephalin-releasing neurons inhibited

○ enkephalin

● morphine

enkephalin receptor

FIG. 17-5 Hypothesized steps in the production of addiction to morphine.

However, another possible cause for tolerance has nothing to do with drug degradation but results from the drug's actions as a neurotransmitter. The drug's effects add to those of the normally occurring neurotransmitter, thereby producing an increased effect that, by feedback mechanisms, decreases the release of the naturally occurring neurotransmitter. For example, tolerance to opiate drugs such as morphine, which react with some types of the receptors normally activated by enkephalin, may occur in the following way (Fig. 17-5). Enkephalin is normally continually released, at least to some degree, by those neurons that use it as their transmitter so that the enkephalin receptors are always exposed to a certain amount of stimulation. When morphine is present, it binds to any receptors not already occupied by enkephalin and thereby increases the analgesic-euphorigenic effects of the enkephalin pathways. As the enkephalin receptors are stimulated more and more strongly (by the morphine-enkephalin combination), a feedback system (still hypothetical) decreases the firing rate of the enkephalin-releasing neurons, thereby decreasing the synthesis and release of the transmitter and the number of receptors activated by it. The receptors, no longer receiving their usual amount of enkephalin, can therefore accept more morphine. In other words, to regain the previous analgesic-euphorigenic effects of full receptor activation, increased amounts of the drug are required.

This type of reasoning can also explain the physical symptoms associated with cessation of drug use, i.e., **withdrawal.** When the drug is stopped, the receptors receive for some time neither the drug nor the normal neurotransmitter (because the latter's synthesis and release had been inhibited by the previous prolonged drug use).

NEURAL DEVELOPMENT AND LEARNING

Evolution of the Nervous System

During an early stage of evolution, the nervous system was probably a simple system with a limited number of interneurons interposed between the afferent and efferent nerve cells. The interneuronal component expanded rapidly until it came to be by far the largest part. The interneurons formed networks of increasing complexity, at first involved mainly with the stability of the internal environment and position of the body in space. Those cells with increasing specialization of function came to be localized at one end of the primitive nervous system, and the brain began to evolve.

The brainstem, which is the oldest part of the brain in this evolutionary sense, retains today many of the anatomical and functional characteristics typical of those most primitive brains. With continued evolution, newer, increasingly complex structures were added on top of (or in front of) the older ones. They developed as paired symmetrical tissues, the cerebral hemispheres, and reached their highest degree of sophistication with the formation of the cerebral cortex.

The newer structures served in part to elaborate, refine, modify, and control already existing functions. For example, it is possible to perceive the somatic stimuli of pain, touch, and pressure with only a brainstem, but the stimulus cannot be localized without a functioning cerebral cortex. Perhaps even more important is that these newer, more anterior parts of the brain came to be involved in the perception of goals, the ordering of goal priorities, and the patterning of behaviors to serve in pursuit of these goals.

Development of the Individual Nervous System

All neurons are present at birth, but during postnatal development many important changes in these neurons occur: The cells enlarge, the numbers of dendrites increase, axons elongate, myelinization increases, the number of synaptic contacts increases, neurons migrate to new locations, etc. Notice the changes in visual cortex cells in just the first three months of life (Fig. 17-6) and the gradual change in electric activity (Fig. 17-7). These developmental patterns are reflected in behavioral changes. It is believed that the cortex is relatively nonfunctional at birth, a notion supported by the fact that infants born without a cortex have almost the same behavior and reflexes as shown by a normal infant. As cortical function develops, it seems gradually to exert an inhibitory control over the lower (and phylogenetically older) structures.

With cortical development, some of the reflexes whose integrating centers are in subcortical structures come under at least a degree of cortical control. Examples are the *rooting* and *sucking reflexes* present in infancy. (In these reflexes, the stimulus is a touch on the infant's cheek; in response, the head turns toward the stimulus, the mouth opens, the stimulating object is taken into the mouth, and sucking begins.) The reflex is essential for the survival of the young, but it is superseded by other eating behaviors. As the cerebral hemispheres develop and autonomy increases, allowing the elaboration of hand feeding, the primitive rooting and

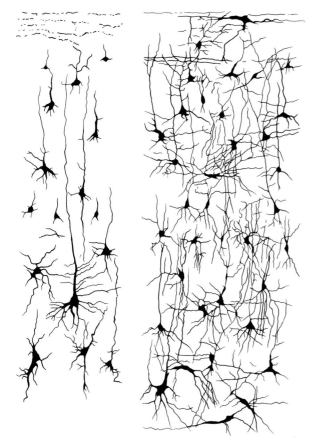

FIG. 17-6 Visual cortex of a newborn (left) and a 3-month-old child (right).

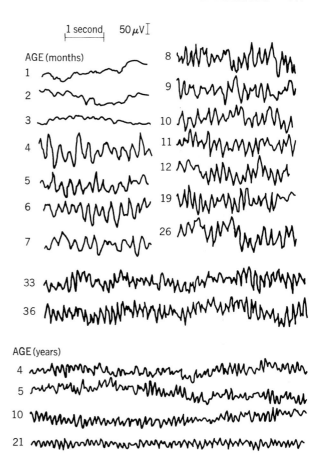

FIG. 17-7 EEGs from one individual, showing the onset of the alpha rhythm at 4 months, the attainment of adult frequencies at 10 years of age, and little change thereafter.

sucking reflexes are inhibited. The basic reflex arc does not disappear; it is simply inhibited. For example, while examining a patient with damaged frontal lobes of the brain, a physician standing behind the patient and out of view quietly reached over and touched the patient's cheek. In response, the patient nuzzled the physician's finger until it was in his mouth and even began sucking on it. Upon realizing what he had unconsciously and automatically done, he was quite embarrassed; but his behavior, released from its normal inhibition because of the brain damage, serves as a perfect illustration of the point made above.

During development of the brain and spinal cord, nerve growth and synaptic contact occur with remarkable precision and selectivity. How do the right nerve connections get established in the first place? The best answer presently available is the **chemoaffinity theory,** in which the complicated

neural circuits are said to grow and organize themselves according to selective attractions between neurons, which are determined by chemical codes under genetic control. Thus, during early stages of their development, neurons acquire individual chemical ''identifications'' by which they can recognize and distinguish each other. The chemical specificity is precise enough to determine not only the particular postsynaptic cell to which the axon tip will grow but even the exact area on the postsynaptic cell that will be contacted. (Recall that synapses closer to the postsynaptic cell's initial segment have greater influence over its activity.)

Several other factors also influence the final neural organization: timing of the development of the individual cells (for example, the axons that reach a postsynaptic neuron earliest have the greatest chance of forming large numbers of synapses with it); the presence of neighboring neurons and glia (which form structural constraints and guides for the developing nerve processes); and chemical events involving neurotransmitters and enzyme systems in the tip of the developing fiber and adjacent cells.

Of course, adequate nutritive environments are required for normal brain development. The developing brain is extremely sensitive to the effects of malnutrition (i.e., a deficiency of calories, protein, trace metals, or vitamins), and some of the damage produced may be permanent. Malnutrition in the prenatal period, during which neurons are dividing, causes a decrease in the rate of cell division, and the final number of cells is less than normal. This change is thought to be permanent and probably cannot be reversed by improving nutrition at a later time after the period for cell division has passed. Postnatally, malnutrition prevents the normal enlargement of cells, but this growth is largely restored with the resumption of an adequate diet.

The level of hormones is also important; for example, decreased thyroid hormone concentrations during development cause a type of mental retardation known as *cretinism.* Although it was formerly believed that the developing fetus "took what it needed" nutritionally from the mother and that, in cases of malnutrition after birth, the brain was "spared" at the expense of the rest of the body, there is increasing evidence that nutritional deprivation both before and after birth has serious and irreversible effects on the chemical and structural maturation of the brain. Intermittent deprivation affects those cells that are still developing at that time more than cells already mature.

Despite the evidence that indicates the selective growth and interconnectivity of nerve cells, the nature-nurture controversy, i.e., how much of development is due to genetically determined patterns and how much to experience and subsequent learning, must also be considered, for, in fact, the course of neural development can be altered. However, the period during which this alteration can occur is genetically built into the neuron's developmental timetable. The modifiability of some neurons is limited to so-called **critical periods** early in development but in others it persists longer. Thus, some neuronal organizational patterns are highly specified early in development and are unmodifiable thereafter. These periods of maximal structural and

functional growth depend upon the availability of proper internal and external environmental conditions for their full development. As an example of such a critical period it was found that the visual systems of sheep dogs deprived of sight in the first 5 weeks of life are anatomically, biochemically, and electrophysiologically retarded; this does not occur if the deprivation occurs from 5 to 10 weeks of age or during a 5-week period in adult dogs.

Other neurons remain uncommitted and are modifiable by function and experience until late in development. But in the case of either limited or lengthy periods of modifiability, it can occur only within the constraints imposed by the neurons' genetic code and its experiential history. The genes specify the capacity, whereas the environmental stimuli determine its specific expression and content.

Learning and Memory

Learning is the increase in the likelihood of a particular response to a stimulus as the consequence of experience. There are many kinds of learning: habituation (defined classically as a reversible decrease in the strength of a response upon repeated stimulation), conditioning, and other forms typical only of higher animals in which learning may occur following a single stimulus and in the absence of direct reinforcement. Generally, rewards or punishments, as mentioned earlier, are crucial ingredients of learning, as is contact with and manipulation of the environment. **Memory** is the relatively permanent form of this change in responsiveness, and the postulated neural correlate of memory is called the **memory trace.**

Memory occurs in two forms, **short-term** and **long-term,** and the process of transferring memories from the short- to the long-term form is called **memory consolidation.** The existence of two forms is borne out by the behavior of patients with bilateral lesions in the hippocampal region of the limbic system. Such patients can recall information immediately after it has been presented, but they cannot retain the material as well as normal people do, particularly if it involves verbal clues. It is as though they were unable to transfer the information from short- to long-term memory storage. Although short-term memory involves cerebral cortex, and long-term memory involves the limbic system, there is no exclusive site for memory storage; removal of various parts of the brain does not remove specific memories, and practically every part of the nervous system is actually capable of undergoing some form of learning.

Short-Term Memory Short-term memory is a limited-capacity storage process that lasts as little as a few seconds and serves as the initial depository of information. After the registration of events in short-term memory, they may either pass away or be retained in long-term memory stores, depending upon factors such as attention and motivation. The distribution is not all-or-none, since some fragments of an event may be retained and others lost.

One theory suggests that the memory trace during the early phases of learning is a reverberating neural circuit in which electric activity passes around and around in closed neural loops until the mechanisms by which it is consolidated into long-term stores are activated. There is certainly evidence for the existence of such neuronal pathways in the brain, and activity, once started in such loops, could be maintained to keep the "memory" of the input for some time. That such reverberating circuits could be responsible for the temporary storage of acquired information in short-term memory is supported by evidence that conditions that interfere with the electric activity of the brain (for example, coma, deep anesthesia, electroconvulsive shock, and insufficient blood supply to the brain) also interfere with the retention of recently acquired information. These same states do not usually interfere with previously laid-down long-term memory. For example, if a man becomes unconscious from a blow on the head, he might not remember anything that happened for about 30 min before he was hit; this is so-called *retrograde amnesia*. As recovery from retrograde amnesia occurs, memories of the most recent events often return last and least completely (although islands of memory can return in a discontinuous fashion). The loss of consciousness in no way interferes with memories of experiences that were learned before the period of amnesia. There are no known chemicals that selectively block the formation of short-term memories.

Memory Consolidation Memory consolidation is a labile process during which the fixation of an experience is susceptible to external interference, the lability lasting for periods of a few seconds to days or even weeks. The amount of time required to consolidate an experience is fairly constant in a given species, although it may be markedly increased or decreased by various drugs that presumably alter the chemical reactions involved in transferring an experience to a more stable form of memory. Agents that enhance learning also generally increase attentiveness, activity, and sensory responsiveness, any of which would increase learning, but these substances are effective even when given after the learning experience and are therefore thought to act directly on consolidation. According to the "reverberating circuit" theory, these drugs facilitate consolidation because they accelerate or prolong the reverberations of activity through the neural circuit that in turn facilitates the permanent changes required for long-term memory of the event.

All experiences do not register in long-term memory. What determines whether consolidation will occur? It is as though, for memory to become fixed, a "fix signal" must follow the event that is to be remembered. Such a signal is probably not specific for a particular event, but rather may indicate "whatever just happened, remember it." Thus, the signal for memory consolidation may be likened to the fixation step in photographic developing; the latent image on the film will fade rapidly unless a chemical fixative is applied. The same fixative is used for photographs of any subject and it itself contains no specific information. We have no real idea as to the nature of the physiological "fix" signal.

Long-Term Memory Behavioral investigations and common experience indicate that memories of past events and well-learned behavior patterns normally can have very long life-spans. They may be changed or suppressed by other experiences but, contrary to popular opinion, memories do not usually fade away or decay with time. This stability and durability, combined with the fact that removing parts of the brain does not remove specific memories, suggests that memory is stored in widespread chemical form or that the memory trace is an alteration in structure of some elements of the brain.

It is generally assumed that the neuronal changes

that underly learning and memory are due to changes in synaptic transmission in the absence of major structural changes in nervous tissue, i.e., without the establishment of new connections by the sprouting and growth of neurons such as occurs during development and repair of injuries. This ability of neural tissue to change its responsiveness to stimulation because of its past history of activation is known as **plasticity.**

The synaptic theory of learning and memory is based on the view that the environmental interactions that result in learning modify communication between neurons at synapses that already exist as a result of heredity and maturation, i.e., preexisting synapses are thought to become more efficient. The critical modifications of synaptic relationships might be a change in the area of synaptic contact, an increase or decrease in the concentration of synaptic vesicles, changes at the presynaptic or postsynaptic membranes, etc. In any case, present evidence indicates that whatever the changes are, they are ultimately ascribable to alterations in the neuron's macromolecules; i.e., that large, stable molecules within neurons are changed during learning and that information is stored in the specific configuration of these molecules.

One of the most interesting new developments concerning learning and memory is the possible role of certain hormones or other neuromodulators in these processes. Peptide hormones such as ACTH and B-lipotropin all contain similar segments seven amino acids in length. Experiments have shown that these substances (or perhaps simply their shared segment) enhance learning of several different kinds of tasks and in some cases protect against amnesia. It has been suggested that these substances may be released in increased amounts during learning situations and affect learning and memory by temporarily increasing the motivational value of certain cues in the environment, perhaps via enhanced arousal of portions of the limbic system. The effects of these substances on learning and memory last 2 to 4 hr and are clearly distinct from their other known effects.

The posterior-pituitary peptide hormones, ADH (vasopressin) and oxytocin, also affect learning. Their effects last longer than those of the ACTH-related peptides and are thought to somehow enhance memory consolidation. These hormones will be discussed in more detail in Chaps. 18 and 24.

Forgetting   Forgetting denotes an inability to retrieve certain memory traces, but it does not necessarily mean that the traces no longer exist in the brain, for memory traces may be inaccessible to certain stimuli whereas they can be reached by others. For example, one may recognize the voice of a telephone caller but not be able to recall his name, which is immediately remembered upon seeing him. Our understanding of the mechanism of forgetting, like that of learning, is still mainly at the level of theories. One, the *interference theory of forgetting,* states that forgetting results from competition between responses at the time of recall. Competition comes from conflicting information stored both before and after the storage of the particular item that is being recalled. Thus we carry with us (in the form of prior learning) the source of much of our forgetting.

CEREBRAL DOMINANCE AND LANGUAGE

The two cerebral hemispheres are basically symmetrical, but they both have functional specializations, the most marked of which is language. All facets of language—the conceptualization of what one wants to say or write, the neuronal control of the act of speaking or writing, and the comprehension of what has been spoken or written—are mediated by structures of the left cerebral hemisphere in over 90 percent of the population. This is also the dominant hemisphere for fine motor control in the large majority of people, and so most people are right-handed (recall that the left hemisphere controls the movements of the right side of the body). Control of both-handedness and language in the left hemisphere suggests that this hemisphere is specialized for controlling successive changes in the precise positioning of muscle groups. However, the correlation between handedness and language function is not perfect, and for many left-handed people, language functions are controlled by the left hemisphere (in the remaining left-handed people, language is represented bilaterally or in the right hemisphere).

Just as the left hemisphere is dominant for language in most people, the right hemisphere appears to be dominant for many nonlanguage processes, particularly those that require spatial ability. The right hemisphere also predominately processes sounds such as musical patterns. Memories are asymmetrically stored, too, verbal memories being greater in the left hemisphere and nonverbal memories, e.g., visual patterns, being greater in the right. It may be that even the emotional responses of the two hemispheres are different; some authors claim that when electroconvulsive shock is administered in the treatment of depression, better effects are obtained when both electrodes are placed over the right hemisphere.

Much of the evidence about the specific functions of the two cerebral hemispheres has been obtained from studies on patients whose main commissures (nerve fiber bundles) joining the hemispheres have been cut to relieve uncontrollable epilepsy (neuronal activity, which starts at a cluster of abnormal cortical neurons, spreads throughout adjacent cortex, and gives rise to seizures and convulsions). In essence, this operation leaves two separate cerebral hemispheres with a single brainstem—two separate mental domains within one head. Events experienced, learned, and remembered by one hemisphere remain unknown to the other because the memory processing of one hemisphere is inaccessible to the other. Because control of language production resides in the left hemisphere, only that hemisphere can communicate orally or in writing about its conscious experience. When vision is experimentally limited so that only that part of the retina whose fibers pass to the right hemisphere is excited, the left hemisphere, which controls speech, is unaware of the visual experience and the patient is unable to describe it either orally or in writing. In contrast, when the portion of retina projecting to the left hemisphere is activated alone, the patient can describe the experience without difficulty. However, the right silent hemisphere does understand some langauge even though it is itself incapable of producing speech, for if the word for an object is flashed only to that hemisphere and the patient is told to point to the object or to demonstrate the use of the object, the patient complies (responding with the left hand since the right hemisphere controls the muscles of the opposite side of the body).

The left hemispheric specialization for processing speech is clearly in evidence in 5-year-olds and, in fact, seems to be present and effectively operating (for speech sounds at least) in the first few months of life. This indicates that the left hemisphere is slated from birth to be the speech hemisphere, a hypothesis supported by language-associated anatomical and physiological asymmetries in the two hemispheres present at birth. This fundamental asymmetry is more flexible in early years than in later life. For example, accidental damage to the left hemisphere of children under the age of two causes no impediment to future language development, and language develops in the intact right hemisphere. Even if the left hemisphere is traumatized after the onset of language, language is reestablished in the right hemisphere after transient periods of loss. The prognosis becomes rapidly worse as the age at which damage occurs increases, so that after the early teens, language is interfered with permanently. The dramatic change in the possibility of establishing language (or a second language) in the teens is possibly related to the fact that the brain attains its final structural, biochemical, and functional maturity at that time. Apparently, with maturation of the brain, language functions are irrevocably assigned and the utilization of language propensities of the right hemisphere is no longer possible. Interestingly, if the left-hemisphere damage occurs early enough in the child's development so that language functions can be transferred to the right hemisphere, defects occur in those functions in which the right hemisphere normally plays a role, i.e., language develops at the expense of other right-hemisphere functions.

Different areas of left cerebral cortex are related to specific aspects of language. Areas in the frontal lobe near motor cortex are involved in the articulation of speech, whereas areas in the parietal and temporal lobe are involved in sensory functions and language interpretations. Such cortical specialization is demonstrated by the *aphasias,* specific language deficits not due to mental retardation or paralysis of the muscles involved in speech. Damage to parietal and temporal lobe areas (Wernicke's area) result in aphasias related to conceptualization—the patients cannot understand spoken or written language even though their hearing and vision are unimpaired. In contrast, damage to the areas near motor cortex (Broca's area) results in expressive aphasias—the patients are unable to carry out the coordinated respiratory and oral movements necessary for language even though they can move their lips and tongue; they understand spoken language and know exactly what they want to say but are unable to speak. Expressive aphasias are often associated with an inability to write.

CONCLUSION

Until recently it was commonly thought that control of most complex behavior such as thinking, remembering, learning, etc., was handled almost exclusively by the cerebral cortex. Actually, damage of cortical areas outside of motor and sensory areas produces behavioral results that are subtle rather than obvious (the one exception, language, is highly sensitive to cortical damage), and stimulation of the cortex causes little change in the orientation or level of excitement of the animal.

In general, it is best to consider that particular behavioral functions are not controlled exclusively by any one area of the nervous system but that the control is shared or influenced by structures in other areas. Cortex and the subcortical regions—particularly limbic and reticular systems—form a highly interconnected system in which many parts contribute to the final expression of a particular behavioral performance. Moreover, the nervous system is so abundantly interconnected that it is difficult to know where any particular subsystem begins or ends.

In the early seventeenth century Descartes taught that all things in nature, including human beings, are machines. The brain's mode of operation was compared to that of a clock. When computers became widely used, the brain was compared to a computer. The most recent analogy compares the brain to a hologram, a photographic process that records specially processed lightwaves themselves, rather than the image of an object. These widely divergent analogies only emphasize how little we know of how the brain really functions.

KEY TERMS

consciousness	reticular activating system	short-term memory
states of consciousness	sleep centers	long-term memory
conscious experience	motivation	memory trace
electroencephalogram	reward	memory consolidation
alpha rhythm	punishment	plasticity
EEG arousal	critical period	forgetting
slow-wave sleep	learning	cerebral dominance
paradoxical sleep	emotion	

REVIEW EXERCISES

1 Define states of consciousness and distinguish from conscious experience.
2 Match electroencephalogram (EEG) patterns with the five states of consciousness listed in Fig. 17-1.
3 Define alpha rhythm and describe the subjective mood often associated with it.
4 Define EEG arousal and describe the subjective mood often associated with it.
5 Compare slow-wave sleep and paradoxical sleep in terms of EEG pattern, duration, ease of arousal, muscle tone, and dream episodes.
6 Describe two opposing systems for maintaining arousal or sleep.
7 Describe the role of sensory input in the maintenance of normal conscious experience.
8 Define motivation in behavioral terms.
9 Describe the relationship between emotional behavior and the limbic system, hypothalamus, and cortex.
10 Define tolerance to a drug; describe a mechanism by which tolerance could occur.
11 Describe the chemoaffinity theory of nerve-connection development.
12 State the role of critical periods in neural development.
13 Compare short-term and long-term memory, including a discussion of memory consolidation.
14 Define plasticity in the nervous system.
15 State the role of the left cerebral hemisphere in language function.
16 Define aphasia.

CHAPTER

18

THE ENDOCRINE SYSTEM

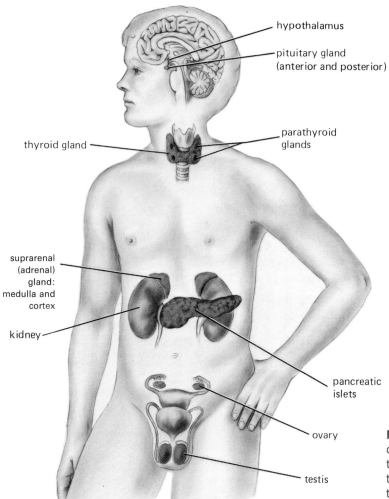

hypothalamus

pituitary gland
(anterior and posterior)

parathyroid
glands

thyroid gland

suprarenal
(adrenal)
gland:
medulla and
cortex

kidney

pancreatic
islets

ovary

testis

FIG. 18-1 Location of the major endocrine glands. Note that the parathyroid glands actually lie on the posterior surface of the thyroid. Also note that this hypothetical bisexual figure has both ovaries and testes; this does not occur in normal human beings.

Although the term **endocrine** broadly denotes any ductless gland, we shall use it in its more usual restricted sense to include only those glands whose secretory products are hormones; thus, for example, the liver, which primarily secretes nonhormonal materials (glucose and other metabolites) into the blood, is generally excluded from this category. A **hormone** is a chemical substance secreted into the blood, which carries it to other sites in the body where its actions are exerted. In terms of chemical structure, hormones generally fall into four categories: *steroids, amino acid derivatives, peptides,* and *proteins.*

Figure 18-1 illustrates the locations of the major endocrine glands. Clearly, the endocrine system differs from most of the other organ systems of the body in that the various glands are not in anatomic continuity with each other; however, they do form a system in the functional sense. It should also be noted that some of the glands are completely distinct organs (the pituitary, for example), whereas others are found within larger organs that have nonendocrine functions as well (the gonads and pancreas, for example).

The endocrine system constitutes the second great communications system of the body (the first being the nervous system), the hormones serving as blood-borne messengers that control and integrate many bodily functions: reproduction (Chap. 27), organic metabolism and energy balance (Chap. 26), and mineral metabolism (Chap. 24).

The nervous and endocrine systems actually

function as a single interrelated system. The central nervous system, particularly the hypothalamus, plays a crucial role in controlling hormone secretion; conversely, hormones markedly alter neural function and strongly influence many types of behavior. These interrelationships form the area of study known as **neuroendocrinology.**

Table 18-1 summarizes the physiology of the major endocrine glands and the hormones they secrete. As emphasized above, the hormones function as components of the body's control systems; accordingly, we have chosen to describe their detailed physiology in later chapters in the context of the

control systems in which they participate. The aim of this chapter is to provide the foundation for these later descriptions by presenting: (1) the general characteristics and principles that apply to almost all hormones; (2) the anatomy of the simple endocrine glands (the anatomy of the hormone-secreting cells of the organs that have multiple functions—gonads, pancreas, kidneys, gastrointestinal tract, and thymus—is described in subsequent chapters with the overall anatomy of these organs); and (3) the types of direct input that act upon endocrine-secreting cells to cause production and release of the hormones into the blood.

TABLE 18-1   SUMMARY OF THE MAJOR HORMONES (*Continued on next page*)

Gland	Hormone	Major Function Is Control of:
Hypothalamus	Releasing hormones Oxytocin Antidiuretic hormone	Secretion by the anterior pituitary (see posterior pituitary) (see posterior pituitary)

Anterior pituitary	Growth hormone (somatotropin, GH, STH)*
	Thyroid-stimulating hormone (TSH, thyrotropin)
	Adrenocorticotropic hormone (ACTH, corticotropin)
	Prolactin
	Gonadotropic hormones:
	  Follicle-stimulating hormone (FSH)
	  Luteinizing hormone (LH)
		Growth; organic metabolism
		Thyroid gland
		
		Adrenal cortex
		
		Breasts (milk synthesis)
		Gonads (gamete production and sex hormone synthesis)

Posterior pituitary†	Oxytocin Antidiuretic hormone (ADH, vasopressin)	Milk secretion; uterine motility Water excretion
Adrenal cortex	Cortisol Androgens Aldosterone	Organic metabolism; response to stresses Growth and, in women, sex drive Sodium and potassium excretion
Adrenal medulla	Epinephrine Norepinephrine	Organic metabolism; cardiovascular function; response to stresses
Thyroid	Thyroxine (T-4) Triiodothyronine (T-3) Calcitonin	Energy metabolism; growth Plasma calcium
Parathyroids	Parathyroid hormone (parathormone, PTH, PH)	Plasma calcium and phosphate

* The names and abbreviations in parentheses are synonyms.
† The posterior pituitary stores and secretes these hormones; they are synthesized in the hypothalamus.

TABLE 18-1	SUMMARY OF THE MAJOR HORMONES *(Continued)*	
Gland	**Hormone**	**Major Function Is Control of:**

Gonads:
  Female:
    ovaries
  Male: testes
	Estrogens
	Progesterone
	Testosterone
		Reproductive system; growth and
		  development; breasts
		Reproductive system; growth and
		  development

Pancreas	Insulin Glucagon Somatostatin	Organic metabolism; plasma glucose
Kidneys	Renin Erythropoietin (ESF) 1,25-Dihydroxyvitamin D_3	Adrenal cortex; blood pressure Erythrocyte production Calcium balance

Gastrointestinal
  tract
	Gastrin
	Secretin
	Cholecystokinin
	Gastric inhibitory peptide
	Somatostatin
		Gastrointestinal tract; liver; pancreas;
		  gallbladder

Thymus	Thymus hormone (thymosin)	Lymphocyte development
Pineal	Melatonin	?Sexual maturity

Note: This table is by no means comprehensive, mainly because we have left out those substances suspected of being hormones but not conclusively established to have a function in human beings. Perhaps the most interesting of these omissions are the endorphins, whose roles as neurotransmitters are described in Chap. 16 but which are also secreted by the pituitary. Also not shown are the hormones secreted by the placenta (see Chap. 27).

HORMONE TARGET-CELL SPECIFICITY

Hormones travel in the blood and are therefore able to reach virtually all tissues. This is obviously very different from the efferent nervous system, which can send messages selectively to specific organs. Yet the body's response to hormones is not all-inclusive but highly specific, in some cases involving only one organ or group of cells. In other words, despite the ubiquitous distribution of a hormone via the blood, only certain cells are capable of responding to the hormone; they are known as **target cells.** Cells respond in a highly characteristic manner only to certain hormones; this ability to respond depends upon the presence of specific receptors for those hormones. As described in Chap. 7, the receptor is the site in the target cell of the first interaction of the hormone with the cell. Specialization of target-cell receptors explains the specificity of action of hormones.

GENERAL FACTORS THAT DETERMINE THE PLASMA CONCENTRATIONS OF HORMONES

Rate of Secretion

With few exceptions, hormones are not secreted at constant rates. Indeed, as emphasized previously, it is essential that any regulatory system be capable of altering its output, and this is the case for the hormonal components of the body's regulatory systems, just as it is true for the neural components. Most, if not all, hormones are released in short bursts with little or no release occurring between bursts; accordingly, the blood concentrations of hormones may fluctuate rapidly over brief time periods. When an endocrine cell is stimulated by appropriate inputs, the bursts of hormone release occur more frequently so that the average plasma concentration of the hormone increases. In contrast, in the absence of stimulation or in the pres-

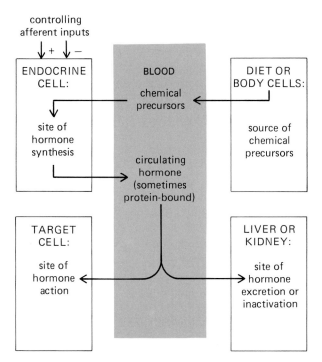

FIG. 18-2 General factors that determine blood concentrations of hormones.

or its metabolites is directly proportional to the rate of glandular secretion.

Transport in the Blood

Most of the nonprotein hormone molecules which circulate in the blood are bound to various plasma proteins; the free moiety may be quite small and is in equilibrium with the bound fraction:

"Free" hormone + protein $\rightleftharpoons$ hormone-protein

It is important to realize that only the free hormone can exert effects on the target cells.

Activation of Hormones after Secretion

Earlier in this section we mentioned metabolic transformation of the hormone in terms of its inactivation. Certain hormones, however, are relatively inactive in the forms released from their endocrine glands and must undergo further molecular alteration before they are able to exert their hormonal effects; i.e., the metabolism of a hormone may yield molecules that are more active, rather than less, compared to the original hormone. An example of this phenomenon is provided by testosterone, which is converted to dihydrotestosterone by certain of its target cells; the latter steroid then elicits the overall response of the target cell.

Particularly interesting (and confusing) is the possibility that different metabolites of the same hormone may be the mediators of different effects of that hormone. For example, several different polypeptide fragments of the protein hormone, growth hormone, circulate simultaneously in the plasma, and it has been extremely difficult to ascertain whether the intact growth-hormone molecule acts on target cells or whether the fragments are the active agents, each perhaps exerting its own unique effect.

Wherever such activation through molecular transformation is required, deficiency of the enzymes that mediate the activation may constitute an important source of malfunction in disease. For example, a deficiency in the relevant target cells of the enzyme that converts testosterone to dihydrotestosterone would result in the failure of testosterone to act on those target cells; the end result for those cells would be the same as if no testosterone were being secreted.

ence of active inhibitory inputs, the bursts decrease in frequency or stop altogether and the plasma concentration of the hormone decreases. This pattern is quite analogous to the phenomena of facilitation and inhibition manifested by neural control mechanisms.

Rates of Inactivation and Excretion

The concentration of a hormone in the plasma depends not only upon the rate of secretion but also upon the rate of removal from the blood by either excretion or metabolic transformation (Fig. 18-2). Sometimes the hormone is inactivated by the cells upon which it acts, but for most hormones, the pathway of removal from the blood is the liver or the kidneys. Accordingly, patients with kidney or liver disease may suffer from excess of certain hormones solely as a result of reduced hormone inactivation.

The rate of urinary excretion is often used as an indicator of secretory rate because for many hormones the rate of urinary excretion of the hormone

MECHANISMS OF HORMONE ACTION

Cell Responses to Hormones

The end result of a hormone's stimulation of its target cells is an alteration in the rate at which the target cells perform a specific activity: muscle cells increase or decrease contraction; epithelial cells (and other cell types as well) alter their rates of water or solute transport; and gland cells secrete more or less of their secretory products. The direct cause of many of these observed changes is a hormone-induced change in the activity of a crucial enzyme in the cell. What is meant by a "crucial enzyme" in this context? Recall that most metabolic reactions are truly reversible, whereas others proceed generally in one direction under the influence of one set of enzymes and in the reverse direction under the influence of a second set of enzymes. The enzymes that catalyze these "one-way" reactions are the primary ones regulated by various hormones. Let us consider, as an example, the relationship between glucose and glycogen in the liver:

$$\text{Glucose} \underset{\text{enzyme B}}{\overset{\text{enzyme A}}{\rightleftharpoons}} \text{glycogen}$$

Although there are actually multiple steps in both pathways, each catalyzed by a different enzyme, two enzymes (A + B) are particularly critical because they catalyze the major irreversible reactions in opposing directions. The hormone insulin increases the activity of enzyme A and reduces that of enzyme B; the result is increased formation of glycogen from glucose. In contrast, the hormone epinephrine increases the activity of enzyme B and decreases that of enzyme A; the result is facilitation of the catabolism of glycogen to glucose.

How is the activity of a particular type of enzyme increased? One way is for the cell to produce more of the enzyme. In Chap. 3 we described the mechanics of protein synthesis and the control of these processes. Hormones may exert effects on the genetic apparatus of their target cells to induce (or repress) the synthesis of RNA and, in turn, the proteins (enzymes) whose synthesis is directed by the particular RNA. This may well be the major biochemical action of most steroid hormones (although it is by no means limited to steroid hormones).

There are, however, other ways by which enzyme activity may be altered without a change in the total number of enzyme molecules in the cells, i.e., with no change in enzyme synthesis. Many enzymes exist in both active and inactive forms; thus, the number of active enzymes can be increased by converting some of the inactive molecules into active ones, and many hormones influence this conversion.

Certain hormones induce both the synthesis of new enzyme molecules and an increased activity of the enzyme molecules already present. The advantages of this dual effect are considerable: Induction of new enzyme synthesis requires hours to days, whereas the activation of molecules already present can occur within minutes. Thus, the hormone simultaneously exerts a rapid effect and sets into motion a long-term adaptation.

As widespread as such hormone-induced enzyme changes are, it is not always possible to explain the cell's response in these terms. For example, the facilitation of glucose uptake induced in its target cells by insulin does not seem to be caused by an enzyme change but rather by an alteration in the carrier process itself. This is true for several other hormone-induced changes in membrane transport. Changes in muscle tone are another type of hormone-induced response not always explainable in terms of enzyme changes.

Hormone Receptors

We have thus far described only the end results of a hormone's stimulation of its target cells. However, this response is only the final event in a sequence triggered when the hormone interacts with the target cell. In all cases, this initial interaction occurs between hormone and target-cell receptors, i.e., molecules of the target cells that have a specific capacity to bind the hormone. As emphasized in earlier chapters, this "recognition" (binding) is made possible by the configuration of portions of the receptor molecules such that a "match" exists between them and the specific hormone. It is the presence of the hormone-receptor combination that initiates the chain of intracellular biochemical events that leads ultimately to the cell's overall response.

Where in the target cells are the hormone receptors located? The receptors for steroid hormones (and possibly several nonsteroid hormones as well) are soluble proteins in the cytoplasm; steroids are quite lipid-soluble and readily cross the plasma membrane to enter the cytoplasm and combine with their specific receptors. In contrast, the receptors for most nonsteroid hormones (proteins, peptides, and amines) are proteins in the plasma membranes of the target cells.

Receptor modulation is a normal component of hormonal control systems. It explains why certain hormones, when present chronically in large amounts, frequently manifest a reduction in their effectiveness. For example, many obese, but otherwise normal, people have high concentrations of insulin in their plasma (for reasons to be described in Chap. 26) but do not manifest any signs of insulin excess; this is, at least in part, because of a loss of receptors on insulin's target cells. One factor that causes this change is insulin itself! Thus, a persistently elevated plasma insulin concentration induces (by unknown biochemical events) a loss of insulin receptors, and this prevents the target cells from overreacting to the high hormone concentration. This clearly constitutes an important negative-feedback control over insulin's actions. Similar controls seem to be exerted by many other hormones over their specific receptors.

Events Elicited by Hormone-Receptor Binding

In a sense, we began this section on the mechanisms of hormone action at the end of the story, i.e., with the end result or overall response of the target cell to the hormone. We then jumped to the beginning of the story, the hormone receptors. Now we must fill in the middle, namely the events that lead from the hormone-receptor combination to the overall response; much of this still remains quite uncertain.

As described above, the receptors for most nonsteroid hormones are on the outer surface of the plasma membrane. For at least 12 of these hormones, the event triggered by hormone-receptor combination is activation of membrane-bound adenylate cyclase with subsequent generation of cyclic AMP inside the cell. Thus, these hormones all utilize the cAMP second-messenger system described in Chap. 7, the overall response representing the final event in a sequence of biochemical reactions. It is clear, however, that all the actions of all nonsteroid hormones are not mediated by activation of this system. In at least one case (insulin), inhibition of adenylate cyclase (with subsequent reduction of cellular cAMP) may be impor-

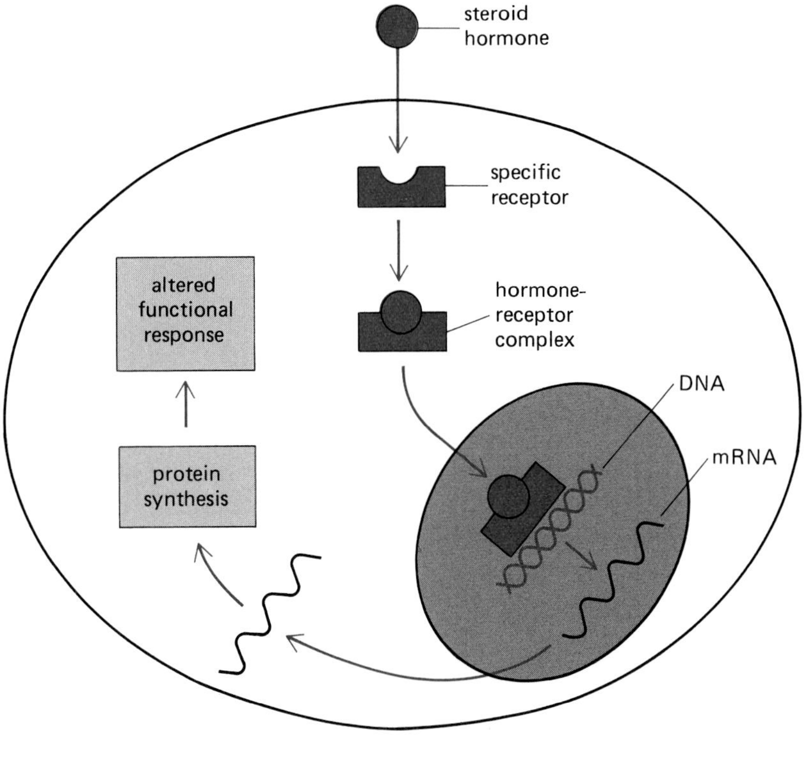

FIG. 18-3 Steroid hormones enter the target cell and bind to a specific cytoplasmic receptor. The hormone-receptor complex is then transported into the nucleus, where it binds to specific sites on the DNA molecule and activates transcription of new RNA, which mediates the response characteristic of the cell by directing protein synthesis.

tant; in other cases, cGMP (Chap. 7) may be involved; while in still others neither cAMP nor cGMP is adequate to explain the response and the explanation remains unknown.

Many polypeptide hormones can gain entry to the interior of cells, apparently by endocytosis. What they do after they have gained entry remains unknown. One hypothesis is that they (or the peptide fragments that are the result of their intracellular degradation) combine with intracellular receptors and exert long-term effects upon the cell.

Steroid Hormones   As mentioned above, the common denominator of the effects of the steroid hormones is an increased synthesis of proteins (enzymes, structural proteins, etc.) by their specific target cells. This increased protein synthesis is the result of stimulation of mRNA synthesis by the hormone-receptor complex. Recall that steroid hormones enter the cytoplasm of their target cells and combine with receptors (Fig. 18-3). This binding activates the receptor so that it moves into the cell's nucleus, carrying with it the bound hormone. There the bound hormone combines with a DNA-associated protein specific for it. This interaction causes the original receptor molecule to split, one fragment remaining bound to the DNA-associated protein while the other interacts with an adjacent segment of DNA. It is this last reaction that triggers transcription of the DNA segment, i.e., synthesis of mRNA, which can then enter the cytoplasm and serve as a template for synthesis of a specific protein. Note that this entire sequence depends upon several "recognitions": hormone with cytoplasmic receptor and receptor with nuclear acceptor. The latter is essential for proper positioning of the receptor-hormone complex along the DNA so that the appropriate gene is transcribed.

Hormone Interactions on Target Cells

Cells are constantly exposed to the simultaneous effects of many hormones. This allows for complex hormone-hormone interactions on the target cells, including inhibition, synergism, and the important phenomenon known as **permissiveness.** In general terms, frequently hormone A must be present for the full exertion of hormone B's effect. In essence, A is "permitting" B to exert its action. For example (Fig. 18-4), the hormone epinephrine causes marked

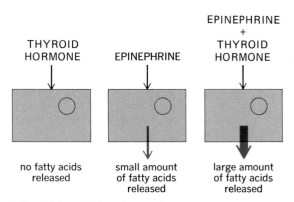

FIG. 18-4   Ability of thyroid hormone to permit epinephrine-induced liberation of fatty acids from adipose-tissue cells.

release of fatty acids from adipose tissue only in the presence of thyroid hormone. Many of the defects seen when an endocrine gland is removed or ceases to function because of disease actually result from loss of the permissive powers of the hormone secreted by that gland.

Pharmacologic Effects

Administration of very large quantities of a hormone may have results that are never seen in a normal person, although these so-called **pharmacologic effects** sometimes occur in endocrine diseases when excessive amounts of hormone are secreted. These effects are of great importance in medicine because hormones in pharmacologic doses are used as therapeutic agents. Perhaps the most famous example is that of the adrenal hormone cortisol, which is highly useful in suppressing allergic and inflammatory reactions. Mental changes, including outright psychosis, may also be induced by large quantities of cortisol and are frequently a striking symptom of patients suffering from hyperactive adrenal glands.

THE ENDOCRINE GLANDS: ANATOMY AND CONTROL OF SECRETION

The immediate inputs that control the secretion of hormones by the various endocrine glands fall into one of five categories, and an appreciation of them should greatly facilitate the understanding of each specific hormone as it is discussed in subsequent chapters. Table 18-2 summarizes the categories to

TABLE 18-2 SUMMARY OF THE CONTROL OF HORMONE SECRETION*

In response to neural, hormonal, or metabolic inputs, hypothalamic neurons themselves release:

Oxytocin ⎱
Antidiuretic hormone ⎰ (from posterior pituitary)
Hypothalamic releasing hormones

Hypothalamic releasing hormones directly control the release from the anterior pituitary of:

Growth hormone
Thyroid-stimulating hormone (TSH)
Adrenocorticotropic hormone (ACTH)
Gonadotropic hormones (FSH and LH)
Prolactin

Anterior pituitary hormones directly control the release of:

Thyroid hormone
Cortisol (from adrenal cortex)
Gonadal hormones
    (Female: estrogen and progesteron)
    (Male: testosterone)

Autonomic neurons directly control the release of:

Epinephrine and norepinephrine (from adrenal medulla)
Renin (from kidney)
Insulin and glucagon (from pancreas)
Gastrointestinal hormones
?Others

Plasma concentrations of ions or nutrients directly control the release of:

Parathyroid hormone
Insulin and glucagon (from pancreas)
Aldosterone (from adrenal cortex)
Calcitonin (from thyroid glands)

* This table does not necessarily list all the controls of each hormone.

be discussed; it must be reemphasized that all the material contained in this table will be presented again in later chapters where relevant. It should also be noted that the table does not constitute a complete inventory of all hormones or of the immediate inputs for any given hormone but includes only those that help to illustrate the common patterns.

The Pituitary

It is evident from Table 18-1 that the **pituitary gland** (or *hypophysis cerebri*) is of major importance in hormone secretion. This gland lies in the *sella turcica,* a pocket in the sphenoid bone just below the hypothalamus (Fig. 18-5) to which it is connected by a stalk (the **infundibulum**) containing nerve fibers and small blood vessels. It consists of three lobes, the **anterior, intermediate,** and **posterior lobes,** each of which is a more or less distinct gland. The term **adenohypophysis** refers to the anterior and intermediate lobes and their part of the stalk; **neurohypophysis** refers to the posterior lobe and its part of the stalk. In human beings, the intermediate lobe is rudimentary, and its function is unclear.

The Anterior Pituitary Hormones Despite its close proximity to the brain, the anterior pituitary is not neural but is composed of true glandular tissues, which produce at least six different polypeptide hormones. Secretion of each of the six hormones may occur independently of the others; i.e., the anterior pituitary comprises, in effect, six endocrine glands anatomically associated in a single structure. Until recently, the cells of the anterior pituitary were grouped mainly according to their propensities for being stained by certain dyes; those cells stained strongly by acid dyes are known as *acidophils,* or α *cells,* and those stained strongly by basic dyes are *basophils,* or β *cells.* These two groups comprise the chromophilic or "color-attracting" cells, in contrast to a third group, the *chromophobes,* which have little affinity for either dye. However, more recent evidence using histologic techniques specific for the different hormones has documented that these cell types are not homogeneous; i.e., there are distinct subgroups within the categories, each group responsible for the secretion of one (or, at most, two) hormones. Thus, different groups of acidophils secrete growth hormone, prolactin, or ACTH, and groups of basophils secrete TSH, FSH, and LH. The precise function of the chromophobes is unknown; they may represent nonsecretory phases of the other cell types.

The major function of two of the anterior pituitary hormones is to stimulate the secretion of other hormones: (1) **Thyroid-stimulating hormone** (TSH) induces secretion of thyroid hormones (thyroxine and triiodothyronine) from the thyroid; and (2) **adrenocorticotropic hormone** (ACTH), meaning "hormone that stimulates the adrenal cortex," is responsible for stimulating the secretion of cortisol. Thus, the important target organs for TSH and ACTH are the thyroid and adrenal cortex, respectively (whether these hormones have other functions elsewhere in the body remains controversial).

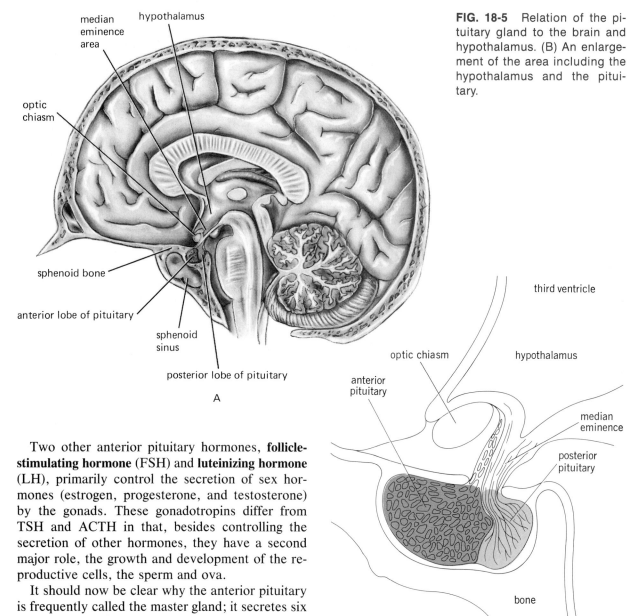

median eminence area

hypothalamus

optic chiasm

sphenoid bone

anterior lobe of pituitary

sphenoid sinus

posterior lobe of pituitary

A

FIG. 18-5 Relation of the pituitary gland to the brain and hypothalamus. (B) An enlargement of the area including the hypothalamus and the pituitary.

third ventricle

optic chiasm

hypothalamus

anterior pituitary

median eminence

posterior pituitary

bone

B

Two other anterior pituitary hormones, **follicle-stimulating hormone** (FSH) and **luteinizing hormone** (LH), primarily control the secretion of sex hormones (estrogen, progesterone, and testosterone) by the gonads. These gonadotropins differ from TSH and ACTH in that, besides controlling the secretion of other hormones, they have a second major role, the growth and development of the reproductive cells, the sperm and ova.

It should now be clear why the anterior pituitary is frequently called the master gland; it secretes six hormones itself[1] and controls the secretion of three or four (depending upon the person's sex) other hormones.

The two remaining anterior pituitary hormones have not been shown to exert major control over the secretion of other hormones. **Prolactin**'s major target organs are the breasts, and **growth hormone**

[1]It has also been found recently that both the intermediate and anterior pituitary contain large amounts of peptides known as *endorphins*. The possible hormonal role of these substances, which are also found in the central nervous system, is unknown.

exerts multiple metabolic effects upon many organs and tissues. The target organs and functions of the anterior pituitary hormones are summarized in Fig. 18-6. It should be noted that the tropic hormones of the anterior pituitary control not only the synthesis and secretion of their target-gland hormones but also the growth and development of the target glands themselves.

ANTERIOR PITUITARY

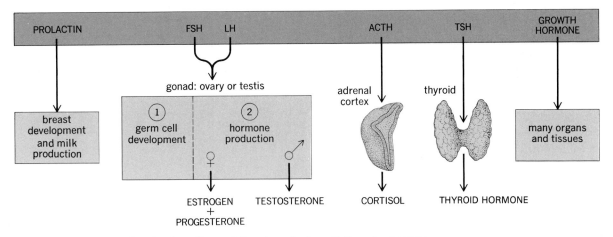

FIG. 18-6   Target and major functions of the six anterior pituitary hormones.

Control of Anterior Pituitary Secretion   What direct inputs control secretion of the anterior pituitary hormones? One important type of input for the tropic hormones is the target-gland hormone itself, a beautiful example of negative feedback (Chap. 7). Thus, ACTH stimulates cortisol secretion and increases the blood concentration of cortisol, which acts upon the anterior pituitary to inhibit ACTH release. In this manner, any increase in ACTH secretion is partially prevented by the resultant increase in cortisol secretion, as illustrated in Fig. 18-7. This same pattern of negative feedback is exerted by the thyroid hormones and the sex hormones on their respective pituitary tropic hormones (the sex hormone effects are actually more complex than those shown in the figure, as will be discussed

in Chap. 27). It is evident that such a system is highly effective in damping hormonal responses, i.e., limiting the extremes of hormone secretory rates; it also serves to maintain the plasma concentration of the hormone relatively constant whenever a primary change occurs in the catabolism of the hormone. If this negative-feedback relationship were the sole source of anterior pituitary control, however, there would be no way of altering anterior pituitary output; some unchanging equilibrium blood concentrations of pituitary and target-gland hormone would always be maintained. Obviously, there must be some other type of input to the anterior pituitary. In reality, this other input is the major controller of anterior pituitary function.

The major inputs that control release of anterior

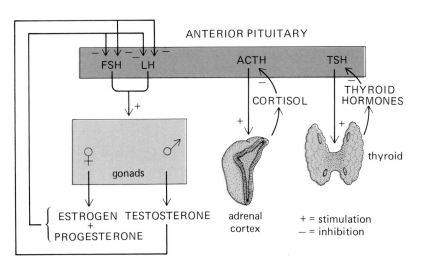

FIG. 18-7   Negative feedback of target-cell hormones on their respective anterior pituitary tropic hormones. The gonadal hormones actually exert complex effects on the pituitary as described in Chap. 27.

pituitary hormones are a group of so-called **releasing hormones** produced in the hypothalamus. An appreciation of the anatomic relationships between the hypothalamus and anterior pituitary is essential for understanding this process. Although the anterior pituitary lies just below the hypothalamus, there are no important neural connections between the two, but there is an unusual capillary-to-capillary connection. The superior hypophyseal arteries end in the base of the hypothalamus (the **median eminence**) as intricate capillary tufts that recombine into the **hypophyseal** (or hypothalamopituitary) **portal veins** (the term *portal* denotes veins that connect two distinct capillary beds). These pass down the stalk that connects the hypothalamus and pituitary and enter the anterior pituitary where they break into a second capillary bed, the *anterior pituitary capillaries,* which provide most of the vascular supply to that organ.

Thus the hypophyseal portal veins offer a local route for flow of capillary blood from hypothalamus to anterior pituitary. The axons of neurons that originate in diverse areas of the hypothalamus terminate in the median eminence around the capillary tufts, i.e., the origins of the portal vessels. These neurons secrete into the capillaries substances that are carried by the portal vessels to the anterior pituitary, where they act upon the various pituitary cells to control hormone secretion (Fig. 18-8).

We are dealing with multiple discrete hypothalamic substances, each secreted by a unique group of hypothalamic neurons and influencing the release of one or, in some cases, two of the six anterior pituitary hormones (Table 18-3 and Fig. 18-9). Because most of these substances stimulate release of their relevant hormones, they were originally termed *hypothalamic releasing factors;* however,

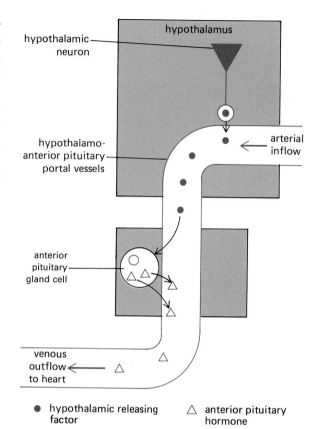

● hypothalamic releasing factor   △ anterior pituitary hormone

FIG. 18-8   Control of anterior pituitary secretion by a hypothalamic releasing hormone.

they fulfill the criteria for the definition of a hormone and are now generally called "hormones" rather than "factors." Each of the **hypothalamic releasing hormones** is named according to the anterior-pituitary hormone whose secretion it controls; for example, ACTH (also known as corticotropin) secretion is stimulated by **corticotropin releasing hormone** (CRH). As shown in Table 18-3

TABLE 18-3   MAJOR HYPOTHALAMIC RELEASING HORMONES AND THEIR EFFECTS ON THE ANTERIOR PITUITARY

Hypothalamic Releasing Hormone	Effect on Anterior Pituitary
Corticotropin releasing hormone (CRH)	Stimulates secretion of ACTH (corticotropin)
Thyrotropin releasing hormone (TRH)	Stimulates secretion of TSH (thyrotropin)
Growth hormone releasing hormone (GRH)	Stimulates secretion of GH
Somatostatin (also known as growth-hormone release inhibiting hormone, GIH)	Inhibits secretion of growth hormone
Gonadotropin releasing hormone (GnRH)	Stimulates secretion of LH and FSH
Prolactin releasing hormone (PRH)	Stimulates secretion of prolactin
Prolactin release inhibiting hormone (PIH)	Inhibits secretion of prolactin

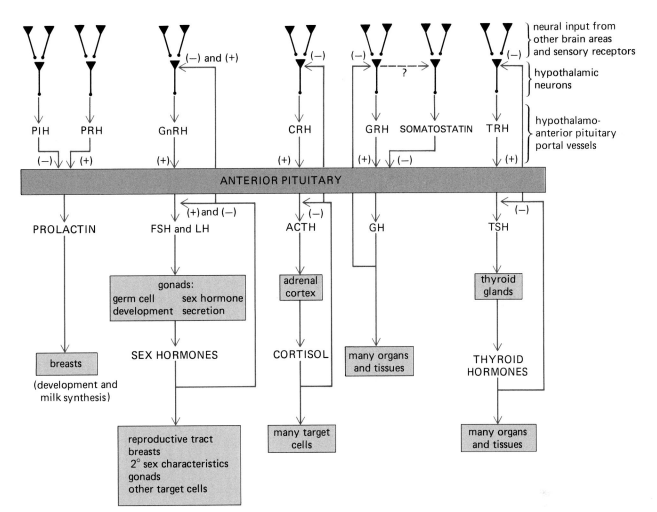

FIG. 18-9 Summary of hypothalamic-anterior-pituitary-target-organ relationships. The neural input to the hypothalamus from other brain areas and sensory receptors is both inhibitory and excitatory. It is not known whether somatostatin release is influenced in a feedback manner by GH (the ? in the diagram). The interactions between gonadal hormones and the hypothalamus–anterior pituitary are complex and will be described in Chap. 27. This model will certainly become more complicated as research progresses. For example, current evidence suggests that prolactin may exert a negative-feedback control over its own secretion via the hypothalamus and that this hormone may also be influenced by TRH as well as by PIH and PRH. Definitions of abbreviations are listed in Table 18-3.

and Fig. 18-9, at least two of the hypothalamic hormones inhibit, rather than stimulate, release of anterior pituitary hormones. One of these inhibits the secretion of prolactin and is termed **prolactin-release inhibiting hormone** (PIH). The other inhibits secretion of growth hormone and is commonly called **somatostatin.**

This system would be much easier to research if there were always a strict one-to-one correspondence of hypothalamic releasing hormone and anterior pituitary hormone, but this is not always the case, as is clear from Table 18-3 and Fig. 18-9. First, at least one of the hypothalamic hormones influences the secretion of more than one anterior-pituitary hormone; thus, **gonadotropin releasing hormone** (GnRH) controls the secretion of both LH and FSH. Secondly, dual systems of hypothalamic hormones, one stimulatory and the other inhibitory, operate on several anterior-pituitary hormones: Prolactin secretion is inhibited by hypothalamic PIH but is stimulated by a second hypothalamic hormone, PRH; the secretion of growth hormone is also controlled by opposing stimulatory and inhibitory hypothalamic releasing hormones. Clearly, in such dual-control cases, the overall anterior-pituitary response depends upon the relative amounts of the two opposing hormones released by the hypothalamic neurons. This is simply one more example of how the output of an endocrine gland cell may reflect the integration of multiple inputs to it.

Before leaving this subject it should be noted that only three of the hypothalamic releasing hormones have thus far been completely isolated, identified, and synthesized—those that have are small peptides. These same peptides have been found in locations other than the hypothalamus, particularly in other areas of the brain and in the gastrointestinal tract. For example, somatostatin is found in the stomach and pancreas (mainly in cells adjacent to other endocrine cells). This has raised the possibility that somatostatin is synthesized in these areas and serves locally to modulate the secretion of hormones and other substances from adjacent endocrine and exocrine cells. Of particular interest is the finding of these peptides in areas of brain other than hypothalamus; this has raised the possibility that they may serve as neurotransmitters or neuromodulators in the central nervous system.

We stated earlier that the nervous and endocrine systems actually function as a single interrelated system; the anterior pituitary may be the master gland, but its function is primarily controlled by the hypothalamus via the releasing hormones. It now appears that many diseases characterized by inadequate secretion of one or more pituitary hormones really are due to hypothalamic malfunction rather than primary pituitary disease.

Our analysis has pushed the critical question one step further: The hypothalamic releasing hormones control anterior pituitary function, but what controls secretion of the releasing hormones? The answer is neural and hormonal input to the hypothalamic neurons that secrete the releasing hormones. The hypothalamus receives neural input, both facilitory and inhibitory, from virtually all areas of the body. The specific type of input that controls the secretion rate of the individual releasing hormones will be described in future chapters when we discuss the relevant anterior pituitary or target-gland hormone. It suffices for now to point out that endocrine disorders may be generated, via alteration of hypothalamic activity, by all manner of

neural activity, such as stress, anxiety, etc. An example is sterility caused by severe emotional upsets.

Hormonal influences upon the hypothalamus are also important. Some of the negative-feedback effect of thyroid hormone, cortisol, and sex hormones upon pituitary tropic-hormone secretion is actually mediated via the hypothalamus, i.e., by inhibition of releasing-hormone secretion. For example, cortisol acts not only directly upon the anterior pituitary to inhibit ACTH secretion but also upon the hypothalamus to inhibit CRH secretion, an event which also reduces ACTH secretion. Presently, there is still controversy over the quantitative importance of these two negative-feedback sites for the various hormones, but the generalization that both sites are involved, albeit to varying extents for each hormone, seems warranted. In addition, it is quite likely that growth hormone and prolactin, the two anterior pituitary hormones that have no target-organ hormones, exert negative-feedback controls via the hypothalamus, over their own secretion. In this regard, recent evidence indicates that a significant portion of the blood flow from the pituitary actually goes back toward the central nervous system before entering the systemic circulation. Therefore, it is quite possible that pituitary hormones may reach the brain in high concentrations to influence the secretion of hypothalamic hormones, behavior, etc. Our description of the basic interrelationships of the hypothalamus, anterior pituitary, and target glands is now complete (Fig. 18-8).

Hypothalamus–Posterior-Pituitary Function   The **posterior pituitary** lies just behind the anterior pituitary in the same bony pocket in the sphenoid bone at the base of the hypothalamus, but its structure is totally different from its neighbor's. The posterior pituitary is actually an outgrowth of the hypothalamus and is true neural tissue. Two well-defined clusters of hypothalamic neurons, the *supraoptic* and *paraventricular nuclei,* send out nerve fibers that pass by way of the connecting stalk to end in the posterior pituitary in close proximity to capillaries (Fig. 18-10). The two hormones, **oxytocin** and **antidiuretic hormone** (ADH, also known as vasopressin), released from the posterior pituitary are actually synthesized in the hypothalamic cells. Enclosed in small vesicles, they move down the cytoplasm of the neuron axons to accumulate at the

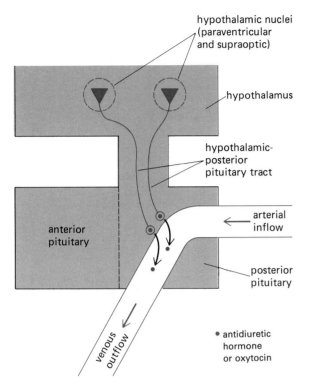

FIG. 18-10 Schematized relationship between the hypothalamus and posterior pituitary

nerve endings. Release into the capillaries occurs in response to generation of action potentials in the axon. Thus, these hypothalamic neurons secrete hormones in a manner quite analogous to that described previously for hypothalamic releasing hormones, the essential difference being that the releasing hormones are secreted into capillaries that empty directly into the anterior pituitary, whereas the posterior pituitary capillaries drain primarily into the general body circulation.

It is evident, therefore, that the term *posterior pituitary hormones* is somewhat of a misnomer because the hormones are actually synthesized in the hypothalamus, of which the posterior pituitary is merely an extension. We must therefore add two more names to our growing list of hormones under direct or indirect control of the hypothalamus.

The Adrenal Glands

The paired adrenal (or *suprarenal*) glands are small triangular organs, one on top of each kidney (Fig. 18-11); each measures only approximately 5 by 3

by 1 cm. Each adrenal comprises two distinct endocrine glands, an inner adrenal medulla and its surrounding adrenal cortex.

The **adrenal cortex,** which forms the larger part of the gland, secretes several steroid hormones, the most important of which are: (1) **aldosterone,** which is essential for the maintenance of sodium and potassium balance and is therefore known as a **mineralocorticoid;** (2) **cortisol,** known as a **glucocorticoid** because it has important effects, among others, on carbohydrate and protein metabolism; and (3) **sex steroids,** produced in small but significant quantities. The adrenal cortex is itself divided into three layers: a thin outer **zona glomerulosa,** a middle **zona fasciculata,** and an inner **zona reticularis** that abuts on the medulla (Fig. 18-11B, C). The cells of all three layers contain the abundant lipids typical of steroid-secreting cells; the zona glomerulosa synthesizes aldosterone, and the other zones produce cortisol (and the sex steroids). As described above, the secretion of cortisol is totally under the control of ACTH from the anterior pituitary, whereas aldosterone secretion is controlled by nonpituitary inputs discussed in Chap. 24.

The **adrenal medulla,** constituting only 10 percent of the total gland, contains *chromaffin cells* arranged in rows along the edges of wide venous sinuses. These cells are under the control of sympathetic preganglionic neurons and, in response to nerve stimulation, secrete their hormones into the extracellular space that surrounds the venous sinuses; the hormones pass through the sinus walls and enter the circulation. In human beings, the hormone released by the medulla is for the most part the amino acid derivative **epinephrine** (a smaller amount of **norepinephrine** is also secreted). In controlling epinephrine secretion, the adrenal medulla behaves just like a sympathetic ganglion and is dependent upon stimulation by the sympathetic preganglionic fibers. Destruction of these incoming nerves causes marked reduction of epinephrine release and failure to increase secretion in response to the usual physiologic stimuli.

The adrenal medulla is best viewed as a general reinforcer of sympathetic activity. Its secretion of epinephrine into the blood serves to increase the overall sympathetic functions of the body. We shall discuss in later chapters the specific reflexes that cause enhanced sympathetic activity and elicit epinephrine secretion; suffice it to say that these re-

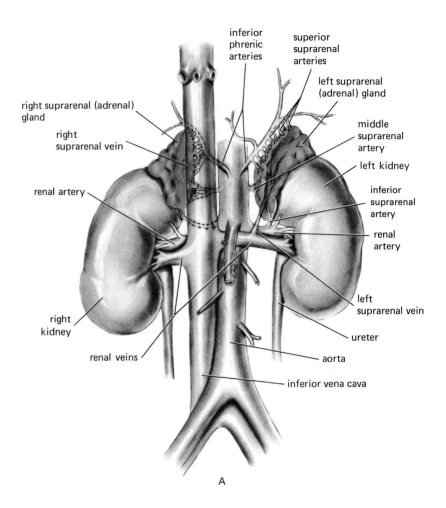

right suprarenal (adrenal) gland

right suprarenal vein

renal artery

right kidney

renal veins

inferior phrenic arteries

superior suprarenal arteries

left suprarenal (adrenal) gland

middle suprarenal artery

left kidney

inferior suprarenal artery

renal artery

left suprarenal vein

ureter

aorta

inferior vena cava

A

flexes are under the strong control of higher brain centers, including the hypothalamus.

Role of the Autonomic Nervous System in Hormone Secretion

We have seen that one hormone, epinephrine, is actually a product of the sympathetic nervous system. This is not the only influence the autonomic nervous system has on the endocrine system. Certain of the endocrine glands receive a rich supply of sympathetic and/or parasympathetic neurons, whose activity influences their rates of hormone secretion. As shown in Table 18-2, examples are the secretion of renin by the kidneys, insulin and glucagon by the pancreas, and the gastrointestinal hormones. In all cases studied, except the adrenal medullary hormones, the autonomic input does not serve as the sole regulator of the hormone's secretion but serves rather as only one of multiple inputs.

The Thyroid Gland

The **thyroid gland** secretes two iodine-containing amino acid derivatives, **thyroxine** and **triiodothyronine** (collectively known as **thyroid hormone**), which have widespread effects on energy metabolism, growth and development, and brain function. The thyroid is located in the lower part of the neck in front of the trachea (Fig. 18-12A). It consists of *right* and *left lobes* connected by a narrow *isthmus,* all of which are covered by a thin fibrous connective-tissue capsule. The basic unit of the thyroid is a **follicle,** which is a sphere of epithelial **follicular cells** surrounding a semifluid material known as **colloid** (Fig. 18-12B). Surrounding the follicles is a loose connective tissue in which **parafollicular cells** (also called *C,* for clear, or light, cells) are embedded. These cells, which are the second major cell type of the thyroid, secrete still another hormone, **calcitonin.**

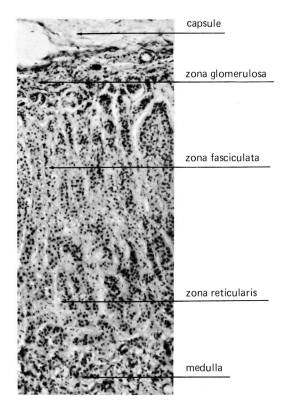

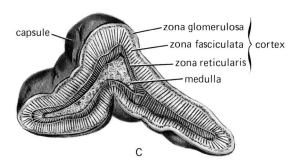

FIG. 18-11 (A) Location of adrenal (suprarenal) glands. (B) Photomicrograph of a section through adrenal cortex (a small section of the medulla can be seen at the bottom). (C) Section through adrenal gland.

Thyroid physiology received its greatest stimulus when it was discovered that the enlarged thyroids (goiters) so common among inland populations were completely preventable by the administration of small quantities of iodine, as little as 4 g/year. The other ingredient for thyroxine synthesis is the amino acid tyrosine, which can be produced from a wide variety of substances in the body and therefore offers no supply problem.

Much of the ingested iodine absorbed by the gastrointestinal tract (which converts it to iodide, the ionized form of iodine) is removed from the blood by the follicular cells of the thyroid, which manifest a remarkably powerful active-transport mechanism for iodide. Once in the cells, the iodide is converted to an active form of iodine, which is then attached to certain tyrosyl residues of a large glycoprotein known as *thyroglobulin,* which is in the colloid of the follicle. Some of the iodinated tryrosyl residues then become linked together to form thyroid hormone residues of this protein. The normal gland may store several weeks' supply of thyroid hormone in this bound form.

Hormone release into the blood occurs by enzymatic splitting of the residues from the thyroglobulin, and the entry of this freed hormone into the blood. Finally, as described above, the overall process is controlled by a pituitary **thyroid-stimulating hormone,** which stimulates certain key rate-limiting steps and thus alters the rate of thyroid hormone secretion. A variety of defects—dietary, hereditary, or disease-induced—may decrease the amount of hormone released into the blood. One such defect results from dietary iodine deficiency. However, this deficiency need not lead to permanent reduction of thyroid hormone secretion because the thyroid gland enlarges (goiter) so as to provide greater utilization of whatever iodine is available. This response is mediated by thyroid-stimulating hormone, which induces thyroid enlargement whenever the blood concentration of thyroid hormone decreases; the blood concentration of thyroid-stimulating hormone increases because of diminished feedback inhibition secondary to lowered blood thyroid hormone.

Once in the blood, most of the thyroid hormone becomes bound to certain plasma proteins and circulates in this form. This protein-bound hormone is in equilibrium with a much smaller amount of free hormone, the latter being the effective hormone. Since most of the iodine in the plasma is in the thyroid-hormone molecules, and since most of the thyroid hormone is bound to protein, a useful estimate of the level of circulating thyroid hormone is gained by measuring the plasma protein-bound iodine (PBI).

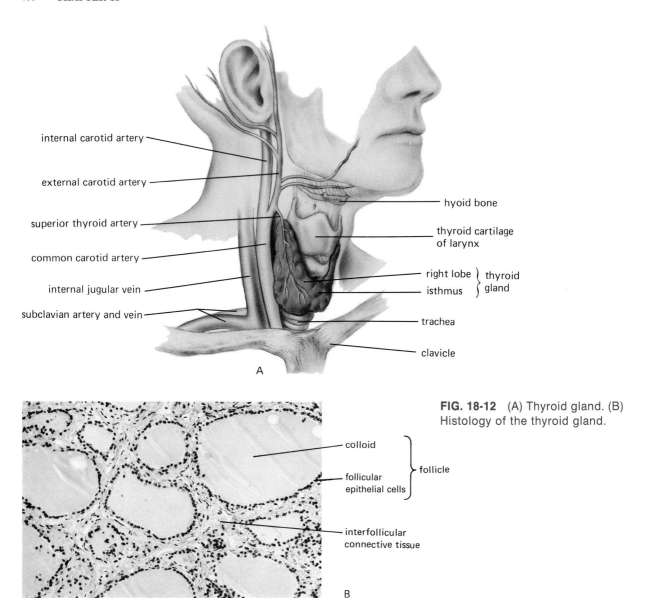

internal carotid artery

external carotid artery

superior thyroid artery

common carotid artery

internal jugular vein

subclavian artery and vein

hyoid bone

thyroid cartilage of larynx

right lobe } thyroid

isthmus } gland

trachea

clavicle

A

FIG. 18-12 (A) Thyroid gland. (B) Histology of the thyroid gland.

colloid

follicular epithelial cells } follicle

interfollicular connective tissue

B

Finally, it is not yet clear whether thyroxine itself acts upon its target cells or whether it must first be metabolized by the cells to a different molecule (probably triiodothyronine, T_3) that then serves as the active form of the hormone; increasing evidence favors the latter possibility.

The Parathyroid Glands

The **parathyroid hormone** is one of the most important regulators of plasma calcium and phosphate concentrations (Chap. 24). The **parathyroid glands** are very small oval structures that lie along the back of the thyroid gland between the gland and its capsule (Fig. 18-13). Usually there are four parathyroids, two behind each lobe of the thyroid. Connective tissue surrounds the parathyroids, forming a thin capsule, and passes into the substance of the gland, carrying with it the large blood vessels, nerves, and lymphatics. The glandular cells themselves (the **principal** or **chief cells**), which secrete parathyroid hormone, are arranged in columns separated by a rich network of sinusoidal capillaries.

Parathyroid hormone is the first hormone in our

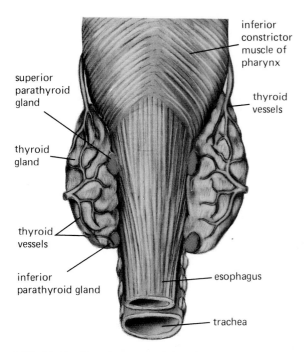

FIG. 18-13 Esophagus and thyroid gland viewed from the back. The parathyroid glands (usually four in number) are on the posterior surface of the thyroid.

discussion whose secretion is not regulated mainly by the pituitary, hypothalamus, or sympathetic nervous system. Rather, the cells of the parathyroid gland respond directly to the calcium concentration of the blood supplying them. Since the major function of parathyroid hormone is to regulate plasma calcium concentration, this control mechanism is appropriate and remarkably simple: The hormone-secreting cells are themselves sensitive to the plasma substance their hormone regulates.

As shown in Table 18-2, this type of pattern exists not just for parathyroid hormone but for other hormones as well. Thus, the secretion of insulin and glucagon is analogous in that the pancreatic cells that produce these hormones are controlled, in large part, directly by the glucose concentration of the plasma supplying them. The adaptive rationale is also analogous in that a major function of insulin and glucagon is regulation of the plasma glucose concentration. Similarly, aldosterone helps to regulate plasma potassium concentration, and the adrenal cells that secrete it are directly controlled, at least in part, by the plasma potassium concentration.

Summary

Table 18-2 should now appear less formidable. The control mechanisms for most of the hormones involve the direct or indirect participation of the hypothalamus and pituitary. The anterior pituitary hormones are controlled primarily by releasing hormones secreted into the hypothalamopituitary portal vessels by neurons in the hypothalamus. In turn, the four anterior pituitary tropic hormones control hormone secretion by their target glands—the thyroid, adrenal cortex, and gonads. These glands exert a negative-feedback control over their own secretion via the effects of their hormones on both the hypothalamus and anterior pituitary. The hypothalamus produces two other hormones, oxytocin and ADH, which are released from nerve endings in the posterior pituitary. And finally, the hypothalamus exerts profound control over the autonomic nervous system, including the adrenal medulla. The hypothalamus is, of course, not the only brain area controlling autonomic outflow. Thus, this small area of brain, which weighs 4 g in the adult human being, acts as a compact integrating center that receives messages, both neural and hormonal, from all areas of the body and sends out efferent messages via both the nerves and hormones. As we shall see, it also regulates body temperature, food intake, water balance, and a host of other autonomic, endocrine, and behavioral activities.

Despite the central role of the hypothalamopituitary system and the autonomic nervous system, there are important hormones whose secretion is controlled, at least in part, by distinct mechanisms. These include aldosterone (from the adrenal cortex), insulin and glucagon (from the pancreas), parathyroid hormone, several kidney hormones, and a group of gastrointestinal hormones.

THE PROBLEM OF MULTIPLE HORMONE SECRETION

The phenomenon of multiple hormone secretion by a single gland is clearly evident in Table 18-1. In certain glands, such as the pancreas, the hormones are secreted by completely distinct cells; in other glands, it is likely that, although some separation of function exists, a single cell may secrete more than one hormone. Even in such cases, it is essential to realize that each hormone has its own unique control mechanism; i.e., there is no massive undifferentiated release of the multiple hormones.

A frequent source of confusion concerning the endocrine system involves not multiple hormone secretion by a single gland but the mixed general functions exhibited by the gonads, pancreas, kidneys, and gastrointestinal tract. All these organs contain endocrine gland cells but also perform other completely distinct nonendocrine functions. For example, most of the pancreas is concerned with the production of digestive enzymes. These enzymes are produced by exocrine glands, i.e., the secretory products are transported not into the blood but into ducts that lead, in this case, into the intestinal tract. The endocrine function of the pancreas is performed by completely distinct nests of endocrine cells, the islets of Langerhans, which are scattered throughout the pancreas. This pattern is true for all the organs of mixed function; the endocrine function is always subserved by gland cells distinct from the other cells that constitute the organ.

KEY TERMS

endocrine gland	pharmacologic effects	adrenal glands
hormone	pituitary gland	adrenal medulla
target cell	anterior pituitary	adrenal cortex
hormone receptor	posterior pituitary	thyroid gland
receptor modulation	hypothalamic releasing hormones	parathyroid gland
permissiveness		

REVIEW EXERCISES

1 State the four general chemical categories of hormones.
2 State the basis of hormone target-cell specificity.
3 List the endocrine glands.
4 Describe the general factors that determine the plasma concentrations of hormones.
5 State the ways in which hormones can be altered following their secretion.
6 Describe the general types of cell responses to hormones.
7 Describe how a hormone can cause changes in an enzyme's concentration and activity.
8 State where the receptors for the different classes of hormones are located.
9 Describe the role of cyclic AMP in the action of hormones.
10 Describe the sequence of events triggered by the combination of a steroid hormone with a receptor for it.
11 Describe and contrast permissive and pharmacologic effects of hormones.
12 List the five general paths involved in the control of hormone secretion.

13 Describe the anatomy of the pituitary gland and list the hormones produced by the anterior and posterior lobes; give a one-sentence description of each hormone's function.
14 Describe the hypothalamic control of anterior pituitary hormone secretion; list the major hypothalamic releasing hormones and their effect on the anterior pituitary.
15 Describe the negative-feedback effects of hormones on the hypothalamic-pituitary system.
16 State the origin of the two posterior-pituitary hormones and the immediate stimulus for their release.
17 State the zones of the adrenal cortex and the hormones they produce.
18 Describe the anatomy of the thyroid glands, the hormones they produce, and the hormones' metabolism.
19 Define the following terms: "one-way" reactions, adenohypophysis, neurohypophysis, acidophils (alpha cells), basophils (beta cells), chromophils, chromophobes, median eminence, hypophyseal portal veins.

PART

4

TRANSPORT AND PROCESSING OF MATERIALS

CHAPTER

19

BLOOD

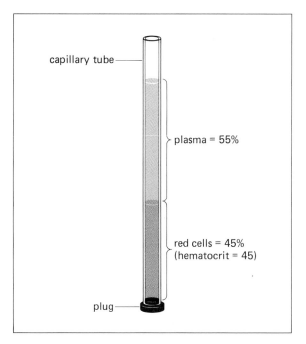

capillary tube

plasma = 55%

red cells = 45%
(hematocrit = 45)

plug

FIG. 19-1 If a blood-filled capillary tube is centrifuged, the red cells become packed in the lower portion, and the percentage of red cells (hematocrit) can be determined.

Blood is composed of specialized cellular elements and the liquid, **plasma,** in which they are suspended. The cells are the **erythrocytes,** or red blood cells; the **leukocytes,** or white blood cells; and the **platelets** (which are really cell fragments).

Ordinarily, the constant motion of the blood keeps the cells well dispersed throughout the plasma, but if a sample of blood is allowed to stand (clotting prevented), the cells slowly sink to the bottom. This process can be speeded up by centrifuging (Fig. 19-1). By this means, the percentage of the total blood volume that is cells, known as the **hematocrit,** can be determined. The normal hematocrit is approximately 45 percent. (Since the vast majority of all blood cells are erythrocytes, the total cell volume is approximately equal to the erythrocyte volume.) The total blood volume of an average person is approximately 8 percent of his or her total body weight. Accordingly, for a 70-kg person:

Total blood weight = 0.08 × 70 kg = 5.6 kg
One kilogram of blood occupies approximately 1 L;
    therefore, total blood volume = 5.6 L

The hematocrit is 45 percent; therefore
    total cell volume = 0.45 × 5.6 L = 2.5 L
Plasma volume = 5.6 − 2.5 L = 3.1 L

PLASMA

Plasma is an extremely complex liquid. It consists of a large number of organic and inorganic substances dissolved in water (Table 19-1). The most abundant of these substances by weight are the proteins, which together compose approximately 7 percent of the total plasma weight. The **plasma proteins** vary greatly in their structure and function, but they can be classified, according to certain physical and chemical reactions, into three broad groups: the **albumins, globulins,** and **fibrinogen.** The albumins are three to four times more abundant than the globulins and usually are of smaller molecular weight. The plasma proteins, with notable exceptions, are synthesized by the liver, the major exception being the group known as **immune globulins,** which are formed in the lymph nodes and other lymphoid tissues.

The plasma proteins serve a host of important functions that will be described in relevant chapters, but it must be emphasized that normally they are *not* taken up by cells and utilized as metabolic fuel. Accordingly, they must be viewed quite differently from most other organic constituents of plasma, such as glucose, which use the plasma as a vehicle for transport but function in cells. The plasma proteins function in the plasma itself or, under certain circumstances, in the interstitial fluid. Finally, plasma is distinguished from **serum,** which is plasma from which fibrinogen has been removed as a result of clotting.

In addition to the organic solutes—proteins, nutrients, and metabolic end products—plasma contains a large variety of mineral electrolytes, the concentrations of which are shown in Table 19-1. Comparison of the concentrations in millimols per liter for these electrolytes and protein may seem puzzling in view of the statement that protein is the most abundant plasma solute by *weight*. Remember, however, that molarity is a measure not of the weight but of the *number* of molcules or ions per unit volume. Protein molecules are so large in comparison to sodium ions that a very small number of them greatly outweighs a much larger number of sodium ions.

TABLE 19-1   CONSTITUENTS OF ARTERIAL PLASMA

Constituent	Amount/Concentration	Major Functions
Water	93% of plasma weight	Medium for carrying all other constituents
Electrolytes (inorganic):	Total < 1% of plasma weight	Keep H_2O in extracellular compartment; act as buffers; function in membrane excitability
Na$^+$	142 mM	
K$^+$	4 mM	
Ca^{2+}	2.5 mM	
Mg^{2+}	1.5 mM	
Cl$^-$	103 mM	
HCO$^-_3$	27 mM	
Phosphate (mostly HPO$_4{}^{2-}$)	1 mM	
SO$_4{}^{2-}$	0.5 mM	
Proteins:	Total = 7.3 g/100 mL (2.5 mM)	Provide nonpenetrating solutes of plasma; act as buffers; bind other plasma constitutents (lipids, hormones, vitamins, metals, etc.); clotting factors; enzymes, enzyme precursors; antibodies (immune globulins); hormones
Albumins	4.5 g/100 mL	
Globulins	2.5 g/100 mL	
Fibrinogen	0.3 g/100 mL	
Gases:		
CO$_2$	2 mL/100 mL plasma	
O$_2$	0.2 mL/100 mL	
N$_2$	0.9 mL/100 mL	
Nutrients:		
Glucose and other carbohydrates	100 mg/100 mL (5.6 mM)	
Total amino acids	40 mg/100 mL (2 mM)	
Total lipids	500 mg/100 mL (7.5 mM)	
Cholesterol	150–250 mg/100 mL (4–7 mM)	
Individual vitamins	0.0001–2.5 mg/100 mL	
Individual trace elements	0.001–0.3 mg/100 mL	
Waste products:		
Urea	34 mg/100 mL (5.7 mM)	
Creatinine	1 mg/100 mL (0.09 mM)	
Uric acid	5 mg/100 mL (0.3 mM)	
Bilirubin	0.2–1.2 mg/100 mL (0.003–0.018 mM)	
Individual hormones	0.000001–0.05 mg/100 mL	

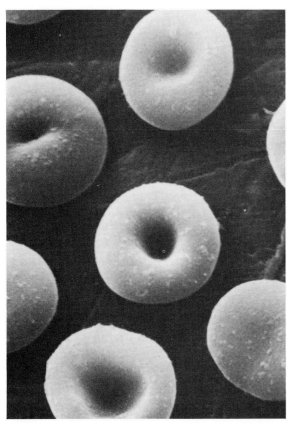

FIG. 19-2 Scanning electron micrograph of erythrocytes.

CELLULAR ELEMENTS OF THE BLOOD

Erythrocytes

Each cubic millimeter of blood contains approximately 5 million **erythrocytes,** and the total number in the human body is about 25 trillion. These cells are shaped like a biconcave disc, i.e., thicker at the edge than in the middle, like a doughnut with a center depression on each side instead of a hole (Fig. 19-2). This shape and their small size (7 μm in diameter) have adaptive value in that oxygen and carbon dioxide, carriage of which is the major function of the erythrocytes, can rapidly diffuse throughout the cell interior. The plasma membranes of erythrocytes contain specific proteins that differ from person to person, and these confer upon the blood immunological characteristics that are classified according to so-called blood type, as will be described in Chap. 28.

The outstanding physiologic characteristic of

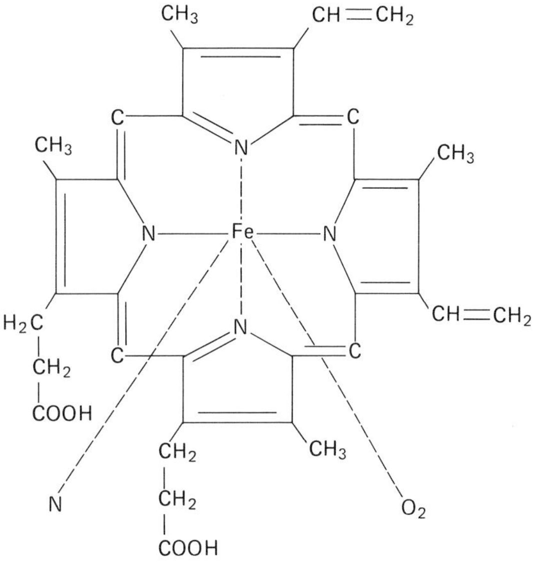

FIG. 19-3 Heme. The nitrogen, N, at the lower left attaches to a polypeptide chain of globin to form a subunit of hemoglobin. Four of these make a single hemoglobin molecule.

erythrocytes is the presence of the iron-containing protein **hemoglobin,** which binds oxygen; hemoglobin constitutes approximately one-third of the total cell weight. Each hemoglobin molecule is made up of four subunits, each consisting of a **heme** portion conjugated to a polypeptide; the four polypeptides are collectively called **globin.** The heme portion of the molecule (Fig. 19-3) contains iron (Fe^{2+}), and it is this iron that binds oxygen.

The amino acid sequences in the polypeptide chains are, of course, genetically determined, and mutant genes cause the production of abnormal hemoglobins, some of which may cause serious malfunction. For example, **sickle cell anemia** is caused by the presence of a particular form of abnormal hemoglobin molecules that, at the low oxygen concentrations that exist in many capillaries, interact with each other to form fiberlike structures; these distort the erythrocyte membrane and cause the cell to form sickle shapes or other bizarre forms. The result is both blockage of the capillaries, with consequent tissue damage, and destruction of the deformed erythrocytes, with consequent anemia. All this because of a change in one of 287 amino acids in the globin molecule!

Another erythrocyte substance of great importance is the enzyme carbonic anhydrase, which, as we shall see, facilitates the transportation of carbon dioxide.

Erythrocyte and Hemoglobin Balance Erythrocyte volume and hemoglobin content are subject to physiologic control. Erythrocytes are incomplete cells in that they lack nuclei and the metabolic machinery to synthesize new proteins. Thus, they can neither reproduce themselves nor maintain their normal structure for any length of time. As essential enzymes in them deteriorate and are not replaced, the cells age and ultimately die. Fortunately, their oxygen-carrying ability is not significantly diminished during the aging period.

The average life-span of an erythrocyte is approximately 120 days, which means that almost 1 percent of the total erythrocytes in the body are destroyed every day. Destruction of erythrocytes is accomplished by a group of large phagocytic cells, the **macrophages,** found in liver, spleen, bone marrow, and lymph nodes, usually lining the blood vessels or lying close to them. These cells, the physiology of which will be described in detail in Chap. 28, ingest and destroy the erythrocytes by breaking down their large complex molecules with lysosomal enzymes.

In the process, hemoglobin molecules are split to yield heme and the polypeptide chains of globin. Iron is removed from the heme, enters the blood, and is transported to the bone marrow (and other cells) to be reused in the synthesis of new heme groups. The remainder of the heme is broken down and converted to a yellow molecule named **bilirubin,** which is released into the blood. The liver cells pick up this substance from the blood and add it to the bile, but if the liver is damaged or overloaded because of an abnormally high rate of erythrocyte destruction, the bilirubin may accumulate in the blood and give a yellow color to the skin. This is called **jaundice.**

Obviously, in a normal person, a quantity of erythrocytes equal to that destroyed must be simultaneously produced and released into the circulatory system. The site of erythrocyte production (**erythropoiesis**) is the bone marrow. The erythrocytes are descended from bone marrow stem cells called **erythrocyte progenitors.** These cells contain no hemoglobin but do have nuclei and are capable of cell division. After several divisions, cells emerge that are identifiable as immature erythro-

cytes because they contain hemoglobin. As maturation continues, these cells accumulate increased amounts of hemoglobin, and their nuclei become progressively smaller until the cells finally divide in two and the fragment that contains the nuclear portion is phagocytized. Even after losing its nucleus, a newly produced erythrocyte still contains some ribosomes that produce a weblike appearance when treated with special stains; this appearance gives the young red blood cells the name **reticulocyte.** The mature erythrocytes leave the bone marrow and enter the general circulation, in which they will travel for some 120 days.

This growth process requires the usual nutrients: amino acids, lipids, and carbohydrates. In addition, certain growth factors, including vitamin B_{12} and folic acid, are essential. Finally, the formation of an erythrocyte requires the materials that go into the making of hemoglobin, such as iron and the amino acids that are incorporated into globin. A lack of any of these growth factors or raw materials results in the failure of normal erythrocyte formation and a decreased quantity of effective circulating erythrocytes. The substances that are most commonly lacking are iron, folic acid, and vitamin B_{12}.

Iron Iron is obviously an essential component of the hemoglobin molecule because it is to this element that the oxygen actually binds. The balance of iron and its distribution in the body are shown schematically in Fig. 19-4. About 70 percent of the

FIG. 19-4 Summary of iron balance. The size of the boxes represents the quantity of iron involved. The magnitude of exchange in a particular direction varies according to conditions.

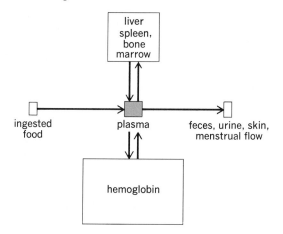

total body iron is in hemoglobin, and the remainder is stored primarily in the liver, spleen, and bone marrow. As erythrocytes are destroyed, most of the iron released from hemoglobin is returned to these depots from which it can again be released for hemoglobin synthesis. Small amounts of iron, however, are lost each day via the urine, feces, sweat, and cells sloughed from the skin. In addition, women lose a significant quantity of iron via menstrual blood. In order to remain in iron balance, the amount of this metal lost from the body must be replaced by ingestion of iron-containing foods. An upset of this balance results either in iron deficiency and inadequate hemoglobin production or in an excess of iron in the body and serious toxic effects.

The homeostatic control of iron balance is unusual in that it resides primarily in the intestinal epithelium, which actively absorbs the iron from ingested food. Only a small fraction of the ingested iron is absorbed and, what is more important, this fraction is increased or decreased depending upon the state of body iron balance. These fluctuations appear to be mediated by changes in the iron content of the intestinal epithelium itself, as described in Chap. 25.

The tendency of iron to form insoluble precipitates has been circumvented in the body by the evolution of iron-binding proteins that act both as carrier molecules for transferring iron in the plasma (and between cells) and as storage molecules for iron; the carrier iron-protein complex is known as **transferrin** and the storage complex as **ferritin.**

Vitamin B_{12} and Folic Acid Normal erythrocyte formation requires extremely small quantities (one-millionth of a gram per day) of a cobalt-containing molecule, vitamin B_{12}, which permits final maturation of the erythrocyte. Vitamin B_{12} is not synthesized in the body, which therefore depends upon dietary intake for the substance. Absorption of vitamin B_{12} from the gastrointestinal tract into the blood is described in Chap. 25. Dietary deficiency of vitamin B_{12} or impaired intestinal absorption of it leads to the failure of erythrocyte proliferation and maturation known as *pernicious anemia*. Since vitamin B_{12} is also needed for normal functioning of the nervous system, specifically for myelin formation, a variety of neurologic symptoms may accompany the anemia.

Folic acid is a vitamin essential for the formation

of DNA because it is involved in the synthesis of purines and thymine. Accordingly, its deficiency results in failure of normal erythrocyte multiplication and maturation, as well as poor growth and development in many other tissues.

Regulation of Erythrocyte Production In a normal person, the total volume of circulating erythrocytes remains remarkably constant. Such constancy is required both for delivery of oxygen to the tissues and for maintenance of the blood pressure. From the preceding paragraphs it should be evident that a constant number of circulating erythrocytes can be maintained only by balancing erythrocyte production and destruction (or loss). Such balance is achieved by controlling the rate of erythrocyte production. During periods of severe erythrocyte destruction or hemorrhage, the normal rate of erythrocyte production can be increased more than sixfold.

It is easiest to discuss the control of erythrocyte production by first naming the mechanisms *not* involved. In the previous section, we listed a group of nutrient substances, such as iron and vitamin B_{12}, which must be present for normal erythrocyte production. However, none of these substances actually *regulates* the rate of production.

The direct control of erythropoiesis is exerted by a hormone called **erythropoietin.** This hormone is produced in the blood, but its formation is controlled by an enzyme that is secreted by the kidneys and probably other sites as well. Normally, a small quantity of erythropoietin is circulating, which stimulates the bone marrow to produce erythrocytes at a certain basal rate. An increase in circulating erythropoietin further stimulates the bone marrow and increases erythropoiesis, whereas a decrease in circulating erythropoietin below the normal level decreases erythropoiesis.

The common denominator of changes that increase circulating erythropoietin is precisely what one would logically expect—a decreased oxygen delivery to the kidneys (and other erythropoietin-controlling tissues). As will be described in Chap. 23, this can result from a decrease in the number of erythrocytes or their hemoglobin content, decreased blood flow, or decreased oxygen delivery from the lungs into the blood. As a result, erythropoietin formation is increased, erythrocyte production is increased, the oxygen-carrying capacity

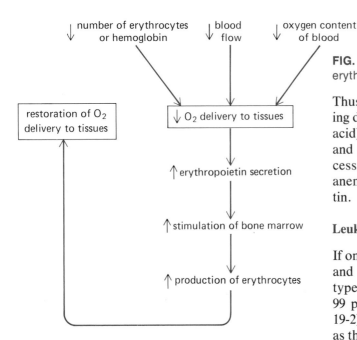

↓ number of erythrocytes or hemoglobin   ↓ blood flow   ↓ oxygen content of blood

↓ O_2 delivery to tissues

restoration of O_2 delivery to tissues

↑ erythropoietin secretion

↑ stimulation of bone marrow

↑ production of erythrocytes

FIG. 19-5   Role of erythropoietin in regulation of erythrocyte production and oxygen delivery.

Thus, anemia has a wide variety of causes, including dietary deficiencies (of iron, vitamin B_{12}, or folic acid), bone-marrow failure (caused by toxic drugs and cancer, for example), excessive blood loss, excessive destruction of erythrocytes (as in sickle cell anemia), and inadequate secretion of erythropoietin.

Leukocytes

If one takes a drop of blood, adds appropriate dyes, and examines it under a microscope, the various types of blood cells can be seen (Fig. 19-6). Over 99 percent of all the cells are erythrocytes (Table 19-2). The remaining cells, the **leukocytes,** all have as their major function defense against foreign cells, such as bacteria, and other foreign matter. Their roles are described in detail in Chap. 28, and we introduce them here to complete our survey of blood. Leukocytes are classified according to their structure and affinity for various dyes. The name **polymorphonuclear granulocytes** refers to the three types of cells with lobulated nuclei and abundant membrane-bound granules in their cytoplasm. The granules of one group show no dye preference, and the cells are therefore called **neutrophils.** The granules of the second group take up the red dye eosin, thus giving the cells their name **eosinophils.** Cells of the third group have an affinity for a basic dye and

of the blood is increased, and oxygen delivery to the tissues is returned toward normal (Fig. 19-5); it is interesting that testosterone in physiologic amounts also stimulates the formation of erythropoietin, and this may account for the fact that hemoglobin concentration in men is 1 to 2 percent higher than the concentration in women.

Anemia   **Anemia** is a reduced total blood hemoglobin. It may be due to a decrease in the total number of erythrocytes (each having a normal quantity of hemoglobin), to a diminished concentration of hemoglobin per erythrocyte, or both.

FIG. 19-6   Normal blood cell types. The two types of lymphocytes, B and T lymphocytes, cannot be distinguished by light or even electron microscopy; therefore we have shown a single cell.

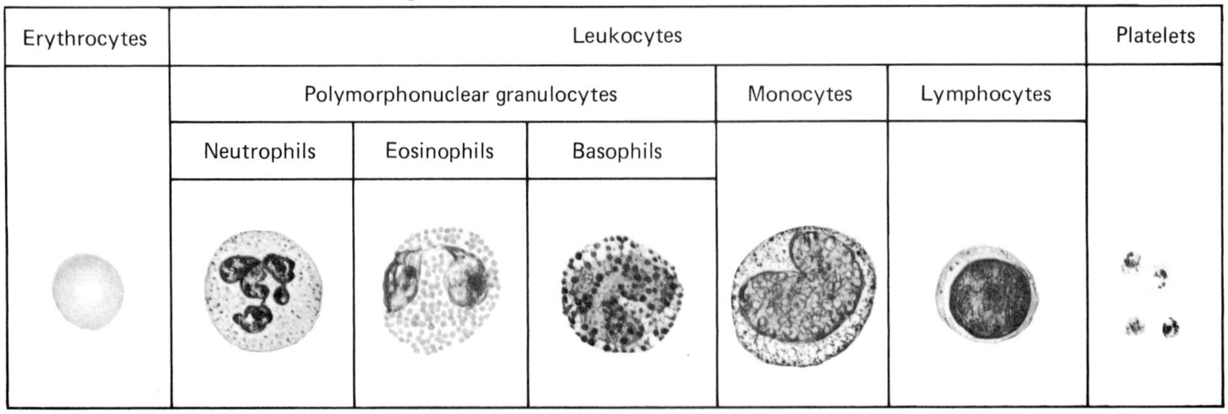

Erythrocytes	Leukocytes					Platelets
	Polymorphonuclear granulocytes			Monocytes	Lymphocytes	
	Neutrophils	Eosinophils	Basophils			

TABLE 19-2	NUMBERS AND DISTRIBUTION OF ERYTHROCYTES, LEUKOCYTES, AND PLATELETS IN NORMAL HUMAN BLOOD

Total erythrocytes = 5,000,000 cells per cubic millimeter of blood
Total leukocytes = 7000 cells per cubic millimeter of blood
Percent of total leukocytes
  Polymorphonuclear granulocytes
    Neutrophils, 50–70
    Eosinophils, 1–4
    Basophils, 0.1
  Mononuclear cells
    Monocytes, 2–8
    Lymphocytes, 20–40
Total platelets = 250,000 cells per cubic millimeter of blood

are called **basophils.** All three types of granulocytes are produced in the bone marrow and released into the circulation. Unlike the erythrocytes, the major functions of granulocytic leukocytes are exerted not in the blood vessels but in the interstitial fluid; i.e., granulocytes utilize the circulatory system only as the route for reaching a damaged or invaded area. Once there, they leave the blood vessels to enter the tissue and perform their functions.

The primary function of the neutrophil is **phagocytosis,** the ingestion and destruction of particulate material. It remains in the circulation for only 8 to 12 hr before entering connective-tissue spaces and is incapable of division, so that its supply must be continuously replenished by the bone marrow (it is estimated that 125 billion neutrophils are produced each day). Eosinophils and basophils are also phagocytic cells. Basophils, in addition, contain powerful chemicals, such as histamine, that they release locally, a response which, as we shall see, may importantly contribute to tissue damage and allergy. The basophil is virtually identical to the group of cells known as **mast cells** that are found in connective tissue throughout the body and do not circulate.

A fourth type of leukocyte quite different in appearance from the three granulocytic types is the **monocyte.** These cells, which are also produced by the bone marrow, are somewhat larger than the granulocytes, with a single oval or horseshoe-shaped nucleus and relatively few cytoplasmic granules. Upon entering an area that has been invaded or wounded, monocytes are transformed into **macrophages,** another type of phagocytic cell. Thus, the function of circulating monocytes is to provide to the issues a source of new macrophages, the function of which is described in Chap. 28.

The final class of leukocyte is the **lymphocyte.** The outstanding structural features of lymphocytes are a relatively large nucleus and scanty surrounding cytoplasm. The circulating pool of lymphocytes continually travels from the blood across capillary walls to the lymph, thence to lymph nodes and, finally, back to the blood. However, circulating lymphocytes constitute only a very small fraction of the total body lymphocytes, most of which are found at any instant in the lymphoid tissues. The lymphocytes (and a daughter-cell line, the **plasma cells**) are responsible for specific defenses against foreign invaders and will be described in Chap. 28.

In an normal person the white blood cells have approximately the distribution shown in Table 19-2.

Platelets

The circulating platelets (Fig. 19-7) are colorless corpuscles much smaller than erythrocytes and containing numerous granules. There are approximately 250 million per cubic millimeter of blood (compare this with the number of white blood cells given in Table 19-2). The platelets, which are actually cell fragments and lack nuclei, originate from certain large cells (**megakaryocytes**) in bone marrow. They are apparently portions of cytoplasm of these cells that are pinched off and enter the circulation, where they have a life-span of about 10 days. They play crucial roles in blood clotting, as described in a subsequent section.

Formation of Blood Cells

Blood cells are formed in discrete clusters, or colonies. Most experts believe that all blood cell types differentiate from a single cell type known as a **colony forming unit,** or **CFU.** The CFUs are able to migrate to and populate various tissues, and the specific organ that contains the bulk of the CFUs and, thereby, serves as the major hemopoietic organ changes throughout fetal life.

In very early fetal life, the CFUs reside in the

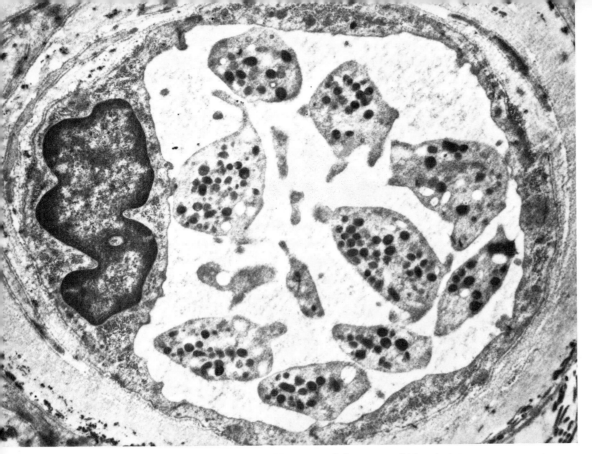

FIG. 19-7 Electron micrograph of a capillary containing several blood platelets in its lumen. Note the numerous granules and lack of nuclei.

yolk sac; next they migrate to the liver. Beginning at about the third fetal month the spleen becomes the major hemopoietic organ, and mainly primitive red blood cells are formed at this time. Finally, in the later portions of gestation the bone marrow becomes increasingly important, and it remains the major hemopoietic organ throughout the remainder of the individual's life. (The spleen and liver in the adult also contain small numbers of CFUs and may, under abnormal conditions, serve as hemopoietic organs.) The bone marrow produces the erythrocytes, granulocytes, monocytes, platelets, and one kind of lymphocyte, the **B lymphocytes.** The thymus produces the other kind of lymphocyte, the **T lymphocytes.**

Despite the fact that the lymphocytes originate in the bone marrow and thymus, most of them are housed in lymphatic (i.e., lymphocyte-containing) tissues such as the lymph nodes, spleen, and tonsils. These tissues function as sieves and accumulate lymphocytes mainly by filtering them from the blood. In addition, the lymphatic tissues are able to form some lymphocytes from precursors that have migrated there from the bone marrow and thymus,

and lymphatic tissues become active lymphocyte producers as part of the immune response, which will be discussed more fully in Chap. 28.

All blood cells derived from bone marrow (i.e., all erythrocytes, monocytes, granulocytes, platelets, and B lymphocytes) are referred to as **myeloid** (Greek *myelos,* marrow), whereas the T lymphocytes, which are derived from CFUs in the thymus, a lymphatic organ, are called **lymphoid.** The relationships between the various blood cell types and their precursors are shown in Fig. 19-8.

HEMOSTASIS: THE PREVENTION OF BLOOD LOSS

All animals with a vascular system must be able to minimize blood loss consequent to vessel damage. In human beings, blood coagulation is only one of several important mechanisms for **hemostasis,** the prevention of blood loss. The mechanism that predominates varies, depending upon the kind and number of vessels damaged and the location of the injury.

412

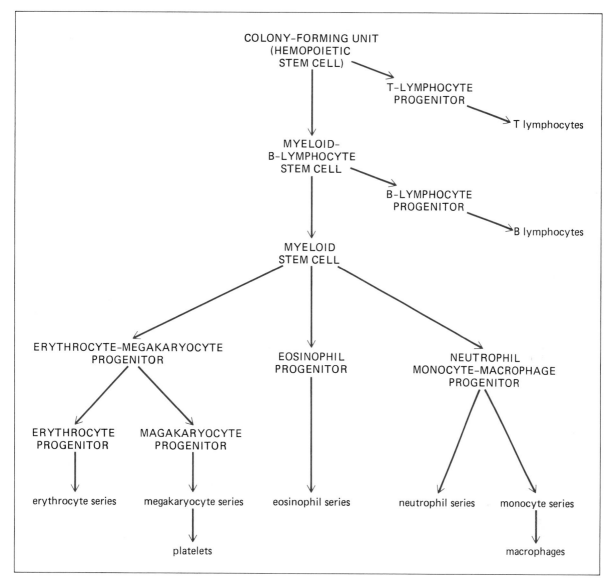

FIG. 19-8 Derivation of the different blood cell types from a colony-forming unit. The origin of basophils is unknown, but presumably they derive from the myeloid stem cell. (A *stem cell* is able both to renew its own type and to differentiate into other cell types; in contrast a *progenitor cell,* while able to differentiate into other cell types, is unable to renew its own type.)

The basic prerequisites for bleeding *(hemorrhage)* are (1) loss of vessel continuity or marked increase in permeability so that cells and fluid can leak out, and (2) a pressure inside the vessel greater than outside, since hemorrhage occurs by bulk flow. Accordingly, bleeding ceases if at least one of two requirements is met: The pressure difference favoring blood loss is eliminated, or the damaged portion of the vessel is sealed. All hemostatic processes accomplish one of these two requirements.

We shall discuss the probable sequence of events in response to damage to small blood vessels—arterioles, capillaries, venules—because they are the most common source of bleeding in everyday life and because the hemostatic mechanisms are most effective in dealing with such injuries. In contrast, the bleeding from a severed artery of medium or large size is not usually controllable by the body and requires radical aids such as application of pressure and ligatures. Venous bleeding is less danger-

TABLE 19-3   SYNOPSIS OF HEMOSTATIC EVENTS IN SMALL BLOOD VESSELS

1 Initial constriction of the damaged vessel
2 Sticking together of injured endothelium
3 Clumping of platelets to form a plug
4 Facilitation of the initial vasoconstriction
5 Blood coagulation, i.e., formation of a fibrin clot
6 Retraction of the clot

ous because veins have low blood pressure; indeed, the drop in pressure induced by simple elevation of the bleeding part may stop the hemorrhage. In addition, if the venous bleeding goes into the tissues, the accumulation of blood (*hematoma*) may increase interstitial pressure enough to eliminate the pressure gradient required for continued blood loss. The hemostatic events in small vessels are listed above in Table 19-3. The first four events (Fig. 19-9) usually occur within seconds, whereas the formation of a clot (step 5) takes several minutes. However, the different events actually do not occur in a neat orderly sequence but overlap in time and are closely interrelated functionally.

Blood Coagulation: Clot Formation

Despite the participation of the first four mechanisms listed in Table 19-3, blood coagulation is the dominant hemostatic defense in human beings, as attested by the fact that, with few exceptions, clotting defects are the cause of abnormal bleeding. Pure vascular defects that interfere with the preclot hemostatic mechanisms do occur but are much less frequently the cause of abnormal bleeding.

The event that transforms blood into a solid gel is the conversion of the plasma protein **fibrinogen** to **fibrin.** Fibrinogen is a soluble, large, rod-shaped protein produced by the liver and always present in the plasma of normal persons. Its conversion to fibrin is catalyzed by the enzyme **thrombin.**

$$\text{Fibrinogen} \xrightarrow{\text{thrombin}} \text{fibrin}$$

In this reaction, several small negatively charged polypeptides are split from fibrinogen, conferring upon the remaining large molecule the ability to bind to other similar molecules. They join each other end to end and side by side to form the polymer known as fibrin (Fig. 19-10). This polymeriza-

FIG. 19-9 Summary of hemostatic mechanisms not dependent upon blood coagulation. The dashed line indicates the positive-feedback effect of ADP on platelet adhesion and agglutination. Prostaglandins, not shown in the figure, may either inhibit or facilitate platelet clumping.

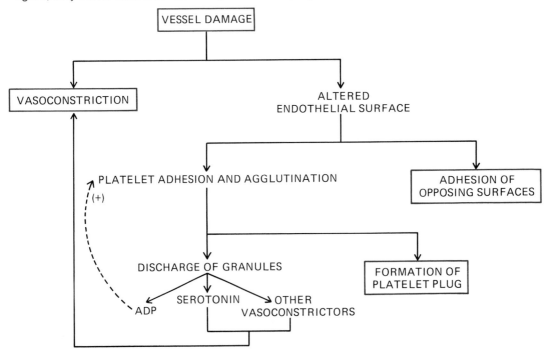

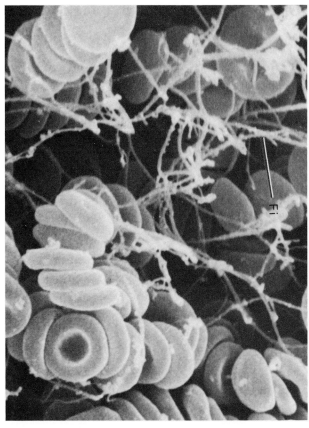

FIG. 19-10 Scanning electron micrograph of fibrin.

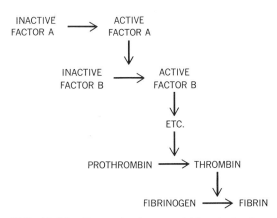

FIG. 19-11 Cascade theory of blood clotting. Each substance left of an arrow is normally present in plasma but requires activation by the action of the previous substance in the sequence. Factor A is Hageman factor; the term *factor* signifies a plasma protein.

tion causes the fluid portion of the blood to gel, rather like a gelatin dessert. In addition, cellular elements of the blood become entangled in the meshwork and contribute to its strength. It must be emphasized that the clot is basically the result of fibrin and can occur in the absence of blood cells (except the platelets).

Since fibrinogen is always present in the blood, the enzyme thrombin must normally be absent and its formation must be triggered by vessel damage. The generation of thrombin follows the same general principle as fibrin, in that an inactive precursor, **prothrombin,** is produced by the liver and is normally present in the blood. Prothrombin is converted to thrombin during clot formation by a specific enzyme that splits off peptide from the inactive prothrombin molecule.

$$\text{Prothrombin} \xrightarrow{\text{Enzyme}} \text{thrombin}$$
$$\downarrow$$
$$\text{Fibrinogen} \longrightarrow \text{fibrin}$$

We have now only pushed the essential question one step further back: Where does the enzyme come from that catalyzes the conversion of the prothrombin to thrombin? The answer is that this enzyme is always present in the plasma in an inactive form and is converted to its active form by yet another enzyme that was itself activated by another enzyme, and so on. Thus, we are dealing with a cascade of plasma proteins, each normally an inactive proteolytic enzyme until activated by the previous one in the sequence (Fig. 19-11). Ultimately, the final enzyme in the sequence is activated and in turn catalyzes the conversion of prothrombin to thrombin. The first factor in the clotting sequence is designated **factor XII** or **Hageman factor** and, as will be described later, it has several other important functions in addition to initiating clotting.

The adaptive value of this type of cascade system lies in the amplification gained at each step, i.e., in the manyfold increase in the number of active molecules produced (recall from Chap. 7 that an analogous cascade of enzyme activations was described for the cyclic AMP-protein kinase system). However, there is at least one disadvantage in such a cascade in that a single defect, either hereditary or disease-induced, anywhere in the system can block the entire cascade and thereby interfere with clot formation. The clotting factors are listed in Table 19-4.

A disruption in the cascade of reactions leading to fibrin formation occurs in bleeding disorders such

TABLE 19-4	NUMERICAL SYSTEM FOR NAMING BLOOD CLOTTING FACTORS.*
Factor Number	**Factor Names**
I	Fibrinogen
II	Prothrombin
III	Thromboplastin
IV	Calcium
V	Proaccelerin, labile factor, accelerator globulin
VII	Proconvertin, SPCA, stable factor
VIII	Antihemophilic factor (AHF), antihemophilic factor A, antihemophilic globulin (AHG)
IX	Plasma thromboplastic component (PTC), Christmas factor, antihemophilic factor B
X	Stuart-Prower factor
XI	Plasma thromboplastin antecedent (PTA), antihemophilic factor C
XII	Hageman factor, glass factor
XIII	Fibrin stabilizing factor, Laki-Lorand factor

* Factor VI is not a separate entity and is no longer used.

as **hemophilia.** Hemophilia occurs when one of the protein factors essential in the cascade, factor VIII (or antihemophilic factor), is defective and inactive. Hemophilia is a sex-linked genetic disorder in which the gene for the defective factor VIII of hemophilia is a gene that is located on the X chromosome. (See Chap. 27 for a discussion of human genetics and X and Y chromosomes.) Thus, a woman who has one normal X chromosome and one with a defective gene for factor VIII will be a genetic carrier of hemophilia, but she herself will not have the disease. However, a male, who has

only one X chromosome, will be a hemophiliac if that chromosome carries the defective gene.

In addition to these plasma proteins, calcium is required as cofactor for several of the enzymatic steps in coagulation: however, calcium deficiency is never a cause of clotting defects in human beings because only very small concentrations are required.

Figure 19-11 reveals that we have still not answered the basic question of what *initiates* clotting. What activates the first enzyme (Hageman factor or factor A in the figure) in the catalytic sequence? The answer is contact of this protein with a damaged vessel surface, most likely with the collagen fibers that underly the damaged endothelium (as was true for platelet aggregation). However, even contact activation of this first factor cannot produce clotting in the absence of platelets. A phospholipid substance exposed on the surface of platelets during their adhesion and agglutination is required as cofactor for several of the steps in the catalytic sequence; this phospholipid seems to provide a surface for interaction of certain of the plasma clotting factors.

At last we have reached the connecting link between vessel injury and initiation of clotting. Thus, for two reasons, the critical event that initiates clot formation is contact of the blood with a damaged surface: (1) It activates the first factor in the activation sequence, and (2) it causes platelet adhesion

FIG. 19-12 Summary of blood-clotting mechanism. The dashed line indicates the positive-feedback effect of thrombin on platelet adhesion (recall that ADP exerts a similar positive feedback) and activation of Hageman factor.

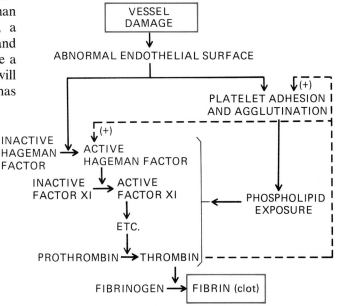

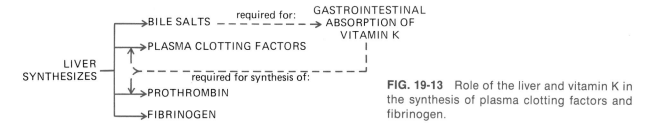

FIG. 19-13 Role of the liver and vitamin K in the synthesis of plasma clotting factors and fibrinogen.

and exposure of the phospholipid cofactor. One final detail is that thrombin markedly enhances the adhesion and agglutination of platelets and the activation of Hageman factor; thus, once thrombin formation has begun, the overall reaction progresses explosively because of the positive-feedback effects of thrombin. Our growing figure can now be completed (Fig. 19-12).

This entire process occurs only locally at the site of vessel damage. Each active component is formed, functions, and is then rapidly inactivated by enzymes in the plasma or local tissue, without spilling over into the rest of the circulation. Otherwise, because of the chain-reaction nature of the response, the appearance of platelet phospholipid and any single activated factor in the overall circulation would induce massive widespread clotting throughout the body.

The reason should also be clear why blood coagulates when it is taken from the body and put in a glass tube. This has nothing whatever to do with exposure to air, as popularly supposed, but happens because the glass surface induces the same activation of Hageman factor and aggregation of platelets as does a damaged vessel surface. A silicone coating markedly delays clotting by reducing the activating effects of the glass surface.

The liver plays several important indirect roles in the overall functioning of the clotting mechanism (Fig. 19-13). First, it is the site of production for many of the plasma factors including prothrombin, and fibrinogen. Second, the bile salts, produced by the liver are required for normal gastrointestinal absorption of the fat-soluble vitamin K, which is an essential cofactor in the liver's synthesis of prothrombin and several other plasma factors. When vitamin K is not present in adequate amounts, these clotting factors are still produced, but they lack the critical binding sites that would permit them to interact with platelet phospholipid and become activated. It should be clear, therefore, why patients with liver disease or defective gastrointestinal fat

absorption frequently have serious bleeding problems.

Finally, a word must be said about the contribution of a tissue (rather than blood) factor to clotting. If one extracts almost any of the body's tissues and injects the extract into unclotted normal blood in a siliconized tube, clotting occurs within seconds. The explanation is that the tissues contain a lipoprotein substance known as **tissue thromboplastin** that can substitute for platelet phospholipid as well as several of the plasma factors; thus, an abnormal surface is no longer required to initiate clotting. This is known as the **extrinsic clotting pathway** to distinguish it from the **intrinsic pathway** described earlier. Its quantitative contribution to normal intravascular clotting is unclear but it may play an essential role in the response to many bacterial infections by initiating fibrin clots that may block further spread of the bacteria.

Clot Retraction

When blood is carefully collected and placed in a glass test tube, clotting usually occurs within 5 to 8 min, and the entire volume of blood appears as a coagulated gel. However, during the next 30 min, a striking transformation occurs; the clot literally retracts, squeezing out the fluid that constituted a large fraction of the gel. The end result is a small hard clot at the bottom of the tube with a large volume of serum floating on top (plasma without fibrinogen is called serum). The fibrin meshwork with its entangled cells has become denser and stronger. This process is known as **clot retraction.** It is due to the platelets. As fibrin strands form around them during clotting, the agglutinated mass of platelets sends out adhering pseudopods along them. The pseudopods then contract, pulling the fibrin fibrils together and squeezing out the serum. This contraction is produced by actomyosin contractile proteins in the platelets. An event similar to clot retraction also occurs in the body. The various

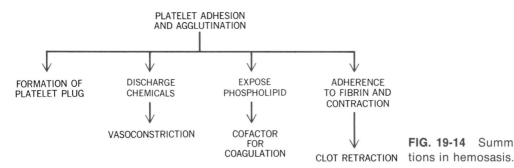

FIG. 19-14 Summary of platelet functions in hemosasis.

roles of platelets in hemostasis are summarized in Fig. 19-14.

The Anticlotting System

It has long been observed that clots frequently disappear after lengthy standing. Blood also fails to clot in a variety of other circumstances. This is usually caused by a proteolytic enzyme called **plasmin,** which is able to decompose fibrin, thereby dissolving a clot. The physiology of plasmin bears some striking similarities to that of the coagulation factors in that it circulates in blood in an inactive form **(plasminogen)** that is enzymatically converted to active plasmin by the action of activated Hageman factor as well as by other substances found in tissue fluid (Fig. 19-15). It may seem paradoxical that the same substance, Hageman factor, triggers off simultaneously the clotting and anticlotting systems. Yet, in fact, this makes sense since the generated plasmin becomes trapped within the newly formed clot and very slowly dissolves it, thereby contributing to tissue repair at a time when the danger of hemorrhage is past.

The anticlotting system no doubt has other functions as well. It may be that small amounts of fibrin are constantly being laid down throughout the vascular tree and that plasmin acts on this fibrin to prevent clotting. The lung tissue, for example, contains a substance that activates plasmin; this probably explains the lung's ability to dissolve the fibrin clumps that its capillaries filter from the blood. Moreover, the uterine wall is extremely rich in a similar activator, and thus normal menstural blood generally does not clot.

A second naturally occurring anticoagulant is **heparin.** This substance, found in various cells of the body, especially mast cells, acts by interfering with the activation of several of the clotting factors

and with the ability of thrombin to split fibrinogen. (Not surprisingly, activated platelets secrete a substance that inhibits heparin.) Despite its presence in the body and the fact that it is the most powerful anticoagulant known, it is not clear whether heparin plays a normal physiologic role in clot prevention. On the other hand, heparin is widely used as an anticoagulant in medicine.

Excessive Clotting: Intravascular Thrombosis

Formation of a clot in a bleeding vessel is obviously a homeostatic physiologic response, but the formation of clots in intact vessels is pathological. Such an intravascular clot is known as a **thrombus.** It may occur in the veins, the microcirculation, or arteries. Coronary arterial occlusion (obstruction of an artery in the heart) secondary to **thrombosis** (formation of a clot) is one of the major killers in the United States today. The sequence of events leading to thrombosis is the subject of intensive study, and numerous theories have been proposed. A brief synopsis of some of them is in order because of the great importance of the subject and its illustration of the basic physiologic processes.

One of the dominant theories today postulates that the clotting mechanism in persons prone to thrombosis is hyperactive, as manifested by the reduced time it takes for withdrawn blood to clot in a test tube. Perhaps one of the plasma factors is present in excessive amounts, or a normally occurring anticoagulant is deficient. This theory emphasizes that the blood itself is the cause of excessive clotting. However, there seems little question that hypercoagulability is not always essential for thrombosis, since hemophiliacs have been known to suffer from coronary thrombosis.

A second category of theories puts the blame on the blood vessels. Since initiation of blood clotting

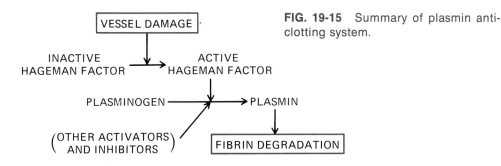

FIG. 19-15   Summary of plasmin anti-clotting system.

is primarily dependent upon the state of the blood vessel lining, even minor transient alterations in the endothelial surface could trigger the cascade sequence that leads to clot formation. These vessel-oriented theories can explain many of the situations associated with an increased probability of vascular thrombosis (Fig. 19-16): (1) Stasis, i.e., decreased movement, of blood in veins, which occurs during quiet standing, valve malfunction, or cardiac insufficiency, may induce damage in the vein wall as a result of oxygen lack. (2) Inflammation of veins (*phlebitis*) and other vessels caused by bacteria, allergic reactions, or toxic substances may cause vessel damage. (3) The presence of abnormal smooth muscle cells and the deposition of lipids in arterial walls (atherosclerosis) causes marked thickening and irregularity of the arterial lining.

Thus, the three major conditions that predispose to clot formation are consistent with the damaged-lining theory. It should be noted that this concept and the hypercoagulability theory are not mutually exclusive. Both probably are valid, depending upon the circumstances.

Regardless of the initiating event, there is no question that a clot, no matter how small, provides a suitable surface upon which more clot can form. Thus the thrombus grows and may eventually occlude the entire vessel, thereby leading to damage of the tissue supplied or drained by the vessel. A second important factor in vessel closure during clot growth is the release of vasoconstrictors from freshly adhered platelets. Finally, the chances are greatly increased of clot fragments breaking off and being carried to the lungs or other organs. These **emboli** not only plug the small blood vessels but in the lung may induce totally inappropriate cardiovascular reflexes that culminate in disturbance of the heart's rhythm, and death.

The prevention of new clot growth, with its associated consequences, is the major reason for giving patients anticoagulant drugs. Anticoagulants now in use include heparin and a group of drugs that interfere with the normal synthesis of several of the plasma clotting factors, including prothrombin, by blocking the action of vitamin K. Indeed, one of the important observations leading to the discovery of vitamin K was that animals on a diet of spoiled sweet clover hay, which contains these anticoagulants, manifested serious bleeding tendencies. Of course, a patient receiving any anticoagulant is prone to bleeding.

FIG. 19-16   Damaged-vessel theory of abnormal intravascular clotting.

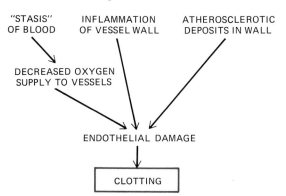

KEY TERMS

plasma	leukocytes	hemorrhage
albumins	neutrophils	fibrin
globulins	eosinophils	factor XII (Hageman factor)
fibrinogen	basophils	extrinsic and intrinsic clotting pathways
serum	monocytes	clot retraction
erythrocytes	lymphocytes	plasmin
hemoglobin	platelets	thrombosis
erythropoiesis	hemostasis	

REVIEW EXERCISES

1 State the types of blood cells.
2 List the types of plasma proteins.
3 Describe the anatomy of erythrocytes and its adaptive significance.
4 State the general structure of hemoglobin and describe its catabolic pathway.
5 Describe the sequence of events, cells involved, and essential chemicals for erythropoiesis.
6 Describe the metabolism of iron and its balance in the body.
7 Describe the control of erythrocyte production.
8 State the types of leukocytes and their approximate numerical distribution in blood; describe the function of each type of leukocyte.
9 Describe the origin of platelets.
10 Describe the sites of formation of the different blood cells.
11 State the two basic prerequisites for bleeding to occur.
12 List the hemostatic events that occur prior to clot formation.

13 List the sequence of events that lead from damage to a vessel to the formation of fibrin; state at which points calcium and platelet phospholipid are important.
14 List the functions of the liver in blood clotting and the role of vitamin K.
15 Contrast the extrinsic and intrinsic clotting pathways and state the function of tissue thromboplastin.
16 Describe the events and mechanism of clot retraction.
17 List the various roles of platelets in hemostasis.
18 Describe the anticlotting system.
19 Describe two theories that explain excessive clotting.
20 Define the following terms: hematocrit, sickle cell anemia, bilirubin, jaundice, pernicious anemia, anemia, erythropoietin, megakaryocytes, colony-forming unit, myeloid cells, lymphoid cells, hemophilia, heparin, embolus.

CHAPTER

THE CARDIOVASCULAR SYSTEM: THE HEART

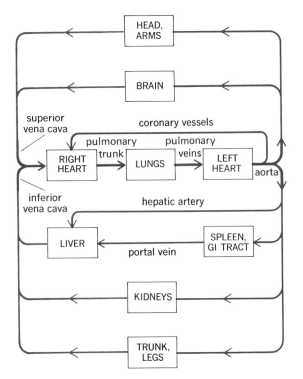

FIG. 20-1 Diagrammatic representation of the cardiovascular system in the adult human being. Darker shading indicates blood with high oxygen content. The commonly used terms *right* and *left heart* refer to the right and left halves of the heart, which is a single organ.

Physiology as an experimental science began in 1628, when William Harvey demonstrated that the entire circulatory system forms a circle, so that blood is continuously being pumped out of the heart through one set of vessels and returned to the heart via a different set. In human beings, as in all mammals, there are actually two circuits, both originating and terminating in the heart, which is divided longitudinally into two functional halves (Fig. 20-1). Blood is pumped via one circuit (the **pulmonary circulation**) from the right half of the heart through the lungs and back to the left half of the heart. It is pumped via the second circuit (the **systemic circulation**) from the left half of the heart through all the tissues of the body, except, of course, the lungs, and back to the right half of the heart. In both circuits, the vessels that carry blood away from the heart are **arteries,** and the vessels that carry blood back to the heart are **veins.** In the systemic circuit, blood leaves the heart via a single large artery, the **aorta,** and returns to the heart via two large veins, the **superior** and **inferior venae cavae.** In the pulmonary circuit, blood leaves the heart via the **pulmonary trunk** and returns to the heart via the four **pulmonary veins.**

In a normal person, blood can pass from the systemic veins to the systemic arteries only by first being pumped through the pulmonary circuit; thus, all the blood returning from the body tissues is oxygenated before it is pumped back to them. The total volumes of blood pumped through the pulmonary and systemic circuits during a given period of time are equal. In other words, the right heart pumps the same amount of blood as the left heart. Only under unusual circumstances, such as malfunction of one half of the heart, do these volumes differ from each other, and then only transiently. In a resting normal person, the amount of blood pumped simultaneously by each half of the heart is approximately 5 L/min. During heavy work or exercise, the volume may increase as much as sevenfold to 35 L.

ANATOMY OF THE HEART

The heart is a muscular organ located in the *mediastinum,* i.e., the central portion of the thoracic cavity between the two lungs. One-third of the heart lies to the right of the median plane and two-thirds to the left (Fig. 20-2). The walls of the heart are composed primarily of muscle (**myocardium**), the structure of which is different from either skeletal or smooth muscle. The inner surface of the myocardium, i.e., the surface in contact with the blood in the heart chambers, is lined by a thin layer of cells (**endothelium**). The heart is enclosed in a closely adhering fibrous sac, the **pericardium.** The inner surface of the pericardium and the outer surface of the heart are lined with serous membrane, which secretes a watery fluid into the narrow cavity between the sac and the heart (Fig. 20-3). This solution serves as a lubricant as the heart moves within the sac.

The heart is divided longitudinally into right and left halves (Fig. 20-4), each consisting of two chambers, an **atrium** and a **ventricle.** The cavities of the atrium and ventricle on each side of the heart communicate with each other, but the right chambers do not communicate directly with those on the left.

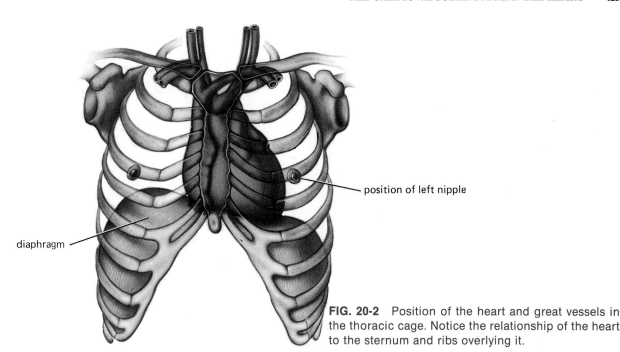

FIG. 20-2 Position of the heart and great vessels in the thoracic cage. Notice the relationship of the heart to the sternum and ribs overlying it.

diaphragm

position of left nipple

They are separated from each other by partitions called the **interatrial** and **interventricular septa.** Thus, right and left atria and right and left ventricles are distinct.

The tip of the left ventricle, which points downward, forward, and toward the left, forms the *apex* of the heart, and the left atrium, which faces backward and toward the right, forms the major part of the *base.* A small ear-like pouch extends from each atrium; these pouches are termed the *right* and *left auricles* (the term auricle is often wrongly used synonymously with atrium).

Perhaps the easiest way to picture the architecture of the heart is to begin with its fibrous skeleton, which comprises four interconnected rings of dense connective tissue (Fig. 20-5). To the tops of these

rings are anchored the muscle masses of the atria, pulmonary trunk, and aorta. To the bottoms are attached the muscle masses of the ventricles. To these rings are attached four sets of valves (Fig. 20-5).

Between the cavities of the atrium and ventricle in each half of the heart are the **atrioventricular**

FIG. 20-3 The pericardium and pericardial cavity.

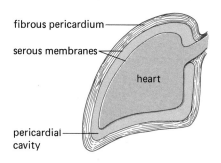

fibrous pericardium

serous membranes

heart

pericardial cavity

FIG. 20-4 Diagrammatic section of the heart. The arrows indicate the direction of blood flow.

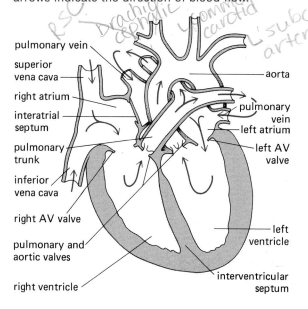

pulmonary vein

superior vena cava

right atrium

interatrial septum

pulmonary trunk

inferior vena cava

right AV valve

pulmonary and aortic valves

right ventricle

aorta

pulmonary vein

left atrium

left AV valve

left ventricle

interventricular septum

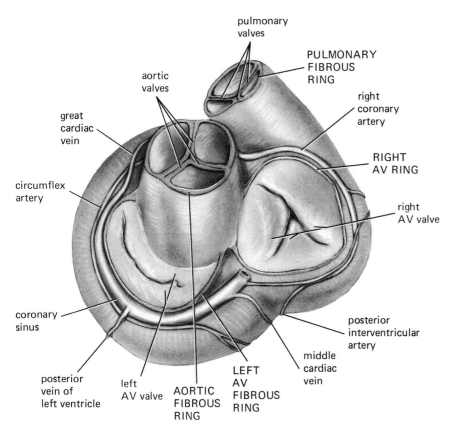

FIG. 20-5 Four connective-tissue rings compose the fibrous skeleton of the heart. To these are attached the valves, the trunks of the aorta and the pulmonary artery, and the muscle masses of the cardiac chambers. The right atrioventricular (AV) valve is formed of three cusps, or leaflets, and is also called the tricuspid valve. The left AV valve, which consists of two cusps, is the bicuspid, or mitral, valve. The opening of the aorta is guarded by the aortic valve and the pulmonary trunk by the pulmonary valve. The aortic and pulmonary valves are called semilunar valves because of the shape of their leaflets.

valves (AV valves), which permit blood to flow from atrium to ventricle but not from ventricle to atrium. The right AV valve is called the **tricuspid** value and the left is the **bicuspid,** or **mitral,** valve. When the blood is moved from atrium to ventricle, the valves lie open against the ventricular wall, but when the ventricles contract, the valves are brought together by the increasing pressure of the ventricular blood, and the atrioventricular opening is closed (Fig. 20-6). Blood is therefore forced into the pulmonary trunk (from the right ventricle) and into the aorta (from the left ventricle) instead of back into the atria. To prevent the valves themselves from being

forced upward into the atrium, they are fastened by thin fibrous strands, the **chordae tendineae,** to the **papillary muscles,** which are projections of the ventricular walls. These muscular projections do *not* open or close the valves; they act only to limit the valves' movements and prevent them from being everted.

The openings of the ventricles into the pulmonary trunk and aorta are also guarded by valves, the **pulmonary** and **aortic valves,** which permit blood to flow into these arteries but close immediately, preventing reflux of blood in the opposite direction. There are no true valves at the entrances of the

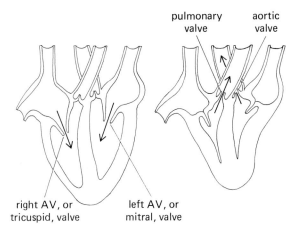

FIG. 20-6 Position of the heart valves and direction of blood flow during ventricular relaxation (left) and ventricular contraction (right).

venae cavae and pulmonary veins into the right and left atrium, respectively.

We can now list the structures through which blood flows in passing from the systemic veins to the systemic arteries (Fig. 20-7). The driving force for this flow of blood, as we shall see, comes solely from the active contraction of the cardiac muscle. The valves play no part at all in initiating flow and only prevent the blood from flowing in the opposite direction.

The Vascular System of the Heart

The blood in the heart chambers does not exchange nutrients and metabolic end products with the cells that constitute the heart walls. The heart, like all other organs, receives its blood supply via arterial branches (the right and left coronary arteries, Figs. 20-8 and 20-9) that arise from the aorta just after its origin at the left ventricle. The **right coronary artery** passes to the right and runs along the groove known as the *coronary sulcus,* between the right atrium and right ventricle and continues around to the posterior surface of the heart as far as the sulcus that runs down the posterior surface between the two ventricles. The main branch of the right coronary artery, the *posterior interventricular artery* (formerly called the posterior descending branch), turns and runs into the sulcus between the two ventricles.

The **left coronary artery** divides shortly after its

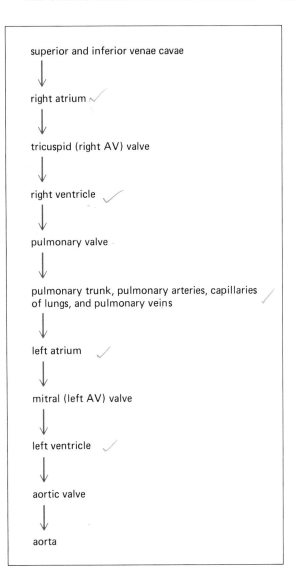

FIG. 20-7 Path of blood flow from the systemic circulation through the pulmonary circulation and back to the systemic circulation.

origin at the aorta into anterior interventricular and circumflex branches. The *anterior interventricular* artery (formerly called the anterior descending artery) runs in the groove between the two ventricles on the anterior surface of the heart and gives off branches to both ventricles. A *left marginal artery* (not officially named) branches from the anterior interventricular artery. The *circumflex artery* runs in the coronary sulcus between the left atrium and left ventricle and continues along the posterior surface of the heart. The largest branch of the circum-

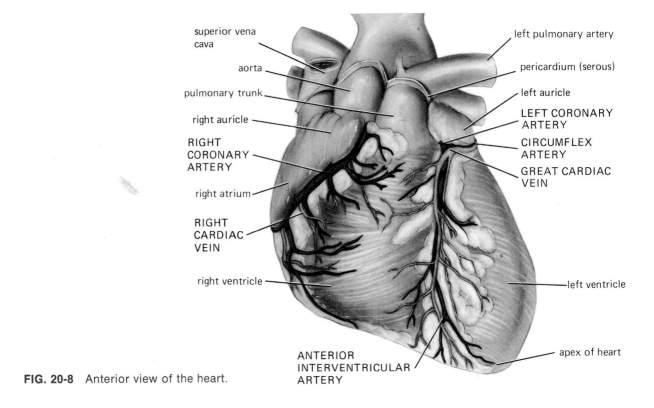

FIG. 20-8   Anterior view of the heart.

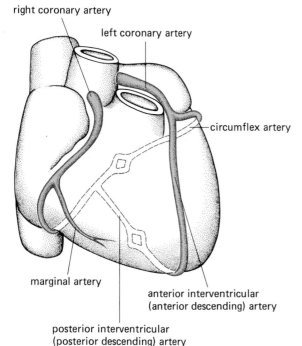

FIG. 20-9   Diagram of the coronary arteries.

flex artery is called (not officially) the *posterior* or *left ventricular branch* and runs along the posterior surface of the left ventricle. A *right marginal artery* (not officially named) also branches from the right coronary artery.

In general, the right coronary artery supplies the right atrium, right ventricle, and the interatrial septum and, therefore, generally supplies the sinoatrial and atrioventricular nodes, two crucial components of the electrical conduction system of the heart. (The conduction system will be discussed in detail later.) The left coronary artery generally supplies the left atrium, left ventricle, and interventricular septum and thus nourishes parts of the electrical conducting system of the ventricles. However, there are many variations in the coronary arteries and, therefore, many exceptions to these generalizations.

Most of the veins of the heart (referred to as **cardiac veins** rather than coronary veins) run parallel to the arteries (Figs. 20-8 and 20-10). The **great cardiac vein** lies in the groove between the ventricles on the anterior surface of the heart (with the anterior interventricular artery) and turns toward

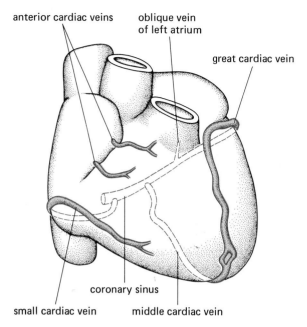

FIG. 20-10 Diagram of the cardiac veins.

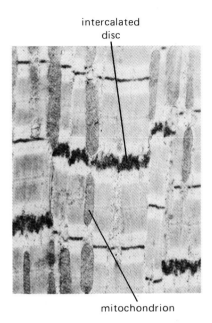

Fig. 20-11 Electron micrograph of cardiac muscle. Note the striations similar to those of skeletal muscle. The wide dark bands, called intercalated disks, occur where adjacent muscle fibers interdigitate.

the left between the left ventricle and left atrium. The great cardiac vein becomes continuous with the **coronary sinus,** which ends in the posterior wall of the right atrium where it empties into the atrial chamber. The major veins that feed into the great cardiac vein and coronary sinus on the posterior surface of the heart (the *oblique vein of the left atrium,* the *middle cardiac vein,* and the *small cardiac vein*) are shown in Fig. 20-10. In addition to these veins, there are several small *anterior cardiac veins* that arise on the wall of the right atrium as well as numerous *least cardiac veins;* these veins empty directly into the heart chambers. The coronary sinus returns the far greater part of the blood from the cardiac muscle to the heart chambers, but some is returned by the other veins as well.

The Heart Walls

The walls of the atria and ventricles are composed of layers of myocardium that are tightly bound together and completely encircle the blood-filled chambers. Thus, when the walls of a chamber contract, they come together like a squeezing fist, thereby exerting pressure on the blood they enclose. Cardiac muscle cells combine certain of the properties of smooth and skeletal muscle. The individual cell is striated (Fig. 20-11) because of the repeating sarcomeres, which contain both the thick myosin and thin actin filaments described for skeletal muscle. However, cardiac cells are considerably shorter than skeletal fibers and have several branching processes. The processes of adjacent cells are joined end to end at **intercalated discs,** within which are two types of membrane junctions: (1) desmosomes, which hold the cells together and to which the myofibrils are attached; and (2) gap junctions, which allow action potentials to be transmitted from one cardiac cell to another, in a manner similar to that in smooth muscle.

Besides the ordinary myocardial fibers, certain areas of the heart contain specialized muscle fibers that are essential for normal excitation of the heart. These muscle fibers constitute a network known as the **conducting system** and are in contact with the ordinary cardiac fibers via gap junctions that permit passage of action potentials from the conducting system to the regular myocardial cells.

The heart receives a rich supply of sympathetic and parasympathetic nerve fibers, the latter contained in the vagus nerves. The postganglionic sympathetic fibers terminate upon the cells of the specialized conducting system of the heart as well as on the ordinary myocardial cells of the atria and

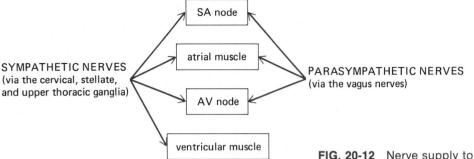

FIG. 20-12 Nerve supply to the heart.

ventricles. The parasympathetic neurons also innervate the conducting system and the atrial myocardial cells but have only sparse distribution to the ventricular myocardium (Fig. 20-12). The sympathetic postganglionic fibers release norepinephrine, and the parasympathetics release acetylcholine. The myocardial receptors for norepinephrine are β-adrenergic and those for acetylcholine are muscarinic.

HEARTBEAT COORDINATION

Contraction of cardiac muscle is triggered by depolarization of the muscle fiber membrane. Before dealing with the mechanisms by which membrane excitation is initiated and spread in the heart we may consider what would happen if all the many muscle fibers in the heart were to contract in a random manner. One result would be lack of coordination between pumping by the atria and ventricles, but this defect is dwarfed by the more serious lack of muscle coordination in the ventricles. Random contraction of the many ventricular muscle cells would cause the blood to be sloshed back and forth in the ventricular cavities instead of being ejected into the aorta and pulmonary trunk. Thus, the complex muscle masses that form the ventricular pumps must contract more or less simultaneously for efficient pumping.

Such coordination is made possible by two factors, already mentioned: (1) The gap junctions allow spread of an action potential from one fiber to the next so that excitation in one muscle fiber spreads throughout the heart; (2) the specialized conducting system in the heart facilitates the rapid and coordinated spread of excitation. Where and how does the action potential first arise, and what is the path and sequence of excitation?

Origin of the Heartbeat

Certain cardiac muscle cells, like certain forms of smooth muscle, are *autorhythmic;* i.e., they are capable of spontaneous, rhythmic self-excitation. When the individual cells of a salamander embryo heart are separated and placed in salt solution, they beat spontaneously, but they beat at different rates. The top half of Fig. 20-13 shows recordings of membrane potentials from two such cells, the most important feature of which is the gradual depolarization that causes the membrane potential to reach threshold, at which point an action potential occurs. Following the action potential, the membrane potential returns to the initial resting value, and the gradual depolarization begins again. This capacity for autonomous depolarization toward threshold makes the rhythmic self-excitation of the muscle cells possible. It is caused by a decreasing membrane permeability to potassium, but just how this change is generated "spontaneously" remains obscure. It should be evident that the slope of this depolarization, i.e., the rate of membrane potential change per unit time, determines how quickly threshold is reached and the next action potential elicited. Accordingly, cell A has a faster rate of firing than cell B.

In the course of the salamander experiment above, many individual cells form gap junctions. When such a junction is formed between two cells previously contracting autonomously at different rates, both the joined cells contract at the faster rate (Fig. 20-13, bottom). In other words, the faster cell sets the pace, causing the initially slower cell to contract at the faster rate. The mechanism is straightforward: The action potential generated by the faster cell causes depolarization, via the gap junctions, of the second cell's membrane to thresh-

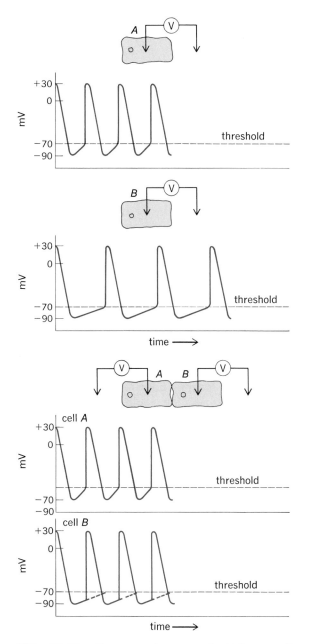

FIG. 20-13 Transmembrane potential recordings from cardiac muscle cells grown in tissue culture. The dashed lines in the bottom recording of cell B indicate the course depolarization would have followed if the cells had not been joined.

old, at which point an action potential occurs in this second cell. The important generalization emerges that, because of the gap junctions between cardiac muscle cells, all the cells are excited at the rate set by the cell with the fastest autonomous

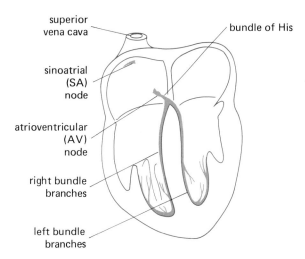

FIG. 20-14   Conducting system of the heart.

rhythm. Precisely the same explanation holds for the origination of the heartbeat in the intact heart.

Several areas in the conducting system of the adult mammalian heart demonstrate these same characteristics of autorhythmicity and pacemaking, the one with the fastest inherent rhythm being a small mass of specialized myocardial cells embedded in the right atrial wall near the entrance of the superior vena cava (Fig. 20-14). Called the **sinoatrial (SA) node,** it is the normal pacemaker for the entire heart. Figure 20-15A is an intracellular recording from an SA node cell; note the slow depolarization toward threshold that initiates the action potential. Compare this SA nodal action potential to that of an unspecialized nonautorhythmic atrial cell, which fails to show the pacemaker potential (Fig. 20-15B). If the activity of the SA node is depressed or conduction from it blocked, another portion of the conducting system may take over as pacemaker. In contrast, ordinary atrial and ventricular muscle fibers, which constitute 99 percent of the total cardiac muscle, are not normally capable of generating pacemaker potentials and do so only under abnormal conditions.

Sequence of Excitation

The sequence of cardiac excitation is illustrated in Fig. 20-16. The cells of the SA node form gap junctions with the surrounding atrial myocardial fibers. From the SA node, the wave of excitation spreads throughout both atria along ordinary atrial myocardial cells, passing from cell to cell by way of the

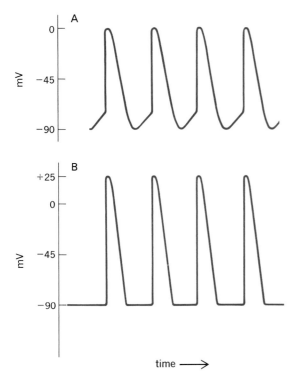

FIG. 20-15 Transmembrane potential recordings from SA node cell (A) and atrial muscle fiber (B).

At the base of the right atrium very near the interventricular septum, the wave of excitation encounters a second small mass of specialized cells, the **atrioventricular (AV) node.** This node, and the bundle of special conducting fibers (the **bundle of His**) that leaves the node, constitute the only conducting link between the atria and ventricles, all other areas being separated by nonconducting connective tissue. This anatomic pattern ensures that excitation will travel from atria to ventricles only through the AV node, but it also means that malfunction of the AV node may completely dissociate atrial and ventricular excitation. The AV node manifests one particularly important characteristic: The propagation of action potentials through the node is delayed for approximately 0.1 sec (mainly because of the small diameter of these cells), allowing the atria to contract and empty their contents into the ventricles before ventricular contraction.

After leaving the AV node, the impulse travels through the bundle of His. These fibers in turn divide into **right** and **left bundle branches,** which run down the interventricular septum and then branch into the **Purkinje fibers** that run throughout much of the right and left ventricular myocardium (Fig. 20-14). Finally, these fibers form gap junctions with ordinary myocardial fibers through which the impulse spreads from cell to cell in the remaining myocardium. The rapid conduction along the special conducting fibers and their highly diffuse distribution cause depolarization of all right and left ventricular cells more or less simultaneously and ensure a single coordinated contraction.

gap junctions. There are also specialized fiber bundles **(internodal tracts)** that conduct the impulse from the SA node directly to the left atrium and the base of the right atrium, thereby ensuring the virtually simultaneous excitation of both atria.

How does the excitation spread to the ventricles?

FIG. 20-16 Sequence of cardiac excitation. Atrial excitation is complete before ventricular excitation begins because of the delay at the AV node.

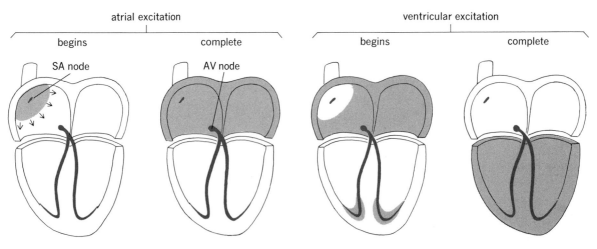

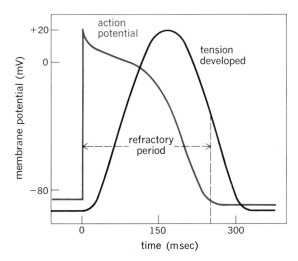

FIG. 20-17 Relationship between membrane potential changes and contraction in a single cardiac muscle cell. The refractory period lasts almost as long as the contraction.

Refractory Period of the Heart

The pumping of blood requires alternate periods of contraction and relaxation. Imagine the result of a prolonged tetanic contraction of cardiac muscle like that described for skeletal muscle. Obviously, pumping would cease and death would ensue. In reality such contractions never occur in the heart because of the long refractory period of cardiac muscle. Recall that in any excitable membrane an action potential is accompanied by a period during which the membrane cannot be reexcited. Follow-

ing this absolute refractory period comes a second period during which the membrane can be depolarized again but only by a stimulus more intense than usual. In skeletal muscle, the absolute refractory periods are very short (1 to 2 msec) compared with the duration of contraction (20 to 100 msec), and a second contraction can be elicited before the first is over. In contrast, the absolute refractory period of cardiac muscle lasts almost as long as the contraction (250 msec), and the muscle cannot be reexcited in time to produce summation (Fig. 20-17).

A common situation explainable in terms of the refractory period is shown in Fig. 20-18. In many people, drinking several cups of coffee causes increased excitability of areas other than the SA node as a result of the action of caffeine. When such an area of the atria or ventricles actually initiates an action potential, it is known as an **ectopic focus.** When one of these areas fires just after completion of a normal contraction but before the next SA nodal impulse, a premature wave of excitation occurs. As a result, the next normal SA nodal impulse occurs during the refractory period of the premature beat and is not propagated because the myocardial cells are refractory (the SA node still fires because it has a shorter refractory period). The second SA nodal impulse after the premature contraction is propagated normally. The net result is an unusually long delay between beats. The contraction after the delay is unusually strong, and the person is aware of his or her heart pounding. If the ectopic focus continued to discharge at a rate higher than that of the SA node, it might then capture the

FIG. 20-18 Effect of an ectopic discharge on ventricular contraction. The arrows indicate the times at which the SA node or ectopic focus fires. The premature beat induced by the ectopic discharge makes the ventricular muscle refractory at the time of the next SA-node impulse. The failure of this impulse to induce a contraction results in a longer period than normal before the next beat.

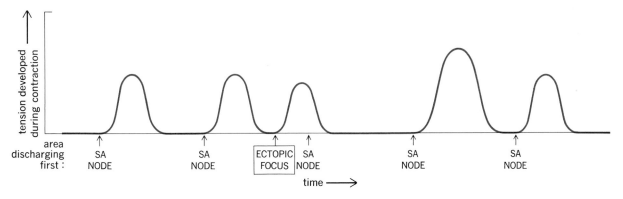

role of pacemaker and drive the heart at rates as high as 200 to 300 beats per minute, compared with a normal rate of 70.

Many similar examples are important clinically and help to clarify the normal physiologic process. For example, disease may damage cardiac tissue and hamper conduction through the AV node so that only a fraction of the atrial impulses are transmitted into the ventricles; thus, the atria may have a rate of 80 beats per minute and the ventricles only 60. If there is complete block at the AV node, none of the atrial impulses get through and the AV node itself begins to initiate excitation at its own spontaneous rate; i.e., the AV node (or a portion of the conducting system just below) is capable of pacemaker action when released from capture by impulses from the SA node. This rate is quite slow, generally 25 to 40 beats per minute, and completely out of synchrony with the atrial contractions, which continue at the normal higher rate. Under such conditions, the atria are totally ineffective as pumps because they are usually contracting against closed AV valves, but atrial pumping, as we shall see, is relatively unimportant for cardiac functioning (except during relatively strenuous exercise). Some patients have transient recurrent episodes of complete AV block signaled by fainting spells (a result of decreased brain blood flow). These spells result because the ventricles do not begin their own impulse generation immediately and their pumping ceases temporarily.

Partial AV block is not always caused by disease and may frequently represent a normal life-saving adaptation. Imagine a patient with an ectopic area driving the atria at 300 beats per minute. Ventricular rates this high are very inefficient because there is inadequate time to fill the ventricles between contractions. Fortunately, the long refractory period of the AV node may prevent passage of a significant fraction of the impulses and the ventricles beat at a slower rate.

Another group of abnormalities apparently is characterized by a prolonged or unusual conduction route so that the impulse constantly meets an area that is no longer refractory and keeps traveling around the heart in a so-called circus movement. This may lead to continuous, completely disorganized contractions (*fibrillation*), which can cause death if they occur in the ventricles. Indeed, ventricular fibrillation is the immediate cause of death

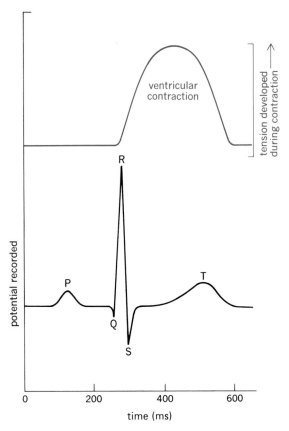

FIG. 20-19 Typical electrocardiogram. P, atrial depolarization; QRS, ventricular depolarization; T, ventricular repolarization.

from electrocution and many instances of heart attack.

The Electrocardiogram

The **electrocardiogram (EKG)**[1] is a tool for evaluating the electrical events in the heart. The action potentials of cardiac muscle produce currents that flow throughout the body fluids, becoming weaker with increasing distances from the heart. These currents can be detected at the surface of the skin by attaching small metal plates (connected to a sensitive voltmeter) at different places on the body; the recording produced by the voltmeter is the EKG.

Figure 20-19 and the top of Fig. 20-20 illustrate a typical normal EKG recorded as the potential difference between the right and left wrists. The

[1] Traditionally, electrocardiogram has been abbreviated EKG, the K derived from the Greek word for heart. ECG is actually the preferred term.

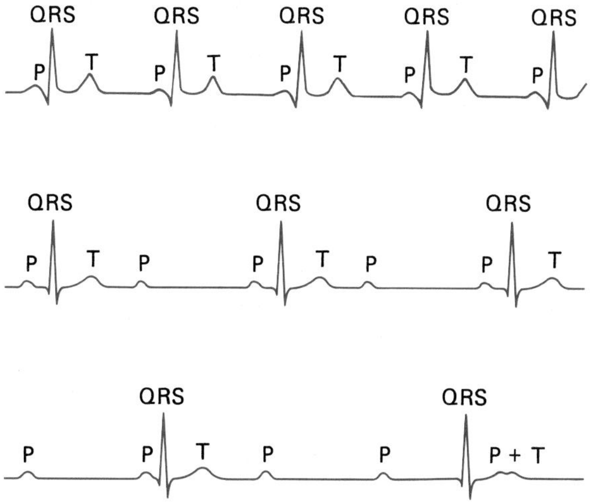

FIG. 20-20 Electrocardiograms from a healthy person and two persons suffering from atrioventricular block. Upper tracing: A normal EKG. Middle tracing: Partial block; one-half of the atrial impulses are transmitted to the ventricles. Lower tracing: Complete block; there is absolutely no synchrony between atrial and ventricular electric activities.

first wave **P** represents atrial depolarization. The second complex **QRS,** occurring approximately 0.1 to 0.2 sec later, represents ventricular depolarization. The final wave **T** represents ventricular repolarization. No manifestation of atrial repolarization is evident because it occurs during ventricular depolarization and is masked by the QRS complex. Figure 20-20, center, gives one example of the clinical usefulness of the EKG. This patient is suffering from partial AV nodal block so that only one-half the atrial impulses are being transmitted. Note that every second P wave is not followed by a QRS and T. Because many myocardial defects alter normal impulse propagation, and thereby the shapes of the waves, the EKG is a powerful tool for diagnosing heart disease.

Excitation-Contraction Coupling

The depolarization ("excitation") of cardiac muscle cells triggers their contraction. As described in Chap. 11, the mechanism that couples excitation and contraction is an increase in the cytosolic concentration of calcium; this calcium combines with the regulator protein, troponin, with the resulting removal of tropomyosin's inhibition of cross-bridge formation between actin and myosin. There are two sources of the increased cytosolic calcium during the action potential: (1) diffusion of calcium from interstitial fluid across the plasma membrane into the cytosol during the plateau of the action potential, and (2) release of calcium from the sarcoplasmic reticulum. During repolarization, cytosolic calcium is restored to its original extremely low level by active transport of calcium back across the plasma membrane or into the sarcoplasmic reticulum, and the muscle relaxes. As we shall see, changes in cytosolic calcium account for many situations in which the strength of cardiac muscle contraction is altered.

MECHANICAL EVENTS OF THE CARDIAC CYCLE

Fluid always flows from a region of higher pressure to one of lower pressure. (This important concept will be developed formally in Chap. 21; at present, we need deal with it only as an intuitively obvious phenomenon.) The sole function of the heart is to generate the pressures that produce blood flow, the heart valves serving to direct the flow. The orderly process of depolarization described in the previous section triggers contraction of the atria followed rapidly by ventricular contraction.

A single contraction-relaxation episode of the heart constitutes a **cardiac cycle.** The cardiac cycle is divided into two major phases: a period of ventricular contraction (**systole**), followed by a period of ventricular relaxation (**diastole**). We start our analysis of the cardiac cycle at the far left of Fig. 20-21 with the events of mid-to-late diastole, considering first only the left heart; events on the right side have an identical sequence, but the pressure changes are smaller.

Mid-to-Late Diastole

The left atrium and ventricle are both relaxed; left atrial pressure is very slightly higher than left ventricular pressure (because blood is entering the atrium from the pulmonary veins); therefore, the AV valve is open, and blood is passing from atrium to ventricle. This is an important point: The ventricle receives blood from the atrium throughout most of diastole, not just when the atrium contracts. Indeed, at rest, approximately 80 percent of ven-

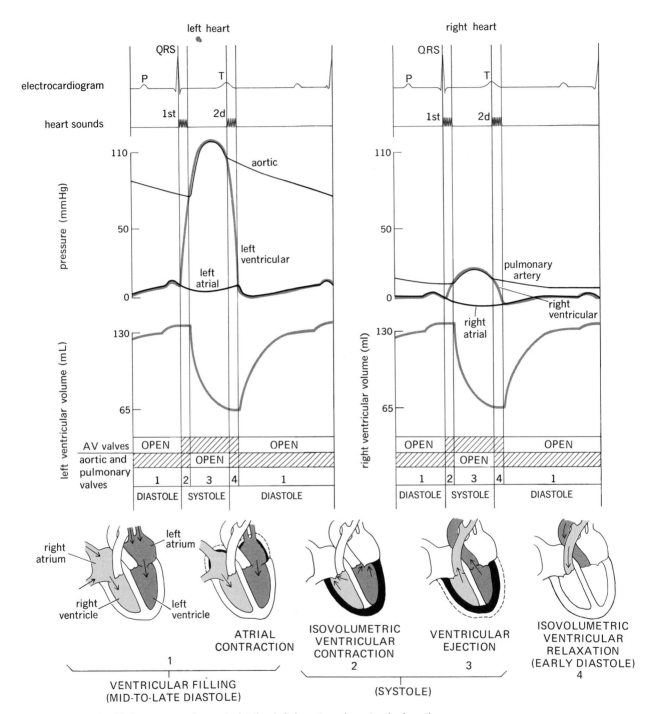

FIG. 20-21 *(Left)* Summary of events in the left heart and aorta during the cardiac cycle. At *c'* the AV valve closes; at *o'* it opens. At *o* the aortic valve opens; at *c* it closes. *(Right)* Summary of events in the right heart and pulmonary arteries during the cardiac cycle At *c'* the AV valve closes; at *o'* it opens. At *o* the pulmonary valve opens; at *c* it closes. In the lower part of the figure, the contracting portions of the heart are shown in black.

tricular filling occurs before atrial contraction.[2] Note that the aortic valve (between aorta and left ventricle) is closed because the aortic pressure is higher than the ventricular pressure. The aortic pressure is slowly falling because blood is moving out of the arteries and through the vascular tree; in contrast, ventricular pressure is rising slightly because blood is entering from the atrium, thereby expanding the ventricular volume. At the very end of diastole, the SA node discharges, the atrium depolarizes (as shown by the P wave of the EKG), the atrium contracts (note the small rise in atrial pressure), and a small volume of blood is added to the ventricle. The amount of blood in the ventricle just prior to systole is called the **end-diastolic volume.**

Systole

The wave of depolarization passes through the ventricle and triggers ventricular contraction. As the ventricle contracts, it squeezes the blood contained in it and ventricular pressure rises steeply. Almost immediately, this pressure exceeds the atrial pressure and closes the AV valve, thus preventing backflow into the atrium. Since, for a brief period, the aortic pressure still exceeds the ventricular, the aortic valve remains closed and the ventricle does not empty despite contraction. This early phase of systole is called **isovolumetric ventricular contraction** because ventricular volume is constant; i.e., the lengths of the muscle fibers remain approximately constant as in an isometric skeletal muscle contraction. This brief phase ends when ventricular pressure exceeds aortic, the aortic valve opens, and **ventricular ejection** occurs. The ventricular volume curve shows that ejection is rapid at first and then tapers off. The ventricle does not empty completely; the amount remaining after ejection is called the **end-systolic volume.** As blood flows into the aorta, the aortic pressure rises with ventricular pressure. Atrial pressure also rises slowly throughout the entire period of ventricular ejection because of continued inflow of blood from the veins. Note

that peak aortic pressure is reached before the end of ventricular ejection; i.e., the pressure actually is beginning to fall during the last part of systole despite continued ventricular ejection. This occurs because the rate of blood ejection during this last part of systole is quite small (as shown by the ventricular volume curve) and is less than the rate at which blood is leaving the aorta (and other large arteries) via the arterioles; accordingly the volume and, therefore, the pressure in the aorta begin to decrease.

Early Diastole

When contraction stops, the ventricular muscle relaxes rapidly because of release of tension created during contraction. Ventricular pressure therefore almost immediately falls below aortic pressure, and the aortic valve closes. However, ventricular pressure still exceeds atrial pressure so that the AV valve remains closed. This phase of early diastole, obviously the mirror image of early systole, is called **isovolumetric ventricular relaxation.** It ends as ventricular pressure falls below atrial pressure, the AV valve opens, and ventricular filling begins. Filling occurs rapidly at first and then slows down. The fact that ventricular filling is almost complete during early diastole is of the greatest importance; it ensures that filling is not seriously impaired during periods of rapid heart rate, e.g., exercise, emotional stress, and fever, despite a marked reduction in the duration of diastole. However, when rates of approximately 200 beats per minute or more are reached, filling time does become inadequate so that the volume of blood pumped during each beat is decreased and cardiac pumping is impaired. Significantly, the AV node in normal adults does not conduct at rates greater than 200 to 250 beats per minute.

Pulmonary Circulatory Pressures

Figure 20-21 also summarizes the simultaneously occurring events in the right heart and pulmonary arteries, the patterns being virtually identical to those just described for the left heart. There is one striking quantitative difference: The pressures in the right heart are considerably lower during systole. The pulmonary circulation is a low-pressure system (for reasons to be described later). This

---

[2] It is for this reason that the conduction defects discussed above, which eliminate the atria as efficient pumps, do not seriously impair cardiac function, at least at rest. Many persons lead relatively normal lives for many years despite atrial fibrillation. Thus, in many respects, the atrium may be conveniently viewed as merely a continuation of the large veins.

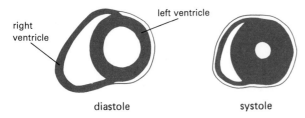

right ventricle

left ventricle

diastole   systole

FIG. 20-22   The relative shape and wall thickness of the right and left ventricles during diastole and systole.

difference is clearly reflected in the ventricular architecture, the right ventricular wall being much thinner than the left (Fig. 20-22). Note, however, that despite the lower pressure, the right ventricle ejects the same amount of blood as the left.

Heart Sounds

Two **heart sounds** (see Fig. 20-21) are normally heard through a stethoscope placed on the chest wall. The first sound, a low pitched *lub,* is associated with closure of the AV valves at the onset of systole; the second, a high-pitched *dup,* is associated with closure of the pulmonary and aortic valves at the onset of diastole. These sounds, which result from vibrations caused by valvular closure, are perfectly normal, but abnormal sounds, known as **heart murmurs,** are frequently (although not always) a sign of heart disease. Murmurs can be produced by blood flowing rapidly in the usual direction through an abnormally narrowed valve (*stenosis*), backward through a damaged leaky valve (*regurgitation*), or between the two atria or two ventricles via a small hole in the septum. The exact timing and location of the murmur provide the physician with a powerful diagnostic clue. For example, a murmur heard throughout systole suggests a narrowed pulmonary or aortic valve or a hole in the interventricular septum, whereas a murmur heard throughout diastole suggests a leaky mitral or tricuspid valve. The diagnosis can then be completed by the use of specialized techniques.

THE CARDIAC OUTPUT

The volume of blood pumped by each ventricle per minute is called the **cardiac output,** usually expressed as liters per minute. Note that the cardiac output is the amount of blood pumped by *each* ventricle, *not* the total amount pumped by both ventricles. The cardiac output is determined by multiplying the number of times the heart beats each minute (**heart rate**) and the volume of blood ejected by each ventricle during each beat (**stroke volume**).

$$\underset{\text{L/min}}{\text{Cardiac output}} = \underset{\text{beats/min}}{\text{heart rate}} \times \underset{\text{L/beat}}{\text{stroke volume}}$$

For example, if each ventricle has a rate of 72 beats per minute and ejects 70 mL with each beat, the cardiac output is:

$$\text{CO} = 72 \text{ beats/min} \times 0.07 \text{ L/beat} = 5.0 \text{ L/min}$$

These values are approximately normal for a resting adult. During periods of exercise, the cardiac output may reach 30 to 35 L/min. Obviously, heart rate or stroke volume or both must have increased. Physical exercise is but one of many situations in which various tissues and organs require a greater flow of blood; e.g., flow through skin vessels increases when heat loss is required, and flow through intestinal vessels increases during digestion. Some of the increased flow can be obtained merely by decreasing blood flow to some other organ, i.e., by redistributing the cardiac output, but most of the supply must come from a greater cardiac output. The following description of the factors that alter the two determinants of cardiac output, heart rate and stroke volume, applies in all respects to both the right and left heart since stroke volume and heart rate are the same for both.

Control of Heart Rate

The rhythmic discharge of the SA node occurs spontaneously in the complete absence of any nervous or hormonal influences; however, it is under the constant influence of both nerves and hormones. As mentioned earlier, a large number of parasympathetic and sympathetic fibers end on the SA node as well as on other areas of the conducting system.

Stimulation of the parasympathetic (vagus) nerves (or local application of acetylcholine) causes slowing of the heart and, if strong enough, may stop the heart completely for some time. The effects of the sympathetic nerves are just the reverse; nerve stimulation (or local application of norepinephrine) increases the heart rate. Both the sympathetic and parasympathetic nerves normally discharge at some finite rate, but in the resting state the parasympa-

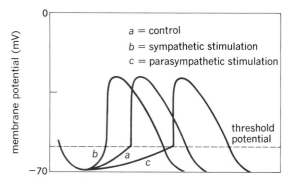

FIG. 20-23 Effects of sympathetic and parasympathetic nerve stimulation on the slope of the pacemaker potential of an SA node cell.

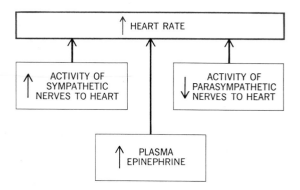

FIG. 20-24 Summary of major factors which influence heart rate. All effects are exerted upon the SA node. The figure, as drawn, shows how heart rate is increased; reversal of all arrows in the boxes would illustrate how heart rate is decreased.

thetic influence is dominant and the normal heart rate is slower than the inherent autonomous discharge rate of the SA node.

Figure 20-23 illustrates the nature of the sympathetic and parasympathetic influence on SA node function. Sympathetic stimulation increases the slope of the pacemaker potential, thus causing the cell to reach threshold more rapidly and the heart rate to increase. Stimulation of the parasympathetics has the opposite effect: The slope of the pacemaker potential decreases, threshold is reached more slowly, and heart rate decreases. How do these autonomic mediators alter the slope of the pacemaker potential? Acetylcholine does so by increasing the permeability of the membrane to potassium; the story for norepinephrine is not yet clear, but the most likely explanation is that it does just the opposite, decreasing the membrane permeability to potassium.

Factors other than the cardiac nerves can alter heart rate. Epinephrine, the hormone liberated from the adrenal medulla, speeds the heart; this is not surprising since epinephrine is similar in structure to norepinephrine (Chap. 13). Temperature, plasma electrolyte concentrations, and hormones other than epinephrine can also affect heart rate. However, these are generally of lesser importance, and the heart rate is primarily regulated very precisely by the balancing of the slowing effects of parasympathetic discharge against the accelerating effects of sympathetic discharge, both operating on the SA node (Fig. 20-24).

Control of Stroke Volume

The second variable that determines cardiac output is stroke volume. It is important to recognize that the ventricles never completely empty themselves of blood during contraction; therefore, a more forceful contraction, i.e., a greater shortening of myocardial fibers, can produce an increase in stroke volume. Changes in the force of contraction can be produced by a variety of factors, but two are dominant under most physiological conditions: (1) changes in the degree of stretching of the ventricular muscle secondary to changes in ventricular volume, and (2) alterations in the magnitude of sympathetic nervous system input to the ventricles.

The Relationship between Stroke Volume and End-Diastolic Volume It is possible to study the completely intrinsic adaptability of the heart by means of the so-called heart-lung preparation. Tubes are placed in the heart and vessels of an anesthetized animal so that blood flows from the very first part of the aorta (just above the exit of the coronary arteries) into a blood-filled reservoir and from there into the right atrium. The blood is then pumped by the right heart as usual via the lungs into the left heart (Fig. 20-25). The net effect is to nourish the heart and lungs normally and to deprive the rest of the animal's body of blood and cause death, thereby abolishing all nervous and hormonal activity. A key feature of this preparation is that the pressure that causes blood flow into the heart (venous pressure)

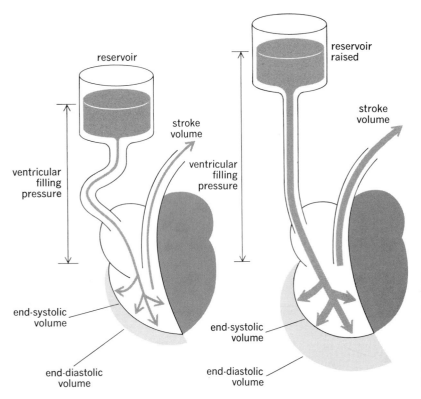

FIG. 20-25 Demonstration of the intrinsic control of stroke volume. By raising the reservoir, the pressure causing ventricular filling is increased. The increased filling distends the ventricle, which responds with an increased strength of contraction.

can be altered simply by raising or lowering the reservoir. This is analogous to altering the venous and right atrial pressures and causes changes in the quantity of blood entering the ventricles during diastole. Thus, when the reservoir is raised, ventricular filling increases, thereby increasing end-diastolic volume. The more distended ventricle responds with a more forceful contraction. The net result is a new steady state, in which the ventricle is distended and diastolic filling and stroke volume are both increased, but equally. The mechanism that underlies this completely intrinsic adaptation is that cardiac muscle, like other muscles, shows a relationship between muscle fiber length and tension development. However, unlike skeletal muscle, cardiac muscle length in the resting state is less than that which yields maximal tension during contraction so that an increase in length produces an increase in contractile tension. Thus, in the experiment above, the increased diastolic volume stretches the ventricular muscle fibers and causes them to contract more forcefully.

The British physiologist Starling observed that there was a direct proportion between the diastolic volume of the heart, i.e., the length of its muscle fibers, and the force of contraction of the following

systole, and this relationship is now referred to as *Starling's law of the heart*. A typical response curve obtained by progressively increasing end-diastolic volume is shown in Fig. 20-26. Note that marked overstretching causes the force of contraction, and thereby the stroke volume, to fall off. Thus, heart muscle manifests a length-tension re-

FIG. 20-26 Relationship between ventricular end-diastolic volume and stroke volume (Starling's law of the heart). The data were obtained by progressively increasing ventricular filling pressure, as in Fig. 20-25. The horizontal axis could have been labelled "sarcomere length," and the vertical "contractile force," i.e., this is a length-tension curve.

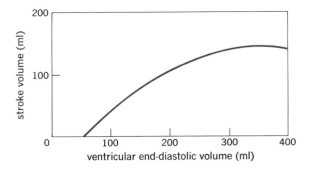

lationship very similar to that described earlier for skeletal muscle (see Fig. 11-19) and explainable in terms of the sliding-filament mechanism of muscle contraction. This intrinsic relationship between end-diastolic volume and stroke volume, originally demonstrated in the heart-lung preparation, applies equally to the intact human being. End-diastolic volume, therefore, becomes a crucial determinant of cardiac output.

What then are the factors that determine end-diastolic volume, i.e., the degree of ventricular distension just before systole? The simplest way to approach this question is to view the ventricle as an elastic chamber, much like a balloon. A balloon enlarges when one blows into it because the internal pressure acting upon the wall becomes greater than the external pressure. The more air blown in, the higher the internal pressure becomes. The degree of distension therefore depends upon the pressure difference across the wall and the distensibility of the wall. This is precisely the situation for the ventricle. We shall ignore the problem of ventricular distensibility (which changes little under physiologic conditions) and concentrate only on the **transmural,** or across-the-wall, difference in pressure.

The internal pressure is the fluid pressure exerted by the blood against the walls. The external pressure surrounding the heart is the pressure in the chest cavity (thorax), or **intrathoracic pressure.** A typical end-diastolic ventricular blood pressure is 4 mmHg. Physiologists state pressures with the atmospheric pressure, i.e., the pressure of the atmospheric air surrounding the body, given as zero. Thus, when we say the end-diastolic pressure is 4 mmHg, we really mean 4 mmHg greater than atmospheric pressure. If we assume the atmospheric pressure to be 760 mmHg, the true internal pressure is 764 mmHg. This distinction must be remembered, especially when one is discussing events within the thoracic cage, because, for reasons to be described in Chap. 23, the intrathoracic pressure of the fluid surrounding the heart, lungs, and all other intrathoracic structures is *less* than atmospheric. This subatmospheric intrathoracic pressure is frequently termed a "negative" pressure, but this terminology should be avoided because there is no such thing in nature as a negative pressure. The intrathoracic pressure averages approximately 5 mmHg *less* than atmospheric pressure (or a true pressure of 755 mmHg); accordingly, the pressure difference acting to distend the ventricles at the end of diastole is $4 + 5 = 9$ mmHg, i.e., $764 - 755 = 9$ mmHg.

We can now answer our question. End-diastolic ventricular volume can be increased either by increasing the intraventricular blood pressure or by decreasing the intrathoracic pressure, or both. The latter occurs during inspiration (see Chap. 23) and accounts, in part, for the increased stroke volume that usually occurs during inspiration. However, it is primarily by changes in the end-diastolic intraventricular pressure that end-diastolic volume is controlled. And, as described earlier, the diastolic ventricular blood pressure is determined by the atrial pressure. As we shall see, the atrial pressure, in turn, is determined by the pressure in the veins that empty into the atria.

The significance of this mechanism should now be apparent; an increased flow of blood from the veins into the heart automatically forces an equivalent increase in cardiac output by distending the ventricle and increasing stroke volume, just like the heart-lung apparatus when we increased "venous return" by elevating the reservoir. This is probably the single most important mechanism for maintaining equality of right and left output. Should the right heart, for example, suddenly begin to pump more blood than the left, the increased blood flow to the left ventricle would automatically produce an equivalent increase in left ventricular output and blood would not be allowed to accumulate in the lungs. Another example of Starling's law has already been described above, namely, the pounding that occurs after a premature contraction. Recall that an unusually long period elapses between the premature contraction and the next contraction; the period for diastolic filling is increased, end-diastolic volume increases, and the force of contraction is increased (Fig. 20-18). It is this strong contraction, which may actually lift the heart against the chest wall, that the person is aware of. In summary, end-diastolic volume and stroke volume are generally increased whenever atrial pressure increases; the factors that determine atrial pressure will be described in Chap. 21.

The Sympathetic Nerves Sympathetic nerves are distributed not only to the SA node and conducting system but to all myocardial cells. The effect of the sympathetic transmitter, norepinephrine, is to increase ventricular (and atrial) strength of contraction at any given initial muscle-fiber length, i.e., end-diastolic volume. This is defined as increased **contractility** and reflects an increased availability of calcium and thereby an increased number of active cross-bridges. Not only is the contraction more powerful but both it and relaxation occur more rapidly. These latter effects are quite important since, as described earlier, increased sympathetic activity to the heart also increases heart rate. As the heart rate increases, the time available for diastolic filling decreases, but the more rapid contraction and relaxation induced simultaneously by the sympathetic nerves partially compensate for this problem. Because the ventricles relax so rapidly after a contraction the intraventricular pressure falls rapidly, thereby creating an enhanced pressure gradient for flow of blood into the ventricles. This is, in essence, a ''sucking'' effect that facilitates ventricular filling. The ability of these effects to maintain diastolic filling in the face of a decreased filling-time is, of course, not unlimited, and diastolic filling is significantly reduced at very high heart rates. The significance of this interplay between diastolic filling time, heart rate, and contractility will be analyzed further in the section on exercise in Chap. 22.

Circulating epinephrine produces changes in contractility similar to those induced by the sympathetic nerves to the heart. They both increase contractility by inducing an increased movement of calcium into the cytosol during excitation, and they speed relaxation by enhancing the rate at which this calcium is transported back out of the cytosol. In contrast to the sympathetic nerves, the parasym-

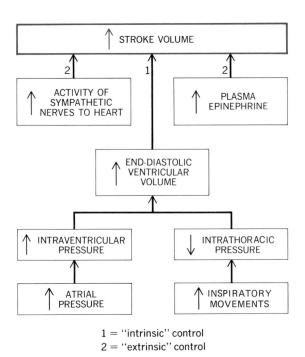

1 = ''intrinsic'' control
2 = ''extrinsic'' control

FIG. 20-28 Major factors which influence stroke volume. The figure as drawn shows how stroke volume is increased; a reversal of all arrows in the boxes would illustrate how stroke volume is decreased. (Refer to the text for details).

pathetic nerves to the heart have relatively little effect on ventricular contractility.

The interrelationship between Starling's law and the cardiac nerves as measured in a heart-lung preparation is illustrated in Fig. 20-27. The dashed line is the same as the line shown in Fig. 20-26 and was obtained by slowly raising ventricular pressure while measuring end-diastolic volume and stroke volume; the solid line was obtained similarly for the same heart but during sympathetic-nerve stimulation. Starling's law still applies, but during nerve stimulation the stroke volume is greater at any given end-diastolic volume. In other words, the increased contractility leads to a more complete ejection of the end-diastolic ventricular volume.

In summary (Fig. 20-28), stroke volume is controlled both by an intrinsic cardiac mechanism dependent only upon changes in end-diastolic volume and by an extrinsic mechanism mediated by the cardiac sympathetic nerves (and circulating epinephrine). The reflexes that control the nerves will be described in Chap. 22.

FIG. 20-27 Effects on stroke volume of stimulating the sympathetic nerves to the heart.

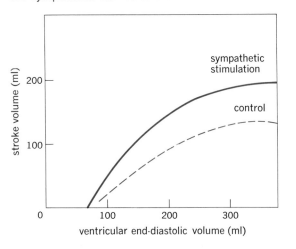

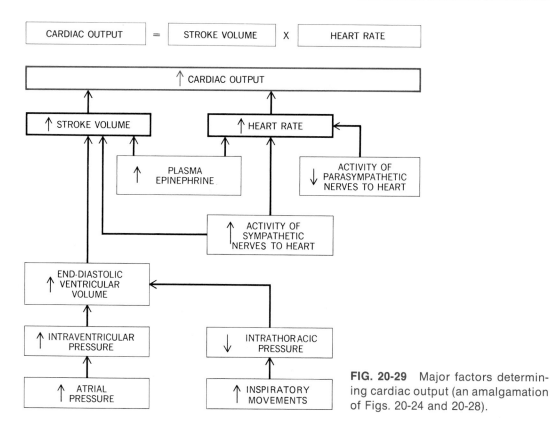

FIG. 20-29 Major factors determining cardiac output (an amalgamation of Figs. 20-24 and 20-28).

Summary of Cardiac Output Control

A summary of the major factors that determine cardiac output is presented in Fig. 20-29, which combines the information of Fig. 20-24 (factors that influence heart rate) and Fig. 20-28 (factors that influence stroke volume).

Cardiac Energetics

During contraction, the heart, like other kinds of muscle, converts chemical energy stored in ATP into mechanical work and heat. The total energy utilized per unit time is determined by a large number of factors, but the most important are the arterial pressure against which the ventricles must pump and the heart rate (i.e., the number of times per minute this pressure must be developed). This means that the energy expended by the heart may vary markedly in different situations having the same cardiac output. For example, if the heart rate is doubled and the stroke volume halved, cardiac output does not change but the energy required increases markedly. To take another example, the hearts of persons with *hypertension* (high arterial blood pressure) consume more energy than normal in pumping blood even though their cardiac output is normal. From these facts it should be clear why people with impaired coronary blood flow (i.e., decreased supply of oxygen and nutrients) are subjected to the greatest danger when their heart rate and arterial blood pressure increase, as during emotional excitement.

KEY TERMS

pulmonary circulation
systemic circulation
pulmonary trunk
pulmonary veins
aorta
venae cavae
myocardium
atrium

ventricle
valve
myocardial fibers
sinoatrial node
autorhythmicity
atrioventricular node
electrocardiogram (EKG)
systole

diastole
isovolumetric
ejection
end-diastolic volume
cardiac output
heart rate
stroke volume
Starling's law of the heart
contractility

REVIEW EXERCISES

1   Distinguish between the pulmonary and systemic circulations.
2   List in order the heart chambers, valves, and blood vessels through which the blood passes as it moves from the venae cavae to the aorta.
3   Explain how the heart beat originates.
4   Describe the spread of excitation throughout the heart.
5   List the areas of the heart nourished by the right coronary artery and its branches and the physiologic significance of these areas. Make a similar list for the left coronary artery.
6   Explain the significance of the delay in conduction of cardiac excitation at the AV node.
7   Explain the functional significance of the long refractory period characteristic of heart muscle.
8   Draw a normal electrocardiogram tracing; name each wave and describe the electrical events that accompany it. Assuming that contraction of the heart follows excitation, indicate on your tracing the duration of systole and diastole.

9   Describe the mechanical events of the heart and the pressure changes in the left ventricle and aorta during the following phases of the cardiac cycle: mid-diastole, atrial contraction, isovolumetric ventricular contraction, ventricular ejection, and isovolumetric ventricular relaxation.
10 Compare the pressures in the pulmonary circulation with those in the systemic circulation.
11 Calculate the cardiac output for a person whose heart rate is 70 beats/min and stroke volume is 35 mL/beat. State whether this is a normal value.
12 Describe the effect of the sympathetic and parasympathetic nerves on the heart rate; on the strength of cardiac contraction.
13 Describe the effect of increasing end-diastolic volume on the strength of cardiac contraction.
14 List the two most important factors that affect energy expenditure by the heart.

CHAPTER

21

THE CARDIOVASCULAR SYSTEM: BLOOD VESSELS AND LYMPHATICS

SECTION A: DESIGN OF THE VASCULAR SYSTEM

BASIC PLAN OF THE CIRCULATORY SYSTEM

As mentioned in the preceding chapter, in the systemic vascular circuit blood leaves the left half of the heart via the **aorta.** All the systemic **arteries,** which supply blood to the various organs and tissues, branch either directly or indirectly from this single large artery. These arteries divide in a highly characteristic manner into progressively smaller branches, much of the branching occurring in the specific organ or tissue supplied. Ultimately the smallest branches, called **arterioles,** divide into a huge number of very small, thin vessels termed **capillaries.** The capillaries unite to form larger vessels, **venules,** which, in turn, unite to form fewer and still larger vessels termed **veins** (Fig. 21-1). The veins from different organs and tissues unite to form two large veins, the **inferior vena cava** (from the lower portion of the body) and the **superior vena cava** (from the upper half of the body). By these two veins blood is returned to the right half of the heart.

There are exceptions to this general pattern of blood flow: **Portal systems** have two capillary beds. Thus, in a portal system blood flows from the left heart through the aorta, a smaller artery, arterioles, one set of capillaries, a portal vein, a second set of capillaries, venules, a vein, and back to the heart. Portal systems serve specialized functions and occur between the hypothalamus and anterior pituitary (the hypothalamohypophyseal portal system, Chap. 18) and between many organs of the digestive system and the liver (the hepatic portal system, Chap. 25).

The pulmonary circulation is similar to the systemic circuit. Blood leaves the right half of the heart via a single large artery, the **pulmonary trunk,** which divides into the right and left **pulmonary arteries** that supply the right and left lungs. In the lungs, the arteries continue to branch, generally following the air passages. Ultimately the arterial branches divide into capillaries. These capillaries unite to form small venules, which unite to form larger and larger veins. The blood leaves the lungs via the largest of these, the **pulmonary veins.** Rather than a single large vein leaving each lung, there are generally two. Thus, four pulmonary veins, two from each lung, empty into the left atrium.

The blood flowing through the systemic veins, right half of the heart, and pulmonary arteries, has a low oxygen content. As this blood flows through the lung capillaries, it picks up large quantities of oxygen; therefore, the blood in the pulmonary veins, left heart, and systemic arteries is high in oxygen. As this blood flows through the capillaries of tissues and organs throughout the body, much of the oxygen leaves the blood, resulting in the low oxygen content of systemic venous blood.

Note that all the blood pumped by the right heart flows through the lungs; in contrast, only a fraction of the total left ventricular output flows through any single organ or tissue. In other words, the systemic circulation comprises numerous different pathways ''in parallel.'' They all originate as large arteries branching directly or indirectly from the aorta. The only significant deviation from this pattern is the blood supply to the liver, much of which is not arterial but venous blood from the portal system, which has just left the pancreas, spleen, and gastrointestinal tract.

Overview of Major Arteries and Veins

The largest arteries are those nearest the heart. Likewise, since the capillaries flow into small veins that, in turn, join to form larger veins, the largest veins are those that empty into the heart. Many of the branches of the arterial (Fig. 21-2) and venous (Fig. 21-3) trees are named, and in general the arteries and veins that run parallel to each other have

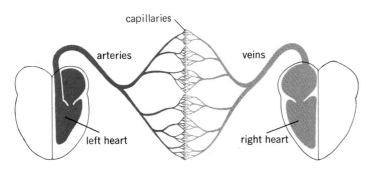

FIG. 21-1 Systemic circulation as two trees connected by capillaries. As indicated by the color change, oxygen leaves the blood during passage through the capillaries.

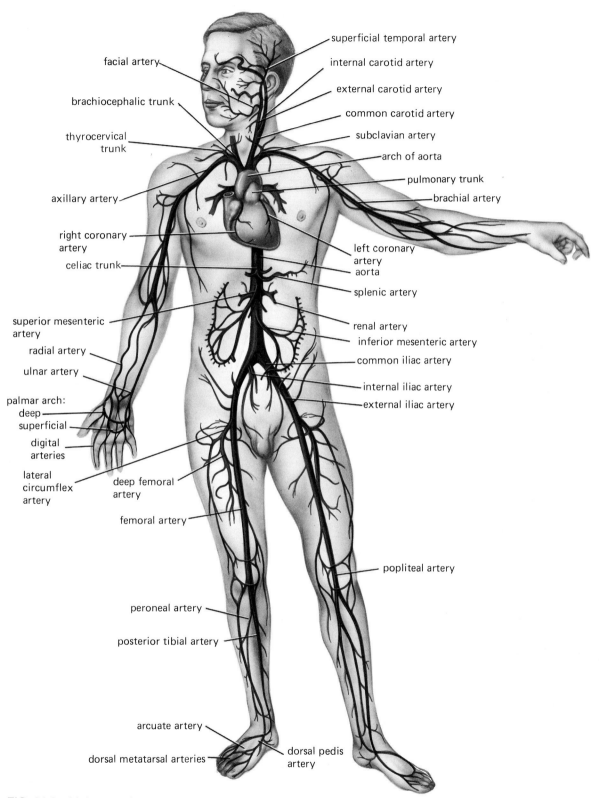

facial artery

brachiocephalic trunk

thyrocervical trunk

axillary artery

right coronary artery

celiac trunk

superior mesenteric artery

radial artery

ulnar artery

palmar arch:
 deep
 superficial

digital arteries

lateral circumflex artery

deep femoral artery

femoral artery

peroneal artery

posterior tibial artery

arcuate artery

dorsal metatarsal arteries

dorsal pedis artery

superficial temporal artery

internal carotid artery

external carotid artery

common carotid artery

subclavian artery

arch of aorta

pulmonary trunk

brachial artery

left coronary artery

aorta

splenic artery

renal artery

inferior mesenteric artery

common iliac artery

internal iliac artery

external iliac artery

popliteal artery

FIG. 21-2 Major arteries.

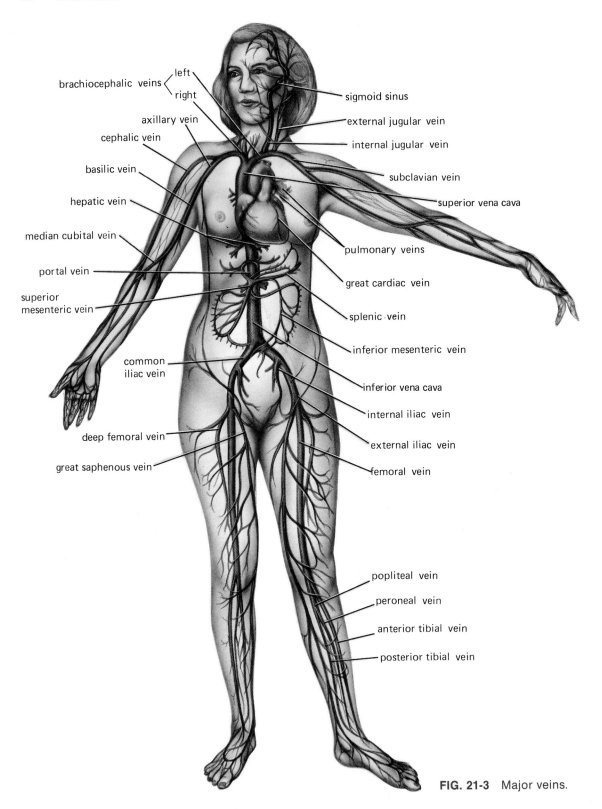

brachiocephalic veins { left
right

axillary vein

cephalic vein

basilic vein

hepatic vein

median cubital vein

portal vein

superior
mesenteric vein

common
iliac vein

deep femoral vein

great saphenous vein

sigmoid sinus

external jugular vein

internal jugular vein

subclavian vein

superior vena cava

pulmonary veins

great cardiac vein

splenic vein

inferior mesenteric vein

inferior vena cava

internal iliac vein

external iliac vein

femoral vein

popliteal vein

peroneal vein

anterior tibial vein

posterior tibial vein

FIG. 21-3 Major veins.

the same name. The vessels that are too small to be dissected easily are usually not named. Variations in the branching patterns of blood vessels occur frequently, but they are much more common in the veins than the arteries, particularly in the superficial veins, i.e., the veins that run just below the skin.

In general, the systemic veins lie in the same connective-tissue beds as the arteries. The major exceptions to this generalization are the veins of the brain, the portal vein, which drains much of the digestive tract, and the superficial veins of the extremities. The superficial veins, while forming a system of their own, unite at points with deeper lying veins.

THE AORTA

As the aorta (Figs. 21-2 and 21-4) leaves the left ventricle, it passes upward and then, at the upper end of the sternum, arches backward and toward the left so that it comes to lie on the left side of the vertebral column. As the aorta descends, it gradually moves back toward the midline and passes through an opening (the *aortic hiatus*) in the diaphragm. The descending aorta is described as *thoracic* or *abdominal,* depending upon whether it lies above or below the diaphragm. As it continues to descend, the diameter of the aorta decreases rapidly as major vessels branch from it. The aorta terminates at the level of the fourth lumbar vertebra by

FIG. 21-4 The aorta and its branches.

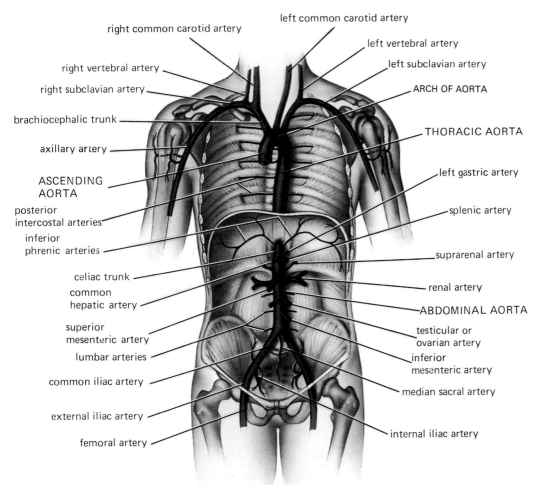

TABLE 21-1   MAJOR ARTERIES THAT BRANCH FROM VARIOUS SEGMENTS OF THE AORTA

Artery	Major Branches	Major Region Supplied
Ascending Aorta		
Right and left coronary arteries		Heart (myocardium, epicardium, endocardium)
	Marginal branch (right)	
	Posterior interventricular branch (right)	
	Anterior interventricular branch (left)	
	Circumflex (left)	
Arch of the Aorta		
Brachiocephalic trunk		
	Right common carotid artery	Head and neck
	Right subclavian artery	Right arm
Left common carotid artery	(Branches of the common carotid arteries are listed in Table 21-2.)	Head and neck
Left subclavian artery	(Branches of the subclavian arteries are listed in Table 21-3.)	Left arm
Thoracic Portion of Descending Aorta		
Bronchial arteries		Bronchial airways, tissues of lungs, esophagus, pericardium
Esophageal arteries		Esophagus
Mediastinal branches		Lymph nodes and fatty tissue of the mediastinum
Superior phrenic branches		Diaphragm
Posterior intercostal arteries		Intercostal spaces
	Dorsal branch	Spinal cord, muscles, and skin of back
	Muscular branch	Chest muscles
	Cutaneous branch	Skin of chest
	Mammary branch	Breasts
Subcostal arteries		Abdominal muscles, muscles and skin of back

dividing into right and left **common iliac arteries** (which supply the pelvic area and the legs). The branches of the aorta are listed in Table 21-1.

Just after the aorta arises from the left ventricle, it gives off the right and left **coronary arteries.** In fact, the openings of these branches lie almost completely behind the cusps of the aortic valve when the valve is open. The coronary arteries are described in Chap. 20.

The **arch of the aorta** gives off three branches: the **brachiocephalic trunk** (formerly called the "innominate artery"), which is the largest branch of the aorta; the **left common carotid artery;** and the **left subclavian artery.** At the base of the neck, the brachiocephalic trunk divides into a **right subclavian**

artery and a **right common carotid artery.** The course of the subclavian arteries is described later under Major Vessels of the Arm and that of the common carotid arteries below under Major Vessels of the Head and Neck.

The **thoracic portion of the descending aorta** runs from the 4th to the 12th thoracic vertebra, where it passes through the diaphragm. The branches of the thoracic aorta are small since the large organs of the thorax, i.e., the lungs and the heart, receive their chief blood supply via the pulmonary and coronary arteries rather than from the thoracic aorta. The small branches that do exist are collectively called **visceral branches** when they go to the lungs, esophagus, or pericardium, as do the *bronchial* and

TABLE 21-1 MAJOR ARTERIES (Continued)

Artery	Major Branches	Major Region Supplied
Abdominal Aorta		
Inferior phrenic arteries		Diaphragm and lower esophagus
	Superior suprarenal branches	Suprareneal (adrenal) glands
Celiac trunk (Fig. 21-12)		
	Left gastric artery	Stomach, esophagus
	Common hepatic artery	Liver, stomach, pancreas, duodenum, gallbladder, and bile duct
	Splenic artery	Pancreas, stomach, omentum, spleen
Middle suprarenal arteries		Suprarenal (adrenal) gland
Superior mesenteric artery (Fig. 21-13)		Most of the small intestine, cecum, ascending colon, part of the transverse colon
	Inferior pancreaticoduodenal artery	Head of pancreas, duodenum
	Jejunal and ileal branches	Small intestine
	Ileocolic artery	Ascending and part of the transverse colon, appendix, lower ileum
	Right colic artery	Ascending colon
	Middle colic artery	Transverse colon
Renal arteries		Kidneys, ureters, surrounding tissue
	Inferior suprarenal arteries	Suprarenal (adrenal) glands
Testicular (or ovarian) arteries		Testes (or ovaries)
Inferior mesenteric artery (Fig. 21-13)	Left colic artery	Transverse and descending colon
	Sigmoid arteries	Descending and sigmoid colon
	Superior rectal artery	Rectum
Lumbar arteries		Muscles and skin of the back, vertebral canal, and its contents
Median sacral artery		Sacral vertebrae, rectum
Common iliac arteries (Fig. 21-14)		Pelvic region and legs
	External iliac	Leg (see Table 21-4)
	Internal iliac (Fig. 21-14)	Viscera and walls of the pelvis, perineum, and gluteal region

esophageal arteries; they are called **parietal branches** when they supply the chest muscles, skin, and breasts, as do the *mediastinal, superior phrenic,* and *posterior intercostal* and *subcostal branches* (see Table 21-1).

The **abdominal portion of the descending aorta** is a direct continuation of the thoracic portion; the name simply changes as the vessel passes through the diaphragm. The major branches that leave the abdominal aorta before its termination supply the digestive organs and the kidneys. The three branches to the digestive tract are unpaired and arise from the front of the aorta. They are the **celiac trunk,** which supplies upper abdominal organs including the stomach, liver, duodenum, pancreas,

and spleen; the **superior mesenteric artery,** which supplies some of the organs just mentioned but especially the small intestine and much of the large intestine; and the **inferior mesenteric artery,** which supplies the more caudal portion of the large intestine. The arteries to the kidneys, the **renal arteries,** leave the sides of the aorta and are paired, one artery supplying each kidney. The branches of the abdominal aorta will be further discussed below under Major Vessels of the Abdomen and Pelvis.

MAJOR VESSELS OF THE HEAD AND NECK

The **right** and **left common carotid arteries** are the principal arteries of the head and neck (Fig. 21-5

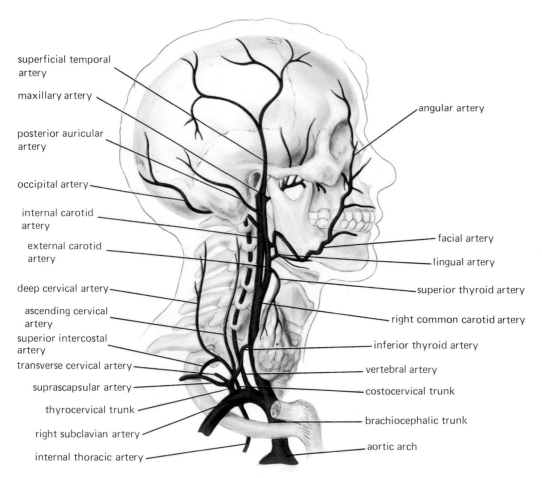

superficial temporal artery

maxillary artery

posterior auricular artery

occipital artery

internal carotid artery

external carotid artery

deep cervical artery

ascending cervical artery

superior intercostal artery

transverse cervical artery

suprascapular artery

thyrocervical trunk

right subclavian artery

internal thoracic artery

angular artery

facial artery

lingual artery

superior thyroid artery

right common carotid artery

inferior thyroid artery

vertebral artery

costocervical trunk

brachiocephalic trunk

aortic arch

FIG. 21-5 The principal arteries of the head and neck.

and Table 21-2). They have different origins, the right common carotid arising from the brachiocephalic trunk and the left arising directly from the aortic arch, but as they ascend in the neck, one on each side of the trachea (the windpipe), their paths become similar. Where they lie against the trachea, their pulse can be easily felt. High in the neck, at the level of the upper border of the thyroid cartilage (the Adam's apple), each common carotid divides into an external and internal carotid artery. Normally, except for this final division, there are no branches of the common carotids.

The **external carotid** gives off eight named branches, which rebranch extensively to supply most of the structures in the head and neck. (The major exception is the brain, which is supplied by the internal carotid and vertebral arteries, as discussed below.) The *facial artery,* a prominent branch, whose pulsations can be felt as it passes over the base of the mandible, is the chief artery of the face; its name, however, changes to *angular artery* as it continues above the lower border of the nose. The external carotid ends by dividing in the region of the parotid gland into its two terminal branches, the *maxillary* and the *superficial temporal arteries.* The latter lies subcutaneously in front of the ear, where its pulse can be felt.

The **internal carotid artery** does not give off any branches in the neck but runs upward and passes into the cranial cavity through the *carotid canal* in the temporal bone. Once in the cranial cavity, it gives off branches to the eye and associated tissues (the *ophthalmic artery,* which accompanies the optic nerve into the orbit), the pituitary, and the dura, but its chief branches, the *anterior* and *middle cerebral arteries,* supply the anterior portion of the brain.

TABLE 21-2   MAJOR ARTERIES OF THE HEAD AND NECK

Artery	Major Branches	Major Region Supplied
External carotid artery		
	Superior thyroid artery	Thyroid gland and adjacent muscles
	Ascending pharyngeal artery	Pharynx, larynx, auditory tube, and cerebral meninges
	Lingual artery	Tongue and floor of the mouth, oropharynx, sublingual gland, and neighboring muscles
	Facial artery	Muscles and tissues of the face below the level of the eyes, the submandibular gland, tonsil, and soft palate, auditory tube, tissue under chin
	Sternocleidomastoid artery	Sternocleidomastoid muscle
	Occipital artery	Muscles, skin, and other tissue in the region behind the ear and the back part of the scalp; cerebral meninges
	Posterior auricular artery	Parotid gland, muscles, skin, and other tissues of the ear and posterior scalp regions
	Superficial temporal artery	Parotid gland, temporomandibular joint, outer ear, forehead, temporal region of the scalp, adjacent muscles
	Maxillary artery	Upper and lower jaws, teeth, muscles of mastication, palate, nose, and dura mater
Internal carotid artery		
	Hypophyseal artery	Pituitary gland
	Ophthalmic artery	Eye, lacrimal gland, ocular muscles, nasal cavity, forehead
Right and left subclavian arteries		
	Vertebral artery	Spinal cord, vertebrae and surrounding tissues, deep neck structures, brain
	Thyrocervical trunk	Thyroid gland, neck muscles, trachea, esophagus
	Costocervical trunk	Muscles at the back of the neck, vertebral canal, first intercostal space

The chief arteries of the arms are the **right** and **left subclavian arteries** but, while passing the base of the neck, these arteries also give off branches to the head and neck. These early branches are on each side the **vertebral artery,** thyrocervical trunk, internal thoracic artery, and costocervical trunk. (The more distal course of the subclavian arteries is included with Major Vessels of the Arm, below.

Each **vertebral artery** ascends through openings in the transverse process of the cervical vertebrae, giving off branches to the deep muscles of the neck before entering the cranial cavity. The *thyrocervical trunk,* after giving off branches to the shoulder, ends as the *inferior thyroid artery.* The *internal thoracic* (formerly the internal mammary) *artery* supplies tissues of the wall of the thoracic cavity. The *costocervical trunk* supplies tissues of the posterior wall of the thoracic cavity and tissues of the back.

The chief veins that drain the head and neck (Fig. 21-6 are the **internal jugular veins,** which arise as continuations of the sigmoid sinuses in the skull. Venous blood from the anterior part of the scalp and skin of the face drains via the *facial vein* and its retromandibular branch to the *common facial vein,* a tributary of the internal jugular vein. The veins that drain deeper structures of the head and neck such as the tongue, pharynx, and thyroid also empty directly or indirectly into the internal jugular vein. The internal jugular veins run straight down either side of the neck in the carotid sheath along with the common carotid artery, the vagus nerve, and many of the deep lymph nodes of the neck. As it passes behind the clavicle, each internal jugular joins the *subclavian vein* from the same side of the body to form a brachiocephalic vein. The two brachiocephalic veins, in turn, enter the thorax and

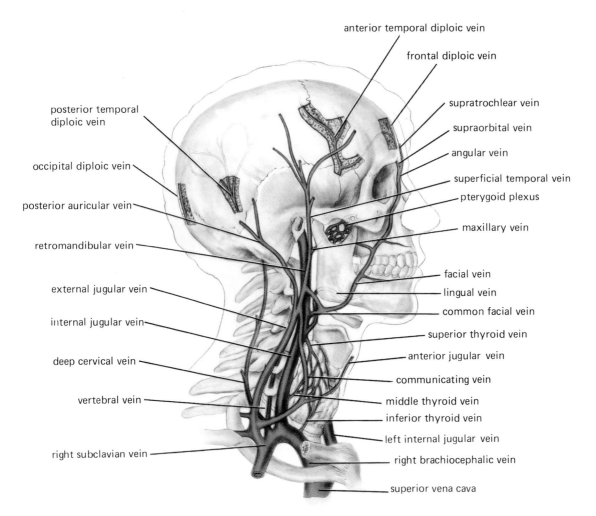

anterior temporal diploic vein

frontal diploic vein

supratrochlear vein

supraorbital vein

angular vein

superficial temporal vein

pterygoid plexus

maxillary vein

facial vein

lingual vein

common facial vein

superior thyroid vein

anterior jugular vein

communicating vein

middle thyroid vein

inferior thyroid vein

left internal jugular vein

right brachiocephalic vein

superior vena cava

posterior temporal diploic vein

occipital diploic vein

posterior auricular vein

retromandibular vein

external jugular vein

internal jugular vein

deep cervical vein

vertebral vein

right subclavian vein

FIG. 21-6 Veins of the head and neck.

unite to form the superior vena cava, which after a short distance enters the heart.

On each side of the head a smaller **external jugular vein** begins near the level of the angle of the mandible by the union of the *posterior auricular* and a portion of the *retromandibular veins*. Blood from the skin and scalp above and behind the ear drains into the external jugular vein as does blood from the front of the neck, via the small *anterior jugular vein*. The external jugular empties into the subclavian vein from the same side before the union of the subclavian vein with the internal jugular; the external jugular is the only named tributary of the subclavian vein.

The *right* and *left vertebral veins* drain the vertebral and spinal cord regions and the deep muscles of the back of the neck. They descend along the

path of the vertebral artery to empty into the brachiocephalic vein from the same side.

Blood Supply of the Brain and Spinal Cord

As mentioned above, the arterial blood supply to the brain comes not only from the internal carotid arteries but also from the paired vertebral arteries (Fig. 21-7A), which supply the posterior circulation of the brain. After entering the skull through the foramen magnum, the vertebral arteries run through the subarachnoid space and pass upward along the brainstem to send branches to the cerebellum and medulla and downward along the anterior surface of the spinal cord to feed the *anterior spinal artery* (Fig. 21-7A).

As the two vertebral arteries reach the level of

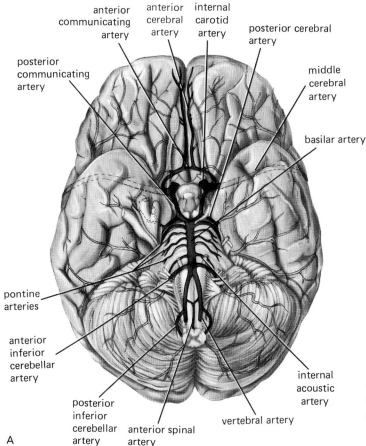

anterior communicating artery

anterior cerebral artery

internal carotid artery

posterior cerebral artery

posterior communicating artery

middle cerebral artery

basilar artery

pontine arteries

anterior inferior cerebellar artery

internal acoustic artery

posterior inferior cerebellar artery

anterior spinal artery

vertebral artery

A

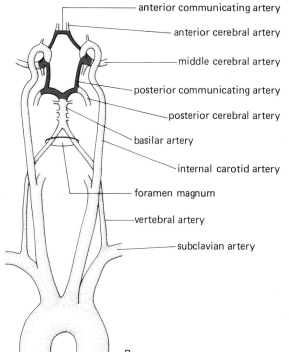

anterior communicating artery

anterior cerebral artery

middle cerebral artery

posterior communicating artery

posterior cerebral artery

basilar artery

internal carotid artery

foramen magnum

vertebral artery

subclavian artery

B

FIG. 21-7   (A) Major arteries at the base of the brain and the circle of Willis. (B) Relationship of the circle of Willis to the major systemic arteries.

the pons, they join to form the **basilar artery,** which sends branches to the pons, midbrain, and parts of the cerebellum before ending at the upper boundary of the pons. As it ends, the basilar artery branches and forms paired *posterior cerebral arteries,* which supply the occipital lobes of the cerebrum and parts of the temporal lobes (Fig. 21-7A and B).

As mentioned above, the blood supply to the anterior part of the brain is furnished by the two **internal carotid arteries,** which enter the cranial cavity through the *carotid canal* of the temporal bone and almost immediately branch into the *anterior* and *middle cerebral arteries* (Fig. 21-7A and B). The anterior branch supplies blood to the medial surface of the cerebral hemispheres, and the middle branch supplies principally the lateral surfaces. Thus, of the three cerebral arteries in each hemisphere, two arise from the internal carotid and one from the vertebral artery. However, the three cerebral arteries anastomose with each other on the

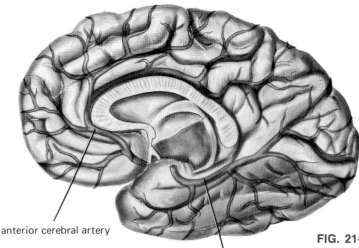

anterior cerebral artery

posterior cerebral artery

FIG. 21-8 Distribution of the anterior and posterior cerebral arteries over the midline surface of the cerebral hemisphere.

hemisphere surface. The distribution of the anterior and posterior cerebral arteries can be seen in Fig. 21.8.

The two blood supplies to the brain (the vertebrals and internal carotids) are also connected by *posterior communicating arteries,* and the two anterior cerebral arteries are connected by an *anterior communicating artery* (Fig. 21-7B). These connecting arteries complete a circle, the **circle of Willis,** which provides an important safeguard to ensure the supply of blood to all parts of the brain despite the blockage of either the vertebral or internal carotid arteries.

The veins of the brain (Fig. 21-9), comprising the cerebral, cerebellar, and brainstem veins, have very thin walls and no valves. The drainage from the outer and medial surfaces of the hemispheres is into superficial veins, whereas deeper tissues drain into *deep cerebral veins* and a large *basal vein,* which in turn empty into the **great cerebral vein** (of Galen). As they leave the substance of the brain, the veins pierce the arachnoid membrane, cross the subarachnoid space, and enter the dura, within which they form large venous sinuses, which drain eventually into the internal jugular veins (Figs. 21-6 and 21-9).

The spinal cord is nourished by several arteries, including the *spinal arteries,* which run along the anterior and posterior surfaces of the cord, and *segmental arteries,* which enter the vertebral canal at irregular intervals along its length. The blood leaves the cord via the *anterior* and *posterior spinal veins.*

MAJOR VESSELS OF THE ARM

The major artery of the arm is the **subclavian artery** (Fig. 21-10 and Table 21-3). The right subclavian artery arises from the brachiocephalic trunk, whereas the left one arises directly from the aortic arch, but they both lie mainly behind the right and left clavicles, respectively. Each subclavian artery gives rise to three vessels, mentioned earlier (the vertebral artery, thyrocervical trunk, and costocervical trunk), that supply the head and neck; it also gives rise to the *internal thoracic artery* (formerly the internal mammary artery), which supplies tissues of the chest wall and mediastinum.

As the subclavian artery passes through the axillary fossa, or armpit, its name is changed to **axillary artery.** This artery supplies structures of the lateral chest wall, shoulder joint, and arm. It's chief branches are the *superior* and *lateral thoracic arteries,* the *thoracoacromial artery,* the *subscapular artery,* and the *anterior* and *posterior humeral circumflex arteries.* The thoracoacromial and lateral thoracic arteries are the major arteries of the chest wall; the subscapular artery supplies the posterior region of the armpit and lateral region of the tho-

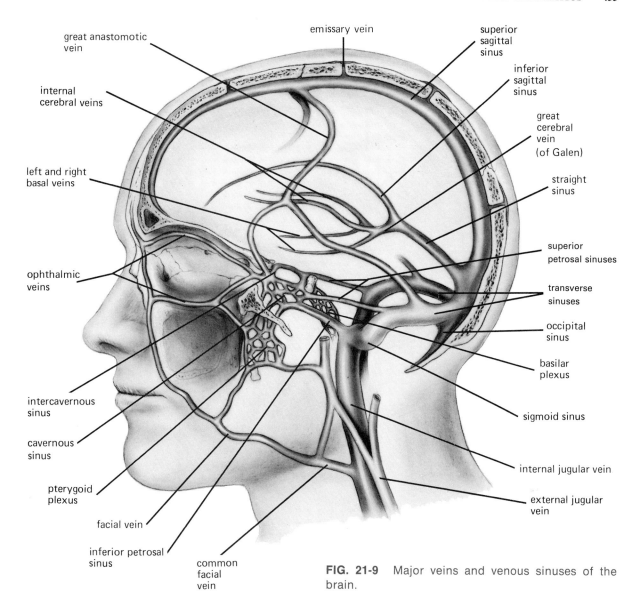

FIG. 21-9 Major veins and venous sinuses of the brain.

racic wall; and the circumflex arteries, as their name suggests, pass around the upper end of the humerus (the bone of the upper arm).

The **brachial artery** is the continuation of the axillary artery; the name change occurs at the lateral border of the teres major muscle, and the brachial artery ends just below the elbow by dividing into the **radial** and **ulnar arteries.** Initially, the brachial artery lies medial to the humerus, but it shifts position until at its termination it lies on the anterior surface of the arm. It is the brachial artery that is commonly used for blood pressure measurements.

The major branches of the artery are the *deep brachial artery,* to the posterior portion of the upper arm, a *nutrient artery* to the humerus, and *superior* and *inferior ulnar collateral arteries* to the tissues on the medial side of the upper arm and the elbow.

The **radial artery** lies on the lateral (thumb) side of the arm. A short distance below its origin, it gives off a *radial recurrent artery,* which supplies muscle and other tissues in the region of the elbow, and as it passes more distally, it gives off several branches to the muscles of the forearm and to the tissues of the wrist. The radial artery approaches

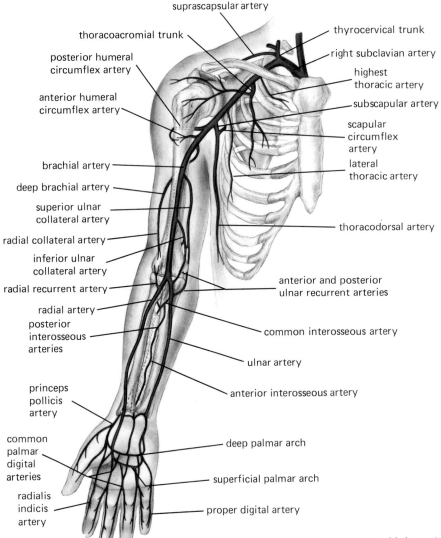

suprascapular artery

thoracoacromial trunk

posterior humeral
circumflex artery

anterior humeral
circumflex artery

brachial artery

deep brachial artery

superior ulnar
collateral artery

radial collateral artery

inferior ulnar
collateral artery

radial recurrent artery

radial artery

posterior
interosseous
arteries

princeps
pollicis
artery

common
palmar
digital
arteries

radialis
indicis
artery

thyrocervical trunk

right subclavian artery

highest
thoracic artery

subscapular artery

scapular
circumflex
artery

lateral
thoracic artery

thoracodorsal artery

anterior and posterior
ulnar recurrent arteries

common interosseous artery

ulnar artery

anterior interosseous artery

deep palmar arch

superficial palmar arch

proper digital artery

FIG. 21-10 Major arteries of the arm.

the surface of the forearm just above the wrist where it is commonly used for taking the pulse.

The **ulnar artery,** which passes down the medial side of the forearm, gives off branches to the elbow region (the two *ulnar recurrent arteries*) and to the anterior and posterior surfaces of the forearm (the *anterior* and *posterior interosseous arteries,* which branch from the *common interosseous artery,* and various *muscular branches*).

Terminal branches of the radial and ulnar arteries join, forming *palmar* and *dorsal carpal arches* that cross the wrist as well as *deep* and *superficial palmar arches* that cross the hand. The radial artery contributes primarily to the deep palmar arch and to the arteries on the back of the hand, and the

ulnar artery is the chief supplier of the superficial palmar arch. Arteries project from the two palmar arches to supply the metacarpals and the fingers.

Both deep and superficial veins drain the arms (Fig. 21-11), although the superficial veins are larger and play a greater role. The two sets of veins interconnect freely with each other, and both sets have valves that direct the flow of blood back to the heart.

The dorsal and palmar surfaces of the hand have complex networks of superficial veins that drain the fingers and the metacarpals. The dorsal network gives rise on its ulnar (little finger) side to the **basilic vein** and on its radial side to the **cephalic vein.** The venous plexus on the palmar surface of the hand

TABLE 21-3   MAJOR ARTERIES OF THE ARM

Artery	Major Branches	Major Region Supplied
Subclavian artery		
	Vertebral artery	(See Table 21-2)
	Internal thoracic (mammary) artery	Thoracic pleura, pericardium, fat and lymph nodes of mediastinum, sternum, intercostal muscles, muscles and skin of chest wall, breast, muscles and skin of abdominal wall
	Thyrocervical trunk	(See Table 21-2)
	Costocervical trunk	(See Table 21-2)
	Dorsal scapular artery	Muscles of the upper back and shoulder
Axillary artery		
	Highest thoracic artery	Pectoralis muscles and chest wall
	Thoracoacromial artery	Muscles of the chest and shoulder
	Lateral thoracic artery	Muscles of the chest
	Subscapular artery	Muscles of the chest and shoulder
	Anterior and posterior circumflex humeral arteries	Head of the humerus, shoulder joint, muscles of shoulder
Brachial artery		
	Deep brachial artery (profunda)	Muscles of the upper arm, humerus
	Main nutrient artery	Muscles and tissues of the upper arm
	Muscular branches	Muscles of the upper arm
	Inferior and superior ulnar collateral arteries	Elbow and upper region of the forearm
Radial artery		
	Radial recurrent artery	Elbow joint and muscles of the forearm
	Muscular branches	Muscles of the radial (thumb) side of the forearm
	Palmar carpal branch	Bones and joints of the wrist
	Superficial palmar branch	Muscles of the thumb
	Dorsal carpal branch	Lower ends of the radius and ulna; gives rise to arteries supplying metacarpals
Ulnar artery		Muscles on the medial side of the forearm and hand
	Anterior ulnar recurrent and posterior arteries	Elbow joint and muscles of the upper forearm
	Common interosseous artery	Wrist and forearm
	Muscular branches	Muscles of the ulnar region of the forearm
	Deep branch	Hypothenar muscles
	Common palmar digital arteries	Fingers

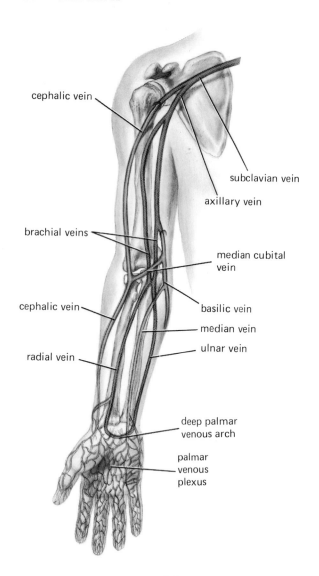

FIG. 21-11 Major veins of the arm.

cephalic vein

subclavian vein

axillary vein

brachial veins

median cubital vein

cephalic vein

basilic vein

median vein

ulnar vein

radial vein

deep palmar venous arch

palmar venous plexus

The deep veins of the arm occur in pairs and accompany the arteries that have corresponding names. Thus, veins from the fingers and metacarpals drain into *deep* and *superficial venous arches,* which in turn drain into paired **radial** and **ulnar veins.** After receiving numerous tributaries from tissues of the forearm, the radial and ulnar veins unite near the elbow to form paired **brachial veins.** The brachial veins join the unpaired axillary vein, which is a continuation of one of the large superficial veins (the basilic vein) and receives the other major superficial vein (the cephalic vein). The axillary vein therefore provides drainage for all the veins of the arm.

MAJOR VESSELS OF THE ABDOMEN AND PELVIS

The digestive tract receives its blood supply from three unpaired branches of the aorta (Fig. 21-4): the **celiac trunk,** the **superior mesenteric artery,** and the **inferior mesenteric artery.** The **celiac trunk** leaves the aorta just below the diaphragm and is typically a short trunk with three branches: the *left gastric artery,* which passes toward the junction of the stomach and esophagus; the *common hepatic artery,* which runs to the right along the upper border of the pancreas; and the *splenic artery,* which runs to the left along the upper border of the pancreas.

The stomach receives its arterial supply from all three branches of the celiac trunk (Fig. 21-12), the major arteries to the stomach being the *right and left gastric arteries* and the *right and left gastroepiploic arteries.* In addition, the upper part of the stomach receives *short gastric arteries.* The spleen receives its blood supply from branches of the *splenic artery,* whereas the pancreas and duodenum, which generally share a blood supply, are nourished by branches of the *common hepatic* and *superior mesenteric* arteries.

The unpaired **superior mesenteric artery** (Fig. 21-13) leaves the aorta about 1 cm below the origin of the celiac trunk. It passes downward behind the pancreas and in front of the duodenum and gives rise to some 20 or so branches that anastomose with each other to form a series of arches. From these arches branch the arteries (the *arteriae rectae*) that actually go to the intestinal wall. The anastomosing arches provide uninterrupted blood

drains into the **median vein.** A little below the elbow, the cephalic vein gives rise to the **median cubital vein,** which is commonly used for venipunctures where it passes across the front of the elbow to join the basilic vein. The basilic vein then continues upward, but its name is changed to **axillary vein** as it approaches the armpit. The axillary vein receives the cephalic vein just below the level of the clavicle. As the axillary vein passes between the clavicle and the first rib to enter the neck, it becomes the **subclavian vein.** The course of the subclavian vein was described above with the Major Vessels of the Head and Neck.

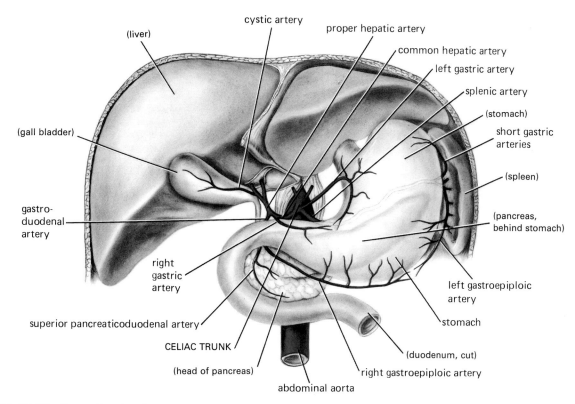

cystic artery

(liver)

proper hepatic artery

common hepatic artery

left gastric artery

splenic artery

(stomach)

short gastric arteries

(gall bladder)

(spleen)

gastro-duodenal artery

(pancreas, behind stomach)

right gastric artery

left gastroepiploic artery

superior pancreaticoduodenal artery

stomach

CELIAC TRUNK

(duodenum, cut)

(head of pancreas)

right gastroepiploic artery

abdominal aorta

FIG. 21-12   The celiac trunk.

supply during twisting movements of the intestines and their supporting tissues. The superior mesenteric artery also supplies the cecum, ascending colon, and most of the transverse colon through its *ileocolic, right colic,* and *middle colic branches.*

The **inferior mesenteric artery** (Fig. 21-13), which lies behind the peritoneum for most of its length, supplies the left one-third of the transverse colon and the entire descending colon, sigmoid colon, and rectum. The inferior mesenteric artery gives off a *left colic artery* as well as *sigmoid* and *rectal branches.*

The remaining branches of the aorta stay behind the peritoneum and are distributed to the abdominal wall and to the kidneys, suprarenal (adrenal) glands, and gonads via paired *suprarenal arteries,* **renal arteries,** and *ovarian* or *testicular arteries* (also called in both sexes the internal spermatic arteries). The large terminal branches of the aorta are the **right** and **left common iliac arteries;** each of these in turn, at the level of the lumbosacral articulation, divides into an **internal** and **external iliac artery.** The internal branch (formerly the "hypo-

gastric artery") supplies the pelvic viscera such as the bladder, rectum, and in the female, the uterus and vagina. It also sends branches to the muscles of the buttock and to the genitals and other perineal structures. The external branch supplies the leg.

Branches of the **internal iliac artery** (Fig. 21-14) are quite variable, but in general a *vesicle branch* supplies the bladder, prostate and seminal vesicles, and ureter; a *rectal artery* supplies the lower rectum and prostate and seminal vesicles; a *uterine branch* supplies the uterus; a *vaginal branch* supplies the vagina; the *obturator artery* supplies muscles and bones of the pelvic region; the *internal pudendal* supplies the genitalia and perineum; a *gluteal artery* supplies the buttocks and back of thighs; a *sacral artery* supplies the sacral vertebrae and their contents; and the *iliolumbar artery* supplies the lumbar vertebrae, ilium, and nearby muscles. The **internal iliac vein** and its tributaries generally follow the course of the arteries and need no special mention (except that the *umbilical artery,* which in its final extent becomes a solid cord and continues to the umbilicus, the navel, as the *medial umbilical liga-*

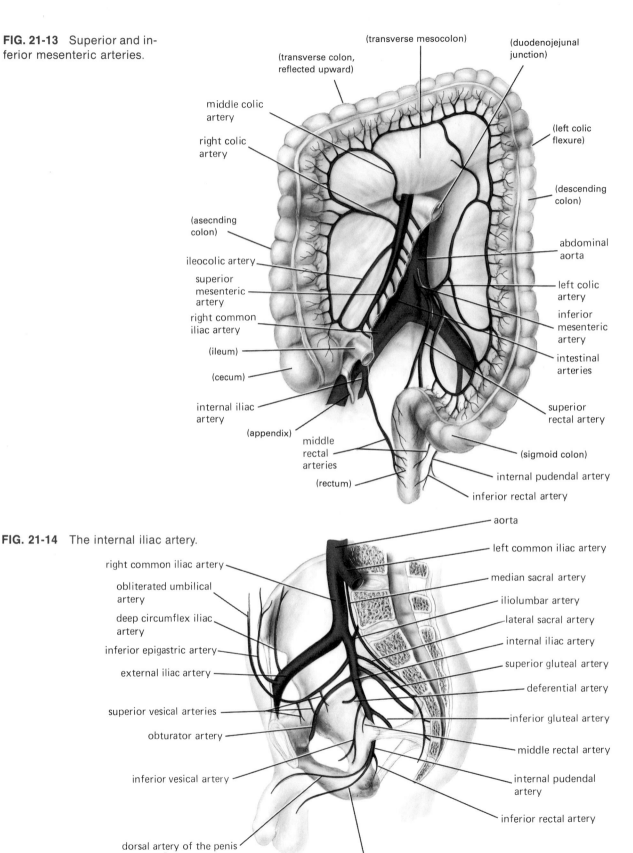

FIG. 21-13 Superior and inferior mesenteric arteries.

(transverse colon, reflected upward)

(transverse mesocolon)

(duodenojejunal junction)

middle colic artery

right colic artery

(left colic flexure)

(descending colon)

(asecnding colon)

abdominal aorta

ileocolic artery

superior mesenteric artery

left colic artery

inferior mesenteric artery

right common iliac artery

(ileum)

intestinal arteries

(cecum)

internal iliac artery

superior rectal artery

(appendix)

middle rectal arteries

(sigmoid colon)

internal pudendal artery

(rectum)

inferior rectal artery

FIG. 21-14 The internal iliac artery.

aorta

left common iliac artery

right common iliac artery

median sacral artery

obliterated umbilical artery

iliolumbar artery

deep circumflex iliac artery

lateral sacral artery

inferior epigastric artery

internal iliac artery

external iliac artery

superior gluteal artery

deferential artery

superior vesical arteries

inferior gluteal artery

obturator artery

middle rectal artery

inferior vesical artery

internal pudendal artery

inferior rectal artery

dorsal artery of the penis

perineal artery

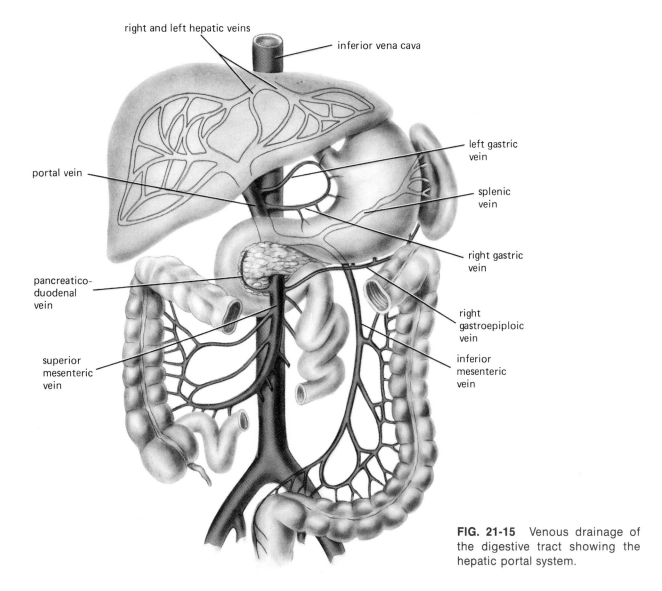

right and left hepatic veins

inferior vena cava

left gastric vein

splenic vein

portal vein

right gastric vein

pancreatico- duodenal vein

right gastroepiploic vein

superior mesenteric vein

inferior mesenteric vein

FIG. 21-15 Venous drainage of the digestive tract showing the hepatic portal system.

ment, is not accompanied by an umbilical vein).

The large vein that returns blood to the right heart from all the regions below the diaphragm that are supplied by the abdominal aorta or its branches is the **inferior vena cava** (Fig. 21-15), which is formed by the union of the **right** and **left common iliac veins** and passes upward next to the aorta. The inferior vena cava receives *lumbar veins,* the *right ovarian* or *testicular vein,* the *right renal* and *suprarenal veins,* and the *left renal vein* (which has already been joined by the *left suprarenal, inferior phrenic,* and *left gonadal veins*); just before passing through the diaphragm, the inferior vena cava is joined by the **hepatic veins** from the liver and the *right inferior phrenic vein.*

The **hepatic veins** (Fig. 21-15), mentioned rather casually above, deserves special attention because through them the venous blood from most of the digestive tract drains into the inferior vena cava. The venous drainage from these organs is not led directly back to the heart through successively larger veins, as is the case for most tissues; instead, the venous blood passes through a second capillary bed, forming a portal system. Thus, venous blood from the abdominal viscera is collected into the **portal vein,** which passes to the liver where it again breaks into capillary beds and thus brings the portal blood into intimate contact with the liver cells. The portal blood and the blood from the liver's own arterial supply leave via the hepatic veins, which

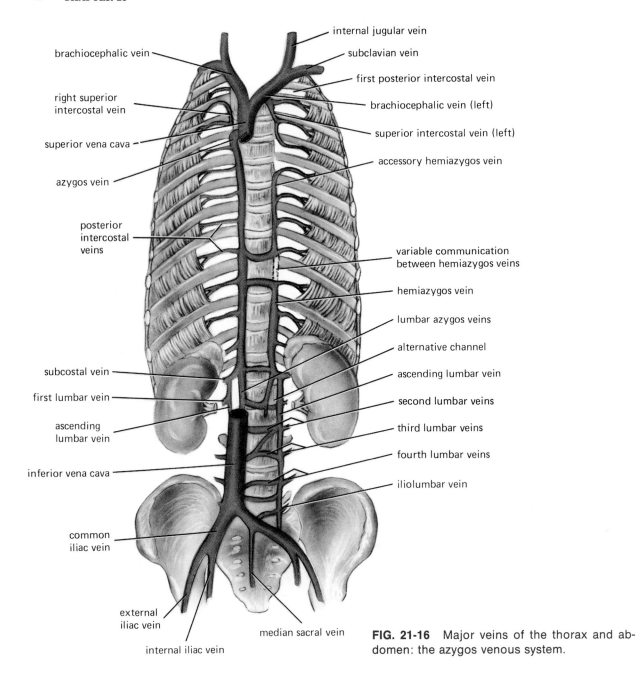

internal jugular vein

brachiocephalic vein

subclavian vein

first posterior intercostal vein

right superior intercostal vein

brachiocephalic vein (left)

superior vena cava

superior intercostal vein (left)

azygos vein

accessory hemiazygos vein

posterior intercostal veins

variable communication between hemiazygos veins

hemiazygos vein

lumbar azygos veins

alternative channel

subcostal vein

ascending lumbar vein

first lumbar vein

second lumbar veins

ascending lumbar vein

third lumbar veins

fourth lumbar veins

inferior vena cava

iliolumbar vein

common iliac vein

external iliac vein

median sacral vein

internal iliac vein

FIG. 21-16 Major veins of the thorax and abdomen: the azygos venous system.

drain into the inferior vena cava. The portal vein is formed by the union of the *superior mesenteric* and *splenic veins;* the splenic vein receives blood from the *pancreatic* and *inferior mesenteric veins.* Even the veins that drain the stomach empty into the portal vein.

The segmental veins that drain the thoracic wall and vertebral column do not drain directly into the venae cavae; they drain instead into the **azygos system of veins** (Fig. 21-16), which lies along the posterior wall of the thoracic cage. This system includes the **azygos vein,** which runs along the right side of the vertebral column, and the *hemiazygos* and *accessory hemiazygos veins,* which run along the left side. The azygos is a continuation of the *right ascending lumbar vein,* and the hemiazygos is a continuation of the left. The hemiazygos and accessory hemiazygos veins cross the midline to empty into the azygos vein, which drains into the superior vena cava. Because of its connections with the ascending lumbar veins (and the inferior vena cava) the azygos system can serve as a bypass for the inferior vena cava if it should become obstructed. In addition to the *intercostal veins,* the *esophageal, pericardial,* and *bronchial veins* drain into the azygos system.

MAJOR VESSELS OF THE LEG

The blood vessels of the legs follow the same general plan as do those of the arms. The arterial supply to the legs (Fig. 21-17 and Table 21-4) is provided by the **external iliac arteries,** which are formed by the division of the common iliac arteries. As the external iliac artery enters the leg, passing behind the inguinal ligament, it becomes the **femoral artery** and lies on the front and medial side of the thigh. While in this location it gives off branches that turn back upward to supply the abdominal wall and inguinal region (the *superficial epigastric, superficial iliac circumflex,* and *inguinal arteries*) and the perineum (the *external pudendal artery*). The major branch, however, of the femoral artery is the **profunda femoris** (*deep femoral*) **artery.** The profunda gives rise to four *perforating arteries,* so called because they pass through certain of the adductor muscles. They form a series of loops, anastomosing with each other and with branches of the iliac artery above and the popliteal artery below. *Lateral* and *medial femoral circumflex arteries,* which nourish tissues of the hip and upper thigh, arise directly from the femoral artery or from the profunda. Near the lower end of the femoral artery, a *descending genicular artery* arises that supplies tissues around the knee.

The femoral artery passes between the extensor and adductor muscles of the thigh as it descends, and about two-thirds of the way down the thigh it emerges at the back of the leg where it is known as the **popliteal artery.** (The region at the back of the knee is the popliteal fossa.) Branches of the popliteal artery, the so-called *genicular arteries,* anastomose to form a ring around the knee and upper ends of the tibia and fibula (bones of the lower leg). The popliteal artery ends just below the knee by

branching to form the **anterior** and **posterior tibial arteries.**

The **anterior tibial artery** arises at the back of the leg but passes forward to run down the front and continues along the upper surface (dorsum) of the foot as the dorsal artery of the foot, the **dorsalis pedis artery.** Before entering the foot, however, the anterior tibial artery gives off branches that supply the knee joint (the *anterior* and *posterior tibial recurrent arteries*), *muscular branches,* and branches that supply the ankle (the *medial* and *lateral anterior malleolar arteries*). The arteries around the ankle, including the dorsalis pedis and malleolars as well as other branches of the anterior and posterior tibial arteries, anastomose freely and form an arterial network around the ankle joint. The course of the anterior tibial artery ends at its most distal part, the dorsalis pedis artery, and turns downward to form part of the *plantar arch.*

The **posterior tibial artery** runs down the back of the leg and onto the plantar surface, or sole, of the foot. After giving rise to a large *nutrient artery* (to the tibia), to the *circumflex fibular artery* (to the muscles in the leg), and to several unnamed branches, the posterior tibial forms its largest branch, the *peroneal artery,* which supplies muscles of the lower leg and the tibia and adds anastomosing branches to the arterial ring that surrounds the ankle. The terminal branches of the posterior tibial artery are the *medial plantar artery,* which passes to the medial side of the foot, and the *lateral plantar artery,* which is the larger of the two branches and passes to the lateral side of the foot. The lateral plantar artery then turns medially to unite with the dorsalis pedis artery to complete the *deep plantar arch.* Branches from this arch bring arterial blood to the toes.

The veins of the leg (Fig. 21-18) consist of superficial and deep veins. *Perforating,* or *communicating, veins* connect the two sets of veins and are valved in such a way that blood is normally conducted from the superficial to the deep veins.

The major superficial veins are the **great** and *small saphenous veins,* both of which originate on the dorsum, or upper surface, of the foot from the *dorsal venous arch.* This arch receives blood from the toes and the irregular network of veins that drains the dorsal surface of the foot. The **great saphenous vein,** the longest vein in the body, arises from the medial end of the arch and receives trib-

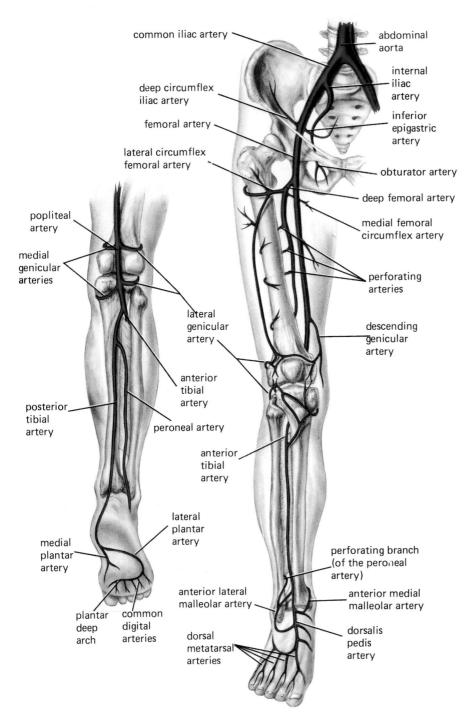

FIG. 21-17   Arterial supply to the leg.

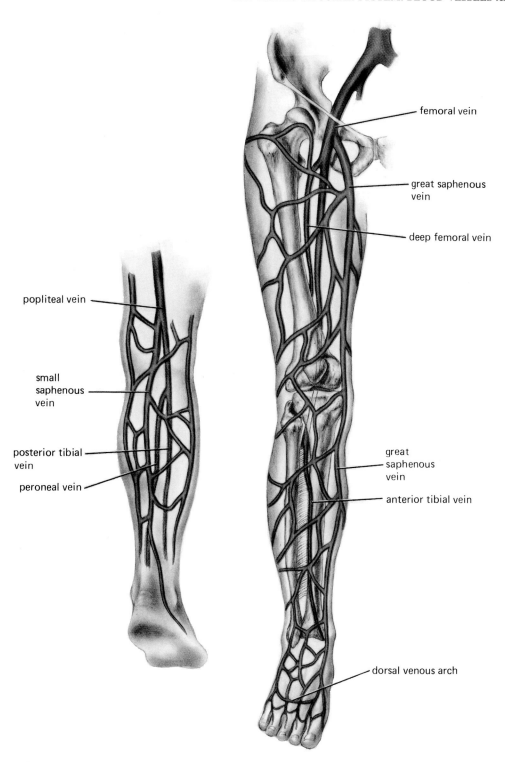

femoral vein

great saphenous vein

deep femoral vein

popliteal vein

small saphenous vein

posterior tibial vein

peroneal vein

great saphenous vein

anterior tibial vein

dorsal venous arch

FIG. 21-18 Venous drainage of the leg.

TABLE 21-4 MAJOR ARTERIES OF THE LEG

Artery	Major Branches	Major Region Supplied
Femoral artery		
	Muscular branches	Muscles of the upper thigh
	Profunda femoris (deep femoral) artery	Principal arterial supply to the adductor, extensor, and flexor muscles
Popliteal artery		
	Sural arteries	Gastrocnemius, soleus, and plantaris muscles
	Superior, middle, and inferior genicular arteries	Knee joint
Anterior tibial artery		
	Anterior and posterior tibial recurrent arteries	Knee joint, tibiofibular joint
	Muscular branches	Adjacent muscles and skin
	Medial anterior malleolar artery	Medial side of the ankle joint
	Lateral anterior malleolar artery	Lateral side of the ankle
Dorsalis pedis artery (dorsal artery of the foot)		
	Tarsal arteries	Tarsal region of foot
	Arcuate artery	Metatarsal region of foot
	Deep plantar artery	Metatarsal region of foot and toes
Posterior tibial artery		
	Nutrient artery	Tibia
	Circumflex fibular artery	Bone and joint structures of the knee, lateral muscles of the leg
	Peroneal artery	Muscles, fibula
	Medial malleolar branches	Ankle joint
	Calcaneal branch	Heel
	Medial plantar artery	Metatarsal region and toes
	Lateral plantar artery	Adjoining muscles, skin, and sole of the foot

utaries from the sole and skin of the foot as well as communications from the small saphenous vein. After passing up along the medial side of the leg and knee, the great saphenous vein swings toward the front of the thigh where it ends in the **femoral vein.** In a standing person, the superficial veins of the leg, like the deep veins, must support the weight of a column of blood that extends to the heart; unlike the deep veins, however, the superficial veins are not subject to periodic emptying by the muscle pump (see Chap. 22) and, therefore, are particularly susceptible to enlargements known as *varicosities.*

The femoral vein, parallel to the femoral artery, passes under the inguinal ligament and enters the abdomen where it is known as the *external iliac vein.* The *internal iliac vein* from the buttock and pelvis joins the external iliac to form the **common iliac vein.**

The **small saphenous vein** begins in the foot at the lateral end of the dorsal venous arch and ascends at the back of the leg. Along its course, it anastomoses frequently not only with the great saphenous, as mentioned above, but also with the deep veins. Most of the tributaries of the great and small saphenous veins are not named.

The small saphenous vein empties into the **popliteal vein** behind the knee. The popliteal vein, which is called the **femoral vein** higher up in the thigh, is formed by the union of the paired **anterior**

tibial and **posterior tibial veins,** deep veins of the leg that accompany the corresponding arteries. The femoral and popliteal veins are also deep veins.

The deep veins lie alongside the arteries and generally have corresponding names. Paired veins accompany the arteries from the foot to the knee, whereas from the knee upward there is usually a single vein with each artery. The deep drainage system of the foot flows into the *plantar venous arch,* which in turn flows into the paired posterior tibial veins. As these veins pass up the leg, they receive tributaries, the largest of which are the *peroneal veins,* and then join the anterior tibial veins to form the popliteal vein. The course of the deep veins after the formation of the popliteal has been mentioned above.

SECTION B: FUNCTION OF THE VASCULAR SYSTEM

Beyond a distance of a few cell diameters, diffusion is not sufficiently rapid to meet the metabolic requirements of cells. In multicellular organisms larger than a microscopic cluster of cells, some mechanism other than simple diffusion is needed to transport molecules rapidly over the long distances between the cells and the body's surface and between the various specialized tissues and organs. This problem is met in the animal kingdom by a **circulatory system,** which comprises the blood; the set of tubes, blood vessels, through which the blood flows; and a pump, the heart, which produces this flow. The heart and blood vessels together are termed the **cardiovascular system.**

Rapid bulk flow of blood through all parts of the body via the blood vessels is produced by pressures created by the pumping action of the heart. The extraordinary degree of branching of these vessels assures that most cells are within a few cell diameters of at least one of the smallest branches, the capillaries. Diffusion across the walls of the capillaries and through the interstitial fluid permits the exchange of nutrients and metabolic end products between the capillary blood and the cells near the capillary. Thus, the circulation utilizes bulk flow to solve the problem of delivering blood to the various organs and tissues but uses diffusion for the actual exchanges between the blood and the cells of these organs and tissues. To comprehend the functional characteristics of the various blood vessel types, one must be familiar with the basic physical principles that underly the bulk flow of fluids.

BASIC PRINCIPLES OF PRESSURE, FLOW, AND RESISTANCE

Fluid flows through a tube in response to a difference (*gradient*) in pressure between the two ends of the tube. The total volume that flows per unit time is directly proportional to the difference in pressure. It is not the absolute pressure in the tube that determines flow but the difference in pressure between the two ends. This direct proportionality of flow F and pressure difference ΔP can be written

$$F = k \, \Delta P$$

where k is the proportionality constant that describes how much flow occurs for a given pressure difference. Knowing only the pressure difference between two ends of a tube is not enough to determine how much fluid will flow; one must also know the numerical value of k. This constant is simply a measure of the ease with which fluid will flow through a tube. Usually we do not deal directly with k but with the reciprocal of k, known as **resistance R.**

$$R = \frac{1}{k}$$

Thus, the larger k is, the smaller R is. R answers the question: How difficult is it for fluid to flow through a tube at any given pressure? Its name, resistance, is therefore quite appropriate. Substituting $1/R$ for k, the basic equation becomes

$$F = \frac{\Delta P}{R}$$

In words, flow is directly proportional to the pressure difference and inversely proportional to the resistance. As we shall see in the next chapter, this equation is directly applicable to fluid dynamics in the cardiovascular system where F becomes cardiac output, ΔP becomes mean arterial blood pressure, and R becomes the resistance to flow offered by the blood vessels.

Determinants of Resistance

Resistance is essentially a measure of friction because basically it is the friction between the tube wall and fluid and between the molecules of the fluid themselves that opposes flow. Resistance depends upon the nature of the fluid and the geometry of the tube.

Nature of the Fluid   Syrup flows less readily than water because there is greater friction between the molecules of syrup as they slide past each other than there is between the molecules of water. This property of fluids is called **viscosity;** the more friction, the higher the viscosity. In abnormal states, changes in blood viscosity, which are generally caused by increases in hematocrit, can have important effects on the resistance to flow, but under most physiologic conditions, the viscosity of blood is relatively constant.

Geometry of the Tube   Both the length and radius of a tube affect its resistance to flow by determining the surface area of the tube in contact with the fluid (Fig. 21-19). Resistance is directly proportional to the length of the tube, but since the lengths of the

blood vessels remain constant in the body, length is not a factor in the control of resistance and we shall not be concerned with this relationship. Because of complex relationships between the tube wall and the fluid, the resistance increases markedly as tube radius decreases. The exact relationship is given by the following formula, which states that the resistance is inversely proportional to the fourth power of the radius (the radius multiplied by itself four times).

$$R \propto \frac{1}{r^4}$$

The extraordinary dependence of resistance upon radius can be appreciated by the fact (shown in Fig. 21-19) that doubling the radius increases the flow sixteen fold. As we shall see, the radius of blood vessels can be changed significantly and constitutes the most important factor in the control of resistance to blood flow.

GENERAL STRUCTURAL CHARACTERISTICS OF BLOOD VESSELS

The functional and structural characteristics of the blood vessels change with successive branching, yet the entire cardiovascular system from the heart to the smallest capillary has one structural component in common, a smooth, low-friction lining of endothelial cells. In addition, all vessels larger than capillaries have supporting layers of tissue that surround the endothelium to withstand the pressure of the contained blood, to dampen pressure pulsations, and to minimize flow variations throughout the cardiac cycle. They also have muscle fibers to control the diameter of the vessel lumen.

These tissue layers from within outward are named the **tunica intima, tunica media,** and **tunica adventitia** (Fig. 21-20 and Table 21-5). In general, the tunica intima comprises a single layer of endothelial cells supported by a thin connective-tissue network. The tunica media consists of smooth muscle and elastic tissue, and the tunica adventitia is composed largely of collagenous connective tissue. The adventitia binds the vessel to the connective tissue through which it runs and contains the nerves and small blood vessels that supply the vessel wall. The thickness and composition of the outer two layers vary with the type and size of vessel, reflecting its functional role. The tunica media is the thick-

FIG. 21-19   Effects of tube length and radius on flow.

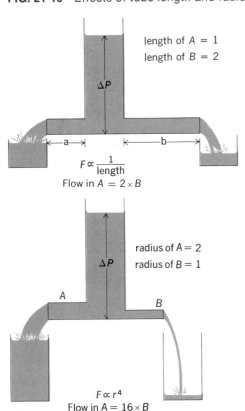

length of A = 1
length of B = 2

$F \propto \dfrac{1}{length}$

Flow in A = 2 × B

radius of A = 2
radius of B = 1

$F \propto r^4$

Flow in A = 16 × B

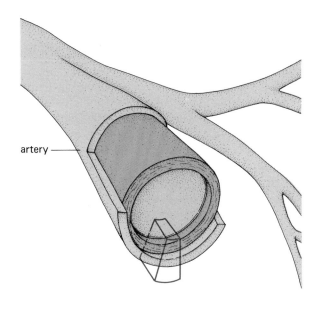

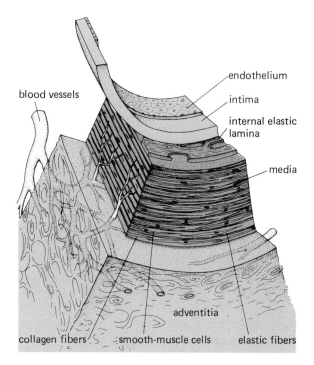

FIG. 21-20 Structure of an artery wall.

TABLE 21-5 CHARACTERISTICS OF HEART AND BLOOD VESSEL WALLS*

	Tunica Intima			Tunica Media		Tunica Adventitia
	Endothelium	Intima	Internal Elastic Lamina	Media	External Elastic lamina	Adventitia
Heart	Present	**Endocardium** Thick; CT† and smooth muscle	Absent	**Myocardium** Very thick cardiac muscle	Absent	**Epicardium** CT, vessels, and nerves
Elastic arteries (> 1 cm diam.)	Present	Thick; CT and smooth muscle	Present	Thick; elastic fibers, some smooth muscle	Indistinct	Thin; CT, smooth muscle, vessels, and nerves
Muscular arteries (< 1 cm diam.)	Present	Thick; CT and smooth muscle	Present	Thickest tunic; smooth muscle in 30–40 concentric layers	Present	Thick; CT, smooth muscle, vessels, and nerves
Arterioles (≤ 0.5 mm diam.)	Present	Thin; CT	Present	Thick; smooth muscle in 1–2 layers	Indistinct	Thin; CT and nerves
Capillaries (8–10 μm diam.)	Present	Absent	Absent	Absent	Absent	Thin; CT
Venules (10 μm–1 mm diam.)	Present	Thin; CT	Absent	Absent or smooth muscle in 1–2 layers	Absent	Thin; CT; thickness increases with vessel size
Veins (≥ 2 mm diam.)	Present	Thin; CT	Present	Smooth muscle in a few thin layers	Absent	Very thick; CT, smooth muscle, vessels, and nerves

†CT = connective tissue

est coat in arteries, whereas the adventitia is the thickest coat in veins; moreover, arteries generally have thicker walls and smaller lumina than veins (Fig. 21-21).

ARTERIES

Arteries are classed as either **elastic** or **muscular arteries,** depending upon the relative amounts of elastic or muscle tissues in their walls. The aorta and other large arteries have thick walls with a relatively thick tunica media (Table 21-5), which is filled with repeating concentric layers of elastic tissue separated by layers of smooth muscle cells and connective-tissue fibers, whereas, in comparison, the tunica adventitia is thin. Smaller arteries and arterioles have much more smooth muscle than large arteries, relative to their thickness.

The walls of the larger arteries themselves receive blood vessels, called the **vasa vasorum,** i.e., vessels of vessels, but nutrients for the smaller arteries simply diffuse in from the blood that is in the vessel itself. The arterial walls are supplied with nerves, the majority of which are postganglionic sympathetic fibers. They course through the adventitia, contacting the smooth muscle cells through breaks in the external elastic lamina lying between the tunica media and adventitia. The arteries are

enclosed in thin connective-tissue sheaths, usually with an accompanying vein and nerve.

Despite the presence of smooth muscle in the vessel wall, there is relatively little fluctuation in vessel diameter. Because the arteries have large radii, they serve as low-resistance pipes that conduct blood to the various organs. Their second major function, related to their elasticity, is to act as a pressure reservoir for maintaining blood flow through the tissues during diastole.

Arterial Blood Pressure

Recall once more the factors that determine the pressure in an elastic container, e.g., a balloon filled with water. The pressure inside the balloon depends upon the volume of water in it and the distensibility of the balloon, i.e., how easily its walls can be stretched. If the walls are very stretchable, large quantities of water can go in with only a small rise in pressure; conversely, a small quantity of water causes a large pressure rise in a balloon with low distensibility.

These principles can now be applied to an analysis of arterial function. The contraction of the ventricles ejects blood into the pulmonary and systemic arteries during systole. If a precisely equal quantity of blood were to flow simultaneously out of the arteries via arterioles, the total volume of blood in the arteries would remain constant and arterial pressure would not change. Such, however, is not the case. As shown in Fig. 21-22, a volume of blood equal to only about one-third the stroke volume leaves the arteries during systole. The excess volume distends the arteries by raising the arterial pressure. When ventricular contraction ends, the stretched arterial walls recoil passively (like a stretched rubber band upon release) and the arterial pressure continues to drive blood through the arterioles. As blood leaves the arteries, the arterial volume and, therefore, the pressure slowly fall, but the next ventricular contraction occurs while there is still adequate blood in the arteries to stretch them partially.

The aortic pressure pattern shown in Fig. 21-23 is typical of the pressure changes that occur in all the large systemic arteries. The pulmonary-artery pressure profile is similar but with all pressures smaller. The maximum pressure is reached during

FIG. 21-21   A scanning electron micrograph in which a muscular artery and medium-sized vein in the same connective-tissue bed can be compared.

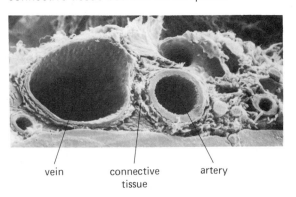

vein   connective   artery
            tissue

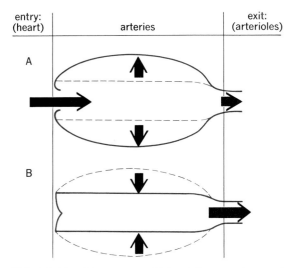

entry: (heart)   arteries   exit: (arterioles)

A

B

FIG. 21-22 Movement of blood into and out of the arteries during the cardiac cycle. The lengths of the arrows denote relative quantities. During systole A, less blood leaves the arteries than enters and the arterial walls stretch. During diastole B, the walls recoil passively, driving blood out of the arteries.

peak ventricular ejection and is called **systolic pressure.** The minimum pressure occurs just before ventricular contraction and is called **diastolic pressure.** They are generally recorded as systolic/diastolic, that is, 125/75 mmHg in our example. The pulse,

FIG. 21-23 Typical aortic pressure fluctuations during the cardiac cycle. The small disturbance in the downward pressure curve (the **dicrotic notch**) occurs because of the following: As the heart relaxes, left ventricular pressure falls below aortic pressure and, for a brief moment, blood flows back toward the heart. This reverse flow snaps the aortic valve shut, blood flows again toward the periphery, and arterial pressure resumes its smooth decline.

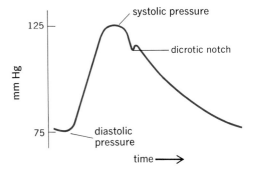

systolic pressure

125

dicrotic notch

mm Hg

75   diastolic pressure

time ⟶

which can be felt in an artery, is a result of the difference between systolic and diastolic pressure. This difference (125 − 75 = 50) is called the **pulse pressure.** The factors that alter pulse pressure are the following: (1) An increased stroke volume tends to elevate systolic pressure because of greater arterial stretching by the additional blood. (2) Decreased arterial distensibility, as in atherosclerosis, the major cause of hardening of the arteries, may cause a marked increase in systolic pressure because the wall is stiffer; i.e., any given volume of blood produces a greater pressure rise.

It is evident from the figure that arterial pressure and, therefore, flow is constantly changing throughout the cardiac cycle and the average pressure **(mean pressure)** throughout the cycle is not merely the value halfway between systolic and diastolic pressure, because diastole usually lasts longer than systole. The true mean arterial pressure can be obtained only by complex methods, but for most purposes it is approximately equal to the diastolic pressure plus one-third of the pulse pressure. Thus, in our example,

$$\text{Mean pressure} = 75 + (\tfrac{1}{3} \times 50) = 92 \text{ mmHg}$$

The mean arterial pressure is the most important of the pressures described because it is the average pressure driving blood into the tissues throughout the cardiac cycle. In other words, if the pulsatile pressure changes were eliminated and the pressure throughout the cardiac cycle were always equal to the mean pressure, the total flow would be unchanged. It is a closely regulated quantity; indeed, the reflexes that accomplish this regulation constitute the basic cardiovascular control mechanisms and will be described in detail in Chap. 22.

One last point: We can refer to "arterial" pressure without specifying to which artery we are referring because the aorta and other arteries have such large diameters that they offer only negligible resistance to flow and the pressures are therefore similar everywhere in the arterial tree.

Measurement of Arterial Pressure

Both systolic and diastolic blood pressure are readily measured in human beings with the use of a sphygmomanometer (Fig. 21-24). A hollow cuff is wrapped around the arm and inflated with air to a

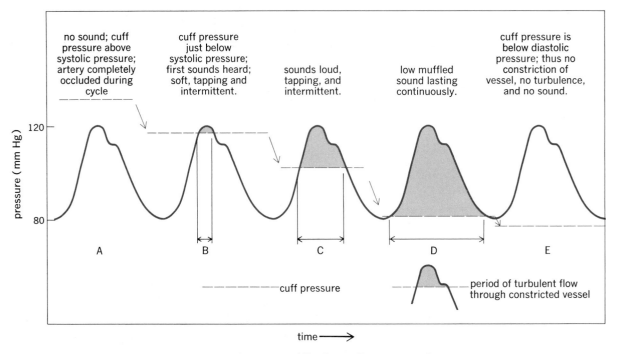

FIG. 21-24 Sounds heard through a stethoscope while the cuff pressure of a sphygmomanometer is gradually lowered. Systolic pressure is recorded at B and diastolic pressure at the point of sound disappearance (D to E).

pressure greater than systolic blood pressure (Fig. 21-24A). The high pressure in the cuff is transmitted through the tissues of the arm and completely collapses the arteries under the cuff, thereby preventing blood flow to the lower arm. The air in the cuff is now slowly released, causing the pressure in the cuff and arm to drop. When cuff pressure has fallen to a point just below the systolic pressure (Fig. 21-24B), the arterial blood pressure at the peak of systole is greater than the cuff pressure, causing the artery to expand and allow blood flow for this brief time. During this interval, the blood flow through the partially occluded artery occurs at a very high velocity because of the small opening for blood passage and the large pressure gradient. The high-velocity blood flow produces turbulence and vibration, which can be heard through a stethoscope placed over the artery just below the cuff. The pressure measured on the manometer attached to the cuff at which sounds are first heard as the cuff pressure is lowered is identified as the systolic blood pressure. These first sounds are soft tapping

sounds, corresponding to the peak systolic pressure reached during ejection of blood from the heart. As the pressure in the cuff is lowered further, the time of blood flow through the artery during each cycle becomes longer (Fig. 21-24C). The tapping sound becomes louder as the pressure is lowered. When the cuff pressure reaches the diastolic blood pressure, the sounds become dull and muffled, as the artery remains open throughout the cycle, allowing continuous turbulent flow (Fig. 21-24D). Just below diastolic pressure all sound stops as flow is now continuous and nonturbulent through the completely open artery. Thus, systolic pressure is measured as the cuff pressure at which sounds first appear and diastolic pressure as the cuff pressure at which sounds first disappear.

ARTERIOLES

Each organ or tissue obviously receives only a fraction of the total left ventricular cardiac output. The

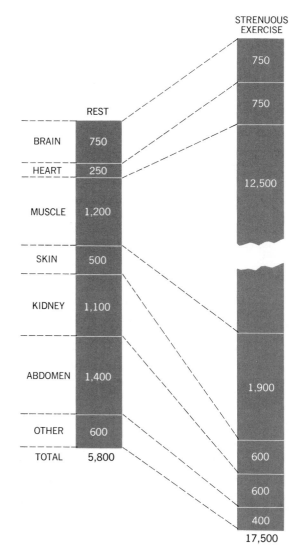

FIG. 21-25 Distribution of blood flow to the various organs and tissues of the body at rest and during strenuous exercise. The numbers show blood flow in milliliters per minute.

changed. In contrast, the energy consumption of muscular tissues of the body (heart, skeletal muscle, uterus, etc.) varies directly with the degree of muscle activity. Now compare the values at rest with those for exercise in Fig. 21-25. There is a large increase in blood flow to the exercising skeletal muscle and heart. Skin blood flow also has increased, kidney flow has decreased, and brain flow is unchanged. Obviously, the total cardiac output has increased; but more important for our present purposes, the distribution of flow has greatly changed. Cardiac output is distributed to the various organs and tissues according to their functions and needs at any given moment. The remainder of this section describes the arterioles, that segment of the vascular tree primarily responsible for control of blood-flow distribution.

As the terminal branches of the smallest arteries reach diameters of 100 μm or so, the arterioles begin. Their endothelial lining is separated from the smooth muscle cells of the tunica media by mere patches of elastic tissue, which become more and more scanty as the arterioles decrease in size until they finally disappear in the finest terminal arterioles. The smooth muscle cells are wound around the arteriole in spiral bands (Fig. 21-26) and are electrically coupled to each other by occasional gap junctions. There are several layers of smooth muscle cells in the larger arterioles, but they too thin out until there is just a single layer in the finest terminal arterioles.

Figure 21-27 illustrates the major principles by which contraction or relaxation of the vascular smooth muscle and the resulting change in vessel diameter control blood-flow distribution. The principles are presented in terms of a simple model, a fluid-filled tank with a series of compressible outflow tubes. What determines the rate of flow through each exit tube? As always,

$$\text{Flow} = \frac{\Delta P}{R}$$

Since the driving pressure is identical for each tube, differences in flow are completely determined by differences in the resistance to flow offered by each tube. The lengths of the tubes are approximately the same, and the viscosity of the fluid is a constant; therefore, differences in resistance offered by the tubes are due solely to differences in their radii.

typical distribution for a resting normal adult is given in Fig. 21-25. The digestive tract (including the liver), the kidneys, and the brain receive the largest supplies. Perhaps the most striking aspect of brain blood flow is its remarkable constancy. Whether we are staring blankly into space or contemplating the theory of relativity, the *total* energy consumption of the brain remains relatively un-

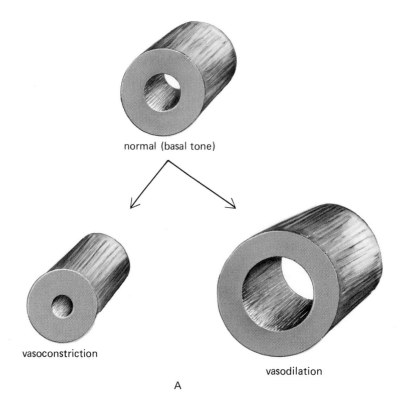

normal (basal tone)

vasoconstriction

vasodilation

A

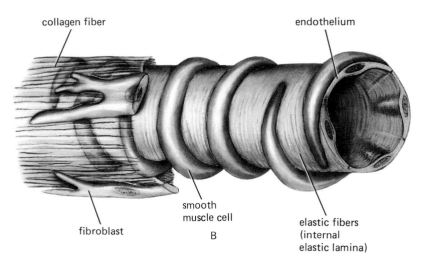

collagen fiber

endothelium

smooth
muscle cell

B

fibroblast

elastic fibers
(internal
elastic lamina)

FIG. 21-26 (A) Change in lumen diameter and wall thickness of a dilated and constricted arteriole as compared to normal. (B) Structure of a small arteriole wall. The endothelium forms the internal coat, the smooth muscle fibers the middle coat, and the connective-tissue elements (fibroblasts and collagen fibers) the external coat.

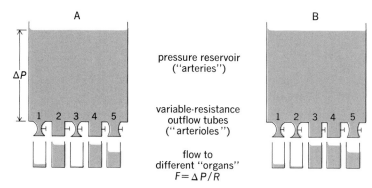

FIG. 21-27 Physical model of the relationship between arterial pressure, arteriolar radius in different organs, and blood-flow distribution. Blood has been shifted from organ 2 to organ 3 (in going from A to B) by constricting the "arterioles" of 2 and dilating those of 3.

Obviously, the widest tube has the greatest flow. If we equip each outflow tube with an adjustable cuff, we can obtain any combination of relative flows we wish.

This analysis can now be applied to the cardiovascular system. The tank is analogous to the arteries, which serve as a pressure reservoir, the major arteries themselves being so large that they contribute little resistance to flow. The smaller terminal arteries begin to offer some resistance, but it is relatively slight. Therefore, all the arteries of the body can be considered a single pressure reservoir. The arteries branch within each organ into the next series of smaller vessels, the arterioles, which are now narrow enough to offer considerable resistance. The arterioles are the major site of resistance in the vascular tree and are therefore analogous to the outflow tubes in the model.

The blood flow through any organ is given by the following equation, assuming for simplicity that the venous pressure, the other end of the pressure gradient, is zero.

$$F_{\text{organ}} = \frac{\text{mean arterial pressure}}{R_{\text{organ}}}$$

Since the arterial pressure, i.e., the driving force for flow through each organ, is identical throughout the body, differences in flows between organs depends entirely on the relative resistance (R_{organ}) offered by the arterioles of each organ. The arteriolar radii are subject to precise physiologic control; their walls consist of relatively little elastic tissue but much smooth muscle, which can relax or contract, thereby changing the radius of the inside (lumen) of the arteriole. Thus, the pattern of blood-flow distribution depends primarily upon the degree of arteriolar smooth muscle contraction in each organ and tissue.

The smooth muscles that surround arterioles possess a large degree of inherent myogenic activity, i.e., "spontaneous" contraction (myogenic means originating within a muscle). This basal tone is responsible for a large portion of the basal resistance offered by the arterioles. However, what is more important is that a variety of physiologic factors play upon these smooth muscles to either increase or decrease their degree of contraction, thereby increasing or decreasing the vessels' resistance. The mechanisms that control these changes in arteriolar resistance fall into general categories: (1) local controls and (2) extrinsic (reflex) controls. The local controls serve to couple blood flow through an organ with the metabolic requirements of that organ, whereas the external controls are efferent pathways of reflexes that serve to integrate the overall activity of the cardiovascular system.

Local Controls

The term **local control** denotes the fact that individual organs, to greater or lesser extent, possess completely inherent mechanisms for altering their own arteriolar resistances, which confers upon them the capacity to self-regulate their blood flows.

Certain organs and tissues, particularly the heart, skeletal muscle, and other muscular organs, manifest an increased blood flow (**hyperemia**) any time their metabolic activity is increased. For example, the blood flow to exercising skeletal muscle increases in direct proportion to the increased activity of the muscle. This phenomenon—the adjustment of blood flow to the level of tissue activity—is known as **functional hyperemia.** It is the direct result of arteriolar dilation in the more active organ. This vasodilation does not depend upon the presence of nerves or hormones but is a locally mediated response. The adaptive value of the phenomenon should be readily apparent; an increased rate of activity in an organ automatically produces an increased blood flow to that organ by dilating its arterioles.

The arterioles dilate because their smooth muscle relaxes, i.e., loses a fraction of its basal contractile tone. The factors that act upon vascular smooth muscle in functional hyperemia to cause it to relax are local chemical changes that result from the increased metabolic cellular activity, but the relative contributions of the various factors implicated seem to vary, depending upon the organs involved and the duration of the increased activity. At present, therefore, we can only name but not quantify some of the factors that appear to be involved: decreased oxygen concentration; increased concentrations of carbon dioxide, hydrogen ion, and metabolic intermediates, such as adenosine; increased concentration of potassium (perhaps as a result of enhanced movement out of skeletal muscle cells during the more frequent action potentials); increased osmolarity (i.e., decreased water concentration, as a result of the increased breakdown of high-molecular-weight substances); and increased concentrations of prostaglandins. Changes in all these variables have been shown to cause arteriolar dilation under controlled experimental conditions, and they all

may contribute to the functional hyperemia response (Fig. 21-28). It must be emphasized that all these chemical changes act locally upon the arteriolar smooth muscle, causing it to relax, which causes the vessel to dilate; no nerves or hormones are involved.

It should not be too surprising that the phenomenon of functional hyperemia is most highly developed in heart and skeletal muscle, which show the widest range of normal metabolic activities of any organs or tissues. It is highly efficient, therefore, that their supply of blood be primarily determined locally by their rates of activity.

Exocrine glands such as the pancreas, salivary glands, and sweat glands also manifest functional hyperemia during periods of increased secretion; so do the stomach and the intestines during digestion. In these cases, a mechanism different from that described above for the muscular organs is responsible. The sequence is as follows: The tissue undergoing increased secretory activity releases an enzyme called **kallikrein,** which acts locally on a protein called **kininogen;** the result is the formation of a peptide, **kallidin,** which, in turn, is split by other locally occurring enzymes to form **bradykinin.** Both kallidin and bradykinin are potent relaxers of vascular smooth muscle and so bring about vasodilation and an increased blood flow. This kallikrein-kinin system has other important functions and will be described more fully in Chapter 28.

The discussion of local metabolic controls has thus far emphasized increased metabolic activity of the tissue or organ as the initial event leading to vasodilation. However, similar changes in the local concentrations of the implicated metabolic factors may occur under a very different set of conditions, namely when a tissue or organ suffers a reduction in its blood supply (**ischemia**). The supply of oxygen to the tissue is diminished and the local oxygen concentration decreased. Simultaneously, the con-

FIG. 21-28 Local control of blood flow in functional hyperemia.

centrations of carbon dioxide, hydrogen ion, and metabolites all increase because they are not removed by the blood as fast as they are produced, and prostaglandin synthesis is increased. In other words, the local metabolic changes that occur during both increased metabolic activity and ischemia are similar (although not identical) because both situations reflect an imbalance of blood supply and level of cellular metabolic activity. The adaptive value of vasodilation in response to ischemia, say, as a result of partial occlusion of the artery that supplies the tissue, is that it automatically tends to maintain the blood flow to the tissue.

Extrinsic (Reflex) Controls

Sympathetic Nerves   Most arterioles receive a rich supply of sympathetic postganglionic nerve fibers. These nerves (with one major exception) release norepinephrine, which combines with α-adrenergic receptors on the vascular smooth muscle to cause vasoconstriction. If almost all the nerves to arterioles are constrictor in action, how can reflex arteriolar dilation be achieved? Since the sympathetic nerves are seldom completely quiescent but discharge at some finite rate, which varies from organ to organ, the nerves always cause some degree of tonic constriction; from this basal position, further constriction is produced by increased sympathetic activity, whereas dilation can be achieved by decreasing the rate of sympathetic activity below the basal level. The skin offers an excellent example of these processes: Skin arterioles of a normal unexcited person at room temperature are already under the influence of a high rate of sympathetic discharge; an appropriate stimulus (fear, loss of blood, etc.) causes reflex enhancement of this activity; the arterioles constrict further, and the skin pales. In contrast, an increased body temperature reflexly inhibits the sympathetic nerves to the skin, the arteroles dilate, and the skin flushes. This generalization cannot be stressed too strongly: Control of the sympathetic constrictor nerves to arteriolar smooth muscle can accomplish either dilation or constriction.

In contrast to the local control processes, the primary functions of these nerves are concerned not with the coordination of local metabolic needs

and blood flow but with reflexes that help maintain an adequate blood supply at all times to vital organs such as the brain and heart.[1] The common denominator of these reflexes is the regulation of arterial blood pressure; they will be described in detail in Chap. 22.

There is, as we said, one exception to the generalization that sympathetic nerves to arterioles release norepinephrine. A group of sympathetic (*not* parasympathetic) nerves to the arterioles in skeletal muscle instead releases acetylcholine, which causes arteriolar dilation and increased blood flow. It still must be emphasized that *most* sympathetic nerves to skeletal muscle arterioles release norepinephrine; thus skeletal muscle arterioles receive a dual set of sympathetic nerves. The only known function of the vasodilator fibers is in the response to exercise or stress and will be described in Chap. 22; the vasoconstrictor fibers mediate all other situations that involve neural control of skeletal muscle arterioles.

Parasympathetic Nerves   With but few major exceptions (the blood vessels of certain areas in the genital tract), there is no significant parasympathetic innervation of arterioles. It is true that stimulation of the parasympathetic nerves to certain glands is associated with an increased blood flow, but this is secondary to the increased metabolic activity induced in the gland by the nerves, with resultant release of kallikrein and the initiation of local functional hyperemia.

Hormones   Several hormones cause contraction or relaxation of arteriolar smooth muscle. One of these is epinephrine, the hormone released from the adrenal medulla. In most vascular beds, epinephrine, like the sympathetic nerves, causes vasoconstriction; surprisingly, in other vascular beds, epinephrine may induce vasodilation. However, it is likely that the effects of circulating epinephrine on arterioles are quantitatively of little significance

---

[1] The vasculature of the brain and heart receives sympathetic nerve fibers whose activity may be altered under certain circumstances. However, present evidence indicates that these nerve fibers are usually of negligible importance when compared with the local control of these vascular beds.

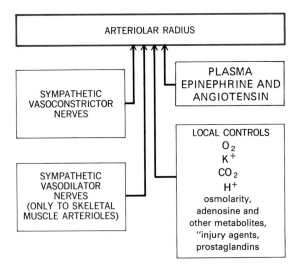

FIG. 21-29 Major factors affecting arteriolar radius.

when compared with those exerted by norepinephrine released from sympathetic nerve endings.

Another hormone that exerts a potent vasoconstrictor action on most arterioles is **angiotensin.** This peptide is generated in the blood as the result of the action of another hormone, the enzyme **renin,** on a plasma protein, **angiotensinogen.** The control and multiple functions of the so-called renin-angiotensin system will be described in Chap. 24; suffice it here to emphasize that an increase in plasma angiotensin will produce a widespread increase in the degree of arteriolar vasoconstriction.

Figure 21-29 summarizes the factors that determine arteriolar radius. Of course, the local factors and extrinsic reflex controls do not exist in isolation from one another but manifest important interactions.

CAPILLARIES

At any given moment, approximately 5 percent of the total circulating blood is in the capillaries. Yet it is this 5 percent that is performing the ultimate function of the entire system, namely, the exchange of nutrients and metabolic end products. All other segments of the vascular tree subserve the overall aim of getting adequate blood flow through the capillaries. The capillaries permeate practically every tissue of the body; most cells are no more than 0.15

mm from a capillary. Therefore, diffusion distances are very small, and exchange is highly efficient. There are thousands of kilometers of capillaries in an adult person, each individual capillary being only about 1 mm long.

Capillaries throughout the body vary somewhat in structure, but the typical capillary (Fig. 21-30) is a thin-walled tube of endothelial cells one layer thick without elastic tissue, connective tissue, or smooth muscle to impede movement of water and solutes. The squamous cells that constitute the endothelial tube interlock like pieces of a jigsaw puzzle. Sinusoids differ from other capillaries in that they are much wider and have irregular lumens, and their walls, particularly in the liver, spleen, and bone marrow, contain fixed macrophages. The permeability of capillaries varies throughout the body, liver sinusoids being the "leakiest" and brain capillaries the "tightest."

Anatomy of the Capillary Network

Figure 21-31 illustrates diagrammatically the general anatomy of the small vessels that constitute the so-called microcirculation. Blood enters the capillary network from the arterioles. Most tissues appear to have two distinct types of capillaries: "true" capillaries and thoroughfare channels. The thoroughfare channels connect arterioles and venules directly. From these channels exit and reenter the network of true capillaries across which materials actually exchange. The site at which a true capillary exits from a thoroughfare channel is protected by a ring of smooth muscle, the **precapillary sphincter,** which continually opens and closes (due to local controls) so that flow through any given capillary is usually intermittent. Generally, the more active the tissue, the more precapillary

FIG. 21-30 Structure of the capillary wall.

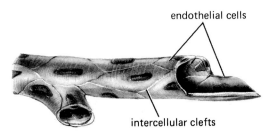

than that of the arterioles. (2) Because capillaries have no smooth muscle, their radius (and, therefore, their resistance) is not subject normally to active control and simply reflects the volume of blood delivered to them via the arterioles (and the volume leaving via the venules).

Velocity of Capillary Blood Flow

Figure 21-32 illustrates a simple mechanical model of a series of 1-cm diameter balls being pushed down a single tube that branches into narrower tubes. Although each tributary tube has a smaller cross section than the wide tube, the sum of the tributary cross sections is much greater than the area of the wide tube. Let us assume that in the wide tube each ball moves 3 cm/min. If the balls are 1 cm in diameter and they move two abreast, six balls leave the wide tube per minute and enter the narrow tubes, and six balls leave the narrow tubes per minute. At what speed does each ball move in the small tubes? The answer is 1 cm/min. This example illustrates the following important generalization: When a continuous stream moves through consecutive sets of tubes, the velocity of flow decreases as the sum of the cross-sectional

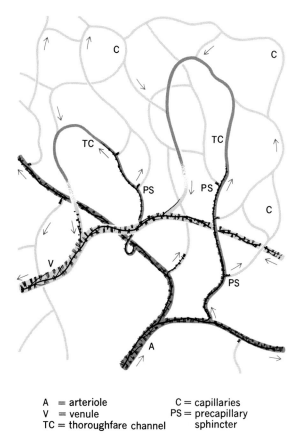

A = arteriole   **C** = capillaries
V = venule   **PS** = precapillary
TC = thoroughfare channel   sphincter

FIG. 21-31 Diagram of microcirculation. Note the thinning of the smooth muscle coat (lines across the vessel) in the thoroughfare channels and its complete absence in the true capillaries. The black lines on the surface of the vessels are nerve fibers leading to smooth muscle cells.

sphincters are open at any moment. The sphincters are best visualized as functioning in concert with arteriolar smooth muscle to regulate not only the total flow of blood through the tissue capillaries but the number of functioning capillaries as well.

Resistance of the Capillaries

Since a capillary is very narrow, it offers a considerable resistance to flow, but for two reasons, the resistance is not of critical importance for cardiovascular function: (1) Despite the fact that the capillaries are actually narrower than the arterioles, the huge total number of capillaries provides such a great cross-sectional area for flow that the *total* resistance of *all* the capillaries is considerably less

FIG. 21-32 Relationship between cross-sectional area and velocity of flow. The total cross-sectional area of the small tubes is three times greater than that of the large tube. Accordingly, velocity of flow is one-third as great in the small tubes.

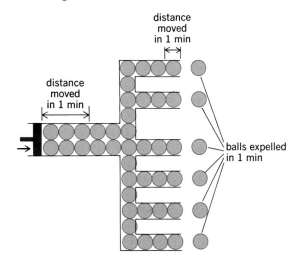

distance moved in 1 min

distance moved in 1 min

balls expelled in 1 min

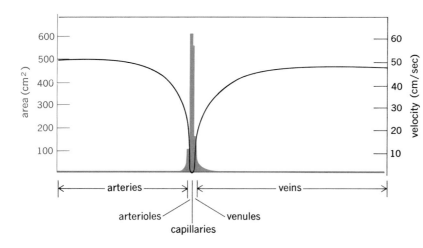

FIG. 21-33 Relation between cross-sectional area and velocity of flow in the systemic circulation. The values are those for a 30-lb dog. Velocity of blood flow through the capillaries is about 0.07 cm/sec.

areas of the tubes increases. This is precisely the case in the cardiovascular system (Fig. 21-33); the blood velocity is very great in the aorta, progressively slows in the arteries and arterioles, and then markedly slows as it passes through the huge cross-sectional area of the capillaries (600 times the cross-sectional area of the aorta). The speed then progressively increases in the venules and veins because the cross-sectional area decreases. The adaptive significance of this phenomenon is very great; blood flows through the capillaries so slowly (0.07 cm/sec) that there is adequate time for exchange of nutrients and metabolic end products between the blood and tissues.

Diffusion across the Capillary Wall: Exchanges of Nutrients and Metabolic End Products

There is no active transport of solute across the capillary wall, and materials cross primarily by diffusion. As described in the next section, there is some movement of fluid by bulk flow, but it is of negligible importance for the exchange of nutrients and metabolic end products. Because fat-soluble substances penetrate cell membranes easily, they probably pass directly through the endothelial capillary cells. In contrast, many ions and molecules are poorly soluble in fat and must pass through as yet unidentified "pores" in individual endothelial cells or through the clefts between adjacent cells. Small amounts of protein pass through most capillaries mainly by endocytosis and exocytosis (see

Chap. 4). (This is an exception to our generalization that most solutes cross by diffusion.) In any case, nearly all nutrients and metabolic end products diffuse across the capillary with great speed.

What is the sequence of events involved in capillary-cell transfers? Tissue cells do not exchange material directly with blood; the interstitial fluid always acts as middleman. Thus, nutrients diffuse across the capillary wall into the interstitial fluid, from which they gain entry to cells. Conversely, metabolic products move first across plasma membranes of cells into interstitial fluid,

FIG. 21-34 Movements of nutrients and metabolic products between plasma and tissue cells. The substance may enter or leave the cell by diffusion or carrier-mediated transport, but it moves across the capillary wall only by diffusion, dependent upon capillary–interstitial fluid concentration gradients. (Two capillaries are shown for convenience; actually, of course, nutrients and products leave and enter the same capillary.)

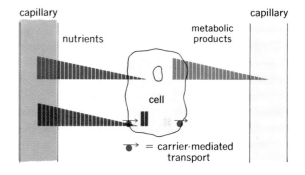

from which they diffuse into the plasma. Thus, two membrane transport processes must always be considered, that across the capillary wall and that across the tissue cell plasma membrane (Fig. 21-34). The plasma membrane step may be by diffusion or by carrier mediated transport, but, as described above, the transcapillary movement is always by diffusion (except in the case of some proteins). Since, to achieve *net* transport of any substance by diffusion, a concentration gradient is required, transcapillary diffusion of nutrients and metabolic products proceeds primarily in one direction because of diffusion gradients for these substances between the blood and interstitial fluid.

How do these diffusion gradients arise? Let us take two examples, those of glucose and carbon dioxide transcapillary movement in muscle. Glucose is continuously consumed after being transported from interstitial fluid into the muscle cells; this removal from interstitial fluid lowers the interstitial fluid glucose concentration below that of plasma and creates the gradient for glucose diffusion out of the capillary. Carbon dioxide is continuously produced by muscle cells, thereby creating an increased intracellular carbon dioxide concentration, which causes diffusion of carbon dioxide into the interstitial fluid. In turn, this causes the interstitial carbon dioxide concentration to be greater than that of plasma and produces carbon dioxide diffusion into the capillary. We chose these particular examples to emphasize that net movement of a substance between interstitial fluid and cells is the event that establishes the transcapillary plasma-interstitial diffusion gradients.

When a tissue increases its rate of metabolism, it must obviously obtain more nutrients from the blood and eliminate more metabolic products. One important mechanism for achieving this is the increase in diffusion gradients between plasma and cells that occurs for two reasons: (1) Increased cellular consumption of oxygen and nutrients lowers their intracellular concentrations, while increased production of carbon dioxide and other end products raises their cell concentrations. Let us return to our example of glucose and muscle. When the muscle increases its activity, it also increases its uptake of glucose, thereby lowering the interstitial glucose concentration below normal. This sets up an increased plasma-interstitial glucose concentra-

tion gradient that causes the *net* diffusion of glucose out of the capillary to be increased. In other words, this change in concentration gradient has allowed a greater fraction of the total blood glucose to be extracted from the blood as it flows through the capillaries. (2) Simultaneously, blood flow to the area increases, caused by the action of local metabolic factors on arterioles, and this additional blood flow prevents the plasma concentrations of oxygen and nutrients from falling rapidly and those of carbon dioxide and other end products from rising rapidly as diffusion proceeds. The increase in local metabolic factors facilitates diffusion by another mechanism as well; it dilates precapillary sphincters, and the resulting increase in the number of open capillaries increases the total surface area available for diffusion and decreases the average distance between cells and capillaries.

Bulk Flow across the Capillary Wall: Distribution of the Extracellular Fluid

Since the capillary wall is highly permeable to water and to almost all the solutes of the plasma with the exception of the plasma proteins, it behaves like a porous filter through which protein-free plasma (**ultrafiltrate**) moves by bulk flow through the clefts between the cells under the influence of a hydrostatic pressure gradient. The magnitude of the bulk flow is directly proportional to the hydrostatic pressure difference between the inside and outside of the capillary, i.e., between the capillary blood pressure and the interstitial-fluid pressure. Normally, the former is much larger than the latter, so that a considerable hydrostatic pressure gradient exists to drive the filtration of protein-free plasma out of the capillaries into the interstitial fluid. Why then does all the plasma not filter out into the interstitial space? The explanation was first elucidated by Starling (the same scientist who expounded the law of the heart that bears his name) and depends upon the principles of osmosis.

In Chap. 4 we described how a net movement of water occurs across a semipermeable membrane from a solution of high water concentration to a solution of low water concentration. Recall that the concentration of water depends upon the concentration of solute molecules or ions dissolved in the water. When two solutions A and B, which are separated by a semipermeable membrane, have identical concentrations of all solutes, the water concentrations are identical and no net water movement occurs. When, however, a quantity of a nonpermeating substance is added to solution A, the water concentration of A is reduced below that of solution B and a net movement of water will occur by osmosis from B into A. Of great importance is that osmotic flow of water (solvent) "drags" along with it any dissolved solutes to which the membrane is highly permeable. Thus, a difference in water concentration can result in the movement of both water and permeating solute in a manner similar to the bulk flow produced by a hydrostatic pressure difference. The difference in water concentration that results from the presence of the nonpenetrating solute can therefore be expressed in units of pressure (millimeters of mercury).

This analysis can now be applied to capillary fluid movements. The plasma in the capillary and the interstitial fluid outside it contain large quantities of low-molecular-weight solutes (crystalloids), e.g., sodium, chloride, or glucose. Since the capillary lining is highly permeable to all these crystalloids, they all have almost identical concentrations in the two solutions. There are small concentration differences occurring for substances consumed or produced by the cells, but these tend to cancel each other and, accordingly, no significant water-concentration difference is caused by the presence of the crystalloids. In contrast, the plasma proteins cross capillary walls only very slightly and therefore have a very low interstitial-fluid concentration. This difference in protein concentration between plasma and interstitial fluid means that the water concentration of the plasma is lower than that of interstitial fluid, inducing an osmotic flow of water from the interstitial compartment into the capillary. Along with the water are carried all the different types of crystalloids dissolved in the interstitial fluid. Thus, osmotic flow of fluid, like bulk flow, does not alter the concentrations of the low-molecular-weight substances of plasma or interstitial fluid.

In summary, two opposing forces act to move fluid across the capillary: (1) The hydrostatic pressure difference between capillary blood pressure and interstitial-fluid pressure favors the filtration of a protein-free plasma out of the capillary; and (2) the water-concentration difference between plasma and interstitial fluid, which results from the protein-

concentration differences, favors the osmotic movement of interstitial fluid into the capillary. Accordingly, the movements of fluid depend directly upon four variables: the capillary hydrostatic pressure, interstitial hydrostatic pressure, plasma protein concentration, and interstitial-fluid protein concentration.

We may now consider quantitatively how these variables act to move fluid across the capillary wall (Fig. 21-35). Much of the arterial blood pressure has already been dissipated as the blood flows through the arterioles, so that pressure at the beginning of the capillary is 35 mmHg. Since the capillary also offers resistance to flow, the pressure continuously decreases to 15 mmHg at the end of the capillary. The interstitial pressure is essentially zero.[2] The difference in protein concentration between plasma and interstitial fluid causes a difference in water concentration (plasma water concentration less than interstitial-fluid water concentration), which induces an osmotic flow of fluid into the capillary equivalent to that produced by a hydrostatic pressure difference of 25 mmHg. It is evident that in the first portion of the capillary the hydrostatic pressure difference is greater than the osmotic forces and a net movement of fluid out of the capillary occurs; in the last portion of the capillary, however, a net force causes fluid movement into the capillary (termed **absorption).** The net result is that the early and late capillary events tend to cancel each other out, and there is little overall net loss or gain of fluid (Fig. 21-35A). In a normal person there is a small net filtration; as we shall see, this is returned to the blood by lymphatics.

This analysis of capillary fluid dynamics in terms of different events that occur at the arterial and venous ends of the capillary is, in fact, oversimplified. It is very likely that many capillaries manifest only net filtration or net absorption along their entire length because the arterioles that supply them are either so dilated or so constricted as to yield a capillary hydrostatic pressure above or below 25 mmHg along the entire length of the capillary. This does not alter the basic concept that, taken as a unit, a capillary bed manifests net absorption or filtration, depending upon the average levels of hy-

---

[2] The exact value of interstitial pressure remains controversial. Many physiologists believe that it is not zero but is actually subatmospheric. The outcome of this controversy will not, however, alter the basic concepts presented here.

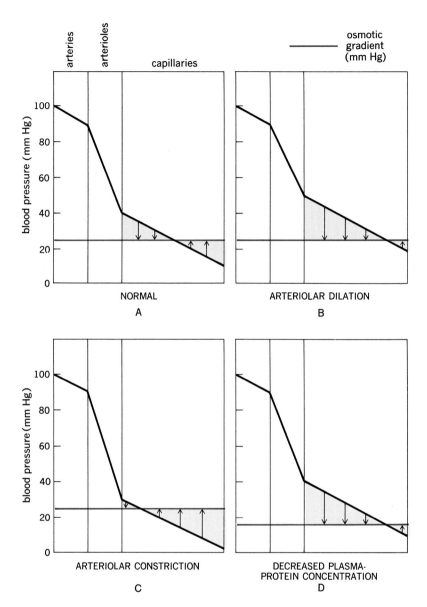

FIG. 21-35 Relevant filtration-absorption forces acting across the capillary wall in several situations. The colored horizontal lines indicate the osmotic gradient in mmHg. Arrows down indicate filtration out of the capillary. Arrows up indicate fluid movement from interstitium into capillary. The shaded areas denote relative magnitudes of the fluid movements.

drostatic pressures in the individual capillaries constituting the bed.

Figure 21-35B and C illustrate the effects on this equilibrium of changing capillary pressure. In Figure 21-35B, the arterioles in the organ have been dilated, and the capillary pressure therefore increases since less of the arterial pressure is dissipated in the passage through the arterioles. Outward filtration now predominates, and some of the plasma enters the interstitial fluid. In contrast, marked arteriolar constriction (Fig. 21-35C) produces decreased capillary pressure and net movement of interstitial fluid into the vascular compart-

ment. Figure 21-35D shows how net absorption or filtration can be produced in the absence of capillary pressure changes whenever plasma protein concentration is altered (lowered in our example). Thus, in liver disease, protein synthesis decreases, plasma protein concentration is reduced, plasma water concentration is increased, net filtration occurs, and fluid accumulates in the interstitial space **(edema).**

The transcapillary protein-concentration difference can also be decreased by a quite different event, namely, the leakage of protein across the capillary wall into the interstitium whenever the capillary lining is damaged. This reduces the protein-concentration difference, and edema occurs as a result of the unchecked hydrostatic pressure difference still acting across the capillary. The fluid accumulation in a blister is an excellent example.

The major function of this capillary filtration-absorption equilibrium should now be evident: It determines the distribution of the extracellular-fluid volume between the vascular and interstitial compartments. Obviously, the ability of the heart to pump blood depends upon the presence of an adequate volume of blood in the system. Recall that the interstitial fluid volume is three to four times larger than the plasma volume; therefore, the interstitial fluid serves as a reservoir that can supply additional fluid to the circulatory system or draw off excess. This equilibrium plays an important role in the physiologic response to many situations, such as hemorrhage.

It should be stressed again that capillary filtration and absorption do not alter *concentrations* of any substance (other than protein) since movement is by bulk flow; i.e., everything in the plasma (except protein) or the interstitial fluid moves together. The reason this process of filtration plays no significant role in the exchange of nutrients and metabolic end products between capillary and tissues is that the total quantity of a substance (such as glucose or oxygen) moving into or out of the capillary during bulk flow is extremely small in comparison with the quantities moving by diffusion. For example, during a single day approximately 20,000 g of glucose crosses the capillary into the interstitial fluid by diffusion but only 20 g enters by bulk flow. (Actually, only a small fraction of this glucose is utilized by the cells, the remainder moving back into the blood, again almost entirely by diffusion.)

VEINS

The walls of the veins are composed of three coats; the main difference between the veins and the arteries is in the relative thinness of the middle coat in the veins (Table 21-5). The tunica media of the veins has less smooth muscle and elastic connective tissue than the same layer in arteries, features related to the lower blood pressures they encounter.

Most of the pressure imparted to the blood by the heart is dissipated as blood flows through the arterioles and capillaries, so that pressure in the small venules is approximately 15 mmHg and only a small pressure remains to drive blood back to the heart. One of the major functions of the veins is to act as low-resistance conduits for blood flow from the tissues back to the heart. This function is performed so efficiently that the total pressure drop from venule to right atrium is only about 10 mmHg, the right atrial pressure being 0 to 5 mmHg. The resistance is low because the veins have a large diameter.

The veins perform a second extremely important function that has only recently been appreciated: They adjust their total capacity to accommodate variations in blood volume so as to maintain venous pressure and, thereby, venous return to the heart. The veins are the last set of tubes through which the blood must flow on its trip to the heart. The force immediately driving this venous return is the venous pressure (more precisely, the pressure gradient between the veins and atria). In turn, the rate of venous return, i.e., inflow to the atria, is one of the most important determinants of atrial pressure. In a discussion of the control of cardiac output in Chap. 20, we emphasized that the atrial pressure is the major determinant of ventricular end-diastolic volume and thereby of intrinsic control of stroke volume. Combining these two statements, we now see that venous pressure is a crucial determinant of stroke volume via the intermediation of atrial pressure and ventricular end-diastolic volume.

Determinants of Venous Pressure

The factors that determine pressure in any elastic tube, as we know, are the volume of fluid in it and the distensibility of its wall. Accordingly, total blood volume is one important determinant of venous pressure. The relatively thin walls of the veins are much more distensible than arterial walls and can accommodate large volumes of blood with a relatively small increase of internal pressure. Approximately 60 percent of the total blood volume is present in the systemic veins at any given moment (Fig. 21-36), but the venous pressure averages less than 10 mmHg. In contrast, the systemic arteries contain less than 15 percent of the blood at a pressure of approximately 100 mmHg. This pressure-volume relationship of the veins allows them to act as a reservoir for blood.

The walls of the veins contain smooth muscle

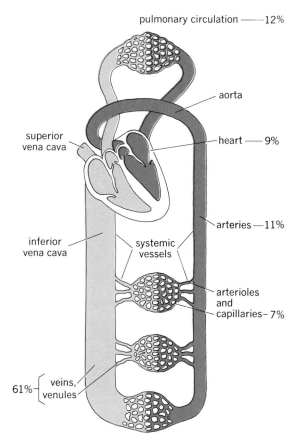

FIG. 21-36 Distribution of blood in the different portions of the cardiovascular system.

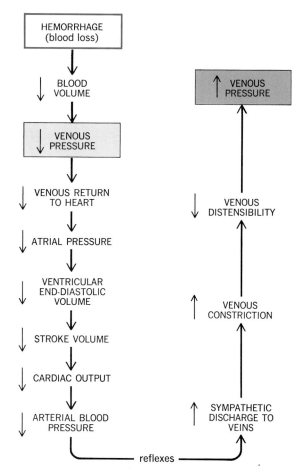

FIG. 21-37 The path by which a decreased blood volume leads to contraction of venous smooth muscle and the role of the resulting venoconstriction in maintaining venous pressure during blood loss. The increased venous smooth muscle contraction returns the decreased venous pressure toward, but not to, normal. The reflexes involved in this response are described in Chap. 22.

richly innervated by sympathetic vasoconstrictor nerves, stimulation of which causes venous constriction, thereby increasing the stiffness of the wall, i.e., making it less distensible, and raising the pressure of the blood in the veins. Increased venous pressure drives more blood out of the veins into the right heart. Thus, venous constriction exerts the same effect on venous return as giving a transfusion.

The great importance of this effect can be visualized by the example in Fig. 21-37. A large decrease in total blood volume initially reduces the pressures everywhere in the circulatory system, including the veins; venous return to the heart decreases, and cardiac output decreases. However, reflexes to be described cause increased sympathetic discharge to the venous smooth muscle, which contracts, thereby returning venous pressure toward normal, restoring venous return and cardiac output.

Two other mechanisms can decrease venous capacity, increase venous pressure, and facilitate venous return; these are the **skeletal muscle "pump"** and the effects of respiration upon thoracic and abdominal veins (**respiratory "pump"**). During skeletal muscle contraction, the veins that run through the muscle are partially compressed, which reduces their diameter and decreases venous capacity (Fig. 21-38). As seen in Chap. 23, during inspiration, the diaphragm descends, pushes on the abdominal contents, and increases abdominal pressure. The large veins that pass through the abdomen

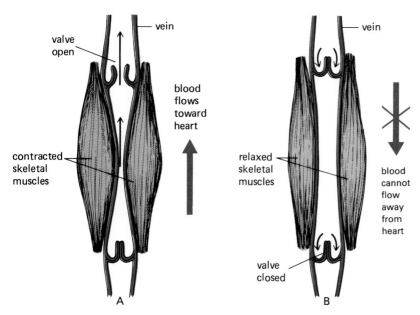

FIG. 21-38   The skeletal muscle pump. (A) During muscle contraction, venous diameter decreases and venous pressure rises. The resulting increase in blood flow can occur only toward the heart because of the valves in the veins. (B) As the muscles relax, blood flow away from the heart is prevented by the venous valves.

are partially compressed by this increased pressure. Simultaneously, the pressure in the chest (thorax) decreases, and this decrease is transmitted passively to the intrathoracic veins and right atrium. The net effect is to increase the pressure gradient between veins outside the thorax and the right atrium; accordingly, venous return to the heart is enhanced during inspiration. The increased movement of blood that results from the muscle and respiratory pumps occurs only toward the heart because venous valves prevent backflow.

In summary (Fig. 21-39), the effects of the venous smooth muscle contraction, the skeletal muscle pump, and the respiratory pump are to facilitate return of blood to the heart. The net result is that atrial pressure and, thereby, cardiac output are determined in large part by these factors.

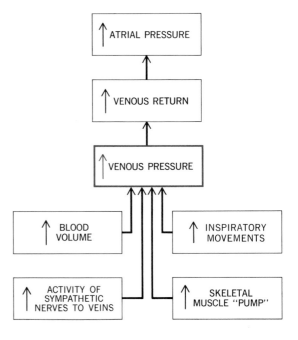

FIG. 21-39   Major factors determining venous pressure and thereby atrial pressure. The figure as drawn shows how venous and atrial pressures are increased; reversing the arrows in the boxes would indicate how these pressures can be reduced.

FIG. 21-40 (A) Normal venous valve. Any tendency toward retrograde flow would immediately push the valve leaflets together. (B) Demonstration of the location of venous valves in the forearm.

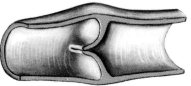

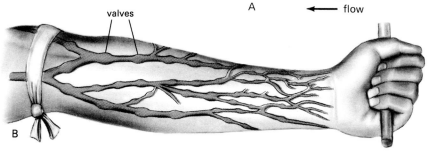

Effects of Venous Constriction on Resistance to Flow

We have seen that decreasing the diameter of the veins increases venous pressure, which increases venous return to the heart. However, this decreased diameter also increases resistance to flow, a phenomenon that would retard venous return if the effect of venous constriction upon resistance were not so slight as to be negligible. The veins have such large diameters that a slight decrease in size (which has great effects on venous capacity) produces little increase in resistance. This is just the opposite of the arterioles, which are so narrow that they contain little blood at any moment (Fig. 21-36) and further decrease has little effect on blood displacement toward the heart but even a slight decrease in diameter produces a marked increase in resistance to flow. Flow back to the heart, therefore, tends to be impaired by arteriolar constriction and enhanced by venous constriction. It should be stressed, however, that abnormally great increases in venous resistance, say from an internal blood clot or a tumor compressing from the outside, may markedly impair blood flow. Under such conditions, blood accumulates behind the lesion, pressures in the small veins and capillaries drained by the occluded vein increase, capillary filtration increases, and the tissue becomes edematous, i.e., swollen with excess interstitial fluid.

Venous Valves

Many veins, particularly in the limbs, have valves that allow flow only toward the heart (Fig. 21-40).

Why are these valves necessary if the pressure gradient created by cardiac contraction always moves blood toward the heart? We have seen how two other forces, the muscle pump and inspiratory movements, facilitate flow of venous blood. When these forces squeeze the veins, blood would be forced in both directions if the valves were not there to prevent backward flow. As we shall see, valves also play a role in countering the effects of upright posture.

This completes our description of the pressure changes throughout the vascular tree. The normal pressure profiles for the systemic and pulmonary circulation are given in Fig. 21-41. Note that the

FIG. 21-41 Summary of pressures in the vascular system. Compare these pressures with the distribution of blood volumes in different portions of the cardiovascular system shown in Fig. 21-36.

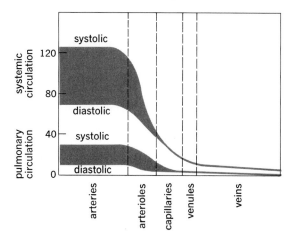

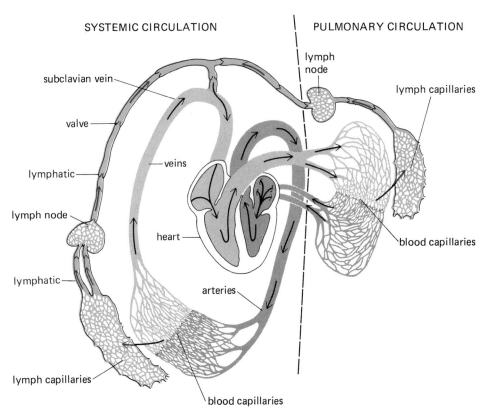

SYSTEMIC CIRCULATION   PULMONARY CIRCULATION

lymph node

subclavian vein

valve

lymphatic

lymph node

heart

lymphatic

veins

arteries

lymph capillaries

blood capillaries

lymph capillaries

blood capillaries

FIG. 21-42 A comparison of the lymphatic circulation (shown here in color), which begins in blind-end lymph capillaries, and the cardiovascular system (shown here in gray), which is a continuous circuit, and the relationship between the lymphatic and cardiovascular systems.

pulmonary pressures are considerably smaller than the systemic pressures for reasons shortly to be described. Note also that the resistance offered by the arterioles effectively damps the pulse; by doing so, the arterioles convert the pulsatile arterial flow into a continuous capillary flow.

THE LYMPHATIC SYSTEM

The lymphatic vessels are not part of the circulatory system *per se* but constitute a one-way route for the movement of interstitial fluid from the tissue spaces to blood. The lymphatic vessels arise as blind-end **lymph capillaries** (Fig. 21-42); they are present in almost all organs and form an extensive network throughout the body. The lymph capillaries resemble systemic capillaries; they are thin walled

and are permeable to virtually all interstitial-fluid constituents, including protein. The **lymph,** as the fluid is called once it has entered the lymphatic vessels, drains from the lymph capillaries into larger vessels called **lymphatics.** Ultimately the lymph is collected into one of the two major lymphatic ducts, the **thoracic duct** and the **right lymphatic duct,** which empty their lymph into the left and right brachiocephalic veins, respectively (Fig. 21-43). (The lower, expanded part of the thoracic duct is the **cisterna chyli;** it lies in the upper part of the abdomen slightly behind the aorta.) Thus, all fluid drained from the tissue spaces is eventually returned to the venous circulation. At some point before returning to the venous bloodstream, the lymph flows through chains of lymph nodes.

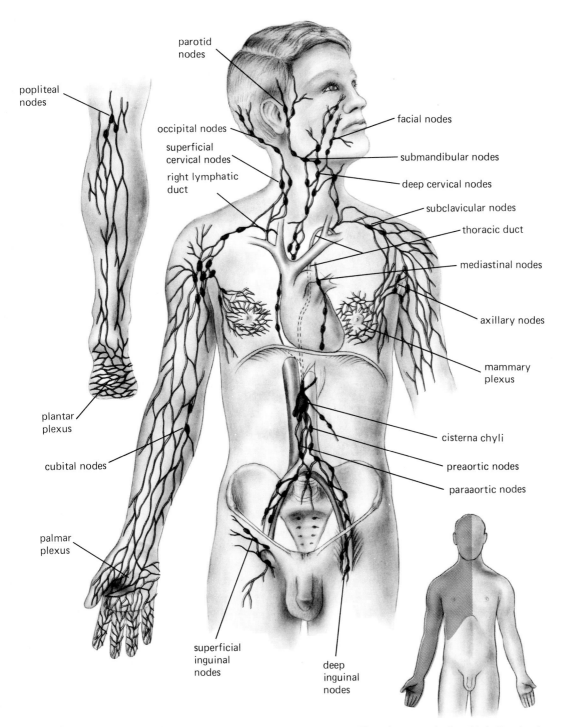

FIG. 21-43 Location of the major lymph nodes and vessels. The cisterna chyli, which lies in the upper part of the abdomen, is the lower expanded end of the thoracic duct. The numerous nodes in the abdominal organs are not shown. The figure at the lower right is shaded to indicate the region drained by the right lymphatic duct; lymphatics from the rest of the body drain into the thoracic duct.

Functions of the Lymphatic System

Return of Excess Filtered Fluid   In a normal person, the fluid filtered out of the capillaries each day exceeds the fluid reabsorbed by approximately 3 L. This excess is returned to the blood via lymphatics. Partly for this reason, lymphatic obstruction leads to edema.

Return of Protein to the Blood   Most capillaries have a slight permeability to protein and, accordingly, there is a small steady loss of protein from the blood into the interstitial fluid. The protein is returned to the circulatory system via the lymphatics. The breakdown of this cycle is the most important cause of the marked edema seen in patients with lymphatic malfunction. Because protein (in

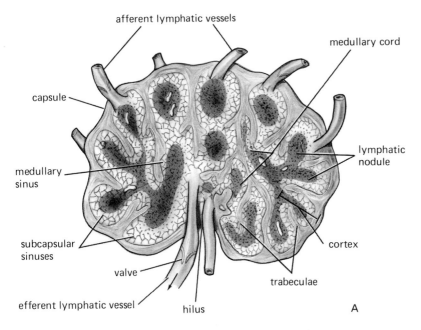

FIG. 21-44   (A) Diagram of a lymph node. Lymph enters the node via the afferent lymphatic vessels and leaves via the efferent lymphatic vessels. The lymphatic nodules are in color. (B) Histologic cross section through a lymph node.

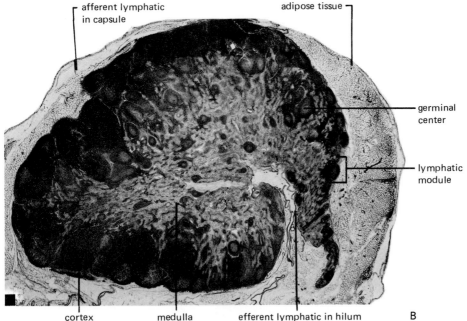

small amounts) is normally lost from the capillaries, failure of the lymphatics to remove it allows the interstitial protein concentration to increase to a level equal to that of the plasma. This failure reduces or eliminates the protein-concentration difference and thus the water-concentration difference across the capillary wall and permits the net movement of quantities of fluid out of the capillary into the interstitial space.

Specific Transport Functions In addition to the nonspecific transport functions, the lymphatics also provide the pathway by which certain substances reach the blood. The most important is fat absorbed from the gastrointestinal tract via lymphatic capillaries called **lacteals.** Certain high-molecular-weight hormones also may reach the blood via the lymphatics.

Defense Mechanisms Besides its transport functions, the lymphatic system plays important roles in the body's defenses against disease. The lymph nodes, located along the larger lymphatic vessels, mediate this function; they remove particles and cells, add lymphocytes to the lymph, and are the

sites of antibody production. (These functions will be described in Chap. 28).

Mechanism of Lymph Flow

How does the lymph move with no heart to push it? The best explanation at present is that lymph flow depends primarily upon forces external to the vessels, i.e., the pumping action of the skeletal muscle through which the lymphatics flow and the effects of respiration on thoracic-cage pressures. Since the lymphatics have valves similar to those in veins, external pressures cause only flow toward the points at which the lymphatics enter the circulatory system.

Lymph Nodes

Lymph enters a node via *afferent lymphatic vessels* on the convex border of the node, trickles through the *medullary sinuses* of the node, and leaves via *efferent lymphatic vessels* on the other side (Fig. 21-44). Each node is enclosed in a fibrous connective-tissue capsule from which partitions, or *trabeculae*, dip into the substance of the node. A re-

FIG. 21-45 Scanning electron micrograph of portions of a lymph node showing medullary sinuses and medullary cords. The cells lining the sinuses can be seen as well as reticular cells, macrophages, and lymphocytes in the sinus.

medullary sinus

medullary cord

cells lining sinus

reticular cells

lymphocytes

macrophages

ticular network extends from the trabeculae into all parts of the node. The spaces of the reticular network are packed with lymphocytes and plasma cells. The substance of the node is divided into an outer *cortex* and an inner *medulla.* Many of the cells in the cortical region of the node are collected in *lymphatic follicles* or *nodules,* which may contain *germinal centers* where B lymphocytes differentiate. The germinal centers also contain some T lymphocytes, macrophages, and plasma cells, although the plasma cells, which are formed from differentiating B lymphocytes, move from the cortex into more central portions of the node as soon as they are formed. The lymphocytes, plasma cells, and macrophages in the medullary portion of the node form *medullary cords* (Fig. 21-45).

After the afferent lymphatics pierce the capsule of the lymph node, they empty into large sinuses that lie directly beneath the capsule. From this *subcapsular sinus,* branches extend into the cortex and on into the medullary region of the node and converge at the concave side of the node where they flow into the efferent lymphatic vessels, which leave the node. As the lymph flows through the lymphatic sinuses, some lymphocytes are removed from it to be stored temporarily in the node and others are added to the lymph. Some of the lymphocytes released into the lymph are those previously stored, and others are newly formed. The lymphatic sinuses are lined with macrophages, which phagocytize particulate matter, such as dust (inhaled into the lungs), cellular debris, and bacteria and other microorganisms.

We have been describing lymph flow through the nodes, but they also receive a blood supply. Most of the blood vessels of the node enter at its concave surface and pass up into the cortex, where they break into a rich capillary network. The veins into which these capillaries drain leave the node near the point at which the arteries entered. As blood flows through the nodal vessels, both B and T lymphocytes pass from the blood into the meshwork of the node where they are stored for variable periods of time before passing into the lymph sinuses to be recirculated via the lymphatic circulation to the blood stream. Thus, lymphocytes pass from the blood to the lymph via the lymph nodes. Some of the major nodes are indicated in Fig. 21-43.

KEY TERMS

cardiovascular system	resistance	absorption
artery	systolic pressure	ultrafiltrate
arteriole	diastolic pressure	hydrostatic pressure
capillary	mean pressure	water-concentration difference
venule	bulk flow	venous capacity
vein	diffusion	lymphatic system
pressure gradient	filtration	lymph
flow		

REVIEW EXERCISES

Section A

1 Diagram the general plan of the circulatory system; label the right and left hearts; systemic and pulmonary circulations; aorta and pulmonary trunk; pulmonary and systemic arteries, capillaries, and veins; inferior and superior venae cavae, and the parallel circuits in the systemic circulation.

2 Indicate whether the oxygen content of the blood is high or low (or increasing or decreasing) in each labeled part of the previous diagram.

3 Differentiate between a portal system and the general plan of the circulation.

4 Diagram the aorta; identify its three major divisions and the principal arteries branching from each division.

5 Describe the principal arteries and veins of the head and neck.

6 Describe the principal blood supply of the brain; include a description of the circle of Willis and explain its significance.

7 Identify the principal arteries and veins of the arms; list in order the vessels through which blood passes as it travels from the left heart to the right thumb and back to the right heart.

8 Describe the principal arteries and veins of the abdomen.

9 Describe the portal system of the liver and its significance.

10 Describe the principal arteries and veins of the legs; list in order the vessels through which blood passes as it travels from the left heart to the left great toe and back to the right heart.

Section B

1 Explain the relationship between flow, pressure gradient, and resistance; write the equation that expresses this relationship.

2 List the factors that determine resistance to flow through a set of tubes; indicate which is the chief factor that determines resistance to flow through the vascular system.

3 Diagram the structure of an arterial wall; label the endothelial cells, intima, internal elastic lamina, media, external elastic lamina, and adventitia; indicate which layers comprise the tunica intima, tunica media, and tunica adventitia.

4 List the principal tissue types that constitute each of the layers of an arterial wall.

5 List the anatomical features that distinguish arteries, arterioles, capillaries, and veins.

6 Define and give normal values for systolic pressure, diastolic pressure, pulse pressure, and mean arterial pressure.

7 Explain how blood pressure is measured using a sphygmomanometer.

8 Differentiate between local and extrinsic control of blood flow to an organ; give two examples of each.

9 Describe functional hyperemia.

10 Define ischemia, kallikrein system, and myogenic response.

11 Explain how sympathetic control of vessels can achieve either an increased or decreased blood flow.

12 Describe the role of the sympathetic nerves, parasympathetic nerves, and hormones in control of blood flow.

13 Explain why the resistance to flow through capillaries is small despite the small radius of the vessels.

14 Compare the velocity of flow through capillaries with flow in other vessels.

15 Differentiate between bulk flow and diffusion; use as examples the movement of molecules in capillaries and across their walls.

16 Define capillary filtration, capillary absorption, ultrafiltrate, and edema.

17 List the four variables that affect movement of molecules across capillary walls; describe the effect of each variable on capillary absorption and filtration.

18 List the determinants of venous pressure.

19 Explain how the muscle pump, respiratory pump, and venous valves affect venous return to the heart.

20 List the general functions of the lymphatic system.

CHAPTER

THE CARDIOVASCULAR SYSTEM: INTEGRATION OF FUNCTION IN HEALTH AND DISEASE

In Chap. 7 we described the fundamental ingredients of all reflex control systems: (1) the internal-environmental variable being regulated, that is, maintained relatively constant, and the receptors sensitive to it; (2) afferent pathways passing information from the receptors to (3) a control center, which integrates the different afferent inputs; (4) efferent pathways controlling activity of (5) effector organs, whose output raises or lowers the level of the regulated variable. The control and integration of cardiovascular function will be described in these terms. The major variable being regulated is the systemic arterial blood pressure. The central role of arterial pressure and the adaptive value of keeping it relatively constant should be apparent from the ensuing discussion.

ARTERIAL PRESSURE, CARDIAC OUTPUT, AND ARTERIOLAR RESISTANCE

Adequate blood flow through the vital organs (brain and heart) must be maintained at all times; the brain, for example, suffers irreversible damage within 3 min of ischemia. In contrast, many areas of the body, e.g., gastrointestinal tract, kidneys, skeletal muscle, and skin, can withstand moderate reductions of blood flow for longer periods of time or even severe reductions if only for a few minutes.

The mean arterial blood pressure is the driving force for blood flow through all the organs. The distribution of flow, i.e., the actual flow through the various organs, at any given arterial pressure depends primarily upon the radii of the arterioles in each vascular bed. However, a critical relationship not emphasized before is implicit in the basic pressure-flow equation: These two factors, arterial pressure and arteriolar resistance, are *not* independent variables; arteriolar resistance is one of the major determinants of arterial pressure. This can be illustrated by the simple mechanical model shown in Fig. 22-1. A pump pushes fluid into a cylinder at the rate of 1 L/min; at steady state, fluid leaves the cylinder via the outflow tubes at 1 L/min, and the height of the fluid column, which is the driving pressure for outflow, remains stable. Assuming that the radii of the adjustable outflow tubes are all equal so that the flows through them are equal, we disturb the steady state by loosening the cuff on the outflow tube 1, thereby increasing its radius, reducing its resistance, and increasing its flow. The total outflow for the system is now greater than 1 L/min, more fluid leaves the reservoir than enters via the pump, and the height of the fluid column begins to decrease. In other words, a change in outflow resistance must produce changes in the pressure of the reservoir (unless some compensatory mechanism is

FIG. 22-1 Model that illustrates the dependency of arterial blood pressure upon arteriolar resistance; shown are the effects of dilating one arteriolar bed upon arterial pressure and organ blood flow if no compensatory adjustments occur. The middle panel is a transient state before the new equilibrium occurs. In one respect, the illustration is misleading in that the arterial reservoir is shown containing very large quantities of blood. As we have seen, the volume of blood in the arteries is, in fact, quite small.

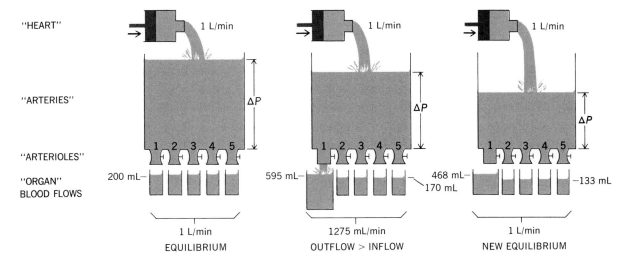

brought into play). As the pressure falls, the rate of outflow via all tubes decreases. Ultimately, in our example, a new steady state is reached when the reservoir pressure is low enough to cause only 1 L/min outflow despite the decreased resistance of tube 1.

This analysis can be applied to the cardiovascular system by equating the pump with the heart, the reservoir with the arteries, and the outflow tubes with various arteriolar beds. An analogy to opening outflow tube 1 is exercise. During exercise, the skeletal muscle arterioles dilate, primarily because of functional hyperemia, which thereby decreases resistance. If the cardiac output and the arteriolar diameters of all other vascular beds remain unchanged, the increased runoff through the skeletal muscle arterioles causes a decrease in arterial pressure. This, in turn, decreases flow through all other organs of the body. Indeed, even the exercising muscles themselves suffer a lessening flow (below that seen immediately after they dilated) as arterial pressure falls. Thus, the only way to guarantee the essential flow to the vital organs and the additional flow to the exercising muscle is to prevent the arterial pressure from falling.

This can be accomplished by changing the radii of the other arteriolar vascular beds or the cardiac output or both. Figure 22-2 demonstrates the first major possibility, simultaneously tightening one or more of outflow tubes 2 to 5. This partially compensates for the decreased resistance of tube 1, and the total outflow resistance of all tubes can be shifted back toward normal and therefore, the total outflow remains near 1 L/min. Of course, the distribution of flow is such that flow in tube 1 is increased and all the others are decreased. If, for some reason, tube 5 is declared a vital pathway the flow through which should never be altered, the adjustments can always be made in tubes 1 to 4.

Applied to the body, this process is obviously analogous to control of total vascular resistance. When the skeletal muscle arterioles dilate during exercise, the total resistance of all vascular beds can still be maintained if arterioles constrict in other organs, such as the kidneys and gastrointestinal tract, which can readily suffer moderate flow reductions for at least short periods of time. In contrast, the arterioles supplying the brain remain unchanged, which assures constant blood flow to the brain.

This type of resistance juggling, however, can compensate only within limits. Obviously if tube 1 opens very wide, even total closure of the other tubes cannot compensate completely. Moreover, if the closure is prolonged, absence of flow will cause severe tissue damage. There must therefore be a second compensatory mechanism to maintain pressure in the reservoir: increasing the inflow by increasing the activity of the pump (Fig. 22-3). Thus, at the new equilibrium, the total outflow and inflow are still equal, the reservoir pressure is unchanged, outflow through tubes 2 to 5 is unaltered, and the entire increase in outflow occurs through tube 1. Applied to the body, it should be evident that, when the blood vessels dilate, arterial pressure can be maintained constant by stimulating the heart to increase cardiac output. Thus, the regulation of arterial pressure not only assures blood supply to the

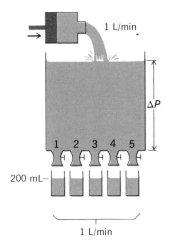

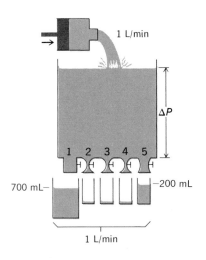

FIG. 22-2 Compensation for dilation in one bed by constriction in others. When outflow tube 1 is opened, outflow tubes 2 to 4 are simultaneously tightened so that the *total* outflow resistance remains constant, total rate of runoff remains constant, and reservoir pressure remains constant.

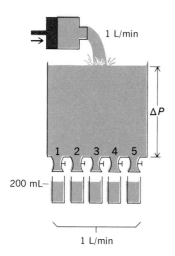 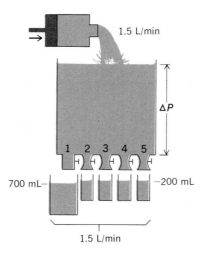

FIG. 22-3 Compensation for dilation by increasing pump output. When outflow tube 1 is dilated, the total resistance decreases and total rate of runoff increases. The pump output is increased by precisely the same amount, so that reservoir pressure remains constant.

vital organs but provides a means for coordinating cardiac output with total tissue requirements. In summary, the regulation of arterial blood pressure is accomplished by control of both cardiac output and arteriolar resistance.

It should now be possible to formalize these relationships for the cardiovascular system, using our basic pressure-flow equation. Flow through any tube is directly proportional to the pressure gradient between the ends of the tube and inversely proportional to the resistance.

$$\text{Flow} = \frac{\Delta P}{R}$$

Rearranging terms algebraically

$$\Delta P = \text{flow} \times R$$

This is simply another way of looking at the same equation, a way that clearly shows the dependence of pressure upon flow and resistance, which we have just described, using our model. Because the vascular tree is a continuous closed series of tubes, this equation holds for the entire system, i.e., from the very first portion of the aorta to the last portion of the vena cava just at the entrance to the heart. Therefore

$$
\begin{aligned}
\text{Flow} &= \text{cardiac output} \\
\Delta &= \text{mean aortic pressure} \\
&\quad - \text{late vena cava pressure} \\
R &= \text{total resistance}
\end{aligned}
$$

where total resistance means the sum of the resistances of all the vessels in the systemic vascular

tree. This usually is termed **total peripheral resistance.**

Since the late vena cava pressure is very close to 0 mmHg, the formula

$$
\begin{aligned}
\Delta P = \ &\text{mean aortic pressure} \\
&- \text{late vena cava pressure}
\end{aligned}
$$

becomes

$$\Delta P = \text{mean aortic pressure} - 0$$

or

$$\Delta P = \text{mean aortic pressure}$$

Moreover, since the mean pressure is essentially the same in the aorta and all large arteries, the pressure term in the equation becomes

$$\Delta P = \text{mean arterial pressure}$$

The pressure-flow equation for the entire vascular tree now becomes

Mean arterial pressure
  = cardiac output × total peripheral resistance

Recall that the arteries and veins are so large that they contribute very little to the total peripheral resistance. The major sites of resistance are the arterioles. The capillaries also offer significant resistance, but they have no muscle and their diameter reflects primarily the diameter of the arterioles that supply them. For these reasons, it is convenient to consider changes in arteriolar radius as virtually the only determinant of variations in total peripheral resistance (a change in blood viscosity will also influence resistance). Thus, the equation

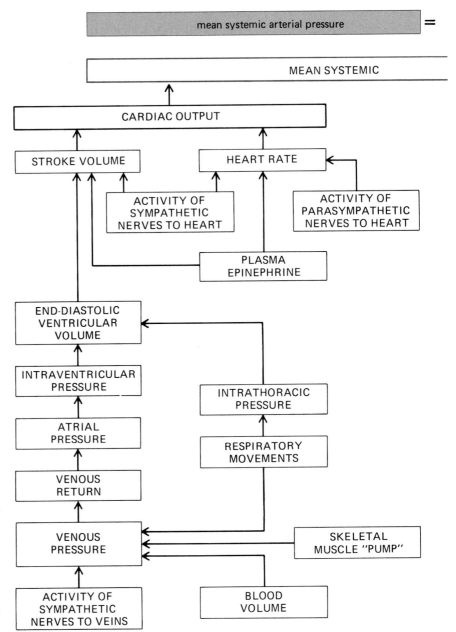

FIG. 22-4 Summary of effector mechanisms and efferent pathways that regulate systemic arterial pressure, an amalgamation of Figs. 20-29, 21-29, and 21-39, with the addition of the effect of hematocrit on resistance.

formally states the basic relationships described earlier, namely, that arterial blood pressure can be altered, either by changing cardiac output or arteriolar resistance, or both.

This equation is the fundamental equation of cardiovascular physiology. Given any two of the variables, the third can be calculated. For example, we can now explain why pulmonary arterial pressure

is much lower than the systemic arterial pressure. The blood flow per minute, i.e., cardiac output through the pulmonary and systemic arteries, is, of course, the same; therefore, the pressures can differ only if the resistances differ. Thus, the pulmonary arterioles must be wider and offer much less resistance to flow than the systemic arterioles. In other words, the total pulmonary vascular resistance is

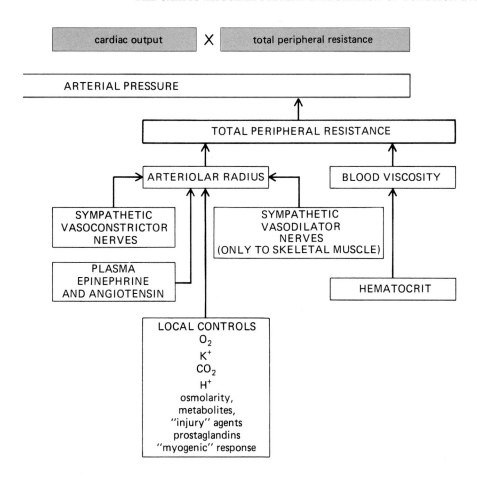

lower than the total systemic peripheral resistance. This permits the pulmonary circulation to function as a low-pressure system.

Figure 22-4 presents the grand scheme of effector mechanisms and efferent pathways that regulate systemic arterial pressure.[1] None of this informa-tion is new, all of it having been presented in pre-vious figures. A change in any single variable shown in the figure will, if all others remain constant, pro-duce a change in mean arterial pressure by altering either cardiac output or total peripheral resistance. Conversely, any such deviation in mean arterial pressure can be minimized by the reflex alterations of some other variable. It should be evident from the figure that the reflex control of cardiac output and peripheral resistance involves primarily (1) sympathetic nerves to heart, arterioles, and veins; and (2) parasympathetic nerves to the heart. In one sense, we have approached arterial pressure regu-lation backward, in that the past chapters have de-scribed the effector sites (heart, arterioles, veins) and motor pathways (autonomic nervous system). Now we must complete the reflexes by describing the monitoring systems (receptors and afferent

---

[1] Any model of a system as complex as the circulatory system must, of necessity, be an oversimplification. One basic defi-ciency in our model is that its chain of causal links is entirely unidirectional (from bottom to top in Figure 22-4), whereas, in fact, there are also important causal interactions in the reverse direction. For example, as we have shown, a change in cardiac output influences arterial pressure, but it is also true that a change in arterial pressure has an important effect on cardiac output. To take another example, atrial pressure has been shown to influence cardiac output, but it is also true that cardiac output influences atrial pressure.

pathways to the brain) and the control centers in the brain.

Before doing so, however, we must emphasize that, despite the preeminent role of the autonomic nervous system in the reflexes that control arterial pressure, it certainly is not the only factor, as shown in Fig. 22-4. Blood volume, in particular, is a crucial determinant, and the regulation of blood volume is, therefore, a critical component of blood pressure control. Short-term regulation of blood volume is achieved mainly by alteration of the so-called Starling forces that determine the distribution of extracellular fluid between the vascular and interstitial compartments. These forces were described in Chap. 21, and an example of their potent role will be given in Hemorrhage and Hypotension later in this chapter. Long-term control is achieved principally by regulation of red cell volume through the erythropoietin system described in Chap. 19 and plasma volume through the control of salt and water excretion by the kidneys, which will be described in Chap. 24.

CARDIOVASCULAR CONTROL CENTERS IN THE BRAIN

The primary cardiovascular control center is in the brainstem medulla. The nerve fibers from this center synapse with the autonomic neurons that control heart and blood vessel activity and exert dominant influence over these neurons. The cardiovascular center is absolutely essential for blood-pressure regulation. The relevant medullary neurons are sometimes divided into cardiac and vasomotor centers, which are then further subdivided and classified, but because these areas actually constitute diffuse networks of highly interconnected neurons, we prefer to call the entire area the **cardiovascular center.**

The input to the medullary neurons and the distribution of their axons are such that when the activity of the parasympathetic nerves to the heart is increased, the activity of the sympathetic nerves to the heart, as well as to the arterioles and the veins, is usually slowed (Fig. 22-5). Conversely, parasympathetic inhibition and sympathetic stimulation are usually elicited simultaneously.

There is always some continuous discharge of the autonomic nerves. Therefore, the heart can be

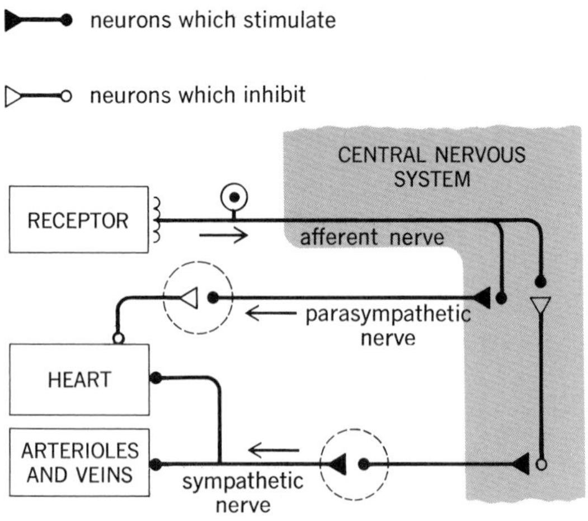

FIG. 22-5 A diagrammatic representation of reciprocal innervation in the control of the cardiovascular system. Afferent input which stimulates the parasympathetic nerves to the heart simultaneously inhibits the sympathetic nerves to the heart, arterioles, and veins.

slowed by two simultaneous events: inhibition of the sympathetic activity to the SA node and enhancement of the parasympathetic activity to the SA node. The converse is also true for accelerating the heart. In contrast, only sympathetic fibers significantly innervate the ventricular muscle itself and the arteriolar and venous smooth muscle. However, the muscle activity can still be decreased below normal by inhibiting the basal sympathetic activity.

Other areas of the brain, particularly in the cerebral cortex and the hypothalamus, have an important influence on blood pressure, but there is good reason to believe that they exert their effects mainly via the medullary centers; i.e., nerve impulses from them descend to the medulla and through synaptic connections alter the discharge of the medullary neurons. It is through these pathways that factors such as pain, anger, body temperature, and many others can alter blood pressure. There is at least one major exception: The sympathetic vasodilator fibers to skeletal muscle arterioles are apparently not controlled by the medullary centers but are under the direct influence of neuronal pathways that originate in the cerebral cortex and hypothalamus. These pathways and the sympathetic vasodilators are activated only during exercise and stress and play no role in the many other cardiovascular responses.

RECEPTORS AND AFFERENT PATHWAYS

We have now to discuss the last (really the first) step in arterial pressure regulation, namely, the receptors and afferent pathways bringing information into the medullary centers. The most important of these are the **arterial baroreceptors.**

Arterial Baroreceptors

It is only logical that the reflexes that homeostatically regulate arterial pressure originate primarily with pressure-sensitive receptors in the arteries. High in the neck where each of the carotid arteries divides into an external and internal carotid artery, the wall of the artery is thinner than usual and contains a large number of branching, vinelike nerve endings (Fig. 22-6). This small portion of the artery is called the **carotid sinus.** There the nerve endings are apparently highly sensitive to stretch or distortion; since the degree of wall stretching is directly related to the pressure within the artery, the carotid sinus actually serves as a pressure receptor (**baroreceptor**). The nerve fibers come together and travel with the glossopharyngeal (IXth cranial) nerve to the medulla, where they eventually synapse upon the neurons of the cardiovascular center. An area functionally similar to the carotid sinuses found in the arch of the aorta constitutes a second important arterial baroreceptor. The afferent fibers from these **aortic arch baroreceptors** travel with the vagus (Xth cranial) nerve.

Action potentials recorded in single afferent fibers from the carotid sinus demonstrate the pattern of response by these receptors. If the arterial pressure in the carotid sinus is artificially controlled at a steady nonpulsatile pressure of 100 mmHg, there is a tonic rate of discharge by the nerve. This rate of firing can be increased by raising the arterial pressure, and it can be decreased by lowering the pressure. The receptors show no acute fatigue or adaptation. The arterial baroreceptors are responsive not only to the mean arterial pressure but to the pulse pressure as well, an increased pulse pressure producing an increased firing. This responsiveness adds a further degree of sensitivity to blood-pressure regulation because small changes in certain important factors (such as blood volume) cause changes in pulse pressure before they become serious enough to affect mean pressure.

Our description of the major blood pressure-regulating reflex is now complete. An increase in arterial pressure increases the rate of discharge of the carotid sinus and aortic arch baroreceptors. These impulses travel up the afferent nerves to the medulla and, via appropriate synaptic connections with the neurons of the medullary cardiovascular centers, induce (1) slowing of the heart because of decreased sympathetic discharge and increased parasympathetic discharge, (2) decreased myocardial contractility because of decreased sympathetic activity, (3) arteriolar dilation because of decreased sympathetic discharge to arteriolar smooth muscle, and (4) venous dilation because of decreased sympathetic dischage to venous smooth muscle. The net result is a decreased venous return, decreased cardiac output (decreased heart rate and stroke volume), decreased peripheral resistance, and return of blood pressure toward normal.

Other Baroreceptors

Other portions of the vascular system contain nerve endings sensitive to stretch, namely, other large arteries, the large veins, the pulmonary vessels, and the cardiac walls themselves. Most of them seem to function like the carotid sinus and aortic arch in that they show increased rates of discharge with increasing pressure. By means of these receptors, the medulla is kept constantly informed about the venous, atrial, and ventricular pressures, and a further degree of sensitivity is gained. Thus, a slight decrease in atrial pressure begins to facilitate the sympathetic nervous system even before the change lowers cardiac output and arterial pressure far enough to be detected by the arterial baroreceptors. As we shall see in Chap. 24, the atrial baroreceptors are particularly important for the control of body sodium and water and, therefore, blood volume.

Chemoreceptors

The aortic and carotid arteries contain specialized structures sensitive primarily to the concentrations of oxygen in arterial plasma. Since these receptors are far more important for the control of respiration, they are described in Chap. 23, but they also send information to the medullary cardiovascular centers, the result being that blood pressure tends

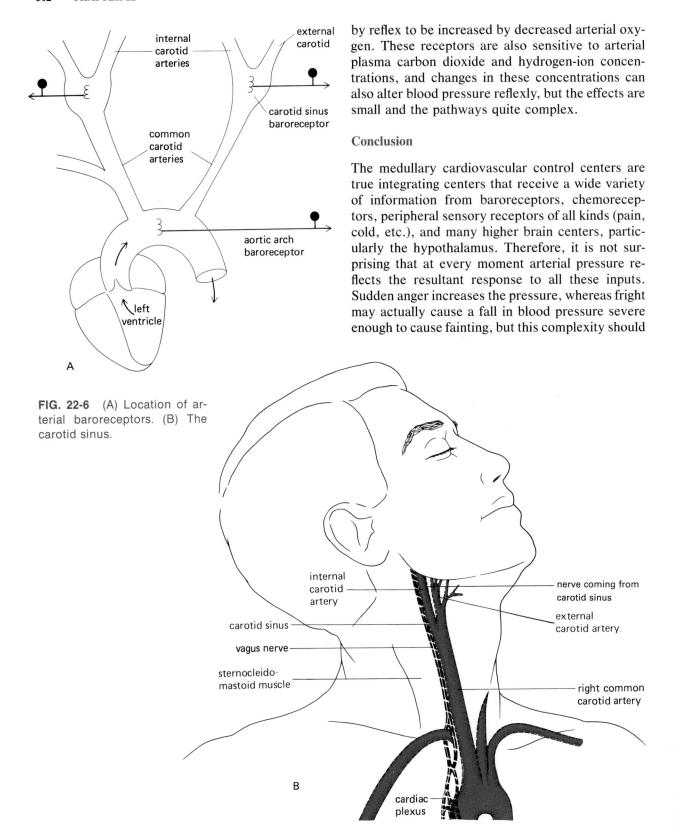

by reflex to be increased by decreased arterial oxygen. These receptors are also sensitive to arterial plasma carbon dioxide and hydrogen-ion concentrations, and changes in these concentrations can also alter blood pressure reflexly, but the effects are small and the pathways quite complex.

Conclusion

The medullary cardiovascular control centers are true integrating centers that receive a wide variety of information from baroreceptors, chemoreceptors, peripheral sensory receptors of all kinds (pain, cold, etc.), and many higher brain centers, particularly the hypothalamus. Therefore, it is not surprising that at every moment arterial pressure reflects the resultant response to all these inputs. Sudden anger increases the pressure, whereas fright may actually cause a fall in blood pressure severe enough to cause fainting, but this complexity should

FIG. 22-6 (A) Location of arterial baroreceptors. (B) The carotid sinus.

not obscure the important generalization that the primary regulation of arterial pressure is exerted by the baroreceptors, particularly those in the carotid sinus and aortic arch. Other inputs may alter the pressure somewhat from minute to minute, but the mean arterial pressure in a normal person is maintained by the baroreceptors within quite narrow limits.

HEMORRHAGE AND HYPOTENSION

The decrease in blood volume caused by bleeding produces a drop in blood pressure (**hypotension**) by

the sequence of events previously shown in Fig. 21-37). The most serious consequences of the lowered blood pressure are the reduced blood flow to the brain and cardiac muscle. Compensatory mechanisms restoring arterial pressure toward normal are summarized in Fig. 22-7; their effects can best be appreciated from the data of Table 22-1. Kidney flow is even lower 5 min after the hemorrhage, despite the improved arterial pressure. We recall, however, that one of the important compensatory mechanisms is increased arteriolar constriction in many organs; thus, kidney blood flow is reduced in order to maintain arterial blood pressure and

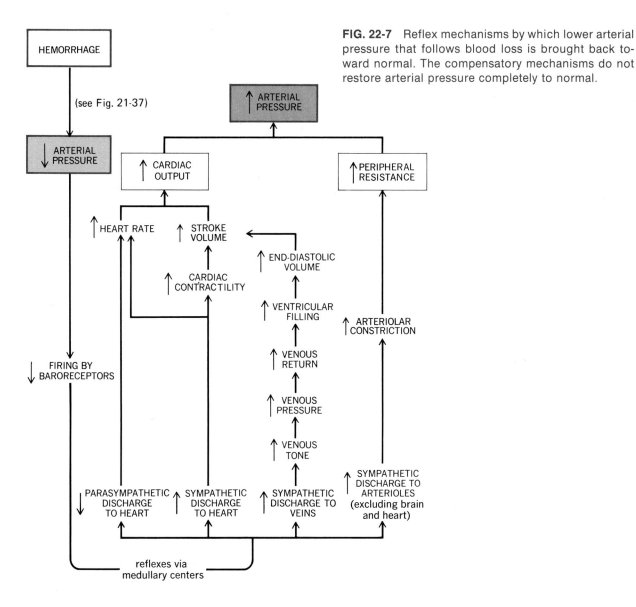

FIG. 22-7 Reflex mechanisms by which lower arterial pressure that follows blood loss is brought back toward normal. The compensatory mechanisms do not restore arterial pressure completely to normal.

TABLE 22-1   CARDIOVASCULAR EFFECTS OF HEMORRHAGE

	Prehemor-rhage	Posthemorrhage	
		Immediate	5 min
Arterial pressure, mmHg	125/75	80/55	115/75
Left atrial pressure, mmHg	4	2	2.5
End-diastolic volume, mL	150	75	90
Stroke volume, mL	75	40	53
Heart rate, beats/min	70	70	91
Cardiac output, mL/min	5250	2800	4775
Kidney blood flow, mL/min	1300	1000	850
Brain blood flow, mL/min	1300	1000	1275

thereby brain and heart blood flow.

A second important compensatory mechanism involves capillary fluid exchange and results from both the decrease in blood pressure and the increase in arteriolar constriction. Both of these changes decrease capillary hydrostatic pressure,

TABLE 22-2   FLUID SHIFTS AFTER HEMORRHAGE

	Normal	Immediately after Hemorrhage	18 hr after Hemorrhage
Total blood volume, mL	5000	4000 (↓ 20%)	4900
Erythrocyte volume, mL	2300	1840 (↓ 20%)	1840
Plasma volume, mL	2700	2160 (↓ 20%)	3060
Plasma albumin mass, g	135	108 (↓ 20%)	125

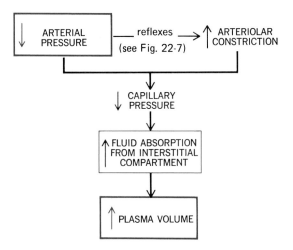

FIG. 22-8 Mechanisms that compensate for blood loss by movement of interstitial fluid into the capillaries. This response is diagrammed in Fig. 21-35C.

thereby favoring absorption of interstitial fluid (Fig. 22-8). Thus, the initial event—blood loss and decreased blood volume—is in large part compensated for by the movement of interstitial fluid into the vascular system. Indeed, as shown in Table 22-2, 12 to 24 hr after a moderate hemorrhage, the blood volume may be restored virtually to normal.

The entire compensation is a result of expansion of the plasma volume; replacement of the lost erythrocytes requires many days. Note that at 18 hr much of the albumin lost in the hemorrhage has been replaced. This phenomenon is of great importance for expansion of plasma volume, as can be seen by considering the following discussion. As capillary hydrostatic pressure decreases as a result of the hemorrhage, interstitial fluid enters the plasma; this fluid, however, contains virtually no protein, so that its entrance dilutes the plasma proteins and increases the plasma water concentration. The resulting reduction of the water-concentration difference between the capillaries and the interstitial fluid hinders further fluid reabsorption and would prevent the full compensatory expansion if it were not that replacement of the plasma protein minimizes this fall in protein concentration, and movement of interstitial fluid into the plasma can continue. Synthesis of albumin contributes to this replacement to only a minor degree during the first 24 hr; most of the replacement albumin that appears in the plasma during this early stage is interstitial protein carried by the lymphatics to the vascular compartment as a result of increased lymph flow. The mechanisms responsible for the increased lymph flow are presently unclear; nor is it known what stimulates the liver to synthesize new protein.

We must emphasize that this capillary mechanism has only redistributed the extracellular fluid; ultimate replacement of the plasma lost from the body involves the control of fluid ingestion and kidney function, both described in Chap. 24. Similarly, replacement of the lost red cells requires stimulation of erythropoiesis (by erythropoietin). Both these replacement processes require days to weeks in contrast to the rapidly occurring reflex compensations described in Fig. 22-7.

Loss from the body of large quantities of extracellular fluid (rather than whole blood) can also cause hypotension. This may occur via the skin, as in severe sweating or burns; via the gastrointestinal tract, as in diarrhea or vomiting; or via unusually large urinary losses. Regardless of the route, the loss decreases circulating blood volume and produces symptoms and compensatory phenomena similar to those seen in hemorrhage.

Hypotension may be caused by events other than blood or fluid loss. Hypotension and fainting in response to strong emotion are fairly common. Somehow the higher brain centers involved with emotions act upon the medullary cardiovascular centers to inhibit sympathetic activity and enhance parasympathetic activity, which results in decreased arterial pressure and brain blood flow. Fortunately, this whole process is usually transient, with no aftereffects, although a weak heart may suffer damage during the period of reduced blood flow to the cardiac muscle.

Other important causes of hypotension seem to have a common denominator in the liberation within the body of chemicals that relax arteriolar smooth muscle. There the cause of hypotension is clearly excessive arteriolar dilation and reduction of peripheral resistance, an important example being the hypotension that occurs during severe allergic responses.

You should now understand the physiologic reasons for not treating patients with hypotension in ways commonly favored by the uninformed, namely, administering alcohol and covering the person with mounds of blankets. Both alcohol and excessive body heat, by actions on the central nervous system, cause profound dilation of skin arterioles, which lowers peripheral resistance and decreases arterial blood pressure still further. As shown below, the worst possible thing is to try to get the person to stand up.

Shock

The compensatory mechanisms described above for hemorrhage are highly efficient, so that losses of as much as 1 to 1.5 L of blood (approximately 20 percent of total blood volume) can be sustained with only slight reductions of mean arterial pressure or cardiac output. In contrast, when much greater losses occur or when compensatory adjustments are inadequate, the decrease in cardiac output may be large enough to impair seriously the blood supply to the various tissues and organs. This is known as **circulatory shock.** The cardiovascular system, itself, suffers damage if shock is prolonged, and as it deteriorates, cardiac output declines even more and shock becomes progressively worse and ultimately irreversible—the person dies even after blood transfusions and other appropriate therapy. Damage to the heart by its low blood flow and by toxins released from other damaged tissues is one factor responsible for irreversible shock; another is progressive decrease in venous return to the heart because of clotting, blood pooling, and leakage of plasma out of capillaries.

Just as hemorrhage is not the only cause of hypotension and decreased cardiac output, so it is not the only cause of shock. Hypovolemia (decreased blood volume) caused by loss of fluid other than blood, excessive release of vasodilators as in allergy and infection, loss of neural tone to the cardiovascular system, trauma—all these can lead to severe reductions of cardiac output and the positive-feedback cycles that culminate in irreversible shock.

Upright Posture

The simple act of getting out of bed and standing up is equivalent to a mild hemorrhage, because the changes in the circulatory system in going from a lying, horizontal position to a standing, vertical position result in a decrease in the effective circulating

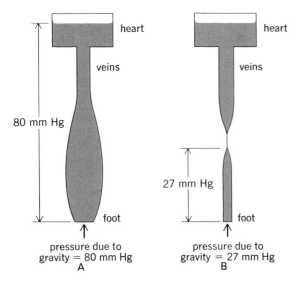

FIG. 22-9 Role of contraction of the leg skeletal muscles in reducing capillary pressure and filtration in the upright position. The skeletal muscle contraction compresses the veins and thus causes intermittent complete emptying so that the column of blood is interrupted.

blood volume. The decrease results from the action of gravity upon the long, continuous columns of blood in the vessels between the heart and the feet. All the pressures we have given in previous sections of this chapter were for the horizontal position, in which all the blood vessels are at approximately the same level as the heart, and the weight of the blood produces negligible pressure. In the vertical position (Fig. 22-9A), the intravascular pressure everywhere becomes equal to the usual pressures that result from cardiac contraction plus an additional pressure equal to the weight of a column of blood from the heart to the point of measurement; in a foot capillary, for example, the pressure increases from 25 mmHg (the pressure that results from cardiac contraction) to 105 mmHg (the extra 80 mmHg pressure is due to the weight of the blood).

The veins are highly distensible structures, as we have seen. The increased hydrostatic pressure in the veins of the legs that occurs upon standing pushes outward upon the vein walls, which causes marked distension with resultant pooling of blood; i.e., much of the blood that emerges from the capillaries simply remains in the expanding veins rather than returning to the heart. Simultaneously, the marked increase in capillary pressure caused by the

gravitational force produces increased filtration of fluid out of the capillaries into the interstitial space (most of us have experienced swollen feet after a day's standing). The combined effects of venous pooling and increased capillary filtration are a significant reduction in the effective circulating blood volume in a manner very similar to a mild hemorrhage. The ensuing decrease in arterial pressure causes reflex compensatory adjustments similar to those shown in Fig. 22-7 for hemorrhage.

Perhaps the most effective compensation is the contraction of skeletal muscles of the leg, which produces intermittent, complete emptying of veins in the leg, so that uninterrupted columns of venous blood from the heart to the feet no longer exist. The result is a decrease in venous distension and pooling and a marked reduction in capillary hydrostatic pressure and fluid filtration out of the capillaries (Fig. 22-9B). An example of the importance of this compensation (really its absence) is when soldiers faint after standing very still, i.e., with minimal contraction of the abdominal and leg muscles, for long periods of time. Here the fainting may be considered adaptive in that venous and capillary pressure changes induced by gravity are eliminated once the person is prone, the pooled venous blood is mobilized, and the previously filtered fluid is reabsorbed into the capillaries. Thus, the wrong thing to do to anyone who has fainted is to hold the person upright.

EXERCISE

In order to maintain muscle activity during exercise, a large increase in blood flow is required to provide the oxygen and nutrients consumed and to carry away the carbon dioxide and heat produced. Thus, cardiac output may increase from a resting value of 5 L/min to the maximal values of 35 L/min obtained by trained athletes. The increased skeletal muscle blood flow results from marked dilation of the skeletal muscle arterioles mediated by local factors associated with functional hyperemia. In addition, in a person just about to begin exercising, the skeletal muscle flow actually increases before the onset of muscular activity and of functional hyperemia. This anticipatory response that provides a rapid initial supply of blood to the muscle is mediated by the sympathetic vasodilator fibers.

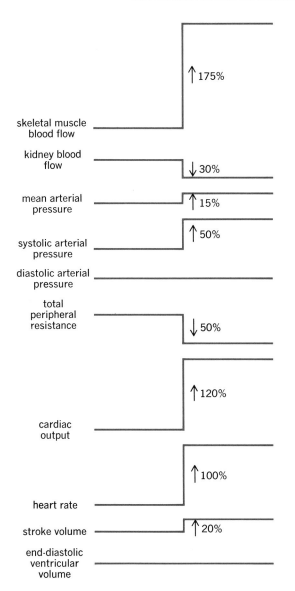

FIG. 22-10 Summary of cardiovascular changes during mild exercise.

Once exercise has begun, however, these sympathetic vasodilator nerves are of little importance and functional hyperemia plays the primary role in producing vasodilation.

As mentioned earlier, the decrease in peripheral resistance that results from dilation of skeletal muscle arterioles is partially offset by constriction of arterioles in other organs, particularly the gastrointestinal tract and kidneys (Fig. 22-10). However, the "resistance juggling" is quite incapable of compensating for the huge dilation of the muscle arte-

rioles, and the net result is a marked decrease in total peripheral resistance.

The cardiac output increase during exercise is caused by an increased heart rate and stroke volume associated with greater sympathetic activity and less parasympathetic activity to the heart. The heart-rate changes are usually much greater than stroke-volume changes. Note (Fig. 22-10) that, in our example, the increased stroke volume occurs without change in end-diastolic ventricular volume; accordingly, the stroke volume increase is ascribable completely to the increased contractility induced by the cardiac sympathetic nerves. The stability of end-diastolic volume in the face of reduced time for ventricular filling (increased heart rate) is, in part, attributable to the fact, described earlier, that the sympathetic nerves increase the speed of relaxation.

It would be incorrect, however, to leave the impression that enhanced sympathetic activity to the heart completely accounts for the elevated cardiac output that occurs in exercise, for such is not the case. The fact is that cardiac output could not be increased if the venous return to the heart were not simultaneously facilitated to the same degree, for otherwise end-diastolic volume would fall and stroke volume would decrease (because of Starling's law). Therefore, factors promoting venous return during exercise are extremely important. They are (1) the marked activity of the skeletal muscle pump, (2) the increased "respiratory-pump" activity that is a result of increased respiratory movements, (3) the sympathetically mediated increase in venous tone, and (4) the ease with which blood flows from arteries to veins through the dilated skeletal muscle arterioles. These factors may be so powerful that venous return is enhanced enough to cause an increase in end-diastolic ventricular volume; under such conditions, stroke volume (and, thereby, cardiac output) is further enhanced.

What happens to arterial blood pressure during exercise? As always, the mean arterial pressure depends upon the cardiac output and peripheral resistance. During most forms of exercise, the cardiac output tends to increase somewhat more than the peripheral resistance decreases so that mean arterial pressure usually increases slightly. However, the pulse pressure may show a marked increase because of greater systolic pressure and a relatively constant diastolic pressure. The systolic pressure rise is due primarily to the faster ejection.

It is evident from this description that the sympathetic nerves (and inhibition of the parasympathetics) play an important role in the cardiovascular response to exercise. However, a problem arises when we try to understand the mechanisms that control the autonomic nervous system during exercise. The pulsatile (and mean) arterial pressure tends to be elevated, which should cause the arterial baroreceptors to signal the medullary centers to decrease cardiac output and dilate arterioles. Obviously, then, the arterial baroreceptors not only cannot be the origin of the cardiovascular changes in exercise but actually oppose these changes. Similarly, other possible inputs, such as oxygen and carbon dioxide, can be eliminated because they show little, if any, change. We shall meet this same problem again when we describe increased respiration during exercise (Chap. 23). The best present working hypothesis is that a center in the hypothalamus acts, via descending pathways, upon the medullary centers (and directly upon the sympathetic vasodilators to skeletal muscle arterioles) to produce the changes in autonomic function so characteristic of exercise. What is the input to the hypothalamus? We do not know for certain, but it is probably information coming from the same motor areas of the cerebral cortex that are responsible for the skeletal muscle contraction. It is likely that, as these fibers descend from cerebral cortex to the spinal-cord motor neurons, they give off branches to the relevant hypothalamic centers. This system would nicely coordinate the skeletal muscle contraction with the blood supply needed to support it. This combination of events is also triggered during times of excitement, stress, and anger and, in such cases, is viewed as a preparation for exercise (''fight or flight'').

HYPERTENSION

Hypertension (high blood pressure) is defined as a chronically increased arterial pressure. In general, the dividing line between normal pressure and hypertension is taken to be 140/90 mmHg. However, the diastolic pressure is usually the more important index of hypertension. Heart failure (see below), brain stroke (occlusion or rupture of a cerebral blood vessel), and kidney damage are all caused by prolonged hypertension and its attendant strain on the various organs.

Theoretically, hypertension could result from an increase in cardiac output or peripheral resistance or both. In fact, at least in well-established hypertension, the major abnormality is increased peripheral resistance as a result of abnormally reduced arteriolar diameter. What causes arteriolar narrowing? In most cases, we do not know, and hypertension of unknown cause is called **essential hypertension.** In a small fraction of cases, the cause of hypertension is known: (1) Certain tumors of the adrenal medulla secrete excessive amounts of epinephrine; (2) certain tumors of the adrenal cortex secrete excessive amounts of hormones, which leads to hypertension by as yet unknown mechanisms; (3) many diseases that damage the kidneys or decrease their blood supply are associated with hypertension, but despite intensive efforts by many investigators, the actual factor(s) that mediate **renal hypertension** remains unknown. Increased release of renin with subsequent increased generation of angiotensin almost certainly plays a role, but it is unlikely that this is the only causal factor. (The renin-angiotensin system will be described in Chap. 24.)

Besides the three known causes of hypertension described above and several others, we are left with the vast majority of patients in the essential, or unknown, category. Many hypotheses have been proposed for increased arteriolar constriction but none proved. At present, much evidence seems to point to excessive sodium ingestion or retention within the body as a common denominator of renal, adrenal, and even essential hypertension, but how the sodium is involved in the increased arteriolar constriction remains unknown.

You might wonder why the arterial baroreceptors do not, by way of the reflexes they initiate, return the blood pressure to normal. The reason seems to be that, in chronic hypertension, the baroreceptors are ''reset'' at a higher level; i.e., they regulate blood pressure but at a greater pressure.

CONGESTIVE HEART FAILURE

The heart may become weakened for many reasons; regardless of cause, however, the failing heart in-

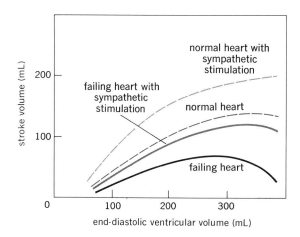

FIG. 22-11 Relationship between end-diastolic ventricular volume and stroke volume in normal and failing hearts. The normal curves are those shown previously in Figs. 20-26 and 20-27. The failing heart can still eject an adequate stroke volume if the sympathetic activity to it is increased or if the end-diastolic volume increases, i.e., if the ventricle becomes more distended.

duces a similar procession of signs and symptoms grouped under the category of **congestive heart failure.** The basic defect in heart failure is a decreased contractility of the heart, but the molecular mechanism is unknown. As shown in Fig. 22-11, the failing heart shifts downward to a lower Starling curve. How can this be compensated for? Increased sympathetic stimulation would help to increase contractility, and this does occur. However, an even more striking compensation is increased ventricular end-diastolic volume. The heart in severe failure is generally engorged with blood, as are the veins and capillaries, the major cause being an increase (sometimes massive) in plasma volume. The sequence of events that leads to the increased plasma volume is as follows: Decreased cardiac output causes a decrease in mean and pulsatile arterial pressure; this triggers reflexes (to be described in Chap. 24) that induce the kidneys to reduce their excretion of sodium and water. The retained fluid then causes expansion of the extracellular volume, which increases venous pressure, venous return, and end-diastolic ventricular volume and thus tends to restore stroke volume toward normal.

Another result of elevated venous and capillary pressure is increased filtration out of the capillaries, with resulting edema. This accumulation of tissue fluid may be the chief feature of ventricular failure; the legs and feet are usually most prominently involved (because of the additional effects of gravity), but the same engorgement is occurring in other organs and may cause severe malfunction. The most serious result occurs when the left ventricle fails; in this case, the excess fluid accumulates in the lung air sacs (**pulmonary edema**) because of increased pulmonary capillary pressure, and the patient may actually drown in his own fluid. This situation usually worsens at night; during the day, because of the patient's upright posture, fluid accumulates in the legs, but it is slowly absorbed when the patient lies down at night, the plasma volume expands, and an attack of pulmonary edema is precipitated.

Moreover, what began as a useful compensation for decreased cardiac contractility becomes potentially lethal because the tension-length relationship for muscle holds only up to a point, beyond which further stretching of the muscle may actually cause decreased strength of contraction. Thus, expansion of plasma volume may so increase end-diastolic volume as to decrease contractility (Fig. 22-11) and produce a rapidly progressing downhill course.

The treatment for congestive heart failure is easily understood in these terms: The precipitating cause should be corrected if possible; contractility can be increased by a drug known as digitalis; excess fluid should be eliminated by the use of drugs that increase excretion of sodium and water by the kidneys; the patient's activity level should be adjusted so as to reduce the cardiac output required to fulfill the body's metabolic needs.

"HEART ATTACKS" AND ATHEROSCLEROSIS

We have seen that the myocardium does not extract oxygen and nutrients from the blood within the atria and ventricles but depends upon its own blood supply via the coronary vessels. The rate of blood flow through the coronary arteries depends primarily upon the arterial diastolic blood pressure (the coronary vessels are completely occluded during systole by the constriction of the surrounding myocardial cells) and the resistance offered by the coronary vessels. The degree of coronary-vessel constriction, or dilation, is normally determined mainly by local metabolic control mechanisms, with some degree of neural control. Insufficient coronary blood flow leads to myocardial damage and, if severe enough, to destruction of the myocardium (*infarction*), a so-called **heart attack.** This may occur as a result of decreased arterial pressure but is more commonly caused by increased vessel resistance following coronary atherosclerosis.

Atherosclerosis is a disease characterized by a thickening of the arterial wall with abnormal smooth muscle cells and deposits of cholesterol. The mechanism by which atherosclerosis reduces coronary blood flow is quite simple; the thickening of the arterial wall narrows the vessel lumen and increases resistance to flow. This usually worsens over time and often leads ultimately to complete occlusion. The mechanisms that initiate wall thickening are not clear, but it is known that a high plasma cholesterol, smoking, hypertension, obesity, and a variety of other factors predispose to this disease. The suspected relationship between atherosclerosis and blood concentrations of cholesterol has probably received the most widespread attention, and many studies have documented that high concentrations of this lipid in the blood increase the rate and the severity of the atherosclerotic process. (Cholesterol metabolism is described in Chap. 26.)

Acute coronary occlusion may occur because of sudden formation of a clot on the vessel surface roughened by atherosclerotic thickening or breaking off of a deposit or clot, which then lodges downstream, completely blocking a smaller vessel. If, on the other hand, the occlusion is only gradual and does not occur in a major coronary vessel, the heart may remain uninjured because of the development, over time, of new accessory vessels that supply the same area of myocardium. Before complete occlusion many patients experience recurrent transient episodes of inadequate coronary blood flow, usually during exertion or emotional tension. The pain associated with this is termed **angina pectoris.**

The cause of death from coronary occlusion and myocardial infarction may be either severe hypotension that results from weakened contractility or disordered cardiac rhythm that results from damage to the cardiac conducting system. Moreover, the severe hypotension that may be associated with a heart attack is frequently worsened by reflex inhibition of the sympathetic nervous system and en-hancement of the parasympathetics. The origin of these totally inappropriate and frequently lethal reflexes is not known. Finally, should the patient survive an acute coronary occlusion, the heart may be left permanently weakened, and a slowly progressing heart failure may ensue. On the other hand, many people lead quite active and normal lives for many years after a heart attack.

We do not wish to leave the impression that atherosclerosis attacks only the coronary vessels, for such is not the case. Most arteries of the body are subject to this same occluding process. For example, cerebral occlusions (**strokes**) are common in the aged and constitute an important cause of sickness and death. Wherever the atherosclerosis becomes severe, the resulting systems always reflect the decrease in blood flow to the specific area.

THE FETAL CIRCULATION AND CHANGES IN THE CIRCULATORY SYSTEM AT BIRTH

During its 9 months in the uterus the fetus, of course, has no direct contact with the outside world; therefore, its lungs cannot exchange gases and its gastrointestinal tract cannot provide nutrients. Rather, the fetus must obtain its nutrients from and excrete its wastes into the mother's blood. One of the dominant features of the fetal circulation, therefore, is the arterial and venous supply to the **placenta,** the structure in the pregnant uterus that performs these nutrient and excretory functions (see Chap. 27).

Blood is carried from the placenta to the fetus by an **umbilical vein,** which enters the abdomen of the fetus through the **umbilicus,** or **navel,** and goes immediately to the liver (Fig. 22-12). This blood carries the oxygen and nutrients required by the fetus. From the liver, the blood enters the inferior vena cava where it mixes with the blood from the inferior half of the fetus. (One of the vessels that carry blood from the liver to the inferior vena cava is the **ductus venosus,** which is obliterated and present as the **ligamentum venosum** in the adult.)

The blood from the inferior vena cava enters the right atrium. Rather than flowing into the right ventricle, most of this blood passes through an opening in the interatrial septum (the **foramen ovale**) directly into the left atrium where it mixes with the small

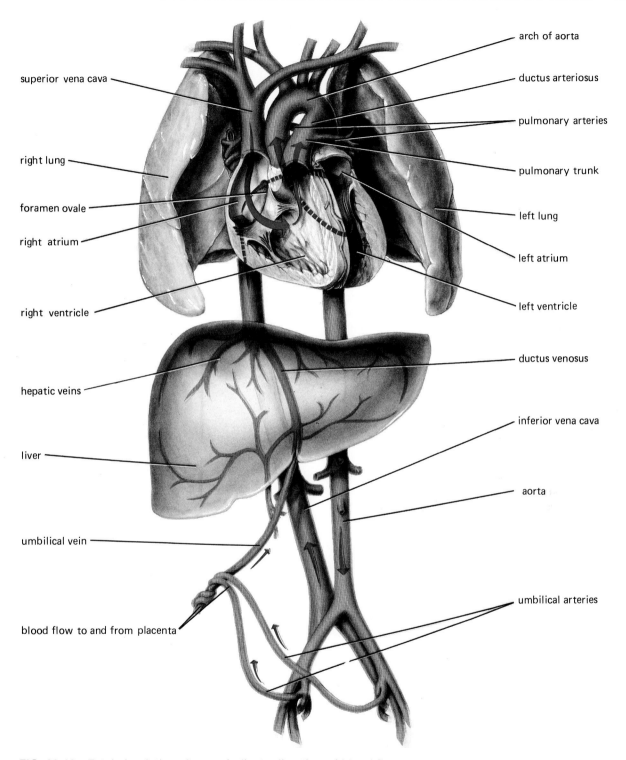

superior vena cava

arch of aorta

ductus arteriosus

pulmonary arteries

pulmonary trunk

right lung

foramen ovale

right atrium

left lung

left atrium

right ventricle

left ventricle

ductus venosus

hepatic veins

inferior vena cava

liver

aorta

umbilical vein

umbilical arteries

blood flow to and from placenta

FIG. 22-12   Fetal circulation. Arrows indicate direction of blood flow.

amount of blood returning to the heart from the pulmonary veins (see below). From the left atrium, the blood passes to the left ventricle and then into the aorta.

Some of the blood from the inferior vena cava and most of the blood from the superior vena cava, upon entering the right atrium, passes not through the foramen ovale but into the right ventricle and on into the pulmonary trunk. However, the resistance to blood flow through the lungs is very high in the fetus and only a small portion of the blood in the pulmonary trunk actually enters the lung tissue. Most of it is shunted through the **ductus arteriosus,** a fetal vessel which carries blood from the pulmonary trunk to the aorta (Fig. 22-12). The ductus arteriosus enters the descending portion of the aortic arch beyond the sites where the coronary arteries and the arteries to the upper limbs and head exit. Therefore, most of the blood from the inferior vena cava (which has passed through the foramen ovale and left heart) flows to the heart muscle, upper limbs, and head of the fetus while the blood from the superior vena cava (which has passed through the right ventricle, pulmonary trunk, and ductus arteriosus) is shunted to the lower limbs, abdominal organs, and the two **umbilical arteries,** which carry it back to the placenta. The significance of this distribution lies in the fact that inferior vena cava blood is much better oxygenated than superior vena cava blood because much of the former has returned from the placenta, the site of the fetus's oxygen uptake; accordingly, the heart and brain receive better-oxygenated blood.

In order for blood to flow from the right atrium into the left atrium or from the pulmonary trunk through the ductus arteriosus into the aorta, the pressures in the right heart must exceed those in the left heart and systemic circulation, a condition that reverses shortly after birth.

At birth the muscular walls of the umbilical vessels constrict in response to various stimuli such as changes in oxygen tension, temperature, and the degree of stretch on the vessels. This constriction markedly increases the total systemic peripheral resistance in the fetal circulation and also decreases venous return to the heart. Simultaneously the infant's first breath is stimulated by temperature or tactile stimulation during birth or, perhaps, by the decrease in blood oxygen content that results from the infant's separation from the placenta. The mechanical actions of the chest movements associated with the first breath and the smooth muscle relaxation induced by the increased blood oxygen that results from the breath cause the resistance in the pulmonary vessels to decrease markedly. This decrease, coupled with the decreased venous return to the heart and the increased total peripheral resistance, lower the pressure in the right heart and raise it in the left heart, which thereby reverses the fetal pressure gradient. This reversal, in turn, abruptly closes the valve over the foramen ovale, and the leaflets of the valve physically fuse within several days. Simultaneously, the decreased pressure in the pulmonary vessels and the increased pressure in the aorta cause a reversal of flow through the ductus arteriosus; this is shortlived, however, because the walls of the ductus begin to constrict and the vessel lumen becomes obliterated, which thereby eliminates all flow through it. Closure of the ductus is stimulated by the increased oxygen content of the blood flowing through it and probably by hormones as well. Failure of either the ductus arteriosus or foramen ovale to close after birth leads to some of the more common congenital heart defects; fortunately, these can usually be corrected surgically.

KEY TERMS

total peripheral resistance	**circulatory shock**	**stroke**
cardiovascular center	**hypertension**	**umbilical veins and arteries**
arterial baroreceptors	**atherosclerosis**	**foramen ovale**
carotid sinus	**heart attack**	**ductus arteriosus**
hypotension	**angina pectoris**	

REVIEW EXERCISES

1 Describe how arteriolar resistance is one of the major determinants of arterial pressure.

2 Describe two general ways in which arterial pressure can be maintained constant when peripheral resistance decreases markedly in one vascular bed.

3 State the equation that relates mean arterial pressure, cardiac output, and total peripheral resistance.

4 Draw a figure to summarize the effector mechanisms and efferent pathways that regulate arterial pressure.

5 State the location of the major cardiovascular control centers in the brain.

6 Describe the locations and functioning of the arterial baroreceptors.

7 Describe the reflex responses (via the baroreceptors) to an elevation of arterial pressure.

8 Describe the reflex responses to hemorrhage or hypotension, including the mechanisms that lead to mobilization of interstitial fluid.

9 Describe how plasma albumin, albumin synthesis, hematocrit, and erythrocyte synthesis change with time after a hemorrhage.

10 State the events that lead to irreversible circulatory shock.

11 Describe the events that occur when a person stands up and the reflex responses that maintain arterial pressure at this time.

12 Describe the control of skeletal muscle arterioles during exercise.

13 State the effects of exercise on heart rate, stroke volume, cardiac output, peripheral resistance, mean arterial pressure, and pulsatile arterial pressure.

14 State the roles of the arterial baroreceptors and brain control centers in the cardiovascular response to exercise.

15 List three known causes of hypertension.

16 Describe the sequence of events that lead to formation of edema in congestive heart failure.

17 Describe fetal circulation and the changes that occur at birth.

18 Define the terms *carotid sinus baroreceptor, aortic arch baroreceptor,* and *hypovolemia.*

CHAPTER

23

THE RESPIRATORY SYSTEM

SECTION A: ORGANIZATION OF THE RESPIRATORY SYSTEM

Most cells in the human body obtain the bulk of their energy from chemical reactions that involve oxygen. In addition, cells must be able to eliminate carbon dioxide, the major end product of these oxidations. A unicellular organism can exchange oxygen and carbon dioxide directly with the external environment, but this is obviously impossible for most cells of a complex organism like the human body, since only a small fraction of the cells (skin, gastrointestinal lining, respiratory lining) is in direct contact with the external environment. Large animals have specialized systems for the supply of oxygen and elimination of carbon dioxide as well as organs of gas exchange with the external environment. Such organs in human beings are **lungs.** There are also specialized blood components that permit the transportation of large quantities of oxygen and carbon dioxide between the lungs and cells.

Before describing the basic processes of oxygen supply, carbon dioxide elimination, and breathing control, we must first define the terms *respiration* and *respiratory system*. **Respiration** has two quite different meanings: (1) the metabolic reaction of oxygen with carbohydrate and other organic molecules and (2) the exchange of gas between the cells of an organism and the external environment. The various steps of the second process form the subject matter of this chapter; the first process was described in Chap. 3. The term **respiratory system** refers only to those structures that are involved in the exchange of gases between the blood and external environment; it does not include the transportation of gases in the blood or gas exchange between blood and the tissues. Admittedly, this definition is arbitrary since it includes only half of the processes involved in respiration, but it has become firmly established by long usage. The respiratory system consists of the lungs, the series of airways that lead to the lungs, and the chest structures responsible for movement of air in and out of the lungs.

In order for air to reach the lungs, it must pass first through a series of air passages that connect the lungs to the nose and mouth (Fig. 23-1) and then through the continuation of these air passages in the lungs themselves (Fig. 23-2). All these air passages together are termed the **conducting portion** of the respiratory system. Their walls are supported by bone or cartilage to ensure that the airways remain open. The smallest tubes of the conducting portion of the airway system lead into the **respiratory portion** where the passages contain on their walls tiny blind sacs, the **alveoli,** which are sites of gas exchange between the airway and the blood. The system of air passages terminates in clusters of alveoli (Fig. 23-3). The air passageways and alveoli receive a rich supply of blood vessels, which con-

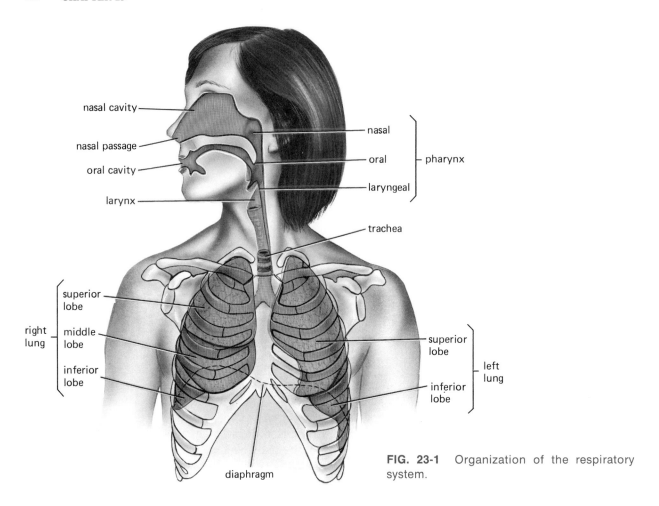

FIG. 23-1 Organization of the respiratory system.

stitute a large portion of the total lung substance (Fig. 23-4). The lungs also contain large quantities of elastic connective tissue, which plays an important role in breathing.

NOSE AND PHARYNX

Air can enter the respiratory passages by either the nose or mouth, although the nose is the normal route. The **nasal cavity** is an irregularly shaped space that extends from the hard palate, which separates the nose and mouth cavities, upward to the base of the cranial cavity (Fig. 23-5). It is divided into right and left nasal cavities by the **nasal septum.** The front of each cavity opens onto the face through the **nares,** or **nostril;** the back opens into the nasal pharynx through the **posterior nasal aperture,** or **choana.**

The slight enlargement of each cavity just inside the nostril is the **vestibule.** It is lined with coarse hairs, which tend to trap foreign substances carried into the nose with the inspired air. The roof of the nasal cavity contains the receptors for the sense of smell and is called the *olfactory region.* The remaining *respiratory region* extends from the nasal septum to the lateral walls, which have three bony projections, the **superior, middle,** and **inferior conchae** (Fig. 23-5). These markedly increase the respiratory surface of the nasal cavity and help to warm and moisten the inspired air. Under each concha is an air space, the **superior, middle,** and **inferior meatus,** which connects freely with the main air passage of the cavity. The lacrimal ducts (tear ducts) empty into the nasal cavities.

After passing through the nose (or mouth), air enters the **pharynx** (throat), a passage common to the routes followed by air and food. The pharynx is a muscular tube lined with mucous membrane; it

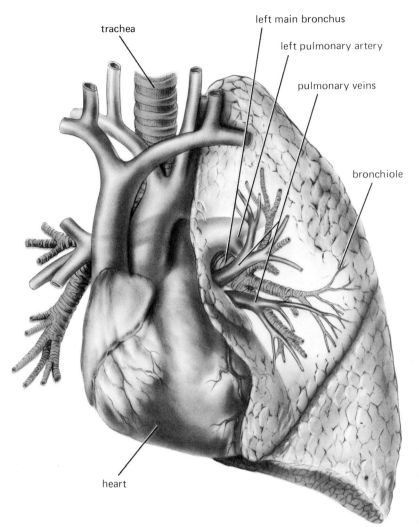

trachea

left main bronchus

left pulmonary artery

pulmonary veins

bronchiole

heart

FIG. 23-2 Large airways of the human lung, starting with the trachea and continuing with its many branches in the lungs. In this figure the substance of the lungs appears transparent so that the relationships between the airways and blood vessels can be seen.

extends from the base of the skull downward behind the nasal cavity, mouth, and larynx where it becomes continuous with the esophagus (Fig. 23-5).

The upper, *nasal part* of the pharynx lies behind the nasal cavities and communicates with them through the posterior nasal apertures. The nasal pharynx also communicates with the middle-ear cavities through the auditory, or eustachian, tubes (Figs. 15-23A and 23-5). Near these openings patches of lymphoid tissue, the *pharyngeal tonsil,* lie in the mucous membrane. At the level of the soft palate, the nasal region of the pharynx ends and the *oral pharynx* begins; it extends downward to the level of the epiglottis. Anteriorly, the oral pharynx opens into the oral (mouth) cavity. The oral pharynx contains the *palatine tonsils,* which

with the pharyngeal tonsils form a partial ring of lymphoid tissue around the opening of the digestive and respiratory tubes; the ring is completed by the *lingual tonsils.* (The histology and function of the tonsils will be discussed in Chap. 28.) The pharynx continues downward as the *laryngeal pharynx,* which opens into the larynx.

LARYNX

The **larynx,** which lies in the front part of the neck, serves three functions: (1) It is the air passageway between the pharynx and the trachea; (2) it acts as a protective sphincter to prevent solids and liquids from passing into the bronchi and lungs; and (3) it is involved in the production of sounds. The frame-

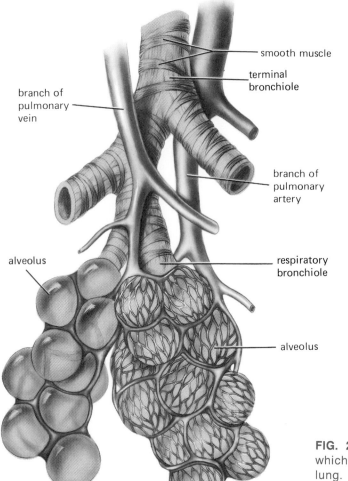

smooth muscle

terminal bronchiole

branch of pulmonary vein

branch of pulmonary artery

alveolus

respiratory bronchiole

alveolus

FIG. 23-3 Clusters of alveoli form alveolar sacs, which are connected to the air passageways of the lung. The intimate relationship between the respiratory system and blood vessels at the level of the alveoli can be seen.

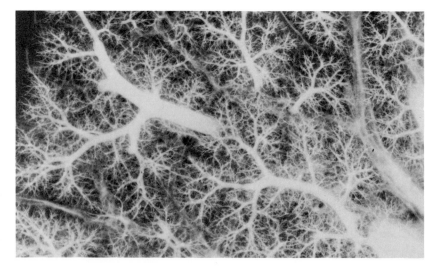

FIG. 23-4 Radiograph of the blood vessels in a slice of inflated and fixed mammalian lung on which a barium arteriogram was performed before slicing.

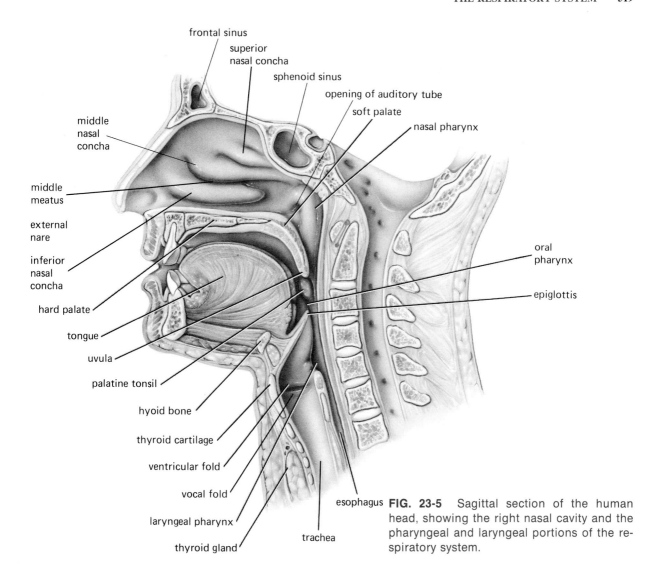

frontal sinus
superior nasal concha
sphenoid sinus
opening of auditory tube
soft palate
nasal pharynx
middle nasal concha
middle meatus
external nare
inferior nasal concha
hard palate
tongue
uvula
palatine tonsil
hyoid bone
thyroid cartilage
ventricular fold
vocal fold
laryngeal pharynx
thyroid gland
trachea
esophagus
oral pharynx
epiglottis

FIG. 23-5 Sagittal section of the human head, showing the right nasal cavity and the pharyngeal and laryngeal portions of the respiratory system.

work of the larynx consists of nine cartilages and the ligaments, membranes, and muscles that connect them. The largest cartilage of the larynx, the **thyroid cartilage** (Fig. 23-6A and B), forms the hard bump in the front of the neck known as the **laryngeal prominence,** or Adam's apple. This prominence is hardly visible in children or adult females, but it is quite marked in males after puberty. The inlet to the larynx lies in the anterior wall of the pharynx; it is bounded by the epiglottis, arytenoid cartilages, and the aryepiglottic folds (Fig. 23-6C).

Continuing from the inlet, the cavity of the larynx expands into a wide *vestibule.* Two sets of thick, membranous ridges protrude from the laryngeal wall at the lower end of the vestibule. The upper pair, the **vestibular folds,** are often called *false vocal cords* and are not involved in the production of sounds. The lower pair, the **vocal folds,** or *true vocal cords,* contain the **vocal ligament** (Fig. 23-7). The vocal folds, ligaments, and intervening space (the **rima glottidis,** or glottic aperture) together form the sound-producing apparatus known as the **glottis.** The rima glottidis extends back between the two arytenoid cartilages and, in its entirety, forms the narrowest part of the laryngeal cavity. The laryngeal muscles (some of which are shown in Fig. 23-7) can bring the vocal folds together for vocalization (during which the vocal folds are made taut and caused to vibrate as air expired from the lungs passes over them) or breathholding. Muscles with

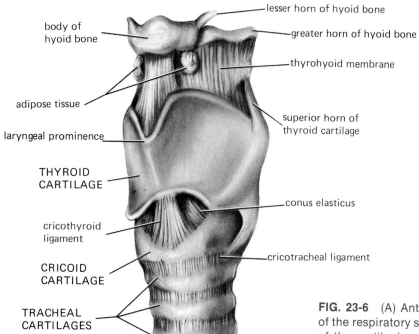

body of
hyoid bone

lesser horn of hyoid bone

greater horn of hyoid bone

thyrohyoid membrane

adipose tissue

laryngeal prominence

superior horn of
thyroid cartilage

**THYROID
CARTILAGE**

conus elasticus

cricothyroid
ligament

cricotracheal ligament

**CRICOID
CARTILAGE**

**TRACHEAL
CARTILAGES**

A

FIG. 23-6   (A) Anterior view of the laryngeal portion of the respiratory system. (B) Relationship of the parts of the cartilaginous skeleton of the larynx, showing the arytenoid corniculate, and part of the cricoid cartilages behind the thyroid cartilage. (C) Sagittal section through the larynx. The anterior part of the larynx is toward the left in B and C.

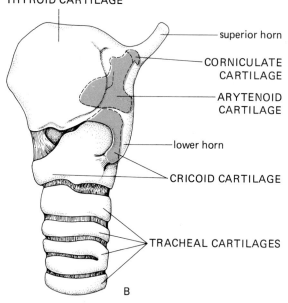

THYROID CARTILAGE

superior horn

**CORNICULATE
CARTILAGE**

**ARYTENOID
CARTILAGE**

lower horn

CRICOID CARTILAGE

TRACHEAL CARTILAGES

B

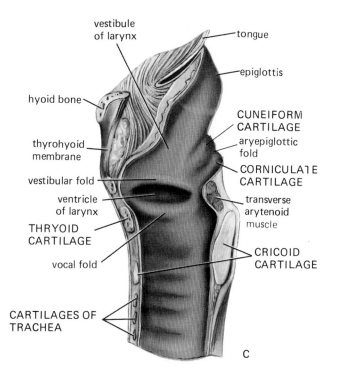

vestibule
of larynx

tongue

epiglottis

hyoid bone

**CUNEIFORM
CARTILAGE**
aryepiglottic
fold

thyrohyoid
membrane

**CORNICULATE
CARTILAGE**

vestibular fold

transverse
arytenoid
muscle

ventricle
of larynx

**THRYOID
CARTILAGE**

**CRICOID
CARTILAGE**

vocal fold

**CARTILAGES OF
TRACHEA**

C

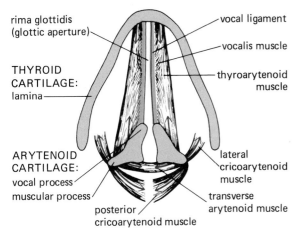

FIG. 23-7   Diagrammatic cross section of the larynx at the glottic aperature, showing the action (colored arrows) of certain of the laryngeal muscles.

opposite actions separate the vocal folds during forced inspirations, such as gasping.

TRACHEA AND MAIN BRONCHI

The **trachea** constitutes the tubular air passageway from the larynx to the lungs (Fig. 23-1), where at the **carina** it divides into two **main** (primary) **bronchi** (sing.: *bronchus*). The main bronchi, after traveling only a short distance beyond this tracheal division, enter the lungs.

The trachea and bronchi are kept open by irregular rings of cartilage embedded in their walls. The

FIG. 23-8   Cross section through the trachea.

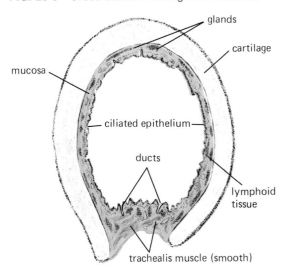

tracheal rings are C-shaped, almost completely circling the air passageway, with their openings at the back where the trachea lies against the esophagus (Fig. 23-8). The gaps are bridged by smooth muscle and connective-tissue bands, which cause changes in the diameter of the tracheal lumen when they contract or are stretched. Adjacent rings are joined by connective tissue and smooth muscle. The structure of the main bronchi is similar to that of the trachea.

LUNGS

The **lungs** are located in the thoracic cavity, one on each side, separated from each other by the mediastinum and its contents (Fig. 23-9). They lie free in their respective halves of the thoracic cavity, except for attachments to the heart and trachea. The heart, which lies in the mediastinum slightly to the left of midline, occupies part of the thoracic cavity on that side, and the left lung is slightly smaller than the right. The left lung is divided into two **lobes,** the right into three (see Fig. 23-1). The medial surface of each lung has a slight depression that surrounds the place where the main bronchus and its accompanying nerves and blood vessels enter the lung. This depression is the **hilum** of the lung; the bronchus, nerves, and vessels form the **root** of the lung. The upper part of the lung, which lies against the top of the thoracic cavity, is the **apex,** and the lower part lying against the diaphragm is the **base.**

The substance of the lungs is light and spongy, and their color, at least at birth, is rose-pink. By adulthood the color changes to a dark slatey gray mottled with patches that can be almost black because of inhaled particulate matter that has been retained and deposited in the lung tissues.

Intrapulmonary Airways

The lungs are not simply hollow balloons but have a highly organized structure that consists of air-containing tubes, blood vessels, and elastic connective tissue. The airway system of a lung has been compared to a tree because the main bronchus entering it undergoes many divisions, the resulting branches becoming smaller and closer together with each division (Fig. 23-2). The large branches are called **bronchi.** Their walls contain cartilage that

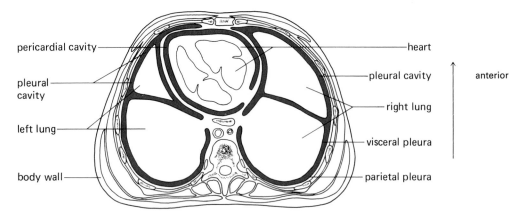

FIG. 23-9   A diagrammatic cross section of the thorax in which the pleural and pericardial cavities (here exaggerated in size) are indicated by shading.

decreases in amount as the bronchi successively branch. As the cartilage declines, the muscle layer increases, and rings of smooth muscle completely surround the bronchi. The two main bronchi divide into *lobar,* or *secondary, bronchi,* one of which enters each of the lobes of the lungs where they divide into *tertiary bronchi,* of which there are 10 in each lung. Major branches such as these are termed *segmental bronchi* if they enter a self-contained, functionally independent unit of the lung. Such independent segments of lung tissue are known as *bronchopulmonary segments.*

Where the cartilage disappears and the diameter is reduced to about 1 mm, the bronchi become **bronchioles.** The rings of smooth muscle persist and are joined together by diagonal muscle fibers. Like the trachea and main bronchi, the smooth muscle of the intrapulmonary airways is richly innervated and sensitive to certain circulating hormones, e.g., epinephrine. Contraction or relaxation of this muscle alters resistance to air flow. The bronchioles continue to divide; their last branches are the *terminal bronchioles.* This marks the ends of the conducting portion of the respiratory system and the beginning of the thinner-walled respiratory portion where the actual gas exchange occurs.

Site of Gas Exchange in the Lungs: The Alveoli

Beyond the terminal bronchioles, the airway walls contain small outpocketings of epithelium. As the airway continues to divide, these outpocketings, or

alveoli, increase in frequency until the airway ends in grapelike clusters of alveoli (Fig. 23-3). The alveoli are tiny cup-shaped, hollow sacs whose open ends are continuous with the lumens of the smallest bronchioles and the alveolar ducts. The alveoli (Fig. 23-10) are lined by a continuous single thin layer of epithelial cells that rest on a thin basement membrane, which in turn rests on a very loose mesh of connective-tissue elements that constitutes the interstitial space of the alveolar walls. Most of the alveolar walls is occupied by capillaries, the endothelial lining of which is separated from the alveolar epithelial lining only by the very thin interstitial space. Indeed, the interstitial space may be absent altogether; i.e., the basement membranes of the epithelium and endothelium may actually fuse. Thus, the blood in a capillary is separated from the air in an alveolus only by an extremely thin barrier (0.2 μm compared with 7 μm, which is the diameter of an average red blood cell). The total area of alveoli in contact with capillaries is 138 m^2 (roughly 80 times greater than the total body surface area). This immense area combined with the thin barrier permits the rapid exchange of large quantities of oxygen and carbon dioxide.

In addition to its thin squamous cells, the alveolar epithelium contains specialized cuboidal cells, **type II cells,** which produce surfactant, which will be discussed below. The alveolar wall also contains other connective-tissue cells and phagocytic cells (macrophages) to defend the lungs against most microorganisms. Finally, there are pores in the al-

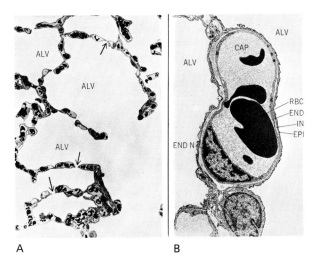

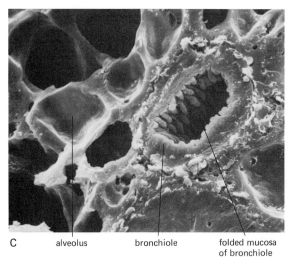

A B

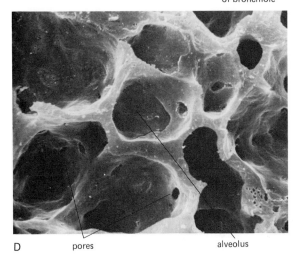

C alveolus bronchiole folded mucosa of bronchiole

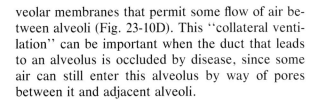

D pores alveolus

FIG. 23-10 (A) Low-power electron micrograph of dog lung alveoli (ALV). Note the capillaries in the walls between alveoli (the dark disclike objects are erythrocytes). The arrows denote pores in the alveolar walls. (B) Higher magnification of a portion of an alveolar wall, showing a single capillary (CAP) surrounded by an alveolar epithelial cell. The nucleus (END N) and cytoplasm (END) of a capillary endothelial cell are visible, as are the nucleus and cytoplasm (EPI) of an alveolar epithelial cell. Note that the blood in the capillary is separated from air in the alveoli (ALV) only by the thin membrane that consists of endothelium, interstitial fluid (IN), and epithelium. (C and D) Scanning electron micrographs of rat and mouse lung.

veolar membranes that permit some flow of air between alveoli (Fig. 23-10D). This "collateral ventilation" can be important when the duct that leads to an alveolus is occluded by disease, since some air can still enter this alveolus by way of pores between it and adjacent alveoli.

RESPIRATORY MUCOSA

The entire airway is lined with epithelial tissue the characteristics of which vary, depending upon the area of the respiratory system (Table 23-1). The oral part of the pharynx, through which food passes, and the surfaces of the larynx around the vocal cords are subject to excessive amounts of abrasion and are lined with stratified squamous epithelium. From the lower border of the larynx to the opening of the alveoli, the airways are lined with columnar epithelium. Many of these cells contain hairlike projections, called **cilia,** which constantly beat toward the pharynx (Fig. 23-11). In addition, the airway lining contains glands and surface cells that secrete a thick substance **(mucus),** which lines the respiratory passages as far down as the bronchioles. Any particulate matter such as dust contained in the inspired air sticks to the mucus, which is constantly moved by the cilia to the pharynx, and then is swallowed and eliminated in the feces. Besides keeping the lungs clean, this mechanism is important in the body's defenses against bacterial infection, since many bacteria enter the body on dust particles. A major cause of lung infection is probably paralysis of ciliary activity by noxious agents, including substances in cig-

TABLE 23-1   COMPOSITION OF THE RESPIRATORY AIRWAYS

Component	Predominant Cell Types in Airway Lining	Cartilage?	Muscle?
Conducting Portion: Trachea	Pseudostratified tall columnar ciliated epithelial cells and goblet cells, which secrete mucus; mucus-secreting exocrine glands are present.	Yes, 16–20 incomplete rings.	Yes, bands bridge the gaps between ends of cartilages.
Bronchi	Large bronchi are similar to trachea, but as the bronchi get smaller, the height of the columnar cells decreases.	Yes, present as large, irregular, often helical plates.	Yes, forms a complete layer in the larger bronchi; present as isolated bundles in the smaller bronchi.
Bronchioles and terminal bronchioles	Cuboidal epithelium with some ciliated cells; mucus-secreting elements are few or absent; the number of ciliated cells gradually decreases in the terminal bronchioles and a new cell type (Clara cell) appears.	No	Yes, present as isolated bundles rather than as a complete layer.
Respiratory Portion: Respiratory bronchioles	Low cuboidal epithelium that changes to flat transitional cells at the entrance of the occasional alveoli that are present.	No	Yes, present as isolated bundles
Alveolar ducts	Lining is so extensively interrupted by alveoli that there is essentially no airway wall between the rings of cells that surround the alveolar entrances.	No	Yes, muscle forms a sphincter at the outlet of the last alveolar duct.
Alveoli	Squamous epithelium (alveolar type I cells), alveolar type II cells, and macrophages (''dust cells'').	No	No

arette smoke. This, coupled with the stimulation of mucus secretion induced by these same agents, may result in partial or complete airway obstruction by the stationary mucus. (A smoker's early-morning cough is the attempt to clear this obstructive mucus from the airways.) A second protective mechanism is provided by the phagocytic cells, which are present in the respiratory-tract lining in great numbers. These cells, which engulf dust, bacteria, and debris, are also injured by cigarette smoke and other air pollutants. Air is also warmed and moistened by contact with the epithelial lining as it flows through the respiratory passages.

BLOOD AND NERVE SUPPLY OF THE RESPIRATORY SYSTEM

The lungs receive blood from two different arterial systems: from the pulmonary arteries, which carry deoxygenated (venous) blood to the lungs to pick up oxygen and release carbon dioxide, and from the bronchial arteries, which carry arterial blood to nourish the lung tissues.

The two **pulmonary arteries** arise from the pulmonary trunk shortly after it leaves the right ventricle, one artery entering each lung. The arteries branch, generally following the branches of the bronchi down to the last respiratory bronchioles, and finally end in the dense capillary networks in the alveolar walls where the oxygen and carbon dioxide of the blood equilibrate with the alveolar air. Blood that is now oxygenated flows from the capillaries into branches of the pulmonary veins, which also receive a small amount of deoxygenated blood directly from arteriovenous shunts. The branches of the veins gradually merge until all the blood is collected into the four large **pulmonary veins,** two from each lung, which pass to the left atrium.

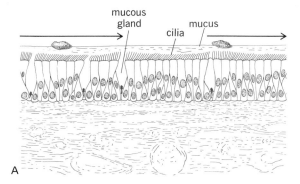

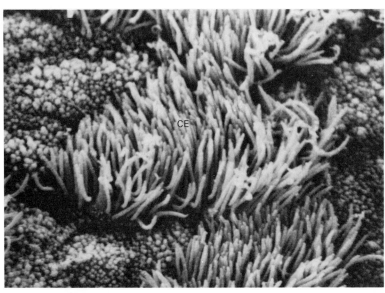

FIG. 23-11 (A) Epithelial lining of the respiratory tract. The arrows indicate the upward direction in which the cilia move the overriding layer of mucus, to which foreign particles are stuck. (B) Scanning electron micrograph of the mucosa of a mammalian trachea. The two cell types of the mucosa, the ciliated epithelial cell (CE) and mucus-secreting goblet cells can be seen.

A

B

The **bronchial arteries** arise from the thoracic portion of the descending aorta and form capillary plexuses that supply the airways, pleura, large blood vessels, and nerves of the lungs. Some of the blood from the bronchial capillary plexuses drains into the pulmonary veins, mixing with the oxygenated blood that has just left the alveolar capillaries; the rest empties into the **bronchial veins,** which eventually return the blood to the superior vena cava. The blood flow through the bronchial vessels is only 1 percent of the pulmonary blood flow.

Pulmonary Capillary Pressure

We wish to reemphasize the fact, described in Chap. 21, that the pulmonary circulation is a *low-pressure* circuit. The normal pulmonary capillary pressure, which is the major force that favors movement of fluid out of the pulmonary capillaries into

the interstitium, is only 15 mmHg. It is below the major force that favors absorption, namely, the plasma osmotic pressure of 25 mmHg. Accordingly, the alveoli normally do not accumulate fluid, a feature essential for normal gas exchange. Should pulmonary capillary pressure increase above the osmotic pressure, as occurs when the left ventricle "fails," fluid accumulates in the interstitial space of the alveoli (**pulmonary edema)** and even in the alveoli themselves; this impairs gas exchange between alveolus and capillary.

Afferent nerve fibers travel from the lungs to the central nervous system from receptor endings, and efferent fibers to the lungs alter the activity of airway smooth muscle and gland cells. The airways receive their principle innervation via the vagus (Xth cranial) nerves; the acetylcholine released from these parasympathetic fibers acts to stimulate the smooth muscle and glands. There are also sympathetic nerve endings in the airway walls that inhibit the glands and cause relaxation of the smooth muscle.

THORACIC CAGE

The thoracic cage is a closed compartment. It is bounded at the neck by muscles and connective tissue, and it is completely separated from the abdomen by the diaphragm. The outer walls of the thoracic cage are formed by the sternum (breastbone), 12 pairs of ribs, intercostal muscles, and other tissues that lie between the ribs. These walls also contain large amounts of elastic connective tissue.

The **diaphragm** is the principal muscle of respiration, but fibers of both external and internal intercostal muscles are also active during the various phases of the respiratory cycle; those firing during inspiration are called the **inspiratory intercostals,** and those active during forced expiration are the **expiratory intercostals.** (Another function of the intercostal muscles is to stiffen the intercostal spaces so that the tissues there are not drawn inward with the decrease in intrathoracic pressure.) Additional muscles can be recruited during forced respiration; some of the accessory muscles of respiration are the scalenes, sternocleidomastoids, posterior neck and back muscles (including the trapezius), muscles of the palate, tongue, and cheeks, and the abdominal muscles.

Movements of the Thorax

The thoracic cage is quite mobile, and this mobility is ultimately responsible for respiration. Its dimensions increase during inspiration and return to resting size during normal expiration. The changes during inspiration are the result of contraction of various muscles, notably the diaphragm and inspiratory intercostals. Expiration is usually passive and occurs simply as a result of cessation of the inspiratory contractions.

During inspiration, the dimensions of the thoracic cage are increased in all three directions. Movement of the ribs, mainly by the inspiratory intercostal muscles, is responsible for the lateral and anteroposterior changes. Each rib can be regarded as a lever whose fulcrum lies near the vertebral column so that a slight movement of the rib at its

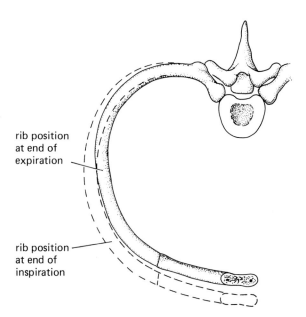

FIG. 23-12   Position of a rib at end inspiration (dashed line) and at end expiration (solid line).

vertebral end is much magnified at its sternal end. When a rib is elevated by muscle contraction, it pushes the sternum forward, which increases the anteroposterior dimensions. Simultaneously, the shaft of the rib moves outward from the medial plane, which increases the lateral dimensions of the thoracic cage (Fig. 23-12).

The vertical (craniocaudal) dimension of the thoracic cavity is increased mainly by contraction of

FIG. 23-13   Movements of chest wall and diaphragm during breathing. The contracting intercostal muscles move the ribs upward and outward during inspiration while the contracting diaphragm moves downward. The dashed lines indicate the inspiratory positions.

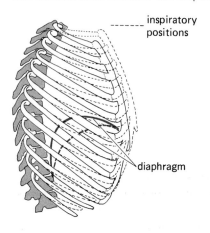

the diaphragm (Fig. 23-13). As the muscle fibers of the diaphragm contract, the central tendinous dome moves downward, pushing on the abdominal viscera, whose movements down and outward are made easier by the simultaneous relaxation of the muscles of the abdominal wall.

Relation of the Lungs to the Thoracic Cage

The preceding section described how the dimensions of the thoracic cage are altered during normal respiration. To understand how these changes produce similar changes in the dimensions of the lungs, the relationship of the lungs to the thoracic cage must be appreciated.

Firmly attached to the entire interior of the thoracic cage is the **pleura,** a thin layer of loose connective tissue covered by a layer of mesothelium. (**Mesothelium** is the simple squamous epithelium that covers the surface of serous cavities such as the peritoneal, pericardial, and pleural cavities.) The pleura forms two completely enclosed sacs in the thoracic cage, one on each side of the midline. The relationship between the lungs and pleura can be visualized by imagining what happens when one punches a fluid-filled balloon (Fig. 23-14). The arm represents the main bronchus that leads to the lung, the fist is the lung, and the balloon is the pleural sac. The outer portion of the fist becomes coated by one surface of the balloon. In addition, the balloon is pushed back upon itself so that its surfaces lie close together. This is precisely the relation between the lung and pleura except that the pleural surface that coats the lung is firmly attached to the lung surface. This layer of pleura (called the **visceral plura**) and the outer layer which lines the interior thoracic wall (the **parietal pleura**) are so close to each other that they are virtually in contact, being separated only by a very thin layer of **intrapleural fluid.**

FIG. 23-14   Relationship of lungs, pleura, and thoracic cage, analogous to pushing one's fist into a fluid-filled balloon. Note that there is no communication between the right and left intrapleural fluids. The volume of intrapleural fluid is greatly exaggerated; normally it consists of an extremely thin layer of fluid between the pleural membrane that lines the inner surface of the thoracic cage and the pleural membrane that lines the surface of the lungs.

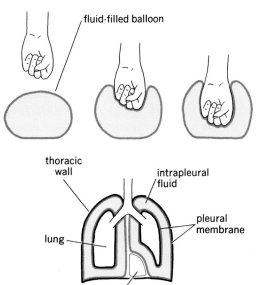

METABOLIC FUNCTIONS OF THE LUNGS

The lungs have a variety of functions in addition to their roles in gas exchange and regulation of H^+ concentration. Most notable are their contributions to altering the blood concentrations of a large number of biologically active substances; some substances are inactivated and others are secreted into the blood as it flows through the lungs.

The lungs also act as a "sieve" that traps and dissolves small blood clots generated in the systemic circulation, which thereby prevents them from reaching the systemic arterial blood where they could occlude blood vessels in other organs.

SECTION B: VENTILATION

The alveoli are such a great distance from the external environment that diffusion could not possibly provide the amount of oxygen and carbon dioxide exchange necessary for metabolism; therefore, a pressure gradient capable of generating bulk flow between the two regions is needed. This pressure gradient is developed by changes in lung volume; the pressure in the air spaces of the lungs decreases as their volume increases, and vice versa. This relationship between the pressure of a gas and its volume is described by Boyle's Law, where V is the volume of a fixed number of mols of gas, k is the proportionality constant, and P the pressure at a constant temperature

$$P = k \frac{1}{V}$$

The lungs, themselves, however, lack muscle and are therefore passive elastic containers with no inherent ability to increase their volume. Lung expansion is accomplished instead by the action of the intercostal muscles, which move the ribs, and the diaphragm muscle.

INVENTORY OF STEPS INVOLVED IN RESPIRATION

1   Exchange of air between the atmosphere (external environment) and alveoli. This process includes the movement of air in and out of the lungs and the distribution of air in the lungs. Not only must a large volume of new air be delivered to the alveoli but it must be distributed proportionately to the millions of alveoli within each lung. This entire process is called **ventilation** and occurs by bulk flow.
2   Exchange of oxygen and carbon dioxide between alveolar air and lung capillaries by diffusion. The volume and distribution of the pulmonary (lung) blood flow are extremely important for normal functioning of this process.
3   Transportation of oxygen and carbon dioxide by the blood. This includes the flow of blood and the forms of the gases within the blood.
4   Exchange of oxygen and carbon dioxide between the blood and tissues of the body by diffusion as blood flows through tissue capillaries.

EXCHANGE OF AIR BETWEEN ATMOSPHERE AND ALVEOLI: VENTILATION

Like blood, air moves by bulk flow from a high pressure to a low pressure. Bulk flow can be described by the equation

$$F = k(P_1 - P_2)$$

That is, flow is proportional to the pressure difference between two points, k being the proportionality constant. For air flow into or out of the lungs, the two relevant pressures are the **atmospheric pressure** and the **alveolar pressure** i.e., the pressure in the alveoli.

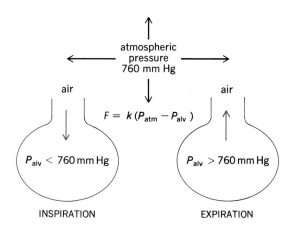

FIG. 23-15   Relationships required for ventilation. When the alveolar pressure P_{alv} is less than atmospheric pressure, air enters the lungs. Flow F is directly proportional to the pressure difference.

$$F = k(P_{atm} - P_{alv})$$

The atmospheric pressure is 760 mmHg at sea level and is obviously not subject to control, short of putting a person in a spacesuit or diving bell. Since atmospheric pressure remains relatively constant, if air is to be moved in and out of the lungs, the air pressure in the lungs, i.e., the alveolar pressure, must be made alternately less than and greater than atmospheric pressure (Fig. 23-15).

Concept of Intrapleural Pressure

When the chest is opened during surgery, if one is careful to cut only the thoracic wall but not the lung, the lung on the side of the incision collapses immediately (Fig. 23-16). Normally the highly elastic lungs are stretched within the intact chest, and the force responsible is eliminated when the chest is opened. This force is the pressure difference across the lung wall. An analogy may help illustrate the mechanism (Fig. 23-17). Imagine two balloons

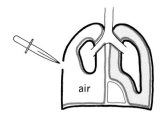

FIG. 23-16   Lung collapse caused by cutting the thoracic cage. Note that the air in the pleural space did *not* come from the lungs since the lung wall is still intact.

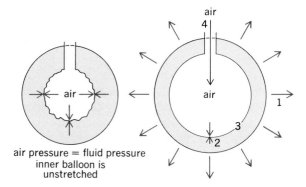

air
4
air
1
3
2
air

air pressure = fluid pressure
inner balloon is
unstretched

FIG. 23-17 How fluid between two balloons causes the inner balloon to expand whenever the outer one does. (1) Outer balloon is expanded by an outside force. (2) Expansion of the fluid space causes the fluid presure to decrease. (3) Fluid pressure is less than internal air pressure; therefore, the wall of the inner balloon is pushed out. (4) As the inner balloon expands, its internal air pressure decreases and air moves in from the atmosphere.

of slightly different size, one inside the other. The space between the balloons is completely filled with water. The inner, smaller balloon is open at the top so that there is free communication between its interior and the atmosphere; therefore, it contains air at atmospheric pressure. The forces that act upon the wall of the inner ballon are the inner air pressure and the water pressure surrounding it. Initially, as shown in the left half of Fig. 23-17, these pressures are approximately equal, and there is no tension in the wall of the inner balloon. When we enlarge the outer balloon by pulling on it in all directions, the inner balloon expands by an almost equal amount and its walls become highly stretched and taut. This occurs because the fluid pressure of the water that surrounds the inner balloon becomes less than the air pressure inside it.

Water is highly indistensible; i.e., any attempt to expand or compress a completely water-filled space causes a marked decrease or increase, respectively, of the fluid pressure in the space (it is much more difficult to compress a water-filled balloon than an air-filled balloon of the same size). Thus the pull on the external balloon produces a drop in the fluid pressure that surrounds the inner balloon, which now becomes less than the air pressure in the inner balloon; this difference in pressure across the wall pushes out the wall of the inner balloon. As the inner wall moves outward, the internal air pressure

falls slightly, but atmospheric air immediately enters through the opening so that atmospheric pressure is restored. Thus, the fluid pressure remains lower than the internal air pressure and the inner balloon expands until the force of its elastic recoil becomes great enough to balance this distending pressure difference.

We can apply this analogy to the lungs (air-filled inner balloon), thoracic cage (outer balloon), and intrapleural fluid (the water between the balloons). At rest, i.e., even when no respiratory-muscle contraction is occurring to cause inspiration, elastic forces in the tissues of the thoracic cage tend to pull it away from the outer surface of the lungs. This drops the *intrapleural* fluid pressure below that of the alveolar air pressure, a pressure difference that forces the lungs to distend. The lungs must expand to virtually the same degree as the thoracic cage, and their elastic walls become stretched. The tendency for the lungs to recoil as a result of this stretch is balanced by the difference between the alveolar air pressure and the intrapleural fluid pressure.

Why the lung collapses when the chest wall is opened should now be apparent. The low intrapleural pressure is significantly less than the pressure of the atmospheric air outside the body; i.e., it is subatmospheric. When the chest wall is pierced, atmospheric air rushes into the intrapleural space, the pressure difference across the lung wall is eliminated, and the stretched lung collapses. Air in the intrapleural space is known as a *pneumothorax*.

The subatmospheric pressure of the intrapleural fluid is normally maintained throughout life (only during a forced expiration does intrapleural pressure exceed atmospheric pressure). Regardless of whether the person is inspiring, expiring, or not breathing at all, the intrapleural pressure is always lower than the air pressure in the lungs, and the lungs are considerably stretched. However, the gradient between intrapleural and alveolar pressures does vary during breathing and directly causes the changes in lung size that occur during inspiration and expiration. The magnitude of the change in lung volume (ΔV) caused by a given change in the pressure difference across the lung wall (ΔP) is defined as the **lung compliance.** The higher this ratio ($\Delta V/\Delta P$), the more compliant (i.e., the more stretchable) is the lung. In the disease *emphysema,* the internal structures of the lung become more stretchable; i.e., their compliance increases. This results in a marked increase in resting lung size because the resting alveolar-intrapleural pressure difference distends the lung much more than usual. This phenomenon has important deleterious consequences for gas exchange, as will be described later.

Since intrapleural pressure is transmitted throughout the fluid that surrounds not only the lungs but the heart and other intrathoracic structures as well, it is frequently termed the **intrathoracic pressure.** Recall that subatmospheric intrathoracic pressure was mentioned in describing the forces that determine end-diastolic ventricular volume (Chap. 21).

Inspiration

The left half of Fig. 23-18 summarizes events which occur during inspiration. Just before the inspiration begins, i.e., at the conclusion of the previous ex-

FIG. 23-18   Summary of alveolar and intrapleural pressure changes and air flow during inspiration and expiration of 500 mL of air. Note that, on the pressure scale at the left, normal atmospheric pressure (760 mmHg) has a scale value of zero.

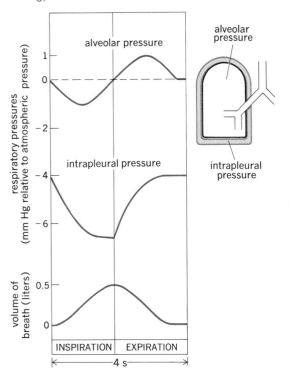

piration, the respiratory muscles are relaxed and no air is flowing. The intrapleural pressure is subatmospheric (for reasons described above). The alveolar pressure, i.e., the air pressure in the alveoli, is exactly atmospheric because the alveoli are in free communication with the atmosphere via the airways. Inspiration is initiated by the contraction of the diaphragm and inspiratory intercostal muscles, which, as described above, enlarge the volume of the thoracic cage.

As the thoracic cage begins to move away from the lung surface, the intrapleural fluid pressure abruptly decreases, i.e., becomes even more subatmospheric. This increases the difference between the alveolar and intrapleural pressures, and the lung wall is pushed out. Thus, when the inspiratory muscles increase the thoracic dimensions, the lungs are also forced to enlarge because of the changes in intrapleural pressure. This further stretching of the lung causes an increase in the volumes of all the air-containing passages and alveoli in the lung. As the alveoli enlarge, the air pressure in them drops to less than atmospheric, and this difference in pressures causes bulk flow of air from the atmosphere through the airways into the alveoli until their pressure again equals atmospheric.

Expiration

The expansion of thorax and lungs produced during inspiration by active muscular contraction stretches both lung and thoracic wall elastic tissue. When inspiratory-muscle contraction ceases and these muscles relax, the stretched tissues recoil to their original length since there is no force left to maintain the stretch. An obvious analogy is the snap of a stretched rubber band when it is released. The tissue recoil causes a rapid and complete reversal of the inspiratory process, as shown in the right side of Fig. 23-18. The thorax and lungs spring back to their original sizes, alveolar air becomes temporarily compressed so that its pressure exceeds atmospheric, and air flows from the alveoli through the airways out into the atmosphere.

Expiration at rest is thus completely passive, depending only upon the cessation of inspiratory-muscle activity and the relaxation of these muscles. Under certain conditions (during heavy exercise, for example) expiration is facilitated by the contraction of expiratory intercostal muscles and abdominal muscles, which actively decrease thoracic dimensions. Contraction of the abdominal muscles helps by increasing intraabdominal pressure to force the diaphragm up higher into the thorax and by depressing the lower ribs and flexing the trunk.

It should be noted that the analysis of Fig. 23-18 treats the lungs as a single alveolus. The fact is that there are significant regional differences in both alveolar and intrapleural pressures throughout the lungs and thoracic cavity. These differences are due, in part, to the effects of gravity and to local differences in the elasticity of the chest structures. They are of great importance in determining the pattern of ventilation.

Quantitative Relationship between Atmosphere-Alveolar Pressure Gradients and Air Flow: Airway Resistance

What is the quantitative relationship between the atmosphere-alveolar pressure gradients and the volume of air flow? It is expressed by precisely the same equation given for the circulatory system.

Flow = pressure gradient/resistance

The volume of air that flows in or out of the alveoli per unit time is directly proportional to the pressure gradient between the alveoli and atmosphere and inversely proportional to the resistance to flow offered by the airways. Normally the magnitude of this pressure gradient is increased by taking deeper breaths, i.e., by increasing the strength of contraction of the inspiratory muscles. This causes, in order, a more rapid expansion of the thoracic cage, a greater pressure gradient across the lung wall, increased expansion of the lungs, and increased flow of air into the lungs (Fig. 23-19).

What factors determine airway resistance? Resistance is (1) directly proportional to the magnitude of the interactions between the flowing gas molecules, (2) directly proportional to the length of the airway, and (3) inversely proportional to the fourth power of the airway radius. These factors, of course, are counterparts of the factors that determine resistance in the circulatory system; and in the respiratory tree, just as in the circulatory tree, resistance is largely controlled by the radius of the airways.

The airway diameters are normally so large that they actually offer little total resistance to air flow, and interaction between gas molecules is also usu-

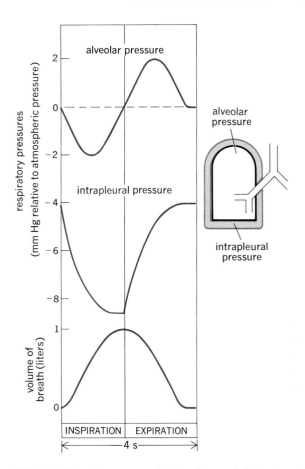

FIG. 23-19 Summary of alveolar and intrapleural pressure changes and air flow during inspiration of 1000 mL of air. Compare these values with those given in Fig. 23-18.

ally negligible, as is the contribution of airway length. Therefore, the total resistance remains so small that minute pressure gradients suffice to produce large volumes of air flow. As we have seen (Fig. 23-18), the average pressure gradient during a normal breath at rest is less than 1 mmHg; yet, approximately 500 mL of air is moved by this tiny gradient.

The diameter of the airways becomes critically important in certain conditions such as asthma, which is characterized by severe airway smooth muscle contraction and constriction and plugging of the airways by mucus. Airway resistance to air flow may become great enough to prevent air flow completely, regardless of the atmosphere-alveolar pressure gradient.

Airway size and resistance may be altered by

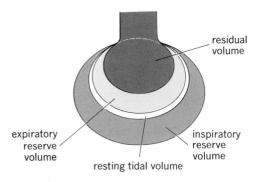

FIG. 23-20   Lung volumes.

physical, nervous, or chemical factors. The most important normal physical factor is simply expansion of the lungs; during inspiration, airway resistance decreases because as the lungs expand, the airways within the lung are also widened. Conversely, during expiration, airway resistance increases. For this reason patients with abnormal airway resistance, as in asthma, have much less difficulty inhaling than exhaling, with the result that air may be trapped in the lungs, the lung volume progressively increasing.

Neural regulation of the airway size is mediated by the autonomic nervous system; the sympathetic neurons cause relaxation of the airway smooth muscle (decreased resistance) and the parasympathetics cause smooth muscle contraction (increased resistance). These reflexes are important in causing air-

way constriction upon inhalation of chemical irritants but their precise contribution to control of airway resistance under normal conditions is unclear.

As would be expected from knowledge of the effects of the sympathetic nerves on airway resistance, circulating epinephrine also causes airway dilation. This is a major reason for administering epinephrine or epinephrine-like drugs to patients suffering from airway constriction, as in an asthmatic attack. In contrast, histamine causes bronchiolar constriction (and increased mucus secretion as well) and may be the cause of the airway constriction observed in allergic attacks. This explains the use of antihistamines to relieve the respiratory symptoms of allergies. Finally, the effects of the prostaglandins on pulmonary airways in asthma and other diseases may be of particular importance since the lungs take up, metabolize, and release various members of the prostaglandin family, some of which are airway constrictors, some dilators.

Contraction of certain skeletal muscles can also influence the airway diameter. For example, the alae nasi widen the nostrils, and the laryngeal muscles alter the diameter of the larynx.

Lung-Volume Changes during Breathing

The volume of air entering or leaving the lungs during a single breath is called the **tidal volume**

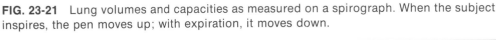

FIG. 23-21   Lung volumes and capacities as measured on a spirograph. When the subject inspires, the pen moves up; with expiration, it moves down.

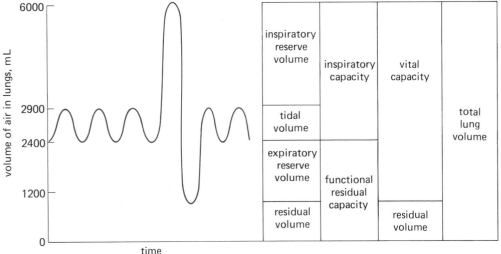

(Figs. 23-20 and 23-21). For a breath under resting conditions, this is approximately 500 mL. We are all aware that the resting thoracic excursion is small compared with a maximal breathing effort. The volume of air that can be inspired over and above the resting tidal volume is called the **inspiratory reserve volume** and amounts to 2500 to 3500 mL of air. At the end of a normal expiration, the lungs still contain a large volume of air, part of which can be exhaled by active contraction of the expiratory muscles; it is called the **expiratory reserve volume** and measures approximately 1000 mL of air. Even after a maximal expiration, some air (approximately 1000 mL) still remains in the lungs and is termed the **residual volume.**

What, then, is the maximum amount of air that can be moved in and out during a single breath? It is the sum of the normal tidal, inspiratory-reserve, and expiratory-reserve volumes. This total volume is called the **vital capacity.** During heavy work or exercise, a person uses part of both the inspiratory and expiratory reserves (particularly the former) but rarely uses more than 50 percent of his or her total vital capacity, because deeper breaths than this require exhausting activity of the inspiratory and expiratory muscles. The greater depth of breathing during exercise greatly increases pulmonary ventilation, and still larger increases are produced by increasing the rate of breathing as well.

The **total pulmonary ventilation** per minute is determined by the tidal volume times the respiratory rate (expressed as breaths per minute). For example, at rest, a normal person moves approximately 500 mL of air in and out of the lungs with each breath and takes 10 breaths each minute. The total minute ventilatory volume is therefore 500 mL $\times$ 10 = 5000 mL of air per minute. However, as we shall see, not all this air is available for exchange with the blood.

Air Distribution within the Lungs

Anatomic Dead Space The respiratory tract, as we have seen, is composed of conducting airways and the alveoli. In the lungs, exchanges of gases with the blood occur only in the alveoli and not in the conducting airways, the total volume of which is approximately 150 mL. Picture, then, what occurs during expiration: 500 mL of air is forced out of the alveoli and through the airways. Approxi-

mately 350 mL of this air is exhaled at the nose or mouth, but approximately 150 mL still remains in the airways at the end of expiration. During the next inspiration, 500 mL of air flows into the alveoli, but the first 150 mL of entering air is not atmospheric but the 150 mL of alveolar air left behind. Thus, only 350 mL of new atmospheric air enters the alveoli during the inspiration. At the end of inspiration, 150 mL of fresh air also fills the conducting airways, but no gas exchange with the blood can occur there. At the next expiration, this fresh air will be washed out and again replaced by old alveolar air, thus completing the cycle. The end result is that 150 mL of the atmospheric air entering the respiratory system during each inspiration never reaches the alveoli but is merely moved in and out of the airways. Because these airways do not permit gas exchange with the blood, the space within them is termed the **anatomic dead space.**

The volume of fresh atmospheric air entering the alveoli during each inspiration, then, equals the tidal volume minus the volume of air in the anatomic dead space. Thus, for a normal breath

$$
\begin{array}{rl}
\text{Tidal volume} = & 500 \text{ mL} \\
\text{Dead space} = & \underline{150 \text{ mL}} \\
\text{Fresh air entering alveoli} = & 350 \text{ mL}
\end{array}
$$

This total is the **alveolar ventilation** *per breath.* This value is somewhat confusing because it seems to indicate that only 350 mL air enters and leaves the alveoli with each breath. This is not true—the total is 500 mL of air, but only 350 mL is fresh air.

What is the significance of the anatomic dead space and alveolar ventilation? Total pulmonary ventilation is equal to the tidal volume of each breath multiplied by the number of breaths per minute. But since only that portion of inspired atmospheric air that enters the alveoli, i.e., the alveolar ventilation, is useful for gas exchange with the blood, the magnitude of the alveolar ventilation is of much greater significance than the magnitude of the total pulmonary ventilation, as can be demonstrated readily by the data in Table 23-2.

In this experiment subject A breathes rapidly and shallowly, B normally, and C slowly and deeply. Each subject has exactly the same total pulmonary ventilation; i.e., each is moving the same amount of air in and out of the lungs each minute. Yet, when we subtract the anatomic-dead-space ventilation from the total pulmonary ventilation, we find

TABLE 23-2   EFFECT OF BREATHING PATTERNS ON ALVEOLAR VENTILATION

Subject	Tidal Volume, mL/breath	×	Frequency, breaths/min	=	Total Pulmonary Ventilation, mL/min	Dead Space Ventilation, mL/min	Alveolar Ventilation, mL/min
A	150		40		6000	150 × 40 = 6000	0
B	500		12		6000	150 × 12 = 1800	4200
C	1000		6		6000	150 × 6 = 900	5100

marked differences in alveolar ventilation. Subject A has no alveolar ventilation (and would become unconsicous in several minutes), whereas C has a considerably greater alveolar ventilation than B, who is breathing normally. The important deduction to be drawn from this example is that increased depth of breathing is far more effective in elevating alveolar ventilation than an equivalent increase of breathing rate. Conversely, a decrease in depth can lead to critical reduction of alveolar ventilation. This is because a fraction of each tidal volume represents anatomic-dead-space ventilation. If the tidal volume decreases, this fraction increases until, as in subject A, it may represent the entire tidal volume. On the other hand, any increase in tidal volume goes entirely toward increasing alveolar ventilation. Alveolar ventilation *per minute* is calculated as follows:

Alveolar ventilation (mL/min) =
        frequency (breaths/min)
        × [tidal volume (mL) − dead space (mL)]

These concepts have important physiologic implications. Most situations, such as exercise, that necessitate an increased oxygen supply (and carbon dioxide elimination) reflexly call forth a relatively greater increase in breathing depth than rate. Indeed, well-trained athletes can perform moderate exercise with very little increase, if any, in respiratory rate. The mechanisms by which rate and depth of respiration are controlled will be described in a later section of this chapter.

Alveolar Dead Space   Some inspired air is not useful for gas exchange with the blood even though it reaches the alveoli because some alveoli, for various reasons, receive too little blood supply for their size. Air that enters these alveoli during inspiration cannot exchange gases efficiently because there is insufficient blood. The inspired air in this **alveolar dead space** must be distinguished from that in the anatomic dead space. It is quite small in normal persons but may reach lethal proportions in several kinds of lung diseases, e.g., emphysema. It is minimized by local mechanisms that match air and blood flows.

Work of Breathing

During inspiration, active muscular contraction provides the energy required to expand the thorax and lungs. What determines how much work these muscles must perform in order to provide a given amount of ventilation? First, there is simply the stretchability of the thorax and lungs. To expand these structures they must be stretched. The easier they stretch, i.e., the more compliant the lung, the less energy is required for a given amount of expansion. Much of the work of breathing goes into stretching the elastic tissue of the lung, but an even larger fraction goes into stretching a different kind of "tissue"—water itself! The air in each alveolus is separated from the alveolar membranes by an extremely thin layer of fluid; in a sense, therefore, the alveoli may be viewed as air-filled bubbles lined with water. At an air-water interface, the attractive forces between water molecules cause them to squeeze in upon the air within the bubble (Fig. 23-22). This force, known as **surface tension,** makes the water lining very like highly stretched rubber that constantly tries to shorten and resists further stretching. Thus inspiration requires considerable energy to expand the lungs because of the difficulty of distending these alveolar bubbles. Indeed, the surface tension of pure water is so great that lung expansion would require exhausting muscular effort and the lungs would tend to collapse. It is extremely

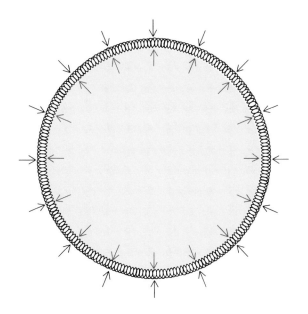

FIG. 23-22 Forces that act on the surface of a bubble. The springs and dark arrows represent the surface tension that results from the cohesive forces of water molecules at the air-water interface. This tension is opposed by the air pressure within the bubble (colored arrows).

important, therefore, that specialized alveolar cells, the type II cells, produce a phospholipoprotein complex, known as **pulmonary surfactant,** which intersperses with the water molecules on the alveolar surface and markedly reduces their cohesive force, which thereby lowers the surface tension.

A striking example of what occurs when insufficient surfactant is present is provided by the disease known as *respiratory-distress syndrome,* which all too frequently afflicts premature infants in whom the surfactant-synthesizing cells are too immature to function adequately. The infant is able to inspire only by the most strenuous efforts that may ultimately cause complete exhaustion, inability to breathe, lung collapse, and death. Normal maturation of the surfactant-synthesizing apparatus is facilitated by the hormone cortisol, the secretion of which is increased late in pregnancy. Accordingly, it has been found that administration of cortisol-like drugs to some pregnant women 48 hours prior to the delivery of a premature infant may provide an important means of combating this disease. Unfortunately, postnatal administration of cortisol to the infant seems to be ineffective.

The second factor that determines the degree of muscular work required for a certain amount of ventilation is the magnitude of the airway resistance. When airway resistance is increased by bronchiolar constriction or by secretions (as in asthma), the usual pressure gradient does not suffice for adequate air inflow and a deeper breath is required to create a larger pressure gradient.

One might imagine from this discussion (and from observing an athlete exercising hard) that the work of breathing uses up a major portion of the energy spent by the body. Not so; in a normal person, even during heavy exercise, the energy needed for breathing is only about 3 percent of the total expenditure. It is only in disease, when the work of breathing is markedly increased by structural changes in the lung or thorax, by loss of surfactant, or by an increased airway resistance, that breathing itself becomes an exhausting form of exercise.

SECTION C: GAS EXCHANGE AND TRANSPORT

Alveolar ventilation is only the first step in the total respiratory process. Oxygen must move across the alveolar membranes into the pulmonary capillaries, be transported by the blood to the tissues, leave the tissue capillaries, and finally cross plasma membranes to gain entry into cells. Carbon dioxide must follow a similar path in reverse (Fig. 23-23).

At rest, during each minute, body cells consume approximately 200 mL of oxygen and produce approximately the same amount of carbon dioxide. The relative amounts depend primarily upon what nutrients are being used for energy; e.g., when glucose is utilized, one molecule of carbon dioxide is produced for every molecule of oxygen consumed.

$$C_6H_{12}O_6 + 6O_2 \longrightarrow 6CO_2 + 6H_2O + energy$$

The ratio (CO_2 produced)/(O_2 consumed) is known as the **respiratory quotient (RQ)** and accordingly is 1 for glucose. When fat is utilized, only 7 molecules of carbon dioxide are produced for every 10 molecules of oxygen consumed, and RQ = 0.7. On a mixed diet the RQ is less than 1.0, and more O_2 is consumed than CO_2 produced; however, for simplicity, Fig. 23-23 assumes that the carbon dioxide and oxygen amounts are equal and the total vol-

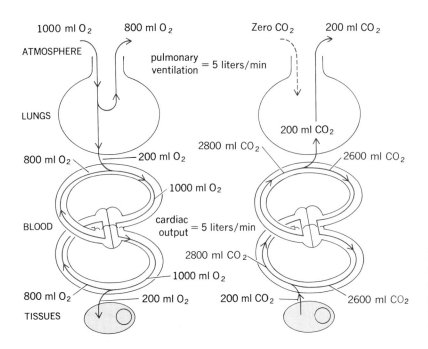

FIG. 23-23 Summary of oxygen and carbon dioxide exchanges between atmosphere, lungs, blood, and tissues during 1 min. RQ is assumed to be 1 (as will be explained latter in this section of text).

umes of air inspired and expired are therefore identical.

At rest, the total pulmonary ventilation equals 5 L of air per minute. Since only 20 percent of atmospheric air is oxygen (most of the remainder is nitrogen), the total oxygen input is $20\% \times 5 L = 1 L$ of O_2 per minute. Of this inspired oxygen, 200 mL crosses the alveoli into the pulmonary capillaries, and the remaining 800 mL is exhaled. This 200 mL of oxygen is carried by 5 L of blood, which is the pulmonary blood flow (cardiac output) per minute. Note, however, that blood entering the lungs already contains large quantities of oxygen, to which this 200 mL is added. This blood is then pumped by the left ventricle through the tissue capillaries of the body, and 200 mL of oxygen leaves the blood to be taken up and utilized by cells. Because only a fraction of the total blood oxygen actually leaves the blood, some oxygen remains in the blood when it returns to the heart and lungs. It is obvious but important that the quantities of oxygen added to the blood in the lungs and removed in the tissues are identical. As shown by Fig. 23-23, the story reads in reverse for carbon dioxide. As we shall see, most of the blood carbon dioxide is actually in the form $HCO_3{}^-$, but we have shown it as CO_2 for simplicity.

The pumping of blood by the heart obviously propels oxygen and carbon dioxide between the lungs and tissues by bulk flow, but what forces induce the net movement of these molecules across the alveolar, capillary, and cell membranes? The answer is diffusion: There is no active membrane transport for oxygen or carbon dioxide. As described in Chap. 4, diffusion can effect the net transport of a substance only when a concentration gradient exists for it. Understanding the mechanisms involved depends upon familiarity with some basic chemical and physical properties of gases, to which we now turn.

BASIC PROPERTIES OF GASES

A gas consists of individual molecules constantly moving at great speeds. Since rapidly moving molecules bombard the walls of any vessel that contains them, they therefore exert a pressure against the walls. The magnitude of the pressure is increased by anything that increases the bombardment. The pressure a gas exerts is proportional to (1) the temperature, because heat increases the speed at which molecules move, and (2) the concentration of the gas, i.e., the number of molecules per unit volume. In other words, when a certain number of molecules are compressed into a smaller volume, there are more collisions with the walls. The pressure of a gas is therefore a measure of the concentration and speed of its molecules.

Of great importance is the relationship between different gases, i.e., different kinds of molecules, such as oxygen and nitrogen, in the same container. In a mixture of gases, the pressure exerted by each gas is independent of the pressure exerted by the others because gas molecules are normally so far apart that they do not interfere with each other. Since each gas behaves as though the other gases were not present, the total pressure of a mixture of gases is simply the sum of the individual pressures. These individual pressures, termed **partial pressures,** are denoted by a P in front of the symbol for the gas, the partial pressure of oxygen thus being represented by P_{O_2}. Thus, the partial pressure of a gas is a measure of the concentration of that gas in a mixture of gases. Net diffusion of a gas will occur from a region where its partial pressure (concentration) is high to a region where it is low. Gas partial pressures are usually expressed in millimeters of mercury (mmHg), the same unit used for the expression of hydrostatic pressure.

BEHAVIOR OF GASES IN LIQUIDS

Several factors determine the uptake of gases by liquids and the behavior of gases dissolved in liquids. When a gas comes into contact with a liquid, the number of gas molecules that dissolve in the liquid is directly proportional to the pressure of the gas. This phenomenon is clear from the basic definition of pressure. Suppose, for example, that oxygen is placed in a closed container half full of water. Oxygen molecules constantly bombard the surface of the water, some entering the water and dissolving. Since the number of molecules striking the surface is directly proportional to the pressure of the oxygen gas P_{O_2}, the number of molecules entering the water is also directly proportional to P_{O_2}.

How many entering molecules actually stay in the water? Since the dissolved oxygen molecules are also constantly moving, some of them strike the water surface from below and escape into the free oxygen above. The rate of escape from the water and the rate of entry into the water are equal when the rates of bombardment are equal. By definition we say that the partial pressure of the oxygen in the liquid is equal to its partial pressure in the gas phase at this point. Thus, we come back to our earlier statement: The number of gas molecules that will dissolve in a liquid is directly proportional to the partial pressure of the gas. When the partial pressure in the gas phase is higher than the partial pressure of the gas in a liquid, there will be a net diffusion into the liquid. Conversely, if a liquid that contains a dissolved gas at high partial pressure is exposed to that same free gas at lower partial pressure, gas molecules will diffuse from the liquid into the gas phase until the partial pressures in the two phases become equal. These are precisely the phenomena that occur between alveolar air and pulmonary capillary blood.

It should also be apparent that dissolved gas molecules diffuse *within* the liquid from a region of higher partial pressure to a region of lower partial pressure, an effect that underlies the exchange of gases between cells, tissue fluid, and capillary blood throughout the body.

This discussion has been in terms of proportionalities rather than absolute amounts. The number of gas molecules that will dissolve in a liquid is *proportional* to the partial pressure, but the *absolute* number also depends upon the *solubility* of the gas in the liquid. Thus, if a liquid is exposed to two different gases at the same partial pressures, the numbers of molecules of each gas that are dissolved at equilibrium are not necessarily identical but reflect the solubilities of the two gases. Nevertheless, doubling the partial pressures doubles the number of gas molecules dissolved.

PRESSURE GRADIENTS OF OXYGEN AND CARBON DIOXIDE IN THE BODY

With these basic gas properties as foundation, we can discuss the diffusion of oxygen and carbon dioxide across alveolar, capillary, and plasma membranes. The pressures of these gases in atmospheric air and in various sites of the body are given in Fig. 23-24 for a resting person at sea level. The rest of this section is devoted to an elaboration of this figure.

Atmospheric air consists primarily of nitrogen and oxygen with very small quantities of water vapor, carbon dioxide, and inert gases such as argon. The sum of the partial pressures of all these gases is termed **atmospheric,** or **barometric pressure.** It

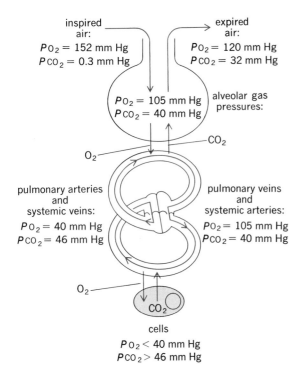

inspired _____ expired
air: air:
P_{O_2} = 152 mm Hg P_{O_2} = 120 mm Hg
P_{CO_2} = 0.3 mm Hg P_{CO_2} = 32 mm Hg

P_{O_2} = 105 mm Hg   alveolar gas
P_{CO_2} = 40 mm Hg   pressures:

CO$_2$

O$_2$

pulmonary arteries   pulmonary veins
       and   and
systemic veins: systemic arteries:
P_{O_2} = 40 mm Hg P_{O_2} = 105 mm Hg
P_{CO_2} = 46 mm Hg P_{CO_2} = 40 mm Hg

O$_2$

CO$_2$

cells
P_{O_2} < 40 mm Hg
P_{CO_2} > 46 mm Hg

FIG. 23-24 Summary of carbon dioxide and oxygen pressures in the inspired and expired air at various places in the body.

varies in different parts of the world as a result of differences in altitude, but at sea level it averages 760 mmHg. Since air is 20 percent oxygen, the P_{O_2} of inspired air is 20% × 760 = 152 mmHg at sea level.

The first question suggested by Fig. 23-24 is why the partial pressures of the constituents of expired air are not identical to those of alveolar air. Recall that approximately 150 mL of the inspired atmospheric air during each breath never gets down into the alveoli but remains in the airways (dead space). This air does not exchange carbon dioxide or oxygen with blood and is expired along with alveolar air during the subsequent expiration. Therefore, the P_{O_2} and P_{CO_2} of the total expired air are higher and lower, respectively, than those of alveolar air.

The next question concerns the alveolar gas pressures themselves. One might logically reason that the alveolar gas pressures must vary considerably during the respiratory cycle, since new atmospheric air enters only during inspiration. In fact, however, the variations in alveolar P_{O_2} and P_{CO_2} during the cycle are so small as to be negligible because, as

explained in the section on lung volumes, a large volume of gas is always left in the lungs after expiration. This remaining alveolar gas contains large quantities of oxygen and carbon dioxide, and when the relatively small volume of new air enters, it mixes with the alveolar air already present, lowering its P_{CO_2} and raising its P_{O_2}, but only by a small amount. For this reason, the alveolar-gas partial pressures remain *relatively* constant throughout the respiratory cycle, and we may use the single alveolar pressures shown in Fig. 23-24 in our subsequent analysis of alveolar-capillary exchange, ignoring the minor fluctuations.

Our next question concerns the exchange of gases between alveoli and pulmonary capillary blood. The blood that enters the pulmonary capillaries is, of course, systemic venous blood pumped to the lungs via the pulmonary arteries. Having come from the tissues, it has a high P_{CO_2} (46 mmHg) and a low P_{O_2} (40 mmHg). As it flows through the pulmonary capillaries, it is separated from the alveolar air only by an extremely thin layer of tissue. The differences in the partial pressures of oxygen and carbon dioxide on the two sides of this alveolar-capillary membrane result in the net diffusion of oxygen into the blood and of carbon dioxide into the alveoli. As this diffusion occurs, the capillary blood P_{O_2} rises above its original value and the P_{CO_2} falls. The net diffusion of these gases ceases when the alveolar and capillary partial pressures become equal. In a normal person, the rates at which oxygen and carbon dioxide diffuse are so rapid and the blood flow through the capillaries so slow (at rest, a red blood cell takes 0.75 sec to pass through the pulmonary capillaries) that complete equilibrium is always reached. Thus, the blood that leaves the lungs to return to the heart has essentially the same P_{O_2} and P_{CO_2} as alveolar air.

The diffusion of gases between alveoli and capillaries may be impaired in a number of ways. The disease *emphysema,* which is intimately related to cigarette smoking, is characterized by the breakdown of the alveolar walls with the formation of fewer but larger alveoli. The result is a reduction in the total area available for diffusion. In a different kind of defect, caused by membrane thickening or by pulmonary edema, the area available for diffusion is normal but the molecules must travel a greater distance. Finally, without becoming thicker the alveolar walls may become denser and less permeable, as for example when beryllium is inhaled and deposited on the walls.

As the arterial blood enters capillaries throughout the body, it becomes separated from the interstitial fluid only by the thin, highly permeable capillary membrane. The interstitial fluid, in turn, is separated from intracellular fluid by plasma membranes that are also quite permeable to oxygen and carbon dioxide. Metabolic reactions that occur in these cells are constantly consuming oxygen and producing carbon dioxide. Therefore, as shown in Fig. 23-24, intracellular P_{O_2} is lower and P_{CO_2} higher than in blood. As a result, a net diffusion of oxygen occurs from blood to cells, and a net diffusion of carbon dioxide from cells to blood. In this manner, as blood flows through capillaries, its P_{O_2} decreases and its P_{CO_2} increases. This accounts for the venous blood values shown in Fig. 23-24. Venous blood returns to the right ventricle and is pumped to the lungs, where the entire process begins again.

In summary, the consumption of oxygen in the cells and the supply of new oxygen to the alveoli create P_{O_2} gradients that produce net diffusion of oxygen from alveoli to blood in the lungs and from blood to cells in the rest of the body. Conversely, the production of carbon dioxide by cells and its elimination from the alveoli via expiration create P_{CO_2} gradients that produce net diffusion of carbon dioxide from cells to blood in the rest of the body and from blood to alveoli in the lungs.

TRANSPORT OF OXYGEN IN THE BLOOD: THE ROLE OF HEMOGLOBIN

Table 23-3 summarizes the oxygen content of arterial blood. Each liter of arterial blood contains the same number of oxygen molecules as 200 mL of pure gaseous oxygen. Oxygen is present in two forms: (1) physically dissolved in the blood water, and (2) reversibly combined with hemoglobin molecules. The amount of oxygen that can be dissolved in blood is directly proportional to the P_{O_2} of the blood, but because oxygen is relatively insoluble in water, only 3 mL of oxygen can be dissolved in 1 L of blood at the normal alveolar and arterial P_{O_2} of 100 mmHg. In contrast, in 1 L of blood, 197 mL of oxygen, more than 98 percent of the total, is carried in the red blood cells bound to hemoglobin.

Recall that the shape of the erythrocyte is a bi-

TABLE 23-3   OXYGEN CONTENT OF ARTERIAL BLOOD
1 liter arterial blood contains:
3 mL O_2 physically dissolved
197 mL O_2 chemically bound to hemoglobin
Total 200 mL O_2
Cardiac output = 5 L/min
O_2 carried to tissues/min = $5 \times 200 = 1000$ mL

concave disc, i.e., thicker at the edge than in the middle, like a doughnut with a center depression on each side instead of a hole. Its shape and small dimensions have adaptive value in that oxygen and carbon dioxide can rapidly diffuse throughout the entire cell interior.

The outstanding physiologic characteristic of erythrocytes is the presence of the iron-containing protein **hemoglobin,** which constitutes approximately one-third of the total cell weight. Hemoglobin is a protein composed of four polypeptide chains, each of which contains a single atom of iron (see also Chap. 19). It is with the iron that the oxygen combines; thus, each hemoglobin molecule can combine with four molecules of oxygen. However, the equation for the reaction of oxygen and hemoglobin is usually written in terms of a single polypeptide chain.

$$O_2 + Hb \rightleftharpoons HbO_2 \qquad (23\text{-}1)$$

Hemoglobin combined with oxygen (**HbO$_2$**) is called **oxyhemoglobin;** not combined (**Hb**), it is called **reduced hemoglobin** or deoxyhemoglobin. When all the iron atoms in hemoglobin are combined with O_2, the hemoglobin solution is said to be *fully saturated.* When hemoglobin exists as mixed Hb and HbO$_2$, it is said to be *partially saturated.* The **percentage saturation** of hemoglobin is a measure of the fraction of total hemoglobin binding sites that are combined with oxygen.

What factors determine the extent to which oxygen will combine with hemoglobin? By far the most important is the P_{O_2} of the blood, i.e., the concentration of dissolved oxygen. The blood hydrogen-ion concentration and temperature also play significant roles as do certain chemicals produced by the red blood cells themselves.

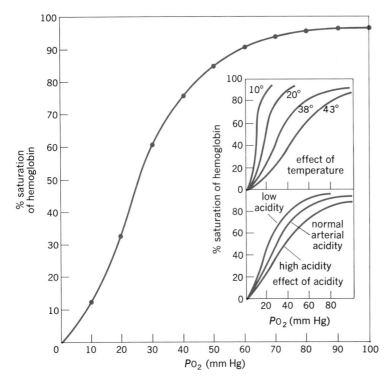

FIG. 23-25 Hemoglobin-oxygen dissociation curve. The large curve applies to blood at 38°C and the normal arterial hydrogen-ion concentration (acidity). The inset curves illustrate the effects of altering temperature and acidity on the relationship between Po_2 and hemoglobin saturation with oxygen.

Effect of Po_2 on Hemoglobin Saturation

From inspection of the equation for the reaction of oxygen and hemoglobin given earlier (Eq. 23-1) and the principle of mass action, it is obvious that raising the Po_2 of the blood should increase the combination of oxygen with hemoglobin. The experimentally determined quantitative relationship between these variables is shown in Fig. 23-25. When a sample of blood is exposed to a large volume of oxygen, a net diffusion of oxygen occurs from the gas into the blood until the blood and gas partial pressures become equal. By this means, we produce in 10 different samples of blood 10 different oxygen partial pressures ranging from 10 to 100 mmHg and analyze the effect of Po_2 on hemoglobin saturation by measuring the fraction of hemoglobin combined with oxygen in each case.

Data from such an experiment are plotted in Fig. 23-25, which is called an **oxygen-hemoglobin dissociation curve,** i.e., saturation curve. It is an S-shaped curve with a steep slope between 10 and 60 mmHg Po_2 and a flat portion between 70 and 100 mmHg Po_2. In other words, the extent to which hemoglobin combines with oxygen increases very rapidly from 10 to 60 mmHg so that, at a Po_2 of 60 mmHg, 90 percent of the total hemoglobin is com-

bined with oxygen. From this point on, a further increase in Po_2 produces only a small increase in oxygen binding. The adaptive importance of this plateau at higher Po_2 values is very great for the following reason. Many situations (severe exercise, high altitudes, cardiac or pulmonary disease) are characterized by a moderate reduction of alveolar and arterial Po_2. Even if the Po_2 fell from the normal value of 100 to 60 mmHg, the total quantity of oxygen carried by hemoglobin would decrease by only 10 percent, since hemoglobin saturation is still close to 90 percent at a Po_2 of 60 mmHg. The plateau therefore provides an excellent safety factor in the supply of oxygen to the tissues.

We now retrace our steps and reconsider the movement of oxygen across the various membranes, this time including hemoglobin in our analysis. It is essential to recognize that the oxygen that is bound to hemoglobin does not contribute directly to the Po_2 of the blood. Only gas molecules which are free in solution, i.e., physically dissolved, do so. Therefore, the diffusion of oxygen is governed only by that portion that is dissolved, a fact that permitted us to ignore hemoglobin in discussing transmembrane pressure gradients. However, the presence of hemoglobin plays a critical role in determining the *total amount* of oxygen that will dif-

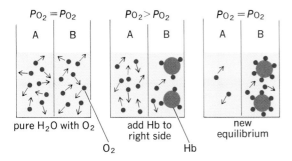

$P_{O_2} = P_{O_2}$ $P_{O_2} > P_{O_2}$ $P_{O_2} = P_{O_2}$

pure H_2O with O_2 add Hb to right side   new equilibrium

O_2 Hb

FIG. 23-26 Effects of adding hemoglobin on the distribution of oxygen between two compartments separated by a semipermeable membrane. At the new equilibrium, the P_{O_2} values are equal to each other but lower than before the hemoglobin was added; however, the total oxygen, i.e., both dissolved and chemically combined with hemoglobin, is much higher on the right side of the membrane.

fuse, as illustrated by a simple example (Fig. 23-26). Two solutions separated by a semipermeable membrane contain equal quantities of oxygen, the gas pressures are equal, and no net diffusion occurs. Addition of hemoglobin to compartment B destroys this equilibrium because much of the oxygen combines with hemoglobin. Despite the fact that the *total quantity of oxygen* in compartment B is still the same, the number of molecules *dissolved* has decreased; therefore, the P_{O_2} of compartment B is less than that of A, and net diffusion of oxygen occurs from A to B. At the new equilibrium, the oxygen pressures are once again equal, but almost

all the total oxygen is in compartment B and is combined with hemoglobin.

Let us now apply this analysis to the lung and tissue capillaries (Fig. 23-27). The plasma and erythrocytes that enter the lungs have a P_{O_2} of 40 mmHg, and the hemoglobin is 75 percent saturated. Oxygen diffuses from the alveoli because of its higher P_{O_2} (100 mmHg) into the plasma; this increases plasma P_{O_2} and induces diffusion of oxygen into the erythrocytes, which elevates erythrocyte P_{O_2} and causes increased combination of oxygen and hemoglobin. Thus, the vast preponderance of the oxygen diffusing into the blood from the alveoli does not remain dissolved but combines with hemoglobin. In this manner, the blood P_{O_2} remains less than that of the alveolar P_{O_2} until hemoglobin is virtually completely saturated, and the diffusion gradient favoring oxygen movement into the blood is maintained despite the very large transfer of oxygen. In the tissue capillaries, the procedure is reversed: As the blood enters the capillaries, plasma P_{O_2} is greater than interstitial fluid P_{O_2} and net oxygen diffusion occurs across the capillary wall; plasma P_{O_2} is now lower than erythrocyte P_{O_2}, and oxygen diffuses out of the erythrocyte into the plasma. The lowering of erythrocyte P_{O_2} causes the dissociation of HbO_2, which thereby liberates oxygen. Simultaneouly, the oxygen that diffused into the interstitial fluid is moving into cells along the concentration gradient generated by cell utilization of oxygen. The net result is a transfer of large quan-

FIG. 23-27 Oxygen movements in the lungs and tissues. Movement of inspired air into the alveoli is by bulk flow; all movements across membranes are by diffusion.

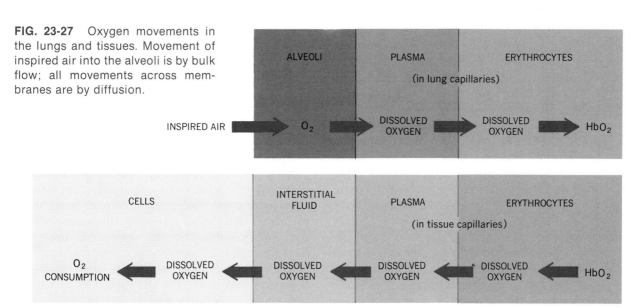

tities of oxygen from HbO_2 into cells purely by passive diffusion.

The fact that hemoglobin is still 75 percent saturated at the end of tissue capillaries under resting conditions underlies an important automatic mechanism by which cells can obtain more oxygen whenever they increase their activity. An exercising muscle consumes more oxygen, which thereby lowers its intracellular P_{O_2}; this increases the overall blood-to-cell P_{O_2} gradient and the diffusion of oxygen out of the blood. In turn, the resulting reduction of erythrocyte P_{O_2} causes additional dissociation of hemoglobin and oxygen. An exercising muscle can thus extract virtually all the oxygen from its blood supply. This process is so effective because it takes place in the steep portion of the hemoglobin dissociation curve. Of course, an increased blood flow to the muscle also contributes greatly to the increased oxygen supply.

Effect of Acidity on Hemoglobin Saturation

The large oxygen-hemoglobin dissociation curve illustrated in Fig. 23-25 is for blood that has a specific hydrogen-ion concentration equal to that found in arterial blood. When the same experiments are performed at a different level of acidity, the curve changes significantly. (The higher the hydrogen-ion concentration, the higher the acidity, and vice versa.) A group of such curves is shown in the inset of Fig. 23-25 for different hydrogen-ion concentrations. It is evident that, regardless of the existing acidity, the percentage saturation of hemoglobin is still determined by the P_{O_2}. However, a change in acidity causes highly significant shifts in the entire curve. An increased hydrogen-ion concentration moves the curve downward and to the right, which means that, at any given P_{O_2}, hemoglobin has less affinity for oxygen when the acidity is high. For reasons to be described later, the hydrogen-ion concentration in the tissue capillaries is greater than in arterial blood. Blood flowing through tissue capillaries becomes exposed to this elevated hydrogen-ion concentration and therefore loses even more oxygen than if the decreased P_{O_2} had been the only factor involved. Conversely, the hydrogen-ion concentration is lower in the lung capillaries than in the systemic venous blood, so that hemoglobin picks up more oxygen in the lungs than if only the P_{O_2} were involved. Finally, the more active a tissue

is, the greater its hydrogen-ion concentration; accordingly, hemoglobin releases even more oxygen during passage through these tissue capillaries, which thereby provides the more active cells with additional oxygen. The hydrogen ion exerts this effect on the affinity of hemoglobin for oxygen by combining with the hemoglobin and altering its molecular structure. The importance of this combination for the regulation of extracellular acidity will be described in Chap. 24.

Effect of Temperature on Hemoglobin Saturation

The effect of temperature on the oxygen-hemoglobin dissociation curve (inset of Fig. 23-25) resembles that of an increase in acidity. The implication is similar: Actively metabolizing tissue, e.g., exercising muscle, has an elevated temperature, which facilitates the release of oxygen from hemoglobin as blood flows through the muscle capillaries.

Effect of DPG on Hemoglobin Saturation

Red cells contain large quantities of the substance **2,3-diphosphoglycerate (DPG),** which is present in only trace amounts in other mammalian cells. DPG, which is produced by the red cells during glycolysis, binds reversibly with hemoglobin and causes it to change its conformation and release oxygen. The effect of increased DPG is to shift the curve downward and to the right (just as does an increased temperature or hydrogen-ion concentration). The net result is that whenever DPG is increased there is enhanced unloading of oxygen as blood flows through the tissues. Such an increase is triggered by a variety of conditions associated with decreased oxygen supply to the tissues and helps to maintain oxygen delivery.

In summary, we have seen that oxygen is transported in the blood primarily in combination with hemoglobin. The extent to which hemoglobin binds oxygen is dependent upon the P_{O_2}, hydrogen-ion concentration, temperature, and DPG. These factors cause the release of large quantities of oxygen from hemoglobin during blood flow through tissue capillaries and virtually complete conversion of reduced hemoglobin to oxyhemoglobin during blood flow through lung capillaries. An active tissue increases its extraction of oxygen from the blood because of its lower P_{O_2} and higher hydrogen-ion concentration and temperature.

FUNCTIONS OF MYOGLOBIN

Myoglobin, an iron-containing protein found in cardiac and skeletal muscle cells, resembles hemoglobin in that it binds oxygen reversibly. Its major function is to act as an intracellular carrier that facilitates the diffusion of oxygen throughout the muscle cell. In addition, it provides a store of oxygen that the cell can call upon during sudden changes in activity.

TRANSPORT OF CARBON DIOXIDE IN THE BLOOD

The quantity of carbon dioxide that is dissolved in blood at physiologic carbon dioxide partial pressures is quite small, certainly much smaller than the large volume of carbon dioxide that must be constantly transported from the tissues to the lungs.

Carbon dioxide can undergo the reaction

$$CO_2 + H_2O \rightleftharpoons H_2CO_3$$
$$\text{carbonic acid}$$

that goes quite slowly unless it is catalyzed by the enzyme **carbonic anhydrase.** The quantities of both dissolved carbon dioxide and carbonic acid are directly proportional to the P_{CO_2} of the plasma. The actual amount of carbonic acid in blood is small

because most of the carbonic acid ionizes according to the equation

$$H_2CO_3 \rightleftharpoons HCO_3^- + H^+$$
$$\text{bicarbonate ion}$$

This reaction proceeds very rapidly and requires no enzyme. Combining these two equations, we find

$$CO_2 + H_2O \overset{\text{carbonic anhydrase}}{\rightleftharpoons}$$
$$H_2CO_3 \rightleftharpoons HCO_3^- + H^+ \qquad \textbf{(23-2)}$$

Thus, the addition of carbon dioxide to a liquid results ultimately in bicarbonate and hydrogen ions.

Carbon dioxide can also react directly with proteins, particularly hemoglobin, to form *carbamino* compounds.

$$CO_2 + Hb \rightleftharpoons HbCO_2 \qquad \textbf{(23-3)}$$

When arterial blood flows through tissue capillaries, oxyhemoglobin gives up oxygen to the tissues and carbon dioxide diffuses from the tissues into the blood, where the following processes occur (Fig. 23-28).

1 A small fraction (8 percent) of the carbon dioxide remains physically dissolved in the plasma and red blood cells.
2 The largest fraction (81 percent) of the carbon dioxide undergoes the reactions described in

FIG. 23-28 Summary of carbon dioxide movements and reactions as blood flows through tissue capillaries. All movements across membranes are by diffusion. Note that most of the CO_2 ultimately is converted to HCO_3^-, this occurs almost entirely in the erythrocytes (because the carbonic anhydrase is located there), but most of the HCO_3^- then diffuses out of the erythrocytes into plasma.

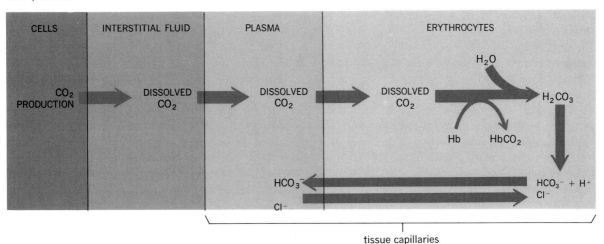

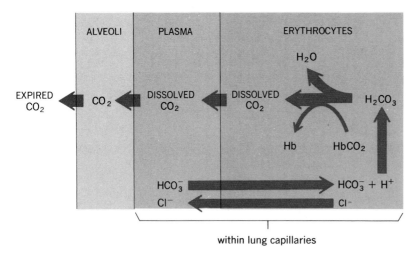

FIG. 23-29 Summary of carbon dioxide movements and reactions as blood flows through the lung capillaries. All movements across membranes are by diffusion. The plasma-erythrocyte phenomena are simply the reverse of those that occur during blood flow through the tissue capillaries, as shown in Fig. 23-28. The breakdown of H_2CO_3 is catalyzed by carbonic anhydrase.

Eq. 23-2 and is converted into bicarbonate and hydrogen ions. This occurs primarily in the red blood cells because they contain large quantities of the enzyme carbonic anhydrase but the plasma does not. However, most of the bicarbonate diffuses out of the red cells and is carried in the plasma; bicarbonate, in contrast to carbon dioxide, is extremely soluble. As the bicarbonate leaves the cells, chloride ions move in (Fig. 23-28), and the balance of negatively charged ions across the red cell membrane is maintained. This exchange of chloride for bicarbonate is called the **chloride shift.** It works in the opposite direction in the lung capillaries; i.e., bicarbonate enters the red blood cells and chloride leaves (Fig. 23-29). The reaction described in Eq. 23-2 explains why tissue capillary hydrogen-ion concentration is higher than that of the arterial blood and increases as metabolic activity increases. The fate of these hydrogen ions will be discussed in Chap. 24.

3   The remaining fraction (11 percent) of the carbon dioxide reacts directly with hemoglobin to form $HbCO_2$, as in Eq. 23-3.

Since these are all reversible reactions, i.e., they can proceed in either direction, depending upon the prevailing conditions, why do they all proceed primarily to the right, toward generation of HCO_3^- and $HbCO_2$, as blood flows through the tissues? Once again, the answer is provided by the law of mass action: It is the increase in carbon dioxide concentration that drives these reactions to the right as blood flows through the tissues.

Obviously, a sudden lowering of blood P_{CO_2} has just the opposite effect. HCO_3^- and H^+ combine to give H_2CO_3, which generates carbon dioxide and water. Similarly, $HbCO_2$ generates hemoglobin and free carbon dioxide. This is precisely what happens as venous blood flows through the lung capillaries (Fig. 23-29). Because the blood P_{CO_2} is higher than alveolar, a net diffusion of carbon dioxide from blood into alveoli occurs. This loss of carbon dioxide from the blood lowers the blood P_{CO_2} and drives these chemical reactions to the left, thus generating more dissolved carbon dioxide. Normally, as fast as this carbon dioxide is generated from HCO_3^- and H^+ and from $HbCO_2$, it diffuses into the alveoli. In this manner, all the carbon dioxide delivered into the blood in the tissues now is delivered into the alveoli, from which it is expired and eliminated from the body.

TRANSPORT OF HYDROGEN IONS BETWEEN TISSUES AND LUNGS

In the previous section, we pointed out (Eq. 23-2) that hydrogen ions and bicarbonate ions are generated by the dissociation of H_2CO_3 after its formation by the hydration of CO_2. However, most of these hydrogen ions do not remain free in the blood during transit from tissues to lungs (if they did, a toxic level of free hydrogen ion would result), but rather become bound to plasma proteins and hemoglobin. This latter protein is quantitatively far more important for the following reason: Reduced hemoglobin has a much greater affinity for hydrogen ion

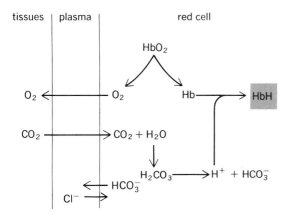

FIG. 23-30 Removal of free hydrogen ions from solution by hemoglobin as blood flows through tissue capillaries.

than oxyhemoglobin does. As blood flows through the tissues, a fraction of oxyhemoglobin loses its oxygen and is transformed into reduced hemoglobin. Simultaneously, a large quantity of carbon dioxide enters the blood and undergoes (primarily in the red blood cells) the reactions that ultimately generate HCO_3^- and H^+. Because reduced hemoglobin has a strong affinity for hydrogen ions, most of these hydrogen ions become bound to hemoglobin (Fig. 23-30). In this manner only a small number of hydrogen ions remain free, and the acidity of venous blood is only slightly greater than that of arterial blood. As the venous blood passes through the lungs, all these reactions are reversed. Hemoglobin becomes saturated with oxygen, and its ability to bind hydrogen ions decreases. The hydrogen ions are released, whereupon they react with HCO_3^- to give CO_2, which diffuses into the alveoli and is expired.

In a previous section, we described how the hydrogen-ion concentration of the blood is an important determinant of the ability of hemoglobin to bind oxygen. Now we have shown how the presence of oxygen is an important determinant of hemoglobin's ability to bind hydrogen ion. The adaptive value of these phenomena is enormous.

SECTION D: CONTROL OF RESPIRATION

In a human being at rest, the body's cells consume approximately 200 mL of oxygen per minute. Under conditions of high oxygen need, e.g., exercise, the rate of oxygen consumption may increase as much as thirtyfold. Equivalent amounts of carbon dioxide are simultaneously eliminated. It is obvious, therefore, that mechanisms must exist that coordinate breathing with metabolic demands. We shall see that the control of breathing also plays an important role in the regulation of the acidity of the extracellular fluid.

In dealing with the mechanisms by which the basic respiratory processes are controlled, we shall be concerned primarily with two questions: By what mechanisms are rhythmic breathing movements generated? What factors control the rate and depth of breathing, i.e., the total ventilatory volume?

NEURAL GENERATION OF RHYTHMIC BREATHING

Like cardiac muscles, the inspiratory muscles normally contract rhythmically; however, the origins of these contractions are quite different. Cardiac muscle has automaticity; i.e., it is capable of self-excitation. The nerves to the heart merely alter this basic inherent rate and are not actually required for cardiac contraction. On the other hand, the diaphragm and intercostal muscles consist of skeletal muscle, which cannot contract unless stimulated by nerves. Thus, breathing depends entirely upon cyclic respiratory muscle excitation by the **phrenic nerves** (to the diaphragm) and the **intercostal nerves** (to the intercostal muscles). The two phrenic nerves arise from the 3rd, 4th, and 5th cervical spinal nerves, one on each side of the vertebral column. They pass downward in the mediastinum and penetrate the diaphragm, where each divides to supply the various areas of that muscle. The 11 pairs of intercostal nerves are formed from the 1st to 11th thoracic spinal nerves. Destruction of the phrenic and intercostal nerves or the spinal cord areas from which they originate (as in poliomyelitis, for example) results in paralysis of the respiratory muscles and death, unless some form of artificial respiration can be rapidly instituted.

At the end of expiration, when the chest is at rest, a few impulses are still passing down these nerves. Like other skeletal muscles, therefore, the respiratory muscles have a certain degree of resting tonus. This muscular contraction is too slight to move the chest but plays a role in maintaining normal posture. Inspiration is initiated by an increased rate of firing of these inspiratory motor units. As more and more new motor units are recruited, thoracic expansion increases. In addition, the firing frequency of the individual units increases. By these two measures, the force of inspiration increases as it proceeds. Then almost all these units stop firing, the inspiratory muscles relax, and expiration occurs as the elastic lungs recoil. In addition, when expiration is facilitated by contraction of expiratory muscles, the nerves to these muscles, having been quiescent during inspiration, begin firing during expiration.

By what mechanism are nerve impulses to the respiratory muscles alternately increased and decreased? Control of this neural activity resides primarily in neurons with cell bodies in the lower portion of the brainstem, the medulla, which also contains the cardiovascular control centers. If the spinal cord is cut at any point between the medulla and the areas of the spinal cord from which the phrenic and intercostal nerves originate, breathing ceases. This experiment demonstrates that these efferent nerves are controlled by synaptic connections with neurons that descend in the spinal cord from the medulla.

By means of tiny electrodes placed in various parts of the medulla to record electric activity, neurons have been found that discharge in synchrony with inspiration and cease discharging during expiration. These neurons are called **inspiratory neurons.** They provide, via either direct or interneuronal connections, the rhythmic input to the motor neurons that innervate the inspiratory muscles.

What are the factors that induce firing of the medullary inspiratory neurons? There are two distinct components of this problem: (1) What factors are essential for the generation of rhythmical bursts of firing, and (2) what factors modify the timing of the bursts and their intensity? This section deals only with the first question, the subsequent section with the second.

It is quite likely, although not definitely proven, that the medullary inspiratory neurons have inherent automaticity and rhythmicity, i.e., the capacity for cyclical self-excitation. This alone could account for alternating cycles of inspiration and expiration. However, synaptic inputs to them from other neurons also play important roles. There are

at least three sources of such input: (1) other neurons in the medulla; (2) neurons with cell bodies in the pons, the area of brainstem just above the medulla; and (3) stretch receptors in the lung airways.

1   The medullary inspiratory neurons give off, in addition to their descending fibers, processes that synapse upon and stimulate adjacent medullary neurons. These neurons, in turn, send processes back to the inspiratory neurons and inhibit them. Thus the firing of the medullary inspiratory neurons generates, via this loop, an inhibitory input to itself, which thereby helps to terminate its own firing.

2   The medullary inspiratory neurons receive a rich synaptic input, either directly or via the inhibitory neurons adjacent to them, from neurons in various areas in pons. The precise contributions of these areas in the generation of the normal respiratory rhythm is still the subject of debate.

3   **Pulmonary stretch receptors** lie in the airway smooth muscle layer and are activated by lung inflation. The afferent nerve fibers from them travel in the vagus nerves to the medulla and inhibit the inspiratory neurons either directly or by way of the adjacent inhibitory neurons. Thus, feedback from the lungs themselves helps to terminate inspiration. However, the threshold of these receptors in human beings is very high and this pulmonary stretch receptor reflex plays a role in setting respiratory rhythm only under conditions of very large tidal volumes, as in exercise.

It must be emphasized that cessation of inspiratory neuronal activity is all that is required for expiration to occur, since expiration is normally a passive process dependent only upon relaxation of the inspiratory muscles. In contrast, during active expiration, cessation of firing of the medullary inspiratory neurons is accompanied by the activation of other descending pathways that synapse upon and stimulate the motor neurons to those muscles whose contractions produce active expiration.

CONTROL OF VENTILATORY VOLUME

In the preceding section, we were concerned with the mechanisms that generate rhythmic breathing.

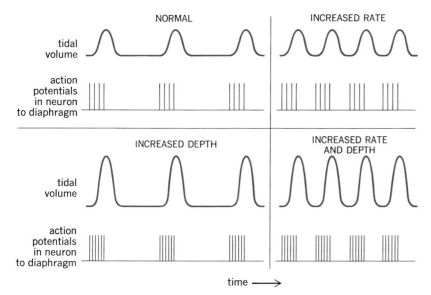

NORMAL

INCREASED RATE

tidal volume

action potentials in neuron to diaphragm

INCREASED DEPTH

INCREASED RATE AND DEPTH

tidal volume

action potentials in neuron to diaphragm

time ⟶

FIG. 23-31 Recordings of depths of respiration (tidal volume) and frequency of action potentials in motor neurons to the diaphragm. Note that the rate of discharge during the inspiratory "burst" determines the tidal volume, whereas the time between bursts determines the respiratory rate.

It is obvious that the actual respiratory rate is not fixed but can be altered over a wide range. Similarly, the depth and force of breathing movements can also be altered. As we have seen, these two factors, rate and depth, determine the alveolar ventilatory volume. Generally, rate and depth change in the same direction, although there may be important quantitative differences. For simplicity, we shall describe the control of total ventilation without attempting to discuss whether rate or depth makes the greatest contribution to the change.

Depth of respiration depends upon the number of motor units firing and their frequency of discharge, whereas respiratory rate depends upon the length of time that elapses between the bursts of motor unit activity (Fig. 23-31). As described above, the respiratory motor units are directly controlled by descending pathways from the medullary respiratory neurons. The efferent pathways for control of ventilation are therefore clear-cut, and the critical question becomes: What is the nature of the afferent input to these centers? In other words, what variables does the control of ventilatory volume regulate?

This may seem a ridiculously complex way of phrasing a question when the answer seems so intuitively obvious. After all, respiration ought to supply oxygen as fast as it is consumed and ought to excrete carbon dioxide as fast as it is produced. But how do the respiratory centers "know" what the body's oxygen requirements are? The logical

way to approach this question is to ask what detectable changes would result from imbalance of metabolism and ventilation. Certainly the most obvious candidates are the blood P_{O_2} and P_{CO_2}. Inadequate ventilation would lower the P_{O_2} because consumption would get ahead of supply and would elevate the P_{CO_2} because production would exceed elimination. Less obvious, perhaps, is the fact that arterial hydrogen-ion concentration also is exquisitely sensitive to changes in ventilation. Recall the equilibrium between H^+, HCO_3^-, and CO_2.

$$CO_2 + H_2O \rightleftharpoons H_2CO_3 \rightleftharpoons HCO_3^- + H^+$$

As described by the principle of mass action, any increase in carbon dioxide concentration drives this reaction to the right, which liberates additional hydrogen ions. Any increase or decrease in blood P_{CO_2} is accompanied by changes in blood hydrogen-ion concentration.

However, these statements of fact in no way prove that any of these blood concentrations actually are involved in respiratory regulation. The next step is to ask whether there are, indeed, receptors that can detect the levels of these variables in plasma and transmit the information to the medullary respiratory centers. If so, where are these receptors located and what contribution do they make to overall control of ventilation? We shall describe the answers to these questions first for oxygen and then for carbon dioxide and hydrogen ion.

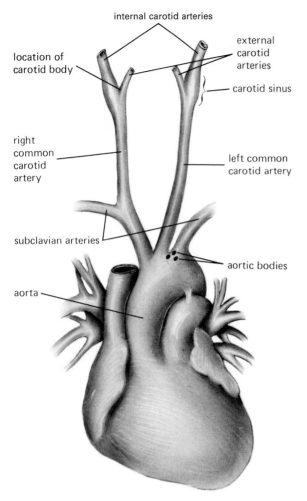

internal carotid arteries

external carotid arteries

location of carotid body

carotid sinus

right common carotid artery

left common carotid artery

subclavian arteries

aortic bodies

aorta

FIG. 23-32 Location of the carotid and aortic bodies. Each common carotid bifurcation contains a carotid sinus and a carotid body.

Control of Ventilation by Oxygen

The rationale of the experiments to be described is quite simple: Alteration of inspired P_{O_2} produces changes in blood P_{O_2}; if blood P_{O_2} is important in controlling ventilation, we should observe definite changes in ventilation. If a normal person takes a single breath of 100 percent oxygen (without being aware of it), after a latent period of about 8 sec, a transient reduction in ventilation occurs of approximately 10 to 20 percent. Such studies have demonstrated that the normal sea-level blood P_{O_2} of 100 mmHg exerts a tonic stimulatory effect adequate to account for approximately 20 percent of total ventilation. Thus, the increase in blood P_{O_2} produced

by the 100 percent oxygen removed this tonic stimulation and reduced ventilation. Conversely, a reduction in blood P_{O_2} below normal stimulates ventilation.

The receptors stimulated by low P_{O_2} are located at the bifurcation of the common carotid arteries and in the arch of the aorta, quite close to, but distinct from, the baroreceptors described in Chap. 22 (Fig. 23-32). Known as the **carotid** and **aortic bodies** or, simply, the **peripheral chemoreceptors,** they are composed of epithelial-like cells and neuron terminals in intimate contact with the arterial blood; of the two groups, the carotid bodies are far more important. The nerve fibers from each carotid body travel with the glossopharyngeal (IXth cranial) nerve and those from the aortic arch travel with the vagus (Xth cranial nerve); they enter the brainstem, where they synapse ultimately with the neurons of the medullary centers. A low P_{O_2} increases the rate at which the receptors discharge, which results in an increased number of action potentials traveling up the afferent nerve fibers and a stimulation of the medullary inspiratory neurons. If an animal is exposed to a low P_{O_2} after the afferent nerves from the carotid and aortic bodies have been cut, the usual rapid increase in alveolar ventilation is not observed, demonstrating that the immediate stimulatory effects of a low P_{O_2} are mediated via these pathways. (Indeed, in the absence of the carotid bodies, low P_{O_2} actually reduces ventilation through a direct depressive effect on the medullary respiratory neurons.)

What is the precise stimulus to these chemoreceptors? It is most likely their own internal P_{O_2}, that is, the concentration of dissolved oxygen within them. Normally, their blood supply is so huge relative to their utilization of oxygen that their internal P_{O_2} is virtually identical to the arterial P_{O_2}. For this reason we can state that, in effect, they monitor arterial P_{O_2}. Thus, any time arterial P_{O_2} is reduced by lung disease, hypoventilation, or high altitude, the internal chemoreceptor P_{O_2} will change similarly, the chemoreceptors will be stimulated and initiate a compensatory increase in ventilation.

Because the chemoreceptors respond to P_{O_2}, not to total blood oxygen content, they will not be stimulated in situations in which there are modest reductions in total oxygen content (e.g., moderate anemia) but no change in arterial P_{O_2}. However, the close correspondence between arterial P_{O_2} and internal-chemoreceptor P_{O_2} is eliminated whenever there is a very marked reduction in the total amount of oxygen that reaches the chemoreceptors, as in severe anemia or severe hypotension (which results in diminished blood supply to the receptors). In these situations, arterial P_{O_2} is still normal but chemoreceptor P_{O_2} falls because the total supply of oxygen to its cells is not adequate to maintain chemoreceptor P_{O_2} at the original arterial value. Accordingly, the chemoreceptors are stimulated and elicit increased ventilation; however, it must be re-emphasized that even in these cases, the receptors are responding directly to their own reduction in internal P_{O_2}, not to the reduction in total oxygen content of the blood. This same analysis holds true when total oxygen delivery is reduced by carbon monoxide, a gas that reacts with the same iron-binding sites on the hemoglobin molecule as does oxygen and has such a high affinity for these sites that even small amounts reduce the amount of oxygen combined with hemoglobin. Since carbon monoxide does not affect the amount of oxygen that can dissolve in blood, the arterial P_{O_2} is unaltered and no increase in peripheral chemoreceptor output occurs (until the carbon monoxide poisoning becomes very severe).

Control of Ventilation by Carbon Dioxide and Hydrogen Ion

The important role that CO_2 plays in the control of ventilation can be demonstrated by a simple experiment. Figure 23-33 illustrates the effects on respiratory volume of increasing the P_{CO_2} of inspired air. Normally, atmospheric air contains virtually no carbon dioxide. In the experiment illustrated, the subject breathed from bags of air that contained variable quantities of carbon dioxide. The presence of carbon dioxide in the inspired air caused an elevation of alveolar P_{CO_2} and thereby an elevation of arterial P_{CO_2} as well. This increased P_{CO_2} markedly stimulated ventilation, an increase of 5 mmHg in alveolar P_{CO_2} causing a 100 percent increase in ventilation. Obviously, ventilation is sensitive to acute changes in the P_{CO_2}.

Why can people hold their breath for only a relatively short time? The lack of ventilation causes an accumulation of carbon dioxide and increased blood P_{CO_2}. The ability of this increased P_{CO_2} to stimulate certain of the respiratory neurons is so

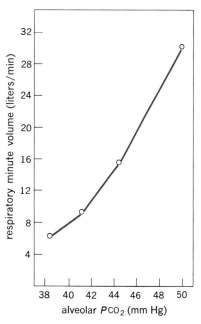

FIG. 23-33 Effects on respiration of increasing alveolar P_{CO_2} by addition of carbon dioxide to inspired air. The increase in respiratory volume is due to an increase in both tidal volume and respiratory rate. This study was performed on normal human subjects.

powerful that it overcomes the voluntary inhibition of respiration. Unfortunately, underwater swimmers have misguidedly made use of these facts. They voluntarily hyperventilate for several minutes before submerging thereby lowering their blood P_{CO_2} and enabling them to prolong their breath-holding time. This is a very dangerous procedure, particularly during exercise, when oxygen consumption is high; because of our relative insensitivity to oxygen deficits, a rapidly decreasing P_{O_2} may cause unconsciousness and drowning.

What is the mechanism and afferent pathway by which the P_{CO_2} controls ventilation? The evidence, to date, indicates that the effects of carbon dioxide on ventilation are due not to carbon dioxide itself but to the associated changes in hydrogen-ion concentration. For example, the stimulant effects of breathing mixtures that contain large amounts of carbon dioxide are probably due not to the effects of molecular carbon dioxide but to those of the increased hydrogen-ion concentration that results from the chemical reactions described above. This is why we stressed the relationship between P_{CO_2} and hydrogen-ion concentration.

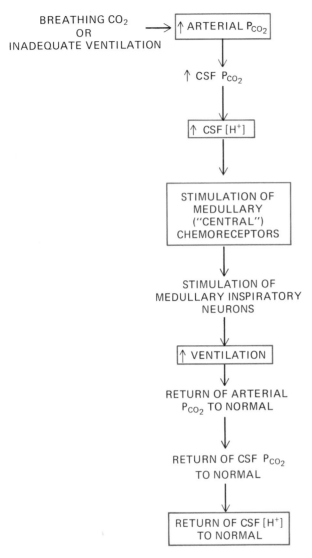

BREATHING CO_2
OR
INADEQUATE VENTILATION $\longrightarrow$ ↑ ARTERIAL P_{CO_2}

↑ CSF P_{CO_2}

↑ CSF [H^+]

STIMULATION OF
MEDULLARY
("CENTRAL")
CHEMORECEPTORS

STIMULATION OF
MEDULLARY INSPIRATORY
NEURONS

↑ VENTILATION

RETURN OF ARTERIAL
P_{CO_2} TO NORMAL

RETURN OF CSF P_{CO_2}
TO NORMAL

RETURN OF CSF [H^+]
TO NORMAL

FIG. 23-34 Negative-feedback control of cerebrospinal fluid hydrogen-ion concentration via regulation of ventilation.

There are two groups of hydrogen-ion receptors that participate in this reflex response. One is the same peripheral chemoreceptors—the carotid bodies—that initiated the hyperventilatory response by low P_{O_2}. Thus, these receptors are stimulated both by low P_{O_2} and high hydrogen-ion concentration. However, whereas they are the only important receptors in the low-oxygen reflex, they play only a minor role in the CO_2-dependent hydrogen-ion reflex. The major receptors for this latter reflex are the so-called **central chemoreceptors,** located in the brainstem medulla; they monitor the hydrogen-ion

concentration of the cerebrospinal fluid. The sequence of events is illustrated in Fig. 23-34: CO_2 diffuses rapidly across the membranes that separate arterial blood and brain tissue so that any increase in arterial P_{CO_2} causes a rapid identical increase in cerebrospinal fluid P_{CO_2}; this increased P_{CO_2}, by mass action, causes the generation of an increased number of hydrogen ions, i.e., an increased hydrogen-ion concentration. The latter stimulates the central chemoreceptors that, via synaptic connections, stimulate the medullary respiratory neurons to increase ventilation; the end result is a return of arterial and cerebrospinal fluid P_{CO_2} and hydrogen-ion concentration to normal.

One is left with the rather startling conclusion that the control of breathing (at least during rest) is aimed primarily at the regulation of *brain hydrogen-ion concentration!* However, this actually makes perfectly good sense, in terms of survival of the organism. As emphasized above, the relationship between carbon dioxide and hydrogen ion ensures that the regulation of hydrogen-ion concentration also produces relative constancy of P_{CO_2} as well. To continue the chain, the close relationship between oxygen consumption and carbon dioxide production ensures that ventilation adequate to maintain P_{CO_2} constant by excreting carbon dioxide as fast as it is produced also suffices to supply adequate oxygen (except in certain kinds of lung disease). Moreover, the shape of the hemoglobin dissociation curve minimizes any minor deficiency of oxygen supply, and any major oxygen deficit induces reflex respiratory stimulation via the carotid and aortic bodies. In contrast, brain function is extremely sensitive to changes in hydrogen-ion concentration so that even small increases or decreases in brain hydrogen-ion concentration could induce serious malfunction.

Throughout this section we have described the stimulatory effects of carbon dioxide on ventilation. It should also be noted that very high levels of carbon dioxide depress the entire central nervous system, including the respiratory centers, and may therefore be lethal. Closed environments, such as submarines and space capsules, must be designed so that carbon dioxide is removed as well as oxygen supplied.

Finally, it should be noted that the effects of increased P_{CO_2} (via changes in hydrogen-ion concentration), on the one hand, and decreased P_{O_2}, on the other, not only exist as independent inputs

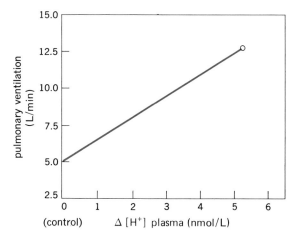

FIG. 23-35 Change in ventilation in response to an elevation of plasma hydrogen-ion concentration, produced by the administration of lactic acid.

to the medulla but manifest synergistic interactions as well; ventilatory response to combined hypoxia and increased P_{CO_2} is considerably greater than the sum of the responses to each alone.

Control of Ventilation by Changes in Arterial Hydrogen-Ion Concentration not due to Altered Carbon Dioxide

In the previous section we described how changes in hydrogen-ion concentration secondary to altered P_{CO_2} influence ventilation. However, there are many situations in which a change in hydrogen-ion concentration occurs as a result of some cause other than a primary change in P_{CO_2}. In contrast to their relative lack of importance in mediating CO_2-induced hydrogen-ion concentration changes, the peripheral chemoreceptors play the major role in stimulating ventilation whenever arterial hydrogen-ion concentration is changed by any means other than by elevated P_{CO_2}. For example, increased addition of lactic acid to the blood causes hyperventilation (Fig. 23-35) almost entirely by way of the peripheral chemoreceptors. The central chemoreceptors do not respond in this case because hydrogen ions penetrate the blood-brain barrier so poorly that cerebrospinal-fluid hydrogen-ion concentration is not increased appreciably by the administration of the lactic acid; this is in contrast to the ease with which changes in arterial P_{CO_2} produce equivalent changes in cerebrospinal-fluid hydrogen-ion concentration.

The converse of the above situation is also true; i.e., when arterial hydrogen-ion concentration is lowered by any means other than by a reduction of P_{CO_2} (for example, by loss of hydrogen ions from the stomach in vomiting), ventilation is inhibited because of decreased peripheral chemoreceptor output. The adaptive value such reflexes have in regulating arterial hydrogen-ion concentration is described in Chap. 24.

CONTROL OF VENTILATION DURING EXERCISE

During heavy exercise, the alveolar ventilation may increase ten to twentyfold to supply the additional oxygen needed and excrete the excess carbon dioxide products. On the basis of our three variables— P_{O_2}, P_{CO_2}, and hydrogen-ion concentration—it might seem easy to explain the mechanism that induces this increased ventilation. Unhappily, such is not the case.

Decreased P_{O_2} as the Stimulus?

It would seem logical that, as the exercising muscles consume more oxygen, blood P_{O_2} would de-

FIG. 23-36 Relation of ventilation, arterial gas pressures, and hydrogen-ion concentration to the magnitude of muscular exercise.

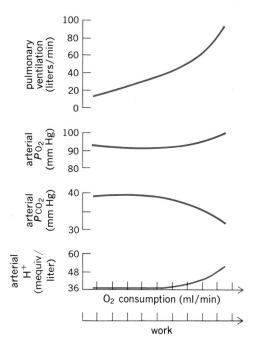

crease and stimulate respiration. But, in fact, arterial P_{O_2} is not significantly reduced during exercise (Fig. 23-36). The alveolar ventilation increases in exact proportion to the oxygen consumption; therefore, P_{O_2} remains constant. Indeed, in exhausting exercise, the alveolar ventilation may actually increase relatively more than oxygen consumption, resulting in an increased P_{O_2}.

Increased P_{CO_2} as the Stimulus?

This is virtually the same story. Despite the marked increase in carbon dioxide production, the precisely equivalent increase in alveolar ventilation excretes the carbon dioxide as rapidly as it is produced, and arterial P_{CO_2} remains constant (Fig. 23-36). Indeed, for the same reasons given above for oxygen, the arterial P_{CO_2} may actually decrease during exhausting exercise.

Increased Hydrogen Ion as the Stimulus?

Since the arterial P_{CO_2} does not change (or decreases) during exercise, there is no accumulation of excess hydrogen ion as a result of carbon dioxide accumulation. Although there is an increase in arterial hydrogen-ion concentration for quite a different reason, namely, generation and release into the blood of lactic acid and other acids during exercise, the changes in hydrogen-ion concentration are not nearly great enough, particularly in only moderate exercise, to account for the increased ventilation.

We are left with the fact that, despite intensive study for more than 100 years by many of the greatest respiratory physiologists, we do not know what input stimulates ventilation during exercise. Our big three—P_{O_2}, P_{CO_2}, and hydrogen ion—appear presently to be inadequate, but many physiologists still believe that they will ultimately be shown to be the critical inputs. They reason that the fact that P_{CO_2} remains constant during moderate exercise is very strong evidence that ventilation is actually controlled by P_{CO_2}. In other words, if P_{CO_2} were not the major controller, how else could it remain unchanged in the face of the marked increase in its production? They reason that there may be a change in sensitivity of the chemoreceptors to hydrogen ion so that changes undetectable by experimental methods might be responsible for the stimulation of ventilation.

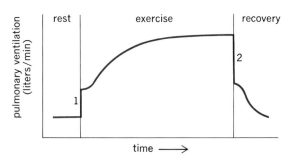

FIG. 23-37 Ventilation changes in time during exercise. Note the abrupt increase (1) at the onset of exercise and the equally abrupt but larger decrease (2) at the end of exercise.

Moreover, the problem is even more complicated; as shown in Fig. 23-37, there is an abrupt increase (within seconds) in ventilation at the onset of exercise and an equally abrupt decrease at the end. Clearly, these changes occur too rapidly to be explained by alteration of chemical constituents of the blood.

In Chap. 22, we described how the cardiovascular responses to exercise were probably mediated by input to the hypothalamus from branches of the neuron pathways that descend from the cerebral cortex to the motor nerves. It is likely that these fibers send branches to the medullary respiratory neurons as well; however, the magnitude of this contribution to stimulation of ventilation is not known.

CONTROL OF RESPIRATION BY OTHER FACTORS

Temperature

An increase in body temperature frequently occurs as a result of increased physical activity and contributes to stimulation of alveolar ventilation. This facilitation of the respiratory centers is probably due both to a direct physical effect of increased temperature upon the respiratory neurons and to stimulation via pathways from thermoreceptors of the hypothalamus.

Reflexes from Joints and Muscles

Many receptors in joints and muscles can be stimulated by the physical movements that accompany muscle contraction. It is quite likely that afferent pathways from these receptors play a significant role in stimulating respiration during exercise. Thus, the mechanical events of exercise help to coordinate alveolar ventilation with the metabolic requirements of the tissues. This input is usually considered to be a stimulus not only during steady-state exercise but to be the cause of the abrupt increase in ventilation at the onset of exercise; however, recent data suggest that the latter actually represents a conditioned (i.e., learned) response.

Protective Reflexes

A group of responses, most familiar being the cough and the sneeze, protect the respiratory tract against irritant materials. These reflexes originate in receptors that line the respiratory tract. When they are excited, the result is stimulation of the medullary respiratory centers in such a manner as to produce a deep inspiration and a violent expiration. In this manner, particles can be literally exploded out of the respiratory tract. The cough reflex is inhibited by alcohol, which may contribute to the susceptibility of alcoholics to choking and pneumonia. Another example of a protective reflex is the immediate cessation of respiration that is frequently triggered when noxious agents are inhaled.

Voluntary Control of Respiration

Although we have discussed in detail the involuntary nature of most respiratory reflexes, it is quite obvious that we retain considerable voluntary control of respiratory movements. This is accomplished by descending pathways from the cerebral cortex to the motor neurons that supply the intercostal muscles and diaphragm. As we have seen, this voluntary control of respiration cannot be maintained when the involuntary stimuli, such as an elevated P_{CO_2} or hydrogen-ion concentration, become intense. Besides the obvious forms of voluntary control, e.g., breath holding, respiration must also be controlled during the production of complex voluntary actions such as speaking and singing.

The Heimlich Maneuver

A sudden, sharp increase in alveolar pressure caused by a forceful elevation of the diaphragm forms the basis of the **Heimlich maneuver,** which is

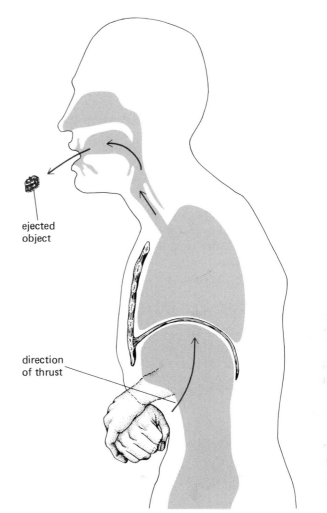

ejected object

direction of thrust

FIG. 23-38 The Heimlich maneuver. The rescuer's fist forms a knob, which is placed against the victim's abdomen. A quick upward thrust causes elevation of the diaphragm and a forceful expiration. As air is forced through the trachea and larynx, the foreign object in the airway is expelled.

used to aid persons who are choking on foreign matter caught in the respiratory tract. The sudden decrease in thoracic volume, which is responsible for the increased alveolar pressure, is produced as the rescuer's fist, which has been placed against the victim's abdomen slightly above the navel and well below the xiphoid process of the sternum, is pressed into the abdomen with a quick upward thrust (Fig. 23-38). The forceful expiration thus produced expels the object caught in the respiratory tract and causing choking. The Heimlich maneuver can be performed on persons who are standing, sitting, or lying down; it can even be performed by the victim.

MAXIMAL OXYGEN UPTAKE

When a person is subject to progressively increasing work loads, there is a linear increase in the amount of oxygen taken up by the muscle cells until a point is reached beyond which a further increase in work load does not result in further oxygen uptake. This is defined as the **maximal oxygen uptake.** Work loads greater than this can be achieved only for very short periods of time because the muscles are operating by anaerobic glycolysis to a large extent, as manifested by a large increase in lactic acid production. What sets this upper limit for oxygen uptake and therefore work? In a normal person it is usually not set by any "saturation" of intracellular oxidative reactions, i.e., by an inability of the muscle cells to use more oxygen. Rather, it is set by the maximal ability of the cardiovascular-respiratory systems to provide oxygen, specifically by the inability to increase cardiac output beyond a certain point.

What limits cardiac output during exercise? The answer is the interaction between heart rate and stroke volume. Stroke volume increases with work load, although not to the same degree as does heart rate, but then decreases from maximal values at work loads beyond maximal oxygen uptake. The major factors responsible for this decrease are the very rapid heart rate (which decreases diastolic filling time) and failure of the peripheral factors favoring venous return (muscle pump, respiratory pump, venoconstriction, arteriolar dilation; see Chap. 21) to elevate venous pressure high enough to maintain adequate ventricular filling during the very short time available. The net result is a decrease in end-diastolic volume that causes, by Starling's law, a decrease in stroke volume.

HYPOXIA

Hypoxia is defined as a deficiency of oxygen or oxygen utilization at the tissue level. Potential

TABLE 23-4   SOME CAUSES OF HYPOXIC HYPOXIA
1 Decreased P_{O_2} in inspired air (high altitude)*
2 Hypoventilation
Increased airway resistance (foreign body, asthma)
Paralysis of respiratory muscles (poliomyelitis)
Skeletal deformities
Inhibition of medullary respiratory centers (morphine)
Decreased "stretchability" of the lungs and thorax (deficient surfactant, thickened lung or chest tissues)
Lung collapse (pneumothorax)
3 Deficient alveolar-capillary diffusion
Decreased area for diffusion (pneumonia)
Thickening of alveolar-capillary membranes (beryliosis)
4 Abnormal matching of ventilation and blood flow (emphysema)

* The phrases in parentheses are specific examples in each category.

causes of hypoxia can be classed in four general categories: (1) *hypoxic hypoxia,* in which the arterial P_{O_2} is reduced; (2) *anemic hypoxia,* in which the arterial P_{O_2} is normal but the total oxygen content of the blood is reduced because of inadequate numbers of red cells, deficient or abnormal hemoglobin, or competition for the hemoglobin molecule by carbon monoxide; (3) *ischemic hypoxia,* in which the basic defect is too little blood flow to the tissues; and (4) *histotoxic hypoxia,* in which the quantity of oxygen that reaches the tissue is normal but the cell is unable to utilize the oxygen because a toxic agent (cyanide, for example) has interfered with the cell's metabolic machinery.

Clearly, anemic hypoxia is mainly the result of abnormal blood formation or destruction, ischemic hypoxia is caused by failure of the cardiovascular system, and histotoxic hypoxia represents malfunction of the tissue cells themselves. In contrast to these three categories, hypoxic hypoxia is a result of either high-altitude exposure or malfunction of the respiratory system. The causes are many, and a list of certain of the major ones provides an excellent review of the interplay between the various components of the respiratory system described in this chapter (Table 23-4).

KEY TERMS

respiration	expiration	alveolar ventilation
respiratory system	atmospheric pressure	surface tension
alveoli	alveolar pressure	respiratory quotient (RQ)
conducting airways	intrapleural pressure	partial pressure
respiratory airways	tidal volume	oxygen-hemoglobin dissociation
surfactant	residual volume	inspiratory neurons
diaphragm	total pulmonary ventilation	carotid bodies
intercostal muscles	vital capacity	(peripheral chemoreceptors)
inspiration	dead space	central chemoreceptors
		maximal oxygen uptake

REVIEW EXERCISES

Section A

1 Define respiration (both meanings), respiratory system, ventilation, inspiration, and expiration.
2 Diagram the respiratory system and label the nasal passages, nasal, oral, and laryngeal portions of the pharynx, larynx, trachea, bronchi, diaphragm, parietal and visceral pleura, and intrapleural space.
3 List the five major functions of the conducting portion of the airways.
4 State the protective roles of the cilia, mucus, and phagocytic cells.
5 Diagram an alveolus and label the alveolar epithelial cell, capillary, capillary endothelial cell, interstitial cell, and erythrocyte.
6 Explain the difference between the pulmonary and bronchial blood supplies to the lung.

Section B

1 Define ventilation, bulk flow, atmospheric pressure, and alveolar pressure.
2 Describe the mechanical events that occur in the respiratory cycle and the air movements that result from them.
3 Describe the relation between bulk flow, atmospheric pressure, alveolar pressure, and airway resistance.
4 Define intrapleural pressure; describe and explain its changing values during the respiratory cycle.
5 Describe the changes in alveolar pressure during the respiratory cycle.
6 Define pneumothorax and intrathoracic pressure.
7 Explain the role of the diaphragm and intercostal muscles during a complete respiratory cycle.
8 List the determinants of airway resistance.

9 Define tidal volume, inspiratory reserve, expiratory reserve, residual volume, and vital capacity.
10 Given the appropriate data, explain how one calculates total pulmonary ventilation and alveolar ventilation.
11 Define anatomic and alveolar dead space.
12 State the determinants of the work of breathing.
13 Define surfactant and describe its role.

Section C

1 Define respiratory quotient, gas pressure, and partial pressure.
2 List the determinants of gas pressure.
3 List the determinants of the uptake of gases by a liquid.
4 Define diffusion; contrast bulk flow and diffusion.
5 State the P_{O_2} of alveolar air, expired air, and atmospheric air; explain why they are different. Do the same for P_{CO_2}.
6 Explain why there is little variation in alveolar gas composition throughout the respiratory cycle.
7 List the P_{O_2} in arterial and venous blood; explain why they are different. Do the same for P_{CO_2}.
8 Define hemoglobin and describe its role in gas transport.
9 Define oxyhemoglobin, reduced hemoglobin, hemoglobin saturation, and dissolved blood gases.
10 State two ways oxygen is carried in the blood and the relative importance of each.
11 Draw an oxygen-hemoglobin dissociation (hemoglobin saturation) curve, label the axes, and mark the normal values for arterial and venous blood.
12 State the effects of changing temperature, acidity, and DPG concentration on the curve.

13 Define maximal oxygen uptake and explain how it affects exercise capacity.

14 Define carbamino compound and carbonic anhydrase.

15 Describe carbon dioxide transport in the blood; include a discussion of the role of carbonic anhydrase.

16 State the relation between carbon dioxide and hydrogen-ion concentration of the blood.

Section D

1 State the type of muscle that constitutes the respiratory musculature and describe its innervation.

2 Describe the medullary respiratory centers, their patterns of activity during the respiratory cycle, and the inputs that influence them.

3 Describe the control of respiration by oxygen, including the receptors, afferent and efferent pathways, integrating center, effector organs, and responses to changes in P_{O_2}.

4 Distinguish between the peripheral and central chemoreceptors.

5 Describe the control of ventilation by carbon dioxide and hydrogen ion, including the receptors, afferent and efferent pathways, integrating center, effector organs, and responses to changes in P_{CO_2} and hydrogen-ion concentration.

6 Describe the relationship between P_{O_2} and P_{CO_2}–hydrogen-ion concentration in the control of ventilation.

7 List additional factors that control respiration; state the effect of each of these stimuli upon respiration.

8 Describe the ventilatory response to exercise, breath holding, altitude, and the Heimlich maneuver.

9 List the four general categories of hypoxia and define hypoxic, anemic, ischemic, and histotoxic hypoxia.

CHAPTER

THE URINARY SYSTEM: REGULATION OF WATER AND ELECTROLYTES

SECTION A: STRUCTURE OF THE URINARY SYSTEM AND BASIC PRINCIPLES OF RENAL PHYSIOLOGY

The **urinary system** is responsible for the formation of urine and its conveyance out of the body. Urine is formed by the **kidneys.** It then drains from each kidney via a large duct, the **ureter,** which leads to the **urinary bladder,** the reservoir for urine (Fig. 24-1). The **urethra,** the last component of the urinary system, carries urine from the bladder to the outside. In males, the urethra also serves as the passageway for semen.

STRUCTURE OF THE KIDNEY AND URINARY SYSTEM

The kidneys are paired organs that lie in the posterior abdominal wall behind the abdominal cavity (retroperitoneal), one on each side of the vertebral column at the level of the T12 to L3 vertebrae. The right kidney usually lies a little lower than the left, possibly because the bulk of the liver is on the right side just above the right kidney. Each kidney is surrounded by a mass of fatty connective tissue, the *perirenal fat.* The medial border of the kidney is indented by a deep fissure called the *hilum,* through which pass the renal vessels and nerves and in which lies the funnel-shaped continuation of the upper end of the ureter, the **renal pelvis** (Fig. 24-2). The outer convex border of the renal pelvis is divided into **major calyces,** each of which subdivides into several **minor calyces.** The minor calyces are cupped around the projecting apexes of the **renal papillae.**

The interior of each kidney is divided into two zones: an inner **renal medulla** and an outer **renal cortex.** The medulla consists of a number of cone-shaped masses (the **renal pyramids**) whose apexes form the papillae (Fig. 24-2). Each pyramid, topped by a region of the renal cortex, forms a **lobe** of the kidney. Approximately eight such lobes are present in each kidney. That part of the renal cortex that dips down between adjacent pyramids forms the **renal columns.**

In human beings, each kidney is composed of approximately 1 million tiny, closely packed units, all similar in structure and function, bound together by small amounts of connective tissue that contains blood vessels, nerves, and lymphatics. One such unit, or **nephron,** is shown in Fig. 24-3. The nephron consists of a vascular component (the **glomerulus**) and a tubular component. The mechanisms by which the kidneys perform their functions depend on the relationship between these two components.

Throughout its course, the **tubule** is composed of a single layer of epithelial cells that differ in structure and function along its length. It originates as a blind sac, known as **Bowman's capsule,** which is lined with thin epithelial cells. On one side, Bowman's capsule is intimately associated with the glomerulus; on the other it opens into the tubule's first portion, which is highly coiled and is known as the **proximal convoluted tubule.** The next portion of the tubule is a sharp hairpinlike loop, called **Henle's loop.** The tubule once more becomes coiled (the **distal convoluted tubule**) and finally runs a straight course as the **collecting duct.** From Bowman's capsule to the beginning of the collecting duct, each of the million tubules is completely separate from its neighbors. The tiny collecting ducts

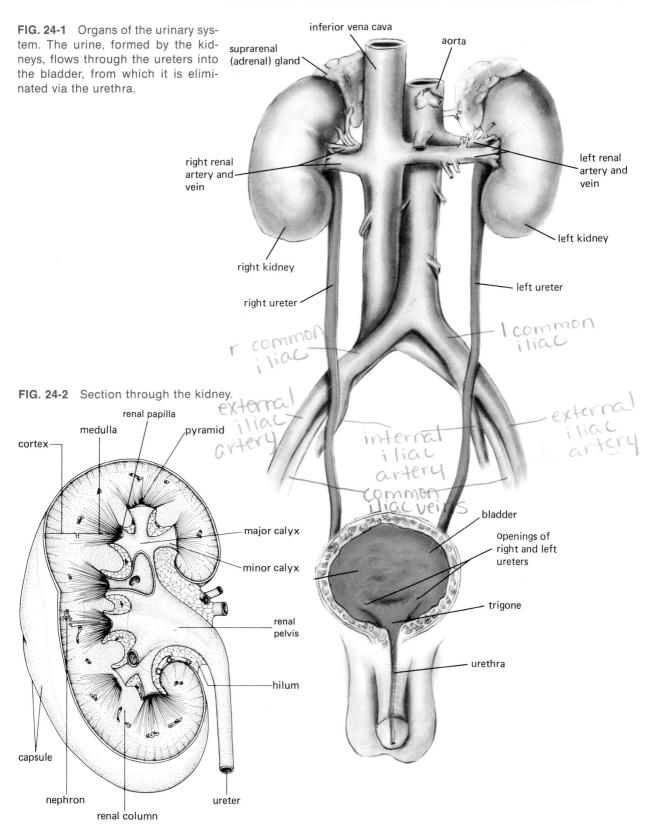

FIG. 24-1 Organs of the urinary system. The urine, formed by the kidneys, flows through the ureters into the bladder, from which it is eliminated via the urethra.

inferior vena cava

aorta

suprarenal (adrenal) gland

right renal artery and vein

left renal artery and vein

left kidney

right kidney

left ureter

right ureter

r common iliac

l common iliac

external iliac artery

external iliac artery

FIG. 24-2 Section through the kidney.

internal iliac artery

common iliac veins

renal papilla

medulla

pyramid

cortex

major calyx

minor calyx

bladder

openings of right and left ureters

renal pelvis

trigone

hilum

capsule

urethra

nephron

renal column

ureter

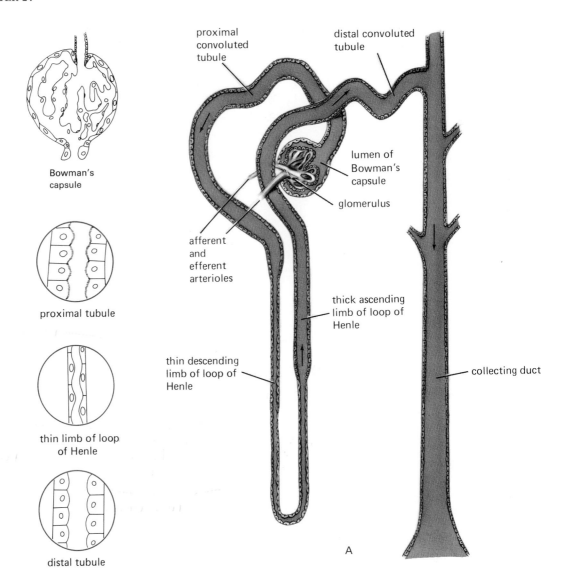

Bowman's capsule

proximal tubule

thin limb of loop of Henle

distal tubule

proximal convoluted tubule

distal convoluted tubule

lumen of Bowman's capsule

glomerulus

afferent and efferent arterioles

thick ascending limb of loop of Henle

thin descending limb of loop of Henle

collecting duct

A

from separate tubules join to form larger ducts, which in turn form even larger ducts, which finally empty into the large central cavities of the renal pelvis (Fig. 24-2). The urine is not altered after it leaves the collecting ducts. From the renal pelvis on, the remainder of the urinary system serves simply as plumbing.

To return to the other component of the nephron: What are the origin and nature of the glomerulus? Blood is supplied to the kidneys through the **renal arteries,** which arise on each side of the aorta. Each renal artery divides before entering the substance of the kidney, forming the *interlobar arteries*. These ascend in the renal columns between the pyramids until they reach the bases of the pyramids, where

they give off many *arcuate arteries,* which run horizontally along the region of the corticomedullary junction (Fig. 24-4). The *interlobular arteries* arise from the arcuates, and each of these gives off, at right angles to itself, a series of arterioles, the **afferent arterioles,** each of which leads to a compact tuft of capillaries. This tuft of capillaries is the glomerulus, which protrudes into Bowman's capsule and is completely covered by the epithelial lining of the capsule (Fig. 24-5). The functional significance of this anatomic arrangement is that blood in the glomerulus is separated from the space within Bowman's capsule by only a thin layer of tissue composed of (1) the single-celled capillary lining, (2) a layer of basement membrane, and (3) the single-celled lining

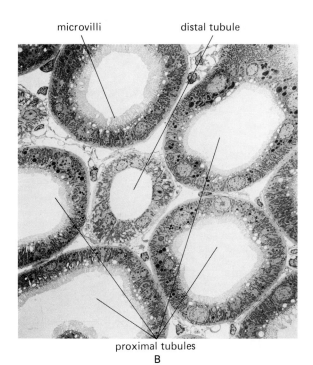

microvilli   distal tubule

proximal tubules

B

FIG. 24-3 (A) Basic structure of a nephron (at left). (B) Cross section through proximal and distal tubules of a mammalian kidney.

of Bowman's capsule. Filtration of fluid occurs from the capillaries through this thin barrier into Bowman's capsule.

The glomerular capillaries instead of leading into veins, recombine to form another set of arterioles, the **efferent arterioles** through which blood leaves the glomerulus. Each efferent arteriole soon divides into a second set of capillaries (Fig. 24-4), the **peritubular capillaries,** which are profusely distributed to, and intimately associated with, all the remaining portions of the tubule. They rejoin to form the interlobular veins that end in the arcuate veins. These drain through the interlobar veins into the **renal veins,** by which blood ultimately leaves the kidney.

There are important regional differences in the location of the various tubular and vascular components, as shown in Fig. 24-4. The cortex contains all the glomeruli, proximal and distal convoluted tubules, and the first portions of the loops of Henle and collecting ducts. In human beings, approximately 85 percent of the nephrons originate in glomeruli located in the outer, or *superficial,* cortex and have relatively short loops of Henle, which extend only into the outer medulla. The remaining 15 percent originate in glomeruli located in the in-

nermost, or *juxtamedullary,* cortex (i.e., "cortex just adjacent to the medulla"). These *juxtamedullary nephrons,* in contrast to the *superficial nephrons,* have long loops that extend deep into the medulla, where they run parallel to the collecting ducts. The vascular organization of the juxtamedullary nephrons also differs in that the efferent arterioles drain not only into the usual peritubular-capillary network but also into thin hairpin-loop vessels (*vasa recta*), which run parallel to the loops of Henle and collecting ducts in the medulla. This arrangement has considerable significance for renal function, as will be described later.

One last anatomic feature should be pointed out. Note that, as each loop of Henle ascends into the cortex to become a distal tubule, the tubule contacts the arterioles that supply its nephron of origin (Fig. 24-3). This area of contact is marked by unique structural changes in both the arterioles and tubules and is known as the **juxtaglomerular apparatus.** Its structure and function will also be discussed later.

THE BALANCE CONCEPT

The major task achieved by the kidneys' formation of urine is regulation of the salt and water composition of the internal environment. A cell's function depends not only upon receiving a continuous supply of organic nutrients and eliminating its metabolic end products but also upon the existence of stable physicochemical conditions in the extracellular fluid that bathes it. Among the most important substances that contribute to these conditions are water, sodium, potassium, calcium, and hydrogen ion. This chapter is devoted to a discussion of the mechanisms by which the total amounts of these substances in the body and their concentrations in the extracellular fluid are maintained relatively constant, mainly as a result of the kidney's activities.

A substance appears in the body either as a result of ingestion or as a product of metabolism. Conversely, a substance can be excreted from the body or consumed in a metabolic reaction. Therefore, if the quantity of any substance in the body is to be maintained at a constant level over a period of time, the total amounts ingested and produced must equal the total amounts excreted and consumed. This is a general statement of the **balance concept.** For water and hydrogen ion, all four possible pathways apply; however, for the mineral electrolytes such

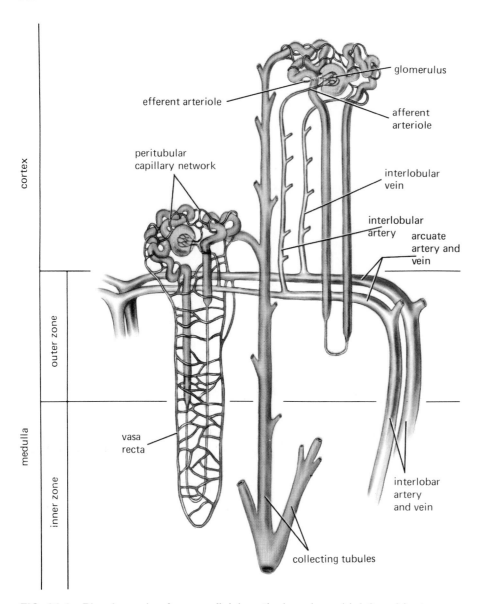

cortex

medulla

outer zone

inner zone

glomerulus

efferent arteriole

afferent arteriole

peritubular capillary network

interlobular vein

interlobular artery

arcuate artery and vein

vasa recta

interlobar artery and vein

collecting tubules

FIG. 24-4   Blood supply of a superficial cortical nephron (right) and juxta-medullary nephron (left). For simplicity, we have shown that the efferent arteriole from a glomerulus supplies only the nephron associated with that glomerulus, but such is not the case in fact; generally each nephron is perfused by capillaries from glomeruli of many other nephrons as well.

as sodium and potassium, balance is simpler because they are neither synthesized nor consumed by cells, and their total body balance thus reflects only ingestion versus excretion.

As an example, let us describe the balance for total body water (Table 24-1). It should be recognized that these are average values, which are subject to considerable variation. The two sources of

body water are metabolically produced water, resulting largely from the oxidation of organic nutrients, and ingestion of water in liquids and so-called solid food (a rare steak is approximately 70 percent water). There are four sites from which water is lost to the external environment: skin, lungs, gastrointestinal tract, and kidneys. (Menstrual flow constitutes a fifth source of water loss.)

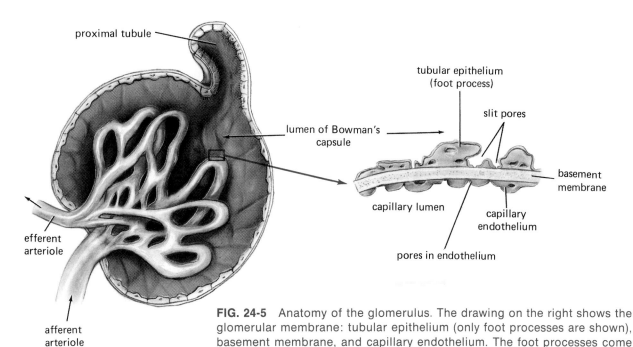

FIG. 24-5 Anatomy of the glomerulus. The drawing on the right shows the glomerular membrane: tubular epithelium (only foot processes are shown), basement membrane, and capillary endothelium. The foot processes come together at the cell body like the tentacles of an octopus.

The loss of water by evaporation from the cells of the skin and the lining of respiratory passageways is a continuous process, often referred to as **insensible loss** because the person is unaware of its occurrence. Additional water can be made available for evaporation from the skin by the production of sweat. The normal gastrointestinal loss of water (in feces) is quite small but can be severe in vomiting or diarrhea.

TABLE 24-1	NORMAL ROUTES OF WATER GAIN AND LOSS IN ADULTS	
		Milliliters per Day
Intake:		
In liquids		1200
In food		1000
Metabolically produced		350
Total		2550
Output:		
Insensible loss (skin and lungs)		900
Sweat		50
In feces		100
Urine		1500
Total		2550

Under normal conditions, as can be seen from Table 24-1, water loss exactly equals water gain, and no net change of body water occurs. This is no accident but the result of precise regulatory mechanisms. The question then is: Which processes involved in water balance are controlled to make the gains and losses balance? The answer, as we shall see, is voluntary intake (thirst) and urinary loss. This does not mean that none of the other processes is controlled but that their control is not primarily oriented toward water balance. Catabolism of organic nutrients, the major source of the water of oxidation, is controlled by mechanisms directed toward regulation of energy balance. Sweat production is controlled by mechanisms directed toward temperature regulation. Insensible loss (in human beings) is truly uncontrolled, and fecal water loss is generally quite small and unchanging.

The mechanism of thirst is certainly of great importance, since body deficits of water, regardless of cause, can be made up only by ingestion of water, but it is also true that our fluid intake is often influenced more by habit and sociological factors than by the need to regulate body water. The control of urinary water loss is the major mechanism by which body water is regulated.

By similar analysis, we find that the body balances of most of the ions in the extracellular fluid are regulated primarily by the kidneys. To appreciate the importance of these kidney regulations, one need only make a partial list of the more important simple inorganic substances that constitute the internal environment and which are regulated by the kidney: water, sodium, potassium, chloride, calcium, magnesium, sulfate, phosphate, and hydrogen ion. It is worth repeating that normal biological processes depend on the constancy of this internal environment, the implication being that the amounts of these substances must be held within very narrow limits, regardless of large variations in intake and abnormal losses that result from disease (hemorrhage, diarrhea, vomiting, etc.). Indeed, the extraordinary number of substances that the kidney regulates and the precision with which these processes normally occur accounted for the kidney's being the last stronghold of the nineteenth-century vitalists, who simply would not believe that the laws of physics and chemistry could fully explain renal function. By what mechanism does urine flow rapidly increase when a person ingests several glasses of liquid? How is it that the patient on an extremely low salt intake and the heavy salt eater both urinate precisely the amounts of salt required to maintain their sodium balance? What mechanisms decrease the urinary calcium excretion of children deprived of milk?

This regulatory role is obviously quite different from the popular conception of the kidneys as glorified garbage disposal units that rid the body of assorted wastes and "poisons." It is true that the complex chemical reactions that occur in cells result ultimately in end products collectively called waste products; e.g., the catabolism of protein produces approximately 30 g of urea per day. Other waste substances produced in relatively large quantities are uric acid (from nucleic acids), creatinine (from muscle creatine), and the end products of hemoglobin breakdown. There are many others, not all of which have been completely identified. Most of these substances are eliminated from the body as rapidly as they are produced, primarily by way of the kidneys. Some of these waste products are harmless, although the accumulation of certain of them during periods of renal malfunction accounts for some of the disordered body functions in severe kidney disease. However, many of the problems that occur in renal disease are due simply to dis-

ordered water and electrolyte balance.

The kidneys have another excretory function that is presently assuming increased importance, namely, the elimination from the body of foreign chemicals (drugs, pesticides, food additives, etc.). A final kidney function is the formation of at least three substances that are components of hormonal systems: erythropoietin (Chap. 19), renin, and the active form of vitamin D (to be discussed later in this chapter).

BASIC RENAL PROCESSES

Urine formation begins with the filtration of essentially protein-free plasma through the glomerular capillaries into Bowman's capsule. This **glomerular filtrate** contains all low-molecular-weight substances in virtually the same concentrations as plasma. The final urine that enters the renal pelvis is quite different from the glomerular filtrate because, as the filtered fluid flows from Bowman's capsule through the remaining portions of the tubule, its composition is altered. This change occurs by two general processes, tubular reabsorption and tubular secretion. The tubule is at all points intimately associated with the peritubular capillaries, a relationship that permits transfer of materials between peritubular plasma and the inside of the tubule (*tubular lumen*). When the direction of transfer is from tubular lumen to peritubular capillary plasma, the process is called **tubular reabsorption.** Movement in the opposite direction, i.e., from peritubular plasma to tubular lumen, is called **tubular secretion.** (This term must not be confused with **excretion;** to say that a substance has been excreted is to say only that it appears in the final urine.) These relationships are illustrated in Fig. 24-6.

The most common relationships between these basic renal processes—glomerular filtration, tubular reabsorption, and tubular secretion—are shown in Fig. 24-7. Plasma that contains substances X, Y, and Z enters the glomerular capillaries. A certain quantity of protein-free plasma that contains these substances is filtered into Bowman's capsule, enters the proximal tubule, and begins its flow through the rest of the tubule. The remainder of the plasma, also containing X, Y, and Z, leaves the glomerular capillaries via the efferent arteriole and enters the peritubular capillaries. The cells that compose the

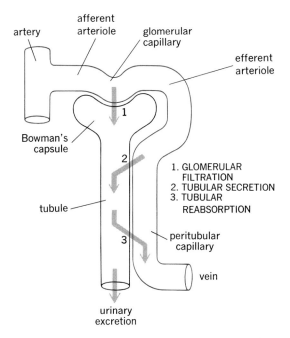

FIG. 24-6   Three basic components of renal function.

tubular epithelium can actively transport X (not Y or Z) from the peritubular plasma into the tubular lumen, but not in the opposite direction. By this combination of filtration and tubular secretion all the plasma that originally entered the renal artery

is cleared of substance X, which leaves the body via the urine, which reduces the amount of X remaining in the body. If the tubule were incapable of reabsorption, the Y and Z originally filtered at the glomerulus would also leave the body via the urine, but the tubule can transport Y and Z from the tubular lumen back into the peritubular plasma. The amount of reabsorption of Y is small, so that most of the filtered material does escape from the body, but for Z the reabsorptive mechanism is so powerful that virtually all the filtered material is transported back into the plasma, which flows through the renal vein back into the vena cava. Therefore no Z is lost from the body. Hence the processes of filtration and reabsorption have canceled each other out, and the net result is as though Z had never entered the kidney at all.

The kidney operates only on plasma (the erythrocytes supply oxygen to the kidney but serve no other function in urine formation). Each substance in plasma is handled in a characteristic manner by the nephron, i.e., by a particular combination of filtration, reabsorption, and secretion. The critical point is that the rates at which the relevant processes proceed for many of these substances are subject to physiologic control. What is the effect, for example, if the Y filtration rate is increased or its reabsorption rate decreased? Either change

FIG. 24-7   Renal manipulation of three substances X, Y, and Z. X is filtered and secreted but not reabsorbed. Y is filtered, and a fraction is then reabsorbed. Z is filtered and is completely reabsorbed.

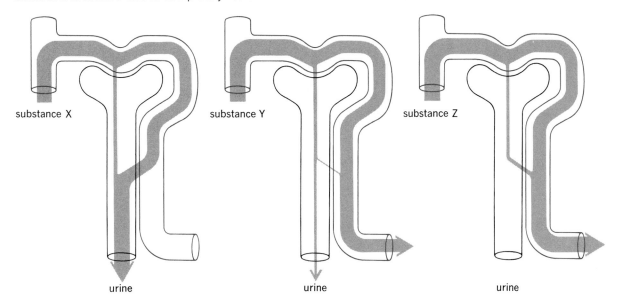

means that more Y is lost from the body via the urine. By triggering such changes in filtration or reabsorption whenever the plasma concentration of Y rises above normal, homeostatic mechanisms regulate plasma Y.

Glomerular Filtration

As described in Chap. 21, capillaries are freely permeable to water and small solutes. They are relatively impermeable to large molecules (colloids), the most important of which are the plasma proteins. The barrier offered by the glomerulus and adherent epithelium of Bowman's capsule behaves qualitatively like any capillary; accordingly, the glomerular filtrate, i.e., the fluid within Bowman's capsule, is essentially protein-free and contains all low-molecular-weight substances in virtually the same concentrations as the plasma. This may seem surprising since the glomerular barrier is structurally different from most other capillaries in that it is composed of three layers: capillary endothelium, basement membrane, and a single-celled layer of epithelial cells. These epithelial cells (*podocytes*) have an unusual octopuslike structure in that they have a large number of extensions, or foot processes, that rest on the basement membrane (Figs. 24-5 and 24-8). Slits exist between adjacent foot processes and probably constitute the path through which the filtrate, once past the endothelial cells

and basement membrane, travels to enter Bowman's capsule. The presence of this epithelial layer may account for certain quantitative differences between the glomerulus and other capillaries.

The hydrostatic pressure of the blood in the glomerular capillaries is less than the mean blood pressure in the large arteries of the body (approximately 100 mgHg), since pressure is dissipated as the blood passes through the arterioles that connect the renal artery branches to the glomeruli. The glomerular capillary pressure is usually about 50 mmHg. This is about half of mean arterial pressure and is considerably higher than in other capillaries of the body because the afferent arterioles that lead to the glomeruli are wider than most arterioles and therefore offer less resistance to flow.

The capillary hydrostatic pressure favoring filtration is not completely unopposed. There is, of course, fluid in Bowman's capsule that results in a capsular hydrostatic pressure of 10 mmHg resisting further filtration into the capsule. A second opposing force results from the presence of protein in the plasma and its absence in Bowman's capsule. As in other capillaries, this unequal distribution of protein causes the water concentration of the plasma to be less than that of the fluid in Bowman's capsule. Again, as in other capillaries, the water-concentration difference is due completely to the plasma protein since all the low-molecular-weight solutes have virtually identical concentrations in the plasma and

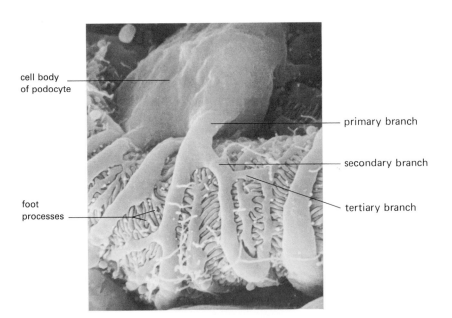

cell body of podocyte

foot processes

primary branch

secondary branch

tertiary branch

FIG. 24-8 A scanning electron micrograph of the cell body of a podocyte and its primary, secondary, and tertiary branches. The terminal branches (foot processes) of adjacent cells interdigitate, separated only by narrow slits through which fluid filters from the plasma to Bowman's capsule.

TABLE 24-2	FORCES INVOLVED IN GLOMERULAR FILTRATION	
Forces		Millimeters of mercury (mmHg)
Favoring filtration:		
Glomerular capillary blood pressure		50
Opposing filtration:		
Fluid pressure in Bowman's capsule		10
Osmotic gradient		30
(water-concentration difference due to protein)		
Net filtration pressure		10

Bowman's capsule. The difference in water concentration induces an osmotic flow of fluid (water plus all the low-molecular-weight solutes) from Bowman's capsule into the capillary, a flow that opposes filtration. Its magnitude is equivalent to the bulk flow produced by a hydrostatic pressure difference of 30 mmHg.

As can be seen from Table 24-2, the net filtration pressure is approximately 10 mmHg. This pressure initiates urine formation by forcing an essentially protein-free filtrate of plasma through glomerular pores into Bowman's capsule and thence down the tubule into the renal pelvis. The glomerular membranes serve only as a filtration barrier and play no active, i.e., energy-requiring, role. The energy that produces glomerular filtration is the energy transmitted to the blood (as hydrostatic pressure) when the heart contracts.

Before leaving this topic, we must point out the reason for use of the term "essentially protein-free" in describing the glomerular filtrate. In reality there is a very small amount of protein in the filtrate since the glomerular membranes are not perfect sieves for protein. Normally, less than 1 percent of serum albumin and no globulin are filtered; whatever protein is filtered is completely reabsorbed so that no protein appears in the final urine. However, in diseased kidneys, the glomerular membranes may become much more leaky to protein so that large quantities are filtered and some of this protein appears in the urine.

Rate of Glomerular Filtration (GFR)   In human beings, the average volume of fluid filtered from the plasma into Bowman's capsule is 180 L/day (ap-

proximately 45 gal)! The implications of this remarkable fact are extremely important. When we recall that the average total volume of plasma is approximately 3 L, it follows that the entire plasma volume is filtered by the kidneys some 60 times a day. It is, in part, this ability to process such huge volumes of plasma that enables the kidneys to excrete large quantities of waste products and to regulate the constituents of the internal environment so precisely. The second implication concerns the magnitude of the reabsorptive process. The average person excretes between 1 and 2 L of urine per day. Since 180 L of fluid is filtered, approximately 99 percent of the filtered water must have been reabsorbed into the peritubular capillaries, the remaining 1 percent escaping from the body as urinary water.

To complete this discussion of glomerular filtration we must consider the magnitude of the **total renal blood flow.** We have seen that none of the red blood cells and only a portion of the plasma that enters the glomerular capillaries are filtered into Bowman's capsule and that the remainder passes via the efferent arterioles into the peritubular cap-

Fig. 24-9 Magnitude of glomerular filtration rate, total renal plasma flow, and total renal blood flow. Only 20 percent of the plasma that enters the kidneys is filtered from the glomerulus into Bowman's capsule. The remaining 80 percent flows through the glomerulus into the efferent arteriole and thence into the peritubular capillaries.

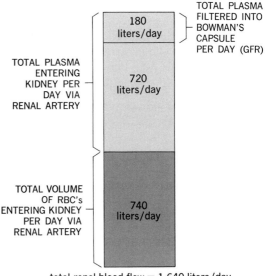

total renal blood flow = 1,640 liters/day

illaries. Normally, the glomerular filtrate constitutes approximately one-fifth of the total plasma that enters the kidney. Thus the total renal plasma flow is equal to 5 × 180 = 900 L/day. Since plasma constitutes approximately 55 percent of whole blood, the total renal blood flow, i.e., erythrocytes plus plasma, must be approximately 1640 L/day (1.1 L/min). Thus, the kidneys receive one-fifth to one-fourth of the total cardiac output (5 L/min) although their combined weight is less than 1 percent of the total body weight! These relationships are illustrated in Fig. 24-9.

Tubular Reabsorption

Many filterable plasma components are either completely absent from the urine or present in smaller quantities than were originally filtered at the glomerulus. This fact alone is sufficient to prove that these substances undergo tubular reabsorption. An idea of the magnitude and importance of these reabsorptive mechanisms can be gained from Table 24-3, which summarizes data for a few plasma components, which are handled by filtration and reabsorption.

These are typical values for a normal person on an average diet. There are at least three important conclusions to be drawn from this table: (1) The quantities of material that enter the nephron via the glomerular filtrate are enormous, generally larger than their total body stores; e.g., if reabsorption of water ceased but filtration continued, the total plasma water would be urinated within 30 min. (2) The quantities of waste products, such as urea, that are excreted in the urine are generally sizable fractions of the filtered amounts; thus, in mammals coupling a large glomerular filtration rate with a limited urea reabsorptive capacity permits rapid ex-

cretion of the large quantities of this substance produced constantly as a result of protein breakdown. (3) In contrast to urea and other waste products, the amounts of most useful plasma components, e.g., water, electrolytes, and glucose, that are excreted in the urine represent much smaller fractions of the filtered amounts because of reabsorption. The rates at which many of these substances are reabsorbed (and therefore the rates at which they are excreted) are constantly subject to physiologic control.

Types of Reabsorption A bewildering variety of ions and molecules is found in the plasma. With the exception of the proteins (and a few ions tightly bound to protein), these materials are all present in the glomerular filtrate, and most are reabsorbed, to varying extents. It is essential to realize that tubular reabsorption is a different process than glomerular filtration. The latter occurs by bulk flow in which water and all low-molecular-weight solutes move together. In contrast, tubular reabsorption of various substances is by more or less discrete mediated transport mechanisms and by diffusion. The phrase "more or less" in the above sentence denotes several facts: (1) The mediated-transport systems for reabsorption of different ions and molecules are frequently coupled, just as they are in other cell membranes (Chap. 4); (2) in many cases a single reabsorptive system transports several different components if they are similar in structure. For example, many of the simple carbohydrates are reabsorbed by the same system.

As described in Chap. 4, transport processes can be categorized broadly as *active* or *passive,* and there are many examples of each in the kidney. The process is passive if no cellular energy is involved in the transport of the substance, i.e., if the substance moves downhill by simple or facilitated diffusion as a result of an electric or chemical concentration gradient. Active transport, on the other hand, can produce net movement of the substance uphill against its concentration or electric gradient and therefore requires energy expenditure by the transporting cells.

In this chapter we shall ignore the fact that multiple membranes lie between the lumina of the tubule and capillary and treat the tubular epithelium as a single membrane that separates tubular fluid and plasma.

	TABLE 24-3	**AVERAGE VALUES FOR SEVERAL COMPONENTS HANDLED BY FILTRATION AND REABSORPTION**	
Substance	Amount Filtered per Day	Amount Excreted per Day	Percent Reabsorbed
Water, L	180	1.8	99.0
Sodium, g	630	3.2	99.5
Glucose, g	180	0	100
Urea, g	54	30	44

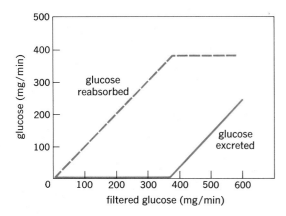

FIG. 24-10 Saturation of the glucose transport system. Glucose is administered intravenously to a person so that plasma glucose and, thereby, filtered glucose are increased.

Most, if not all, of the active reabsorptive systems in the renal tubule can transport only limited amounts of material per unit time because the membrane carrier responsible for the transport becomes saturated. The classic example is the tubular transport process for glucose. As we know, normal persons do not excrete glucose in their urine because all filtered glucose is reabsorbed; but it is possible to produce urinary excretion of glucose in a completely normal person merely by administering large quantities of glucose directly into a vein. The filtered quantity of any freely filterable plasma component such as glucose is equal to its plasma concentration multiplied by the glomerular filtration rate. If we assume that, as we give our subject intravenous glucose, the glomerular filtration remains constant, the filtered load of glucose will be directly proportional to the plasma glucose concentration (Fig. 24-10). We shall find that even after the plasma glucose concentration has doubled, the urine will still be glucose-free, indicating that the **maximal tubular capacity,** T_m, for reabsorbing glucose has not yet been reached. But as the plasma glucose and the filtered load continue to rise, glucose finally appears in the urine. From this point on any further increase in plasma glucose is accompanied by an increase in excreted glucose, because the T_m has now been exceeded. The tubules are now reabsorbing all the glucose they can, and any amount filtered in excess of this quantity cannot be reabsorbed and appears in the urine. This is precisely what occurs in the patient with diabetes mellitus. In this disease the hormonal control of plasma glucose is defective and glucose may rise to extremely high values. The filtered load of glucose becomes great enough to exceed the T_m and glucose appears in the urine. There is nothing wrong with the tubular transport mechanism for glucose, it is simply unable to reabsorb the huge filtered load.

The plasma glucose in normal persons never becomes high enough to cause urinary excretion of glucose because the reabsorptive capacity for glucose is much greater than necessary for normal filtered loads. However, for certain other substances, e.g., phosphate, the reabsorptive T_m is very close to the normal filtered load; thus small increases in plasma concentrations of such substances will result in increased excretion.

Just as glucose provides an excellent example of an actively transported solute, urea provides an example of passive transport. Since urea is filtered at the glomerulus, its concentration in the very first portion of the tubule is identical to its concentration in peritubular capillary plasma. Then, as the fluid flows along the tubule, water reabsorption occurs, increasing the concentration of any intratubular solute not being reabsorbed at the same rate as the water, with the result that intratubular urea concentration becomes greater than the peritubular plasma urea concentration. Accordingly, urea is able to diffuse passively down this concentration gradient from tubular lumen to peritubular capillary. Urea reabsorption is thus a passive process and completely dependent upon the reabsorption of water (itself passive, as we shall see), which establishes the diffuse gradient. In human beings, urea reabsorption varies between 40 and 60 percent of the filtered urea, the lower figure holding when water reabsorption is low and the higher when it is high.

Passive reabsorption is also of considerable importance for many foreign chemicals. The renal tubular epithelium acts in many respects as a lipid barrier; accordingly, lipid-soluble substances, like urea, can penetrate it fairly readily. Recall that one of the major determinants of lipid solubility is the polarity of a molecule—the less polar, the more lipid-soluble. Many drugs and environmental pollutants are nonpolar and, therefore, highly lipid-soluble. This makes their excretion from the body via the urine difficult because they are filtered at the glomerulus and then reabsorbed, like urea, as water reabsorption causes their intratubular concentrations to increase. Fortunately, the liver transforms most of these substances to progressively more polar metabolites that, because of their reduced lipid solubility, are poorly reabsorbed by the tubules and can therefore be excreted.

Tubular Secretion

Tubular secretory processes, which transport substances from peritubular capillaries to the tubular lumen, i.e., in the direction opposite to tubular reabsorption, constitute a second pathway into the tubule, the first being glomerular filtration. Like tubular reabsorption processes, secretory transport may be either active or passive. Of the large number of different substances transported into the tubules by tubular secretion, only a few are normally found in the body, the most important being hydrogen ion and potassium. We shall see that most of the excreted hydrogen ion and potassium enters the tubules by secretion rather than filtration. Thus, renal regulation of these two important substances is accomplished primarily by mechanisms that control the rates of their tubular secretion. The kidney is also able to secrete a large number of foreign chemicals, thereby facilitating their excretion from the body; penicillin is an example.

In the remainder of this chapter we shall see how the kidney functions in a variety of homeostatic processes and how renal function is coordinated with that of other organs. Before turning to the individual variables being regulated, we complete our basic story by describing the mechanisms for eliminating urine from the body.

MICTURITION (URINATION)

From the kidneys, urine flows to the bladder through the ureters (Fig. 24-1), propelled by peristaltic contractions of the smooth muscle that makes up the ureteral wall. Under normal rates of urine formation, the downward-progressing peristaltic contractions of these muscles propel several jets of urine each minute into the bladder where it is temporarily stored. The ureters, which are 25 to 30 cm long and about 3 mm in diameter, pass from the renal pelvis downward behind the peritoneum, approach the posterior wall of the bladder, pass obliquely through a gap in the bladder's muscular wall, and open into the bladder interior. The walls

of the ureters have three coats: an inner *mucosa* of connective-tissue fibers covered by transitional epithelium; a middle *muscular coat* of circular and longitudinal layers of smooth muscle fibers and an outer connective-tissue *fibrous coat,* which blends with the kidney capsule and bladder wall at the two ends of the ureters.

The bladder is a balloonlike structure whose size, shape, and location vary with the amount of urine it contains (and with the degree of distension of nearby viscera such as the large intestine and uterus). The highly folded wall of the bladder flattens out to accommodate the increased volume as the bladder fills with urine.

Like the ureters, the bladder has a three-layered wall: an inner *mucosa,* lined with transitional epithelium and continuous with the mucosa of the ureters above and the urethra below; a middle muscular layer, which forms the **detrusor muscle;** and an external fibrous layer. The detrusor muscle consists of three layers of smooth muscle fibers, an external and internal layer of longitudinally arranged fibers, and a middle layer of fibers oriented in a circular direction (although there is considerable intermingling of the three layers). Upon contraction, the walls squeeze inward, increasing the pressure of the urine in the bladder. The smooth muscle layer at the base of the bladder constitutes the first portion of the urethra and is called the **vesicle,** or **internal, sphincter.** It is not a distinct muscle but the last portion of the bladder; however, when the bladder is relaxed, this circular muscle is closed and functions as a sphincter. As the smooth muscle of the bladder contracts, this sphincter is pulled open simply by changes that occur in bladder shape during contraction. In other words, no special mechanism is required for its relaxation.

Urine flows from the bladder whenever the pressure in the bladder exceeds the resistance in the urethra. In an infant, micturition is basically a local spinal reflex. The bladder muscle receives a rich supply of parasympathetic fibers via the pelvic splanchnic nerves, the stimulation of which causes bladder contraction. (The role of the sympathetic nerves in micturition and even their distribution to the bladder are disputed.) The bladder wall contains stretch receptors whose afferent fibers enter the spinal cord and eventually synapse with these parasympathetic pathways and stimulate them. When the bladder contains only small amounts of urine,

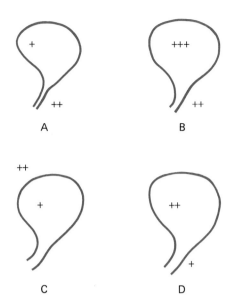

FIG. 24-11 Micturition does not occur when sphincter tension exceeds bladder pressure (A); it does occur when the reverse is true. The relationship between the sphincter and bladder may be altered so micturition occurs by contraction of the detrusor muscle (B), an increase in external pressure (C), a decrease in sphincter tension (D), or a combination of the three.

its internal pressure is low, there is little stimulation of the bladder stretch receptors, and the parasympathetics are relatively quiescent; urethral resistance exceeds bladder pressure and micturition does not occur (Fig. 24-11A). As the bladder fills with urine, it becomes distended, and the stretch receptors are gradually stimulated until their output becomes great enough to contract the bladder so that bladder pressure exceeds urethral pressure and urination occurs (Fig. 24-11B). Thus, the entire process is quite analogous to any other spinal reflex.

Voluntary control of micturition is by way of skeletal muscles related to the bladder, particularly the muscles of the pelvic diaphragm. This group of muscles fills the outlet of the pelvis and helps support some of the pelvic viscera, including that part of the bladder that opens into the urethra. Voluntary relaxation of the pelvic diaphragm allows the neck of the bladder to move downward; this decreases the resistance in the urethra while simultaneously stretching the walls of the bladder, an event that leads to reflex contraction of the smooth muscle of the bladder via the parasympathetic nerves. If the increased pressure that results from

contraction of the bladder smooth muscle is insufficient to overcome urethral resistance, pressure is further increased by voluntary contraction of the abdominal muscles and the diaphragm (i.e., the respiratory diaphragm) (Fig. 24-11C). In contrast, micturition is stopped by contraction of the pelvic diaphragm. The urethral sphincter, a circular layer of skeletal muscle that surrounds the urethra and relaxes and contracts with the pelvic diaphragm, is not thought to play a major role in voluntary control of micturition.

SECTION B: REGULATION OF SODIUM AND WATER BALANCE

Table 24-1 is a typical balance sheet for water; Table 24-4 for sodium chloride. As with water, the excretion of sodium via the skin and gastrointestinal tract is normally quite small but may increase markedly during severe sweating, vomiting, or diarrhea. Hemorrhage, of course, can result in loss of large quantities of both sodium and water.

Control of the renal excretion of sodium and water constitutes the most important mechanism for the regulation of body sodium and water. The excretory rates of these substances can be varied over an extremely wide range; e.g., a gross consumer of salt may ingest 20 to 25 g of sodium chloride per day, whereas a patient on a low-salt diet may ingest only 50 mg. The normal kidney can readily alter its excretion of salt over this range. Similarly, urinary water excretion can be varied physiologically from approximately 400 mL/day to 25 L/day, depending upon whether one is lost in the desert or participating in a beer-drinking contest.

TABLE 24-4 NORMAL ROUTES OF SODIUM CHLORIDE INTAKE AND LOSS	
	Grams per Day
Intake:	
Food	10.5
Output:	
Sweat	0.25
Feces	0.25
Urine	10.0
Total output	10.5

BASIC RENAL PROCESSES FOR SODIUM, CHLORIDE, AND WATER

Sodium, chloride, and water do not undergo tubular secretion; each is freely filterable at the glomerulus and approximately 99 percent is reabsorbed as it passes down the tubules (Table 24-3). Indeed, the vast majority of all renal energy production is used to accomplish this enormous reabsorptive task. The tubular mechanisms for reabsorption of these substances can be summarized by two generalizations: (1) The reabsorption of sodium and/or chloride is an active process; i.e., it is carrier mediated, requires an energy supply, and can occur against an electric or concentration gradient. In most segments of the tubule, sodium is the actively transported ion with the reabsorption of chloride coupled to it by several mechanisms; in at least one tubule segment, the situation is reversed; i.e., chloride is the actively transported ion, sodium movement being coupled to it. The end result is the same, and for simplicity we shall usually refer simply to "sodium reabsorption." (2) The reabsorption of water is by diffusion (osmosis) and depends upon the reabsorption of sodium chloride.

How does the active reabsorption of sodium lead to the passive reabsorption of water? Since the concentrations of all low-molecular-weight solutes are virtually identical in plasma and in Bowman's capsular fluid, no significant transtubular concentration gradients exist initially for sodium, chloride, or water. As the fluid flows down the tubule, sodium is actively reabsorbed into the peritubular capillaries. What happens to the water concentration of the tubular fluid as a result of this reabsorption? Obviously the removal of solute lowers osmolarity, i.e., raises water concentration relative to that of plasma. Thus, a water concentration gradient is created between tubular lumen and peritubular plasma that constitutes a driving force for water reabsorption via osmosis. If the water permeability of the tubular epithelium is very high, water molecules are reabsorbed passively almost as rapidly as the actively transported sodium ions, so that the tubular fluid is only slightly more dilute than plasma. In this manner almost all the filtered sodium and water could theoretically be reabsorbed and the final urine would still have approximately the same osmolarity as plasma; however, this reabsorption of water can occur only if the tubular epithelium is highly perme-

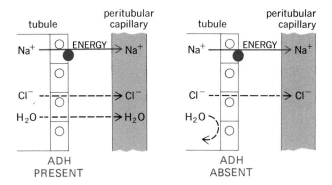

FIG. 24-12 Effect of ADH on water reabsorption. ADH causes the tubule to be highly permeable to water so that water movement can accompany sodium and chloride reabsorption. Without ADH, the tubule becomes quite impermeable to water, and water reabsorption is impeded. Note that ADH does not alter reabsorption of sodium or chloride ions.

able to water. No matter how great the water concentration gradient, water cannot move if the epithelium is impermeable to it.

The permeability of the last portions of the tubules (the late distal tubules and collecting ducts) to water is subject to physiologic control. The major determinant of this permeability is a peptide hormone known as **antidiuretic hormone (ADH),** or **vasopressin.** In the absence of ADH (Fig. 24-12) the water permeability of the late distal tubule and collecting duct is very low; sodium reabsorption proceeds normally because ADH has little or no effect on sodium reabsorption, but water is unable to follow and thus remains in the tubule to be excreted as a large volume of urine. On the other hand, in the presence of ADH, the water permeability of these last nephron segments is very great, water reabsorption is able to keep up with sodium reabsorption, and the final urine volume is small.

Several crucial aspects of sodium and water reabsorption should be emphasized:

1 The tubular response to ADH shows graded changes as the concentration of ADH is altered, which permits fine adjustments of water permeability and excretion.
2 Excretion of large quantities of sodium always results in the excretion of large quantities of water. This follows from the passive nature of water reabsorption, since water can be reab-

sorbed only if sodium is reabsorbed first. As we shall see, this relationship has considerable importance for the regulation of extracellular volume.

3 In contrast, large quantities of water can be excreted even though the urine is virtually free of sodium. This process we shall find critical for the renal regulation of extracellular osmolarity.

CONTROL OF SODIUM EXCRETION: REGULATION OF EXTRACELLULAR VOLUME

The renal compensation for increased body sodium is excretion of the excess sodium. Conversely, a deficit in body sodium is prevented by a reduction of urinary sodium to an absolute minimum, thus retaining within the body the amount that would otherwise have been lost via the urine.

Since sodium is freely filterable at the glomerulus and actively reabsorbed but not secreted by the tubules, the amount of sodium excreted in the final urine represents the resultant of two processes, glomerular filtration and tubular reabsorption.

$$\text{Sodium excretion} = \text{sodium filtered} - \text{sodium reabsorbed}$$

It is possible, therefore, to adjust sodium excretion by controlling one or both of these two variables (Fig. 24-13). For example, what happens if the quantity of filtered sodium increases (as a result of a higher GFR) but the rate of reabsorption remains constant? Clearly, sodium excretion increases. The same final result could be achieved by lowering sodium reabsorption while holding the GFR constant. Finally, sodium excretion could be raised greatly by elevating the GFR and simultaneously reducing reabsorption. Conversely, sodium excretion could be decreased below normal levels by lowering the GFR or raising sodium reabsorption or both. Control of GFR and sodium reabsorption is in fact the mechanism by which renal regulation of sodium balance is accomplished.

The reflex pathways by which changes in total body sodium balance lead to changes in GFR and sodium reabsorption include (1) "volume" receptors (the reasons for the use of quotation marks will soon be apparent) and the afferent pathways that lead from them to the central nervous system and endocrine glands; (2) efferent neural and hormonal

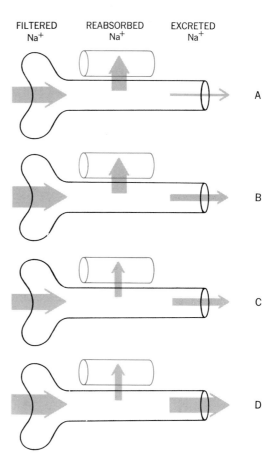

FILTERED Na$^+$ REABSORBED Na$^+$ EXCRETED Na$^+$

FIG. 24-13 Sodium excretion is determined by the amount filtered at the glomerulus (the GFR) and the amount reabsorbed. Sodium excretion is increased by increasing the GRF (B), by decreasing reabsorption (C), or by a combination of both (D). The arrows indicate relative magnitudes of filtration, reabsorption, and excretion.

pathways to the kidneys; and (3) renal effector sites: the renal arterioles and tubules.

Rather than extracellular volume, per se, the variables monitored in sodium-regulating reflexes are those that are altered as a result of changes in this volume, specifically intravascular and intracardiac pressure. For example, a decrease in plasma volume tends to lower the hydrostatic pressures in the veins, cardiac chambers, and arteries. These changes are detected by receptors (for example, the carotid sinus baroreceptor) in these structures and initiate the reflexes that lead to the renal sodium retention that helps restore the plasma volume toward normal.

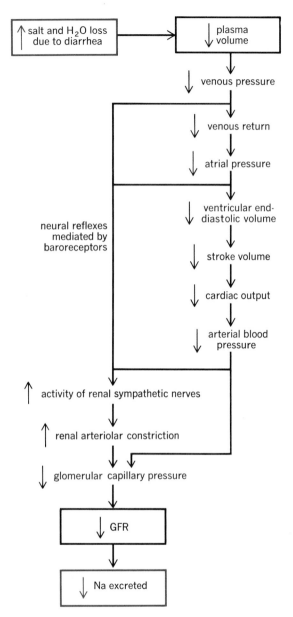

FIG. 24-14 Pathway by which the GFR is decreased when plasma volume decreases. The baroreceptors that initiate the sympathetic reflex are probably located in large veins and the walls of the heart, as well as in the carotid sinuses and aortic arch.

Control of GFR

Figure 24-14 summarizes the reflex pathway by which negative sodium balance (as, for example, in diarrhea) causes a decrease in GFR. Note that this is simply the basic baroreceptor reflex described in Chap. 22, where it was pointed out that decrease in cardiovascular pressure causes reflex vasoconstriction in many areas of the body. Conversely, an increased GFR can result from increased plasma volume and contribute to the increased renal sodium loss that returns extracellular volume to normal.

Control of Tubular Sodium Reabsorption

So far as long-term regulation of sodium excretion is concerned, the control of tubular sodium reabsorption is more important than that of GFR. One major controller of the rate of tubular sodium reabsorption is aldosterone.

Aldosterone and the Renin-Angiotensin System The adrenal cortex produces a hormone, **aldosterone,** that stimulates sodium reabsorption, specifically by the distal tubules and collecting ducts. In the complete absence of this hormone, a person may excrete 25 g of salt per day, whereas excretion may be virtually zero when aldosterone is present in large quantities. In a normal person, the amounts of aldosterone produced and salt excreted lie somewhere between these extremes, varying with the amount of salt ingested. (It is interesting that aldosterone also stimulates sodium transport by other epithelia in the body, namely, by sweat and salivary

FIG. 24-15 The juxtaglomerular apparatus. The macula densa is in lighter color.

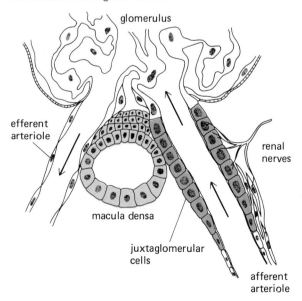

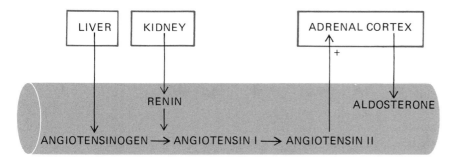

FIG. 24-16 Summary of the renin-angiotensin-aldosterone system. The plus sign denotes the stimulatory effect of angiotensin II on aldosterone secretion.

glands and the intestine. The net effect is the same as that exerted on the renal tubules—a movement of sodium out of the luminal fluid into the blood. Thus aldosterone is an "all-purpose" stimulator of sodium retention.)

Aldosterone secretion (and thereby tubular sodium reabsorption) is controlled by reflexes that involve the kidneys themselves. Specialized cells (the **juxtaglomerular cells,** Fig. 24-15) lining portions of the afferent arterioles in the kidneys synthesize and secrete into the blood an enzyme known as **renin** that splits off a small polypeptide, **angiotensin I,** from a large plasma protein, **angiotensinogen** (Fig. 24-16). Angiotensin I then undergoes further enzymatically mediated cleavage to form **angiotensin II,** the active molecule in this sequence. Angiotensin II is a profound stimulator of aldosterone secretion and constitutes the major input, to the adrenal gland, that controls the production of this hormone.

Angiotensinogen is synthesized by the liver and is always present in the blood; therefore, the rate-limiting factor in angiotensin formation is the concentration of plasma renin that, in turn, depends upon the rate of renin secretion by the kidneys. The critical question now becomes: What controls the rate of renin secretion? There are multiple inputs to the renin-secreting cells and it is not yet possible to assign quantitative roles to each of them. It is likely that the renal sympathetic nerves and decreased renal arterial pressure constitute important inputs in normal persons; this makes excellent sense, teleologically, since a reduction in body sodium and extracellular volume lowers blood pressure and triggers off increased sympathetic discharge to the kidneys (as shown in Fig. 24-14), which sets off the hormonal chain of events that restores sodium balance and extracellular volume to normal (Fig. 24-17).

Clearly, by helping to regulate sodium balance and thereby plasma volume, the renin-angiotensin system contributes to the control of arterial blood pressure; however, this is not the only way in which it influences arterial pressure. Recall from Chap. 22

FIG. 24-17 Pathway by which aldosterone secretion is increased when plasma volume is decreased. (The pathways by which the GFR is reduced are shown in Fig. 24-14.

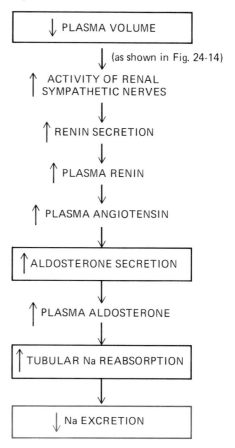

that angiotensin is a potent constrictor of arterioles and that this effect on peripheral resistance increases arterial pressure. Abnormal increases in the activity of the renin-angiotensin system contribute to the development of hypertension (high blood pressure) via both the sodium-retaining and arteriole-constricting effects.

There are a large number of factors other than aldosterone that are capable of altering tubular sodium reabsorption, and renal physiologists are presently attempting to quantitate their relative contributions to the control of sodium excretion in both health and disease.

Conclusion

The control of sodium excretion depends upon the control of two variables of renal function, the GFR and sodium reabsorption. The latter is controlled, at least in part, by the renin-angiotensin–aldosterone hormone system and, in part, by other less well-defined factors. The reflexes that control both GFR and sodium reabsorption are essentially blood pressure–regulating reflexes since they are probably most frequently initiated by changes in arterial or venous pressure (or cardiac output). This is only fitting since cardiovascular function depends upon an adequate plasma volume, which as a component of the extracellular fluid volume, normally reflects the mass of sodium in the body. In normal persons, these regulatory mechanisms are so precise that sodium balance does not vary by more than 2 percent despite marked changes in dietary intake or in losses due to sweating, vomiting, or diarrhea. In several types of diseases, however, sodium balance becomes deranged by the failure of the kidneys to excrete sodium normally. Sodium excretion may fall virtually to zero despite continued sodium ingestion, and the person may retain huge quantities of sodium and water, which leads to abnormal expansion of the extracellular fluid and edema. The most important example of this phenomenon is congestive heart failure (Chap. 22). A person with a failing heart manifests a decreased GFR and increased aldosterone secretion rate, both of which contribute to the virtual absence of sodium from the urine. In addition to aldosterone, the other less well-defined factors also cause decreased sodium excretion by enhancing tubular sodium reabsorption. The net result is expansion of plasma volume, increased capillary pressure, and filtration of fluid into the interstitial space, i.e., edema. Why are these sodium-retaining reflexes all stimulated despite the fact that the expanded extracellular volume should call forth sodium-losing responses (increased GFR and decreased aldosterone)? We do not know, although it seems fairly certain that the lower cardiac output as a result of cardiac failure is somehow responsible. In addition to treating the basic heart disease, physicians use diuretics, or drugs that inhibit tubular sodium or chloride reabsorption, which leads to greater sodium and water excretion.

ADH Secretion and Extracellular Volume

Although we have discussed extracellular-volume regulation only in terms of the control of sodium excretion, it is clear that, to be effective in altering extracellular volume, the changes in sodium excretion must be accompanied by equivalent changes in water excretion. We have already pointed out that the ability of water to be reabsorbed when sodium is reabsorbed depends upon ADH. Accordingly, a decreased extracellular volume must reflexly call forth increased ADH secretion as well as increased aldosterone secretion. What is the nature of this reflex? As described in Chap. 18, ADH is produced by a discrete group of hypothalamic neurons whose axons terminate in the posterior pituitary, from which ADH is released into the blood. These hypothalamic cells receive input from several vascular baroreceptors, particularly a group located in the left atrium. The baroreceptors are stimulated by increased atrial blood pressure, and the impulses resulting from this stimulation are transmitted via the vagus nerves and ascending pathways to the hypothalamus, where they inhibit the ADH-producing cells. Conversely, decreased atrial pressure causes less firing by the baroreceptors and stimulation of ADH synthesis and release (Fig. 24-18). The adaptive value of this baroreceptor reflex, one more in our expanding list, should require no comment.

RENAL REGULATION OF EXTRACELLULAR OSMOLARITY

We turn now to the renal compensation for pure water losses or gains, e.g., a person drinking 2 L of

FIG. 24-18 Pathway by which ADH secretion is increased when plasma volume decreases.

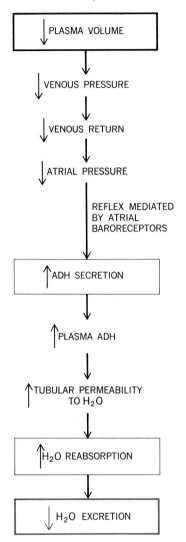

plasma or one that has a lower osmolarity than plasma (hypoosmotic urine), the latter occurring whenever water reabsorption lags behind solute reabsorption, i.e., when plasma ADH is reduced. Clearly the formation of a hypoosmotic urine is a good compensation for an excess of water in the body.

At the other end of the spectrum, the final urine may indeed reach a concentration considerably greater than that of the plasma; moreover, this concentrated urine is produced without violating the generalization that water reabsorption is always passive. But how can the kidneys produce a hyperosmotic urine, i.e., a urine having an osmolarity greater than that of plasma? (Recall that osmolarity refers to the total solute concentration of a solution.) For this to occur, does not water reabsorption have to "get ahead" of solute reabsorption? How can this happen if water reabsorption is always secondary to solute (particularly sodium chloride) reabsorption? To this question we now turn.

Urine Concentration: The Countercurrent System

The ability of the kidneys to produce concentrated urine is not merely an academic problem. It is a major determinant of one's ability to survive without water. The human kidney can produce a maximal urinary concentration of 1400 mosmol/L. The urea, sulfate, phosphate, and other waste products (plus the smaller number of nonwaste ions) that are excreted each day amount to approximately 600 mosmol; therefore, the water required for their excretion constitutes an obligatory water loss and equals

$$\frac{600 \text{ mosmol/day}}{1400 \text{ mosmol/L}} = 0.444 \text{ L/day}$$

As long as the kidneys are functioning, excretion of this volume of urine will occur, despite the absence of water intake. In a sense, persons who lack access to water may literally urinate to death (due to fluid depletion). If the body could produce a urine with an osmolarity of 6000 mosmol/L then only 100 mL of obligatory water need be lost each day and survival time would be greatly expanded. A desert rodent, the kangaroo rat, does just that; this animal never drinks water, for the water produced in its body by oxidation of foodstuffs is ample for maintenance of water balance.

water, where no change in total salt content of the body occurs, only total water. The most efficient compensatory mechanism is for the kidneys to excrete the excess water without altering their usual excretion of salt, and this is precisely what they do. ADH secretion is inhibited by a reflex, as will be described below, tubular water permeability of the collecting ducts becomes very low, sodium reabsorption proceeds normally but water is not reabsorbed, and a large volume of extremely dilute urine is excreted. In this manner, the excess pure water is eliminated.

Thus it is easy to see how the kidneys produce a final urine that has the same osmolarity as that of

The kidneys produce concentrated urine by a complex interaction of events involving the so-called **countercurrent multiplier system** that resides in the loop of Henle. Recall that the loop of Henle, which is interposed between the proximal and distal convoluted tubules, is a hairpin loop that extends into the inner portions of the kidney (the renal medulla). The fluid flows in opposite directions in the two limbs of the loop, thus the name countercurrent. Let us list the critical characteristics of this loop.

1   The ascending limb of the loop of Henle (i.e., the limb leading to the distal tubule) actively transports sodium chloride out of the tubular lumen into the surrounding interstitium. (In the ascending limb of the loop of Henle chloride is the actively transported ion with sodium following passively; for simplicity, we shall refer to the process as "sodium chloride transport.") It is fairly impermeable to water, so that water cannot follow the sodium chloride.

2   The descending limb of the loop of Henle (i.e., the limb into which drains fluid from the proximal tubule) does not actively transport sodium chloride; it is the only tubular segment that does not. Moreover, it is highly permeable to water but relatively impermeable to sodium chloride.

Given these characteristics, imagine the loop of Henle filled with a stationary column of fluid supplied by the proximal tubule. At first, the concentration everywhere would be 300 mosmol/L, since fluid leaving the proximal tubule is isosmotic to plasma, equivalent amounts of sodium chloride and water having been reabsorbed by the proximal tubule.

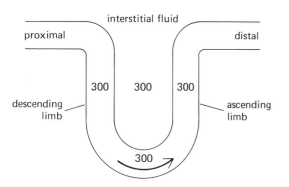

Now let the active pump in the ascending limb transport sodium chloride into the interstitium until a limiting gradient (say 200 mosmol/L) is established between ascending-limb fluid and interstitium.

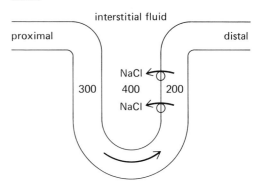

A limiting gradient is reached because the ascending limb is not completely impermeable to sodium and chloride; accordingly, passive backflux of ions into the lumen counterbalances active outflux, and a steady-state limiting gradient is established.

Given the relatively high permeability of the descending limb to water, what net flux now occurs between interstitium and descending limb? There is a net diffusion of water out of the descending limb until the osmolarities are equal. The interstitial osmolarity is maintained at 400 mosmol/L during this equilibration because of continued active sodium chloride transport out of the ascending limb.

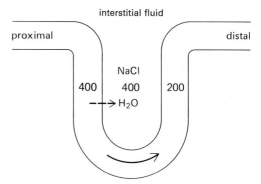

Note that as a result of this purely passive water movement across the descending limb, the osmolarities of the descending limb and interstitium bcome equal and both are higher than that of the ascending limb.

So far we have held the fluid stationary in the loop, but of course it is actually continuously flowing. Let us look at what occurs under conditions of

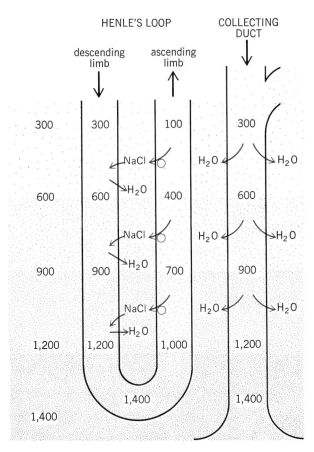

FIG. 24-19 Operation of countercurrent multiplication of concentration in the formation of hypertonic urine.

flow (Fig. 24-19). Note that the fluid is progressively concentrated as it flows down the descending limb and then is progressively diluted as it flows up the ascending limb. Whereas only a 200-mosmol/L gradient is maintained across the ascending limb at any given *horizontal level* in the medulla, there exists a much larger osmotic gradient from the top of the medulla to the bottom (300 mosmol/L versus 1400 mosmol/L). In other words, the 200-mosmol/L gradient established by active sodium chloride transport has been *multiplied* because of the *countercurrent* flow in the loop (i.e., flow in opposing directions through the two limbs of the loop). It should be emphasized that the active sodium chloride transport mechanism in the ascending limb is the essential component of the entire system; without it, the countercurrent flow would have no effect whatever on concentrations.

But what has this system really accomplished? Certainly, it concentrates the loop fluid to 1400 mosmol/L, but then it immediately redilutes it so that the fluid entering the distal tubule is actually more dilute than the plasma. Where is the final urine concentrated and how?

The site of final concentration is the collecting ducts that pass through the renal medulla parallel to the loops of Henle and are bathed by the interstitial fluid of the medulla. In the presence of maximal levels of ADH, fluid in the late distal tubules reequilibrates with peritubular plasma and is isosmotic to plasma (that is, 300 mosmol/L) when it enters the collecting ducts. As this fluid then flows through the collecting ducts it equilibrates with the ever-increasing osmolarity of the interstitial fluid. Thus, the real function of the loop countercurrent multiplier system is to concentrate the *medullary interstitium*. Under the influence of ADH, the collecting ducts are highly permeable to water, which diffuses out of the collecting ducts into the interstitium as a result of the osmotic gradient (Fig. 24-19). The net result is that the fluid at the end of the collecting duct has equilibrated with the interstitial fluid at the tip of the medulla. By this means, the final highly concentrated urine contains relatively less of the filtered water than solute, which is precisely the same as adding pure water to the extracellular fluid, and thereby compensating for a pure water deficit.

In contrast, in the presence of low plasma ADH concentration, the collecting ducts become relatively impermeable to water and the interstitial osmotic gradient is ineffective in inducing water movement out of the collecting ducts; therefore a large volume of dilute urine is excreted, thereby compensating for a pure water excess.

Osmoreceptor Control of ADH Secretion

To reiterate, pure water deficits or gains are compensated for by partially dissociating water excretion from that of salt through changes in ADH secretion. What variable controls ADH under such conditions? The answer is changes in extracellular osmolarity. The adaptive rationale should be obvious, since osmolarity is the variable most affected by pure water gains or deficits. Osmoreceptors are located in the supraoptic and paraventricular nuclei

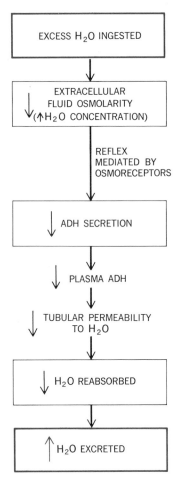

FIG. 24-20 Pathway by which ADH secretion is lowered and water excretion raised when excess water is ingested.

of the hypothalamus (Chap. 18), and information from them is transmitted by neurons to the hypothalamic cells that secrete ADH. Via these connections an increase in osmolarity stimulates the ADH-secreting neurons and increases the rate of ADH secretion; conversely, decreased osmolarity inhibits ADH secretion (Fig. 24-20).

We have now described two different afferent pathways that control the ADH-secreting hypothalamic cells, one from baroreceptors and one from osmoreceptors. These hypothalamic cells are therefore true integrating centers whose rate of activity is determined by the total synaptic input. To add to the complexity, these cells receive synaptic input from many other brain areas, so that ADH secretion (and therefore urine flow) can be altered by pain, fear, and a variety of other factors. However, these effects are usually short-lived and should not obscure the generalization that ADH secretion is determined primarily by the states of extracellular volume and osmolarity. Alcohol is a powerful inhibitor of ADH release, which probably accounts for much of the large urine flow that accompanies the ingestion of alcohol.

The disease *diabetes insipidus,* which is distinct from diabetes mellitus, i.e., sugar diabetes, illustrates what happens when the ADH system is disrupted. This disease is characterized by the constant excretion of a large volume of highly dilute urine (as much as 25 L/day). In most cases, the flow can be restored to normal by the administration of ADH. These patients apparently have lost the ability to produce ADH, usually as a result of damage to the hypothalamus. Thus, renal tubular permeability to water is low and unchanging regardless of extracellular osmolarity or volume. The very thought of having to urinate (and therefore to drink) 25 L of water per day underscores the importance of ADH in the control of renal function and body water balance.

THIRST

Now we must turn to the other component of the balance, control of intake. It should be evident that large deficits of salt and water can be only partly compensated by renal conservation and that ingestion is the ultimate compensatory mechanism. The subjective feeling of **thirst,** which drives one to obtain and ingest water, is stimulated both by a lower extracellular volume and a higher plasma osmolarity, the adaptive significance of both being self-evident. Note that these are precisely the same changes that stimulate ADH production. Indeed, the centers that mediate thirst are also located in the hypothalamus very close to those areas that produce ADH. Damage to these centers abolishes thirst completely. Conversely, electric stimulation of them may induce profound and prolonged drinking (these water-intake centers are very close to, but distinct from, the food-intake centers to be described in a later chapter). Because of the similarities between the stimuli for ADH secretion and thirst, it is tempting to speculate that the receptors (osmoreceptors and atrial baroreceptors) that initiate the ADH-controlling reflexes are identical to those for thirst. Much evidence indicates that this is, indeed, the case.

SECTION C: REGULATION OF POTASSIUM, CALCIUM, AND HYDROGEN-ION CONCENTRATIONS

POTASSIUM REGULATION

The potassium concentration of extracellular fluid is a closely regulated quantity. The importance of maintaining this concentration in the internal environment stems primarily from the role of potassium in the excitability of nerve and muscle. Recall that the resting-membrane potentials of these tissues are directly related to the ratio of intracellular to extracellular potassium concentration. Raising the external potassium concentration depolarizes membranes, which increases cell excitability and causes abnormalities of cardiac muscle conduction and rhythmicity. Conversely, lowering the external potassium hyperpolarizes cell membranes and reduces their excitability, and early manifestations of potassium depletion are weakness of skeletal muscles, particularly in the limbs, and abnormalities of cardiac function.

Since most of the body's potassium is found in cells, primarily as a result of active ion-transport systems located in plasma membranes, even a slight alteration of the rates of ion transport across cell membranes can produce a large change in the amount of extracellular potassium. Unfortunately, little is known about the physiologic control of these transport mechanisms, and our understanding of the regulation of extracellular potassium concentration will remain incomplete until further data are obtained on this critical subject.

The normal person remains in total body potassium balance (as is true for sodium balance) by daily excreting an amount of potassium in the urine equal to the amount ingested minus the small amounts normally eliminated in the feces and sweat. Normally potassium losses via sweat and the gastrointestinal tract are small (although large quantities can be lost by the latter during vomiting or diarrhea). Again the control of renal function is the major mechanism by which body potassium is regulated.

Renal Regulation of Potassium

Potassium is freely filterable at the glomerulus. The amounts of potassium excreted in the urine are gen-

erally a small fraction (10 to 15 percent) of the quantity filtered, thus establishing the existence of tubular potassium reabsorption. However, under certain conditions the excreted quantity may actually exceed that filtered, thus establishing the existence of tubular potassium secretion. The subject is therefore complicated by the fact that potassium can be both reabsorbed and secreted by the tubule. However, normally most of the filtered potassium is reaborbed (by active transport) regardless of changes in body potassium balance. In other words, the reabsorption of potassium does not seem to be controlled so as to achieve potassium homeostasis. The important result of this phenomenon is that changes in potassium *excretion* are due to changes in potassium *secretion* (Fig. 24-21). Thus, during potassium depletion when the homeostatic response is to reduce potassium excretion to a minimal level, there is no significant potassium secretion, and only the small amount of potassium escaping reabsorption is excreted. In all other situations, to this same small amount of unreabsorbed potassium is added a variable amount of secreted potassium. Thus, in describing the homeostatic control of potassium excretion we may ignore changes in GFR or potassium reabsorption and

FIG. 24-21 Basic renal processing of potassium. Since virtually all the filtered potassium is reabsorbed, potassium excreted in the urine results from tubular secretion.

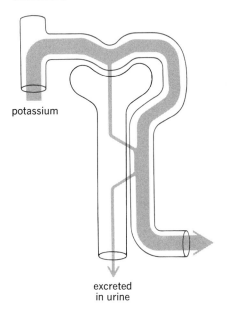

potassium

excreted
in urine

focus only on the factors that alter the rate of tubular potassium secretion.

One of the most important of these factors is the potassium concentration of the renal tubular cells themselves. When a high-potassium diet is ingested, potassium concentration in most of the body's cells increases, including the renal tubular cells. This higher concentration facilitates potassium secretion into the lumen and raises potassium excretion. Conversely, a low-potassium diet or a negative potassium balance, e.g., from diarrhea, lowers renal-tubular-cell potassium concentration; this reduces potassium entry into the lumen and decreases potassium excretion, thereby helping to reestablish potassium balance.

A second important factor that controls potassium secretion is the hormone aldosterone, which besides assisting tubular sodium reabsorption enhances tubular potassium secretion. The reflex by which changes in extracellular volume control aldosterone production is completely different from the reflex initiated by an excess or deficit of potassium. The former constitutes a complex pathway that involves renin and angiotensin; the latter, however, seems to be much simpler (Fig. 24-22). The aldosterone-secreting cells of the adrenal cortex are

FIG. 24-22 Pathway by which an increased potassium intake induces greater potassium excretion mediated by aldosterone.

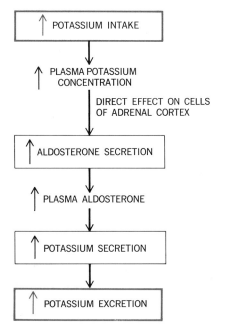

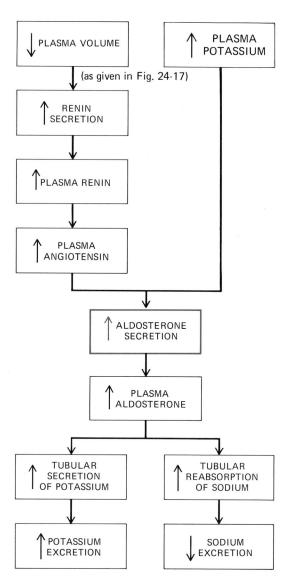

FIG. 24-23 Summary of the control of aldosterone and its functions.

apparently themselves sensitive to the potassium concentration of the extracellular fluid that bathes them. For example, an increased intake of potassium leads to an increased extracellular potassium concentration, which in turn directly stimulates aldosterone production by the adrenal cortex. This extra aldosterone circulates to the kidney, where it increases tubular potassium secretion and thereby eliminates the excess potassium from the body. Conversely, a lowered extracellular potassium concentration would decrease aldosterone production and thereby inhibit tubular potassium secretion. Less potassium than usual would be excreted in the urine, thus helping to restore the normal extracellular potassium concentration. Again, complete compensation depends upon the ingestion of additional potassium. The control and renal tubular effects of aldosterone are summarized in Fig. 24-23.

CALCIUM REGULATION

Extracellular calcium concentration is also normally held relatively constant, and the requirement for precise regulation stems primarily from the profound effects of calcium on neuromuscular excitability. A low calcium concentration increases the excitability of nerve and muscle cell membranes so that patients with diseases in which low calcium occurs suffer from *hypocalcemic tetany,* characterized by skeletal muscle spasms. Hypercalcemia is also dangerous in that it causes cardiac arrhythmias as well as depressed neuromuscular excitability.

Effector Sites for Calcium Homeostasis

At least three effector sites are involved in the regulation of extracellular calcium concentration: bone, the kidneys, and the gastrointestinal tract.

Bone   Approximately 99 percent of total body calcium is contained in bone, which consists primarily of a framework of organic molecules upon which calcium phosphate crystals are deposited. Contrary to popular opinion, bone is not an absolutely fixed, unchanging tissue but is constantly being remolded and, what is more important, is available for either the withdrawal or deposit of calcium from extracellular fluid.

Gastrointestinal Tract   Under normal conditions considerable amounts of ingested calcium are not absorbed from the intestine but are eliminated in the feces. Accordingly, control of the active transport system that moves this ion from gut lumen to blood can result in large increases or decreases in the rate of absorption. In Chap. 25 we will discuss the indiscriminate nature of most gastrointestinal absorptive processes. This is certainly not true for calcium absorption, which is subject to quite precise hormonal control.

Kidneys   The kidneys handle calcium by filtration and reabsorption. In addition, as we shall see, the renal handling of phosphate also plays an important role in the regulation of extracellular calcium.

Parathormone

All three of the effector sites described above are subject directly or indirectly to control by a protein hormone called **parathormone** (also called *parathyroid hormone*), produced by the parathyroid glands. Parathormone production is controlled by the calcium concentration of the extracellular fluid that bathes the cells of these glands. Lower calcium concentration stimulates parathormone production and release, and a higher concentration does just the opposite. It should be emphasized that extracellular calcium concentration acts directly upon the parathyroids (just as was true of the relation between extracellular potassium and aldosterone production) without any intermediary hormones or nerves.

Parathormone exerts at least four distinct effects on the bone, gastrointestinal tract, and kidneys (Fig. 24-24A).

1   It increases the movement of calcium (and phosphate) from bone into extracellular fluid, making available this immense store of calcium for the regulation of the extracellular calcium concentration. Theories about these mechanisms are still controversial.
2   It increases gastrointestinal absorption of calcium. This is not a direct action of parathormone on the intestine but is the indirect result of the hormone's effect on vitamin D, as discussed below.
3   It increases renal-tubular calcium reabsorption, which decreases urinary calcium excretion.
4   It reduces the renal-tubular reabsorption of phosphate, which raises its urinary excretion and lowers extracellular phosphate concentration; this facilitates release of more calcium from bone.

The adaptive value of these effects should be obvious: They all result in a higher extracellular calcium concentration, which compensates for the lower concentration that originally stimulated par-

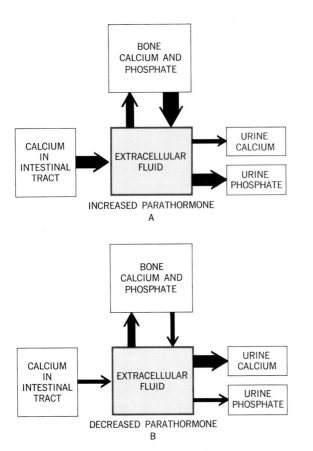

FIG. 24-24 Effects of parathormone on the gastrointestinal tract, kidneys, and bone; the arrows signify relative magnitudes. Note that when parathormone is decreased there is net movement of calcium and phosphate into bone, urine calcium is raised, and gastrointestinal absorptions of calcium is reduced.

athormone production. Conversely, an increase in extracellular calcium concentration inhibits normal parathormone production, which produces increased urinary and fecal calcium loss and net movement of calcium from extracellular fluid into bone (Fig. 24-24B).

Parathormone has other functions, notably its role in milk production, but the four effects discussed above constitute the major mechanisms by which it integrates various organs and tissues in the regulation of extracellular calcium concentration.

Vitamin D

Vitamin D plays an important role in calcium metabolism, as attested by the fact that its deficiency results in poorly calcified bones. Vitamin D can be produced by the skin in the presence of sunlight, but since inadequate sunlight and people's clothing often prevent this reaction, they are dependent on dietary intake for its supply, and it is classed as a vitamin. The major action of vitamin D is to stimulate active calcium absorption by the intestine. Thus, the major event in vitamin D deficiency is decreased gut calcium absorption, which results in decreased plasma calcium. In children, the newly formed bone protein matrix fails to be calcified normally because of the low plasma calcium, which leads to the disease *rickets.*

The molecular form in which vitamin D is ingested or formed in the skin is relatively inactive and must undergo several transformations in the liver and kidneys before it is able to stimulate transport of calcium by the intestine. (The active form of vitamin D is *1,25 dihydroxyvitamin D₃.*) The molecular alteration that occurs in the kidneys seems to be particularly critical and is, itself, subject to physiologic regulation by several inputs. One of these is parathormone. Accordingly, a decreased plasma calcium stimulates the secretion of parathormone that, in turn, enhances the renal alteration of vitamin D that, by its actions on the intestine, helps to restore plasma calcium to normal. Thus, the feedback loops for parathormone and vitamin D activity are closely intertwined.

Calcitonin

Another hormone, known as **calcitonin,** has significant effects on plasma calcium. Calcitonin is secreted by the parafollicular (or clear) cells in the thyroid gland that surround but are completely distinct from the thyroxine-secreting follicles (for anatomy of the thyroid gland, see Chap. 18). Calcitonin lowers plasma calcium primarily by inhibiting bone resorption. Its secretion is controlled directly by the calcium concentration of the plasma that supplies the thyroid gland; increased calcium causes increased calcitonin secretion. Thus, this system constitutes a second feedback control over plasma calcium concentration, one that is opposed to the parathormone system. However, its overall contribution to calcium homeostasis is minor compared with that of parathormone.

HYDROGEN-ION REGULATION

Most metabolic reactions are highly sensitive to the hydrogen-ion concentration[1] of the fluid in which they occur. This sensitivity is due primarily to the influence on enzyme function exerted by the hydrogen ion. Accordingly, the hydrogen-ion concentration of the extracellular fluid is closely regulated.

Basic Definitions

The hydrogen ion is an atom of hydrogen that has lost its only electron. When dissolved in water, many compounds dissociate reversibly to produce negative charged ions (anions) and hydrogen ions, e.g.,

$$\text{Lactic acid} \rightleftharpoons H^+ + \text{lactate}^-$$
$$H_2CO_3 \rightleftharpoons H^+ + HCO_3^-$$

Carbonic acid   Bicarbonate

Any compound capable of liberating a hydrogen ion in this manner is called an **acid.** Conversely, any substance that can accept a hydrogen ion is termed a **base.** Thus, in the reactions above, lactate and bicarbonate are bases because they can bind hydrogen ions. The hydrogen-ion concentration often is referred to in terms of **acidity:** the higher the hydrogen-ion concentration, the greater the acidity (it must be understood that the hydrogen-ion concentration of a solution refers only to the hydrogen ions that are *free* in solution).

Strong and Weak Acids

Strong acids dissociate completely when they dissolve in water. For example, hydrochloric acid added to water gives

$$HCl \longrightarrow H^+ + Cl^-$$

Hydrochloric
acid

Virtually no HCl molecules exist in the solution, only free hydrogen ions and chloride ions. On the other hand, **weak acids** are those which do not dissociate completely when dissolved in water. For example, when dissolved, only a fraction of lactic acid molecules dissociate to form lactate and hydrogen ions, and the other molecules remain intact. This characteristic of weak acids underlies an important chemical and physiologic phenomenon, *buffering*.

Buffer Action and Buffers

Figure 24-25A pictures a solution made by dissolving lactic acid and sodium lactate in water. The sodium lactate dissociates completely into sodium ions and lactate ions, but only a very small fraction of the lactic acid molecules dissociate to form hydrogen ions and lactate ions. Accordingly, the solution has a relatively low concentration of hydrogen ion and relatively high concentrations of undissociated lactic acid molecules, sodium ions, and lactate ions. Lactic acid, hydrogen ion, and lactate are in equilibrium with each other.

$$\text{Lactic acid} \rightleftharpoons \text{lactate}^- + H^+$$

By the mass-action law, an increase in the concentration of any substance on one side of the arrows forces the reaction in the opposite direction. Conversely, a decrease in the concentration of any substance on one side of the arrows forces the reaction toward that side, i.e., in the direction that generates more of that substance.

What happens if we add hydrochloric acid to this solution (Fig. 24-25A)? Hydrochloric acid, a strong acid, completely dissociates and liberates hydrogen ions. This excess of hydrogen ions drives the lactic acid reaction to the left, and many of the hydrogen ions liberated from the dissociation of hydrochloric acid thus combine with lactate to give undissociated lactic acid. As a result, many of the hydrogen ions generated by the dissociation of hydrochloric acid do not remain free in solution but become incorporated into lactic acid molecules. The final hydrogen-ion concentration is therefore smaller than if the hydrochloric acid had been added to pure water.

Conversely, if instead of adding HCl, we add a chemical that *removes* hydrogen ions, the lactic acid reaction is driven to the right, lactic acid molecules dissociate to generate more hydrogen ions, and the original fall in hydrogen-ion concentration is minimized. This process that prevents large changes in hydrogen-ion concentration when hydrogen ion is added or removed from a solution is

---

[1] Hydrogen-ion concentration is frequently expressed in terms of pH, which is defined as the negative logarithm to the base 10 of the hydrogen-ion concentration: $pH = -\log H^+$. This can be confusing for several reasons, not the least of which is that pH decreases as H^+ concentration increases. We have chosen not to use pH in this text.

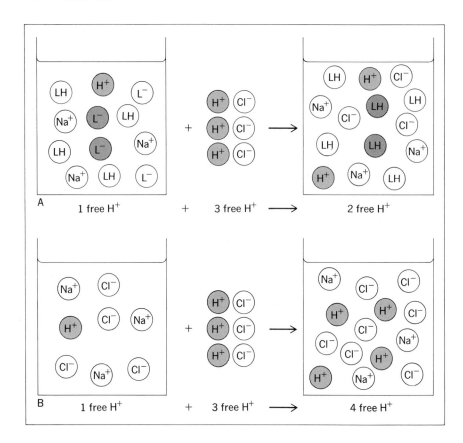

FIG. 24-25 (A) Example of a buffer system. LH = lactic acid, L⁻ = lactate. When HCl is added to the beaker, two of the hydrogen ions react with lactate to give lactic acid; therefore, only one of the three added hydrogen ions remains free in solution. Contrast this to B, in which no buffer is present and all three of the added hydrogen ions remain free in solution.

termed **buffering,** and the chemicals (in this case lactate and lactic acid) are called **buffers** or **buffer systems.**[2]

Generation of Hydrogen Ions

There are three major sources of hydrogen ion in the body:

1   Phosphorus and sulfur are present in large quantities in many proteins and other biologically important molecules. The catabolism of these molecules releases phosphoric and sulfuric acids into the extracellular fluid. These acids, to a large extent, dissociate into hydrogen ions and anions (phosphate and sulfate). For example,

$$H_2PO_4^- \longrightarrow H^+ + HPO_4^{2-}$$
$$H_2SO_4 \longrightarrow 2\ H^+ + SO_4^{2-}$$

---

[2] Of necessity, this description of acids and buffering is oversimplified and incomplete; e.g., no mention has been made of the ionization of water itself or of the importance of the hydroxyl ion. The interested reader can find descriptions of acid-base chemistry in any textbook of biochemistry.

2   Many organic acids, e.g., fatty acids and lactic acid, are produced in metabolic reactions and also liberate hydrogen ion by dissociation.

3   We have already described (Chap. 23) the major source of hydrogen ion, namely, liberation of hydrogen ion from metabolically produced carbon dioxide via the reactions

$$CO_2 + H_2O \longrightarrow H_2CO_3 \longrightarrow H^+ + \underset{\text{Bicarbonate}}{HCO_3^-}$$

As described in Chap. 23, the lungs normally eliminate carbon dioxide from the body as rapidly as it is produced in the tissues. As blood flows through the lung capillaries and carbon dioxide diffuses into the alveoli, the chemical reactions which originally generated HCO_3^- and H^+ from carbon dioxide and water in the venous blood are reversed.

$$CO_2 + H_2O \longleftarrow H_2CO_3 \longleftarrow H^+ + HCO_3^-$$

As a result, all the hydrogen ion generated from carbonic acid is reincorporated into water molecules. Therefore, there is normally no net gain or loss of hydrogen ion in the body from this source, but what happens when lung disease prevents adequate elimination of carbon dioxide? The retention of carbon dioxide means an elevated extracellular P_{CO_2}. Some of the hydrogen ions generated from the reaction of this carbon dioxide with water are also retained and raise the extracellular hydrogen-ion concentration. This hydrogen ion must be eliminated from the body (via the kidneys) in order for hydrogen-ion balance to be maintained.

Buffering of Hydrogen Ions in the Body

Between the generation of hydrogen ions in the body and their excretion (to be described) what happens? The hydrogen-ion concentration of the extracellular fluid is extremely small, approximately 0.00004 mmol/L. What would happen to this concentration if just 2 mmol of hydrogen ion remained free in solution following its dissociation from an acid? Since the total extracellular fluid is approximately 12 L, the hydrogen-ion concentration would increase by 2/12, or approximately 0.167 mmol/L. Since the original concentration was only 0.00004, this would represent more than a 4000-fold increase. Obviously, such a rise does not really occur, for we have already observed that the extracellular hydrogen ion is kept remarkably constant. Therefore, of the 2 mmol of hydrogen ion liberated in our example, only an extremely small portion can have remained free in solution. The vast majority has been bound (buffered) by other ions. The kidneys ultimately eliminate excess hydrogen ions, but it is buffering that minimizes hydrogen-ion concentration changes until excretion occurs.

The most important body buffers are bicarbonate–CO_2, large anions such as plasma protein and intracellular phosphate complexes, and hemoglobin. Recall that only free hydrogen ions contribute to the acidity of a solution. These buffers all act by binding hydrogen ions according to the general reaction

$$\text{Buffer}^- + H^+ \rightleftharpoons \text{H-buffer}$$

It is evident that H-buffer is a weak acid in that it can exist as the undissociated molecule or can dissociate to buffer$^-$ + H^+. When hydrogen-ion concentration increases, the reaction is forced to the right and more hydrogen ion is bound. Conversely,

when hydrogen-ion concentration decreases, the reaction proceeds to the left and hydrogen ion is released. In this manner, the body buffers stabilize hydrogen-ion concentration against changes in either direction.

Bicarbonate–CO_2   In this section and in Chap. 23 we have already described the relationships between HCO_3^-, H^+, and CO_2. Let us once more write the pertinent equations in their true forms as reversible reactions.

$$H^+ + HCO_3^- \rightleftharpoons H_2CO_3 \rightleftharpoons H_2O + CO_2$$

The basic mechanism by which this system acts as a buffer should be evident: An increased extracellular hydrogen-ion concentration drives the reaction to the right, H^+ and HCO_3^- combine, hydrogen ion is thereby removed from solution, and the hydrogen-ion concentration returns toward normal. Conversely, a decreased extracellular hydrogen-ion concentration drives the reaction to the left, CO_2 and H_2O combine to generate hydrogen ion, and this additional hydrogen ion returns hydrogen-ion concentration toward normal. One reason for the importance of this buffer system is that the extracellular bicarbonate concentration is normally quite high and is closely regulated by the kidneys. A second reason stems from the relationship between extracellular hydrogen-ion concentration and carbon dioxide elimination from the body.

When additional hydrogen ion is added to the extracellular fluid, i.e., when the H^+ combines with HCO_3^-, the extent to which this reaction can restore hydrogen-ion concentration to normal depends upon precisely how much additional H^+ actually combines with HCO_3^- and is thereby removed from solution. Complete compensation (which never actually occurs) can be obtained only if the HCO_3^-–CO_2 reaction proceeds to the right until all the additional H^+ has combined with HCO_3^-. However, this reaction obviously generates CO_2, which seriously hinders the further buffering ability of HCO_3^- because, as can be seen from the equation, any increase in the concentration of CO_2, by the mass-action law, tends to drive the reaction back to the left, which thus prevents further net combination of $H^+ + HCO_3^-$. In reality, however, this expected increase in extracellular CO_2 does not occur. Indeed, during periods of increased body hydrogen-ion production from or-

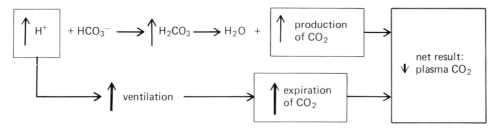

FIG. 24-26 Effects of excess hydrogen ions on plasma carbon dioxide. The direct effect, by mass action, is to increase the production of carbon dioxide, but the indirect effect is to lower carbon dioxide by reflexly stimulating breathing. Since the latter effect predominates, the net effect is a reduction of plasma carbon dioxide.

ganic, phosphoric, or sulfuric acids, the extracellular CO_2 is actually *decreased!* What causes this? We have already studied the mechanism in Chap. 23: A greater extracellular hydrogen-ion concentration stimulates the respiratory centers to increase alveolar ventilation and thereby causes greater elimination of carbon dioxide from the body (Fig. 24-26). Thus, although an increased combination of H^+ and HCO_3^- generates more carbon dioxide, respiratory stimulation produced by the higher extracellular hydrogen-ion concentration results in the elimination of carbon dioxide even faster than it is generated. As a result, the extracellular carbon dioxide decreases and, by the mass-action law, the further combination of H^+ and HCO_3^- is actually facilitated.

A lower extracellular hydrogen-ion concentration that results from either decreased hydrogen-ion production or increased hydrogen-ion loss from the body is compensated for by just the opposite buffer reactions: (1) The lower hydrogen-ion concentration drives the HCO_3^-–CO_2 reaction to the left, carbon dioxide and water combine to generate H^+ + HCO_3^-, and this additional hydrogen ion returns hydrogen-ion concentration toward normal; and (2) the lower hydrogen-ion concentration decreases alveolar ventilation and carbon dioxide elimination by inhibiting the medullary respiratory centers. The elevated extracellular carbon dioxide that is the result of this process also serves to drive the HCO_3^-–CO_2 reaction to the left, which allows further generation of hydrogen ion.

When lung disease is itself the cause of the increased hydrogen-ion concentration, the efficiency of this buffer system is obviously greatly impaired. Actually a large fraction of any hydrogen-ion excess or deficit is always buffered by the other buffers listed.

Hemoglobin as a Buffer   In Chap. 23 we pointed out that most metabolically produced carbon dioxide is carried from the tissues to the lungs in the form of HCO_3^-. These bicarbonate ions are generated by the hydration of CO_2 to form H_2CO_3 and the dissociation of H_2CO_3 to HCO_3^- and H^+. As blood flows through the lungs, the reaction is reversed and H^+ and HCO_3^- recombine. However, these hydrogen ions must be buffered while they are in transit from the tissues to the lungs, a function performed primarily by hemoglobin.

Renal Regulation of Extracellular Hydrogen-Ion Concentration

It should be emphasized that none of the buffer systems described above eliminates hydrogen ion from the body; instead, they combine with the hydrogen ion. Binding thus removes the hydrogen ion from solution and prevents it from contributing to the free-hydrogen-ion concentration; the actual elimination of hydrogen ion is normally performed only by the kidneys.

The quantity of phosphoric, sulfuric, and organic acids formed depends primarily upon the type and quantity of food ingested. A high-protein diet, for example, results in increased protein breakdown and release of large quantities of sulfuric acid. The average American diet results in the liberation of 40 to 80 mmol of hydrogen ion each day. If hydrogen-ion balance is to be maintained, the same quantity must be eliminated. This loss occurs via the kidneys. In addition, the kidneys must be capable of *altering* their hydrogen-ion excretion in response to changes in body hydrogen-ion production, regardless of whether the source is carbon dioxide (in lung disease) or phosphoric, sulfuric, or organic acids. The kidneys must also be able to compensate for any gastrointestinal loss or gain of hydrogen ion resulting from disease.

Virtually all the hydrogen ion excreted in the urine enters the tubules via tubular secretion, the mechanism and its control being quite complex. Suffice it to say that the presence of excess acid in the body induces an increased urinary hydrogen-ion excretion. Conversely, the kidneys respond to a decreased amount of acid in the body by lowering urinary hydrogen-ion excretion. The controlling effect of the acid appears to be primarily exerted directly upon the tubular cells, with no nerve or hormone intermediates. Obviously, such a system is effective in stabilizing hydrogen-ion concentration at the normal value.

The ability of the kidneys to excrete H^+ depends both upon tubular hydrogen-ion secretion and upon the presence of buffers in the urine to combine with the hydrogen ions. The major urinary buffers are HPO_4^{2-} and ammonia NH_3.

$$HPO_4^{2-} + H^+ \longrightarrow H_2PO_4^-$$
$$NH_3 + H^+ \longrightarrow NH_4^+$$

The HPO_4^{2-} in the tubular fluid has been filtered and not reabsorbed. In contrast, the ammonia of the tubular fluid is formed by the tubular cells themselves by the deamination of certain amino acids transported into the renal tubular cells from the peritubular capillary plasma. From the cells the ammonia diffuses into the lumen. The amount that remains in the lumen depends upon the hydrogen-ion concentration there, since the tubular cell membrane is quite permeable to ammonia but not to NH_4^+. Accordingly, the more hydrogen ion there is in the lumen, the greater the conversion of NH_3 to NH_4^+. Thus increased secretion of hydrogen ion automatically induces greater excretion of NH_4^+. Another important feature of this system is that the rate of ammonia production by the renal tubular cells increases whenever extracellular hydrogen-ion concentration remains elevated for more than 1 to 2 days; this extra ammonia provides the additional buffers required for combination with the increased hydrogen ions secreted.

Classification of Disordered
Hydrogen-Ion Concentration

Acidosis refers to any situation in which the hydrogen-ion concentration of arterial blood is elevated; **alkalosis** denotes a reduction. Acidosis and alkalosis are the results of an imbalance between hydrogen-ion gain and loss, and all such situations fit into two distinct categories: (1) **respiratory acidosis** or **alkalosis,** and (2) **metabolic acidosis** or **alkalosis.**

As its name implies, the first category results either from failure of the lungs to eliminate CO_2 as fast as it is produced (acidosis) or from elimination of CO_2 faster than it is produced (alkalosis). As described earlier, the imbalance of arterial hydrogen-ion concentration in such cases is completely explainable in terms of mass action. Thus, the hallmark of a respiratory acidosis is an elevated arterial CO_2 and H^+, that of respiratory alkalosis is a reduction in both values.

The second category, metabolic acidosis or alkalosis, includes all situations other than those in which the primary problem is respiratory. Some common forms of metabolic acidosis are excessive production of lactic acid (during severe exercise or hypoxia) or of ketone acids (in uncontrolled diabetes mellitus or in fasting, Chap. 26). Metabolic acidosis can also result from excessive loss of bicarbonate, as in diarrhea. A frequent cause of metabolic alkalosis is persistent vomiting, with its associated loss of hydrogen ions (as HCl) from the stomach.

What is the arterial CO_2 in metabolic, as opposed to respiratory, acidosis or alkalosis? Since by definition metabolic acidosis and alkalosis must be caused by something other than excess retention or loss of CO_2, one might have predicted that arterial CO_2 would be unchanged, but such is not the case. As described previously, the elevated hydrogen-ion concentration associated with the metabolic acidosis, say as a result of retention of lactic acid, stimulates ventilation by a reflex via the peripheral chemoreceptors and lowers arterial CO_2; by mass action this helps restore the hydrogen-ion concentration toward normal. Conversely, a person with metabolic alkalosis, say caused by vomiting, will by a reflex have ventilation inhibited; the result is a rise in arterial CO_2 and, by mass action, an associated elevation of hydrogen-ion concentration. To reiterate, the CO_2 changes in metabolic acidosis and alkalosis are not the cause of the acidosis or alkalosis but rather are compensatory reflex responses to primary nonrespiratory abnormalities.

KIDNEY DISEASE

The term kidney disease is no more specific than "car trouble," since many diseases affect the kidneys. Bacteria cause kidney infections, most of which are collectively called *pyelonephritis*. A common type of kidney disease, *glomerulonephritis*, results from an allergy incident to throat infection by a specific group of bacteria. Congenital defects, stones, tumors, and toxic chemicals are possible sources of kidney damage. Obstruction of the urethra or a ureter may cause injury as a result of a buildup of pressure and may predispose the kidneys to bacterial infection.

Disease can attack the kidney at any age. Experts estimate that there are at present more than 3 million undetected cases of kidney infection in the United States and that 25,000 to 75,000 Americans die of kidney failure each year.

Early symptoms of kidney disease depend greatly upon the type of disease involved and the specific part of the kidney affected. Although many diseases are self-limited and produce no permanent damage, others progress if untreated. The end state of progressive diseases, regardless of the nature of the damaging agent (bacteria, toxic chemical, etc.), is a shrunken, nonfunctioning kidney. Similarly, the symptoms of profound renal malfunction are independent of the damaging agent and are collectively known as *uremia*, literally "urine in the blood."

The severity of uremia depends upon how well the impaired kidneys are able to preserve the constancy of the internal environment. Assuming that the patient continues to ingest a normal diet that contains the usual quantities of nutrients and electrolytes, what problems arise? The key fact to keep in mind is that the kidney destruction has markedly reduced the number of functioning nephrons. Accordingly, the many substances that gain entry to the tubule primarily by filtration are filtered in diminished amounts. This category includes sodium chloride, water, calcium, and a number of waste products. In addition, the excretion of potassium, hydrogen ion, and certain other substances is im-

paired because there are too few nephrons capable of normal tubular secretion. Thus the buildup of many substances in the blood causes the symptoms and signs of uremia. Abnormal secretion of the various hormones produced by the kidneys also plays a role.

Artificial Kidney

The artificial kidney is an apparatus that eliminates the excess ions and wastes that accumulate in the blood when the kidneys fail. Blood is pumped from one of the patient's arteries through tubing that is bathed by a large volume of fluid. The tubing then conducts the blood back into the patient by way of a vein. The tubing is generally made of a cellophane, which is highly permeable to most solutes but relatively impermeable to protein—characteristics quite similar to those of capillaries. The bath fluid, which is constantly replaced, is a salt solution similar in ionic concentrations to normal plasma. The basic principle is simply that of dialysis, or diffusion. Because the cellophane is permeable to most solutes, as blood flows through the tubing, solute concentrations tend to equilibrate in the blood and bath fluid. Thus, if the plasma potassium concentration of the patient is above normal, potassium diffuses out of the blood into the bath fluid. Similarly, waste products and excesses of other substances diffuse across the cellophane tubing and are eliminated from the body.

At first patients were placed on the artificial kidney only when there was reason to believe that their kidney damage was only temporary and that recovery would occur if the patient could be kept alive during temporary renal failure. However, technical improvements have permitted many patients to utilize the artificial kidney several times a week for unlimited periods, and thus patients with permanent kidney failure can be kept alive.

The other major hope for patients with permanent renal failure is kidney transplantation. Although great strides have been made, the major problem remains the frequent rejection of the transplanted kidney by the recipient's body (see Chap. 28).

KEY TERMS

ureter	balance concept	aldosterone
urinary bladder	glomerular filtration	renin-angiotensin system
urethra	tubular reabsorption	countercurrent multiplier system
nephron	tubular secretion	parathormone
glomerulus	excretion	vitamin D (1,25 dihydroxyvitamin D_3)
Bowman's capsule	total renal blood flow	calcitonin
proximal convoluted tubule	micturition	acid
loop of Henle	antidiuretic hormone (ADH)	base
distal convoluted tubule	glomerular filtration rate (GFR)	buffer system
collecting duct		

REVIEW EXERCISES

Section A

1 Draw a diagram of the urinary system and label the kidneys, ureters, bladder, and urethra.
2 Draw a section of the kidney and label the hilum, renal pelvis, major and minor calyces, a renal pyramid, a renal column, a lobe, and the renal artery and vein.
3 Draw a nephron and label the glomerulus, Bowman's capsule, proximal convoluted tubule, loop of Henle, distal convoluted tubule, collecting duct, afferent and efferent arterioles, and peritubular capillaries.
4 Explain the balance concept; use water as an example.
5 Define and differentiate between glomerular filtration, tubular reabsorption, tubular secretion, and excretion.
6 Give the approximate volume of water filtered, reabsorbed, and excreted by the kidneys each day.
7 Describe the factors that affect glomerular filtration.
8 List the plasma substances that are filtered at the glomerulus, and on a second list indicate the substances that are not filtered.
9 Define glomerular filtration rate and distinguish between glomerular filtration rate and total renal blood flow.
10 Explain how plasma composition can be altered by varying the filtration, reabsorption, or secretion of a given substance.
11 Explain the relationship between bladder pressure and sphincter tension in micturition.

Section B

1 Describe the way sodium, chloride, and water are handled in the proximal tubule.
2 Describe the control(s) of plasma ADH concentration and the effect of ADH on the nephron.

3 Explain how total body sodium is related to extracellular fluid volume.
4 Explain the effect of aldosterone.
5 Describe the relationship between aldosterone and the renin-angiotensin system.
6 Describe the reflex pathways by which decreased plasma volume leads to a decreased sodium excretion; decreased water excretion.
7 Describe the conditions under which the osmolarity of the medullary interstitium can affect the osmolarity of the urine.
8 Describe the effect of hypotonic plasma on ADH secretion.

Section C

1 Describe the way potassium is handled by the kidneys.
2 Describe the effect of potassium on aldosterone secretion and the effect of aldosterone on the renal handling of potassium.
3 List the major organs that play a role in calcium homeostasis.
4 List the hormones that control plasma calcium concentration and the factors that control their release.
5 Describe the role of vitamin D in calcium balance.
6 Distinguish between an acid and a base.
7 Describe the relationship between acidity and hydrogen-ion concentration.
8 Explain buffering using HCO_3^-–H_2CO_3 as an example.
9 Differentiate between a strong acid and a weak acid.
10 Describe the way in which the kidneys function to regulate extracellular hydrogen-ion concentration.
11 Distinguish between respiratory and metabolic acidosis and give an example of each.
12 Distinguish between respiratory and metabolic alkalosis and give an example of each.

CHAPTER

THE DIGESTIVE SYSTEM

SECTION A: OVERVIEW OF THE GASTROINTESTINAL SYSTEM AND ITS REGULATION

The **gastrointestinal system** includes the gastrointestinal tract (mouth, esophagus, stomach, small and large intestines), salivary glands, and portions of the liver and pancreas (Fig. 25-1). Its function is to transfer nutrients, salts, and water from the external environment to the internal environment, where they can be distributed to the cells by the circulatory system. Most food is taken into the mouth as large particles that consist of high-molecular-weight substances, such as proteins and polysaccharides, which are unable to cross cell membranes. Before these substances can be absorbed, they must be broken down into smaller molecules, such as amino acids and monosaccharides. This breaking-down process, **digestion,** is accomplished mainly by the action of acid and digestive enzymes secreted into the gastrointestinal tract or contained in its lining cells. The small molecules that result from digestion cross the cells of the intestinal tract (**absorption**) and enter the blood or lymph. While these processes are taking place, smooth muscle contractions move the luminal contents through the tract.

Contrary to popular belief, the gastrointestinal system is not a major excretory organ for eliminating wastes from the body. It is true that small amounts of certain end products, such as the breakdown products of hemoglobin, are normally elimi-

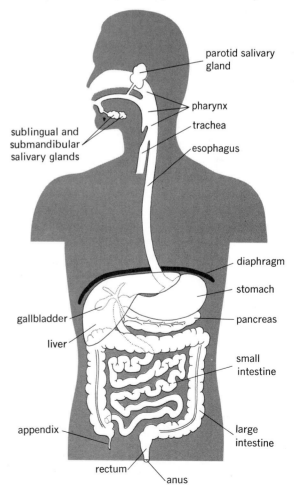

FIG. 25-1   Anatomy of the gastrointestinal system.

parotid salivary gland

pharynx

trachea

esophagus

sublingual and submandibular salivary glands

diaphragm

stomach

pancreas

gallbladder

liver

small intestine

appendix

large intestine

rectum

anus

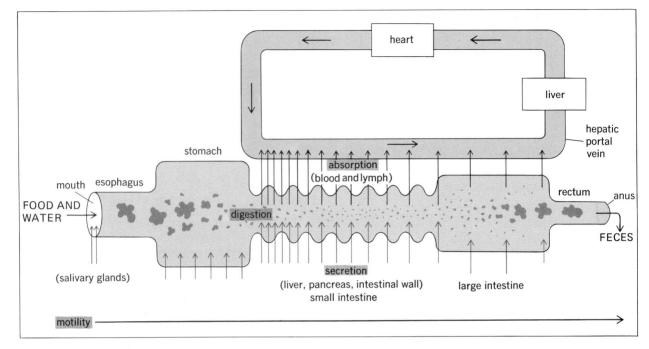

FIG. 25-2 Summary of gastrointestinal activity involving motility, secretion, digestion, and absorption.

nated in the feces, but the elimination of most wastes from the internal environment is performed by the lungs and kidneys. Feces consist primarily of bacteria and ingested material which failed to be digested and absorbed during its passage along the gastrointestinal tract, i.e., material that was never actually part of the internal environment.

In this chapter we examine the anatomy of the gastrointestinal system and four aspects of gastrointestinal function: (1) secretion, (2) digestion, (3) absorption, and (4) motility (Fig. 25-2). In Chap. 26 the mechanisms that control the distribution and utilization of absorbed food products as well as hunger and thirst will be considered.

GENERAL STRUCTURE OF THE GASTROINTESTINAL TRACT

The **gastrointestinal tract** is a tube of variable diameter, 4.5 m (15 ft) in length, that begins with the mouth and ends with the anus. The lumen of this tube is continuous with the external environment, which means that its contents are technically out-

side the body. This fact is relevant to an understanding of some of the tract's properties. For example, the lower portion of the intestinal tract is inhabited by millions of bacteria, most of which are harmless and even beneficial in this location; however, if the same bacteria enter the body proper, as may happen, for example, in the case of a ruptured appendix, they are extremely harmful and even lethal.

Although there is some regional variation, the gastrointestinal tract throughout most of its length has the same general structure as the segment of intestine illustrated in Fig. 25-3. From the esophagus to the anus, its wall is made up of four layers that, beginning with the innermost, are as follows: (1) mucosa, (2) submucosa, (3) muscularis externa, the main muscle coat, and (4) serosa.

The **mucosa** consists of three layers: epithelium, lamina propria, and muscularis mucosa. The luminal surface of the mucosa is generally not flat and smooth, but highly convoluted, with many ridges and valleys that greatly increase the total surface area available for interaction with the luminal contents. From the stomach on, the mucosa is lined by a single layer of **epithelium,** across which

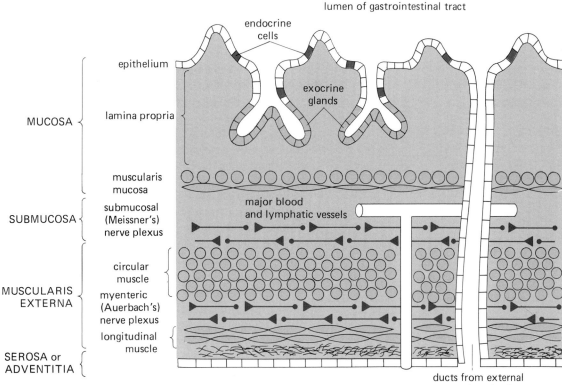

FIG. 25-3 Representative longitudinal section of wall of the gastrointestinal tract. Not shown are the smaller blood vessels, neural connections between the two plexuses, and neural terminations on the muscles and glands.

absorption takes place. Included in this epithelial layer are cells that secrete mucus and other cells that release hormones into the blood. Invaginations of the epithelial layer into the underlying tissue form tubular exocrine glands that secrete mucus, acid, enzymes, water, and ions into the lumen. Just below the epithelial surface is a layer of connective tissue, the **lamina propria,** through which pass small blood vessels, nerve fibers and lymphatic ducts. The lamina propria is separated from the underlying tissues by a thin layer of smooth muscle, the **muscularis mucosa.**

Beneath the mucosa is a second connective-tissue layer, the **submucosa,** that contains the larger blood vessels and lymphatics from which branches penetrate into the overlying mucosal layer and the underlying muscular layer. Also located in this layer is a network of nerve cells known as the **submucosal,** or **Meissner's, plexus.**

Next is the **muscularis externa,** which in most regions of the gastrointestinal tract consists of two smooth muscle layers: an outer layer with fibers that run almost longitudinally along the tube and an inner layer where the muscles lie with a more circular orientation. Contractions of the longitudinal layer shorten the gastrointestinal tract, and contractions of the circular layer constrict its lumen, which exerts pressure on the contents of the lumen and causes them to advance. Between these two layers of smooth muscle is another, more extensive, plexus of nerve cells, the **myenteric,** or **Auerbach's, plexus.**

The outer layer of the gastrointestinal wall is called the **serosa** or **adventitia** (fibrosa), depending on whether it is covered by peritoneum or by adjacent tissues, as is the case with the esophagus, for example. *Serosal membranes* consist of a connective-tissue layer and a sheet of mesothelium, and they secrete a watery substance that allows the organs to move past each other and the walls of the

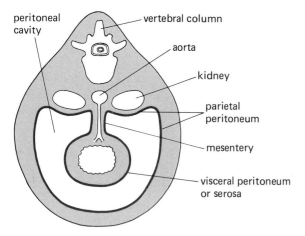

FIG. 25-4 Diagram of the peritoneum and peritoneal cavity.

cavity. Similar membranes line the thoracic and pericardial cavities.

Support of the Abdominal Portion of the Gastrointestinal Tract

The stomach and intestines are suspended in the abdominal cavity by thin sheets of serous mem-

brane-covered connective tissue, the **peritoneum,** which are attached to the body wall. The peritoneum, like the pleura of the thoracic cavity, can be viewed as a membranous bag that has been invaginated by organs (Fig. 25-4). The invaginating organs in the thorax are, of course, the lungs; those in the abdominal, or peritoneal, cavity are the stomach and intestines. The uninvaginated part of the peritoneum that adheres to the abdominal wall is the **parietal peritoneum,** and the part pushed inward by the organs (and, therefore, adhering to their surfaces) is the **visceral peritoneum** (Fig. 25-4). Thus, visceral peritoneum forms the outer layer (the serosal layer) of the gastrointestinal tract.

As the invaginating organs moved away from the abdominal wall during development of an individual, the serosal membranes that were dragged behind them fused together and formed a sheet of connective tissue covered on both sides by secretory mesothelium. The sheets that attach the stomach to the body wall or to other organs are the **greater** and **lesser omenta** (sing.: *omentum*); those that attach the intestines to the body wall are the **mesenteries** (Fig. 25-5). (The mesentery for the large intestine is the **mesocolon.**) The omenta and mesenteries contain numerous fat deposits and transmit

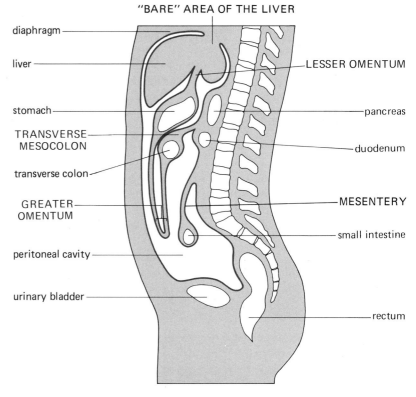

FIG. 25-5 Midsagittal section of the abdominal (peritoneal) cavity.

the blood vessels, lymphatics, and nerves to the digestive tract.

The parietal and visceral layers of the peritoneum are virtually in contact with each other so that the space between them, the **abdominal,** or **peritoneal, cavity,** is actually only a potential space. The small amount of fluid that fills the slitlike space of the cavity is secreted by the mesothelial cells of the peritoneum.

ROLE OF THE VARIOUS ORGANS OF THE GASTROINTESTINAL SYSTEM

The role of the various organs of the gastrointestinal system is summarized in Fig. 25-6. Digestion begins in the **mouth** with chewing, which breaks up large food particles into small particles that can be swallowed.

The salivary glands secrete into the mouth a mucus solution **(saliva)** that moistens and lubricates the food particles prior to swallowing. (Mucus is secreted by gland cells throughout the gastrointestinal tract. In addition to its lubricating function, mucus, in some regions, provides a layer at the epithelial surface that protects underlying cells from abrasion and from the enzymes and other chemicals in the lumen.) Saliva also contains a carbohydrate-digesting enzyme, **amylase,** which begins the breakdown of large polysaccharides into smaller fragments. A third function of saliva is to dissolve some of the molecules in the food particle; only in this dissolved state can they react with the chemoreceptors in the mouth and give rise to the sensation of taste.

The next segments of the tract, the **pharynx** and the **esophagus,** contribute nothing to digestion but simply provide the connection between the mouth and the stomach.

The **stomach** is a hollow structure that initially stores the contents of a meal. Some of the glandular cells in the epithelium that lines the stomach secrete a very strong acid, **hydrochloric acid.** This acid alters the protein structure in the ingested food so as to break up connective tissue and cells, forming a solution of molecules known as **chyme.** The acid in the stomach (gastric acid) thus continues the process begun by chewing, namely reducing large particles of food to smaller particles. Acid, however, has little ability to break down proteins and polysaccharides all the way to amino acids and glucose,

and it has virtually no digestive action at all on fats. A second function of gastric acid is to kill most of the bacteria that enter along with food. This process is not 100 percent effective, and some bacteria normally survive to take up residence and multiply in the intestinal tract, particularly the large intestine.

Also secreted by the stomach glands are the precursors of several protein-digesting enzymes, collectively known as **pepsin,** that split proteins into small peptide fragments. Although polysaccharides and proteins are partially digested while in the stomach by the actions of salivary amylase and gastric pepsin, the end products are large and electrically charged, so that they cannot diffuse across the epithelium. The stomach lining also lacks carrier-mediated transport systems for them. The fats in the gastric chyme, which are not soluble in water, tend to aggregate into large droplets, and are not available for absorption. Thus, little absorption of nutrients occurs across the wall of the stomach, and its major functions are to store, dissolve, and partially digest the contents of a meal, and to deliver chyme to the small intestine in amounts optimal for digestion and absorption in that segment.

Most digestion and absorption occurs in the **small intestine** where large molecules of intact or partially digested carbohydrate, fat, and protein are broken down by enzymes into monosaccharides, fatty acids and glycerol, and amino acids, all of which are able to cross the membranes of the epithelial cells either by diffusion or mediated transport and pass into the blood or lymph. Other organic nutrients, as well as salts and water, are also absorbed in this segment. Some of the enzymes that carry out this digestive process are on the luminal membranes of the epithelial cells of the small intestine, but most are secreted by the pancreas.

The exocrine portion of the **pancreas** secretes enzymes specific for each of the major classes of organic molecules. These enzymes enter the small intestine through a duct that leads from the pancreas to the first portion of the small intestine, the **duodenum.** The exocrine pancreas also secretes large amounts of sodium bicarbonate, which neutralizes the high acidity of the chyme that comes from the stomach.

Because fat is insoluble in water, the digestion of fat in the small intestine requires special processes to solubilize these molecules. Fat solubilization is brought about by a group of detergent molecules, known as **bile salts,** that are secreted in a solution

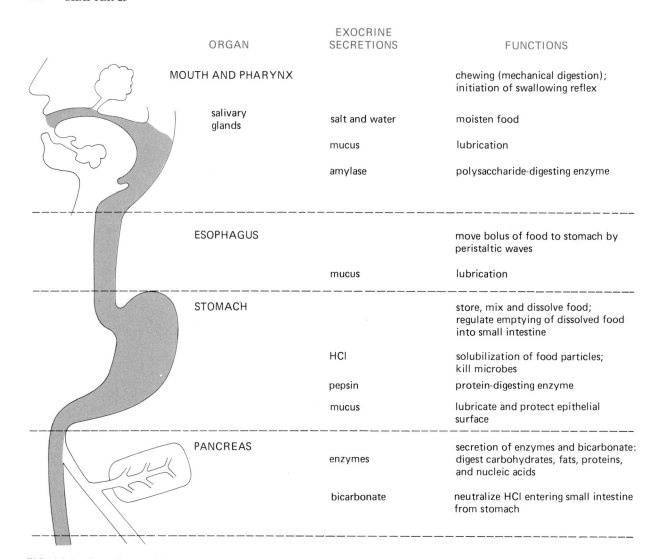

ORGAN	EXOCRINE SECRETIONS	FUNCTIONS
MOUTH AND PHARYNX		chewing (mechanical digestion); initiation of swallowing reflex
salivary glands	salt and water	moisten food
	mucus	lubrication
	amylase	polysaccharide-digesting enzyme
ESOPHAGUS		move bolus of food to stomach by peristaltic waves
	mucus	lubrication
STOMACH		store, mix and dissolve food; regulate emptying of dissolved food into small intestine
	HCl	solubilization of food particles; kill microbes
	pepsin	protein-digesting enzyme
	mucus	lubricate and protect epithelial surface
PANCREAS	enzymes	secretion of enzymes and bicarbonate: digest carbohydrates, fats, proteins, and nucleic acids
	bicarbonate	neutralize HCl entering small intestine from stomach

FIG. 25-6 Functions of the gastrointestinal organs.

known as **bile** by the liver into bile ducts that eventually join the pancreatic duct and empty into the duodenum. In addition to bile salts, bile contains sodium bicarbonate, which helps to neutralize acid from the stomach. Between meals, secreted bile is stored in a small sac underneath the liver, the **gallbladder,** which concentrates the bile by absorbing salts and water. During a meal, the gallbladder contracts, which causes a concentrated solution of bile to be injected into the small intestine.

Monosaccharides and amino acids are absorbed across the wall of the small intestine, mainly by specific carrier-mediated transport processes in the epithelial membranes, whereas fatty acids enter the epithelial cells by diffusion. Salts are actively absorbed, and water follows passively down osmotic gradients. Most digestion and absorption has been completed by the middle portion of the small intestine.

The motility of the small intestine mixes the contents of the lumen and slowly advances the contents toward the **large intestine.** Only a small volume of salts, water, and undigested material is passed on to the large intestine, which stores the undigested material prior to defecation and concentrates it by absorbing salts and water.

It must be recognized that although the average adult consumes about 800 g of solid food and 1200

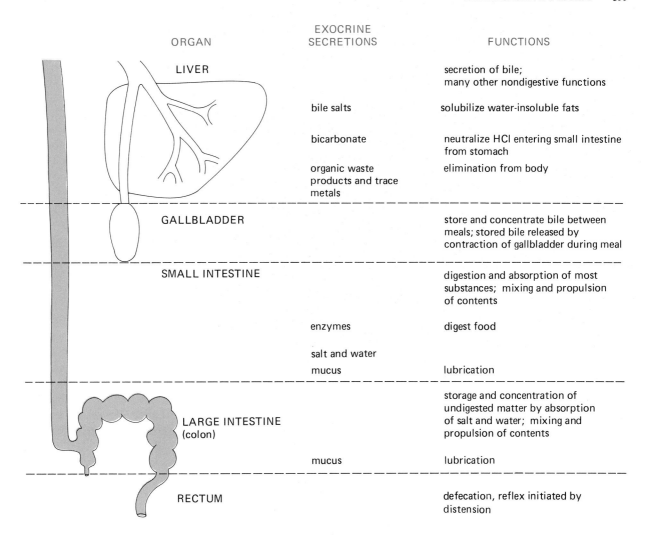

ORGAN	EXOCRINE SECRETIONS	FUNCTIONS
LIVER		secretion of bile; many other nondigestive functions
	bile salts	solubilize water-insoluble fats
	bicarbonate	neutralize HCl entering small intestine from stomach
	organic waste products and trace metals	elimination from body
GALLBLADDER		store and concentrate bile between meals; stored bile released by contraction of gallbladder during meal
SMALL INTESTINE		digestion and absorption of most substances; mixing and propulsion of contents
	enzymes	digest food
	salt and water	
	mucus	lubrication
LARGE INTESTINE (colon)		storage and concentration of undigested matter by absorption of salt and water; mixing and propulsion of contents
	mucus	lubrication
RECTUM		defecation, reflex initiated by distension

FIG. 25-7 Average amounts of fluid and food ingested, secreted, absorbed, and excreted from the gastrointestinal tract daily.

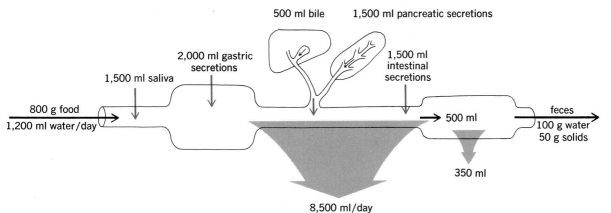

mL of water per day, this is only a fraction of the total material that enters the gastrointestinal tract. To the approximately 2000 mL of ingested food and drink are added about 7000 mL of fluid from the salivary glands, stomach, pancreas, liver, and intestinal tract (Fig. 25-7). Of this total volume, only about 100 mL is lost in the feces and the rest is absorbed across the walls of the intestine. Almost all the secreted salts and enzymes (the latter after digestion to amino acids) are also absorbed back into the circulation. Thus, little of the huge volume of secreted fluid is actually lost from the body.

REGULATION OF GASTROINTESTINAL PROCESSES

Basic Principles

A special environment must be created in the lumen of the gastrointestinal tract for digestion and absorption to occur. Accordingly, glandular secretions and smooth muscle contractions must be controlled to provide this optimal environment. Unlike other control systems, which regulate some variable in the extracellular fluid, the control mechanisms of the gastrointestinal system regulate conditions in the lumen of the tract. In other words, they are not governed to any great extent by the nutritional state of the body but rather by the volume and composition of the luminal contents. Most of the gastrointestinal control systems are reflexes that often contain both neural and endocrine components. As reflexes, they can be analyzed in terms of stimulus, receptor, afferent and efferent components, integrating center, effector organ, and response. On the other hand, some gastrointestinal cells respond directly to substances in the lumen or to a chemical messenger (paracrine agent) that has diffused from its cell of origin to the target cell.

It is appropriate that the majority of the receptors for the control systems are located in the wall of the gastrointestinal tract itself. Most gastrointestinal reflexes are initiated by a relatively small number of luminal stimuli: (1) distension of the wall by the luminal contents; (2) osmolarity (total solute concentration) of the chyme; (3) acidity (H^+ concentration) of the chyme; and (4) products formed by the enzymatic digestion of carbohydrates, fats, and proteins—monosaccharides, fatty acids, peptides, and amino acids. Signals initiated by these stimuli act on mechanoreceptors, osmoreceptors, and chemoreceptors and trigger neuroendocrine reflexes that influence the effectors (the muscle layers in the wall of the tract and the exocrine glands that secrete substances into its lumen); thus, both the receptors and effectors are in the gastrointestinal system.

These reflexes, like other negative-feedback systems, prevent large changes in the variables that are the stimuli for the reflexes; in so doing, they maintain optimal conditions for digestion and absorption. For example, increased acidity in the duodenum, a result of acid emptying into it from the stomach, reflexly inhibits the secretion of acid by the stomach, which prevents over-acidification of the duodenum (the adaptive value of this reflex is that digestive enzymes in the duodenum are inhibited by high acidity). To take another example, distension of the small intestine, caused by the presence of material not yet absorbed, reflexly increases its contractile activity, which thereby facilitates mixing of the chyme with the intestinal secretions and brings it into contact with the epithelium. At the same time, the contractions of the stomach are reflexly inhibited, which thereby decreases the rate at which additional material is emptied into the intestine.

Let us now look at the general types of neural and hormonal pathways that mediate these reflexes.

Neural Regulation

The gastrointestinal tract has in its walls its own local nervous system, in the form of the two major nerve plexuses mentioned earlier, the myenteric and submucosal plexuses (Fig. 25-3). These two plexuses are found throughout the length of the gastrointestinal tract beginning with the esophagus. They are composed of neurons that form synaptic junctions with other neurons in the plexus or that end near smooth muscles and glands. Moreover, many nerve processes pass between the two plexuses so that neural activity in one plexus influences the activity in the other. The nerve processes in both plexuses branch profusely, and stimulation at one point in the plexus can lead to impulses conducted both up and down the tract. Thus, activity initiated in the plexus in the upper part of the small intestine may affect smooth muscle and gland activity in the stomach as well as in the lower part of

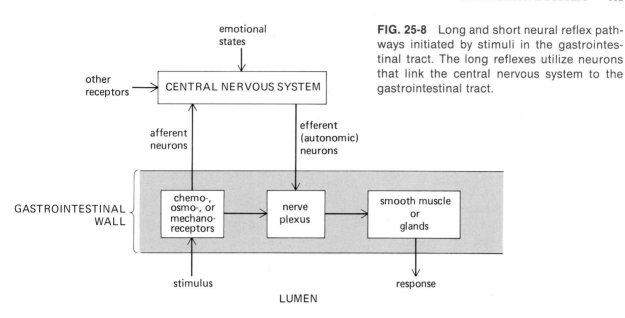

FIG. 25-8 Long and short neural reflex pathways initiated by stimuli in the gastrointestinal tract. The long reflexes utilize neurons that link the central nervous system to the gastrointestinal tract.

the intestinal tract. Many of the receptors mentioned in the previous section are, in fact, on the endings of the plexus neurons themselves.

The neural connections in the plexuses permit reflexes that are independent of the central nervous system. This is not to say that the tract is not subject to control from the central nervous system; nerve fibers from both the sympathetic and parasympathetic branches of the autonomic nervous system enter the intestinal tract and synapse with neurons in the plexuses. Via these pathways, the central nervous system can influence the motor and secretory activity of the gastrointestinal tract. The major autonomic nerves that supply the gastrointestinal tract are the **vagus nerves** (Xth cranial), which send branches to the stomach, small intestine, and upper portion of the large intestine. These nerves are composed of efferent parasympathetic fibers and many afferent fibers from receptors and nerve plexuses in the gastrointestinal wall.

Thus, two types of neural reflex pathways link a stimulus to a response (Fig. 25-8): so-called **short reflexes,** which utilize the receptors, nerve plexuses, and effector cells of the gastrointestinal tract itself; and **long reflexes,** which may use either gastrointestinal receptors or other receptors (such as those stimulated by the sight or smell of food) and which incorporate autonomic nerve fibers from the central nervous system or autonomic ganglia as part of their efferent pathway. Complex behavioral influences, e.g., emotions, also operate, in large part, through the central nervous system. Some controls are mediated solely by short pathways or long pathways, whereas others use both simultaneously.

Hormonal Regulation

The first hormone discovered (in 1902) was a substance extracted from the wall of the small intestine. When injected into the blood of a dog, this substance caused the pancreas to secrete a solution that contained a high concentration of sodium bicarbonate. This gastrointestinal hormone was named **secretin,** and with its discovery the science of endocrinology was born.

Unlike the endocrine glands of the gonads, adrenals, pituitary, or thyroid, the hormone-secreting cells of the gastrointestinal tract are not clustered into discrete organs but consist of individual cells scattered throughout the epithelium of the stomach and intestine. One surface of the cell is exposed to the lumen of the gastrointestinal tract; at this surface, various substances in the chyme stimulate the cell and thus cause hormonal release from the other surface, the basal portion of the cell. After its release, the hormone enters capillaries, and the circulation provides the route by which the hormone gets back to its target cells in the gastrointestinal system.

| TABLE 25-1 MAJOR PROPERTIES OF GASTROINTESTINAL HORMONES |||||

Hormone	Major Location of Hormone-Producing Cells	Stimuli for Hormone Release	Major Target Cell Responses
Gastrin	Stomach	Peptides in stomach; distension of stomach; nerve stimulation; (increased gastric acid inhibits release)	Stimulates *gastric* acid secretion
Secretin	Small intestine	Acid in small intestine	Stimulates *pancreatic* bicarbonate secretion; potentiates CCK's stimulation of *pancreatic* enzyme secretion; in the *liver,* stimulates fluid and bicarbonate secretion into the bile
Cholecystokinin (CCK)	Small intestine	Amino acids or fatty acids in small intestine	Inhibits *gastric* motility; potentiates secretin's stimulation of *pancreatic* bicarbonate secretion; stimulates *pancreatic* enzyme secretion; in the *liver.* potentiates secretin's stimulation of fluid and bicarbonate secretion into the bile; stimulates *gallbladder* contraction
Gastric-inhibitory peptide (GIP)	Small intestine	Fatty acids or monosaccharides in small intestine	Inhibits *gastric* acid secretion

Over the years a large number of extracts from various regions of the gastrointestinal tract have been injected into animals and found to produce changes in gastrointestinal secretory and contractile activity. However, as described in Chap. 18, tissue extracts may contain, in addition to hormones, other chemicals such as neurotransmitters and paracrine substances that are capable of stimulating glandular and muscular activity. Over a dozen different substances are currently being investigated for possible hormonal activity, but only four have met all the criteria required for acceptance as a proven hormone. These are **secretin, cholecystokinin (CCK), gastrin,** and **gastric inhibitory peptide (GIP),** all of which are small polypeptides. Further investigations are almost certain to add other hormones to this list..

Table 25-1 summarizes the major characteristics of the four established hormones. The table not only serves as a reference for future discussions but can be used to illustrate two generalizations that we wish to emphasize: (1) each hormone participates in a feedback system for regulating the luminal environment and, in doing so, may affect multiple target organs; and (2) each target organ is usually affected by more than one of the hormones.

The first generalization can be seen by looking at cholecystokinin, for example. The presence of fatty acids in the duodenum triggers the release of this hormone from the duodenal wall, and it, in turn, slows gastric emptying (so as not to overload the small intestine), stimulates release from the pancreas of enzymes (including the enzyme that digests fat), and causes the gallbladder to contract (delivering to the intestine bile salts that are also required for fat digestion and absorption).

The second generalization becomes apparent when one looks in the last column for effects on the pancreas; secretin and CCK both act to stimulate pancreatic bicarbonate secretion. To take another example, gastrin and GIP both affect the stomach, but in this case the controlling hormones have opposite effects; gastrin stimulates gastric acid secretion and GIP inhibits it. (The term *potentiation* signifies the fact that, in the presence of CCK, secretin stimulates the release of a bicarbonate secretion that is greater than that produced by the sum of the effects of either hormone alone.)

It should be emphasized that Table 25-1 lists only the major effects produced by physiologic concentrations of these hormones. Many additional responses have been observed following the injection of these hormones, often at very high concentrations, into animals and people, and it remains to be determined whether these responses play any role in the normal regulation of the gastrointestinal tract.

Phases of Gastrointestinal Control

The neural and hormonal control of the gastrointestinal system is, in large part, divisible into three phases—cephalic, gastric, and intestinal phases—according to the location of the stimuli that initiate a reflex. The **cephalic phase** is initiated by stimulation of receptors in the head (cephalic)—sight, smell, taste, and chewing, as well as various emotional states. The efferent pathways for the reflex changes elicited by these stimuli involve the parasympathetic fibers in the vagus nerves that pass to almost all regions of the gastrointestinal tract.

Stimuli applied to the wall of the stomach initiate reflexes that constitute the **gastric phase** of regulation. These stimuli are peptides formed during the digestion of protein, distension, and acid. The responses to these stimuli are mediated by the nerve plexuses (short reflexes), by long reflex pathways that involve the external nerves to the gastrointestinal tract, and by the release of gastrin.

Finally, the **intestinal phase** is initiated by stimuli in the lumen of the intestinal tract—distension, acidity, osmolarity, and the various digestive products of carbohydrates, fats, and proteins. Like the gastric phase, the intestinal phase is mediated by both long and short neural reflexes and by the release of hormones, in this case those from the intestine—secretin, CCK, and GIP.

Note that each phase is named for the site at which the stimuli initiate the reflex and not for the sites of effector activity, since each phase is characterized by efferent output to virtually all organs in the gastrointestinal tract. These phases do not occur in temporal sequence except at the very beginning of a meal. Rather, during ingestion and the much longer absorptive period, reflexes characteristic of all three phases occur together.

SECTION B: THE ORAL CAVITY, PHARYNX, AND ESOPHAGUS

THE ORAL CAVITY

The **oral cavity** (mouth) consists of a small outer portion, the **vestibule,** between the teeth (and gums) and the lips, and a larger portion behind the teeth and gums; the latter extends back to the **oropharyngeal isthmus** where the oral cavity joins the oral portion of the pharynx (Fig. 25-9).

The lips are highly muscular folds covered externally by skin and internally by the stratified squamous epithelial mucous membrane that lines the oral cavity. The midpoint of each lip is connected to the corresponding gum by a fold of mucous membrane called a **frenulum.**

The roof of the mouth consists of a hard and a soft palate; the **hard palate** is formed by the *maxilla* and *palatine bones* (Fig. 9-13).

The mucous membrane that overlies the palate is covered by stratified squamous epithelium, the surface cells of which are continually sloughed off and replaced by cells that divide in deeper layers of the epithelium. The **soft palate** is a flexible sheet of densely packed collagen fibers that projects back from the hard palate and dips downward. Its chief function is to close off the opening between the respiratory and oral regions of the pharynx (the **pharyngeal isthmus**) (Fig. 25-9) during swallowing, sucking, blowing, and the production of some speech sounds. This closure is achieved by the action of skeletal muscles (the *levator veli palatini*), which elevate the soft palate and draw it toward the posterior wall of the pharynx as the posterior wall is being drawn forward by contraction of the *palatopharyngeal muscles*.

The **uvula** hangs down from the soft palate (Fig. 25-10). Two folds of mucous membrane pass from the sides of the uvula to the pharyngeal walls and tongue; they form the **palatoglossal** and the **palatopharyngeal arches** (so named because they contain the palatoglossus and palatopharyngeal muscles, respectively). Between these two arches on each

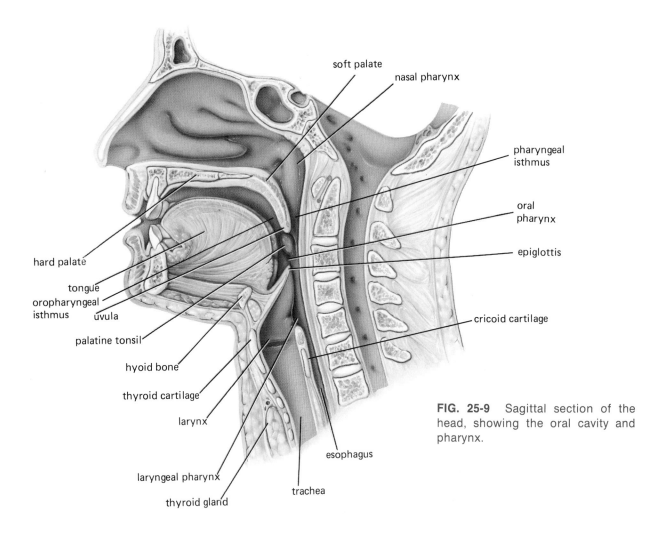

FIG. 25-9 Sagittal section of the head, showing the oral cavity and pharynx.

side of the oral cavity are the palatine tonsils. The region between the two palatoglossal arches is the **isthmus of the fauces** (Fig. 25-10).

The Tongue

The **tongue** is a muscular organ associated with chewing, swallowing, speech, and taste. Its anterior two-thirds is in the oral cavity, and the posterior one-third lies in the pharynx, where the **root** of the tongue is attached to the hyoid bone and mandible. A V-shaped groove, the **sulcus terminalis,** marks the division between these two parts. The surface of the tongue is dotted with **papillae,** elevations in the epithelium and underlying connective tissue

that give the tongue its rough appearance (Fig. 25-11); they often contain taste buds within their walls. The inferior surface of the tongue is attached to the floor of the oral cavity by the frenulum of the tongue **(frenulum linguae).**

The right and left halves of the tongue contain extrinsic muscle fibers (the *genioglossus, hyoglossus,* and *styloglossus*) (Fig. 25-12), which have their attachments outside the tongue, and intrinsic fibers, which are contained totally within it; these fibers together constitute the **lingual muscles.** The muscle fibers cross each other in intricate patterns that account for the tongue's extremely varied and delicate movements. The extrinsic muscles act with the intrinsic ones to flatten the tongue, pull it up-

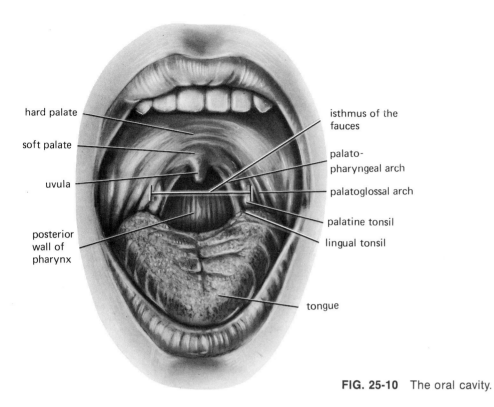

hard palate

soft palate

uvula

posterior
wall of
pharynx

isthmus of the
fauces

palato-
pharyngeal arch

palatoglossal arch

palatine tonsil

lingual tonsil

tongue

FIG. 25-10   The oral cavity.

FIG. 25-11   Upper surface of the tongue, showing the distribution of the
papillae and the palatine tonsils.

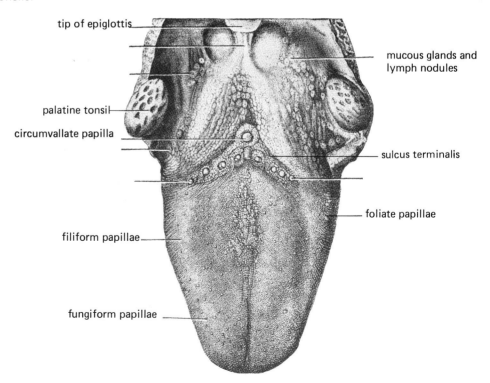

tip of epiglottis

palatine tonsil

circumvallate papilla

filiform papillae

fungiform papillae

mucous glands and
lymph nodules

sulcus terminalis

foliate papillae

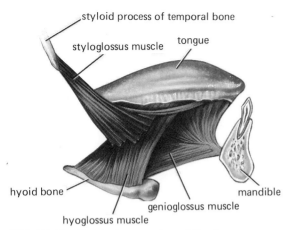

FIG. 25-12   Extrinsic muscles of the tongue.

ward and backward or downward and forward, or protrude it. **Lingual glands** on the tongue provide mucus and watery (serous) secretions.

The Teeth

The teeth are the hardest and most chemically stabilized tissues in the body. A tooth consists of two parts, the **crown,** which is covered by hard, whitish **enamel,** and the **root,** which is covered by yellow-

ish, bonelike **cement** (Fig. 25-13). Most of the tooth substance consists of a calcified connective tissue called **dentin,** which contains a **pulp cavity** expanded in the crown region into a *pulp chamber,* narrowed in the root into a *pulp,* or *root, canal,* and open at the *apical foramen.* **Dental pulp,** from which the cavity takes its name, is a connective tissue that contains the vascular and nervous supply for the dentin. The pulp cavity is lined with a layer of cells, the **odontoblasts,** which are involved in the production of dentin.

The teeth are seated in their sockets in the maxilla (upper jaw) or mandible (lower jaw) bone by bundles of connective-tissue fibers, the *periodontal ligaments* (Fig. 25-13), which allow slight movements of the teeth in response to forces generated in chewing. Near the junction of the root and crown, the periodontal ligament is covered by the **gingiva,** or *gum,* which is continuous with the oral mucosa that lines most of the oral cavity. The periodontal tissues contain receptors that provide in-

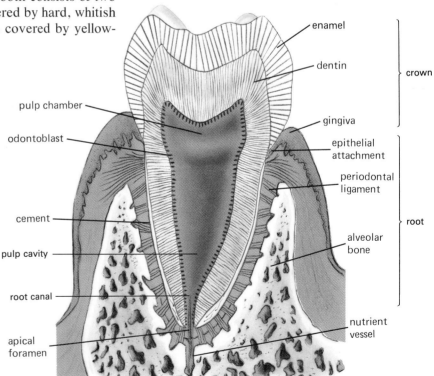

first permanent
mandibular molar

FIG. 25-13   Section of a tooth in situ.

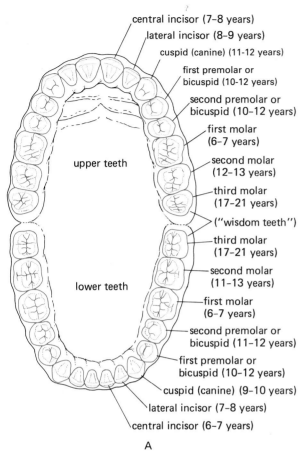

central incisor (7–8 years)
lateral incisor (8–9 years)
cuspid (canine) (11-12 years)
first premolar or
bicuspid (10-12 years)
second premolar or
bicuspid (10-12 years)
first molar
(6–7 years)
second molar
(12–13 years)
third molar
(17–21 years)
("wisdom teeth")
third molar
(17–21 years)
second molar
(11–13 years)
first molar
(6–7 years)
second premolar or
bicuspid (11–12 years)
first premolar or
bicuspid (10–12 years)
cuspid (canine) (9–10 years)
lateral incisor (7–8 years)
central incisor (6–7 years)

upper teeth

lower teeth

A

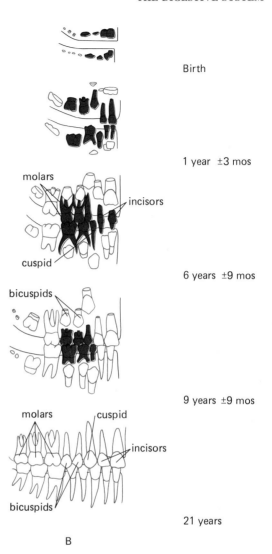

Birth

1 year ±3 mos

molars
incisors
cuspid

6 years ±9 mos

bicuspids

9 years ±9 mos

molars   cuspid
incisors

bicuspids

21 years

B

FIG. 25-14 (A) The permanent teeth and their times of eruption. (B) Development of the teeth. Deciduous teeth are shown in gray and permanent teeth in color. Note the roots of the permanent teeth. The incisors, cuspids, and bicuspids have one root, although the bicuspids of the upper jaw may have two. The molars of the lower jaw have two roots, and the molars of the upper jaw have three. The upper cuspids have the longest roots.

formation concerning the pressure gernerated during chewing to the brain centers that control chewing movements.

People, like other mammals, have two sets of teeth during their lifetime, a set of small **deciduous teeth** (the "baby teeth") and a set of larger **permanent teeth.** There are 20 deciduous teeth (10 in the upper and 10 in the lower jaw) and 32 permanent teeth. The first two teeth on each side of the midline of the upper and lower jaws are the **central** and **lateral incisors,** respectively (Fig. 25-14A, B). The deciduous incisors appear at about 6 months of age. Next to each lateral incisor is a single **canine,** or

cuspid, tooth. (The term *cusp* refers to the conical projections on the tooth, a cuspid tooth having one projection, a bicuspid having two.) Next to the canine tooth are two **molars,** which appear in children of about 2 years of age and are replaced by age 12 by two permanent teeth, which are called the first and second **premolars,** or **bicuspids.** As part of the permanent dentition, three **molars** emerge to lie next to the premolars; the first, the "6 year molar," erupts about age 6, the second at about age 12, and the third ("wisdom teeth") appears much later, if at all.

Chewing

The primary function of the teeth is to bite off and grind down chunks of food into pieces small enough to be swallowed. The incisors of an adult man can exert forces of 25 to 50 lb, and the molars the 200 lb required to crack a walnut or support trapeze artists by their teeth. The rhythmic act of chewing is controlled by the somatic nerves to the skeletal muscles of the mouth and jaw (the masseter, temporal, and lateral and medial pterygoid muscles; see Chap. 12). In addition to the voluntary control of these muscles, rhythmic chewing motions are reflexly activated by the pressure of food against the gums, teeth, hard palate, and tongue. Activation of these pressure receptors leads to inhibition of the muscles that hold the jaw closed; the resulting relaxation of the jaw reduces the pressure and allows a new cycle of contractile activity.

Prolonged chewing, so characteristic of human beings, does not appear to be essential to the digestive process for most foods (many animals, such as the dog and cat, swallow their food almost immediately). Although chewing prolongs the subjective pleasure of taste, it does not appreciably alter the rate at which the food is digested and absorbed. On the other hand, attempting to swallow a particle of food too large to enter the esophagus may lead to choking, if the particle lodges over the trachea and blocks the entry of air into the lungs. A surprising number of preventable deaths occur each year from choking, the symptoms of which are often confused with the sudden onset of a heart attack so that no attempt is made to remove the obstruction from the airway. The Heimlich maneuver, described in Chap. 23, will often generate sufficient pressure in the lungs to dislodge the obstructing particle from the airways.)

The Salivary Glands and Secretion of Saliva

Saliva is secreted by three pairs of exocrine glands, the parotid, submandibular, and sublingual glands (Fig. 25-1), as well as by numerous small glands distributed throughout the lining of the oral cavity. The **parotid glands,** which are the largest of the salivary glands, lie on the sides of the face, in front of the ears (Fig. 25-15); their ducts (the parotid ducts, formerly Stensen's ducts) open into the mouth opposite the second molar teeth in the upper jaw. The **submandibular glands** lie in front of the angles of the mandible, and their ducts (formerly Wharton's ducts) empty at the sides of the frenulum of the tongue. The **sublingual glands,** the smallest of the three pairs, are beneath the mucous membrane of the floor of the mouth. Their secretions

FIG. 25-15 The salivary glands and their ducts.

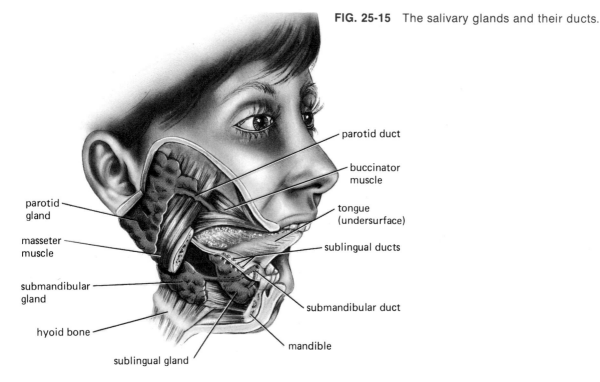

parotid duct

buccinator muscle

tongue (undersurface)

sublingual ducts

submandibular duct

mandible

parotid gland

masseter muscle

submandibular gland

hyoid bone

sublingual gland

flow through many sublingual ducts, most of which open onto the floor of the mouth.

These glands contain secretory cells arranged as acini at the ends of ducts. The salivary glands are classified as compound tubuloacinar glands, containing both tube-shaped and flask-shaped (acinar) components (see Fig. 6-6). The epithelial cells which form the acini, are of two types: serous cells (which produce a watery secretion) and mucous cells. The parotid glands have acini with only serous cells, whereas the sublingual and submandibular acini are mixed; i.e., they contain cells of both types. The glands also contain myoepithelial cells, the contractions of which help to move the secretions along the ducts. Water accounts for 99 percent of the secreted fluid; the remaining 1 percent consists of various salts and proteins. The major proteins of saliva are the *mucins,* which, when mixed with water, form highly viscous mucus. Mucins are also secreted by gland cells throughout the gastrointestinal tract. In the mouth the watery solution of mucus moistens and lubricates the food particles, which allows them to pass easily along the esophagus into the stomach.

Another protein secreted by the salivary glands is the enzyme **amylase,** which catalyzes the breakdown of polysaccharides into disaccharides. Although salivary amylase starts the digestive process in the mouth, food does not remain there long enough for much digestion to occur; however, amylase continues its digestive activity in the stomach until inhibited by the hydrochloric acid there.

The secretion of saliva is controlled by sympathetic and parasympathetic nerves to the glands. However, unlike their antagonistic activity in most organs, both divisions of the autonomic nervous system stimulate secretion although the parasympathetic branch causes by far the greatest increase in fluid volume.

During sleep very little saliva is secreted, but in the awake state a basal rate of about 0.5 mL/min keeps the mouth moist. Food in the mouth increases the rate of salivary secretion. This reflex response is initiated by stimulation of chemoreceptors and pressure receptors in the walls of the mouth and tongue. Afferent fibers from the receptors pass to the brainstem medulla, which contains the integrating centers that control the autonomic output to the salivary glands. The most potent stimuli for salivary secretion are acid solutions, e.g.,

fruit juices and lemons, which may lead to a maximal secretion of 4 mL of saliva per minute.

During the course of a day, between 1 and 2 L of saliva is secreted, most of which is swallowed. The proteins in saliva are broken down into amino acids by the digestive enzymes in the stomach and intestinal tract, and the amino acids, salts, and water absorbed into the circulation. This is a typical pattern for most of the secretions of the gastrointestinal tract. Although large amounts of fluid may be secreted into the tract during the course of a day, most of it, together with its salt and protein content, is digested and reabsorbed. If these secretions are not reabsorbed, large quantities of fluid and salt can be lost through the gastrointestinal tract.

THE PHARYNX

The **pharynx** is a tube 12 to 14 cm long, whose wall consists largely of skeletal muscle. It lies behind the nasal cavities, oral cavity, and larynx. In Chap. 23, we concentrated on the respiratory portion of the pharynx; here we discuss the digestive portions. The **oral pharynx** (that portion bounded by the soft palate above and the upper border of the epiglottis below) is separated from the nasal pharynx by the **pharyngeal isthmus** (Fig. 25-9). During swallowing this isthmus is closed by elevation of the soft palate and movement forward of the posterior pharyngeal wall so that food cannot pass up into the nose.

The openings of the pharynx into the nasal, oral, and laryngeal cavities are on its anterior surface, and the pharyngeal muscle masses are largely on its lateral and posterior sides. Three of the four paired muscles of the pharynx are constrictors (the *superior, middle,* and *inferior constrictors,* Fig. 25-16), whereas the fourth pair of muscles, the *stylopharyngeus,* runs more longitudinally. The lower fibers of the inferior constrictors blend with the circular muscle fibers of the esophagus.

The interior of the pharynx contains the posterior third of the tongue as it bends back and down toward the root (Fig. 25-9); the **epiglottis** is attached to the tongue by three *glossoepiglottic folds.* At the level of the upper border of the epiglottis, the **laryngeal part of the pharynx** begins. It is wide above but narrows rapidly below to become continuous with the esophagus at the lower border of the cricoid cartilage of the larynx (Fig. 25-9).

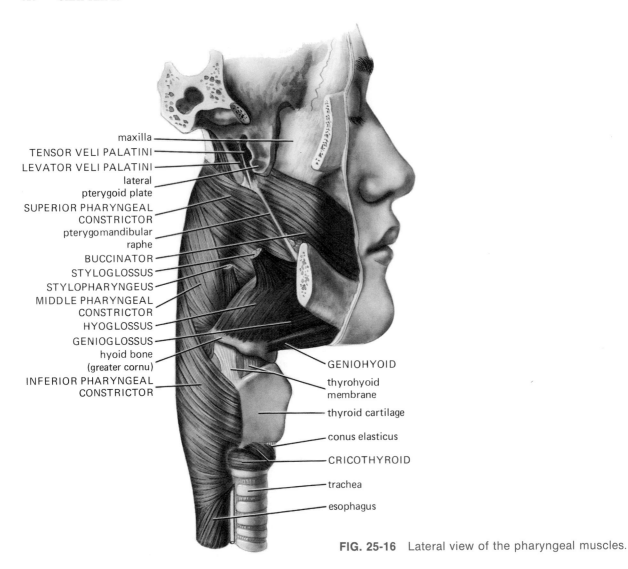

maxilla
TENSOR VELI PALATINI
LEVATOR VELI PALATINI
lateral
pterygoid plate
SUPERIOR PHARYNGEAL
CONSTRICTOR
pterygomandibular
raphe
BUCCINATOR
STYLOGLOSSUS
STYLOPHARYNGEUS
MIDDLE PHARYNGEAL
CONSTRICTOR
HYOGLOSSUS
GENIOGLOSSUS
hyoid bone
(greater cornu)
INFERIOR PHARYNGEAL
CONSTRICTOR

GENIOHYOID
thyrohyoid
membrane
thyroid cartilage
conus elasticus
CRICOTHYROID
trachea
esophagus

FIG. 25-16   Lateral view of the pharyngeal muscles.

THE ESOPHAGUS

The **esophagus** is a flaccid tube about 25 cm long. Its layers blend with those of the pharynx above and the stomach below. Its epithelium is similar to that of the pharynx, being of the stratified squamous type. Mucous glands occur in the mucosa and deeper in the submucosa. The muscularis externa of the upper third of the esophagus in human beings is formed of skeletal muscle, the lower third of smooth muscle, and the middle third of both types of muscle fibers. At rest, the upper opening into the esophagus is closed, and this region of the esoph-

agus forms the **superior esophageal,** or *hypophar-yngeal,* **sphincter** (Fig. 25-17). The skeletal muscles in this region are so arranged that, when they are relaxed, the sphincter is closed by passive elastic tensions in the walls.

The esophagus descends through the thorax and enters the abdominal cavity by passing through the **esophageal hiatus** (or opening) of the diaphragm. Normally some 2 cm of the esophagus lie below the diaphragm, but in cases of *hiatal hernia* the terminal portions of the esophagus as well as part of the stomach slide upward through an abnormally en-larged opening and lie above the diaphragm in the

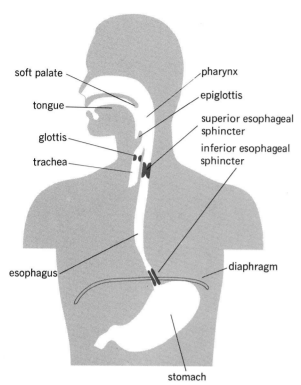

soft palate

tongue

glottis

trachea

pharynx

epiglottis

superior esophageal sphincter

inferior esophageal sphincter

esophagus

diaphragm

stomach

FIG. 25-17 Location of superior and inferior esophageal sphincters.

thoracic cavity.

The last 4 cm of the esophagus before it enters the stomach is known as the **inferior esophageal,** or *gastroesophageal,* **sphincter.** Anatomically, this region appears no different from adjacent regions, but its functional activity is distinct. Like the superior esophageal sphincter, this sphincter remains tonically contracted, and thus forms a barrier between the contents of the stomach and the esophagus. When swallowing is not taking place most of the esophagus is relaxed and flaccid.

Most of the esophagus lies in the thoracic cavity and is subject to a subatmospheric intrathoracic pressure of about -5 to -10 mmHg. The stomach, however, lying below the diaphragm, has an internal pressure slightly above atmospheric, $+5$ to $+10$ mmHg, due to its compression by the contents of the abdominal cavity. Thus, without the inferior esophageal sphincter, the pressure gradient from the stomach to the esophagus would tend to force the contents of the stomach into the esophagus.

The ability of the inferior esophageal sphincter to maintain a barrier between the stomach and esophagus is helped by the fact that its last portion lies below the diaphragm and is subject to the same pressures in the abdominal cavity as the stomach. Thus, if the pressure of the abdominal cavity is raised, e.g., during cycles of respiration or by contraction of the abdominal muscles, the pressures of both the stomach contents and this terminal segment of the esophagus are raised together and there is no change in the pressure gradient between the two.

During pregnancy the growth of the fetus increases the pressure on the abdominal contents and can displace the terminal segment of the esophagus through the diaphragm into the thoracic cavity. The inferior esophageal barrier is therefore no longer assisted by changes in abdominal pressure, and during the last 5 months of pregnancy there is a tendency for the increased pressures in the abdominal cavity to force some of the contents of the stomach up into the esophagus. The hydrochloric acid from the stomach contents irritates the walls of the esophagus and causes contractile spasms of the smooth muscle, both of which are associated with the pain known as *heartburn* because the sensation appears to be located in the region over the heart. Heartburn often subsides in the last weeks of pregnancy as the uterus descends prior to delivery and decreases the pressure on the abdominal organs. Another potential cause of heartburn is a large meal, which may raise the pressure in the stomach enough to produce reflux of acid into the esophagus. A newborn child has no intraabdominal segment of the esophagus and thus has a tendency to regurgitate.

In the abnormal condition called *achalasia,* the inferior esophageal sphincter fails to relax, and a whole meal may become lodged in the esophagus and enter the stomach very slowly. Distension of the esophagus results in pain in the chest that is often confused with pain that originates from the heart. Achalasia appears to be the result of damage to, or absence of, the myenteric nerve plexus in the region of the inferior esophageal sphincter.

During swallowing, air is trapped in the pharynx ahead of the bolus, and most of this passes into the trachea before the glottis closes. Air mixed with the bolus or saliva is swallowed and either expelled by belching or passed on into the intestine.

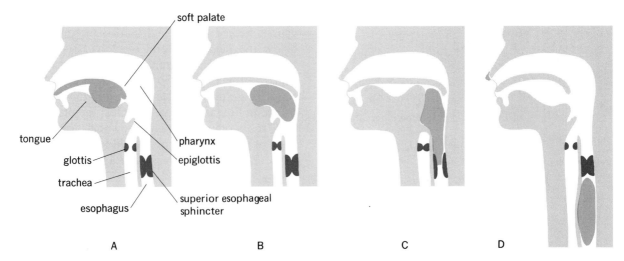

FIG. 25-18 Movement of a bolus of food through the pharynx and upper esophagus during swallowing.

SWALLOWING

Swallowing is a complex reflex initiated when the tongue forces a bolus of food into the rear of the mouth and pressure receptors in the walls of the pharynx are stimulated. They send afferent impulses to the swallowing center in the medulla, which coordinates the sequence of events during the swallowing process via efferent impulses to the muscles in the pharynx, larynx, and esophagus and to the respiratory muscles.

As the bolus of food moves into the pharynx, the soft palate is elevated and lodges against the back wall of the pharynx, which seals off the nasal cavity and prevents food from entering it (Fig. 25-18A and B). The swallowing center generates a precisely timed sequence of nerve impulses that inhibit respiration and draw the pharyngeal walls upward and inward as the tongue pushes the bolus to the back of the pharynx. The larynx is raised and the glottis (the area around the vocal cords) is closed (Fig. 25-18B and C) keeping food from getting into the trachea. Successive contractions of the pharyngeal constrictor muscles and gravity (when the body is upright) move the bolus past the epiglottis into the lower part of the pharynx and on into the esophagus. As the bolus passes the epiglottis, it tips the epiglottis backward (Fig. 25-18C), and thus forms a trough down which the bolus slides into the esophagus. It is the closure of the glottis, however, and not the folding of the epiglottis that is primarily

responsible for preventing food from entering the trachea. This **pharyngeal phase of swallowing** lasts about 1 sec.

The **esophageal phase of swallowing** begins with the relaxation of the superior esophageal sphincter (Fig. 25-18C). Immediately after the bolus has passed (Fig. 25-18D), the sphincter closes, the glottis opens, and breathing resumes. Once in the esophagus, the bolus is moved toward the stomach by a progressive wave of muscle contractions that proceeds along the esophagus at a steady rate, compressing the lumen and forcing the bolus ahead of it. Such waves of contraction in the muscle layers that surround a tube are known as **peristaltic waves.** An esophageal peristaltic wave takes about 9 sec to reach the stomach. Swallowing can occur while a person is upside down because it is not primarily gravity but the peristaltic wave that moves the bolus to the stomach.

The inferior esophageal sphincter is relaxed throughout the period of swallowing, which allows the bolus to enter the stomach. After the bolus has passed, the sphincter closes and thus reseals the junction between the esophagus and the stomach.

Once swallowing has been initiated, it cannot be stopped voluntarily even though it involves skeletal muscles. This reflex is a stereotyped all-or-none response, the entire coordination of which resides in the swallowing center in the medulla. Swallowing is an example of a reflex in which multiple responses occur in a regular temporal sequence that is predetermined by the synaptic connections between neurons in the coordinating center. Since both skeletal and smooth muscles are involved, the swallowing center must direct efferent activity in both the somatic nerves (to the skeletal muscle) and autonomic nerves (to the smooth muscle) of the esophagus (the latter nerves being the parasympathetic fibers that reach the plexuses in the esophageal wall by way of the vagus nerves). Simultaneously, afferent fibers from receptors in the wall of the esophagus send information to the swallowing center that can then alter the efferent activity. For example, if a large particle of food does not reach the stomach during the initial peristaltic wave, the distension of the esophagus activates receptors that initiate repeated waves of peristaltic activity (*secondary peristalsis*) that are not preceded by the pharyngeal phase of swallowing.

SECTION C: THE STOMACH

STRUCTURE OF THE STOMACH

The **stomach** is a chamber located between the end of the esophagus and the beginning of the small intestine. Figure 25-19 indicates the major parts of the stomach and Fig. 25-20 the folds, or **rugae,** present on its inner surface when it is empty. (These gradually flatten out as the stomach becomes filled with food.)

At the gastroesophageal junction the stratified squamous epithelium of the esophagus changes abruptly to the simple columnar epithelium characteristic of the stomach. Some of these cells secrete mucus, which lubricates the stomach wall and protects it to some extent from the abrasive contents of the meal and from the digestive enzymes and acid in the gastric juice. The cells have a very short life-span and are replaced every 3 days or so by the division of undifferentiated epithelial cells in the stomach lining.

The lining of the stomach is indented by numerous **gastric pits** (Fig. 25-20), which are the openings to the tubular **gastric glands** that lie in the mucosa (Fig. 25-21). The cells in the gastric pits secrete mucus. Deeper down in the tubular glands themselves are **parietal** (wall) **cells,** also known as **oxyntic cells,** which secrete hydrochloric acid (and intrinsic factor). Yet a third group of cells, **chief cells,** secrete the enzyme precursor pepsinogen. Finally, the endocrine cells that secrete the hormone gastrin are scattered throughout the epithelium of the **antrum** (the lower portion of the stomach that does not

FIG. 25-19   Interior of the stomach.

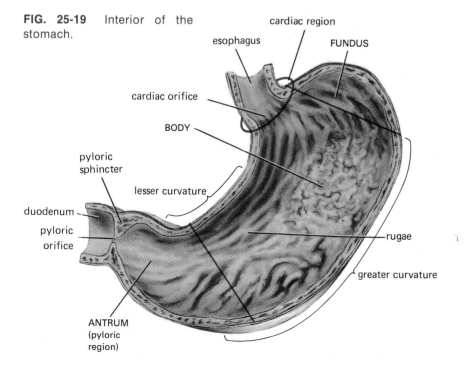

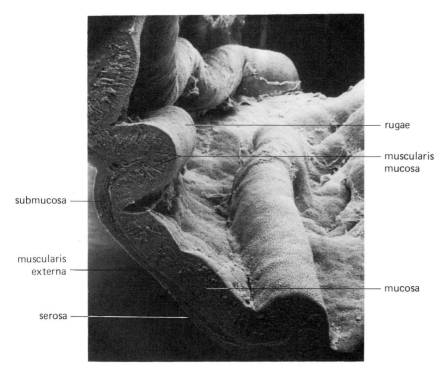

FIG. 25-20 A scanning electron micrograph through the stomach wall. Note the four basic components of the gastrointestinal tract wall (the mucosa, submucosa, muscularis externa, and serosa) as well as the thin muscularis mucosa. Note also the folds (rugae) in the mucosal and submucosal layers. The many small openings on the surface of the rugae are the openings into gastric pits.

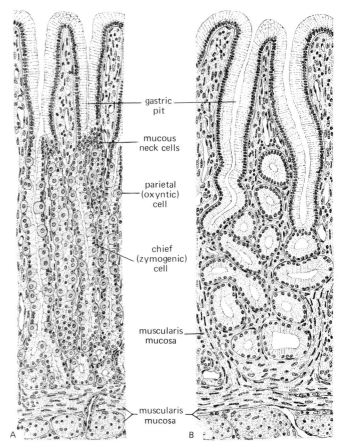

FIG. 25-21 Drawings of the mucosa from the fundic (A) and antral (B) regions of the stomach. Note the relative absence of parietal (acid-secreting) and chief (pepsinogen-secreting) cells in the antrum of the stomach.

secrete acid, Fig. 25-19). Thus, each of the four major secretions of the stomach—mucus, acid, pepsinogen, and the hormone gastrin—is secreted by a distinct cell type.

Digestion in the stomach is limited to the breakdown of large lumps of food into a solution of individual molecules and fragments of large molecules, most of which are still too large to be absorbed. This mixture of fluids, partially digested food particles, and enzymes has the consistency of a thick soup and is known as **chyme.**

The ability of acid and pepsin to break down particles of food depends upon their coming in contact with the food in the stomach. There is very little mixing of solid food in the stomach until it reaches the antrum, where strong peristaltic waves compress and mix the contents. Food in the body of the stomach remains as a semisolid mass that is attacked by acid and pepsin only at its surface, the interior of the mass remains free of acid. On the other hand, although salivary amylase is inactivated by hydrochloric acid, it may continue to act upon starch for long periods of time in the stomach if it is inside the bolus, where acid has not yet reached. Bacteria in the same location may also escape the sterilizing action of hydrochloric acid.

The most important function of the stomach is to regulate the rate at which chyme enters the small intestine, where most of the process of digestion and absorption takes place. In the absence of the stomach, a normal-sized meal moves so rapidly through the small intestine that only a fraction of the food has time to be digested and absorbed.

ACID SECRETION

The human stomach secretes about 2 L of hydrochloric acid per day. Hydrochloric acid is a strong acid that completely dissociates in water into hydrogen and chloride ions. The concentration of hydrogen ions in the lumen of the stomach may reach 150 mM, three million times greater than the concentration in the blood.

Hydrochloric acid is secreted by the *parietal cells* that lie in tubular glands in the body of the stomach (Fig. 25-21). These cells have many large mitochondria, and their cell membranes are indented at the luminal surface to form minute intracellular channels (*canaliculi*) that penetrate deep into the cell. The canaliculi greatly enlarge the surface area of the cell available for secretion, and their intracellular distribution brings them into close association with the mitochondria that produce the ATP necessary for operating the active-transport systems involved in acid secretion.

The secretion of hydrochloric acid involves the active transport of hydrogen and chloride ions. The two ions appear to be transported by separate pumps in the luminal membrane of the parietal cells (Fig. 25-22, right side). The enzyme carbonic anhydrase, similar to the enzyme in red blood cells,

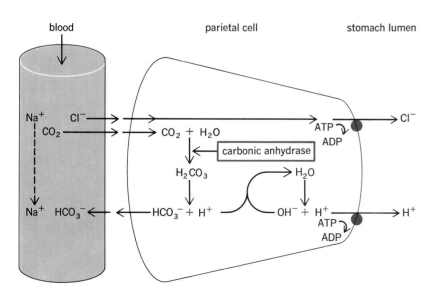

FIG. 25-22 Secretion of HCl by parietal cells and the release of bicarbonate ions into the blood. (Begin at the left side of the figure and follow the fate of the NaCl and CO_2 delivered in the blood.)

is present in the gastric mucosa and is believed to be involved in the secretion of acid, probably by way of the general pathway illustrated in Fig. 25-22. Note that for each hydrogen ion secreted into the lumen of the stomach, a bicarbonate ion is released into the blood in exchange for chloride; therefore, the venous blood that leaves the stomach is more alkaline than the arterial blood that enters it.

The Control of Gastric Acid Secretion

The parietal cells secrete acid in response to chemical stimuli. Some chemicals, such as amino acids and caffeine (a substance found in coffee, tea, cocoa, and cola drinks), can stimulate the parietal cells directly, but more generally the secretory cells are influenced by the body's own chemical mediators, i.e., by certain neurotransmitters, hormones, and paracrines. The best defined of these chemical messengers are acetylcholine, gastrin, and histamine, all three of which act to increase acid secretion.

Acetylcholine is released in the gastric mucosa by the postganglionic endings of parasympathetic nerves and by neurons that are part of the plexuses in the stomach walls. Acetylcholine release can, therefore, be increased by activation of long (vagal) or short reflex arcs. **Gastrin,** the only one of the three chemical messengers that is a hormone, is secreted into the circulation by glands in the pyloric region of the stomach. It is released in response to neural input to the gastrin-secreting cells and to peptides and amino acids that bathe these cells. Gastrin secretion is inhibited by highly acid stomach contents and it may be inhibited as well by other hormones and by paracrine agents. **Histamine** is known to play a significant role in normal acid secretion by the parietal cells, but neither its site of origin nor mechanism of action is known.

The secretion of acid falls into two distinct phases, a **basal,** or **fasting, phase,** and a **stimulated phase,** which occurs during a meal and generally exceeds the basal rate by ten- to twenty-fold. The stimulated phase of acid secretion is in turn divided into three phases: the cephalic, gastric, and intestinal phases. These three phases of stimulated secretion actually overlap; they are summarized in Table 25-2.

The Cephalic Phase Gastric acid secretion begins with the sight, smell, taste, or chewing of appetizing food. This cephalic phase of acid secretion is mediated by vagal (parasympathetic) fibers to the stomach, and the parietal cells are stimulated by the acetylcholine released by the postganglionic neurons and by the small amounts of gastrin that are also secreted during this phase.

The Gastric Phase The two factors responsible for stimulating acid secretion during the gastric

---

TABLE 25-2   CONTROL OF HCl SECRETION DURING A MEAL

Stimuli	Pathways to the Parietal Cells	Result
Cephalic Stimuli:		
Sight	Vagus nerves via	↑ HCl secretion
Smell	gastric nerve	
Taste	plexuses and gastrin	
Chewing		
Gastric Contents:		
Distension	Long and short neural reflexes;	↑ HCl secretion
↓ H⁺ concentration (protein buffering)	gastrin;	
Peptides	direct action of peptides	
Intestinal Contents:		
Distension	*Long and short neural reflexes;	↓ HCl secretion
↑ H⁺ concentration	gastric-inhibitory peptide;	
Nutrients	other hormones (?)	
↑ Osmolarity		

* The neural and hormonal pathways triggered during the intestinal phase have as one of their final common pathways a decrease in gastrin secretion.

phase, the period during which food is present in the stomach, are distention of the stomach and chemical stimulation of the parietal cells by gastrin or directly by substances in the chyme. Gastric distension stimulates mechanoreceptors in the stomach walls; the integrating centers of these reflexes are in the neural plexuses of the gastric walls, and their efferent components stimulate the parietal cells.

The ability of food in the stomach to stimulate acid secretion depends on the chemical nature of the food, and most commonly this stimulation occurs by way of gastrin. Amino acids and protein fragments directly stimulate gastrin release, but glucose and fat do not. In addition, proteins bind (buffer) hydrogen ion, and this reduced hydrogen-ion concentration results, as we have mentioned, in stimulation of gastrin release. As a result, acid secretion during a meal is proportional to the protein content of the meal. Alcohol, contrary to popular belief, does not stimulate gastric-acid secretion in people. High acidity in the lumen of the stomach is a potent inhibitor of gastrin release and, therefore, of acid secretion.

The Intestinal Phase Substances in the intestine may lead to the stimulation or inhibition of gastric-acid secretion, again depending on their nature. For example, fat, hypertonic solutions, and acid in the duodenum inhibit gastric-acid secretion, whereas amino acids tend to increase it. The message from the intestine is carried to the acid- (or gastrin-) secreting cells of the stomach by both neural and hormonal messengers. For want of a better name, the inhibitory hormone (or hormones) is called **"enterogastrone,"** and the neural pathway, the **"enterogastric reflex."** Hormones released by the intestine and suspected of acting as enterogastrones include cholecystokinin (CCK), gastric inhibitory peptide (GIP), and secretin.

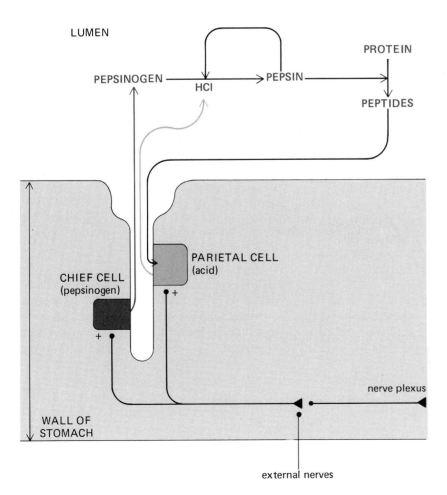

FIG. 25-23 Pathways that regulate the secretion and activation of pepsinogen in the stomach. The peptides affect the parietal cells mainly via gastrin.

PEPSIN SECRETION

Pepsin is secreted by the chief cells in an inactive form known as **pepsinogen.** The high acidity in the lumen of the stomach converts pepsinogen into the active enzyme **pepsin** by breaking off a small fragment of the molecule and thereby changing the shape of the protein and exposing its active site.

Pepsin catalyzes the splitting of specific peptide bonds in proteins, which produces peptide fragments composed of several amino acids. Since high acidity is required for pepsin activation, it is appropriate that, as we have seen, these peptide fragments stimulate gastrin secretion, which in turn increases acid secretion. Once activated, pepsin acts autocatalytically upon pepsinogen to form more pepsin (Fig. 25-23). The synthesis and secretion of inactive pepsinogen followed by its activation to pepsin provides an example of a process that occurs frequently in enzyme secretion: By synthesizing an inactive form of the enzyme, the cell is protected from internal digestion by the enzyme.

The primary pathway for stimulating pepsinogen secretion is input to the chief cells from the nerve plexuses. In general, pepsinogen secretion parallels the secretion of acid.

GASTRIC MOTILITY

The empty stomach has a volume of about 50 mL, and the diameter of its lumen is little larger than that of the small intestine. When a meal is swallowed, the smooth muscle in the body of the stomach relaxes and thus allows the volume of the lumen to increase with little increase in pressure. This **receptive relaxation** is mediated by the vagus (parasympathetic) nerves and coordinated by the swallowing center.

Peristaltic waves proceed along the walls of the stomach at the rate of about 3 per minute. Each wave begins near the entry of the esophagus and produces only a weak ripple as it proceeds over the body of the stomach, one too weak to produce much mixing of the luminal contents with the acid and pepsin secreted by the stomach. As the wave approaches the larger mass of muscle that surrounds the antrum, it speeds up and produces a more powerful wave of contraction that does mix the contents of the antrum and expels them into the duodenum.

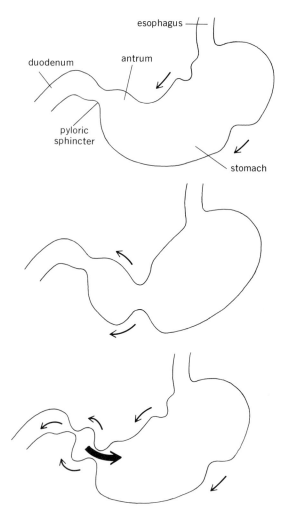

FIG. 25-24 Peristaltic waves that pass over the stomach empty a small amount of material into the duodenum. Most of the material is forced back into the antrum.

The **pyloric sphincter** is a ring of smooth muscle and connective tissue between the terminal antrum and the duodenum. It exerts only a slight pressure at the junction and is open most of the time. When a strong peristaltic wave arrives at the antrum, the pressure of the antral contents is increased; however, the antral contraction also closes the pyloric sphincter, with the result that the pressure forces most of the antral contents back into the body of the stomach, and only small amounts pass into the duodenum (Fig. 25-24).

What is responsible for the characteristics of the peristaltic waves? Its rhythmicity (3 per minute) is

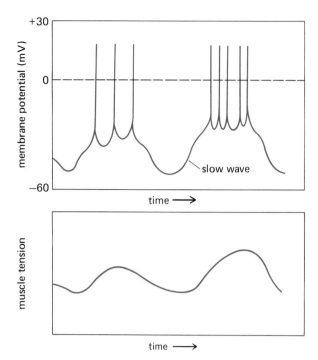

FIG. 25-25 Slow wave oscillations in membrane potential trigger bursts of action potentials. The mechanical activity of smooth muscle correlates with the frequency of the slow wave potential changes and the magnitude of the mechanical response to the frequency of action potentials in each burst.

due to the fact that pacemaker cells in the longitudinal muscle layer near the esophagus undergo spontaneous depolarization-repolarization cycles, termed the **basic electrical rhythm,** at this rate. The depolarizations are propagated through gap junctions along the longitudinal muscle layer of the stomach and induce similar waves in the overlying circular muscle layer (perhaps via gap junctions between the longitudinal and circular layers). In the absence of other input, however, these depolarizations are too small to cause the muscle membranes to reach threshold and fire the action potentials required to elicit peristaltic contractions.

Excitatory neurotransmitters and hormones act upon the muscle to help depolarize the membrane and thereby bring it closer to threshold, so that action potentials are generated at the peak of the basic electrical rhythm cycle. The number of spikes fired with each wave determines the strength of the elicited smooth muscle contraction (Fig. 25-25). Thus, whereas the frequency of contraction is determined by the basic electrical rhythm and remains essentially constant, the force of contraction (and therefore, the amount of gastric emptying per contraction) is determined by neural and hormonal input, i.e., by regulatory reflexes.

There are important gastric and intestinal phases to these reflexes. Distension of the stomach increases the rate of gastric emptying; therefore, the larger a meal the faster the initial rate of emptying, and as the volume of the stomach decreases, so does the rate of emptying. Long and short neural reflexes triggered by mechanoreceptors in the stomach mediate this response to distention. In contrast, distension of the duodenum as well as the presence in it of fat, acid, or hypertonic solutions all produce inhibition of gastric emptying mediated by release of intestinal hormones (and to a lesser extent by neural reflexes); fat in the duodenum is the most potent of the chemical stimuli that lead to this reflex inhibition of gastric motility.

In addition to the control of gastric emptying by the gastric volume and duodenal contents, motility is influenced by external nerve fibers to the stomach. Inhibition of motility occurs when parasympathetic activity is decreased or sympathetic activity increased. Via these pathways, pain and emotions, such as sadness, depression, and fear tend to decrease motility, whereas aggression or anger tend to increase it. These relationships are not always predictable, however, and different people show different responses to apparently similar emotional states.

ABSORPTION BY THE STOMACH

Very little food is absorbed from the stomach into the blood. There are no special transport systems for salts, amino acids, and sugars in the stomach walls like those found in the intestine, and most of the digestion products in the stomach are large, highly charged, and ionized so that they cannot diffuse across cell membranes. The lipids in the food that reaches the stomach are not soluble in water and tend to separate into large droplets that do not mix with the acid contents of the stomach. Since little of this lipid comes into contact with the membranes that line the stomach, little is absorbed.

Several classes of molecules, however, can be absorbed directly by the stomach. A prime example is ethyl alcohol, CH_3CH_2OH. Alcohol is water-soluble, but since it is not ionized, it also has some degree of lipid solubility, which allows it to diffuse across lipid membranes and reach the bloodstream through the walls of the stomach. Although alcohol is absorbed from the stomach, it is absorbed more rapidly from the intestinal tract, which has a greater membrane surface area available for absorption. Accordingly, drinking a glass of milk before a cocktail party or eating high-fat hors d'oeuvres inhibits the rate of gastric emptying through the reflexes from the duodenum and slows down the rate of alcohol absorption but does not stop it.

Weak acids, most notably acetylsalicyclic acid (aspirin), are also absorbed directly across the walls of the stomach. A weak acid is fully ionized in solutions of low hydrogen-ion concentration and in its ionized form is unable to diffuse through cell membranes. In the highly acid environment of the stomach, however, a weak acid is almost totally converted to its un-ionized form, which is lipid-soluble and can cross the cell membrane. Once in the low-acid environment of the cell, the weak acid again ionizes, and liberates a hydrogen ion, which tends to make the cell interior more acid. Therefore, if enough weak acid enters in a short time, the intracellular acidity may rise sufficiently to damage the cell. Although most people show few ill effects from the normal dosage of aspirin, small amounts of blood can be found in the stomachs of most individuals after taking aspirin, and in some individuals it can produce severe hemorrhaging. Since alcohol damages the gastric mucosa, the combination of aspirin and alcohol increases the damage and is more likely to produce bleeding than either substance alone.

VOMITING

Vomiting is the forceful expulsion of the contents of the stomach and upper intestinal tract through the mouth. Like swallowing, vomiting is a complex reflex coordinated by the region of the brainstem medulla, in this case the *vomiting center*. Neural input to this center from receptors in many different regions of the body can initiate this reflex. For example, excessive distension of the stomach or duodenum, various substances that act upon chemoreceptors in the wall of the small intestine or the brain, increased pressure in the skull, rotating movements of the head (motion sickness), intense pain or tactile stimuli applied to the back of the throat can all initiate vomiting. What is the adaptive value of the vomiting reflex? Obviously, the removal of ingested toxic substances before they can be absorbed is of benefit. Moreover, the nausea that usually accompanies vomiting may have the adaptive value of conditioning the individual to avoid the future ingestion of foods that contain the same toxic substances. Why other types of stimuli, such as those that produce intense pain or motion sickness, have become linked to the vomiting center is less clear.

Vomiting is usually preceded by increased salivation, sweating, faster heart rate, and feelings of nausea—all characteristic of a general discharge of the autonomic nervous system. Vomiting begins with a deep inspiration, closure of the glottis, and elevation of the soft palate. The abdominal and thoracic muscles contract and thus raise the abdominal pressure, which is transmitted to the contents of the stomach; the inferior esophageal sphincter relaxes, and the high abdominal pressure forces the contents of the stomach into the esophagus. This initial sequence of events may occur repeatedly without vomiting and is known as *retching*. Vomiting occurs when the pressure applied to the contents of the stomach becomes great enough to force the contents past the superior esophageal sphincter and into the mouth. Vomiting is also accompanied by strong contractions of the upper portion of the small intestine that tend to force some of the intestinal contents back into the stomach. Thus, some bile may be present in the vomitus.

Excessive vomiting can lead to large losses of fluids and salts that normally would be reabsorbed in the small intestine. This can result in severe dehydration, upset the salt balance of the body, and produce circulatory problems as a result of decrease in plasma volume. In particular, the loss of acid results in a lowering of body acidity. If the amount of vomiting is slight, the control mechanisms that act by way of the lungs and kidneys adjust the acidity toward normal levels by hypoventilation and by excreting bicarbonate in the urine.

ULCERS

Considering the high concentration of acid and pepsin secreted by the stomach, it is natural to wonder why the stomach does not digest itself. Several factors protect the walls of the stomach (and duodenum). The cells of the luminal surface of the mucosa secrete a slightly alkaline mucus, which forms a layer 1.0 to 1.5 mm thick over the stomach surface. The protein content of mucus and its alkalinity tend to neutralize hydrogen ions in the immediate area of the epithelium, which leads to formation of a chemical barrier between the highly acid contents of the lumen and the cell surface. In addition, the membranes of cells that line the stomach have a very low permeability to hydrogen ions, and this prevents their entry into the underlying mucosa. Moreover, as we have seen, the lateral surfaces of the epithelial cells are joined by tight junctions so that there is no passage between the cells by which material could diffuse from the lumen into the mucosa. Finally, the epithelial cells that line the walls of the stomach are continually replaced every few days by cell division. The mucous layer, the cell membranes, and cell replacement all contribute to maintaining a barrier between the contents of the lumen and the underlying tissues.

Yet, in some people these protective mechanisms are inadequate, and erosions (**ulcers**) of the gastric wall occur. Ulcers may also occur in the lower part of the esophagus as a result of the reflux of acid into the esophagus from the stomach, and in the upper part of the small intestine that is bathed by the acid from the stomach. If severe enough, the ulcer may damage the underlying blood vessels and cause bleeding into the lumen. On occasion, the ulcer may penetrate the entire gastric or duodenal wall, with the leakage of luminal contents into the abdominal cavity. About 10 percent of the population of the United States are found at autopsy to have ulcers, which are about 10 times more frequent in the walls of the duodenum than in the stomach.

Ulcer formation requires that the mucosal barrier be broken, which exposes the underlying tissue to the corrosive action of acid and pepsin; however, it is not clear what produces the initial damage to the barrier. It is likely that many factors are involved; genetic susceptibility, drugs, decreased blood flow, bile salts, excess acid and pepsin are just some of the possibilities.

SECTION D: THE PANCREAS AND LIVER

THE PANCREAS AND PANCREATIC EXOCRINE SECRETIONS

The **pancreas** is a soft, lobulated organ enclosed in a thin capsule (Fig. 25-26A); it lies along the posterior abdominal wall behind the stomach. The pancreas is a mixed gland that contains endocrine and exocrine portions. The exocrine cells of the pancreas secrete two solutions involved in the digestive process, one that contains a high concentration of sodium bicarbonate and the other a large number of digestive enzymes. These solutions are secreted into ducts from the individual lobules, which converge into a single duct, called simply the **pancreatic duct** that runs from left to right through the pancreas (Fig. 25-26B). Before emptying into the duodenum, the pancreatic duct may join the common bile duct from the liver in a **hepatopancreatic ampulla,** or *ampulla of Vater.* The opening of the ampulla into the duodenum is surrounded by a small raised area, the *major duodenal papilla.* A smaller **accessory pancreatic duct** usually also drains the pancreas (Fig. 25-26B); it enters the duodenum at the *minor duodenal papilla.*

The enzyme portion of the exocrine secretions is released from **acinar cells** of the exocrine glands (Fig. 25-27). These cells have a high density of granular endoplasmic reticulum and many zymogen granules, which indicates a high capacity for synthesizing and secreting proteins. The bicarbonate solution appears to be secreted by the cells that line the early portions of the ducts that lead from the acinar cells.

The endocrine cells of the pancreas occur in clusters known as the **islets of Langerhans** or, simply, the *pancreatic islets,* interspersed between the acini. The islets have neither capsules nor ducts; their secretory products are released directly into the circulatory system. The islet cells and the hormones they secrete (insulin, glucagon, somatostatin, and possibly others) will be discussed in Chap. 26.

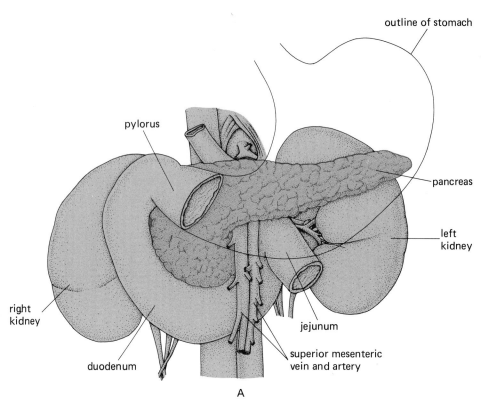

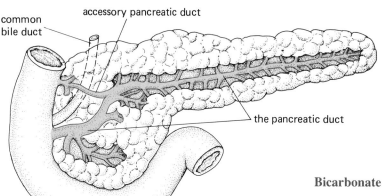

FIG. 25-26 (A) Location of the pancreas. Most of the stomach is indicated only in outline. (B) The ducts of the pancreas.

Bicarbonate Secretion

During the course of a day, 1.5 to 2.0 L of solution is secreted by the pancreas into the duodenum, most of which is reabsorbed. The concentration of sodium bicarbonate in the pancreatic secretions rises with the rate of secretion and may approach values of 150 mM at maximal rates of secretion (6 to 7 mL/min). Since the concentration of bicarbonate in the blood is about 27 mM, the secretion of bicarbonate by the pancreas is an active process.

The mechanism of bicarbonate secretion is similar to the process of hydrochloric acid secretion by the stomach, the crucial difference being a reverse orientation of the transport systems, such that the bicarbonate ions are released into the lumen rather

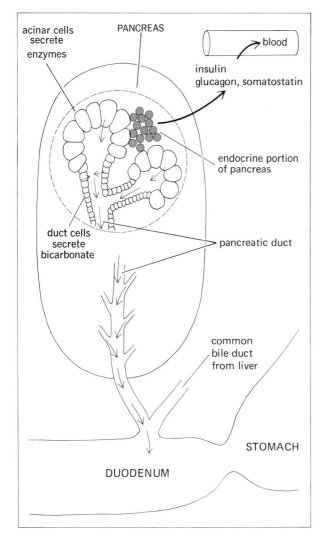

FIG. 25-27 Diagram of the structure of the pancreas. The gland areas are greatly enlarged relative to the entire pancreas.

than the blood. The pancreas, like the stomach, contains a high concentration of carbonic anhydrase, and inhibition of this enzyme reduces bicarbonate secretion, as it does hydrogen-ion secretion in the stomach.

Secretion of an alkaline solution of bicarbonate ions necessitates the formation of an equivalent amount of acid, and the blood that leaves the pancreas is therefore more acid than the blood that enters it. Accordingly, loss of large quantities of bicarbonate ions from the intestinal tract during periods of prolonged diarrhea leads to acidification of the blood, just as loss of acid from the stomach by vomiting leads to alkalinization of the blood.

Normally there is no change in the acidity of the blood since the increase in blood bicarbonate by the stomach is balanced by the increase in blood acidity in the pancreas, and the acid secreted by the stomach is neutralized by the bicarbonate secreted by the pancreas.

Enzyme Secretion

The enzymes secreted by the pancreas digest triacylglycerols, polysaccharides, and proteins to fatty acids, sugars, and amino acids, respectively. A partial list of these enzymes and their activity is given in Table 25-3. Most are secreted in an inactive form, similar to the secretion of pepsinogen by the stomach, and are then activated in the duodenum by another enzyme. For example, **enterokinase,** which is embedded in the luminal plasma membrane of the intestinal epithelium, is a proteolytic enzyme that splits off a peptide from pancreatic **trypsinogen** and then forms the active enzyme **trypsin.** Trypsin, like enterokinase, is a proteolytic enzyme and once

TABLE 25-3 PANCREATIC ENZYMES

Enzyme	Substrate	Action
Trypsin; chymotrypsin	Proteins	Breaks amino acid bond in the interior of proteins to form peptide fragments
Carboxypeptidase	Proteins	Splits off terminal amino acid from end of protein that contains a free carboxyl group
Lipase	Lipids (triacylglycerols)	Splits off fatty acids from positions 1 and 3 of triacylglycerols and thus forms free fatty acids and 2-monoglycerides
Amylase	Polysaccharides	Similar to salivary amylase; splits polysaccharides into a mixture of glucose and maltose
Ribonuclease; deoxyribonuclease	RNA; DNA	Splits nucleic acids into free mononucleotides

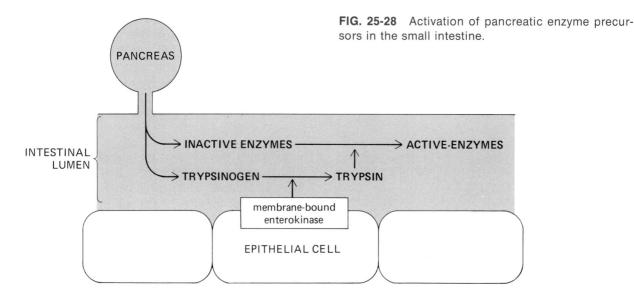

FIG. 25-28   Activation of pancreatic enzyme precursors in the small intestine.

activated it then proceeds to activate the other enzyme precursors from the pancreas in a similar manner (Fig. 25-28), as well as performing its major role of digesting protein. As we have mentioned, the secretion of enzymes in an inactive form is one mechanism for preventing these potent enzymes from digesting the cells in which they are formed.

Control of Pancreatic Exocrine Secretion

Increased pancreatic secretion during a meal is mediated mainly by the vagus (parasympathetic) nerves to the pancreas and by the intestinal hormones secretin and CCK.

The stimulus for release of the hormone, **secretin,** from the walls of the duodenum is the presence of acid in the duodenum. Secretin elicits a marked increase in the amount of bicarbonate and the volume of fluid secreted by the pancreas but only a slight stimulation of enzymatic secretion. Thus, the action of secretin is primarily on the duct cells that secrete bicarbonate. Since bicarbonate neutralizes the acid that enters the intestine from the stomach, it is appropriate that the major stimulus for secretin release is acid in the duodenum; this provides a negative-feedback control system to maintain the neutrality of the intestinal contents. As the acid is neutralized, the stimulus for secretin release is decreased and less bicarbonate is secreted by the pancreas.

The second duodenal hormone, **CCK,**[1] produces a marked increase in pancreatic enzyme secretion but little increase in bicarbonate secretion. Recall that the stimulus for CCK release is the presence of the organic components in chyme—amino acids and fatty acids—rather than acid, and it is thus appropriate that CCK stimulates primarily enzyme secretion that leads to the digestion of fat and protein.

In spite of the multiple number of hormones, nerves, and varieties of interactions between the stomach, duodenum, and pancreas, the overall adaptive significance of these interactions should be reemphasized (Fig. 25-29). As chyme empties from the stomach into the duodenum, its content of acid, amino acids, and fatty acids stimulates the release of secretin and CCK from the wall of the duodenum; these hormones stimulate the secretion of bicarbonate and enzymes from the pancreas. The pancreatic enzymes act upon the large molecules in the chyme to produce amino acids, sugars, and fatty acids that can be absorbed and also maintain the chemical stimuli responsible for hormone release until such time as the nutrients are absorbed. The bicarbonate from the pancreas neutralizes the acid

---

[1] Prior to its isolation, the duodenal hormone that stimulates the secretion of enzymes by the pancreas was called pancreozymin; however, upon purification it was found to be identical to the hormone cholecystokinin.

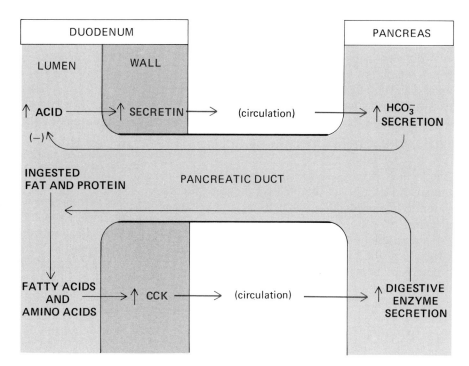

FIG. 25-29 Feedback loops that control pancreatic secretion of bicarbonate and enzymes. Only the major effects of CCK and secretin on the pancreas are shown. Hormonal pathways are colored. (Begin examination of this figure with events in the duodenum that lead to changes in secretion of secretin and CCK.)

FIG. 25-30 Pathways that regulate pancreatic secretion.

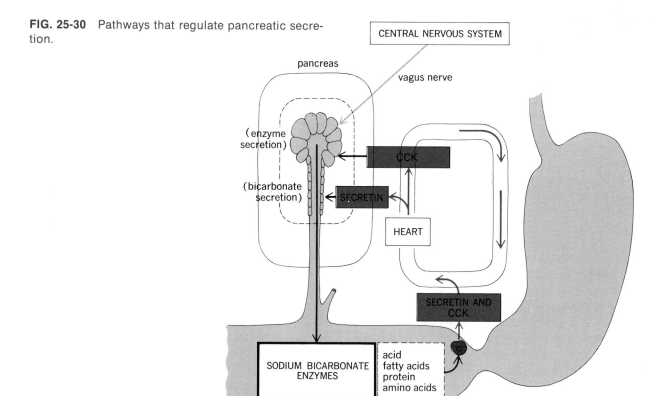

from the stomach and thus prevents ulcerative damage to the walls of the intestine; it also provides an environment in which the enzymes from the pancreas, which are inactive in a highly acid environment, can be active. Figure 25-30 summarizes these pathways that control pancreatic secretion.

THE LIVER AND SECRETION OF BILE

Bile, which is essential for the digestion of fat in the small intestine, is secreted by the liver. In addition to bile production, the cells of the liver (**hepatocytes**) play a role in iron metabolism, plasma-protein production, detoxication of substances that circulate in the plasma, metabolism of breakdown products of hemoglobin, and numerous other biochemical pathways, some of which will be discussed in Chap. 26.

Structure of the Liver

The liver, weighing from 1 to 2.5 kg, is the largest gland in the body. It lies just under the diaphragm in the upper and upper right regions of the abdominal cavity (see Fig. 25-1). The liver comprises two lobes; the right lobe is much the larger and accounts for about five-sixths of the size of the organ. The under, or visceral, surface of the liver is marked by a deep fissure, the **porta hepatis,** through which pass four main tubes: the *portal vein* and *hepatic artery,* which are entering the substance of the liver, and the *common hepatic duct* and *lymphatic,* which are leaving.

The liver is held in place by a complex series of ligaments, the most important of which are the **falciform ligament,** which connects the anterior surface of the liver to the diaphragm and anterior abdominal wall; the **round ligament** (*ligamentum teres*), which extends to the umbilicus and is, in fact, the obliterated umbilical vein of the fetus; and the **coronary ligaments,** which also join the liver to the diaphragm. The **lesser omentum,** a mesentary (see Fig. 25-5), passes from the lower surface of the liver to the lesser curvature of the stomach and to small segments of the esophagus and duodenum.

The blood supply to the liver is unusual in that it comes from two sources: the **hepatic artery** (a branch of the celiac trunk) and the **portal vein,** which receives blood from the stomach, intestines,

pancreas, gallbladder, and spleen before passing to the liver (Fig. 25-31). Thus, blood that returns to the heart from the abdominal portion of the digestive system first passes through the liver and thus gives the liver cells first crack at the foodstuffs absorbed into the bloodstream, a concept that will be important in the discussions in Chap. 26. (In contrast, fats absorbed in the intestinal wall pass into lymph vessels and are released into the general circulation via the thoracic duct; they do not go directly to the liver.)

The portal vein empties into sinusoids in the liver (Fig. 25-32A) that carry the blood from the portal vein to the **central veins,** which in turn drain the blood from the liver into the **hepatic vein** and inferior vena cava. The hepatic arteries also feed into the sinusoids, which therefore carry mixed venous and arterial blood (Fig. 25-32A). It is this blood that nourishes the hepatic cells. The liver sinusoids are wider than ordinary capillaries, but they are capillarylike in that their endothelial lining allows the passage of substances from the blood to the interstitium. The sinusoid lining also contains phagocytic cells (the **hepatic macrophages,** or *Kupffer cells*).

The basic unit of the liver substance is the **hepatic lobule.** In the center of the lobule is the **central vein,** a small tributatry of the hepatic vein (Fig. 25-32B). The hepatocytes are arranged in sheets one cell thick that radiate out from the central vein to the periphery of the lobule to meet a cluster of three tubes that consist of a branch of the hepatic artery, a branch of the portal vein, and a small bile duct, all enclosed in a connective-tissue sheath. This cluster is called the *portal area, portal triad,* or *portal tract.* One surface of the liver cells faces the sinusoids; the other surface faces small **bile canaliculi,** which drain the bile produced by the hepatic cells into larger bile ducts (Fig. 25-32A). Blood-borne materials have free access to the hepatic cells but are prevented from passing directly into the bile canaliculi by tight junctions between the cells that form their boundaries. Material does pass from the hepatic cells into the canaliculi.

The many bile ducts from the right lobe of the liver converge to form a single duct, as do those from the left lobe. These ducts, the **right** and **left hepatic ducts,** join shortly after leaving the liver to form the **common hepatic duct** (Fig. 25-33).

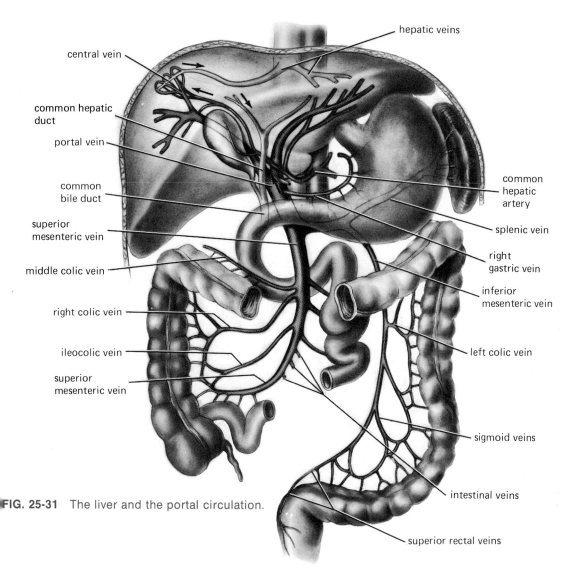

central vein

hepatic veins

common hepatic
duct

portal vein

common
bile duct

superior
mesenteric vein

middle colic vein

right colic vein

ileocolic vein

superior
mesenteric vein

common
hepatic
artery

splenic vein

right
gastric vein

inferior
mesenteric vein

left colic vein

sigmoid veins

intestinal veins

superior rectal veins

FIG. 25-31 The liver and the portal circulation.

The Gallbladder

In human beings, a side branch (the **cystic duct**) from the common hepatic duct leads to the **gallbladder.** This somewhat elongated sac is lined with a layer of mucosa that is thrown into many folds when the organ is contracted. The tall columnar epithelial cells of the mucosa have many microvilli like the absorptive cells of the small intestine. A thin layer of smooth muscle cells surrounds the mucosa and the layer of connective tissue that underlies the mucosa.

The gallbladder stores bile, concentrates it, and eventually releases it into the intestinal tract. The duct that extends from the juncture of the cystic and common hepatic ducts is the **common bile duct** (Fig. 25-33), and as described earlier, this duct joins the pancreatic duct in the ampulla of Vater before emptying into the small intestine. The ampulla is surrounded by smooth muscle known as the *sphincter of the hepatopancreatic ampulla,* or more commonly, the *sphincter of Oddi.* A part of this sphincter surrounds the common bile duct before it enters the ampulla and is particularly well developed. When this sphincter is closed, the bile produced by the liver is prevented from entering the duodenum and flows instead through the cystic duct and into the gallbladder for storage and concentration.

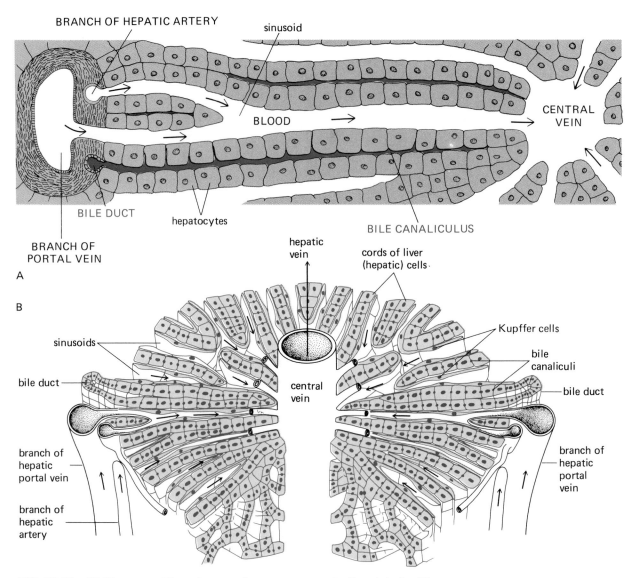

FIG. 25-32 (A) Diagram of the microscopic appearance of a liver lobule. (B) Portion of the liver, showing the location of the bile ducts with respect to the liver cells.

The gallbladder can be surgically removed without impairing the function of bile in the intestinal tract; in fact many animals which secrete bile do not even have a gallbladder; e.g., rats and horses.

Bile

Bile is essential for the digestion of fat in the small intestine. Fat digestion presents special problems because, as fat is released from the breakdown of food particles in the stomach, the molecules of tri-acylglycerol aggregate together to form large globules that are immiscible with the chyme. The function of bile is to break down these large globules into a suspension of very fine droplets about 1 μm in diameter, a process known as **emulsification.** This emulsion of fat can then be digested and absorbed. We shall discuss the details of this process in Section E of this chapter.

Bile consists of a salt solution that contains five primary ingredients: (1) bile salts, (2) cholesterol, (3) lecithin (a phospholipid), (4) bile pigments and

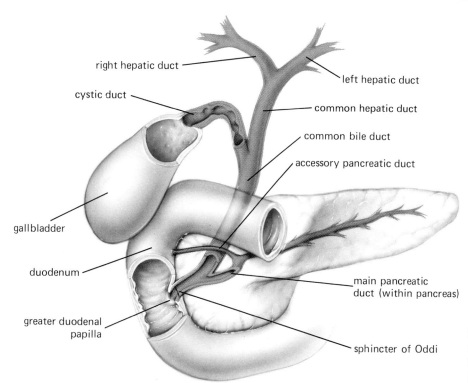

right hepatic duct

left hepatic duct

cystic duct

common hepatic duct

common bile duct

accessory pancreatic duct

gallbladder

duodenum

main pancreatic
duct (within pancreas)

greater duodenal
papilla

sphincter of Oddi

FIG. 25-33 The normal arrangement of the duct system that drains bile from the liver.

certain other end products of organic metabolism, and (5) certain trace metals. The first three of these are synthesized by the hepatocytes and are involved in the emulsificaiton of fat in the small intestine. The fourth, bile pigments, is one of the few substances that are normally excreted from the body through the intestinal tract. The major bile pigment is bilirubin, which is a breakdown product of hemoglobin formed in the macrophages of the spleen and bone marrow and, to a lesser extent, in the Kupffer cells of the liver sinusoids (Chap. 19). The cells of the liver extract bilirubin from the circulation and secrete it into the bile by an active process. Bile pigments are yellow and give bile its golden color. These pigments are modified in the intestinal tract by digestive enzymes to form pigments that give feces their brown color. In the absence of bile secretion, feces are grayish white. Some of the pigments are reabsorbed during their passage through the intestinal tract and are eventually excreted in the urine, which gives urine its yellow color.

From the standpoint of gastrointestinal function, bile salts are the most important components of bile because they are involved in the digestion and absorption of fats. The total amount of bile salt in the body is about 3.6 g; yet during the digestion of a single fatty meal, as much as 4 to 8 g of bile salts may be emptied into the duodenum. This is possible because most of the bile salts that enter the intestinal tract are reabsorbed in the lower part of the small intestine and returned to the liver, where they are again secreted into the bile. This "recycling" circulatory pathway from the intestine to the liver (via the portal vein) and back to the intestine (via the common bile duct) provides the route generally known as the *enterohepatic circulation.*

Control of Bile Secretion The volume of bile secreted depends upon the active transport into the bile canaliculi of solutes, which produce an osmotic gradient between the bile and the blood and thus cause water to follow by diffusion. Inorganic ions (sodium, chloride, and bicarbonate) and bile salts constitute the majority of the solutes that produce this water flow. Since bicarbonate ions in the bile help to neutralize acid in the duodenum, just as is true for bicarbonate ions that come from the pancreas, it is appropriate that bile flow, like pancreatic flow, is stimulated by secretin in response to the presence of acid in the duodenum; secretin does so by stimulating the liver's ion secretion.

The secretion of bile salts is stimulated by the bile salts themselves; the greater the plasma concentration of of bile salts, the greater is the rate of their secretion (and the greater the volume of bile flow). Thus, between meals, when there is little bile salt in the intestine and, therefore, little absorbed, the plasma concentration of bile salts is low and little is secreted (i.e., the enterohepatic cycling of bile salts is minimal). During a meal, the higher plasma concentration of absorbed bile salts leads to an increased rate of bile secretion. This, along with the secretin-induced stimulation of ion secretion, increases bile flow. The liver also synthesizes new bile salts to replace the 5 percent of the total bile salt pool that is not absorbed in the small intestine.

Although, as we have seen, the greatest volume of bile is secreted during and just after a meal, some bile is always being secreted by the liver. When the

sphincter of Oddi is closed, as we have mentioned, the bile secreted by the liver is shunted into the gallbladder. The cells that line the gallbladder actively transport sodium from the bile back into the plasma. As solute is pumped out of the bile, water follows by osmosis. The net result is a five- to tenfold concentration of the constituents of the bile. Shortly after the beginning of a meal, the sphincter of Oddi relaxes and the gallbladder contracts, which discharges concentrated bile into the duodenum. The signal for gallbladder contraction and sphincter relaxation is the intestinal hormone CCK, appropriately so because the stimulus for this hormone's release is the presence of fatty acids and amino acids in the duodenum. (It is from this ability to cause contraction of the gallbladder that cholecystokinin received its name: *chole,* bile; *kystis,* bladder; *kinin,* to move.) Figure 25-34 summarizes the factors that control the release of bile.

FIG. 25-34 Pathways that regulate bile secretion. Start following the loops beginning with ingested fat, ingested protein, and acid. Note the interaction with the pancreatic loop, diagrammed in Fig. 25-29. Hormonal pathways are colored. The thick outer lines denote the enterohepatic circulation.

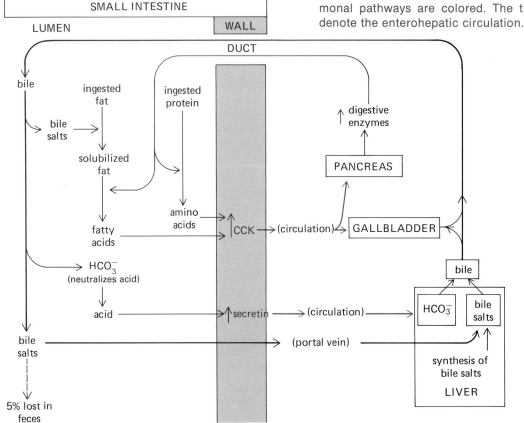

Gallstones

Bile contains not only bile salts but also cholesterol and phospholipids, which are water-insoluble and are maintained in the bile fluid as micelles in association with the bile salts. However, if the proportions of cholesterol, phospholipids, and bile salts do not fall within certain limited ranges, only part of the cholesterol will be solubilized in the micelle and the rest precipitates in the bile duct or gallbladder, and thus *gallstones* are formed. The concentration of secreted cholesterol, in proportion to bile salts and phospholipid, is normally very near the point at which cholesterol precipitates; therefore, small increases in the concentration of biliary cholesterol are sufficient to initiate gallstone formation.

If the gallstone is small, it may pass through the bile duct into the intestine with no complications. A larger stone may become lodged in the gallbladder at its connection with the cystic duct and cause contractile spasms of the smooth muscle, and pain. A more serious complication arises when a gallstone lodges in the common bile duct and thereby prevents bile from entering the intestine. The absence of bile markedly decreases the rate of fat digestion and absorption.

The buildup of pressure in a blocked bile duct inhibits further secretion of bile. As a result, bilirubin accumulates in the blood and tissues to produce the yellowish coloration in the skin known as *jaundice*. Jaundice can also occur, in the absence of a blocked bile duct, if the hepatocytes are diseased and therefore fail to secrete bilirubin as fast as it is normally produced or if an accelerated rate of red blood cell breakdown results in an increased production of the pigment.

Because, in most individuals, the duct from the pancreas joins the common bile duct just before it enters the duodenum, a gallstone that becomes lodged at this point prevents both bile and pancreatic secretions from entering the intestine, and the result is failure both to neutralize acid from the stomach and to digest adequately most organic nutrients, not just fat.

Why some individuals develop gallstones and others do not is still unclear. Women, for example, have about twice the incidence of gallstone formation as do men, and American Indians have a very high incidence compared to other ethnic groups in the United States. Increased cholesterol secretion by the liver may be one of the contributing factors in these susceptible individuals, but the factors responsible for the increased secretion are unknown. Attempts to lower blood cholesterol (in the hopes of decreasing the incidence of coronary artery disease) by drugs that increase the liver's secretion of cholesterol into the bile have resulted in an increased incidence of gallstone formation.

SECTION E: THE INTESTINES

THE SMALL INTESTINE

The **small intestine** consists of about 3 m (9 ft) of tubing, coiled in the central and lower parts of the abdominal cavity, that leads from the stomach to the large intestine. It is encircled by the large intestine (Fig. 25-1) and separated from the anterior abdominal wall by the greater omentum. The first portion of the small intestine is the **duodenum,** a short (20-cm) section that forms an incomplete circle around the head of the pancreas. The hepatopancreatic ampulla (combined bile-pancreatic duct) enters the duodenum about 1.5 cm from the gastric-duodenal junction, i.e., the pylorus. The remainder of the small intestine, which comprises the **jejunum** and **ileum,** is longer and more highly coiled than the duodenum. Although there is a gradual change in morphology from the beginning of the jejunum to the end of the ileum, there is no sharp distinction between them and the division is somewhat arbitrary. The jejunum occupies approximately two-fifths of the length of the combined jejunal-ileal segment (i.e., about 1 m, or 3 ft), and the ileum accounts for the remaining 2 m (6 ft). Normally, most digestion and absorption has occurred by the middle of the jejunum. Thus, the intestine has a considerable functional reserve, making it almost impossible to exceed its absorptive capacity even when exceedingly large quantities of food are ingested.

Structure of the Intestinal Mucosa

The mucosa of the small intestine is highly folded (Fig. 25-35). The folds (**plicae circulares,** or *valves of Kerckring*) are permanent; i.e., they are not flat-

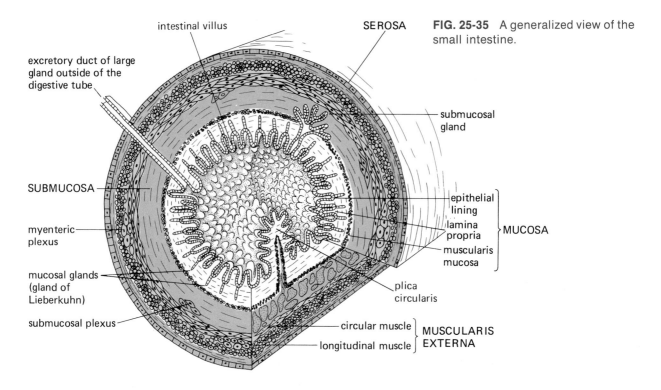

intestinal villus

SEROSA

FIG. 25-35 A generalized view of the small intestine.

excretory duct of large gland outside of the digestive tube

submucosal gland

SUBMUCOSA

epithelial lining

lamina propria

muscularis mucosa

MUCOSA

myenteric plexus

mucosal glands (gland of Lieberkuhn)

plica circularis

submucosal plexus

circular muscle

longitudinal muscle

MUSCULARIS EXTERNA

tened when the intestine is distended. The surface of the plicae is further convoluted by fingerlike projections known as **villi** (Figs. 25-35 and 25-36). The surface of each villus is covered with a single layer of epithelial cells whose surface membranes form small projections known as **microvilli** (Fig. 25-37). These microvilli contain several enzymes important for digestion. The combination of folded mucosa, villi, and microvilli increases the total surface area of the small intestine available for absorption about 600-fold over that of a flat-surfaced tube of the same length and diameter. The total surface area of the human small intestine has been estimated to be about 300 m², or equivalent to the area of a tennis court.

Villi are illustrated in Fig. 25-36, although they change in structure and number along the length of the small intestine; they are longer and have a larger surface area at the duodenal end of the intestine than at the ileal end. (This is also true of the plica circulares, which are large and close together in the duodenum, smaller and sparcer in the jejunum, and nonexistent by mid–ileum.) The center of each villus is occupied by a capillary network that branches from an arteriole and drains into a venule and by a single blind ended lymph vessel, a **lacteal.** Nerve fibers and smooth muscle cells are also present in the villus.

The villi of the intestine move back and forth independently of each other, propelled by the contraction of the small numbers of smooth muscle in each of them. The motion of the villi is increased following a meal, and associated with the increased motion is a greater flow of lymph as the lacteals are "pumped," i.e., alternately compressed. Irritation of the mucosal wall by lumps of food or stimulation of the sympathetic nerves to the intestine causes contraction of the muscularis mucosae and greater mucosal folding.

Associated with the intestines are three types of glands. One type, the large glands outside the intestine but connected to it by a duct, e.g., the pancreas and liver, have already been discussed. A second type, the **glands of Brunner,** are in the submucosa of the duodenum. They are mainly mucussecreting and expel their products into the bottom of the third type of intestinal gland, the **glands of Lieberkuhn,** which open into the intestine at the base of the villi.

The cells that line the intestinal tract are contin-

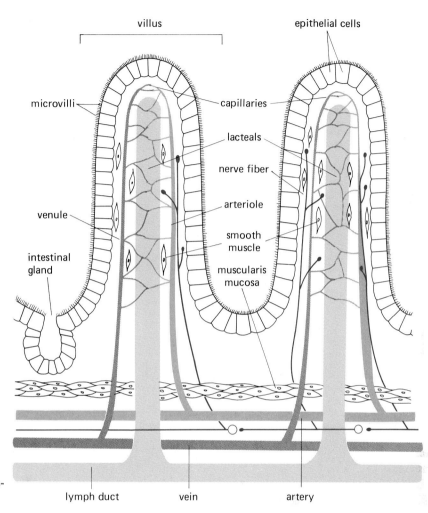

FIG. 25-36 Diagram of two intestinal villi.

ually replaced by new cells that arise from the mitotic activity of the cells at the base of the villi. The new cells differentiate and migrate to the top of the villi and thus replace older cells, which disintegrate and are discharged into the lumen of the intestine. These disintegrating epithelial cells release into the lumen their intracellular enzymes, which contribute to the digestive process. The entire epithelium is replaced approximately every 5 days, and this continuous discharge of cells amounts to about 250 g per day.

The surface of the intestinal tract (and the blood-forming regions of the bone marrow and other tissues that have a high rate of cell division) is very sensitive to damage by x-rays, atomic radiation, and anticancer drugs during the period of the mitotic cycle when DNA is being replicated.

The epithelial cell layer forms the barrier that separates molecules in the lumen from the interstitial fluid that bathes the capillaries and lacteals. Movement across this layer of cells occurs both by diffusion and by a number of specific carrier-mediated transport systems. The digestive enzymes embedded in the plasma membranes of the epithelial cells at the luminal surface are not only frequently responsible for the final states of enzymatic digestion but also are associated closely with the membrane carrier mechanisms for the digestive end products.

The blood flow to the small intestine at rest averages about 1 L/min (one-fifth the resting cardiac output), but increases during periods of digestive activity as a result of functional hyperemia (secondary to the increased metabolic activity of the intestinal cells) and of reflexes triggered by mechanical distension of the lumen by chyme.

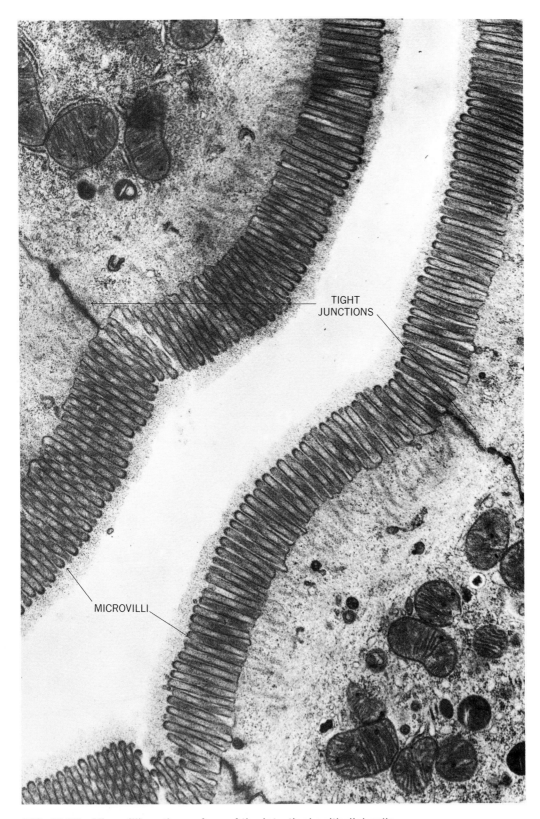

TIGHT
JUNCTIONS

MICROVILLI

FIG. 25-37 Microvilli on the surface of the intestinal epithelial cells.

TABLE 25-4 TYPICAL FOOD AND WATER INTAKE PER DAY		
	Average per Day	Percent Solids
Carbohydrate, g	500	62.5
Protein, g	200	25
Fat, g	80	10
Salt, g	20	2.5
Water, mL	1200	

As described previously, the venous drainage from the intestines (as well as from the pancreas and portions of the stomach) is unusual in that the blood does not empty directly into the vena cava, but passes first to the liver via the hepatic portal vein; there it flows through a second capillary network before leaving the liver to return to the heart. Thus, material absorbed into the capillaries of the intestine may be processed by the liver before entering the general circulation.

Digestion and Absorption

Almost all digestion and absorption of food and water occur in the small intestine. During the course of a day, the average adult consumes about 900 g of solid food and 1100 g of water (Table 25-4), but this is only a fraction of the total material that enters the gastrointestinal tract. To the approximately 2000 mL of ingested food and drink is added about 7000 mL of fluid from the salivary glands, stomach, pancreas, liver, and intestinal tract (Fig. 25-7). Of this total volume of 9000 mL, only about 500 mL passes into the large intestine each day; 94 percent is absorbed across the walls of the small intestine.

Carbohydrates The daily intake of carbohydrate varies considerably, ranging from 250 to 800 g/day in a typical American diet. Most of the carbohydrate (Table 25-5) is in the form of the plant polysaccharide, starch, with smaller amounts of the disaccharides sucrose (table sugar) and lactose (milk sugar). In infants, lactose from milk makes up the majority of the carbohydrate in the diet. Only very small amounts of monosaccharides are normally present. The polysaccharide cellulose, which is present in plant cells, cannot be broken down by the enzymes secreted by the gastrointestinal tract,

so cellulose passes through the small intestine in undigested form and enters the large intestine, where it is partially digested by bacteria.

As mentioned earlier, starch is partially digested by salivary amylase during passage through the stomach, and this digestion is continued in the small intestine by pancreatic amylase. The products formed by these amylases are the disaccharide maltose and short chains composed of about eight glucose molecules. Enzymes in the plasma membranes of the epithelial cells then split these molecules, as well as the disaccharides sucrose and lactose, into monosaccharides.

The monosaccharides, glucose and galactose, liberated from this breakdown of polysaccharides and disaccharides, are actively transported across the intestinal epithelium into the blood. The carriers that transport them into the cells are located very near the membrane disaccharidases, such that the products released from these enzymes immediately combine with the transport carriers. This active-transport system requires sodium ions; the latter combine with the same carrier that transports the sugar, and both are moved into the cell simultaneously.

Protein An intake of about 50 g of protein each day is required by an adult to supply essential amino acids and replace the amino acid nitrogen converted to urea and lost in the urine. A typical American diet contains about 125 g of protein. In addition to these dietary proteins, 10 to 30 g of protein (mostly as enzymes) is secreted into the gastrointestinal tract by the various glands, and about 25 g of protein is derived from the disintegration of epithelial cells. All this protein is normally

TABLE 25-5 CARBOHYDRATES		
Class	Examples	Made up of
Polysaccharides	Starch	Glucose
	Cellulose	Glucose
	Glycogen	Glucose
Disaccharides	Sucrose	Glucose-fructose
	Lactose	Glucose-galactose
	Maltose	Glucose-glucose
Monosaccharides	Glucose	
	Fructose	
	Galactose	

broken down into amino acids and absorbed by the small intestine.

Proteins are broken down to peptide fragments by pepsin in the stomach and by trypsin and chymotrypsin secreted by the pancreas. The peptide fragments are further digested to free amino acids by carboxypeptidase from the pancreas and aminopeptidase located in the intestinal epithelial membranes. (These enzymes' names designate the fact that they split off amino acids from the carboxyl and amino ends of the peptide chains, respectively.) The free amino acids are actively transported across the walls of the intestine. Several different carrier systems are available for transporting different classes of amino acids. Some of these require sodium, just as is true for carbohydrate transport. In addition to single amino acids, short chains of two or three amino acids are also actively absorbed.

Small amounts of intact proteins are able to cross the epithelium and gain access to the interstitial fluid without being digested. These proteins are engulfed by the plasma membrane of the epithelial cells (endocytosis), move through the cytoplasm, and are released on the opposite side of the cell by the reverse process, exocytosis. The capacity to absorb intact proteins is greater in a newborn infant than in adults; antibodies secreted in the mother's milk may be absorbed by the infant in this manner and thus provide a short-term passive immunity until the child begins to produce his or her own antibodies. Moreover, scattered throughout the mucosa of the small intestine are small clusters of unencapsulated lymphoid tissue, such as **Peyer's patches,** that respond to foreign proteins that reach them. These cells provide a local defense against certain foreign proteins that may enter the body across this large surface area exposed to the external environment.

Fat   The amount of fat in the diet varies from about 25 to 160 g/day. Most of this fat is in the form of triacylglycerols; the remainder is primarily phospholipids and cholesterol. There are very few free fatty acids in the average diet. Almost all the fat that enters the digestive tract is absorbed, and the 3 to 4 g of fat in the feces is derived primarily from the bacteria in the large intestine.

Most fat digestion occurs in the small intestine under the influence of pancreatic **lipase,** which cat-

alyzes the splitting of the bonds that link the fatty acids to the first and third carbon atoms of glycerol and thus produces free fatty acids and 2-monoglycerides (glycerol with a single fatty acid attached at the second carbon atom) as products. To digest fat, the water-soluble lipase molecule must come into contact with a molecule of triacylglycerol, but fats are not soluble in water and tend to separate from the water phase into droplets of lipid. In this form they enter the small intestine from the stomach. If the fat remained aggregated in droplets, the only area for lipase action would be the surface of the droplet, which contains only a small fraction of all the molecules in the droplet. Digestion of fat in this state is very slow. Furthermore, the products of lipase action, fatty acids and 2-monoglycerides, are themselves insoluble in water. The role of bile salts, supplied by the liver, is to eliminate both these problems: They break up the large lipid droplets into a number of smaller lipid droplets, a fat-solubilizing process known as **emulsification,** and they combine with the fatty acids and 2-monoglycerides produced by the action of lipase to form the water-soluble particles known as **micelles.**

Bile salts consist of a large steroid portion that is lipid-soluble and a small carbon chain that contains an ionized carboxyl group (Fig. 25-38). The steroid portion dissolves in the large lipid droplets and thus leaves the polar hydroxyl groups and ionized side chain exposed at the surface where they can interact with polar water molecules. As mechanical agitation in the intestine breaks up the large droplets, the resulting smaller droplets become coated in this manner with bile salts. The polar and ionized groups on the bile salts at the surface of these droplets prevent their coalescing back into larger droplets because of the repulsion of the charged surfaces. The resulting suspension of small lipid droplets, each about 1 μm in diameter, is known as an **emulsion** (Fig. 25-38). Because the surface area of lipid accessible to lipase in an emulsion is much greater than in a single large droplet, the rate of lipid digestion is increased.

Although digestion is speeded up by emulsification, absorption of the insoluble products would be very slow if it were not for the second action of bile salts, facilitation of the formation of micelles. Micelles are similar to the droplets in an emulsion, but they are much smaller. (Whereas a lipid emulsion

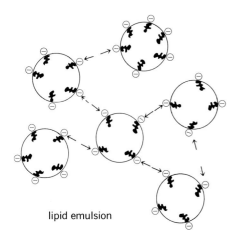

large lipid droplet

$C-NH-CH_2-COO^- + Na^+$

bile salt (glycocholic acid)

FIG. 25-38 Emulsification of fat by bile salts.

lipid emulsion

appears cloudy because of the relatively large size of its emulsion droplets, a solution of micelles is perfectly clear.) Micelles are only 3 to 10 nm in diameter and consist of bile salts, fatty acids, phospholipids, and 2-monoglycerides all clustered together with the polar ends of each molecule oriented to the micelles' surface, where they interact with polar water molecules.

Although free fatty acids and monoglycerides have an extremely low solubility in water, a few individual molecules can exist free in solution, and it is in this form that they are absorbed across the cell membrane by simple diffusion because of their high degree of solubility in the structural lipids of the cell membrane. As the free fatty acids and monoglycerides diffuse into the epithelial cells, they are replaced by others that are leaving the micelles because the lipids in the micelles are in equilibrium with those free in solution. In this manner, the luminal concentration of free lipids is maintained and diffusion into cells continues. In the absence of micelles, fat absorption occurs so slowly that it is

not completely absorbed and some fat is passed on to the large intestine and excreted in the feces.

Fatty acids are the primary form of fat entering the epithelial cells, but very little free fatty acid is released from the cell because during their passage through the epithelial cells the products of fat digestion are resynthesized into triacylglycerols. Release of the lipid droplet is believed to be similar to the release of protein secretory granules, i.e., by exocytosis. The small droplets released are 0.1 to 3.5 μm in diameter and known as **chylomicrons.** They contain about 90 percent triacylglycerol and small amounts of phospholipid, cholesterol, free fatty acids, and protein.

The released chylomicrons pass into the lacteals (lymph vessels) located in each villus. Entrance of the chylomicrons into the lacteals rather than into the capillaries is explainable by the relative permeabilities of the two vessel types. Thus, fat is absorbed into the lymphatic system, which eventually empties into the large veins; in contrast, most other substances absorbed from the intestinal tract enter the capillaries and pass through the liver by way of the hepatic portal vein before reaching the general circulation. Figure 25-39 summarizes the pathway taken by fat in moving from the intestinal lumen into the lymphatic system. In the next chapter we shall discuss how the lipids in the chylomicron are made available to the cells of the body.

Vitamins The fat-soluble vitamins, A, D, E, and K, require no digestion (i.e., molecular alteration) for their absorption. They are released from food by the digestive juices of the stomach and become

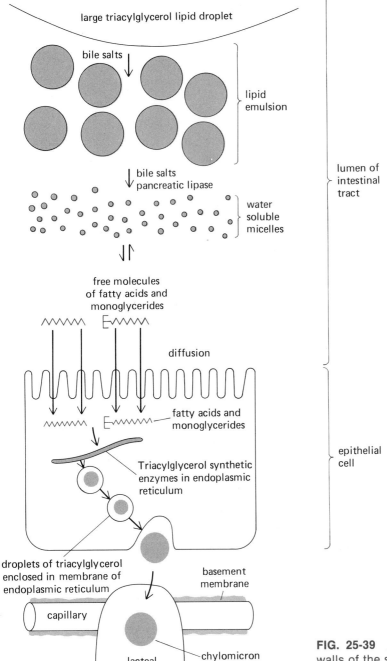

FIG. 25-39 Summary of fat absorption across the walls of the small intestine.

dissolved in the fat droplets that pass into the small intestine. As the fat is solubilized by the action of bile salts, these vitamins become incorporated into the micelles along with the products of fat digestion; the free vitamin molecules, in equilibrium with the vitamins in the micelles, diffuse across the mem-

branes of the epithelial cells as do the products of fat digestion. Any interference with the secretion of bile or the action of bile salts in the intestine decreases the absorption of fat-soluble vitamins.

Most water-soluble vitamins are absorbed by diffusion or carrier-mediated transport across the wall of the intestinal tract; one, however, vitamin B_{12}, is a very large, charged molecule that is unable to move through plasma membranes. In order to be absorbed, vitamin B_{12} must combine with a special protein, known as **intrinsic factor.** This protein, secreted in the stomach, binds to sites on the luminal surface of epithelial cells in the lower portion of the ileum, and this binding initiates membrane endocytosis, which leads to the absorption of the vitamin. Vitamin B_{12} is required for red blood cell formation; when the vitamin is deficient, a form of anemia known as *pernicious anemia* develops. This may occur when the stomach has been removed (as a treatment for ulcers or gastric cancer) or because intrinsic factor is not being secreted by the stomach. Since the absorption of vitamin B_{12} occurs specifically in the lower part of the ileum, removal of this segment of the intestine can also result in pernicious anemia. If the absorption of this vitamin suddenly ceases for any of the above reasons, it may be months or even years before the symptoms of pernicious anemia develop, because the liver stores considerable quantities of vitamin B_{12}.

Nucleic Acids Small amounts of nucleic acids, DNA and RNA, are present in food and are broken down into nucleotides by enzymes secreted by the pancreas. Enzymes in the intestinal epithelial membranes act upon these nucleotides to release free bases and monosaccharides, which are then actively transported across the epithelium by specific carrier systems.

Water and Minerals Water is the most abundant substance in the intestinal chyme, and the membranes of the epithelial cells, as well as the channels between the cells, are very permeable to it. Therefore, a net diffusion of water (osmosis) occurs across the epithelium whenever a water concentration gradient is established as a result of differences in the total solute concentration (osmolarity) on the two sides. Such differences are established by the active transport of solutes. In other words, active solute transport establishes the osmotic gradients that lead to a net movement of water (this mechanism of coupling water movement to the transport of solutes has already been described for the kidney in Chap. 24, and the same basic principles apply to water reabsorption in the intestinal tract).

The net absorption of water has an important effect upon the absorption of other substances that cross the epithelium by diffusion. As water is absorbed, the volume of the luminal contents decreases, which thereby concentrates any solutes not absorbed at the same rate. This rise in concentration secondary to water reabsorption provides the concentration gradient for the net diffusion of these substances across the intestinal wall.

In addition to sodium and chloride, other minerals present in smaller concentrations, such as potassium, magnesium, and calcium, are also absorbed, as are trace elements such as iron, zinc, and iodide. Consideration of the transport processes associated with all of these is beyond the scope of this book, and we shall briefly consider as examples the absorption of only two of them, calcium and iron.

Calcium We have repeatedly emphasized that for most substances, especially the major organic nutrients, the total amount of the substance delivered to the small intestine gets absorbed, i.e., the amount of nutrient absorbed is not regulated by the nutritional state of the body. The story for calcium is quite different in that the amount absorbed is controlled by reflexes that respond to changes in plasma calcium concentration. As described in Chap. 24, this regulation is mediated by the active form of vitamin D; via this pathway decreases in the plasma concentration of calcium lead to increases in active vitamin D and to enhancement of calcium active transport across the intestinal epithelium.

Iron Iron absorption depends in part on the type of food in which it is contained and which is ingested along with it. This is because the iron becomes bound to negatively charged ions or complexes, which can either retard or enhance its transport. For example, iron in liver is much more absorbable than is iron in egg yolk, since the latter contains phosphates that bind the iron to form an insoluble complex.

Iron ions are actively transported into intestinal epithelial cells where most are incorporated into **ferritin,** a protein-iron complex that functions as an intracellular store of iron. Some of the absorbed iron is released from the cell and becomes bound to a specific protein in the plasma, **transferrin,** in which form it circulates throughout the body.

The absorption of iron, like that of calcium, varies, depending upon the iron content of the body. When body stores are ample, the amount of iron bound to ferritin is increased and this somehow reduces further iron absorption. When the body stores of iron drop, as a result of hemorrhage, for example, so does the amount of iron bound to intestinal ferritin, and this increases iron absorption. The absorption of iron is typical of that of other trace metals in several respects: Cellular storage proteins and plasma carrier proteins are involved, and control of absorption is a major mechanism for the homeostatic control of the body's content of the metal.

Secretions

In addition to the fluids that enter the small intestine from the stomach, liver, and pancreas, 2000 mL of fluid moves across the wall of the small intestine from the blood into the lumen each day. Still larger volumes of fluid are simultaneously moving in the opposite direction—from the lumen into the blood—so that normally there is an overall net reabsorption of fluid from the small intestine. The study of intestinal secretion is difficult, and the factors that control these secretions are less understood than those that control other gastrointestinal processes.

A second source of materials that enter the small intestine is the continuous disintegration of the intestinal epithelium. These cells divide, differentiate, function for a few days, and then are discharged from the surface of the intestine. Their breakdown releases enzymes, other organic materials, ions, and water into the lumen of the intestine.

The secretion of mucus throughout the length of the intestine provides yet a third source of organic materials and fluid.

Motility

When the motion of the small intestine is observed by x-ray fluoroscopic examination, the contents of the lumen are seen to move back and forth with little apparent net movement toward the large intestine. Thus, in contrast to the waves of peristaltic contraction that sweep over the surface of the stomach, the primary motion of the small intestine is a stationary, oscillating contraction and relaxation of rings of smooth muscle (Fig. 25-40). Each contract-

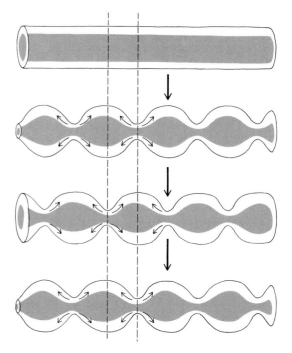

FIG. 25-40 Segmentation movements of the small intestine. The small arrows indicate the movements of the luminal contents, the large arrows the passage of time.

ing segment is about 1 to 4 cm long and the contraction lasts about 5 sec. The chyme in the lumen of a contracting segment is forced both up and down the intestine. This rhythmical contraction and relaxation of the intestinal sections, known as **segmentation,** produces a continuous division and subdivision of the intestinal contents that thoroughly mixes the chyme in the lumen and brings it into contact with the intestinal wall.

These segmenting movements of the small intestine are initiated by electrical activity generated by pacemaker cells in the longitudinal smooth muscle. The frequency of segmentation is set by the frequency of the intestinal basic electrical rhythm but unlike the stomach, which normally has a single basic electrical rhythm, the intestinal rhythm varies from segment to segment, each successive pacemaker having a slightly lower frequency than the one above. Segmentation in the duodenum occurs at a frequency of about 12 contractions per minute, whereas in the terminal portion of the ileum segmentation occurs at a rate of only 9 contractions per minute. Since the frequency of contraction is greater in the upper portion of the small intestine than in the lower, more chyme is forced, on the average, downward than is forced upward.

The magnitude of segmentation can be altered by the nerve plexuses, external nerves (parasympathetic simulation of the intestine increases contractile activity, and sympathetic stimulation decreases it), and hormones. As is true for the stomach, these reflexes produce changes in the intensity of smooth muscle contraction but do not change the natural frequency of the pacemakers. Following a meal, distension increases the intensity of segmenting contractions (paradoxically, the actual propulsion achieved is decreased, because the contracting segments narrow the lumen of the intestine and thus produce a greater resistance to flow). These changes in contractile activity are mediated by both short and long reflexes.

After most of a meal has been absorbed, the segmenting type of contractions declines and, instead, **peristaltic waves** begin to sweep over the intestine, beginning in the duodenum. These waves travel about 60 to 70 cm and then die out; however, the site at which the wave is initiated slowly moves down the small intestine. By the time the waves reach the end of the ileum, new waves are beginning in the duodenum and the process is repeated. These peristaltic waves sweep any material still remaining in the lumen of the small intestine on into the large intestine. Upon the arrival of a new meal in the stomach, they cease and are replaced by the segmenting type of contractile actitivy. The mechanism that controls the generation and progression of these waves is unknown.

Local distension of the intestine produces yet another characteristic response. The region just above the distended portion contracts and the region just below (toward the large intestine) relaxes. The contracted segment then progresses several centimeters down the intestine to produce peristaltic waves much shorter than those described in the previous paragraph. The coordination of this response is mediated by the nerve plexuses. These short peristaltic waves always proceed in the direction of the large intestine (a property known as the **law of the intestine**).

Several other specific reflexes are worth mentioning. Contractile activity in the ileum increases during periods of gastric emptying, and this is known as the **gastroileal reflex** (conversely, distension of the ileum produces decreased gastric motil-ity, the **ileogastric reflex**); and large distensions of the intestine, injury to the intestinal wall, and various bacterial infections in the intestine lead to a complete cessation of motor activity, the **intestino-intestinal reflex.** These last three reflexes appear to have as their efferent pathways the external nerves.

A person's emotional state can also affect the contractile activity of the intestine and, thus, the rate of propulsion of chyme and the time available for digestion and absorption. Fear tends to decrease motility, whereas hostility increases it, although these responses vary greatly in different individuals. These responses are all mediated by the external nerves.

As much as 500 mL of air may be swallowed along with a meal. Most of this air travels no farther than the esophagus and is eventually expelled by belching. Some of the air, however, reaches the stomach and is passed on to the intestines, where its percolation through the chyme, as the intestinal contents are mixed, produces gurgling sounds that are often quite loud.

THE LARGE INTESTINE

The **large intestine,** a tube about 6 cm (2.5 in) in diameter, forms the last 120 cm (4 ft) of the gastrointestinal tract (Fig. 25-41). The first portion of the large intestine, the **cecum,** forms a blind-ended pouch from which the **appendix,** a small (about 3½ in) fingerlike projection, extends; the appendix has no known digestive function. The large intestine is not coiled like the small intestine but consists largely of three relatively straight segments: the **ascending, transverse,** and **descending** colon. The terminal portion of the descending colon is S-shaped and forms the **sigmoid colon,** which empties into a short section, the **rectum.** As the large intestine turns sharply downward and passes through the pelvic diaphragm, the rectum becomes the **anal canal,** which terminates in the opening, the **anus.**

A distinguishing feature of the large intestine is found in the arrangement of the longitudinal muscle layer; rather than surrounding the circular muscle as it does in the rest of the intestinal tract, the longitudinal muscle lies in three narrow bands, the **teniae coli.** These bands, which are shorter than the large intestine, pucker the walls of the intestine into

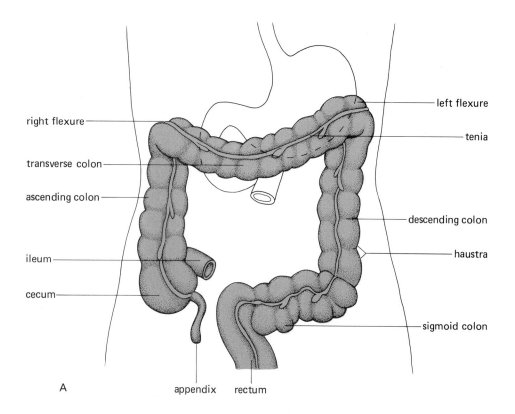

right flexure

transverse colon

ascending colon

ileum

cecum

left flexure

tenia

descending colon

haustra

sigmoid colon

A

appendix   rectum

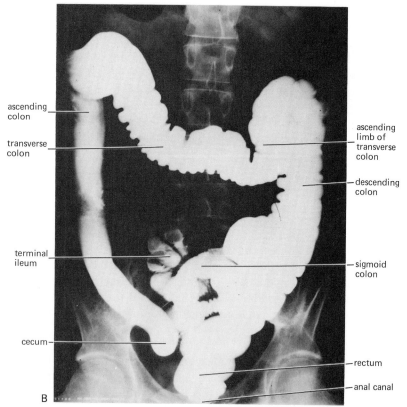

ascending colon

transverse colon

terminal ileum

cecum

ascending limb of transverse colon

descending colon

sigmoid colon

rectum

anal canal

B

FIG. 25-41   (A) The large intestine in situ. (B) X-ray of the large intestine. Haustra are apparent, especially in the transverse colon.

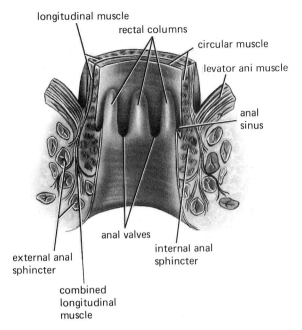

longitudinal muscle

rectal columns

circular muscle

levator ani muscle

anal sinus

anal valves

internal anal sphincter

external anal sphincter

combined longitudinal muscle

FIG. 25-42 Drawing of a longitudinal section through the anal canal and anus.

crosswise outpocketings called **haustra.** The mucosal layer also differs from that of the small intestine in that it has no villi and is considerably thicker, having deeper tubular glands, or crypts of Lieberkuhn.

The mucosa in the anal canal forms a series of longitudinal folds, the **rectal columns,** which at their lower ends are joined by thin membranes, the **anal valves,** and are separated by small pockets, the **anal sinuses** (Fig. 25-42). Many small veins are present in this area and, if they become dilated and bulge inward to encroach on the intestinal lumen, form *internal hemorrhoids. External hemorrhoids* result from the dilation of veins at or close to the anus. Crosswise projections of folds of mucosa, submucosa, and circular muscle, the **transverse rectal folds,** (not shown in Fig. 25-42), extend into the lumen of the rectum and help support its contents.

In the lower rectal segment of the large intestine, the bands of tenia gradually become wider and finally meet to form a complete layer of longitudinal muscle around the rectum. At this point, the haustra disappear. The circular muscle thickens to form the **internal anal sphincter** and is largely surrounded by the external sphincter (Fig. 25-42). The internal sphincter has the same innervation as the smooth muscle of the rest of the large intestine and probably does not function as a sphincter, despite its name. The **external anal sphincter,** which is skeletal muscle, consists of superficial and deep parts. The superficial part passes between the coccyx and either skin or tendon anterior to the anus, where it separates into two bundles that run on either side of the anus; the deep part circles the anal canal as a true sphincter. Part of the *levator ani muscle* functions with the external sphincter in controlling defecation.

Although the large intestine has a greater diameter than the small intestine and is about half as long, its epithelial surface area is only about 1/30 that of the small intestine. The large intestine secretes no digestive enzymes and is responsible for the absorption of only about 4 percent of the total intestinal contents per day. Its primary function is to store and concentrate fecal material prior to defecation.

Chyme enters the colon through the **ileocecal valve** that separates the ileum from the cecum (Fig. 25-43). This valve, which is made up of the last part of the musculature of the ileum, is normally closed. After a meal, however, when the gastroileal reflex increases the contractile activity of the ileum, the

FIG. 25-43 The cecum, ileocecal valve, and appendix. Part of the intestinal wall has been removed.

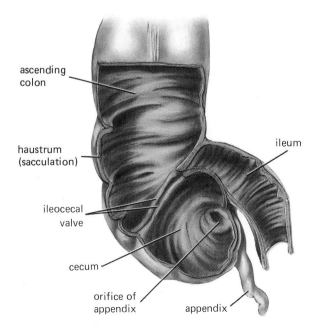

ascending colon

haustrum (sacculation)

ileocecal valve

cecum

orifice of appendix

ileum

appendix

valve opens each time the terminal portion of the ileum contracts and allows chyme to enter the large intestine. Distension of the colon, on the other hand, produces a reflex contraction of the valve that prevents further material from entering.

About 500 mL of chyme enters the colon from the small intestine each day. Most of this material is derived from the secretions of the lower small intestine, since most of the ingested food has been absorbed before reaching the large intestine. The secretions of the colon are scanty and consist mostly of mucus.

The primary absorptive process in the large intestine is the active transport of sodium from the lumen to blood with the accompanying osmotic reabsorption of water. If fecal material remains in the large intestine for a long time, almost all the water is reabsorbed and dry fecal pellets are left behind. There is a small net leakage of potassium into the colon, and severe depletion of total body potassium can result from diarrhea when large volumes of fluid are excreted in the feces.

The large intestine also absorbs some of the products (vitamins, for example) synthesized by the bacteria that inhabit this region. Although this source of vitamins generally provides only a small part of the normal daily requirements, it may make a significant contribution when dietary intake of vitamins is low.

Other bacterial products contribute to the production of intestinal gas (*flatus*). This gas is a mixture of nitrogen and carbon dioxide with small amounts of the inflammable gases hydrogen, methane, and hydrogen sulfide. Bacterial fermentation produces gas in the colon at the rate of about 400 to 700 mL/day.

Motility and Defecation

Contractions of the circular smooth muscle of the colon produce a segmentation motion considerably slower than that in the small intestine; a contraction may occur only once every 30 min. Because of this slow movement, material that enters the colon from the small intestine remains for 18 to 24 hr, which provides time for bacteria to grow and multiply. Three to four times a day, generally after meals, a marked increase in motility occurs (this usually coincides with the gastroileal reflex, described earlier). This increased motility, in which large segments of the ascending and transverse colon contract simultaneously, is a **mass movement;** it pro-

pels fecal material one-third to three-fourths of the length of the colon in a few seconds.

The sudden distension of the walls of the rectum produced by the mass movement of fecal material is the normal stimulus for defecation. It initiates the **defecation reflex,** which is mediated primarily by the nerve plexuses but can be reinforced by external nerves to the terminal end of the large intestine. The reflex response consists of a contraction of the rectum, relaxation of the anal sphincters, and increased peristaltic activity in the sigmoid colon. This activity is sufficient to propel the feces through the anus. Defecation is normally assisted by a deep inspiration followed by closure of the glottis and contraction of the abdominal and chest muscles; this causes a marked increase in intraabdominal pressure, which is transmitted to the contents of the large intestine and assists in the elimination of feces. This maneuver also causes a rise in intrathoracic pressure that leads to a sudden rise in blood pressure followed by a fall as the venous return to the heart is decreased (Chap. 21). In the elderly the cardiovascular stress that results from the strain of defecation may precipitate a stroke or heart attack.

The internal anal sphincter is composed of smooth muscle, but the external anal sphincter, as we have mentioned, is skeletal muscle under voluntary control. Brain centers may, via descending pathways, override the afferent input from the rectum and thereby keep the external sphincter closed and allow a person to delay defecation. The conscious urge to defecate accompanies the initial distension of the rectum. If defecation does not occur, the tension in the walls of the rectum decreases and the urge to defecate, mediated by stretched mechanoreceptors in the wall, subsides until the next mass movement propels more feces into the rectum, which increases the volume and again initiates the defecation reflex.

About 150 g of feces, consisting of about 100 g of water and 50 g of solid material, is eliminated from the body each day. The solid matter is made up mostly of bacteria, undigested cellulose, debris from the turnover of the intestinal epithelium, bile pigments, and small amounts of salts.

MALFUNCTIONING OF THE INTESTINAL TRACT

Since the function of the gastrointestinal system is to digest and absorb nutrients along with salts and water, most malfunctions of this organ system affect either the nutritional state of the body or its salt and water content. The effects of gastrointestinal malfunctioning on nutrition should be obvious, but the consequences for maintaining body fluid balance may require amplification. We have seen that the gastrointestinal system secretes 7000 mL into the lumen each day; the salt and water of these secretions are derived from plasma whose volume is only 3000 mL. Therefore, any significant decrease in the absorption of fluid by the intestinal tract leads to a decrease in plasma volume. Listed below are only a few familiar examples of disordered gastrointestinal function.

Lactase Deficiency (Milk Intolerance)

Lactose is the major carbohydrate of milk. It cannot be absorbed until digested into its components, glucose and galactose, which are readily absorbed by carrier-mediated active-transport processes. This digestion is performed by the enzyme *lactase,* embedded in the plasma membrane of the intestinal epithelium. Lactase is present at birth (except in those rare cases when an individual has a genetic defect and is unable to form the active enzyme), and then its concentration declines in many individuals between 18 and 36 months of age. (This does not occur in approximately 75 percent of white Americans and in most northern Europeans.) In the absence of lactase, lactose remains in the lumen of the small intestine, where it partially prevents water absorption (since the absorption of water requires prior absorption of solute to provide an osmotic gradient for osmosis). This unabsorbed lactose-containing fluid is passed on to the large intestine where bacteria, which do have the enzymes capable of metabolizing lactose, produce from it large quantities of gas (which distends the colon and thus produces pain) and organic products that inhibit active-transport processes and increase osmolarity. The result is the accumulation of fluid in the lumen of the large intestine. Thus, the diarrhea associated with lactase deficiency is a consequence both of diminished fluid absorption in the small and large intestine and of fluid secretion into the colon. Fluid loss in infants with congenital lack of lactase may be so large as to be fatal if they are not transferred to a diet that lacks lactose. The response to milk ingestion by adults whose lactase levels have diminished during development varies from mild discomfort to severely dehydrating diarrhea, according to the volume of milk ingested and the amount of lactase present in the intestine.

Constipation and Diarrhea

Many people have a mistaken belief that unless they have bowel movements every day the absorption of toxic substances from fecal material in the large intestine will somehow poison them. Attempts to identify such toxic agents in the blood following prolonged periods of fecal retention have been unsuccessful. In unusual cases where defecation has been prevented for a year or more by blockage of the rectum, no serious ill effects were noted except for the discomfort of carrying around the extra weight of 50 to 100 lb of feces. There appears to be no physiologic necessity for having bowel movements regulated by a clock; whatever maintains a person in a comfortable state is physiologically adequate, whether this means a bowel movement after every meal, once a day, or only once a week.

There are, on the other hand, symptoms—headache, loss of appetite, nausea, and abdominal distension—that may arise when defecation has not occurred for long periods of time. The symptoms of constipation appear to be caused not by toxins but by the distension of the rectum and large intestine; in fact, inflating a balloon in the rectum of a normal individual produces similar sensations. In addition, the longer fecal material remains in the large intestine, the more water is absorbed and the harder and drier the feces become, which makes defecation more difficult and sometimes painful.

Decreased motility of the large intestine is the primary factor that causes constipation. This often occurs in old age or may result from damage to the nerve plexuses of the colon that coordinate motility. Emotional stress can also affect the motility of the large intestine. One of the factors that increase the motility of the colon, and thus oppose the development of constipation, is distension. Certain materials in the diet that are not absorbed and thus remain in the lumen perform this function; these include cellulose and fiber (certain undigested substances that bind water).

Diarrhea is characterized by an increased water content in the feces and is usually accompanied by an increased frequency of defecation stimulated by the distension of the colon by the increased fluid in the lumen. Diarrhea is ultimately the result of decreased fluid absorption or increased fluid secretion. It is popularly thought that the increased motility that usually accompanies diarrhea actually causes the diarrhea (by decreasing the time available for fluid absorption). However, there is no good evidence for this, and it may be that the increased motility is simply the result of distension produced by increased fluid secretion.

A number of bacterial, protozoan, and viral diseases of the intestinal tract cause diarrhea. In many cases they do so by releasing toxic agents that either alter various ion transport processes in the epithelial membranes or injure the epithelium by direct penetration into the mucosa (thereby leading to increased fluid secretion). As we have described earlier, the presence of unabsorbed solutes in the lumen, as a result of decreased digestion or absorption, also results in retained fluid and diarrhea.

The major consequence of prolonged severe diarrhea is the effect of the loss of fluid upon the circulatory system. Accompanying this loss of fluid is also a loss of various ions, especially potassium and bicarbonate, the latter ions derived from the liver and pancreatic secretions. Since a net loss of bicarbonate leaves behind an equivalent amount of acid, the extracellular fluid becomes more acid during periods of diarrhea.

KEY TERMS

digestion	pepsin	gastric phase
secretion	small intestine	intestinal phase
absorption	pancreas	peristaltic wave
mucosa	liver	antrum
peritoneum	bile	basic electrical rhythm
peritoneal cavity	gallbladder	enterohepatic circulation
saliva	large intestine	villi
amylase	secretin	lacteal
esophagus	cholecystokinin (CCK)	emulsification
stomach	gastrin	segmentation
chyme	cephalic phase	defecation

REVIEW EXERCISES

Section A

1 Draw a diagram of the gastrointestinal tract and label the mouth, esophagus, stomach, small and large intestines, salivary glands, liver, and pancreas.
2 List the functions of the gastrointestinal tract.
3 Define digestion.
4 Draw a diagram of a cross section of the gastrointestinal "tube" and label the mucosa, epithelium, lamina propria, muscularis mucosa, submucosa, submucus nerve plexus, muscularis externa, circular muscle layer, myenteric nerve plexus, longitudinal muscle layer, serosa, and mesothelium.
5 Define parietal and visceral peritoneum, peritoneal cavity, greater and lesser omenta, mesentery, and mesocolon.
6 List three functions of saliva.
7 Name the enzyme present in saliva and state its function.
8 List two functions of hydrochloric acid in the stomach.

9 Define chyme and pepsin.
10 Define duodenum, jejunum, ileum, villi, microvilli, and lacteal.
11 List the four classes of stimuli that initiate gastrointestinal reflexes.
12 Define long and short neural reflexes.
13 List the four established gastrointestinal hormones, the major site of production of each, the stimuli that initiate their release, their target cells, and the effects of their release.
14 List the three phases of gastrointestinal control and describe the types of stimuli that initiate each phase.

Section B

1 Draw a sagittal section of the mouth and pharynx and label the oral cavity, oropharyngeal isthmus, pharyngeal isthmus, and hard and soft palates.

2 Draw a frontal view of the oral cavity as seen with the mouth open, and label the uvula, palatine tonsils, and isthmus of the fauces.

3 Draw a tooth and label the crown, root, enamel, dentin, pulp cavity, pulp canal, apical foramen, and periodontal ligaments.

4 Draw the upper and lower jaw of an adult and show the teeth; label the central and lateral incisors, canine or cuspids, premolars or bicuspids, and molars, and give the average time of eruption of each.

5 List the stimuli for salivary secretion.

6 State approximately how much saliva is secreted each day and explain what happens to it.

7 Draw the pharynx and esophagus and label the oral pharynx, epiglottis, laryngeal pharynx, superior esophageal sphincter, esophagus, and inferior esophageal sphincter.

8 Explain the statement that swallowing is both voluntary and reflex.

9 Describe the act of swallowing; include discussions of peristalsis and secondary peristalsis and the actions of the superior and inferior esophageal sphincters.

Section C

1 List the secretions of parietal and chief cells.

2 List the four major secretions of the stomach.

3 Describe the processes of digestion that occur in the stomach.

4 Draw a diagram of the stomach and label the esophagus, fundus, body, antrum, pyloric sphincter, and duodenum.

5 Describe the chemical processes involved in HCl secretion by the stomach.

6 Describe the factors that control HCl secretion by the stomach.

7 Describe the pathways that regulate the secretion and activation of pepsinogen and describe the relationship between pepsinogen and pepsin.

8 Define basic electrical rhythm.

9 Describe gastric emptying and the factors that control it.

10 Describe vomiting and the vomiting center.

11 Describe the consequences of vomiting on acid-base balance of the interstitial fluid of the body.

12 Describe ulcers and their relation to gastric-acid secretion.

Section D

1 Contrast the exocrine and endocrine functions of the pancreas.

2 Name the two exocrine secretions of the pancreas and describe their roles in digestion.

3 Define enterokinase, trypsinogen, trypsin, chymotrypsin, carboxypeptidase, lipase, amylase, ribonuclease, and deoxyribonuclease and state their roles in digestion.

4 Describe the neural and hormonal control of pancreatic exocrine secretion.

5 List the four main tubes that enter or leave the liver and give the function of each.

6 Draw a diagram of the microscopic structure of the liver and show the relationship between its blood supply, hepatocytes, and secretory ducts for bile.

7 Describe the structure of the gallbladder and explain its function.

8 Describe bile and explain its role in digestion.

9 State where bile is formed, stored, and released into the duodenum.

10 Describe the enterohepatic circulation of bile salts.

11 Describe the factors that control bile secretion.

12 Describe gallstones and explain their cause.

Section E

1 Draw a diagram of an intestinal villus and label the microvilli, capillaries, arterioles, venules, and lacteal.

2 Describe the turnover of the intestinal epithelium.

3 Describe active and passive transport across the intestinal epithelium.

4 Describe the role of tight junctions in the exchange of materials across the intestinal wall.

5 Describe the digestion of proteins, fats, and carbohydrates in the small intestine.

6 Describe the absorption of salts and water in the intestine.

7 Define segmentation and describe its control.

8 Describe the roles of segmentation and peristalsis in digestion and in moving food through the small intestine.

9 Define gastroileal reflex, ileogastric reflex, intestinointestinal reflex, and law of the intestine.

10 Describe the absorption of vitamin B_{12}; include a discussion of intrinsic factor.

11 Draw a diagram of the large intestine and label the cecum; appendix; ileocecal valve; ascending, transverse, and descending colon; sigmoid colon; rectum; anal canal; and anus.

12 List the substances that are absorbed or secreted in the large intestine.

13 Define mass movement and describe its role in triggering the defecation reflex.

14 State the facts and fallacies regarding constipation.

CHAPTER

26

ENERGY
BALANCE

SECTION A: CONTROL AND INTEGRATION OF CARBOHYDRATE, PROTEIN, AND FAT METABOLISM

In Chap. 3, we described the basic chemistry of living cells and their need for a continuous supply of nutrients. Although a certain fraction of these organic molecules is used in the synthesis of structural cell components, enzymes, coenzymes, hormones, antibodies, and other molecules that serve specialized functions, most of the molecules in the food we eat are used by cells to provide the chemical energy required to maintain cell structure and function.

Essential for an understanding of organic metabolism is awareness of the remarkable ability of most cells, particularly those of the liver, to convert one type of molecule into another. These interconversions permit the human body to utilize the wide range of molecules found in different foods; there are limits, however, and certain molecules must be present in the diet in adequate amounts. Enough protein must be ingested to provide the nitrogen needed for synthesis of protein and other nitrogenous substances, and this protein must contain an adequate quantity of specific amino acids. These amino acids are called **essential** because they cannot

be formed in the body by conversion from another molecule type; eight of the 20 amino acids fit this category. The other essential organic nutrients are a small group of fatty acids and the vitamins (Table 26-1). This last group was discovered when diets adequate in both total calories and essential amino acids and fatty acids were found to be incapable of maintaining health.

The concept of a dynamic catabolic-anabolic steady state is also a critical component of organic metabolism. With few exceptions, e.g., DNA, virtually all organic molecules are continuously broken down and rebuilt, often at a rapid rate. The turnover rate of body protein is approximately 100 g/day; i.e., this quantity is broken down into amino acids and resynthesized each day. Few of the atoms present in a person's skeletal muscle a month ago are still there today.

With these basic concepts of **molecular interconvertibility** and **dynamic steady state** as foundation, we can discuss organic metabolism in terms of total body interactions. Figure 26-1 summarizes the major pathways of protein metabolism. The **amino acid pools,** which constitute the body's total free amino acids, are derived primarily from ingested protein (which is degraded to amino acids during digestion) and from the continuous breakdown of body protein. These pools are the source of amino acids for

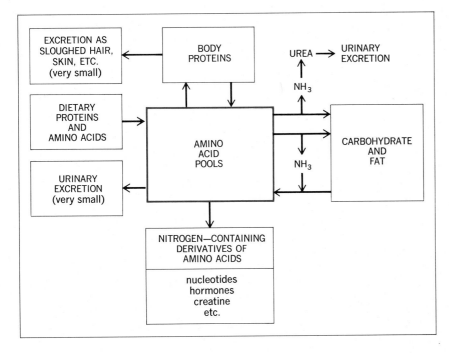

FIG. 26-1 Amino acid pools and major pathways of protein metabolism.

TABLE 26-1 ESSENTIAL ORGANIC NUTRIENTS

Essential Nutrient	RDA[c] for Healthy Adult Male, mg	Dietary Sources	Major Body Functions	Deficiency	Excess
Amino acids: Isoleucine Leucine Lysine Methionine Phenylalanine Threonine Tryptophan Tyrosine Valine	700 1000 800 1100 1100 500 250 1100 800	From proteins: Good sources: Legume grains, dairy products, meat, fish Adequate sources: Rice, corn, wheat Poor sources: Cassava, sweet potato	Precursors of structural protein, enzymes and coenzymes, antibodies, hormones, metabolically active compounds, neurotransmitters, and porphyrins. Certain amino acids have specific functions: (a) Methionine is required for methyl group metabolism (b) Tryptophan is a precursor of serotonin (c) Tyrosine is a precursor of epinephrine and other catecholamines and thyroxine (d) Arginine is a precursor of polyamines and urea	Deficient protein intake leads to development of kwashiorkor and, coupled with low energy intake, to marasmus	Excess protein intake possibly aggravates or potentiates chronic disease states
Arginine[a] Cystine[b] Histidine[c]	0 1100 0				
Fatty Acids: Arachidonic Linoleic Linolenic	6000 6000 6000	Vegetable fats (corn, cottonseed, soy oils); wheat germ; vegetable shortenings	Involved in cell membrane structure and function; precursors of prostaglandins (regulation of gastric function, release of hormones, smooth muscle activity)	Poor growth; skin lesions	Not known

TABLE 26-1 ESSENTIAL ORGANIC NUTRIENTS (*Continued on following page*)

Essential Nutrient	RDA[c] for Healthy Adult Male, mg	Dietary Sources	Major Body Functions	Deficiency	Excess
Vitamins:					
Water-soluble:					
Vitamin B₁ (thiamine)	1.5	Pork, organ meats, whole grains, legumes	Coenzyme (thiamine pyrophosphate) in reactions that involve the removal of carbon dioxide	Beriberi (peripheral nerve changes, edema, heart failure)	None reported
Vitamin B₂ (riboflavin)	1.8	Widely distributed in foods	Constituent of two flavin nucleotide coenzymes involved in energy metabolism (FAD and FMN)	Reddened lips, cracks at corner of mouth (cheilosis), lesions of eye	None reported
Niacin	20	Liver, lean meats, grains, legumes (can be formed from tryptophan)	Constituent of two coenzymes involved in oxidation-reduction reactions (NAD and NADP)	Pellagra (skin and gastrointestinal lesions; nervous, mental disorders)	Flushing, burning and tingling around neck, face, and hands
Vitamin B₆ (pyridoxine)	2	Meats, vegetables, whole-grain cereals	Coenzyme (pyridoxal phosphate) involved in amino acid metabolism	Irritability, convulsions, muscular twitching, dermatitis near eyes, kidney stones	None reported
Pantothenic acid	5–10	Widely distributed in foods	Constituent of coenzyme A, which plays a central role in energy metabolism	Fatigue, sleep disturbances, impaired coordination, nausea (rare in human beings)	None reported
Folacin	0.4	Legumes, green vegetables, whole-wheat products	Coenzyme (reduced form) involved in transfer of single-carbon units in nucleic acid and amino acid metabolism	Anemia, gastrointestinal disturbances, diarrhea, red tongue	None reported

TABLE 26-1 ESSENTIAL ORGANIC NUTRIENTS (Continued)

Essential Nutrient	RDA[c] for Healthy Adult Male, mg	Dietary Sources	Major Body Functions	Deficiency	Excess
Vitamin B₁₂	0.003	Muscle meats, eggs, dairy products (not present in plant foods)	Coenzyme involved in transfer of single-carbon units in nucleic acid metabolism	Pernicious anemia, neurological disorders	None reported
Biotin	Not established; usual diet provides 0.15–0.3	Legumes, vegetables, meats	Coenzyme required for fat synthesis, amino acid metabolism, and glycogen formation	Fatigue, depression, nausea, dermatitis, muscular pains	None reported
Choline	Not established; usual diet provides 500–900	All foods containing phospholipids (egg yolk liver, grains, legumes)	Constituent of phospholipids; precursor of neurotransmitter acetylcholine	Not reported in human beings	None reported
Vitamin C (ascorbic acid)	45	Citrus fruits, tomatoes, green peppers, salad greens	Maintains intercellular matrix of cartilage, bone and dentine; important in collagen synthesis	Scurvy (degeneration of skin, teeth, blood vessels; epithelial hemorrhages)	Relatively nontoxic; possibility of kidney stones
Fat-soluble:					
Vitamin A (retinol)	1	Provitamin A (beta-carotene) widely distributed in green vegetables; retinol present in milk, butter, cheese, fortified margarine	Constituent of rhodopsin (visual pigment); maintenance of epithelial tissues; role in mucopolysaccharide synthesis	Xerophthalmia (keratinization of ocular tissue), night blindness, permanent blindness	Headache, vomiting, peeling of skin, anorexia, swelling of long bones
Vitamin D[d]	0.001	Cod-liver oil, eggs, dairy products, fortified milk, and margarine	Promotes growth and mineralization of bones; increases absorption of calcium	Rickets (bone deformities) in children; osteomalacia in adults	Vomiting, diarrhea, loss of weight, kidney damage
Vitamin E (tocopherol)	15	Seeds, green leafy vegetables, margarines, shortenings	Functions as an antioxidant to prevent cell-membrane damage	Possibly anemia	Relatively nontoxic

TABLE 26-1 ESSENTIAL ORGANIC NUTRIENTS (Continued)

Essential Nutrient	RDA[f] for Healthy Adult Male, mg	Dietary Sources	Major Body Functions	Deficiency	Excess
Vitamin K[e] (phylloquinone)	0.03	Green leafy vegetables; small amount in cereals, fruits, and meats	Important in blood clotting (involved in formation of clotting factors)	Hemorrhage	Relatively nontoxic; synthetic forms at high doses may cause jaundice

[a] Nonessential if sufficient dietary phenylalanine.
[b] Nonessential if sufficient dietary methionine.
[c] Unnecessary in adults in short-term studies but probably essential for normal growth of children.
[d] If adequate sunlight is available, no dietary intake is required since skin synthesizes this "vitamin," which is more properly termed a "hormone" (see Chap. 18).
[e] Synthesized by intestinal bacteria; dietary intake normally unnecessary.
[f] RDA: Recommended Daily Allowance

Source: N. S. Scrimshaw and V. R. Young, "The Requirements of Human Nutrition," *Sci. Amer.*, 235(3):50, Sept. 1976.

resynthesis of body protein and a host of specialized amino acid derivatives, such as nucleotides, epinephrine, etc. A very small quantity of amino acid and protein is lost from the body via the urine, skin, hair, and fingernails.

The interactions between amino acids and the other nutrient types, carbohydrate and fat, are extremely important: Amino acids may be converted into carbohydrates or fat by removal of ammonia (deamination); one type of amino acid may participate in the formation of another by passing its nitrogen group to a carbohydrate. Both these processes were described in greater detail in Chap. 3 and are mentioned here to emphasize the interconvertibility of protein, carbohydrate, and fat. The ammonia, NH_3, formed during deamination is converted by the liver into urea, which is then excreted by the kidneys as the major end product of protein metabolism. Not all the events that relate to amino acid metabolism occur in all cells; for example, urea is formed in one organ (the liver) but excreted by another (the kidneys), but the concept of a pool is valid because all cells are interrelated by the vascular system and blood.

If any of the essential amino acids is missing from the diet, negative nitrogen balance (i.e., output greater than intake) always results. Apparently, the proteins for which that amino acid is essential cannot be synthesized, and the other amino acids that would have been incorporated into the proteins are deaminated and their nitrogen excreted as urea. It should be obvious, therefore, why a dietary requirement for protein cannot be specified without regard to the amino acid composition of that protein. Protein is graded in terms of how closely its ratio of essential amino acids approximates the ideal, which is their relative proportions in body protein. The highest-quality proteins are those found in animal products, whereas the quality of most plant proteins is lower. Nevertheless, it is quite possible to obtain adequate quantities of all essential amino acids from a mixture of plant proteins alone although the total quantity of protein ingested must be larger.

Figure 26-2 summarizes the metabolic pathways for carbohydrate and fat, which are considered together because of the high rate of conversion of carbohydrate to fat (see below and Chap. 3). The similarities between Figs 26-1 and 26-2 are obvious, but there are several critical differences: (1) The major fate of both carbohydrate and fat is catabo-

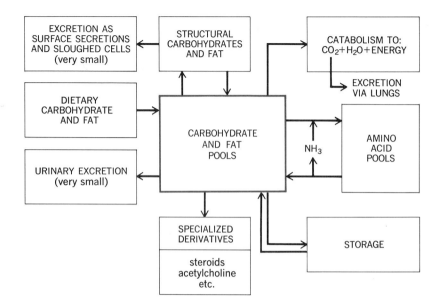

FIG. 26-2 Major pathways of carbohydrate and fat metabolism. "Carbohydrate and fat pools" include the simple unspecialized carbohydrates and fats dissolved in the body fluids. Note that structural carbohydrate and fat are constantly being broken down and resynthesized.

lism to yield energy, whereas amino acids can supply energy only after they are converted to carbohydrate or fat; and (2) excess carbohydrate and fat can be stored as such, whereas excess amino acids are not stored as protein but are converted to carbohydrate and fat.

In discussing the mechanisms that regulate the magnitude and direction of these molecular interconversions, we shall see that the liver, adipose tissue (the storage tissue for fat), and muscle are the dominant effectors and that the major controlling inputs to them are a group of hormones and the sympathetic nerves to adipose tissue and the liver. At this point, the reader should review the biochemical pathways described in Chap. 3, particularly those that deal with glucose and the interconversions of carbohydrate, protein, and fat.

EVENTS OF THE ABSORPTIVE AND POSTABSORPTIVE STATES

When food is readily available, human beings can get along by eating small amounts of food all day long if they wish. This, however, is clearly not true for most other animals, for early man, or for most persons today, and mechanisms exist for survival during alternating periods of plenty and fasting. We speak of two functional states: the **absorptive state,** during which ingested nutrients enter the blood from the gastrointestinal tract, and the **postabsorptive state,** during which the gastrointestinal tract is

empty and energy must be supplied by the body's endogenous stores. Since an average meal requires approximately 4 hr for complete absorption, our usual three-meal-a-day pattern places us in the postabsorptive state during the late morning and afternoon and almost the entire night. The average person can easily withstand a fast of many weeks (so long as water is provided), and extremely obese patients, who were given only water and vitamins, were able to fast for many months.

The absorptive state can be summarized as follows: During absorption of a normal meal: (1) glucose provides the major energy source; (2) only a small fraction of the absorbed amino acids and fat is utilized for energy; (3) another fraction of amino acids and fat is used to resynthesize the continuously degraded body proteins and structural fat, respectively; and (4) most of the amino acids and fat as well as carbohydrate not oxidized for energy are transformed into adipose-tissue fat.

In the postabsorptive state: (1) carbohydrate is synthesized in the body, but its utilization for energy is greatly reduced; (2) the oxidation of endogenous fat provides most of the body's energy supply; and (3) fat and protein synthesis are curtailed and net breakdown occurs.

Absorptive State

Figure 26-3 summarizes the major pathways to be described. Although the figure may appear formidable at first glance, it should give little difficulty after

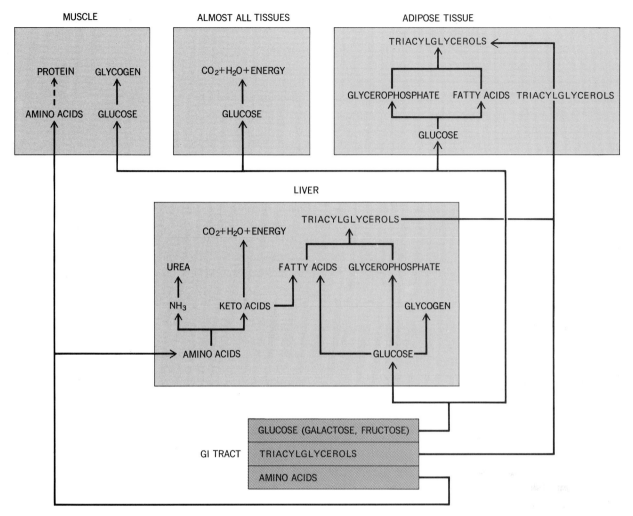

FIG. 26-3 Major metabolic pathways of the absorptive phase. Glycerophosphate is the form of glycerol used to synthesize triacylglycerol.

we have described the component parts, and it should be referred to constantly during the following discussion.

We shall assume an average meal to contain approximately 65 percent carbohydrate, 25 percent protein, and 10 percent fat. Recall from Chap. 25 that these nutrients enter the blood and lymph from the gastrointestinal tract primarily as monosaccharides, amino acids, and triacylglycerols, respectively. (As mentioned earlier, triacylglycerols were termed triglycerides until recently). The first two groups enter the blood, which leaves the gastrointestinal tract to go directly to the liver by way of the hepatic portal vein, allowing this remarkable biochemical factory to alter the composition of the

blood before it returns to the heart to be pumped to the rest of the body. In contrast, the fat droplets are absorbed into the lymph and not into the blood; the lymph drains via the thoracic ducts into the subclavian veins in the neck, and therefore, the liver does not get first crack at the absorbed fat.

Glucose Some of the absorbed carbohydrate is galactose and fructose, but because the liver converts most of these carbohydrates immediately into glucose (and because fructose enters essentially the same metabolic pathways as does glucose), we shall simply refer to these sugars as glucose. As shown in Fig. 26-3, much of the absorbed carbohydrate enters the liver cells, but little of it is oxidized for energy; instead, it is built into the polysaccharide glycogen or transformed into fat. The importance of glucose as a precursor of fat cannot be overemphasized; note that glucose provides both the glycerol (as glycerophosphate) and the fatty acid moieties of triacylglycerols. Some of this fat synthesized in the liver may be stored there, but most is released into the blood and then stored in adipose tissue cells. Much of the absorbed glucose that did not enter liver cells but remained in the blood enters adipose tissue cells, where it is transformed into fat; another fraction is stored as glycogen in skeletal muscle and certain other tissues; and a very large fraction enters the various cells of the body and is oxidized to carbon dioxide and water and thereby provides the cells' energy requirements. Glucose is usually the body's major energy source during the absorptive state.

Triacylglycerols   Almost all ingested fat is absorbed into the lymph as fat droplets (chylomicrons) that contain mainly triacylglycerols, which are hydrolyzed into their component molecules—glycerol and free fatty acids—by enzymes located on the inner surface of the capillary endothelium. These products can then diffuse across the capillary wall and enter the tissue cells. For simplicity's sake we have shown in Fig. 26-3 only the eventual overall result of this processing, namely, the deposition of most of the ingested fat, once again combined into triacylglycerol, in adipose tissue. There are three major sources of adipose tissue triacylglycerol: (1) ingested fat, (2) triacylglycerol synthesized in adipose tissue, and (3) triacylglycerol synthesized in the liver and transported via the blood to the adipose tissue. For simplicity, we have not shown in Fig. 26-3 that a fraction of fat is also oxidized during the absorptive state by various organs to provide energy. The actual amount utilized depends upon the content of the meal and the person's nutritional status.

Amino Acids   Many of the absorbed amino acids enter liver cells and are entirely converted into carbohydrate (keto acids) by removal of the NH_3 portion of the molecule. The ammonia is converted by the liver into urea, which diffuses into the blood and is excreted by the kidneys. The keto acids that are formed can enter the Krebs tricarboxylic acid cycle and be oxidized to provide energy for the liver cells. Finally, the keto acids can also be converted to fatty acids and thereby participate in fat synthesis by the liver. Most of the ingested amino acids are not taken up by the liver cells, and these enter other cells of the body (Fig. 26-3). Although virtually all cells require a constant supply of amino acids for protein synthesis, we have simplified the diagram by showing only muscle because it constitutes the great preponderance of body mass and therefore contains the most important store, quantitatively, of body protein. Other organs of course participate, but to a lesser degree, in the amino acid exchanges that occur during the absorptive and postabsorptive states. After entering the cells, the amino acids may be synthesized into protein. This process is represented by the dotted line in Fig. 26-3 to call attention to an important fact: Excess amino acids are not stored as protein, in the sense that glucose and fat are stored as fat and to a lesser degree as glycogen. Eating large amounts of protein does not significantly increase body protein; the excess amino acids are merely converted into carbohydrate or fat. On the other hand, a minimal supply of ingested amino acids is essential to maintain normal protein stores by preventing net protein breakdown in muscle and other tissues. During the usual alternating absorptive and postabsorptive states, the fluctuations in total body protein are relatively small. This discussion applies only to the adult; the growing child, of course, manifests a continuous increase of body protein.

Conclusion   During the absorptive period anabolism exceeds catabolism; energy is provided primarily by glucose, body protein is maintained, and excess calories (regardless of source) are stored mostly as fat. Glycogen constitutes a quantitatively less important storage form of carbohydrate. The use of fat to store excess calories is an excellent adaptation for mobile animals, because 1 g of triacylglycerol contains more than twice as many calories as 1 g of protein or glycogen and because there is very little water in adipose tissue.

Postabsorptive State

The essential problem during this period is that no glucose is being absorbed from the intestinal tract, yet the plasma glucose concentrations must be maintained because the nervous system is an obligatory glucose utilizer; i.e., it is unable to oxidize any other nutrient for energy (there is an important exception that will be described subsequently). Perhaps the most convenient way of viewing the events of the postabsorptive state is in terms of how the blood glucose concentration is maintained. These events fall into two categories: (1) sources of glucose, and (2) glucose sparing (fat utilization).

Sources of Blood Glucose   The sources of blood glucose during the postabsorptive period (Fig. 26-4) are as follows:

1   Glycogen stores in the liver are broken down to liberate glucose but are adequate only for a short time. After the absorptive period is completed, the normal liver contains less than 100 g of glycogen; at 4 kcal/g, this provides 400 kcal, enough to fulfill the body's total caloric need for only 4 hr.
2   Glycogen in muscle (and to a lesser extent other tissues) provides approximately the same amount of glucose as the liver. A complication arises because muscle lacks the necessary enzyme to form free glucose from glycogen.[1] But glycolysis breaks the glycogen down into pyruvate and lactate, which are then liberated into the blood, circulate to the liver, and are converted into glucose. Thus, muscle glycogen contributes to the blood glucose via the liver.
3   As shown in Fig. 26-4, the catabolism of triacylglycerols yields glycerol and fatty acids. The former can be converted into glucose by the liver, but the latter cannot. Thus, a potential source of glucose is adipose tissue triacylglycerol breakdown, in which glycerol is liberated into the blood, circulates to the liver, and is converted into glucose.
4   The major source of blood glucose during prolonged periods of fasting comes from protein. Large quantities of protein in muscle and to a lesser extent other tissues are not absolutely essential for cell function; i.e., a sizable fraction of cell protein can be catabolized without serious cellular malfunction. There are, of course,

limits to this process, and continued protein loss ultimately means functional disintegration, sickness, and death. Before this point is reached, protein breakdown can supply large quantities of amino acids that are converted into glucose by the liver.

In conclusion, for survival of the brain, plasma glucose concentration must be maintained. Glycogen stores, particularly in the liver, form the first line of defense, are mobilized quickly, and can supply the body's needs for several hours, but they are inadequate for longer periods. Under such conditions, protein and, to a much lesser extent, fat, supply amino acids and glycerol, respectively, for production of glucose by the liver. Hepatic synthesis of glucose from pyruvate, lactate, glycerol, and amino acids is known as **gluconeogenesis,** i.e., new formation of glucose. During a 24-hr fast, it amounts to approximately 180 g of glucose. The kidneys are also capable of glucose synthesis from the same sources, particularly in a prolonged fast (several weeks) at the end of which they may be contributing as much glucose as the liver.

Glucose Sparing (Fat Utilization)   A simple calculation reveals that even the 180 g of glucose per day produced by the liver during fasting cannot possibly supply all the body's energy needs: 180 g/day × 4 kcal/g = 720 kcal/day, whereas normal total energy expenditure equals 1500 to 3000 kcal/day. The following essential adjustment must therefore take place during the transition from absorptive to postabsorptive state: The nervous system continues to utilize glucose normally, but virtually all other organs and tissues markedly reduce their oxidation of glucose and depend primarily on fat as their energy source, and thus spare the glucose produced by the liver to serve the obligatory needs of the nervous system. The essential step is the catabolism of adipose tissue triacylglycerol to liberate fatty acids into the blood. These fatty acids are picked up by virtually all tissues (excluding the nervous system), enter the Krebs cycle, and are oxidized to carbon dioxide and water, and thereby provide energy. The liver, too, utilizes fatty acids for its energy source, but its handling of fatty acids during fasting is unique. It oxidizes them to acetyl CoA, which is processed into a group of compounds called **ketone bodies,** instead of being oxidized fur-

---

[1] Muscle glycogen is broken down in several steps to glucose 6-phosphate rather than free glucose, which is then catabolized via glycolysis and the Krebs cycle for energy.

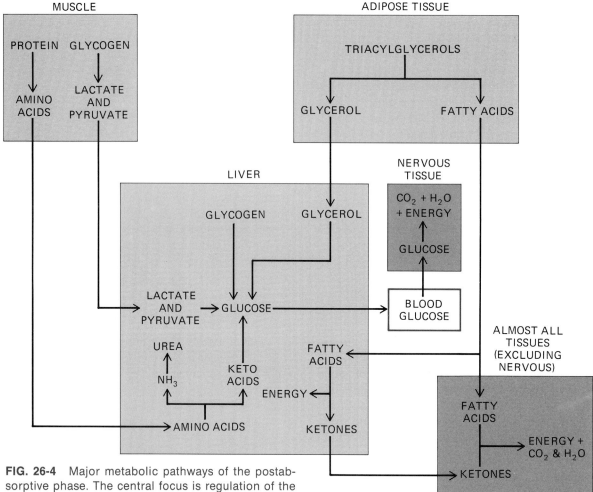

FIG. 26-4 Major metabolic pathways of the postabsorptive phase. The central focus is regulation of the blood glucose concentration.

ther via the Krebs cycle. (One of these substances is acetone, some of which is exhaled and accounts for the distinctive breath odor of persons who are undergoing prolonged fasting or suffering from severe untreated diabetes mellitus.) These ketone bodies are released into the blood and provide an important energy source for the many tissues capable of oxidizing them via the Krebs cycle.

The net result of fatty acid utilization during fasting (as much as 160 g/day) is provision of energy for the body and sparing of glucose for the brain. The combined effects of gluconeogenesis and the switch over to fat utilization are so efficient that, after several days of complete fasting, the plasma glucose concentration is reduced only by a few percent. After one month, it is decreased only 25 percent.

There also occurs an important change in brain metabolism with prolonged starvation. Apparently, after 4 to 5 days of fasting, the generalization that the brain is an obligatory glucose utilizer is no longer valid, for the brain begins to utilize large quantities of ketone bodies, as well as glucose, for its energy source. The survival value of this phenomenon is very great; if the brain significantly reduces its glucose requirement (by utilizing ketones instead of glucose), much less protein need be broken down to supply the amino acids for gluconeogenesis. Accordingly, the protein stores will last longer, and the ability to withstand a long fast without serious tissue disruption is enhanced.

Thus far, our discussion has been purely descriptive; we now turn to the endocrine and neural factors that so precisely control and integrate these

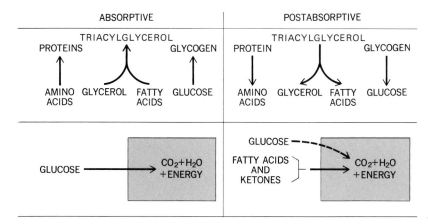

FIG. 26-5 Summary of critical shifts in transition from absorptive to postabsorptive states.

metabolic pathways and transformations. Without question, the most important single factor is insulin. As before, the reader should constantly refer to Figs. 26-3 and 26-4. We shall focus primarily on the following questions (Fig. 26-5) raised by the previous discussion: (1) What controls the shift from the net anabolism of protein, glycogen, and triacylglycerol to net catabolism? (2) What induces primarily glucose utilization during absorption and fat utilization during postabsorption; i.e., how do cells "know" they should start oxidizing fatty acids and ketones instead of glucose? (3) What drives net hepatic glucose synthesis (gluconeogenesis) and release during postabsorption?

ENDOCRINE AND NEURAL CONTROL OF THE ABSORPTIVE AND POSTABSORPTIVE STATES

Insulin

Insulin is a protein hormone secreted by the islets of Langerhans, clusters of endocrine cells in the pancreas (Fig. 26-6). Each of the 1 million or so islets in the human pancreas consists of polyhedral cells clustered together around a rich capillary network. Appropriate staining techniques reveal several types of islet cells, the most prominent being the β, or *beta,* and α, or *alpha,* cells. The beta cells, which account for 75 percent of the islet cells, are the source of insulin, whereas the alpha cells secrete glucagon. A third cell type, the *D,* or *delta,* cells contain somatostatin, a substance found also in the hypothalamus where it functions as growth hormone inhibitory hormone; the role of pancreatic somatostatin in normal physiology remains unclear.

Insulin acts directly or indirectly on most tissues of the body, with the notable exception of most brain areas. The effects of insulin are so important and widespread that an injection of this hormone into a fasting person duplicates the absorptive state pattern of Fig. 26-3 (except, of course, for the absence of gastrointestinal absorption). Conversely, persons who suffer from insulin deficiency (diabetes mellitus) manifest the postabsorptive pattern of Fig. 26-4. From these statements alone, it might appear (correctly) that secretion of insulin is stimulated by eating and inhibited by fasting.

Insulin induces in its target cells a large number of changes (Fig. 26-7) that fall into two general categories: alteration of either (1) membrane transport or (2) enzyme function.

Effects on Membrane Transport Glucose enters most cells by the carrier-mediated mechanism described as facilitated diffusion in Chap. 4. Insulin stimulates this facilitated diffusion of glucose into cells, particularly muscle and adipose tissue; greater glucose entry into cells increases the availability of glucose for all the reactions in which glucose participates. It is important to note that insulin does not alter glucose uptake by the brain, nor does it influence the active transport of glucose across the renal tubule and gastrointestinal epithelium. Insulin also stimulates the active transport of amino acids into most cells and thereby makes more amino acids available for protein synthesis.

Effects on Enzymes The enhanced membrane transport of glucose and amino acids induced by insulin favors, by mass action, glucose oxidation

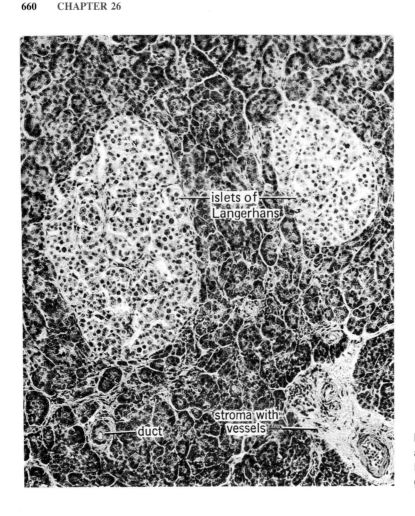

FIG. 26-6 Section of the pancreas of a rhesus monkey, showing two islets of Langerhans and surrounding exocrine glandular tissue.

FIG. 26-7 Major effects of insulin upon organic metabolism. Each solid arrow ($\rightarrow$) represents a process enhanced by insulin, whereas ($-\times\rightarrow$) denotes a reaction inhibited by insulin. Conversely, when insulin concentration is low, the activity of the dashed pathways is enhanced and that of the solid pathways is decreased. The nontransport effects are mediated by insulin-dependent enzymes. All these effects are not necessarily exerted upon all cells.

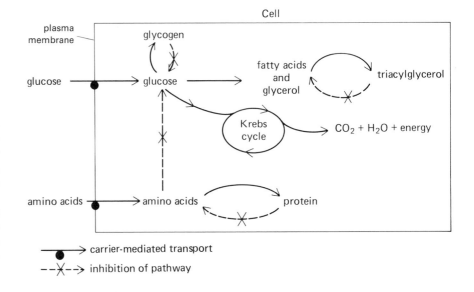

and net synthesis of glycogen, triacylglycerol, and protein. However, in addition, insulin alters the activities or concentrations of many of the intracellular enzymes involved in the anabolic and catabolic pathways of these substances so as to achieve the same result. Glycogen synthesis provides an excellent example: Increased glucose uptake per se stimulates glycogen synthesis but, in addition, insulin increases the activity of the rate-limiting step in glycogen synthesis and inhibits the enzyme that catalyzes glycogen catabolism; thus, insulin favors glucose transformation into glycogen by a triple-barrelled effect!

Insulin also inhibits almost all the critical liver enzymes that catalyze gluconeogenesis; the net result is that insulin abolishes glucose release by the liver (and causes net uptake instead).

Effects of Decreased Insulin   We have thus far dealt with the positive effects of insulin. Clearly insulin deficit will have just the opposite results. When the blood concentration of insulin decreases, the metabolic pattern is shifted toward decreased glucose entry and oxidation and a net catabolism of glycogen, protein, and triacylglycerol. In other words, these metabolic pathways are in a dynamic state, capable of proceeding, in terms of net effect, in either direction. For this reason, energy metabolism can be shifted from the absorptive to the postabsorptive pattern merely by lowering the rate of insulin secretion: (1) Glucose entry and oxidation decrease; (2) glycogen breakdown increases; (3) net protein catabolism liberates amino acids into the blood; (4) net triacylglycerol catabolism liberates glycerol and fatty acids into the blood, and the resulting higher fatty acid concentration in blood facilitates cellular uptake of fatty acids, which, in turn, stimulates fatty acid oxidation; and (5) gluconeogenesis is stimulated not only by the increased availability of precursors (amino acids and glycerol) but by enzyme changes in the liver itself. The glucose released by the liver can be utilized by the brain since its glucose uptake is not insulin dependent.

Control of Insulin Secretion   Insulin secretion is directly controlled by the glucose concentration of the blood that flows through the pancreas, a simple system that requires no participation of nerves or other hormones. An increase in blood glucose con-

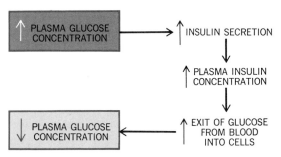

FIG. 26-8   Negative-feedback nature of plasma glucose control over insulin secretion.

centration stimulates insulin secretion; conversely, a reduction inhibits secretion. The feedback nature of this system is shown in Fig. 26-8. A rise in plasma glucose stimulates insulin secretion; insulin induces rapid entry of glucose into cells (as well as cessation of glucose output by the liver); this transfer of glucose out of the blood reduces the blood concentration of glucose and thereby removes the stimulus for insulin secretion, which returns to its previous level.

Figure 26-9 illustrates typical changes in plasma glucose and insulin concentrations following a normal carbohydrate-rich meal. Note the close association between the rising blood glucose (resulting from gastrointestinal absorption) and the plasma insulin increase induced by the glucose rise. The

FIG. 26-9   Blood concentrations of glucose and insulin following ingestion of 100 g of glucose. Study performed on normal human subjects.

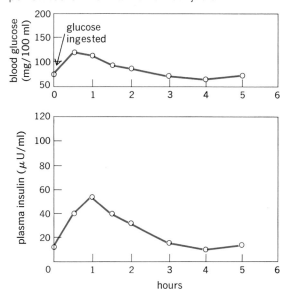

lower postabsorptive values for plasma glucose and insulin concentrations are not the lowest attainable, and prolonged fasting induces even further reductions of both variables until insulin is barely detectable in the blood.

Plasma glucose is not the sole control over insulin secretion, for insulin secretion is sensitive to numerous other types of input. One of the most important is the plasma concentration of certain amino acids, an elevated amino acid concentration causing enhanced insulin secretion. This is easily understandable because amino acid concentrations increase after eating, particularly after a high-protein meal. The increased insulin stimulates cell uptake of these amino acids. There are also important neural and hormonal controls over insulin secretion. For example, one or more of the hormones secreted by the gastrointestinal tract (see Chap. 25) in response to the presence of food stimulates the release of insulin; this provides an "anticipatory" (or "feedforward") component to glucose regulation.

Diabetes Mellitus   The name diabetes, meaning syphon or running through, was used by the Greeks over 2000 years ago to describe the striking urinary volume excreted by certain people. Mellitus, meaning sweet, distinguishes this urine from the large quantities of insipid urine produced by persons suffering from ADH deficiency (Chap. 24). This sweetness of the urine was first recorded in the seventeenth century, but in England the illness had long been called the "pissing evil." Because of the marked weight loss despite huge food intake, the body's substance was believed to be dissolving and pouring out through the urinary tract, a view not far from the truth. In 1889, experimental diabetes was produced in dogs by surgical removal of the pancreas, and 32 years layer, in 1921, Banting and Best discovered insulin.

A tendency toward diabetes can be inherited. We say tendency because diabetes often is not an all-or-none disease but may develop slowly, and overt signs may all but disappear with appropriate measures, e.g., weight reduction. The cause of diabetes is relative insulin deficiency. We have described how a lowered insulin concentration induces virtually all the metabolic changes characteristic of the postabsorptive state (Fig. 26-4). The picture presented by an untreated diabetic is a gross caricature

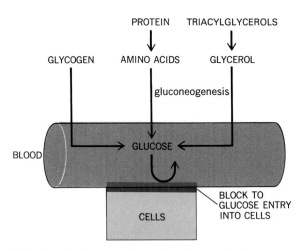

FIG. 26-10   Factors that elevate blood glucose concentration in insulin deficiency.

of this state (Fig. 26-10). The catabolism of triacylglycerol with resultant elevation of plasma fatty acids and ketones is an appropriate response because these substances must provide energy for the body's cells, which are prevented from taking up adequate glucose by the insulin deficiency. In contrast, glycogen and protein catabolism and the marked gluconeogenesis so important to maintain plasma glucose during fasting are completely inappropriate in the diabetic since plasma glucose is already high because it cannot enter into cells. These reactions serve only to raise the plasma glucose still higher, with disastrous consequences. Only the brain is spared glucose deprivation because its uptake of glucose is not insulin-dependent.

The elevated plasma glucose of diabetes induces changes of serious consequence in renal function. In Chap. 24, we pointed out that a normal person does not excrete glucose because all glucose filtered at the glomerulus is reabsorbed by the tubules. However, the elevated plasma glucose of diabetes may so increase the filtered load of glucose that the maximum tubular reabsorptive capacity is exceeded and large amounts of glucose may be excreted. For the same reasons, large amounts of ketones may also appear in the urine. These urinary losses, of course, only aggravate the situation by further depleting the body of nutrients and leading to weight loss. Far worse, however, is the effect of these solutes on sodium and water excretion. In Chap. 24, we saw how tubular water reabsorption is a passive process induced by active solute reab-

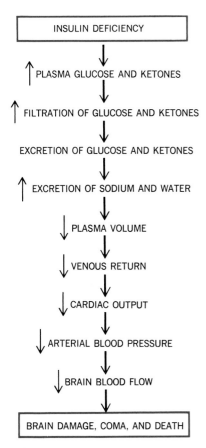

FIG. 26-11   Effects of severe untreated insulin deficiency on renal function.

centration have on respiration? A marked increase in ventilation occurs in response to stimulation of the medullary respiratory centers by hydrogen ion, and the resulting overexcretion of carbon dioxide further helps to keep the hydrogen-ion concentration below lethal limits.

The grim picture painted above is seen only in persons with severe uncontrolled diabetes, those with almost total inability to secrete insulin; the great majority of diabetics never develop this so-called *ketoacidosis,* but they do suffer from a variety of chronic abnormalities that may include atherosclerosis, small vessel and nerve disease, susceptibility to infection, and blindness. The mechanisms by which insulin deficiency contributes to the development of these malfunctions is the subject of much current investigation.

A surprising discovery was that many diabetics actually have increased plasma concentrations of insulin! Their elevated plasma glucose is a result, therefore, not of an absolute deficiency of insulin but of a failure of insulin to influence target cells normally. This insulin insensitivity apparently has multiple causes, including a reduction in the number of insulin receptors.

Glucagon

Insulin unquestionably plays a central role in controlling the metabolic adjustments required for feasting or fasting, but other pathways are also involved. **Glucagon,** a protein hormone produced by the alpha cells of the pancreatic islets,[2] is probably the most important of these. The major effects of glucagon on organic metabolism (mediated by generation of cyclic AMP in its target cells) are all opposed to those of insulin (Fig. 26-12): (1) increased glycogen breakdown in the liver, (2) increased synthesis of glucose (gluconeogenesis) by the liver, and (3) increased breakdown of adipose tissue triacylglycerol (fat mobilization). Thus, the overall results of glucagon's effects are to increase the plasma concentrations of glucose and fatty acids, all important events of the postabsorptive period.

sorption. In diabetes, the osmotic force exerted by unreabsorbed glucose and ketones holds water in the tubule, and reabsorption of the water is thereby prevented. For several reasons (the mechanisms are beyond the scope of this book) sodium reabsorption is also retarded. The net result is marked excretion of sodium and water, which leads, by the sequence of events shown in Fig. 26-11, to hypotension, brain damage, and death.

Another serious abnormality in diabetes is a markedly increased hydrogen-ion concentration, a result primarily of the accumulation of ketone bodies, which, as moderately strong acids, generate large amounts of hydrogen ion by dissociation. The kidneys respond to this increase by excreting more hydrogen ion and are generally able to maintain balance fairly well, at least until the volume depletion described above interferes with renal function. What effect does the increased hydrogen-ion con-

---

[2] Glucagon and glucagonlike substances are also secreted by cells in the lining of the gastrointestinal tract morphologically identical to the alpha cells; the significance of this nonpancreatic glucagon is unclear.

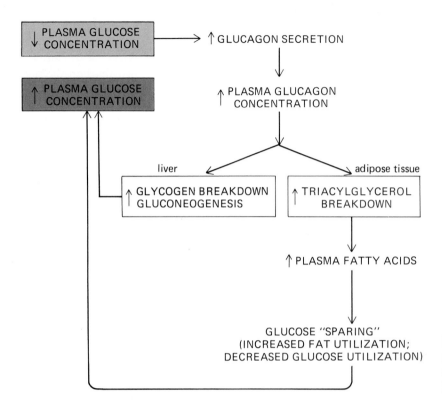

FIG. 26-12 Negative-feedback nature of plasma glucose control over glucagon secretion (compare to Fig. 26-8).

From a knowledge of these effects, one would logically suppose that the secretion of glucagon should be increased during the postabsorptive period and prolonged fasting, and such is the case. The stimulus is a decreased or decreasing plasma glucose concentration. The adaptive value of such a reflex is obvious; a decreasing plasma glucose induces increased release of glucagon, which, by its effects on metabolism, serves to restore normal blood glucose levels and at the same time supply fatty acids for cell utilization (Fig. 26-12). Conversely, an increased or increasing plasma glucose inhibits glucagon's secretion, which thereby helps to return plasma glucose toward normal. The pathway for this reflex is quite simple: The alpha cells in the pancreas respond directly to changes in the glucose concentration of the blood that perfuses the pancreas.

Thus far, the story is quite uncomplicated; the beta and alpha cells of the pancreatic islets constitute a push-pull system for regulating plasma glucose by producing two hormones (insulin and glucagon, respectively) whose actions and controlling inputs are just the opposite of each other. However, we must now point out a complicating feature: A

second major control of glucagon secretion is the plasma amino acid concentration (acting directly on the alpha cells), and in this regard the effect is identical rather than opposite to that for insulin; glucagon secretion, like that of insulin, is strongly stimulated by a rise in plasma amino acid concentration such as occurs following a protein-rich meal. Thus, during absorption of a carbohydrate-rich meal that contains little protein, there occurs an increase in insulin secretion alone, caused by the rise in plasma glucose, but during absorption of a low-carbohydrate–high-protein meal, glucagon also increases, under the influence of the increased plasma amino acid concentration. The usual meal is somewhere between these extremes and is accompanied by a rise in insulin and relatively little change in glucagon since the simultaneous increases in blood glucose and amino acids counteract each other so far as glucagon secretion is concerned. Of course, regardless of the type of meal ingested, the postabsorptive period is always accompanied by a rise in glucagon secretion (and a decrease in insulin secretion).

What is the adaptive value of the amino acid–glucagon relationship? Imagine what might occur were glucagon not part of the response to a high-protein meal: Insulin secretion would be increased by the amino acids but, since little carbohydrate was ingested and therefore available for absorption, the increase in plasma insulin could cause a marked and sudden drop in plasma glucose. In reality, the rise in glucagon secretion caused by the amino acids permits the hyperglycemic effects of this hormone to counteract the hypoglycemic actions of insulin, and the net result is a stable plasma glucose. Thus, a high-protein meal virtually free of carbohydrate can be absorbed with little change in plasma glucose despite a marked increase in insulin secretion.

Because glucagon induces an elevation of plasma glucose, an important question is whether certain forms of diabetes might involve not only deficiency of insulin (either absolute or relative) but also an excess of glucagon. One approach to finding an answer to this question has been to measure the concentration of glucagon in the plasma of persons with diabetes. Recall that, since glucagon secretion is inhibited by an elevated plasma glucose, one would expect to find a low plasma glucagon in diabetic persons (because they have increased plasma glucose) if glucagon had nothing to do with their disease. However, the actual finding is that most diabetic persons have plasma glucagon concentrations that are increased (either absolutely or relative to their elevated plasma glucose). The failure of glucagon to decrease seems to result from a partial loss (for unknown reasons) of glucose-sensing ability by the alpha cells. How much does this absolute or relative glucagon excess contribute to the carbohydrate and fat abnormalities typical of diabetes? This question is being intensively investigated via the use of agents capable of inhibiting glucagon secretion, but no clear answer is yet available. The question is of profound clinical importance because if glucagon were an important contributory factor, the therapy of diabetes might focus on inhibition of glucagon secretion rather than (or, much more likely, in addition to) insulin administration.

Sympathetic Nervous System

Insulin and glucagon are the major hormones involved in the minute-to-minute control over the metabolic adjustments associated with feasting and fasting, but there is a third efferent pathway that may play an important role, particularly when the transition to fasting is associated with a rapidly falling plasma glucose concentration. This is the sympathetic nervous system, and the specific components of this system involved are: (1) the adrenal medulla, the endocrine gland that is an integral part of the sympathetic nervous system and that secretes mainly epinephrine into the blood; and (2) the sympathetic neurons to adipose tissue cells.

The major actions of epinephrine are to stimulate glycogenolysis (by both liver and skeletal muscle) and to stimulate the breakdown of adipose tissue triacylglycerol (fat mobilization); this latter effect is also strongly stimulated by the sympathetic neurons to adipose tissue. Thus, the overall effects on organic metabolism of enhanced activity of the sympathetic nervous system are the opposite of those of insulin and similar to those of glucagon: increased plasma concentrations of glucose, glycerol, and fatty acids (Fig. 26-13).

As might be predicted from these effects, an important stimulus that leads to increased secretion of epinephrine and activity of the neurons to adipose tissue is a decreased or decreasing plasma glucose. This is the same stimulus that, as described above, leads to increased secretion of glucagon (although the receptors and pathways are totally different), and the adaptive value of the response is also the same: Blood glucose is restored toward normal and fatty acids are supplied for cell utilization. The pathway for this reflex is illustrated in Fig. 26-13. As described in Chap. 14, epinephrine release is controlled entirely by the preganglionic sympathetic fibers to the adrenal medulla; the receptors, glucose receptors in the brain (probably in the hypothalamus), initiate the reflexes that lead ultimately to increased activity in these neurons and in the sympathetic pathways to adipose tissue. This reflex is triggered when the plasma glucose concentration is rapidly decreasing, and many of the symptoms associated with such a state are accounted for by the actions of the large quantity of epinephrine, particularly as it affects the cardiovascular and nervous systems—palpitations, tremor, headache, nervousness, cold sweating, dizziness, etc.

It is not yet known whether the sympathetic nervous system plays any role in the metabolic adjustments to prolonged fasting (as contrasted with acute reductions in plasma glucose described above). Recent evidence suggests that the overall activity of

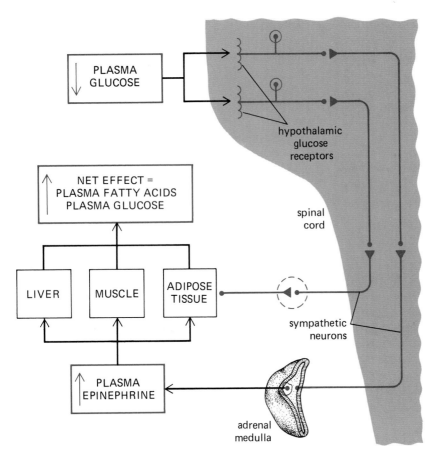

FIG. 26-13 Control of epinephrine secretion and sympathetic nerves to adipose tissue by plasma glucose concentration. A decrease in plasma glucose stimulates the hypothalamic glucose receptors and, via the reflex chain shown, restores plasma glucose to normal while at the same time increasing plasma fatty acids.

the sympathetic nervous system is, contrary to expectations, actually decreased under such conditions; the possible adaptive significance of this change is discussed later in this chapter.

Conclusion

To a great extent insulin may be viewed as the "hormone of plenty"; it is increased during the absorptive period and decreased during postabsorption. The effects on organic metabolism of glucagon and the sympathetic nervous system are, in various ways, opposed to those of insulin and these systems are activated (the latter at least acutely) during the postabsorptive period. The influence of plasma glucose concentration is paramount in this regard; it produces opposite effects on insulin secretion and on the secretion of glucagon and epinephrine and the activity of the sympathetic nerves to adipose tissue.

It should also be noted that the secretion of glucagon and the activity of the sympathetic nervous

system are stimulated not only when plasma glucose is reduced but during exercise and in a variety of nonspecific "stresses," both physical and emotional. This constitutes a mechanism for the mobilization of energy stores for coping with a fight-or-flight situation (see Chap. 14).

Finally, in addition to the three hormones described in this section, there are others—growth hormone, cortisol, thyroxine, and the sex steroids—that have important effects (described elsewhere) on organic metabolism and the flow of nutrients. However, the secretion of these hormones is not usually keyed to the absorptive-postabsorptive phases but is controlled by other factors. Growth hormone is somewhat of an exception in that its rate of secretion does respond to prolonged fasting or rapidly occurring decreases in plasma glucose concentration; however, changes in response to the three-meals-a-day pattern are minimal, and its short-term effects on carbohydrate and fat metabolism are relatively minor compared to its long range effects on protein metabolism and growth.

REGULATION OF PLASMA CHOLESTEROL

In the previous sections we described the flow of lipids to and from adipose tissue in the form of fatty acids and triacylglycerols. These lipids do not circulate as free molecules in the plasma, for their insolubility in water makes that impossible. Rather, they are bound to particular plasma proteins, and the complexes are known as **lipoproteins.** The various lipoproteins differ considerably in the relative amounts of fat and protein that they contain. Those that have little protein and much fat are categorized as **low-density lipoproteins** (since fat is less dense than protein), whereas those with the opposite distribution are **high-density lipoproteins.**

Another very important lipid, **cholesterol,** was not mentioned in earlier sections because it, unlike the fatty acids and triacylglycerols, does not serve as a metabolic fuel but rather as a precursor for bile salts, steroid hormones, plasma membranes, and other specialized molecules. It, too, circulates in combination with lipoproteins, particularly the low-density fraction. A huge number of studies have been devoted to the factors that govern its plasma concentration because it is generally accepted that high plasma cholesterol predisposes to the development of atherosclerosis, the arterial wall thickening that leads to strokes, heart attacks, and other forms of damage (see Chap. 21 for a discussion of the etiology of this thickening). (Interestingly, a high concentration of high-density lipoproteins is associated with a decreased incidence of atherosclerosis.)

Figure 26-14 summarizes the major pathways involved in the metabolism of plasma cholesterol. There are two major sources of cholesterol gain, diet and synthesis in the body; there are two major routes of loss, excretion in the feces and catabolism to bile acids. There are multiple homeostatic control points in this system. Most important is that the synthesis of cholesterol by the liver is controlled in a negative-feedback manner by dietary cholesterol; i.e., ingestion of cholesterol inhibits hepatic synthesis. Conversely, when dietary cholesterol is reduced, hepatic synthesis is increased. For this reason, reduction of dietary cholesterol alone produces only very small reductions in plasma cholesterol. Thus, the explanation for the elevated plasma cholesterol exhibited by persons who ingest large amounts of animal fat is not simply an increase in cholesterol ingestion. However, another component of such a diet does seem to be of great importance, namely, the amount of **saturated fatty acids** ingested. These substances, found mainly in animal sources, somehow stimulate the synthesis of cholesterol while at the same time reducing its excretion. Accordingly, reducing the saturated fat content of the diet may lower plasma cholesterol 15 to 20 percent. In contrast, **unsaturated fatty acids** (found mainly in vegetable oils and other plant products) enhance cholesterol excretion and catabolism so that an increase in their dietary consumption helps to reduce plasma cholesterol. Just how the various fatty acids exert these effects on cholesterol metabolism is not completely known.

Factors other than diet also influence cholesterol

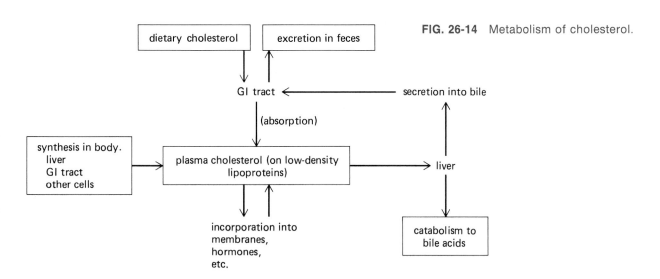

FIG. 26-14 Metabolism of cholesterol.

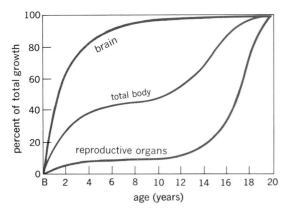

FIG. 26-15   Rate of growth.

metabolism. For example, stress and cigarette smoke have been implicated as factors that elevate plasma cholesterol, but the quantitative contributions of such factors have not been fully documented.

CONTROL OF GROWTH

A simple gain in body weight does not necessarily mean true growth since it may represent retention of either excess water or fat. In contrast, true growth usually involves lengthening of the long bones and increased cell division and enlargement, but the real criterion is increased accumulation of protein. Human beings manifest two periods of rapid growth (Fig. 26-15), one during the first 2 years of life, which is actually a continuation of rapid fetal growth, and the second during adolescence.

Note that total body growth may be a poor indicator of the rate of growth of specific organs (Fig. 26-15). Another important implication of differential growth rates is that the so-called critical periods of development vary from organ to organ. Thus, a period of severe malnutrition during infancy, when the brain is growing extremely rapidly, may produce stunting of brain development, which is irreversible, whereas reproductive organs would be little affected.

External Factors That Influence Growth

An individual's growth capacity is genetically determined, but there is no guarantee that the maximum capacity will be attained. Adequacy of food supply and freedom from disease are the primary external factors that influence growth. Lack of sufficient amounts of any of the essential amino acids, essential fatty acids, vitamins, or minerals interferes with growth, and total protein and total calories must be adequate. No matter how much protein is ingested, growth cannot be normal if caloric intake is too low, since the protein is simply oxidized for energy.

On the other hand, it is important to realize that one cannot stimulate growth beyond the genetically determined maximum by eating more than adequate vitamins, protein, or total calories; this produces obesity, not growth.

Sickness can stunt growth, but if the illness is temporary, upon recovery a child manifests a remarkable growth spurt that rapidly brings him or her up to a normal growth curve. The mechanisms that control this important phenomenon are unknown, but the process illustrates the strength and precision of a genetically determined sequence.

Hormonal Influences on Growth

The hormones most important to human growth are growth hormone, thyroid hormone, insulin, androgens, and estrogen, which exert widespread effects. In addition, ACTH, TSH, prolactin, and FSH and LH selectively influence the growth and development of their target organs, the adrenal cortex, thyroid gland, breasts, and gonads, respectively.

Growth Hormone   Removal of the pituitary in young animals arrests growth. Conversely, administration of large quantities of growth hormone to young animals causes excessive growth. When excess growth hormone is given to adult animals after the actively growing cartilaginous areas of the long bones have disappeared, it cannot lengthen the bones further, but it does produce the disfiguring bone thickening and overgrowth of other organs known as *acromegaly*. These experiments have spontaneously occurring human counterparts, as shown in Fig. 26-16. Thus growth hormone is essential for normal growth and in abnormally large amounts can cause excessive growth.

Growth hormone has a variety of effects on organic metabolism, but its growth-promoting effects are due primarily to its ability to stimulate protein synthesis in most tissues and organs. This it does

FIG. 26-16 Progression of acromegaly. (Top left) Normal, age 9 years; (top right) age 16 years, with possible early coarsening of features; (bottom left) age 33 years, well-established acromegaly; (bottom right) age 52 years, end stage, acromegaly with gross disfigurement.

by increasing membrane transport of amino acids into cells and by increasing the activity of ribosomes, events essential for protein synthesis. Growth hormone also causes large increases of mitotic activity and cell division, the other major components of growth.

The effects on cartilage and bone are dramatic. Growth hormone promotes bone lengthening by stimulating protein synthesis in both the cartilaginous center and bony edge of the epiphyseal plates as well as by increasing the rate of osteoblast mitosis (Chap. 8). It is presently uncertain as to whether growth hormone itself exerts these growth-promoting effects on cartilage, bone, and other tissues, or whether growth hormone causes the release from the liver of so-called **somatomedins** that then are carried by the blood to the target tissues and induce the effects.

Thyroid hormone   Infants and children with deficient thyroid function have retarded growth, which can be restored to normal by administration of physiologic quantities of **thyroid hormone (TH)** (this term refers collectively to two hormones, secreted by the thyroid gland, thyroxine and triiodothyronine, which have very similar effects). Administration of excess TH, however, does not cause excessive growth (as was true of growth hormone) but rather marked catabolism of protein and other nutrients, as will be explained in the section on energy balance. The essential point is that normal amounts of TH are necessary for normal growth, the most likely explanation apparently being that TH promotes the effects of growth hormone on protein synthesis; certainly the absence of TH significantly reduces the ability of growth hormone to stimulate amino acid uptake, ribosomal activation, and RNA synthesis. TH also plays a crucial role in the closely related area of organ development, particularly that of the central nervous system. Hypothyroid infants (*cretins*) are mentally retarded, a defect that can be completely repaired by adequate treatment with TH; if the infant is untreated for long, however, the developmental failure is largely irreversible. The defect is probably a result of failure of nerve myelination that occurs as a result of thyroid deficiency.

Insulin   It should not be surprising that adequate amounts of insulin are necessary for normal growth

since insulin is, in all respects, an anabolic hormone. Its stimulatory effects on amino acid uptake and protein are particularly important in favoring growth.

Androgens and Estrogen   In Chap. 27 we shall describe in detail the various functions of the sex hormones in directing the growth and development of the sex organs and the obvious physical characteristics that distinguish male from female. Here we are concerned only with the effects of these hormones on general body growth.

Sex hormone secretion begins in earnest at about the age of 8 to 10 years old and progressively increases to reach a plateau within 5 to 10 years. The testicular hormone, testosterone, is the major male sex hormone, but other androgens similar to it are also secreted in significant amounts by the adrenal cortex of both sexes. Females manifest a sizable increase of adrenal androgen secretion during adolescence. However, the adrenal androgens are not nearly so potent as testosterone. During adolescence the large increases in secretion of estrogen, the dominant female sex hormone, are virtually limited to the female. Thus the relative quantities of androgen and estrogen are very different between the sexes.

Androgens strongly stimulate protein synthesis in many organs of the body, not just the reproductive organs, and the adolescent growth spurt in both sexes is due, at least in part, to these anabolic effects. Similarly, the increased muscle mass of men compared with that of women may reflect their greater amount of more potent androgen. Androgens stimulate bone growth but also ultimately stop bone growth by inducing complete conversion of the epiphyseal plates to bone. This accounts for the pattern seen in adolescence, i.e., rapid lengthening of the bones culminating in complete cessation of growth for life, and explains several clinical situations: (1) Unusually small children treated before puberty with large amounts of androgens may grow several inches very rapidly but then stop completely; (2) eunuchs may be very tall because bone growth, although slower, continues much longer because of the persistence of the epiphyseal plates.

Estrogen profoundly stimulates growth of the female sex organs and sexual characteristics during adolescence (Chap. 27) and is also required for normal growth of bone and tissues.

SECTION B: REGULATION OF TOTAL BODY ENERGY BALANCE

BASIC CONCEPTS OF ENERGY EXPENDITURE AND CALORIC BALANCE

The breakdown of organic molecules liberates the energy locked in their intramolecular bonds (Chap. 3). This is the source of energy utilized by cells in their preformance of the various forms of biological work (muscle contraction, active transport, synthesis of molecules, etc.). The first law of thermodynamics states that energy can be neither created nor destroyed but can be converted from one form to another. Thus, internal energy liberated (ΔE) during breakdown of an organic molecule can either appear as heat (H) or be used for performing work (W).

$$\Delta E = H + W$$

In all animal cells, most of the energy appears immediately as heat, and the rest is used for work. (As described in Chap. 3, the energy used for work must first be incorporated into molecules of ATP, the subsequent breakdown of which serves as the immediate energy source for the work.) It is essential to realize that the body is not a heat engine since it is totally incapable of converting heat into work. The heat is, of course, valuable for maintaining body temperature.

It is customary to divide biological work into two general categories: (1) **external work,** i.e., movement of external objects by contracting skeletal muscles, and (2) **internal work,** which constitutes all other forms of biological work, including skeletal muscle activity not moving external objects. As we have seen, much of the energy liberated from the catabolism of nutrients appears immediately as heat; only a small fraction is used for performance of external or internal work. What may not be obvious is that all internal work is ultimately transformed into heat except during periods of growth (Fig. 26-17). Several examples will illustrate this essential point.

1   Internal work is performed during cardiac contraction, but this energy appears ultimately as heat generated by the resistance (friction) to flow offered by the blood vessels.
2   Internal work is performed during secretion of

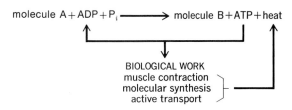

FIG. 26-17   General pattern of energy liberation in a biological system. Most of the energy released when nutrients, such as glucose, are broken down appears immediately as heat. A smaller fraction goes to form ATP, which can be subsequently broken down and the released energy coupled to biological work. Ultimately, the energy that performs this work is also completely converted into heat.

HCl by the stomach and $NaHCO_3$ by the pancreas, but this work appears as heat when the H^+ and HCO_3^- react in the small intestine.
3   The internal work performed during synthesis of a plasma protein is recovered as heat during the inevitable catabolism of the protein, since, with few exceptions, all bodily constituents are constantly being built up and broken down. However, during periods of net synthesis of protein, fat, etc., energy is stored in the bonds of these molecules and does not appear as heat.

Thus, the total energy liberated when organic nutrients are catabolized by cells may be transformed into body heat, appear as external work, or be stored in the body in the form of organic molecules, the latter occurring only during periods of growth (or net fat deposition in obesity). The total energy expenditure of the body is therefore given by the equation

$$\text{Total energy expenditure} =$$
heat produced + external work + energy shortage

The units for energy are **kilocalories** (Chap. 3), and total energy expenditure per unit time is called the **metabolic rate.** (In the field of nutrition, 1 Calorie implies, by convention, 1 large calorie, which is actually 1 kilocalorie.)

Determinants of Metabolic Rate

Since many factors cause the metabolic rate to vary (Table 26-2), when one wishes to compare metabolic rates of different people, it is essential to control as many of the variables as possible. The test used clinically and experimentally to find the **basal** metabolic rate (BMR) is designed to accomplish this by standardizing conditions: The subject is at mental and physical rest, in a room at comfortable temperatures, and has not eaten for at least 12 hr. These conditions are arbitrarily designated basal, the metabolic rate during sleep being actually less than the BMR. The measured BMR is then compared with previously determined normal values for a person of the same weight, height, age, and sex.

BMR is often appropriately termed the metabolic *cost of living*. Under these conditions, most of the energy is expended, as might be imagined, by the heart, liver, kidneys, and brain. Its magnitude is related not only to physical size but to age and sex as well. A growing child's resting metabolic rate, relative to size, is considerably higher than an adult's because the child expends a great deal of energy in net synthesis of new tissue. On the other end of the age scale, the metabolic cost of living gradually decreases with advancing age, for unknown reasons. The female's resting metabolic rate is generally less than that of the male (even taking into account size differences) but increases markedly, for obvious reasons, during pregnancy and lactation. The greater demands upon the body by infection or other disease generally increase total energy expenditure; moreover, the presence of fever increases metabolic rate.

The ingestion of food also increases the metabolic rate, as shown by measurement of the oxygen consumption or heat production of a resting man before and after eating; the metabolic rate is 10 to 20 percent higher after eating. This effect of food on metabolic rate is known as the **specific dynamic action (SDA).** Protein gives the greatest effect, carbohydrate and fat less. The cause of SDA is not what one might expect, namely, the energy expended in the digestion and absorption of ingested food. These processes account for only a small fraction of the increased metabolic rate (intravenous administration of amino acids produces almost the same SDA effect as oral ingestion of the same material). Most of the increased heat production appears to be secondary to the processing of exogenous nutrients by the liver, since it does not occur in an animal whose liver has been removed. In contrast to eating, prolonged fasting causes a decrease in metabolic rate. This is due, in part, simply to reduction of body mass, but even when expressed on a per weight basis, metabolic rate is reduced. The mechanism is unclear, but the adaptive value of this change is considerable since it decreases the amount of nutrient stores that must be catabolized each day. One possible mechanism may be a reduction in sympathetic nervous system activity.

All these influences on metabolic rate are small compared with the effects of muscular activity (Table 26-3). Even minimal increases in muscle tone significantly increase metabolic rate, and severe exercise may raise heat production more than fifteenfold. Changes in muscle activity also explain part of the effects on metabolic rate of sleep (decreased

TABLE 26-2   FACTORS THAT AFFECT THE METABOLIC RATE
Age
Sex
Height, weight, and surface area
Growth
Pregnancy, menstruation, lactation
Infection or other disease
Body temperature
Recent ingestion of food (SDA)
Prolonged fasting
Muscular activity
Emotional state
Sleep
Environmental temperature
Circulating levels of various hormones, especially epinephrine and thyroid hormone

TABLE 26-3 ENERGY EXPENDITURE DURING DIFFERENT TYPES OF ACTIVITY FOR A 70 kg MAN	
Form of Activity	kcal/hr
Awake, lying still	77
Sitting at rest	100
Typewriting rapidly	140
Dressing or undressing	150
Walking level at 2.6 mi/hr	200
Sexual intercourse	280
Bicycling on level, 5.5 mi/hr	304
Walking 3 percent grade at 2.6 mi/hr	357
Sawing wood or shoveling snow	480
Jogging (5.3 mi/hr)	570
Rowing 20 strokes/min	828
Maximal activity (untrained)	1440

muscle tone), reduced environmental temperature (increased muscle tone and shivering), and emotional state (unconscious changes in muscle tone).

Metabolic rate is strongly influenced by the hormones epinephrine and thyroid hormone. The intravenous injection of epinephrine may promptly increase heat production by more than 30 percent. As we have seen, epinephrine has powerful effects on organic metabolism, and its calorigenic, i.e., heat-producing, effect may be related to its stimulation of glycogen and triacylglycerol catabolism since ATP splitting and energy liberation occur in both the breakdown and the subsequent resynthesis of these molecules. Regardless of the mechanism, whenever epinephrine secretion is stimulated, the metabolic rate rises. This probably accounts for part of the greater heat production associated with emotional stress, although increased muscle tone is also contributory.

Thyroid hormone (TH) also increases the oxygen consumption and heat production of most body tissues, a notable exception being the brain. The mechanism of this calorigenic effect remains controversial. Long-term excessive TH, as in persons with hyperthyroidism, induces a host of effects secondary to the hypermetabolism that well illustrate the interdependence of bodily functions: The increased metabolic demands markedly increase hunger and food intake; the greater intake frequently remains inadequate to meet the metabolic needs, and net catabolism of endogenous protein and fat stores leads to loss of body weight; excessive loss of skeletal muscle protein results in muscle weakness; catabolism of bone protein weakens the bones and liberates large quantities of calcium into the extracellular fluid, which results in increased plasma and urinary calcium; the hypermetabolism increases the requirement for vitamins, and vitamin deficiency diseases may occur; respiration is increased to supply the required additional oxygen; cardiac output is also increased, and, if prolonged, the enhanced cardiac demands may cause heart failure; the greater heat production activates heat-dissipating mechanisms, and the person suffers from marked intolerance to warm environments. These are only a few of the many results induced by the calorigenic effect of TH. The important effects of TH that relate to growth and development, described earlier, appear to be quite distinct from the calorigenic effect.

Determinants of Total Body Energy Balance

Using the basic concepts of energy expenditure and metabolic rate as a foundation, we can consider total body energy balance in much the same way as any other balance, i.e., in terms of input and output. The laws of thermodynamics dictate that, in the steady state, the total energy expenditure of the body equals total body energy intake. We have already identified the ultimate forms of energy expenditure: internal heat production, external work, and net molecular synthesis (energy storage). The source of input, of course, is the energy contained in ingested food. Therefore the energy balance equation is

$$
\begin{matrix}
\text{Energy} \\
\text{(from} \\
\text{food} \\
\text{intake)}
\end{matrix}
=
\begin{matrix}
\text{internal} \\
\text{heat} \\
\text{produced}
\end{matrix}
+
\begin{matrix}
\text{external} \\
\text{work}
\end{matrix}
+
\begin{matrix}
\text{energy} \\
\text{storage}
\end{matrix}
$$

Our equation includes no term for loss of fuel from the body via urinary excretion of nutrients. In a normal person, almost all the carbohydrate and amino acids filtered at the glomerulus are reabsorbed by the tubules, so that the kidneys play no significant role in the regulation of energy balance. In certain diseases, however, the most important being diabetes, urinary losses of organic molecules may be quite large and would have to be included in the equation. In all normal persons very small losses occur via the urine, feces, and as sloughed hair and skin, but we can ignore them as being negligible.

As predicted by this energy balance equation, three states are possible:

Energy intake = internal heat production
                        + external work
                  (body weight constant)
Energy intake > internal heat production
                        + external work
                  (body weight increases)
Energy intake < internal heat production
                        + external work
                  (body weight decreases)

In most adults, body weight remains remarkably constant over long periods of time, implying that precise physiologic regulatory mechanisms operate to control (1) food intake or (2) internal heat production plus external work or (3) both. Actually, all these variables are subject to control in human beings, but the amount of food intake is the dominant factor. Control mechanisms for heat production are aimed primarily at regulating body temperature, rather than total energy balance. For example, when someone is cold, her body produces additional heat by shivering even if she is starving; conversely, a fat person is not automatically impelled by his hypothalamus to run around the block—quite the reverse in most cases.

It is essential to understand that as shown by the energy balance equation, an individual's degree of activity, i.e., heat production plus external work, is one of the essential determinants of total body energy balance, but its automatic physiologic control is not aimed at achieving such a balance. Moreover, a person's total activity generally reflects the kind of work that person does, his or her inclination toward sports, etc. The important generalization is that caloric intake is the major factor that is automatically controlled to maintain energy balance and constant body weight. To alter the two examples cited above: When exposure to cold or running around the block causes increased energy expenditure, the individually automatically increases food intake by an amount sufficient to match the additional energy expended (this example, however, will be qualified later in the section on obesity).

CONTROL OF FOOD INTAKE

The control of food intake can be analyzed in the same way as any other biological control system, i.e., in terms of its various components (regulated variable, receptor sensitive to this variable, afferent pathways, etc.). As our previous description emphasized, the variable that is maintained relatively constant in this system is total body energy content. Here we have a problem similar to that described in Chap. 24 for total body sodium balance: What kind of receptors could possibly detect the total body content of a particular variable, in this case calories? It is very unlikely that any such receptors exist, and so the system must instead depend on signals that are intimately related to total energy storage—the plasma concentrations of glucose, fatty acids, glycerol, and amino acids.

These signals provide both short-term and long-term information concerning the body's energy stores. For example (Fig. 26-18), plasma glucose

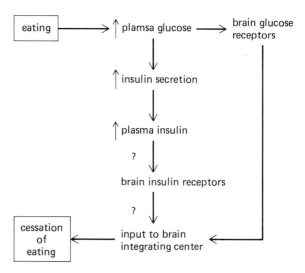

FIG. 26-18 Hypothesized negative-feedback inhibition of eating by glucose and insulin.

concentration (and the rate of cellular glucose utilization) rises during eating; detection of this increase by glucose receptors (in the brain) constitutes a signal to those portions of the brain that control eating behavior and leads to cessation of eating. Conversely, fasting would decrease glucose utilization, remove this signal, and promote eating. Plasma concentrations of fatty acids or glycerol could serve as a long-term indicator of the body's total adipose tissue mass (itself an excellent indicator of total body energy stores) if their basal rates of release from adipose tissue were in direct proportion to the size of the adipose tissue cells; this seems to be the case for glycerol.

In addition to the plasma concentrations of these metabolic fuels, the concentrations of the hormones that regulate them may be important negative-feedback signals to the brain's integrating centers. For example, insulin (which increases during food absorption) has been shown to suppress hunger (Fig. 26-18), whereas glucagon (which increases during fasting) is known to stimulate hunger. Epinephrine and the gastrointestinal hormones released during meal ingestion may also contribute important signals.

Finally, a nonchemical correlate of food ingestion—body temperature—may be yet another signal. The specific dynamic action of food, i.e., the increase in metabolic rate induced by eating, tends to raise body temperature, and this constitutes a signal inhibitory to eating. Such a mechanism would also be consistent with the fact that people eat more in colder climates than in warm ones.

It would certainly be simpler if we could determine the precise contributions of each of these signals to the control of food intake and describe the location and nature of the receptors stimulated by each, but this information is simply not available at present. The confusion does not lessen when we consider the brain areas that integrate all these afferent inputs and cause the individual to either seek out and ingest food or desist from doing so. Areas of the hypothalamus are involved, but there are other brain areas that play important roles as well; the ways in which these various neuronal collections interact remain to be determined.

Thus far we have described the control of food intake in a manner identical to that for any other homeostatic control system, such as those that regulate sodium or potassium. Thus, the automatic "involuntary" nature of the system was emphasized. However, although total energy balance unquestionably reflects in large part the reflex input from some combination of glucose receptors, glycerol receptors, thermostats, etc., it also is strongly influenced by the reinforcement (both positive and negative) of such things as smell, taste, texture, psychologic associations, etc. Thus, the behavioral concepts of reinforcement, drive, and motivation described in Chap. 17 must be incorporated into any comprehensive theory of food-intake control. Obviously, these psychologic factors, which have little to do with energy balance, are of very great importance in obese persons. It should be emphasized, however, that most people whose obesity is ascribed to psychologic factors do not continuously gain weight; their automatic homeostatic control mechanisms are operative but maintain total body energy content at supranormal levels.

Obesity

Obesity has been called the most common disease in the United States. Despite adverse effects on health, the social stigma of being overweight, and a bewildering array of new diets and treatments, the prevalence of obesity in the United States is actually increasing. The term *disease* is perfectly justified because obesity is correlated with illness and premature death from a multitude of causes. The seriousness of being overweight is underlined by statistics that show a mortality rate more than 50 percent greater than normal in overweight persons in the same age groups.

Ultimately, all obesity represents failure of normal food-intake control mechanisms, although what causes this failure (and how best to treat it) is unknown for the great majority of obese persons. It is widely assumed that psychologic factors, habit, and social custom are the predominant causes of obesity, but many investigators believe that this view does not explain many cases of the disease. The relationship between food intake and low levels of physical activity has also received recent attention.

As an exercise in energy balance, let us calculate how rapidly a person can expect to lose weight on a reducing diet. Suppose a woman, whose metabolic rate per 24 hr is 2000 kcal, goes on a 1000 kcal/day diet. How much of her own body fat will be required to supply this additional 1000 kcal/day? Almost all of the organic nutrients lost are fat, and fat contains 9 kcal/g; therefore, 1000 kcal/day $\div$ 9 kcal/g = 111 g/day, or 777 g/week. Approximately another 77 g of water is lost from the adipose tissue along with this fat (adipose tissue is 10 percent water) so that the grand total for 1 week's loss equals 854 g, or 1.8 lb. Thus, during her diet she can reasonably expect to lose approximately this amount of weight each week. Actually, the amount of weight lost during the first week might be considerably greater because, for reasons poorly understood, a large amount of water may be lost early in the diet, particularly when the diet contains little carbohydrate. This early loss, which is really of no value so far as elimination of excess fat is concerned, often underlies the wild claims made for fad diets (indeed, to enhance this effect, drugs that cause the kidneys to excrete even more water are sometimes included in the diet). Clearly, weight loss is a slow process that requires patience and a meaningful reshaping of eating patterns.

REGULATION OF BODY TEMPERATURE

Animals capable of maintaining their body temperatures within very narrow limits are termed **homeothermic.** The adaptive significance of this ability

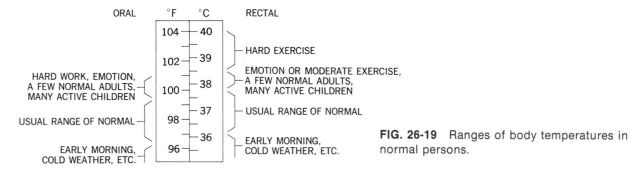

FIG. 26-19 Ranges of body temperatures in normal persons.

stems primarily from the marked effects of temperature upon the rate of chemical reactions in general and enzyme activity in particular. Homeothermic animals are spared the slowdown of all bodily functions that occurs when the body temperature falls. However, the advantages obtained by a relatively high body temperature impose a great need for precise regulatory mechanisms since even moderate elevations of temperature begin to cause nerve malfunction, protein denaturation, and death. Most people suffer convulsions at a body temperature of 41°C (106 to 107°F), and 43°C (110°F) is the absolute limit for life. In contrast, most body tissues can withstand marked cooling (to less than 7°C, or 45°F), which has found an important place in surgery when the heart must be stopped, since the dormant cold tissues require little nourishment.

Figure 26-19 illustrates several important generalizations about normal human body temperature: (1) Oral temperature averages about 0.5°C (1.0°F) less than rectal; thus, all parts of the body do not have the same temperature. (2) Internal temperature is not absolutely constant but varies several degrees in perfectly normal persons in response to activity pattern and external temperature; and, in addition, there is a characteristic diurnal fluctuation, so that temperature is lowest during sleep and slightly higher during the awake state even if the person remains relaxed in bed. An interesting variation in women is a higher temperature during the last half of the menstrual cycle (Chap. 27).

If temperature is viewed as a measure of heat "concentration," temperature regulation can be studied by our usual balance methods. In this case, the total heat content of the body is determined by net difference between heat produced and heat lost from the body. Maintaining a constant body temperature implies that, overall, heat production must equal heat loss. Both these variables are subject to precise physiologic control.

Temperature regulation offers a classic example of a biological control system; its generalized com-

FIG. 26-20 Summary of temperature regulation. Heat loss from the body depends directly upon the external environment and upon changes controlled by temperature-regulating reflexes. In certain environments, heat gain rather than heat loss may actually occur.

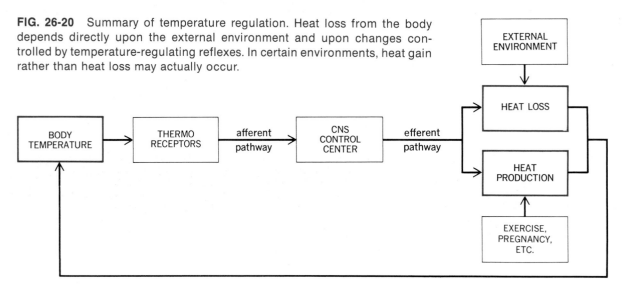

ponents are shown in Fig. 26-20. The balance between heat production and heat loss is continuously being disturbed, either by changes in metabolic rate (exercise being the most powerful influence) or by changes in the external environment that alter heat loss. The resulting small changes in body temperature reflexly alter the output of the effector organs, which drive heat production or heat loss and restore normal body temperature.

Heat Production

The basic concepts of heat production have already been described. Recall that heat is produced by virtually all chemical reactions that occur in the body and that the cost-of-living metabolism by all organs sets the basal level of heat production, which can be increased as a result of skeletal muscular contraction or the action of several hormones.

Changes in Muscle Activity   The first muscle changes in response to cold are a gradual and general increase in skeletal muscle tone. This soon leads to shivering, the characteristic muscle response to cold, which consists of oscillating rhythmic muscle tremors that occur at the rate of about 10 to 20 sec. So effective are these contractions that body heat production may be increased severalfold within seconds to minutes. Because no external work is performed, all the energy liberated by the metabolic machinery appears as internal heat. As always, the contractions are directly controlled by the efferent motor neurons to the muscles. During shivering these nerves are controlled by descending pathways under the primary control of the hypothalamus. It is important to note that this "shivering pathway" can be suppressed, at least in part, by input from the cerebral cortex because a cold man ceases to shiver when he starts to perform voluntary activity. Besides increased muscle tone and shivering, which are completely reflex in nature, human beings also use voluntary heat-production mechanisms such as foot stamping, hand clapping, etc.

Thus far, our discussion has focused primarily on the muscular response to cold; the opposite reactions occur in response to heat. Muscle tone is reflexly decreased and voluntary movement is also diminished ("It's too hot to move."). However, these attempts to reduce heat production are rela-

tively limited in capacity both because muscle tone is already quite low normally and because of the direct effect of a body temperature increase on metabolic rate.

Nonshivering ("Chemical") Thermogenesis   In most experimental animals chronic cold exposure induces an increase in metabolic rate, which is not a result of increased muscle activity. Indeed, as this so-called nonshivering thermogenesis increases over time, it is associated with a decrease in the degree of shivering. The cause of nonshivering thermogenesis has been the subject of considerable controversy; present evidence suggests that it is mainly caused by an increased secretion of epinephrine. (Thyroid hormone may also be involved, but this is much less clear.) Equally controversial is the question of whether nonshivering thermogenesis is a significant phenomenon in human beings. Regardless of the outcomes of these debates, it seems clear that human hormonal changes and nonshivering thermogenesis are of secondary importance and that changes in muscle activity constitute the major control of heat production for temperature regulation, at least in the early response to temperature changes.

Heat-Loss Mechanisms

The surface of the body exchanges heat with the external environment by radiation, conduction and convection (Fig. 26-21), and water evaporation.

Radiation, Conduction, Convection   The surface of the body constantly emits heat, in the form of electromagnetic waves (**radiation**). Simultaneously, all other dense objects are radiating heat. The rate of emissions is determined by the temperature of the radiating surface. Thus, if the surface of the body is warmer than the *average* of the various surfaces in the environment, net heat is lost; the rate is directly dependent upon the temperature difference. The sun, of course, is a powerful radiator, and direct exposure to it greatly decreases heat loss by radiation or may reverse it.

Conduction is the exchange of heat not by radiant energy but simply by transfer of thermal energy from atom to atom or molecule to molecule. Heat, like any other quantity, moves down a concentration gradient, and thus the body surface loses or

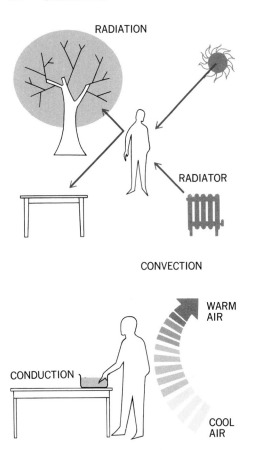

RADIATION

RADIATOR

CONVECTION

WARM
AIR

CONDUCTION

COOL
AIR

FIG. 26-21 Mechanisms of heat transfer. By radiation, heat is transferred by electromagnetic waves (solid arrows); in conduction, heat moves by direct transfer of thermal energy from molecule to molecule (dashed arrow).

gains heat by conduction only through direct contact with cooler or warmer substances, including, of course, the air.

Convection is the process whereby air (or water) next to the body is heated, moves away, and is replaced by cool air (or water), which in turn follows the same pattern. It is always occurring because warm air is less dense and therefore rises, but it can be greatly facilitated by external forces such as wind or fans. Thus, convection aids conductive heat exchange by continuously maintaining a supply of cool air. In the absence of convection, negligible heat would be lost to the air, and conduction would be important only in such unusual circumstances as immersion in cold water. (Because of the great importance of air movement in

aiding heat loss, attempts have been made to quantitate the cooling effect of combinations of air speed and temperature; the most useful tool has been the wind-chill index.) Henceforth we shall also imply convection when we use the term conduction.

It should now be clear that heat loss by radiation and conduction is largely determined by the temperature difference between the body surface and the external environment. It is convenient to view the body as a central core surrounded by a shell that consists of skin and subcutaneous tissue (for convenience, we shall refer to the complex shell of tissues simply as skin) whose insulating capacity can be varied. It is the temperature of the central core that is being regulated at approximately 37°C (99°F); in contrast, as we shall see, the temperature of the outer surface of the skin changes markedly. If the skin were a perfect insulator, no heat would ever be lost from the core; the outer surface of the skin would equal the environmental temperature (except during direct exposure to the sun), and net conduction or radiation would be zero. The skin, of course, is not a perfect insulator, so that the temperature of its outer surface generally lies somewhere between that of the external environment and the core. Of great importance for temperature regulation of the core is that the skin's effectiveness as an insulator is subject to physiologic control by changing the blood flow to the skin. The more blood that reaches the skin from the core, the more closely the skin's temperature approaches that of the core. In effect, the blood vessels diminish the insulating capacity of the skin by carrying heat to the surface (Fig. 26-22). These vessels are controlled primarily by vasoconstrictor sympathetic nerves. Vasoconstriction may be so powerful that the skin of the finger, for example, may undergo a 99 percent reduction in blood flow during exposure to cold.

Exposure to cold increases the difference between core and environment; in response, skin vasoconstriction increases skin insulation, reduces skin temperature, and lowers heat loss. Exposure to heat decreases (or may even reverse) the difference between core and environment; in order to permit the required heat loss, skin vasodilation occurs, the gradient between skin and environment increases, and heat loss increases. Although we have spoken of skin temperature as if it were uniform throughout the body, certain areas participate

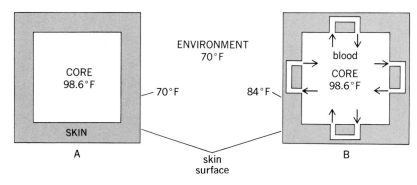

FIG. 26-22 Relationship of skin's insulating capacity to its blood flow. (A) Skin as a perfect insulator, i.e., with zero blood flow, the temperature of the skin outer surface equals that of the external environment. When the skin blood vessels dilate (B), the increased flow carries heat to the body surface, i.e., reduces the insulating capacity of the skin, and the surface temperature becomes intermediate between that of the core and the external environment.

much more than others in the vasomotor responses; accordingly, skin temperatures vary with location.

What are the limits of this type of process? The lower limit is obviously the point at which maximal skin vasoconstriction has occurred; any further drop in environmental temperature increases the environment-skin temperature difference and causes excessive heat loss. At this point, the body must increase its heat production to maintain temperature. The upper limit is set by the point of maximal vasodilation, the environmental temperature, and the core temperature itself. At high environmental temperatures, even maximal vasodilation cannot establish a core-environment temperature difference large enough to eliminate heat as fast as it is produced. Another heat-loss mechanism, sweating, is therefore brought strongly into play. Thus, the vasomotor contribution to temperature regulation is effective in the midrange of environmental temperature (20 to 28°C, or 70 to 85°F), but the major burden is borne by increased heat production at lower temperatures and by increased heat loss via sweating at higher temperatures.

Two other important mechanisms for altering heat loss by radiation and conduction remain: changes in surface area and clothing. Curling up into a ball, hunching the shoulders, and similar maneuvers in response to cold reduce the surface area exposed to the environment and thereby decreases radiation and conduction. For human beings, clothing is also an important component of temperature regulation; it substitutes for the insulating effects of feathers in birds and fur in other mammals. The principle is similar in that the outer surface of the clothes now forms the true "exterior" of the body surface. The skin loses heat directly to the air space trapped by the clothes; the clothes in turn pick up heat from the inner air layer and transfer it to the

external environment. The insulating ability of clothing is determined primarily by its type as well as by the thickness of the trapped air layer. We have spoken thus far only of the ability of clothing to reduce heat loss; the converse is also desirable when the environmental temperature is greater than body temperature, since radiation and conduction then produce heat gain. Human beings therefore insulate themselves against temperatures that are greater than body temperature by wearing clothes. The clothing, however, must be loose so as to allow adequate movement of air to permit evaporation, the only source of heat loss under such conditions. White clothing is cooler because it reflects more radiant energy, which dark colors absorb. Contrary to popular belief, loose-fitting light-colored clothes are far more cooling than going nude during direct exposure to the sun.

Evaporation   Evaporation of water from the skin and lining membranes of the respiratory tract is the second major process for loss of body heat. Thermal energy is required to transform water from the liquid to the gaseous state. Thus, whenever water vaporizes from the body's surface, the heat required to drive the process is conducted from the surface, and the surface is thereby cooled. Even in the absence of sweating, there is still loss of water by diffusion through the skin, which is not completely waterproof. A like amount is lost during expiration from the respiratory lining. This **insensible water loss** amounts to approximately 900 mL/day in human beings and accounts for a significant fraction of total heat loss.

In contrast to this passive water diffusion, sweating requires the active secretion of fluid by sweat glands and its extrusion into ducts, which carry it to the skin surface. The sweat is pumped to the surface by periodic contraction of cells that resemble smooth muscle in the ducts. Production and delivery of sweat to the surface are stimulated by the sympathetic nerves. Sweat is a dilute solution that contains primarily sodium chloride. The loss of this salt and water during severe sweating can cause diminution of plasma volume adequate to provoke hypotension, weakness, and fainting. It has been estimated that there are over 2.5 million sweat glands spread over the adult human body, and production rates of over 4 L/hr have been reported. This is 9 lb of water, the evaporation of which would eliminate almost 2400 kcal from the body!

It is essential to recognize that sweat must evaporate in order to exert its cooling effect. The most important factor that determines evaporation is the water vapor concentration of the air, i.e., the humidity. The discomfort suffered on humid days is caused by the failure of evaporation; the sweat glands continue to secrete, but the sweat simply remains on the skin or drips off. Most other mammals differ from human beings in lacking sweat glands. They increase their evaporative losses primarily by panting, which thereby increases pulmonary air flow and increases water losses from the lining of the respiratory tract, and they deposit water for evaporation on their fur or skin by licking.

Heat loss by evaporation of sweat gradually dominates as environmental temperature rises because radiation and conduction decrease linearly as the body-environment temperature difference diminishes. At environmental temperatures above that of the body, heat is actually gained by radiation and conduction, and evaporation is the sole mechanism for heat loss. A person's ability to survive such temperatures is determined by the humidity and by the maximal sweating rate. For example, when the air is completely dry, human beings can survive a temperature of 130°C (266°F) for 20 min or longer, whereas very moist air at 46°C (115°F) is not bearable for even a few minutes.

Changes in sweating determine human beings' chronic adaptation to high temperatures. A person newly arrived in a hot environment has poor ability to do work initially, body temperature rises, and severe weakness and illness may occur. After several days, there is great improvement in work tolerance with little increase in body temperature, and the person is said to have **acclimatized** to the heat. Body temperature is kept low because there is an earlier onset of sweating and because of increased rates of sweat production. The sodium content of the sweat is reduced, which minimizes the loss of sodium from the body. The mechanisms remain unknown, although heightened aldosterone secretion plays an important role in reducing the sodium.

Table 26-4 summarizes the mechanisms that regulate temperature, none of which is an all-or-none response but calls for a graded progressive increase or decrease in activity. As we have seen, heat production via skeletal muscle activity becomes extremely important at the cold end of the spectrum, whereas increased heat loss via sweating is critical at the hot end.

Brain Centers Involved in Temperature Regulation

Neurons in the hypothalamus and other brain areas, via descending pathways, control the output of somatic motor nerves to skeletal muscle (muscle tone and shivering) and of sympathetic nerves to skin arterioles (vasoconstriction and dilation), sweat glands, and the adrenal medulla. In animals in

TABLE 26-4	SUMMARY OF EFFECTOR MECHANISMS IN HEAT REGULATION
Stimulated by Cold	
Decrease heat loss	Vasoconstriction of skin vessels; reduction of surface area (curling up, etc.); behavioral (put on warmer clothes, raise thermostat setting, etc.)
Increase heat production	Shivering and increased voluntary activity; (?) increased secretion of thyroid hormone and epinephrine; increased appetite
Stimulated by Heat	
Increase heat loss	Vasodilation of skin vessels; sweating; behavioral (put on cooler clothes, turn on fan, etc.)
Decrease heat production	Decreased muscle tone and voluntary activity; (?) decreased secretion of thyroid hormone and epinephrine; decreased appetite

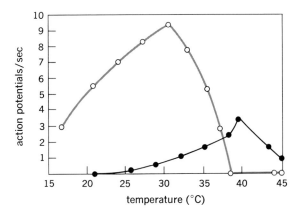

FIG. 26-23 Discharge rates of a typical skin cold receptor (color) and warm receptor (black) in response to changes in temperature.

which thyroid hormone is an important component of the response to cold, these centers also control the output of hypothalamic TSH-releasing hormone (TRH).

Afferent Input to the Integrating Centers

The final component of temperature-regulating systems we must describe is really the first component, i.e., the afferent input. Obviously, these temperature-regulating reflexes require receptors capable of detecting changes in the body temperature. There are two groups of receptors, one in the skin (**peripheral thermoreceptors**) and the other in deeper body structures (**central thermoreceptors).**

Peripheral Thermoreceptors In the skin (and certain mucous membranes) are nerve endings usually categorized as *cold* and *warm receptors*. In one sense, these are misleading terms because cold is not a separate entity but a lesser degree of warmth. Really there are two populations of temperature-sensitive skin receptors, one stimulated by a lower and the other by a higher range of temperatures (Fig. 26-23). Information from these receptors is transmitted via the afferent nerves and ascending pathways to the hypothalamus and other integrating areas, which respond with appropriate efferent output. In this manner, the firing of cold receptors stimulates heat-producing and heat-conserving mechanisms, whereas enhanced firing of warmth receptors accomplishes just the opposite.

Central Thermoreceptors It should be clear that the skin thermoreceptors alone would be highly inefficient regulators of body temperature for the simple reason that it is the core temperature, not the skin temperature, that is actually being regulated. On theoretical grounds alone it was apparent that core, i.e., central, receptors had to exist somewhere in the body, and numerous experiments have localized them to the hypothalamus, spinal cord, abdominal organs, and other internal locations. In unanesthetized dogs, local warming of hypothalamic neurons through previously implanted probes causes them to fire rapidly (Fig. 26-24) and reproduces the entire picture of the dog's usual response to a warm environment: he becomes sleepy, stretches out to increase his surface area, pants heavily, salivates, and licks his fur. Conversely, local cooling induces vasoconstriction, intensive shivering, fluffing out of the fur, and curling up. These hypothalamic thermoreceptors have synaptic connections with hypothalamic and other integrating centers, which also receive input from other central thermoreceptors as well as from the skin thermoreceptors. The precise relative contributions of the various thermoreceptors remain the subject

FIG. 26-24 Effect of local heating of a discrete area of hypothalamus on the discharge rate of a single thermosensitive hypothalamic neuron.

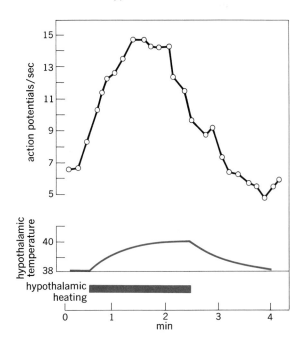

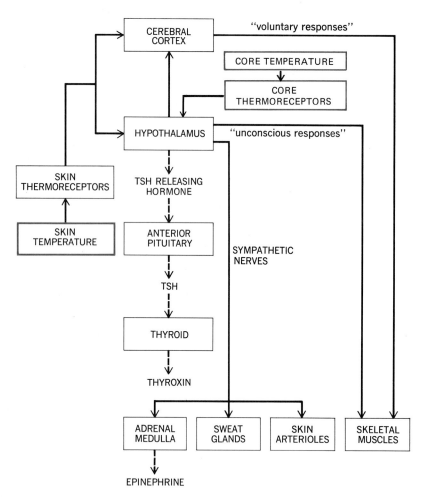

FIG. 26-25 Summary of temperature-regulating mechanisms. The dashed lines are hormonal pathways, which are probably of minor importance in human beings. Not shown are other nonhypothalamic integrating areas.

of considerable debate. Temperature-regulating reflexes are summarized in Fig. 26-25.

Fever

The elevation of body temperature so commonly induced by infection is due not to a breakdown of temperature-regulating mechanisms but to a "resetting of the thermostat" in the hypothalamus or other brain area. Thus, a person with a fever regulates his or her temperature in response to heat or cold but at a higher set point. The onset of fever is frequently gradual, but it is most striking when it occurs rapidly in the form of a chill. It is as though the thermostat were suddenly raised. The person suddenly feels cold, and marked vasoconstriction and shivering occur; the person also curls up and puts on more blankets because of feeling cold. This

association of heat conservation and increased heat production serves to drive body temperature up rapidly. Later, the fever breaks as the thermostat is reset to normal; the person feels hot, throws off the covers, and manifests profound vasodilation and sweating.

What is the basis for the resetting? As is described in Chap. 28 on resistance to infection, certain endogenously produced chemicals known as **pyrogens** are released from white blood cells in the presence of infection or inflammation. These pyrogens act directly upon the thermoreceptors in the hypothalamus (and perhaps other brain areas) to alter their rate of firing and their input to the integrating centers. This effect of pyrogens may be mediated via local release of prostaglandins, which then directly alter thermoreceptor function. Consistent with this hypothesis is the fact that aspirin, which reduces fever by restoring thermoreceptor activity toward normal, inhibits the synthesis of prostaglandins.

In addition to infection and inflammation, in which fever is induced by pyrogens, as described above, there are other situations in which hyperthermia is produced by quite different mechanisms. Excessive blood levels of epinephrine or thyroid hormone that result from diseases of the adrenal medulla or thyroid gland elevate the body temperature by direct actions on heat-producing metabolic reactions rather than by altering the hypothalamic thermoreceptor setting. Certain lesions of the brain do not reset the hypothalamus but rather completely destroy its normal regulatory capacity; under such conditions lethal hyperthermia may occur very rapidly. *Heat stroke* is also characterized by a similar breakdown in function of the regulatory centers. It is frequently a positive-feedback state in which, because of inadequate balancing of heat loss and production, body temperature becomes so high that the brain's regulatory centers are put out of commission and body temperature therefore rises even higher. Thus a patient suffering from heat

stroke manifests a dry skin (absence of sweating) despite a markedly elevated body temperature.

Heat stroke, the attainment of a body temperature at which vital bodily functions are endangered, should be distinguished from *heat exhaustion*. The former is caused by a breakdown in heat-regulating mechanisms, whereas the latter is not the result of failure of heat regulation but rather is the inability to meet the price of heat regulation. Heat exhaustion is a state of collapse as a result of hypotension brought on by depletion of plasma volume (secondary to sweating) and by extreme dilation of skin blood vessels, i.e., by decreases in both cardiac output and peripheral resistance. Thus, heat exhaustion occurs as a direct consequence of the activity of heat-loss mechanisms; because these mechanisms have been so active, the body temperature is only modestly elevated. In a sense, heat exhaustion is a safety valve that, by forcing cessation of work when heat-loss mechanisms are overtaxed, prevents the larger rise in body temperature that would precipitate the far more serious condition of heat stroke.

KEY TERMS

dynamic steady state	**glucagon**	**work**
absorptive state	**epinephrine**	**metabolic rate**
postabsorptive state	**cholesterol**	**total body energy balance**
glucose sparing	**growth hormone**	**conduction**
insulin	**thyroid hormone**	**thermoreceptor**
gluconeogenesis		

REVIEW EXERCISES

Section A

1 Define essential nutrient, molecular interconvertibility, and dynamic steady state.
2 Define metabolic pool; diagram the major inputs and outputs of the amino acid and carbohydrate-lipid pools.
3 Define absorptive and postabsorptive state.
4 Diagram and explain the major events of the absorptive period that concern the fate of ingested carbohydrate, protein, and fat; include the role of the liver in fat synthesis.
5 Diagram and explain the major events of the postabsorptive period; list the sources of blood glucose during this period.
6 Define gluconeogenesis.

7 Describe the metabolism of the brain during early and long-term phases of the postabsorptive period.
8 Describe insulin and its site of production.
9 Describe the major effects of insulin on organic metabolism.
10 State the effects of insulin deficit.
11 Describe the control of insulin secretion.
12 Describe the metabolic events that occur in a person with untreated diabetes mellitus.
13 Describe and explain the renal function and respiratory state of a person with untreated diabetes mellitus.
14 Describe glucagon and its site of production.
15 Describe the major effects of glucagon on organic metabolism.
16 Describe the control of glucagon secretion.

17 Describe the role of the sympathetic nervous system on plasma glucose, glycerol, and fatty acid concentrations.

18 Define cholesterol, high-density lipoprotein, and low-density lipoprotein.

19 List the sources of cholesterol gain and loss.

20 Differentiate between true growth and a simple gain in body weight.

21 State how growth hormone's effects on metabolism stimulate growth.

22 Describe the effects of thyroid hormone on growth.

23 Describe the effects of insulin on growth.

24 Describe the effects of the androgens and estrogen on growth.

Section B

1 State the first law of thermodynamics; describe the equation $E = H + W$.

2 Distinguish between internal and external work.

3 State the relation between heat produced, external work, energy storage, and total energy expenditure in terms of the first law of thermodynamics.

4 Define metabolic rate; describe how it is measured.

5 Define basal metabolic rate; distinguish between metabolic rate and basal metabolic rate.

6 Define metabolic cost of living and specific dynamic action.

7 List the factors that influence metabolic rate; state the mechanisms by which they exert their effects.

8 Use the energy balance equation (Energy = internal heat produced + external work + energy storage) to describe the results of a food intake greater than energy expenditure.

9 Describe the signals that provide short-term and long-term information about the body's energy stores.

10 Describe the relation between caloric intake and exercise in the maintenance of body weight.

11 Define obesity; state why it occurs.

12 Describe the normal variations in body temperature.

13 Identify the physiologic sources of heat production.

14 Describe the mechanisms of heat loss; include discussions of radiation, conduction, and convection and their relation to blood flow, sweating, and behavioral changes.

15 Describe how evaporation from the body surface and linings of the respiratory tract results in heat loss.

16 Describe the information pathways that coordinate temperature regulation; include a discussion of central and peripheral temperature receptors and the integrating centers for temperature regulation.

17 Define fever, heat stroke, and heat exhaustion; use the concept of temperature set point.

PART

5

REPRODUCTION
AND
MAINTENANCE
OF "SELF"

CHAPTER

THE REPRODUCTIVE SYSTEM

Before beginning detailed descriptions of the male and female reproductive systems, it may be worthwhile to summarize some important terminology. The primary reproductive organs are known as the **gonads,** the **testes** in the male and the **ovaries** in the female. In both sexes, the gonads serve dual functions: (1) production of the reproductive cells, **sperm** or **ova;** and (2) secretion of the so-called **sex hormones.** The systems of ducts through which the sperm or ova are transported and the glands that line or empty into the ducts are termed the **accessory reproductive organs** (in the female the breasts are also usually included in this category). The region that contains the external genitalia and openings of the urinary system is the **perineum,** or perineal region (this term is sometimes used to include the anal region as well). Finally, the **secondary sexual characteristics** comprise the many external differences (hair, body contours, etc.) between male and female that are not directly involved in reproduction.

The gonads and accessory reproductive organs are present at birth but remain relatively small and nonfunctional until the onset of **puberty,** at about 10 to 14 years of age. The secondary sexual characteristics are virtually absent until puberty. The term puberty signifies the attainment of sexual maturity in the sense that conception becomes possible; as commonly used, it refers to the 3 to 5 years of sexual development that culminate in the attainment of sexual maturity. The term **adolescence** has a much broader meaning and includes the total period of transition from childhood to adulthood in all respects, not just sexual.

SECTION A: THE MALE REPRODUCTIVE SYSTEM

ANATOMY OF THE MALE REPRODUCTIVE TRACT

The essential male reproductive functions are the manufacture of sperm (**spermatogenesis**) and the deposition of the sperm in the female. The organs that carry out these functions are the testes, epididymides (sing., epididymis), ductus deferens, ejaculatory ducts, and penis together with accessory glands: the seminal vesicles, prostate, and bulbourethral glands (Fig. 27-1). The **testes** are surrounded by a tough, connective-tissue capsule and are suspended outside the body in a sac, the **scrotum,** in which each testis lies in a separate pouch. The skin of the scrotum differs from other skin in that it contains much smooth muscle and, in addition, is tightly fused to an underlying sheet of smooth muscle, the *dartos.*

During embryonic development, the testes lie near the kidneys in the posterior wall of the abdominal cavity; i.e., they are retroperitoneal, but they eventually descend into the scrotum and carry with them the arteries, veins, and nerves that supply them and the ducts (ductus deferens) that connect them with the urethra. This descent usually occurs during the eighth month of fetal life, but it may be as late as a year or so after birth. These structures form the core of the *spermatic cord.* Early in its development, each testis is connected to the inguinal (groin) region by a ligament, which deter-

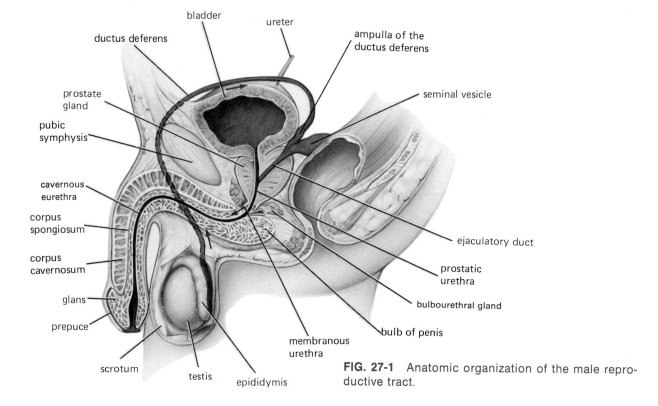

FIG. 27-1 Anatomic organization of the male reproductive tract.

mines the course along which the testis descends. As the testis approaches the groin, the ligament thickens and becomes known as the *gubernaculum* (Fig. 27-2). The testes descend between the peritoneum and the outer facial layers of the abdominal wall; the facial layers form a pouch, the inguinal bursa, and eventually become the coverings of the testes and spermatic cords. Each gubernaculum is attached to the floor of its respective bursa, and the testis follows its gubernaculum down into the scrotum. This descent is essential for normal sperm production during adulthood because sperm for-

FIG. 27-2 Sequence of events during descent of the testes. (A) The scrotum begins to swell. (B) The testis has almost completed its descent as the enfolding of peritoneum (the vaginal process) becomes prominent. (C) Descent is complete, and the tunica vaginalis represents the vestigial lower portion of the vaginal process.

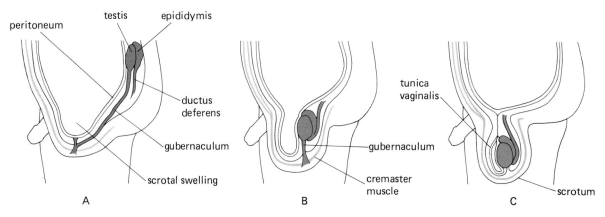

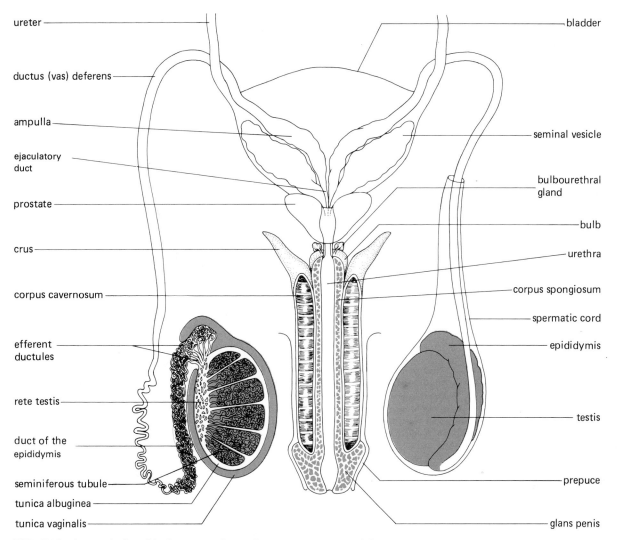

ureter

ductus (vas) deferens

ampulla

ejaculatory
duct

prostate

crus

corpus cavernosum

efferent
ductules

rete testis

duct of the
epididymis

seminiferous tubule

tunica albuginea

tunica vaginalis

bladder

seminal vesicle

bulbourethral
gland

bulb

urethra

corpus spongiosum

spermatic cord

epididymis

testis

prepuce

glans penis

FIG. 27-3 Interrelationship between the various components of the urogenital tract of the male as seen from the rear. The spermatozoa are formed in the highly coiled seminiferous tubules. Part of the left testis has been cut away to reveal internal structures.

mation requires a temperature lower than normal internal body temperature. The actual mechanism of descent is unknown.

Each testis (Fig. 27-3) is somewhat oval and is enclosed in a dense connective-tissue sheath, the *tunica albuginea,* from which projections pass into the substance of the testis and form septa, which incompletely divide the gland into some 200 to 300 lobules. Each lobule contains one or more tiny convoluted **seminiferous tubules,** which are lined with the sperm-producing **spermatogenic cells.** The testes

also serve an endocrine function, the manufacture of the primary male sex hormone testosterone. The endocrine-secreting **interstitial cells** (Leydig cells) lie in the small connective-tissue spaces that surround the seminiferous tubules (Fig. 27-4). Thus, the sperm-producing and testosterone-producing functions of the testes are carried out by different cells.

The seminiferous tubules unite as they leave the lobules of the testis to form larger ducts. As they join, they pack tightly together and interchange

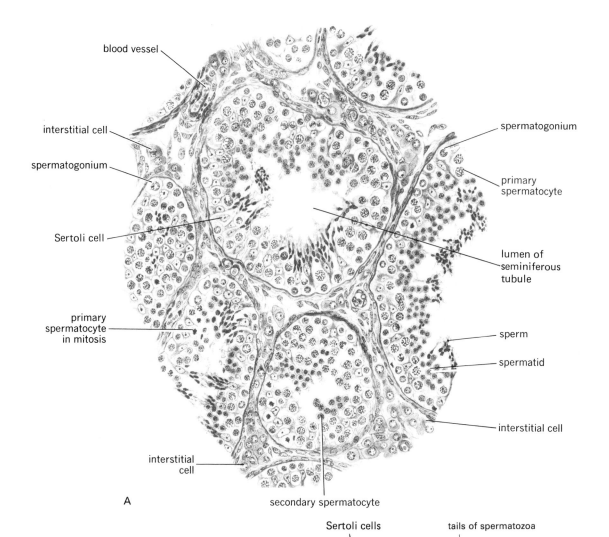

blood vessel

interstitial cell

spermatogonium

Sertoli cell

primary
spermatocyte
in mitosis

interstitial
cell

spermatogonium

primary
spermatocyte

lumen of
seminiferous
tubule

sperm

spermatid

interstitial cell

A

secondary spermatocyte

Sertoli cells

tails of spermatozoa

B

primary spermatocyte

spermatogonium

spermatids

FIG. 27-4 (A) Section of human testis. The seminiferous tubules show various stages of spermatogenesis. (B) A scanning electron micrograph of a cross section of a seminiferous tubule that shows spermatogonia, the least differentiated of the cells close to the outer lining membrane of the tubule. Primary spermatocytes, spermatids, and Sertoli cells can also be seen in the wall of the tubule, and the tails of spermatids undergoing transition into mature spermatozoa can be seen extending into the tubule lumen.

branches to form the **rete testis** (Fig. 27-3). **Efferent ductules** leave the rete testis and pass to the **epididymis** where they unite to form a highly convoluted single duct, the **duct of the epididymis.** In the tail of the epididymis, the duct of the epididymis becomes the **ductus (vas) deferens,** which widens and follows a relatively straight course. After the ductus leaves the epididymis, it joins the blood vessels and nerves of the testis and passes upward via the spermatic cord. The two ductus deferens, one from each testis, pass upward, cross the inguinal canal, and enter the pelvis (although they do not enter the abdominal cavity); each duct then passes behind the urinary bladder, widens to form the **ampulla,** narrows again, receives the opening of a seminal vesicle, and enters the substance of the prostate gland where it empties into the urethra. The narrowed portion of the ductus, between the ampulla and the urethra, is the **ejaculatory duct.** The two **seminal vesicles** are not typical glands but are long coiled tubes whose inner epithelial lining has a high secretory capacity, whose folded surface permits distension for the storage of secreted fluid, and whose adjacent convoluted surfaces are bound together by connective tissue to form a relatively compact structure.

The **prostate** surrounds the urethra as the urethra leaves the bladder (therefore, this portion of the urethra is known as the prostatic urethra). Enlargement of the prostate, relatively common in men past middle age, may obstruct the urethra and interfere with urination. The prostate is surrounded by a capsule, and its substance consists of many individual glands that open by ducts into the urethra. These glands not only produce secretions but are able to accommodate large quantities of stored secretions. Testosterone is required to bring about full development of the prostate and seminal vesicles.

The **penis** consists chiefly of three cylindrical cores of *erectile (cavernous) tissue,* the paired *corpora cavernosa* and a single *corpus spongiosum* through which the urethra passes (Fig. 27-3). The corpus spongiosum extends beyond the corpora cavernosa at the tip of the penis and is expanded to form the **glans.** Each core of erectile tissue is surrounded by a dense connective-tissue tunica albuginea except in the glans, where the skin itself is continuous with the corpus spongiosum beneath it. The three cylinders of erectile tissue are bound together by loose connective tissue, and it is to this that the skin of the main part of the penis (the body of the penis) is bound. A circular fold of skin, the **prepuce,** extends forward to cover the glans (Fig. 27-3) unless the prepuce has been removed surgically, a process known as **circumcision.** At its proximal end, the penis enters the perineum, and the two corpora cavernosa separate from each other and from the corpus spongiosum to form the *crura* (sing., crus) of the penis. Between the two crura the corpus spongiosum continues backward and enlarges to form the *bulb of the penis* (Fig. 27-3).

The urethra is known as the **membranous urethra** as it leaves the prostate; it retains this name only until it enters the bulb of the penis, when it becomes the **cavernous urethra.** In the membranous portion, the urethra is surrounded by skeletal muscle fibers (of the urogenital diaphragm) that form the sphincter muscle of the urethra, sometimes known as the *external sphincter of the bladder.* Two small bodies, each about the size of a pea, lie close to the membranous urethra; these glands, the **bulbourethral,** or **Cowper's, glands,** whose cells are mainly of the mucus-secreting type, empty via ducts into the first portion of the cavernous urethra (Fig. 27-3). In addition to the seminal vesicles, prostate, and bulbourethral glands, many small mucus-secreting glands lie in the wall of the urethra itself and add their secretions to the urethral contents.

SPERMATOGENESIS

The testis is composed primarily of the many highly coiled seminiferous tubules. In Fig. 27-4B, a microscopic section of an adult human testis, it can be seen that the tubules contain many cells, the vast majority of which are in various stages of division; these are the spermatogenic cells.

Each seminiferous tubule is surrounded by a spermatogenic epithelium underlain by a basement membrane, and only the outermost layer of spermatogenic cells is in contact with this membrane; these are undifferentiated germ cells termed **spermatogonia,** which, by dividing mitotically, provide a continuous source of new cells. Some spermatogonia move away from the basement membrane and increase markedly in size. Each of these large cells, now termed a **primary spermatocyte,** divides to form two **secondary spermatocytes,** each of which in turn divides into two **spermatids;** the latter ultimately are transformed into mature **spermatozoa** (Fig. 27-5).

The division of the primary spermatocytes by meiosis differs from the ordinary mitotic division. During mitosis (Chap. 3) each daughter cell receives

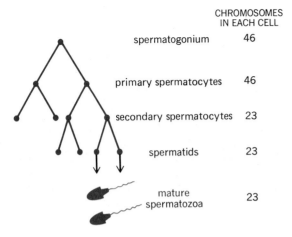

FIG. 27-5 Summary of spermatogenesis. Each spermatogonium yields eight mature sperm, each of which contains 23 chromosomes.

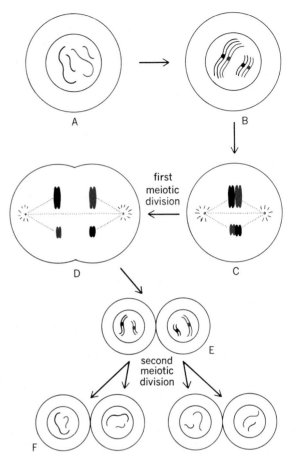

FIG. 27-6 Separation of chromosomes during cell divisions of meiosis. The maternal chromosomes are colored.

the full number of chromosomes; in **meiosis,** which is really two divisions in succession, each final cell receives only *half* of the chromosomes present in the original cell. The primary spermatocyte contains 46 chromosomes (the diploid number), 23 from each parent. Prior to its division each of these 46 chromosomes becomes duplicated, and during the condensation into chromosomes the duplicated maternal and paternal chromatin threads pair with each other to form a four-stranded chromosome (Fig. 27-6). Thus, there are 23 four-stranded chromosomes instead of 46 two-stranded chromosomes. When the primary spermatocyte divides, the two maternal strands pass into one of the secondary spermatocytes and the paternal strands into the other. Since each of the 23 four-stranded chromosomes becomes attached to the spindle fibers at random, one chromosome may have its maternal strands oriented toward one pole of the cell and the next chromosome have its maternal strands oriented toward the opposite pole. Thus, when the maternal and paternal strands separate, the two secondary spermatocytes, in general, receive a mixture of maternal and paternal chromosomes. It would be extremely improbable that all 23 maternal chromosomes would end up in one cell and all 23 paternal chromosomes in the other.

The random distribution of maternal and paternal chromosomes during meiosis provides the basis for much of the variability in the genetic constitution of the offspring from a single set of parents. Over 8 million (2^{23}) different combinations of maternal and paternal chromosomes can result from the distribution of the 23 chromosomes during meiosis. This is only the minimum number of possible combinations of genetic material because segments of the maternal and paternal chromosomes may exchange with each other in a process known as **crossing over,** which occurs during chromosomal pairing before the first meiotic division. This forms chromosomes that contain both maternal and paternal genes in the same chromosome. Thus, all the genes initially present in a paternal or maternal chromosome do not necessarily pass into the same cell during meiosis.

The final phase of spermatogenesis, the transformation of the spermatids into mature spermatozoa, involves no further cell divisions. The head of a

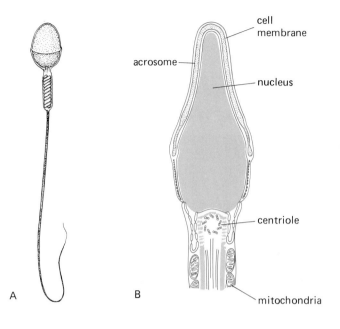

A   B

cell membrane

acrosome

nucleus

centriole

mitochondria

mature spermatozoon (Fig. 27-7) consists almost entirely of the nucleus, a dense mass of DNA that bears the sperm's genetic information. The tip of the nucleus is covered by the **acrosome,** a protein-filled vesicle that contains several lytic enzymes that enable the sperm to enter the ovum. The tail comprises a group of actomyosinlike contractile filaments, which produces a whiplike movement of the tail capable of propelling the sperm at a velocity of 1 to 4 mm/sec. The mitochondria of the spermatid form the midpiece of the tail and probably provide the energy for its movement.

Throughout the entire process of maturation, the developing germ cells remain intimately associated

FIG. 27-7   (A) Human mature sperm seen in frontal view. (B) The close-up is a side view.

FIG. 27-8   A drawing of the seminiferous epithelium, showing the relation of the Sertoli cells and germ cells.

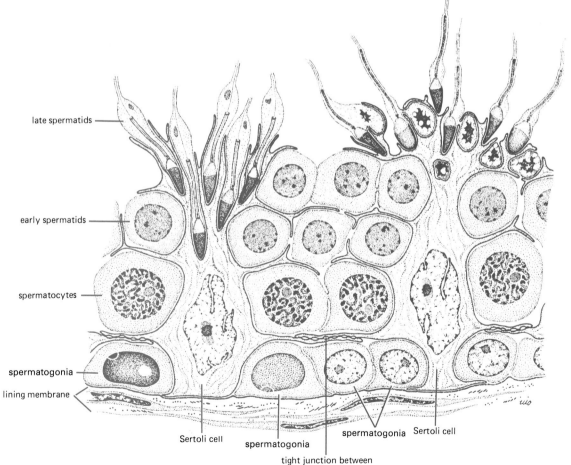

late spermatids

early spermatids

spermatocytes

spermatogonia

lining membrane

Sertoli cell

spermatogonia

tight junction between Sertoli cells

spermatogonia   Sertoli cell

with another type of cell, the **Sertoli cell** (Fig. 27-8). Each Sertoli cell extends from the basement membrane of the seminiferous tubule all the way to the fluid-filled lumen and is joined to adjacent Sertoli cells by means of tight junctions. Thus, the Sertoli cells form an unbroken ring around the outer circumference of the seminiferous tubule and divide the tubule into two compartments: a basal compartment (between the basement membrane and the tight junction) and a central (adluminal) compartment that includes the lumen. This arrangement has several very important results. First, the ring of interconnected Sertoli cells forms a ''barrier'' that limits the movement of chemicals from blood into the lumen of the seminiferous tubule (the membranes that surround the entire tubule form a second component of the so-called blood-testis barrier). Second, mitosis of the spermatogonia takes place entirely in the basal compartment, and the primary spermatocytes must move through the tight junctions of the Sertoli cells (which open to make way for them) to gain entry into the central compartment. In this compartment, the spermatids are contained in recesses formed by invaginations of the luminal membranes of the Sertoli cells until mature enough for release. Apparently, the Sertoli cells serve as the route by which nutrients and chemical signals reach the developing germ cells. As we shall see, they may also produce hormones and other chemicals important for the control of spermatogenesis.

The mechanisms that guide the remarkable cellular transformation from spermatid to mature sperm remain uncertain. In any small segment of seminiferous tubules, the entire process of spermatogenesis proceeds in a regular sequence. For example, at any given time, virtually all the primary spermatocytes in one portion of the tubule are undergoing division, whereas in an adjacent segment, the secondary spermatocytes may be dividing. The entire process in a single area takes approximately 72 days. In mammals that breed seasonally, spermatogenesis is periodic, activity being followed by degeneration of the spermatogonia and shrinking of the seminiferous tubules. In contrast, nonseasonal breeders, such as human beings, manifest continuous spermatogenesis. Perhaps the most amazing characteristic of spermatogenesis is its sheer magnitude: the normal human male may manufacture several hundred million sperm per day.

DELIVERY OF SPERM

Besides serving as a route for sperm exit, the duct system performs several important functions: (1) The epididymis and first portion of the ductus deferens store sperm prior to ejaculation; (2) during passage through the epididymis or storage there (approximately 9 to 14 days are required for a sperm to traverse the epididymis), a maturation process occurs, without which the sperm would be nonmotile and infertile when they enter the female tract; (3) during ejaculation, the sperm are expelled from the epididymis and ductus deferens by strong contractions of the smooth muscle that lines the duct walls.

Movement of sperm through these ducts is accomplished by two means (the sperm themselves are nonmotile at this time): (1) the pressure created by the continuous formation of sperm and fluid, and (2) a peristalticlike action exerted by the smooth muscle cells in the duct walls.

The prostate, seminal vesicles, and bulbourethral glands secrete small quantities of fluid continuously and much larger quantities during sexual intercourse (coitus). The secretions constitute the bulk of the ejaculated fluid, the **semen,** which contains a large number of different chemical substances, the functions of which are presently being worked out. For example, the seminal vesicles secrete large quantities of the carbohydrate fructose, utilized by the sperm contractile apparatus for energy. We shall describe later the possible contributions of the prostaglandins, present in very large concentrations in seminal fluid. Another function of the seminal fluid is that of sheer dilution of the sperm (in human beings, sperm constitute only a few percent of the total ejaculated semen); without such dilution, motility is impaired.

Erection

The primary components of the male sexual act are **erection** of the penis, which permits entry into the female vagina, and **ejaculation** of the sperm-containing semen into the vagina. Erection is a vascular phenomenon that can be understood from the structure of the penis (Fig. 27-9). Normally the arterioles that supply the bodies of erectile tissue in the penis are constricted so that they contain little blood and the penis is flaccid. During sexual excitation, the arterioles dilate, the chambers become engorged

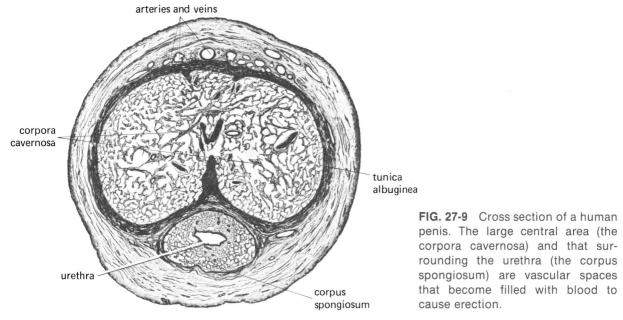

arteries and veins

corpora
cavernosa

tunica
albuginea

urethra

corpus
spongiosum

FIG. 27-9 Cross section of a human penis. The large central area (the corpora cavernosa) and that surrounding the urethra (the corpus spongiosum) are vascular spaces that become filled with blood to cause erection.

with blood, and the penis becomes rigid; moreover, as the erectile tissues expand, the veins that empty them are passively compressed, thus minimizing outflow and contributing to the engorgement. This entire process occurs rapidly; complete erection sometimes takes only 5 to 10 sec. The vascular dilation is accomplished by stimulation of the nervi erigentes (parasympathetic nerve) and inhibition of the sympathetic nerves to the arterioles of the penis (Fig. 27-10). This appears to be one of the few cases of direct parasympathetic control over high-resistance blood vessels. In addition to these vascular effects, the parasympathetic nerves stimulate urethral glands to secrete a mucuslike material that aids in lubrication of the head of the penis. What receptors and afferent pathway initiate these reflexes? The primary input comes from highly sensitive mechanoreceptors in the tip of the penis. The afferent fibers that carry the impulses synapse in the lower spinal cord and trigger the efferent outflow. It must be stressed, however, that higher brain centers, via descending pathways, may exert profound facilitative or inhibitory effects upon the efferent neurons. Thus, thoughts or emotions can cause erection in the complete absence of mechanical stimulation of the penis; conversely, failure of erection (**impotence**) may frequently be due to psychologic factors. The ability of alcohol to interfere with erection is probably a result of its effects on higher brain centers.

FIG. 27-10 Reflex pathway for erection. The reflex is initiated by mechanoreceptors in the penis. Input from higher centers can facilitate or inhibit this reflex.

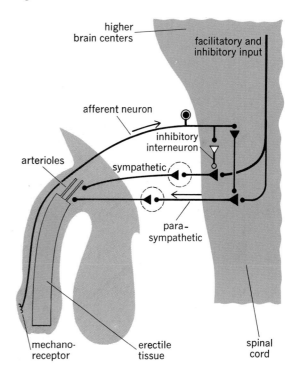

higher
brain centers

facilitatory and
inhibitory input

afferent neuron

inhibitory
interneuron

arterioles

sympathetic

para-
sympathetic

mechano-
receptor

erectile
tissue

spinal
cord

Ejaculation

This process is basically a spinal reflex, the afferent pathways apparently being identical to those described for erection. When the level of stimulation reaches a critical level, a patterned automatic sequence of efferent discharge is elicited to both the smooth muscle of the duct system and the skeletal muscle at the base of the penis. The precise contribution of various pathways is complex, but the overall response can be divided into two phases: (1) The genital ducts and glands constrict, as a result of sympathetic stimulation to them, and empty their contents into the urethra (**emission**); (2) the semen is then expelled from the penis by a series of rapid muscle contractions. During ejaculation the sphincter at the opening of the bladder is closed so that sperm cannot enter the bladder nor can urine be expelled. The rhythmic contractions of the muscles that occur during ejaculation are associated with intense pleasure and many systemic physiologic changes; the entire event is termed the **orgasm.** A marked skeletal muscle contraction throughout the body is followed by the rapid onset of muscular and psychologic relaxation, and there is also a marked increase in heart rate and blood pressure. Once ejaculation has occurred, there is a so-called latent period during which a second erection is not possible; this period is quite variable but may be hours in normal men.

The average volume of fluid ejaculated is 3 mL that contains approximately 300 million sperm. However, the range of normal values is extremely large, and older ideas of the minimal concentration of sperm required for fertility are now being re-evaluated. Although quantity is important, it is obvious that the quality of the sperm is a critical determinant of fertility.

HORMONAL CONTROL OF MALE REPRODUCTIVE FUNCTIONS

Virtually all aspects of male reproductive functions are either directly controlled or indirectly influenced by testosterone or the anterior pituitary gonadotropins, follicle-stimulating hormone (FSH) and luteinizing hormone (LH). These pituitary hormones were named for their effects in the female, but their molecular structures are precisely the same in both sexes. (LH, in the male, is frequently called interstitial-cell-stimulating hormone, ICSH.) FSH and LH exert their effects only upon the testes, whereas testosterone manifests a broad spectrum of actions not only on the testes but on the accessory reproductive organs, the secondary sexual characteristics, sexual behavior, and organic metabolism in general.

Effects of Testosterone

Testosterone, the hormone produced by the interstitial cells of the testes, is the major male sex hormone. Other steroids with actions similar to those of testosterone are produced by the adrenal cortex and are, together with testosterone, collectively known as **androgens.** Although the adrenally produced androgens constitute a large fraction of total blood androgens in men, they are much less potent than testosterone and are unable to maintain testosterone-dependent functions should testosterone secretion be decreased or eliminated by disease or castration (removal of the gonads). Accordingly, we shall not discuss further the adrenal androgens in our description of male reproductive function.

Spermatogenesis Adequate amounts of testosterone are essential for spermatogenesis, and sterility is an invariable result of testosterone deficiency. The cells that secrete testosterone are the interstitial cells; as shown in Fig. 27-4, they lie scattered between the seminiferous tubules. It seems likely that the stimulatory effects of testosterone on spermatogenesis are exerted locally by the hormone that diffuses from the interstitial cells into the seminiferous tubules. Testosterone is not the only hormone required for spermatogenesis; the pituitary gonadotropins are also required, and their relationship with testosterone will be described subsequently.

It must be emphasized that although testosterone is required for the process of spermatogenesis, testosterone production does not depend upon spermatogenesis. In other words, testosterone deficiency produces sterility by interrupting spermatogenesis, but interference with the function of the seminiferous tubules does not alter normal testosterone production by the interstitial cells. The great importance of this relationship is that a simple, effective method of sterilizing the male is **vasectomy** (surgical removal of a segment of the ductus, i.e., vas, deferens, which carries sperm from the testes). This procedure prevents the delivery (though not the production) of sperm but does not appear to alter secretion of testosterone. Whether vasectomy produces any nonreproductive effects on health is presently being extensively studied.

Accessory Reproductive Organs The morphology and function of the entire male duct system, glands, and penis all depend upon testosterone. Following removal of the testes (castration) in the adult, all the accessory reproductive organs decrease in size, the glands markedly reduce their rates of secretion, and the smooth muscle activity of the ducts is inhibited. Erection and ejaculation may become diminished. These defects disappear upon the administration of testosterone.

Secondary Sex Characteristics In the animal world secondary sex characteristics range from the exotic courtship dances of salamanders to the mane of the lion to the stag's antlers. In many species they are important for normal sexual functions; antlers used in fighting for the female and the deep attracting voice of the tree toad are two examples from literally thousands.

In men, virtually all the obvious masculine secondary characteristics are testosterone-dependent. For example, a male castrated before puberty does not develop a beard or axillary or pubic hair. A strange and unexplained finding is that baldness, although genetically determined in part, does not occur in castrated men (castration will not, of course, reverse baldness). Other secondary sexual characteristics that depend on testosterone are the deepening of the voice (resulting from growth of the larynx), skin texture, thick secretion of the skin oil glands (predisposing to acne), and the masculine pattern of muscle and fat distribution. This leads us into an area of testosterone effects usually described as general metabolic effects but very difficult to separate from the secondary sex characteristics. It is obvious that the bodies of men and women (even excepting the breasts and external genitals) have very different appearances; a woman's curves are due in large part to the feminine distribution of fat, particularly in the region of the hips and lower abdomen but in the limbs as well. A castrated male gradually develops this pattern; conversely, a woman treated with testosterone

loses it. A second very obvious difference is that of skeletal muscle mass; testosterone exerts a profound effect on skeletal muscle to increase its size. The overall relationship of testosterone to general body growth was described in Chap. 26.

Behavior Most of our information comes from experiments on animals other than human beings, but even from our fragmentary information about human beings, there is little doubt that the development and maintenance of normal sexual drive and behavior in men are dependent upon testosterone and may be seriously impaired by castration. However, it is a mistake to assume that deviant male sexual behavior must therefore be due to testosterone deficiency or excess. For example, most (but not all) male homosexuals have normal rates of testosterone secretion; although administration of exogenous testosterone may sometimes increase sexual activity in these men, they remain homosexual. To date, no clear-cut correlation has been established between homosexuality or hypersexuality and hormonal status in either men or women.

A question that has recently become the subject of enormous controversy is whether testosterone influences other human behavior in addition to sex; i.e., are there any inherent male-female differences or are the observed differences in behavior all socially conditioned. There is little doubt that behavioral differences based on sex do exist in other mammals; for example, in mice aggression is clearly greater in males and is dependent on testosterone. Obviously, it will be difficult to answer such questions with respect to human beings but attempts are now being made to study them in a controlled scientific manner.

Mechanism of Action of Testosterone Testosterone, like other steroid hormones, crosses cell membranes readily and, within target cells, combines with specific cytoplasmic receptors. The hormone-receptor complex then moves into the nucleus where the testosterone-receptor complex influences the transcription of certain genes into mRNA. The result is a change in the rate at which the target cell synthesizes the proteins coded for by these genes, and it is the changes in the concentrations of these proteins that underlie the cell's overall response to the hormone. For example, testosterone induces increased synthesis of enzymes in the prostate

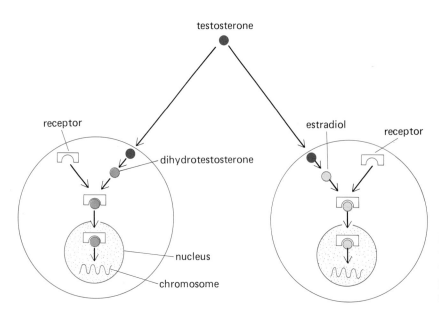

FIG. 27-11 Transformation of testosterone either to dihydrotestosterone or to estradiol is accomplished by enzymes in many target cells.

gland, which then catalyze the formation of the gland's secretion.

In Chap. 18 we mentioned that hormones sometimes must undergo transformation in their target cells in order to be effective, and this is true of testosterone in certain of its target cells. In some, after its entry to the cytoplasm, testosterone undergoes an enzyme-mediated transformation into another steroid, *dihydrotestosterone,* and it is this molecule that then combines with a specific receptor and moves into the nucleus to exert its effect (Fig. 27-11). There are men whose cells lack the enzyme required for formation of dihydrotestosterone, and these men display normal masculinization of those cells that respond directly to testosterone (skeletal muscles, for example) but failure of normal growth of those in which dihydrotestosterone is the actual messenger (the prostate, for example).

Quite startling has been the recent discovery that, in still other target cells, notably neurons in certain areas of the brain, testosterone is transformed not to dihydrotestosterone but to estrogen, which then combines with cytoplasmic receptors, moves into the nucleus, etc. (Fig. 27-11). Thus, a "male" sex hormone must first be transformed into a "female" sex hormone to be able to influence those neurons that, in turn, mediate so-called male behavior!

Testosterone is not a uniquely male hormone since it is found in very low concentration in the blood of normal women (as a result of production by the ovaries and adrenal glands). Of greater importance is the fact that androgens other than testosterone are found in quite significant concentrations in the blood of women. The sites of production are mainly the adrenal glands, which contribute these same nontestosterone androgens in the male. In contrast to their lack of significance in the male, adrenal androgens do play several important roles in the female, specifically stimulation of general body growth (Chap. 26) and maintenance of sexual drive. In several disease states, the female adrenals may secrete abnormally large quantities of androgen and produce virilism: the female fat distribution disappears, a beard appears along with the male body-hair distribution, the voice lowers in pitch, the skeletal muscle mass enlarges, the clitoris (homologue of the male penis) enlarges, and the breasts diminished in size. These changes illustrate the sex-hormone dependency of secondary sex characteristics.

Anterior Pituitary and Hypothalamic Control of Testicular Function

Follicle-stimulating hormone (FSH) and **luteinizing hormone** (LH) are essential for male reproductive function. (Following removal of the anterior pituitary, the testes decrease greatly in weight, and spermatogenesis and testosterone secretion almost cease.) FSH stimulates spermatogenesis by an ac-

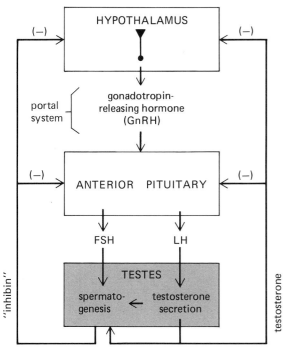

FIG. 27-12 Summary of hormonal control of testicular function. The negative signs indicate that testosterone inhibits LH secretion via both the hypothalamus and the anterior pituitary. Testosterone reaches the seminiferous tubules to stimulate spermatogenesis both by local diffusion and by release into the blood and recirculation to the testes. FSH influences spermatogenesis via an action on the Sertoli cells and is, in turn, inhibited by a substance, inhibin, released by the Sertoli cells. Prolactin, another anterior pituitary hormone (not shown in the figure) potentiates the effect of LH on the secretion of testosterone.

No discussion of the anterior pituitary is complete without inclusion of the hypothalamus. As described in Chap. 18, all anterior pituitary hormones are controlled by releasing hormones secreted by discrete areas of the hypothalamus and reaching the pituitary via the hypothalamopituitary portal blood vessels. This input is essential for sexual function since destruction of the relevant hypothalamic areas stops spermatogenesis and markedly reduces testosterone secretion. Figure 27-12 summarizes the hypothalamic–anterior pituitary–testicular relationships; note that the hypothalamus exerts no direct effects on the testes but rather influences them indirectly via its control of the anterior pituitary gonadotropins.

Several of the paths shown in Fig. 27-12 remain controversial. A major unsettled question is whether there exist two distinct hypothalamic hormones, one for each gonadotropin. Although the evidence is not yet conclusive, it favors the hypothesis that there is only a single releaser of both LH and FSH, appropriately called *gonadotropin-releasing hormone* (GnRH).

Given that GnRH controls the secretion of both FSH and LH, we must now ask what controls GnRH. Recall from Chap. 18 that the hypothalamic cells that produce releasing hormones are neurons, which secrete the releasing hormones into the hypothalamopituitary portal vascular system in response to action potentials generated in them. This electrical activity in the neurons that secrete GnRH may be in part spontaneous, but it also reflects both synaptic input from other brain areas (these neural inputs are not indicated in Fig. 27-12) and inhibitory influences of testicular hormones that reach the hypothalamus via the blood.

The precise nature of this latter negative-feedback inhibition also remains unsettled. The observed fact is that testosterone, at physiologic levels, inhibits the secretion of LH but has little effect on FSH secretion. This is not what would be expected were testosterone to exert its major inhibitory effect on the hypothalamic secretion of GnRH since this latter hormone influences the secretion of both FSH and LH. Therefore, it is likely that testosterone's major negative-feedback effect is directly on the anterior pituitary rather than on the hypothalamus; i.e., testosterone directly inhibits the pituitary's LH-releasing mechanism but has little or no effect on FSH.

tion on the seminiferous tubules. However, this effect is not exerted directly on the spermatogenic cells but via FSH stimulation of Sertoli cells, which as we have seen, are in intimate contact with the spermatogenic cells at all stages of development. LH stimulates testosterone secretion by a direct action on the interstitial cells, and because testosterone is required for spermatogenesis, LH is indirectly involved in this latter process as well (Fig. 27-12).

Prolactin, a hormone released from the anterior pituitary, significantly potentiates both the effect of LH on the secretion of testosterone and the effect of testosterone on the reproductive tract and other androgen sensitive tissues.

How then do the testes inhibit FSH secretion, if not via testosterone? It is most likely that the inhibitory chemical signal to the hypothalamus and anterior pituitary is a protein hormone termed *inhibin,* which is released into the blood from the seminiferous tubules themselves, probably from the Sertoli cells. That the Sertoli cells are the source of the negative-feedback inhibition of FSH secretion makes sense, since, as described above, the stimulatory effect of FSH on spermatogenesis is exerted via the Sertoli cells; thus, the Sertoli cells are, in two ways, the link between FSH and spermatogenesis. The solution to this problem of how the seminiferous tubules exert an inhibitory influence on FSH secretion is not merely of academic interest, since an agent that inhibits FSH only and not LH would constitute an ideal male contraceptive; spermatogenesis, but not testosterone secretion, would be eliminated.

Despite all this complexity, one should not lose sight of the fact that the secretion of LH and FSH in the male normally proceeds at a rather fixed, continuous rate during adult life; accordingly, sper-matogenesis and testosterone secretion also occur at relatively unchanging rates. This is completely different from the large cyclical swings of activity so characteristic of the female reproductive hormones. A word of caution may be in order here, however: Recent work in nonprimate mammals has indicated that testosterone levels can be made to vary in response to various sexual and social stimuli; the relevance of these studies for men is unknown.

SECTION B: THE FEMALE REPRODUCTIVE SYSTEM

ANATOMY OF THE FEMALE REPRODUCTIVE TRACT

Internal Genitalia

The female reproductive system's primary organs—the ovaries and duct system (uterine tubes, uterus, and vagina)—are below the peritoneum in the lower

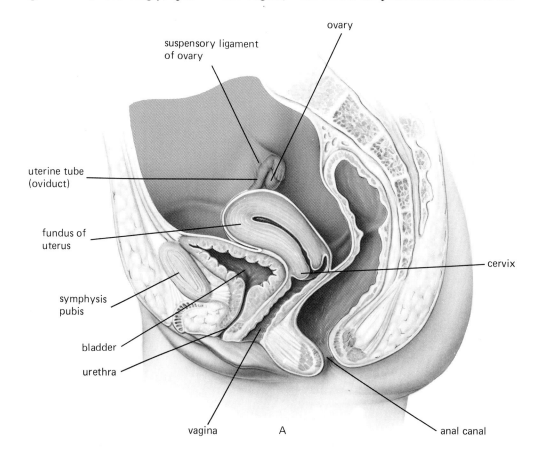

suspensory ligament of ovary

ovary

uterine tube (oviduct)

fundus of uterus

symphysis pubis

bladder

urethra

cervix

vagina   A   anal canal

pelvis and constitute the internal genitalia. As shown in Fig. 27-13A, the urinary bladder lies immediately behind the symphysis pubis (part of the bony pelvic girdle), and the vagina is immediately behind the bladder; the uterus, turned forward, lies above the vagina and bladder. In the female, the urinary and reproductive duct systems are entirely separate.

Each **ovary** lies in an *ovarian fossa,* a depression on the lateral pelvic wall. The **uterine tubes (oviducts, Fallopian tubes)** are not directly connected to the ovaries but open into the peritoneal cavity close

to them. This opening of each uterine tube is a trumpet-shaped expansion surrounded by long fingerlike projections (the **fimbriae**) that are lined with ciliated epithelium.

The other ends of the uterine tubes empty directly into the cavity of the **uterus** (womb), which is a hollow thick-walled muscular organ lying in the pelvis between the bladder and the rectum (Fig. 27-13A). The uterus can be divided approximately in its middle into the **corpus,** or **body,** above and the **cervix** below. The cervix terminates at the small opening into the vagina. The portion of the corpus

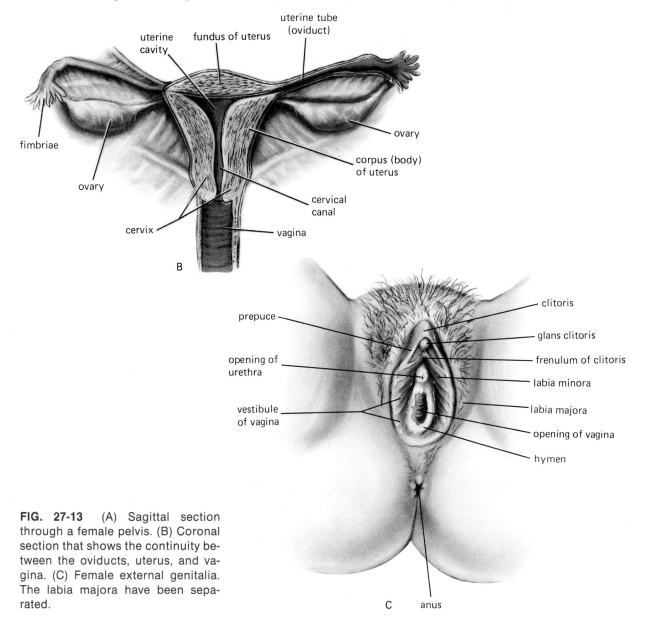

FIG. 27-13 (A) Sagittal section through a female pelvis. (B) Coronal section that shows the continuity between the oviducts, uterus, and vagina. (C) Female external genitalia. The labia majora have been separated.

above the entrance of the uterine tubes is the **fundus** (Fig. 27-13B). The wall of the uterus has three layers: an external serous peritoneal layer (*perimetrium*), a middle smooth muscle layer (*myometrium*), and an inner mucous membrane (*endometrium*).

The **vagina** is a fibromuscular tube that extends from the uterine cervix to the vestibule and is lined with stratified epithelium. Its walls are much thinner than those of the uterus and consist of an internal mucous membrane and a muscular coat.

These internal genitalia are supported by ligaments covered by folds of peritoneum. The ovaries are supported by the *suspensory ligaments* (Fig. 27-13A), which contain the ovarian nerves and vessels, whereas two peritoneal folds, the *broad ligaments*, connect the uterus to the lateral walls of the pelvis. The uterine tubes lie on the upper border of the broad ligaments, i.e., on the side facing the peritoneal cavity, and the ovaries are attached to the posterior surfaces. In other words, the broad ligaments act as mesenteries of the uterus and uterine tubes, and they contain the blood vessels and proper ligaments, such as the *ligamentum teres* (round ligament), of the uterus as well as the *ligament of the ovary*.

External Genitalia

The external genitalia include the mons pubis, labia majora and minora, the clitoris, the vestibule of the vagina, and the greater vestibular glands. The term **vulva** includes all these parts. The **mons pubis** is a rounded elevation of adipose tissue in front of the symphysis pubis. The **labia majora,** the female analogue of the scrotum, are two prominent skin folds that overlie subcutaneous fat (Fig. 27-13C). The **labia minora** are small skin folds that lie between the labia majora; between the two labia minora is the vaginal **vestibule,** which contains anteriorly the opening of the urethra and, behind that, the vaginal opening. This opening is generally partly closed by a thin fold of mucous membrane, the **hymen,** prior to the initial episode of sexual intercourse. A number of glands open into the vestibule, the largest of which are the paired **greater vestibular glands.**

Anteriorly, each labium minorus divides in two; the upper part passes above the clitoris, the lower part below. The **clitoris** is an erectile structure homologous to the penis. Finally, the floor that underlies the pelvic girdle is known as the perineum; it is not part of the external genitalia but is stretched and may be torn or surgically incised (*episiotomy*) during childbirth.

OVARIAN FUNCTION

Unlike the continuous sperm production of the male, the maturation and release of the female germ cell, the **ovum,** are cyclic. This pattern is true not only for ovum development but for the function and structure of virtually the entire female reproductive system. These cycles, which last an average of 28 days, are called **menstrual cycles.**

The ovary, like the testis, serves a dual purpose: (1) production of ova; (2) secretion of the female sex hormones, estrogen and progesterone.

Ovum and Follicle Growth

In Fig. 27-14A, a cross section of an ovary, note the discrete cell clusters known as **primary, or primordial, follicles.** Each is composed of one ovum surrounded by a single layer of **follicular cells.** (As will be described in a subsequent section, the ovum is referred to by different names at each stage of development, but for clarity we shall simply use here the term ovum.) At birth, each ovary contains about 1 million such ova, and no new ones appear after birth. Thus, in marked contrast to the male, the newborn female already has all the germ cells she will ever have. Only a few, perhaps 400, are destined to reach full maturity during her active reproductive life. All the others degenerate over the years, so that none remain by the time she reaches menopause at approximately 50 years of age. One result of this is that the ova that are released (ovulated) near menopause are 30 to 35 years older than those ovulated just after puberty; it has been suggested that certain congenital defects much commoner among children of older women than those of younger women are the result of aging changes in the ovum.

The development of the follicle is characterized by an increase in size of the ovum and a proliferation of the surrounding follicular cells (Fig. 27-14B, C, and D). (Follicles with more than one layer of cells in their wall are **secondary follicles.**) The ovum becomes surrounded by a thick membrane, the *zona pellucida;* however, despite the presence of

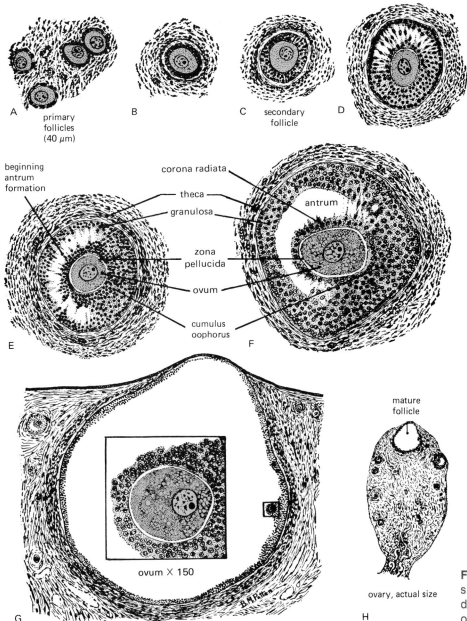

primary
follicles
(40 μm)

secondary
follicle

beginning
antrum
formation

corona radiata

theca

granulosa

antrum

zona
pellucida

ovum

cumulus
oophorus

ovum × 150

mature follicle (1.5 cm)

mature
follicle

ovary, actual size

FIG. 27-14 Drawings that show a series of stages in development of the human ovum and ovarian follicle.

the zona pellucida, the inner layer of follicular cells (the so-called *corona radiata*) remains intimately associated with the ovum by means of cytoplasmic processes that traverse the zona pellucida and form gap junctions with the ovum. It is uncertain whether estrogen or other chemical messengers cross these gap junctions (which become uncoupled at the time of ovulation), but it may be that the follicular cells play a role in mediating the actions of gonadotropic hormones on the ovum analogous to the role of the Sertoli cells in the male.

As the follicle grows, new cell layers are formed, not only from mitosis of the original follicle cells but from the growth of specialized ovarian connective-tissue cells. Thus, the follicle, originally composed only of the ovum and its surrounding layers of follicular cells, becomes invested with additional outer layers of cells known as the **theca** (Fig. 27-

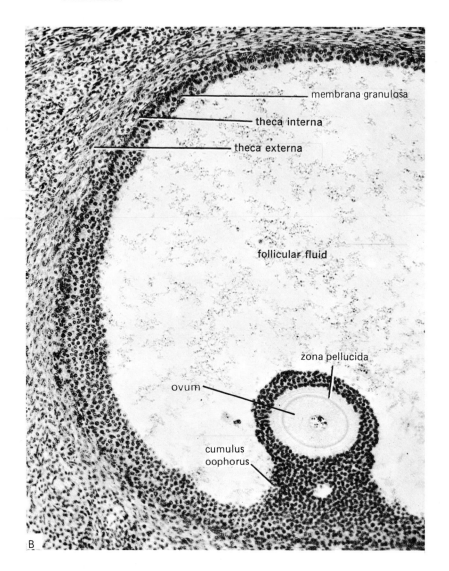

FIG. 27-15 Section through an ovarian follicle.

14E). When the follicle reaches a certain diameter, a fluid-filled space, the **antrum,** begins to form in the midst of the follicular cells as a result of fluid they secrete (Fig. 27-14E and F).

By the time the antrum begins to form, the ovum has reached full size. From this point on, the follicle grows in part because of continued follicular-cell proliferation but largely because of the expanding antrum. Ultimately, the ovum, surrounded by the zona pellucida and several layers of follicular cells (the *cumulus oophorus*), occupies a little hill that projects into the antrum (Fig. 27-14G).

As the follicle develops, the theca, which surrounds it, differentiates into two layers, a *theca interna* and *externa*. Lining the antrum are the follicular cells that were separated from the ovum by formation of the antrum; these cells form the *membrana granulosa* (Fig. 27-15). Eventually, the antrum becomes so large (about 1.5 cm) that the completely mature follicle actually balloons out on the surface of the ovary. **Ovulation** occurs when the wall at the site of ballooning ruptures and the ovum, surrounded by its tightly adhering zona pellucida and cumulus, is carried out of the ovary by the antral fluid. Many women experience varying degrees of abdominal pain at approximately the mid-

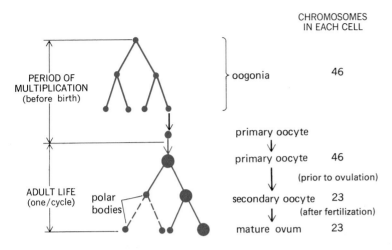

CHROMOSOMES
IN EACH CELL

oogonia   46

primary oocyte
    ↓
primary oocyte   46

    ↓   (prior to ovulation)

secondary oocyte   23

    ↓   (after fertilization)

mature ovum   23

FIG. 27-16 Summary of ovum development. Compare with the male pattern of Fig. 27-5. Each primitive ovum produces only one mature ovum, which contains 23 chromosomes.

point of their menstrual cycles, which had generally been presumed to represent abdominal irritation induced by the entry of follicular contents at ovulation. However, precise timing of ovulation has indicated that this time-honored concept may be wrong, and the cause of discomfort remains unclear.

In the adult ovary, there are always several antrum-containing follicles of varying sizes, but during each cycle, normally only one follicle reaches the complete maturity just described; the process requires approximately 2 weeks. All the other partially matured antral follicles undergo degeneration at some stage in their growth but the mechanism is unknown. On occasion (1 to 2 percent of all cycles), two or more follicles reach maturity, and more than one ovum may be ovulated. This is the commonest cause of multiple births; in such cases the siblings are fraternal, i.e., not identical.

Oogenesis The ova present at birth (**primary oocytes**) are the result of numerous mitotic divisions of the primitive ova, the **oogonia** (a term analogous to spermatogonia in the male), which occurred during early fetal development (Fig. 27-16). At some point the oogonia cease dividing in the fetus and then differentiate into **primary oocytes** (analogous to primary spermatocytes). These primary oocytes all begin the first division of meiosis but do not complete it; accordingly, all the ova present at birth are primary oocytes that contain 46 replicated DNA strands and are in a state of meiotic arrest. This division is completed only just before an ovum is

about to be ovulated and is analogous to the division of the primary spermatocyte because each daughter cell receives 23 replicated chromosomes. However, in this division one of the two resulting cells, the **secondary oocyte,** retains virtually all the cytoplasm; the other, the **polar body,** is very small. In this manner, the already full-size ovum loses half of its chromosomes but almost none of its nutrient-rich cytoplasm. The second cell division of meiosis occurs in the uterine tube after ovulation (indeed, only after penetration by a sperm) and the daughter cells each retain 23 chromosomes. Once again, one daughter cell, now the fertilized ovum, retains nearly all the cytoplasm. The net result is that each primary oocyte is capable of producing only one mature fertilizable ovum (Fig. 27-16); in contrast, each primary spermatocyte produces four viable spermatozoa.

Formation of Corpus Luteum

After rupture of the follicle and discharge of the antral fluid and the ovum, a transformation occurs in the follicle, which collapses, and the antrum fills with partially clotted fluids, ultimately to be replaced with connective tissue. The granulosa and theca interna cells enlarge greatly; the entire gland-like structure is known as the **corpus luteum** (Fig. 27-17). If the discharged ovum is not fertilized, i.e., if pregnancy does not occur, the corpus luteum reaches its maximum development within approximately 10 days and then rapidly degenerates. If pregnancy does occur, the corpus luteum grows and persists until near the end of pregnancy.

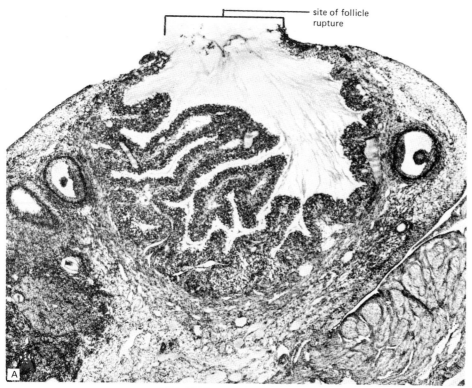

site of follicle
rupture

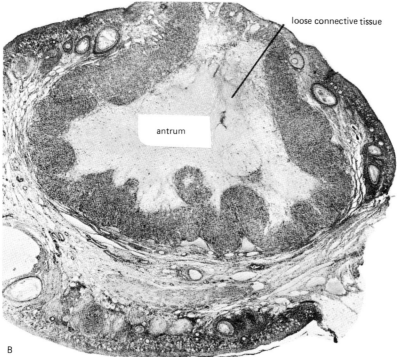

loose connective tissue

antrum

FIG. 27-17 Photomicrograph of two stages in the development of the corpus luteum of the cat. (A) The recent site of rupture as well as the remarkable foldings of the stratum granulosum. (B) Several days later the luteal cells appear glandular and the antrum is being invaded by a loose connective tissue.

Ovarian Hormones

Just as *androgen* refers to a group of hormones with similar actions, the term *estrogen* denotes not a single specific hormone but rather a group of steroid hormones that have similar effects on the female reproductive tract. These include *estradiol, estrone,* and *estriol,* of which the first is the major estrogen secreted by the ovaries. It is common to refer to any of them simply as **estrogen,** and we shall follow this practice. Estrogen is secreted mainly by the granulosa cells (not the ovum) and corpus luteum. **Progesterone,** the second ovarian hormone, may be secreted in very minute amounts by the granulosa cells, but its major source is the corpus luteum. The detailed physiology of these hormones will be described subsequently.

Cyclic Nature of Ovarian Function

The length of a menstrual cycle varies considerably from woman to woman but averages about 28 days. Day 1 is the first day of menstrual bleeding, and in a typical 28-day cycle, ovulation occurs around day 14. In terms of ovarian function, therefore, the menstrual cycle may be divided into two approximately equal phases: (1) the **follicular phase,** during which a single follicle and oocyte develop to full maturity; (2) the **luteal phase,** during which the corpus luteum is the active ovarian structure. It must be stressed that the day of ovulation varies from woman to woman and frequently in the same woman from month to month.

CONTROL OF OVARIAN FUNCTION

The basic factors that control ovum development, ovulation, and formation of the corpus luteum are analogous to the controls described for testicular function in that the anterior pituitary gonadotropins, FSH and LH, and the gonadal sex hormone, estrogen, play primary roles. However, the overall schema is more complex in the female since it includes a second important gonadal hormone (progesterone) and a hormonal cycling, quite different from the more stable continuous rates of hormone secretion in the male.

For purposes of orientation, let us look first at the changes in the blood concentrations of all four participating hormones during a normal menstrual cycle (Fig. 27-18). Note that FSH is slightly elevated in the early part of the follicular phase of the menstrual cycle and then steadily decreases throughout the remainder of the period except for a transient midcycle peak. LH is quite constant during most of the follicular phase but then shows a very large midcycle surge (peaking approximately 24 hr before ovulation) followed by a progressive slow decline during the luteal phase.

The estrogen pattern is more complex. After remaining fairly low and stable for the first week (as the follicle develops), it rises to reach a peak just before LH starts off on its surge. This peak is followed by a dip, a second rise (due to secretion by the corpus luteum), and finally, a rapid decline during the last days of the cycle. The progesterone pattern is simplest of all; virtually no progesterone is secreted by the ovaries during the follicular phase, but very soon after ovulation, the developing corpus luteum begins to secrete progesterone, and from this point the progesterone pattern is similar to that for estrogen. It is hoped that, after the following discussion, the reader will understand how these changes are all interrelated to yield a self-cycling pattern.

Control of Follicle and Ovum Development

Growth and development of the follicles depend upon follicle-stimulating hormone (FSH) and luteinizing hormone (LH)—and on estrogen, which may act, in large part, locally within the ovary. Because estrogen is secreted mainly by cells of the follicle, its secretion rate progressively increases as the follicle enlarges. This secretion requires stimulation by FSH and LH in a fairly complex way. LH acts upon the cells of the theca interna to stimulate the synthesis of androgens that diffuse across the basement membrane that separates the thecal and granulosa layers and enter the granulosa cells; there they are converted into estrogen by enzymes whose activity is under the control of FSH. Thus, both FSH and LH are directly required for normal follicle and ovum development as well as for estrogen secretion (Fig. 27-19).

As in the male, in the female the secretion of these pituitary hormones, in turn, requires stimulation by hypothalamic GnRH. The rate at which this releasing hormone is secreted by specific hypothalamic neurons during the first part of the cycle

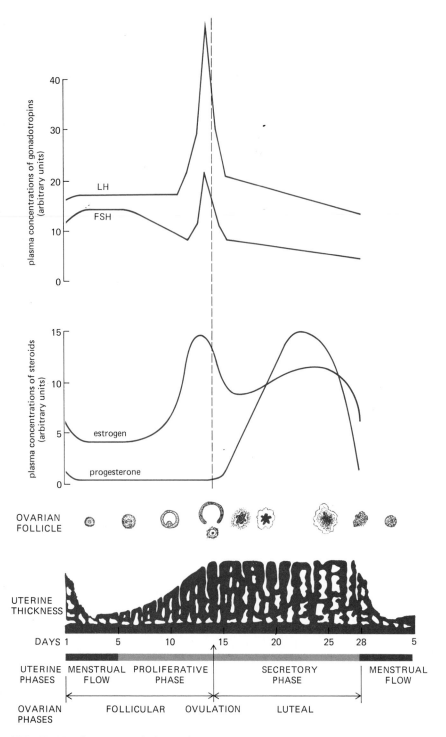

FIG. 27-18 Summary of plasma hormone concentrations, ovarian events, and uterine changes during the menstrual cycle.

↓ ↓ in FSH & LH

Uterus

E_2 — ↑P_4 Recept.

↑ Secretory endometria

↓ Myometrium activity

Cervix

Firm neck

thick + acidic mucus

Vagina

Block estrogen action

Breast

Lobular - alveolar tissue

↑ in growth

Body Temp
↑ 1.5 °F at ovulation

Kidneys

Block aldosterone
action ~ ↓ Na from
body 7 days - Ovulation

---

3 Role the —. Human invertebrate
: 1 : 1 : as   . Gonadotropin (HCG)
    — Placenta lactogen (HPL)
    — Placental thyrotropin (HPT)

2 Steroids — Estrogen (Estriol & Estradiol)
                Progesterone

Provides androgen

Placenta produces
Estradiol + Estriol
↑ progesterone recept.
↑ growth of uterus
     ↑ musle layer

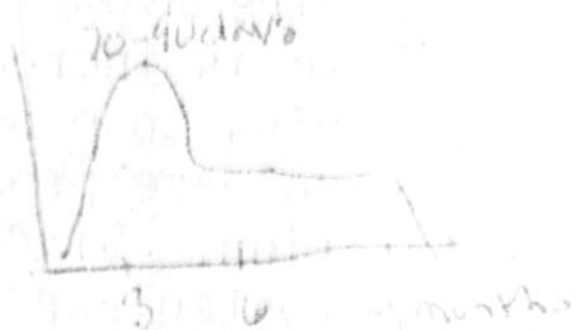

Progesterone - need to prevent abortion

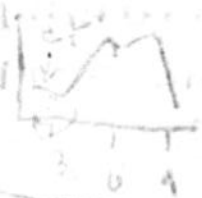

#CG
10 divisions
3   6   9 months

Synctytrophoblast ... cells — produces
with the placenta + serum (hormones)

#HPL
3   6   9 months
actually fetal wk.

Action
Body - allows corpus luteum to
       cont. to secrete prog est.
       ↑ maternal glucose
       ↑↑ O2 to fetus
wks - stimulates growth
       secrete hormones

If no stimulation of secretion
the fetus will be formed.

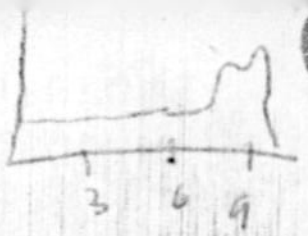

#HPT

1 3 6 9

Action
↑ breast growth
↑ a.a. from to cross
   placenta for fetus
↑ glucose transport
   across placenta

#HPT

1 3 6 9

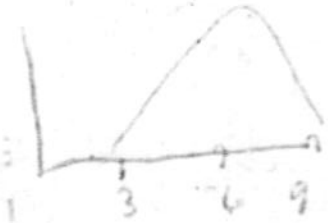

action
↑ thyroid acti.

1. produce due
2. ↓ vagine con...
3. mm nat logic

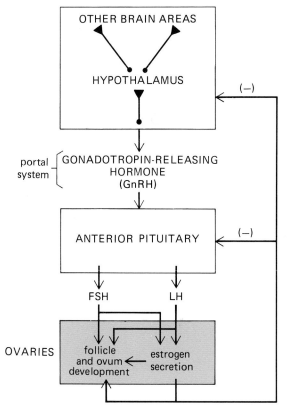

FIG. 27-19 Summary of hormonal control of follicle and ovum development and estrogen secretion during the follicular phase of the menstrual cycle. Compare with the analogous pattern for the male (Fig. 27-12). The negative signs indicate that estrogen inhibits both the hypothalamus and the anterior pituitary. Estrogen reaches the developing ovum and follicle both by local diffusion and by release into the blood and recirculation to the ovaries.

is probably determined by a combination of spontaneous neuronal automaticity, synaptic input from other brain areas, and negative-feedback inhibition by estrogen that reaches the brain via the blood. This latter input is responsible for the decline in FSH secretion seen during the second week of the cycle (Fig. 27-18).

Control of Ovulation

If one administers small quantities of FSH and LH each day to a woman whose pituitary has been removed (because of disease), she manifests normal follicle development, ovum maturation, and estrogen secretion, but she does not ovulate. If, on the other hand, after approximately 14 days of this ther-

apy, she is given one or two larger injections of LH, ovulation occurs. This is precisely what happens in the normal woman; the follicle and ovum mature for 2 weeks under the influence of FSH, LH, and estrogen, and ovulation is triggered by a rapid brief outpouring from the pituitary of larger quantities of LH (Fig. 27-18). This LH then causes ovulation by inducing increased synthesis of enzymes that catalyze the dissolution of the thin ovarian membrane at the bulge of the mature follicle.

Thus, the midcycle surge of LH emerges as perhaps the single most decisive event of the entire menstrual cycle; indeed, it is the presence of this surge that most distinguishes the female pattern of pituitary secretion from the relatively unchanging pattern of the male. What causes the LH surge?

In the previous section we described how estrogen exerts a negative-feedback inhibition on the hypothalamus and pituitary; the fact is that this inhibitory effect occurs only when blood estrogen concentration is relatively low, as during the first part of the follicular phase. In contrast, high blood

FIG. 27-20 Ovulation and corpus luteum formation are induced by the markedly increased LH, itself induced by the stimulatory effects of high levels of estrogen.

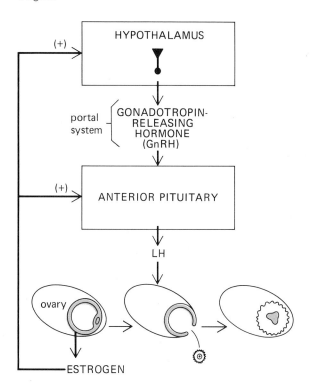

concentrations of estrogen, as exist during the estrogen peak of the late follicular phase, stimulate the hypothalamus to secrete GnRH and also act directly upon the pituitary to enhance the sensitivity of LH-releasing mechanisms to this GnRH (Fig. 27-20); these stimulatory effects are often called "positive-feedback" effects. The net result is that, as estrogen secretion rises rapidly during the last half of the follicular phase, its blood concentration eventually becomes high enough to cause an outpouring of LH, which, in turn, induces ovulation.

Why is there only one surge per cycle; i.e., why does the elevated estrogen that exists throughout most of the luteal phase not keep inducing LH surges? The answer is that, as a result of ovulation and succeeding corpus luteum formation, both induced by the LH surge, progesterone secretion begins and the progesterone-estrogen combination (unlike estrogen alone) exerts only inhibitory effects on the hypothalamopituitary mechanisms for LH release.

Control of the Corpus Luteum

The LH surge not only induces ovulation but also transforms the remaining follicular cells into the corpus luteum. Once the corpus luteum has been formed, its maintenance requires some stimulatory support from LH, but the amount of LH needed is quite small. What causes the corpus luteum to degenerate if no pregnancy results? The blood concentration of LH shows no sudden decrease during the late luteal phase, so it is difficult to blame regression on any sudden withdrawal of LH. In some mammalian species, it seems that regression is actively induced by some substance (perhaps prostaglandin) produced by the nonpregnant uterus, but such seems not to be the case in women. There are other hypotheses, including the idea that the corpus luteum has a "built-in" life-span of approximately 10 to 14 days and that it "self-destructs" unless prevented from doing so by the onset of pregnancy. How pregnancy does this will be described subsequently.

During its short life, the corpus luteum secretes large quantities of estrogen and progesterone. These hormones, particularly estrogen, exert a powerful negative-feedback inhibition (via the hypothalamus and anterior pituitary) of tonic FSH and LH secretion. Accordingly, during the luteal phase of the cycle, pituitary gonadotropin secretion is reduced, which explains the diminished rate of follicular maturation during this second half of the cycle. With degeneration of the corpus luteum, blood estrogen and progesterone concentrations decrease, FSH and LH increase, and a new follicle is stimulated to mature.

Summary of Changes during the Menstrual Cycle

The events described thus far in this section are summarized in Fig. 27-18, which shows the ovarian and hormonal changes during a normal menstrual cycle.

1 Under the influence of FSH, LH, and estrogen, a single follicle and ovum reach maturity at about 2 weeks.
2 During the second week, under the influence of LH and FSH, estrogen secretion by the follicle progressively increases.
3 For several days near midperiod, production of LH (and FSH, to a lesser extent) increases sharply as a result of the stimulatory effects of high levels of estrogen on the brain and the pituitary.
4 The high concentration of LH induces rupture of the mature follicle, and ovulation occurs (it is not known what role if any the increased FSH plays).
5 The ruptured follicle is rapidly transformed into the corpus luteum, which secretes large quantities of both estrogen and progesterone.
6 The high blood concentrations of these hormones inhibits the release of LH and FSH, which thereby lowers their blood concentrations and prevents the development of a new follicle or ovum during the last 2 weeks of the period; in addition, they prevent any additional LH surge.
7 Failure of ovum fertilization is associated with the degeneration of the corpus luteum during the last days of the cycle.
8 The disintegrating corpus luteum is unable to maintain its secretion of estrogen and progesterone, and their blood concentrations drop rapidly.
9 The marked decrease of estrogen and progesterone removes the inhibition of FSH (and LH) secretion.

10   The blood concentration of FSH (and LH) begins to rise, follicle and ovum development are stimulated, and the cycle begins anew.

UTERINE CHANGES IN THE MENSTRUAL CYCLE

Profound changes in uterine morphology occur during the menstrual cycle and are completely attributable to the effects of estrogen and progesterone. Estrogen stimulates growth of the uterine smooth muscle (myometrium) and the endometrium that lines its cavity, and, in addition, induces the synthesis of receptors for progesterone. The **endometrium** consists of a layer of epithelium and an underlying connective-tissue layer, the endometrial *stroma,* which is filled with simple tubular glands that open onto the epithelial surface. The stroma itself consists of two layers, a superficial one whose character changes greatly with the menstrual cycle and, indeed, is almost completely shed at menstruation, and a thin, deep layer whose character and thickness do not change (Fig. 27-21). It is this deeper layer that regenerates another superficial layer after the menstrual flow has stopped. Progesterone acts upon this estrogen-primed endometrium to convert it to an actively secreting tissue: The glands become coiled and filled with secreted glycogen; the blood vessels become spiral and more numerous; various enzymes accumulate in the glands and connective tissue of the lining (Fig. 27-21B). All these changes are ideally suited to provide a hospitable environment for implantation of a fertilized ovum.

Estrogen and progesterone also have important effects on the mucus secreted by the cervix. Under the influence of estrogen alone, this mucus is abundant, clear, and nonviscous; all these characteristics are most pronounced at the time of ovulation and facilitate penetration by sperm. In contrast, progesterone causes the cervical mucus to become thick and sticky, in essence a "plug" that may constitute an important blockade against the entry of bacteria from the vagina—a further protection for the fetus should conception occur.

The endometrial changes throughout the normal menstrual cycle should now be readily understandable (Fig. 27-18).

1   The fall in blood progesterone and estrogen, which results from regression of the corpus luteum, deprives the highly developed endometrial lining of its hormonal support; the immediate result is profound constriction of the uterine blood vessels (Fig. 27-22), which leads to diminished supply of oxygen and nutrients. Disintegration starts, the entire lining begins to slough, and the **menstrual flow** begins, marking the first day of the cycle.

2   After the initial period of vascular constriction, the endometrial arterioles dilate, which results in hemorrhage through the weakened capillary walls; the menstrual flow consists of this blood mixed with endometrial debris. (Average blood loss per period equals 50 to 150 mL.)

3   The **menstrual phase** continues for 3 to 5 days, during which time blood estrogen levels are low.

4   The menstrual flow ceases as the endometrium repairs itself and then grows under the influence of the rising blood estrogen concentration; this period, the **proliferative phase,** lasts for the 10 days between cessation of menstruation and ovulation.

5   Following ovulation and formation of the corpus luteum, progesterone, acting in concert with estrogen, induces the secretory type of endometrium described above.

6   This period, the **secretory phase,** is terminated by disintegration of the corpus luteum, and the cycle is completed.

It is evident that the phases of the menstrual cycle can be named either in terms of the ovarian or uterine events. Thus, the ovarian follicular phase includes the uterine menstrual and proliferative phases; the ovarian luteal phase is the same as the uterine secretory phase. The essential point is that the uterine changes simply reflect the effects of varying blood concentrations of estrogen and progesterone throughout the cycle. In turn, the secretory pattern of these hormones reflects the complex hypothalamic–anterior pituitary–ovarian interactions described previously.

NONUTERINE EFFECTS OF ESTROGEN AND PROGESTERONE

The uterine effects of the sex hormones described above represent only one set of a wide variety of effects exerted by estrogen and progesterone (Table

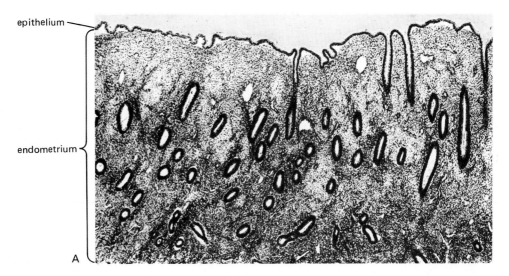

epithelium

endometrium

A

FIG. 27-21 (A) Estrogen-primed endometrium (follicular phase). (B) Progesterone-primed endometrium (luteal phase). Parts A and B (at right) are the same magnification.

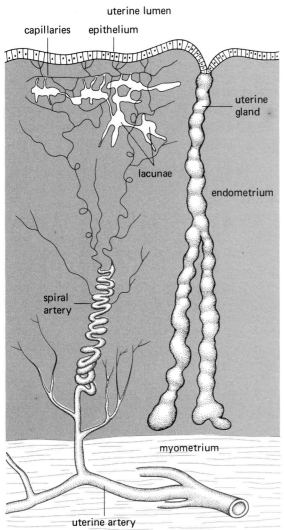

FIG. 27-22 (at left) Diagram of the endometrium. The dominant structures are the uterine glands and the spiral arteries.

27-1). The effects of estrogen in the female are analogous to those of testosterone in the male in that estrogen exerts dominant control over all the accessory sex organs and secondary sex characteristics. Estrogenic stimulation maintains the entire female genital tract—uterus, oviducts, vagina—the glands that line the tract, the external genitalia, and the breasts. It is responsible for the female body-hair distribution and the general female body configuration: narrow shoulders, broad hips, and the characteristic female "curves," the result of fat deposition in the hips, abdomen, and other places.

Estrogen has much less general anabolic effect on nonreproductive tissues than testosterone but probably contributes to the general body growth spurt at puberty. Finally, as described above, estrogen is required for follicle and ovum maturation; its increased secretion at puberty, in concert with that of the pituitary gonadotropins, permits ovulation and the onset of menstrual cycles. For the rest of the woman's reproductive life, estrogen continues to support the ovaries, accessory organs, and secondary sex characteristics. Because its blood concentration varies so markedly throughout the cycle, associated changes in all these dependent functions

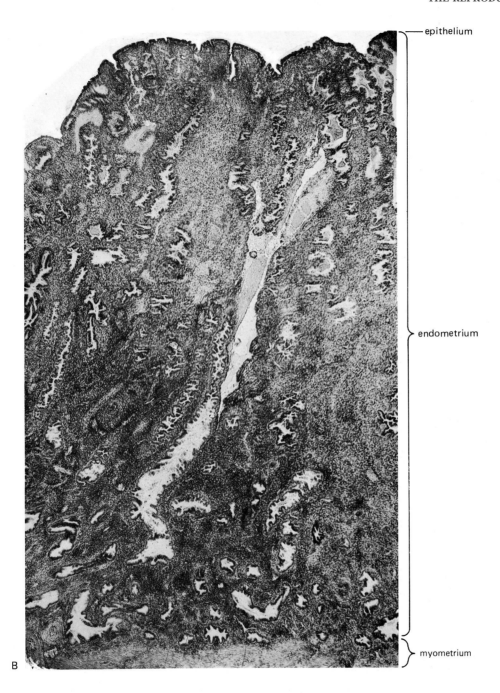

epithelium

endometrium

myometrium

B

occur; uterine manifestations are the most striking.

As was true for testosterone, estrogen acts on the cell nucleus, and its biochemical mechanism of action appears to be at the level of the genes themselves. It should be reemphasized that estrogen is not a uniquely female hormone. Small and usually insignificant quantities of estrogen are secreted by the male adrenal and the testicular interstitial cells, and the latter probably are responsible for the breast enlargement so commonly observed in pubescent boys; apparently, the rapidly developing interstitial cells release significant quantities of estrogen along with the much larger amounts of testosterone.

TABLE 27-1   EFFECTS OF FEMALE SEX STEROIDS

Effects of Estrogens:

1 Growth of ovaries and follicles
2 Growth and maintenance of the smooth muscle and epithelial linings of the entire reproductive tract. Also:

Oviducts:	Increased motility and ciliary activity
Uterus:	Increased motility
	Secretion of abundant, clear cervical mucus
Vagina:	Increased "cornification" (layering of epithelial cells)

3 Growth of external genitalia
4 Growth of breasts (particularly ducts)
5 Development of female body configuration; narrow shoulders, broad hips, converging thighs, diverging arms
6 Stimulation of fluid sebaceous gland secretions ("antiacne")
7 Pattern of pubic hair (actual growth of pubic and axillary hair is androgen-stimulated)
8 Stimulation of protein anabolism and closure of the epiphyses (? due to stimulation of adrenal androgens)
9 ?Sex drive and behavior
10 Reduction of blood cholesterol
11 Vascular effects (deficiency → "hot flashes")
12 Feedback effects on hypothalamus and anterior pituitary
13 Fluid retention

Effects of Progesterone:

1 Stimulation of secretion by endometrium; also induces thick, sticky cervical secretions
2 Stimulation of growth of myometrium (in pregnancy)
3 Decrease in motility of oviducts and uterus
4 Decrease in vaginal "cornification"
5 Stimulation of breast growth (particularly glandular tissue)
6 Inhibition of effects of prolactin on the breasts
7 Feedback effects on hypothalamus and anterior pituitary

Progesterone is present in significant amounts only during the luteal phase of the menstrual cycle, and its effects are less widespread than those of estrogen; the endometrial changes are the most prominent. Progesterone also exerts important effects on the breasts, the oviducts, and the uterine smooth muscle, the significance of which will be described later. Progesterone also causes a transformation of the cells that line the vagina (decreased cornification), and the microscopic examination of some of these cells provides an indicator that ovulation has or has not occurred. Note that in this regard, and others in Table 27-1, progesterone exerts an "antiestrogen effect" (probably by inhibiting the synthesis of estrogen receptors).

Another indicator that ovulation has occurred is a small rise (approximately 0.5°C) in body temperature that usually occurs at this time and persists throughout the luteal phase. The cause of this change is not known.

FEMALE SEXUAL RESPONSE

The female response to sexual intercourse is very similar to that of the male in that it is characterized by marked vasocongestion and muscular contraction in many areas of the body. For example, increasing sexual excitement is associated with engorgement of the breasts and erection of the nipples as a result of contraction of muscle filaments in them. The clitoris (Fig. 27-13), which is a homologue of the penis and is composed primarily of erectile tissue and endowed with a rich supply of sensory nerve endings, also becomes erect. During intercourse, the vaginal epithelium becomes highly congested and secretes a mucuslike lubricant.

The final stage of female sexual excitement may be the process of orgasm, as in the male; if no orgasm occurs, there is a slow resolution of the physical changes and sexual excitement. The female has no counterpart to male ejaculation, but with this exception, as mentioned above, the physical correlates of orgasm are very similar in the sexes: the heart rate and blood pressure increase; the female counterpart of male genital contraction is transient rhythmic contraction of the vagina and uterus. Orgasm seems to play no essential role in assuring fertilization.

A final question related to the female sexual response is sex drive. Incongruous as it may seem, sexual desire in women is more dependent upon androgens than estrogen. Thus sex drive is usually not altered by removal of the ovaries (or its physiologic analog, menopause). In contrast, sexual desire is greatly reduced by adrenalectomy, since these glands are the major source of androgens in women. Finally, women who receive large doses of testosterone (for the treatment of breast cancer) generally report a large increase in sexual desire.

This completes our survey of normal reproductive physiology in the nonpregnant female. In weaving one's way through this maze, it is all too easy to forget the prime function subserved by this entire system, namely, reproduction. Accordingly, we must now return to the mature ovum we left free in the abdominal cavity, find it a mate, and carry it through pregnancy, delivery, and breast feeding.

PREGNANCY

Following ejaculation into the vagina, the sperm live approximately 48 hr; after ovulation, the ovum remains fertile for 10 to 15 hr. The net result is that for pregnancy to occur, sexual intercourse must be performed no more than 48 hr before or 15 hr after ovulation (these are only average figures, and there is probably considerable variation in the survival time of both sperm and ovum). However, even these short time limits are probably too generous since, although fertile, the older ova manifest a variety of malfunctions after fertilization that frequently result in their rapid death.

Ovum Transport

At ovulation, the ovum is extruded from the ovary, and its first mission is to gain entry into the oviduct. The end of the oviduct has long, fingerlike projections lined with ciliated epithelium (Fig. 27-13). At ovulation, the smooth muscle of these projections causes them to pass over the ovary while the cilia beat in waves toward the interior of the duct; these motions sweep in the ovum as it emerges from the ovary and start it on its trip toward the uterus.

Once in the oviduct, the ovum moves rapidly for several minutes, propelled by cilia and by contractions of the duct's smooth muscle coating; the contractions soon diminish, and ovum movement becomes so slow that it takes several days to reach the uterus. Thus, fertilization must occur in the oviduct because of the short life-span of the unfertilized ovum.

Sperm Transport

Transport of the sperm to the site of fertilization in the oviduct is so rapid that the first sperm arrive within 30 min of ejaculation. This is far too rapid to be accounted for by the sperm's own motility; indeed, the movement produced by the sperm's tail is probably essential only for the final stages of approach and penetration of the ovum. The act of intercourse itself provides some impetus for transport out of the vagina into the uterus because of the fluid pressure of the ejaculate and the pumping action of the penis during orgasm. After intercourse, the primary transport mechanism may be the contraction of the uterine and oviduct musculature. The factors that control these muscular contractions remain obscure but may involve prostaglandins in the semen. The mortality rate of sperm during the trip is huge; of the several hundred million deposited in the vagina, only a few thousand reach the oviduct. This is one of the major reasons that there must be so many sperm in the ejaculate to permit pregnancy.

In addition to aiding transport of sperm, the female reproductive tract exerts a second critical effect on them, namely, the conferring upon them of the capacity for fertilizing the egg. Although, as we have mentioned, sperm gain some degree of maturity during their stay in the epididymis, they are still not able to penetrate the zona pellucida that surrounds the ovum until they have resided in the female tract for some period of time. The mechanism by which this process, known as **capacitation,** occurs is still very poorly understood but the result is a change in the sperm membrane that permits release of the acrosomal enzymes upon contact.

Entry of the Sperm into the Ovum

The sperm makes initial contact with the cells that surround the ovum, presumably by random motion (Fig. 27-23), because there is no good evidence of the existence of "attracting" chemicals. Having made contact, the sperm rapidly moves between these adhering cells of the cumulus and through the zona pellucida by releasing from its acrosomal cap enzymes that break down cell connections and intermolecular bonds.

Once through the zona pellucida, the sperm makes contact with the ovum plasma membrane, fuses with it, and slowly passes through into the cytoplasm (frequently losing its tail in the process). Upon penetration by the sperm, the ovum completes its second meiotic division, and the one daughter cell with practically no cytoplasm—the second polar body—is extruded. The nuclei of the

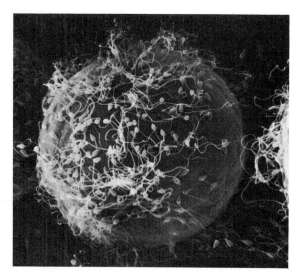

FIG. 27-23 Thousands of sperm bind to the surface of this sea-urchin egg but, as with human beings, only one sperm will fertilize the egg.

sperm and ovum then unite; the cell now contains 46 chromosomes, and fertilization is complete. However, viability depends upon stopping the entry of additional sperm. The mechanism of this "block to polyspermy" is as follows. The fusion of the sperm and ovum plasma membrane causes secretory vesicles located around the ovum's periphery to release their contents into the space between the plasma membrane and zona pellucida. Some of these molecules are enzymes that break down binding sites for sperm in the zona pellucida and thereby prevent additional sperm from moving through the zona; others may cause the zona or plasma membrane of the ovum, itself, to become impenetrable. The immediate trigger for the release of these vesicles is a transient increase in cytoplasmic calcium that results from membrane changes induced by fusion of sperm and ovum; i.e., calcium is acting as a "second messenger" in this system (the sperm being the "first messenger"). This calcium very likely also triggers activation of ovum enzymes required for the ensuing cell divisions and embryogenesis.

The fertilized egg is now ready to begin its development as it continues its passage down the oviduct to the uterus. If fertilization had not occurred, the ovum would slowly disintegrate and usually be phagocytized by the lining of the uterus. Rarely, a fertilized ovum remains in the oviduct,

where implantation may take place. Such tubal pregnancies cannot succeed because of lack of space for the fetus to grow, and surgery may be necessary. Even more rarely a fertilized ovum may move in the wrong direction and be expelled into the abdominal cavity where implantation may take place and (rarely) proceed to term, the infant being delivered surgically.

Early Development, Implantation, and Placentation

During the 3- to 4-day passage through the oviduct, the fertilized ovum undergoes a number of cell divisions (identical twins result when a single fertilized ovum at a very early stage of development becomes completely divided into two independently growing cell masses). After reaching the uterus, it floats free in the intrauterine fluid (from which it receives nutrients) for several more days, all the while undergoing cell division. This entire time span corresponds to days 14 to 21 of the typical menstrual cycle. Thus, while the ovum is undergoing fertilization and early development, the uterine lining is simultaneously being prepared by estrogen and progesterone to receive it. On approximately the twenty-first day of the cycle, i.e., 7 days after ovulation, **implantation** occurs. The fertilized ovum has by now developed into a ball of cells that surround a recently formed central fluid-filled cavity and is known as a **blastocyst.**

A section of the blastocyst is shown in Fig. 27-24. Note the disappearance of the zona pellucida,

FIG. 27-24 A blastocyst.

inner cell mass or embryoblast

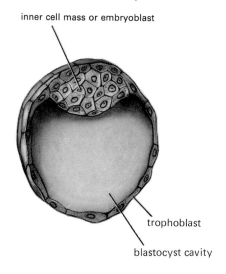

trophoblast

blastocyst cavity

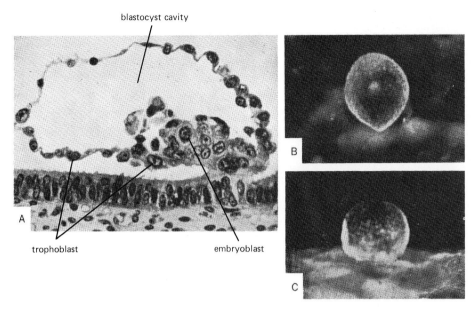

blastocyst cavity

A

trophoblast   embryoblast

B

C

FIG. 27-25 Photomicrographs that show initial implantation of 9-day monkey blastocyst. (A) Cross section of the embryo and uterine lining. (B) An entire blastocyst attached to the uterus, viewed from above. (C) The same embryo viewed from the side.

an event necessary for implantation. The inner cell mass is destined to develop into the fetus itself, whereas the outer lining of cells is already differentiating into the specialized cells, **trophoblasts,** which will form the nutrient membranes for the fetus, as described below. Once the zone pellucida has disintegrated, the trophoblastic layer rapidly enlarges and makes contact with the uterine wall. The trophoblast cells of blastocysts recovered from the uterus have been found to be quite sticky, particularly in the region that overlies the inner cell mass; it is this portion that adheres to the endometrium upon contact and initiates implantation (Fig. 27-25). This initial contact somehow induces rapid development of the trophoblasts, tongues of which penetrate between endometrial cells. By this means, the blastocyst completely embeds within the endometrium by the eleventh day (Fig. 27-26); the nutrient-rich cells of the endometrium will provide the metabolic fuel and raw materials for the developing embryo. This system, however, is adequate to provide for the embryo only during the first few weeks when it is very small. The system that takes over after this is the fetal circulation and placenta (after the end of the second month, the embryo is known as a **fetus**).

The **placenta** is a combination of interlocking fetal and maternal tissues that serves as the organ of exchange between mother and fetus. The expanding trophoblastic layer breaks down endometrial capillaries, which thus allows maternal blood to ooze into spaces called *lacunae,* in the outer layers of the trophoblast (Fig. 27-26); clotting is prevented by the presence of some anticoagulant substance produced by the trophoblasts.

As the lacunae enlarge, cores of trophoblasts, or **villi,** are left extending from the blastocyst, through the lucanae, and into the underlying endometrium. By day 15, the lacunae connect with endometrial arteries, and maternal circulation through the lacunae is fully established. The tips of the extending villi unite to form a shell of trophoblast around the blastocyst, which is the boundary of the embryonic tissue. At this time the various germ layers are developing in the embryo (a more complete discussion of embryonic and fetal development is delayed until later in the chapter), and embryonic blood vessels appear in the cores of the villi and connect, through the developing umbilical vessels, with the embryonic heart. By the third week, fetal blood begins to circulate in the capillaries of the villi; thus, the villi are supplied with fetal and maternal

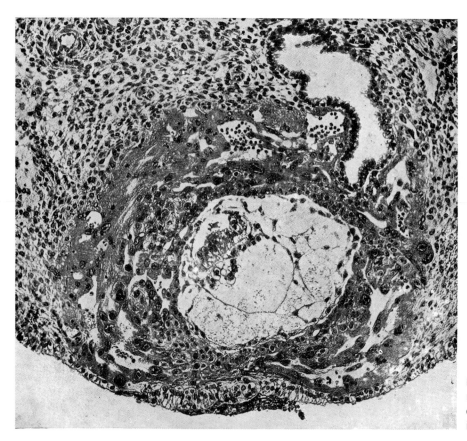

FIG. 27-26 Eleven-day human embryo, completely embedded in the uterine lining.

blood, exchange between the two blood supplies is facilitated, and the entire mechanism for nutrition of the fetus is in operation. Waste products move from the fetal blood across the villi walls of the placenta into the maternal blood; nutrients move in the opposite direction. Many substances, such as oxygen and carbon dioxide, move by simple passive diffusion, whereas other substances are carried by active-transport mechanisms in the placental membranes.

At first, as described above, the trophoblastic projections simply lie in endometrial spaces filled with blood, lymph, and some tissue debris. This basic pattern (Fig. 27-27) is retained throughout pregnancy, but many structural alterations have the net effect of making the system more efficient; e.g., the trophoblastic layer thins, and the distance between maternal and fetal blood is thereby reduced. It must be emphasized that there is exchange of materials between the two blood streams but no actual mingling of the fetal and maternal blood. The maternal blood enters the placenta via the *uterine artery,* percolates through the spongelike endome-

trium, and then exits via the *uterine veins;* similarly, the fetal blood never leaves the fetal vessels. In several ways, the system is analogous to the artificial kidney described in Chap. 24, with the endometrial vascular spaces serving as the bath through which the fetal vessels course like the dialysis tubing. A major difference is that active transport is an important component of placental function. Moreover, the placenta must serve not only as the embryo's kidney but as its gastrointestinal tract and lungs as well. Finally, as will be discussed below, the placenta (probably the trophoblasts) secretes several hormones of crucial importance for maintenance of pregnancy.

The fetus, floating in its completely fluid-filled cavity and attached by the umbilical cord to the placenta, develops into a viable infant during the next 9 months (Fig. 27-28) (babies born premature, often as early as 7 months, frequently survive). A point of great importance is that during these 9 months the developing embryo and, later, fetus is subject to considerable influence by a host of factors (noise, chemicals, etc.) that affect the mother.

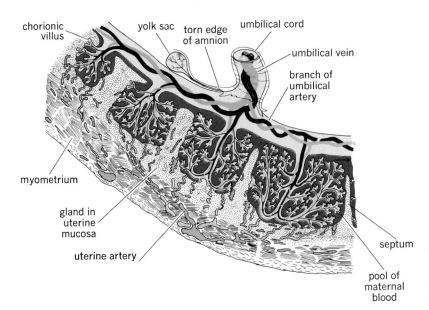

chorionic villus

yolk sac

torn edge of amnion

umbilical cord

umbilical vein

branch of umbilical artery

myometrium

gland in uterine mucosa

uterine artery

septum

pool of maternal blood

FIG. 27-27 Schematic diagram of the interrelations of fetal and maternal tissues in the placenta. The placenta becomes progressively more developed from left to right.

Via the placenta, drugs taken by the mother can reach the fetus and influence its growth and development. The thalidomide disaster was our major reminder of this fact in recent years, as is the growing number of babies who suffer heroin withdrawal symptoms after birth as a result of their mothers' drug use during pregnancy. Two other examples: Lead and DDT cross the placenta very easily. We do not know the potential effects, if any, on the fetus of many agents in the environment.

Hormonal Changes during Pregnancy

Throughout pregnancy, the specialized uterine structures and functions depend upon high concentrations of circulating estrogen and progesterone (Fig. 27-29). During approximately the first 2 months of pregnancy, almost all these steroid hormones are supplied by the extremely active corpus luteum formed after ovulation. Recall that if pregnancy had not occurred, this glandlike structure would have degenerated within 2 weeks after ovulation; in contrast, continued corpus luteum growth and steroid secretion occur when pregnancy does occur. Persistence of the corpus luteum is essential since continued secretion of estrogen and progesterone is required to sustain the uterine lining and prevent menstruation (which does not occur during pregnancy). We have mentioned our lack of knowledge of the factors that cause corpus luteum degen-

eration during a nonpregnant cycle; its persistence during pregnancy is due, at least in part, to a hormone from the blastocyst and placenta called **chorionic gonadotropin** (CG). Almost immediately after beginning their endometrial invasion, the trophoblastic cells start to secrete CG into the maternal blood. This protein hormone has properties very similar to those of LH (although it is chemically different), and it strongly stimulates steroid secretions by the corpus luteum.

Recall that secretion of both LH and FSH is powerfully inhibited by estrogen and progesterone; therefore, the blood concentrations of these pituitary gonadotropins remain extremely low throughout pregnancy. By this means, further follicle development, ovulation, and menstrual cycles are eliminated for the duration of the pregnancy.

The detection of CG in urine or plasma is the basis of most pregnancy tests. The secretion of CG increases rapidly during early pregnancy to reach a peak at 60 to 80 days after the end of the last menstrual period; it then falls just as rapidly, so that by the end of the third month it has reached a low but definitely detectable level that remains relatively constant for the duration of the pregnancy. Associated with this falloff of CG secretion, the placenta itself begins to secrete large quantities of estrogen and progesterone. The very marked increases in blood steroids during the last 6 months of pregnancy are due almost entirely to the placen-

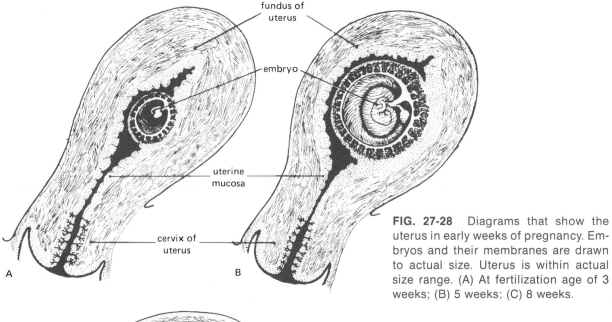

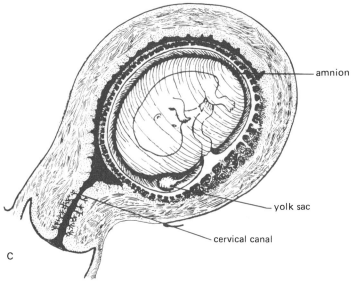

FIG. 27-28 Diagrams that show the uterus in early weeks of pregnancy. Embryos and their membranes are drawn to actual size. Uterus is within actual size range. (A) At fertilization age of 3 weeks; (B) 5 weeks; (C) 8 weeks.

FIG. 27-29 Urinary excretion of estrogen, progesterone, and chorionic gonadotropin during pregnancy. Urinary excretion rates are an indication of blood concentrations of these hormones.

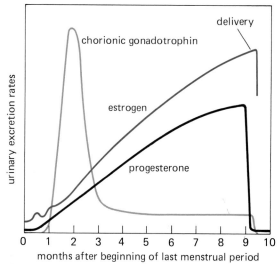

tal secretion. The corpus luteum remains, but its contribution is dwarfed by that of the placenta; indeed, removal of the ovaries during the last 7 months has no effect at all upon the pregnancy, whereas removal during the first 2 months causes immediate loss of the fetus (**abortion**).

An important and clinically useful aspect of placental steroid secretion is that the placenta does not have the enzymes required for the complete synthesis of the major estrogen of pregnancy, **es-**

triol. However, the enzymes it lacks are present in the adrenal cortex of the fetus; therefore, the placenta and fetal adrenals working together, with intermediates transported between them via the fetal circulation, produce the estriol required to maintain the pregnancy. Since the fetus is necessary for estriol production, measurement of this hormone in maternal blood provides a means for monitoring the well-being of the fetus.

Finally, the placenta not only produces steroids and CG but several other hormones as well. The best-documented one at present is a hormone that has effects very similar to those of growth hormone and prolactin. This hormone, **chorionic somatomammotropin,** may play an important role in the mother by maintaining a positive protein balance, mobilizing fats for energy, stabilizing plasma glucose at relatively high levels to meet the needs of the fetus, and facilitating development of the breasts.

Of the numerous other physiologic changes, hormonal and nonhormonal, in the mother during pregnancy, many, such as increased metabolic rate and appetite, are obvious results of the metabolic demands placed upon her by the growing fetus. Of great importance is salt and water metabolism. During a normal pregnancy, body sodium and water increase considerably; the extracellular volume alone rises by approximately 1 L. At present, it appears that important factors that cause fluid retention are renin, aldosterone, ADH, and estrogen, all of which act upon the kidneys. Some women retain abnormally great amounts of fluid and manifest protein in the urine and hypertension, which if severe enough may cause convulsions. These are the symptoms of the disease known as *toxemia of pregnancy.* Since it can usually be well controlled by salt restriction, in the United States it is seen primarily in the lower socioeconomic segments of the population that do not obtain adequate medical care during pregnancy. Despite the obvious association with salt retention, all attempts to determine the factors responsible for the disease have failed.

Parturition (Delivery of the Infant)

A normal human pregnancy lasts approximately 38 weeks, although many babies are born 1 to 2 weeks earlier or later. Delivery of the infant, followed by the placenta, is produced by strong rhythmic contractions of the uterus. Actually, beginning at approximately 30 weeks, weak and infrequent uterine contractions occur and gradually increase in strength and frequency. During the last month, the entire uterine contents shift downward so that the baby is brought into contact with the cervix (the outlet of the uterus). In over 90 percent of births, the baby's head is downward and acts as the wedge to dilate the cervical canal. By the onset of labor, the uterine contractions have become coordinated and quite strong (although usually painless at first) and occur at approximately 10- to 15-min intervals. Usually during this period or before, the membrane that surrounds the fetus ruptures and the intrauterine fluid escapes out the vagina. As the contractions, which begin in the upper portion and sweep down the uterus, increase in intensity and frequency, the cervical canal is gradually forced open to a maximum diameter of approximately 10 cm (Fig. 27-30). Until this point, the contractions have not moved the fetus out of the uterus but have served only to dilate the cervix. Now the contractions move the fetus through the cervix and vagina. At this time the mother, by bearing down to increase abdominal pressure, can help the uterine contractions to deliver the baby. The umbilical vessels and placenta are still functioning, so that the baby is not yet on its own, but within minutes of delivery both the infant's and mother's placental vessels completely contract, the entire placenta becomes separated from the underlying uterine wall, and a wave of uterine contractions delivers the placenta (the afterbirth). Ordinarily, the entire process from beginning to end proceeds automatically and requires no real medical intervention, but in a small percentage of cases, the position of the baby or some maternal defect can interfere with normal delivery. The baby's position is important for several reasons: (1) If the baby is not oriented head first, another portion of the body is in contact with the cervix and is generally a far less effective wedge; (2) because of its large diameter compared with the rest of the body, if the head went through the canal last, it might be obstructed by the cervical canal, which could lead to obvious problems when the baby attempts to breathe; (3) if the umbilical cord becomes caught between the birth canal and the baby, mechanical compression of the umbilical vessels can result. Despite these potential difficulties, however, most babies who are not oriented head first are born normally and with little difficulty.

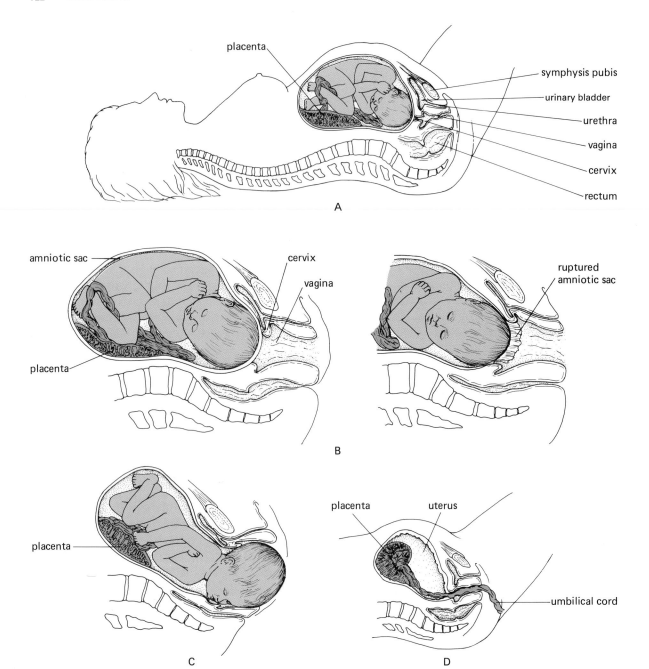

FIG. 27-30   Stages of parturition. (A) Parturition has not yet begun. (B) Dilation of the cervix. (C) The fetus is moving through the birth canal. (D) The placenta is coming loose from the uterine wall preparatory to its expulsion.

What mechanisms control the events of parturition? Let us consider a set of fairly well-established facts.

1   The uterus is composed of smooth muscle that is capable of autonomous contractions and has inherent rhythmicity, both of which are facilitated by stretching the muscle.
2   The efferent neurons to the uterus are of little importance in parturition since anesthetizing them in no way interferes with delivery.
3   Progesterone exerts a powerful inhibitory effect upon uterine contractility (apparently by decreasing its sensitivity to estrogen and oxytocin, which stimulate uterine contractions). Shortly before delivery, the secretion of progesterone sometimes drops, perhaps due to "aging" changes in the placenta.
4   Oxytocin, one of the hormones released from the posterior pituitary, is an extremely potent uterine-muscle stimulant. Oxytocin is reflexly released as a result of input into the hypothalamus from receptors in the uterus, particularly the cervix.
5   The pregnant uterus contains several prostaglandins, at least one of which is a profound stimulator of uterine smooth muscle; an increase in the release of this substance has been demonstrated during labor.

These facts can now be put together in a unified pattern, as shown in Fig. 27-31. The precise contributions of each of these factors are unclear; moreover, we cannot answer the crucial question: Which factor (if any) actually initiates the process? Once started, the uterine contractions exert a pos-

itive-feedback effect upon themselves via reflex stimulation of oxytocin and local facilitation of the muscle's inherent contractility; but what *starts* the contractions? The decrease in progesterone cannot be essential since it simply does not occur in most women. Nor is uterine distension or the presence of a fetus a requirement, as attested by the remarkable fact that typical "labor" begins at the expected time in some animals from which the fetus has been removed weeks previously. A primary role for prostaglandins has recently been touted but the evidence is far from conclusive. Regardless of their relative contributions to normal parturition, both prostaglandins and oxytocin are useful clinically in artificially inducing labor.

Another candidate for initiator of parturition is the hormone cortisol secreted by the adrenal cortex. There is a marked rise in cortisol secretion during the last days of pregnancy, and cortisol is known to enhance the action on the myometrium of several of the factors shown in Fig. 27-31.

Lactation

Perhaps no other process so clearly demonstrates the intricate interplay of various hormonal control mechanisms as milk production. The endocrine control has been established by numerous investigations and observations, none more striking than that in 1910 of Siamese twins: when one twin became pregnant, both women lactated after delivery.

The breasts are formed of compound tubuloalveolar glandular tissue (the **mammary glands**); fatty and fibrous connective tissue between the lobes of the glands; and blood vessels, nerves, and lymph vessels (Fig. 27-32). The epithelium-lined ducts, which drain the many separate glands, converge at the nipples where they open onto the surface. These ducts, the **lactiferous ducts,** branch all through the breast tissue and terminate in saclike structures (**alveoli**) typical of exocrine glands. The alveoli, which secrete the milk, look like bunches of grapes with stems that terminate in the ducts. The clusters of alveoli and the single duct into which they drain form a lobule of the glandular tissue. Beneath the **areola** (the colored area of skin that surrounds the nipple) the ducts are dilated and form **lactiferous sinuses,** which serve as reservoirs of milk in lactating women. The alveoli and ducts (for much of their length) are surrounded by specialized contractile

FIG. 27-31   Factors that stimulate uterine contractions during parturition. Note the positive-feedback nature of several of the inputs. What initiates parturition is not known.

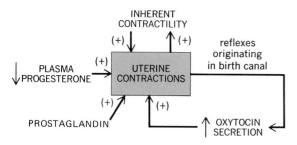

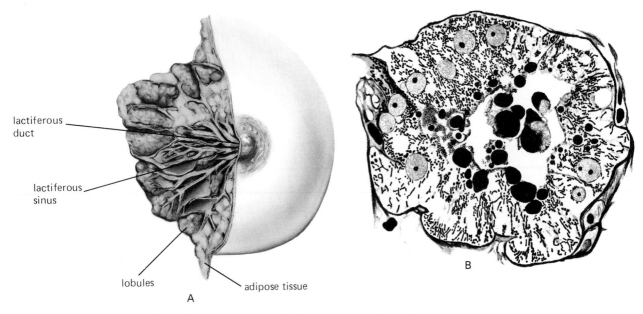

lactiferous duct

lactiferous sinus

lobules

adipose tissue

A

B

FIG 27-32 (A) Dissection of the female breast, showing secretory lobules during lactation. (B) Alveolus of a lactating mammary gland of a rabbit. The cells contain fat droplets (stained black), which, together with the adjacent cytoplasm, are extruded into the lumen.

cells called **myoepithelial cells.** Before puberty, the breasts are small, with little internal glandular structure. With the onset of puberty, the increased estrogen causes a marked enhancement of duct growth and branching, but little development of the alveoli, and much of the breast enlargement at this time is a result of a fat deposition. Progesterone secretion also commences at puberty (during the luteal phase of each cycle), and this hormone also contributes to breast growth.

During each menstrual cycle, breast morphology fluctuates in association with the changing blood concentrations of estrogen and progesterone, but these changes are small when compared with the marked breast enlargement that occurs during pregnancy as a result of the prolonged stimulatory effects of high plasma concentrations of estrogen, progesterone, prolactin, and chorionic somatomammotropin. This last hormone, as described earlier, is secreted by the placenta, whereas prolactin is secreted by the anterior pituitary. Under the influence of all these hormones, the alveolar structure becomes fully developed.

Prolactin secretion is low prior to puberty (presumably because of a high rate of release of prolactin-inhibiting hormone (PIH) from the hypothala-

mus) but increases considerably at puberty in girls (but not in males). This increase is caused by the stimulatory effects of estrogen (it is not known whether the estrogen acts on the hypothalamus to inhibit PIH secretion or on the anterior pituitary to reduce the sensitivity of the prolactin-secreting cells to PIH). During pregnancy, there is a marked increase in prolactin secretion (due to the elevation in plasma estrogen) that begins at about 8 weeks and rises throughout the remainder of the pregnancy.

Prolactin is the single most important hormone that promotes milk production. Yet, despite the fact that prolactin is elevated and the breasts markedly enlarged and fully developed as pregnancy progresses, there is no secretion of milk. This is because estrogen and progesterone, in large concentration, prevent milk production by inhibiting the action of prolactin on the breasts (Table 27-1). Thus, although estrogen causes an increase in the secretion of prolactin and acts with it in promoting breast growth and differentiation, it (along with progesterone) is antagonistic to prolactin's ability to induce milk secretion. Parturition removes the sources of the large amounts of sex steroids and, thereby, the inhibition of milk production.

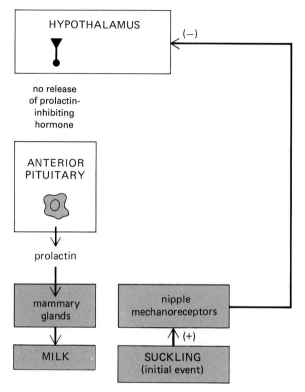

FIG. 27-33 Nipple suckling reflex. The neural pathway is schematic; actually multiple interneurons are involved. The initial event is stimulation of the nipple mechanoreceptors.

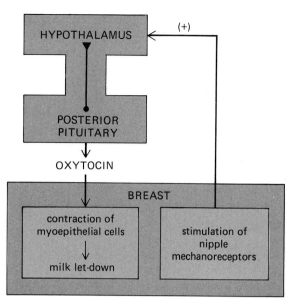

FIG. 27-34 Suckling-reflex control of oxytocin secretion and milk let-down. As with Fig. 27-33, begin following the figure with the nipple mechanoreceptors.

The major factor that maintains the secretion of prolactin during lactation is reflex input to the hypothalamus from receptors in the nipples that are stimulated by suckling (Fig. 27-33). After several months, however, plasma prolactin concentrations return to prepregnancy levels and lactation is no longer dependent upon this hormone. Milk production ceases soon after the mother stops nursing her infant but continues uninterrupted for years if nursing is continued.

One final reflex process is essential for nursing. Milk is secreted into the lumen of the alveoli, but because of their structure the infant cannot suck the milk out. It must first be moved into the ducts, from which it can be sucked. This process is called **milk let-down** and is accomplished by contraction of the myoepithelial cells that surround the alveoli; the contraction is directly under the control of oxytocin, which is reflexly released very rapidly (within 1 min) by suckling (Fig. 27-34), just like

prolactin. Higher brain centers also exert important influence over oxytocin release, and a nursing mother may actually leak milk when she hears her baby cry. In view of the central role of the nervous system in lactation reflexes, it is no wonder that psychologic factors can interfere with a woman's ability to nurse.

The end result of all these processes, the milk, contains four major constituents: water, protein, fat, and the carbohydrate lactose. The mammary alveolar cells must be capable of extracting the raw materials—amino acids, fatty acids, glycerol, glucose, etc.—from the blood and building them into the higher-molecular-weight substances. These synthetic processes require the participation of prolactin, insulin, growth hormone, cortisol, and probably other hormones.

Another important neuroendocrine reflex triggered by suckling (and mediated, in part, by prolactin) is inhibition of the hypothalamic-pituitary-ovarian chain at a variety of steps, with resultant block of ovulation. This inhibition apparently is relatively short-lived in many women, and approximately 50 percent begin to ovulate despite continued nursing. Pregnancy is common in women lulled into false security by the mistaken belief that failure to ovulate is always associated with nursing.

TABLE 27-2   POSSIBLE MEANS OF PREVENTING FERTILITY

Possible Means of Preventing Fertility in Men*:

I   Interference with sperm survival
   A   Prevention of maturation process in epididymis
   B   Prevention of function of accessory glands
      1   Prevention of androgen action on accessory glands
      2   Prevention of formation of accessory-gland secretion
      3   Prevention of accessory-gland secretion from entering urethra
   C   Creation of hostile environment to sperm in ductus deferens or urethra

II   Interference with testicular function at testicular level
   A   Prevention of androgen action on seminiferous tubules
   B   Prevention of action of FSH on seminiferous tubules
   C   Prevention of sperm division

III   Interference with pituitary function
   A   Prevention of FSH secretion at a pituitary level
   B   Prevention of action of GnRH on FSH-secreting cells in pituitary
   C   Prevention of secretion of GnRH
   D   Abolishment of extrahypothalamic central nervous factors required for normal reproductive function

Possible Means of Preventing Fertility in Women†:

I   Interference with ovarian function at ovarian level
   A   Prevention of initiation of follicle growth (no response to early FSH rise)
   B   Prevention of response to ovulatory LH surge
   C   Prevention of maturation of follicle or maturation of ova (e.g., inhibition of meiotic division)
   D   Prevention of corpus luteum formation
   E   Prevention of ovarian estrogen secretion
   F   Prevention of ovarian progesterone secretion
   G   Prevention of the maintenance of the corpus luteum during early pregnancy

II   Interference with pituitary function
   A   Prevention of FSH or LH secretion, or both, at a pituitary level
   B   Prevention of action of releasing hormones on pituitary
   C   Prevention of secretion of releasing hormones
   D   Alteration of estrogen or progesterone action on tonic center (too much or too little)
   E   Prevention of estrogen action on cyclic center
   F   Abolishment of extrahypothalamic central nervous factors required for normal reproductive hypothalamic function

III   Interference with sperm action
   A   Prevention of sperm entrance to vagina
   B   Prevention of sperm entrance to uterus
   C   Prevention of sperm entrance to uterine tubes
   D   Creation of hostile environment to sperm in vagina, uterus, or uterine tubes
   E   Prevention of sperm capacitation
   F   Prevention of sperm penetration of ovum by action on sperm

IV   Interference with ovum action
   A   Prevention of ova release from ovary
   B   Prevention of ova entrance to uterine tubes
   C   Creation of hostile environment for ova in uterine tubes
   D   Prevention of sperm penetration of ovum by action on ova

V   Prevention of survival of fertilized ova
   A   Prevention of fertilized ova from undergoing mitosis
   B   Alteration of migration along uterine tube (too fast or too slow)
   C   Prevention of implantation of blastocyst in uterine wall
   D   Destruction or expulsion of embryo after implantation in uterine wall
   E   Prevention of CG secretion by placenta
   F   Prevention of sex steroid secretion by placenta

---

* Not listed are surgical interventions removing the source of sperm, such as castration, or preventing egress of sperm, such as vasectomy.
† Not listed is surgical removal of any of the component organs (e.g., uterus, uterine tubes, ovaries).

Fertility Control

Table 27-2 summarizes possible means of preventing fertility. We present it because it should serve to summarize a majority of the information presented in this chapter (there are no new facts in the table). For each possibility listed there are many possible preventive aspects; for example, implantation is extremely complex and requires a long sequence of events. Thus, for each possibility, scientific investigation revolves around the question: What are the normal physiologic events that make the process possible and how can we intervene to prevent them?

Until recently, techniques of birth control (**contraception**) were primarily those that prevent sperm from reaching the ovum: vaginal **diaphragms,** sperm-killing **jellies,** and male **condoms.** Each of these methods has drawbacks, and more important, they are sometimes ineffective (Table 27-3). Another widely used method is the so-called **rhythm method,** in which couples merely abstain from sexual intercourse near the time of ovulation. Unfortunately, it is difficult to time ovulation precisely even with laboratory techniques; e.g., the small rise in body temperature or change in cervical mucus and vaginal epithelium, all of which are indicators of ovulation, occur only *after* ovulation. This problem, combined with the marked variability of the time of ovulation in many women, explains why this technique is only partially effective.

Since 1950, an intensive search has been made for a simple, effective contraceptive method; the first fruit of these studies was "the pill," the **oral contraceptive.** Its development was based on the knowledge that combinations of estrogen and progesterone inhibit pituitary gonadotropin release and thereby prevents ovulation. The most commonly used agents, at least at first, were combinations of an estrogen- and a progesteronelike substance (progestogens). Each month, this type of pill is taken for 20 days, then discontinued for 5 days; this steroid withdrawal produces menstruation, and the net result is a menstrual cycle without ovulation. The monthly withdrawal is required to avoid "breakthrough" bleeding that would occur if the steroids were administered continuously. Another type of regimen is the so-called sequential method in which estrogen is administered alone for 15 days followed by estrogen plus progestogen for 5 days, followed by withdrawal of both steroids. As with the combination pills, this regimen interferes with the orderly secretion of gonadotropins and prevents ovulation; no LH surge occurs, at least in part because of the constant estrogen levels, i.e., the absence of a rising estrogen level capable of exerting positive feedback.

It has now become clear that certain of the oral contraceptives do not always prevent ovulation yet still are effective because they have multiple antifertility effects. In other words, the hormonal milieu required for normal pregnancy is such that these exogenous steroids interfere with many of the steps between intercourse and implantation of the blastocyst. Taken correctly, they are almost 100 percent effective. Serious side effects, such as intravascular clotting, have been reported but only in a small number of women; nonetheless, only time can show how many undesirable effects will ultimately appear as a result of chronic alteration of normal hormonal balance.

Another type of contraceptive that is highly effective (although not 100 percent) and which illustrates the dependency of pregnancy upon the right conditions is the **intrauterine device** (the IUD) (Table 27-3). Placing one of these small objects in the uterus prevents pregnancy, perhaps by somehow interfering with the endometrial preparation for acceptance of the blastocyst.

The search goes on for an effective method that will reduce even further the possibility of unwanted side effects and will still be easy to use. Almost every possible process shown in Table 27-2 is worth following up. Two recent developments are postin-

TABLE 27-3	EFFECTIVENESS OF CONTRACEPTIVE METHODS

Method	Pregnancies per 100 Women per Year
None	115
Douche	31
Rhythm	24
Jelly alone	20
Withdrawal	18
Condom	14
Diaphragm	12
Intrauterine device	5
Oral contraceptive	1
correctly used	0

tercourse medications: prostaglandins and the estrogenlike diethylstilbestrol (DES). Both cause increased contractions of the female genital tract and may, therefore, cause expulsion of the fertilized ovum or failure to implant. Enthusiasm for DES has been tempered by the possibility that it may have cancer-producing properties.

Fertility prevention is of great importance, but the other side of the coin is the problem of unwanted infertility. Approximately 10 percent of the married couples in the United States are infertile. There are many reasons—some known, some unknown—for infertility; indeed, Table 27-2 also serves as a list of possible causes of infertility. Careful investigation of infertile couples frequently permits diagnosis and therapy of the basic problem.

EMBRYONIC AND FETAL DEVELOPMENT

As we have seen, the development of a human body from a single fertilized egg takes place over a period of 9 months in the mother's uterus. The major shaping of the body occurs during the first 2 months following fertilization. This early stage of development is divided into two periods: the first 3 weeks, the **period of germ disk formation,** followed by the

embryonic period, weeks 4 through 8, during which the germ disk gives rise to the various organs (Fig. 27-35). By the end of 2 months, the embryo has assumed a human form with a face, arms, legs, fingers, and toes, although it is only about 3 cm long, weighs about 10 g, and is incapable of independent survival outside the mother's womb. The embryo now is known as a **fetus.** During the remaining 7 months of pregnancy—the **fetal period**—the organ systems continue to mature and develop the detailed organized structures that will enable them to function independently after birth.

Formation of the Trilaminar Germ Disk (Weeks 1 to 3)

While still in the uterine tube (oviduct), the fertilized egg cell (**zygote**) undergoes four cell divisions, which results in a spherical cluster of 16 cells known as the **morula.** During these initial cell divisions, the cleavage divisions, there is no increase in total volume as compared to that of the original fertilized egg.

On the third day following fertilization, fluid begins to accumulate between the cells of the morula and thus transforms it into a **blastocyst.** As mentioned earlier, the outer cell layer of the blastocyst, the trophoblast, attaches to the uterus and eventually forms the placenta, and the inner cell mass, or embryoblast, forms the embryo.

By the eighth day the embryoblast becomes organized into two distinct cell layers: large columnar cells, the **ectodermal layer,** and smaller cuboidal cells, the **entodermal layer.** Together, these two layers form the **bilaminar germ disk** (Fig. 27-36). The ectodermal layer is separated from the trophoblast by a small fluid-filled cavity, the **amniotic cavity.** As the embryo develops, the amniotic cavity surrounds it and thereby encases the embryo in a fluid-filled, protective environment in which it will live during its intrauterine development. (It is the breaking of this amniotic sac that produces the watery discharge at the time of birth.) On the entodermal side of the germ disk is the **blastocele,** which will develop into the **yolk sac** and eventually form the lumen of the gastrointestinal tract.

By the end of the second week (Fig. 27-37), a third fluid-filled cavity, the **chorionic cavity,** appears between the trophoblast layer and a thin layer of cells that line the yolk sac. As the chorionic cavity

FIG. 27-35 Periods of development from fertilization to birth.

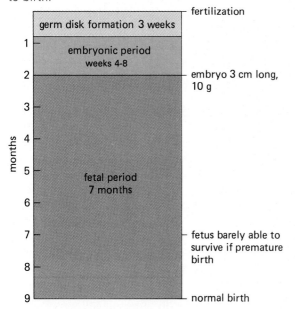

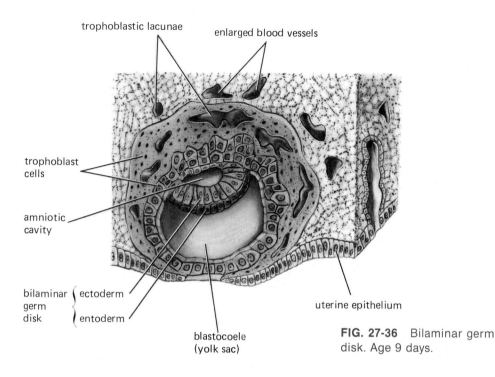

trophoblastic lacunae

enlarged blood vessels

trophoblast cells

amniotic cavity

bilaminar germ disk { ectoderm / entoderm

uterine epithelium

blastocoele (yolk sac)

FIG. 27-36 Bilaminar germ disk. Age 9 days.

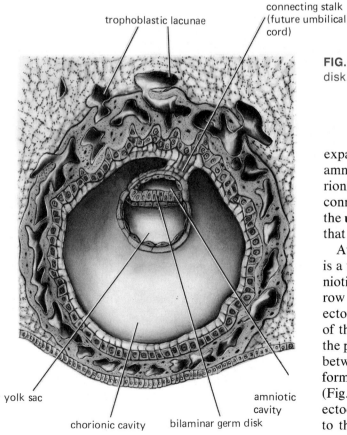

trophoblastic lacunae

connecting stalk (future umbilical cord)

FIG. 27-37 Bilaminar germ disk. Age 13 days.

yolk sac

chorionic cavity

bilaminar germ disk

amniotic cavity

expands, it leaves the germ disk and its associated amniotic cavity and yolk sac suspended in the chorionic cavity, connected to the trophoblast by a thin connecting stalk that will eventually develop into the **umbilical cord,** which contains the blood vessels that connect the fetus to the placenta (Fig. 27-37).

At the beginning of the third week, the germ disk is a two-layered oval disk situated between the amniotic cavity and the yolk sac. At this time, a narrow groove, the **primitive streak,** appears in the ectodermal layer and extends about half the length of the germ disk. Ectodermal cells migrate toward the primitive streak, move into the groove, and then between the ectodermal and entodermal layers to form a third layer of cells, the **mesodermal layer** (Fig. 27-38). As cells continue to move between the ectoderm and entoderm, they migrate laterally and to the front of the primitive streak until the germ

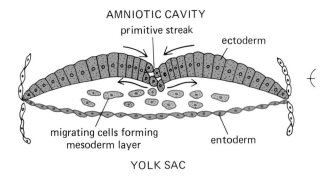

FIG. 27-38   Migration of ectodermal cells through the primitive streak to form the mesodermal layer.

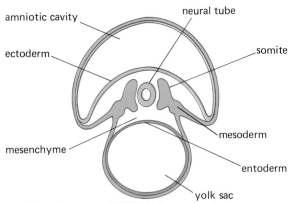

FIG. 27-39   Organization of the mesodermal layer during the third to fifth weeks into somite segments along the neural tube.

disk becomes a three-layered structure. By the end of the third week, having formed the third of the three layers of the germ disk, the primitive streak begins to regress.

The Embryonic Period (Weeks 4 to 8)

During the 5-week embryonic period the three germ layers give rise to the various tissues and organs of the body. In the following sections we shall examine the contributions of each of the three germ layers to the morphological development of the embryo.

Ectoderm   A portion of the ectodermal layer folds into a hollow tube, the **neural tube,** which forms the brain, spinal cord and associated nerves, and the central mass of cells of the adrenal gland (the adrenal medulla). The special sense organs of the eyes, ears, nose, and mouth are formed by foldings of ectodermal cells at the appropriate locations during various stages of development. In addition to these structures associated with the nervous system, the ectodermal layer, which eventually covers the entire outer surface, gives rise to the epidermis, which forms the outer layer of the skin, and to various structures associated with it: the hair, nails, and sweat and sebaceous (oil) glands, and in the female, the glandular portion of the breast. The ectoderm also lines the oral cavity where it gives rise to the salivary glands and the enamel layer of the teeth. The pituitary is formed from ectodermal tissue as an invagination of a pouch from the roof of the mouth meets an invagination from the overlying neural tube to form the anterior and posterior portions of the gland, respectively.

Mesoderm   By the end of the third week the mesoderm layer along the neural tube begins to break up into distinct segments to form pairs of **somites** (Fig. 27-39). By the middle of the fifth week there are 42 to 44 pairs of somites located along the neural tube. As development proceeds, cells migrate from the somites to the region that underlies the neural tube where they form a loosely woven tissue known as **mesenchyme.** These mesenchymal cells give rise to the cells that form the connective tissues. Additional mesodermal cells from the somites migrate toward the surface where they give rise to the dermal layer of the skin. The remaining cells of the somites multiply, forming masses of tissue known as **myotomes,** which will form the skeletal muscles.

The paired, segmented organization of the somites in the early embryo has important implications for understanding the distribution of sensory and motor functions in the adult. In association with each somite there develops a pair of nerves that extend from each side of the neural tube and become the spinal nerves in the fully developed organism. Each of these nerves eventually innervates the muscles and skin that develop from its corresponding somite. As the embryo develops, the simple geometric relation between the spinal nerves and corresponding somites becomes distorted because different regions of the embryo grow at different rates, but the link between the segmented nerves and their somite of origin remains.

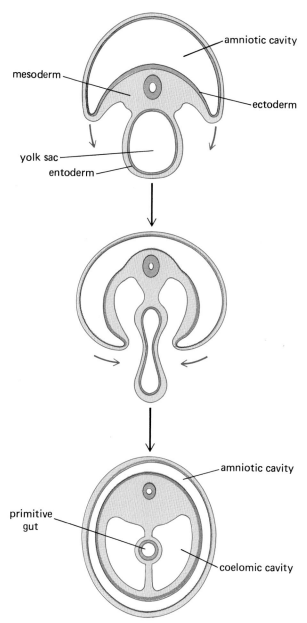

FIG. 27-40 Folding of the embryonic layers during the fourth week to form the coelomic cavity and primitive gut.

traembryonic spaces: the **coelomic cavity,** which is lined with mesoderm and becomes the thoracic and abdominal cavities, and the **primitive gut,** formed by sealing off a portion of the yolk sac, which is lined with entoderm. A portion of the yolk sac remains outside the embryo, in the region of the midgut, connected to it by a thin tube that contains the blood vessels of the umbilical cord that links the embryo to the placenta. The gut, which becomes the gastrointestinal tract, thus begins as a simple tube suspended in the center of the embryo. Initially the ends of this tube, in the region of the mouth and anus, are sealed. As development progresses, the ends of the tube open and are continuous with the amniotic fluid in the cavity surrounding the embryo.

Arising from the entodermal lining of the gut wall are a number of organs, all of which begin as small pouches or ducts that extend out from the surface of the gut into the surrounding coelom. These include the organs directly associated with gastrointestinal function such as the liver, gallbladder, bile ducts, and pancreas, as well as organs with other functions related to the external environment: the lungs, urinary bladder, and the auditory tubes that link the middle chamber of the ears to the oral pharynx.

Some of the organs that arise from the entoderm lose their association with the lumen of the gut and become separate glands; these include the thyroid and parathyroid glands that arise in the neck as outpouchings of the gut wall.

A summary of the organs and tissues that arise from the three germ layers is provided in Table 27-4.

During the 5 weeks of the embryonic period many congenital malformations may arise. A **congenital malformation** is a gross structural defect present at the time of birth, such as a cleft palate, absence of fingers or toes, blindness, deafness, defects in the structure of the heart and major blood vessels, etc. Approximately 5 percent of all infants are born with one or more such malformations. A number of factors, including certain viral and bacterial infections, radiation, a variety of drugs and chemical pollutants, the nutritional state of the mother, and genetic and chromosomal abnormalities, have been implicated as possible causes of congenital malformations. The type of malformation induced depends upon the stage of development at the time the embryo is exposed to the agent. If exposed

Entoderm   The third layer, the entoderm, eventually grows to surround the surface of the yolk sac. During the fourth week, the embryo undergoes extensive folding, both in the longitudinal direction in which the head and the tail regions fold toward each other and laterally in which the edges of the germ disk fold downward until they meet underneath the embryo (Fig. 27-40) and form two in-

TABLE 27-4   ORGANS AND TISSUES THAT ARISE FROM THE THREE GERM LAYERS

Ectoderm	Mesoderm	Entoderm
Nervous System:	**Muscle:**	**Epithelium of:**
Brain	Cardiac	Larynx
Spinal cord	Smooth	Trachea
Peripheral nerves	Skeletal	Lungs
Ganglia		Esophagus
Special sensory receptors of eye,	**Connective Tissue:**	Stomach
ear, nose, and mouth	Fibrous connective tissue	Intestines
General sensory receptors	Adipose tissue	Liver
Posterior pituitary (neurohypo-	Bone and bone marrow	Gallbladder
physis)	Blood cells	Pancreas
Adrenal medulla	Lymphatic tissue	Bladder
	Reticuloendothelial system	Urethra
Skin (epithelium of the integument):		Vagina
Epidermis	**Skin** (fibrous connective tissue):	Inner ear
Hair	Dermis	Auditory tubes
Nails		Thyroid
Sweat glands	**Epithelium of:**	Parathyroid
Sebaceous glands	Thoracic and abdominal cavities (pleura,	Thymus
Mammary glands	peritoneum, pericardium)	
	Kidneys	
Epithelium of:	Ureters	
Salivary glands	Gonads and associated ducts	
Nasal cavity	Adrenal cortex	
Mouth	Lining of heart (endocardium) and ves-	
Anal cavity	sels (endothelium)	
Enamel of teeth		
Anterior pituitary (adenohypo-		
physis)		

early in the embryonic period, a large number of major defects may result, whereas at later stages the same agent may affect only those organs that are undergoing a critical step in their development at the time of exposure. Environmental agents, which may cause significant anatomic malformations during the embryonic period, often produce less obvious morphologic damage during the fetal period but still institute functional impairment, particularly in the developing nervous system, where various forms of mental retardation can result.

The Fetal Period (Weeks 9 to 38)

By the end of the eighth week the embryo has acquired a human form and is now known as a fetus (Fig. 27-41). The remaining 30 weeks of pregnancy, the **fetal period,** is a period of major growth and maturation (Fig. 27-42). Rarely will the fetus survive outside the mother until it has undergone about 28 weeks of development, at which time the lungs have become sufficiently developed to function in gas exchange, and the appropriate neural connections have been made to the respiratory muscles to regulate the rhythmic patterns of breathing. Although most of the initial events associated with the formation of the various organs occur during the embryonic period, the tissues continue to differentiate and shape the fine structures of the organs during the fetal period. The brain, in particular, has a very long period of development, which continues, in part, even after birth.

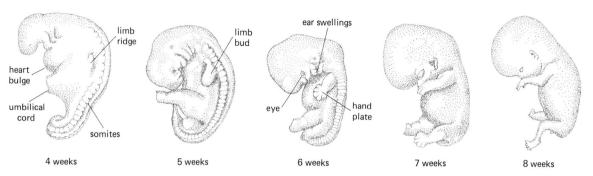

FIG. 27-41   Changes in embryonic form during the last 4 weeks of the embryonic period.

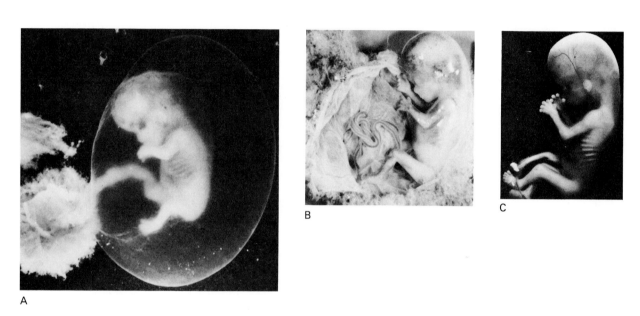

FIG. 27-42   Human fetuses at different stages of development. (A) 10 weeks; (B) 12 weeks; (C) 14 weeks; (D) 16 weeks; and (E) 20 weeks.

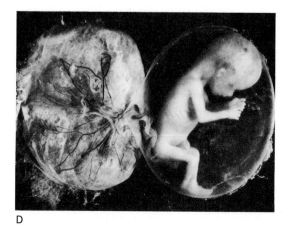

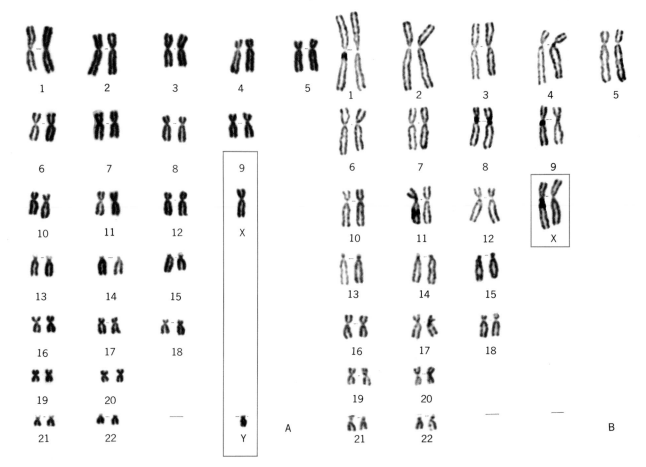

FIG. 27-43 Normal chromosomes in male (A) and female (B) cells. Each chromosome other than the X and Y chromosomes has an identifying number assigned to it.

SECTION C: THE CHRONOLOGY OF SEX DEVELOPMENT

SEX DETERMINATION

Sex is determined by the genetic inheritance of the individual, specifically by two chromosomes called the **sex chromosomes.** All human cells have 46 chromosomes, 22 pairs of somatic (nonsex) chromosomes and 1 pair of sex chromosomes. The larger of the sex chromosomes is the X chromosome and the smaller, the Y chromosome. Genetic males possess one X and one Y, whereas females have two X chromosomes (Fig. 27-43). Thus the genetic difference between male and female is simply the difference in one chromosome.

The reason for the approximately equal sex distribution of the population should be readily apparent (Fig. 27-44): The female can contribute only an X chromosome, whereas the male, during meiosis, produces sperm, half of which are X and half of which are Y. When the sperm and ovum join, 50 percent should have XX and 50 percent XY. Interestingly, however, sex ratios at birth are not 1:1. Rather, there tends to be a slight preponderance of male births (in England, the ratio of male to female births is 1.06). Even more surprising, the ratio at the time of conception seems to be much higher. From various types of evidence, it has been estimated that there may be 30 percent more male conceptions than female. There are several implications of these facts. First, there must be a considerably larger *in utero* death rate for males. Sec-

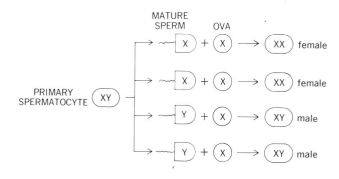

FIG. 27-44 Basis of genetic sex determination.

ond, the "male," that is, the Y bearing, sperm must have some advantage over the "female," or X bearing, sperm in reaching and fertilizing the egg. It has been suggested, for example, that since the Y chromosome is lighter than the X, the "male" sperm might be able to travel more rapidly. There are numerous other theories, but we are far from an answer. Moreover, it should be pointed out that conception and birth ratios show considerable variation in different parts of the world and, indeed, in rural and urban areas of the same country.

An easy method for distinguishing between the sex chromosomes was found quite by accident; the cells of female tissue (scrapings from the cheek mucosa are convenient) contain a readily detected nuclear mass that is a condensed X chromosome (in the normal female only one X chromosome functions in each somatic cell). This is not usually found in male cells. The method has proved valuable when genetic sex was in doubt. Its use and that of the more exacting tissue-culture visualization have revealed a group of genetic sex abnormalities characterized by such bizarre chromosomal combinations as XXX, XXY, X, and many others. The end result of such combinations is usually the failure of normal anatomic and functional sexual development. For example, patients with only one sex chromosome, an X, show no gonadal development.

SEX DIFFERENTIATION

It is not surprising that persons with abnormal genetic endowment manifest abnormal sexual development, but careful study has also revealed people with normal chromosomal combinations but abnormal sexual appearance and function. For example, a genetic male (XY) may have testes and female internal genitalia (vagina and uterus); such people are termed male *pseudohermaphrodites*. This kind of puzzle leads us into the realm of sex differentiation, i.e., the process by which the fetus develops the male or female characteristics. The genes *directly* determine only whether the individual will have testes or ovaries; virtually all the rest of sexual differentiation depends upon hormones secreted by this genetically determined gonad.

Differentiation of the Gonads

The primordial germ cells that will reside in the gonad of the embryo do not originate in the embryo itself but rather in the yolk sac entoderm. These cells migrate to the genital ridge of the embryo and there join mesenchyme cells, which will cover the gonad. For the first two months, the gonads, whether testes in a male embryo or ovaries in a female embryo, develop at similar sites, and the two types of gonads cannot be distinguished from each other; but shortly after that time in the genetic male the testes develop rapidly and begin to secrete testosterone. In the genetic female, ovaries begin to develop only several weeks later. Secretions of the embryonic gonad, testis or ovary, then regulate the remainder of the individual's sexual development.

Differentiation of Internal and External Genitalia

The very early fetus is sexually bipotential so far as its internal duct system and external genitalia are concerned. The primitive reproductive tract consists, in addition to the primitive gonad, of a double genital duct system (**wolffian and mullerian ducts**) and a common opening for the genital ducts and urinary system to the outside. The internal reproductive organs develop from different portions of the duct system: In the male, the wolffian ducts

FIG. 27-45 Stages in the development of the internal genitalia.

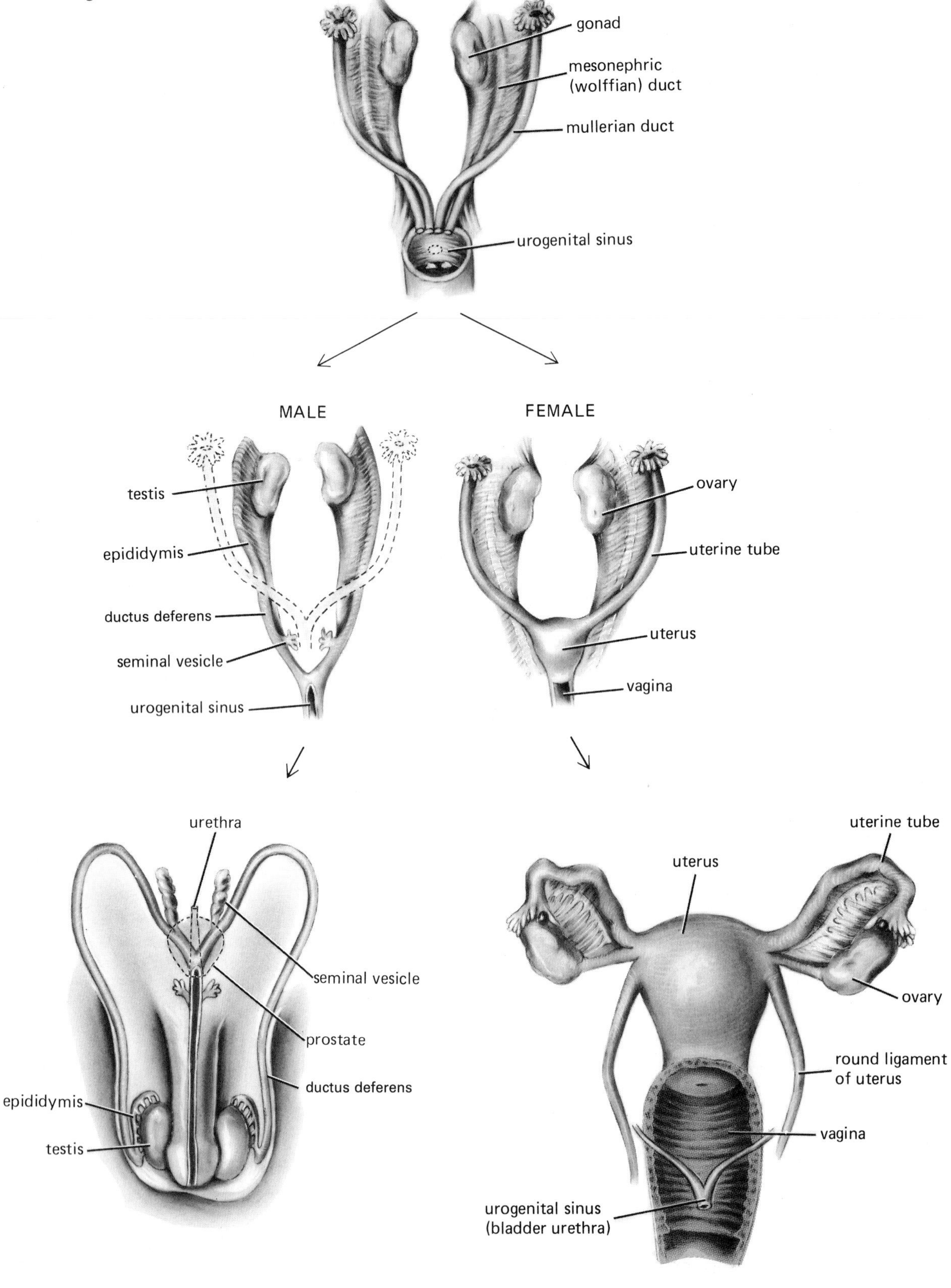

INDIFFERENT STAGE

gonad

mesonephric (wolffian) duct

mullerian duct

urogenital sinus

MALE

testis

epididymis

ductus deferens

seminal vesicle

urogenital sinus

FEMALE

ovary

uterine tube

uterus

vagina

urethra

seminal vesicle

prostate

ductus deferens

epididymis

testis

uterus

uterine tube

ovary

round ligament of uterus

vagina

urogenital sinus (bladder urethra)

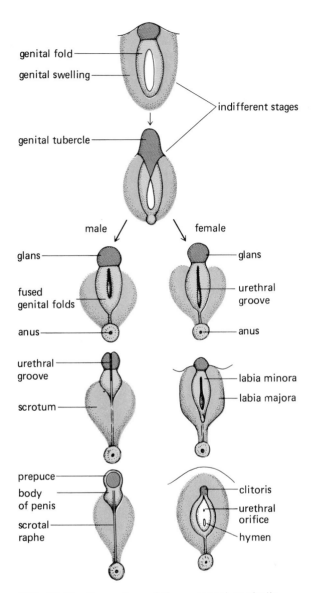

genital fold

genital swelling

indifferent stages

genital tubercle

male   female

glans   glans

fused
genital folds   urethral
groove

anus   anus

urethral
groove   labia minora
labia majora

scrotum

prepuce   clitoris
body
of penis   urethral
orifice
scrotal
raphe   hymen

FIG. 27-46 Formation of the external genitalia.

persist and the mullerian ducts regress, whereas in the female, the opposite happens; i.e., the mullerian duct system persists and the wolffian ducts degenerate (Fig. 27-45). The external genitalia in the two sexes develop from the genital tubercle, folds, and swelling (Fig. 27-46).

Which of the two duct systems develops depends not on the genetic makeup of the duct cells but rather on the presence or absence of two substances secreted by the fetal testes—testosterone and the **mullerian inhibiting substance.** These cause the mul-

lerian duct system to degenerate and the wolffian ducts to differentiate into epididymis, ductus deferens, ejaculatory duct, and seminal vesicle. Externally, the genital folds elongate and fuse to form the penis and urethra, and the urogenital swellings form the scrotum. Anatomical development of the male reproductive organs is complete by the end of the first trimester of pregnancy, and only growth of the genitalia and descent of the testes remain to be accomplished. The female embryo, lacking the two secretory products of a testis, develops (from the mullerian system) uterine tubes, uterus, and part of the vagina. Thus, the system is basically quite simple: If a functioning male gonad is present to secrete testosterone and mullerian inhibiting substance, a male duct system and external genitalia will form, and if these "male" secretions are absent, a female reproductive system will develop (a female gonad need not be present for the female organs to develop).

PUBERTY

Puberty is the period, usually occurring sometime between the ages of 10 and 14, during which the reproductive organs mature and reproduction becomes possible. For example, in the male, the seminiferous tubules begin to produce sperm; the genital ducts, glands, and penis enlarge and become functional; the secondary sex characteristics develop; and sexual drive is initiated. All these phenomena are effects of testosterone, and puberty in the male is the direct result of the onset of testosterone secretion by the testes (Fig. 27-47). Although significant testosterone secretion occurs during late fetal life, within the first days of birth testosterone secretion becomes very low and remains so until puberty.

The critical question is: What stimulates testosterone secretion at puberty? Or conversely: What inhibits it before puberty? Experiments with testis or pituitary transplantation in other mammals and studies of hormone injections in human beings have suggested the hypothalamus as the critical site of control. Before puberty, the hypothalamus fails to secrete significant quantities of GnRH to stimulate secretion of FSH and LH (Fig. 27-47B); deprived of these gonadotropic hormones, the testes fail to produce sperm or large amounts of testosterone.

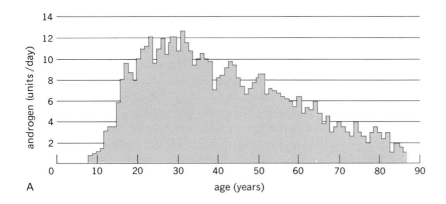

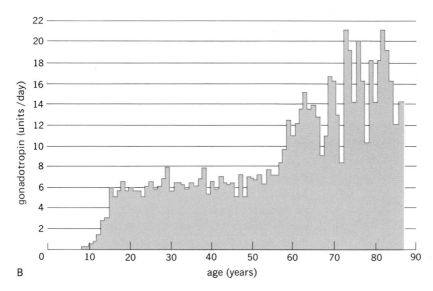

FIG. 27-47 (A) Excretion of androgen in the urine of normal boys and men, an indicator of the blood concentration of androgen. (B) Excretion of anterior pituitary gonadotropins (FSH and LH) in the urine of normal boys and men, an indicator of the blood concentration of gonadotropins.

Puberty is initiated by an unknown alteration of brain function that permits increased secretion of the hypothalamic GnRH. The mechanism of this change remains unknown, but one likely hypothesis is that prior to puberty the hypothalamus is so sensitive to the negative-feedback effects of testosterone that even the extremely low blood concentratino of this hormone in prepubescent boys is adequate to block secretion of the releasing hormone. Of course, this simply raises the question of what brain change occurs at puberty to lower responsiveness to the negative feedback. In any case, the process is not abrupt but develops over several years, as evidenced by slowly rising plasma concentrations of the gonadotropins and testosterone.

The situation in the female is analogous to that for the male. Throughout childhood, estrogen is secreted at very low levels (Fig. 27-48). Accordingly, the female accessory sex organs remain small and nonfunctional; there are minimal secondary sex characteristics, and ovum maturation does not occur. As for the male, prepuberal dormancy is probably due mainly to deficient secretion of the hypothalamic GnRH, and the onset of puberty is occasioned by an alteration in brain function that raises secretion of this releasing hormone, which in turn stimulates secretion of pituitary gonadotropins. The increased estrogen stimulated by the gonadotropins then induces the striking changes associated with puberty and, through its stimulatory effect on the brain and pituitary, permits menstrual cycling to begin. "Maturation" of the positive-feedback mechanism also occurs at puberty, for it is not possible to elicit an LH surge in prepubescent girls

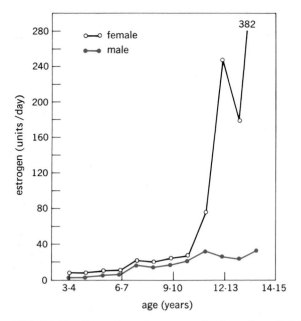

FIG. 27-48 Excretion of estrogen in the urine of children, an indicator of the blood concentration of estrogen.

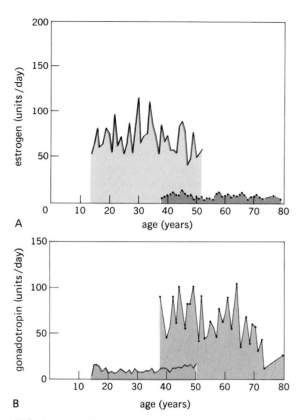

FIG. 27-49 (A) Excretion of estrogen in the urine of women from puberty to senescence, an indicator of blood concentration of estrogen: before menopause (color) and after menopause (gray). (B) Excretion of gonadotropins in the urine of nonpregnant women, an indicator of blood concentrations of gonadotropins: before menopause (color) and after menopause (gray).

by giving estrogen.

It should be recognized that the maturational events of puberty usually proceed in an orderly sequence but that the ages at which they occur may vary among individuals. In boys, the first sign of puberty is acceleration of growth of testes and scrotum; pubic hair appears a trifle later, and axillary and facial hair still later. Acceleration of penis growth begins on the average at 13 (range 11 to 14.5 years) and is complete by 15 (13.5 to 17). But note that, because of the overlap in ranges, some boys may be completely mature whereas others at the same age (say 13.5) may be completely prepubescent. Obviously, this can lead to profound social and psychological problems. In girls, appearance of "breast buds" is usually the first event (average age, 11 years) although pubic hair may, on occasion, appear first. **Menarche,** the first menstrual period, is a later event (average, 12.3 years in the United States) and occurs almost invariably after the peak of the total body growth spurt has passed. The early menstrual cycles are usually not accompanied by ovulation so that conception is generally not possible for 12 to 18 months after menarche.

MENOPAUSE

Ovarian function declines gradually from a peak usually reached before the age of 30. Figure 27-49A and B demonstrate that the cause of the decline is decreasing ability of the aging ovaries to respond to pituitary gonadotropins, in large part because of the diminished number of follicles. Estrogen secretion drops despite the fact that the gonadotropins, partially released from the negative-feedback inhibition by estrogen, are secreted in greater amounts. Ovulation and the menstrual periods become irregular and ultimately cease completely; **menopause** is the period during the cessation of menstruation. Some ovarian secretion of estrogen generally continues beyond these events but gradually diminishes until it is inadequate to maintain the estrogen-dependent tissues: the breasts and genital organs gradually atrophy; the decrease in protein anabolism causes thinning of the skin and bones; however, sexual drive is frequently not diminished and may even be increased. The hot flashes, so typical of menopause, result from dilation of the skin arterioles that causes a feeling of warmth and marked sweating; why estrogen deficiency causes this is unknown. Many of these symptoms of menopause can be reduced by the administration of estrogen, but the safety of such administration is being seriously questioned because of the possibility that estrogen may facilitate development of breast or uterine cancers.

In the male, the changes that occur with aging are much less drastic than those in the female. Once testosterone and pituitary gonadotropin secretions are initiated at puberty, they continue throughout adult life (Fig. 27-49A and B). A steady decrease in testosterone secretion in later decades apparently reflects slow deterioration of testicular function. The mirror-image rise in gonadotropin secretion is due to diminishing negative-feedback inhibition from the decreasing plasma testosterone concentration. Along with the decreasing testosterone levels, both libido and potency diminish with advancing age, although many men continue to enjoy active sex lives into their seventies and eighties and fertility has been documented in men in their eighties. Thus, there is usually no complete cessation of reproductive function analogous to menopause.

KEY TERMS

gonads	follicle-stimulating hormone (FSH)	estrogen
sperm	luteinizing hormone (LH)	progesterone
spermatogenesis	androgens	menstrual cycle
testes	ovaries	implantation
penis	uterus	embryo
Sertoli cells	vagina	fetus
semen	clitoris	placenta
erection	ovum	parturition
ejaculation	oogenesis	contraception
orgasm	follicle	ectoderm, entoderm, and mesoderm
testosterone	corpus luteum	menopause

REVIEW EXERCISES

Section A

1 State the primary reproductive organs and their dual function.
2 Describe the anatomy of the testes, including their descent.
3 Describe the duct system that leads from the seminiferous tubules to the penis, including the glands that empty into it.
4 List the stages of spermatogenesis and describe the structure of the mature sperm.
5 Describe the anatomy and function of the Sertoli cells.
6 Describe the physiologic events responsible for erection and ejaculation.
7 List the effects of testosterone and its mechanism of actions.
8 Describe the role of gonadotropin-releasing hormone, FSH, LH, prolactin, and inhibin in the control of testicular function; describe the negative-feedback interactions in this system.
9 Define the following terms: accessory reproductive organs, secondary sexual characteristics, puberty, adolescence, interstitial (Leydig) cells, and dihydrotestosterone.

Section B

1 Describe the anatomy of the female internal and external genitalia.
2 List the stages of ovum and follicle growth and describe the anatomical characteristics of each.
3 Describe the formation of the corpus luteum.
4 List the ovarian hormones.
5 Name the stages of the menstrual cycle in two ways.
6 Draw a figure that illustrates the plasma concentrations of FSH, LH, estrogen, and progesterone during the menstrual cycle.
7 Describe the hormonal control of follicle and ovum development, ovulation, and formation of the corpus luteum.
8 List the feedback interactions between the hormones at various times in the cycle.
9 Describe the uterine changes that occur in the menstrual cycle and their hormonal control.
10 List the nonuterine effects of estrogen and progesterone; state various ways of determining what stage of the cycle is present.
11 List the sequence of events in the female response to sexual intercourse.
12 State the survival times of sperm and ovum.
13 Describe the mechanisms of ovum and sperm transport and the entry of sperm into the ovum.
14 Explain how identical and fraternal twins occur.
15 Describe implantation and placentation, including the layers of the placenta and their functions.
16 List the hormonal changes that occur during pregnancy.
17 State the factors that stimulate uterine contractions during parturition.
18 Describe the structure of the breasts and the hormonal control of breast development and lactation.
19 Describe the formation of the trilaminar germ disk.
20 List the contributions of each of the three germ layers to morphological development of the embryo.
21 Define the following terms: theca interna, theca externa, membrana granulosa, estrone, estriol, cornification, block to polyspermy, tubal pregnancy, zygote, blastocyst, milk let-down, somites, and myotomes.

Section C

1 Distinguish sex determination from sex differentiation; contrast the differentiation of the gonads from that of the internal and external genitalia.
2 Describe the present theory for the initiation of puberty.
3 State the physiologic events that lead to menopause.

CHAPTER

DEFENSE MECHANISMS OF THE BODY: IMMUNOLOGY, FOREIGN CHEMICALS, AND STRESS

SECTION A: IMMUNOLOGY: THE BODY'S DEFENSE AGAINST FOREIGN MATERIALS

Immunity constitutes all the physiologic mechanisms that allow the body to recognize materials as foreign or abnormal and to neutralize or eliminate them; in essence, these mechanisms maintain uniqueness of "self." Classically, immunity referred to the resistance of the body to microbes (viruses, bacteria, and other unicellular and multicellular organisms). It is now recognized, however, that the immune system has more diverse functions than this. It is involved both in the elimination of "worn-out" or damaged body cells (such as old erythrocytes) and in the destruction of abnormal or . mutant cell types that arise in the body. This last function, known as **immune surveillance,** constitutes one of the body's major defenses against cancer.

It has also become evident that immune responses are not always beneficial but may result in serious damage to the body. In addition, the immune system seems to be involved in the process of aging. Finally, it constitutes the major obstacle to successful transplantation of organs.

Immune responses may be classified into two categories: specific and nonspecific. **Specific immune responses,** the function of lymphocytes (and cells derived from lymphocytes), depend upon prior exposure to a specific foreign material, recognition of it upon subsequent exposure, and reaction to it. In contrast, the **nonspecific immune responses** do not require previous exposure to the particular foreign substance; rather they nonselectively protect against foreign substances without having to recognize their specific identities. They are particularly important during the initial exposure to a foreign organism before the specific immune responses have been activated.

Immune responses can be viewed in the same way as other homeostatic processes in the body, i.e., as stimulus-response sequences of events. In such an analysis, the groups of cells that mediate the final responses are effector cells. Before we begin our detailed description of immunology by introducing the various effector cells involved, let us first mention their major opponents: microbes (or microorganisms), i.e., bacteria, viruses, and fungi.

Bacteria are unicellular organisms that have not only a plasma membrane but also an outer coating, the cell wall. Most bacteria are self-contained complete cells in that they have all the machinery required to sustain life and to reproduce themselves. In contrast, the **viruses** are essentially nucleic acid cores surrounded by a protein coat. They lack both the enzyme machinery for energy production and the ribosomes essential for protein synthesis. Thus, they cannot survive by themselves but must "live" inside other cells whose biochemical apparatus they make use of; i.e., they are obligatory parasites. Other types of microorganisms and multicellular parasites are potentially harmful to human beings, but we shall devote most of our attention to the body's defense mechanisms against bacteria and viruses.

How do microorganisms cause damage and endanger health? There are many answers to this question, depending upon the specific bacterium or virus involved. Some bacteria themselves cause cellular destruction by locally releasing enzymes that break down cell membranes and organelles. Others give off toxins that disrupt the functions of organs and tissues. The effect of viral habitation and replication in a cell depends upon the type of virus. Some viral particles, upon entering a cell, multiply very rapidly and kill the cell by depleting it of essential components or by directing it to produce toxic substances; with the death of the host cell, the viral particles leave and move on to another cell. In contrast, other viral particles replicate very slowly, and the viral nucleic acid may even become associated with the cell's own DNA molecules to replicate along with them and be passed on to the daughter cells during cell division; such a virus may remain in the cell or its offspring for many years. As we shall see, cells infected with these so-called ''slow'' viruses may be damaged by the body's own defense mechanisms turned against the cells because they are no longer "recognized" as "self." Finally, it is likely that certain viruses cause transformation of their host cells into cancer cells.

EFFECTORS OF THE IMMUNE SYSTEM

The major effector-cell types (Table 28-1) are the white blood cells (leukocytes, described in Chap. 19), plasma cells, and macrophages. **Plasma cells** are derived from lymphocytes and are the cells that secrete antibodies. **Macrophages** are derived from monocytes, have as their major function phagocytosis, and are found scattered throughout the tissues of the body. The structure of these cells varies from place to place, but their common distinctive features are numerous cytoplasmic granules and the ability to ingest almost any kind of foreign particle. Tissue macrophages are capable of mitotic activity, but the major means of repletion of tissue macrophages is by influx of blood monocytes and their differentiation locally. The effector cells of the immune system are distributed throughout the organs and tissues, but many are housed in the lymphoid organs (Table 28-2), which contain dense clusters of lymphocytes and comprise the lymph nodes, thymus, spleen, and several unencapsulated aggregations of lymphocytes.

TABLE 28-1   MAJOR EFFECTOR-CELL TYPES OF THE IMMUNE SYSTEM

| | White Blood Cells (Leukocytes) | | | | | Plasma Cells | Macrophages |
| | Polymorphonuclear Granulocytes | | | Lymphocytes | Monocytes | | |
	Neutrophils	Eosinophils	Basophils				
Percent of total leukocytes	50–70%	1–4%	0.1%	20–40%	2–8%		
Primary site of production	Bone marrow	Bone marrow	Bone marrow	Bone marrow, thymus, and other lymphoid tissues	Bone marrow	Derived from B lymphocytes in lymphoid tissue	Most are formed from monocytes
Primary known function	Phagocytosis of bacteria, cell debris, and antibody-primed foreign matter; release of chemicals (endogenous pyrogen, chemotaxic factors, etc.)	? ? ?	Release of histamine and other chemicals (similar to tissue mast cells)	*B cells:* production of antibodies (after transformation into plasma cells) *T cells:* responsible for cell-mediated immunity; influence B cells	Transformed into tissue macrophages with high phagocytic activity; release of granulopoietin	Production of antibodies	Phagocytosis of bacteria, cell debris, and antibody-primed foreign matter; assist in antibody formation and T-cell sensitization; release of granulopoietin

TABLE 28-2	SOME FUNCTIONS OF LYMPHOID ORGANS	
Lymph nodes (including unencapsulated lymphoid tissues)	1 2 3	Remove and store lymphocytes Form and add new lymphocytes to lymph flow through nodes Remove particulate matter by macrophage phagocytosis
Spleen	1 2 3 4	Produce red blood cells in fetus but not adult Produce new lymphocytes Remove products of red cell degradation and other foreign matter by macrophage phagocytosis Store red blood cells, which can be added to circulation by contraction of the spleen
Thymus	1 2	Produce T lymphocytes Secrete hormones (thymosin)

Lymph Nodes

The lymph nodes were described in Chapter 21. They are the only lymphoid organs that have lymphatic vessels emptying into them.

Thymus

The **thymus** is an encapsulated lymphoid organ that lies in the mediastinum behind the sternum but in front of the pulmonary vessels, aorta, and trachea; it consists of two unequally sized lobes connected by fatty tissue (Fig. 28-1). The size of the thymus varies with age; it is relatively large at birth and grows until puberty when it gradually atrophies and is partly replaced by fatty and fibrous tissue. The thymus is pinkish-gray and highly lobulated; each lobule has an outer cortex that contains densely packed lymphocytes and an inner medulla with lymphocytes in looser array. The thymus produces the T lymphocytes of the body from stem cells that emigrate to it from the bone marrow. However, there are normally no well developed germinal centers in the thymus similar to those of the lymph nodes. Following differentiation, the T cells leave

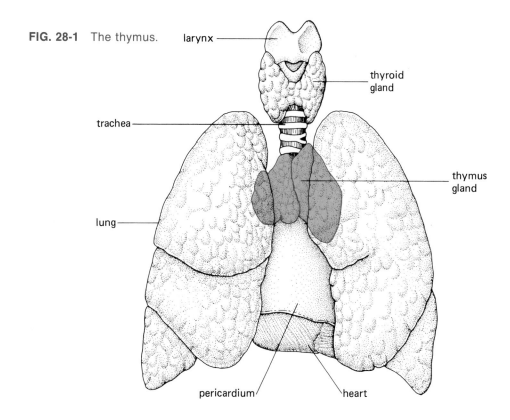

FIG. 28-1 The thymus.

larynx

thyroid gland

trachea

thymus gland

lung

pericardium   heart

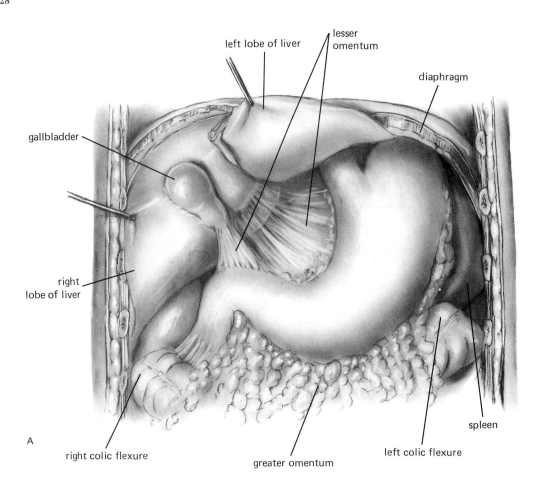

left lobe of liver

lesser omentum

diaphragm

gallbladder

right lobe of liver

spleen

A

right colic flexure

greater omentum

left colic flexure

the thymus and supply the T-lymphocyte regions of the lymph nodes, spleen, and other lymphatic tissues. In addition to the T lymphocytes, the thymus also contains macrophages, epithelial cells, and fat cells. Aggregations of epithelial cells in the medulla of the thymic nodules form **thymic corpuscles,** whose function is unknown. There are no afferent lymphatic vessels to the thymus, but efferent lymphatics pass from the organ to lymph nodes in the mediastinum. The thymus also secretes a group of hormones that probably promote the differentiation of T lymphocytes; they are presently known collectively as **thymosin.** The role of the thymus in immunity will be described below.

Spleen

The **spleen** is the largest of the lymphoid organs. It lies in the left part of the abdominal cavity between the stomach and diaphragm (Fig. 28-2A). It is a highly vascular organ, purplish in color. The two

most important functions of the spleen are: (1) removal from the bloodstream of worn-out or damaged erythrocytes and other cellular debris, a task accomplished by the macrophages of the spleen; and (2) production of antibodies (discussed below with the immunologic responses of the lymphoid tissues). A third, less important function of the spleen is the storage of blood cells and of the iron and bile pigments that are formed during the breakdown of hemoglobin by the macrophages.

Partitions, or **trabeculae,** from the connective-tissue capsule extend into the spleen and divide it into lobules; they also carry arteries, veins, and nerves into the spleen. The interior of the spleen is filled with a reticular meshwork, the red pulp and the white pulp.

The **red pulp** is made up of large branching, thin-walled blood vessels, the **splenic sinusoids,** and the cellular **splenic cords,** which lie between them (Fig. 28-2B). (Note that the splenic sinusoids are blood vessels, whereas the sinuses of the lymph nodes

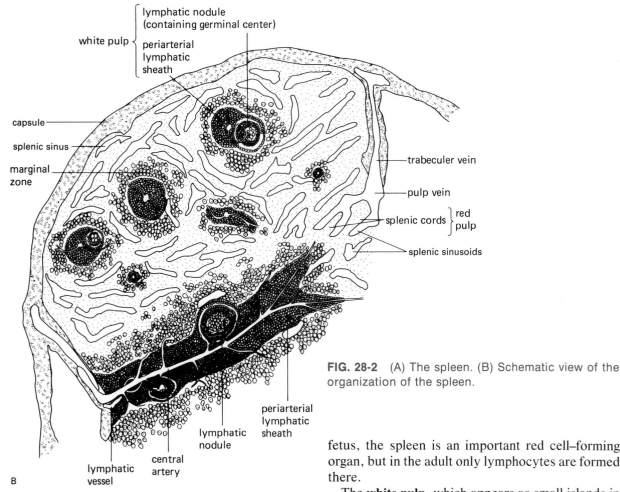

white pulp { lymphatic nodule (containing germinal center)
periarterial lymphatic sheath

capsule

splenic sinus

marginal zone

trabeculer vein

pulp vein

splenic cords } red pulp

splenic sinusoids

periarterial lymphatic sheath

lymphatic nodule

central artery

lymphatic vessel

B

FIG. 28-2 (A) The spleen. (B) Schematic view of the organization of the spleen.

are lymphatics.) The splenic cords contain erythrocytes, macrophages, platelets, granulocytes, and reticular cells. Blood, rather than lymph, percolates through the red pulp, but knowledge about the path of blood flow is uncertain. The most widely accepted theory, the *open-circulation theory,* states that blood can flow through the spleen by several paths, one of which passes from the terminal arteries, out into the reticular meshwork of the red pulp, back into the splenic sinuses, and on to the veins that ultimately drain from the spleen. Thus, while some of the arterial vessels connect directly with the sinuses to form a closed circulation, most empty into the splenic substance where the cords act as filters to pass the cells through quickly, delay them for variable periods, or if they are damaged, destroy and phagocytize them. The red pulp accounts for much of the substance of the spleen. In the human

fetus, the spleen is an important red cell–forming organ, but in the adult only lymphocytes are formed there.

The **white pulp,** which appears as small islands in the red pulp, is the actual lymphoid tissue of the spleen. This pulp may occur as cylindrical sheaths that surround arteries as they pass through the substance of the spleen (Fig. 28-2B) or as oval clusters, known as **lymphatic nodules.** The sheaths contain predominantly T lymphocytes, whereas the nodules consist of B lymphocytes and may even contain germinal centers where B lymphocyes proliferate and differentiate. A **marginal zone** lies between the red and white pulp; it is here that much of the filtering of the blood brought to the spleen is initiated.

Unencapsulated Lymphoid Tissues

Deposits of unencapsulated lymphatic tissue occur at various sites throughout the body. They may be small and appear as single lymph nodules scattered throughout loose connective tissue or the mem-

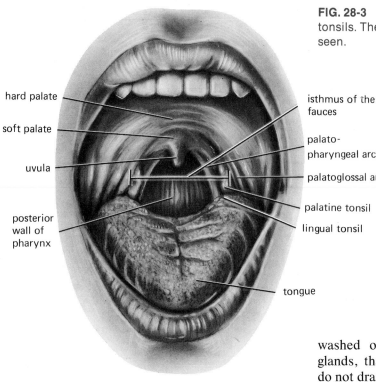

FIG. 28-3 Cavity of the mouth, showing tonsils. The pharyngeal tonsil cannot be seen.

hard palate

soft palate

uvula

posterior wall of pharynx

isthmus of the fauces

palato-pharyngeal arch

palatoglossal arch

palatine tonsil

lingual tonsil

tongue

branes lining the respiratory, digestive, urinary, or reproductive tracts; or they may be aggregated into easily discernable clusters such as the tonsils in the nasopharynx, pharynx, and tongue or Peyer's patches in the lower part of the small intestine.

The periphery of **lymph nodules,** in contrast to lymph nodes, is not sharply defined because they have no limiting capsule. The nodules sometimes develop a germinal center where B lymphocytes proliferate and differentiate. The nodules produce lymphocytes and sometimes plasma cells and serve as part of the immune mechanism discussed below.

Three groups of tonsils, the **palatine, pharyngeal,** and **lingual tonsils,** form a circular band of lymphoid tissue around the opening of the digestive and respiratory tubes. It is the palatine tonsils (Fig. 28-3) that are commonly called "the tonsils" and the single pharyngeal tonsil that is called "the adenoid," particularly when it is enlarged. The surfaces of the lingual and palatine tonsils are indented by infoldings of the overlying epithelium to form **crypts,** which extend into the substance of the tonsil. Debris can accumulate in the crypts and lead to infection (*tonsillitis*), although this is less apt to happen in the lingual tonsils, whose crypts are

washed out by the secretions from underlying glands, than in the palatine tonsils, whose crypts do not drain glands. The tonsilar substance consists of aggregates of lymphatic nodules, many of which contain lymphocyte-forming germinal centers and numerous leukocytes. The germinal centers are especially prominent in children and young adults.

The **Peyer's patches,** usually 20 to 30 in number, occur most commonly in the lower part of the small intestine (the lower ileum) although they may be present in other parts of the small intestine as well. Like the tonsils, they are aggregates of lymphatic nodules. The wall of the appendix (the **vermiform appendix**) also contains lymphatic tissues that consist of clumped nodules.

NONSPECIFIC IMMUNE MECHANISMS

External Anatomic and Chemical "Barriers"

The body's first lines of defense against infection are the barriers offered by surfaces exposed to the external environment. Very few microorganisms can penetrate the intact skin. The sweat, sebaceous, and lacrimal glands secrete chemical substances that are highly toxic to certain forms of bacteria. The mucous membranes also contain antimicrobial chemicals, but more important, mucus is sticky. When particles adhere to it, they can be swept away by ciliary action, as in the upper respiratory tract, or engulfed by phagocytic cells. Other specialized surface "barriers" are the hairs at the entrance to the nose, the sensitive epithelium that initiates the cough reflex, and the acid secretion of the stomach. Finally, a major barrier to infection is the normal microbial flora of the skin and other linings exposed to the external environment; these microbes suppress the growth of other potentially more virulent microorganisms.

Inflammatory Response

Despite the effectiveness of the external barriers, small numbers of microorganisms penetrate them every day. Once the invader has gained entry, it triggers off **inflammation,** the basic response to injury. The local manifestations of the inflammatory response are a complex sequence of highly interrelated events, the overall functions of which are to bring phagocytes into the damaged area so that they can destroy (or inactivate) the foreign invaders and set the stage for tissue repair. The sequence of events that constitute the inflammatory reaction varies, depending upon the injurious agent (bacteria, cold, heat, trauma, etc.), the site of injury, and the state of the body, but the similarities are in many respects more striking than the differences. It should be emphasized that, in this section, we describe inflammation in its most basic form, i.e., the nonspecific innate response to foreign material. As we shall see, inflammation remains the basic scenario for the acting out of specific immune responses as well; the difference is that the entire process is amplified and made more efficient by the participation of antibodies and sensitized lymphocytes, i.e., agents of specific immune responses.

Using bacterial infection as our example, we briefly present the following sequence of events in an inflammatory response.

1   Initial entry of microbes
2   Vasodilation of the vessels of the microcirculation that lead to increased blood flow
3   A marked increase in vascular permeability to protein
4   Filtration of fluid into the tissue with resultant swelling
5   Exit of neutrophils (and, later, monocytes) from the vessels into the tissues

6   Phagocytosis and destruction of the microbes
7   Tissue repair

All these events, whether part of the nonspecific or specific mechanisms, are mediated by chemical substances released or generated locally, and these will be described subsequently. The familiar gross manifestations of this process are redness, swelling, heat, and pain; pain is the result both of distension and the direct effect of released substances on afferent nerve endings.

Vasodilation and Increased Permeability to Protein
Immediately upon microbial entry, chemical mediators dilate most of the vessels of the microcirculation in the area and somehow alter the material between the endothelial cells so as to make it quite leaky to large molecules. Tissue swelling is directly related to these changes: The arteriolar dilation increases capillary hydrostatic pressure and thereby favors filtration of fluid out of the capillaries; in addition, the protein that leaks out of the vessels as a result of increased permeability builds up locally in the interstitium and thereby diminishes the difference in protein concentration between plasma and interstitium (recall from Chap. 21 that this difference is mainly responsible for fluid movement from interstitium into capillaries).

The adaptive value of all these vascular changes is twofold: (1) The increased blood flow to the inflamed area increases the delivery of phagocytic leukocytes and plasma proteins crucial for immune responses; and (2) the increased capillary permeability to protein ensures that the relevant plasma proteins—all normally restrained by the capillary membranes—can gain entry to the inflamed area.

Chemotaxis   Within 30 to 60 min after the onset of inflammation, a remarkable interaction occurs between the vascular endothelium and circulating neutrophils. First, the blood-borne neutrophils begin to stick to the inner surface of the endothelium. The process is quite specific since erythrocytes show no tendency to stick and other leukocytes do so only later, if at all. How the endothelium is made sticky by injury remains unknown.

Following their surface attachment, the neutrophils begin to manifest considerable amoeba-like activity. Soon a narrow amoeboid projection is inserted into the space between two endothelial cells, and the entire neutrophil then squeezes into the interstitium. The alterations of vessel structure described above may facilitate this process by loosening the intercellular connections, or the neutrophil may simply pry the connection apart by the force of its amoeboid movement. By this process (known as neutrophil exudation), huge numbers of neutrophils migrate into the inflamed areas of tissue and move toward the microbes. This entire response of the neutrophils is known as **chemotaxis** and is induced by chemical mediators that are generated in the inflamed area.

Movement of leukocytes into the tissue is usually not limited only to neutrophils. Monocytes follow, but much later, and once in the tissue are transformed into macrophages; meanwhile, some of the macrophages normally present in the tissue have begun to multiply by mitosis and to become motile. Thus, all the phagocytic cell types are present in the inflamed area. Usually the neutrophils predominate early in the infection but tend to die off more rapidly than the others, which thereby yields a predominantly macrocytic picture later. In contrast, in certain types of allergies and inflammatory responses to parasites the eosinophils are in striking preponderance. Thus, one of the tests for allergically induced "runny nose" is to study a small amount of nasal discharge for the presence of eosinophils.

Phagocytosis   Phagocytosis is the primary function of the inflammatory response, and the increased blood flow, vascular permeability, and leukocyte exudation serve only to ensure the presence of adequate numbers of phagocytes and to provide the milieu required for the performance of their function. Having arrived at the site, phagocytes must "pick out" what to attack by discerning certain characteristics on the surfaces of the foreign cells (or damaged native cells); in the case of nonspecific inflammation, what these characteristics are and how they trigger phagocytosis is not clear.

The process of ingestion is illustrated in Fig. 28-4. The phagocyte engulfs the organism by endocytosis, i.e., membrane invagination and pouch (**phagosome**) formation. Once inside, the microbe remains in the phagosome with a layer of phagocyte plasma membrane separating it from the phagocyte cytoplasm. The next step is known as **degranulation** (Fig. 28-4): The membrane that surrounds the phagosome makes contact with one of the phagocyte's lysosomes (these appear microscopically as "granules") that are filled with a variety of digestive enzymes. Once contact between the phagosome and lysosome is made, the membranes fuse (the disappearance of the lysosome granules during this fusion leads to the term "degranulation"), and the microbe is exposed to the lysosomal enzymes capable of killing it by breaking down its macromolecules.

The lysosomal enzymes are not the only mechanism in the phagosome for killing bacteria. The phagosome produces a high concentration of hydrogen peroxide that is extremely destructive to macromolecules.

The lysosomal enzymes and hydrogen peroxide not only kill the microorganism but also catabolize it into low-molecular-weight products that can then

FIG. 28-4. Large molecules and particulate substances enter the cell by endocytosis, forming a membrane-bound phagosome (at left in diagram). A phagosome may then merge with a lysosome, which brings together the digestive enzymes of the lysosome and the contents of the phagosome. After digestion has taken place, the contents may be released to the outside of the cell by exocytosis.

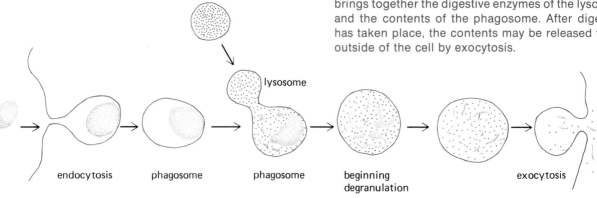

lysosome

lysosome

endocytosis   phagosome   phagosome   beginning
degranulation   exocytosis

be safely released from the phagocyte or actually utilized by the cell in its own metabolic processes. This entire process need not kill the phagocyte, which may repeat its function over and over before dying. Nondegradable foreign particles (such as wood, tattoo dyes, or metal) and certain species of microorganisms may be retained indefinitely in macrophages.

The neutrophils may also release lysosomal granules into the extracellular fluid; the enzymes released from these granules attack extracellular debris at the injury site, which makes it easier for the macrophages to phagocytize it at the battle's end and thus paves the way for repair of the damaged area.

Thus far we have presented the role of the phagocytes, but what are the microorganisms doing all this time to protect themselves? Most of them do very little. Certain kinds, however, release substances that diffuse into the phagocyte cytoplasm and disrupt the membranes of the lysosomes to thereby allow these potent chemicals to destroy the phagocyte itself.

Finally, it should be noted that, by mechanisms described below, some microbes are killed during inflammation without having been previously phagocytized.

Tissue Repair   The final stage of the inflammatory process is tissue repair. Depending upon the tissue involved, regeneration of organ-specific cells may or may not occur (for example, regeneration occurs in skin and liver but not in muscle or the central nervous system; the latter two tissues are incapable of cell division in adults). In addition, fibroblasts in the area divide rapidly and begin to secrete large quantities of collagen.

The end result may be complete repair (with or without a scar), abscess formation, or granuloma formation. An *abscess* is basically a bag of pus (microbes, leukocytes, and liquefied debris) walled off by fibroblasts and collagen. This occurs when tissue breakdown is very severe and when the microbes cannot be eliminated but only contained. When this stage has been reached, the abscess must be drained for it will not resolve spontaneously.

A *granuloma* occurs when the inflammation has been caused by certain microbes (such as the bacteria causing tuberculosis) that are engulfed by phagocytes but survive in them. It also occurs when

the inflammatory agent is a nonmicrobial substance that cannot be digested by the phagocytes. The granuloma consists of numerous layers of phagocytic-type cells, the central ones of which contain the offending material. The whole thing is, itself, usually surrounded by a fibrous capsule. Thus, a person may harbor live tuberculosis-producing bacteria for many years and show no ill effects as long as the microbes are contained in the granuloma and not allowed to escape.

Chemical Mediators of the Nonspecific Inflammatory Response

The events described in the previous section are elicited by chemical substances of varying origins. Some are released from tissue cells present in the area prior to the invasion; others are brought into the area by circulating leukocytes from which they are then released; still others are newly generated in the interstitial fluid of the area as a result of enzyme-mediated reactions. There are a large number of these chemical mediators and no quantitative assessment of their roles is presently available. We describe here three of the mediators that have been most studied to date: histamine, the kinin system, and the complement system. (See Table 28-3.)

Histamine   **Histamine** is present in many tissues but is particularly concentrated in mast cells, circulating basophils, and platelets. Release of histamine is induced by a variety of factors, and the released histamine acts directly on vascular smooth muscle and endothelial cells to cause vasodilation and increased permeability, respectively. In addition, histamine has profound effects on nonvascular smooth muscle, of which perhaps the most significant is its constriction of the respiratory airways.

TABLE 28-3 SUMMARY OF MAJOR MEDIATORS IN NONSPECIFIC INFLAMMATION	
Effect	**Mediators**
Vasodilation and increased permeability	Histamine
	Kinins
	Complement
Chemotaxis	Kinins
	Complement
Enhancement of phagocytosis	Complement
Direct microbial killing	Complement

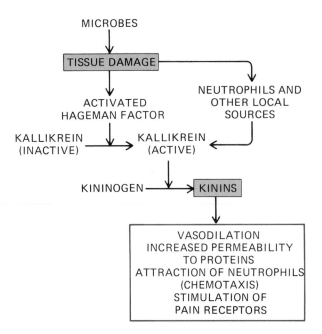

FIG. 28-5 Pathways for kinin generation in nonspecific inflammatory reactions.

The Kinin System   The kinin system was described briefly in Chap. 21 as a local regulator of blood flow in glandular structures. The **kinins** are small polypeptides whose vascular effects are similar to those of histamine and which, in addition, are powerful chemotaxic agents. The kinins are generated in plasma in the following ways (Fig. 28-5). There is present in plasma an enzyme, known as *kallikrein,* which exists normally in an inactive form but which, when activated, catalyzes the splitting-off of the kinins from another normally occurring plasma protein known appropriately as *kininogen.* In the absence of specific immune responses, the most important activator of kallikrein is a chemical substance we have already met in the section of Chap. 19 on hemostasis: the activated Hageman factor. Here is one of many connections between blood clotting and the immune system. Recall that the Hageman factor is, itself, inactive until activated locally by contact with altered vascular surfaces; once activated, it catalyzes the first steps in the cascade sequences that lead to both clotting and plasmin formation. Now we see that, in addition, it leads to the activation of the key enzyme of the kinin-generating system and thereby contributes to the first phases of the inflammatory process.

In addition to being generated from its inactive form in blood, active kallikrein is also found in many different tissues as well as in neutrophils. Once inflammation has begun, the kallikrein present in these sources contributes to the generation of kinin. It should be noted that, by their effects on afferent neuron terminals, the kinins may account for much of the pain associated with inflammation.

The Complement System   The **complement** system is yet another example (the clotting, anticlotting, and kinin systems are others) of a "system" that consists of a group of plasma proteins that normally circulate in the blood in an inactive state. Upon activation of the first protein of the group, there occurs a sequential cascade in which active molecules are generated from inactive precursors. In each of the other cascade systems we have described, only the final activated molecule (fibrin in clotting, plasmin in anticlotting, kinins in the kinin system) is the direct mediator of the biological response of that system; in contrast, in the complement system, certain of the different proteins activated along the cascade may function as mediators of a particular response (vasodilation, chemotaxis, etc.) as well as activators of the next step. Since this system consists of eleven distinct proteins (as well as several active protein fragments split during their enzymatic activation), the overall system is enormously complex and we will make no attempt to identify the roles of individual complement proteins except for purposes of illustration or clarity.

As summarized in Fig. 28-6, one or more of the activated complement molecules is capable of mediating virtually every step of the inflammatory response. Certain of them enhance vasodilation and increased permeability by direct effects on the vasculature as well as by stimulating release of histamine from mast cells (and platelets) and activating plasma kallikrein. They also facilitate neutrophil exudation by acting as powerful chemotaxic agents. Another complement component enhances phagocytosis by coating the microbe and permitting the phagocyte to ingest it more efficiently, most likely by acting as receptors to which the phagocyte can bind. This is a particularly important function because many virulent bacteria have a thick polysaccharide capsule that strongly resists engulfment by phagocytes. Finally, yet another complement pro-

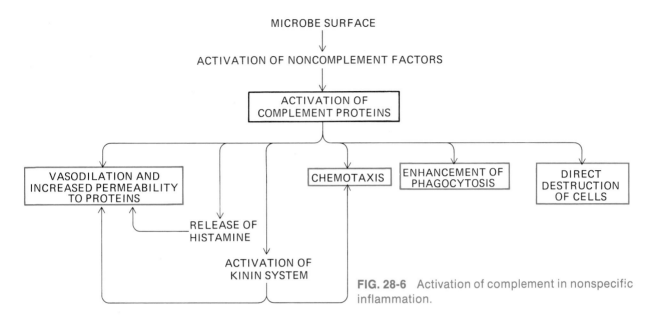

FIG. 28-6 Activation of complement in nonspecific inflammation.

tein mediates direct killing of microbes without prior phagocytosis. The complement becomes embedded in the microbial surface and somehow forms a tunnel; this makes the microbe leaky and eventually kills it.

In describing the actions of the complement system we have emphasized the sequential or cascade nature of the reactions required for the generation of active mediators, but we have so far ignored the problem of just how the sequence is initiated. As we shall see in a later section, antibody is required to activate the very first molecules in the complete complement sequence, but antibody is not present in nonspecific inflammatory responses. Fortunately, the first (antibody-dependent) steps can be bypassed; i.e., alternative mechanisms exist for initiating the sequence at a point beyond these first steps. Little biological activity is lost in bypassing the first steps because no inflammatory mediators are generated by them.

What are the initiators in this alternate antibody-independent complement pathway? The major one is a normally circulating plasma protein that is not itself a member of the complement family and is capable of binding to certain types of carbohydrate chains commonly present on bacterial or viral surfaces (but not present on the body's own cells). This binding changes the conformation so as to confer upon it the capacity to initiate the enzymatic splitting of complement C_3.

Systemic Manifestations of Inflammation

We have thus far described the *local* inflammatory response. What are the systemic, overall responses? Probably the single most common and striking systemic sign of injury is fever. The substance primarily responsible for the resetting of the hypothalamic thermostat, as described in Chap. 26, is a protein (**endogenous pyrogen**) released by the neutrophils (and perhaps other cells) that participate in the inflammatory response.

One would expect fever, being such a consistent concomitant of infection, to play some important protective role, and recent evidence, at least for nonmammalian species, strongly suggests that such is the case (this is a very important question in light of the widespread use of aspirin to suppress fever). In contrast to the possible benefits of fever, there is no question that an extremely high fever may be quite harmful, particularly its effects on the functioning of the central nervous system, and convulsions are not infrequent in highly febrile young children.

Endogenous pyrogen has effects in addition to eliciting fever; in particular, it causes movement of iron (and other trace metals) out of the plasma and into the liver. The net result is to reduce the extracellular concentration of iron, a phenomenon that is beneficial since bacteria require the presence of high concentrations of iron to multiply. Thus, endogenous pyrogen may exert a spectrum of effects that have the common denominator of enhancing resistance to infection.

Another systemic manifestation of many bacterial diseases is a marked increase in the synthesis and release of neutrophils by the bone marrow (in contrast, viral infections frequently are associated with decreased numbers of circulating neutrophils). This stimulation of neutrophil (''granulocyte'') production is mediated by a circulating chemical named **granulopoietin** (just as the hormone that stimulates erythrocyte production is called erythropoietin). Granulopoietin seems to be produced and released mainly by monocytes and macrophages, and its release is induced by the presence of bacterial products. Thus, the mononuclear phagocytic cells not only participate in the destruction of bacteria but, in the process, release chemicals that elicit increased production of the other major class of phagocytes, the neutrophils.

Interferon

Interferon is a nonspecific defense mechanism against viral infection. It is a protein that inhibits viral multiplication and is produced by several different cell types in response to a viral infection. Its production (or lack of production) can be understood from basic principles of protein synthesis (Chap. 3). When there is no virus in the host cell, the potential for interferon synthesis exists but the actual synthesis is repressed. Entry of a virus into the cell induces interferon synthesis by the usual DNA-RNA-ribosomal mechanisms; the inducer seems to be viral nucleic acid. Once synthesized, interferon may leave the cell, enter the circulation, and bind to the plasma membrane of other cells. (Thus, interferon produced by one cell can protect other cells from viral infection.) The interferon molecule does not, itself, have antiviral activity; rather its binding to the plasma membrane of a cell triggers the synthesis of ''antiviral'' proteins by the cell to which it is bound. These antiviral proteins act by blocking synthesis in the cell of macromolecules the virus requires for its multiplication. The fact that interferon itself is not antiviral but rather stimulates the cells to make their own antiviral proteins is another example of amplification—a single molecule of interferon can trigger the synthesis of a lot of antiviral protein so that only a few interferon

molecules seems to be required to protect a cell.

To reiterate, interferon is *not* specific; all viruses induce the same kind of interferon synthesis, and interferon in turn can inhibit the multiplication of many different viruses. For many years the hope has been to administer interferon to patients who suffer from viral infections in order to enhance their resistance. Several studies now completed with exogenous interferon have been successful in conferring increased resistance against different viruses and against a form of cancer.

SPECIFIC IMMUNE RESPONSES

Traditionally, specific immune responses are placed in one of two categories, according to the nature of the effector mechanisms employed in the response. One category has been termed **humoral,** or **antibody-mediated, immunity,** in recognition of the central role of circulating antibodies in the destructive process; the antibodies are secreted by *B lymphocytes* (to be more precise, by the *plasma cells* into which B lymphocytes differentiate upon being stimulated). The second category of specific immune defenses is **cell-mediated immunity,** and it is mediated not by antibodies but by intact cells, in this case the *T lymphocytes*.

The origin of B and T lymphocytes was described in Chap. 19. At some point in their development, cells destined to be T cells enter (or arise within) the thymus (thus the name T cell), which in some manner confers upon them the ability to differentiate and mature into cells competent to act as effectors in cell-mediated immunity. The T cells then leave the thymus and take up residence in the other lymphoid tissues, but the thymus continues to stimulate them and their offspring by means of thymosin, which its epithelial cells secrete. In contrast, the precursors of the B cells are derived from the bone marrow prior to their taking up residence in lymph nodes and other lymphoid tissues.

Unlike the nonspecific defense mechanisms, B-cell and T-cell responses depend upon the cells' ''recognizing'' the specific foreign matter to be attacked or neutralized. This recognition is made possible by the fact that molecular components of foreign cells (and other foreign matter) combine specifically with receptor sites on the surface of B cells and/or T cells and thereby trigger off the attack. These foreign molecular components that stimulate a specific immune response are known as **antigens.** The B-cell system is characterized by an ability to recognize an enormous variety of different specific antigens, and its major function is to confer, via its secreted antibodies, resistance against most bacteria (and their toxins) and some viruses. By contrast, the T-cell system appears to recognize a more limited number of antigens, but these antigens, as we shall see, are the crucial cell-surface markers that label cells "self" or "non-self"; the T cells confer major resistance against fungi, viruses (once they have become intracellular), parasites, the few bacteria that must live inside cells to survive, cancer cells, and solid-tissue transplants.

B cells and T cells influence each other in a variety of ways. On the one hand, antibodies (a B-cell product) may either facilitate or decrease the ability of an attacking T cell to destroy a foreign cell. On the other hand, T cells may either enhance or suppress the secretion of antibody by B cells. This last phenomenon led to the discovery that T cells do not constitute a homogenous population but are of three kinds: (1) **cytolytic T cells,** which upon activation perform the role classically ascribed to T cells—the killing of those foreign cells listed in the previous paragraph; (2) **"helper" T cells,** which enhance antibody production; and (3) **"suppressor" T cells,** which reduce antibody production. Thus, so interrelated are T cells and B cells that two of the three classes of T cells have as their function the regulation of B cells.

Humoral Immunity

Table 28-4 summarizes the sequence of events that result in antibody-directed destruction of bacteria. We withhold discussion of step 1, induction of antibody synthesis, until we have described the nature of antibodies and their mechanisms of action.

Antibodies   An **antibody** is a specialized protein capable of combining with the specific antigen that stimulated its production. The word *specific* is essential in the definition since an antigenic substance reacts only with the type of antibodies elicited by its own kind or an extremely closely related kind of molecule. Specificity is thus related to the structure of the antigen and its antibody.

Antibodies are all proteins, each composed of

TABLE 28-4	SEQUENCE OF EVENTS IN HUMORAL IMMUNITY AGAINST BACTERIA
1	Bacterial antigen is carried to lymphoid tissues where it binds to surface of specific B lymphocytes. (Macrophages "help" this process; T cells may "help" or "suppress" it.)
2	B lymphocyte differentiates into plasma cell and secretes into the blood antibody specific for that antigen.
3	Antibody circulates to site of infection and combines with antigen on surface of bacteria.
4	Presence of antibody bound to antigen facilitates phagocytosis and activates complement system that further enhances phagocytosis and also directly kills bacteria.
5	Certain of the B lymphocytes differentiate into memory cells capable of responding very rapidly should the bacteria be encountered again.

four interlinked polypeptide chains (Fig. 28-7). The two long chains are called *heavy chains,* and the two short ones are *light chains* (these chains are all joined by disulfide bridges). The amino acid sequences along the chains are constant for each of the five classes of antibodies from one antibody to the next except for relatively short sequences in portions of opposing light and heavy chains (Fig. 28-7). These sequences are unique for each of the extremely large number of antibodies that an individual can produce and constitute the binding sites (two per antibody) for the antigen specific for that antibody. The constant sequences in the "stem" of the heavy chains also contain binding sites for molecules and cells that function as the effectors for antibody action.

Antibodies belong to the family of proteins known as **immunoglobulins,** which may be subdi-

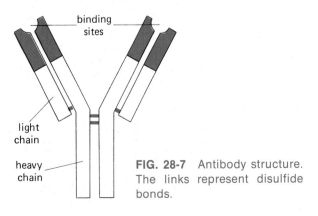

FIG. 28-7 Antibody structure. The links represent disulfide bonds.

vided into five classes according to the types of light and heavy chains they contain. It is very easy to misunderstand this concept, which states that there are five *classes* of immunoglobulins (antibodies), not five antibodies; each class contains many thousands of unique antibodies. These classes are designated by the letters G, A, M, D, and E after the symbol Ig (for immunoglobulin). IgG (the G stands for gamma; thus "gamma globulin" is a synonym for immunoglobulin G, the most abundant plasma antibodies) and IgM antibodies provide the bulk of specific immunity against bacteria and viruses. The other class we shall be concerned with is IgE, for these antibodies mediate certain allergic responses. IgA antibodies are produced by lymphoid tissue that lines the gastrointestinal, respiratory, reproductive, and urinary tracts and exert their major activities in the secretions of these tracts. The function of the IgD class is presently uncertain.

Many small chemicals injected into the body do not induce antibody formation because an essential determinant of a molecule's capacity to serve as an antigen is size (most antigens have molecular weights greater than 10,000). However, many smaller molecules can act as antigens after first attaching themselves to one of the host's proteins, forming a complex large enough to induce antibody formation. Still other low-molecular-weight substances incapable of inducing antibody synthesis because of their small size can combine with antibodies induced by another antigen; in such cases the structural unit of the large true antigen that was critical in the induction process must have been similar or identical to that of the small molecule. These last two phenomena explain why many small molecules can cause allergic attacks.

The antigens we shall be most concerned with are large protein or carbohydrate molecules that are in the outer surface of microbes or are microbial products, such as bacterial toxins. However, components of almost any foreign cell or molecule not normally present in the body may act as an antigen (moreover, even normal body components can induce antibody formation under unusual circumstances). Once the antibodies have been formed (the mechanism will be discussed in a subsequent section), they are released into the blood, reach the site where the antigen is located, and combine with it.

Functions of Antibodies   In a previous section we described how a local inflammatory response is induced nonspecifically by any tissue damage. Now we shall see that the presence of antigen-antibody complexes triggers off events that profoundly amplify the inflammatory response. In other words, the major function of humoral immune mechanisms is to enhance and make more efficient the inflammatory response and cell elimination already initiated in a nonspecific way by the invaders. Thus, vasodilation, increased vascular permeability to protein, neutrophil exudation, phagocytosis, and killing of microbes without prior phagocytosis are all markedly enhanced.

Activation of complement system   There are several mechanisms by which the presence of antigen-antibody complexes enhances inflammation, but the single most important involves the complement system. As described earlier, this system, which provides mediators for virtually every process in inflammation, is involved to some extent in the nonspecific inflammatory response, but the presence of antibody attached to antigen profoundly increases its participation because the antibody-antigen complex is a powerful activator of the first step in the complement sequence. Precisely how this activation occurs is not completely clear, but it is certain that the critical event is combination of the first complement molecule in the sequence to the antibody of the antibody-antigen complex.

It is important to note that the complement molecule binds not to the unique antigen-specific binding sites in the antibody's "prongs," but rather to binding sites in the antibody's stem. Since the latter are the same in virtually all antibodies of the IgG and IgM classes, the complement molecule will bind to *any* antibodies that belong to these classes. In other words, there is only one set of complement molecules, and once activated, they do essentially the same thing, regardless of the specific identity of the invader. In contrast, the formation of antibodies to antigens on the invader and their subsequent combination are highly specific.

To reiterate, the function of the antibodies is to "identify" the invading cells as foreign by combining with antibody-specific antigens on the cell's surface; the complement system is subsequently activated when the first complement molecules in the sequence combine with this antigen-bound antibody, and the cascade of activated complement molecules then mediate the actual attack. Once activated, the complement components that facilitate phagocytosis or kill the microbes outright are so nonspecific in their ability to bind to cells that they are potentially capable of attacking the body's own cells as well; this does not normally occur because their period of activation is extremely short-lived and they do not have time to get to an "innocent" host cell from the microbe on whose surface the original activation had occured.

Direct enhancement of phagocytosis   Activation of complement is not the only mechanism by which antibodies enhance phagocytosis. Merely the presence of antibody attached to antigen on the microbe's surface has some enhancing effect. This combination of antibody with microbial antigen increases the activity of phagocytes bound to the same microbe by somehow enhancing release of membrane-bound calcium into the phagocyte's cytosol and thereby triggers the interactions that lead to membrane invagination.

Direct neutralization of bacterial toxins and viruses
Bacterial toxins and certain viral components act as antigens to induce antibody production. The antibodies then combine chemically with the toxins and viruses to "neutralize" them. Neutralization in reference to a virus means that the combined antibody prevents attachment of the virus to host cell membranes and thereby prevents virus entry into the cell. Similarly, antibodies neutralize bacterial toxins by combining chemically with them and thus prevent the interaction of the toxin with susceptible cell membrane sites. In both cases, since each antibody has two binding sites for combination with antigen, chains of antibody-antigen complexes are formed (see Fig. 28-8 below) and are then phagocytized.

Antibody Production   When a foreign antigen reaches a lymphoid tissue, it triggers off antibody synthesis (Table 28-4 and Fig. 28-9), by certain cells there. The antigen may reach the spleen via the blood, but much more commonly it is carried from its site of entry into the body via the lymphatics to lymph nodes. There it stimulates a tiny fraction of the B lymphocytes to enlarge and undergo rapid cell division, most of the progeny of which then differentiate into plasma cells, which are the active antibody producers. The most striking aspect of this transformation is a marked expansion of the

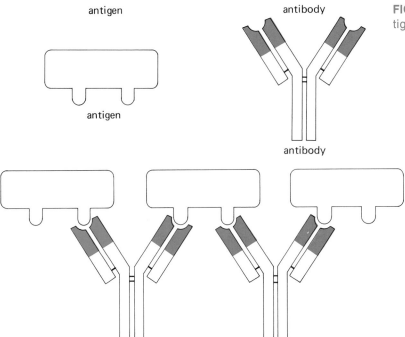

antigen

antibody

antigen

antibody

FIG. 28-8   Interlocking complex of antigens and antibodies.

FIG. 28-9 Induction of antibody synthesis by a microbe. The "processing" of antigen by macrophage is still not clear. The possible role of helper or suppressor T cells is not shown.

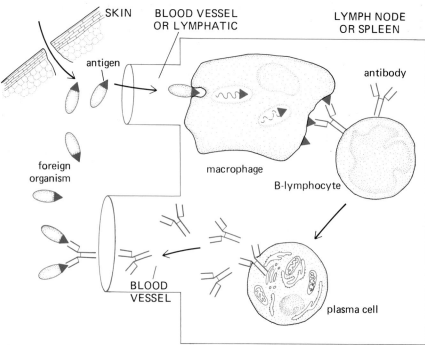

SKIN

BLOOD VESSEL OR LYMPHATIC

LYMPH NODE OR SPLEEN

antigen

antibody

macrophage

B-lymphocyte

foreign organism

plasma cell

BLOOD VESSEL

FIG. 28-10 Electron micrograph of a guinea pig plasma cell. Note the extensive endoplasmic reticulum.

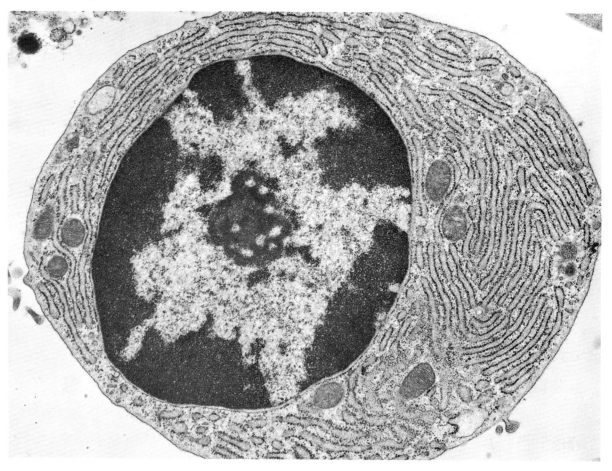

cytoplasm, which consists almost entirely of the granular type of endoplasmic reticulum (Fig. 28-10) found in other cells that manufacture protein for export; after synthesis, the antibodies are released into the blood or lymph. Some of the B-cell progeny do not fully differentiate into plasma cells but rather into "memory" cells, ready to respond rapidly should the antigen ever reappear at a future time; thus, the presence of the "memory" cells avoids much of the delay that occurred during the initial infection.

Note that we stated that only a tiny fraction of the total B cells respond to any given antigen. It can be demonstrated that different antigens stimulate entirely different populations (**clones**) of B cells. This is because the cells of any one lymphocyte clone (and the plasma cells it gives rise to) are capable of secreting only one kind of antibody. Thus, according to this clonal theory, different antigens do not direct a single cell to produce different antibodies; rather each specific antigen triggers activity in the clone of cells already predetermined to secrete only antibody specific to that antigen. The antigen selects this particular clone and no other because the B cell displays on its surface immunoglobulin binding sites identical to those of the antibodies that it is capable of producing (after differentiation into a plasma cell). These surface immunoglobulins act as receptor sites with which the antigen can combine to thereby trigger off the entire process of division, differentiation, and antibody secretion just described. The staggering but statistically possible implication is that there must exist millions of different clones, one for each of the possible antigens an individual *might* encounter during life.

In concluding this section on antibody synthesis, it should be pointed out that interaction between antigen and B-cell receptor site is considerably more complex than that just described above, for it seems clear that macrophages also play a crucial role. During antigen activation of the B cells, there occurs a clustering of macrophages around the relevant B-cell clone. It is likely that the macrophages "process" and "present" the antigen in some way so as to allow it to activate the B-cell receptor sites, but the actual events remain unknown. As mentioned earlier, T cells also play a role, for they may facilitate or inhibit B-cell antibody formation in response to an antigenic stimulus.

Active and Passive Humoral Immunity   We have been discussing antibody formation without regard to the course of events in time. The response of the antibody-producing machinery to invasion by a foreign antigen varies enormously, depending upon whether it has previously been exposed to that antigen. Antibody response to the first contact with a microbial antigen occurs slowly over several days, with some circulating antibody remaining for long periods of time, but a subsequent infection elicits an immediate and marked outpouring of additional antibody. It is evident that this type of "memory" confers a greatly enhanced resistance toward subsequent infection with a particular microorganism. This resistance, built up as a result of actual contact with microorganisms and their toxins or other antigenic components, is known as **active immunity.**

Until modern times, the only way to develop active immunity was actually to suffer an infection, but now a variety of other medical techniques are used, i.e., the injection of vaccines or microbial derivatives. The actual material injected may be small quantities of living or weakened microbes, e.g., polio vaccine, small quantities of toxins, or harmless antigenic materials derived from the microorganism or its toxin. The general principle is always the same: Exposure of the body to the agent results in the induction of the antibody-synthesizing machinery required for rapid, effective response to possible future infection by that particular organism. However, for many microorganisms the memory component of the antibody response does not occur, and antibody formation follows the same time course regardless of how often the body has been infected with the particular microorganism.

A second kind of immunity, known as **passive immunity,** is simply the direct transfer of actively formed antibodies from one person (or animal) to another, by which the recipient thereby receives preformed antibodies. This exchange normally occurs between fetus and mother across the placenta and is an important source of protection for the infant during the first months of life when the antibody-synthesizing capacity is relatively poor. The same principle is used clinically when specific antibodies or pooled gamma globulin are given to a person exposed to or actually suffering from certain infections, such as measles, hepatitis, or tetanus. The protection afforded by this transfer of antibodies is relatively short-lived, usually lasting only a few weeks. The procedure is not without danger since the injected antibodies (often of non-human origin) may themselves serve as antigens and elicit antibody production by the recipient and possibly severe allergic responses.

Summary We may now summarize the interplay between nonspecific and specific immune mechanisms in resisting a bacterial infection. When a bacterium is encountered for the first time, nonspecific defense mechanisms resist its entry and, if entry is gained, attempt to eliminate it by phagocytosis (and, to some extent, by nonphagocytic killing).

Simultaneously, the bacterial antigens induce the differentiation of specific B-cell clones into plasma cells capable of antibody production. If the nonspecific defenses are rapidly successful, these specific immune responses may never play an important role. If only partly successful, the infection may persist long enough for significant amounts of antibody to reach the scene; antibody activates its chemical amplification system—complement—that both enhances phagocytosis and directly destroys the foreign cells. In either case all subsequent encounters with that type of bacteria will be associated with the same sequence of events, with the crucial difference that the specific immune re-

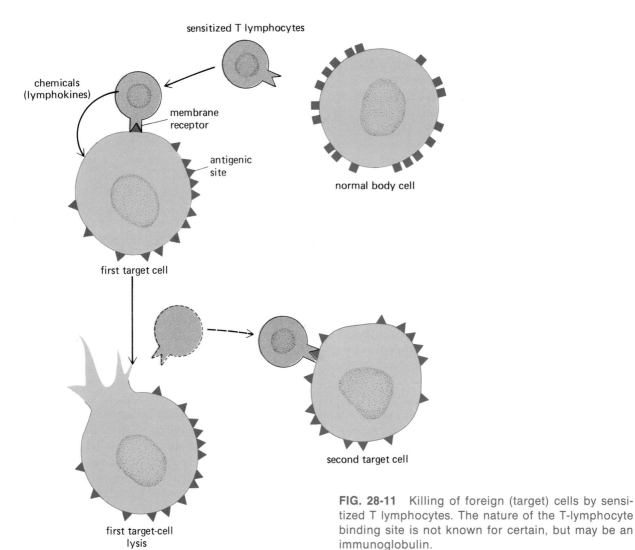

FIG. 28-11 Killing of foreign (target) cells by sensitized T lymphocytes. The nature of the T-lymphocyte binding site is not known for certain, but may be an immunoglobulin.

sponses are brought into play much sooner and with greater force; i.e., the person would enjoy active immunity against that type of bacteria.

Cell-Mediated Immunity

The T lymphocytes are responsible for cell-mediated immunity. T cells, like B cells, are clonal in that each clone of T cells bears on its cell surface genetically determined receptor sites for a specific foreign antigen. Upon exposure to and combination with the appropriate antigen (in the presence of macrophages), the T cell becomes "sensitized," i.e., undergoes enlargement and mitosis, with differentiation of the daughter cells. So far, the story sounds similar to that described earlier for B cells, but the crucial difference has to do with the activities of these differentiated daughter cells: Instead of secreting antibodies (as do the plasma-cell progeny of B cells), these cells actually combine with the antigen on the surface of the foreign cell, and this triggers the release locally of a powerful battery of chemicals (known collectively as **lymphokines**) (Fig. 28-11).

The chemicals released, when sensitized lymphocytes combine with a specific antigen, kill cells directly (Fig. 28-11) and, in addition, act as an amplification system for the facilitation of the inflammatory response and phagocytosis. Thus, cell-mediated immunity is analogous to humoral immunity in that it serves, in large part, to enhance and make more efficient the nonspecific defense mechanisms already elicited by the foreign material. The major difference is that humoral immunity utilizes a circulating group of plasma proteins (the complement system) as its major amplification system whereas the T cells literally produce and secrete their own chemical amplification system.

As might be predicted, some of these chemicals are chemotaxic factors. These serve to attract monocytes and some neutrophils to the area. The monocytes are converted to macrophages and begin their job of phagocytosis. Not only do the T cells secrete chemotaxic factors to attract the macrophages-to-be; they secrete another substance that keeps the macrophages in the area and stimulates them to greater phagocytic activity (indeed, such revved-up macrophages are known as "angry" macrophages). In addition to facilitating the killing of target cells by phagocytosis, the lymphocytes

secrete so-called cytotoxins that are able to kill target cells directly, i.e., without phagocytosis. Here is another analogy to the complement system, with its ability to destroy cells directly as well as to facilitate phagocytosis. Interferon is also released in large quantity from T cells.

The cytolytic T cells we have been describing in this section belong to only one of three distinct T-cell populations mentioned earlier. Cells of the other two classes play totally different roles in specific immunity, namely the facilitation or suppression of B-cell production of antibody. At present we know little about how these "helper" and "suppressor" cells perform their functions but it is likely that they combine with antigenic sites on the microbe, and release chemicals that act upon the B cells to influence their differentiation into plasma cells. Their possible roles in both the excessive and deficient antibody production seen in many human diseases are presently being intensively studied.

One aspect of T-lymphocyte function against viruses requires emphasis (Table 28-5). As stated several times, viruses take up residence in the body's cells; accordingly, in order to attack a virus once it has gained cellular entry, the lymphocyte must destroy the cell, itself. Such destruction occurs because the viral nucleic acid codes for a protein foreign to the body, and once this protein has been synthesized by the host cell, it becomes located on the cell's surface where it can act as an antigen to elicit a T-cell attack. The result is destruction of the host cell with release of the viruses that can then be directly attacked. Generally, only a few host cells must be sacrificed in this way, but once viruses have had a chance to replicate and spread from cell to cell, so many host cells may be attacked by the body's own defenses that serious malfunction may result.

As is true for the B system, some of the sensitized T cells do not actually participate in the immune response but serve as a "memory bank" that greatly speeds up and enhances the immune response if the person is ever exposed to the specific antigen again. Thus, active immunity exists for cell-mediated immune responses just as for antibody responses. Passive immunity also exists in this system and can be conferred by administering sensitized lymphocytes taken from a previously infected person (or animal).

TABLE 28-5	SUMMARY OF HOST RESPONSES TO VIRUSES	
	Main Cells Involved	Comment on Action
Nonspecific Responses:		
Anatomic barriers	Surface linings of body	Simple barrier; antiviral chemicals
Inflammation	Tissue macrophages	Phagocytosis of extracellular virus
Interferon	Multiple cell types	Prevention of viral replication inside host cells
Specific Responses:		
Humoral	Plasma cells derived from B lymphocytes secrete antibodies that:	Neutralize virus and thus prevent entry to cell
		Activate complement that leads to both enhanced phagocytosis and direct destruction of extracellular virus
Cell-mediated	Sensitized T lymphocytes secrete chemicals that:	Destroy host cell and thus induce release of virus so that it can be phagocytized
		Prevent viral replication (this secreted chemical is interferon)

Immune Surveillance: Defense Against Cancer   A major function of cell-mediated immunity is to recognize and destroy cancer cells. This is made possible by the fact that virtually all cancer cells have some surface antigens different from those of other body cells and can, therefore, be recognized as "foreign." It is likely that cancer arises as a result of genetic alteration (by viruses, chemicals, radiation, etc.) in previously normal body cells. One manifestation of the genetic change is the appearance of the new surface antigens. Circulating T cells encounter and become sensitized to these foreign cells, combine with the antigens on their surface, and release the effector chemicals that destroy the cells by the mechanisms described above. It is presently believed that such transformations occur very frequently, i.e., that we may "get cancer once a day" (one expert's estimate), but that the cells are destroyed as fast as they arise. According to this view, only when the cell-mediated system is ineffective in either recognizing or destroying the cells do they multiply and produce clinical cancer.

Rejection of Tissue Transplants   The cell-mediated immune system is also mainly responsible for the recognition and destruction, i.e., **rejection,** of tissue transplants (Fig. 28-12). On the surfaces of all nucleated cells of an individual's body are antigenic protein molecules known as **histocompatibility antigens.** The genes that code for these proteins are, of course, inherited from one's parents, so that the offspring's group of antigens are, in part, similar to the parents' but not identical. Clearly, the more closely related two people are the more similar these antigens will be, but no two people (other than identical twins) have identical groups.

When tissue is transplanted from one individual to another, those surface antigens that differ from the recipient's are recognized as foreign and are destroyed by sensitized circulating T cells.

Some of the most valuable tools aimed at reducing graft rejection are radiation and drugs that kill actively dividing lymphocytes and, thereby, decrease the T-cell population. Unfortunately, this also results in depletion of B cells as well so that antibody production is diminished and the patient becomes highly susceptible to infection. A more discriminating method presently being tried is to prepare and inject into the recipient antibodies against the T cells; by this means, the T cells would be destroyed but not the B cells. As might be predicted from the previous section, graft recipients who receive medication to reduce their T-cell activity manifest an increased tendency to develop certain types of cancer.

Related to the general problem of graft rejection is one of the major unsolved questions of immunology: How does the body avoid producing antibodies or sensitized lymphocytes to its own cells; i.e., how does it distinguish self-antigens from non-

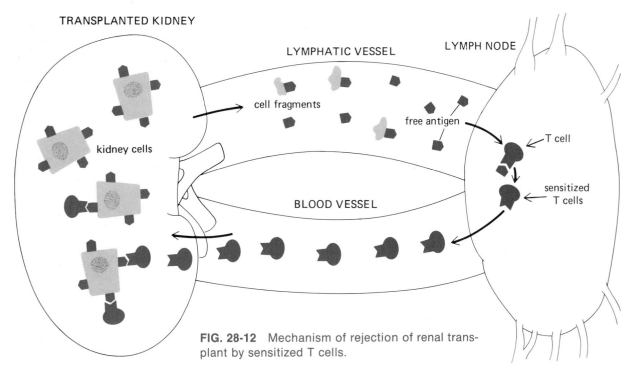

FIG. 28-12 Mechanism of rejection of renal transplant by sensitized T cells.

self-antigens? In general, it appears that any antigens present during embryonic and very early neonatal life are recognized as self and no antibodies or sensitized lymphocytes are formed against them later in life, following maturation of specific immune mechanisms. This can be shown by fooling the embryo in the following manner: Foreign mouse cells are injected into an embryo mouse during intrauterine life; months later, when the mature mouse is given a graft from the same foreign mouse, the graft is not rejected.

Of course, the generalization stated above is empirical fact, not explanation. Present evidence warrants the generalization that the thymus, as might be predicted, is very much involved in this "imprinting" of self-recognition but just how is simply not known. The question is clearly of more than academic interest since, if we understood the mechanism by which tolerance for one's own tissues is established, then we might be able to confer tolerance for transplants.

TRANSFUSION REACTIONS AND BLOOD TYPES

Transfusion reactions are a special example of tissue rejection, which illustrates the fact that antibodies rather than sensitized T cells can sometimes be the major factor in leading to the destruction of nonmicrobial cells. Among the large numbers of erythrocyte membrane antigens, we still recognize those designated A, B, and O as most important. These antigens are inherited, A and B being dominant. Thus, an individual with the genes for either A and O or B and O will develop only the A or B antigen. Accordingly, the possible blood types are A, B, O, and AB.

If the typical pattern of antibody induction were followed, one would expect that a type A person would develop antibodies against type B cells only if the B cells were introduced into the body. However, what is atypical of this system is that even without initial exposure the type A person always has a high plasma concentration of anti-B antibodies. The sequence of events during early life that leads to the presence of the so-called natural antibodies in all type A persons is unknown. Similarly, type B persons have high levels of anti-A antibodies; type AB persons obviously have neither anti-A nor anti-B antibodies; type O persons have both; anti-O antibodies are usually not present in anyone.

With this information as background, what will happen if a type A person is given type B blood? There are two incompatibilities: (1) The recipient's

TABLE 28-6	SUMMARY OF ABO BLOOD-TYPE INTERACTIONS	
Recipient	Donor	Compatible ?
A	O, A	yes
B	O, B	yes
AB	O, A, B, AB	yes
O	O	yes
A	AB, B	no
B	AB, A	no
AB	– – –	– – –
O	AB, A, B	no

anti-B antibodies cause the transfused cells to be attacked; and (2) the anti-A antibodies in the transfused plasma cause the recipient's cells to be attacked. The latter is generally of little consequence, however, because the transfused antibodies become so diluted in the recipient's plasma that they are ineffective. It is the destruction of the transfused cells that produces the problems. The range of possibilities is shown in Table 28-6. It should be evident why type O people are frequently called universal donors whereas type AB people are universal recipients. These terms, however, are misleading and dangerous since there are a host of other incompatible erythrocyte antigens and plasma antibodies besides those of the ABO type. Therefore, except in dire emergency, the blood of donor and recipient must be carefully matched.

Another antigen of medical importance is the so-called **Rh factor** (because it was first studied in rhesus monkeys) now known to be a group of erythrocyte membrane antigens. The Rh system follows the classic immunity pattern in that no one develops anti-Rh antibodies unless exposed to Rh-type cells (usually termed Rh-positive cells) from another person. Although this can be a problem in an Rh-negative person, i.e., one whose cells have no Rh antigen, subjected to multiple transfusions with Rh-positive blood, its major importance is in the mother-fetus relationship.

When an Rh-negative mother carries an Rh-positive fetus, some of the fetal erythrocytes may cross the placental barriers into the maternal circulation and thus induce her to synthesize anti-Rh antibodies. Because the movement of fetal erythrocytes

into the maternal circulation occurs mainly during separation of the placenta at delivery, a first Rh-positive pregnancy rarely offers any danger to the fetus, since delivery occurs before the antibodies can be made. In future pregnancies, however, these anti-Rh antibodies will already be present in the mother and can cross the placenta to attack the erythrocytes of an Rh-positive fetus. The risk increases with each Rh-positive pregnancy as the mother becomes more and more sensitized. Fortunately, Rh disease can be prevented by giving any Rh-negative mother gamma globulin against Rh erythrocytes within 72 hr after she has delivered an Rh-positive infant. These exogenous antibodies bind to the antigenic sites on any Rh erythrocytes that enter the mother's blood during delivery and prevent them from inducing antibody synthesis by the mother (both the cells and exogenous antibody are soon destroyed).

HYPERSENSITIVITY AND TISSUE DAMAGE

Immune responses obviously evolved to protect the body against invasion by foreign matter. Unfortunately, they frequently cause malfunction or damage to the body itself. The term **hypersensitivity** refers to an acquired reactivity to an antigen that can result in bodily damage upon subsequent exposure to that particular antigen. Hypersensitivity responses may be due to activation of either the humoral or cell-mediated system. There are a variety of types of hypersensitivity responses, and we shall describe only two categories: allergy and autoimmune disease.

Allergy

A certain portion of the population is susceptible to sensitization by environmental antigens such as pollen, dusts, foods, etc. Initial exposure to the antigen leads to some antibody synthesis but, more important, to the memory storage that characterizes active immunity. Upon reexposure, the antigen elicits a more powerful antibody response. So far, none of this is unusual. What is it then that leads to body damage? The fact is that these particular antigens stimulate the production of the IgE class of antibodies that, upon their release from plasma cells, circulate to various parts of the body, attach themselves to mast cells (and basophils), and remain there. When exposure occurs to the appropriate antigen, the antigen combines with the IgE attached to the mast cell, the complex triggers entry of calcium into the cell and a resulting release of the mast cell's histamine and other vasoactive chemicals (complement is not involved in IgE effects). These chemicals then initiate a local inflammatory response. Thus, the symptoms of allergy are due to the various effects of these chemicals and the body site in which the antigen–IgE–mast cell combination occurs. For example, when a previously sensitized person inhales ragweed pollen, the antigen combines with IgE-mast cells in the respiratory passages. The chemicals released cause increased mucus secretion, increased blood flow, leakage of protein, and contraction of the smooth muscle that lines the airways. Thus, there follow the symptoms of congestion, running nose, sneezing, and difficulty in breathing that characterize hayfever.

In this manner, the symptoms of allergy may be localized to the site of entry of the antigen. However, sometimes systemic symptoms may result if very large amounts of the vasoactive chemicals released enter the circulation and cause severe hypotension and bronchiolar constriction. This sequence of events (*anaphylactic shock*) can actually cause death and can be elicited in some sensitized people by the antigen in a single bee sting.

A major puzzle to biologists is the inappropriate nature of most allergic responses, which are usually far more damaging to the body than the antigen that triggers them. In other words, we clearly see the maladaptive nature of antigen-IgE reactions but we do not know what normal physiologic function is subsumed by IgE antibodies.

Autoimmune Disease

We must qualify a generalization made previously by pointing out that the body does, all too often, produce antibodies or sensitized T cells against its own tissues, and the result is cell damage or alteration of function. A growing number of human diseases (certain thyroid diseases, for example) are being recognized as **autoimmune** in origin. There are many causes, all of which are characterized by a breakdown in self-recognition that results in turning the body's immune mechanisms against its own tissues.

SECTION B: METABOLISM OF FOREIGN CHEMICALS

The body is exposed to a huge number of environmental chemicals, including inorganic nonnutrient elements, naturally occurring fungal and plant toxins, and synthetic chemicals. This last category is by far the largest, since there are now more than 10,000 foreign chemicals being commercially synthesized (over 1 million have been synthesized at one time or another); these are "foreign" in the sense that they are not normally found in nature. These foreign chemicals inevitably find their way into the body, either because they are purposely administered as drugs (medical or "recreational"), or simply because they are in the air, water, and food we use.

As described in Section A of this chapter, foreign materials can induce inflammation and specific immune responses. However, these defenses are directed mainly against foreign cells and, although noncellular foreign chemicals can also elicit certain of them (as in allergy, for example), such immune responses do not constitute the major defense mechanisms against most foreign chemicals. Rather, molecular alteration (**biotransformation**) and excretion constitute the primary mechanisms.

Of prime concern for the body's handling of any foreign chemical are those factors that determine the effective concentration of the chemical at its sites of action (Fig. 28-13). First, the chemical must gain entry to the body through the gastrointestinal tract, lungs, or skin (or placenta in the case of a fetus). Accordingly, its ability to move across these barriers will have an important influence on its blood concentration. As Fig. 28-13 illustrates, however, the rate of entry into the body is only one of many factors that determine the concentration of the chemical at its site of action. Once in the blood, the chemical may become bound reversibly to plasma proteins or to erythrocytes; this lowers its free concentration and, thereby, its ability to alter cell function. It may be transported into storage depots (for example, DDT into fat tissue), or it may undergo biotransformation. The metabolites that result from these latter enzyme-mediated reactions themselves enter the blood and are subject to the same fates as the parent molecules. Finally, the foreign chemical and/or its metabolites may be elim-

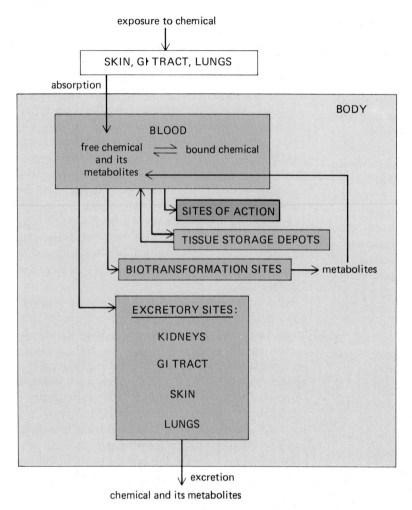

FIG. 28-13 Metabolic pathways for foreign chemicals.

inated from the body in the urine, expired air, skin secretions, or feces (having been deposited in the latter by biliary secretion). All these processes serve to determine the free blood concentration of the chemical, which in turn is a major determinant of the amount of chemical that reaches its sites of action. We will now discuss each of these processes.

ABSORPTION

In practice, most organic molecules move through the lining of some portion of the gastrointestinal tract fairly readily, either by simple diffusion or by carrier-mediated transport. This should not be sur-

prising, since the gastrointestinal tract evolved to favor absorption of the wide variety of nutrient molecules in the environment; the nonnutrient synthetic organic chemicals are the beneficiaries of these relatively nondiscriminating transport mechanisms.

The lung alveoli are also highly permeable to most organic chemicals and therefore offer an easy entrance route for airborne chemicals. They are also an important entry site for airborn metals, which generally penetrate the gastrointestinal tract very poorly. One important aspect of absorption through the lungs is that the liver does not get first crack at the chemical (as it does when entry is via the gastrointestinal tract); by the same token, any chemical that is toxic to the liver is not as dangerous when it enters via the lungs.

Lipid solubility is all important for entry through the skin, so that this route is of little importance for charged molecules but can be used by oils, steroids, and other lipids.

The penetration of the placental membranes by foreign chemicals is one of the most important fields in toxicology, since the effects of environmental agents on the fetus during critical periods of development may be quite marked and, in many cases, irreversible (thalidomide offers a tragic example). We are still relatively ignorant of placental transport mechanisms, but it is clear that diffusion is an important mechanism for lipid-soluble substances and that carrier-mediated systems (which evolved for the carriage of endogenous nutrients) may be usurped by foreign chemicals to gain entry into the fetus.

STORAGE DEPOTS

The major storage depots for foreign chemicals are cell proteins, bone, and fat. The chemical bound in these sites is in equilibrium with the free chemical in the blood so that an increase in blood concentration causes more movement into storage. Conversely, as the chemical is eliminated from the body and its blood concentration falls, movement occurs out of storage depots. It sometimes happens that the storage sites accumulate so much chemical that they themselves become damaged. Finally, there is the possibility of rapid release with potentially toxic effects.

EXCRETION

To appear in the urine, a chemical must be either filtered through the glomerulus or secreted across the tubular epithelium (Chap. 24). Glomerular filtration is, as emphasized in Chap. 24, a bulk flow process so that all low-molecular-weight substances in plasma undergo filtration; accordingly, there is considerable filtration of most environmental chemicals except for those that are mainly bound to plasma proteins (note that protein binding is, therefore, a mixed blessing in that it reduces toxicity but impedes excretion). In contrast, tubular secretion is by discrete transport processes, and many envi-

ronmental chemicals (penicillin is a good example) utilize the mediated transport systems available for naturally occurring substances.

Once in the tubular lumen, either via filtration or tubular secretion, the foreign chemical may still not be excreted for it may be reabsorbed back across the tubular epithelium into the blood. This is a major problem because so many foreign chemicals are highly lipid-soluble; as the filtered fluid moves along the renal tubules, these molecules passively diffuse along with reabsorbed water through the tubular epithelium and back into the blood. The net result is that little is excreted in the urine, and the chemical is retained in the body. If these chemicals could be transformed into more polar (and, therefore, less lipid-soluble) molecules, their passive reabsorption from the tubule would be retarded and they would be excreted more readily. This type of transformation is precisely what occurs in the liver, as described in the next section.

An analogous problem exists for those foreign molecules (and trace metals) excreted into the bile. Many of these substances, having reached the lumen of the small intestine, are absorbed back into the blood and thereby escape excretion in the feces. This cyclic enterohepatic circulation was described in Chap. 25.

BIOTRANSFORMATION

The metabolic alteration of foreign molecules occurs mainly in the liver (but to some extent also in the kidneys, skin, placenta, and other organs). A large number of distinct enzymes and pathways are involved, but the common denominator of most of them is that they transform chemicals into more polar, less lipid-soluble substances. One consequence of this transformation is that the chemical may be rendered less toxic, but this is not always so. The second, more important, consequence is that its tubular reabsorption is diminished and urinary excretion facilitated. Similarly, for substances handled by biliary excretion, gut absorption of the metabolite is less likely so that fecal excretion is enhanced.

The hepatic enzymes that perform these transformations were first discovered in the context of drug metabolism and were called drug-metabolizing enzymes, but their spectrum of action is much wider than this. The preferred name is **microsomal enzyme system (MES)** to denote the fact that they are found mainly in the smooth endoplasmic reticulum. One of the most important facts about this enzyme system is that it is easily inducible; i.e., the activity of the enzymes can be greatly increased by exposure to a chemical that acts as a substrate for the system. However, all is not really so rosy, for the hepatic biotransformation mechanisms vividly demonstrate how an adaptive response may, under some circumstances, turn out to be maladaptive. These enzymes all too frequently "toxify" rather than "detoxify" a drug or pollutant; in fact many foreign chemicals are quite nontoxic until the liver enzymes biotransform them.

Of particular importance is the likelihood that many, if not all, chemicals that cause cancer do so only after biotransformation. The microsomal enzymes can also cause problems in another way, because they evolved primarily not to defend against foreign chemicals (which were much less prevalent during our evolution) but rather to metabolize endogenous substrates, particularly steroids and other fat-soluble molecules. Therefore, their induction by a drug or pollutant increases metabolism not only of that drug or pollutant but of the endogenous substrates as well. The result is a decreased concentration in the body of that normal substrate.

Another fact of great importance concerning the microsomal enzyme system is that, just as certain chemicals induce it, others inhibit it. The presence of such chemicals in the environment could have deleterious effects on the system's capacity to protect against those chemicals it transforms. (Just to illustrate how complex this picture can be, note that any chemical that inhibits the microsomal enzyme system may actually confer protection against those other chemicals that must undergo transformation in order to become toxic.)

Alcohol: An Example of Biotransformation

Alcohol (specifically one type, *ethanol*) offers an excellent example of the role biotransformation plays in determining a substance's toxicity and its influence on the body's responses to other chemicals. The overuse of alcohol is associated with liver damage, and for many years it was thought that this damage was due to the malnutrition that so fre-

quently accompanies alcoholism. It is now clear that, although severe malnutrition may play some role, the toxic damage to the liver is caused mainly by the metabolites of alcohol, produced by the liver cells themselves.

Alcohol is initially broken down by liver cells to hydrogen and acetaldehyde, and these seem to be the major culprits. Acetaldehyde is a very toxic chemical that damages mitochondria. The excess hydrogen exerts its damaging effect more subtly by causing increased accumulation of fat (in the form of triacylglycerol) in the liver cells by two pathways: (1) By mass action it drives the synthesis of fatty acids and glycerol (hydrogen is a substrate in these reactions), the building blocks of triacylglycerol; and (2) the hydrogen is "burned" in the mitochondria (see oxidative phosphorylation, Chap. 3) to supply the liver's energy requirement, and this allows fat, the usual source of hydrogen for the liver's mitochondria, to accumulate. The accumulation of fat leads to enlargement of liver cells and damage to them. (We have restricted our discussion to the toxic effects of hydrogen and acetaldehyde on liver cells; other organs and tissues are also damaged by them in a variety of ways.)

The metabolism of alcohol not only damages the liver cells but leads to marked changes in the metabolism of other drugs and foreign chemicals, all explainable by the fact that these other agents share certain of alcohol's metabolic pathways. When alcohol is taken at the same time as another drug, such as a barbiturate, enhanced effects of the two drugs are observed. This is because the alcohol and the barbiturate compete for the same hepatic microsomal enzyme, resulting in a decreased rate of catabolism (and, therefore, increased blood concentrations) of both.

The situation just described was for simultaneous administration of the two drugs and is independent of whether the individual is a chronic overuser of alcohol or not. Let us look now at the long-term effects of chronic overuse. Alcohol is a powerful inducer of the microsomal enzyme system so that its chronic presence causes an increase in the system's activity. The result is an increase in the rate of catabolism of other chemicals, such as barbiturates, that share the enzymes. Thus, the chronic use of alcohol decreases the potency of any given dose of barbiturate by causing its blood concentration to fall more rapidly.

Putting these last two paragraphs together, one can see that the effect of alcohol on the blood concentrations of other chemicals depends both on whether the two drugs are taken simultaneously and whether the person is a chronic overuser of alcohol. But the situation is even more complex, for if the person has suffered serious liver damage due to chronic alcoholism, then the microsomal enzyme system, instead of being induced, may be inadequate simply because the cells have been damaged. At this stage, then, the metabolism of the other agents will be diminished and their effects will be exaggerated.

Finally, we have presented the chronic effects of alcohol only in terms of the effects on the metabolism of other chemicals. It should be clear, however, that similar effects are exerted on the metabolism of alcohol itself. Thus, chronic overuse, but before significant damage has occurred, results in increased catabolism (because the induction of the microsomal enzymes), and this accounts for much of the "tolerance" to alcohol, i.e., the fact that increasing doses must be taken to achieve a given effect. Once severe damage has occurred, however, the decline in rate of catabolism may lead to increased sensitivity to a given dose, just the opposite of tolerance.

INORGANIC ELEMENTS

Thus far our discussion of environmental chemicals has dealt mainly with organic molecules. Many potentially dangerous substances are not organic chemicals but are normally occurring inorganic elements present in excess because of human activity. Examples are lead and cadmium. As is true for organic drugs and pollutants, the concentrations of inorganic elements at their sites of action depend on rates of entry, excretion, and storage. The gastrointestinal tract offers an important first line of defense since absorption of inorganic elements is usually quite limited. In contrast, airborne inorganic elements gain entry to the blood readily through the lungs. Little is known of the mechanisms by which the kidneys and liver handle most potentially harmful inorganic elements; specifically, it is not known whether adaptive increases in excretion are induced by exposure to the element.

SECTION C: RESISTANCE TO STRESS

Much of this book has been concerned with the body's response to stress in its broadest meaning of an environmental change that must be adapted to if health and life are to be maintained. Thus, any change in external temperature, water intake, etc., sets into motion mechanisms designed to prevent a significant change in some physiologic variable. In this section, however, we describe the basic stereotyped response to stress in the more limited sense of noxious or potentially noxious stimuli. These comprise an immense number of situations, including physical trauma, prolonged heavy exercise, infection, shock, decreased oxygen supply, prolonged exposure to cold, as well as pain, fright, and other emotional stresses. It is obvious that the overall response to cold exposure is very different from that to infection, but in one respect the response to all these situations is the same: Invariably secretion of *cortisol* is increased; indeed, the term stress has come to mean any event that elicits increased cortisol secretion. Also, sympathetic nervous activity is usually increased.

Historically, activation of the sympathetic nervous system was the first overall response to stress to be recognized and was labeled the fight-or-flight response. Only later did further work clearly establish the contribution of the adrenal cortical response. The increased cortisol secretion is mediated entirely by the hypothalamus–anterior pituitary system (Fig. 28-14) and does not occur in animals that lack a pituitary. Thus, afferent input to the hypothalamus induces secretion of ACTH-releasing hormone, which is carried by the hypothalamopituitary portal vessels to the anterior pituitary and stimulates ACTH release. The ACTH, in turn, circulates to the adrenal and stimulates cortisol release.

As described in Chap. 18, the hypothalamus receives input from virtually all areas of the brain and receptors of the body, and the pathway involved in any given situation depends upon the nature of the stress; e.g., ascending pathways from the arterial baroreceptors carry the input during hypotension, whereas pathways from other brain centers mediate the response to emotional stress. The destination is always the same, namely, synaptic connection with the hypothalamic neurons that secrete ACTH-re-

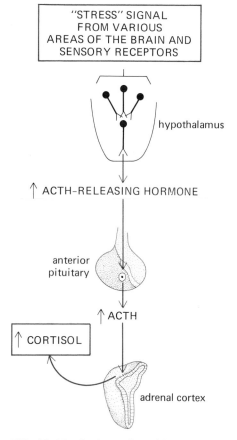

FIG. 28-14 Pathway by which stressful stimuli elicit increased cortisol secretion.

leasing hormone. These same pathways also converge on the hypothalamic areas that control sympathetic nervous activity (including release of epinephrine from the adrenal medulla).

FUNCTIONS OF CORTISOL IN STRESS

Many of cortisol's most important effects are on organic metabolism. Cortisol (1) stimulates protein catabolism, (2) stimulates liver uptake of amino acids and their conversion to glucose (gluconeogenesis), (3) is permissive for stimulation of gluconeogenesis by other hormones (glucagon, growth hormone, etc.), and (4) inhibits glucose uptake and oxidation by many body cells ("insulin antagonism") but not by the brain. Indeed, so striking are these effects that cortisol is often called a **glucocorticoid** (to distinguish it from the other major adrenal steroid, aldosterone, called a **mineralcorticoid** be-

cause its major effects are on sodium and potassium metabolism).

These effects are ideally suited to meet a stressful situation. First, an animal faced with a potential threat is usually forced to forgo eating, and these metabolic changes are essential for survival during fasting—indeed, an adrenalectomized animal rapidly dies of hypoglycemia and brain dysfunction during fasting. Second, the amino acids liberated by catabolism of body protein stores not only provide energy, via gluconeogenesis, but also constitute a potential source of amino acids for tissue repair should injury occur.

A few of the many medically important implications of these cortisol-induced metabolic effects associated with stress are as follows: (1) Any patient ill or subjected to surgery catabolizes considerable quantities of body protein; (2) a diabetic who suffers an infection requires much more insulin than usual; and (3) a child subjected to severe stress of any kind manifests retarded growth. The explanations for these phenomena should be evident.

Cortisol has important effects other than those on organic metabolism. One of the most important is that of enhancing vascular reactivity. A patient who lacks cortisol, faced with even a moderate stress, may develop hypotension and die if untreated. This is due primarily to a marked decrease in total peripheral resistance. For unknown reasons, stress induces widespread arteriolar dilation, despite massive sympathetic nervous system discharge, unless large amounts of cortisol are present. A large part of its counteracting effect is ascribable to the fact that moderate amounts of cortisol permit norepinephrine to induce vasoconstriction, but this can be only part of the story because considerably large amounts of cortisol are required to prevent stress-induced hypotension completely. In other words, the normal cardiovascular response to stress requires *increased* cortisol secretion, not just permissive quantities.

Thus far we have presented the adaptive value of the stress-induced cortisol increase mainly in terms of its role in preparing the body physically for fight or flight, and there is no doubt that cortisol does function importantly in this way. However, cortisol may have other important functions. Table 28-7 is a partial list of the large variety of psychosocial situations associated with increased cortisol secretion; common denominators of many of them are

TABLE 28-7 PSYCHOSOCIAL SITUATIONS
SHOWN TO BE ASSOCIATED WITH
INCREASED PLASMA
CONCENTRATION OR URINARY
EXCRETION OF ADRENAL
CORTICAL STEROIDS

TABLE 28-7 PSYCHOSOCIAL SITUATIONS SHOWN TO BE ASSOCIATED WITH INCREASED PLASMA CONCENTRATION OR URINARY EXCRETION OF ADRENAL CORTICAL STEROIDS

Experimental Animals:

1. Any "first experience" characterized by novelty, uncertainty, or unpredictability.
2. Conditioned emotional responses; anticipation of something previously experienced as unpleasant.
3. Involvement in situations in which the animal must master a difficult task in order to avoid or forestall aversive stimuli. The animal must really be "trying."
4. Situations in which long-standing rules are suddenly changed so that previous behavior is no longer effective in achieving a goal.
5. Socially subordinate animals. (Dominant animals have decreased cortisol.)
6. Crowding (increased social interactions).
7. Fighting or merely observing other animals fighting.

Human Beings:

I Normal persons
  a Acute situations
    1 Aircraft flight
    2 Awaiting surgical operation
    3 Final exams (college students)
    4 Novel situations
    5 Competitive athletics
    6 Anticipation of exposure to cold
    7 Workdays, compared to weekends
    8 Many job experiences
  b Chronic life situations
    1 Predictable personality-behavior profile: aggressive, ambitious, time-urgency
    2 Discrepancy between levels of aspiration and achievement
  c Experimental techniques
    1 "Stress" or "shame" interview
    2 Many motion pictures
II Psychiatric patients
  a Acute anxiety
  b Depression, but only when patient is aware of and involved in a struggle with it

novelty and challenge. Of great interest, therefore, are recent experiments that suggest that cortisol affects memory in experimental animals, most likely through direct actions on the brain. Even more striking, ACTH (independent of its stimulation of cortisol secretion) facilitates learning and memory. (ACTH seems to be only one of many peptides that exert such actions on the brain. See Chap. 13.) Thus, it may well be that the rise in cortisol secretion induced by psychosocial stress helps one to cope with the stress by facilitating the learning of appropriate responses.

Cortisol's Pharmacologic Effects and Disease

There are several situations in which adrenal corticosteroid levels in human beings become abnormally elevated. Patients with excessively hyperactive adrenals (there are several causes of this disease) represent one such situation, but the common occurrence is that of steroid administration for medical purposes. When cortisol is present in very high concentration, the previously described effects on organic metabolism are all magnified, but in addition there may appear one or more new effects, collectively known as the **pharmacologic effects** of cortisol. The most obvious is a profound reduction in the inflammatory response to injury or infection (indeed, reducing the inflammatory response to allergy, arthritis, or other diseases is the major reason for administering the cortisol to patients). Large amounts of cortisol inhibit almost every step of inflammation (vasodilation, increased vascular permeability, phagocytosis) and may decrease antibody production as well. As might be expected, this decreases the ability of the person to resist infections. In addition, large amounts of cortisol may accelerate development of hypertension, atherosclerosis, and gastric ulcers, and may interfere with normal menstrual cycles.

As emphasized above, these pharmacologic effects are known to be elicited when cortisol levels are extremely elevated. Yet an unsettled question of great importance is whether long-standing lesser elevations of cortisol may do the same thing, albeit more slowly and less perceptibly. Put in a different way, do the psychosocial stresses, noise, etc., of everyday life contribute to disease production via increased cortisol?

FUNCTIONS OF THE SYMPATHETIC NERVOUS SYSTEM IN STRESS

A list of the major effects of increased general sympathetic activity almost constitutes a guide on how to meet emergencies. Since all these actions have been discussed in other sections of the book, they are listed here with little or no comment.

1   Increased hepatic and muscle glycogenolysis (provides a quick source of glucose)
2   Increased breakdown of adipose tissue triacylglycerol (provides a supply of glycerol for gluconeogenesis and of fatty acids for oxidation)
3   Increased central nervous system arousal and alertness
4   Increased skeletal muscle contractility and decreased fatigue
5   Increased cardiac output secondary to increased cardiac contractility and heart rate
6   Shunting of blood from viscera to skeletal muscles by means of vasoconstriction in the former beds and vasodilation in the latter
7   Increased ventilation
8   Increased coagulability of blood

The adaptive value of these responses in a fight-or-flight situation is obvious. But what purpose do they serve in the psychosocial stresses so common to modern life when neither fight nor flight is appropriate? As for cortisol, the question yet to be answered is whether certain of these effects, if prolonged, might not enhance the development of certain diseases, particularly atherosclerosis and hypertension. For example, one can easily imagine the increased blood fat concentration and cardiac work contributing to the former disease. Considerable work remains to be done to evaluate such possibilities.

OTHER HORMONES RELEASED DURING STRESS

Other hormones that are usually released during many kinds of stress are aldosterone, antidiuretic hormone, and growth hormone. The increases in ADH and aldosterone ensure the retention of sodium and water, an important adaptation in the face of potential losses by hemorrhage or sweating. (ADH may also influence learning by an action on the brain.) Growth hormone reinforces the insulin antagonism effects of cortisol and the fat-mobilizing effects of epinephrine. Moreover, it probably stimulates the uptake of amino acids by an injured tissue and thereby facilitates tissue repair if needed; but since it cannot counteract the generalized protein catabolic effects of the increased cortisol, gluconeogensis is not hampered.

Finally, recent evidence suggests that this list of hormones whose secretion rates are altered by stress is by no means complete. It is likely that the secretion of almost every known hormone may be influenced by stress. For example, prolactin, thyroid hormone, and glucagon are often increased, whereas the pituitary gonadotropins (LH and FSH), insulin, and the sex steroids (testosterone or estrogen) are decreased. The adaptive significance of many of these changes is unclear, but their possible contribution to stress-induced disease processes may be very important.

KEY TERMS

immunity	antigen	hypersensitivity
immune response	antibody	autoimmune
lymphoid organ	B cells	biotransformation
inflammatory response	T cells	stress
complement system	Rh factor	cortisol

REVIEW EXERCISES

Section A

1 Define immunity and immune surveillance.
2 Distinguish between specific and nonspecific immune responses.
3 State some mechanisms by which bacteria and viruses harm people.
4 List the major effector cells of the immune system, their primary sites of production, and their primary functions.
5 List the lymphoid tissues and their locations.
6 State the functions of the lymph nodes, thymus, and spleen; diagram the structure of each type of tissue.
7 List the external anatomic and chemical defenses against infection.
8 List the sequence of events in an inflammatory response.
9 Describe the adaptive value of the vasodilation and increased capillary permeability that occur with inflammation.
10 Describe the chemotactic response of neutrophils in inflammation.
11 Describe phagocytosis, including the process of degranulation.
12 Define abscess and granuloma.
13 List three chemical mediators of the nonspecific inflammatory response; list the stimuli for their release and the effects of these mediators.
14 Describe the systemic manifestations of inflammation.
15 Define interferon; describe its function.
16 Name the two categories of specific immune responses; list their sites of origin.
17 Define antigen.
18 Describe the functions and interactions of the B- and T-cell systems.
19 List the sequence of events in humoral immunity against bacteria.
20 Define antibody; describe antibody specificity and its structural basis.
21 List the five classes of immunoglobulins and the general functions of each class.
22 State the major functions of antigen-antibody complexes.
23 Describe antibody formation.
24 Contrast active immunity with passive immunity.
25 Describe the mechanisms of cell-mediated immunity.
26 Contrast cell-mediated immunity with humoral immunity.
27 Describe the host responses to a virus infection.
28 Describe immune surveillance.
29 Define histocompatibility antigen.
30 Describe the process of tissue-implant rejection and the attempts made to combat it.
31 List the four blood types and the types of antibodies a person from each of these blood-type groups will have.
32 Describe the Rh-factor antigens.
33 Define hypersensitivity; contrast allergy with autoimmune disease.
34 Describe the process of desensitization.

Section B

1 List the factors that determine the effective concentration of any foreign chemical at its site of action.
2 List the sites of absorption of foreign chemicals.
3 List the major storage sites for foreign chemicals.
4 List the routes of excretion of foreign chemicals.
5 Define biotransformation; describe its harmful and beneficial effects.
6 Define microsomal enzyme system.
7 Describe the metabolic pathways of alcohol; describe the harmful effects of alcohol on cell metabolism.

Section C

1 State the responses common to almost all types of stress.
2 List cortisol's effects on organic metabolism; state their adaptive value.
3 State the relation between cortisol and the sympathetic nervous system in stress.
4 List the eight major effects of increased sympathetic activity during stress.
5 List the other hormones released during stress; list the adaptive value of each.

GLOSSARY

A band One of the transverse bands that make up the repeating striated pattern of cardiac and skeletal muscle; located in the middle of a sarcomere; corresponds to the length of the thick filaments.

abdomen The portion of the body that lies between the diaphragm and the pelvis.

abducens Cranial nerve VI; innervates the lateral rectus muscle of the eyeball.

abduct To draw away from the midline.

abductor A muscle that performs the function of abduction.

abortion Removal or expulsion of a fetus (or embryo) from the uterus before it is viable.

abscess Microbes, leukocytes, and liquified tissue debris walled off by fibroblasts and collagen.

absolute refractory period The time during which an excitable membrane is unable to generate an action potential in response to any stimulus.

absorption Movement of materials across a layer of epithelial cells from a body cavity or compartment toward the blood; e.g., movement from interstitial space into a capillary, from the lumen of the diges-

tive tract into the interstitial space, or from the lumen of a nephron into the interstitial space.

absorptive state The period during which nutrients are entering the bloodstream from the gastrointestinal tract.

acceptor A protein associated with DNA; the site at which a receptor-hormone complex binds and influences genetic transcription.

accessory reproductive organs The system of ducts through which the sperm or ova are transported and the glands that empty into these ducts (in the female the breasts are usually also included in this category).

accommodation Adjustment of the eye for viewing various distances by changing the shape of the lens.

acetone A ketone body produced from acetyl CoA during prolonged fasting or untreated severe diabetes mellitus.

acetyl coenzyme A A metabolic intermediate that transfers acetyl groups to the Krebs cycle and to various synthetic pathways, such as those for fatty acid synthesis.

acetyl group $CH_3\overset{\displaystyle O}{\overset{\|}{C}}—$; transferred

from one reaction to another by way of coenzyme A.

acetylcholine A chemical transmitter released from many peripheral nerve endings (e.g., from postganglionic parasympathetic fibers and at neuromuscular junctions) and from some neurons in the central nervous system.

ACh: *See* acetylcholine.

Achilles tendon The tendon that connects the muscles of the calf of the leg (gastrocnemius and soleus) to the calcaneus bone of the heel.

acid Any compound capable of releasing hydrogen ions to a solution; *see also* strong acid, weak acid.

acidity A description of the hydrogen-ion concentration of a solution; the greater the acidity, the higher the hydrogen-ion concentration.

acidosis Any situation in which the hydrogen-ion concentration of arterial blood is elevated.

acinus A small saclike dilation, as in glands.

acoustic Pertaining to the sense of hearing or to sound.

acrosome A vesicle that contains digestive enzymes located at the head of a sperm.

ACTH *See* adrenocorticotropic hormone.

actin　The globular contractile protein to which the myosin cross bridge binds; located in the thin filaments of muscle.

action potential　The electric signal propagated over long distances by excitable cells, e.g., nerve and muscle; characterized by an all-or-none reversal of the membrane potential in which the inside of the cell temporarily becomes positive relative to the outside; has a threshold and is conducted without decrement.

activation energy　The energy necessary to disrupt existing chemical bonds that hold atoms or molecules together so that new chemical bonds can be formed.

active immunity　A resistance to reinfection acquired as the result of contact with microorganisms, their toxins, or other antigenic components.

active site　The region of an enzyme to which the ligand (substrate) binds; a binding site.

active state　The tension at any instant generated by the muscle sliding filaments as a result of cross-bridge activity.

active transport　A carrier-mediated transport system in which molecules can be transported across a membrane from a region of low to a region of high concentration at the expense of energy provided by metabolism.

acuity　Acuteness or clarity of perception.

adduct　To move toward the midline of the body.

adductor　A muscle that adducts.

adenine　One of the purine bases found in the nucleotides that constitute the subunits of DNA and RNA.

adenohypophysis　The anterior portion of the pituitary gland (hypophysis); synthesizes, stores, and releases the hormones FHS, LH, ACTH, TSH, GH, and prolactin.

adenosine diphosphate (ADP)　The two-phosphate product of ATP breakdown; formed when ATP releases its stored energy, ATP → ADP + P_i + Energy.

adenosine monophosphate (AMP)　The monophosphate derivative of ATP; one of the nucleotides in DNA and RNA.

adenosine triphosphate (ATP)　The major energy carrier that transfers chemical potential energy from one molecule to another.

adenylate cyclase　The enzyme that catalyzes the transformation of ATP to cyclic AMP.

ADH　*See* antidiuretic hormone.

adipocytes　Cells specialized for the synthesis and storage of triacylglycerol (neutral fat) during periods of food uptake; fat cells.

adipose　Fatty; e.g., adipose tissue is tissue composed largely of fat-storing cells.

adolescence　The period of transition from childhood to adulthood in all respects, not just sexual ones.

ADP　*See* adenosine diphosphate.

adrenal glands　Also called suprarenal glands; a pair of endocrine glands located just above each kidney; each gland consists of two regions: an outer layer, the *adrenal cortex,* which secretes mainly cortisol, aldosterone, and androgens, and an inner core, the *adrenal medulla,* which secretes mainly epinephrine.

adrenaline　The British name for epinephrine.

adrenergic　Pertaining to those nerve fibers that release norepinephrine or epinephrine from their synaptic terminals; a compound that acts like norepinephrine or epinephrine.

adrenocorticotropic hormone (ACTH)　A polypeptide hormone secreted by the anterior pituitary; stimulates some of the cells of the adrenal cortex; also called *corticotropin.*

adventitia　The outermost connective-tissue covering of a structure.

aerobic　Requiring air or free oxygen.

afferent arterioles　Vessels in the kidney that convey blood from arteries to glomeruli.

afferent neuron　A neuron whose cell body lies outside the brain or spinal cord; carries information to the central nervous system from the receptors at its peripheral endings; sometimes called primary afferent, or first-order, neuron.

afferent pathway　The component of a reflex arc that transmits information from a receptor to an integrating center; any pathway that conveys information toward the central nervous system (or toward the brain).

affinity　The degree of attraction between a ligand and its binding site; at high-affinity sites the binding is tight, whereas at low-affinity sites the binding is weak.

after-hyperpolarization　A temporary increase in the transmembrane potential that follows an action potential, so that the inside of the cell is more negative than its resting state.

afterbirth　The placenta and associated membranes expelled from the uterus after delivery of an infant.

aggregation factor　A protein that contains several binding sites capable of interacting with specific binding sites on plasma membranes; aggregation factors form cross links between specific types of cells and thus hold them together to form tissues.

albumin　One of the major proteins found in blood plasma; more abundant than the globulins and usually of smaller molecular weight.

aldosterone　A mineralocorticoid steroid hormone secreted by the adrenal cortex; regulates electrolyte balance.

alkalosis　Any situation in which the hydrogen-ion concentration of arterial blood is reduced.

all-or-none　Pertaining to an event that either occurs maximally or does not occur at all, e.g., an action potential.

allergy　The responses to IgE antibodies (and the histamine and other vasoactive chemicals released by them) induced by an antigen.

allosteric enzyme　An enzyme whose activity can be altered by modulator molecules that act at a site other than the substrate binding site.

allosteric regulation Control of the properties of a protein binding site by modulator molecules that act at locations other than the binding site.

alpha-adrenergic receptor A type of receptor for norepinephrine and epinephrine that responds to certain drugs and can be distinguished from a beta-adrenergic receptor.

alpha helix A coiled conformation of a polypeptide chain found in many proteins, resulting from the electrical attraction between repeating peptide bonds along the chain.

alpha motor neuron A large motor neuron that innervates skeletomotor muscle fibers.

alpha rhythm The prominent 8- to 13-Hz oscillation in the EEG of awake, relaxed adults with their eyes closed; associated with decreased levels of attention.

alveolar dead space The volume of inspired air that reaches the alveoli but does not undergo gas exchange with the blood.

alveolar duct A thin-walled airway that branches from the end of the terminal bronchiole; alveoli extend from the end and sides of this duct.

alveolar pressure The air pressure in the pulmonary alveoli.

alveolar ventilation The volume of fresh atmospheric air that enters the alveoli each minute (respiratory rate times the tidal volume minus the anatomic dead space); also called *alveolar ventilatory volume*.

alveolus A small cavity; the thin-walled, air-filled outpocketing from the alveolar duct in the lungs; the cluster of cells at the end of a duct in secretory glands.

amino acid A molecule that contains an amino group, a carboxyl group, and a side chain all attached to a single carbon atom; 20 different amino acids (distinguished by their side chains) form the structural subunits of proteins.

amino acyl synthetase One of 20 different enzymes, each of which catalyzes the linkage of one type of amino acid to its particular type of tRNA during protein synthesis.

amino group $-NH_2$.

aminopeptidases Enzymes in the intestinal epithelial membrane; they break the peptide bond at the end of a polypeptide that has a free amino group.

ammonia NH_3; substance produced during the breakdown of amino acids; converted in the liver to urea.

amnesia Loss of memory; *see also* retrograde amnesia.

AMP *See* adenosine monophosphate.

amphetamine A drug that increases transmission at catecholamine-mediated synapses in the brain.

amphiarthrosis A slightly movable joint.

ampulla A flasklike dilation of a tubular structure.

amphipathic molecule A molecule that contains polar or ionized groups in one region and nonpolar groups in another.

amylase An enzyme that breaks down starch to produce disaccharides and small polysaccharides; secreted by the salivary glands and exocrine portion of the pancreas; also called *alpha amylase* or, if secreted by the salivary glands, *ptyalin*.

anabolic reactions Synthetic chemical reactions; reactions that put small molecular fragments together to form larger molecules.

anaerobic Without oxygen.

anastomosis A connection between two structures, e.g., between two blood vessels.

anatomic dead space The space within the airways of the respiratory tract whose walls do not permit gas exchange.

androgen Any chemical with actions similar to those of testosterone.

anemia A decreased concentration of hemoglobin in the blood.

anemic hypoxia Hypoxia in which the arterial P_{O_2} is normal but the total oxygen content of the blood is reduced because of inadequate numbers of erythrocytes, deficient or abnormal hemoglobin, or competition for the hemoglobin molecule by carbon monoxide.

angina pectoris Chest pain associated with inadequate blood flow to the heart muscle.

angiotensin I A small polypeptide (10 amino acid residues) hormone generated in the blood by the action of renin on a plasma protein, angiotensinogen.

angiotensin II An octapeptide formed by the enzymatic removal of two amino acids from angiotensin I; stimulates the secretion of aldosterone from the adrenal cortex and the contraction of vascular smooth muscle.

angiotensin III A heptapeptide formed by the enzymatic removal of an amino acid from angiotensin II.

angiotensinogen A plasma protein synthesized in the liver; the precursor of the hormone angiotensin.

"angry" macrophages Macrophages whose phagocytic activity has been facilitated by chemicals released from T cells.

anion A negatively charged ion, e.g., Cl^-, SO_4^{2-}.

anoxia The absence of oxygen.

antagonist A muscle whose action opposes the intended movement; e.g., the finger flexors are antagonists to the finger extensors.

anterior The front of the erect human body or toward the front of the body.

anterior pituitary The anterior portion of the pituitary gland; synthesizes, stores, and releases ACTH, GH, TSH, prolactin, FSH, and LH.

antibody A specialized protein secreted by plasma cells and capable of combining with the specific antigen that stimulated its production.

antibody-mediated immunity *See* humoral immunity.

anticodon The three-nucleotide sequence in tRNA that is exposed and able to base-pair with a complementary codon in mRNA during protein synthesis.

antidiuretic hormone (ADH)   A peptide hormone synthesized in the hypothalamus and released from the posterior pituitary; also called *vasopressin;* increases the permeability of the kidney collecting ducts to water.

antigen   A foreign protein or protein-polysaccharide complex that stimulates a specific immune response against itself when introduced into the body.

antihistamine   A chemical that blocks the action of histamine; e.g., it relieves the airway constriction and increased mucus secretion caused by histamine release during allergic atacks.

antrum   (gastric) The lower portion of the stomach, i.e., the region closest to the pyloric sphincter; (ovarian) the fluid-filled cavity in the maturing ovarian follicles.

anus   The distal opening of the rectum through which the end of the gastrointestinal tract opens to the exterior.

aorta   The largest artery in the body; carries blood from the left ventricle to the upper portion of the thoracic cage where it turns and descends to the lower portion of the abdomen.

aortic body   Location of the aortic chemoreceptor.

aortic chemoreceptor   Chemoreceptor near the arch of the aorta; sensitive to oxygen, carbon dioxide, and hydrogen-ion concentrations.

aortic valve   The valve between the left ventricle of the heart and the aorta.

aperture   An opening or orifice.

apex   The top of an organ or part.

aphasia   A specific language deficit not due to mental retardation; deficiency can be in understanding or formulating oral or written language.

apocrine   A type of secretion in which the secretory products accumulate at the apical end of the cell; the entire free end of the cell is then pinched off, which leaves the remainder of the cell to repeat the process.

appendicular skeleton   The portion of the skeleton that consists of the limbs (appendages) and their girdles.

aponeurosis   A flat sheet of fibrous connective tissue that serves as an attachment for muscles.

appendix   Appendix vermiformis; a small, fingerlike projection from the cecum in the large intestine; has no known digestive function in human beings.

aqueous humor   The watery fluid in the anterior and posterior chambers of the eye.

arachnoid   Resembling a spider web; the middle of the three meningeal membranes that cover the surface of the brain and spinal cord.

areola   The ringlike pigmented area around the nipple of the breast.

arrhythmia   Any variation from the normal hearbeat rhythm.

arterial baroreceptors   Nerve endings sensitive to stretch or distortion of the blood vessels (carotid sinus and arch of the aorta) in whose walls they lie; respond to blood pressure.

arteriolar resistance   The resistance offered by the arterioles to blood flow from arteries to capillaries.

arteriole   A narrow vessel, surrounded by several layers of smooth muscle, that regulates the flow of blood from an artery into a capillary network; the primary site at which vascular resistance is regulated.

artery   A thick-walled, elastic vessel that carries blood away from the heart.

arthritis   The inflammation of a joint.

articular cartilage   The cushioning layer of cartilage that covers the ends of bones that form synovial joints.

articulation   A joint; the union between two or more bones.

artificial kidney   An apparatus that eliminates by dialysis the excess ions and wastes that accumulate in the blood when the kidneys fail.

ascending limb (of Henle's loop)   The portion of the renal nephron that leads to the distal convoluted tubule.

association area   *See* cortical association areas.

asphyxia   A decrease in the amount of oxygen in the blood accompanied by an increase in carbon dioxide, which leads to suffocation.

asthma   A disease characterized by airway smooth muscle contraction, airway constriction, and plugging of the airways with mucus.

astigmatism   A defect in vision that occurs because of irregularities in the shape of the cornea.

atherosclerosis   A disease characterized by a thickening of the arterial wall with abnormal smooth muscle cells, deposits of cholesterol, and connective tissue resulting in narrowing of the vessel lumen.

atlas   The first cervical vertebra.

atmospheric pressure   The air pressure of the external environment (760 mmHg at sea level); barometric pressure.

atom   The smallest unit of matter that has chemical characteristics; consists of a complex arrangement of electrons that move around a positively charged nucleus, but the numbers of positive and negative charges are equal so an atom has no net charge; atoms combine to form molecules.

ATP   *See* adenosine triphosphate.

atrioventricular (AV) node   A region, at the base of the right atrium near the interventricular septum, that contains specialized cardiac muscle cells through which electrical activity must pass to reach the ventricles from the atria.

atrioventricular (AV) valve   A valve between an atrium and a ventricle of the heart; the AV valve on the right side of the heart is the *tricuspid valve,* and that on the left is the *mitral valve.*

atrium   A chamber of the heart that receives blood from the veins and passes it on to the ventricle on the same side of the heart.

auditory Pertaining to the sense of hearing.

auditory cortex The region of cerebral cortex that receives the nerve fibers from the auditory pathways.

auricle A portion of the external ear; also a projection from the atria of the heart.

auditory tube The tube that connects the middle-ear cavity to the pharynx; eustachian tube.

autoimmune disease A disease produced by antibodies or T cells sensitized against the body's own cells, which results in damage or alteration of cell function.

automaticity Capable of self-excitation.

autonomic nervous system A component of the efferent division of the peripheral nervous system; innervates cardiac muscle, smooth muscle, and glands; consists of sympathetic and parasympathetic subdivisions.

autoregulation (of blood flow) The ability of individual organs to alter their vascular resistance and, thereby, to control (self-regulate) their blood flows independent of external neural and hormonal influence.

AV node *See* atrioventricular node.

AV valve *See* atrioventricular valve.

axial skeleton Pertaining to the central line of the body determined by the spinal cord; opposite of appendicular.

axilla The armpit.

axon A single process that extends from the cell body of a neuron; propogates action potentials away from cell body; also called *nerve fiber.*

axon collateral A branch of an axon that leads to a set of axon terminals that differ in location from the main axon terminals.

axon terminals The network of fine branches at the end of the axon; each branch ends at a synaptic or neuroeffector junction.

azygos An unpaired structure, e.g., the azygos system of veins.

B lymphocytes Lymphocytes capable of becoming antibody-secreting plasma cells; also called *B cells.*

bacteria Unicellular organisms that have an outer coating, the cell wall, in addition to a cell membrane; are self-contained complete cells in that they have all the machinery required to sustain life and to reproduce themselves.

balance concept The statement that if the quantity of any substance in the body is to be maintained at a constant level over a period of time, the total amounts ingested and produced must equal the total amounts excreted and consumed.

ballismus Sudden, violent, purposeless movements of the distal part of the arms or legs.

barometric pressure *See* atmospheric pressure.

baroreceptor A receptor sensitive to pressure and rate of change in pressure; e.g., the arterial baroreceptors respond to mean arterial pressure and pulse pressure.

Bartholin's glands Two mucus-secreting glands on either side of the vaginal opening in the female.

basal Resting level.

basal ganglia Several nuclei in the cerebral hemispheres that code and relay information associated with the control of muscle movements; specifically, the caudate nucleus, globus pallidus, and putamen.

basal metabolic rate (BMR) A measurement by a clinical test in which the metabolic rate is determined under standardized conditions; i.e., the subject is at mental and physical rest (but not sleeping), at comfortable temperature, and fasting for at least 12 hr; findings are compared with values for a normal person of the same weight, height, age, and sex; the metabolic cost of living.

base Any molecule that can combine with a hydrogen ion; *see also* purine bases, pyrimidine bases.

basement membrane A thin, proteinaceous layer of extracellular material associated with the plasma membranes of many cells, especially epithelial cells, where it is located on the basal side of the cell.

basic electrical rhythm The spontaneous depolarization-repolarization cycles of pacemaker cells in the longitudinal smooth muscle layer of the stomach and intestines, which coordinate the muscular activity of the GI tract.

basilar membrane The membrane in the inner ear that separates the cochlear duct and scala tympani; supports the organ of Corti.

basophil A polymorphonuclear granulocytic leukocyte whose granules stain with basic dyes.

best frequency The sound frequency that causes the greatest response in neurons in the auditory pathway; different neurons have different best frequencies.

beta-adrenergic receptor A type of receptor for norepinephrine and epinephrine that responds to certain drugs and can be distinguished from an alpha-adrenergic receptor.

bicarbonate HCO_3^-; the anion that is the result of the ionization of carbonic acid.

biceps A muscle that has two heads.

bicuspid Having two cusps or tapering projections, e.g., a bicuspid tooth.

bicuspid valve The left atrioventricular valve of the heart; also called the *mitral valve.*

bile A yellow-green fluid that contains bile salts, cholesterol, lecithin (a phospholipid), bile pigments and other end products of organic metabolism, and certain trace metals; secreted by the liver.

bile canaliculi Small ducts adjacent to the liver cells; receive bile secreted by these cells.

bile pigment A pigmented substance (bilirubin) derived from the breakdown of hemoglobin, secreted in bile; gives bile its color.

bile salts A group of steroid molecules secreted by the liver and passed via bile ducts into the intestinal lumen where they promote the solubilization and digestion of fat.

bilirubin A yellow substance that results from the breakdown of heme; excreted in bile as a bile pigment.

binding site A region on a protein that has chemical groups that are able to interact with a specific ligand to bind it to the surface of the protein.

biotin One of the B-complex vitamins.

biotransformation Alteration of foreign molecules by an organism's metabolic pathways.

bipolar cells The neurons in the retina that are postsynaptic to the rods and cones.

blastocyst A ball of developing cells that surround a central cavity; stage of embryonic development.

blood-brain barrier A complex group of anatomic barriers and physiologic transport systems that closely controls the kinds of substances that enter the extracellular space of the brain from the blood and the rate at which they enter.

blood-testis barrier A barrier, formed by Sertoli cells and the tight junctions between them, that limits the movements of chemicals between the blood and the lumen of the seminiferous tubules.

blood type Specific antigenic proteins on the plasma membranes of erythrocytes, e.g., the ABO system.

blue-sensitive cones Cones that contain photopigment most sensitive to light of 445-nm wavelengths.

BMR *See* basal metabolic rate.

bolus A lump of food that has been chewed and mixed with saliva.

bone marrow A highly vascular, cellular substance in the central cavity of some bones; site of the synthesis of erythrocytes, some types of leukocytes, and platelets.

Bowman's capsule The blind sac at the beginning of the tubular component of a kidney nephron;

closely associated with the glomerulus of the nephron.

brachial Pertaining to the arm.

brachial plexus The network of nerve fibers formed from the ventral branch of nerves C5 to C8 and T1.

brainstem A subdivision of the brain that consists of medulla, pons, and midbrain; located between the spinal cord and the cerebrum.

bronchiole A small division of a bronchus (lung airway).

bronchus The branch of the air passage that enters the substance of the lung (the *main bronchus*); a second or third-order branch of the main bronchus (*secondary* or *tertiary bronchi*).

buffering The reversible binding of hydrogen ions by various compounds, e.g., bicarbonate; excess hydrogen ions are bound to the buffer compounds thereby decreasing the hydrogen-ion concentration, whereas a low hydrogen-ion concentration results in a release of hydrogen ions from the buffer compounds; these reversible reactions tend to maintain the acidity of a buffered solution nearly constant.

bulbourethral glands A pair of glands, also known as *Cowper's glands*, located below the prostate gland; secrete an alkaline fluid into the urethra.

bulk flow Movement of fluids or gasses down a pressure gradient.

Ca²⁺ Calcium ion.

calcification The process of depositing calcium salts.

calmodulin A protein present in all nucleated cells that acts as the major calcium binding site that mediates many of the intracellular effects of calcium.

calorie (cal) The unit in which heat energy is measured; the amount of heat needed to raise the temperature of 1 g of water 1°C; one *large calorie,* used to represent the caloric equivalent of various nu-

trients, equals 1000 calories and is denoted one Calorie.

Calorie (Cal) A kilocalorie.

cAMP *See* cyclic AMP.

canaliculus A small passage or canal.

cancellous Spongy, or trabecular, bone tissue.

cancer An uncontrolled malignant growth of cells; neoplasm.

canine A fanglike or pointed tooth.

capacitation Final maturation of the sperm.

capillary Smallest of the blood vessels; water, nutrients, dissolved gases, and other molecules are able to diffuse across capillary walls; composed of a single layer of endothelial cells.

capsule A fibrous or membranous connective-tissue covering.

carbamino compound A compound, e.g., $HbCO_2$, that is the result of the combination of carbon dioxide with protein amino groups, particularly hemoglobin.

carbohydrate A substance composed of atoms of carbon, hydrogen, and oxygen according to the general formula $C_n(H_2O)_n$, where n is any whole number, e.g., glucose, $C_6H_{12}O_6$.

carbon monoxide CO; a gas that reacts with the same iron binding sites on hemoglobin as does oxygen but with a much greater affinity than oxygen.

carbonic acid H_2CO_3; an acid formed from H_2O and CO_2.

carbonic anhydrase The enzyme that catalyzes the reversible reaction by which CO_2 and H_2O combine to form carbonic acid (H_2CO_3).

carboxyl group -COOH.

carboxypeptidases Two enzymes secreted by the exocrine pancreas as precursors, the procarboxypeptidases; break the peptide bond at the end of a protein that has a free carboxyl group.

cardiac Pertaining to the heart.

cardiac cycle One contraction-relaxation episode of the heart.

cardiac output The volume of blood

pumped by each ventricle per minute (not the total output pumped by both ventricles).

cardiovascular Pertaining to the heart and blood vessels.

cardiovascular center A group of neurons in the brainstem medulla that serves as a major integrating center for reflexes that affect heart rate, cardiac contractility, vascular resistance, and vascular capacity in the control of blood pressure.

carotid Pertaining to the two major arteries (the carotid arteries) in the neck that convey blood to the head.

carotid baroreceptor Mechanoreceptors stimulated by distortion or stretch; lie in the walls of the carotid sinus; sensitive to blood pressure.

carotid bifurcation The point at which the common carotid artery divides to form the internal and external carotid arteries.

carotid body *See* carotid chemoreceptors.

carotid chemoreceptors Chemoreceptors near the carotid bifurcation; sensitive to oxygen, carbon dioxide, and hydrogen-ion concentrations.

carotid sinus A dilation of the internal carotid artery just above the carotid bifurcation; location of the carotid baroreceptors.

carpal Pertaining to the wrist.

carrier A protein binding site in a membrane capable of combining with a transported molecule and enabling it to pass through the membrane.

cartilage A semisolid type of connective tissue found in the external ear, nose, trachea, and joints in adults; forms the major skeletal tissue of the embryo prior to replacement by bone tissues.

castration Removal of the gonads, i.e., testes or ovaries.

catabolic reaction A degradative chemical reaction that results in the fragmentation of a molecule into smaller and smaller parts.

catalyst A substance that accelerates chemical reactions but does not itself undergo any net chemical change during the reaction.

cataract An opaque area of the lens of the eye.

cation A positively charged ion.

caudal Toward the feet; inferior.

CCK *See* cholecystokinin.

cecum The dilated pouch at the beginning of the large intestine into which open the ileum, colon, and appendix.

cell The basic structural and functional unit of living things, i.e., the smallest structural unit into which an organism can be divided and still retain the characteristics we associate with life.

cell division The process by which a single cell forms two daughter cells; consists of nuclear division with replication of chromosomes (mitosis) followed by division of the entire cell.

cell-mediated immunity A type of specific immune response mediated by T lymphocytes.

cell membrane *See* membrane.

cellulose A straight-chain polysaccharide composed of thousands of glucose subunits; found in plant cells.

center of gravity The point in a body at which the mass of the body is in perfect balance; if the body were suspended from a string attached to this point, there would be no movement.

central chemoreceptors Receptors in the brainstem medulla that respond to changes in the H^+ and CO_2 concentrations of the cerebrospinal and brain extracellular fluid.

central dogma (of molecular biology) The doctrine that in all organisms genetic information flows from DNA to RNA and then to proteins.

central nervous system (CNS) Brain plus spinal cord.

central thermoreceptors Temperature receptors in the hypothalamus, spinal cord, abdominal organs, and other internal locations.

centrioles Two small bodies composed of nine fused sets of microtubules located in the cell cytoplasm; participate in nuclear and cell division.

centromere The point in a chromosome at which the two chromatin threads are attached to each other.

cephalad Toward the head; anterior.

cephalic Pertaining to the head.

cephalic phase Initiation of the neural and hormonal gastrointestinal reflexes by stimulation of receptors in the head (cephalic)—sight, smell, taste, and chewing.

cerebellum A subdivision of the brain behind the forebrain and above the brainstem; deals with control of muscle movements.

cerebral Pertaining to the brain.

cerebral cortex The 3-mm-thick, cellular layer that covers the cerebrum.

cerebrospinal fluid (CSF) The clear fluid that fills the cerebral ventricles and the subarachnoid space that surrounds the brain and spinal cord.

cerebrum The two hemispheres that form the upper and larger portions of the forebrain.

cervix The lower portion of the uterus; connects the lumen of the uterus with that of the vagina.

CG *See* chorionic gonadotropin.

cGMP *See* cyclic GMP.

channel A small passage through a membrane through which small-diameter molecules and ions can diffuse; a membrane pore.

chemical bonds The interactions between the electrons of adjacent atoms that hold the atoms together in a molecule.

chemical element A given type of atom; e.g., carbon atoms form one type of chemical element, oxygen atoms another; there are over 100 different chemical elements, each characterized by a distinct number of subatomic particles.

chemical equilibrium A situation in which the rates of the forward and reverse reactions of a chemical reaction are equal, and thus there is no net change in the concentrations of the reactants or products.

chemical formula   A term that lists the numbers and kinds of atoms in a given molecule; e.g., H_2O is the chemical formula for water, the molecules of which contain two hydrogen atoms and one oxygen atom.

chemical specificity   Only certain ligands (molecules, ions) can bind with a given type of binding site; the fewer the kinds of ligands capable of binding to that site, the greater its specificity; determined by the shape of the ligand and binding site.

chemical synapse   A synapse at which chemical mediators (neurotransmitters) released in response to action potentials in one neuron diffuse across an extracellular gap to influence the activity of the second neuron.

chemoaffinity theory   The suggestion that developing neural circuits form the proper connections because of selective attractions, determined by individual chemical codes, between the neurons.

chemoreceptors   Afferent nerve endings or cells associated with them that are sensitive to concentrations of certain chemicals.

chemotaxis   The orientation and movement of cells in a specific direction in response to a chemical stimulus.

chiasm   An X-shaped crossing, e.g., as of the optic nerves.

chief cells   Gastric gland cells that secrete pepsinogen, precursor of the enzyme pepsin.

cholecystokinin (CCK)   A polypeptide hormone secreted by cells in the upper small intestine; regulates several gastrointestinal activities including gastric motility and secretion, gall bladder contraction, and pancreatic enzyme secretion; formerly also called pancreozymin.

cholesterol   A steroid molecule; precursor of the steroid hormones and bile acids.

cholinergic   Pertaining to a nerve fiber that releases acetylcholine; a compound that acts like acetylcholine.

chondrocyte   A cartilage-forming cell.

chorion   A membranous layer composed of trophoblast and mesoderm layers; forms the extraembryonic membrane that makes contact with the uterus and forms the placenta.

chorionic gonadotropin (CG)   A protein hormone secreted by the trophoblastic cells of the blastocyst and placenta; maintains the secretory activity of the corpus luteum during the first three months of pregnancy.

chorionic somatomammotropin   A hormone secreted by the placenta with effects (in the mother) similar to those of growth hormone and prolactin.

choroid plexus   A highly vascular, epithelial structure that lines portions of the cerebral ventricles; responsible for the formation of much of the cerebrospinal fluid.

chromatid   The duplicated filament that constitutes half a chromosome; during cell division, each chromatid of the chromosome goes to a different pole of the dividing cell.

chromatin   A substance in the nucleus that consists of fibrous threads of DNA, nuclear proteins, and small amounts of RNA.

chromatin protein   *See* nuclear protein.

chromophore   The retinal component of a photopigment molecule.

chromosome   A highly coiled, condensed form of chromatin that is formed in the cell nucleus during the process of mitosis and meiosis.

chyle   The fatty material taken up by the central lacteals in the villi of the duodenum after a fatty meal.

chylomicron   A small (0.1- to 3.5-μm diameter) lipid droplet, consisting of triacylglycerol, phospholipid, cholesterol, free fatty acids, and protein, that is released from the intestinal epithelial cells and enters the lacteals during fat absorption.

chyme   The solution of partially digested food in the lumen of the stomach and intestines.

chymotrypsins   A family of enzymes secreted by the exocrine pancreas as precursors, the chymotrypsinogens; break certain peptide bonds in proteins and polypeptides.

cilia   Hairlike projections from the surface of specialized epithelial cells; sweep back and forth in a synchronized way to propel material along the cell surface.

circular muscle layer   The layer of smooth muscle that surrounds the submucosal layer in the stomach and intestinal walls, and whose muscle fibers are oriented circumferentially around these organs.

circulatory shock   A condition that results from a decrease in cardiac output severe enough to impair seriously the blood supply to various organs and tissues.

"circus" movement   The abnormal traveling of waves of cardiac excitation via circular routes through cardiac muscle that results in fibrillation.

cisterna   A fluid-filled space or reservoir.

citric acid   A six-carbon molecule that is an intermediate in the Krebs cycle.

citric acid cycle   *See* Krebs cycle.

clitoris   A small body of erectile tissue in the female external genitalia; homologous to the penis.

clone   A group of genetically identical objects.

clot retraction   The tightening up or contraction of a fibrin meshwork that squeezes serum out of a blood clot and strengthens it.

CNS   *See* central nervous system.

CO_2   Carbon dioxide.

coactivation   Almost simultaneous stimulation of alpha and gamma motor neurons that maintains tension on the spindle stretch receptors during skeletal muscle fiber shortening.

cobalamine   Vitamin B$_{12}$; cyanocobalamine.

cochlea   The inner ear; the fluid-filled, spiral-shaped compartment that contains the organ of Corti.

cochlear duct   A fluid-filled, membranous bag that extends almost the entire length of the inner ear and thus divides it into two compartments, the scala vestibuli and the scala tympani; contains the organ of Corti.

code word   The three-nucleotide sequence in DNA that signifies a given amino acid.

codon   The sequence of three nucleotide bases in mRNA that corresponds to a given code word in DNA.

coenzyme   Nonprotein organic molecule that functions as a cofactor; many coenzymes are derived from vitamins; generally serves as a carrier molecule that transfers atoms (such as hydrogen) or small molecular fragments from one reaction to another; is not consumed in the reaction in which it participates and can be reused until it is degraded.

coenzyme A   A derivative of the B-vitamin pantothenic acid; the coenzyme that transfers acetyl groups from one reaction to another.

cofactor   Nonprotein substance (e.g., metal ions or coenzymes) that binds to specific regions of an enzyme and thereby either maintains the shape of the active site or participates directly in the binding of substrate to the active site; is not consumed in the reaction in which it participates and can be reused until it is degraded.

collagen   A fibrous protein that has great strength but no active contractile properties; functions as a structural element in the interstitium of various types of connective tissue, e.g., tendons and ligaments.

collateral ventilation   Movement of air between alveoli through pores in the walls that separate the alveoli.

collecting duct   The portion of the tubular component of a kidney neph-

ron between the distal convoluted tubule and the renal pelvis.

colloid   A large molecule to which the capillaries are relatively impermeable, e.g., a plasma protein.

colon   The large intestine, specifically that part that extends from the cecum to the rectum.

colony (of neurons)   A group of crebral-cortex neurons with related function that lie in a column perpendicular to the cortical surface.

color blindness   A defect in color vision caused by the absence of one or even two of the three cone photopigments or to a defect in color-processing neurons in the visual pathway.

colostrum   The thin, milky fluid secreted by the mammary glands for a few days after birth.

"command" neurons   Neurons or groups of neurons whose activity initiates the series of neural events that result in a voluntary action; the identity of these neurons is unknown.

commissure   A large bundle of myelinated nerve fibers that connects the two cerebral hemispheres.

compensatory growth   A type of regeneration present in many organs after tissue damage.

complement system   A group of plasma proteins that normally circulate in the blood in an inactive state but, upon activation of the first protein, take part in a "cascade" of reactions; the active proteins generated during the cascade mediate certain events of the immune response.

compliance   *See* lung compliance.

concentration   The amount of material present in a unit volume; expressed as mass/volume, e.g., g/L or mmol/L.

concentration gradient   A graded concentration between two regions, e.g., across a cell membrane or epithelial layer of cells.

conducting portion (of the respiratory airways)   Those air passageways that have walls too thick to allow

the exchange of gas between blood and air.

conducting system (of the heart)   A network of cardiac muscle fibers specialized to conduct electrical activity to different areas of the heart.

conductor   Material that has a low resistance to current flow.

condyle   A rounded projection on a bone.

cone   One of the two receptor types for photic energy in the retina.

congenital   Existing at or before birth.

congestive heart failure   A set of signs and symptoms associated with a decreased contractility of the heart and engorgement of the heart, veins, and capillaries with blood.

conjunctiva   The membrane that covers the surface of the eyeball and eyelids except for the cornea.

conscious experience   The things of which a person is aware; the thoughts, feelings, perceptions, ideas, and reasoning during any of the states of consciousness.

conscious movements   *See* voluntary actions.

consciousness   *See* conscious experience, state of consciousness.

constipation   Infrequent or difficult defecation.

contractility   The component of contractile force that depends directly on cross-bridge activity rather than on the extent of filament overlap (fiber length).

contraction   The active process of generating force in a muscle.

contraction time   The time interval between the stimulus and the development of peak tension by a muscle.

control system   A collection of interconnected components that functions to keep a physical or chemical parameter within a predetermined range of values.

convection   The process by which air (or water) next to a warm body is heated, moves away, and is replaced by colder air, which, in turn, follows the same cycle.

convergence   Synapses from many presynaptic neurons act upon a single postsynaptic neuron; the turning of the eyes inward, i.e., toward the nose, to view near objects.

core temperature   The temperature of the inner organs as opposed to that of the skin.

cornea   The transparent structure covering the front of the eye; forms part of the optical system of the eye and helps focus an image on the retina.

cornification   Layering of flat (squamous) epithelial cells.

coronary   Pertaining to the arteries that carry blood to the heart muscle.

corpus callosum   A large, wide band of myelinated nerve fibers that connects the two cerebral hemispheres; one of the brain commissures.

corpus luteum   An ovarian structure formed from the follicle after ovulation; secretes estrogen and progesterone.

cortex   *See* adrenal cortex, cerebral cortex.

cortical association areas   Regions of the parietal and frontal lobes of cerebral cortex that receive input from the different sensory modalities, memory stores, etc., and perform further perceptual processing.

corticospinal pathway   A descending motor pathway that has its nerve cell bodies of origin in the cerebral cortex; the axons pass without synapsing to the region of the motor neurons; also called the *pyramidal tract*.

corticotropin   *See* adrenocorticotropic hormone.

corticotropin releasing hormone (CRH)   The hypothalamic hormone that stimulates ACTH (corticotropin) secretion by the anterior pituitary.

cortisol   A glucocorticoid steroid hormone secreted by the adrenal cortex; regulates various aspects of metabolism.

countercurrent   Flow in opposing directions through the two limbs of a hairpin loop.

countercurrent multiplier system   The mechanism associated with the loops of Henle that creates in the medulla of the kidney a region that has a high interstitial-fluid osmolarity.

covalent bond   A chemical bond between two atoms in a molecule in which each atom shares one of its electrons with the other.

CP   *See* creatine phosphate.

cranial nerve   One of the 24 (12 pairs) of peripheral nerves that extend from the brainstem or forebrain.

craniosacral division   *See* parasympathetic nervous system.

cranium   The bones of the skull that enclose and protect the brain.

creatine phosphate   A molecule that transfers phosphate and energy to ADP to generate ATP.

creatinine   A waste product derived from muscle creatine.

cretin   A mentally retarded person whose defect is the result of insufficient thyroid hormone during the critical periods of brain development.

CRH   *See* corticotropin releasing hormone.

critical period   A time during the developmental history when a system is most readily influenced by both favorable and adverse environmental factors.

cross bridge   A projection that extends from a thick filament in muscle; a portion of the myosin molecule capable of exerting force on the thin filament and causing it to slide past the thick one.

crossed-extensor reflex   Increased activation of the extensor muscles on the side opposite the limb flexion.

crossing over   A process in which segments of maternal and paternal chromosomes exchange with each other during chromosomal pairing before the first meiotic division.

crystalloid   A low-molecular-weight solute, e.g., Na^+, glucose, or urea.

CSF   *See* cerebrospinal fluid.

cubital   Pertaining to the space anterior to the elbow joint.

cumulus   The follicle cells that surround the ovum as it projects into the antrum of the ovarian follicle.

curare   A drug that binds strongly to acetylcholine receptors at neuromuscular junctions, which thus prevents their activation by acetylcholine and thereby causes paralysis.

current   The movement of electric charge; in biological systems, this is achieved by the movement of ions.

cusp   A tapering, pointed projection.

cutaneous   Pertaining to the skin.

cyanide   A poison that works by reacting with the final cytochrome in the cytochrome system associated with oxidative phosphorylation, thereby preventing electron transfer to oxygen and the production of ATP.

cyclic AMP (cAMP)   Cyclic 3',5'-adenosine monophosphate; a cyclic nucleotide that serves as second messenger for many nonsteroid hormones, neurotransmitters and paracrine agents.

cyclic GMP (cGMP)   Cyclic 3',5'-guanosine monophosphate; a cyclic nucleotide that acts as second messenger in some cells, possibly in opposition to cyclic AMP.

cytochrome   One of a group of iron-containing proteins that transfers electrons from one molecule to another during oxidation phosphorylation.

cytochrome system   A chain of enzymes that transfers hydrogen from coenzyme hydrogen carriers, such as NADH, to oxygen to form water; the enzymes are in the inner membranes of mitochondria and are associated with oxidative phosphorylation.

cytokinesis   The division of the cytoplasm during cell division.

cytology   The science that concerns itself with the structure of cells.

cytolytic T cells   A type of T cell that upon activation kills fungi, intracellular viruses, parasites, the few bacteria that can survive only inside cells, cancer cells, and solid-tissue transplants.

cytoplasm The region of a cell located outside of the nucleus.

cytosine One of the pyrimidine bases present in DNA and RNA.

cytosol The watery medium inside cells but outside of cell organelles.

dark adaptation The improvement in the sensitivity of vision after one is in the dark for some time.

daytime vision The vision possible only under conditions of high illumination; color vision.

deamination The removal of an amino ($-NH_2$) group from a molecule, such as an amino acid.

decremental Decreasing in amplitude.

decussation A crossing over, especially of symmetrical parts.

defecation Expulsion of feces from the rectum.

degradative reaction *See* catabolic reaction.

degranulation The fusion of the phagosome and lysosome membranes; the lysosomes, which appear as microscopic "granules," disappear.

dendrite A highly branched process of a neuron that receives synaptic inputs from other neurons.

denervation atrophy A decrease in the size of muscle fibers whose nerve supply is destroyed.

dentin The hard substance of a tooth that surrounds the pulp and is covered by enamel.

deoxyhemoglobin *See* reduced hemoglobin.

deoxyribonucleic acid (DNA) The nucleic acid that stores and transmits genetic information; consists of a double strand (double helix) of nucleotide subunits; the sequence of nucleotides in DNA (three nucleotides specify one amino acid) determines the sequence of amino acids in the proteins synthesized by a cell.

depolarize To change the value of the membrane potential toward zero so the inside of the cell becomes less negative.

dermis The deep, thick, inner layer of skin that underlies the epidermis.

descending colon The part of the large intestine on the left side of the abdomen.

descending limb (of Henle's loop) The segment of a renal nephron into which the proximal tubule drains.

desmosome A type of cell junction that holds cells together; consists of two opposed plasma membranes that remain separated by a 20-nm extracellular space; fibers extend into the cytoplasm from the inner surface of the desmosome and appear to be linked to other desmosomes of the cell.

diabetes insipidus A disease that results from defective ADH control of urine concentration; marked by great thirst and the excretion of a large volume of urine.

diabetes mellitus A disease in which the hormonal control of plasma glucose is defective because of an absolute or relative deficiency of insulin.

diaphragm The dome-shaped sheet of skeletal muscle that separates the thoracic and abdominal cavities; the principle muscle of respiration.

diaphysis The shaft of a long bone.

diarrhea The passage of watery feces.

diarthrosis A freely movable joint.

diastole The period of the cardiac cycle when the ventricles are not contracting.

diastolic pressure The minimum blood pressure during the cardiac cycle, i.e., the pressure just prior to ventricular ejection.

diencephalon The interior part of the brain that consists of the hypothalamus, thalamus, epithalamus, and metathalamus.

differentiation The process by which cells undergo alteration and ac-

quire specialized structural and functional properties during growth and development.

diffuse thalamic projection system Fibers of thalamic neurons that synapse in the cortex and are concerned with maintaining the behavioral characteristics of the awake state and with direction of attention to selected events.

diffusion The movement of molecules from one location to another because of random thermal molecular motion; net diffusion always occurs from a region of higher concentration to a region of lower concentration.

diffusion coefficient (k_D) A proportionality constant that determines the rate at which a substance diffuses through a solution.

diffusion equilibrium The state during which the diffusion fluxes in opposite directions are equal (i.e., the net flux = 0 and there is no concentration gradient).

diffusion potential A voltage created by the separation of charge that results from the diffusion of charged particles in a solution.

digestion The process of breaking down large particles and high-molecular-weight substances into small molecules.

dihydrotestosterone A steroid formed by the enzyme-mediated alteration of testosterone; the active form of testosterone in certain of its target cells.

dipeptide A peptide that, upon hydrolysis, yields two amino acids.

2,3-diphosphoglycerate (DPG) A substance produced by erythrocytes during glycolysis; binds reversibly with hemoglobin and causes it to change conformation and release oxygen.

disaccharidase One of a family of enzymes in the plasma membrane of the intestinal epithelium capable of splitting disaccharides into monosaccharides.

disaccharide A carbohydrate molecule composed of two monosaccharides; e.g., the disaccharide sucrose (table sugar) is composed of glucose and fructose.

dishabituation   Restoration of an orienting response when a novel stimulus (or the same stimulus at a different intensity) replaces a stimulus to which an animal has become habituated.

disinhibition   The removal of inhibition from a neuron that thereby allows its activity to increase.

distal convoluted tubule   A portion of the tubular component of a kidney nephron; situated between Henle's loop and the collecting duct.

disulfide bond   R-S-S-R, where R is the remaining portion of the molecule.

disuse atrophy   A decrease in the size of muscle fibers that are not used for a long period of time.

diuretic   Any substance that causes an increase in volume of urine excreted.

diurnal   Daily; occurring in a 24-hr cycle.

divergence (neuronal)   The process in which a single presynaptic neuron synapses upon and thereby influences the activity of many postsynaptic neurons; (eyes) the movement in which the eyes are turned outward to view distant objects.

DNA   *See* deoxyribonucleic acid.

DNA-dependent-RNA-polymerase *See* transcriptase.

DNA polymerase   The enzyme that during replication joins together nucleotides base paired with the nucleotides of one strand of DNA to form a new double-stranded molecule of DNA.

dominant hemisphere   The cerebral hemisphere that controls the hand used most frequently for intricate tasks, e.g., the left hemisphere in a right-handed person.

dopa   Dihydroxyphenylalanine; an intermediary in the catecholamine synthetic pathway.

dopamine   A catecholamine neurotransmitter; a precursor of epinephrine and norepinephrine.

dorsal   Pertaining to or toward the back.

dorsal root   A group of afferent nerve fibers that enters the left and right sides of the region of the spinal cord that faces the back of the body.

double bond   Two chemical bonds formed between the same two atoms; symbolized by $=$, e.g., the carbon dioxide molecule (CO_2) has two double bonds, $O=C=O$.

DPG   *See* 2,3-diphosphoglycerate.

dual innervation   Innervation of an organ or gland by both sympathetic and parasympathetic nerve fibers.

duct   A passage or tube.

ductus deferens   One of the paired male reproductive ducts that connect the epididymis to the urethra (pl., ductus deferentia); also called *vas deferens*.

duodenum   First portion of the small intestine; that portion of the intestine between the stomach and jejunum.

dura mater   The outermost fibrous membrane covering the brain.

dyspnea   Inability to breathe adequately.

dystrophy   The degeneration of an organ or tissue.

ear canal   The passageway that leads from the outside of the skull inward to end at the tympanic membrane.

eardrum   *See* tympanic membrane.

ECG   *See* electrocardiogram.

ectoderm   The outermost of the three germ layers of the embryo.

ectopic focus   A region of the heart (other than the SA node) that assumes the role of cardiac pacemaker.

edema   Accumulation of excess fluid in the interstitial space.

EDV   *See* end-diastolic volume.

EEG   *See* electroencephalogram.

EEG arousal   The transformation of the EEG pattern from alpha rhythm to a lower, faster oscillation; associated with increased levels of attention.

effector   A cell or cell collection that changes its activity in response to a neural or hormonal signal; spe-

cifically, muscle and gland cells, but technically all cells, are effectors.

efferent   To convey away from a central point.

efferent arteriole   A vessel that conveys blood from a renal glomerulus to a set of peritubular capillaries.

efferent neuron   A neuron that carries information away from the central nervous system to muscle cells, gland cells, or postganglionic neurons.

efferent pathway   That component of a reflex arc that transmits information from the integrating center to the effector; any pathway that conveys information out of the central nervous system (or away from the brain within the central nervous system).

ejaculation   A reflex that consists of contraction of the genital ducts and skeletal muscle at the base of the penis with resulting expulsion of semen from the penis.

ejaculatory duct   The continuation of the ductus deferens after it receives the duct from the seminal vesicle; joins the urethra within the prostate gland.

EKG   *See* electrocardiogram.

electric charge   One of the fundamental units of measurement; labeled positive or negative depending upon its interaction with other, known electric charges.

electric force   The force that causes the movement of charged particles toward regions that have an opposite charge and away from regions that have a similar charge.

electric potential (or electric potential difference)   *See* potential.

electrical synapse   A synapse at which local currents that result from action potentials in one neuron flow through gap junctions, joining two neurons, to influence the activity of the second neuron.

electrocardiogram (ECG, EKG)   A recording of the electric currents generated by action potentials of cardiac muscle cells.

electrochemical gradient   The force

that determines the magnitude of the net diffusion of charged particles; a combination of the electric gradient (as determined by the voltage difference between two points) and the chemical gradient (as determined by the concentration difference between the same two points).

electrode   A probe to which electric charges can be added (or from which they can be removed) to cause changes in the electric current that flows to a recorder that usually transforms this input into terms of voltage.

electroencephalogram (EEG)   A recording of the electric potential differences between two points on the scalp (or between one point on the scalp and a distant electrode).

electrogenic pump   An active transport system that directly separates electric charge to produce a potential difference.

electrolyte   Mineral ions, e.g., Na$^+$ and K$^+$.

electromagnetic radiation   Radiation composed of waves with electric and magnetic components that travel through matter and vacuum; generally caused by the motion of charged particles and includes gamma rays, x-rays, ultraviolet, visible, and infrared light, and television and radio waves.

electron   A subatomic particle that carries one negative charge and revolves in an orbit around the nucleus of the atom; may be shared between two atoms in the formation of a chemical bond or captured by another atom in the formation of ions.

embryo   An organism during the early stages of development; in human beings, the first two months of intrauterine life.

emission   Movement of the contents of the male genital ducts into the urethra.

emotion   Subjective feelings such as fear, joy, anxiety, amusement, etc.

emphysema   A disease in which the internal structures of the lung become more stretchable, which results in a markedly increased resting-lung size but a decreased surface area for gas exchange with the blood.

emulsification   A fat-solubilizing process in which large lipid droplets are broken into smaller droplets.

emulsion   A suspension of small (1-μm diameter) lipid droplets in an aqueous medium.

enamel   The white, hard substance that covers the dentin and crown of a tooth.

end-diastolic volume (EDV)   The amount of blood in the ventricle of the heart just prior to systole.

end-plate potential (EPP)   A depolarization of the motor end plate of skeletal muscle fibers in response to acetylcholine released from the overlying axon terminal in response to an action potential in the motor neuron; initiates an action potential in skeletal muscle plasma membrane.

end-product inhibition   Inhibition of a metabolic pathway by the action of the final product upon an allosteric site of some enzyme (usually the rate-limiting enzyme) in the pathway; in this case, the end product is the modulator molecule.

end-systolic volume (ESV)   The amount of blood that remains in the ventricle after ejection.

endocardium   The endothelial lining inside the heart.

endocrine   Pertaining to a ductless (i.e., endocrine) gland.

endocrine gland   A group of epithelial cells that are specialized for secretion but have no ducts, so they secrete their products into the extracellular space around the gland cells from which the products diffuse into the bloodstream; a ductless gland whose secretory products are hormones.

endocrine system   All the hormone-secreting glands of the body.

endocytosis   The process in which the plasma membrane of a cell invaginates and the invaginations become pinched off to form small, intracellular, membrane-bound vesicles that enclose a volume of material; when the extracellular substance is fluid, the process is called *fluid endocytosis* (pinocytosis); when specific molecules bind to sites on the plasma membrane and are carried into the cells as the membrane invaginates, the process is called *adsorptive endocytosis* (phagocytosis).

endolymph   The fluid that fills the membranous sacs of the vestibular apparatus and the cochlear duct of the inner ear.

endomysium   The connective tissue that surrounds muscle fibers.

endometrium   Glandular epithelium that lines the uterine cavity.

endoplasmic reticulum   A cell organelle consisting of an interconnected network of membrane-bound tubules, vesicles, and flattened sacs located in the cytoplasm; two types are distinguished: granular, which has ribosomes embedded in its membranes, and agranular, which is smooth surfaced.

endorphins   A group of peptides; some are probably neurotransmitters at the synapses activated by opiate drugs; "endogenous morphines."

endothelium   The thin layer of cells that lines the cavities of the heart and the blood vessels.

energy   The ability to perform work; energy is redistributed in all physical and chemical change, and its presence is revealed only when change is occurring.

enkephalin   A peptide that possibly functions as a neurotransmitter at the synapses activated by opiate drugs.

enterohepatic circulation   A "recycling" pathway for bile salts by reabsorption from the intestines, passage to the liver (via the hepatic portal vein), and back to the intestines (via the bile duct).

enterokinase   An enzyme embedded in the luminal plasma membrane of the intestinal epithelial cells; activates pancreatic trypsinogen.

entoderm   The innermost germ layer of the embryo.

enzymatic activity   The rate at which an enzyme converts substrate to product.

enzyme   A protein that accelerates specific chemical reactions but does not itself undergo any net change during the reaction; a biochemical catalyst.

eosinophil   Polymorphonuclear granulocytic leukocyte whose granules take up the red dye eosin.

epicardium   The outer layer of the heart wall.

epidermis   The outer epithelial layer of the skin that overlies the dermis.

epididymis   A portion of the duct system of the male reproductive system located between the seminiferous tubules and the ductus deferens.

epiglottis   A thin, triangular flap of cartilage that folds down and back to cover the trachea; possibly deflects food away from the respiratory passages during swallowing.

epilepsy   Neuronal activity that starts at a cluster of abnormal cortical neurons and spreads to adjacent neural tissue and thus gives rise to convulsions.

epimysium   The fibrous sheath around muscle.

epinephrine   A hormone secreted by the adrenal medulla and involved in the regulation of organic metabolism; a catecholamine neurotransmitter; also called *adrenaline* (Brit.).

epineurium   The sheath of connective tissue that surrounds a peripheral nerve.

epiphenomenon   An event that accompanies some action but is not causally related to it.

epiphyseal plate   An actively proliferating zone of cartilage near the ends of bones; the region of bone growth.

epithelial transport   *See* transcellular transport.

epithelium   Tissue that covers all body surfaces and lines all body cavities; most glands are formed from epithelial tissue.

EPP   *See* end-plate potential.

EPSP   *See* excitatory postsynaptic potential.

equilibrium   A situation in which no net change occurs within a system with time; no energy input is required to maintain the state of equilibrium.

equilibrium potential   The voltage difference that causes a movement of a particular ion species equal in magnitude but opposite in direction to the concentration gradient that causes net diffusion of that same ion species; at the equilibrium potential there is no net movement of the ion species.

erectile tissue   Spongy vascular tissues in the penis and clitoris that become engorged with blood and cause the tissue to become erect.

erection   A vascular phenomenon in which the penis or clitoris becomes stiff and erect.

erythrocyte   A red blood cell.

erythropoiesis   Formation of erythrocytes.

erythropoietin   A glycoprotein hormone secreted mainly by cells in the kidney; stimulates red-blood-cell production.

esophagus   The portion of the digestive tract that connects the mouth (pharynx) and stomach.

essential amino acids   The amino acids that cannot be formed by the body at a rate adequate to meet metabolic requirements and, therefore, must be obtained from the diet.

essential hypertension   Hypertension of unknown cause.

essential nutrients   Substances that are required for normal or optimal body function but are synthesized by the body either not at all or in amounts inadequate to achieve balance.

estradiol   A steroid hormone of the estrogen family; the major estrogen secreted by the ovaries.

estriol   A steroid hormone of the estrogen family; the major estrogen of pregnancy.

estrogen   A group of steroid hormones that have similar effects on the female reproductive tract.

estrone   A steroid hormone of the estrogen family.

ESV   *See* end-systolic volume.

euchromatin   Uncoiled region of chromatin; site of most gene transcription.

euphorigen   A drug that elevates mood.

eustachian tube   *See* auditory tube.

excitability   The ability to produce action potentials.

excitation-contraction coupling   The mechanisms in muscle fibers that link depolarization of the plasma membrane with generation of force by the cross bridges.

excitatory postsynaptic potential (EPSP)   A depolarizing graded potential that arises in the postsynaptic neuron in response to the activation of excitatory synaptic endings upon it.

excitatory synapse   A synapse that, when activated, either increases the likelihood that the membrane potential of the postsynaptic neuron will reach threshold and undergo action potentials or increases the firing frequency of existing action potentials.

excretion   The appearance of a substance in the urine or feces.

exocrine gland   A group of epithelial cells that are specialized for secretion and have ducts, which lead to a specific compartment or surface, into which the products are secreted.

exocytosis   The process in which the membrane of an intracellular vesicle fuses with the plasma membrane, the vesicle opens, and the vesicle contents are liberated into the extracellular fluid.

expiration   The bulk flow of air out of the lungs to the atmosphere.

expiratory reserve volume   The volume of air that can be exhaled by maximal active contraction of the expiratory muscles at the end of a normal expiration.

expressive aphasia A language deficit, not due to mental retardation, in which persons cannot perform the coordinated movements necessary for speech even though they are not paralyzed, understand what is spoken to them, and know what they want to say; often involves writing as well.

extension The straightening of a joint.

extensor muscle A muscle whose activity causes straightening of a joint.

external anal sphincter A band of skeletal muscle around the lower end of the rectum; generally under voluntary control.

external environment The environment that immediately surrounds the external surfaces of an organism.

external genitalia Female: the mons pubis, labia majora and minora, clitoris, vestibule of the vagina, and vestibular glands; male: the penis and scrotum.

external urethral sphincter A ring of skeletal muscle that surrounds the lower end of the urethra; prevents micturition when contracted.

external work Movement of external objects by the contraction of skeletal muscles.

extracellular fluid The watery medium that surrounds each of the body's cells; the environment in which the cells live; comprises the interstitial fluid and plasma; accounts for about one-third of total body water.

extrafusal fibers *See* skeletomotor fibers.

extrapyramidal system *See* multineuronal pathways.

extrasystole A heartbeat that occurs before the normal time in the cardiac cycle.

extrinsic Originating from the outside.

extrinsic clotting pathway The formation of fibrin clots by a pathway using tissue thromboplastin.

facet A smooth plane or surface for articulation.

facilitated diffusion A carrier-mediated transport system in which binding sites on carrier proteins are accessible to extra- and intracellular fluids, which thereby permits the transported molecules to be moved in either direction across the membrane; net movement always occurs from high to low concentration, and at equilibrium the concentrations on the two sides of the membrane become equal.

facilitation The subthreshold depolarization of a nerve cell membrane potential when excitatory synaptic input to the cell exceeds inhibitory input.

factor XII *See* Hageman factor.

fallopian tube *See* uterine tubes.

farsighted Pertaining to the defect in vision that occurs because the eyeball is too short for the lens, so that near objects are focused behind the retina.

fascia The thin layers of connective tissue that form sheets of supporting tissue to which other tissues are attached.

fascicle A small bundle of nerve or muscle fibers.

fat mobilization The increased breakdown of triacylglycerols and release of glycerol and fatty acids into the blood.

fat-soluble vitamins Vitamins A, D, E, and K, which are soluble in nonpolar solvents and insoluble in water.

fatty acid A chain of carbon atoms 14 to 22 carbons long (16 and 18 are most common) with a carboxyl group at one end through which it can be linked to glycerol in the formation of triacylglycerol (neutral fat); *see also* polyunsaturated fatty acid, saturated fatty acid, unsaturated fatty acid.

feces Waste material that consists primarily of water, bacteria, and ingested matter that failed to be digested and absorbed in the intestines; the material expelled from the large intestine during defecation.

feedback A situation in which the ultimate effects of a system themselves influence the system; *see also* negative feedback, positive feedback.

feedforward An aspect of some control systems that allows the system to anticipate changes of stimuli to its sensor.

ferritin An iron-binding protein that is the storage form for iron.

fertilization Union of a sperm with an ovum.

fetus An organism during the later stages of development prior to birth, in human beings the period from the second month of gestation until birth.

fever An increased body temperature that is the result of the setting of the ''thermostat'' of the temperature-regulating mechanisms at a higher-than-normal level.

fibrillation Extremely rapid contractions of cardiac muscle in an unsynchronized, repetitive way that prevents effective pumping of blood.

fibrin The polymer formed when protein fragments that result from the enzymatic cleavage of fibrinogen bind to other similar fragments, i.e., to polymerize; has the ability to turn blood into a solid gel (clot).

fibrinogen The plasma protein precursor of fibrin.

fibroblast A connective-tissue cell that synthesizes and secretes the protein that forms extracellular collagen.

filtration Movement of essentially protein-free plasma across the walls of a capillary out of its lumen as a result of a pressure gradient across the capillary wall.

fimbria Fingerlike projections that surround the trumpet-shaped expansion at the ovarian end of each uterine tube.

first law of thermodynamics The maxim that energy is neither created nor destroyed during any chemical or physical process; law of the conservation of energy.

"first" messenger The extracellular hormone, paracrine agent, or neurotransmitter that combines with a plasma-membrane or cytosol receptor.

first-order neuron The afferent neuron.

fissure Any fold, cleft, or groove.

flatus Intestinal gas.

flexion The bending of a joint.

flexion reflex Flexion of those joints that withdraw an injured part away from a painful stimulus.

flexor muscle A muscle whose activity causes bending of a joint.

fluid mosaic model (membrane structure) Molecular structure of cell membranes that consists of proteins embedded in a bimolecular layer of phospholipids; the phospholipid layer has the physical properties of a fluid, which thus allows the membrane proteins to move laterally within the lipid layer.

flux The amount of material that crosses a surface in a unit of time; *see also* net flux, unidirectional flux.

folic acid Pteroylglutamic acid; a water-soluble vitamin of the B-complex group essential for the formation of DNA; also called *folacin*.

follicle *See* ovarian follicle.

follicle-stimulating hormone (FSH) A polypeptide hormone secreted by the anterior pituitary in both males and females; one of the gonadotropins whose target organs are the gonads.

follicular cell A cell of the ovarian follicle.

follicular phase That portion of the menstrual cycle during which a single follicle and ovum develop to full maturity prior to ovulation.

fontanelle An unossified area of the cranium of an infant (soft spot).

foramen A small opening.

forebrain A subdivision of the brain that consists of the right and left cerebral hemispheres (the cerebrum) and the diencephalon.

fossa A pit or hollow area.

fovea The area near the center of the retina in which cones are most concentrated and which, therefore, gives rise to the most acute vision.

frontal lobe A region of the anterior cerebral cortex, where motor cortex, speech centers, and some association cortices are located.

fructose A five-carbon sugar found in the dissaccharide sucrose (table sugar).

FSH *See* follicle-stimulating hormone.

functional hyperemia The increased blood flow through a tissue that results from an increase in its metabolic activity.

galactose A six-carbon monosaccharide found in the dissaccharide lactose (milk sugar).

gall bladder A small sac underneath the liver; concentrates bile and stores it between meals; contraction of the gall bladder ejects bile into the common bile duct and small intestine.

gallstone A precipitate of cholesterol (and occasionally other substances as well) in the gall bladder of bile duct.

gametes The germ cells or reproductive cells; the sperm cells in a male and the ova in a female.

gametogenesis Formation of the male (sperm) and female (ovum) sex cells.

gamma globulin Immunoglobulin G (IgG), the most abundant plasma antibody; a plasma protein.

gamma loop An indirect pathway that leads to the facilitation of alpha motor neurons; the route from descending motor pathway to gamma motor neuron to spindle muscle fiber to stretch receptor and afferent neuron to alpha motor neuron.

gamma motor neuron A small motor neuron that controls spindle (intrafusal) muscle fibers.

ganglion A cluster of neuronal cell bodies located outside the central nervous system (pl., ganglia).

ganglion cells The neurons in the retina that are postsynaptic to the bipolar cells; their axons form the optic nerve.

gap junction A type of cell junction that allows direct electric current flow between adjacent cells; the two opposing plasma membranes come within 2 to 4 nm of each other and are joined by small channels or tubes through which ions and small molecules can flow between the cytoplasms of the two cells.

gas equilibration The condition in which the partial pressure of a gas in the liquid phase and the partial pressure of that gas in the gaseous phase are equal.

gastric Pertaining to the stomach.

gastric-inhibitory peptide (GIP) A polypeptide hormone secreted by cells of the upper small intestine; inhibits gastric motility.

gastric phase (of gastrointestinal control) The initiation of neural and hormonal reflexes by stimulation of the wall of the stomach by distension, acidity, and peptides.

gastrin A polypeptide hormone secreted by the antral region of the stomach; stimulates gastric acid secretion.

gastroilial reflex A reflex increase in contractile activity of the ileum during periods of gastric emptying.

gastrointestinal system Gastrointestinal tract plus the salivary glands and portions of the liver and pancreas.

gastrointestinal tract Mouth, esophagus, stomach, small and large intestines.

gene A unit of hereditary information; the portion of a DNA molecule that contains the information, coded in its nucleotide sequence, required to determine the amino acid sequence of a single polypeptide chain.

generator potential *See* receptor potential.

germ cells Cells that give rise to the male and female gametes, the sperm and ova.

GFR *See* glomerular filtration rate.

GH *See* growth hormone.

GIP *See* gastric-inhibitory peptide.

gland A group of epithelial cells specialized for the function of secretion; *see also* endocrine gland, exocrine gland.

glial cell A nonneuronal cell type in the brain and spinal cord; helps regulate the extracellular environment of the central nervous system; also *neuroglial cell.*

globin The polypeptide chains of a hemoglobin molecule.

globulin One of the types of protein found in blood plasma, less abundant than albumin and usually of larger molecular weight.

glomerular filtration The movement of an essentially protein-free plasma through the capillaries of the renal glomerulus into Bowman's capsule.

glomerular filtration rate (GFR) The volume of fluid filtered through the renal glomerular capillaries into the renal tubules per unit time.

glomerulus A capillary tuft that forms the vascular component at the beginning of a kidney nephron; intimately associated with Bowman's capsule.

glossal Pertaining to the tongue.

glottis The opening between the vocal cords.

glucagon A polypeptide hormone secreted by the A (or alpha) cells of the islets of Langerhans of the pancreas; its action on several target cells leads to a rise in plasma glucose.

glucocorticoid One of several hormones produced by the adrenal cortex that have major effects on glucose metabolism, e.g., cortisol.

gluconeogenesis Synthesis by the liver of glucose from pyruvate, lactate, glycerol, or amino acids; new formation of glucose.

glucose The major monosaccharide (carbohydrate) in the body; a six-carbon sugar, $C_6H_{12}O_6$; the catabolism of glucose during the absorptive state provides most of the energy for the cells of the body.

glucose-6-phosphate A carbohydrate formed by the transferring of a phosphate from ATP to glucose; the first step in the pathways of glycolysis and glycogen synthesis.

glucose sparing The switch from glucose to fat utilization by most cells during the postabsorptive state.

glycerol A three-carbon carbohydrate molecule; forms the backbone of triacylglycerol (neutral fat).

glycerol phosphate An important intermediate in glycolysis and triacylglycerol synthesis.

glycogen A highly branched polysaccharide composed of thousands of glucose subunits; the major form of carbohydrate storage in the body; also called *animal starch.*

glycogenesis The formation of glycogen from carbohydrates.

glycogenolysis The breakdown of glycogen.

glycolipid Lipid with a covalently bound carbohydrate group; present in small amounts in the plasma membrane oriented with the carbohydrate at the extracellular surface.

glycolysis The breakdown of glucose to form two molecules of lactic acid; occurs in the absence of oxygen and gives rise to the net synthesis of two molecules of ATP; enzymes located in the cytoplasm.

glycoprotein Protein with a covalently linked carbohydrate group; present in plasma membranes with the carbohydrate facing the extracellular surface.

GnRH *See* gonadotropin releasing hormone.

goblet cell A type of epithelial cell that secretes mucus.

goiter An enlarged thyroid.

Golgi apparatus A cellular organelle that consists of membranes and vesicles, usually located near the nucleus; processes newly synthesized proteins for secretion.

Golgi tendon organs Mechanoreceptor endings of afferent nerve fibers wrapped around collagen bundles in tendons; activated by distortion when tension is applied to the tendon.

gonad Gamete-producing reproductive organ, i.e., the testes in the male and the ovaries in the female.

gonadotropin-releasing hormone (GnRH) The hypothalamic hormone that controls LH and FSH secretion by the anterior pituitary.

gonadotropins Hormones secreted by the anterior pituitary that control the function of the gonads; FSH and LH.

granulocyte Any of several types of white blood cells that contain granules in their cytoplasm—eosinophils, neutrophils, and basophils.

granulopoietin The hormone, produced and released mainly by macrophages, that stimulates neutrophil ("granulocyte") production.

granulosa cells Cells that line the antrum in the ovarian follicle.

gray matter That portion of the brain and spinal cord that appears gray in unstained specimens and consists mainly of cell bodies and the unmyelinated portions of nerve fibers.

green-sensitive cones Cones that contain photopigment most sensitive to light of 535-nm wavelengths.

GRH *See* growth hormone–releasing hormone.

groin The junction of the abdomen and thighs.

growth An increase in body size; i.e., lengthening of the long bones, increased cell division and cell volume, and accumulation of protein; *see also* compensatory growth.

growth hormone (GH) A polypeptide hormone secreted by the anterior pituitary; also called *somatotropin* or *somatotropic hormone* (STH); stimulates body growth by means of its actions on carbohydrate and protein metabolism.

growth hormone–releasing hormone (GRH, GHRH) The hypothalamic hormone that stimulates the secretion of growth hormone by the anterior pituitary.

GTP *See* guanosine triphosphate.

guanine One of the purine bases in DNA and RNA.

guanosine triphosphate (GTP) A high-energy molecule similar to ATP except that it contains the base guanine rather than adenine.

gustatory Pertaining to the sense of taste.

gyrus A convolution of the brain's surface.

H⁺ Hydrogen ion; the concentration of hydrogen ions in a solution determines the acidity of a solution.

H zone One of the transverse bands that make up the repeating striated pattern of cardiac and skeletal muscle; a light region that bisects the A band of striated muscle; corresponds to the space between the ends of the thin filaments; thus, it consists only of thick filaments.

habituation A reversible decrease in the strength of a response upon repeatedly administered stimulation.

Hageman factor Factor XII; the initial factor in the sequence of reactions that results in fibrin formation.

hair cells Mechanoreceptors in the organ of Corti and vestibular apparatus.

haploid Having a chromosome number equal to that of the gametes, i.e., half the diploid number found in most somatic cells.

haustra The sacculations of the wall of the large intestine.

haversian canal A canal in bone tissue that contains nerves and blood vessels; surrounded by concentric layers of calcified bone tissue.

Hb *See* reduced hemoglobin.

HbCO₂ *See* carbamino compound.

HbO₂ *See* oxyhemoglobin.

H₂CO₃ Carbonic acid.

heart attack Damage to or death of cardiac muscle, i.e., a myocardial infarct.

heart-lung preparation An experimental setup in which the performance of a heart that lacks any hormonal or neural input can be determined in response to varying venous pressures.

heart murmur The abnormal heart sounds caused by the turbulent flow of blood through narrowed or leaky valves or a hole in the interventricular or interatrial septum.

heart rate The number of contractions of the heart per minute.

heart sounds The noises that result from vibrations caused by closure of the atrioventricular valves (first heart sound) and pulmonary and aortic valves (second heart sound).

heartburn Pain that seems to occur in the region of the heart but is due either to spasms of the esophageal muscle following stimulation by acid stomach contents refluxed into the esophagus or to the direct stimulation of pain receptors in the esophageal wall by the acid.

heat exhaustion A state of collapse due to hypotension brought about both by depletion of plasma volume secondary to sweating and by extreme dilation of skin blood vessels; the thermoregulatory centers are still functioning.

heat stroke A positive-feedback situation in which a heat gain greater than heat loss causes body temperature to become so high that the brain thermoregulatory centers do not function, allowing temperature to rise even closer to the point at which vital bodily functions are endangered.

"helper" T cells A type of T cell that regulates B cells in a way that leads to enhanced antibody production.

hematocrit The percentage of the total blood volume occupied by blood cells.

hematoma A local, extravascular accumulation of blood.

hematopoiesis *See* hemopoiesis.

heme The iron-containing organic molecule bound to each of the four polypeptide chains of a hemoglobin molecule; the iron-containing portion of a cytochrome.

hemoglobin A red-colored protein, located in erythrocytes, that transports most of the oxygen in the blood; an oligomeric protein composed of four polypeptide chains, each of which contains a heme group that has a single atom of iron, with which oxygen reversibly combines.

hemopoiesis The formation of blood cells; also called hematopoiesis.

hemorrhage Bleeding.

hemostasis Prevention of the loss of blood; stopping the flow of blood through a particular vessel.

Henle's loop The hairpinlike segment of the tubular component of a kidney nephron; situated between the proximal and distal tubules.

heparin An anticlotting agent found in various cells and tissues; an anticoagulant drug.

hepatic Pertaining to the liver.

hepatic portal vein The vein that conveys blood from the capillary beds in the intestines (as well as portions of the stomach and pancreas) to the capillary beds in the liver.

hertz (Hz) The measure used for wave frequencies; cycles per second.

heterochromatin Region of chromatin that remains tightly coiled; stains more darkly than the uncoiled regions; genes located in these regions are inactive.

hiatus A gap or opening in an organ.

hilus A depression or pit, especially where vessels exit or enter an organ.

hippocampal "arousal" *See* theta rhythm.

hippocampus A portion of the limbic system in the brain.

histamine A paracrine agent found in many tissues of the body but particularly in mast cells, circulating basophils, and platelets; a possible neurotransmitter.

histocompatibility antigens Antigenic protein molecules on the surface of all nucleated cells; differ from person to person.

histone One of a group of positively charged proteins associated with DNA to form chromatin; interacts with the phosphate groups of the nucleotides to cause supercoiling of the chromatin threads.

histotoxic hypoxia Hypoxia in which the quantity of oxygen that reaches the tissues is normal but the cell is unable to utilize the oxygen because a toxic agent has interfered with its metabolic machinery.

holocrine gland A gland whose secretion results from the disintegration of the entire gland cell.

homeostasis The relatively stable physical and chemical composition of the internal environment of the body that results from the actions of compensating regulatory systems.

homeostatic system A control system that consists of a collection of interconnected components and functions to keep a physical or chemical parameter of the internal environment relatively constant.

homeotherm An animal capable of maintaining its body temperature within very narrow limits.

homozygous Possessing an identical pair of genes that specify the same genetically determined component.

hormone A chemical substance synthesized by an endocrine gland and secreted into the blood that carries it to other cells in the body where its actions are exerted.

HR *See* heart rate.

humoral immunity A type of specific immune response in which circulating antibodies play a central role in the destructive process.

hyaline Glassy and transparent, as hyaline cartilage.

hydrocephalus A disease in which blockage of the outflow paths for cerebrospinal fluid cause the fluid to build up, resulting in considerable damage to the brain tissue.

hydrochloric acid HCl; a strong acid secreted into the stomach lumen by gland cells that line the stomach.

hydrogen bond A weak bond between two molecules in which a negative region of one polarized substance is electrostatically attracted to a positively polarized hydrogen atom in the other.

hydrogen peroxide H_2O_2; a chemical produced by the phagosomes; highly destructive to the macromolecules within them.

hydrolysis The breakdown of a chemical bond with the addition of the elements of water (—H and —OH) to the products formed.

hydrostatic pressure The pressure exerted by a fluid.

hydroxyl group R—OH, a polar chemical group where R is the remaining portion of the molecule.

hymen A thin fold of mucous membrane that partially overlies the vaginal opening.

hyperemia Increased blood flow; *see also* active hyperemia.

hyperglycemia Increased plasma glucose concentration above normal levels.

hyperopia *See* farsighted.

hyperosmotic Having an osmolarity greater than normal plasma.

hyperplasia An increase in the number of cells in a tissue or organ.

hyperpolarize The change in the transmembrane potential so that the inside of the cell becomes more negative than its resting state.

hypersensitivity (to an antigen) An acquired reactivity to an antigen that can result in bodily damage upon subsequent exposure to that particular antigen, e.g., allergies.

hypertension Chronically increased arterial blood pressure.

hyperthyroidism The secretion of excess thyroid hormone.

hypertonic solution A solution that contains a higher concentration of membrane-impermeable solute particles than cells contain; cells placed in a hypertonic solution will shrink because of the diffusion of water (osmosis) out of the cell.

hypertrophy An increase in size; enlargement of a tissue or organ that results from an increase in cell size rather than in cell number.

hyperventilation Ventilation greater than that needed to maintain plasma P_{CO_2} normal.

hypocalcemic tetany A disease characterized by skeletal muscle spasms caused by a low plasma calcium concentration.

hypoosmotic Having an osmolarity less than normal plasma.

hypophysis The pituitary gland.

hypotension Decreased blood pressure.

hypothalamic releasing factors *See* hypothalamic releasing hormones.

hypothalamic releasing hormones Hormones released from hypothalamic neurons into the hypothalamopituitary portal vessels to control the release of the anterior pituitary hormones; formerly called hypothalamic releasing factors.

hypothalamopituitary portal vessels Vessels that transport blood from a capillary network in the hypothalamus to one in the anterior pituitary.

hypothalamus A brain region below the thalamus responsible for the integration of many basic behavioral patterns that involve correlation of neural and endocrine function, especially those concerned with regulation of the internal environment.

hypothyroid Insufficient thyroid hormone.

hypotonic solution A solution that contains a lower concentration of membrane-impermeable solute particles that cells contain; cells placed in a hypotonic solution will swell due to the diffusion of water (osmosis) into the cell.

hypoventilation Ventilation insufficient to maintain plasma P_{CO_2} normal.

hypoxia A deficiency of oxygen at the tissue level.

hypoxic hypoxia Hypoxia due to decreased arterial P_{O_2}.

Hz *See* hertz.

I band One of the transverse bands that make up the repeating striated pattern of cardiac and skeletal muscle; located between the A bands of adjacent sarcomeres and bisected by the Z line; extends the length of the thin filaments up to but not including regions of thick and thin filament overlap.

ICSH *See* luteinizing hormone.

ileocecal sphincter A ring of smooth muscle that separates the ileum from the cecum, i.e., the small and large intestines.

ileum The final, longest segment of the small intestine.

iliogastric reflex A reflex decrease in gastric motility in response to distension of the ileum.

immune responses *See* nonspecific immune responses, specific immune responses.

immune surveillance The recognition and destruction of abnormal or mutant cell types that arise within the body.

immunity The physiologic mechanisms that allow the body to recognize materials as foreign or abnormal and to neutralize or eliminate them; the mechanisms that maintain the uniqueness of "self."

immunization The process of rendering a person immune.

immunoglobulins A family of plasma proteins to which the antibodies belong; subdivided into five classes: IgG, IgA, IgD, IgM, and IgE (Ig stands for immunoglobulin).

implantation The event during which the fertilized ovum adheres to the uterine wall and becomes embedded in it.

impotence Failure of erection of the penis, thus, inability of the male to perform sexual intercourse.

inborn errors of metabolism Inherited diseases, often due to a single gene, in which there is failure to produce an active enzyme or else an abnormally active one is produced.

incisure A cut, notch, or incision.

inclusion Any thing that is enclosed.

inducer The molecule that, when combined with the repressor protein, inactivates the repressor to thereby free the operator gene from inhibition and initiate synthesis of the proteins coded by the operon.

induction Stimulation of enzyme synthesis by action of the enzyme substrate on the genetic mechanisms that control enzyme synthesis.

infarct An area of dead tissue that results from local ischemia caused by diminished circulation supplying the area.

inferior esophageal sphincter Smooth muscles of the last 4 cm of the esophagus that can act as a sphincter and close off the opening of the esophagus into the stomach.

inferior vena cava The large vein that carries blood from the lower half of the body to the right atrium of the heart.

inflammation The local response to injury characterized by swelling, pain, and increased temperature and redness in the region of injury due to increased local blood flow.

inflammatory response *See* inflammation.

ingestion The act of taking material into the body via the mouth.

inguinal Pertaining to the groin area.

inhibin A protein hormone secreted from seminiferous tubules (probably by the Sertoli cells).

inhibitory postsynaptic potential (IPSP) A hyperpolarizing graded potential that arises in the postsynaptic neuron in response to the activation of inhibitory synaptic endings upon it.

inhibitory synapse A synapse that, when activated, decreases the likelihood that the membrane potential of the postsynaptic neuron will reach threshold and fire an action potential or decreases the firing frequency of existing action potentials.

initial segment The first portion of the axon of a neuron plus the part of the cell body where the axon is joined.

inner ear The cochlea; contains the organ of Corti.

innervate To supply with nerves.

inorganic ion Any ion that does not contain carbon, e.g., Na^+, Cl^-, and K^+.

inosine A nucleotide base present in tRNA; base-pairs with cytosine.

insensible loss The water loss of which a person is unaware, i.e., the loss of evaporation from the cells of the skin and the lining of the respiratory passages.

inspiration The bulk flow of air from the atmosphere into the lungs.

inspiratory muscles Those muscles whose contraction contributes to inspiration; the diaphragm is the major inspiratory muscle.

inspiratory neuron A neuron whose cell body is in the brainstem medulla and which fires in synchrony with inspiration and ceases firing during expiration.

inspiratory reserve volume The volume of air that can be inspired over and above the resting tidal volume.

insulin A polypeptide hormone secreted by the B (or beta) cells of the islets of Langerhans of the pancreas; stimulates the carrier-mediated movement of glucose into most cells and the enzymes that transform glucose into fat and glycogen.

integral proteins Membrane proteins that are embedded in the phospholipid bilayer of the membrane and cannot be removed without disrupting the lipid components of the membrane; may span the membrane or be at only one side.

integrator A cell whose output is determined by many often-conflicting bits of incoming information.

integument The skin.

intercostal muscles Skeletal muscles that lie between the ribs; contraction results in elevation or compression of the rib cage during inspiration or expiration.

interference theory The theory that things are forgotten because of competition with conflicting information stored before or after the storage of the particular item being recalled.

interferon A protein that stimulates cells to synthesize "antiviral" proteins, which in turn interfere with the multiplication of viruses in those cells.

internal anal sphincter A ring of smooth muscle around the lower end of the rectum.

internal environment The extracellular fluid that immediately surrounds cells in the body, thus, the environment in which these cells live; includes the interstitial fluid and blood plasma.

internal urethral sphincter Part of the smooth muscle wall of the urinary bladder that, by its contraction and relaxation, opens and closes the outlet from the bladder to the urethra.

internal work Energy-requiring activities (work) in the body, e.g., cardiac contraction, internal secretions, membrane active transport, synthetic chemical reactions, smooth muscle contractions, and the muscular movement associated with respiration.

interneuron A neuron whose cell body and axon lie entirely within the central nervous system.

interphase The period of the cell-division cycle between the end of one division and the beginning of the next; DNA replication occurs in this phase.

interstitial Lying between; the extracellular space of a tissue.

interstitial cell–stimulating hormone (ICSH) *See* luteinizing hormone.

interstitial cells (of the testes) Testosterone-secreting cells that lie between the seminiferous tubules in the testes.

interstitial fluid The extracellular fluid that surrounds the cells of a tissue; plasma, which surrounds blood cells, is not considered interstitial fluid.

interstitium Interstitial space; the space between tissue cells.

interventricular septum The partition in the heart that separates the cavities of the right and left ventricles.

intestinal phase (of gastrointestinal control) The initiation of neural and hormonal reflexes by stimulation of the walls of the intestinal tract by distension, acidity, osmolarity, and the various products of carbohydrate, fat, and protein digestion.

intestinointestinal reflex A reflex cessation of contractile activity of the intestines in response to large distensions of the intestine, injury to the intestinal wall, or various bacterial infections in the intestine.

intima The innermost layer of a blood vessel.

intraabdominal pressure The pressure within the abdominal cavity.

intracellular fluid The fluid in the cells.

intrafusal fibers *See* spindle fibers.

intramembranous ossification The process of bone formation whereby bone is formed directly in membranous tissue.

intrapleural fluid The thin film of fluid in the thoracic cavity between the pleura that lines the inner wall of the thoracic cage and the pleura that covers the lungs.

intrapleural pressure The pressure within the pleural space generated by the tendency of the lung and chest wall to pull away from each other; also called *intrathoracic pressure*.

intrathoracic pressure The pressure within the intrapleural fluid of the chest cavity; intrapleural pressure.

intrinsic clotting pathway The intravascular sequence of fibrin formation initiated by Hageman factor.

intrinsic factor A glycoprotein secreted by the epithelial lining of the stomach and necessary for absorption of vitamin B_{12}.

ion Any atom or small molecule that contains an unequal number of electrons and protons and, therefore, carries a net positive or negative electric charge, e.g., Cl^-, Na^+, and NH_4^+.

ionic hypothesis An explanation of the voltage changes during an action potential in terms of membrane permeability changes and the movements of sodium and potassium ions across the membrane.

ionization The dissociation of a substance in solution into atoms or molecules that carry a net electric charge, i.e., the formation of ions.

IPSP *See* inhibitory postsynaptic potential.

iris The ringlike structure that surrounds the pupil of the eye; consists of pigmented epithelium and muscle which by contraction adjusts the size of the pupil.

irreversible reaction A chemical reaction that proceeds predominantly in only one direction because large quantities of energy are released in the reaction; may be reversed by an alternate route that involves a different enzyme and an additional substrate to supply the large quantity of energy required.

ischemia A reduced blood supply to an organ or tissue.

ischemic hypoxia Hypoxia in which the basic defect is too little blood flow to the tissues to meet the metabolic demands.

islets of Langerhans The clusters of endocrine cells in the pancreas that produce insulin, glucagon, and somatostatin.

isometric contraction Contraction of a muscle in such a way that it develops tension but is unable to shorten because the opposing load is greater than the tension developed.

isotonic *See* isotonic contraction, isotonic solution.

isotonic contraction Shortening of a muscle that results in the moving of a load.

isotonic solution A solution that contains the same number of nonpenetrating solute particles as cells contain; not synonymous with isoosmotic; a solution in which a cell will neither shrink nor swell.

isovolumetric ventricular contraction The early phase of systole when both the atrioventricular and aortic valves are closed.

isovolumetric ventricular relaxation The early phase of diastole when both the atrioventricular and aortic valves are closed.

isthmus A narrow connection between two larger bodies.

jaundice A condition in which the skin is yellowish because of a buildup in the blood and tissues of bilirubin as a result of its failure to be excreted by the liver into the bile.

jejunum The middle segment of the small intestine.

joint The union between two or more bones of the skeleton.

k_D *See* diffusion coefficient.

k_p *See* membrane permeability constant.

kallikrein The enzyme in plasma and tissues that generates kinins from their inactive precursors.

kcal Kilocalorie; *see also* calorie.

KCl Potassium chloride.

keratin A scleroprotein found in hair and nails on the surface of the epidermis.

keto acid A molecule that contains a carbonyl group

$$(-\overset{\overset{\textstyle O}{\|}}{C}-)$$

and a carboxyl group

$$(-\overset{\overset{\textstyle O}{\|}}{C}-OH);$$

a product of amino acid deamination.

ketoacidosis A form of metabolic acidosis due to increased concentration of ketone bodies; a symptom of severe untreated diabetes mellitus.

ketone bodies Products of fatty acid oxidation, e.g., acetoacetic acid, acetone, and β-hydroxybutyric acid, that accumulate in the blood during starvation and in untreated diabetes mellitus.

kilocalorie (kcal) A Calorie; 1000 calories; *see also* calorie.

kininogen The protein from which the kinins are generated.

kinins Small polypeptides generated in plasma from kininogen by the enzyme kallikrein; act as paracrines.

knee jerk An example of the stretch reflex; stretch of the quadriceps femoris (thigh) muscle is elicited by depressing its tendon as the tendon passes over the knee; the reflex response (contraction of the muscle) causes extension of the knee.

Krebs cycle A metabolic pathway that utilizes fragments derived from the breakdown of carbohydrate, protein, and fat and produces carbon dioxide and hydrogen; the hydrogen, via oxidative phosphorylation, ultimately combines with oxygen to form water; also known as the *citric acid cycle* or the *tricarboxylic acid cycle*.

labia majora Two prominent skin folds that are part of the female external genitalia; the female analogue of the scrotum.

labia minora Two small skin folds that lie between the labia majora and surround the urethral and vaginal openings.

labyrinth A system of connecting cavities or canals, e.g., the bony labyrinth in the temporal bone of the skull.

lacrimal glands The glands that secrete tears.

lactase The enzyme that breaks lactose (milk sugar) into glucose and galactose; located in the luminal plasma membrane of the small-intestinal epithelia.

lactate The anion formed by the loss of hydrogen ion from lactic acid.

lactation The production and secretion of milk.

lacteal A blind-ended lymph vessel in the center of each intestinal villus.

lactic acid A three-carbon molecule; the end product of the anaerobic breakdown of glucose.

lactose Milk sugar; a disaccharide composed of glucose and galactose.

lacuna A small, hollow cavity.

lamella A thin leaf or plate, e.g., a narrow concentric ring that surrounds the haversian canal in compact bone.

lamina propria A layer of connective tissue that underlies the epithelial layer of the stomach and intestines.

larynx The part of the air passageway that connects the pharynx and trachea and contains the vocal cords.

latent period The period (several milliseconds) between initiation of an action potential in a muscle fiber and the beginning of mechanical activity; the period during which the events of excitation-contraction coupling occur.

lateral inhibition A method of refining information in neural pathways whereby the fibers inhibit each other, the most active causing the greatest inhibition of adjacent fibers.

lateral sac The enlarged region at the end of each segment of sarcoplasmic reticulum adjacent to the transverse tubule.

law of conservation of energy *See* first law of thermodynamics.

law of the intestine The fact that the short peristaltic waves (i.e., those that travel only several centimeters), occurring in response to local distension of the intestine, always proceed in the direction of the large intestine.

law of mass action The maxim according to which an increase in the concentration of a reactant of a chemical reaction causes the reaction to proceed and thus form more product; also called the principle of mass action.

learning The increase in the likelihood of a particular response to a stimulus as a consequence of experience.

lens The adjustable part of the optical system of the eye that helps focus an image on the retina regardless of the distance the object is from the eye.

leukocyte A white blood cell.

LH *See* luteinizing hormone.

ligament A band of connective tissue that connects bones or supports the viscera.

ligand Any molecule or ion that binds to the surface of any other molecule, such as protein, by noncovalent bonds.

ligase The enzyme that forms recombinant DNA by linking the fragmented ends of two DNA molecules previously split by a restriction enzyme.

limbic system An interconnected group of brain structures within the cerebrum involved with emotions and learning.

lipid Any of several types of molecules composed primarily of carbon and hydrogen atoms; characterized by insolubility in water; include fatty acids, triacylglycerols, phospholipids, and steroids.

lipoprotein A complex of fat and protein, the amounts of which vary; *low-density plasma lipoproteins* have little protein and much fat, whereas *high-density plasma lipoproteins* have the opposite distribution.

load The force exerted on a muscle by an external object.

lobule A small lobe.

local-circuit neuron A neuron that connects with cells in its immediate vicinity and has a short axon or, sometimes, none at all.

local current flow The movement of positive ions from a membrane region with a high positive charge through the cytoplasm or extracellular fluid toward a membrane region of more negative charge, and the simultaneous movement of negative ions in the opposite direction; leads to depolarization or hyperpolarization of adjacent segments of a membrane.

local potential A small, graded potential difference between two points; conducted decrementally,

has no threshold and no refractory period; EPSPs and IPSPs are examples of local potentials.

local response A biological response in the immediate vicinity of the stimulus (without the involvement of nerves or hormones) with the net effect of counteracting the stimulus.

local spinal reflex A reflex whose afferents, efferents, and integrating center are located within only a few segments of the spinal cord.

lockjaw A disease produced by the toxin of the tetanus bacillus; spasm of the jaw muscles occurs early in the disease.

locus A specific site; in genetics, the location of a gene on a chromosome.

loin That portion of the back between the thorax and pelvis.

long reflex A type of neural reflex in which the integrating center lies within the central nervous system.

long-term memory A memory that has a relatively long life-span.

longitudinal muscle layer The smooth muscle layer that surrounds the circular muscle layer in the stomach and intestinal walls and whose fibers are oriented longitudinally along these organs.

loop of Henle *See* Henle's loop.

lumen The space within a hollow tube or organ.

luminal Pertaining to the lumen of an organ.

lung compliance The magnitude of the change in lung volume caused by a given change in pressure difference across the lung wall; i.e., the greater the lung compliance, the more stretchable the lung wall.

luteal phase The last half of the menstrual cycle during which the corpus luteum is the active ovarian structure.

luteinizing hormone (LH) A polypeptide hormone secreted by the anterior pituitary that acts upon the gonads; one of the gonadotropins; sometimes called interstitial cell–stimulating hormone (ICSH) in the male.

lymph nodes Small organs, containing lymphocytes, located along the course of the lymph vessels; site of lymphocyte formation and storage.

lymphatic Pertaining to the system of vessels that conveys lymph from the tissues to the blood.

lymphocyte A type of leukocyte; mainly responsible for the "specific defenses" of the body against foreign invaders.

lymphoid tissue Lymph nodes, spleen, thymus, tonsils, and aggregates of lymphoid follicles such as those that line the gastrointestinal tract.

lymphokines Chemicals released from a T cell after combination of the T cell with its specific antigen.

lysosome Spherical or oval cell organelle surrounded by a single membrane; contains digestive enzymes that break down bacteria and large molecules that have entered a cell by endocytosis, and can digest damaged components of the cell itself.

M line One of the transverse bands that make up the repeating striated pattern of cardiac and skeletal muscle; a thin dark band in the middle of the sarcomere; produced by linkages between the thick filaments.

macrophage A cell type that has as its major function the phagocytosis of foreign matter.

malignant tumor A mass of cancer cells that rapidly grow, divide, and invade surrounding tissues, disrupt the structure and function of organs, and eventually lead to death of the organism.

maltase An enzyme that catalyzes the hydrolysis of the disaccharide maltose into two molecules of glucose.

maltose A disaccharide composed of two glucose molecules; formed during the digestion of starch.

mammary glands The milk-secreting glands in the breasts.

mass movement Contraction of large segments of the ascending and transverse colon that propels fecal material into the rectum.

mast cell A connective-tissue cell similar to a basophil except that it does not circulate in the blood; local injury releases histamine from mast cells.

mastication Chewing.

matrix The extracellular substance that surrounds connective-tissue cells; has some structural organization, such as connective-tissue fibers, as opposed to being a simple ionic solution.

maximal oxygen uptake The maximal amount of oxygen that can be utilized by the body per unit time during intense exercise.

maximal tubular capacity (T_m) The maximal rate of mediated transport of a substance across the wall of the kidney nephrons.

mean arterial pressure The mean blood pressure during the cardiac cycle; approximately equal to diastolic pressure plus one-third of the pulse pressure.

mechanoreceptor A sensory receptor that responds preferentially to mechanical stimuli such as bending, twisting, or compressing.

mediastinum The extrapulmonary space between the two lungs that extends from the sternum to the backbone.

mediated transport Movement of molecules across membranes by binding to protein carrier molecules in the membrane; characterized by specificity, competition, and saturation; includes both facilitated diffusion and active transport.

medulla The central portion of an organ, as opposed to the periphery or cortex.

meiosis A process of nuclear division that leads to the formation of the gametes (sperm and ova); consists of two cell divisions in succession, the first and second meiotic divisions; the final cells receive only half of the chromosomes present in the original cell.

membrane A structural barrier composed of phospholipids and proteins associated with the cell surface and its organelles, e.g., endoplasmic reticulum, mitochondria, plasma membrane, etc., which provides a selective barrier to the movement of molecules and ions across the membrane and provides a structural framework to which enzymes, fibers, and a variety of ligands can bind.

membrane permeability constant (k_p) A proportionality constant that defines the rate at which a substance diffuses through a membrane given the value of the concentration gradient across the membrane.

membrane potential The voltage difference between the inside and outside of the cell; also called *transmembrane potential*.

memory The relatively permanent form of a change in responsiveness to a particular stimulus.

"memory" cells Lymphocytes that were exposed to a certain antigen but did not differentiate fully into plasma cells or lymphokine-secreting T cells; rather, they serve as a rapidly responding pool of cells should that antigen be encountered again.

memory consolidation The processes by which memory is transferred from its short-term to its long-term form.

memory trace The postulated neural substrate of memory.

menarche The onset, at puberty, of menstrual cycling.

meninges The three membranes that cover the brain and spinal cord—the dura mater, pia mater, and arachnoid.

menopause The cessation of menstrual cycling, usually in the midforties.

menstrual cycle The cyclic rise and fall in female reproductive hormones; a period of approximately 28 days in female human beings.

menstrual fluid Blood mixed with debris from the disintegrating endometrium.

menstrual phase That stage of the menstrual cycle during which menstruation is occurring; also called the *menstrual period*.

menstruation The flow of menstrual fluid from the uterus.

merocrine A type of gland in which the secreting cells remain intact throughout the process of formation and discharge of the secretory products; in contrast to apocrine and holocrine secretion.

MES *See* microsomal enzyme system.

mesencephalon The midbrain.

mesenchyme The network of embryonic connective tissues in the mesoderm.

mesentery A sheet of connective tissue that connects the outer surface of the stomach and intestines to the abdominal wall.

messenger RNA (mRNA) The form of ribonucleic acid that transfers the genetic information corresponding to one gene (or, at most, a few genes) from DNA, in the nucleus, to the ribosomes, in the cytoplasm where proteins are synthesized.

metabolic acidosis An acidosis due to a metabolic production of acids other than carbon dioxide, e.g., the production of lactic acid during severe exercise or of ketones in diabetes mellitus.

metabolic alkalosis An alkalosis resulting from the removal of hydrogen ions from the body by mechanisms other than the respiratory removal of carbon dioxide, e.g., the loss of HCl from the stomach with persistent vomiting.

metabolic cost of living *See* basal metabolic rate.

metabolic pathway The sequence of enzyme-mediated chemical reactions that lead to the formation of a particular product.

metabolic rate The total energy expenditure of the body per unit time.

metabolism All the chemical reactions that occur within a living organism.

metastasis The breaking away of cancer cells from the parent tumor and their spread by way of the circulatory or lymphatic system to other parts of the body.

methyl group —CH_3.

Mg^{2+} Magnesium ion.

micelle A soluble cluster of amphipathic molecules (fatty acids, 2-monoglycerides, and bile salts), formed during the digestion of fat in the small intestine, in which the polar regions of the molecules line the outer surface of the micelle, and the nonpolar regions are within the micelle.

microbes Minute organisms, including bacteria, protozoans, and fungi.

microcirculation Circulation in blood vessels 100 μm or less in diameter.

microliter A unit of volume equal to 0.001 mL, or 0.000001 L.

micrometer (μm) 10^{-6} m; a micron; equivalent to 0.000039 in; formerly abbreviated μ.

microsomal enzyme system (MES) Enzymes, found in the smooth endoplasmic reticulum of liver cells, that transform molecules into more polar, less lipid-soluble substances; formerly called "drug-metabolizing enzymes."

microtubules Tubular filaments in the cytoplasm of cells that provide internal support for cells; may act in maintaining and changing cell shape and in producing the movements of organelles within the cell.

microvilli Small fingerlike projections from the surface of epithelial cells that greatly increase the absorptive surface area of a cell; a characteristic of the epithelium that lines the small intestine and kidney nephrons.

micturition Urination.

midbrain The part of the brain between the pons and forebrain.

middle-ear cavity An air-filled space deep within the temporal bone; contains the three ossicles (ear bones) that conduct the sound waves from the tympanic membrane to the cochlea.

milk intolerance The inability to digest milk sugar (lactose) because of lack of the intestinal enzyme lactase.

milk let-down The process by which milk is moved from the alveoli of the mammary glands into the ducts, from which it can be sucked.

milliliter (mL) A unit of volume equal to 0.001 L.

millimeter (mm) 10^{-3} m; equivalent to 0.03937 in.

millivolt (mV) 0.001 V.

mineralocorticoids Steroid hormones, produced by the adrenal cortex, that have their major effects on sodium and potassium balance; aldosterone is the major mineralocorticoid.

mitochondria Rod-shaped or oval organelles in cell cytoplasm that produce most of the ATP used by the cells; site of Krebs cycle and oxidative-phosphorylation enzymes; site of carbon dioxide production and oxygen utilization.

mitosis A process of nuclear division in which DNA is duplicated and an identical set of chromosomes is passed to each daughter cell at the time of cell division.

mitral valve The valve between the left atrium and left ventricle of the heart; also called the *bicuspid valve*.

mm *See* millimeter.

modality (of a stimulus) The quality or kind of stimulus; e.g., touch and hearing are different modalities.

modulator A molecule that, by acting at a protein's regulatory site, alters the properties of other binding sites on the protein and thus regulates the functional activity of the protein.

molarity A unit of concentration; the number of moles of solute per liter of solution.

mol (also mole) A unit that indicates the number of molecules of a substance present; the number of moles = weight in grams/molecular weight; one mole of any substance contains the same number of molecules as one mole of any other substance (the number is 6 × 10^{23}, Avagadro's number).

molecular weight A number equal to the sum of the atomic weights of all the atoms in a molecule; e.g., glucose ($C_6H_{12}O_6$) has a molecular weight of 180 [(6C × 12) + (12H × 1)(6 O × 16) = 180].

molecule A chemical structure formed by linking atoms together.

monocyte A type of leukocyte; leaves the bloodstream and is transformed into a tissue macrophage.

monosaccharide A carbohydrate that consists of a single sugar molecule; contains generally five or six carbon atoms, e.g., glucose ($C_6H_{12}O_6$).

monosynaptic reflex A reflex in which the afferent limb of the reflex arc (the afferent neuron) directly activates the efferent limb (the motor neuron); the stretch reflex is the only known monosynaptic reflex in human beings.

morula A solid, globular mass of cells.

motor Pertaining to muscles and movement.

motor cortex A strip of cerebral cortex along the posterior border of the frontal lobe; gives rise to many (but certainly not all) of the axons that descend in the corticospinal pathway to the motor neurons.

motor end plate The specialized region of a muscle-cell plasma membrane that lies directly under the axon terminal of a neuron; contains the receptor sites for the neurotransmitter acetylcholine; analogous to the subsynaptic membrane of postsynaptic neuron.

motor neuron An efferent neuron that innervates skeletal muscle fibers.

motor potential Electrical activity that can be recorded over motor cortex about 50 to 60 msec before a movement begins.

motor unit A motor neuron plus the muscle fibers it innervates.

mRNA *See* messenger RNA.

mucin A protein that forms mucus when mixed with water.

mucopolysaccharide A polysaccharide that contains an amino sugar, found in mucus and various connective-tissue secretions.

mucosa (of the stomach and intestines) The three layers of the stomach and intestinal wall nearest the lumen, i.e., the epithelium, lamina propria, and muscularis mucosa.

müllerian duct A part of the embryo that, in a female, develops into the ducts of the reproductive system but in a male degenerates.

müllerian regression factor A protein secreted by the Sertoli cells of the primitive testes that causes the müllerian ducts to regress.

multineuronal pathways The descending motor pathways that are made up of chains of neurons synapsing in the basal ganglia, several other subcortical nuclei, and the brainstem; also called the *multisynaptic pathways* and the *extrapyramidal system*.

multiunit smooth muscle Smooth muscle that exhibits little, if any, propagation of electrical activity from fiber to fiber and whose contractile activity is closely coupled to its neural input; *see also* single-unit smooth muscle.

muscarinic receptor An acetylcholine receptor that can be stimulated by the mushroom poison muscarine; most of these receptors are on smooth muscle and gland cells.

muscle A number of muscle fibers bound together by connective tissue.

muscle fatigue A decrease in the mechanical response of muscle with prolonged stimulation.

muscle fiber A muscle cell.

muscle spindle A specialized receptor organ in skeletal muscle that responds to stretch.

muscle-spindle stretch receptors The stretch receptors within the muscle spindles that are activated by stretch and provide information about muscle length.

muscle tension The force exerted by a contracting muscle on an object.

muscle tone The small amount of tension produced by skeletal muscles under resting conditions as a result of a low-frequency discharge of the motor neurons; in smooth muscle, tone may be maintained by spontaneous electrical activity within the muscle itself.

muscularis mucosa A thin layer of smooth muscle that underlies the lamina propria in the stomach and intestines.

mutation Any alteration in the base sequence of DNA that alters the genetic information stored in DNA.

mV *See* millivolt.

myasthenia gravis A neuromuscular disease associated with fatigue and skeletal muscle weakness; due to decreased numbers of ACh receptors at the motor end plate.

myelin Insulating material that covers the axons of many neurons; consists of multiple layers of myelin-forming cell plasma membranes that are wrapped concentrically around the axon.

myeloid Pertaining to bone marrow.

myenteric plexus A network of nerve cells that lies between the circular and longitudinal muscle layers in the esophagus, stomach, and intestinal walls; Auerbach's plexus.

myoblast The embryological cell that gives rise to muscle fibers.

myocardium Cardiac muscle.

myoepithelial cells Specialized contractile cells that surround the alveoli of mammary glands.

myofibrils Longitudinal bundles of thick and thin contractile filaments arranged in a repeating sarcomere pattern; located in the cytoplasm of striated muscles.

myogenic Originating within a muscle.

myoglobin A protein in muscle fibers, similar to hemoglobin, that binds oxygen.

myometrium Uterine smooth muscle.

myopia *See* nearsighted.

myosin The contractile protein that forms the thick filaments of muscle.

myosin ATPase An enzymatic site on the globular head of myosin that catalyzes the breakdown of ATP with the release of chemical energy that is used to produce the force of muscle contraction.

NaCl Sodium chloride.

NAD *See* nicotinamide adenine dinucleotide.

Na$^+$—K$^+$ ATPase An enzyme in the plasma membranes of most cells; splits ATP in the presence of sodium and potassium ions and releases energy that is used to transport sodium out of the cell and potassium into the cell.

nanometer (nm) 10^{-9} m; equivalent to 0.000000039 in; a millimicron; formerly abbreviated mμ.

nares The openings of the nasal cavity.

nasal Pertaining to the nose.

nearsighted Pertaining to the defect in vision that occurs because the eyeball is too long for the lens, so that images of distant objects are focused in front of the retina.

negative feedback. An aspect of control systems in which the ultimate effects of the system counteract the original stimulus.

nephron The basic functional unit of the kidney whose function is the formation of urine; consists of a glomerulus, Bowman's capsule, proximal tubule, Henle's loop, distal tubule, and collecting duct.

nerve A collection of nerve fibers (axons) in the peripheral nervous system.

nerve fiber *See* axon.

net flux The net amount of a substance that diffuses across a surface in a unit of time; equal to the difference between two opposite unidirectional fluxes.

neural Pertaining to the nervous system.

neuroeffector junction Functional relationship between a nerve fiber and a muscle or gland cell, e.g., a neuromuscular junction.

neuroendocrinology The study of the interrelated functions of the nervous and endocrine systems.

neurofibril Thin fibers found in nerve cells.

neuroglia The supportive, nonneural cells in the nervous system; glia.

neurohypophysis The posterior lobe of the pituitary gland.

neuromodulator A chemical that amplifies or dampens neuronal activity by altering the neurons directly or by influencing the effectiveness of a neurotransmitter; does not act as a specific neurotransmitter.

neuromuscular junction An anatomically specialized junction between an axon terminal of a motor neuron and a skeletal muscle cell where the chemical mediator (acetylcholine), released in response to action potentials in the nerve, diffuses across an extracellular gap to initiate action potentials in the muscle membrane.

neuron A nerve cell.

neurotransmitter A chemical agent released by one neuron that acts upon a second neuron or upon a muscle or gland cell and alters its electrical state or activity.

neutral fat Triacylglycerol; formerly called triglyceride; consists of a molecule of glycerol to which three fatty acids are attached.

neutral molecule Any molecule that has no polar or ionized groups and thus has no electric charge.

neutrophil A polymorphonuclear granulocytic leukocyte whose granules show preference for neither eosin nor basic dyes.

neutrophil exudation The amebalike movement of neutrophils from the lumen of capillaries out into the extracellular space of tissues.

NH₃ Ammonia.

NH₄⁺ Ammonium ion.

niacin One of the B-complex vitamins; an essential ingredient of the coenzyme NAD.

nicotinamide adenide dinucleotide (NAD) A coenzyme that transfers hydrogen from one reaction to another.

nicotinic receptors Those acetylcholine receptors that respond to the drug nicotine; primarily, the receptors at the motor end plate and the receptors on postganglionic neurons of the autonomic nervous system.

night vision The only vision possible under conditions of low illumination, i.e., rod vision.

nm *See* nanometer.

nociceptor A sensory receptor whose stimulation causes the sensation of pain.

node of Ranvier Regions located at intervals along the sheath of myelinated nerve fibers where the axon membrane is in contact with the extracellular fluids.

nonpolar molecule A molecule that contains predominantly neutral chemical bonds and thus has no polar or ionized groups.

nonspecific immune responses Responses that do not require previous exposure to a particular foreign material but nonselectively protect against them without having to recognize their specific identities.

nonspecific pathway A chain of synaptically connected neurons in the brain or spinal cord that are activated by sensory units of several different modalities.

nonspecific thalamic nuclei Nuclei in the cerebrum whose axons project to widely dispersed areas of cerebral cortex.

norepinephrine A catecholamine neurotransmitter released at most sympathetic postganglionic endings, from the adrenal medulla, and in many regions of the central nervous system.

novocaine A local anesthetic that works by preventing the increased membrane permeability to sodium ions required to produce an action potential in the axon of a neuron.

noxious Poisonous; harmful to health.

nuclear envelope The double membrane that surrounds the nucleus of a cell.

nuclear pores Openings in the nuclear envelope.

nuclear protein The protein portion of chromatin; involved in the control of gene activity.

nucleic acid A straight-chain polymer of nucleotides in which the phosphate of one of the repeating nucleotide subunits is linked to the sugar of the adjacent one; functions in the storage and transmission of genetic information; includes both DNA and RNA.

nucleolus A prominent, densely staining nuclear structure that consists of granular and filamentous elements; contains the portions of DNA that code for rRNA.

nucleoside A molecule that consists of a purine or pyrimidine base and a sugar (ribose or deoxyribose).

nucleotide The molecular subunit of nucleic acids that consists of a purine or pyrimidine base, a sugar, and phosphoric acid.

nucleus A large spherical or oval membrane-bound cell organelle present in most cells; contains most of the cell's DNA and some of its RNA; (neural) a cluster of neuronal cell bodies within the CNS.

obesity An excessive accumulation of fat by the adipose tissues of the body.

occipital cortex The posterior region of the cerebral cortex where the primary visual cortex is located.

Ohm's law The relationship between current (I), voltage (E), and resistance (R) such that $I = E/R$.

olfactory Pertaining to the sense of smell.

olfactory mucosa The region of mucous membrane in the upper part of the nasal cavity that contains the receptors for the sense of smell.

olfactory nerve Cranial nerve I; consists of the axons of bipolar cells whose peripheral terminals contain the chemoreceptors for the sense of smell.

oligodendroglia A type of glial cell; responsible for the generation of myelin around nerve fibers in the central nervous system.

oligomeric protein A protein that consists of more than one polypeptide chain.

omentum A fold of peritoneum (mesentery) attached to the stomach.

oocyte The ovum, or egg cell.

oogonia Primitive ova that upon mitotic division before birth, give rise to the primary oocytes.

operator gene The initial gene of an operon that controls the transcription of the genes in the operon; the site of repressor binding.

operon A collection of genes that can be induced or repressed as a unit.

ophthalmic Pertaining to the eye.

opsin The protein component of a photopigment molecule.

optic nerve Cranial nerve II; formed of the axons of the retinal ganglion cells.

oral Pertaining to the mouth.

orbit The bony cavity of the skull that contains the eyeball.

organ A collection of tissues joined in a structural unit to serve a common function; e.g., the heart is an organ that comprises muscle, nerve, connective, and epithelial tissues that together serve as a pump for blood.

organ of Corti The collection of structures in the inner ear capable of transducing the energy of waves in a fluid medium into the electrochemical energy of action potentials in the auditory nerve.

organ system A collection of organs that together serve an overall function; e.g., the kidneys, ureters, urinary bladder, and urethra are the organs that constitute the urinary system.

organelles Multimolecular structural components of cells that perform specialized functions; e.g., the mitochondria and endoplasmic reticulum are cell organelles.

organic Pertaining to molecules that contain the element carbon.

organic acid An acid molecule that contains carbon atoms, e.g., carbonic and lactic acids.

osmolarity The total solute concentration of a solution; a measure of water concentration in that the higher the osmolarity of a solution, the lower its water concentration.

osmoreceptor A neural receptor that responds to changes in the osmolarity of the surrounding fluid.

osmosis The net diffusion of water due to a difference in water concentration on the two sides of a membrane impermeable to solute molecules; the water always moves from the region of low solute concentration (high water concentration) to the region of high solute concentration (low water concentration).

osmotic equilibrium A condition in which the solute (osmolar) concentrations on the two sides of a membrane are equal; thus, the water concentrations are also equal.

osteoblast The type of cell responsible for laying down the protein matrix of bone.

osteoclast A cell that degrades the extracellular matrix of bone; osteoblasts and osteoclasts may be different functional stages of the same cell.

osteocyte An osteoblast that has become embedded in bone matrix.

otoliths Calcium-carbonate ''stones'' in the gelatinous mass of the utricle and saccule.

oval window The membrane-covered opening that separates the middle-ear cavity from the scala vestibuli of the inner ear; receives the footplate of the stapes (one of the three middle-ear bones).

ovarian follicle The ovum and its encasing follicular, granulosa, and theca cells at any stage of its development prior to ovulation.

ovary The gonad in the female.

oviducts *See* uterine tubes.

ovulation The release of an ovum, surrounded by its zona pellucida and cumulus, from the ovary.

ovum The female gamete that is fertilized by a sperm cell at the time of conception (pl., ova).

oxidative phosphorylation The process by which the energy released from the combination of hydrogen with molecular oxygen is used to form ATP from ADP and inorganic phosphate; occurs in the mitochondria through a highly specialized coupling process that involves cytochromes embedded in the inner mitochondrial membrane.

oxyhemoglobin (HbO_2) hemoglobin combined with oxygen.

oxyntic cells *See* parietal cells.

oxytocin A peptide hormone of nine amino acids synthesized in the hypothalamus and released from the posterior pituitary; stimulates the mammary glands to release milk and the uterus to contract.

P wave The component of the electrocardiogram that reflects depolarization of the atria.

pacemaker potential The spontaneous depolarization to threshold potential of the plasma membrane of some nerve and muscle cells.

pacinian corpuscle A mechanoreceptor whose structure, which resembles onionlike layers around a central nerve fiber, is specialized to respond to vibrating stimuli.

pain A sensation that results from tissue damage (or threatened tissue damage) and often accompanied by an emotional component such as fear, anxiety, or a sense of unpleasantness.

palpitation A rapid, often irregular beating of the heart that can be felt by the individual.

pancreas A gland in the abdomen near the stomach and connected by a duct to the small intestine; contains both endocrine gland cells, which secrete the hormones insulin and glucagon into the bloodstream, and exocrine gland cells, which secrete digestive enzymes and bicarbonate into the intestine.

pancreatic duct The large duct that carries the exocrine secretion of the pancreas into the duodenum.

pancreatic lipase An enzyme secreted by the exocrine pancreas;

acts on triacylglycerols to form 2-monoglycerides and free fatty acids.

pancreozymin *See* cholecystokinin.

pantothenic acid One of the B-complex vitamins.

paracrine agent A chemical agent that, when released, exerts its effects on tissues located near the site of its secretion, in contrast to a hormone that travels by way of the blood to exert its effects on cells far removed from the site of its secretion; excludes neurotransmitters.

paradoxical sleep The state of sleep associated with small, rapid, EEG oscillations (indistinguishable from those of EEG arousal) and dreaming and with complete loss of tone in the postural muscles, and irregular respiration and heart rate.

parasympathetic nervous system That portion of the autonomic nervous system whose preganglionic fibers leave the central nervous system at the brainstem and sacral portion of the spinal cord.

parasympathomimetic A chemical that produces effects similar to those of the parasympathetic nervous system.

parathormone *See* parathyroid hormone.

parathyroid glands The four parathyroid hormone–secreting glands on the back of the thyroid gland.

parathyroid hormone (PTH) A peptide hormone secreted by the parathyroid glands; also called *parathormone;* regulates calcium and phosphate concentrations of the internal environment.

parietal cells Gastric gland cells that secrete hydrochloric acid and intrinsic factor; also called *oxyntic cells.*

parietal lobe The region of the cerebral cortex where sensory cortex and some association cortex are located.

parkinsonism A disease characterized by tremor, rigidity, and a delay in the initiation of movement; a result of degeneration of the dopamine-liberating axons from the substantia nigra.

parotid One of the three pairs of salivary glands.

partial pressure The pressure exerted by one molecular species of a gas; e.g., P_{O_2}, the partial pressure of oxygen, is that part of the total pressure that is due to oxygen molecules; in a pure gas, the partial pressure equals the total pressure; a measure of the concentration of a gas in a mixture of gases.

parturition Birth; delivery of an infant.

passive immunity A resistance to infection that results from the direct transfer of antibodies or T cells from one person (or animal) to another.

P_{CO_2} The partial pressure of carbon dioxide.

pepsin An enzyme formed in the stomach from the precursor pepsinogen; breaks protein down to peptide fragments.

pepsinogen The inactive precursor of the enzyme pepsin; secreted by the chief cells of the gastric mucosa.

peptide A short polypeptide chain.

peptide bond The chemical bond

$$(\text{—NH—}\overset{\displaystyle O}{\overset{\displaystyle \|}{\text{C}}}\text{—})$$

that joins two amino acids; forms the backbone of protein molecules.

perception The conscious experience of objects and events of the external world that we acquire from the neural processing of sensory information.

perfusion The passage of fluid through a vessel or organ.

pericardium The fibrous connective-tissue sac that surrounds the heart.

perichondrium The connective-tissue membrane that surrounds cartilage.

perimysium The connective-tissue layers that bind a number of muscle fibers into bundles.

perineum The floor of the pelvis; the anatomical space bounded by the scrotum in the male and the mons pubis in the female in front, by the medial surface of the thighs laterally, and by the buttocks behind.

periosteum The connective-tissue membrane that covers the outer surface of bones.

peripheral chemoreceptors The carotid and aortic bodies that respond to changes in blood oxygen, carbon dioxide, and hydrogen ions.

peripheral nervous system The nerve fibers that extend from the brain and spinal cord.

peripheral proteins Membrane proteins that are water soluble and can be removed from the membrane by solutions that alter the electric charge on the proteins.

peripheral thermoreceptors Cold and warm receptors in the skin and certain mucous membranes.

peristaltic wave A progressive wave of muscle contraction that proceeds along the wall of a tube, compressing the lumen of the tube and causing the contents of the tube to move.

peritoneum The membrane that lines the abdominal cavity and covers the viscera.

peritubular capillaries Capillary network closely associated with the tubular components of the kidney nephrons.

permeability The degree to which molecules are able to move through a barrier such as a cell membrane or epithelial layer of cells.

permissiveness (of a hormone) The situation whereby small quantities of one hormone are required in order for a second hormone to exert its full effect upon a cell.

pernicious anemia A disease in which erythrocytes fail to proliferate and mature because of a deficiency of vitamin B_{12}.

pH The negative logarithm to the base 10 of the hydrogen-ion concentration; a measure of the acidity of a solution: the pH decreases as the acidity increases.

phagocytosis A form of endocytosis in which large multimolecular particles, such as bacteria, are taken into the cell by invagination of the plasma membrane.

phagosome   The membrane-enclosed pouch formed when a phagocyte engulfs a microbe or other particulate matter.

pharmacological effects (of a hormone)   The effects produced by much larger quantities of a hormone in the body than are normally present.

pharynx   Throat; a passage common to the routes followed by food and air.

phasic   Intermittent, as opposed to tonic.

phosphatase   An enzyme that removes phosphate from organic molecules.

phosphocreatine   A compound used to store small amounts of metabolic energy that can be rapidly transfered to ATP; found in large amounts in muscle tissue.

phosphodiesterase   The enzyme that catalyzes the breakdown of the second messenger cyclic AMP to AMP (adenosine monophosphate).

phospholipid   A subclass of lipid molecules with a glycerol backbone to which are attached two fatty acids and a phosphate group ($-PO_4-$) plus a small polar or ionized nitrogen-containing molecule; a major component of cell membranes.

phosphoric acid   An acid generated during the catabolism of phosphorus-containing compounds; dissociates to form phosphate ions and hydrogen ions.

phosphorylation   Addition of a phosphate group to an organic molecule.

photon   A quantum of electromagnetic radiation.

photopigment   A light-sensitive molecule altered by the absorption of photic energy of certain wavelengths; consists of an opsin bound to a chromophore.

phrenic nerve   A nerve that passes from the cervical region of the spinal cord to the diaphragm muscle.

pia mater   The meningeal covering closest to the brain or spinal cord.

PIH   *See* prolactin inhibiting hormone.

pinocytosis   *See* endocytosis.

pitch   The degree of how low or high a sound is perceived; related to the frequency of the sound wave.

pituitary   An endocrine gland that lies in a pocket of bone just below the hypothalamus; formerly called the hypophyseal gland, or hypophysis.

placenta   A combination of interlocking fetal and maternal tissues in the uterus that serves as the organ of molecular exchange between maternal and fetal circulation.

plasma   The fluid, noncellular portion of the blood; a component of the extracellular fluid compartment.

plasma cells   Cells that are derived from lymphocytes and secrete antibodies.

plasma membrane   The cell membrane that forms the outer surface of the cell and separates its contents from the surrounding extracellular fluid.

plasmin   A proteolytic enzyme able to decompose fibrin and, thereby, dissolve blood clots.

plasminogen   The inactive precursor of the enzyme plasmin.

plasticity   The ability of neural tissue to change its responsiveness because of its past history of activation.

platelet   A cell fragment that is present in the blood and plays a role in blood clotting; formed in bone marrow from megakaryocytes.

pleura   A thin sheet of cells firmly attached to the interior of the thoracic cage and folding back upon itself to form two completely enclosed sacs within the thoracic cavity; each sac is also attached to the surface of the lung.

plexus   A network of interlacing nerve fibers or blood vessels.

pneumothorax   Air in the intrapleural space.

Po_2   The partial pressure of oxygen.

polar body   The small cell that results from the unequal distribution of cytoplasm during the divisions of the primary and secondary oocytes.

polar molecule   Substance that contains a number of electrically polarized chemical bonds in which electrons are shared unequally between atoms, the region of the molecule to which the electrons are drawn carrying a negative charge and the region from which they are drawn carrying a positive charge; polar molecules are soluble in water.

polarized   Having two electric poles, one negative and one positive.

polarized covalent bond   A chemical bond in which two electrons are shared unequally between two atoms; the atom to which the electrons are drawn carries a negative charge and the other atom carries a positive charge.

polymer   A very large molecule formed by the linking together of many smaller similar subunits.

polymodal   A neuron that responds to stimuli of more than one modality.

polymorphonuclear granulocyte   A subclass of leukocytes; there are three types: neurophils, eosinophils, and basophils.

polypeptide   Polymer that consists of amino acid subunits joined in sequence by peptide bonds between the amino group of one amino acid and the carboxyl group of the adjacent one.

polyribosome   One strand of mRNA to which are attached a number of ribosomes.

polysaccharide   A large carbohydrate molecule formed by linking many monosaccharide subunits together; for example, glycogen and starch.

polysynaptic reflex   A reflex that employs one or more interneurons in its reflex arc.

polyunsaturated fatty acid   A fatty acid that contains more than one double bond.

pons The region of the brain between the medulla and the midbrain.

pool The body's readily available quantity of a particular substance; frequently identical to the quantity present in the extracellular fluid.

portal vessels Blood vessels that link two capillary networks.

positive feedback An aspect of control systems in which the ultimate effects of the system influence it in such a way that these effects are increased, i.e., that the original stimulus is strengthened; an unstable situation because the output of the system will rapidly and progressively increase in magnitude.

postabsorptive state The period during which nutrients are not being absorbed by the gastrointestinal tract and energy must be supplied by the body's endogenous stores.

posterior pituitary That portion of the pituitary from which oxytocin and antidiuretic hormone are released.

postganglionic neuron A neuron of the autonomic nervous system whose cell body lies in a ganglion and whose axon terminals form neuroeffector junctions with smooth muscle, cardiac muscle, or glands; a neuron that conducts impulses away from a ganglion toward the periphery.

postsynaptic neuron A neuron that conducts information away from a synapse; a neuron acted upon by a synapse.

postsynaptic potential A local potential that arises in the postsynaptic neuron in response to the activation of synapses upon it; *see also* excitatory postsynaptic potential, inhibitory postsynaptic potential.

postural reflexes Those reflexes that maintain or restore stable, upright posture.

potential The voltage difference between two points; also called *potential difference*.

prandial drinking Drinking associated with a meal.

precapillary sphincter A ring of smooth muscle around a "true" capillary at the point at which it exits from a thoroughfare channel or arteriole.

preganglionic neuron A neuron of the atonomomic nervous system whose cell body lies in the central nervous system and whose axon terminals lie within a ganglion; a neuron that conducts impulses from the central nervous system toward a ganglion.

preprogram A sequence of neuronal activity that results in a movement independent of afferent input that may occur during the course of the movement.

presbyopia The impairment in vision that results from stiffening of the lens, which makes accommodation difficult.

pressure The force exerted on a surface.

presynaptic inhibition A relationship between two neurons in which the axon terminals of one neuron end on the synaptic knob of a second presynaptic neuron; action potentials in the first neuron decrease the release of neurotransmitter from the second neuron and thus decrease the magnitude of the postsynaptic potential produced in the third (postsynaptic) neuron.

presynaptic neuron A neuron that conducts information toward a synapse.

PRH *See* prolactin releasing hormone.

primary afferent *See* afferent neuron.

primary follicle Any cell cluster in the ovaries that consists of one ovum surrounded by a single layer of follicular cells.

primary oocyte Germ cell of the female; undergoes the first meiotic division to form a secondary oocyte (and a polar body).

primary spermatocyte A male germ cell derived from the spermatogonia; undergoes meiotic division to form two secondary spermatocytes.

primary structure (of a protein) The sequence of amino acids along the polypeptide chain of the molecule.

principle of mass action *See* law of mass action.

progenitor cell A cell that can differentiate into other cell types but is unable to renew its own type.

progesterone A steroid hormone secreted primarily by the corpus luteum and placenta; stimulates secretion by uterine glands, inhibits contraction of uterine smooth muscle, and stimulates breast growth.

program A related sequence of neural activity.

projection neuron A neuron that has a long axon that connects it with distant parts of the nervous system or with muscles or glands.

prolactin A polypeptide hormone secreted by the anterior pituitary; stimulates milk secretion by the mammary glands.

prolactin inhibiting hormone (PIH) The hypothalamic hormone that inhibits prolactin secretion by the anterior pituitary.

prolactin releasing hormone (PRH) The hypothalamic hormone that stimulates prolactin secretion by the anterior pituitary.

proliferative phase That stage of the menstrual cycle between menstruation and ovulation during which the endometrium repairs itself and grows.

prostaglandins A group of unsaturated, modified fatty acids that function as chemical messengers (paracrine agents); synthesized in most—possibly all—cells of the body.

prostate gland A large gland that encircles the urethra in the male; secretes fluid into the urethra.

protease An enzyme that breaks the peptide bonds between amino acids in polypeptide chains.

protein A large polymer that consists of one or more linear sequences of amino acid subunits joined by peptide bonds.

protein binding site *See* binding site.

protein kinase A group of enzymes,

each of which catalyzes the addition of a phosphate group to a specific protein.

proteolytic Causing the breakage of the peptide bonds that link amino acids in proteins.

prothrombin The inactive precursor to the enzyme thrombin; produced by the liver and normally present in the plasma.

protomer Each of the single polypeptide chains in an oligomeric protein.

proton A subatomic particle in the nucleus of atoms; carries one positive charge.

protoplasm An old term used to refer to the material in cells.

proximal tubule The first tubular component of a nephron after Bowman's capsule.

psychological fatigue The mental factors that cause individuals to stop exercising even though their muscles are not fatigued and are still able to contract.

PTH *See* parathyroid hormone.

ptyalin *See* amylase.

puberty The attainment of sexual maturity in the sense that conception becomes possible; as commonly used, it refers to the 3 to 5 years of sexual development that culminate in the attainment of sexual maturity.

pulmonary Pertaining to the lungs.

pulmonary circulation The portion of the cardiovascular system between the pulmonary trunk (as it leaves the right ventricle) and the pulmonary veins (as they enter the left atrium); the circulation through the lungs.

pulmonary edema Accumulation of fluid in the lung interstitium and air sacs.

pulmonary stretch receptor An afferent nerve ending (or cell closely associated with it) in the airway smooth muscle and activated by lung inflation.

pulmonary surfactant *See* surfactant.

pulmonary trunk The large artery that carries blood from the right ventricle of the heart to the lungs.

pulmonary valve The valve between the right ventricle of the heart and the pulmonary trunk.

pulmonary ventilation *See* total pulmonary ventilation.

pulse pressure The difference between systolic and diastolic blood pressures.

pupil The black-appearing opening in the iris of the eye through which light passes to reach the retina.

purine bases Adenine and guanine; double-ring, nitrogen-containing subunits of nucleotides and nucleosides.

Purkinje fibers The muscle fibers in the ventricles of the heart that rapidly conduct action potentials throughout the venticles.

pyloric sphincter A ring of smooth muscle and connective tissue between the terminal antrum of the stomach and the first segment of the small intestine.

pyramidal tract *See* corticospinal pathway.

pyridoxine Vitamin B_6.

pyrimidine bases Cytosine, thymine, and uracil; single-ring, nitrogen-containing subunits of nucleotides and nucleosides.

pyrogen An endogenously produced chemical released from white blood cells in the presence of infection or inflammation; acts on thermoreceptors in the hypothalamus (and perhaps other brain areas) to alter their rate of firing and their input to the thermoregulatory integrating centers.

pyruvate The anion formed when the carboxyl group of pyruvic acid dissociates, i.e., loses a hydrogen ion.

pyruvic acid The three-carbon keto-acid in the glycolytic pathway that, in the absence of oxygen, forms lactic acid or, in the presence of oxygen, enters the Krebs cycle.

QRS complex The component of the electrocardiogram that corre-

sponds to the depolarization of the ventricles.

quadriceps A muscle that has four heads.

radiation (of heat) The emission of heat from the surfaces of objects.

ramus A branch.

rarefaction A region in a sound wave where the density of air molecules is low.

RAS *See* reticular activating system.

rate-limiting enzyme The enzyme in a metabolic pathway most easily saturated with substrate; activity of this enzyme determines the flow of substrate through the entire metabolic pathway; usually the site of allosteric regulation in the pathway.

reactant A molecule that enters a chemical reaction; in enzyme-catalyzed reactions, the reactant is called the substrate molecule.

readiness potential Electrical activity that can be recorded over wide areas of association cortex during the ''decision-making period,'' about 0.8 sec before a movement begins.

receptive field (of a neuron) The region surrounding the receptor endings that, if stimulated, causes activity in that neuron.

receptive relaxation The decrease in smooth muscle tension in the walls of hollow organs in response to distention.

receptor *In sensory systems:* the specialized peripheral ending of an afferent neuron or a separate cell intimately connected to it that detects changes in some aspect of the environment; *in chemical communication:* specific proteins either in the plasma membrane or cytosol of a target cell with which a chemical mediator combines to exert its effects.

receptor density The number of receptors per unit area.

receptor potential A graded potential that arises in the ending of an afferent neuron or in a specialized cell intimately associated with it, in response to stimulation of the receptor.

reciprocal innervation Inhibition of motor neurons that activate those muscles whose contraction would oppose the intended movement; e.g., inhibition of flexor motor neurons during extensor motor neuron activation.

recombinant DNA DNA experimentally formed by joining portions of two DNA molecules previously fragmented by a restriction enzyme.

recruitment The activation of additional cells in response to a stimulus of increased magnitude.

rectum A short segment of the large intestine between the sigmoid colon and anus.

reduced hemoglobin (Hb) Deoxyhemoglobin; hemoglobin not combined with oxygen.

reflex The sequence of events elicited by a stimulus; a biological control system mediated by a reflex arc composed of neural, hormonal, or neural and hormonal elements.

reflex arc The components that mediate a reflex; usually, but not always, includes a receptor, afferent pathway, integrating center, efferent pathway, and effector.

reflex response That change brought about by the action of a stimulus upon a reflex arc; effector response.

refractory period The time during which an excitable membrane does not respond to a stimulus whose magnitude is normally sufficient to trigger an action potential; *see also* absolute refractory period, relative refractory period.

regulator site The site on a protein at which interaction with a modulator allosterically alters the properties of other binding sites on the protein.

"regulatory" appetite The desire for a substance in proportion to the deficiency of that substance.

relative refractory period The time during which an excitable membrane will produce an action potential only in response to stimuli of greater strength than the usual threshold strength.

relaxation time The time interval from the development of peak tension by a muscle until the tension has decreased to zero.

REM sleep *See* paradoxical sleep.

renal Pertaining to the kidneys.

renal blood flow The volume of blood delivered to the kidneys per unit of time.

renal medulla The inner region of the kidney.

renal pelvis A large central cavity at the base of each kidney; receives urine from the collecting ducts and empties it into the ureter.

renin An enzyme secreted by the kidney; it catalyzes the splitting off of angiotension I from angiotensionogen in the blood.

replication Formation of a new molecule of DNA identical to the original one.

repolarize To restore the value of the transmembrane potential toward its resting level.

repression Inhibition of enzyme synthesis in the presence of product molecules synthesized by that enzyme (or the enzyme chain of which it is a part); product molecules inhibit enzyme synthesis by effects upon the genes that code for the protein structure of the enzymes.

repressor A protein molecule that can bind to an operator gene and inhibit the transcription of the gene into mRNA.

repressor gene A portion of DNA that contains the genetic instructions for synthesis of a repressor protein.

reproductive duct system Seminiferous tubules, epididymis, ductus deferens, ejaculatory duct, and urethra in the male; uterine tubes, uterus, and vagina in the female.

residual volume The volume of air that remains in the lungs after a maximal expiration.

resistance The hinderance to movement through a particular substance or tube, e.g., the hinderance to the movement of blood through the blood vessels.

respiration The consumption of molecular oxygen during metabolism; the exchange of gas between the cells of an organism and the external environment.

respiratory acidosis An acidosis that results from failure of the lungs to eliminate carbon dioxide as fast as it is produced.

respiratory alkalosis An alkalosis that results from the elimination of carbon dioxide from the lungs faster than it is produced.

respiratory distress syndrome (of the newborn) A disease that afflicts premature infants in whom the surfactant-producing cells are too immature to function adequately.

respiratory "pump" The effect of the changing intrathoracic and intraabdominal pressures associated with respiration on blood flow through vessels in the thoracic and abdominal cavities.

respiratory quotient (RQ) The ratio of CO_2 produced to O_2 consumed during metabolism.

respiratory rate The number of breaths per minute.

respiratory system The structures involved in the exchange of gases between the blood and the external environment, i.e., the lungs, the series of airways that lead into the lungs, and the chest structures responsible for movement of air in and out of the lungs.

resting membrane potential The voltage difference between the inside and outside of a cell in the absence of excitatory or inhibitory stimulation; also called *resting potential*.

restriction enzyme A bacterial enzyme that splits DNA into a number of fragments, acting at different loci in the two strands of the polymer and at points that contain a particular sequence of nucleotides.

retching   The strong involuntary effort to vomit, i.e., deep inspiration, closure of the glottis, elevation of the soft palate, and contraction of abdominal and thoracic muscles, without the actual expulsion of the stomach's contents through the pharynx.

reticular   Resembling a network.

reticular activating system   A collection of neurons in the brainstem reticular formation and its thalamic extension whose activity is concerned with alertness and direction of attention to selected events.

reticular formation   An extensive network of finely branched neurons that extend through the core of the brainstem and are linked functionally to a similar group of cells in the midline of the thalamus.

retina   A thin layer of neural tissue that contains the receptors for vision and lines the back of the eyeball.

retinal   A derivative of vitamin A; forms the chromophore component of a photopigment.

retrograde amnesia   Loss of memory for a time immediately preceding a memory-disturbing trauma such as a blow on the head.

reversible reaction   A chemical reaction that can readily proceed in either direction because only small exchanges of energy are involved in the reaction.

Rh factor   Erythrocyte plasma-membrane antigens used in blood typing that may (Rh +) or may not (Rh −) be present.

rhodopsin   The photopigment in rods; a light-absorbing glycoprotein.

riboflavin   Vitamin B_2.

ribonucleic acid (RNA)   A nucleic acid involved in the decoding of the genetic information and its transfer to the site of protein synthesis; its nucleotides contain the sugar ribose to which is attached one of four bases (adenine, guanine, cytosine, or uracil); exists in three forms: messenger RNA, ribosomal RNA, and transfer RNA.

ribose   A monosaccharide found in RNA.

ribosomal RNA (rRNA)   A polymer of ribose-containing nucleotides that is synthesized in the nucleus but moves to the cytoplasm where it forms part of the ribosome.

ribosome   A cytoplasmic particle that mediates the linking together of amino acids to form proteins.

rickets   A disease in which new bone matrix is inadequately calcified as a result of a deficiency of 1,25-dihydroxyvitamin D_3.

rigor mortis   Death rigor; a stiffness of the muscles that results from a loss of ATP, which is necessary for the dissociation of myosin cross bridges from the actin thin filaments; occurs approximately 3 to 12 hr after death.

RNA   *See* ribonucleic acid.

rod   One of the two receptor types for photic energy; contains the photopigment rhodopsin.

round window   The membrane-covered opening that separates the scala tympani of the inner ear from the middle-ear cavity.

RQ   *See* respiratory quotient.

rRNA   *See* ribosomal RNA.

rugae   The ridges or folds that appear in the wall of the stomach.

SA node   *See* sinoatrial node.

saccade   A very fast, short, jerking movement of the eyeball.

saliva   The secretory product of the salivary glands; a watery solution of various salts and proteins, including the mucins and amylase (ptyalin).

salivary gland   One of three paired glands that secretes into the mouth a solution that contains mucus and the carbohydrate-digesting enzyme amylase.

sarcomere   The repeating structural unit of a muscle myofibril; the region of a myofibril that extends between two adjacent Z lines.

sarcoplasmic reticulum   The endoplasmic reticulum in muscle fibers; forms sleevelike structures around each sarcomere in striated muscle; site of storage and release of calcium ions that bind to troponin to initiate contractile activity.

saturated fatty acid   A fatty acid that does not contain any double bonds.

saturation (of binding sites)   The degree to which binding sites are occupied by ligand; if all are occupied, the binding sites are fully saturated, if half are occupied, the saturation is 50 percent.

scala tympani   The fluid-filled inner-ear compartment that receives sound waves from the basilar membrane and transmits them to the round window.

scala vestibuli   The fluid-filled inner-ear compartment that receives sound waves from the oval window and transmits them to the basilar membrane and cochlear duct.

Schwann cell   A nonneural cell that surrounds all peripheral nerve fibers; the plasma membrane of this cell forms the myelin sheath around a myelinated axon; its analogous cell in the central nervous system is the oligodendroglia.

sclera   The white fibrous tissue that forms the outer protective covering of the eyeball.

scrotum   The sac that contains the testes and epididymides.

SDA   *See* specific dynamic action.

sebaceous glands   Glands in the skin that secrete a fatty substance.

SEC   *See* series elastic component.

second messenger   An intracellular substance formed or released as a result of combination of the original extracellular chemical messenger (i.e., the "first" messenger) with a receptor in the plasma membrane; alters some aspect of the cell's function and thereby mediates the cell's response to the first messenger.

second-order neuron   The interneuron directly activated by the afferent neuron.

secondary peristalsis Esophageal peristaltic waves not immediately preceded by the pharyngeal phase of a swallow.

secondary sexual characteristics The many external differences (hair, body contours, etc.) between male and female that are not directly involved in reproduction.

secondary spermatocyte A male germ cell derived from a primary spermatocyte.

secondary structure (of a protein molecule) The alpha-helical and random-coil conformations of portions of a polypeptide chain.

secretin A peptide hormone secreted by cells in the upper part of the small intestine; the first hormone discovered (in 1902); stimulates the pancreas to secrete bicarbonate into the small intestine.

secretory phase That stage of the menstrual cycle that follows ovulation during which a secretory type of endometrium develops.

segmentation A series of rhythmic contractions and relaxations of rings of intestinal smooth muscle that mix the luminal contents of the intestine.

self-stimulation An experimental technique in which an animal can perform an action that results in the stimulation of electrodes implanted in its brain.

semen The sperm-containing fluid of the male ejaculate.

semilunar valves (of the heart) The aortic and pulmonary valves.

seminal vesicles A pair of large glands that secrete fluid into the two ductus deferentia.

seminiferous tubule A tubule in the testis that contains the cells that differentiate into sperm cells.

semipermeable membrane A membrane permeable to some substances but not to others.

sensor First component of a control system; detects changes in some aspect of the environment; a receptor.

sensory information Afferent information that has a conscious correlate.

sensory unit A single afferent neuron plus all the receptors it innervates.

septum A dividing partition.

series elastic component (SEC) The elastic structures associated with a muscle fiber through which the force generated by the cross bridges is transmitted to the load on the muscle; includes the cross bridges, thick and thin filaments, Z lines, and tendons.

serosa A connective-tissue layer covered by a layer of mesothelium that secretes fluid; surrounds the outer surface of organs such as the stomach and intestinal walls and lines the thoracic, peritoneal, and pericardial cavities.

serotonin A probable neurotransmitter; also functions as a paracrine agent in blood platelets and specialized cells of the digestive tract; 5-hydroxytryptamine, 5-HT.

Sertoli cell A type of cell intimately associated with the developing germ cells in the seminiferous tubules; creates a "blood-testis barrier" and probably mediates hormonal effects on the tubules.

serum Blood plasma from which the fibrinogen has been removed as a result of clotting.

sex chromatin A readily detected nuclear mass not usually found in male cells; consists of a condensed X chromosome.

sex chromosomes The two chromosomes, X and Y, that determine the genetic sex of an individual; genetic males have one X and one Y, and genetic females have two X chromosomes.

sex determination The genetic basis of an individual's sex; XY determines male, and XX female.

sex differentiation The development of male or female reproductive organs.

sex hormones Estrogen, progesterone, and testosterone.

shivering Oscillating rhythmic muscle tremors that occur at a rate of about 10 to 20 per second.

shock *See* circulatory shock.

short reflex A neural reflex in which the afferent and efferent pathways and integrating center are within the plexuses of the gastrointestinal tract.

short-term memory A limited-capacity storage process, possibly lasting only a few seconds, that serves as the initial depository of information from which the memory may be transferred to long-term storage or forgotten.

sickle cell anemia A disease in which one of the 287 amino acids in the polypeptide chains of the hemoglobin molecule is abnormal and, at the low oxygen concentrations that exist in many capillaries, causes the erythrocytes to assume sickle shapes or other bizarre forms that block the capillaries.

sigmoid Shaped like the letter S.

sigmoid colon The S-shaped terminal portion of the colon.

single-unit smooth muscle Smooth muscle in which the fibers are joined by gap junctions that allow electrical activity to be conducted from cell to cell; i.e., the muscle responds to stimulation as a single unit; has pacemaker activity leading to the generation of action potentials in the absence of external stimulation.

sinoatrial (SA) node A region in the right atrium of the heart that contains specialized cardiac muscle cells that depolarize spontaneously at a rate faster than that of other spontaneously depolarizing cells of the heart; the pacemaker of the heart.

sinus A hollow cavity or channel.

skeletal muscle "pump" The pumping effect of contracting skeletal muscles on the blood flow through vessels that underlie them.

skeletomotor fibers The primary skeletal muscle fibers, as opposed to the modified fibers within the muscle spindle.

sleep *See* paradoxical sleep, slow-wave sleep.

sleep centers Clusters of neurons in the brainstem whose activity periodically opposes the tonic activity of the reticular activating system and induces the cycling of slow-wave and paradoxical sleep.

sliding filament model The process of muscle contraction in which shortening occurs as a result of the sliding of thick and thin filaments past each other.

slow viruses Viruses that replicate very slowly and may even become associated with the cell's own DNA and be passed along to the daughter cells upon cell division and that alter the cells so therefore they are no longer recognized as "self."

slow-wave potential A slow oscillation of the membrane potential of some smooth muscle cells that arises from spontaneous cyclical changes in the transport of ions across the membrane.

slow-wave sleep The state of sleep associated with large, slow EEG waves, considerable tonus in the postural muscles, and little change in cardiovascular and respiratory activity.

sodium inactivation The turning off of the increased sodium permeability at the peak of an action potential.

soft palate The nonbony region at the posterior part of the roof of the mouth.

solar plexus The celiac plexus, located behind the stomach; contains sympathetic, parasympathetic, and afferent components.

solubility Ability of the molecules of a substance to go into solution when placed in a given solvent.

solute Molecules or ions dissolved in a liquid (solvent).

solution A liquid (solvent) that contains dissolved molecules or ions (solutes).

solvent The liquid in which molecules or ions (solutes) are dissolved.

somatic Pertaining to the body; related to the framework or outer walls of the body, including skin, skeletal muscle, tendons, and joints.

somatic chromosomes The 22 pairs (in human beings) of nonsex chromosomes.

somatic nervous system A component of the efferent division of the peripheral nervous system distinguished from the autonomic nervous system; innervates skeletal muscle.

somatic receptor A neural receptor that responds to mechanical stimulation of the skin or hairs and underlying tissues, to rotation or bending of joints, to temperature changes, or possibly to some chemical changes.

somatomedins Hormones released mainly by the liver in response to growth hormone.

somatosensory cortex A strip of cerebral cortex in the parietal lobe just behind the junction between the parietal and frontal lobes in which nerve fibers that transmit somatic sensory modalities (e.g., touch, temperature, etc.) synapse.

somatostatin The hypothalamic hormone that inhibits growth hormone and TSH secretion by the anterior pituitary; a possible neurotransmitter; found also in stomach and D cells of the pancreatic islets of Langerhans.

somatotropin *See* growth hormone.

sound wave A disturbance in the air that results from variations in densities of air molecules from regions of high density (compression) to low density (rarefaction).

spatial summation The concept that the effects of simultaneous stimuli (or synaptic inputs) to different points on a cell can be added together to produce a potential change greater than that caused by a single stimulus (or synaptic input).

specific dynamic action (SDA) The increase in metabolic rate induced by eating.

specific immune responses Responses of lymphocytes and cells derived from them that depend upon prior exposure to a specific foreign material, recognition of it upon subsequent exposure, and reaction to it; *see also* cell-mediated immunity, humoral immunity.

specific pathway A chain of synaptically connected neurons in the brain or spinal cord, all of which are activated by sensory units of the same modality, e.g., a pathway that conveys information about only touch or only temperature.

specific thalamic nuclei Nuclei in the cerebrum whose axons project to localized, clearly defined regions of cerebral cortex, e.g., the nuclei that relay precise information about somatic stimuli.

specificity Selectivity; the ability of a protein binding site to react only with one type (or a limited number of types) of molecule.

sperm (spermatozoa) Reproductive cells in the male; the male gamete.

spermatid An immature sperm.

spermatogenesis The manufacture of sperm.

spermatogonia The undifferentiated germ cell that gives rise ultimately to sperm.

spermatozoa *See* sperm.

sphincter of Oddi A ring of smooth muscle that surrounds the common bile duct at its point of entrance into the duodenum.

sphygmomanometer A device that consists of an inflatable cuff and a pressure gauge for measuring blood pressure.

spinal nerve One of the 31 pairs of peripheral nerves that extend from the spinal cord.

spindle apparatus A cell structure that appears during nuclear division and consists of a number of microtubules that pass from one side of the cell to the other between the centrioles or from a centriole to the centromere region of a chromosome.

spindle fibers The modified skeletal muscle fibers within the muscle

spindle receptor organs; also called *intrafusal fibers.*

splanchnic Pertaining to the viscera.

spleen The largest of the lymphoid organs; located in the left part of the abdominal cavity between the stomach and diaphragm.

split brain A condition that results from cutting of the nerve-fiber, bundles that connect the two cerebral hemispheres, which prevents communication between the two halves of the cerebrum.

squamous Scaly or platelike.

stapes The third of the three middle-ear bones; situated so that its footplate rests in the oval window and transmits sound waves to the scala vestibuli of the inner ear.

starch A moderately branched polysaccharide composed of thousands of glucose subunits; found in plant cells.

Starling's law of the heart The law that states that, within limits, an increase in the end-diastolic volume of the heart, i.e., increasing the length of the muscle fibers, increases the force of cardiac contraction during the following systole.

stasis Cessation of blood flow in a vessel.

state of consciousness The degree of mental alertness; e.g., sleep and wakefulness are two states of consciousness.

steady state A situation in which no net change occurs in a system other than the continual energy input to the system that is required to prevent a change; *see also* equilibrium.

stem cell A cell that can renew its own cell type as well as differentiate into other types.

stenosis An abnormally narrowed opening.

sternum Breastbone.

steroid A subclass of lipid molecules; the molecule consists of a skeleton of four interconnected carbon rings to which a few polar groups may be attached.

STH *See* growth hormone.

stimulus A change in the environment detectable by a receptor.

stress Any environmental change that must be adapted to if health and life are to be maintained; any event that elicits increased cortisol secretion.

stretch receptor Afferent nerve endings that are depolarized by stretching, e.g., the nerve endings in the muscle spindle.

stretch reflex A monosynaptic reflex in which stretch of a muscle causes contraction of that muscle; mediated by muscle spindles.

striated muscle Muscle that demonstrates a characteristic transverse banding pattern when viewed with a microscope; skeletal and cardiac muscle.

stroke Damage to a portion of the brain caused by stoppage of the blood supply to that region because of occlusion or rupture of a cerebral vessel.

stroke volume The volume of blood ejected by a ventricle during one beat of the heart.

strong acid An acid whose hydrogen atoms are completely dissociated to form hydrogen ions when the molecules of acid are dissolved in water.

subatmospheric pressure A pressure less than the air pressure of the external environment, e.g., a pressure less than 760 mmHg at sea level.

subatomic particles The units that together make up an atom; e.g., electrons, protons, and neutrons are subatomic particles.

subcortical nuclei Clusters of neuronal cells buried deep in the substance of the cerebrum, includes the basal ganglia.

sublingual glands One of the three pairs of salivary glands.

submandibular glands One of the three pairs of salivary glands.

submucosa A connective-tissue layer that underlies the mucosa in the walls of the stomach and intestines.

submucous plexus A network of nerve cells in the submucosal layer of the esophagus, stomach, and intestinal walls; Meissner's plexus.

subset program The sequence of neural activity that controls a component of a complete action.

substance P A substance thought to be the neurotransmitter at the synaptic endings of the afferent neurons in the pain pathway.

substantia nigra A subcortical nucleus whose cells are deeply pigmented.

substrate The ligand that binds to the specific binding site on an enzyme in an enzyme-catalyzed chemical reaction and undergoes a chemical reaction to form product molecules.

substrate phosphorylation One of the two major ways by which energy is transferred from carbohydrates, lipids, and proteins to other molecules (the other way is oxidative phosphorylation); occurs in the cytoplasm when a phosphate group is transferred directly from an intermediate in the glycolytic pathway to ADP to form ATP; also occurs during one of the Krebs-cycle reactions.

subsynaptic membrane That part of a postsynaptic neuron's plasma membrane that lies directly under the synaptic knob.

subthreshold potentials Any depolarization less than the threshold potential.

subthreshold stimulus Any agent capable of depolarizing the membrane a little but not enough to reach threshold.

sucrose A disaccharide carbohydrate molecule composed of glucose and fructose; table sugar.

sulcus A groove, trench, or depression such as found on the surface of the brain, separating the gyri.

sulfhydryl group R-SH, where R is the remaining portion of the molecule.

sulfuric acid An acid generated during the catabolism of sulfur-containing compounds; dissociates to form sulfate and hydrogen ions.

summation   The increase in the mechanical response of a muscle to action potentials that occur in rapid succession.

superior esophageal sphincter   A ring of skeletal muscle that surrounds the esophagus just below the pharynx that, when contracted, closes the entrance to the esophagus.

superior vena cava   The large vein that carries blood from the upper half of the body to the right atrium of the heart.

supine   Lying on one's back.

"suppressor" T cells   A type of T cell that regulates B cells in a way that leads to reduced antibody production.

suprathreshold stimulus   Any agent capable of depolarizing the membrane beyond its threshold.

surface tension   The unequal attractive forces between water molecules at a surface that result in a net force that acts to reduce the surface area.

surfactant   A phospholipoprotein complex produced by the type II cells of the pulmonary alveoli; markedly reduces the cohesive force of the water molecules that line the alveoli, i.e., reduces the surface tension.

suture   An immovable joint between bones of the cranium.

sweating   The secretion of a hypotonic salt solution onto the skin surface where it can evaporate and thereby lower body temperature.

sympathetic nervous system   That portion of the autonomic nervous system whose preganglionic fibers leave the central nervous system at the thoracic and lumbar portions of the spinal cord.

sympathetic trunk   One of the paired chains of interconnected sympathetic ganglia that lie on either side of the vertebral column.

sympathomimetic   Producing effects similar to those of the sympathetic nervous system.

synapse   An anatomically specialized junction between two neurons where the activity in one neuron influences the excitability of the second; *see also* chemical synapse, electrical synapse, excitatory synapse, inhibitory synapse.

synaptic cleft   The narrow extracellular space that separates the pre- and postsynaptic neurons at a chemical synapse.

synaptic delay   The less than 0.001-sec period between excitation of the presynaptic axon terminal and membrane-potential changes in the postsynaptic neuron.

synaptic knob   The slight swelling at the end of each axon terminal; also called *synaptic terminal*.

synergistic muscle   A muscle whose action aids the intended motion.

synovial   Pertaining to capsular joints.

synthetic reactions   *See* anabolic reactions.

systemic circulation   The circulation through all organs except the lungs; the portion of the cardiovascular system between the aorta (as it leaves the left ventricle) and the superior and inferior venae cavae (as they enter the right atrium).

systole   The period when the ventricles of the heart are contracting.

systolic pressure   The maximum blood pressure reached during the cardiac cycle.

T$_m$ *See* maximal tubular capacity.

T$_3$ *See* triiodothyronine.

T$_4$ *See* thyroxine.

T cells   Lymphocytes that arise in spleen, lymph nodes, or other lymphoid tissue from precursors that at one time were in the thymus where in some way they acquired the ability to differentiate and mature into cells competent to act as effectors in cell-mediated immunity; *see also* cytolytic T cells, "helper" T cells, "suppressor" T cells.

T lymphocytes   *See* T cells.

t tubule   *See* transverse tubule.

T wave   The component of the electrocardiogram that corresponds to repolarization of the ventricles.

tactile   Pertaining to the touch.

taste buds   The sense organs for taste; contain chemoreceptors and supporting cells and are on the tongue, roof of the mouth, pharynx, and larynx.

taste sensations   Sensations traditionally divided into four categories: sweet, sour, salt, and bitter.

tectorial membrane   A structure in the organ of Corti that is in contact with the hairs on the receptor cells.

telencephalon   The anterior portion of the brain.

teleology   The explanation of events in terms of the ultimate purpose served by them, e.g., the statement "urea is excreted in the urine because the body needs to get rid of such waste products."

temporal lobe   The region of cerebral cortex where primary auditory and one of the speech centers are located.

temporal summation   The concept that the effects of two or more stimuli (or synaptic inputs) that occur at different times can be added together to produce a potential change grater than that caused by a single stimulus (or synaptic input).

tendon   A bundle of collagen fibers that connects muscle to bone and transmits the contractile force of the muscle to the bone.

tension   *See* muscle tension.

termination code word   One of the three-nucleotide sequences in DNA that signifies the end of a gene; a "punctuation word" in the genetic code.

termination codon   The sequence of three nucleotides in mRNA that corresponds to a given termination code word in DNA.

tertiary structure (of a protein)   The three dimensional shape of a polypeptide chain.

testis   The gonad in the male.

testosterone   A steroid hormone produced in the interstitial cells of the testes; the major male sex hormone; maintains the growth and development of the reproductive organs and is responsible for the secondary sexual characteristics of males.

tetanus The sustained mechanical response of a muscle to high-frequency stimulation.

tetanus bacillus A bacterium (*Clostridium tetani*) that produces a toxin that blocks inhibitory synapses on motor neurons and causes the disease lockjaw.

TH *See* thyroid hormone.

thalamus A subdivision of the diencephalon that is a way station and integrating center for all sensory (except smell) input on its way to cerebral cortex; also contains motor nuclei and a central core that functions as an extension of the brainstem reticular formation.

thalidomide A sedative drug previously used particularly in pregnant women for morning sickness and subsequently found to cause severe fetal malformations.

theca A cell layer formed from follicle cells and specialized ovarian connective-tissue cells; surrounds the granulosa cells of the ovarian follicle.

thermogenesis Heat generation.

thermoreceptor A sensory receptor that responds preferentially to temperature (and changes in temperature) particularly in a low (cold receptors) or high (warm receptors) range.

theta rhythm (of the hippocampus) Low-frequency high-amplitude electrical activity in recordings (hippocampal EEGs) during the orienting response; electrical manifestation of hippocampal "arousal."

thiamine Vitamin B_1.

thick filament The 12- to 18-nm filaments in muscle cells; located in the central region of a sarcomere in striated muscle fibers; consists of myosin molecules.

thin filament The 5- to 8-nm filaments in muscle cells; attached to the Z line at either end of a sarcomere in striated muscle fibers; consists of actin, troponin, and tropomyosin molecules.

thoracic cage Wall of the thoracic cavity; chest wall.

thoracic cavity Chest cavity.

thoracolumbar division (of the autonomic nervous system) *See* sympathetic nervous system.

thoroughfare channel (in a capillary network) A capillary that connects an arteriole and venule directly and from which branch the "true" capillaries.

threshold potential The membrane potential to which an excitable membrane must be depolarized in order to initiate an action potential; also called *threshold*.

threshold stimulus Any agent capable of depolarizing the membrane to threshold potential.

thrombin The enzyme that catalyzes the conversion of fibrinogen into fibrin.

thrombosis Occlusion of a blood vessel by a clot.

thromboxanes Substances closely related to the prostaglandins; synthesized from arachidonic acid and other essential fatty acids.

thrombus An intravascular blood clot.

thymine One of the pyrimidine bases; the one nucleotide base present in DNA but not in RNA.

thymosin A group of hormones secreted by the thymus that have as their target cells T lymphocytes.

thymus A lymphoid organ that lies in the upper part of the chest; site of T lymphocyte formation and thymosine secretion.

thyroglobulin A glycoprotein to which thyroid hormones bind; the storage form of the thyroid hormones in the thyroid.

thyroid A paired endocrine gland located in the neck.

thyroid hormone The collective term for the hormones released from the thyroid, i.e., thyroxine and triiodothyronine; increases the metabolic rate of most cells in the body.

thyroid-stimulating hormone (TSH) A glycoprotein hormone secreted by the anterior pituitary; also called *thyrotropin*.

thyrotropin *See* thyroid-stimulating hormone.

thyrotropin-releasing hormone (TRH) The hypothalamic hormone that stimulates TSH (thyrotropin) secretion by the anterior pituitary.

thyroxine (T_4) Tetraiodothyronine; an iodine-containing amino acid hormone secreted by the thyroid.

tidal volume The volume of air entering or leaving the lungs during a single breath during any state of respiratory activity.

tight junction A type of cell junction found in epithelial tissues; extends around the circumference of the cell; consists of an actual fusing of the two adjacent plasma membranes and an obliteration of the extracellular space.

tissue Formerly defined as an aggregate of differentiated cells of a similar type; commonly used to denote the general cellular fabric of a given organ, e.g., kidney tissue, brain tissue; the four general classes of tissues are epithelial tissue, connective tissue, nerve tissue, and muscle tissue.

tissue thromboplastin An extravascular enzyme capable of initiating the formation of fibrin clots.

tolerance The state in which increasing doses of a drug are required to achieve effects that initially occurred in response to a smaller dose, i.e., when it takes more drug to do the same job.

tone A state of continuous low-level activity, e.g., the sustained tension in a "resting" muscle.

tonic A continuous activity, as opposed to a phasic one.

total energy expenditure The energy stored plus the energy expended doing external work plus the energy represented in heat production by the body.

total peripheral resistance (TPR) The total resistance of all the vessels in the systemic vascular tree to the flow of blood from the very first portion of the aorta to the very last portion of the venae cavae.

total pulmonary ventilation The volume of air that passes into or out of the respiratory system per minute; equal to the tidal volume times the number of breaths per minute.

total vascular resistance   *See* total peripheral resistance.

toxemia of pregnancy   A disease that occurs in pregnant women and is associated with retention of abnormally great amounts of fluid, the appearance of protein in the urine, hypertension, and possibly convulsions.

toxicology   The study of substances that are harmful to the body.

TPR   *See* total peripheral resistance.

trace element   A mineral present in the body in extremely small quantities; at present 13 different essential trace elements are known that are required for normal growth and function.

trachea   The single airway that connects the larynx with the two bronchi that enter the right and left lungs.

tract   A large bundle of myelinated nerve fibers within the central nervous system.

transamination   A reaction in which an amino group ($-NH_2$) is transferred from an amino acid to a keto acid and the keto acid thus becomes an amino acid.

transcellular transport   The movement of molecules from an extracellular compartment into an epithelial cell and out the opposite side of the cell into a second extracellular compartment.

transcriptase   DNA-dependent-RNA-polymerase, the enzyme that joins the nucleotides of mRNA during its formation.

transcription   Formation of mRNA so that it contains in the linear sequence of its nucleotides the genetic information contained in the DNA gene from which the mRNA is transcribed; the first stage in the transfer of information from DNA to protein synthesis.

transfer RNA (tRNA)   A polymer of ribose-containing nucleotides; dif-

ferent tRNAs combine with different amino acids and contain an anticodon that binds to the codon in mRNA specific for that amino acid and thus arranges the amino acids in the appropriate sequence to form the protein specified by the codons in mRNA; smallest of the three forms of RNA.

transferrin   An iron-binding protein that is the carrier molecule for iron in the plasma.

translation   Assembly of the proper amino acids in the correct order during the formation of a protein according to the genetic instructions carried in mRNA; occurs on the ribosomes of a cell.

transmural pressure   The difference between the pressure exerted on one side of a structure and the pressure exerted on the other side.

transverse tubule   A tubule that extends from the plasma membrane of a striated muscle fiber into the fiber interior and passes between the opposed lateral sacs of adjacent sarcoplasmic-reticulum segments; conducts the muscle action potential into the muscle fiber; *also t tubules.*

TRH   *See* thyrotropin-releasing hormone.

triacylglycerol   A subclass of lipid molecules; neutral fat; composed of glycerol and three fatty acids; formerly called triglyceride.

tricarboxylic acid cycle   *See* Krebs cycle.

tricuspid valve   The valve between the right atrium and the right ventricle of the heart.

"trigger zone"   That low-threshold region of a neuron at which action potentials are initiated; sometimes occurs at the initial segment of an axon.

triglyceride   *See* triacylglycerol.

triiodothyronine (T_3)   An iodine-containing amino acid hormone secreted by the thyroid.

tRNA   *See* transfer RNA.

trophoblast   The outer layer of cells of the blastocyst; contributes to the

placental tissues across which nutrients and wastes are exchanged between the maternal and fetal circulations.

tropomyosin   A protein capable of reversibly covering the binding sites on actin, thereby preventing their combination with myosin; associated with the thin filaments in muscle fibers.

troponin   A protein bound to the actin and tropomyosin molecules of the thin filaments of striated muscle; site of calcium binding that initiates contractile activity in striated muscles.

trypsin   An enzyme secreted by the exocrine pancreas as the precursor trypsinogen; breaks certain peptide bonds in proteins.

TSH   *See* thyroid-stimulating hormone.

tubular reabsorption   The process by which materials are transferred from the lumen of a kidney tubule to the peritubular capillaries.

tubular secretion   The process by which materials are transferred from the peritubular capillaries to the lumen of the kidney tubules.

tunica   A covering.

twitch   The mechanical response of a muscle to a single action potential.

tympanic membrane   The membrane stretched across the end of the ear canal.

ulcer   An erosion, or sore, as in the stomach or intestinal wall.

umbilical cord   The flexible structure that contains the umbilical arteries and vein to convey them between the umbilicus of the fetus and the placenta.

umbilical vessels   The arteries and veins that transport blood between the fetus and placenta.

umbilicus   The depression in the middle of the abdomen that marks the place where the umbilical cord was attached to the fetus; the navel or "belly button."

unidirectional flux The amount of material that crosses a surface in one direction in a unit of time; e.g., in the case of diffusion, $f_{1-2} = k_D C_1$ where f_{1-2} is the flux from compartment 1 to 2, k_D is the diffusion coefficient, and C_1 is the concentration in compartment 1.

universal donor A person who has type O blood, which can be transfused into individuals with any other blood type.

universal recipient A person who has type AB blood; such an individual can receive blood of any other blood type.

unsaturated fatty acid A fatty acid that contains one or more double bonds.

uracil One of the pyrimidine bases; present in RNA but not DNA.

urea NH_2—$\overset{\overset{\displaystyle O}{\|}}{C}$—$NH_2$; synthesized in the liver by linking two molecules of ammonia with carbon dioxide; the major nitrogenous waste product of protein breakdown.

uremia A general term for the symptoms of profound kidney malfunction.

ureter A tube, whose walls contain smooth muscle, that connects the renal pelvis to the urinary bladder.

urethra The tube that conveys urine from the urinary bladder to the outside of the body.

uric acid A waste product derived from nucleic acids.

urinary bladder A thick-walled sac that temporarily stores urine; receives urine from the ureters and empties it into the urethra.

uterine tubes Paired structures that transport the ovum from the ovary to the uterus; also called *oviducts, fallopian tubes.*

uterus Womb; a hollow, thick-walled, muscular pear-shaped organ in the pelvic region of females; consists of the corpus (or body) above and the cervix below; the fertilized ovum implants in the uterine wall and develops within the uterus during the nine months of pregnancy.

V *See* volt.

vagus nerve Cranial nerve X; consists of approximately 90 percent afferent and 10 percent parasympathetic preganglionic efferent fibers.

valve A structure in a tube or passage that prevents the backflow of its contents.

varicose vein An irregularly swollen superficial vein of the lower extremities.

varicosity A swollen region that contains neurotransmitter-filled vesicles along the terminal portions of the axons of autonomic nerve fibers; analogous to a synaptic knob.

vas deferens *See* ductus deferens.

vascular Pertaining to vessels.

vasectomy The cutting and tying off of the ductus (vas) deferentia, which results in sterilization of the male without loss of testosterone secretion by the testes.

vasoconstriction A decrease in the diameter of blood vessels as a result of vascular smooth muscle contraction.

vasodilation An increase in the diameter of blood vessels as a result of vascular smooth muscle relaxation.

vasopressin *See* antidiuretic hormone.

vein A wide, thin-walled vessel that carries blood back to the heart from the venules.

vena cava One of two large veins that return systemic blood to the heart.

venous return The volume of blood that flows to the heart from the veins per unit of time.

ventilation The bulk-flow exchange of air between the atmosphere and the alveoli.

ventral root A group of efferent fibers that leaves the left and right side of the region of the spinal cord that faces the front of the body.

ventricle A cavity, e.g., a ventricle of the heart or the brain.

venule A small, thin-walled vessel that carries blood from a capillary network to a vein.

vesicle A small, membrane-bound intracellular organelle.

vestibular apparatus Membranous sacs in the bony semicircular canals and in the bony cavity adjacent to them (which contains the utricle and saccule).

vestibular nuclei Neuron clusters in the brainstem that receive primary afferent nerve fibers from the vestibular apparatus.

vestibular receptors Hair cells in the gelatinous mass in the semicircular canals, utricle, and saccule.

vestibular system A sense organ (often called the sense organ of balance) deep within the temporal bone of the skull; consists of the three semicircular canals, utricle, and saccule.

vestibule (of the female external genitalia) The area enclosed by the labia minora.

villi Fingerlike projections from the highly folded surface of the small intestine; covered with a single layer of epithelial cells.

virus An "organism" that is essentially a nucleic acid core surrounded by a protein coat and lacks both the enzyme machinery for energy production and the ribosomes for protein synthesis; thus, it cannot reproduce except inside other cells whose biochemical apparatus it makes use of; *see also* slow viruses.

viscera The organs in the thoracic and abdominal cavities.

viscosity The property of a fluid caused by frictional interactions between its molecules that makes it resist flow.

vital capacity The maximal amount of air that can be moved into the lungs following a maximal expiration; the sum of the inspiratory-reserve, normal tidal, and expiratory-reserve volumes.

vitalist One who believes that life processes occur only because of the presence of a "life force" rather than by physicochemical processes alone.

vitamin An organic molecule that must be present in trace amounts to maintain normal health and growth but is not manufactured by the organism's metabolic pathways; over 20 vitamins are presently known; classified as water soluble (vitamins C and the B complex) and fat soluble (vitamins A, D, E, and K).

vitamin D A fat-soluble vitamin; a group of closely related sterols produced by the action of ultraviolet light on the skin or ingested in the diet; 1,25-dihydroxyvitamin D_3, which is formed in the kidneys, is the active form.

vocal cords Two strong bands of elastic tissue stretched across the opening of the larynx and caused to vibrate by the movement of air past them providing the tones of speech.

volt (V) The unit of measurement of the electric potential between two points.

voltage A measure of the potential of separated electric charges to do work; the amount of work done by an electric charge when moving from one point in a system to another; a measure of the electric force between two points.

voluntary actions Actions we think about and are aware of; actions to which our attention is directed.

vulva The mons pubis, labia majora and minora, clitoris, vestibule of the vagina, and vestibular glands.

wavelength The distance between two wave peaks in an oscillating medium.

weak acid An acid whose hydrogen atoms are not completely dissociated to form hydrogen ions when the molecules of acid are dissolved in water.

white matter The portion of the brain and spinal cord that appears white in unstained specimens and consists mainly of myelinated nerve fibers.

withdrawal Physical symptoms (usually unpleasant) associated with cessation of drug use.

wolffian duct A part of the embryonic duct system that, in a male, remains and develops into the ducts of the reproductive system, but degenerates in a female.

work A measure of the amount of energy required to produce a physical displacement of matter; formally defined as the product of force (F) times distance (X), $W = FX$; *see also* external work, internal work.

xylocaine A local anesthetic that works by preventing the increase in membrane permeability to sodium ions required for the generation of action potentials.

yellow-sensitive cones Cones that contain photopigment most sensitive to light of 570-nm wavelengths; often called red-sensitive cones because the range of pigment sensitivity extends closer to the red range than do the sensitivities of the two other cone photopigments.

yolk The part of the egg that provides nutrients for the developing embryo.

Z line The structure that runs across the myofibril of a striated muscle at each end of a sarcomere; anchors one end of the thin filaments.

zona pellucida A thick membrane that separates the ovum from the surrounding follicular cells.

zymogen granule Membrane-bound cytoplasmic vesicle that contains inactive precursor enzymes prior to their secretion from a cell; the protein granules secreted by the exocrine portion of the pancreas.

ILLUSTRATION CREDITS

2-2 Courtesy of Keith R. Porter.
2-3A Courtesy of Keith R. Porter. From T. W. Goodwin and O. Lindberg (eds.), *Biological Structure and Function,* vol. 1, Academic Press, Inc., New York, 1961.
2-4A From W. Bloom and D. W. Fawcett, *Textbook of Histology,* 9th ed., W. B. Saunders Company, Philadelphia, 1968, Fig. 2-14.
2-5A Courtesy of Keith R. Porter.
2-9 Adapted from S. Bernhard, *The Structure and Function of Enzymes,* W. A. Benjamin, Inc., New York, 1968, Fig. 8-2.

4-12D Courtesy of Marilyn G. Farquhar. From M. G. Farquhar and G. E. Palade, *J. Cell Biol.,* **17:**375–412 (1963).

6-3 Photomicrographs courtesy of Raymond H. Kahn.
6-4 Photomicrographs courtesy of Raymond H. Kahn.
6-7 Adapted from T. L. Lentz, *Cell Fine Structure,* W. B. Saunders Company, Philadelphia, 1971, pp. 195 and 287.
6-8 Adapted from W. Bloom and D. W. Fawcett, *A Textbook of Histology,* 9th ed., W. B. Saunders Company, Philadelphia, 1968, Fig. 6-14.
6-9 From R. O. Greep and L. Weiss, *Histology,* 3d ed., McGraw-Hill Book Company, New York, 1973, Fig. 4-2. Used with permission of McGraw-Hill Book Company.
6-10 Adapted from R. Warwick and P. L. Williams, *Gray's Anatomy,* 35th British edition, W. B. Saunders Company, Philadelphia, 1973, Fig. 1.41.
6-11 Courtesy of Raymond H. Kahn.
6-12 (A) Adapted from T. L. Lentz, *Cell Fine Structure,* W. B. Saunders Company, Philadelphia, 1971, p. 71. (B) Courtesy of Raymond H. Kahn.
6-13 Courtesy of Raymond H. Kahn.
6-14 Adapted from G. J. Tortora and N. P. Anagnostakos, *Principles of Anatomy and Physiology,* 3d ed., Harper & Row, Publishers, Inc., New York, 1981, Fig. 5-4.

8-1 From L. Weiss and R. O. Greep, *Histology,* 4th ed., McGraw-Hill Book Company, New York, 1974, Fig. 6-12. Used with permission of McGraw-Hill Book Company.
8-3 Courtesy of Richard G. Kessel and R. H. Kardon, *Tissues and Organs,* W. H. Freeman and Company, San Francisco, 1979. Used with permission.
8-4 Adapted from R. G. Kessel and R. H. Kardon, *Tissues and Organs,* W. H. Freeman and Company, San Francisco. Used with permission.
8-6 Adapted from R. Warwick and P. L. Williams, *Gray's Anatomy,* 35th British ed., W. B. Saunders Company, Philadelphia, 1973, Fig. 3.22.
8-8 Adapted from J. W. Hole, Jr., *Human Anatomy and Physiology,* 2d ed., © 1978, 1981, Wm. C. Brown Company, Publishers, Dubuque, Iowa, Fig. 8-11. Reprinted by permission.

10-1 Adapted from W. H. Hollinshead, *Textbook of Anatomy,* 3d ed., Harper & Row, Publishers, Inc., New York, 1974, Fig. 3-10.
10-5 Adapted from W. H. Hollinshead, *Textbook of Anatomy,* 3d ed., Harper & Row, Publishers, Inc., New York, 1974, Fig. 3-12.

11-2 From E. K. Keith and M. H. Ross, *Atlas of Descriptive Histology,* Harper & Row, Publishers, Inc., New York, 1968, Fig. 1 on Figure Plate 17.
11-3 (A) from L. Weiss and W. O. Greep, *Histology,* 4th ed., McGraw-Hill Book Company, New York, 1977, Fig. 7-4. Used with permission of McGraw-Hill Book Company. (B) From H. E. Huxley, *J. Biophys. Biochem. Cytol.,* **3:**631–648, Fig. 8, Plate 210 (1957).
11-13 Adapted from H. Elias and J. E. Pauly, *Human Microanatomy,* F. A. Davis Company, Philadelphia, 1966, Fig. 7-9.
11-21 Courtesy of John A. Faulkner.
11-22 Courtesy of A. P. Somylo, Pennsylvania Muscle Institute. From A. P. Somylo et al., *Phil. Trans. R. Soc. Lond. B.,* **265:**223–229 (1973).

12-1 Adapted from R. Warwick and P. L. Williams, *Gray's Anatomy,* 35th British ed. W. B. Saunders Company, Philadelphia, 1973, Fig. 5-16.
13-3 Adapted from S. W. Kuffler and J. G. Nicholls, *From Neuron to Brain,* Sinauer Associates, Sunderland, Massachusetts, 1976, p. 9.
13-11 Adapted from R. Poritsky, *J. Comp. Neurol.,* **135:**423 (1969), Plate 8, Fig. 15.
13-13 From W. F. Windle, *Textbook of Histology,* 5th ed., McGraw-Hill Book Company, New York, 1976, Fig. 14-12B. Used with permission of McGraw-Hill Book Company.

14-1 Adapted from R. T. Woodburne, *Essentials of Human Anatomy,* 3d ed., Oxford University Press, New York, 1965, Fig. 18.
14-7 Adapted from R. Warwick and P. L. Williams, *Gray's Anatomy,* 35th British ed., W. B. Saunders Company, Philadelpha, 1973, Fig. 7.22.
14-20 Adapted from W. H. Hollinshead, *Textbook of Anatomy,* 3d ed., Harper & Row, Publishers, Inc., New York, 1974, Fig. 4-5.
14-28 (A) Adapted from S. W. Kuffler and J. G. Nicholls, *From Neuron to Brain,* Sinauer Associates, Sunderland, Massachusetts, 1976, p. 292. (B) Adapted from J. W. Hole, Jr., *Human Anatomy and Physiology,* 2d ed., © 1978, 1981, Wm. C. Brown Company, Publishers, Dubuque, Iowa, Fig. 10-24. Reprinted by permission.

15-5 Adapted from H. Hensel and K. K. A. Bowman, *J. Neurophysiol.,* **23:**564, Fig. 8 (1960).
15-16 Adapted from R. L. Gregory, *Eye and Brain: The Psychology of Seeing,* 2d ed., McGraw-Hill Book Company, New York, 1973, Fig. 4-9. Used with permission of McGraw-Hill Book Company.
15-18 Adapted from R. L. Gregory, *Eye and Brain: The Psychology of Seeing,* 2d ed., McGraw-Hill Book Company, New

York, 1973, Fig. 8-2. Used with permission of McGraw-Hill Book Company.

15-19   Adapted from C. R. Michael, *New Engl. J. Med.*, **288:**724 (1973).

15-25   (A)–(D) Adapted from N. Y. S. Kiang in D. B. Tower (ed.), *The Nervous System,* vol. 3, Raven Press, New York, 1975, p. 82.

15-26   Adapted from M. Alpern, M. Lawrence, and D. Wolsk, *Sensory Processes,* Brooks/Cole Publishing Company, Monterey, California, 1967, p. 94, Fig. 13.

15-27   Adapted from D. O. Kim and C. E. Molnar in D. B. Tower (ed.), *The Nervous System,* vol. 3, Raven Press, New York, 1975, p. 59.

15-28   Scanning electron micrographs by Robert E. Preston; courtesy of Joseph E. Hawkins, Kresge Hearing Research Institute.

15-30   (A, B) Adapted from D. E. Parker, "The Vestibular Apparatus," *Scientific American,* **243:**118 (November 1980), figure on p. 120. Used with permission.

15-31   Adapted from J. Wersall, L. Gleisner, and P. G. Lundquist in A. V. S. de Reuck and J. Knight (eds.), *Myotactic Kinesthetic and Vestibular Mechanisms,* Ciba Foundation Symposium, Little, Brown and Company, Boston, 1967, p. 109, Fig. 10.

15-32   Adapted from D. E. Parker, "The Vestibular Apparatus," *Scientific American,* **243:**118 (November 1980), figure on p. 125. Used with permission.

15-33   (A) Adapted from L. Weiss and R. O. Greep, *Histology,* 4th ed., McGraw-Hill Book Company, New York, 1977, Fig. 18-4. Used with permission of McGraw-Hill Book Company.

16-8   Adapted from J. C. Eccles, *The Understanding of the Brain,* 2d ed., McGraw-Hill Book Company, New York, 1977, Fig. 4-2. Used with permission of McGraw-Hill Book Company.

17-2   Adapted from R. B. Livingston in G. C. Quarton, T. Melnechuk, and F. O. Schmitt (eds.), *The Neurosciences: A Study Program,* The Rockefeller University Press, New York, 1967, p. 509, Figs. 6 and 7.

17-6   Adapted from J. L. Conel, *The Postnatal Development of the Human Cerebral Cortex,* vol. 1, Fig. 104, and vol. 3, Fig. 116, Harvard University Press, Cambridge, Massachusetts, 1939 and 1947.

17-7   Adapted from D. B. Lindsley in J. Field (ed.), *Handbook of Physiology,* sec. 1, vol. 3, American Physiological Society, Bethesda, Maryland, 1960, p. 1574, Fig. 9.

18-11   (B) From and (C) adapted from L. L. Langley et al., *Dynamic Anatomy and Physiology,* 4th ed., McGraw-Hill Book Company, New York, 1974, Fig. 31-13. Used with permission of McGraw-Hill Book Company.

18-12   (B) From W. F. Windle, *Textbook of Histology,* 5th ed., McGraw-Hill Book Company, New York, 1976, Fig. 24-14. Used with permission of McGraw-Hill Book Company.

19-2   From R. G. Kessel and C. Y. Shih, *Scanning Electron Microscopy in Biology,* Springer-Verlag New York, Inc., New York, 1974, p. 265.

19-6   From R. O. Greep and L. Weiss, *Histology,* 3d ed., McGraw-Hill Book Company, New York, 1973. Used with per-

mission of McGraw-Hill Book Company.

19-7   From W. Bloom and D. W. Fawcett, *A Textbook of Histology,* 9th ed., W. B. Saunders Company, Philadelphia, 1968, Fig. 5-15.

19-8   Adapted from A. W. Ham and D. H. Cormack, *Histology,* 8th ed., J. B. Lippincott Company, Philadelphia, 1979, Fig. 12-3.

19-10   From R. G. Kessel and C. Y. Shih, *Scanning Electron Microscopy in Biology,* Springer-Verlag New York, Inc., New York, 1974, p. 265.

20-11   Courtesy of D. W. Fawcett.

20-16   Adapted from R. F. Rushmer, *Cardiovascular Dynamics,* 3d ed., W. B. Saunders Company, Philadelphia, 1970, Fig. 2-12.

20-21   Adapted from W. F. Ganong, *Review of Medical Physiology,* 9th ed., Lange Medical Publications, Los Altos, California, 1979, Figs. 29-1 and 29-2.

20-22   Adapted from E. Braunwald, *New Engl. J. Med.,* **290:**1124 (1974), Fig. 1.

20-23   Adapted from B. E. Hoffman and P. E. Cranefield, *Electrophysiology of the Heart,* McGraw-Hill Book Company, New York, 1960, Fig. 5-2. Used with permission of McGraw-Hill Book Company.

21-7   (B) Adapted from W. H. Hollinshead, *Textbook of Anatomy,* 3d ed., Harper & Row, Publishers, Inc., New York, 1974, Fig. 26-15.

21-20   Adapted from E. P. Benditt, "The Origin of Atherosclerosis," *Scientific American,* **236:**74 (February 1977), figure on p. 76. Used with permission.

21-21   Courtesy of Richard G. Kessel. From R. G. Kessel and R. H. Kardon, *Tissues and Organs,* W. H. Freeman and Company, San Francisco, 1979. Used with permission.

21-31   Adapted from B. W. Zweifach, "The Microcirculation of the Blood," *Scientific American,* January (1959). Used with permission.

21-33   Adapted from R. F. Rushmer, *Cardiovascular Dynamics,* 3d ed., W. B. Saunders Company, Philadelphia, 1969, Fig. 1-5.

21-36   Adapted from A. G. Guyton, *Functions of the Human Body,* 3d ed., W. B. Saunders Company, Philadelphia, 1969, Fig. 121.

21-44   (B) From W. F. Windle, *Textbook of Histology,* 5th ed., McGraw-Hill Book Company, New York, 1976, Fig. 12-5. Used with permission of McGraw-Hill Book Company.

21-45   Courtesy of Richard G. Kessel. From R. G. Kessel and R. H. Kardon, *Tissues and Organs,* W. H. Freeman and Company, San Francisco, 1979. Used with permission.

23-4   Courtesy of E. Robert Heitzman.

23-6   (B) Adapted from J. E. Anderson, *Grant's Atlas of Anatomy,* 7th ed., Williams & Wilkins Company, Baltimore, 1978, Fig. 9-69.

23-7   Adapted from R. T. Woodburne, *Essentials of Human Anatomy,* 3d ed., Oxford University Press, New York, 1965, Fig. 140.

23-8   Adapted from W. Bloom and D. W. Fawcett, *A Textbook of Histology,* 9th ed., W. B. Saunders Company, Philadelphia, 1968, Fig. 29-9.

23-10   (A, B) from E. R. Weibel, *Physiol. Rev.,* **53:**419 (1973), figure on p. 424. (C, D) From R. G. Kessel and C. Y. Shih,

Scanning Electron Microscopy in Biology, Springer-Verlag New York, Inc., New York, 1974, p. 293.

23-11   (B) From R. G. Kessel and C. Y. Shih, *Scanning Electron Microscopy in Biology,* Springer-Verlag New York, Inc., New York, 1974, p. 291.

23-25   Adapted from J. H. Comroe, *Physiology of Respiration,* Year Book Medical Publishers, Chicago, 1965, Fig. 96.

23-33   Adapted from C. J. Lambertsen in V. B. Mountcastle (ed.), *Medical Physiology,* 14th ed., The C. V. Mosby Company, St. Louis, 1980; modified from C. J. Lambertsen, *Anesthesiology,* vol. 21, p. 642, 1980.

23-35   Adapted from C. J. Lambertson et al., *J. Appl. Physiol.,* **16:**480 (1961).

23-36   Adapted from A. Holmgren and M. B. McIlroy, *J. Appl. Physiol.,* **19:**243 (1964), Fig. 108.

23-39   Adapted from H. J. Heimlich and M. H. Uhley, *Ciba Clinical Symposia,* vol. 31, no. 3, 1979.

24-2   Adapted from R. G. Kessel and R. H. Kardon, *Tissues and Organs,* W. H. Freeman and Company, San Francisco, 1979, p. 222. Used with permission.

24-3   (A) Circular inserts adapted and reproduced with permission from R. F. Pitts, *Physiology of the Kidney and Body Fluids,* 3d ed, Year Book Medical Publishers, Inc., Chicago, 1963. (Previously adapted from H. W. Smith, *The Kidney,* Oxford University Press, New York, 1935.) (B) Courtesy of A. B. Maunsbach. From A. B. Maunsbach, *J. Ultrastruct. Res.,* **15:**242–282 (1966).

24-8   Courtesy of Richard G. Kessel. From R. G. Kessel and R. H. Kardon, *Tissues and Organs,* W. H. Freeman and Company, San Francisco, 1979, p. 231. Used with permission.

24-11   Adapted from W. H. Hollinshead, *Textbook of Anatomy,* 3d ed., Harper & Row, Publishers, Inc., New York, 1974, Fig. 21-8.

24-15   Adapted from J. O. Davis, *Am. J. Med.,* **55:**333 (1973), Fig. 1.

25-11   Adapted from L. Weiss and R. O. Greep, *Histology,* 4th ed., McGraw-Hill Book Company, New York, 1977, Fig. 18-4. Used with permission of McGraw-Hill Book Company.

25-14   (A) From J. E. Anderson, *Grant's Atlas of Anatomy,* 7th ed., The Williams & Wilkins Company, Baltimore, 1978, Fig. 7-99.

25-15   Adapted from A. P. Spence and E. B. Mason, *Human Anatomy and Physiology,* Benjamin/Cummings, Menlo Park, California, 1979, Fig. 24-3.

25-16   (B) Adapted from I. Schour and M. Massler, *J. Am. Dental Assoc.,* **28:**1154 (1941).

25-20   Courtesy of Richard G. Kessel. From R. G. Kessel and R. H. Kardon, *Tissues and Organs,* W. H. Freeman and Company, San Francisco, 1979, p. 165. Used with permission.

25-21   (A, B) Adapted from W. F. Windle, *Textbook of Histology,* 5th ed., McGraw-Hill Book Company, New York, 1976, p. 384. Used with permission of McGraw-Hill Book Company. (C) Adapted from S. Ito and R. J. Winchester, *J. Cell Biol.,* **16:**541 (1963), Figs. 10 and 18.

25-32   (A) Adapted from A. Kappas and A. P. Alvares, "How the Liver Metabolizes Foreign Substances," *Scientific American,* **232:**22 (July 1975), figure on p. 26. Used with permission. (B) Adapted from G. J. Tortora and N. P. Anagnostakos, *Principles of Anatomy and Physiology,* 3d ed., Harper & Row, Publishers, Inc., New York, 1981, Fig. 24-13C.

25-37 Courtesy of Susumu Ito. From D. W. Fawcett, *J. Histochem. Cytochem.,* **13:**75 (1965), Fig. 19.

26-6 From W. F. Windle, *Textbook of Histology,* 5th ed., McGraw-Hill Book Company, New York, 1976, Fig. 21-11. Used with permission of McGraw-Hill Book Company.

26-9 Adapted from W. H. Daughaday and D. M. Kipnis, *Recent Prog. Horm. Res.,* **22:**49 (1966), figure on p. 72.

26-16 From W. H. Daughaday in R. H. Williams (ed.), *Textbook of Endocrinology,* 4th ed., W. B. Saunders Company, Philadelphia, 1968, p. 74, Fig. 2-28.

26-19 Adapted from E. F. Dubois, *Fever and the Regulation of Body Temperature,* Charles C Thomas, Publisher, Springfield, Illinois, 1948, p. 8, Fig. 4.

26-23 Adapted from E. Dodt and Y. Zotterman, *Acta Physiol. Scand.,* **26:**345, Fig. 7 (1952), figure on p. 354.

26-24 Adapted from T. Nakayama et al., *Am. J. Physiol.,* **204:**1122 (1963), Fig. 4.

27-4 (A) Adapted from W. Bloom and D. W. Fawcett, *A Textbook of Histology,* 9th ed., W. B. Saunders Company, Philadelphia, 1968, Fig. 31-4. (B) Courtesy of Richard G. Kessel. From R. G. Kessel and R. H. Kardon, *Tissues and Organs,* W. H. Freeman and Company, San Francisco, 1979, p. 261. Used with permission.

27-8 Courtesy of Y. Clermont. Adapted from L. Weiss and R. O. Greep, *Histology,* 4th ed. McGraw-Hill Book Company, New York, 1977. Used with permission of McGraw-Hill Book Company.

27-9 Adapted from W. Bloom and D. W. Fawcett, *A Textbook of Histology,* 9th ed., W. B. Saunders Company, Philadelphia, 1968, Fig. 31-51.

27-11 Adapted from B. S. McEwen, "Interactions between Hormones and Nerve

Tissue," *Scientific American,* **235:**48 (July 1976), figure on p. 57. Used with permission.

27-14 Adapted from B. M. Patten and B. M. Carlson, *Human Embryology,* 3d ed., McGraw-Hill Book Company, New York, 1968, Fig. II-13. Used with permission of McGraw-Hill Book Company.

27-15 From W. F. Windle, *Textbook of Histology,* 5th ed., McGraw-Hill Book Company, New York, 1976, Fig. 26-8. Used with permission of McGraw-Hill Book Company.

27-17 From L. Weiss and R. O. Greep, *Histology,* 4th ed., McGraw-Hill Book Company, New York, 1977. Used with permission of McGraw-Hill Book Company.

27-21 (A, B) From W. F. Windle, *Textbook of Histology,* 5th ed., McGraw-Hill Book Company, New York, 1976, Figs. 26-16 and 26-17. Used with permission of McGraw-Hill Book Company.

27-22 Adapted from L. Weiss and R. O. Greep, *Histology,* 4th ed., McGraw-Hill Book Company, New York, 1977, Fig. 23-27. Used with permission of McGraw-Hill Book Company.

27-23 Courtesy of Dr. Mia Tegner, Scripps Institute of Oceanography.

27-25 From C. H. Heuser and G. L. Streeter, *Carnegie Contrib. Embryol.,* **29:**15 (1941).

27-26 From A. T. Hertig and J. Rock, *Carnegie Contrib. Embryol.,* **29:**127 (1941).

27-27 Adapted from B. M. Patten and B. M. Carlson, *Human Embryology,* 3d ed., McGraw-Hill Book Company, New York, 1968, Fig. VI-19. Used with permission of McGraw-Hill Book Company.

27-28 Adapted from B. M. Patten and B. M. Carlson, *Human Embryology,* 3d ed., McGraw-Hill Book Company, New York, 1968, Fig. VI-11. Used with permission of McGraw-Hill Book Company.

27-30 Adapted from G. J. Tortora and N. P. Anagnostakos, *Principles of Anatomy and Physiology,* 3d ed., Harper & Row,

Publishers, Inc., New York, 1981, Fig. 29-12.

27-36 Adapted from J. Langman, *Medical Embryology,* 3d ed., The Williams & Wilkins Company, Baltimore, 1975, Fig. 3-3.

27-37 Adapted from J. Langman, *Medical Embryology,* 3d ed., The Williams & Wilkins Company, Baltimore, 1975, Fig. 3-6.

27-41 Adapted from J. Langman, *Medical Embryology,* 3d ed., The Williams & Wilkins Company, Baltimore, 1975, Figs. 5-5, 5-18, 5-21.

27-42 Courtesy of Landrum B. Shettles, M.D.

27-43 Courtesy of J. Lejeune, Chaire de Genetique Fondamentale, Paris.

27-46 Adapted from J. D. Wilson, F. W. George, and J. E. Griffin, *Science,* **211:**1278 (1981), Fig. 2B. Copyright © 1981 American Association for the Advancement of Science.

27-47 Adapted from K. Pederson-Biergaard and M. Ionnesen, *Acta Med. Scand.,* **131** (suppl. 213):284 (1948), Figs. I and II.

27-48 Adapted from I. T. Nathanson, L. E. Towne, and L. C. Aub, *Endocrinology,* **28:**851, Fig. 10 (1941).

27-49 Adapted from K. Pederson-Biergaard and M. Ionnesen, *Acta Endocrinol.,* **1:**38 (1948), Figs. 2 and 4.

28-2 (B) Adapted from L. Weiss and R. O. Greep, *Histology,* 4th ed., McGraw-Hill Book Company, New York, 1977, Fig. 15-4. Used with permission of McGraw-Hill Book Company.

28-9 Adapted from S. Singer and H. R. Hilgard, *The Biology of People,* W. H. Freeman and Company, San Francisco, 1978, Fig. 13-3. Used with permission.

28-10 From W. Bloom and D. W. Fawcett, *A Textbook of Histology,* 9th ed., W. B. Saunders Company, Philadelphia, 1968, Fig. 6-20.

28-11 Adapted from J. C. Cerrottini, *Hosp. Pract.,* **12:**57, (November 1977), figure on p. 62.

28-12 Adapted from D. M. Hume, *Hosp. Pract.,* **3:**31 (1968).

INDEX

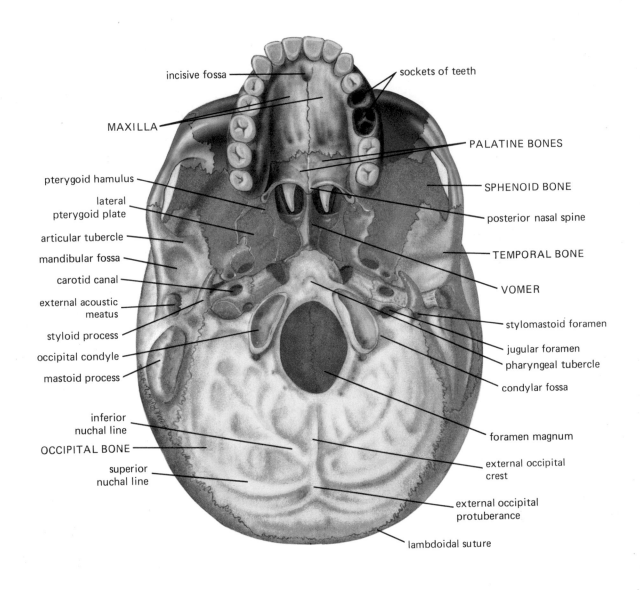

incisive fossa

sockets of teeth

MAXILLA

PALATINE BONES

pterygoid hamulus

SPHENOID BONE

lateral pterygoid plate

posterior nasal spine

articular tubercle

TEMPORAL BONE

mandibular fossa

VOMER

carotid canal

external acoustic meatus

stylomastoid foramen

styloid process

jugular foramen

occipital condyle

pharyngeal tubercle

mastoid process

condylar fossa

inferior nuchal line

OCCIPITAL BONE

foramen magnum

superior nuchal line

external occipital crest

external occipital protuberance

lambdoidal suture

External view of the base of the skull.
(See Fig. 9-13 on page 148 of the text.)